Modeling, Analysis and Optimization of Process and Energy Systems

Modeling, Analysis and Optimization of Process and Energy Systems

F. Carl Knopf
Louisiana State University
Baton Rouge, LA

WILEY
A JOHN WILEY & SONS, INC., PUBLICATION

Published by John Wiley & Sons, Inc., Hoboken, New Jersey.
Published simultaneously in Canada.

Library of Congress Cataloging-in-Publication Data:

Knopf, F. Carl, 1952-
Modeling, analysis and optimization of process and energy systems / F. Carl Knopf.
p. cm.
Includes bibliographical references and index.
ISBN 978-0-470-62421-0 (hardback)
1. Factories–Energy conservation. 2. Manufacturing industries–Energy conservation. 3. Industrial efficiency–Simulation methods. 4. Manufacturing processes–Evaluation. 5. Electric power-plants–Efficiency. I. Title.
TJ163.5.F3K66 2012
658.2'6–dc23

2011015221

Printed in the United States of America.

10 9 8 7 6 5 4 3 2 1

I dedicate this book to my wife Donna and our daughter Megan.

Contents

Preface

Energy costs affect the profitability of virtually every process. This book provides a unified platform for process improvement through the analysis of both the energy demand side—the processing plant—and the energy supply side—available heat and power resources. Emphasis is placed on first quantifying the material and energy flows in a process. The energy needs of the process guide the optimal design of the utility system. Techniques are also presented to ensure that the most cost-effective operation of the utility system is maintained.

Both practicing engineers and engineering students can use the information presented here. For practicing engineers, the book provides a systematic and self-contained approach for minimizing energy use and cost at an operational facility. For chemical, mechanical, petroleum, and energy engineering students, the book provides a detailed evaluation of energy analysis, design, and optimization.

FEATURES OF THE BOOK

There are a number of features of this book that we hope will encourage its use.

The installation of example files, problem solution files, and compiled and source versions of all developed software is detailed in Chapter 1, Section 1.7.

Energy costs, basic economic calculations, and economic uncertainty using Monte Carlo simulations are introduced in Chapters 1 and 2. Levelized utility costing is also developed (Chapter 16).

A systematic approach using either sequential modular (Chapters 3 and 5) or simultaneous-based (Chapters 4 and 5) methodologies is developed for the solution of process material and energy balances. Necessary numerical methods are developed naturally as part of the solution process.

Data reconciliation and gross error detection are introduced (Chapter 6) and applied to an actual cogeneration system (Chapter 11).

Cogeneration system performance and design and off-design calculations are developed using both ideal gas (Chapter 7) and real fluid (Chapter 9) properties.

An open source thermodynamics package (~7000 lines of code) for cogeneration, combustion, and steam calculations is provided in Chapter 8. Codes are provided for problems with field or SI units. The code is used to solve cogeneration design, data reconciliation, and power dispatching problems. Details are provided on how this or any code (written in C, C++, Fortran, etc.) can be seamlessly incorporated into Excel (Appendix A).

Optimal power dispatching for an actual cogeneration system is developed in Chapter 12.

A unified approach to process heat integration and site utility system integration is provided in Chapters 13 and 14. An open source software package is provided to help in the understanding of the basic concepts of heat exchanger network synthesis.

Site emissions are addressed and gas turbine systems are modeled as a series of stirred tank and plug flow reactors (Chapter 15). The ordinary differential equation solver CVODE (from Lawrence Livermore National Laboratory) is made available as a callable routine from Excel, and a reduced kinetics set based on GRI-Mech 3.0 is used to predict emissions from gas turbines.

The economics of carbon dioxide capture in conventional coal-fired utility plants, including steam turbine design and off-design calculations, is addressed in Chapter 16.

Many of the concepts used throughout the text are brought together for the economic analysis of an organic Rankine cycle in Chapter 17.

For several of the heat and power generation topics discussed in the text, "self-contained" Web-based downloadable videos (~30 minutes) with self-study guides and additional problems are available at our Web site, www.cogened.lsu.edu. This site also provides real-time data from the Louisiana State University (LSU) cogeneration system; these data can be used to enhance cogeneration problems and discussions.

There are over 160 completely worked chapter examples. Virtually every example includes a computer-aided (Excel-based) solution. The chapter problems provide an additional 140 problems, with most having computer-aided solutions. A detailed solution manual for the chapter problems is available at the Wiley Web site. A faculty member or practicing engineer can request a copy by sending a letter on a company letterhead.

BACKGROUND

I have assumed that the reader has some knowledge of Excel and programming and has been introduced to basic material and energy balance calculations. Enough detail is provided to help a reader without detailed knowledge of Visual Basic for Applications (VBA) and C.

Throughout this text, a "just in time" approach has been taken to the development of the necessary solution techniques. Developing needed solution techniques often provides the opportunity to improve engineering computer skills. Excel was used as the starting platform for problem solution; however, enhancements made possible by the use of VBA and C programs within Excel are emphasized. The reader is shown, step-by-step, how VBA and C programs can be incorporated into Excel sheets as callable functions and subroutines. The user is given access to all source codes used in this text, which will promote improvements and widespread use.

In the text, I often use both field and SI units. I appreciate that many faculty prefer the sole use of SI units; however, too often I have found that starting engineers make unit mistakes. One solution is practice, which this text provides. As virtually all examples and problems are solved using computer-aided techniques, it is straightforward for the user to change units in the provided solutions. Several examples carry extra significant figures in intermediate calculations to allow direct comparison with Excel sheet values. Some chapter problems are especially important for reinforcing and extending presented materials; for these problems, detailed solutions are provided as part of the text. A detailed solution manual is available for chapter problems.

The optimal design and operation of energy systems can involve the solution of linear programming (LP), nonlinear programming (NLP), mixed-integer linear programming (MILP), or mixed-integer nonlinear programming (MINLP) problems. We utilize both Excel Solver and What's Best for the solution of these problems. What's Best is an Excel add-in for solving optimization problems; a version of What's Best has been supplied by LINDO Systems for use with this text.

USE OF THE BOOK

For engineering students, this book provides a logical progression to allow a better understanding of energy flows in a processing plant. Topics of importance to energy engineering calculations occur naturally. This should prove to be an interesting way of improving skills in coding, using numerical methods to solve engineering problems, and formulating and solving process and utility energy optimization problems.

The book can be integrated into engineering curricula by following one of the following paths.

ENERGY SUSTAINABILITY COURSE

This book can be used in a one-semester special topic course to introduce energy sustainability to third- and fourth-year engineering students. Here the first three-quarters of the course focuses on understanding energy flows in processing plants and how cogeneration and energy efficiency are important aspects of a national energy portfolio; these topics are directly covered in this text. Then using this text as a basis, and combined with outside reading materials, the final quarter of the course can be devoted to detailed analysis of key emerging energy technologies—I suggest including biomass gasification, solar thermal/organic Rankine power plants, and integrated gasification combined cycle and other advanced clean coal processes. A reasonable question is, Why study these topics when there are so many emerging energy technologies? There are several answers to this question: First, biomass and solar thermal/organic Rankine plants represent the breadth of the emerging technologies; second, the best currently available large-scale conservation technology is cogeneration; and finally, coal usage must be addressed since ~50% of the electricity generation in the United States is from coal. In addition, these technologies all share several process units. The energy inputs to these processes (from chemicals, fuel, or radiation from the sun) can be used to produce steam (or to vaporize an organic compound) in a Rankine power cycle; chemical energy can be converted into a synthesis gas and can be used in gas turbines; or some combination of these may be used. This allows the students to see the common features of these processes and allows for a discussion of optimal process designs dependent on the energy source. Students can explore other alternative energy technologies through team-oriented term projects that are suggested in the text.

NUMERICAL METHODS AND CAPSTONE DESIGN COURSES: ESTABLISHING AN "ENERGY THREAD"

Another alternative is to use ~50% of this text in an applied numerical analysis course and to use the remaining chapters as part of a capstone design sequence and within other courses in the curriculum. This is actually how I originally developed the text; we wanted to establish an "energy thread" in our engineering courses without adding a new course. For a sophomore-level applied numerical method course, topics included engineering programming (Chapter 2), solution of linear and nonlinear equations (Chapter 3), solution of linear and nonlinear equation sets (Chapter 4), data analysis and curve fitting (Chapter 6), ordinary differential equations (Chapter 5, Sections 5.4–5.6, and Chapter 15, Sections 15.3 and 15.4), partial differential equations (Chapter 10, Section 10.7), and advanced engineering programming (Appendix A). As part of the capstone design sequence taught to fourth-year engineering students, Chapters 3–5 were quickly reviewed, highlighting the structure of computer-aided solutions to material and energy balances, and then emphasis was placed on optimizing energy resources in processing plants using material from Chapters 13 and 14. Chapters 16 and 17 were used to detail levelized economics. Data reconciliation and gross error detection (Chapters 6 and 11) were used as a lab for third-year engineering students. The chapters on determining gas turbine performance (Chapters 7 and 9) and developing physical property packages (Chapter 8) were used within engineering thermodynamics courses. Modeling gas turbine combustors (Chapter 15, Sections 15.2–15.4) was used within our kinetics and reactor design course.

PROCESS SYSTEMS COURSE (ENGINEERING CURRICULUM)

In an engineering curriculum, this book can be used to help provide an integrated introduction to process synthesis. Following the introductory material and energy balance course, a process perspective of energy costs and basic economics, data reconciliation, gross error detection, heat and power systems, utility system dispatch, heat integration, and cogeneration can be taught using the materials in this book.

INDUSTRIAL USE

One strength of this book will be its use for practicing engineers. Heat and power systems involve large flows that can magnify inaccuracies in physical properties. A major coding effort in the text has been the development of accurate physical properties for utility systems. In Appendix A, we show the user how these thermodynamic codes, or any user-written code, can be seamlessly incorporated into Excel. The thermodynamic properties (~7000 lines of code) for cogeneration, combustion, and steam calculations are described in Chapter 8. Emphasis is also placed on data reconciliation and gross error detection, cogeneration system design and off-design operation, utility system dispatching, heat exchanger network synthesis and site energy integration, and predicting emissions.

SUPPLEMENTARY MATERIALS

For several heat and power generation topics discussed in the text, Web-based downloadable videos (~30 minutes) with documentation and additional problems are available at our Web site, www.cogened.lsu.edu. These materials have been designed for student use and have been tested at LSU, Florida A&M University–Florida State University (FAMU-FSU) (Dr. John Telotte), University of Alabama (Dr. Heath Turner), and University of Florida (Dr. Peng Jiang).

ACKNOWLEDGMENTS

The codes provided here would not have been possible without the efforts of graduate students and postdoctoral research associates with whom I have been fortunate to work. The cogeneration thermodynamics code (Chapter 8) was initially developed by Dr. Shane Stafford and was later modified and completed by Dr. Derya B. Orzyurt. There has been a long collaboration with Dr. Janardhana R. Punuru in developing techniques that allow bridging between Excel and C/C++ code. Dr. Punuru developed the Excel interface for CVODE, which is provided in Chapter 15 (CVODE is the ordinary differential equation solver available from Lawrence Livermore National Laboratory). The initial version of the heat exchanger network synthesis program THEN (Chapter 13) was developed by Sanjay P. Bhargava, Sanjay G. Pethe, and Rajiv Singh and was later coupled to Excel with the help of Dr. Punuru. Lina M. Bustami worked on the heat recovery steam generator problem and Robert Buckley worked on both the initial cogeneration data reconciliation problem and the energy dispatching model.

I also want to thank my colleagues who have made significant contributions to this book. I especially thank Dr. Kerry M. Dooley (LSU) who read and provided corrections for the first draft of each chapter in this text. I have had many discussions about energy systems and the cost of energy generation with Louis Braquet (LB Services) and Dr. David Dismukes (LSU), both of whom reviewed Chapter 1. Dr. Dismukes prepared the table of levelized costs for alternative energy systems in Chapter 17. Richard McKinney reviewed Appendix A and helped provide the needed modifications to move from Microsoft Visual C++ 6.0 to Visual C++ 2008 Express Edition. Peter Davidson and Tony Cupit (LSU Facility Services) helped provide data and cogeneration operational strategies for the optimal energy dispatching model. Dr. Oscar Jimenez Cabeza (GEPROP) and Dr. Roger Nordman (SP Technical Research Institute of Sweden) provided critical reviews of the energy integration chapters.

Dr. John Telotte (FAMU-FSU) reviewed the thermodynamic aspects of Chapters 8 and 15.

I am especially indebted to Dr. Frank Madron (ChemPlant) and Dr. Michael Erbes (Enginomix). Dr. Madron provided many corrections and clarifications to the chapters on data reconciliation. Dr. Erbes provided his expertise on energy sustainability and modeling, cogeneration systems, and turbine performance in both design and off-design operation, and also reviewed Chapters 7, 9, 10, and 16.

Dr. Ralph Pike (LSU) and Dr. G.V. Reklaitis (Purdue University) helped focus the goals of the text and provided suggestions for improvement. Mohammed Syed read the final draft of the text and Vamshi Kandula helped assemble all the materials in the text.

I would especially like to thank Professor Don Freshwater for his suggestion for the cover painting *ICI Wilton Works* by Tom Gamble. I would also like to thank Professor Chris D. Rielly of Loughborough University in Leicestershire, United Kingdom, for allowing its use and for providing the copy.

I acknowledge the financial assistance of the National Science Foundation Phase I and Phase II grants, "Integrating a Cogeneration Facility into Engineering Education," NSF Awards 0535560 (Phase I) and 0716303 (Phase II).

F. CARL KNOPF
Baton Rouge, Louisiana

To view color versions of the figures in this book, please visit: ftp://ftp.wiley.com/public/sci_tech_med/energy_system.

Conversion Factors

Mass and Density	1 kg = 2.2046 lb	1 lb = 0.4536 kg
	1 g/cm^3 = 10^3 kg/m^3	1 lb/ft^3 = 0.016018 g/cm^3
	1 g/cm^3 = 62.428 lb/ft^3	1 lb/ft^3 = 16.018 kg/m^3
Length	1 cm = 0.3937 in.	1 in = 2.54 cm
	1 m = 3.2808 ft	1 ft = 0.3048 m
Velocity	1 km/h = 0.62137 mile/h	1 mile/h = 1.6093 km/h
Volume	1 cm^3 = 0.061024 in.^{3}	1 in^3 = 16.387 cm^3
	1 m^3 = 35.315 ft^3	1 ft^3 = 0.028317 m^3
	1 L = 10^{-3} m^3	1 gal = 0.13368 ft^3
	1 L = 0.0353 ft^3	1 gal = 3.7854 × 10^{-3} m^3
Force	1 N = 1 kg-m/s^2	1 lbf = 32.174 lb-ft/s^2
	1 N = 0.22481 lbf	1 lbf = 4.4482 N
Pressure	1 Pa = 1 N/m^2 = 1 kg/m-s^2	1 lbf/in^2 = 6894.757 Pa
	1 Pa = 1.4504 × 10^{-4} lbf/in^2	1 lbf/in^2 = 144 lbf/ft^2
	1 bar = 10^5 N/m^2 = 10^5 Pa	1 atm = 14.696 lbf/in.^{2}
		1 atm = 1.01325 bars
Energy and Specific Energy	1 J = 1 N-m = 0.73756 ft-lbf	1 ft-lbf = 1.35582 J
	1 kJ = 737.56 ft-lbf	1 Btu = 778.17 ft-lbf
	1 kJ = 0.9478 Btu	1 Btu = 1.0551 kJ
	1 kJ/kg = 0.42992 Btu/lb	1 Btu/lb = 2.326 kJ/kg
	1 kcal = 4.1868 kJ	
Energy Transfer Rate or Power	1 W = 1 J/s = 3.4121 Btu/h	1 Btu/h = 0.29307 W
	1 kW = 1.341 hp	1 hp = 2544.4 Btu/h
		1 hp = 550 ft-lbf/s
		1 hp = 0.7457 kW
Specific Heat or Entropy	1 kJ/kg-K = 0.238846 Btu/lb-R	1 Btu/lb-R = 4.1868 kJ/kg-K
	1 kcal/kg-K = 1 Btu/lb-R	

(Continued)

Temperature	K = °C + 273.15	R = °F + 459.67
	K = (5/9) R	R = (9/5) K = (1.8) K
	°C = (5/9) (°F − 32)	°F = (1.8) °C + 32
	Δ K = Δ°C	Δ R = Δ°F
	Δ K = (5/9) Δ R	Δ R = (1.8) Δ K
Notation Common to the U.S. Power Industry	k (kilo) = 10^3; kW (kilowatt)	
	m = 10^3; mSCF (1000 standard cubic feet)	
	M (mega) = 10^6 ; MW (megawatt)	
	MM = 10^6 ; MMBtu (million British thermal unit)	
Universal Gas Constant, *R*	8.314 J/mol-K	
	0.082057 L-atm/mol-K	
	1.987 cal/mol-K	
	1.987 Btu/(lb-mol-R)	
	82.06 cm^3-atm/mol-K	

List of Symbols

$a_{e,i}$	number of atoms of element e in species i, Chapter 3
A_c	plug flow reactor (PFR) tube cross-sectional area
$C_{p,i}$	isobaric molar heat capacity of species i
$\hat{C}_{P,i}$	isobaric heat capacity of species i per unit mass (specific heat capacity)
$\hat{C}_{P,j}$	isobaric specific heat capacity of stream j (Btu/lb-R, kJ/kg-K)
$C_P^{\text{vapor}}, C_P^{\text{liquid}}, C_P^{ig}$	isobaric molar heat capacity vapor phase, liquid phase, ideal gas
C_{Total}	total cost rate (\$/time), Chapter 10
C_V	isochoric molar heat capacity
d_0, d_i	tube diameter outside, inside, Chapter 10
e	escalation rate for economics calculation, Chapter 16
f_i	fugacity of species i, Chapter 15
f_i^0	standard state fugacity of species i, Chapter 15
f	friction factor, Chapter 16
F_j	mass flow rate of stream j
F_i^j, $F_{t,j}$	mass flow rate of species i in stream j
F_{unit}^j	mass flow rate stream j in given unit, Chapter 16
G	molar Gibbs free energy, Chapter 15
G_i^0	standard molar Gibbs free energy of pure species i, Chapter 15
G_j	gas mass velocity of stream j, Chapter 10
h_i	molar enthalpy of species i
$h_j, h_{(T_j,P_j)}, h(T_j, P_j)$	molar enthalpy of specific stream j
$h_j^{\text{vapor}}, h_j^{\text{liquid}}, h_j^{\text{sat liquid}}$	molar enthalpy of stream j with state indicated: vapor, liquid, saturated liquid
$h_{j,\text{isen}}, h_j^{\text{isen}}, h_{j,a}, h_j^a$	molar enthalpy of stream with thermodynamic path indicated: isentropic, actual
$h_{i,j}$	molar enthalpy of species i in stream j
Δh	change in molar enthalpy, $\Delta h = h_{(T_{\text{out}},P_{\text{out}})} - h_{(T_{\text{in}},P_{\text{in}})}$
h_{ref}, h_0	reference molar enthalpy for stream, generally for field units (77°F or 535.67 R, 14.696 psia) and for SI units (25°C or 298.15 K, 0.101326 MPa)
h_0, $h_{\text{gas side}}$	heat transfer coefficient outside, gas side , Chapter 10
$\hat{h}_i$, $\hat{h}_j$, $\hat{h}_{i,j}$	enthalpy per unit mass of species i; enthalpy per unit mass of stream j (specific enthalpy) (Btu/lb, kJ/kg)
$\hat{h}_{LP}^{\text{sat vapor}}$, $\hat{h}_{\text{HPT out}}^{\text{Real}}$	specific enthalpy of low pressure (steam) at saturated vapor conditions, high-pressure turbine out at actual conditions
$\hat{h}_{i,\text{ref}}, \hat{h}_{j,\text{ref}}, h_{i,\text{ref}}, h_{j,\text{ref}}$	reference specific enthalpy for species i or stream j, molar enthalpy, generally for field units (77°F or 535.67 R, 14.696 psia) and for SI units (25°C or 298.15 K, 0.101326 MPa)
$\Delta\hat{h}_{\text{MP Steam}}, \Delta\hat{h}_{\text{LP Steam}}$	change in specific molar enthalpy for medium pressure, low-pressure steam
H	total enthalpy (Btu, kJ)
$\dot{H}$	total enthalpy rate (Btu/s, kW)

ΔH_r, $\Delta H_{\text{combustion}}$, $\Delta H_{\text{vaporization}}$	heat of reaction, combustion, vaporization
i	equivalent discount rate with escalation for economic calculations, Chapters 16 and 17
i, i_{eff}	discount rate or interest rate per period or time value of money, effective interest rate
k_f, k_b	reaction rate constant forward, backward (reverse)
k	reaction rate constant, thermal conductivity of the material
$[k]$	kth iteration; brackets indicate iteration
k_c	water/steam thermal conductivity, Chapter 10
K_i	equilibrium distribution coefficient for species i
K_{eq}	equilibrium constant of chemical reaction
K_h	exhaust gas thermal conductivity, Chapter 10
$L(x, \lambda)$	Lagrangian function
L_{tube}	tube length, Chapter 10
MW_i	molecular weight of species i
n, n_i	number of periods (economics), number of moles of species i
$N_{\text{tubes\_wide}}$	number of tubes in the selected width, Chapter 10
N_j	molar flow rate of stream j
$N_{i,j}$, N_i^j	molar flow rate of species i in stream j
N_i^k	molar flow rate of species i at iteration k, Runge–Kutta method, Chapter 5
$N_i^{Z_j}$	molar flow rate of species i at position Z_j, Euler's method or Runge–Kutta method, Chapter 5
P_j, P_{ref}	pressure stream j, reference pressure
$\Delta P_{\text{Unit}}^{\text{species}}$, $\Delta P_{\text{Unit}}^{\text{stream}}$	pressure drop in given unit , stream side
q	heat quantity per mole
$\hat{q}$	heat quantity per unit mass
$q_1^{\text{Isentropic}}$, q_1^{real}	steam quality isentropic, real
Q	variance–covariance matrix, Chapters 6 and 11
$\dot{Q} \equiv \dfrac{dQ}{dt}$	rate of heat transfer (Btu/s, kW)
$\dot{Q}_{\text{Unit}}$	rate of heat transfer in specified unit (Btu/s, kW)
$\dot{Q}_{\text{MP}}^{\text{Process}}$, $\dot{Q}_{\text{LP}}^{\text{Process}}$	process heat load satisfied by medium pressure (MP), low pressure (LP) steam, Chapter 14
r, r_i	rate of reaction, rate of reaction species i (single reaction)
r_m	rate of reaction for reaction m, Chapter 15
R	universal gas constant per mole
$\hat{R}$	universal gas constant per unit mass = R/MW_i
R_i	production or generation rate of species i
R_{Thermal}	thermal resistance, Chapter 10
s_i	molar entropy of species i
s_j, $s_{(Tj,Pj)}$, $s(T_j, P_j)$	molar entropy of stream j
s_i^0	molar entropy of species i, at reference T and P
Δs	change in molar entropy
$\hat{s}_i$, $\hat{s}_j$	entropy per unit mass of species i, entropy per unit mass of stream j (specific entropy)
$\hat{s}_j^{\text{isentropic}}$, $\hat{s}_j^{\text{real}}$, $\hat{s}_j^{\text{sat vapor}}$	entropy per unit mass stream j at isentropic, real, and saturated vapor conditions
S	species column vector, Chapter 15
S_{tube}	tube spacing, Chapter 10
T_j	temperature stream j
$T_{j,\text{location}}$	temperature stream j, at specific location
T_{ref}, T_{isen}, T_{actual}, T_a	reference temperature, isentropic temperature, actual temperature, acrtual temperature
T_j^{isen}, $T_j^{\text{sat vap}}$	temperature stream j, isentropic thermodynamic path, saturated vapor condition
T_{zj}	temperature at position Z_j, Euler's method or Runge–Kutta method, Chapter 5
ΔT_{Pinch}, $\Delta T_{\text{approach}}$, $\Delta T_{\text{LMTD,Evaporater}}$	pinch temperature difference, approach temperature difference, log mean temperature difference, unit
$\hat{u}_{\text{in}}$, $\hat{u}_{\text{out}}$	specific internal energy
u_j	estimates of nonmeasured process variables for data reconciliation, Chapters 6 and 11
U	overall heat transfer coefficient

υ	molar volume
$\hat{\upsilon}$, $\hat{\upsilon}_j$	volume per unit mass (specific volume), of stream j
V_c	water velocity in tube (ft/s), Chapter 10
V, V_r	total volume, volume of the reactor
V_{an}	annulus velocity (ft/s), Chapter 16
w_i^j	weight fraction of species i in stream j
w	work per mole
$\hat{w}$	work per unit mass (Btu/lb)
w_c	water/steam flow rate in the single tube (lb/h), Chapter 10
W_{GT}, W_{PT}	gas turbine work, power turbine work
$\dot{W}_T$, $\dot{W}_{Flow}$, $\dot{W}_{shaft}$	rate of work total, rate of work due to fluid flow, rate of shaft work
$x^{[k]}$	kth iteration; brackets indicate iteration
$x_{i,j}$	mole fraction of species i in stream j
x_i^+, x_i	measured value, reconciled value in data reconciliation, Chapters 6 and 11
Δx	insulation thickness, Chapter 10
y_i^j, $y_{i,j}$	mole fraction of species i in stream j
z_j	purchase cost of the ith component ($) in economic calculations, Chapter 10

GREEK LETTERS

α_i^j, $\alpha_{i,j}$	species i split fraction in separator output stream $j = N_{i,j}/N_{i,in}$, Chapters 3 and 5
α_j	species split fraction in splitter output stream $j = N_j/N_{in}$, Chapter 3
α_{site_PHR}, α_{cogen_PHR}	base power to base heat ratio, Chapter 10
$\alpha = k/\rho\hat{C}_p$	thermal diffusivity, Chapter 10
δ_i	denotes whether equipment i is on = 1/off = 0
ε	convergence parameter or tolerance
γ, γ_i	ratio of heat capacities $\gamma = C_P/C_V$, ratio of heat capacities for species i
η, η_{unit}	efficiency, efficiency of given unit
$\eta_{unit}^{isentropic}$	isentropic efficiency of given unit, Chapter 16
λ_k	Lagrange multipliers
μ_i	chemical potential
μ_h	exhaust gas viscosity, Chapter 10
ν_i, $\nu_{m,i}$	stoichiometric coefficient for species i (single reaction), stoichiometric coefficient for species i in reaction m
ν'	reactant stoichiometric coefficient matrix, Chapter 15
ν''	product stoichiometric coefficient matrix, Chapter 15
ξ, ξ_m	extent of reaction (single reaction), extent of reaction for reaction m
ρ	molar density, mole per unit volume
$\hat{\rho}$	specific density, mass per unit volume
σ_i	standard deviation, Chapters 6 and 11
τ	mean residence time
ϕ_i	fugacity coefficient for species i
$\chi^2_{(1-\alpha)}(\nu)$	chi-squared distribution, Chapters 6 and 11
$\frac{1}{2}\vartheta^2$	kinetic energy

Chapter 1

Introduction to Energy Usage, Cost, and Efficiency

1.1 ENERGY UTILIZATION IN THE UNITED STATES

Let us begin by examining energy use in the United States. Figures 1.1 and 1.2 are U.S. energy utilization in 2000 and 2008 as determined by the Energy Information Administration (EIA) and Lawrence Livermore National Laboratory (LLNL). Here, the energy unit used in the figures is quadrillion or quad, and 1 quad = 10^{15} Btu.

Key results from Figures 1.1 and 1.2 as well as energy utilization data from 2000–2008 are summarized in Table 1.1.

Table 1.1 indicates that total energy use in the United States has remained relatively flat for the years 2000–2008. Also, our reliance on "traditional" hydrocarbon energy sources (natural gas, coal, and petroleum) has remained little changed over this near decade. In the future, there is every expectation that alternative energy technologies will grow in importance. For example, wind energy use has increased some fivefold from 2003 to 2008. But in the near term, we should give renewed consideration to the most efficient use of existing energy resources, especially our traditional hydrocarbons. Based on LLNL estimates, Figures 1.1 and 1.2 indicate that overall U.S. energy efficiency is ~42%. The ~58% rejected energy is primarily waste heat from combustion. A motivation for this book is to support existing and planned processing sites and energy facilities to help ensure energy is being used as efficiently as possible.

1.2 THE COST OF ENERGY

We next ask the following question, "Just how important are process energy costs?" Energy costs are often one of the largest single expenses at an industrial processing site along with labor and raw materials. These industrial processing sites would include chemicals and petrochemicals, agrochemicals, pharmaceuticals, plastics, paper and pulp, metal and mining. In these industries, energy costs typically range from about 10% to over 50% of the operating expenses for the site.

To assign an actual dollar cost, we need to pick a price for energy and to estimate a typical process energy load. For the price of energy, we can use the cost of natural gas to industry as a benchmark. Many industries in the United States were built on "\$2" natural gas—this "\$2" refers to \$2 per million British thermal unit (MMBtu) or \$2 per thousand standard cubic foot (mSCF). Roughly 1 mSCF of natural gas contains 1 MMBtu. Figure 1.3 shows the average annual industrial price for natural gas from 1973 until 2009. Figure 1.4 shows the average monthly industrial price for natural gas from January 2000 to November 2010 (EIA). Costs in Figures 1.3 and 1.4 are not adjusted for inflation. Starting in about 2000, there has been an upward trend in natural gas price coupled with large short-term price swings. For example, in 2005, Hurricanes Katrina and Rita contributed to spot market natural gas prices climbing to \$15+ per MMBtu for short periods of time.

Next, we can address, "What is a typical process energy load?" A large industrial (chemical plant or refinery) may use 50–100 × 10^9 Btu/day for heating and power generation. This is the energy required to produce steam and electricity for process use. A medium-sized industrial or a large university may use 10–20 × 10^9 Btu/day for heating and power generation.

Using a long-term projected price of \$8 natural gas (\$8 per MMBtu) as a fuel cost, the utility cost for a large industrial may be on the order of \$200 × 10^6 per year, and a

Modeling, Analysis and Optimization of Process and Energy Systems, First Edition. F. Carl Knopf.

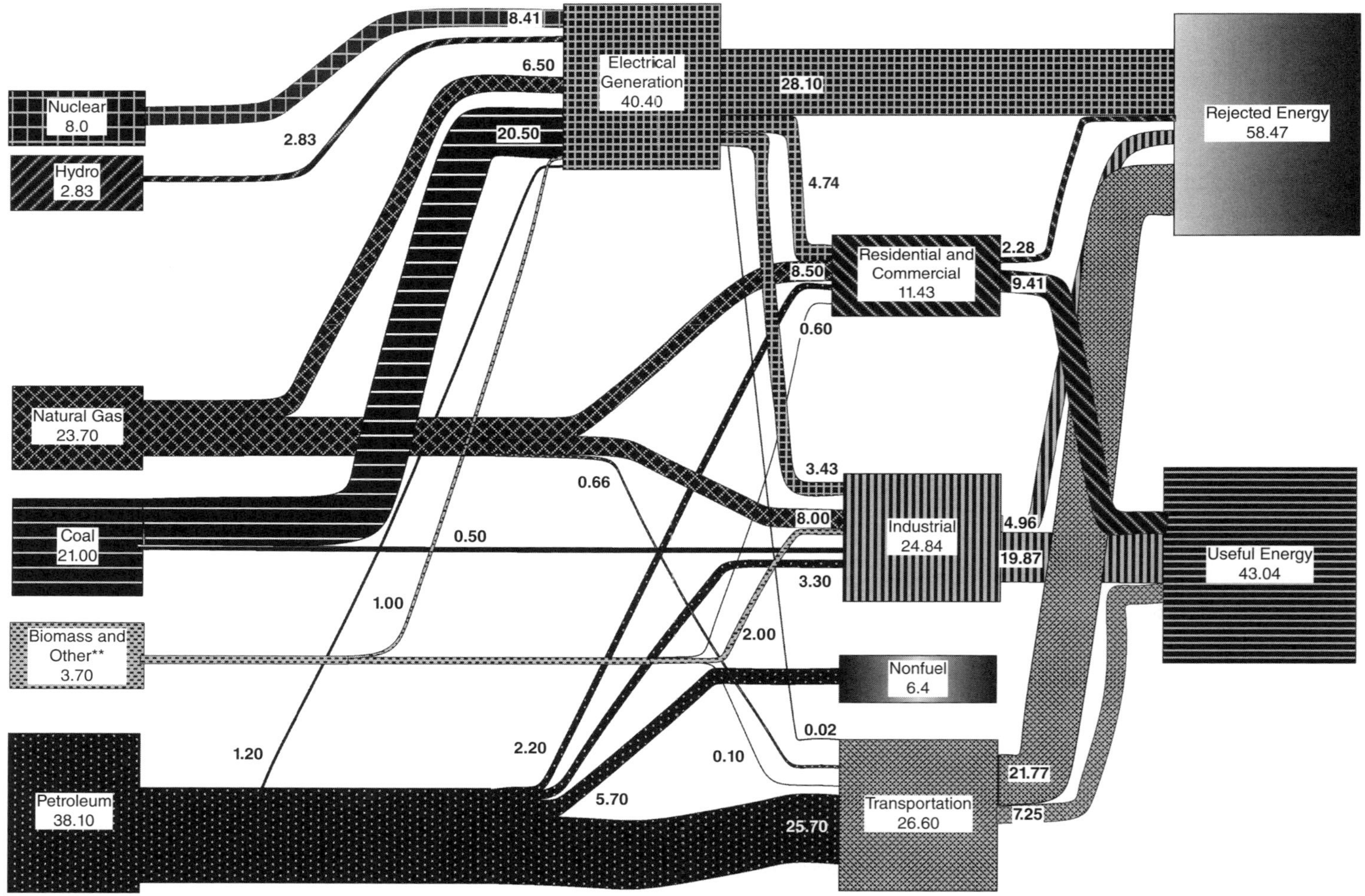

Figure 1.1 Energy use in the United States (EIA and LLNL; see the References for the Web site) in 2000. In this figure, nonfuel from petroleum feed is, for example, petroleum used as a chemical feedstock.

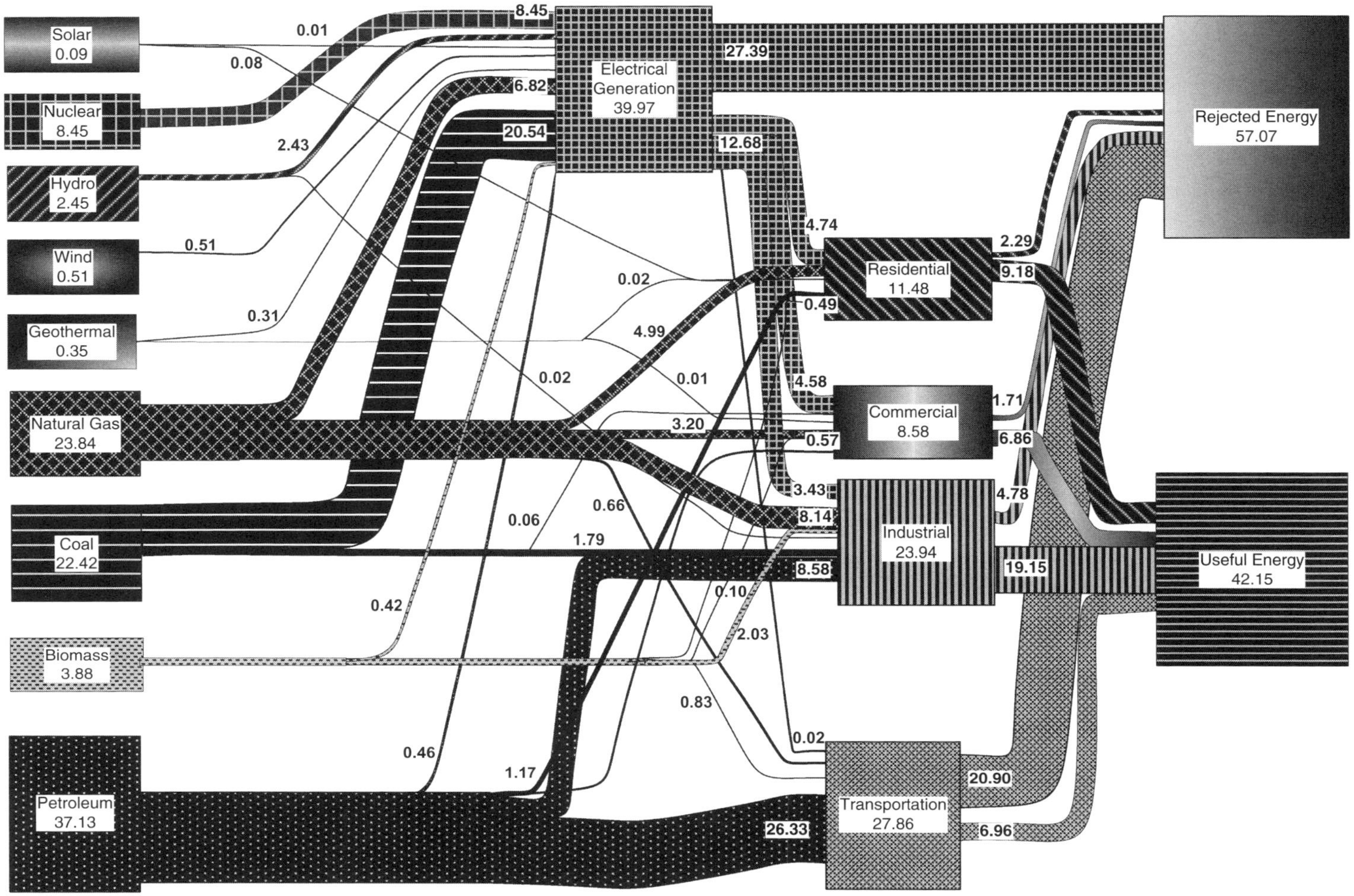

Figure 1.2 Energy use in the United States in 2008 (EIA and LLNL).

Table 1.1 Summary of Energy Use (Quadrillion British Thermal Unit) in the United States from 2000 to 2008 (EIA)

	2000	2001	2002	2003	2004	2005	2006	2007	2008
Total energy use	98.5	97.0	97.0	98.1	100.2	100.4	99.8	101.5	99.2
Energy from									
Solar				0.06	0.06	0.06	0.07	0.08	0.09
Nuclear	8.0	8.0	8.10	7.95	8.22	8.16	8.21	8.41	8.45
Hydro	2.83	2.30	2.60	2.82	2.69	2.70	2.86	2.46	2.45
Wind				0.11	0.14	0.17	0.26	0.31	0.51
Geothermal				0.33	0.34	0.34	0.34	0.35	0.35
Natural gas	23.70	23.20	23.20	22.90	22.93	22.50	22.19	23.63	23.84
Coal	21.0	21.90	22.30	22.32	22.46	22.79	22.44	22.76	22.42
Biomass	3.7*	3.3*	3.2*	2.81	3.02	3.15	3.37	3.61	3.88
Petroleum	38.10	38.0	38.10	38.80	40.29	40.39	39.95	39.81	37.13
Energy from natural gas, coal, and petroleum	82.28	83.10	83.60	84.05	85.68	85.68	84.58	86.30	83.39

*The biomass data for 2000–2002 include wood and waste, geothermal, solar, and wind.

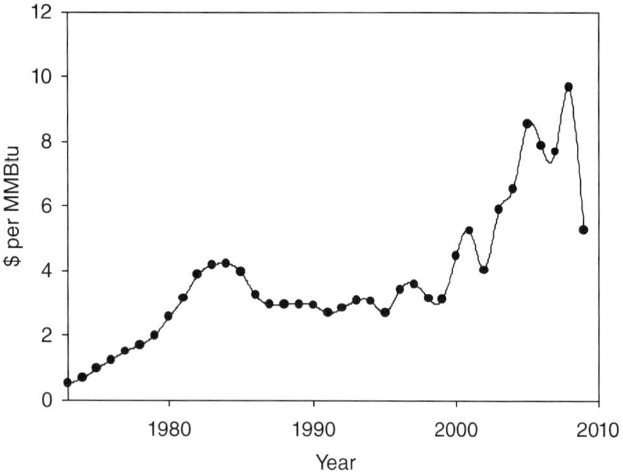

Figure 1.3 Industrial price for delivered natural gas (dollar per MMBtu) and yearly average price from 1970 to 2009 (EIA).

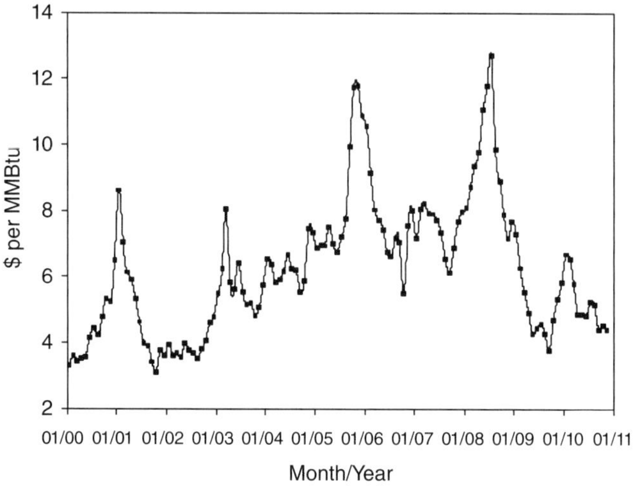

Figure 1.4 Industrial price for delivered natural gas (dollar per MMBtu) and monthly average price from January 2000 to November 2010 (EIA).

medium-sized industrial or large university may see utility costs near $\$40 \times 10^6$ per year. Utility costs may be higher if electricity is purchased from the local utility and steam is generated in boilers, as opposed to taking advantage of combined heat and power generation opportunities.

We can see that energy costs are a significant consideration in most processes. Increasing worldwide demand for energy is expected to keep utility costs on a positive slope. There is a need for careful assessment of the energy currently being used in a process coupled with a systematic approach for reducing this energy use. In many processing plants, electricity is both purchased and generated, and in deregulated areas, plants may sell electricity to the open market. Fluctuating energy costs (see Figure 1.4) necessitate a careful coupling of process energy needs with utility production planning and utility purchase.

1.3 ENERGY EFFICIENCY

There are two commonly used definitions of efficiency, η, one based on the first law of thermodynamics and a second definition based on the second law of thermodynamics. From the first law of thermodynamics, we can write

$$\eta_{1st} = \frac{\text{Usable energy output from the system}}{\text{Energy supplied to the system}}. \quad (1.1)$$

Here, we account for the energy supplied to the system and the usable energy from the system. From the second law of thermodynamics, we can write

$$\eta_{2\text{nd}} = \frac{\text{Minimum theoretical energy required}}{\text{Energy actually used}}. \quad (1.2)$$

Both definitions will find use in the energy and power calculations performed in this text. As Equation (1.2) requires additional development, examples and problems in this chapter will just focus on Equation (1.1).

In many efficiency calculations, Equation (1.1) can be directly used. In some cases, especially when a specific operation is using or producing electricity (e.g., a turbine), Equation (1.2) proves more beneficial. For power (electricity) generating systems, overall performance is often provided by an alterative form of Equation (1.1)—the plant net heat rate,

$$\text{Plant net heat rate} = \frac{\text{Energy supplied to the system, Btu}}{\text{Usable electrical energy output from the system, kW-h}}. \quad (1.3)$$

EXAMPLE 1.1 *Utility Company Power Plant Energy Efficiency*

A coal-fired utility company power plant is shown in Figure 1.5. Here, time is not considered—we are just looking at the energy flows resulting from 1000 lb of coal being added to the boiler. The values provided in Figure 1.5 represent energy inputs to each operation; for example, energy to the boiler from the coal feed is 12,720,000 Btu. Part of this 12,720,000 Btu will be sent to the turbine (as steam) and part will be lost to the stack. Of the 11,194,000 Btu sent to the turbine, 5,261,000 Btu is used for electricity generation. The turbine here is a "condensing steam turbine." This is also typically referred to as a bottoming cycle as all the available energy in the steam for power generation is extracted in the turbine and then the steam is condensed. The 4,933,000 Btu of thermal energy rejected to the environment is required to condense low-pressure steam from the turbine. This condensing allows the boiler feedwater to be recycled as it allows the boiler feedwater pressure to be increased using a pump. The return boiler feedwater contains 1,000,000 Btu.

It is possible to directly convert the energy flows in Figure 1.5 to energy transfer rates, for example, 1000 lb of coal per hour would generate 12,720,000 Btu/h. It is also possible to consider the energy flows in Figure 1.5 to be directly scalable; for example, 2000 lb of coal would generate 25,440,000 Btu. For now, we can use the energy values in Figure 1.5 to determine the efficiency of the boiler, the turbine, and the generator and also the overall process thermal efficiency (sometimes termed the plant net thermal efficiency) and the power plant net heat rate (British thermal unit per kilowatt-hour).

In this example, we are just accounting for energy flow within the utility plant fence line. We are not accounting for transportation and distribution losses. Exported power to the grid often experiences 5–8% transmission and distribution losses before that power can be utilized at the end application (see Problem 1.5).

SOLUTION It can sometimes be confusing trying to consistently determine unit by unit and overall system efficiencies using Equations (1.1) and (1.3). It can be helpful to think of each unit (or system) as a cost center. For example, the boiler "purchases" 12,720,000 Btu of coal and "sells" 11,194,000 Btu to the turbine and takes back "credit" 1,000,000 Btu as condensate return. The efficiency of the boiler is

$$\eta_{\text{Boiler}} = \frac{(11{,}194{,}000 - 1{,}000{,}000)}{12{,}720{,}000} = 80.1\%.$$

The turbine is needed to generate electricity, and here the turbine purchases 11,194,000 Btu and "delivers" 5,261,000 Btu for electricity generation. The efficiency of the turbine is

$$\eta_{\text{Turbine}} = \frac{5{,}261{,}000}{11{,}194{,}000} = 47.01\%.$$

For the overall thermal efficiency, 12,720,000 Btu of coal is supplied to the plant, and 4,844,000 Btu of electricity is delivered for transmission and sale. Here 365,000 Btu of electricity for coal handling, including coal conveyors and coal crushers or pulverizers, is used internally and is not available for "sale"—internal use of electricity is often termed a parasitic load. The utility plant net heat rate is defined as the British thermal unit supplied to the plant divided by the kilowatt-hour of electricity delivered for transmission:

$$\frac{12{,}720{,}000\ \text{Btu}}{1420\ \text{kW-h}} = 8957.7\frac{\text{Btu}}{\text{kW-h}}.$$

Efficiency results for the coal utility plant are summarized in Table 1.2.

It is often convenient to represent the energy flows in Figure 1.5 on a Sankey diagram, which is shown in Figure 1.6a. From the Sankey diagram, where the energy flow is normalized to 100 input units (Figure 1.6b), it is easy to see the overall plant thermal efficiency is 38.08%, here

$$\frac{4{,}844{,}000\ \text{Btu}}{12{,}720{,}000\ \text{Btu}} = 38.08\%.$$

■

EXAMPLE 1.2 *Topping Cycle Cogeneration Plant Energy Efficiency*

A natural gas-fired topping cycle cogeneration facility is shown in Figure 1.7. It is termed a topping cycle because electricity is generated from the steam before the steam is used by the process. The topping cycle uses a back-pressure or "let-down" steam turbine. Unlike the condensing steam turbine of Example 1.1, steam exhaust from a back-pressure turbine still has energy that can be useful in

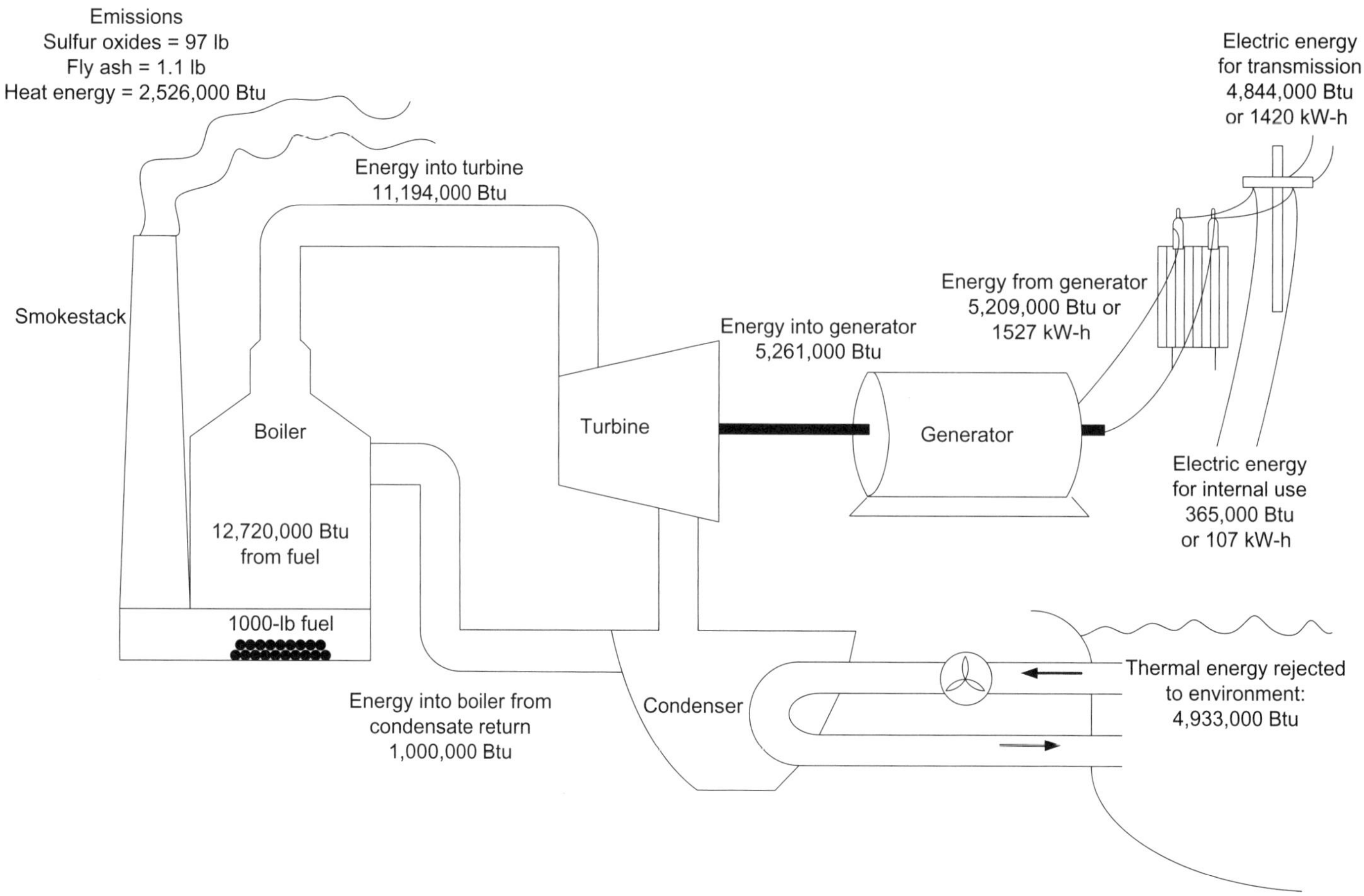

Figure 1.5 Coal-fired electric power plant (adapted from Priest, 1973).

the process. The thermal energy (steam) from the turbine is used for heating within the processing plant and the condensate is returned to the boiler. In a topping cycle cogeneration facility, the ratio of the process steam demand to the process electricity demand is generally large; this is typical of the needs found in a brewery or distillery. The values provided in Figure 1.7 again represent the energy input to each operation. Determine the efficiency of the boiler, the turbine, the generator, and also the overall process thermal efficiency and the cogeneration plant heat rate (British thermal unit per kilowatt-hour). Finally, draw the Sankey diagram with energy flows normalized to 100 input units (British thermal unit).

SOLUTION The efficiencies for the topping cycle are provided in Table 1.2. For the overall thermal efficiency, 12,720,000 Btu of natural gas is supplied to the system, 1,720,000 Btu is available for transmission as electricity (to the process), and 8,439,000 Btu is used as heat in the process. The Sankey diagram is shown in Figure 1.8. ■

EXAMPLE 1.3 *Gas Turbine Cogeneration Plant Energy Efficiency*

A natural gas-fired turbine cogeneration facility typical of a processing plant or university is shown in Figure 1.9. The values provided in Figure 1.9 are the energy input to each operation. Compressed air and natural gas are mixed and combusted in the turbine system. The hot-pressurized exhaust gas loses pressure as it drives the turbine. Part of the energy from the turbine is used to compress the incoming air; part is used to generate electricity; and the remainder used for steam generation. The figure shows that the turbine shaft (here a single shaft) turns both the generator and the air compressor. The energy from the turbine that drives the air compressor (6,400,000 Btu) will add to the energy of the incoming air stream. In later chapters, we will account for energy losses in the compression process. If we examine the mechanical energy from the turbine (6,400,000-Btu compression + 4,250,000-Btu electricity), ~60% of this mechanical energy is used to drive the compressor. The compression process can be viewed as a recycle energy stream—this energy (6,400,000 Btu) from the turbine is actually returned to the turbine by increasing the air feed stream pressure and temperature.

The hot and nearly atmospheric pressure exhaust gas from the turbine (8,470,000 Btu) passes to a waste heat boiler (WHB), which recovers energy, as steam, for use in the process. The steam from the boiler is used for heating within the processing plant and the condensate is returned to the boiler. Determine the efficiency of the boiler, the turbine, and the generator in this power plant. Also determine the overall plant thermal efficiency and the cogeneration plant net heat rate (British thermal unit per kilowatt-hour). Finally, draw the Sankey diagram with energy flows normalized to 100 input units (British thermal unit).

Table 1.2 Efficiencies for Different Energy System Configurations

Operation	Coal Utility Plant	Topping Cycle	Gas Turbine
η_{Boiler}, boiler efficiency	$\frac{10,194,000}{12,720,000} = 80.1\%$	$\frac{10,176,000}{12,720,000} = 80.0\%$	See discussion* $\frac{6,070,000}{8,470,000} = 71.7\%$
η_{Turbine}, turbine efficiency	$\frac{5,261,000}{11,194,000} = 47.0\%$	See discussion following this table*	See discussion following this table*
$\eta_{\text{Generator}}$, generator efficiency	$\frac{5,209,000}{5,261,000} = 99.0\%$	$\frac{1,720,000}{1,737,000} = 99.0\%$	$\frac{4,208,000}{4,250,000} = 99.0\%$
Overall plant thermal efficiency or plant net thermal efficiency	$\frac{4,844,000}{12,720,000} = 38.08\%$	$\frac{10,159,000}{12,720,000} = 79.86\%$	$\frac{10,278,000}{12,720,000} = 80.8\%$
Plant net heat rate	$\frac{12,720,000 \text{ Btu}}{1420 \text{ kW-h}} = 8957.7 \frac{\text{Btu}}{\text{kW-h}}$	See discussion* $\frac{12,720,000 \text{ Btu}}{504 \text{ kW-h}} = 25,238 \frac{\text{Btu}}{\text{kW-h}}$	See discussion* $\frac{12,720,000 \text{ Btu}}{1233 \text{ kW-h}} = 10,316 \frac{\text{Btu}}{\text{kW-h}}$
Incremental heat rate		$\frac{4,281,000 \text{ Btu}}{504 \text{ kW-h}} = 8494 \frac{\text{Btu}}{\text{kW-h}}$	$\frac{6,650,000 \text{ Btu}}{1233 \text{ kW-h}} = 5393.3 \frac{\text{Btu}}{\text{kW-h}}$

*Calculation of these quantities requires additional discussion, which is provided in the text below.

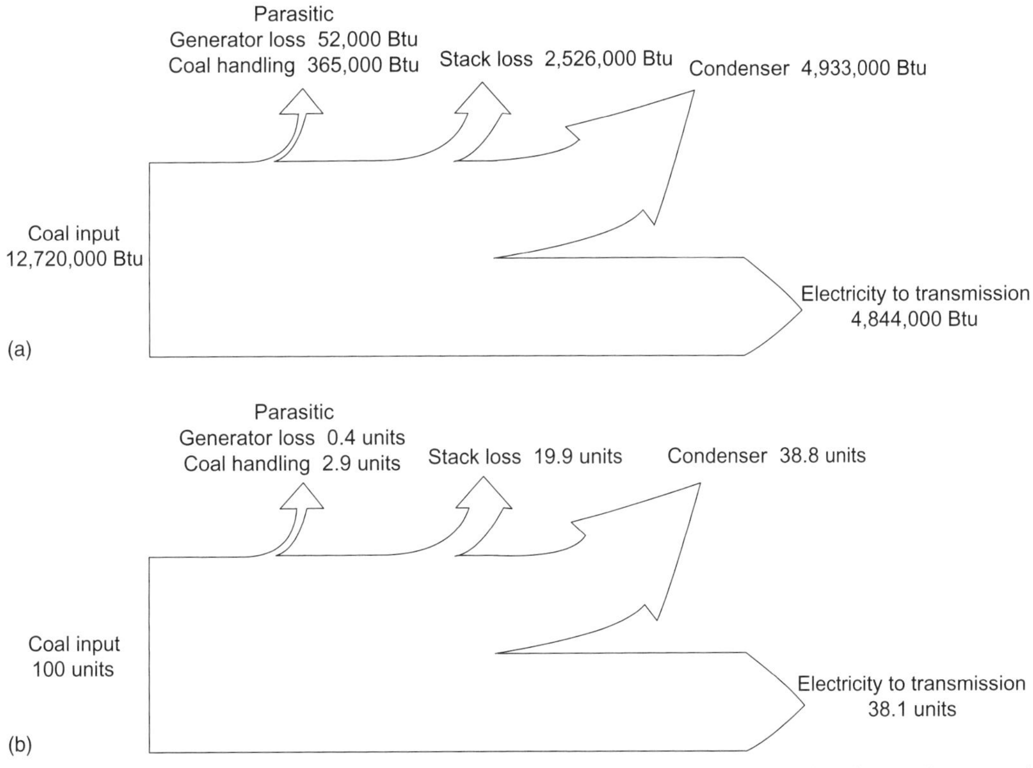

Figure 1.6 (a) Sankey diagram for conventional electricity generation from a coal-fired utility. (b) Sankey diagram for conventional electricity generation from a coal-fired utility with input normalized to 100 units.

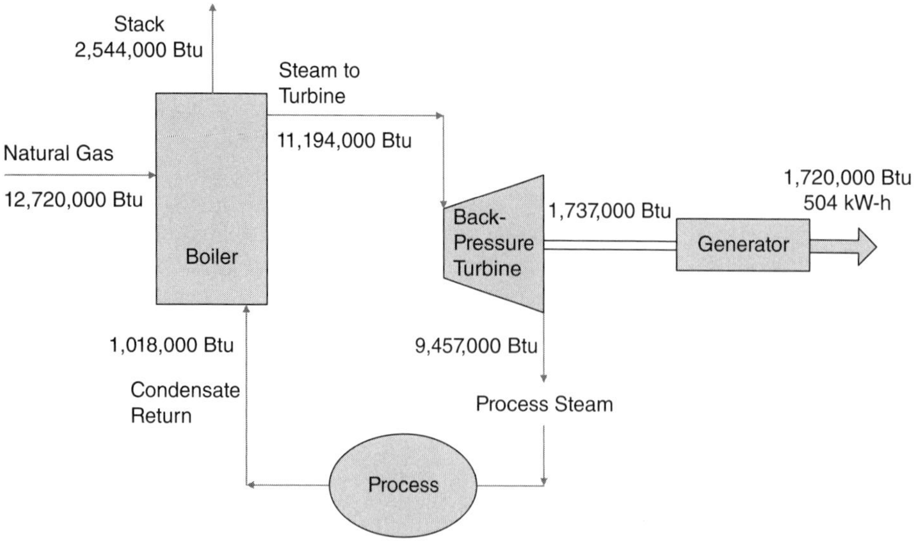

Figure 1.7 Industrial topping cycle cogeneration system.

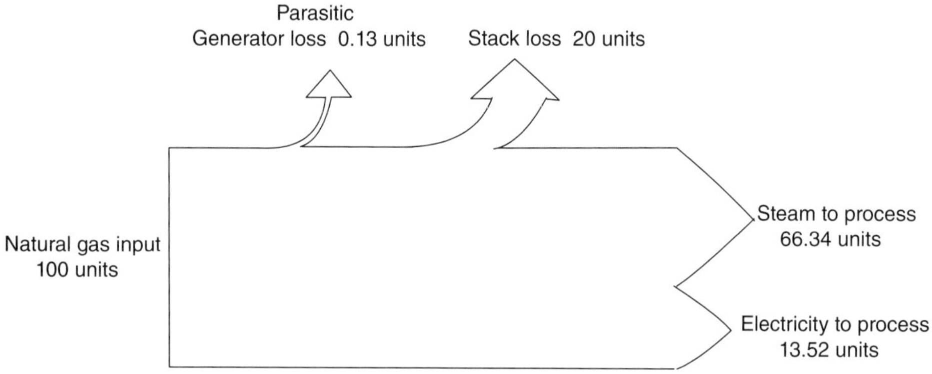

Figure 1.8 Sankey diagram for topping cycle cogeneration facility.

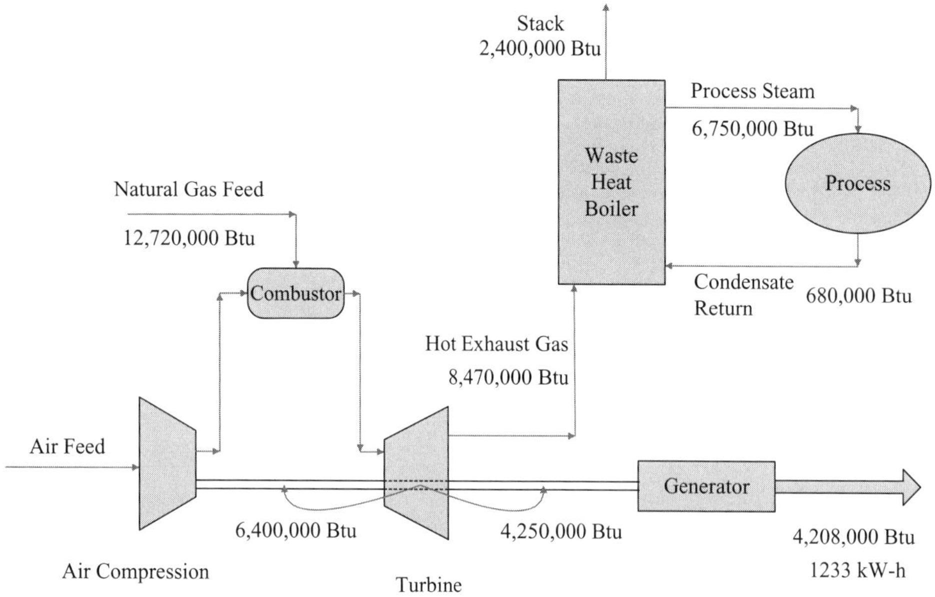

Figure 1.9 Gas turbine cogeneration system.

SOLUTION The efficiencies for the gas turbine-based cogeneration system are provided in Table 1.2. For the overall thermal efficiency, 12,720,000 Btu of natural gas is supplied to the system; 4,208,000 Btu is available for transmission as electricity (to the process); and 6,070,000 Btu is used as heat in the process. The Sankey diagram for the gas turbine system is shown in Figure 1.10. ■

From Table 1.2, the use of the first law efficiency is reasonable for boiler calculations where the fuel to the boiler is fired and the combustion gas is immediately used to raise steam as in Examples 1.1 and 1.2. The calculated efficiency of the boiler in Example 1.3 (the waste heat boiler) can be misleading—we must appreciate that energy (and temperature) has been extracted from the combustion gas prior to the waste heat boiler, which reduces the apparent first law efficiency. In comparison, when additional fuel is added to the hot exhaust gas entering the waste heat boiler, in order to raise additional steam for the process, the efficiency of this "supplemental firing" approaches 99%.

The turbine efficiency for the coal plant is based on all available energy entering the turbine and on the complete extraction of all useful energy to electricity. For both the topping cycle and the cogeneration system, part of the available energy to the turbine is used to produce electricity, but part of the energy remains, which is used as heat within the process. Here, for turbine calculations, efficiencies based on the second law of thermodynamics (Eq. (1.2)) will be developed.

The overall plant thermal efficiency also requires discussion. The thermal efficiency is

$$\text{Thermal efficiency} = \frac{\begin{pmatrix}\text{Usable electrical energy output}\\ + \text{Heat energy used by the process}\end{pmatrix}, \text{Btu}}{\text{Energy supplied to the system, Btu}}. \quad (1.4)$$

The thermal cycle efficiencies in Table 1.2 range from 38.1% for the simple cycle (power only) coal plant to 80.8% for the process cogeneration facility (combined power and process heat use). The underlying justification for much of the energy optimization in this chapter is the efficiency gains made possible when moving from utility plant simple cycles to combined cycle processing facilities. But to be fair, we do need to appreciate that in these efficiency calculations, we are treating electrical energy the same as thermal energy. Clearly, electrical energy is more valuable than thermal energy. Electrical energy can be transported long distances, and low-temperature, low-pressure thermal energy is often of little value. For cogeneration applications to be economical, there must be a home (or need) for all the generated thermal energy.

Plant net heat rates are often used to compare utility plants to process cogeneration facilities. This is especially true when inefficient utility plants with heat rates >10,000 Btu/kW-h are compared to process cogeneration facilities. For cogeneration facilities, a better maker is an incremental heat rate (also shown in Table 1.2), which is defined as

$$\text{Incremental heat rate} = \frac{\begin{pmatrix}\text{Energy supplied to the system}\\ - \text{Heat energy used by the process}\end{pmatrix}, \text{Btu}}{\begin{matrix}\text{Usable electrical energy output from the}\\ \text{cogeneration system, kW-h}\end{matrix}}. \quad (1.5)$$

For example, for the cogeneration system,

$$\text{Incremental heat rate} = \frac{(12{,}720{,}000 - (6{,}750{,}000 - 680{,}000))}{1233} = \frac{6{,}650{,}000}{1233} = 5393.3\frac{\text{Btu}}{\text{kW-h}}.$$

What is beneficial from these examples is to observe from the Sankey diagrams that different cogeneration configurations can produce different amounts of steam and electricity for use in a process. Also, cogeneration systems show a higher thermal efficiency when compared with utility plants, as the steam generated in cogeneration systems is condensed in the process. In utility plants, low-temperature, low-pressure steam is condensed by cooling water. Plant net heat rate values from Table 1.2 can facilitate the analysis of self-generated versus purchased power (electricity) as we will see in the next examples.

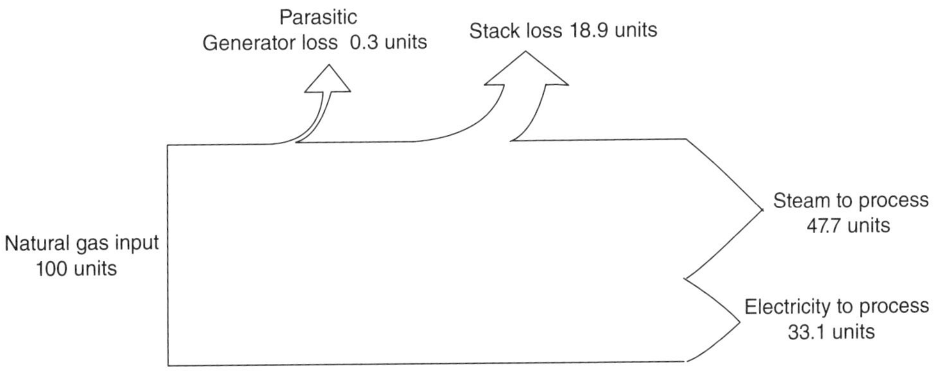

Figure 1.10 Sankey diagram for gas turbine cogeneration system.

1.4 THE COST OF SELF-GENERATED VERSUS PURCHASED ELECTRICITY

Overall process efficiency can often be improved by self-generating electricity and recovering combustion energy for use as heat in the process—cogeneration. Understanding the costs of self-generated electricity and utility company supplied electricity is important in order to accurately access cogeneration opportunities. Based on the heat rate determined in Example 1.1, we can determine the break-even cost for the utility company to supply electricity. The efficiencies in Example 1.1 can be considered relatively independent of the size of the utility plant.

EXAMPLE 1.4 ***The Break-Even Cost of Purchased Electricity***

Using our results from Example 1.1, we can discuss the cost of electricity purchased from the utility company. For example, if coal is \$5 per MMBtu, the fuel cost to generate electricity would be 8957.7 Btu/kW-h × \$5 per MMBtu = 4.48¢ per kilowatt-hour.

Also from Example 1.1, if the fuel was switched from coal to natural gas, we would anticipate a lower plant net heat rate as the parasitic electricity load for coal handling would be eliminated. With natural gas fuel, the expected utility plant heat rate would be

$$\frac{12{,}720{,}000 \text{ Btu}}{1527 \text{ kW-h}} = 8330 \frac{\text{Btu}}{\text{kW-h}}.$$

Using this heat rate and with \$8 natural gas (\$8 per MMBtu), the cost for the utility company to generate electricity is 6.66¢ per kilowatt-hour.

In addition to fuel costs, there would be an expected operation and maintenance (O&M) cost at roughly 0.5¢ per kilowatt-hour and a 0.2¢ per kilowatt-hour transmission cost. So, for a 8957.7 heat rate coal plant, a "break-even" cost for electricity generation would be ~5.2¢ per kilowatt-hour and for a 8330 heat rate natural gas plant ~7.4¢ per kilowatt-hour. A coal plant will have additional costs associated with environmental cleanup compared to a natural gas-fired plant. For example, from Figure 1.5, a 20-MW coal-fired utility plant must address the release of about 1275 lb/h of sulfur oxides to the atmosphere. In fact, for a new construction coal plant, the costs associated with meeting environmental regulations can easily wipe out any savings when compared to a natural gas-fired plant. The effect of potential long-term emission fees on existing coal-fired plants (including CO_2 capture and sequestration costs) is detailed in Chapter 16. ■

We can next discuss the cost of self-generated electricity.

EXAMPLE 1.5 ***The Cost of Self-Generated Electricity without Cogeneration Credit***

Consider the cost of generating electricity from the gas turbine-based cogeneration system of Example 1.3. With \$8 natural gas, the fuel cost to generate electricity is 10,316 Btu/kW-h × \$8 per MMBtu = 8.25¢ per kilowatt-hour. With O&M cost again at roughly 0.5¢ per kilowatt-hour, the cost for self-generation would be ~8.8¢ per kilowatt-hour, which is 68% more than the break-even cost of electricity from a coal-fired utility and 18% more than the break-even cost of electricity from a natural gas-fired utility. Of course, here we have not yet valued the generated steam used in the process. ■

A utility company will produce and often purchase electricity from a number of sources using fuels including coal, natural gas, hydro, and nuclear energy (see Figures 1.1 and 1.2). Coal accounts for ~50% of utility electricity generation here in the United States. Generally, based solely on electricity generation costs, a cogeneration facility cannot compete with electricity generation from a utility company. To keep the economics favorable for a cogeneration system, there must be a home for the generated steam. This "home" is typically the industrial process as indicated in Figures 1.7 and 1.9. In some cases, a portion of the generated steam can also be sent to a steam turbine to produce additional electricity in a combined cycle cogeneration plant. Most cogeneration facilities use natural gas-fired turbines as shown in Figsure 1.9. We must also appreciate that as natural gas fuel costs rise, electricity purchased from a utility company becomes increasingly favorable as their fuel mix (coal, natural gas, hydro, and nuclear energy) and longer-term fuel contracts help stabilize electricity cost.

Finally, in Examples 1.4 and 1.5, we did not include capital costs, which we will do in Example 1.6. The costs in Examples 1.4 and 1.5 represent "incremental costs," assuming the generation resource is already in place.

EXAMPLE 1.6 ***The Cost of Self-Generated Electricity with Cogeneration Credit***

In Example 1.5, cogeneration electricity costs were skewed because we did not value the generated steam. This example considers replacing an existing process utility system—a steam boiler and utility company electricity purchase, with a gas turbine cogeneration system. The calculations in the example can be found in the Excel file **Example 1.6.xls**; see Section 1.7 for details on installing Excel example files and developed software.

SOLUTION ***Current Plant***

A processing plant is currently purchasing 1000 kW from the local utility and is obtaining some 4800 lb/h of steam from a stand-alone boiler. The process uses ~1000 Btu from each pound of steam (British thermal unit per pound steam from the boiler—British thermal unit per pound in the condensate return). Within the processing plant, steam is thermally valued at \$9 per 1000 lb and purchased utility costs are 7¢ per kilowatt-hour plus a demand charge of \$7 per kilowatt-month. The plant operates 8000 h/year.

We first determine the hourly utility cost:

$$\text{Electricity cost:}\left(\frac{\$0.07}{\text{kW-h}}\right)(1000\ \text{kW}) = 70\frac{\$}{\text{h}},$$

Electricity demand charge:

$$\left(\frac{\$7}{\text{kW-month}}\right)(1000\ \text{kW})\left(\frac{12\ \text{months}}{\text{year}}\right)\left(\frac{1\ \text{year}}{8000\ \text{h}}\right) = 10.50\frac{\$}{\text{h}},$$

$$\text{Steam cost:}\left(\frac{\$9}{1000\ \text{lb}}\right)\left(\frac{4800\ \text{lb}}{\text{h}}\right) = 43.20\frac{\$}{\text{h}}.$$

The total utility cost is \$123.70 per hour or \$989,600 per year.

Cogeneration Facility

We want to evaluate replacing our boiler/purchased electricity with a gas turbine-based cogeneration facility with performance values based on Example 1.3. A gas turbine cogeneration system (turbine and heat recovery boiler) will have a total installed cost of \$1100 per kilowatt, and natural gas costs are \$8 per 10^6 Btu:

$$\text{Installed cost:}\left(\frac{\$1100}{\text{kW}}\right)(1000\ \text{kW})\left(\frac{0.16\ \text{CRF}}{\text{year}}\right)\left(\frac{1\ \text{year}}{8000\ \text{h}}\right)$$

$$= 22.00\frac{\$}{\text{h}}.$$

In the installed cost calculation, we are using a capital recovery factor (CRF) of 0.16, which converts the total installed cost (including tax and depreciation considerations) to a uniform annual cost. The CRF is developed in Chapters 2 and 16:

$$\text{Natural gas cost:}\left(\frac{10{,}316\ \text{Btu}}{\text{kW-h}}\right)(1000\ \text{kW})\left(\frac{\$8}{10^6\ \text{Btu}}\right)$$

$$= 82.53\frac{\$}{\text{h}}.$$

We must also check if the cogeneration system will supply enough usable heat in the waste heat boiler to meet our steam needs:

$$\text{Usable heat:}\left(\frac{10{,}316\ \text{Btu}}{\text{kW-h}}\right)(1000\ \text{kW})\left(\frac{47.7\ \text{steam units}}{100\ \text{units in}}\right)$$

$$= 4{,}920{,}732\frac{\text{Btu}}{\text{h}}.$$

The factor

$$\left(\frac{47.7\ \text{steam units}}{100\ \text{units in}}\right)$$

comes from the gas turbine Sankey diagram (Figure 1.10). And the steam generation would be

$$\text{Steam generation:}\left(4{,}920{,}732\frac{\text{Btu}}{\text{h}}\right)\left(\frac{1\text{-lb steam}}{1000\ \text{Btu}}\right)$$

$$= 4921\frac{\text{lb steam}}{\text{h}}.$$

The steam generated from the cogeneration system would exceed the 4800 lb/h currently needed, and some design changes in the cogeneration system may be considered:

O&M (operation and maintenance) costs:

$$\left(\frac{\$0.005}{\text{kW-h}}\right)(1000\ \text{kW}) = 5\frac{\$}{\text{h}}.$$

The cost of the cogeneration system is \$109.53 per hour. The profit if the cogeneration system was installed would be \$14.17 per hour or \$113,376 per year. This is profit as we have accounted for the installed cost of the cogeneration system. The profit from a cogeneration can be very sensitive to fuel costs and purchased electricity costs (see Problem 1.3). Do note that nearly 75% of the cogeneration profit can be traced to the utility demand charge. The demand charge is explained in Chapter 12. ■

1.5 THE COST OF FUEL AND FUEL HEATING VALUE

Before we finish our discussion of energy, the cost of energy, and energy efficiency, it is important to understand the energy content or the heating value of a fuel and how fuel is priced.

EXAMPLE 1.7 *A Special Caution: The Fuel Heating Value*

Let us start our discussion of the fuel heating value by developing the efficiency equation for the boiler shown in Figure 1.11. You may want to return to this example again after reading Chapter 5, where we develop energy balances, but please do read this example before moving to the other chapters.

We define

F_{fuel} = fuel flow rate, lb/h;
HV = fuel heating value, Btu/lb;
$F_{\text{boiler feedwater}}$ = boiler feedwater flow rate, lb/h;
F_{steam} = steam flow rate leaving the boiler, lb/h;
$\hat{h}_1$ = specific enthalpy of boiler feedwater at boiler inlet, Btu/lb; and
$\hat{h}_2$ = specific enthalpy of steam leaving boiler, Btu/lb.

Here, fuel to the boiler supplies heat to create a steam stream, which is at a flow rate F_{steam}. Condensate and makeup water are returned to the boiler at a flow rate, $F_{\text{boiler feedwater}}$. The F_{steam} may or may not equal $F_{\text{boiler feedwater}}$ depending on boiler blowdown. Blowdown is used to control solid buildup in the boiler. The boiler efficiency is

$$\eta_{\text{boiler}} = \frac{\text{Energy in steam exiting boiler} - \text{Energy in feedwater}}{\text{Energy in fuel}}$$

$$= \frac{F_{\text{steam}}\left(\hat{h}_2 - \hat{h}_1\right)}{F_{\text{fuel}}(HV)}, \tag{1.6}$$

where we have neglected any boiler blowdown, so $F_{\text{steam}} = F_{\text{boiler feedwater}}$. ■

The fuel heating value HV deserves special comment, and it will be discussed in more detail in later chapters. The fuel heating value can be defined as the amount of heat liberated when 1 lb of fuel is completely burned and the combustion products are returned to the initial fuel temperature. When the water formed in the combustion process is in vapor form, the fuel heating value is the lower heating value (LHV). When the water formed in the combustion process is condensed, the fuel heating value is the higher heating value (HHV). Values for fuel LHV and HHV are provided in Table 1.3.

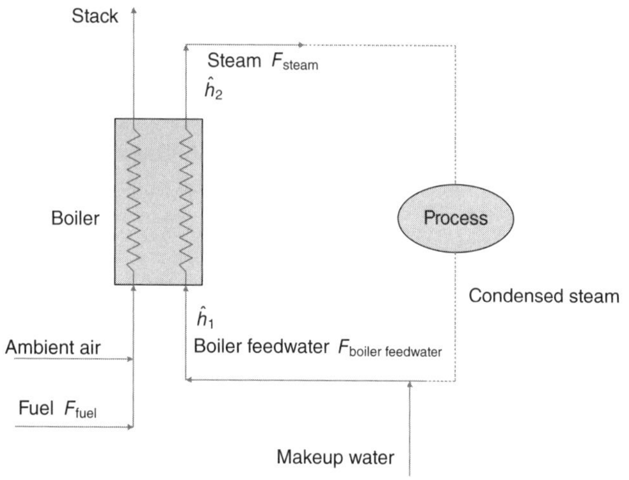

Figure 1.11 Boiler schematic for efficiency calculation.

If we now look at Equation (1.6), for a given set of steam and boiler feedwater conditions (flow rates and enthalpies), the fuel flow rate will be uniquely determined with a known boiler efficiency. However, the boiler efficiency in Equation (1.6) depends on whether the fuel *LHV* or *HHV* is used for *HV* in the equation.

There can be serious consequences when a different fuel heating value (*HHV* vs. *LHV*) is used in efficiency calculations and in subsequent costing calculations. For example, in Table 1.3, the difference in the *HHV* and *LHV* for methane fuel is 2359 Btu/lb. This problem of using *HHV* versus *LHV* is not always obvious. For example, turbine gas manufacturers generally report efficiency based on methane *LHV*. However, methane is priced and purchased on the open market based on its *HHV*.

Generally throughout this chapter, I prefer to use the fuel *LHV* in efficiency calculations—when combustion products exit the stack, water is in the vapor state. Here, then, fuel costs must also be on (or converted to) an *LHV* basis. One exception is optimal cogeneration dispatching, discussed in Chapter 12. In this chapter, all efficiencies and costs were determined based on the fuel *HHV* as this proved more convenient in the development of the economics-driven optimization model. To avoid confusion, power system calculations should explicitly note the use of *HHV* or *LHV* values.

1.6 TEXT ORGANIZATION

In this chapter, we want to provide a platform to explore the connection between the processing side and the utility side of an industrial plant. The processing side is often designed using commercial simulators including Aspen PLUS by Aspen Technologies, HYSYS (as Aspen HYSYS by Aspen Technologies or UniSim Design by Honeywell); and PRO/II by Simulation Sciences Inc. A key feature of these programs is the inclusion of rigorous thermodynamics, which allows the solution of unit operation material and energy balances and quantification of chemical mixture separation in distillation columns and other staged devices.

There are also commercial programs including GateCycle™ (from General Electric) and IPSEpro (from

Table 1.3 Fuel Lower Heating Value (*LHV*) and Higher Heating Value (*HHV*) at 77°F, 1 atm (Bathie, 1996)

Compound	State	*LHV* (Btu/lb)	*LHV* (kJ/kg)	*HHV* (Btu/lb)	*HHV* (kJ/kg)	*HHV–LHV* (Btu/lb)	*HHV–LHV* (kJ/kg)
Methane	Gas	21,501	50,012	23,860	55,499	2359	5487
Ethane	Gas	19,141	44,521	20,911	48,638	1770	4117
Propane	Liquid	19,927	46,351	21,644	50,343	1717	3992
n-Butane	Liquid	19,493	45,342	21,121	49,128	1628	3786

SimTech Simulation Technology) that allow utility system calculations. These programs focus on gas turbine system and steam turbine system design and off-design performance calculations. For off-design turbine calculations, manufacturer-provided performance curves are often utilized.

However, these process design programs and utility design programs do not generally share information. Clearly, an important problem is matching the design energy needs of the process to the best utility system design. Even more challenging is matching the changing energy needs of the actual operational process to the off-design performance of the plant utility system. This book addresses these important issues and hopefully lays the foundation for improved and optimal utility management.

Here we examine how computer-based solutions to process material and energy balances are obtained, allowing the user to construct a simplified Excel-based process simulation for energy tracking and energy forecasting. For the utility plant, we provide rigorous thermodynamics with all thermodynamic functions callable from Excel, Visual Basic for Applications (VBA), or C programs. This approach—an Excel-based process plant simulation and rigorous utility side calculations—allows the design of a utility system that matches the processing plant needs. Most important is that we show both design and off-design performance calculations for utility systems, which are necessary for the optimal operation of the process utility system.

Throughout this chapter, Excel is utilized in problem solutions. Straightforward problems can be solved directly in Excel. As problem complexity grows, Excel with VBA and Excel with VBA and C/C++ are utilized for problem solution. Here, Excel serves as a very convenient interface between the user and these low-level languages.

Chapter 2 provides an introduction to process economics. Here, Excel and Excel with VBA are used to solve economics problems. We show how variables and calculation results can be moved between the Excel sheet and VBA macro subroutines, VBA subroutines, and VBA functions. The concept of economic uncertainty is introduced with the use of Monte Carlo simulations.

Chapter 3 discusses computer-aided solutions of material balance problems using the sequential modular approach and Excel spreadsheets. We need to calculate material flow rates in an industrial process in order to determine energy requirements as well as for design and control purposes. Recycle loops in the process require iterative calculations, and here convergence strategies are explored. The concept of elementary material balance modules, with both natural and alternative specifications, is introduced. Alternative specifications often require the use of single-variable search techniques for problem solution. VBA provides a convenient platform for both recycle acceleration schemes and single-variable search strategies.

Chapter 4 presents computer-aided solutions of material balance problems using the simultaneous or equation-based approach and Excel spreadsheets. We provide a general discussion of the Gauss–Jordan matrix elimination method to solve linear equation sets. Linear material balance problems are then solved using the Gauss–Jordan method. The Newton–Raphson method is developed to allow solution of nonlinear material balance equation sets. These solution methods (Gauss–Jordan and Newton–Raphson) are provided both as VBA subroutine procedures or C callable dynamic link libraries (DLLs).

Appendix A (suggested reading after Chapter 4, Section 4.1) serves to introduce C coding, and a tutorial on C programming is provided in Appendix A—Tutorial. In Appendix A, we show the power in connecting Excel to low-level languages—specifically C code. Here, we construct DLLs of C/C++ code to allow the seamless transfer of variables, vectors, and matrices between Excel $\leftrightarrow$ VBA $\leftrightarrow$ C programs. In Appendix A, we focus on simple examples to show that the transfer of variables and vectors between Excel and C is straightforward. The transfer of matrices is more difficult and requires understanding of column-major and row-major matrix storages. To help understand the transfer of matrices, the Gauss–Jordan method (from Chapter 4, Section 4.1) is developed in C. In Appendix A—Tutorial, we repeat several examples from Chapter 2 to allow the comparison of VBA code and C code.

Chapter 5 incorporates energy balances into elementary process modules and investigates the impact these energy balances have on the sequential and equation-based solution approaches. Energy balances can necessitate the use of inner (energy balances) and outer loops (recycle loops) when the modular approach is used. Energy balances are shown to cause no additional computational difficulties when the equation-based approach (Newton–Raphson method) is used. The solution of reactor energy balances is particularized to plug flow reactors where both the Euler method and the Runge–Kutta fourth-order method are employed to size tubular (plug flow) reactors.

Chapter 6 introduces data reconciliation and gross error detection. Data from an actual plant are not "perfect"—for example, these data will not close material balances or energy balances. For data reconciliation problems with linear (material balances) or nonlinear (energy balances) constraints, the Solver optimization routine available in Excel allows direct solution. For data reconciliation problems with linear constraints (or linearized nonlinear constraints), we also develop Lagrange multipliers for problem solution. We introduce gross error detection (the global test method) and gross error identification (the measurement test method).

Chapter 7 begins our examination of cogeneration systems and system performance in both design and off-design operations. Solution of cogeneration system performance is explored using ideal gas physical properties. We also introduce our Excel callable sheet functions for steam properties. Callable steam properties allow the heat recovery steam generator (HRSG) performance to be determined under off-design operation including supplemental firing.

Chapter 8 explains the construction of the thermodynamics package (*TPSI+*) for cogeneration and combustion calculations. Equations of state for calculating needed thermodynamic properties are based on the work of Reynolds (1979) and developed within a C program. We extend Reynolds's work with the inclusion of refrigerants. Thermodynamic properties from accurate equations of state for pure components can be combined to allow prediction of thermodynamic properties for combustion mixtures. The C program is linked to Excel sheets as a DLL, and all source code for (*TPSI+*) is provided to allow the user to make modifications.

Chapter 9 examines cogeneration system performance in both design and off-design operations using real fluid properties; real fluid properties (*TPSI+*) are developed in Chapter 8. The examples and problems in Chapter 9 parallel those in Chapter 7 allowing comparison of ideal gas and real fluid calculations.

Chapter 10 explores the optimal design of cogeneration systems with known and fixed steam and electrical demands—the CGAM problem. First, the solution of the CGAM cogeneration design problem with ideal gas properties is examined. We next use our thermodynamics package (*TPSI+*) to solve the cogeneration design problem with real fluid properties. *TPSI+* is shown to produce results equivalent to commercial utility design codes such as GateCycle from General Electric Corporation. The use of *TPSI+* also allows incorporation of more realistic cogeneration system constraints in the design problem when compared to the original CGAM problem. Steam generation in the HRSG is explored for both the design and off-design (supplemental firing) cases. Heat conduction problems and performance analysis in the HRSG naturally lead to partial differential equations, and here, a solution using finite difference approximations is developed.

Chapter 11 applies data reconciliation and gross error detection techniques developed in Chapter 6 to data from an operational cogeneration system. Here we can take advantage of the provided thermodynamics package (*TPSI+*) to solve material and energy flows. We also show the difficulties in determining the exact location of a gross error in large energy systems.

Chapter 12 provides for multiperiod energy management and energy dispatching in a cogeneration system. Data from an operational gas turbine cogeneration system were collected for over a 1-year period and steady-state equipment efficiencies were determined. Forecasting energy costs is discussed. A mixed-integer linear programming (MILP) is developed to solve for multiperiod optimum energy management strategies. The Excel add-in What's Best is used to solve the MILP energy dispatch problem; What's Best has been supplied by LINDO Systems for use with this chapter. The model developed here has been successfully used to lower utility costs at an operational cogeneration facility.

Chapter 13 demonstrates how an energy integration analysis in a processing plant can be performed. Minimum hot and cold utility requirements can be determined. In existing plants, these values can be used as energy targets and compared to current usage. Improvements and changes in the existing heat exchanger network can then be evaluated. We provide a heat exchanger network synthesis program *THEN* (~3000 lines of code), which aids in the construction of an optimal heat recovery network. The heat exchanger network synthesis program is a legacy Fortran code, and we show how this code, or any executable code, can be incorporated into Excel with Excel serving as both pre- and postprocessors.

Chapter 14 extends the energy analysis of Chapter 13 to include the integration of the utility system, which supplies hot and cold utilities as well as electrical power. Here, the real fluid cogeneration system developed in Chapters 9 and 10 is modified to meet the energy requirements of a process. The thermodynamics program (*TPSI+*) is also used to develop a steam-based turbine system to meet process energy needs. In a large complex, individual processing plants typically share a common utility system. The potential for energy exchange throughout the total site is explored and the optimal design and synthesis of the site utility system is discussed.

Chapter 15 explains how emissions from heat and power generating systems can be predicted; this is especially important in nonattainment areas. Emissions based on simple stoichiometric and equilibrium calculations are introduced. Equilibrium calculations require the use of thermochemical data. Rigorous modeling of methane combustion in gas turbines is explored by modeling the turbine as a series of stirred tank and plug flow reactors. A reduced kinetics set based on GRI-Mech 3.0 is solved using the ordinary differential equation (ODE) solver CVODE (from LLNL). CVODE is made available as a callable routine from Excel.

In Chapter 16, we explore the impact of CO_2 capture on a conventional coal-fired utility plant. Our Excel-based thermodynamics functions (*TPSI+*) are utilized to predict the performance of a steam turbine-based utility plant operating in the design or off-design mode. Off-design operation is required as the capture process will extract steam from the utility plant for regeneration. We introduce levelized economics, which allows utility pricing; levelized economics is explored for comparison of energy systems. The chapter concludes with examples showing how utility plant performance and economics are altered if CO_2 capture is implemented.

In Chapter 17, we discuss levelized costing of alternative energy systems. We show the connection between levelized costs and the price we must pay for utilities if the alternative technology is implemented. There are many factors that go into determining levelized costs, and we detail the levelized cost determination process for an organic Rankine cycle. In the final example, we introduce the next generation of nuclear reactors.

In this book, all examples and chapter problems are fully worked and available at the Wiley Web site. All computer codes including the rigorous combustion thermodynamics programs and heat exchanger network synthesis are provided as open software. I have been fortunate to teach the material in this book in undergraduate classes and in areas where the students had difficulty. I have tried to supply additional explanation and discussion. I have tried to keep in mind that practicing engineers may be using this book. I appreciate the time demands on practicing engineering, and I have tried to keep the book as self-contained as possible.

1.7 GETTING STARTED

Solutions to all example problems and chapter problems as well as all developed software programs (including all source codes) are available at the Wiley Web site. The POEA folder from the Web site should be copied onto the C drive of your computer as C:\POEA; all provided materials should be in a single folder as shown in Figure 1.12.

Figure 1.12 shows that within the folder POEA, you should find "subfolders": Chapter 1–Chapter 17 holding

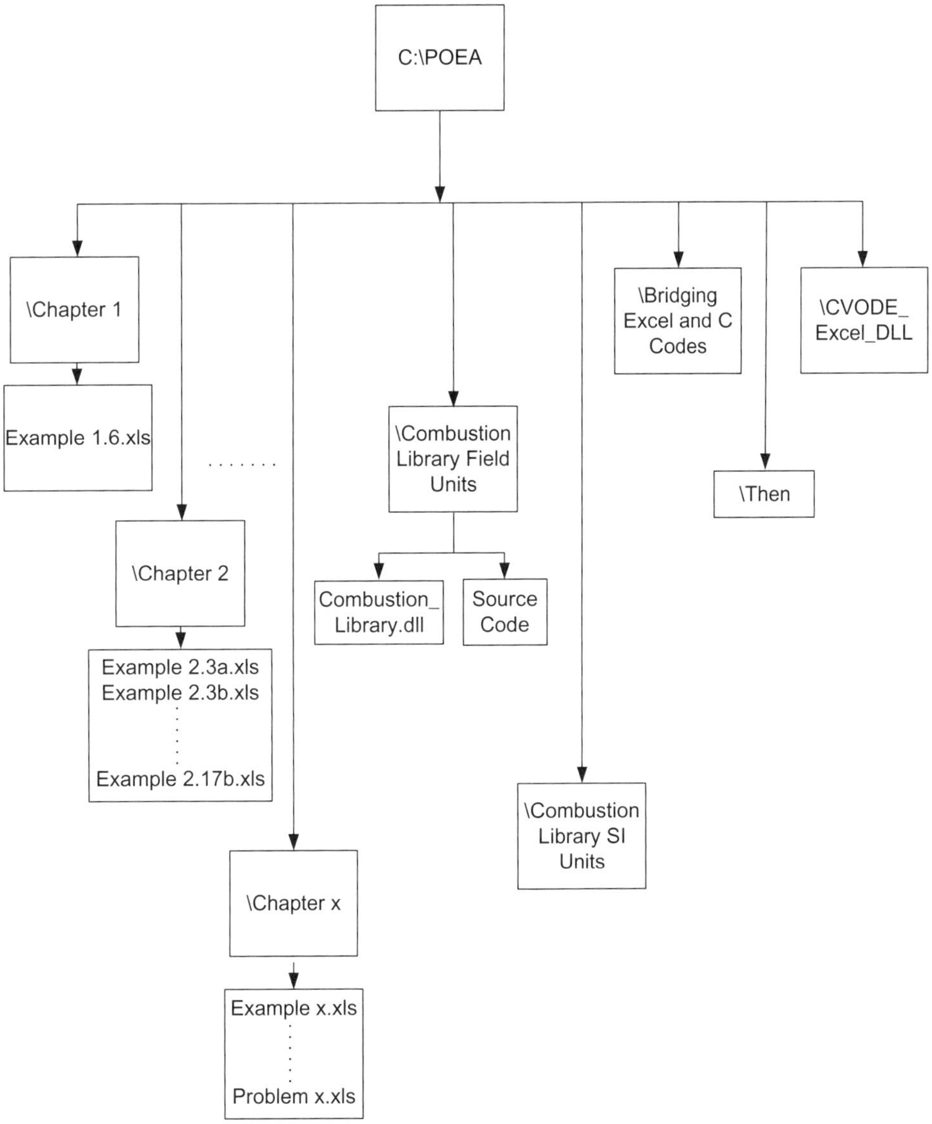

Figure 1.12 Required folder locations for examples, problems, and all provided codes.

solutions to all the chapter examples and some chapter problems. Within the folders Combustion Library Field Units and Combustion Library SI Units, you will find the source code for the thermodynamics program *TPSI+* as well as dynamic links to the source code. The source code for *THEN* Teaching Heat Exchanger Network analysis is provided in the subfolder Then. The executable version of *THEN* (Then.exe) is found in the folder Chapter 13; you should initially run Then.exe from the Excel files provided in Chapter 13. The subfolder CVODE_Excel_DLL holds an Excel callable version of the ODE solver CVODE available from LLNL. CVODE is part of the Sundials suite at LLNL, which also provides other advanced numerical methods; these programs (source code) are available at no cost for noncommercial use. Bridging Excel and C codes holds programs developed in the chapter including the Gauss–Jordan and Newton–Raphson methods. This folder also holds examples and a tutorial on how to connect Excel to C code and how to use C++ 2008 Express. Many of the example files and problem solutions utilize programs within various folders in POEA; in other words, examples and problems in one subfolder may include paths to programs in other subfolders. It is important that POEA be assembled as shown in Figure 1.12. Once you become familiar with paths between programs, it is easy to combine files and programs in any folder that you may want to create.

I used the acronym POEA to hold all work related to process optimization and energy analysis. As you work on examples and chapter problems, it may be best to keep them in a separate work folder within \POEA. In many cases, it will be convenient to copy an example problem and to use it as a starting point for the next example. For some of the more complicated examples and chapter problems, there are template Excel sheets provided in \POEA. These Excel template files can be saved to your work folder. In \POEA, you will also find template Excel sheets that allow connection (established path) to all provided software.

We developed all programs in Excel 2003 and C++ 2008 Express. All programs were also tested in Excel 2007 and Excel 2010. When using Excel 2007 or 2010, there can be a warning that some settings are no longer supported. This will not affect any calculations in the Excel sheets and you can simply press *continue*.

In this text, the standard add-ins provided with Excel—Goal Seek and Solver—are used in many single-variable and multivariable nonlinear optimization problems. Mixed-integer linear problems (Chapter 12—Optimal Power Dispatch in a Cogeneration Facility and Chapter 14—Utility System Superstructure) are solved using the Excel add-in What's Best from LINDO Systems. LINDO Systems has agreed to supply the needed version of What's Best at no cost—details for downloading this software can be found at the Wiley site.

When installing What's Best 10.0, the default path will be to Microsoft Office. Users will have different versions of Microsoft Office, so you will have to establish the path between our provided Excel files and your installed version of What's Best in order to link the WBA.xls file. This is accomplished when you first open the Excel file containing What's Best (this will only occur in files utilizing What's Best) → click on update → edit links → change source → here you will need to give the path to wba.xla; for example, in Excel 2003 C:\Program Files\Microsoft Office\Office 11\Library\Lindo WB\WBA.xla; in Excel 2007\Office 12\; and in Excel 2010\Office 14\ → click on check status → OK (the status should now appear as source is open).

1.8 CLOSING COMMENTS

In Chapter 1, we introduced the cost to produce electric power from utility-based coal or natural gas plants as well as natural gas-fired turbine cogeneration facilities. In later chapters, we will develop more complete economics as well as detailed design and off-design performance calculations for these energy systems. There are opportunities to improve efficiency by understanding and modeling these generation facilities.

The most common mistake in power output calculations and energy costing analyses is a lack of consideration of fuel *LHV* versus *HHV*. Project mistakes on the order of 20% can occur when *LHV* efficiency and power output values from vendors are mixed with fuel sales based on *HHV* (L.J. Braquet, pers. comm.).

REFERENCES

Bathie, W.W. 1996. *Fundamentals of Gas Turbines* (2nd edition). John Wiley and Sons, New York.

Dorf, R.C. 1978. *Energy, Resources, & Policy*. Addison-Wesley Publishing, Reading, MA.

Dorf, R.C. 2001. *Technology, Humans and Society: Toward a Sustainable World*. Academic Press, San Diego, CA.

EIA, Data in Table 1.1 and Figures 1.3 and 1.4 can be found at the EIA Web site http://www.eia.doe.gov/dnav/ng/hist/n3020us3M.htm. Natural gas costing data are in dollar per 1000 ft^3 converted to dollar per MMBtu (accessed July 2010).

EIA and LLNL, Energy flow data in Figures 1.1 and 1.2 can be found at the LLNL Web site https://publicaffairs.llnl.gov/news/energy/archive.html. In these two figures, we have added energy imports to the appropriate energy source and we have subtracted energy exports from the provided numbers to help maintain the overall energy consumption balance. (accessed July 2010).

Hobson, A. 2004. Energy flow diagrams for teaching physics concepts. *Phys. Teach.* 42: 113–117.

Priest, J. 1973. *Problems of our Physical Environment: Energy Transportation Pollution*. Addison-Wesley Publishing, Reading, MA.

Reynolds, W.C. 1979. *Thermodynamic Properties in SI: Graphs, Tables, and Computational Equations for Forty Substances*. Department of Mechanical Engineering, Stanford University, Stanford, CA.

PROBLEMS

1.1 ***Automobile Efficiency*** Energy transfer rates (power) in a typical automobile at normal highway speed are indicated in Figure P1.1. Determine the efficiency of the automobile with power values as shown in Figure P1.1.

1.2 ***Efficiency and Plant Net Heat Rate*** Determine the relation between the overall plant thermal efficiency as given by Equation (1.1) and the plant net heat rate as given by Equation (1.3).

1.3 ***The Effect of Natural Gas Pricing on Cogeneration Profitability*** In Example 1.6, we determined the profit if our

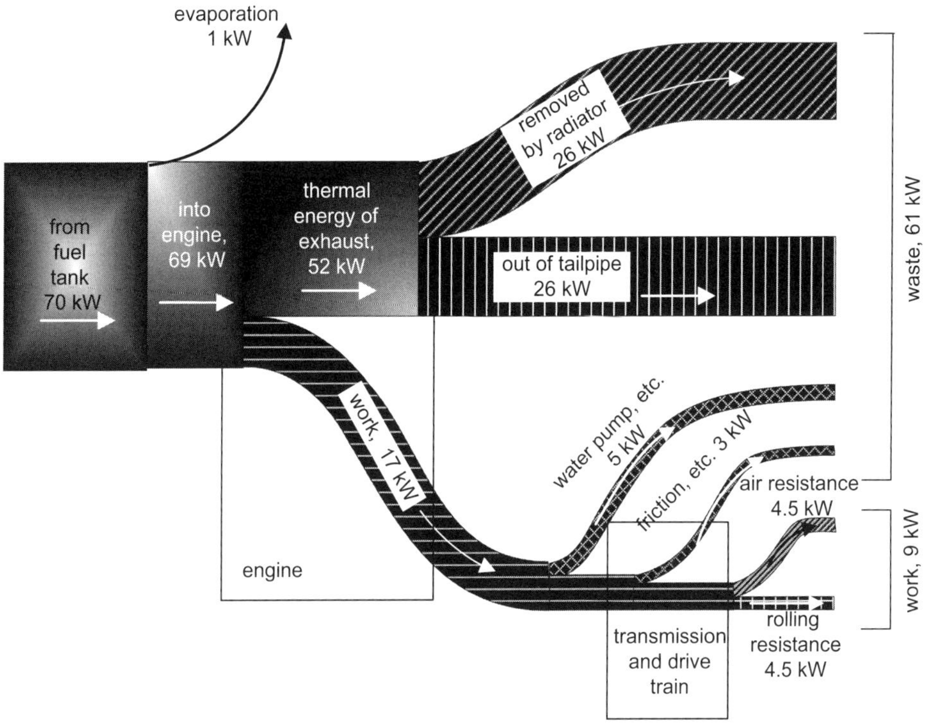

Figure P1.1 Automobile energy transfer rates at steady state and at normal highway speed (adapted from Hobson, 2004).

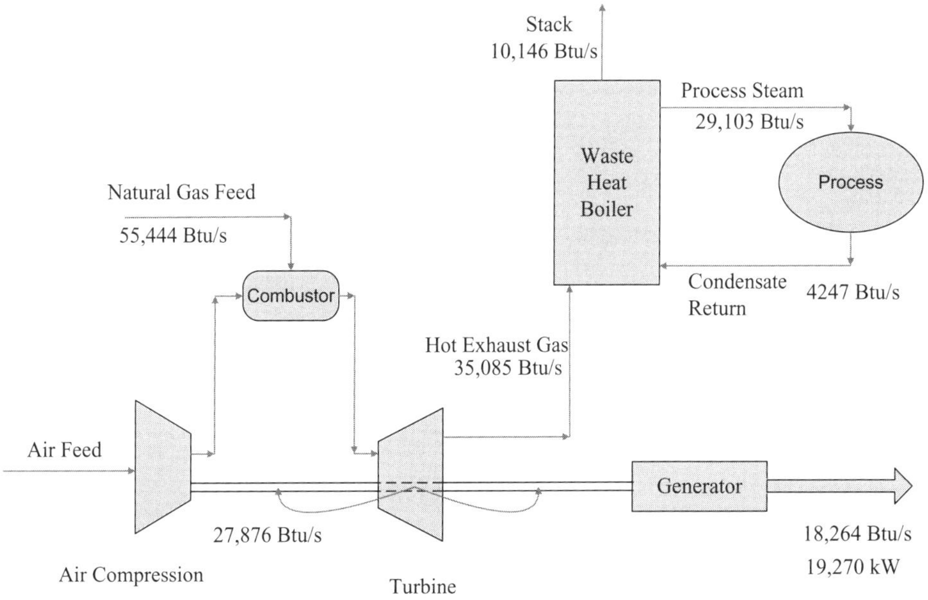

Figure P1.4 Gas turbine cogeneration system.

existing system, consisting of a stand-alone boiler for steam generation and electricity purchase, was replaced with a cogeneration system. As noted in the discussion of Example 1.6, cogeneration profitability is affected by natural gas costs. Determine a price for natural gas when the cogeneration system is no longer profitable.

Solution: The solution can be found using the Excel file **Example 1.6.xls**. If you are not familiar with Excel, you may want to wait until after reading Chapter 2 before using the Excel file. Otherwise, simply vary the natural gas cost (or use Goal Seek) until the profit = 0; at \$9.37 per 10^6 Btu, cogeneration is no longer profitable assuming all other costs remain constant.

We can consider the costs as provided in Example 1.6 to be long-term or levelized costs. Levelized utility costs are explained in Chapter 16. Cogeneration profitability is also strongly affected by the net heat rate, and cogeneration heat rate quickly erodes in part load operation as detailed in Chapters 7 and 9.

1.4 *Gas Turbine Cogeneration Plant Energy Efficiency* A gas turbine-based cogeneration facility is shown in Figure P1.4. The values provided in Figure P1.4 are the energy input rate to each operation. You will note that energy balances do not close "perfectly." There are thermal losses in the process that are not directly shown in the figure. Determine the cogeneration plant net heat rate as British thermal unit per kilowatt-hour and also develop the Sankey energy flow diagram.

1.5 *Coal Energy Efficiency* Determine how much of the energy originally available in coal is ultimately delivered as electricity to the end user. Account for the energy requirements at each stage including raw coal → extraction → processing → transportation → conversion to electricity (see Example 1.1) → electricity grid transmission → end use. Here, you will need to use outside references; see, for example, Dorf (1978, 2001) or other sources.

1.6 ***Who Owns Possible Environmental Credits from a Cogeneration System?*** In Example 1.6, we developed the cost savings that may be possible if a cogeneration system is used to replace an existing stand-alone boiler with electricity purchase from a utility plant. Here we want to explore the possibility of emission reduction, specifically carbon dioxide reduction, if this cogeneration system is installed. An interesting issue with a cogeneration system may be: Who should receive environmental credit for any emission reduction?

Assume the cogeneration system in Example 1.6 is installed, thereby replacing the existing natural gas-fired stand-alone boiler and electricity purchase from a coal-fired utility plant (as in Figure 1.5). Allow that the combustion of 1 MMBtu (or 1 mSCF) of natural gas releases about 120 lb of carbon dioxide. Burning 1000 lb of coal releases about 2400 lb of carbon dioxide. Determine the total carbon dioxide emissions at the processing site (within the fence line) and the utility plant (within the fence line) both before and after cogeneration installation.

Chapter 2

Engineering Economics with VBA Procedures

2.1 INTRODUCTION TO ENGINEERING ECONOMICS

Engineering economics provides for the proper accounting of cash flows (both positive and negative) over the life of a project. An economic analysis can be applied to a single project or to multiple independent projects; generally, for multiple independent projects, a yes or no decision is made on each project. The analysis can also be applied to mutually exclusive projects to help determine the best option.

Engineering economics can be broadly classified as either plant process economics or plant design economics. Plant design economics often involves accounting for, or estimating, a large number of factors including equipment capital and installation costs, profits, as well as taxes and depreciation. For plant operation economics, a less detailed approach is often used, where, for example, costs such as taxes may be ignored.

There are many available methods for economic analyses, but when properly applied, each method should produce the same result. Here we will introduce several methods that are commonly used in engineering economics calculations.

EXAMPLE 2.1 ***Mutually Exclusive Purchases***

A feedwater pump in our process is leaking and it must be replaced. Two pumps are available for purchase. Pump A costs \$1000 and has an anticipated lifetime of 2 years, and pump B costs \$1500 with an expected lifetime of 3 years. Which pump should we install?

To solve this problem, we must first account for the pumps having different lifetimes. Here, we could pick the least common multiple number of years—6 years. Over this 6-year period, we could buy pump A now, then buy a second pump A at the end of year 2 and a third pump A at the end of year 4. In contrast, we could buy a pump B now and buy a second pump B at the end of year 3. As shown in Figure 2.1, both these options carry us out to year 6, where both pumps will need replacement. ■

The next question may be: Do we expect the price of these pumps to increase with time? Notice that future pumps are not purchased at the same time. Are the installation costs the same for both pumps? Pump A requires three installations, while pump B requires two installations to move us out 6 years. Are the costs of operation, for example, electrical efficiency, the same for both pumps? What are the disposal costs associated with each pump? These costs are part of the pump life cycle (from cradle to grave). The tools necessary to make the best selection are developed in the next section.

2.2 THE TIME VALUE OF MONEY: PRESENT VALUE (*PV*) AND FUTURE VALUE (*FV*)

The time value of money is the basic principle that \$100 today is more valuable than \$100 at any time in the future. We are familiar with compound interest. Here, for example, if \$100 is invested at 10% per year, the gain on the investment will be (0.10)(100) = \$10, and funds returned after 1 year will be \$110.

Here, the \$100 represents the present value (PV) of the investment. The \$110 is the future value (FV) and the interest rate per year or discount rate (i) is 10%. The equation relating present value to future value for year 1 is $FV = PV + (PV)(i) = PV(1 + i)$.

If the funds were allowed to remain for a second year, the gain on the investment would be (0.10)(110) = \$11 and the funds returned will be \$121. Here, the equation relating PV to FV would be $FV = PV + (PV)(i) + (PV + (PV)(i))(i) = PV(1 + i)^2$, where 2 is the number of compounding periods (here 2 years).

Modeling, Analysis and Optimization of Process and Energy Systems, First Edition. F. Carl Knopf.

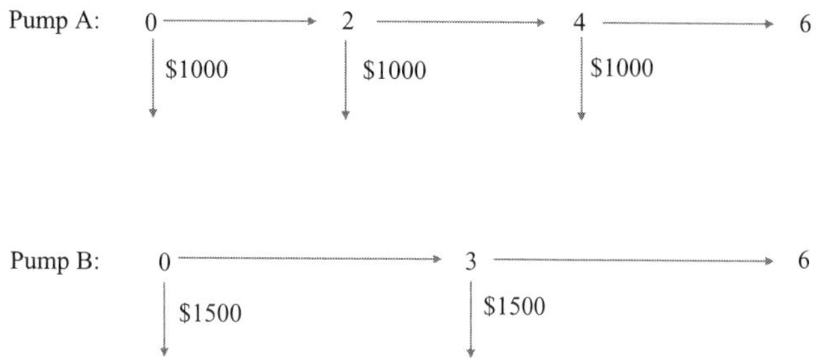

Figure 2.1 Cash flow diagram for pumps A and B (Example 2.1).

In general,

$$FV = PV(1+i)^n, \tag{2.1}$$

where n is the number of compounding periods and i is the interest rate per period.

Economic analysis requires us to accept that different cash flows in time can be equivalent, at a given discount rate. For example, at 10% annual compound interest, a *PV* of $100 is exactly equivalent to an *FV* of $110 in 1 year, or $121 in 2 years. This equivalence allows us to account for the time value of money. In other words, the selected interest rate allows us to bring cash flows occurring at different times to a common point in time, and then an economic evaluation can be performed.

EXAMPLE 2.2 *Effect of the Compounding Period*

What is the effect of a shorter compounding period? Here, allow that the nominal 1-year interest rate is 10%, but the compounding period is monthly. If the same $100 is invested, what is the future value after 1 year?

A nominal interest rate of 10% per year would provide a monthly interest rate of 10%/12 = 0.8333%. The $100 investment would gain 100 (0.008333) = $0.8333 at the end of the first month. The total gain at the end of the second month would be $100.8333 (0.008333) = $0.8402.

Using Equation (2.1),

$$FV = PV(1+i)^n = \$100\left(1+\frac{0.10}{12}\right)^{12} = \$110.47.$$

■

We can also define an effective annual interest rate (i_{eff}), which is the equivalent annual interest rate that results from compounding with time periods shorter than 1 year:

$$FV = PV(1+i_{\text{eff}})^{n=1} \tag{2.2}$$

$$\$110.47 = \$100(1+i_{\text{eff}})^1 \rightarrow \text{solving gives } i_{\text{eff}} = 10.47\%.$$

In other words, a nominal interest rate of 10% per year compounded monthly yields an effective annual interest rate of 10.47%. For most engineering economic problems, the annual compound interest is generally used. For a discussion of simple and continuous interests, see Peters et al. (2003) or Eschenbach (2003).

We can now revisit Example 2.1. Assuming an interest or discount rate of 10% per year, the present value for each pump can be determine by rearranging Equation (2.1) and using Figure 2.1:

$$PV = \frac{FV}{(1+i)^n} \tag{2.3}$$

$$PV_{\text{Pump A}} = \$1000 + \frac{\$1000}{(1+0.1)^2} + \frac{\$1000}{(1+0.1)^4} = \$2509.46$$

$$PV_{\text{Pump B}} = \$1500 + \frac{\$1500}{(1+0.1)^3} = \$2626.97.$$

With this analysis, we would recommend the purchase of pump A; it has a lower present value cost. We have indicated pump cash flows (costs) as positive quantities. We could have used negative cash flows for costs with the results $PV_{\text{Pump A}} = -\$2509.46$ and the $PV_{\text{Pump B}} = -\$2626.97$; again, pump A has the lower *PV* cost. The use of positive or negative terms in cash flow diagrams is the choice of the user. Generally, if all the cash flows are costs, these cash flows are taken as positive quantities. If profits as well as costs are involved, then profits are taken as positive and costs as negative quantities. We have not yet considered installation costs, operational costs, or disposal costs, which may change our recommendation.

At this point, we want to begin using Excel and Visual Basic for Applications (VBA) to solve problems.

EXAMPLE 2.3 *Present Value Analysis Using Excel Sheet and VBA Macro Subroutine Procedure*

Solve Example 2.1, the two-pump selection problem with i = 10% per year, first by using Excel and then a VBA macro subroutine procedure.

SOLUTION ***Solution Excel Sheet***

The Excel sheet solution is shown in **Example 2.3a.xls**. There are two major drawbacks to Excel, both of which can be overcome to some extent. Creating solutions by simply manipulating cells on the sheet makes it difficult for anyone to understand what was actually done. Here we want to emphasize the need to use named cells or named variables, which improves sheet logic. A second problem is that equations are hidden. This problem can be overcome by using the reveal equations option—use keys crtl ~, which is especially effective if named cells have been used.

The Excel sheet and the equations (crtl ~) are shown in Figures 2.2 and 2.3. To name a cell, highlight the cell where the variable value is to be stored. Then, in the Name Box (upper left hand side of the Excel sheet), type in the variable name followed by enter ↵ or the return key. The user then creates equations in the normal fashion, except variable names as opposed to cell locations will appear in the equations. The cells should be named before the equations are entered. ■

> ### *Named Cells*
>
> Cells without names make it extremely difficult for the user to understand your spreadsheet. Even the developer, within a very short period of time, can become confused as to the intent of the cell when cell names are not used.
>
> The naming of cells can be somewhat time-consuming, and Excel is unforgiving. Do be aware that while the contents of named cells can be deleted, this will not delete the name connected to the cell. Once a name has been entered for a cell, name changes are possible only by deleting the existing name; this is done via Insert → Name → Define Names in the workbook. Highlight the names no longer needed and delete.
>
> It is possible to name cells after the spreadsheet has been created, exactly as described above. This will not initially affect any equations; however, the sequence Insert → Name → Apply → and selecting names will allow cell names to appear in equations.

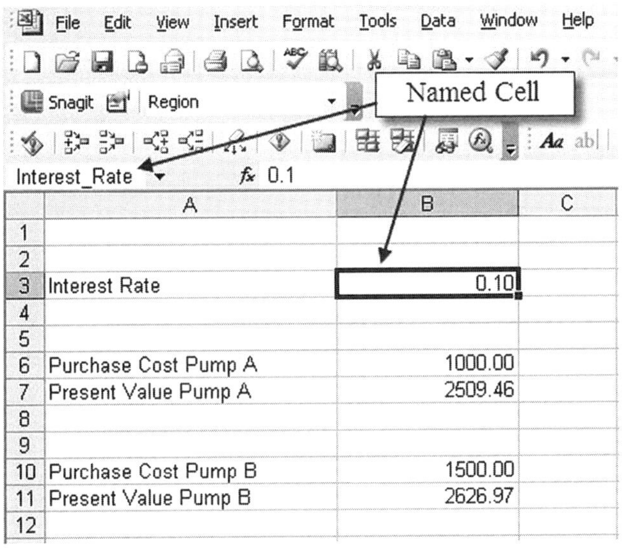

Figure 2.2 Excel sheet showing problem solution with named cells (Example 2.3).

***SOLUTION** Excel Sheet and VBA Macro Subroutine Procedure*

At this point, we are ready to obtain the solution to Example 2.1 by using VBA within the Excel sheet. Here let us read the data: cost of pump A, cost of pump B, and interest rate from the sheet, and calculate the present value of each pump using VBA. We can then write these present value results on the sheet next to the Excel solution.

The Excel solution using VBA is shown in **Example 2.3b.xls**. If you are independently working these examples, the best way to begin is by copying your current solution, **Example 2.3a.xls**, and renaming it **Example 2.3b.xls**. Then, from the Excel toolbar, click on Tools → Macro → Visual Basic Editor and a new screen will appear. On the new toolbar, click on Insert → Module and the screen should appear similar to Figure 2.4. At this point, we are ready to add the needed VBA code, which is shown in Figure 2.5. Note that the line numbers on the right-hand side (RHS) (`line 1, etc.`) are not to be included in the code; these are simply for discussion purposes.

`Option Explicit` is good coding practice requiring that all variables be defined; variables must be defined by type (e.g., `As Double` or `As Single`). `Public Sub Calc_PV()` creates a macro Sub procedure (Sub is short for subroutine). A macro can be called from an Excel sheet; however, parameters cannot be passed in the argument list (the argument list would be variables found within the parentheses of `Public Sub Calc_PV()`)—more on this later.

`Public` is a key word allowing the procedure to be accessed by all other procedures in the module and by all other procedures in all other modules in all projects—more on this later. The key word `Private` limits access only to other procedures within the same module.

`Lines 4-8` define variables as `Double` type (real, double precision). Here we use "variables" as a general term to include variables, data, vectors, matrices, and constants. When important, precise terms will be used.

Variable values are read from the sheet in `lines 10-12`. For example, `line 10, Cost_Pump_A = Sheet1.Cells(6, 2)` reads the cost of pump A from the sheet in row 6, column 2. In `lines 14` and `15,` PV_Pump_A and PV_Pump_B are calculated. Notice that these equations are exactly the same equations we used in the Excel sheet. The

	A	B
1		
2		
3	Interest Rate	0.1
4		
5		
6	Purchase Cost Pump A	1000
7	Present Value Pump A	=Cost_Pump_A + Cost_Pump_A/(1+Interest_Rate)^2 + Cost_Pump_A/(1+Interest_Rate)^4
8		
9		
10	Purchase Cost Pump B	1500
11	Present Value Pump B	=Cost_Pump_B + Cost_Pump_B/(1+Interest_Rate)^3
12		

Figure 2.3 Using (crtl ~) reveals equations in the Excel sheet. Note that the use of named cells makes the equations easier to interpret.

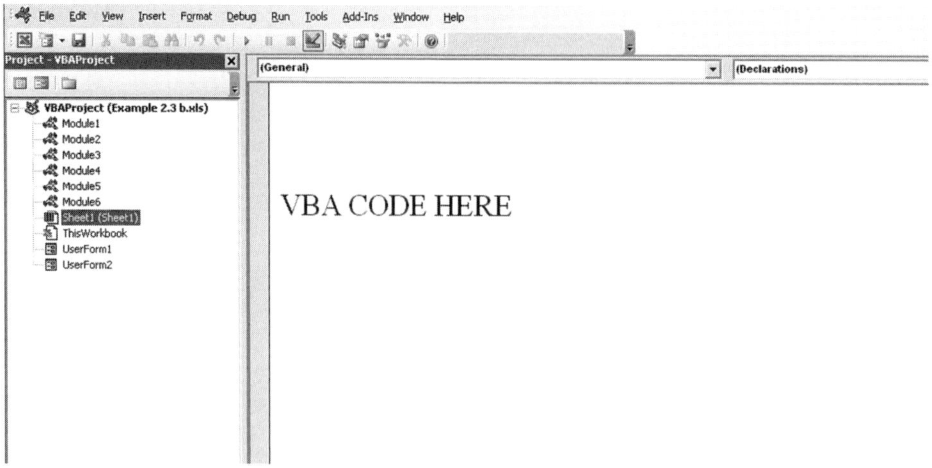

Figure 2.4 Module allowing VBA code addition.

```
line 1   Option Explicit

line 2   Public Sub Calc_PV()
line 3       ' Dimension and declare variable type
line 4       Dim Cost_Pump_A As Double
line 5       Dim Cost_Pump_B As Double
line 6       Dim Interest_Rate As Double
line 7       Dim PV_Pump_A As Double
line 8       Dim PV_Pump_B As Double

line 9       'Read data from sheet
line 10      Cost_Pump_A = Sheet1.Cells(6, 2)
line 11      Cost_Pump_B = Sheet1.Cells(10, 2)
line 12      Interest_Rate = Sheet1.Cells(3, 2)

             'Calulate Present Values
             'A line break or contiuation requires the use of underscore _
line 13      '
             PV_Pump_A = Cost_Pump_A + Cost_Pump_A / (1 + Interest_Rate) ^ 2 _
line 14      + Cost_Pump_A / (1 + Interest_Rate) ^ 4

line 15      PV_Pump_B = Cost_Pump_B + Cost_Pump_B / (1 + Interest_Rate) ^ 3

line 16      'Write Present Values to sheet
line 17      Sheet1.Cells(7, 4) = PV_Pump_A
line 18      Sheet1.Cells(11, 4) = PV_Pump_B
line 19  End Sub
```

Figure 2.5 VBA macro Sub (subroutine) procedure code (Example 2.3—two-pump selection problem).

result for `PV_Pump_A` is then placed on the sheet in row 7, column 4 by `line 18`. The same process is used for `PV_Pump_B`.

In order to run the VBA code, return to the Excel sheet, then click on Tools → Macro → Macros . . . → Macro Name → Calc_PV → Run. The results should appear on the Excel sheet. If the VBA program does not run, the VBA interpreter will stop at the line causing the error. You will not be able to run the Macro until you have corrected the error and reset the VBA program; here use Run → Reset or use the ■ symbol on the toolbar. ■

2.3 ANNUITIES

EXAMPLE 2.4 ***Present Value Analysis Single Investment***

You are the plant manager at the ABC processing plant. A company, PCNew, offers to develop a new plantwide control system for the ABC plant for a one-time fee of \$2MM and yearly updating fees of \$0.25MM (taken as end-of-the-year payments). The cost to upgrade the existing computer system and to install additional control points (required for the PCNew control system) is estimated at \$1MM. PCNew claims that if their control system is adopted yearly, profits should increase by \$0.9MM. PCNew is requesting a 5-year agreement. Is this a good investment if the time value of money is 10% per year?

Here, the cash flow diagram would appear as shown in Figure 2.6. We can evaluate the present value of this investment, where the $PV = PV$ of the increased profit – PV of the costs =

$$\begin{aligned} PV_{\text{investment}} &= \left(\frac{\$0.9\text{M}}{(1+0.1)^1}\right)+\left(\frac{\$0.9\text{MM}}{(1+0.1)^2}\right)+\left(\frac{\$0.9\text{MM}}{(1+0.1)^3}\right) \\ &+\left(\frac{\$0.9\text{MM}}{(1+0.1)^4}\right)+\left(\frac{\$0.9\text{MM}}{(1+0.1)^5}\right) \\ &-\left(3\text{MM}+\left(\frac{\$0.25\text{MM}}{(1+0.1)^1}\right)+\left(\frac{\$0.25\text{MM}}{(1+0.1)^2}\right)\right. \\ &\left.+\left(\frac{\$0.25\text{MM}}{(1+0.1)^3}\right)+\left(\frac{\$0.25\text{MM}}{(1+0.1)^4}\right)+\left(\frac{\$0.25\text{MM}}{(1+0.1)^5}\right)\right) \\ &= 3{,}411{,}708.09 - 3{,}947{,}696.69 = -\$535{,}988.60. \end{aligned} \tag{2.4}$$

The PV of this investment is negative at –\$535,988.60, which indicates this is not a good investment at a time value of money of 10% per year. ■

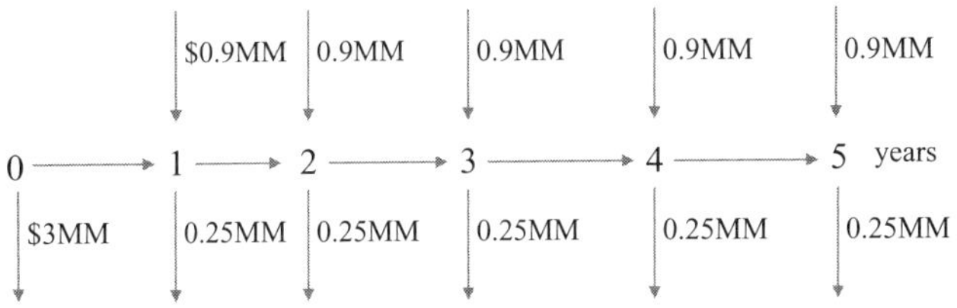

Figure 2.6 Cash flow diagram for Example 2.4—proposed process control system.

The cash flows (both +\$0.9MM and –\$0.25MM) are occurring uniformly over a constant time period, which is termed an annuity. We can develop an equation to determine the present value of uniformly occurring cash flows. Allowing A to be the payment or annuity amount, n the number of uniform periods (e.g., years), and i the interest per period,

$$PV = \left(\frac{A}{(1+i)^1}\right)+\left(\frac{A}{(1+i)^2}\right)+\left(\frac{A}{(1+i)^3}\right)+\left(\frac{A}{(1+i)^4}\right) + \ldots\ldots + \left(\frac{A}{(1+i)^n}\right). \tag{2.5}$$

Multiply both sides by $(1 + i)$:

$$PV(1+i) = (A)+\left(\frac{A}{(1+i)^1}\right)+\left(\frac{A}{(1+i)^2}\right)+\left(\frac{A}{(1+i)^3}\right) + \ldots\ldots + \left(\frac{A}{(1+i)^{n-1}}\right). \tag{2.6}$$

Subtracting Equation (2.6) from Equation (2.5) and rearranging yields

$$A = PV\left(\frac{i(1+i)^n}{(1+i)^n - 1}\right), \tag{2.7}$$

which is called the annuity equation. It can be written as

$$PV = A\left(\frac{(1+i)^n - 1}{i(1+i)^n}\right). \tag{2.8}$$

Do note that annuities are end-of-the-period payments.

EXAMPLE 2.5 *Annuity Present Value Analysis Using Excel and VBA Functions*

Solve Example 2.4, the process control investment opportunity, first by using an Excel sheet and then using a VBA function—use Equation (2.8).

SOLUTION Excel. The Excel sheet solution is shown in **Example 2.5a.xls**. Do note the use of named cells in the solution. Both Equations (2.4) and (2.8) are used in the sheet.

SOLUTION VBA Function Procedure. The annuity equation is useful in solving many economic problems. Frequently used equations that generate a single answer (number) are ideal candidates for VBA functions. VBA functions allow variables in the argument list and they are available anywhere in the Excel sheet.

In contrast to a VBA function procedure, recall that in Example 2.3, we developed a macro subroutine procedure, and this macro was called from the Excel sheet. A macro will not allow variables in the argument list, and in Example 2.3, the necessary variables were read from the Excel sheet.

The solution using a VBA function is provided in **Example 2.5b.xls**. If you are independently working this example, the best way to begin is by copying your current solution, **Example 2.5a.xls**, and renaming it **Example 2.5b.xls**. From the Excel sheet, click on Tools → Macro → Visual Basic Editor and a new screen will appear. On the toolbar, click on Insert → Module. We are now ready to add the needed VBA code, which is shown in Figure 2.7. Again, line numbers of the RHS (`line 1, etc.`) are not to be included in the code; these are simply for discussion purposes.

The function procedure accepts three variables: `Annuity, Interest_Rate, n`, and the present value of the annuity is calculated in `line 4`. The value returned to the sheet must be calculated as the function name `Annuity_PV`. There are no dimension statements required for the three variables in the argument list. These variables are actually sent to the function by reference `ByRef`. The VBA default for passing single variables, vectors, and matrices in argument lists is `ByRef` (by reference). Passing variables by reference `ByRef` or by value `ByVal` is discussed in greater detail later.

From the Excel sheet, this function can be accesed from any cell using f_x → category → User Defined → Annuity_PV. The

```
line 1 Option Explicit

line 2 Public Function Annuity_PV(Annuity, Interest_Rate, n)
           'Calulate Present Value of the Annuity
           'A line break or continuation requires the use of underscore _
           '
line 3     Annuity_PV = Annuity * ((1 + Interest_Rate) ^ n - 1) _
line 4                 / ((Interest_Rate) * ((1 + Interest_Rate) ^ n))
line 5 End Function
```

Figure 2.7 VBA function code for Example 2.5—proposed process control system.

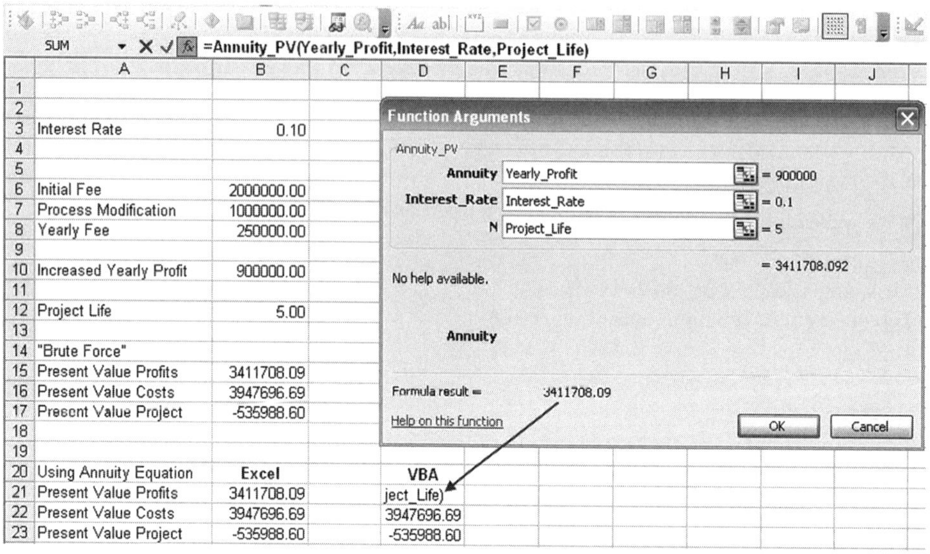

Figure 2.8 Excel sheet screen shot (Example 2.5).

screen should appear as shown in Figure 2.8, where sheet cells containing values for the annuity amount A, interest rate i, and number of years n are generally copied and pasted into the appropriate box. The annuity present value calculated in the cell (here cell D21) will be automatically updated if changes are made in any input cell containing the annuity, interest rate, or number of years. ■

A few words of caution before we move on—Excel/VBA reserves some function names. This will cause problems if we use the same names in our VBA code. The pre-existing function will take priority. For example, in **Example 2.5b.xls**, change the VBA function name from `Annuity_PV` to `PV` in `line 2` and `line 4`. Following the same procedure to access the function (now user-defined `PV`), all will initially appear fine. Input annuity amount, interest rate, and number of years exactly the same as above; however, an answer of –4.14E–06 (or a similar wrong answer) will result. These kinds of errors can be extremely difficult to track down. Do not save this modified version of **Example 2.5b.xls**—we do not want to keep these changes.

The problem here is that *PV* is already a defined function. Go to $f_x \rightarrow$ Financial $\rightarrow$ *PV* and the definition *PV* (rate, nper, pmt, fv, type) appears. You can check your VBA variables against reserved variables by using $f_x \rightarrow$ All. My personal preference is to clarify variable and function names with underscores whenever practical. Currently, Microsoft advocates the use of a capital letter to help clarify variable names, for example, `Interest_Rate` would be `InterestRate`; the Microsoft approach will not eliminate potential problems with reserved VBA functions.

We want to further explore the use of *VBA macro subroutine procedures* (no calling parameters, and directly callable from the Excel sheet), *VBA subroutine procedures* (calling parameters, but cannot be directly called from the Excel sheet), and *VBA function procedures*. Example 2.6 is a straightforward economics problem that we solve using Excel. Then, in four examples (Examples 2.7–2.10), we will explore various solution strategies to Example 2.6 using macros, subroutines, and functions with VBA vectors and VBA matrices.

EXAMPLE 2.6 *Excel Annuity*

A supplemental retirement fund of $500,000 is desired in 20 years. How much money should be set aside each month if the nominal interest rate is 10% per year? Solve this problem on an Excel sheet using the annuity equation. Also, on the Excel sheet, show the future value of these investments at the end of every month (240 months).

The solution is given in **Example 2.6.xls** and is shown in Figure 2.9.

The annuity equation, Equation (2.8), combined with the definition of the future value, Equation (2.1), gives

$$A = FV\left(\frac{i}{(1+i)^n - 1}\right) = \$500{,}000\left(\frac{\frac{0.10}{12}}{\left(1+\frac{0.10}{12}\right)^{20\times 12} - 1}\right) = \$658.44.$$

Therefore, with end-of-the-month payments of $658.44 for 240 months, the target future value of $500,000 will be reached. This calculation is accomplished in cell B6.

Also, in Figure 2.9, recursive calculations were performed to determine the investment future value at the end of every month. This was accomplished in the Excel sheet by copying row 11 (month 2) down the sheet (until a total of 240 months was reached). We could not copy row 10 (month 1) down the sheet. For month 1, the initial annuity payment of $658.44 is made at the end of the month; therefore, no interest calculation is performed. The "Available" column is the future value of the payments and interest at the end of each month.

EXAMPLE 2.7 *VBA Macro Subroutine and VBA Vectors (One-Dimensional Arrays)*

Solve Example 2.6 using a VBA macro with VBA vectors to store the monthly future value amounts. When coding, vectors are generally referred to as one-dimensional arrays.

SOLUTION The solution is provided in **Example 2.7.xls**. If you are independently working this example, the best way to begin is by copying your current solution, **Example 2.6.xls**, and renaming it **Example 2.7.xls**. The possible VBA code is shown in Figure 2.10.

`Lines 3-9` declare variable types and read values from the sheet. The annuity amount A_A ($658.44) is calculated in `line 10` and written to the sheet in row 8, column 2. The variable `Payment` is set = A_A `(line 17)`. Two one-dimensional arrays are defined in `lines 13` and `14`. These vectors are then redimensioned to the number of periods +1 (`n_Periods + 1` here = 240 + 1 = 241) in `lines 15` and `16`. The first value can be stored in the zero element of each vector, and we will take advantage of this fact. The actual elements in these one-dimensional arrays will be stored 0–240. Note that 0–240 will contain 241 elements. VBA is flexible and forgives if the user overruns the index.

The array `Future_Value_Interest(i)` will hold the monthly interest generation, and the array `Future_Value_Annuity_Payout(i)` will contain the monthly future value of the annuity. `Line 18` sets the zero element in the vector

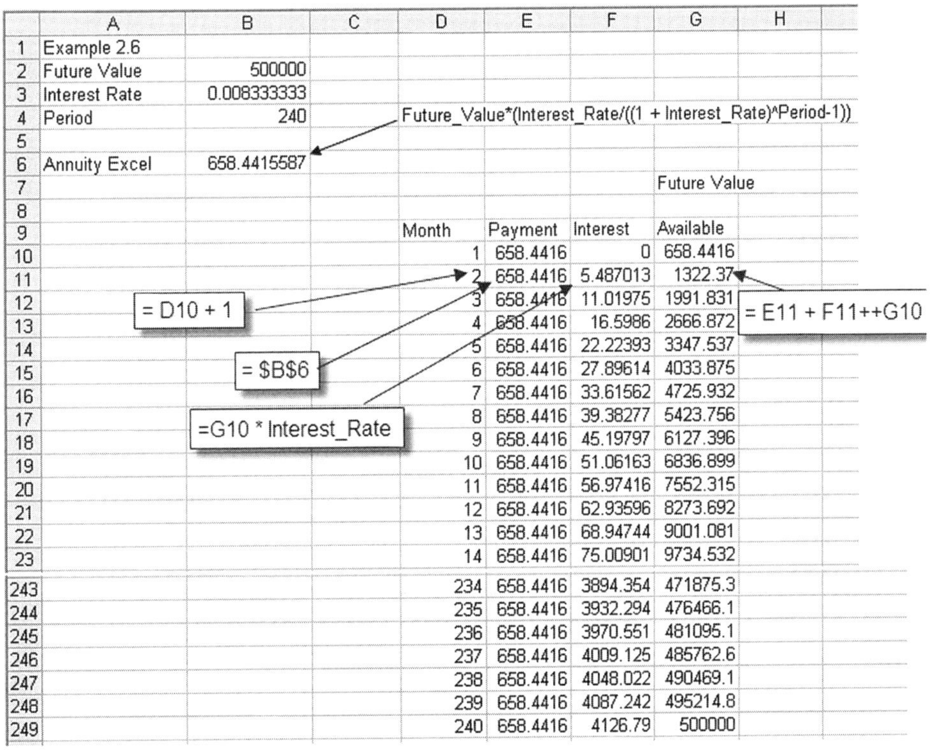

	A	B	C	D	E	F	G	H
1	Example 2.6							
2	Future Value	500000						
3	Interest Rate	0.008333333						
4	Period	240						
5								
6	Annuity Excel	658.4415587						
7							Future Value	
8								
9				Month	Payment	Interest	Available	
10				1	658.4416	0	658.4416	
11				2	658.4416	5.487013	1322.37	
12				3	658.4416	11.01975	1991.831	
13				4	658.4416	16.5986	2666.872	
14				5	658.4416	22.22393	3347.537	
15				6	658.4416	27.89614	4033.875	
16				7	658.4416	33.61562	4725.932	
17				8	658.4416	39.38277	5423.756	
18				9	658.4416	45.19797	6127.396	
19				10	658.4416	51.06163	6836.899	
20				11	658.4416	56.97416	7552.315	
21				12	658.4416	62.93596	8273.692	
22				13	658.4416	68.94744	9001.081	
23				14	658.4416	75.00901	9734.532	
243				234	658.4416	3894.354	471875.3	
244				235	658.4416	3932.294	476466.1	
245				236	658.4416	3970.551	481095.1	
246				237	658.4416	4009.125	485762.6	
247				238	658.4416	4048.022	490469.1	
248				239	658.4416	4087.242	495214.8	
249				240	658.4416	4126.79	500000	

Figure 2.9 Excel solution for Example 2.6—supplemental retirement annuity.

```
Option Explicit                                                    line 1

Public Sub Get_Annuity()                                           line 2
    Dim Future_Value As Double                                     line 3
    Dim Interest_Rate As Double                                    line 4
    Dim n_Periods As Integer                                       line 5
    Dim A_A As Double                                              line 6

    Future_Value = Sheet1.Cells(2, 2)                              line 7
    Interest_Rate = Sheet1.Cells(3, 2)                             line 8
    n_Periods = Sheet1.Cells(4, 2)                                 line 9
    A_A = Future_Value * (Interest_Rate / ((1 + Interest_Rate) ^ n_Periods - 1))
                                                                   line 10
    Sheet1.Cells(8, 2) = A_A                                       line 11

    Dim Payment As Double                                          line 12
    Dim Future_Value_Interest() As Double                         line 13
    Dim Future_Value_Annuity_Payout() As Double                   line 14
    ReDim Future_Value_Interest(n_Periods + 1 ) As Double          line 15
    ReDim Future_Value_Annuity_Payout(n_Periods + 1) As Double     line 16

    Payment = A_A                                                  line 17

    Future_Value_Annuity_Payout(0) = 0.0#                          line 18
    Dim i As Integer                                               line 19
    For i = 1 To n_Periods                                         line 20
        Future_Value_Interest(i) = Future_Value_Annuity_Payout(i - 1) _
                                 * Interest_Rate                   line 21

        Future_Value_Annuity_Payout(i) = Future_Value_Annuity_Payout(i - 1) _
                                 + Future_Value_Interest(i) + Payment      22

        Sheet1.Cells(i + 9, 9) = Future_Value_Interest(i)          line 23
        Sheet1.Cells(i + 9, 10) = Future_Value_Annuity_Payout(i)   line 24

    Next i                                                         line 25
End Sub                                                            line 26
```

Figure 2.10 VBA macro subroutine and VBA vectors (one-dimensional arrays) code for Example 2.7—supplemental retirement annuity.

Future_Value_Annuity_Payout(0) = 0.0#. The # in the code indicates the decimal point in the double-type variable. Lines 19–25 set up a For/Next loop. Here, the counter i represents the month, and it is incremented from 1 to 240 (with increments of one). From line 21,

```
Future_Value_Interest(i) = Future_Value_
Annuity_Payout(i - 1) _
 * Interest_Rate.
```

For I = 1 (month 1),

```
Future_Value_Interest(1) = Future_Value_
Annuity_Payout(0) * Interest_Rate
Future_Value_Interest(1) = 0. * 0.008333 = 0,
```

and from line 22,

```
Future_Value_Annuity_Payout(i) = Future_Value_
Annuity_Payout(i - 1) _
 + Future_Value_Interest(i) + Payment
Future_Value_Annuity_Payout(1) =
Future_Value_Annuity_Payout(0)
 + Future_Value_Interest(1) + Payment = 0. +
0. + 658.44 = 658.44.
```

These values are then placed on the Excel sheet by lines 23 and 24, and then the counter on i is incremented. For i = 2 (month 2),

```
Future_Value_Interest(2) = Future_Value_
Annuity_Payout(1) * Interest_Rate
Future_Value_Interest(2) = 658.44 * 0.008333
= 5.487
```

and

```
Future_Value_Annuity_Payout(2) =
Future_Value_Annuity_Payout(1)
 + Future_Value_Interest(2) + Payment = 658.44
+ 5.487 + 658.44 = 1322.37.
```

Again, values are placed on the Excel sheet and the counter i is incremented.

The construction of the VBA code is the same as the construction on the Excel sheet. In the VBA code, we are able to take advantage of setting the zero element in the vector Future_Value_Annuity_Payout(0) = 0.0#; this allows the iteration scheme to begin immediately in the VBA code (month 1). In comparison with the Excel sheet, we had to wait until the second row (month 2) on the Excel sheet before we could copy the equations down the sheet; this copying establishes the iteration scheme on the Excel sheet.

To run the code from the Excel sheet, click on Tools → Macro → Macros . . . → Macro Name → Get_Annuity → Run. The results should appear on the Excel sheet. ■

EXAMPLE 2.8 *VBA Macro Subroutine and VBA Subroutine*

In Example 2.7, we used a single VBA macro sub (subroutine) procedure to solve the annuity problem Example 2.6. Here we want to use VBA sub (subroutine) procedures. A VBA Sub procedure allows an argument list, and we will utilize this argument list to move information between subroutines. Vectors are again used to store monthly future value amounts.

SOLUTION The possible VBA code is shown in Figure 2.11a and it is provided in **Example 2.8.xls**.

Here, `lines 1-17` are the same as the code developed in Example 2.7. `Line 18` calls Sub procedure `Sub Annuity_Payout` by `Call Annuity_Payout(Interest_Rate, n_Periods, Payment, Future_Value_Interest, Future_Value_Annuity_Payout)`. The variables passed in the argument list are `Interest_Rate, n_Periods, Payment, Future_Value_Interest, Future_Value_Annuity_Payout`. The last two variables of the parameter list are both one-dimensional arrays.

In `lines 19-23`, we take advantage of a For/Next loop. In this example, we are simply indexing on `i`. The general construction of the `For/Next` loop is proved in Figure 2.11b.

Control is then transferred to `line 25` where the passed variables are renamed `I_R, n_P, P, F_V_I, F_V_A_P`. Note that it is not necessary to declare these variables. The variable type in the parameter list of the Sub procedure `Sub Annuity_Payout` is the same variable type in the parameter list of `Call Annuity_Payout`. Even though the variables are renamed in Sub procedure `Sub Annuity_Payout`, the variables have not changed memory location—this is the definition of passing variables by reference (recall the VBA passing variable default is `ByRef`). For example, the variable `I_R` shares the same memory location with `Interest_Rate` and the value in the memory location = `0.008333`. The `0 To 240` elements of `Future_Value_Annuity_Payout` share the same memory location with the `0 To 240` elements of `F_V_A_P`, and currently, this is a vector with `0` in each element.

Given that the passed variables share the same memory location, it is straightforward to see that the code in Example 2.8, `lines 26-31`, is identical to the code in Example 2.7, `lines 18-25`. In Example 2.8, we have moved the two "print statements" (`lines 23-24` in Example 2.7) back into the initial Sub procedure `Sub Get_Annuity()` as `lines 21-22`. After the calculations in the Sub procedure are completed, control is transferred back to `line 18`. The `For` loop in `lines 20-23` prints the vectors `Future_Value_Interest(i), Future_Value_Annuity_Payout(i)` on the Excel sheet. The values in these two vectors were calculated in `Sub Annuity_Payout`. ■

We do want to discuss passing variables in a little more detail before we leave this example. In Example 2.8, we have used the VBA default for passing parameters between Sub procedures. The VBA default is by reference (`ByRef`), which means that when a variable is passed, the memory location for that variable, not the variable actual value, is passed. This is easy to explain for a single variable. Say $x = 5$ and this value is stored in memory address 330307; the memory location is passed. For a vector or matrix, only the memory location for the first element in the vector or matrix is passed; the subsequent memory locations are determined by VBA.

We can also pass passing variables by value (`ByVal`). Here, a temporary memory location is established for each variable in the parameter list. A copy of the value for each variable in the calling Sub procedure is stored in these temporary memory locations. Upon return to the calling Sub procedure, the initial values of the passed variables will be unchanged. To pass variables by value, we can either enclose the variable in parentheses or use the key word `ByVal` variable. See Problem 2.6 for an example of passing variables using by value and by reference.

So far, we have discussed *macro subroutine procedures* (no calling parameters, and directly callable from the Excel sheet) and *subroutine procedures* (calling parameters, but cannot be directly called from the Excel sheet). Both *macro subroutine procedures* and *subroutine procedures* allow values to be read from the Excel sheet and used for calculations. In addition, both allow values to be placed on the Excel sheet, and these values are active; in other words, the Excel sheet will be updated with the values.

We have examined VBA *function procedures* in **Example 2.5b.xls**. A function procedure can be called either from an Excel sheet or a Sub procedure. A function procedure allows calling parameters. In general, a function procedure is written so that it returns one value from the calculations. However, the allowed operations by a function procedure depend on how it is called.

When a VBA function procedure is called directly from an Excel sheet (e.g., from f_x on the toolbar), values for parameters in the argument list can be read from the Excel sheet. Inside the VBA function, it can read additional values from anywhere on the sheet. But the function can only return (write back) a single value. It cannot, for example, change other values on the sheet or write to the sheet in other locations.

When a VBA function procedure is called from a Sub procedure within VBA, it will accept values in the parameter list from the calling Sub procedure. The function can read additional values from the Excel sheet. It can change values anywhere on the sheet. When a VBA function is called from a subroutine, it behaves like a subroutine. This is demonstrated in Example 2.9.

```
Option Explicit                                                    line 1

Public Sub Get_Annuity()                                           line 2
    Dim Future_Value As Double                                     line 3
    Dim Interest_Rate As Double                                    line 4
    Dim n_Periods As Integer                                       line 5
    Dim A_A As Double                                              line 6


    Future_Value = Sheet1.Cells(2, 2)                              line 7
    Interest_Rate = Sheet1.Cells(3, 2)                             line 8
    n_Periods = Sheet1.Cells(4, 2)                                 line 9
    A_A = Future_Value * (Interest_Rate / ((1 + Interest_Rate) ^ n_Periods - 1))
                                                                   line 10
    Sheet1.Cells(8, 2) = A_A                                       line 11

    Dim Payment As Double                                          line 12
    Dim Future_Value_Interest() As Double                          line 13
    Dim Future_Value_Annuity_Payout() As Double                    line 14
    ReDim Future_Value_Interest(n_Periods + 1)                      line 15
    ReDim Future_Value_Annuity_Payout(n_Periods + 1)               line 16

    Payment = A_A                                                  line 17

    Call Annuity_Payout(Interest_Rate, n_Periods, Payment, _
                Future_Value_Interest, Future_Value_Annuity_Payout) line 18

    Dim i As Integer                                               line 19
    For i = 1 To 240                                               line 20

        Sheet1.Cells(i + 9, 9) = Future_Value_Interest(i)          line 21
        Sheet1.Cells(i + 9, 10) = Future_Value_Annuity_Payout(i)   line 22
    Next i                                                         line 23

End Sub                                                            line 24

Sub Annuity_Payout(I_R, n_P, P, F_V_I, F_V_A_P)                    line 25

    F_V_A_P(0) = 0.0#                                              line 26
    Dim i As Integer                                               line 27
    For i = 1 To n_P                                               line 28
        F_V_I(i) = F_V_A_P(i - 1) * I_R                            line 29
        F_V_A_P(i) = F_V_A_P(i - 1) + F_V_I(i) + P                 line 30

    Next i                                                         line 31
End Sub                                                            line 32
```
(a)

For / Next: The general construction is,

For counter = start to finish Step {counter increment, default = 1}
        [code]
        If {condition A met} Then Exit For
        [code for condition A not met]
Next counter

(b)

Figure 2.11 (a) VBA macro subroutine and VBA subroutine (Example 2.8—supplemental retirement annuity). (b) For/Next loop in VBA.

EXAMPLE 2.9 *VBA Macro Subroutine and VBA Functions*

We solved a simple annuity problem using a function in Example 2.5. Here, we want to call an annuity function from a VBA macro. The solution is given in **Example 2.9.xls** and the VBA code is provided in Figure 2.12.

The subroutine macro Sub Get_Annuity() has been discussed in Example 2.7. The construction in line 10 is new. Line 10 calls a function procedure Function A_A_Calc with a parameter list (Future_Value, Interest_Rate,

n_Periods). The variable A_A will be assigned the value returned from the function procedure A_A_Calc.

Lines 13–15 create a public function Public Function A_A_Calc. In line 13, A_A_Calc accepts (Future_Value, Interest_Rate, n) as calling parameters. The calling parameters here may be the same names as in the macro subroutine; this is allowed, but in general, this is not considered good coding practice. A_A_Calc is declared a double type, As Double, at the end of line 13; this type declaration is actually not required, but it is good coding practice.

`Line 14` calculates the annuity amount, which is assigned to `A_A_Calc`, the function name. The value assigned to `A_A_Calc` will be returned to `A_A`.

The function `A_A_Calc` can be called directly from the Excel sheet or indirectly through the macro `Get_Annuity`. Currently, if the macro is run from the Excel sheet, all the needed values, `Future_Value`, `Interest_Rate`, and `n_Periods` are read from the sheet (`lines 7-9`) and the annuity amount is placed in row 8, column 2 (by `line 11`). If the function `A_A_Calc` is directly called from the Excel sheet (f_x → User Defined → `A_A_Calc`), the user must specify the needed values of `Future_Value`, `Interest_Rate`, and `n`, which are generally pasted from cells on the Excel sheet. These can be any values. The function is independent of the macro when the function is directly called from the sheet.

It is straightforward to check the limitations of functions as detailed above by making the following additions to **Example 2.9.xls**. Run the code with these additions, but do not save the changes. First, let us examine the writing capabilities of a function. Add the line of code `Sheet1.Cells(11, 1) = 50.` or `50#` after `line 13`, and the value 50 will be placed in row 11, column 2 if the function is called from the macro. However, if we try and call the function `A_A_Calc` directly from the Excel sheet, the following error will result (#Value!, a value used in the formula, is of the wrong data type).

Next, examine the reading capabilities of a function. Let us add the line of code `Interest_Rate = Sheet1.Cells (3, 4)` after `line 13`, and be sure that in the Excel sheet, row 3, column 4, we have the value 0.02. Whether `A_A_Calc` is called from the macro or directly from the Excel sheet as a function, the value for the interest rate in the annuity calculation will be 0.02. ■

EXAMPLE 2.10 *VBA Macro Subroutine and VBA Matrices (Multidimensional Arrays)*

Here, we want to solve Example 2.6 by using a VBA macro subroutine with a VBA matrix. The solution is given in **Example 2.10.xls** and the VBA code is provided in Figure 2.13. When coding, matrices are generally referred to as multidimensional arrays.

`Lines 1-12` have been discussed in Example 2.7. The annuity amount `A_A` ($658.44) is calculated in `line 10` and written to the sheet in row 8, column 2. The variable `Payment` is set = `A_A` (`line 15`). A multidimensional array (matrix) is defined in `line 13`. The matrix is redimensioned as a two-dimensional array in `line 14`. The first dimension is set as the number of periods + 1 (`n_Periods + 1` here = 240 + 1 = 241), and the second dimension is set to two. A value can be stored in the (zero, zero) element of the two-dimensional array. The actual elements in the matrix will be stored from (0 → 240, 0 → 1). Note that 0–240 will contain 241 elements and 0–1 will contain 2 elements. VBA is flexible and forgives if the user overruns these array indices.

`Line 16` calls Sub procedure `Sub Annuity_Payout`, by, `Call Annuity_Payout(Interest_Rate, n_Periods, Payment, Matrix_Interest_Annuity_Payout)`. The variables passed in the argument list are `Interest_Rate, n_Periods, Payment, Matrix_Interest_Annuity_Payout`. The last variable of the parameter list is a two-dimensional array.

Control is then transferred to `line 23`, where the passed variables are renamed `I_R, n_P, P, M_I_A_P`. It is not necessary to declare these variables. The variable type in the parameter list of the Sub procedure `Sub Annuity_Payout` is the same variable type in the parameter list of `Call Annuity_Payout`. Even though the variables are renamed in Sub procedure `Sub Annuity_Payout`, the variables have not changed memory location—this is the definition of passing variables by reference (recall the VBA passing variable default is `ByRef`).

The first column of the array `M_I_A_P(i, 0)` will hold monthly interest generation. The second column of the array `M_I_A_P(i, 1)` will hold the monthly future value of the annuity. `Line 24` sets the (0, 1) element in the array `M_I_A_P (0, 1) = 0.0#`. The # in the code indicates the decimal point in the double-type variable. `Lines 26-29` set up a `For/Next` loop. Here, the counter `i` represents the month, and it is incremented from 1 to 240 (with increments of one). From `lines 27 and 28`,

For month 1, `i` = 1,

```
M_I_A_P(1, 0) = M_I_A_P(0, 1) * I_R = 0
M_I_A_P(1, 1) = M_I_A_P(0, 1) + M_I_A_P(1,
0) + P = 0 + 0 + 658.44 = 658.44.
```

Then for month 2, `i` = 2,

```
M_I_A_P(2, 0) = M_I_A_P(1, 1) * I_R = 658.44
* 0.008333 = 5.487
M_I_A_P(2, 1) = M_I_A_P(1, 1) + M_I_A_P(2,
0) + P =
658.44 + 5.487 + 658.44 = 1322.37.
```

The construction of the VBA code is the same as the construction on the Excel sheet. In the VBA code, we are able to take advantage of setting the (0, 1) element in the array `M_I_A_P (0, 1) = 0.0#`; this allows the iteration scheme to begin immediately in the VBA code (month 1). When the index `i` reaches 241 (`n_P + 1`), control is returned to the main program with updated values for the array for `Matrix_Interest_Annuity_Payout`. `Lines 17-21` write these values to the Excel sheet.

To run the code from the Excel sheet, click on Tools → Macro → Macros .. → Macro Name → Get_Annuity → Run. The results should appear on the Excel sheet. ■

2.4 COMPARING PROCESS ALTERNATIVES

We have been exploring the basic concepts of the time value of money including present value, future value, annuities, and interest rate or discount cash flow. We have already used a present value analysis and annuities to determine the best economic selection.

There are several commonly used methods to compare alternatives. These include present value, future value,

```
Option Explicit                                                      line 1
Public Sub Get_Annuity()                                             line 2
    Dim Future_Value As Double                                       line 3
    Dim Interest_Rate As Double                                      line 4
    Dim n_Periods As Integer                                         line 5
    Dim A_A As Double                                                line 6

    Future_Value = Sheet1.Cells(2, 2)                                line 7
    Interest_Rate = Sheet1.Cells(3, 2)                               line 8
    n_Periods = Sheet1.Cells(4, 2)                                   line 9

    A_A = A_A_Calc(Future_Value, Interest_Rate, n_Periods)          line 10

    Sheet1.Cells(8, 2) = A_A                                        line 11
End Sub                                                             line 12

Public Function A_A_Calc(Future_Value, Interest_Rate, n) As Double  line 13

    A_A_Calc = Future_Value * (Interest_Rate / ((1 + Interest_Rate) ^ n - 1))
                                                                   line 14
End Function                                                        line 15
```

Figure 2.12 VBA macro subroutine and VBA function code for Example 2.9—supplemental retirement annuity.

```
Option Explicit                                                      line 1

Public Sub Get_Annuity()                                             line 2
    Dim Future_Value As Double                                       line 3
    Dim Interest_Rate As Double                                      line 4
    Dim n_Periods As Integer                                         line 5
    Dim A_A As Double                                                line 6


    Future_Value = Sheet1.Cells(2, 2)                                line 7
    Interest_Rate = Sheet1.Cells(3, 2)                               line 8
    n_Periods = Sheet1.Cells(4, 2)                                   line 9
    A_A = Future_Value * (Interest_Rate / ((1 + Interest_Rate) ^ n_Periods - 1))
                                                                    line 10
    Sheet1.Cells(8, 2) = A_A                                        line 11

    Dim Payment As Double                                           line 12
    Dim Matrix_Interest_Annuity_Payout() As Double                 line 13
    ReDim Matrix_Interest_Annuity_Payout(n_Periods + 1, 2)         line 14

    Payment = A_A                                                   line 15

    Call Annuity_Payout(Interest_Rate, n_Periods, Payment, _
                   Matrix_Interest_Annuity_Payout)                 line 16

    Dim i As Integer                                               line 17
    For i = 1 To 240                                               line 18

        Sheet1.Cells(i + 9, 9) = Matrix_Interest_Annuity_Payout(i, 0)   line 19
        Sheet1.Cells(i + 9, 10) = Matrix_Interest_Annuity_Payout(i, 1)  line 20
    Next i                                                         line 21


End Sub                                                            line 22

Sub Annuity_Payout(I_R, n_P, P, M_I_A_P)                           line 23

    M_I_A_P(0, 1) = 0.0#                                           line 24
    Dim i As Integer                                              line 25

    For i = 1 To n_P                                              line 26

        M_I_A_P(i, 0) = M_I_A_P(i - 1, 1) * I_R                   line 27
        M_I_A_P(i, 1) = M_I_A_P(i - 1, 1) + M_I_A_P(i, 0) + P     line 28

    Next i                                                        line 29

End Sub                                                           line 30
```

Figure 2.13 VBA macro subroutine and VBA multidimensional arrays for Example 2.10—supplemental retirement annuity.

capitalized costs, rate of return (ROR), and others. For a more complete examination of alternatives, there are a number of sources available—see Peters et al. (2003) and Eschenbach (2003).

Often, alternative projects for engineering processes fall into one of three general categories. In the first category, we are presented with a single project and we must make a yes or no decision based on the economic return required by the company. In a second category, we have several possible solutions to the problem and we need to make the best selection—these are considered mutually exclusive projects. In the third category, we have the situation of many unrelated projects but with a finite budget. Here we may select a limited number of these projects.

An example of the first category is our Example 2.4, where a yes or no decision was made on the purchase of a new control system. An example of the second category is our pump selection problem—we must pick one pump from among several. An example of the final category is when a company is evaluating several process improvement opportunities, say, for example, the modification of a distillation train to allow increased throughput, the purchase of a new cooling tower for improved operations, and the upgrading of the boiler and condensate return system. Financial constraints do not allow all three projects. However, in order to be considered, each project must first pass the yes/no test meeting the economic return required by the company.

In Example 2.11, we use present value for the analysis of mutually exclusive projects. Rate of return is used for a yes/no single investment decision in Example 2.12. Equivalent annual cost (EAC) analysis is applied to mutually exclusive alternatives in Example 2.13. We do not specifically address the third category above; however, the first step here is a yes/no decision on each individual project. The reader is referred to a number of excellent texts to further explore economic alternatives—see Peters et al. (2003) and Eschenbach (2003).

2.4.1 Present Value

EXAMPLE 2.11 ***Present Value Analysis with Variable Time Value of Money***

Different interest rates can be incorporated into an economic analysis. This is especially important if one component of an economic analysis, for example, fuel costs, is seeing costs changing rapidly. Here we want to reexamine Example 2.1 and its solution in Example 2.3. Again, let the required interest rate for investment be 10% per year. But now, assume that the pump installation cost is \$200 and this cost increases 5% per year. Also, pump costs are expected to increase 15% per year and pump disposal costs are \$50.

Pump A

Here the cash flow diagram would appear as

0	1	2	3	4	5	6
↓		↓		↓		↓
Purchase		Disposal		Disposal		Disposal
Install		Purchase		Purchase		
		Install		Install		

$$PV_{\text{Pump A}} = \$1000 + \$200 + \left(\frac{\$50}{(1+0.1)^2}\right) + \left(\frac{\$1000(1+0.15)^2}{(1+0.1)^2}\right) + \left(\frac{\$200(1+0.05)^2}{(1+0.1)^2}\right) + \left(\frac{\$50}{(1+0.1)^4}\right) + \left(\frac{\$1000(1+0.15)^4}{(1+0.1)^4}\right) + \left(\frac{\$200(1+0.05)^4}{(1+0.1)^4}\right) + \left(\frac{\$50}{(1+0.1)^6}\right) = \$3939.52. \quad (2.9)$$

Pump B

Here the cash flow diagram would appear as

0	1	2	3	4	5	6
↓			↓			↓
Purchase			Disposal			Disposal
Install			Purchase			
			Install			

$$PV_{\text{Pump B}} = \$1500 + \$200 + \left(\frac{\$50}{(1+0.1)^3}\right) + \left(\frac{\$1500(1+0.15)^3}{(1+0.1)^3}\right) + \left(\frac{\$200(1+0.05)^3}{(1+0.1)^3}\right) + \left(\frac{\$50}{(1+0.1)^6}\right) = \$3653.71. \quad (2.10)$$

Now pump B is a better selection. The concept of using different interest rates (time values of money) allows adjustment for rapidly increasing costs such as energy. The solution to this example is also provided in **Example 2.11.xls**.

2.4.2 Rate of Return (ROR)

Rate of return or internal rate of return (IROR) is the interest rate that makes the present value of the project = 0. Generally, rate of return problems require a trial-and-error or iterative solution. The equation summing the present value for each cash flow is written with the interest rate left as an unknown. The interest rate is then varied until the present value = 0. This is an ideal application for the Goal Seek feature in Excel.

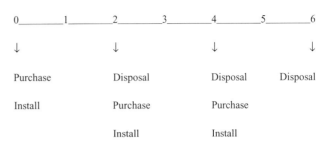

EXAMPLE 2.12 ***Rate of Return Analysis Single Investment and Use of Goal Seek***

Let us revisit Example 2.4 and determine the internal rate of return for the process control system proposed by PCNew. If your company requires a minimum acceptable rate of return (MARR) of 10% per year on any investment, should the control system be built?

SOLUTION We calculate the rate of return for this possible investment by utilizing Equation (2.4) with the present value = 0 and the interest rate as unknown:

$$PV = \left(\frac{\$0.9\text{M}}{(1+i)^1}\right) + \left(\frac{\$0.9\text{MM}}{(1+i)^2}\right) + \left(\frac{\$0.9\text{MM}}{(1+i)^3}\right) + \left(\frac{\$0.9\text{MM}}{(1+i)^4}\right) + \left(\frac{\$0.9\text{MM}}{(1+i)^5}\right) - \left(3\text{MM} + \left(\frac{\$0.25\text{MM}}{(1+i)^1}\right) + \left(\frac{\$0.25\text{MM}}{(1+i)^2}\right) + \left(\frac{\$0.25\text{MM}}{(1+i)^3}\right) + \left(\frac{\$0.25\text{MM}}{(1+i)^4}\right) + \left(\frac{\$0.25MM}{(1+i)^5}\right)\right) = 0? \tag{2.11}$$

In Equation (2.11), different values for i are tried until the $PV = 0$. This value is then compared to the required MARR of 10%. These types of trial-and-error problems are straightforward in Excel.

The Excel solution is given in **Example 2.12.xls**. The equations to determine the present value of each cash flow are entered on the spreadsheet as shown in Figure 2.14a. Cell D6 contains the interest rate (Interest_Rate). The sum of present values is given in F18.

In Excel, on the toolbar, use Tools → Goal Seek. Goal Seek can be used to vary the interest rate to make the present value = 0 as shown in Figure 2.14b.

The internal rate of return is 2.73% per year. This investment falls well below the MARR of 10% per year, so the proposed control system should not be installed. ■

An equivalent alternative procedure is to set the interest rate in Equation (2.11) to the MARR (here 10%) and to determine the present value of the investment. If the present value ≥ 0, then the MARR requirement has been met and the investment should be made. This alternative procedure is correct, but it does not provide an indication of the strength of the investment; the strength of the investment is provided by the rate of return.

The rate of return analysis developed here can be applied to a single investment where a yes/no decision is being made. If mutually exclusive investments are being considered (e.g., choose one among several pumps), then an incremental rate of return analysis must be performed. This topic is discussed in Peters et al. (2003) and in Eschenbach (2003).

2.4.3 Equivalent Annual Cost/ Annual Capital Recovery Factor (CRF)

An equivalent annual cost (or worth) analysis is a convenient method for bringing costs to an annual basis. This is especially useful if the alternatives have different lives. A straightforward way to determine the equivalent annual cost is to determine the present value of the investment and then to convert this present value to an annuity over the life of each project.

Recalling Equation (2.7),

$$A = PV\left(\frac{i(1+i)^n}{(1+i)^n - 1}\right),$$

the connection between the present value and an annuity is clear. If n is in years, then A (the annuity amount) will be the equivalent annual cost. The equivalent annual cost can be determined for the entire project or any component of the project. For example, sometimes it is useful to convert just the capital equipment costs to an equivalent annual cost. In Equation (2.7),

$$\left(\frac{i(1+i)^n}{(1+i)^n - 1}\right)$$

is termed the capital recovery factor. As discussed later in this chapter and in Chapter 10 (The cogeneration design problem—CGAM), the capital recovery factor is very useful in energy optimization problems.

EXAMPLE 2.13 ***Equivalent Annual Cost Analysis***

Let us reexamine the two-pump decision process of Example 2.3. Recall that pump A cost \$1000 with a 2-year life and pump B cost \$1500 with a 3-year life, and the annual interest rate was 10%. In Example 2.3, a present value analysis was performed over a 6-year time period; 6 years was used to bring the alternatives to the same common year. With equivalent annual cost, a common year analysis is not required when costs are not undergoing inflation. The purchase cost of each pump is its PV.

The CRF for n = 2 years; i = 10% is

$$\left(\frac{i(1+i)^n}{(1+i)^n - 1}\right) = \left(\frac{0.1(1+0.1)^2}{(1+0.1)^2 - 1}\right) = 57.62\%.$$

The CRF for n = 3 years; i = 10% is

$$\left(\frac{i(1+i)^n}{(1+i)^n - 1}\right) = \left(\frac{0.1(1+0.1)^3}{(1+0.1)^3 - 1}\right) = 40.21\%.$$

The $EAC_{\text{Pump A}}$ = (\$1000)(0.5762) = \$576.20.

The $EAC_{\text{Pump B}}$ = (\$1500)(0.4021) = \$603.15.

On an annual cost basis, pump A is a better selection, which is the same result we obtained in Example 2.3. ■

We can also perform an equivalent annual cost analysis of Example 2.11. But here, because pump capital costs and installation costs are undergoing inflation, we must again carry the alternative comparison out to a common year.

The CRF for n = 6 years; i = 10% is

$$\left(\frac{i(1+i)^n}{(1+i)^n - 1}\right) = \left(\frac{0.1(1+0.1)^6}{(1+0.1)^6 - 1}\right) = 22.96\%,$$

and using the PV for each pump determined in Example 2.11,

$$EAC_{\text{Pump A}} = (\$3939.52)(0.2296) = \$904.51$$

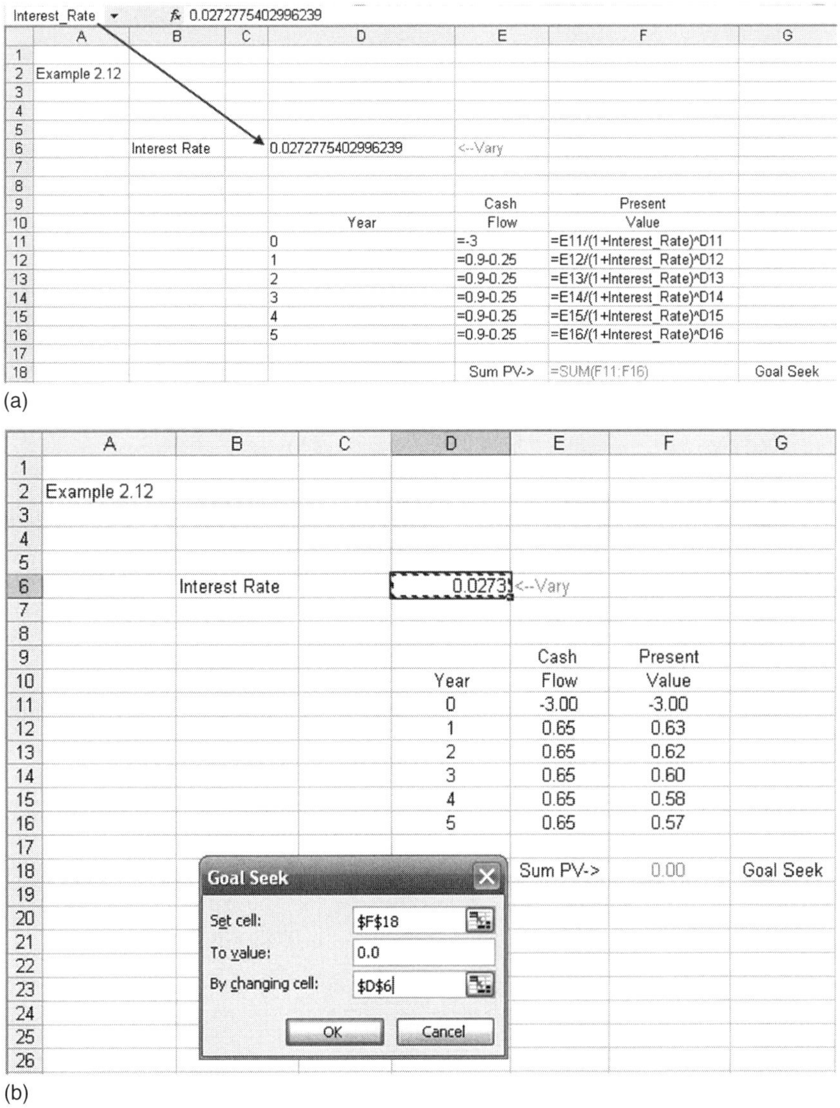

Figure 2.14 (a) Spreadsheet equations for present value calculation. (b) Excel Goal Seek to find the rate of return.

and

$$EAC_{\text{Pump B}} = (\$3653.71)(0.2296) = \$838.89.$$

Here, on an annual cost basis, pump B is a better selection.

2.5 PLANT DESIGN ECONOMICS

We very briefly introduce here the concept of plant design economics. Generally, this economic analysis is applied to new construction. For example, a new chemical plant would have costs for the purchase and installation of equipment, instrumentation, piping, and electrical facilities; these costs are often termed the fixed capital investment (FCI). There would also be costs associated with the start-up of the plant (SU). Working capital (WC) must be set aside to pay expenses including salaries, raw materials, and operating costs. WC is often considered a constant dollar "pot of money" needed to keep the plant operational. The sum of all FCI + SU + WC is termed the total capital investment (TCI).

Once the plant is producing a product, there will be operating costs associated with raw materials, labor costs, maintenance and repair, insurance, and utilities; these are often called operation and maintenance (O&M) costs. Because of income taxes, capital depreciation must be considered in the economic analysis. For simple straight-line depreciation, the depreciation cost per year is

$$\text{Straight line depreciation/year} = \left(\frac{\text{FCI} + \text{SU} - \text{Equipment salvage value}}{\text{Years of useful equipment life}}\right). \quad (2.12)$$

Taxes are paid to local, state, and federal governments, and here, a lumped tax rate of 40% is often used. Taxes are paid on (annual revenues—expenses), where both operating expenses and depreciation are allowable expenses. We can illustrate our discussion with an example.

EXAMPLE 2.14 ***Plant Design Economic Analysis***

A new chemical plant is proposed with an FCI of $25MM. The key economic parameters are the following:

Fifty percent of the FCI is spent 2 years before start-up.

Fifty percent of the FCI is spent 1 year before start-up.

Start-up costs (SU), which are spent at start-up, are 10% of the FCI.

WC is charged at start-up and is 20% of the FCI.

The plant life is anticipated at 10 years.

Depreciation is calculated using a 7-year straight-line method with no salvage value.

The tax rate is 40%.

Annual operating costs are $30MM per year.

Determine the present value of the process, calculated at start-up, if the annual revenue is $45MM and the MARR is 21% on an after-tax basis. The annual revenue and operating costs are assumed to occur at the end of the year of production. Is this project economically viable?

At this point, it is useful to define terms associated with plant design cash flows.

Description	Calculation
FCI—fixed capital investment	
SU—start-up costs	
WC—working capital	
SV—salvage value	
Depre—depreciation	= (FCI + SU − SV)/Equipment life; straight line
BTCF (years n = 1 to N − 1)—before-tax cash flow	= revenue − operating expenses
BTCF—(prior to start-up)	= −(FCI)
BTCF—(time = 0)	= −(WC + SU)
BTCF—(final year, N)	= revenue − operating expenses + SV + WC
TaxInc (years n = 1 to N − 1)—taxable income	= revenue − operating expenses − Depre
TaxInc—(prior to start-up)	= 0
TaxInc—(final year, N)	= revenue − operating expenses − Depre + SV
Tax—tax paid	= (0.40)(TaxInc); with tax rate at 40%
ATCF—after-tax cash flow	= BTCF − Tax
PV ATCF—present value after-tax cash flow	= ATCF/(1 + i)^n

Note that WC is not included as a positive cash flow in taxable income (final year). The taxable income and BTCF columns differ in that the taxable income allows depreciation as a cost. WC cannot be depreciated, but it is recovered along with salvage value at the end of the project. WC is included in the final year BTCF but not in the TaxInc (see Eschenbach, 2003). By using these definitions, we can construct the after-tax analysis, which is done in the Excel file **Example 2.14.xls** and is shown in Figure 2.15.

The MARR is 21%. Here, the depreciation is (FCI + SU)/7 = $3.929MM. The FCI occurs over 2 years with a present value of −(12.5MM)(1 + 0.21)^2 = −$18.30MM and −(12.5MM)(1 + 0.21)^1 = −$15.13MM.

Examining year 1, the revenues are $45MM per year and the operating costs are $30MM per year, giving a BTCF of $45MM − $30MM = $15MM. The taxable income is the BTCF − depreciation = $15MM − $3.929MM = $11.07MM. With a tax rate of 40%, the tax paid would be (0.40)($11.07MM) = $4.429MM. The after-tax cash flow would be $15MM − $4.429MM = $10.57MM; this would be the after-tax profit realized by the process. The present value of the ATCF would be $10.57/(1 + 0.21)^1 = $8.737MM.

The present value of the after-tax cash flow for each year is determined. The present value of the process using a MARR of 21% is $1.82MM. As this is a positive quantity, the MARR is exceeded and the plant should be constructed. We can determine the exact rate of return by using Goal Seek to determine the interest rate (cell D6), which makes the sum of the present values (cell H28) = 0. Show yourself the rate of return would be 22%, which exceeds the required MARR.

It is also useful to plot the cumulative cash position, which helps identify when the project becomes profitable. The cumulative cash position diagram for this problem, using the ATCF column in Figure 2.15, is shown in Figure 2.16. ■

2.6 FORMULATING ECONOMICS-BASED ENERGY OPTIMIZATION PROBLEMS

Economic optimization problems typically involve minimizing the cost of a process. The cost may include all the factors discussed in Example 2.14. However, often the cost is simplified to include just the purchased cost of any new equipment and utilities. These two costs, equipment and utilities, are sometimes referred to as the cost of operation (not to be confused with operating costs discussed above).

Capital costs in optimization problems are often represented by a power law expression,

$$\text{Capital cost}_{E,Q_i} = \text{Base cost}_{E,Q_E}\left(\frac{I}{I_{\text{Base}}}\right)\left(\frac{Q_i}{Q_E}\right)^{\alpha}(f_{\text{Material}})(f_{\text{Pressure}})(f_{\text{Temperature}}), \tag{2.13}$$

where Capital cost$_{E,Q_i}$ is the purchased cost of equipment type E of needed size Q_i (e.g., a 200-m^2 heat exchanger). Base cost$_{E,Q_E}$ is the known cost of equipment type E with known size Q_E. The cost exponent α is a constant for each equipment type; when the value for α is not known, α = 0.6 is often used. To account for different materials of construction, pressure rating, and temperature rating between the base equipment and the needed equipment, factors f_{Material}, f_{Pressure}, and $f_{\text{Temperature}}$ are used (Guthrie, 1969; Ulrich,

	A	B	C	D	E	F	G	H	I	J
1	**Example 2.14**									
2										
3	25	Fixed Capital Investment								
4	2.5	Start Up								
5	5	Working Capital								
6	0	Salvage		0.21	MARR					
7	45	Annual Revenue								
8	30	Operating Costs								
9	3.928571	S.L. Dep								
10	0.4	Tax Rate								
11										
12										
13	Year	BTCF	Deprec	TaxInc	Tax	ATCF		PV ATCF		
14	-2	-12.5			0	-12.50		-18.3013		
15	-1	-12.5			0	-12.50		-15.125		
16	0	-7.5			0	-7.50		-7.5		
17	1	15.00	3.928571	11.07	4.428571	10.57		8.736718		
18	2	15.00	3.928571	11.07	4.428571	10.57		7.220428		
19	3	15.00	3.928571	11.07	4.428571	10.57		5.967296		
20	4	15.00	3.928571	11.07	4.428571	10.57		4.931649		
21	5	15.00	3.928571	11.07	4.428571	10.57		4.075743		
22	6	15.00	3.928571	11.07	4.428571	10.57		3.368383		
23	7	15.00	3.928571	11.07	4.428571	10.57		2.783788		
24	8	15.00		15.00	6	9.00		1.958662		
25	9	15.00		15.00	6	9.00		1.618729		
26	10	20.00		15.00	6	14.00		2.081011		
27										
28							Sum	1.82	Present Value	
29										
30							IRR	0.22	Goal Seek	
31									Set Sum =0	
32									Vary MARR	

Figure 2.15 Excel solution for Example 2.14—plant design economics.

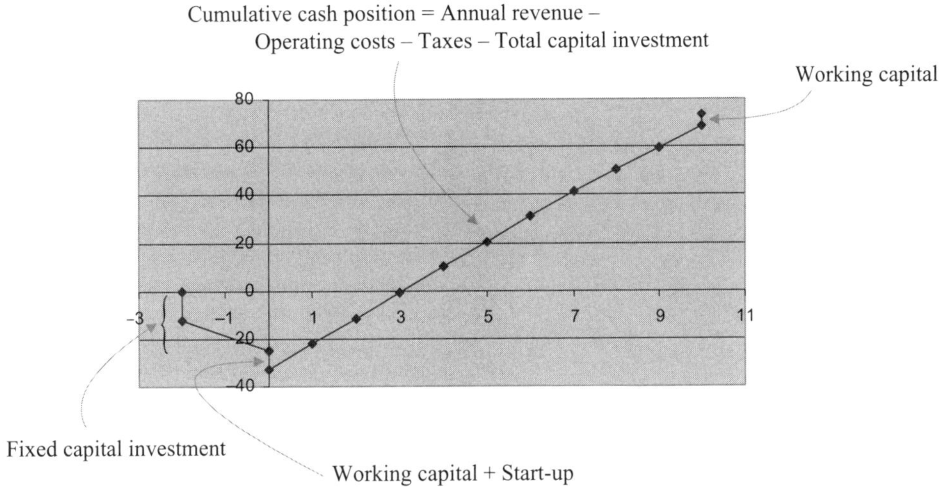

Figure 2.16 Cumulative cash position flow diagram for Example 2.15, time value of money not included.

1984). Capital costs can be brought to the current year, from the base cost year, using cost index I. Values for (I/I_{Base}) are often found using either the Chemical Engineering (CE) Plant Cost Index or the Marshall and Swift (MS) Equipment Cost Index, which are both published monthly in *Chemical Engineering* magazine.

EXAMPLE 2.15 ***Heat Exchanger Cost Estimation***

Estimate the cost of a 200-m^2 shell and tube heat exchanger when the known cost for a 50-m^2 heat exchanger of similar construction and design is \$25,000. The cost exponent for shell and tube heat exchangers is 0.7 over a range of 50–4000 m^2.

SOLUTION Using Equation (2.13),

$$\text{Capital cost}_{200\text{m}^2} = \$25{,}000\left(\frac{200}{50}\right)^{0.7} = \$65{,}975.$$

In an economics optimization problem, the cost of a shell and tube heat exchanger may appear as

$$\text{Capital cost}_{\text{Shell and tube HX}} = \$25{,}000\left(\frac{\text{Area}}{50}\right)^{0.7} = \$1616.82(\text{Area})^{0.7}.$$

This or a similar equation allows the area of the heat exchanger, Area, to be a variable in the optimization problem. The area would be bounded by 50 m^2 (lower limit) and 4000 m^2 (upper limit). ■

A similar procedure is used for all equipment costs, which are then summed to give the total capital cost. To combine the total capital equipment costs and utility costs, they must be placed on the same time basis. Equipment costs, modified using a capital recovery factor as shown in Example 2.13, can be adjusted to the same time basis as utility costs.

2.7 ECONOMIC ANALYSIS WITH UNCERTAINTY: MONTE CARLO SIMULATION

All the examples in this chapter involved "known quantities." There was no uncertainty in reported annual revenues, profits, inflation factors, or any economic parameter. This is not realistic. It is possible to treat any economic parameter as a random variable whose probability distribution will affect the economic analysis. Savage (2003) describes an add-in to Excel, XLSim®, to perform Monte Carlo simulations; Monte Carlo simulations allow variable values to be sampled from distributions. We can produce similar results by coding simulation procedures in VBA or by direct calculations on the Excel sheet.

For our discussions in Section 2.7, profit may be defined as

$$\text{Profit} = \text{Revenues} - \text{Expenses (before taxes)}$$

or

$$\text{Profit} = (\text{Total sales in units}) \times (\text{Selling price/Unit} - \text{Cost/Unit}) - \text{Fixed costs}.$$

EXAMPLE 2.16 ***Profit with Uncertainty***

A new product is being introduced to the market. If there is a low-volume market, estimates are that 60,000 units can be sold at $10 per unit. If there is a high-volume market 100,000 units can be sold, but competition will lower the selling price to $8 per unit (modified from Savage, 2003).

It is also estimated that production costs will be $7.50 per unit. However, with changes in labor rates and raw material costs, this production cost may range from $6 to $9 per unit. Fixed costs for overhead and advertising are $30,000. Determine the profitability.

SOLUTION Often the solution to this problem is formed by using average values for sales, selling price, and production costs. This solution is given in **Example 2.16a.xls** and is shown in Figure 2.17.

As opposed to using average values in economic calculations, we can apply probability distributions to any economic parameter. For example, from the problem statement, there is an implied 50/50 probability that sales may be 60,000 units with a selling price of $10 or 100,000 units with a selling price of $8. This probability can be created in Excel by first taking advantage of the built-in random number (RN) generator =RAND( ); RAND( ) generates an RN between 0 and 1. We could next add an If statement in the Excel cells for Sales and Price by using cells in Figure 2.17 above:

= If(logical test, if true, if false)
for sales = If(RAND( ) > 0.5, F5, E5)
for price = If(RAND ( ) > 0.5, F6, E6)

This construction is not quite correct. As two different RNs are being generated, the low sales/price or high sales/price values will not remain partnered. To overcome this, the RN generator can be moved to a single cell, say $D$14 (cell D14 = RAND( )). Now we can use

for sales = If($D$14 > 0.5, F5, E5)
for price = If($D$14 > 0.5, F6, E6)

and this will keep the low and high scenarios partnered.

Next, we consider unit cost. Let us assume that this cost is normally distributed around $7.5 per unit. We are familiar with the bell-shaped normal distribution, and it will be discussed in detail in Chapter 6 ("Data Reconciliation"). Simply put, if μ is our mean value and σ is our standard deviation, then 68.26% of our unit costs should fall within $\mu \pm \sigma$; 95.44% of our unit costs should fall within $\mu \pm 2\sigma$; and 99.71% of our unit costs (basically all of our results) should fall within $\mu \pm 3\sigma$. For the unit cost, $\mu = 7.5$ and $\sigma = 0.5$; we choose σ such that $\mu \pm 3\sigma$ spans the low and high unit costs.

Excel provides a built-in function, NORMINV (RN, μ, σ), to determine a value (here unit cost) given an RN, mean, and standard deviation. Here, we do not want to use the same RN as for the sales/price, as this would bias the results:

For unit cost (cell $B$7) = NORMINV($E$10, $F$14, 0.5).

The Excel sheet with these two distributions is given in **Example 2.16b.xls** and is shown in Figure 2.18.

After you open this file, press F9, which forces the Excel sheet to calculate one time. You will notice that the two RNs in cells D14 and F14 change value as do the sales/price, unit cost, and calculated profit.

In order to take advantage of our distributions and to generate meaningful economic results, we could press the F9 key many times (>1000) and record the profit result each time. Alternatively, there are several add-in packages to Excel (e.g., the Monte Carlo simulation package XLSim), which will automatically perform this task as well as provide additional probability distributions not found in Excel. Or we can write a VBA code, which will run the

Excel sheet and record the profit results. The VBA macro `Run_Simulation()` to accomplish this task is given in Figure 2.19 and is provided in **Example 2.16b.xls**. Here we are creating what would be considered a simple Monte Carlo simulation to determine profitability.

The code is straightforward. We are creating a For/Next loop, which will index a counter on the Excel sheet (`Sheet1.Cells(13, 2)`), here from 1 to 1000. Each time the counter is indexed, all the values on the sheet will be calculated. This includes the two RN generations, which will then generate a new profit value. This profit value is found from the sheet using, `Profit = Sheet1.Cells(10, 2)`. In the VBA code, we next generate a histogram that keeps track of the number of profit occurrences between two successive values. Here these successive values (and tracking index) are between \$150,000 and ∞ (`HistA`); \$100,000 and \$150,000 (`HistB`); \$50,000 and \$100,000 (`HistC`); \$0 and \$50,000 (`HistD`); −\$50,000 and \$0 (`HistE`); −\$100,000 and −\$50,000 (`HistF`); and −∞ and −\$100,000 (`HistG`). The histogram tracking indexes are normalized and written on the Excel sheet.

If we look back at Figure 2.18, we can see that the results from the histogram (on the sheet in cells B17–B23) indicate that by using the current probabilities, there is a 40% chance the new product will make over \$100,000 in profit, but there is also almost a 20% probability the new product will not make any profit. These probabilities are an interesting way of assessing a potential project. ■

As noted in Example 2.16, one advantage of add-in Monte Carlo simulation packages to Excel is the additional distribution functions they provide. Generally, the user can create these distributions either directly in Excel or by using a combination of Excel and VBA. For example, the original problem in Example 2.16 (Savage, 2003) used a triangular distribution, as opposed to a normal distribution, for the unit cost.

A triangular distribution is not available in Excel. However, following Pritsker (1974), a sample value (here the unit cost) from a triangular distribution can be found as

$$\text{Triag} = \begin{cases} a + \sqrt{(b-a)(c-a)(RN)} & 0 \le RN \le \dfrac{b-a}{c-a} \quad (2.14a) \\ c - \sqrt{(c-b)(c-a)(1-RN)} & \dfrac{b-a}{c-a} \le RN \le 1, \quad (2.14b) \end{cases}$$

where a is the minimum value, b is the mode, and c is the maximum value. For a triangular unit cost distribution in Example 2.16, $a = 6$, $b = 7.5$, and $c = 9$.

	A	B	C	D	E	F	G
1	Profit Start.xls						
2					Scenarios		
3					Low	High	Average
4				Probability	50%	50%	
5	Sales in Units	80000		Units	60000	100000	80000
6	Price per Unit	9		Price	10	8	9
7	Unit Cost	7.5					
8	Fixed Costs	30000		Unit Cost			
9				Low	Likely	High	Average
10	Profit	90000		6	7.5	9	7.5
11							
12		Profit = Sales * (Price - Unit_Cost) - Fixed_Cost					

Figure 2.17 Profitability (Example 2.16) using average values for sales, selling price, and production costs.

	A	B	C	D	E	F	G
1	Profit ND VBA.xls						
2					Scenarios		
3	Financials				Low	High	Average
4				Prob	50%	50%	
5	Sales in Units	100000		Units	60000	100000	80000
6	Price per Unit	8		Price	10	8	9
7	Unit Cost	7.263025546					
8	Fixed Costs	30000		Unit Cost			
9				Low	Likely	High	Average
10	Profit	43697.45		6	7.5	9	7.5
11							
12							
13	Counter	1000		Random Number Generation			
14				0.779534		0.317768	
15							
16		Histogram %					
17	Profit >= \$150,000	6.80					
18	Profit >= \$100,000 but < \$150,000	34.20					
19	Profit >= \$50,000 but < \$100,000	21.50					
20	Profit >= \$0 but < \$50,000	18.30					
21	Profit >= -\$50,000 but < \$0	13.90					
22	Profit >= -\$100,000 but < -\$50,000	4.60					
23	Profit <= -\$100,000	0.70					

Figure 2.18 Profitability (Example 2.16) using distribution values for sales, selling price, and production costs.

EXAMPLE 2.17 ***Profit with Uncertainty User-Generated Distribution***

We can now solve the original problem posed by Savage (2003). In Example 2.16, change the distribution for the unit cost from a normal distribution to a triangular distribution.

The Excel file is given in **Example 2.17a.xls** and results shown in Figure 2.20.

Equation (2.14 a) is used in cell I10 and Equation (2.14 b) is used in cell I11. Again, two RN generators are used to prevent bias in sales/price and unit costs. The VBA code is identical in Examples 2.16 and 2.17. The histogram results indicate that changing from a normal distribution for unit costs to a triangular distribution for unit costs had little effect. Using the probabilities from Figure 2.20, there is again about a 40% chance the new product will make over $100,000 in profit, but there is also almost an 18% probability (down slightly from the normal distribution) the new product will not make any profit. A check of the triangular distribution generation is provided in **Example 2.17b.xls**.

2.8 CLOSING COMMENTS

This chapter introduced basic engineering economics. There are many methods available to evaluate alternative investments, and here we showed the use of present value, annuities, rate of return, and equivalent annual cost. We also introduced plant design economics and showed the development of Monte Carlo simulations for economic evaluation. An important outcome of this chapter was the development of the skill set necessary to utilize Excel and VBA within Excel. Economic evaluation, Excel, and Excel/VBA will prove important throughout this text.

Additional background on engineering economics can be found in the texts by Peters et al. (2003) and Eschenbach (2003). Plant design economics and equipment sizing are detailed by Douglas (1988). For Excel/VBA, the texts by Chapra (2003, 2010) and Albright (2001) provide additional

```
Option Explicit
Public Sub Run_Simulation()

    Dim i As Integer
    Dim HistA As Double, HistB As Double, HistC As Double
    Dim HistD As Double, HistE As Double, HistF As Double
    Dim HistG As Double
    HistA = 0.0#
    HistB = 0.0#
    HistC = 0.0#
    HistD = 0.0#
    HistE = 0.0#
    HistF = 0.0#
    HistG = 0.0#

    For i = 1 To 1000
        Dim Profit As Double

        ' We must make a change on the sheet to have the sheet calculate
        Sheet1.Cells(13, 2) = i

        Profit = Sheet1.Cells(10, 2)

        If Profit >= 150000 Then
            HistA = HistA + 1
ElseIf Profit >= 100000 Then HistB = HistB + 1
ElseIf Profit >= 50000 Then HistC = HistC + 1
ElseIf Profit >= 0 Then HistD = HistD + 1
ElseIf Profit >= -50000 Then HistE = HistE + 1
ElseIf Profit >= -100000 Then HistF = HistF + 1
        Else : HistG = HistG + 1
        End If
    Next i

    ' Remember i is now one more than counter limit in For statement
    Sheet1.Cells(17, 2) = (HistA / (i - 1)) * 100
    Sheet1.Cells(18, 2) = (HistB / (i - 1)) * 100
    Sheet1.Cells(19, 2) = (HistC / (i - 1)) * 100
    Sheet1.Cells(20, 2) = (HistD / (i - 1)) * 100
    Sheet1.Cells(21, 2) = (HistE / (i - 1)) * 100
    Sheet1.Cells(22, 2) = (HistF / (i - 1)) * 100
    Sheet1.Cells(23, 2) = (HistG / (i - 1)) * 100
End Sub
```

Figure 2.19 VBA code for Example 2.16—histogram generation.

	A	B	C	D	E	F	G	H	I	J	K
1	Profit TD VBA.xls										
2					Scenarios						
3	Financials				Low	High	Average				
4				Prob	50%	50%					
5	Sales in Units	100000		Units	60000	100000	80000				
6	Price per Unit	8		Price	10	8	9				
7	Unit Cost	6.550273033									
8	Fixed Costs	30000		Unit Cost					0.5	(b-a)/(d-a)	
9				Low (a)	Likely (b)	High (d)	Average				
10	Profit	114972.70		6	7.5	9	7.5		6.550273	0<= RN<= (b-a)/(d-a)	
11									6.951293	(b-a)/(d-a)<RN<=1	
12											
13	Counter	1000		Random Number Generation							
14				0.885662		0.067289					
15											
16		Histogram %									
17	Profit >= $150,000	9.80									
18	Profit >= $100,000 but < $150,000	29.60									
19	Profit >= $50,000 but < $100,000	26.00									
20	Profit >= $0 but < $50,000	17.10									
21	Profit >= -$50,000 but < $0	10.60									
22	Profit >= -$100,000 but < -$50,000	5.50									
23	Profit <= -$100,000	1.40									

Figure 2.20 Profitability (Example 2.17) using triangular distribution.

reading, and the text by Walkenbach (2002) serves as an excellent reference guide.

REFERENCES

ALBRIGHT, S.C. 2001. *VBA for Modelers: Developing Decision Support Systems with Microsoft Excel*. Duxbury Thompson Learning, Pacific Grove, CA.

CHAPRA, S.C. 2003. *Power Programming with VBA/Excel*. Prentice Hall Peterson Education, Upper Saddle River, NJ.

CHAPRA, S.C. 2010. *Introduction to VBA for Excel* (2nd edition). Prentice Hall Peterson Education, Upper Saddle River, NJ.

DOUGLAS, J.M. 1988. *Conceptual Design of Chemical Processes*. McGraw Hill, Boston.

ESCHENBACH, T.G. 2003. *Engineering Economy: Applying Theory to Practice* (2nd edition). Oxford University Press, New York.

GUTHRIE, K.M. 1969. Data and techniques for preliminary capital cost estimation. *Chem. Eng.* 114–142.

PETERS, M.S., K. TIMMERHAUS, and R.E. WEST 2003. *Plant Design and Economics for Chemical Engineers* (5th edition). McGraw-Hill, Boston.

PRITSKER, A.A.B. 1974. *The GaspIV Simulation Language*. John Wiley & Sons, New York.

REDLICH, O. and J.N.S. KWONG 1949. On the thermodynamics of solutions. V. An equation of state for fugacities of gaseous solutions. *Chem. Rev.* 44: 233–244.

SAVAGE, S.L. 2003. *Decision Making with Insight: Includes Insight.xla 2.0*. Thomas Learning, Belmont, CA.

ULRICH, G.D. 1984. *A Guide to Chemical Engineering Process Design and Economics*. John Wiley and Sons, New York.

WALKENBACH, J. 2002. *Excel 2002 Power Programming with VBA. M and T Books*. Wiley Publishing, New York.

PROBLEMS

2.1 ***Economics and the Annuity Equation*** A recent newspaper headline reads, "Woman Wins $1,000,000 in State Lottery." Under the rules of the lottery, the winner receives $100,000 immediately, with nine additional annual payments of $100,000 each. In other words, the winner receives $100,000 at the beginning of each year for 10 years. If the interest rate is 10% per year, and we ignore income taxes, what is the present value of the $1,000,000 lottery payout?

2.2 ***Nominal and Effective Interest Rates*** A car dealer offers you a "special rate loan" of 5% for 12 months on a $2000 car. The car dealer calculates your monthly payments as $2000 + (5%)($2000) = $2100/12 = $175. Are you actually getting a 5% per year loan? What are the nominal and effective interest rates?

2.3 ***Rate of Return*** A processing facility is considering shutting down or undertaking a major overhaul. The major overhaul will require $2MM. If the overhaul is accomplished, it will extend the life of the plant for five more years with anticipated profits of $500,000 per year. The processing facility requires a minimum rate of return on investment of 10%. Does this investment meet the required MARR?

2.4 ***Working with Multidimensional Arrays*** For Example 2.10, modify the matrix `Matrix_Interest_Annuity_Payout(n_Periods + 1, 2)` so that it contains a third column. Now in the matrix, allow the first column to store the month, the second column the future value of the interest payment, and the third column the future value of the funds available. Write the results on the Excel sheet.

2.5 ***Understanding Sheet Access and For Loops*** Open a worksheet and type = rand ( ) into cell A:1. Here, rand ( ) is an RN generator; it will generate numbers between 0 and 1. Press the F9 key to see it work.

Set up a simple VBA program that will access rand ( ) 1000 times and then compute and report the average value.

Can this same task be accomplished without using VBA? If no, explain why not.

```
Option Explicit

Public Sub Simple_Addition()
Dim x As Double
Dim y As Double
Dim z As Double

'By Reference
x = 1
y = 2
z = 0
Call Add_1(x, y, z)
Sheet1.Cells(2, 2) = x
Sheet1.Cells(3, 2) = y
Sheet1.Cells(4, 2) = z

'By Value Method 1
x = 1
y = 2
z = 0
Call Add_1((x), (y), (z))
Sheet1.Cells(2, 6) = x
Sheet1.Cells(3, 6) = y
Sheet1.Cells(4, 6) = z

'By Value Method 2
x = 1
y = 2
z = 0

Call Add_2(x, y, z)
Sheet1.Cells(2, 8) = x
Sheet1.Cells(3, 8) = y
Sheet1.Cells(4, 8) = z
End Sub

Sub Add_1(a, b, c)
c = a + b
End Sub

Sub Add_2(ByVal a, ByVal b, ByVal c)
c = a + b
End Sub
```

Figure P2.6 VBA code for Problem 2.6.

2.6 ***Variable Transfer By Value and By Reference*** In the simple program shown in Figure P2.6, we add $x + y = z$ in a callable subroutine as $c = a + b$ (in either `Sub Add_1` or `Sub Add_2`) using the by value and by reference options in VBA. Examine the code and determine the value of z from the indicated methods: `'By Reference`, `'By Value Method 1`, `'By Value Method 2`.

2.7 ***Working With Vectors*** This problem does not involve economics, but it provides a good example of working with vectors. We will see similar problems in combustion calculations in later chapters.

For gas or liquid systems, the relation between pressure, temperature, and molar volume can be given by a cubic equation of state (EOS), for example, the Redlich–Kwong EOS (Redlich and Kwong, 1949):

$$P = \frac{RT}{\upsilon - b} - \frac{a}{\upsilon(\upsilon - b)T^{1/2}},$$

where P is the pressure (in atmosphere), T is the temperature (in kelvin), υ is the species molar volume in (liters per mole), R is the universal gas constant (0.08205 atm-L/mole-K), and a and b are species-dependent constants given as

$$a = 0.427\frac{R^2 T_c^{2.5}}{P_c} \quad b = 0.0866R\frac{T_c}{P_c}.$$

T_c is the species critical temperature and P_c is the species critical pressure. T_c and P_c values for selected species are provided in the following table:

Species	T_c (K)	P_c (atm)
Methane	190.061	45.8
Ethane	305.561	48.3
Propane	369.71	42.01
Carbon dioxide	304.201	72.84
Nitrogen	126.271	33.54
Water	647.301	218.2

Develop a VBA macro procedure to calculate P if $T = 600$ K and $\upsilon = 100$ L/mole. Here, place T_c and P_c values for each species on the Excel sheet. Read these values into two VBA vectors and then solve for P for each species (using the vectors) at the given T and υ. Write the P value for each species on the Excel sheet. For comparison purposes, also calculate the species pressure using the ideal gas EOS, $P\upsilon = RT$.

2.8 ***Working with Multidimensional Arrays*** Solve Problem 2.7 by reading T_c and P_c values from the Excel sheet into a single array and then by solving for P (using the array) at the given T and v. Write the P value for each species on the Excel sheet.

2.9 ***Sheet Active for Macro Subroutine Procedures and Subroutine Procedures*** See the impact of adding `Sheet1.Cells(3, 2) = 0.02` after `line 21` in **Example 2.8.xls**.

2.10 ***Project Financing (This Problem Is Taken Directly from Eschenbach, 2003, pp. 348–350)*** A plant has a first cost (capital investment at time = 0) of $900K. After 10 years, the equipment salvage value is $100K; straight-line depreciation is used. Annual revenues will be $300K, and annual operating and maintenance costs are $125K. The firm requires an after-tax rate (MARR) of $i = 9\%$. The effective tax rate is 40%.

Financing is available through a 5-year loan at 10%. The uniform payments are made annually. Assume that 60% of the project will be financed through a loan and 40% internally. The before-tax cash flow, BTCF = Revenue – O&M – Loan payments. The taxable income, TaxInc = Revenue – O&M – Depreciation – Interest payments. Year 0 will include the first cost paid by the company ($–360K), and year 10 will include the salvage value ($100K).

Develop a spreadsheet to find the project after tax: present value and internal rate of return. Allow the fraction of the project to be financed to be a variable.

2.11 ***Project Financing (This Problem Is Taken Directly from Eschenbach, 2003, pp. 348–350)*** Determine how the solution found in Problem 2.10 would change if the Modified Accelerated Cost Recovery System (MACRS) for yearly depreciation was used. Here, depreciation charges would be year 1 = \$128.6K; year 2 = \$220.4K; year 3 = \$157.4K; year 4 = \$112.5; year 5 = \$80.3K; year 6 = \$80.3K; year 7 = \$80.3K; year 8 = \$40.2K; year 9 = year 10 = \$0.

2.12 ***Project Financing with Negative Cash Flows (This Problem Is Taken Directly from Eschenbach, 2003, pp. 347–348)*** Plant construction in an environmentally sensitive area necessitates that a company must undertake a wetland restoration project. The wetland reclamation project would cost \$6M, 80% of which would be financed by a 10-year loan at 9% (with uniform annual payments). Using a 10-year horizon and an after-tax rate (MARR) of $i = 8\%$, what is the present value and equivalent annual cost of this project? The effective tax rate is 40%.

2.13 ***Working with RNs*** Open a worksheet and type =rand ( ) in a cell; it will generate an RN between 0 and 1. The F9 key will update the sheet and change the value in the cell. Set up a simple VBA macro Sub procedure, which will access rand ( ) from the sheet 1000 times and then compute and report the average value to the Excel sheet.

Chapter 3

Computer-Aided Solutions of Process Material Balances: The Sequential Modular Solution Approach

Conceptually, a chemical processing plant may be described as a series of processing units where, first, materials are fed to a reactor to make value-added products, then reactants and products are separated, and unreacted reactants are recycled. Material balances are solved to determine species flows in every stream. The most common computer-aided solution approach, the sequential modular solution technique, builds separate models for each processing unit. Each individual processing unit is solved and information is passed to the downstream units. It is also possible to assemble and solve the material balance equations for all the processing units simultaneously—the simultaneous solution approach. In this chapter, we will explore the sequential modular approach using both Excel and Excel with Visual Basic for Applications (VBA). The simultaneous approach is detailed in Chapter 4.

3.1 ELEMENTARY MATERIAL BALANCE MODULES

We begin with the general material balance equation for an open (flowing) system. The material balance for species i can be written as

$$\begin{pmatrix}\text{The rate of change of the moles}\\ \text{of species } i \text{ in the system}\end{pmatrix} = \begin{pmatrix}\text{The rate at which the moles}\\ \text{of species } i \text{ enter the system}\end{pmatrix} - \begin{pmatrix}\text{The rate at which the moles}\\ \text{of species } i \text{ leave the system}\end{pmatrix} + \begin{pmatrix}\text{The rate at which the moles}\\ \text{of species } i \text{ are generated}\\ \text{in the system}\end{pmatrix}, \quad (3.1)$$

which is often simply stated as accumulation = input − output + generation. At steady state, the left-hand side (LHS) = 0 and, accounting for all input streams, $n_{\text{streams in}}$, and output streams, $n_{\text{streams out}}$, we can write

$$0 = \sum_{j=1}^{n_{\text{streams in}}} N_{i,j} - \sum_{j=1}^{n_{\text{streams out}}} N_{i,j} + (R_i) \quad i = 1, 2, \ldots, n_{\text{species}}, \quad (3.2)$$

where $N_{i,j}$ is the molar flow rate of species i in stream j, and R_i is the species net generation rate. Equation (3.2) can be written for each species where the number of species is n_{species}.

As a first step toward developing a general sequential modular solution approach for material balance problems,

Modeling, Analysis and Optimization of Process and Energy Systems, First Edition. F. Carl Knopf.

we consider a processing plant to be an assembly of four elementary modules or processing units (see Myers and Seider, 1976; Reklaitis, 1983): mixers, separators, splitters, and stoichiometric reactors. The natural specifications to these elementary units would include the flow rates of all species in feed streams, species split fractions for separators and splitters, and stoichiometric coefficients and conversion of key reactants in the reactors. In general, if recycle loops are present, an iterative solution is required. An additional problem is that natural specifications are often missing and are replaced with alternative specifications. Material balance problems with alternative specifications are commonly termed "constrained material balance problems," although we will see these problems are solved as unconstrained optimization problems. We will develop search techniques to solve these problems.

The material balance equations for these four elementary modules are provided below. Here, we assume the streams are single phase, and we apply Equation (3.2) to each of these modules. We will extend these modules to include energy balances in Chapter 5.

3.1.1 Mixer

In a mixer as shown in Figure 3.1, input streams are summed to produce a single output stream. Each input stream may have a different composition.

Species Balances

The n_{species} material balances based on the species molar flow rates in each stream are

$$\sum_{j=1}^{n_{\text{streams}}} N_{i,j} - N_{i,\text{out}} = 0 \quad i = 1, \ldots, n_{\text{species}}, \tag{3.3}$$

or equivalently utilizing the total molar flow rate in each stream,

$$\sum_{j=1}^{n_{\text{streams}}} N_j x_{i,j} - N_{\text{out}} x_{i,\text{out}} = 0 \quad i = 1, \ldots, n_{\text{species}}, \tag{3.4}$$

with the mole fraction normalization equations:

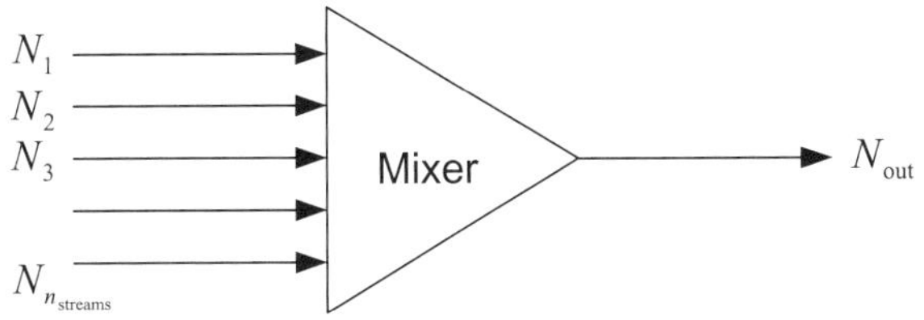

Figure 3.1 Mixer module.

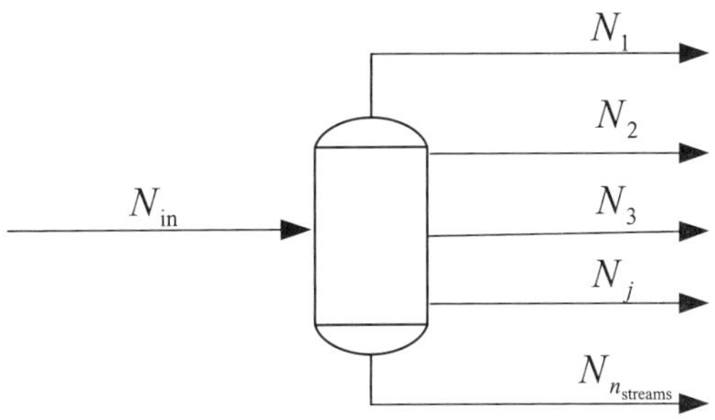

Figure 3.2 Separator module.

$$\sum_{i=1}^{n_{\text{species}}} x_{i,j} = 1 \quad j = 1, \ldots, n_{\text{streams}} \tag{3.5}$$

and

$$\sum_{i=1}^{n_{\text{species}}} x_{i,\text{out}} = 1. \tag{3.6}$$

Total Flow Balance

The sum of the total molar flow rates in equals the total molar flow rate out:

$$\sum_{j=1}^{n_{\text{streams}}} N_j - N_{\text{out}} = 0. \tag{3.7}$$

Equations (3.3)–(3.7) could also have been written in terms of species mass flow rates or the total mass flow rate with weight fraction normalization equations. For brevity, we just list the species material balances utilizing the molar flow rates.

3.1.2 Separator

In a separator, as shown in Figure 3.2, one input stream is separated into n_{streams} output streams—this is the reverse operation of the mixer.

Species Balances

The n_{species} material balances based on the species molar flow rates in each stream are

$$N_{i,\text{in}} - \sum_{j=1}^{n_{\text{streams}}} N_{i,j} = 0 \quad i = 1, \ldots, n_{\text{species}}. \tag{3.8}$$

The species material balances utilizing the total molar flow rate for each stream are

$$N_{\text{in}} x_{i,\text{in}} - \sum_{j=1}^{n_{\text{streams}}} N_j x_{i,j} = 0 \quad i = 1, \ldots, n_{\text{species}}, \tag{3.9}$$

with the mole fraction normalization equations:

$$\sum_{i=1}^{n_{\text{species}}} x_{i,j} = 1 \quad j = 1, \ldots, n_{\text{streams}} \tag{3.10}$$

and

$$\sum_{i=1}^{n_{\text{species}}} x_{i,\text{in}} = 1. \tag{3.11}$$

Separator material balances can be written in terms of the separator's species split fraction $\alpha_{i,j}$:

$$\alpha_{i,j} = \frac{N_{i,j}}{N_{i,\text{in}}} \quad i = 1, \ldots, n_{\text{species}}; \quad j = 1, \ldots, n_{\text{streams}}. \tag{3.12}$$

Here, $\alpha_{i,j}$ is the flow rate of species i in output stream j divided by the flow rate of species i in the input stream. Utilizing species split fractions, the material balances are

$$N_{i,\text{in}} = \sum_{j=1}^{n_{\text{streams}}} N_{i,j} = \sum_{j=1}^{n_{\text{streams}}} \alpha_{i,j} N_{i,\text{in}} \quad i = 1, \ldots, n_{\text{species}} \tag{3.13}$$

and

$$\sum_{j=1}^{n_{\text{streams}}} \alpha_{i,j} = 1 \quad i = 1, \ldots, n_{\text{species}}. \tag{3.14}$$

The species material balance utilizing the total molar flow rate for each stream is

$$N_j x_{i,j} - \alpha_{i,j} N_{\text{in}} x_{i,\text{in}} = 0 \quad i = 1, \ldots, n_{\text{species}}; \quad j = 1, \ldots, n_{\text{streams}} \tag{3.15}$$

with the mole fraction normalization equations for the input and output given by Equations (3.10) and (3.11).

Total Flow Balance

The total molar flow rate in equals the sum of the total molar flow rates out:

$$N_{\text{in}} - \sum_{j=1}^{n_{\text{streams}}} N_j = 0. \tag{3.16}$$

3.1.3 Splitter

A splitter, as shown in Figure 3.3, divides an input stream into two or more output streams such that the species composition in every stream is the same.

Species Balances

As the species compositions are equal in all streams,

$$x_{i,\text{in}} - x_{i,j} = 0 \quad i = 1, \ldots, n_{\text{species}}; \quad j = 1, \ldots, n_{\text{streams}}. \tag{3.17}$$

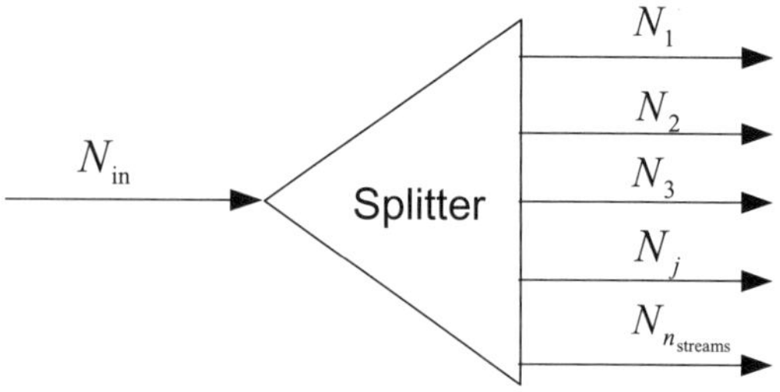

Figure 3.3 Splitter module.

The mole fraction normalization equation is

$$\sum_{i=1}^{n_{\text{species}}} x_{i,\text{in}} = 1. \tag{3.18}$$

Mole fraction normalization equations are generally not written on the output streams as the inlet and outlet species compositions are identical.

Splitter balances can also be written in terms of split fractions, α_j:

$$\alpha_j = \frac{N_j}{N_{\text{in}}} \quad j = 1, \ldots, n_{\text{streams}}. \tag{3.19}$$

Here, α_j is the flow rate of the output stream j divided by the flow rate of the input stream. Utilizing split fractions, the material balances are

$$N_{i,\text{in}} = \sum_{j=1}^{n_{\text{streams}}} N_{i,j} = \sum_{j=1}^{n_{\text{streams}}} \alpha_j N_{i,\text{in}} \quad i = 1, \ldots, n_{\text{species}} \tag{3.20}$$

and

$$\sum_{j=1}^{n_{\text{streams}}} \alpha_j = 1. \tag{3.21}$$

The species material balances utilizing the total molar flow rate for each stream are

$$N_j x_{i,j} - \alpha_j N_{\text{in}} x_{i,\text{in}} = 0 \quad i = 1, \ldots, n_{\text{species}}; \quad j = 1, \ldots, n_{\text{streams}}, \tag{3.22}$$

with the mole fraction normalization equation for the input given by Equation (3.18).

Total Flow Balance

The total molar flow rate in equals the sum of the total molar flow rates out:

$$N_{\text{in}} - \sum_{j=1}^{n_{\text{streams}}} N_j = 0. \tag{3.23}$$

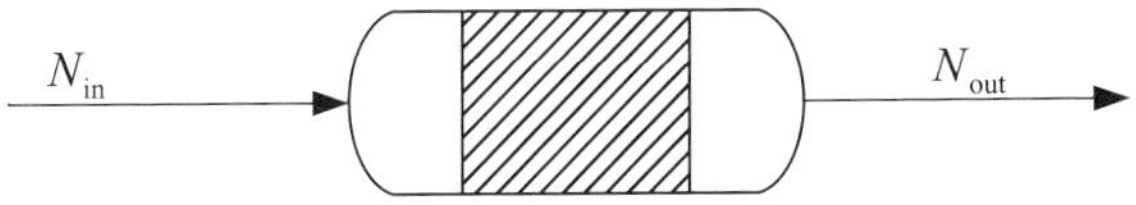

Figure 3.4 Reactor module.

3.1.4 Reactors

The reactor module is shown in Figure 3.4.

Species Balances

For the reactor, we must account for species reacting and products being created. Here, the species material balance, molar rate out = molar rate in, is no longer valid and must be modified to include the molar production or depletion rate for each species (R_i):

$$N_{i,\text{out}} = N_{i,\text{in}} + R_i. \tag{3.24}$$

The molar production rate R_i requires accounting for all reactions, which is conveniently done by defining an extent of reaction (ξ_m) for each independent reaction. For $n_{\text{reactions}}$ independent reactions,

$$R_i = \sum_{m=1}^{n_{\text{reactions}}} \nu_{m,i}\xi_m, \tag{3.25}$$

and here, ($\upsilon_{m,i}$) is the stoichiometric coefficient for each species in the reaction m. The stoichiometric coefficient is generally taken as a positive quantity for products and a negative quantity for reactants. For a reactor with a single reaction, $\xi = (N_{i,\text{out}} - N_{i,\text{in}})/\upsilon_i$. For example, for the single reaction of nitrogen and hydrogen to produce ammonia, $N_2 + 3H_2 = 2NH_3$, we can write the extent of reaction in terms of NH_3 as $\xi = (N_{NH_3,\text{out}} - N_{NH_3,\text{in}})/2$.

For a reactor with multiple reactions (see Problem 3.22), the species material balances are

$$N_{i,\text{out}} = N_{i,\text{in}} + \sum_{m=1}^{n_{\text{reactions}}} (\xi_m \nu_{m,i}) \quad i = 1, \ldots, n_{\text{species}}. \tag{3.26}$$

The species material balances utilizing the total molar flow rate for each stream is

$$N_{\text{out}} x_{i,\text{out}} = N_{\text{in}} x_{i,\text{in}} + \sum_{m=1}^{n_{\text{reactions}}} (\xi_m \nu_{m,i}) \quad i = 1, \ldots, n_{\text{species}}, \tag{3.27}$$

with the mole fraction normalization equations for the input and output:

$$\sum_{i=1}^{n_{\text{species}}} x_{i,\text{out}} = 1 \tag{3.28}$$

and

$$\sum_{i=1}^{n_{\text{species}}} x_{i,\text{in}} = 1. \tag{3.29}$$

In Chapter 5, we develop R_i when provided with the expression for a single rate of reaction, and in Chapter 16, we develop the general expression for R_i accounting for rates of reaction r_m for multiple reactions.

An alternative formulation for reactor material balances, which is especially useful for data reconciliation (Chapter 6), involves the use of elemental balances (Carbon in = Carbon out, Oxygen in = Oxygen out, etc.). The law of mass conservation states that in an open steady-state system, the mass (or number of moles) of each element (C, H, O, etc.) in the system is conserved in both reacting and nonreacting systems. Here, let $a_{e,i}$ represent the number of atoms a of element e in species i; for example, $a_{\text{carbon, ethane}} = 2$. If there are n_{elements}, then the elemental material balances based on the species molar flow rate in each stream are

$$\sum_{i=1}^{n_{\text{species}}} (a_{e,i} N_{i,\text{in}} - a_{e,i} N_{i,\text{out}}) = 0 \quad e = 1, \ldots, n_{\text{elements}}. \tag{3.30}$$

The elemental material balances utilizing the total molar flow rate for each stream are

$$\sum_{i=1}^{n_{\text{species}}} (a_{e,i} N_{\text{in}} x_{i,\text{in}} - a_{e,i} N_{\text{out}} x_{i,\text{out}}) = 0 \quad e = 1, \ldots, n_{\text{elements}}, \tag{3.31}$$

with the mole fraction normalization in Equations (3.28) and (3.29).

The obvious advantage of the elemental balances is the elimination of both the extent of reaction and the requirement that the stoichiometric coefficients be known. However, as we will see in Chapter 5, if a reactor energy balance is required, the extent of the reaction formulation is sometimes preferred.

The remainder of this chapter examines the solution of material balance problems using the elementary modules we developed in this section. A sequential modular solution approach is used with Excel as the calculation platform. Recycle loops in the process necessitate an iterative solution, which is addressed using direct substitution and acceleration strategies.

We also examine material balance problems where a natural process specification has been replaced by an alternative specification. Here we develop single-variable optimization strategies to solve the material balance problem. We will find that the VBA code can be useful in implementing optimization strategies.

3.2 SEQUENTIAL MODULAR APPROACH: MATERIAL BALANCES WITH RECYCLE

The sequential modular solution approach for the process material balance problem treats each processing unit as a "stand-alone" problem. The necessary material balance equations have been developed in Sections 3.1.1–3.1.4. If the species flows into each processing unit are known, and if unit parameters (e.g., split fractions or extents of reaction) are known, the species flows exiting each unit can be determined. We want to examine the sequential modular approach in more detail by solving for species flows in a simplified chemical processing plant—styrene production from ethylbenzene.

EXAMPLE 3.1 ***Styrene Production Material Balances***

Styrene (C_8H_8) and hydrogen (H_2) are produced from ethylbenzene (C_8H_{10}) by catalytic dehydrogenation. Styrene production from ethylbenzene requires a large amount of steam, both to heat the reactor and to prevent coking; coking is the reaction that produces coke, a C_xH_y polymer (where often, $x \sim y$). In the simplified process shown in Figure 3.5, ethylbenzene is reacted, products are separated, and unreacted ethylbenzene and some styrene are recycled.

The fresh feed consists of 3000 mol/h of steam (H_2O) and 100 mol/h of ethylbenzene. Ethylbenzene conversion in the reactor is 65%. The reactor out stream contains styrene, unreacted ethylbenzene, steam, and hydrogen. The reactor out stream is cooled in separator 1, creating three phases: a water phase, an organic phase, and a vapor phase. The water phase is 100% water; the vapor phase is 100% hydrogen; and the organic phase contains an unreacted ethylbenzene and styrene product. In separator 2 (distillation), the organic phase is separated into a styrene-rich product stream and an ethylbenzene-rich recycle stream. The product stream, on a molar flow basis, contains 99% of the styrene produced and 1% of the ethylbenzene in the Organic Out stream. Solve for all species flow rates using a sequential modular approach on an Excel sheet.

SOLUTION The first step is to assemble the material balance equations for this system. The process consists of three of the four elementary modules we developed in Sections 3.1.1–3.1.4, a mixer, a reactor, and two separators.

Mixer

For the mixer, we write, molar rate out = molar rate in, for each species using Equation (3.3):

$$C_8H_{10}\text{ balance: } N_{C_8H_{10}}^{\text{reactor in}} = N_{C_8H_{10}}^{\text{feed}} + N_{C_8H_{10}}^{\text{recycle}}, \tag{3.32}$$

$$C_8H_8\text{ balance: } N_{C_8H_8}^{\text{reactor in}} = N_{C_8H_8}^{\text{feed}} + N_{C_8H_8}^{\text{recycle}}, \tag{3.33}$$

$$H_2\text{ balance: } N_{H_2}^{\text{reactor in}} = N_{H_2}^{\text{feed}} + N_{H_2}^{\text{recycle}}, \tag{3.34}$$

and

$$H_2O\text{ balance: } N_{H_2O}^{\text{reactor in}} = N_{H_2O}^{\text{feed}} + N_{H_2O}^{\text{recycle}}. \tag{3.35}$$

The overall balance (Eq. (3.7)), $N_{\text{reactor in}} = N_{\text{feed}} + N_{\text{recycle}}$, is the sum of the four species balances. The molar feed rate of each species would constitute natural specifications, and here, $N_{C_8H_{10}}^{\text{feed}} = 100$ mol/h, $N_{C_8H_8}^{\text{feed}} = 0$ mol/h, $N_{H_2}^{\text{feed}} = 0$ mol/h, and $N_{H_2O}^{\text{feed}} = 3000$ mol/h. We also know from the problem statement that $N_{H_2}^{\text{recycle}} = 0$ mol/h and $N_{H_2O}^{\text{recycle}} = 0$ mol/h, but for now, we want to write Equations (3.32)–(3.35) in their general form.

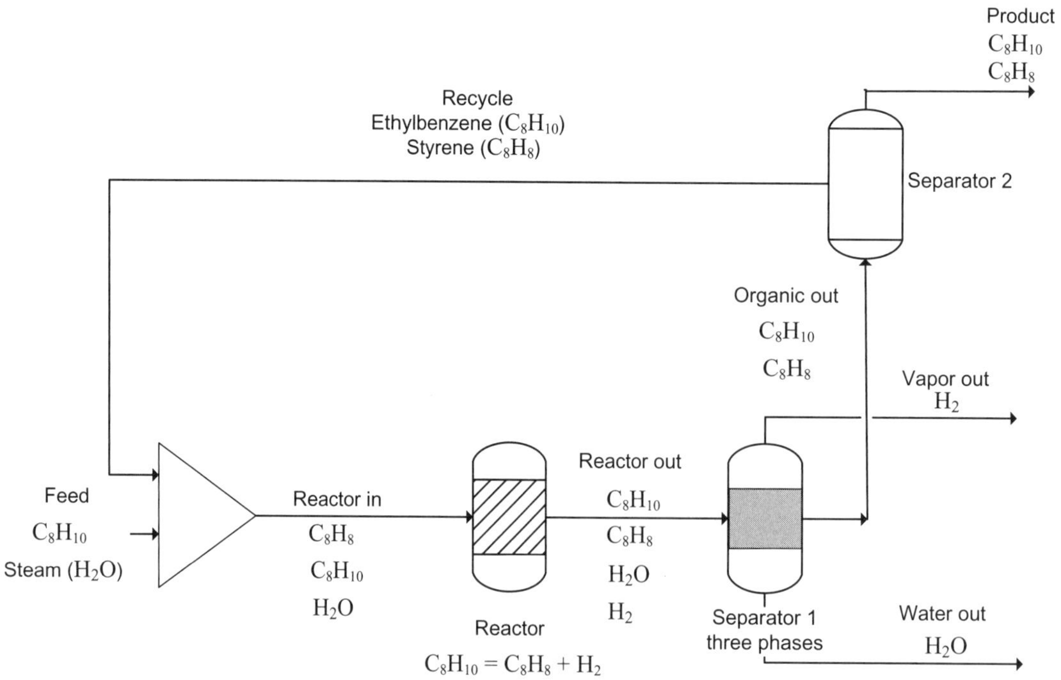

Figure 3.5 Flow sheet for styrene production.

Reactor

Here, there is only one reaction and, for each mole of ethylbenzene reacted, one mole of styrene and one mole of hydrogen will be produced. The stoichiometric coefficient for depletion of ethylbenzene is −1 and the stoichiometric coefficients for the production of styrene and hydrogen are both +1. The molar extent of reaction (ξ) is given as 65% of the ethylbenzene fed to the reactor, $\xi = 0.65(N_{C_8H_{10}}^{\text{reactor in}})$. Also from Equation (3.26), $\xi = (N_{C_8H_{10}}^{\text{reactor out}} - N_{C_8H_{10}}^{\text{reactor in}})/\nu_{C_8H_{10}}$; $\xi = (N_{C_8H_8}^{\text{reactor out}} - N_{C_8H_8}^{\text{reactor in}})/\nu_{C_8H_8}$; and $\xi = (N_{H_2}^{\text{reactor out}} - N_{H_2}^{\text{reactor in}})/\nu_{H_2}$, so the material balances for the reactor are

$$C_8H_{10}\text{ balance: } N_{C_8H_{10}}^{\text{reactor out}} = N_{C_8H_{10}}^{\text{reactor in}} + (-1)(0.65)N_{C_8H_{10}}^{\text{reactor in}}, \quad (3.36)$$

$$\text{Styrene balance: } N_{C_8H_8}^{\text{reactor out}} = N_{C_8H_8}^{\text{reactor in}} + (+1)(0.65)N_{C_8H_{10}}^{\text{reactor in}}, \quad (3.37)$$

$$H_2\text{ balance: } N_{H_2}^{\text{reactor out}} = N_{H_2}^{\text{reactor in}} + (+1)(0.65)N_{C_8H_{10}}^{\text{reactor in}}, \quad (3.38)$$

and

$$H_2O\text{ balance: } N_{H_2O}^{\text{reactor out}} = N_{H_2O}^{\text{reactor in}}. \quad (3.39)$$

Note that in these balances, the production of both styrene and hydrogen is accounted for by the amount of ethylbenzene reacting, $(0.65)N_{C_8H_{10}}^{\text{reactor in}}$. Because the downstream separator removes 100% of the hydrogen produced, $N_{H_2}^{\text{reactor in}} = 0$.

Separator 1

Separator 1 operates so that all steam fed to the reactor and all hydrogen produced in the reaction are removed. The remaining styrene and unreacted ethylbenzene comprise the organic stream from the separator. Here, the separator species split fraction specifications are the following: for C_8H_{10}, $\alpha_{C_8H_{10}}^{\text{organic out}} = 1.0$; for styrene, $\alpha_{C_8H_8}^{\text{organic out}} = 1.0$; for H_2, $\alpha_{H_2}^{\text{vapor out}} = 1.0$; and for H_2O, $\alpha_{H_2O}^{\text{water out}} = 1.0$. The material balances using Equation (3.12) are

$$C_8H_{10}\text{ balance: } N_{C_8H_{10}}^{\text{organic out}} = \left(\alpha_{C_8H_{10}}^{\text{organic out}}\right)\left(N_{C_8H_{10}}^{\text{reactor out}}\right) = N_{C_8H_{10}}^{\text{reactor out}}, \quad (3.40)$$

$$\text{Styrene balance: } N_{C_8H_8}^{\text{organic out}} = \left(\alpha_{C_8H_8}^{\text{organic out}}\right)\left(N_{C_8H_8}^{\text{reactor out}}\right) = N_{C_8H_8}^{\text{reactor out}}, \quad (3.41)$$

$$H_2\text{ balance: } N_{H_2}^{\text{vapor out}} = \left(\alpha_{H_2}^{\text{vapor out}}\right)\left(N_{H_2}^{\text{reactor out}}\right) = N_{H_2}^{\text{reactor out}}, \quad (3.42)$$

and

$$H_2O\text{ balance: } N_{H_2O}^{\text{water out}} = \left(\alpha_{H_2O}^{\text{water out}}\right)\left(N_{H_2O}^{\text{reactor out}}\right) = N_{H_2O}^{\text{reactor out}}. \quad (3.43)$$

Separator 2

Separator 2 operates so that 1% of the ethylbenzene and 99% of styrene (on a molar flow basis) in the organic stream are removed in a product stream, and the remaining species are recycled.

Using the species split fraction specifications and Equation (3.12), the species flow rates for the product stream are

$$C_8H_{10}\text{ specification: } N_{C_8H_{10}}^{\text{product}} = (0.01)\left(N_{C_8H_{10}}^{\text{organic out}}\right) \quad (3.44)$$

and

$$C_8H_8\text{ specification: } N_{C_8H_8}^{\text{product}} = (0.99)\left(N_{C_8H_8}^{\text{organic out}}\right). \quad (3.45)$$

There is no connection between the value 0.01 in Equation (3.44) and 0.99 in Equation (3.45); these equations provide two independent specifications.

The species flow rates for the recycle stream can be found using Equations (3.12) and (3.14) or Equation (3.8); using Equation (3.8),

$$C_8H_{10}\text{ balance: } N_{C_8H_{10}}^{\text{product}} + N_{C_8H_{10}}^{\text{recycle}} = N_{C_8H_{10}}^{\text{organic out}} \quad (3.46)$$

$$C_8H_8\text{ balance: } N_{C_8H_8}^{\text{product}} + N_{C_8H_8}^{\text{recycle}} = N_{C_8H_8}^{\text{organic out}}. \quad (3.47)$$

If we examine Figure 3.5, there are 17 unknown flow rates and one unknown extent of reaction for a total of 18 unknowns. We assembled 13 independent material balances (Eqs. (3.32)–(3.34), (3.36)–(3.43), and (3.46)–(3.47)). The problem description provides two feed flow rates (two independent specifications), a known extent of reaction, and two product specifications (Eqs. (3.44) and (3.45)). In general terms then, we have 18 unknowns and 18 equations, so a solution is possible.

We can assemble this information on an Excel sheet and if we keep the material balance equation sets grouped on the sheet (first the mixer, then the reactor, then separator 1, and then separator 2), we will solve the system by a sequential modular solution approach. In the sequential modular approach, calculations are designed to produce the output species flow rates from the known input species flow rates for each unit operation.

The start of the Excel solution is provided in **Example 3.1a.xls** and is shown in Figure 3.6a. Named cells follow the material balance equations we developed above (use the crtl and ~ keys to reveal the equations used in the construction of the sheet). For example, the first four equations in column Reactor_In can be directly compared to Equations (3.32)–(3.35). Equation (3.35) does not supply independent information, but it is useful to determine the total molar flow rate in a stream if mole fractions are needed. Notice here we have not yet calculated the species flow rates in the recycle stream. Initially, we set the species flow rates in the recycle stream to be zero to allow the column labeled Reactor_In (which is the output of the mixer) to be solved.

Next, the reactor, separator 1, and then separator 2 are solved. This allows new values for the recycle stream, labeled as New Recycle, to be determined. The component flow rates should initially appear as in Figure 3.6a.

We can next bring the material balances for the recycle stream (Eqs. 3.46–3.47) from the New Recycle column to the Recycle column as shown in **Example 3.1b.xls**, and we will obtain the converged values for the flow sheet as shown in Figure 3.6b.

In Figure 3.6b, there is no difference in the equations for the Recycle column and the New Recycle column. We could now eliminate the New Recycle column, but it is convenient to retain this column for later discussions. What is important to see right now is that the recycle loop creates a circular logic problem as values in the Recycle column depend on values in the New Recycle column. Excel allows iterations to be performed to solve these circular logic equations. If a circular reference error occurs, go to the toolbar; click the tabs Tools → Options → Calculation. Then, be sure *Calculation* is set to Automatic, *Iteration* is checked, the *Maximum number of iterations* is 2000 and *Maximum change* is 0.000001. The addition of the recycle stream equations to the Recycle column creates a calculation loop, which will continue solving the sheet until values on the sheet stop changing to within the sheet convergence criterion parameter (Maximum change). ■

	Feed	Recycle	Reactor_In	Reactor_Out	Water_Out	Vapor_Out
N(C8H10)	100	=0	=EBZ_Feed+EBZ_Recycle	=EBZ_R_In - (Conversion)*EBZ_R_In	=EBZ_R_Out*0	=EBZ_R_Out*0
N(C8H8)	=0	=0	=Styrene_Feed+Styrene_Recycle	=Styrene_R_In+(Conversion)*EBZ_R_In	=Styrene_R_Out*0	=Styrene_R_Out*0
N(H2)	=0	=0	=H2_Feed+H2_Recycle	=H2_R_In+(Conversion)*EBZ_R_In	=H2_R_Out*0	=H2_R_Out*1
N(H2O)	3000	=0	=Water_Feed+Water_Recycle	=Water_R_In	=Water_R_Out*1	=Water_R_Out*0
N(total)	=SUM(B11:B14)	=SUM(C11:C14)	=SUM(D11:D14)	=SUM(E11:E14)	=SUM(F11:F14)	=SUM(G11:G14)

	Organic_Out	Product	New Recycle
N(C8H10)	=EBZ_R_Out*1	=EBZ_O_O*0.01	=EBZ_O_O-EBZ_Product
N(C8H8)	=Styrene_R_Out*1	=Styrene_O_O*0.99	=Styrene_O_O-Styrene_Product
N(H2)	=H2_R_Out*0	=H2_O_O*0	=H2_O_O-H2_Product
N(H2O)	=Water_R_Out*0	=Water_O_O*0	=Water_O_O-Water_Product
N(total)	=SUM(H11:H14)	=SUM(I11:I14)	=SUM(J11:J14)

	Feed	Recycle	Reactor_In	Reactor_Out	Water_Out	Vapor_Out	Organic_Out	Product	New Recycle
N(C8H10)	100	0	100	35	0	0	35	0.35	34.65
N(C8H8)	0	0	0	65	0	0	65	64.35	0.65
N(H2)	0	0	0	65	0	65	0	0	0
N(H2O)	3000	0	3000	3000	3000	0	0	0	0
N(total)	3100	0	3100	3165	3000	65	100	64.7	35.3

(a)

	Feed	Recycle	Reactor_In	Reactor_Out	Water_Out	Vapor_Out
N(C8H10)	100	=0	=EBZ_Feed+EBZ_Recycle	=EBZ_R_In - (Conversion)*EBZ_R_In	=EBZ_R_Out*0	=EBZ_R_Out*0
N(C8H8)	=0	=0	=Styrene_Feed+Styrene_Recycle	=Styrene_R_In+(Conversion)*EBZ_R_In	=Styrene_R_Out*0	=Styrene_R_Out*0
N(H2)	=0	=0	=H2_Feed+H2_Recycle	=H2_R_In+(Conversion)*EBZ_R_In	=H2_R_Out*0	=H2_R_Out*1
N(H2O)	3000	=0	=Water_Feed+Water_Recycle	=Water_R_In	=Water_R_Out*1	=Water_R_Out*0
N(total)	=SUM(B11:B14)	=SUM(C11:C14)	=SUM(D11:D14)	=SUM(E11:E14)	=SUM(F11:F14)	=SUM(G11:G14)

	Organic_Out	Product	New Recycle
N(C8H10)	=EBZ_R_Out*1	=EBZ_O_O*0.01	=EBZ_O_O-EBZ_Product
N(C8H8)	=Styrene_R_Out*1	=Styrene_O_O*0.99	=Styrene_O_O-Styrene_Product
N(H2)	=H2_R_Out*0	=H2_O_O*0	=H2_O_O-H2_Product
N(H2O)	=Water_R_Out*0	=Water_O_O*0	=Water_O_O-Water_Product
N(total)	=SUM(H11:H14)	=SUM(I11:I14)	=SUM(J11:J14)

	Feed	Recycle	Reactor_In	Reactor_Out	Water_Out	Vapor_Out	Organic_Out	Product	New Recycle
N(C8H10)	100	53.02218824	153.0221882	53.55776588	0	0	53.55776588	0.5355777	53.02218823
N(C8H8)	0	1.004691135	1.004691135	100.4691135	0	0	100.4691135	99.464422	1.004691135
N(H2)	0	0	0	99.46442236	0	99.46442236	0	0	0
N(H2O)	3000	0	3000	3000	3000	0	0	0	0
N(total)	3100	54.02687938	3154.026879	3253.491302	3000	99.46442236	154.0268794	100	54.02687936

(b)

Figure 3.6 (a) Styrene production, revealed equations, and initial Excel material balances. (b) Styrene production, revealed equations, and converged Excel material balances.

Iterations and Change in Excel

You will also notice on the Excel sheets that we have supplied a cell, C4, to keep track of the number of iterations required to converge the flow sheet. Using the sequence Tools → Options → (Calculation tab) check to be sure the Maximum iterations is ≥2000 and that the Maximum change is ≤0.000001 (E-6). In order to compare different solution methods, it is often very useful to keep track of the number of iterations taken. The values internal to Excel are not readily available, but iteration counting can be accomplished directly on the spreadsheet.

An external iteration counter has been placed in cell C4 by using the equation =IF(C3=0,0,C4 + 0.00000001). If cell C3 = 0, then cell C4 is set to zero and the number of iterations is not tracked on the spreadsheet. However, if cell C3 ≠ 0, then the number of iterations is tracked on the spreadsheet by increasing C4 by E-8 each time the spreadsheet is calculated. It is tempting to increment the counter cell by 1 each time the spreadsheet is calculated. However, the counter (in C4) must be increased by a value less than the Maximum change (E-6) we specified, or the spreadsheet will continue calculations until the Maximum iterations value is reached, something clearly not desired.

The Excel sheet convergence criterion parameter is met when the value in every active sheet cell, Value in cell$_n$, for two successive iterations (k and $k+1$) and $\left|\left(\text{Value in cell}_n^{[k+1]}\right)-\left(\text{Value in cell}_n^{[k]}\right)\right| \le \varepsilon$, the Maximum change.

In the next section, we formally develop the sequential modular direct substitution method, which provides a clear picture of how this type of iteration is performed. Following this, the Wegstein method to accelerate the iteration procedure is discussed.

3.3 UNDERSTANDING TEAR STREAM ITERATION METHODS

To initially solve the first unit (the mixer) in the flow sheet of Figure 3.5, we need values for species flow rates in both the feed stream and the recycle loop. However, species flow rates in the recycle loop are not known; units downstream of the mixer must be solved for to provide species flow rates in the recycle stream. Initially in Example 3.1, we set $N_{C_8H_{10}}^{recycle}$, $N_{C_8H_8}^{recycle}$, $N_{H_2}^{recycle}$, and $N_{H_2O}^{recycle}$ all to zero. This allows the mixer to be solved, which provides the needed input stream values for the next unit, the reactor. Unit by unit calculations continue until updated values for $N_{C_8H_{10}}^{recycle}$, $N_{C_8H_8}^{recycle}$, $N_{H_2}^{recycle}$, and $N_{H_2O}^{recycle}$ are determined. The mixer is then again solved directly using these updated values for species flows in the recycle stream. The iteration process on the recycle stream continues until values in the flow sheet are converged. This process of direct substitution of the recycle stream flow rate values is termed successive substitution.

Tear Stream and Tear Variables

If we consider the first mixer calculations in Figure 3.6 in a little more detail, we have actually used four equations for the mixer output/reactor input (Eqs. (3.32)–(3.35)):

$$C_8H_{10} \text{ balance: } N_{C_8H_{10}}^{\text{reactor in}} = N_{C_8H_{10}}^{\text{feed}} + N_{C_8H_{10}}^{\text{recycle}}$$
$$C_8H_8 \text{ balance: } N_{C_8H_8}^{\text{reactor in}} = N_{C_8H_8}^{\text{feed}} + N_{C_8H_8}^{\text{recycle}}$$
$$H_2 \text{ balance: } N_{H_2}^{\text{reactor in}} = N_{H_2}^{\text{feed}} + N_{H_2}^{\text{recycle}}$$
$$H_2O \text{ balance: } N_{H_2O}^{\text{reactor in}} = N_{H_2O}^{\text{feed}} + N_{H_2O}^{\text{recycle}}.$$

There are known/fixed values for: $N_{C_8H_{10}}^{feed}$, $N_{C_8H_8}^{feed}$, $N_{H_2}^{feed}$, and $N_{H_2O}^{feed}$, leaving eight unknowns: $N_{C_8H_{10}}^{\text{reactor in}}$, $N_{C_8H_8}^{\text{reactor in}}$, $N_{H_2}^{\text{reactor in}}$, $N_{H_2O}^{\text{reactor in}}$, $N_{C_8H_{10}}^{recycle}$, $N_{C_8H_8}^{recycle}$, $N_{H_2}^{recycle}$, and $N_{H_2O}^{recycle}$. We "tear" or remove the variables $N_{C_8H_{10}}^{recycle}$, $N_{C_8H_8}^{recycle}$, $N_{H_2}^{recycle}$, and $N_{H_2O}^{recycle}$ from the equation set, supply values for these tear variables, and then solve the equations for $N_{C_8H_{10}}^{\text{reactor in}}$, $N_{C_8H_8}^{\text{reactor in}}$, $N_{H_2}^{\text{reactor in}}$, and $N_{H_2O}^{\text{reactor in}}$. As the tear variables comprise the entire recycle stream, this is referred to as a "tear stream."

In Section 3.3, we examine in greater detail both single-variable and multivariable iteration methods, specifically two methods commonly used in flow sheeting problems: successive substitution and Wegstein methods.

3.3.1 Single-Variable Successive Substitution Method

In a successive substitution for single-variable problems (also called the fixed-point iteration method), we rearrange the equation $f(x) = 0$ to $x = g(x)$. An initial guess or estimate, $x^{[1]}$, is supplied and $g(x^{[1]})$ is solved. The solution for $g(x^{[1]})$ becomes the next estimate, $x^{[2]} = g(x^{[1]})$. In general, $x^{[k+1]} = g(x^{[k]})$.

EXAMPLE 3.2 ***Single-Variable Successive Substitution Redlich–Kwong Equation of State (EOS)***

In Chapter 8, we discuss thermodynamic EOS to predict fluid properties. The Redlich–Kwong EOS (Redlich and Kwong, 1949) was introduced in Chapter 2, Problem 2.7:

$$P = \frac{RT}{\upsilon - b} - \frac{a}{T^{0.5}\upsilon(\upsilon + b)}. \tag{3.48}$$

This EOS is a cubic equation in υ, which is the molar volume (cubic foot per pound-mole). Here, R is the gas constant 0.7302 ft^3-atm/lb-mol-R, T is temperature (R), P is the pressure (atmosphere), and a and b are species-dependent constants. Often, the pressure and temperature are known and we need to determine the molar volume. We can rearrange Equation (3.48) to the form $x = g(x)$:

$$\upsilon = \frac{RT}{P} - \frac{a(\upsilon - b)}{T^{0.5}\upsilon(\upsilon + b)P} + b. \tag{3.49}$$

Let us determine the molar volume for carbon dioxide at 80 atm and 700 R. The constants a and b for carbon dioxide are a = 21,972 atm-R$^{0.5}$(ft^3/lb-mol)^2 and b = 0.4755 ft^3/lb-mol.

The successive substitution solution, with an initial guess of $\upsilon^{[1]}$ = 300 ft^3/lb-mol, is given in **Example 3.2.xls** and shown in Figure 3.7. Generally, it is important to supply a good initial guess for iterative solutions. Here for example, the ideal gas assumption could be used for

$$\upsilon^{[1]} = \frac{RT}{P} = 6.39 \text{ ft}^3/\text{lb-mol}.$$

However, in this example, we want to see how and if successive substitution converges from an initial guess away from the solution.

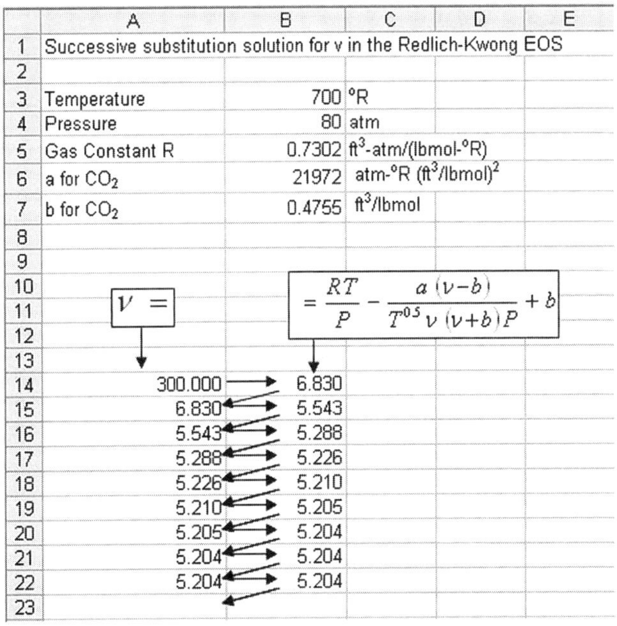

Figure 3.7 Successive substitution solution for υ in the Redlich–Kwong EOS.

The successive substitution solution converges to 5.204 ft^3/lb-mol (from an initial guess of $v^{[1]} = 300$ ft^3/lb-mol) with the iteration progress shown in Figure 3.8a. The successive substitution will converge if, for point $x^{[k]}$, over the region from $x^{[1]}$ to $x^{[\text{solution}]}$,

$$\left|\frac{dg(x^{[k]})}{dx^{[k]}}\right| < 1.0. \tag{3.50}$$

■

If needed, Equation (3.50) can be checked by recalling that

$$\frac{dg(x^{[k]})}{dx^{[k]}}$$

can be written as

$$\frac{\Delta g(x)}{\Delta x} \quad \text{or} \quad \frac{g(x^{[k]}+\varepsilon)-g(x^{[k]})}{\varepsilon}$$

when using a forward difference approximation; ε is a small change in x values. For a problem where successive substitution diverges, see Problem 3.9.

3.3.2 Multidimensional Successive Substitution Method

We can extend successive substitution to multiple dimensions (also called the Gauss–Jacobi method) where we are solving a system of N equations with N unknowns (variables). Here, an initial guess, $x_n^{[1]}$ with $n = 1, 2, \ldots N$ (labeled as $x^{[1]}$), is made for each variable, and then each equation $g_n(x^{[1]})$ is solved. Here, $g_n(x^{[1]}) = g_n(x_1^{[1]}, x_2^{[1]}, x_3^{[1]}, \ldots, x_N^{[1]})$ and $n = 1, 2, \ldots N$. The next estimate is $x_n^{[2]} = g_n(x^{[1]})$, and in general, $x_n^{[k+1]} = g_n(x^{[k]})$.

EXAMPLE 3.3 ***Successive Substitution Method for Recycle Stream of Example 3.1***

We used successive substitution to solve Example 3.1; here, we want to show the procedure more clearly.

Computer-aided design programs often default to successive substitution to solve the material balance equations. The need for

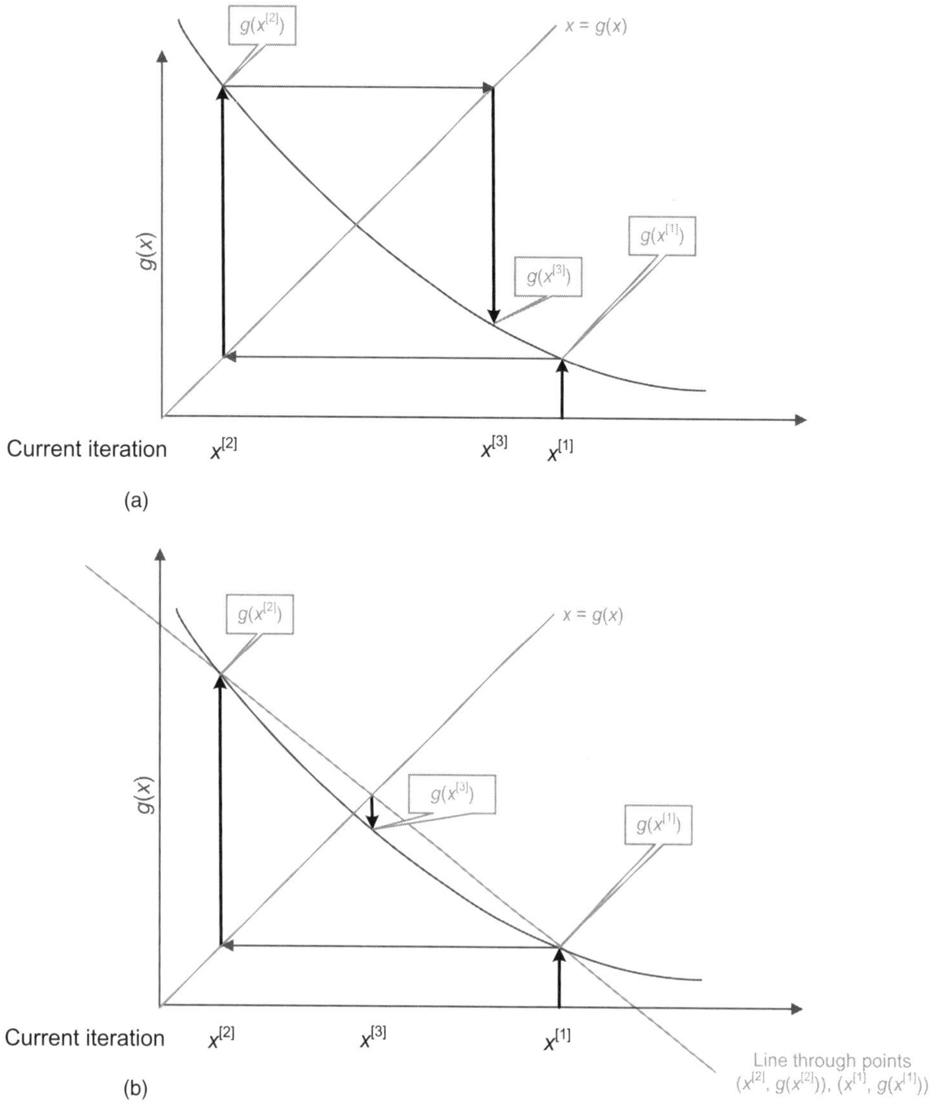

Figure 3.8 (a) Plot of the successive substitution method. (b) Plot of the Wegstein method (discussed in Section 3.3.3).

successive substitution arises in Example 3.1 because the mixer material balances require values for the species molar flow rates in the recycle stream. The successive substitution process can be started by first assuming that the species flow rates in the recycle stream are zero, $x_n^{[1]} = 0$, $n = 1, 2, \ldots N$ (where N = the number of species). This assumption allows the mixer to be solved. Subsequently, the equations for the reactor, separator, and purge are solved, giving new values for the recycle stream $g_n(x^{[1]})$. These new values for the recycle stream are substituted ($x_n^{[2]} = g_n(x^{[1]})$) into the balance equations for the mixer, and the process is repeated until convergence.

See the Excel solution in **Example 3.3a.xls**. When the spreadsheet opens, use Crtl ~ to reveal the equations used to solve the material balance problem.

As shown on Figure 3.9, in the first iteration ($k = 1$), the flow rates of all the components in the recycle loop (Recycle column) are set to zero; this column is ($x_n^{[1]}$). The material balance equations are sequentially solved, and this gives us new values for the flow rates in recycle loop. These values are found in the column New Recycle, $k = 1$; this column is $g_n(x^{[1]})$.

The values $g_n(x^{[1]})$ are then transferred to the Recycle column in the second iteration ($k = 2$); these values are then ($x_n^{[2]}$). The material balance equations are sequentially solved again, and this gives us new values for the flow rates in the recycle loop. These values are found in the column New Recycle, $k = 2$; this column is $g_n(x^{[2]})$.

The process of creating material balances could be repeated down the Excel sheet or we can simply copy $g_n(x^{[2]})$back to the current Recycle column ($k = 1$); these values are then ($x_n^{[3]}$) (see Figure 3.6). When this substitution for $x_n^{[3]}$ is made, a continuous loop is created, which stops when the maximum iterations or the convergence criterion parameter (maximum change), as set in Excel, is reached.

The results should appear as in **Example 3.3b.xls**. ■

The converged values in Figure 3.10 are identical to the values found in Example 3.1. Note that with the current structure of our material balance equations, successive substitution requires that all natural specifications (e.g., the species feed rates) be known.

In Figure 3.10, the $g_n(x^{[2]})$ (cells J21:J25) values, which were copied back to $x_n^{[1]}$ to create $x_n^{[3]}$, and the successive substitution loop, use *absolute addressing ($row $column).*

	New Recycle
$N_{C_8H_{10}}$	=$H$21 – $I$21
$N_{C_8H_8}$	=$H$22 – $I$22
N_{H_2}	=$H$23 – $I$23
N_{H_2O}	=$H$24 – $I$24
N_{total}	=SUM(K21:K24)

The use of absolute addressing, which is analogous to using named cells, ensures that values from these specific cells are used in creating the subsequent $x_n^{[k]}$ values.

EXAMPLE 3.4 *Successive Substitution Directly on Flow Sheet*

It is often convenient to simply create the successive substitution loop directly on the process flow sheet. The process is started in the Excel file **Example 3.4a.xls**, which is shown in Figure 3.11.

As practice, the reader should add the needed recycle balances. As a check, the final result is given in the Excel file **Example 3.4b.xls**. In **Example 3.4b.xls**, the two needed balances for the recycle (EBZ_Recycle = EBZ_Organic – EBZ_Product; Styrene_Recycle = Styrene_Organic – Styrene_Product) have been added. The flow sheet will iterate and converge via successive substitution. ■

	A	B	C	D	E	F	G	H	I	J
1	*Styrene Production - Successive Substitution for Recycle Stream*									
2										
3		Trigger	0	(either 0 to reset or 1 to iterate)						
4		Iterations	0.00000000	iteration count x 10^{-8}						
5										
6		Reaction	C8H10 ---> C8H8 + H2		Conversion	0.65				
7										
8	**Iteration #1**		(k=1)							
9			(x_n^1)							$g_n(x_n^1)$
10		**Feed**	**Recycle**	**Reactor_In**	**Reactor_Out**	**Water_Out**	**Vapor_Out**	**Organic_Out**	**Product**	**New Recycle**
11	N(C8H10)	100	0	100	35	0	0	35	0.35	34.65
12	N(C8H8)	0	0	0	65	0	0	65	64.35	0.65
13	N(H2)	0	0	0	65	0	65	0	0	0
14	N(H2O)	3000	0	3000	3000	3000	0	0	0	0
15	N(total)	3100	0	3100	3165	3000	65	100	64.7	35.3
16										
17										
18	**Iteration #2**		(k=2)							
19			(x_n^2)							$g_n(x_n^2)$
20		**Feed**	**Recycle**	**Reactor_In**	**Reactor_Out**	**Water_Out**	**Vapor_Out**	**Organic_Out**	**Product**	**New Recycle**
21	N(C8H10)	100	34.65	134.65	47.1275	0	0	47.1275	0.471275	46.656225
22	N(C8H8)	0	0.65	0.65	88.1725	0	0	88.1725	87.290775	0.881725
23	N(H2)	0	0	0	87.5225	0	87.5225	0	0	0
24	N(H2O)	3000	0	3000	3000	3000	0	0	0	0
25	N(total)	3100	35.3	3135.3	3222.8225	3000	87.5225	135.3	87.76205	47.53795

Figure 3.9 Styrene production—first two iterations of successive substitution.

	A	B	C	D	E	F	G	H	I	J
1	*Styrene Production - Successive Substitution for Recycle Stream*									
2										
3		Trigger	0	(either 0 to reset or 1 to iterate)						
4		Iterations	0.00000000	iteration count x 10^{-8}						
5										
6		Reaction	C8H10 ---> C8H8 + H2		Conversion	0.65				
7										
8	**Iteration #k**									
9			(x_n^k)							$g_n(x_n^k)$
10		**Feed**	**Recycle**	**Reactor_In**	**Reactor_Out**	**Water_Out**	**Vapor_Out**	**Organic_Out**	**Product**	**New Recycle**
11	N(C8H10)	100	53.02218822	153.0221882	53.55776588	0	0	53.55776588	0.5355777	53.02218822
12	N(C8H8)	0	1.004691135	1.004691135	100.4691135	0	0	100.4691135	99.464422	1.004691135
13	N(H2)	0	0	0	99.46442234	0	99.46442234	0	0	0
14	N(H2O)	3000	0	3000	3000	3000	0	0	0	0
15	N(total)	3100	54.02687935	3154.026879	3253.491302	3000	99.46442234	154.0268794	100	54.02687935
16										
17										
18	**Iteration #k+1**									
19			(x_n^{k+1})							$g_n(x_n^{k+1})$
20		**Feed**	**Recycle**	**Reactor_In**	**Reactor_Out**	**Water_Out**	**Vapor_Out**	**Organic_Out**	**Product**	**New Recycle**
21	N(C8H10)	100	53.02218822	153.0221882	53.55776588	0	0	53.55776588	0.5355777	53.02218822
22	N(C8H8)	0	1.004691135	1.004691135	100.4691135	0	0	100.4691135	99.464422	1.004691135
23	N(H2)	0	0	0	99.46442234	0	99.46442234	0	0	0
24	N(H2O)	3000	0	3000	3000	3000	0	0	0	0
25	N(total)	3100	54.02687935	3154.026879	3253.491302	3000	99.46442234	154.0268794	100	54.02687935

Figure 3.10 Styrene production—successive substitution loop converged.

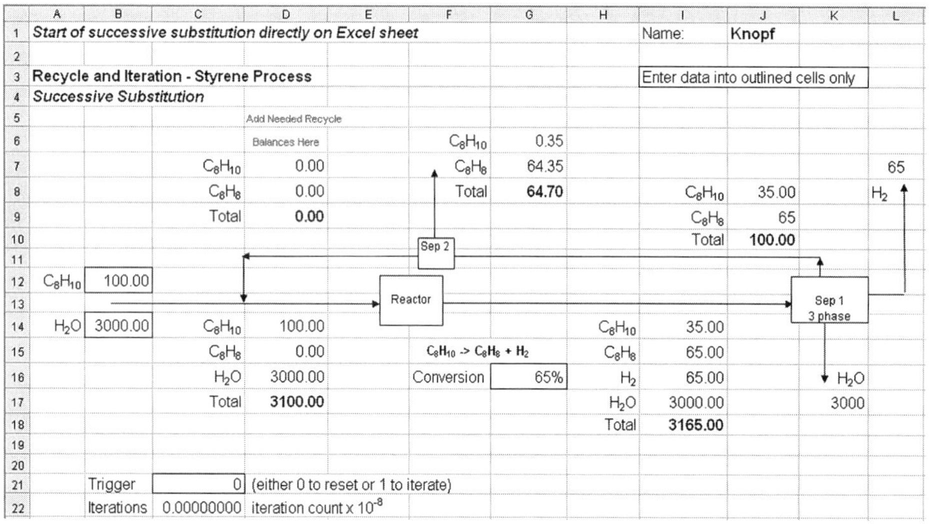

Figure 3.11 Styrene production—start of successive substitution on flow sheet.

Successive substitution generally works well for the closure of recycle loops. It is possible to accelerate recycle loop convergence by utilizing information from past iterations. Next, we discuss one such method, the Wegstein acceleration method.

The Wegstein Solution Method

The Wegstein acceleration method is commonly used to accelerate the closure of recycle loops in computer-aided material balances—it can often be more stable, more likely to reach to the final solution, when compared with successive substitution. The technique takes advantage of information from previous trials to reduce the number of iterations needed to converge the recycle loop. At least two trials must be made before the method can be applied.

3.3.3 Single-Variable Wegstein Method

The Wegstein method generally uses successive substitution for the first two iterations with $x^{[1]} \rightarrow g_n(x^{[1]}) \rightarrow x^{[2]} \rightarrow g_n(x^{[1]})$. If convergence is not satisfied, the third trial point is found by constructing a linear approximation of $g(x)$, whose slope is found by using the two available trial points and their results. This procedure is shown in Figure 3.8b. Here the slope is

$$\text{Slope} = \left(\frac{g(x^{[2]}) - g(x^{[1]})}{x^{[2]} - x^{[1]}}\right), \tag{3.51}$$

giving

$$g(x) = \left(\frac{g(x^{[2]}) - g(x^{[1]})}{x^{[2]} - x^{[1]}}\right)x + \text{intercept}. \tag{3.52}$$

At the point ($x^{[2]}$, $g(x^{[2]})$), the linear function ("$y = mx + b$") is

$$g(x^{[2]}) = \left(\frac{g(x^{[2]}) - g(x^{[1]})}{x^{[2]} - x^{[1]}}\right)x^{[2]} + \text{intercept}; \tag{3.53}$$

solving for the intercept,

$$\text{intercept} = g(x^{[2]}) - \left(\frac{g(x^{[2]}) - g(x^{[1]})}{x^{[2]} - x^{[1]}}\right)x^{[2]}. \tag{3.54}$$

Therefore, the linear approximation of $g(x)$ is given by

$$g(x) = \left(\frac{g(x^{[2]}) - g(x^{[1]})}{x^{[2]} - x^{[1]}}\right)x + g(x^{[2]}) - \left(\frac{g(x^{[2]}) - g(x^{[1]})}{x^{[2]} - x^{[1]}}\right)x^{[2]}. \tag{3.55}$$

At our converged solution, we require $x = g(x)$, and for that reason, we use $x^{[3]} = g(x^{[3]})$, giving

$$g(x^{[3]}) = x^{[3]} = \left(\frac{g(x^{[2]}) - g(x^{[1]})}{x^{[2]} - x^{[1]}}\right)x^{[3]} + g(x^{[2]}) - \left(\frac{g(x^{[2]}) - g(x^{[1]})}{x^{[2]} - x^{[1]}}\right)x^{[2]}. \tag{3.56}$$

Solving for $x^{[3]}$,

$$x^{[3]} = \left(\frac{1}{1 - \left(\frac{g(x^{[2]}) - g(x^{[1]})}{x^{[2]} - x^{[1]}}\right)}\right)\left(g(x^{[2]}) - \left(\frac{g(x^{[2]}) - g(x^{[1]})}{x^{[2]} - x^{[1]}}\right)x^{[2]}\right) \tag{3.57}$$

or

$$x^{[3]} = \left(\frac{1}{1 - \text{Slope}}\right)g(x^{[2]}) - \left(\frac{\text{Slope}}{1 - \text{Slope}}\right)x^{[2]}. \tag{3.58}$$

Setting

$$\theta = \left(\frac{1}{1 - \text{Slope}}\right),$$

the Wegstein method is often written as

$$x^{[3]} = (\theta)g(x^{[2]}) + (1 - \theta)x^{[2]}, \tag{3.59}$$

and, in general,

$$x^{[k+1]} = (\theta)g(x^{[k]}) + (1 - \theta)x^{[k]} \quad (k = 2, 3, \ldots, n_{\text{iterations}}). \tag{3.60}$$

The degree of extrapolation is often limited by setting $\theta = \theta_{max}$ if $\theta > \theta_{max}$, and setting $\theta = \theta_{min}$ if $\theta < \theta_{min}$. This iteration process is then termed a bounded Wegstein method with typical values for θ_{max} and θ_{min} being $\theta_{max} = 10$ and $\theta_{min} = -10$.

EXAMPLE 3.5 ***Single-Variable Wegstein Method for Redlich–Kwong EOS***

Solve the Redlich–Kwong EOS problem of Example 3.2 using the bounded Wegstein method. The solution, with an initial guess of $\upsilon = 300$ cm^3/gmol, is given in **Example 3.5.xls** and shown in Figure 3.12.

The Wegstein method converges to 5.204 ft^3/lb-mol. For the Wegstein method, we also need a series of calculation checks to prevent division by zero. Using the If construction, IF(calculation, calculation true, calculation false):
for the slope,

$$IF\left(x^{[2]} - x^{[1]} = 0, 0, \left(\frac{g(x^{[2]}) - g(x^{[1]})}{x^{[2]} - x^{[1]}}\right)\right);$$ see IF statement column C in the example.
for θ,

$$IF\left(1 - \text{Slope} = 0, 1, \left(\frac{1}{1 - \text{Slope}}\right)\right);$$ see IF statement column D in the example and checks on θ_{max} and θ_{min}; see columns E and F in the example.

3.3.4 Multidimensional Wegstein Method

In the single-dimensional Wegstein method, a linear approximation of $g(x)$ was made by using values at the previous two iteration points. This procedure can be extended to multidimensions, giving for n variables:

$$\text{Slope}_n = \left(\frac{g_n(x^{[2]}) - g_n(x^{[1]})}{x_n^{[2]} - x_n^{[1]}}\right). \tag{3.61}$$

Following the single-variable development, letting

$$\theta_n = \left(\frac{1}{1 - \text{Slope}_n}\right), \tag{3.62}$$

$$x_n^{[k+1]} = (\theta_n)g_n(x^{[k]}) + (1 - \theta_n)x^{[k]} \quad (k = 2, 3, \ldots, n_{\text{iterations}}). \tag{3.63}$$

For the multidimensional case, the degree of extrapolation is limited by setting $\theta_n = \theta_{max}$ if $\theta > \theta_{max}$ and by setting $\theta_n = \theta_{min}$ if $\theta_n < \theta_{min}$. Typical values are $\theta_{max} = 10$ and $\theta_{min} = -10$.

EXAMPLE 3.6 ***Wegstein Method for Recycle Stream Using Excel Sheet***

Solve Example 3.1 using the bounded Wegstein method to converge the recycle loop.

	A	B	C	D	E	F	G	H	I	J
1	Wegstein solution for v in the Redlich-Kwong EOS									
2										
3	Temperature	700	°R		$\theta_{min} =$	-10				
4	Pressure	80	atm		$\theta_{max} =$	10				
5	Gas Constant R	0.7302	ft³-atm/(lbmol-°R)							
6	a for CO_2	21972	atm-°R (ft³/lbmol)²							
7	b for CO_2	0.4755	ft³/lbmol							
8										
9										
10										
11-13	$v =$	$= \dfrac{RT}{P} - \dfrac{a(v-b)}{T^{0.5}v(v+b)P} + b$				Slope $= \left(\dfrac{g(x^2)-g(x^1)}{x^2-x^1}\right)$	with division check			
14										
15	300.000	6.830					$\theta = \left(\dfrac{1}{1-Slope}\right)$			
16	6.830	5.543	0.004	1.004	1.004	1.004	with bounds			
17	5.537	5.287	0.198	1.247	1.247	1.247				
18	5.225	5.209	0.247	1.328	1.328	1.328				
19	5.204	5.204	0.259	1.350	1.350	1.350				
20	5.204	5.204	0.260	1.352	1.352	1.352				
21	5.204	5.204	0.260	1.352	1.352	1.352				
22	5.204	5.204	0.260	1.352	1.352	1.352				
23	5.204	5.204	0.000	1.000	1.000	1.000				

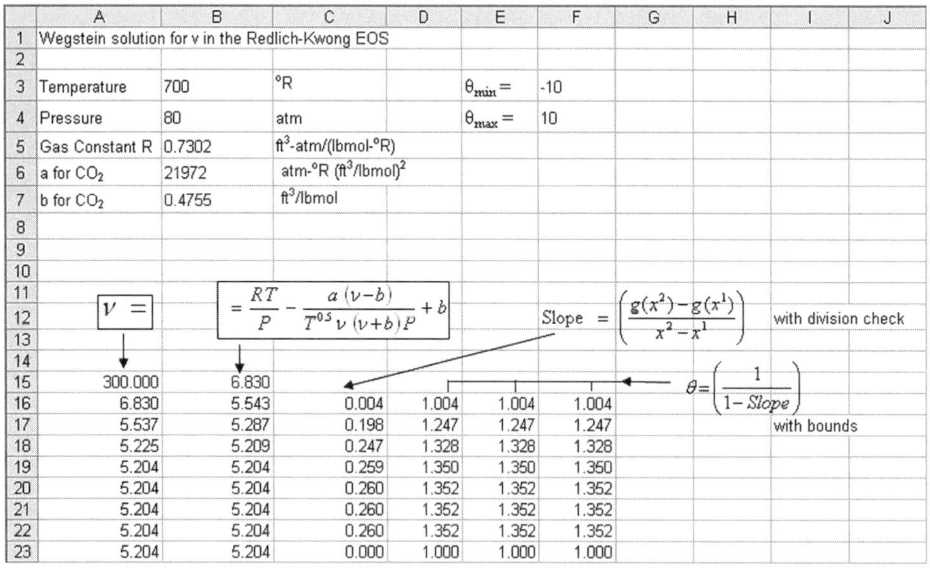

Figure 3.12 Bounded Wegstein solution for v in the Redlich–Kwong EOS.

We can construct the Wegstein method directly on the Excel sheet. Before the Wegstein method can be initiated, we complete two iterations of the flow sheet; these iterations are generally accomplished using successive substitution. This will require two repetitions of the material balance equations. For each iteration of the Wegstein method, we will also need the material balance equations. For example, for four iterations of the Wegstein method, we would need to repeat the material balance set six times on the Excel sheet, the first two material balance equation sets for successive substitution and the next four sets for the Wegstein method. Recall we must look back at the previous two iterations to construct the slope in the current iteration of the Wegstein method. There is really no way to conveniently look back at the two iterations on the Excel sheet without repeating the material balance set. This solution is provided in **Example 3.6.xls** and the first three iterations are shown in Figure 3.13.

In Figure 3.13, the first two iterations use successive substitution: $x_n^{[1]} = 0 \rightarrow g_n(x^{[1]}) \rightarrow x_n^{[2]} \rightarrow g_n(x^{[2]})$. These points are used to solve for $x_n^{[3]}$ by first calculating the slope for each variable in C32:C35 using Equation (3.61). We first check if $(x_n^{[2]} - x_n^{[1]}) = 0$ for each variable to prevent division by 0. If $(x_n^{[k+1]} - x_n^{[k]}) = 0$, the slope is set to zero, otherwise the slope is calculated. Because we are dealing with real numbers, it is generally safer to check if the absolute value of $(x_n^{[k+1]} - x_n^{[k]})$ is less than some precision; we used a precision of seven digits after the decimal.

	Slope
N(C_8H_{10})	=IF(ABS(C21-EBZ_Recycle)<precision,0,(J21-EBZ_New_Recycle)/(C21-EBZ_Recycle))
N(C_8H_8)	=IF(ABS(C22-Styrene_Recycle)<precision,0,(J22-Styrene_New_Recycle)/(C22-Styrene_Recycle))
N(H_2)	=IF(ABS(C23-H2_Recycle)<precision,0,(J23-H2_New_Recycle)/(C23-H2_Recycle))
N(H_2O)	=IF(ABS(C24-Water_Recycle)<precision,0,(J24-Water_New_Recycle)/(C24-Water_Recycle))
N(total)	

Then, θ_n is calculated using Equation (3.62); here, we also check if $(1 - Slope_n) = 0$, in which case θ_n is set = 1 (see cells D32:D35), which results in successive substitution. Bounds are checked, and $x_n^{[3]}$ is calculated (Wegstein Recycle column) using Equation (3.63). This Wegstein Recycle column is then copied into iteration #3, as the $x_n^{[3]}$ Recycle column and the material balances are again solved. The Wegstein method continues, using values from $(x_n^{[2]}, g_n(x^{[2]}))$ and $(x_n^{[3]}, g_n(x^{[3]}))$, to solve for $x_n^{[4]}$, and so on, for subsequent $x_n^{[k]}$ values. In the spreadsheet solution, the material balance set is solved seven times. This process works, but it is not very appealing—we really do not know how many times we need to copy the material balance equation set and Wegstein method down the sheet before convergence is obtained. ∎

We do have several alternatives to the Excel sheet solution shown in **Example 3.6.xls**. One alternative is to solve the material balances with the Wegstein method entirely in VBA. A second alternative is to solve the material balances on the Excel and to use VBA to provide the Wegstein method. This second alternative is shown in Example 3.7.

EXAMPLE 3.7 *Wegstein Method for Recycle Stream Using VBA with the Excel Sheet*

Solve Example 3.1 using the bounded Wegstein method to converge the recycle loop.

Here, we want to use the Excel sheet to solve our material balance set and to use VBA code to update the tear stream values. The solution is provided in **Example 3.7.xls**. The process begins with the construction of our material balances on the Excel sheet as we did in Figure 3.6a; Figure 3.6a is repeated in Figure 3.14. For any set of values for x_n, we can calculate $g_n(x_n)$ for the material balance set.

The VBA code to control the tear stream iterations is shown in Figure 3.15.

The macro subroutine procedure `Sub Flowsheet_Iteration()` is defined in `line 2`. `Lines 4–11` define

	A	B	C	D	E	F	G	H	I	J
1	*Styrene Production - Wegstein Method for Recycle Stream*									
2										
3		Trigger	0	(either 0 to reset or 1 to iterate)						
4		Iterations	0.00000000	iteration count x 10^{-8}						
5		Precision	0.0000001							
6		Reaction	C8H10 ---> C8H8 + H2		Conversion	0.65				
7										
8	**Iteration #1**		(k=1)							
9			(x_n^1)							$g_n(x_n^1)$
10		**Feed**	**Recycle**	**Reactor_In**	**Reactor_Out**	**Water_Out**	**Vapor_Out**	**Organic_Out**	**Product**	**New Recycle**
11	N(C8H10)	100	0	100	35	0	0	35	0.35	34.65
12	N(C8H8)	0	0	0	65	0	0	65	64.35	0.65
13	N(H2)	0	0	0	65	0	65	0	0	0
14	N(H2O)	3000	0	3000	3000	3000	0	0	0	0
15	N(total)	3100	0	3100	3165	3000	65	100	64.7	35.3
16										
17										
18	**Iteration #2**		(k=2)							
19			(x_n^2)							$g_n(x_n^2)$
20		**Feed**	**Recycle**	**Reactor_In**	**Reactor_Out**	**Water_Out**	**Vapor_Out**	**Organic_Out**	**Product**	**New Recycle**
21	N(C8H10)	100	34.65	134.65	47.1275	0	0	47.1275	0.471275	46.656225
22	N(C8H8)	0	0.65	0.65	88.1725	0	0	88.1725	87.290775	0.881725
23	N(H2)	0	0	0	87.5225	0	87.5225	0	0	0
24	N(H2O)	3000	0	3000	3000	3000	0	0	0	0
25	N(total)	3100	35.3	3135.3	3222.8225	3000	87.5225	135.3	87.76205	47.53795
26										
27	Calculate Wegstein parameters - Must solve for every species					Theta max	10			
28						Theta min	-10			
29										
30								(x_n^3)		
31			**Slope**	**Theta Calc**	**Theta Lower**	**Theta Upper**		**Wegstein Recycle**		
32		N(C8H10)	0.3465	1.530221882	1.530221882	1.530221882		53.02218822		
33		N(C8H8)	0.3565	1.554001554	1.554001554	1.554001554		1.01010101		
34		N(H2)	0	1	1	1		0		
35		N(H2O)	0	1	1	1		0		
36		N(total)								
37										
38										
39	**Iteration #3**		(k=3)							
40			(x_n^3)							$g_n(x_n^3)$
41		**Feed**	**Recycle**	**Reactor_In**	**Reactor_Out**	**Water_Out**	**Vapor_Out**	**Organic_Out**	**Product**	**New Recycle**
42	N(C8H10)	100	53.02218822	153.0221882	53.55776588	0	0	53.55776588	0.5355777	53.02218822
43	N(C8H8)	0	1.01010101	1.01010101	100.4745234	0	0	100.4745234	99.469778	1.004745234
44	N(H2)	0	0	0	99.46442234	0	99.46442234	0	0	0
45	N(H2O)	3000	0	3000	3000	3000	0	0	0	0
46	N(total)	3100	54.03228923	3154.032289	3253.496712	3000	99.46442234	154.0322892	100.00536	54.02693345

Figure 3.13 Styrene production—Wegstein method on Excel sheet.

9			(x_n^1)							$g_n(x_n^1)$
10		**Feed**	**Recycle**	**Reactor_In**	**Reactor_Out**	**Water_Out**	**Vapor_Out**	**Organic_Out**	**Product**	**New Recycle**
11	N(C8H10)	100	0	100	35	0	0	35	0.35	34.65
12	N(C8H8)	0	0	0	65	0	0	65	64.35	0.65
13	N(H2)	0	0	0	65	0	65	0	0	0
14	N(H2O)	3000	0	3000	3000	3000	0	0	0	0
15	N(total)	3100	0	3100	3165	3000	65	100	64.7	35.3

Figure 3.14 Material balances on Excel sheet (also Figure 3.6a).

one-dimensional arrays: `x_old(n_components)` will hold $x_n^{[k-1]}$ values; `g_old(n_components)` will hold $g_n\left(x_n^{[k-1]}\right)$; `x_current(n_components)` will hold $x_n^{[k]}$ values; and `g_current(n_components)` will hold $g_n\left(x_n^{[k]}\right)$. Line 12 calls `Sub Successive(x_old, g_old, x_current, g_current, n_components)`, where successive substitution iterations are performed.

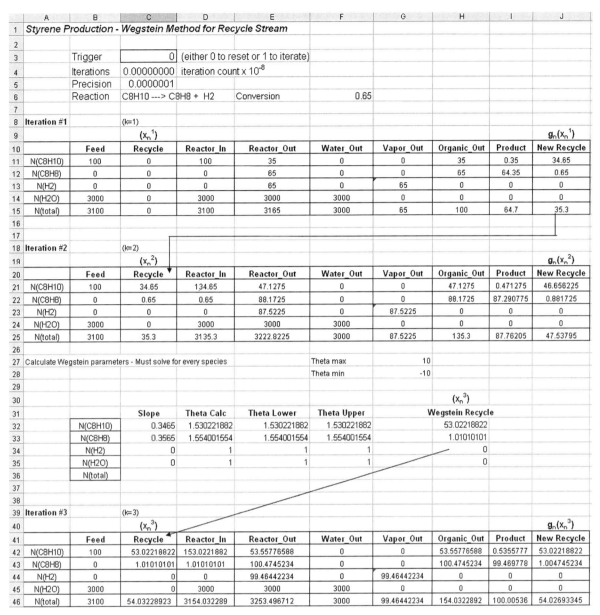

In `Sub Successive`, in `lines 18 and 19`, the current $x_n^{[1]}$ and $g_n\left(x_n^{[1]}\right)$ values are now defined as `x_old` and `g_old`. The values for `g_old` are then placed on the sheet as $x_n^{[2]}$ in `lines 21–23`, and the Excel sheet solves the material balances for $g_n\left(x_n^{[2]}\right)$. In `lines 24–27`, the $x_n^{[2]}$ and $g_n\left(x_n^{[2]}\right)$ values from the Excel sheet become `x_current` and `g_current`. For loops, `For i = 0 To n_components - 1` are used to step through each component. C_8H_{10} flow rate values are stored in the zero element of each one-dimensional array; C_8H_8 flow rate values are stored in the first element, and so on. Control is returned to `Sub Flowsheet_Iteration` where `Sub Wegstein(x_old, g_old, x_current, g_current, n_components)` is called in `line 13`.

In `Sub Wegstein lines 30-35`, we define one-dimensional arrays: `Slope(n_components), Theta(n_components)`, and `x_new(n_components)`. Just as we did in Example 3.6, the values available in `x_old, g_old, x_current, g_current` are used to solve for $x_n^{[3]}$. First, the slope (Eq. (3.61)) for each variable is calculated, and in `lines 41-46`, we use an `If Then Else` statement to prevent division by zero. `If` $(x_n^{[2]} - x_n^{[1]}) = 0$, coded as `x_current(i) - x_old(i) = 0.0#` in `line 42`, `Then` we set the `Slope(i) = 0.0#` in `line 43`, `Else` the Slope_n are calculated in `line 45` using Equation (3.61). A For loop, `For i = 0 To n_components - 1`, is again used to step through each component. As part of the For loop, a second `If Then Else` statement is used in `lines 47-51` to calculate θ_n by Equation (3.62); here, we also check if $(1 - \text{Slope}_n) = 0$, in which case θ_n is set = 1, giving successive substitution. Bounds are checked in `lines 52-53`, and x_n^3 is calculated by Equation (3.63) in `line 54` as `x_new(i) = (1.0# - Theta(i)) * x_current(i) + Theta(i) * g_current(i)`.

```
Option Explicit                                                          line 1

    Public Sub Flowsheet_Iteration()                                     line 2
        Dim n_components As Integer                                      line 3
        n_components = 4

        ' Dimension x and g values
        Dim x_old() As Double                                            line 4
        ReDim x_old(n_components)                                        line 5

        Dim g_old() As Double                                            line 6
        ReDim g_old(n_components)                                        line 7

        Dim x_current() As Double                                        line 8
        ReDim x_current(n_components)                                    line 9

        Dim g_current() As Double                                        line 10
        ReDim g_current(n_components)                                    line 11


        ' Call Successive Substitution for first two iterations
        Call Successive(x_old, g_old, x_current, g_current, n_components) line 12

        ' Call Wegstein Method
        Call Wegstein(x_old, g_old, x_current, g_current, n_components)  line 13

    End Sub                                                              line 14

    Sub Successive(x_old, g_old, x_current, g_current, n_components)     line 15

        ' Read x1 and resulting g(x1) values from sheet
        Dim i As Integer                                                 line 16
        For i = 0 To n_components - 1                                    line 17
            x_old(i) = Sheet1.Cells(i + 11, 3)                           line 18
            g_old(i) = Sheet1.Cells(i + 11, 10)                          line 19
        Next i                                                           line 20

        ' Place g(x1) values on sheet as new x2 values
        For i = 0 To n_components - 1                                    line 21
            Sheet1.Cells(i + 11, 3) = g_old(i)                           line 22
        Next i                                                           line 23

        ' Read x2 and resulting g(x2) values from sheet
        For i = 0 To n_components - 1                                    line 24
            x_current(i) = Sheet1.Cells(i + 11, 3)                       line 25
            g_current(i) = Sheet1.Cells(i + 11, 10)                      line 26
        Next i                                                           line 27

    End Sub                                                              line 28

    Sub Wegstein(x_old, g_old, x_current, g_current, n_components)       line 29

        Dim Slope() As Double                                            line 30
        ReDim Slope(n_components)                                        line 31

        Dim Theta() As Double                                            line 32
        ReDim Theta(n_components)                                        line 33
```

Figure 3.15 VBA code for the Wegstein method with material balances on Excel sheet.

```
        Dim x_new() As Double                                               line 34
        ReDim x_new(n_components)                                           line 35

        ' Set up Wegstein Method iteration loop using j as counter

        Dim j As Integer                                                    line 36
        Dim i As Integer                                                    line 37

        Dim Max_Iteration As Integer                                        line 38
        Max_Iteration = 10                                                  line 39
        For j = 1 To Max_Iteration                                          line 40

            ' Then calculate Slope and Theta
            For i = 0 To n_components - 1                                   line 41
                If (x_current(i) - x_old(i) = 0.0#) Then                    line 42
                    Slope(i) = 0.0#                                         line 43
                Else                                                        line 44
                    Slope(i) = (g_current(i) - g_old(i)) / (x_current(i) - x_old(i))
                                                                            line 45
                End If                                                      line 46

                If ((1 - Slope(i)) = 0.0#) Then                            line 47
                    Theta(i) = 1                                           line 48
                Else                                                        line 49
                    Theta(i) = 1 / (1 - Slope(i))                          line 50
                End If                                                      line 51

                ' Check extrapolation bounds
                If Theta(i) >= 10.0# Then Theta(i) = 10.0#                 line 52
                If Theta(i) <= -10.0# Then Theta(i) = -10.0#               line 53

                ' Calculate new x values for sheet
                x_new(i) = (1.0# - Theta(i)) * x_current(i) + Theta(i) * g_current(i)
                                                                            line 54
            Next i                                                          line 55

            ' Place new x values on sheet

            For i = 0 To n_components - 1                                   line 56
                Sheet1.Cells(i + 11, 3) = x_new(i)                        line 57
            Next i                                                          line 58

            ' Set x and g values
            For i = 0 To n_components - 1                                   line 59
                x_old(i) = x_current(i)                                    line 60
                g_old(i) = g_current(i)                                    line 61
                x_current(i) = Sheet1.Cells(i + 11, 3)                     line 62
                g_current(i) = Sheet1.Cells(i + 11, 10)                    line 63

            Next i                                                          line 64

            ' Here we could check a convergence criterion parameter
            ' (Max_Change) to allow code exit or
            ' Continue to loop until Max_Iterations

        Next j                                                              line 65
    End Sub                                                                 line 66
```

Figure 3.15 (*Continued*)

9			(x_n)						g_n(x_n)
10		Feed	Recycle	Reactor_In	Reactor_Out	Water_Out	Vapor_Out	Organic_Out	Product	New Recycle
11	N(C8H10)	100	53.02218822	153.0221882	53.55776588	0	0	53.55776588	0.5355777	53.02218822
12	N(C8H8)	0	1.004691135	1.004691135	100.4691135	0	0	100.4691135	99.464422	1.004691135
13	N(H2)	0	0	0	99.46442234	0	99.46442234	0	0	0
14	N(H2O)	3000	0	3000	3000	3000	0	0	0	0
15	N(total)	3100	54.02687935	3154.026879	3253.491302	3000	99.46442234	154.0268794	100	54.02687935

Figure 3.16 Styrene production—bounded Wegstein solution method.

The `x_new(i) values` are then placed on the Excel sheet in `lines 56–58`, and the sheet calculates $g_n(x_n^{[3]})$. This completes the first iteration of the Wegstein method. We update our values for `x_old`, `g_old`, `x_current and, g_current` in `lines 59–64`. At this point, we could check if the absolute value $|x_n^{[k-1]} - x_n^{[k]}| \leq$ Max_Change, a convergence criterion parameter, and if true, we could exit the Wegstein subroutine; this additional coding effort is left as a chapter problem. In the current code, we continue the Wegstein iteration method for 10 iterations, and then we exit the For loop. Control is returned to `Sub Flowsheet_Iteration` and the final results are shown in Figure 3.16.

3.4 MATERIAL BALANCE PROBLEMS WITH ALTERNATIVE SPECIFICATIONS

We have examined the solution of a simple styrene material balance problem with recycle. We focused on converging the recycle loop or tear stream using successive substitution or the Wegstein method. Successive substitution could be accomplished directly on the Excel sheet. For the Wegstein method, a combination of VBA (to calculate $x_n^{[k+1]}$) and Excel (to solve the material balances) was found useful.

For the simple styrene process of Example 3.1, all of the natural specifications were provided. However, in many problems, natural specifications are missing but are replaced with other information about the process. The degree of freedom remains zero. A process condition that is not a natural specification is an alternative specification, often termed a process constraint. Solutions of material balance problems with alternative specifications are explored here.

In Example 3.1, the fresh feed to the process consists of 3000 mol/h water (steam) and 100 mol/h of ethylbenzene; these feed rates are both natural specifications. Consider the following modification to Example 3.1. Let us replace the known water feed rate with the following alternative specification. To prevent coking (product conversion to carbon) in the reactor, water at 90% (molar basis) is maintained in the stream entering the reactor.

Eliminating the specification on the water feed rate will reduce the process (and mixer) degree of freedom by one. However, the degree of freedom of the process will remain zero as the molar feed rate of water, $N_{H_2O}^{feed}$, has been replaced with the alternative mole fraction specification. This alternative specification is

$$\frac{N_{H_2O}^{reactor\ in}}{N_{reactor\ in}} = 0.9. \qquad (3.64)$$

The immediate problem is that in order to solve the mixer balances, species flow rates in both the recycle and feed are needed. A direct iterative solution of the styrene flow sheet is no longer possible as a natural specification for the mixer has been eliminated. One solution approach is to assume a value for the missing natural specification (the water feed rate), which then allows a sequential flow sheet solution. Based on the value for the water mole fraction into the reactor, the water feed rate can be adjusted (trial and error) to give the desired mole fraction.

EXAMPLE 3.8 *Styrene Process with Alternative Specification: Trial and Error*

Use trial and error to solve the styrene flow sheet problem of Example 3.1, where the feedwater flow rate has been replaced with the alternative specification given by Equation (3.64).

Following the same logic we used to solve the recycle loop by successive substitution, we can set the water feed rate = 0; this allows the flow sheet to be solved. The file **Example 3.8a.xls** is just a modification to our solution of Example 3.3 in which we add a check on the water composition entering the reactor. The successive substitution solution (for the recycle stream) converges for $N_{H_2O}^{feed} = 0$ to the results shown in Figure 3.17.

We have solved the problem where the water molar feed rate is zero, but this does not meet the specification (constraint) that to prevent coking in the reactor, water at 90% (molar basis) is required in the stream entering the reactor. We can vary the water molar feed rate and check if the required molar concentration into the reactor is met. Here, we are looking for the water feed rate that makes $f(x) = 0$, where, by using Equation (3.64),

$$f(x) = f(\text{water feed rate}) = \frac{N_{H_2O}^{reactor\ in}}{N_{reactor\ in}} - 0.9 = 0.$$

The trial-and-error solution is accomplished in two steps. We first bound Equation (3.64) with two water feed rates, one giving a value for $N_{H_2O}^{reactor\ in} / N_{reactor\ in}$ above 0.9 and the other giving a value below 0.9. We next refine or reduce the interval on $N_{H_2O}^{feed}$ until we are sufficiently close to the correct answer; in this refinement step, we generally do try and keep Equation (3.64) bounded. The answer should be near $N_{H_2O}^{feed} = 1400$ mol/h. ■

This trial-and-error approach of bounding the water feed rate and then reducing the interval until the water mole fraction into the reactor is sufficiently close to our desired result works. As we will see, bounding with interval refinement serves as the basis for most generalized computer-aided solutions to single-variable problems.

EXAMPLE 3.9 *Styrene Process with Alternative Specification: Goal Seek*

Use Excel Goal Seek to solve Example 3.8, the styrene flow sheet problem, where the feedwater flow rate has been replaced with the alternative specification given by Equation (3.64).

One alternative to trial and error would be to use Excel Goal Seek, the single-variable search method for unconstrained problems (introduced in Example 2.12), to vary the water molar feed rate to meet the required composition specification. The use of Goal Seek is shown in Figure 3.18.

Unfortunately, Goal Seek will fail. Goal Seek will guess new water flow rates, but Goal Seek has no provision to allow the flow sheet to converge before it tries another water flow rate—see **Example 3.9.xls**.

EXAMPLE 3.10 *Styrene Process with Alternative Specification: Solver*

Use Excel Solver to solve the styrene flow sheet problem of Example 3.8.

A third solution approach would be to use Excel Solver, the single-variable/multivariable search method for constrained problems, to vary the water molar feed rate to meet the required composition specification. In the Excel toolbar, use Tools → Solver. Here, there would be no constraints—see **Example 3.10.xls**.

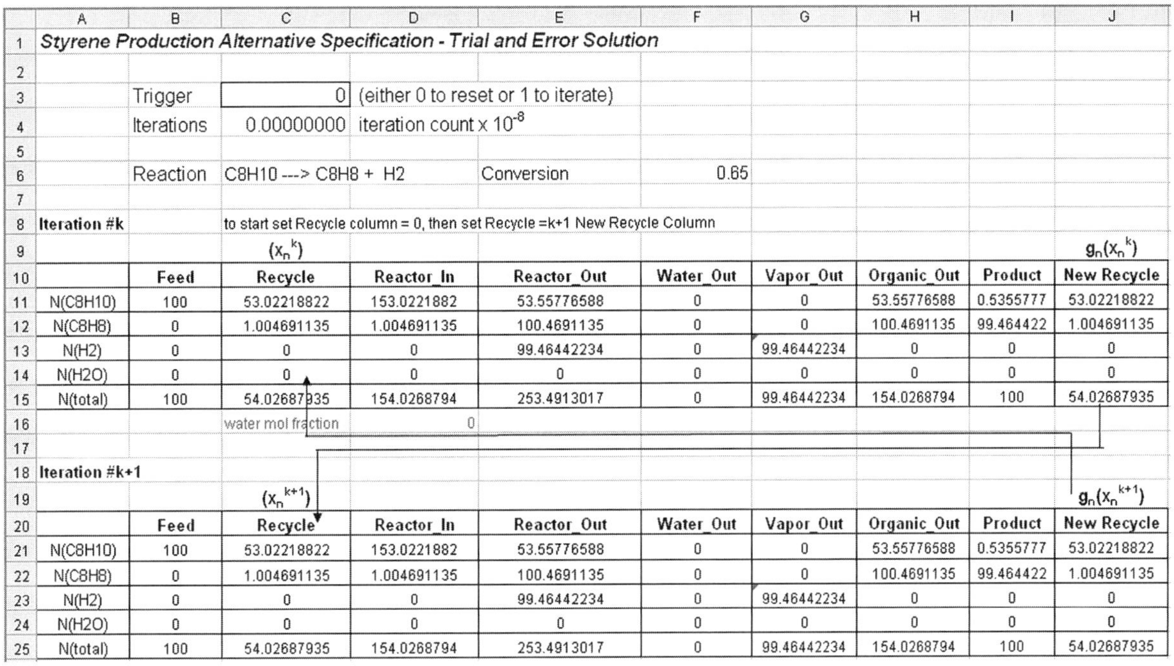

	A	B	C	D	E	F	G	H	I	J
1	Styrene Production Alternative Specification - Trial and Error Solution									
2										
3		Trigger	0	(either 0 to reset or 1 to iterate)						
4		Iterations	0.00000000	iteration count x 10^{-8}						
5										
6		Reaction	C8H10 ---> C8H8 + H2		Conversion	0.65				
7										
8	Iteration #k		to start set Recycle column = 0, then set Recycle =k+1 New Recycle Column							
9			(x_n^k)							$g_n(x_n^k)$
10		Feed	Recycle	Reactor_In	Reactor_Out	Water_Out	Vapor_Out	Organic_Out	Product	New Recycle
11	N(C8H10)	100	53.02218822	153.0221882	53.55776588	0	0	53.55776588	0.5355777	53.02218822
12	N(C8H8)	0	1.004691135	1.004691135	100.4691135	0	0	100.4691135	99.464422	1.004691135
13	N(H2)	0	0	0	99.46442234	0	99.46442234	0	0	0
14	N(H2O)	0	0	0	0	0	0	0	0	0
15	N(total)	100	54.02687935	154.0268794	253.4913017	0	99.46442234	154.0268794	100	54.02687935
16			water mol fraction	0						
17										
18	Iteration #k+1									
19			(x_n^{k+1})							$g_n(x_n^{k+1})$
20		Feed	Recycle	Reactor_In	Reactor_Out	Water_Out	Vapor_Out	Organic_Out	Product	New Recycle
21	N(C8H10)	100	53.02218822	153.0221882	53.55776588	0	0	53.55776588	0.5355777	53.02218822
22	N(C8H8)	0	1.004691135	1.004691135	100.4691135	0	0	100.4691135	99.464422	1.004691135
23	N(H2)	0	0	0	99.46442234	0	99.46442234	0	0	0
24	N(H2O)	0	0	0	0	0	0	0	0	0
25	N(total)	100	54.02687935	154.0268794	253.4913017	0	99.46442234	154.0268794	100	54.02687935

Figure 3.17 Styrene production—trial and error for constraint.

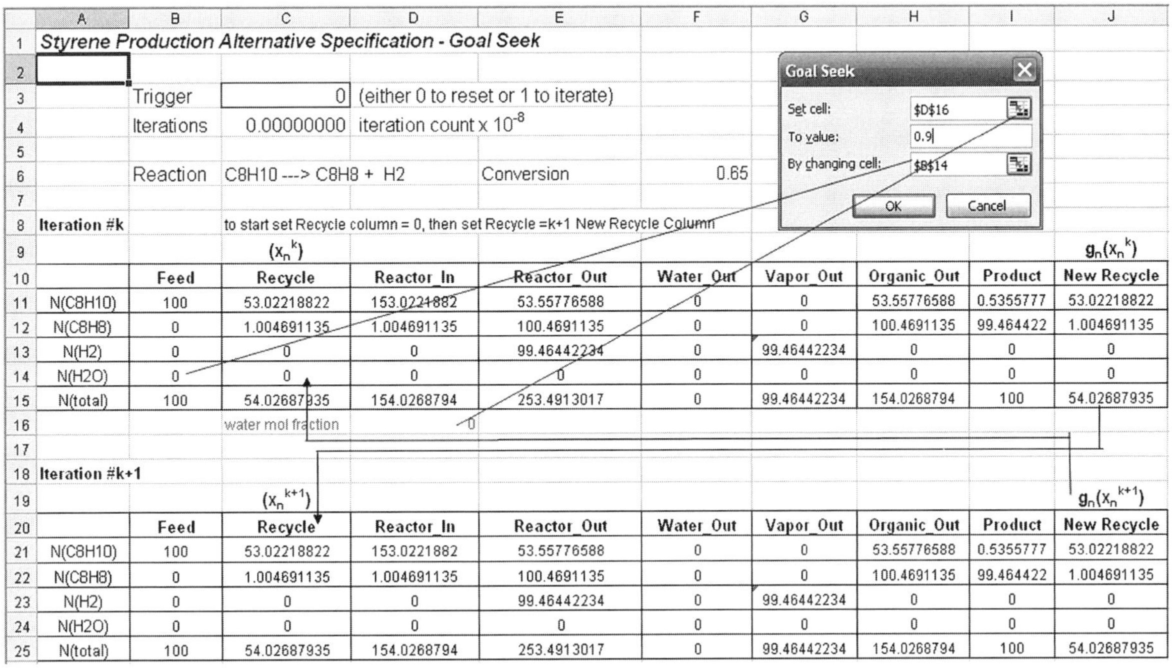

	A	B	C	D	E	F	G	H	I	J
1	Styrene Production Alternative Specification - Goal Seek									
2										
3		Trigger	0	(either 0 to reset or 1 to iterate)						
4		Iterations	0.00000000	iteration count x 10^{-8}						
5										
6		Reaction	C8H10 ---> C8H8 + H2		Conversion	0.65				
7										
8	Iteration #k		to start set Recycle column = 0, then set Recycle =k+1 New Recycle Column							
9			(x_n^k)							$g_n(x_n^k)$
10		Feed	Recycle	Reactor_In	Reactor_Out	Water_Out	Vapor_Out	Organic_Out	Product	New Recycle
11	N(C8H10)	100	53.02218822	153.0221882	53.55776588	0	0	53.55776588	0.5355777	53.02218822
12	N(C8H8)	0	1.004691135	1.004691135	100.4691135	0	0	100.4691135	99.464422	1.004691135
13	N(H2)	0	0	0	99.46442234	0	99.46442234	0	0	0
14	N(H2O)	0	0	0	0	0	0	0	0	0
15	N(total)	100	54.02687935	154.0268794	253.4913017	0	99.46442234	154.0268794	100	54.02687935
16			water mol fraction	0						
17										
18	Iteration #k+1									
19			(x_n^{k+1})							$g_n(x_n^{k+1})$
20		Feed	Recycle	Reactor_In	Reactor_Out	Water_Out	Vapor_Out	Organic_Out	Product	New Recycle
21	N(C8H10)	100	53.02218822	153.0221882	53.55776588	0	0	53.55776588	0.5355777	53.02218822
22	N(C8H8)	0	1.004691135	1.004691135	100.4691135	0	0	100.4691135	99.464422	1.004691135
23	N(H2)	0	0	0	99.46442234	0	99.46442234	0	0	0
24	N(H2O)	0	0	0	0	0	0	0	0	0
25	N(total)	100	54.02687935	154.0268794	253.4913017	0	99.46442234	154.0268794	100	54.02687935

Figure 3.18 Styrene production with alternative specification—Goal Seek solution.

This approach is successful. Solver allows the Excel sheet to converge, via successive substitution, for each trial value of water molar feed rate. After the sheet has converged, control is returned to Solver and a new trial point for the water feed rate is determined. The converged value for the water feed rate is 1386.25 mol/h as shown in Figure 3.19. The iteration counter indicates that the flow sheet was solved 166 times as Solver moved from an initial water feed rate of 0 lb-mol/h to a final value of 1386.25 lb-mol/h to meet the alternative specification of Equation (3.64) in cell D16. ■

Unfortunately, Solver, as provided in the standard Excel installation, will not work in conjunction with user-written VBA subroutine procedures. For example, we can create a subroutine procedure, which will call the Wegstein method to converge the flow sheet. But we cannot, in this subroutine,

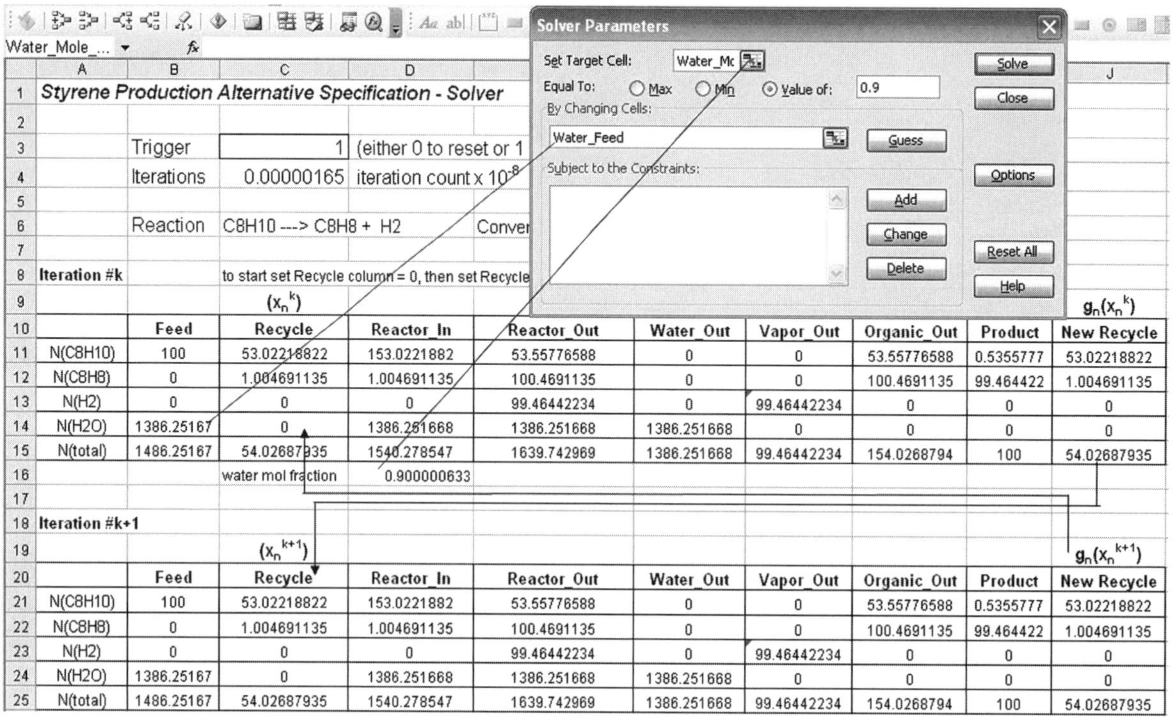

	Feed	Recycle	Reactor_In	Reactor_Out	Water_Out	Vapor_Out	Organic_Out	Product	New Recycle
N(C8H10)	100	53.02218822	153.0221882	53.55776588	0	0	53.55776588	0.5355777	53.02218822
N(C8H8)	0	1.004691135	1.004691135	100.4691135	0	0	100.4691135	99.464422	1.004691135
N(H2)	0	0	0	99.46442234	0	99.46442234	0	0	0
N(H2O)	1386.25167	0	1386.251668	1386.251668	1386.251668	0	0	0	0
N(total)	1486.25167	54.02687935	1540.278547	1639.742969	1386.251668	99.46442234	154.0268794	100	54.02687935

	Feed	Recycle	Reactor_In	Reactor_Out	Water_Out	Vapor_Out	Organic_Out	Product	New Recycle
N(C8H10)	100	53.02218822	153.0221882	53.55776588	0	0	53.55776588	0.5355777	53.02218822
N(C8H8)	0	1.004691135	1.004691135	100.4691135	0	0	100.4691135	99.464422	1.004691135
N(H2)	0	0	0	99.46442234	0	99.46442234	0	0	0
N(H2O)	1386.25167	0	1386.251668	1386.251668	1386.251668	0	0	0	0
N(total)	1486.25167	54.02687935	1540.278547	1639.742969	1386.251668	99.46442234	154.0268794	100	54.02687935

Figure 3.19 Styrene production with alternative specification—Solver solution.

call Solver to vary the water feed rate until the alternative specification given by Equation (3.64) is satisfied.

There are add-ins for Excel, for example, What's Best, from LINDO Systems, which provide for the inclusion of VBA subroutine procedures in the optimization process. A similar add-in is available from the developers of the Excel Solver program—Frontline Systems.

EXAMPLE 3.11 *Styrene Process with Alternative Specification: VBA Wegstein and What's Best Optimizer*

Use the optimizer found in the What's Best add-in to solve the styrene flow sheet problem of Example 3.8. Converge the flow sheet using the Wegstein method (subroutine written in VBA). To solve Example 3.11, the user is provided the student edition of What's Best from LINDO Systems; this program will also be useful in solving the energy management problems of Chapter 12 (see Section 1.7 for installation instructions).

The solution using What's Best combined with a Wegstein method subroutine is given in **Example 3.11.xls**. Here, the Wegstein method VBA subroutines of Example 3.7 can be used without change. We only need to add three lines of code to identify the cells we want to vary, identify the objective function, and call the optimizer:

```
' Identify the cells we want to vary
      wbadjust(Sheet1.Cells(14, 2))
 ' Call our successive substitution /
Wegstein Method to converge the flowsheet
      Call Successive(x_old, g_old, x_
current, g_current, n_components)
      Call Wegstein(x_old, g_old, x_current,
g_current, n_components)
 ' Identify the objective function
      wbBest(Sheet1.Cells(17, 4), "Minimize")
 ' Call the optimization solver
      Application.Run(macro:="WBUsers.
wbSolve")
```

The optimization results are shown in Figure 3.20. The form of the objective function used is discussed in the next section. ■

In Section 3.6, we will write our own VBA-based optimization procedure to solve Example 3.8. We will find it useful to be able to control the optimization sequence, especially in the more advanced flow sheeting problems presented in later chapters. For Example 3.8, VBA code can be used to provide a water molar feed rate to the Excel sheet. The sheet will converge and then the value for the water mole fraction entering the reactor can be read by the VBA code. A new trial for the water molar feed rate can be formulated by the VBA code and the process repeated until the water mole fraction entering the reactor is determined with sufficient accuracy—this will involve bounding and interval refinement as we discussed previously in Example 3.8.

Example 3.8 can be considered a single-variable optimization problem. The single variable is the water molar feed rate. The desired result or objective function is to minimize the difference between the water mole fraction entering the reactor (resulting from the trial water molar feed rate) and 0.9. We discuss solution approaches to single-variable optimization problems in the next section.

	A	B	C	D	E	F	G	H	I	J
1	*Styrene Production Wegstein Method / VBA / WB Solver*									
2										
3		Trigger	1	(either 0 to reset or 1 to iterate)						
4		Iterations	0.00000097	iteration count x 10^{-8}						
5										
6		Reaction	C8H10 ---> C8H8 + H2		Conversion	0.65				
7										
8	**Iteration**									
9			(x_n)							$g_n(x_n)$
10		**Feed**	**Recycle**	**Reactor_In**	**Reactor_Out**	**Water_Out**	**Vapor_Out**	**Organic_Ou**	**Product**	**New Recycle**
11	N(C8H10)	100	53.02218822	153.022188	53.55776588	0	0	53.557766	0.535577659	53.02218822
12	N(C8H8)	0	1.004691135	1.00469113	100.4691135	0	0	100.46911	99.46442234	1.004691135
13	N(H2)	0	0	0	99.46442234	0	99.4644223	0	0	0
14	N(H2O)	1377.0512	0	1377.05116	1377.05116	1377.05116	0	0	0	0
15	N(total)	1477.0512	54.02687935	1531.07804	1630.542462	1377.05116	99.4644223	154.02688	100	54.02687935
16				0.89939972	H_2O mole fraction					
17				3.6034E-07	Objective Function					

Figure 3.20 Styrene production with alternative specification—VBA Wegstein and What's Best optimization.

3.5 SINGLE-VARIABLE OPTIMIZATION PROBLEMS

Rather than trying to solve $f(x) = 0$, it is often more convenient in engineering design problems to solve an equivalent problem to minimize $f(x)$. When minimizing $f(x)$, a plot of the objective function $f(x)$ versus x should be smooth and continuous. The objective function should have a single minimum and the function must be convex. A convex function implies that the line connecting any two points on the objective function will overestimate or lie above the objective function curve; the line connecting two points may also fall on the objective function.

3.5.1 Forming the Objective Function for Single-Variable Constrained Material Balance Problems

To see how we can formulate an objective function for Example 3.8, we can first supply different water feed rates to the Excel sheet in **Example 3.8a.xls** and record the resulting water mole fraction into the reactor. Results are tabulated in the Excel spreadsheet in **Example 3.8b.xls** and shown in Figure 3.21. The first plot in Figure 3.21 shows there is a monotonic increase in the water mole fraction entering the reactor with increasing water feed rate (our single variable). There is no minimum in this plot.

An objective function for this problem can be formed as

$$\text{Minimize } f(x) = \text{Minimize(water mole fraction entering reactor} - 0.9)^2. \quad (3.65)$$

In the second plot in Figure 3.21, (water mole fraction into the reactor $- 0.9)^2$ is plotted against the water feed rate. Here, the objective function is convex and a single minimum exists. The minimum of this plot occurs at a specific water feed rate, and this water feed rate will result in a water mole fraction into the reactor of 0.9.

As we have discussed, the computer solution to a single-variable optimization problem involves two steps: bounding the variable and then interval refinement (reduction) of the bound. For example, the bounding phase of this problem (using the data in Figure 3.21) results in a water feed rate between 1250 and 1750 mol/h; notice that the correct water feed rate falls between these values. This 1250- to 1750-mol/h interval is then refined to a single (optimum) flow rate.

3.5.2 Bounding Step or Bounding Phase: Swann's Equation

Figure 3.22 provides a generic objective function, $f(x)$, plotted against a single variable, x; here, we are trying to minimize the objective function.

Bounding requires an initial guess ($x^{[0]}$) and a step size (Δx) or delta_x. The *direction to move* is determined by comparing $f(x^{[0]})$ to $f(x^{[0]} + \Delta x)$ and $f(x^{[0]} - \Delta x)$; one of three cases will result.

Case A If $f(x^{[0]} - \Delta x) \geq f(x^{[0]}) \geq f(x^{[0]} + \Delta x)$, then the minimum must be to the right of $x^{[0]}$.

Case B If $f(x^{[0]} - \Delta x) \geq f(x^{[0]}) \leq f(x^{[0]} + \Delta x)$, then x has been bounded and the interval refinement phase can begin.

Case C If $f(x^{[0]} - \Delta x) \leq f(x^{[0]}) \leq f(x^{[0]} + \Delta x)$, then the minimum must be to the left of $x^{[0]}$.

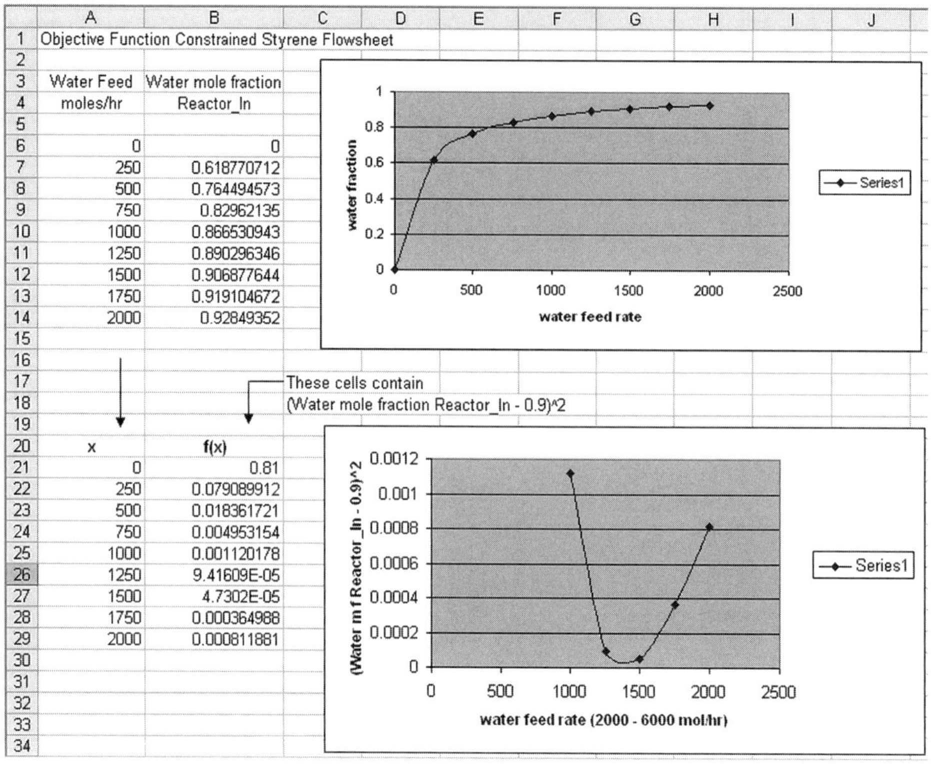

Water Feed moles/hr	Water mole fraction Reactor_In
0	0
250	0.618770712
500	0.764494573
750	0.82962135
1000	0.866530943
1250	0.890296346
1500	0.906877644
1750	0.919104672
2000	0.92849352

x	f(x)
0	0.81
250	0.079089912
500	0.018361721
750	0.004953154
1000	0.001120178
1250	9.41609E-05
1500	4.7302E-05
1750	0.000364988
2000	0.000811881

Figure 3.21 Styrene production—objective function formation for alternative specification.

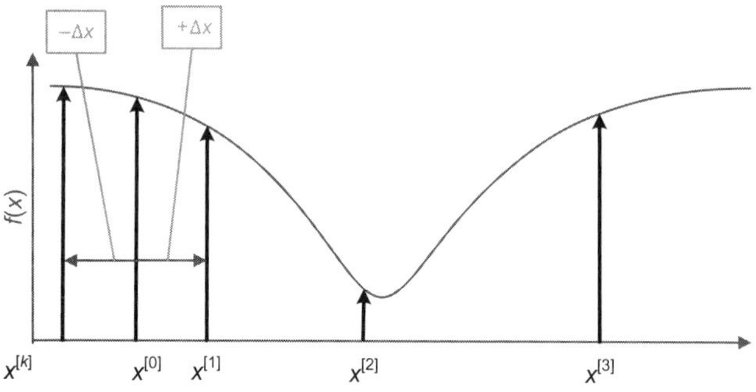

Figure 3.22 Bounding—objective function $f(x)$ versus single variable, x.

Figure 3.22 shows the minimum to the right ($+\Delta x$ direction) of $x^{[0]}$. Trial points can be found using Swann's equation (see Ravindran et al., 2006):

$$x^{[k+1]} = x^{[k]} \pm 2^k (\Delta x), \tag{3.66}$$

and for Figure 3.22, the ± term is set to +. For example, if $x^{[0]} = 200$ and a step size of 50 was used, then the trial points in the figure would be $x^{[0]} = 200$, $x^{[1]} = 250$, $x^{[2]} = 350$, and $x^{[3]} = 550$.

The objective function is *bounded* when $f(x^{[k-1]}) \geq f(x^{[k]}) \leq f(x^{[k+1]})$, and in the figure, this would occur with $x^{[1]}$, $x^{[2]}$, and $x^{[3]}$.

EXAMPLE 3.12 *Bounding Phase for Styrene Process with Alternative Specification "by Hand"*

Before we code the needed bounding phase for the styrene process of Example 3.8, let us confirm our understanding by bounding the water feed rate by hand.

Here, we can create **Example 3.12.xls**, which is simply **Example 3.8a.xls** with the objective function, Equation (3.65) added on the sheet. Using an initial guess for the water feed rate of $x^{[0]} = 200$ and a step size of 50 mol/h, bounding results are shown in Table 3.1.

From the initial guess ($x^{[0]} = 200$), we move in both the negative (−50) and positive (+50) step size directions and first determine if x has been bounded. If x is not bounded, the direction to move is then determined. Here, Case A results, $(f(x^{[0]} - \Delta x) \geq f(x^{[0]}) \geq f(x^{[0]} + \Delta x))$, so we move in the positive step size direction as shown in Table 3.1. At $k = 5$, the objective function is bounded (minimized) between water feed rates of 950 and 3350 mol/h; interval refinement would then occur. The lower bound flow rate of 950 is sometimes noted as $x^{[a]}$ or $x^{[left]}$or $x^{[L]}$. The upper bound flow rate of 3350 may be noted as $x^{[b]}$ or x[right] or $x^{[R]}$.

Table 3.1 Bounding Results for Single-Variable Styrene Production

k and x Values	x Value, Water Feed Rate	$f(x)$, Equation (3.65)	Comments
Initial guess, $x^{[0]} = 200$ mol/h, step size $\Delta x = 50$ mol/h	200	$f(x^{[0]}) = 0.1123$	Initial guess
$k = 0,\ x^{[0]} - 2^k\,(50) = 150$	150	$f(x^{[0]} - \Delta x) = 0.1653$	Not bounded as
$k = 0,\ x^{[0]} + 2^k\,(50) = 250$	250	$f(x^{[0]} + \Delta x) = 0.079$	$f(x^{[0]} - \Delta x) \geq f(x^{[0]}) \geq f(x^{[0]} + \Delta x)$ and correct direction is $+\Delta x$
$k = 0,\ x^{[1]} = x^{[0]} + 2^k\,(50) = 250$	250	$f(x^{[1]}) = 0.079$	
$k = 1,\ x^{[2]} = x^{[1]} + 2^1\,(50) = 350$	350	$f(x^{[2]}) = 0.042$	$f(x^{[0]}) \geq f(x^{[1]}) \geq f(x^{[2]})$
$k = 2,\ x^{[3]} = x^{[2]} + 2^2\,(50) = 550$	550	$f(x^{[3]}) = 0.014$	$f(x^{[1]}) \geq f(x^{[2]}) \geq f(x^{[3]})$
$k = 3,\ x^{[4]} = x^{[3]} + 2^3\,(50) = 950$	950	$f(x^{[4]}) = 0.0016$	$f(x^{[2]}) \geq f(x^{[3]}) \geq f(x^{[4]})$
$k = 4,\ x^{[5]} = x^{[4]} + 2^4\,(50) = 1750$	1750	$f(x^{[5]}) = 0.0004$	$f(x^{[3]}) \geq f(x^{[4]}) \geq f(x^{[5]})$
$k = 5,\ x^{[6]} = x^{[5]} + 2^5\,(50) = 3350$	3350	$f(x^{[6]}) = 0.0031$	$f(x^{[4]}) \geq f(x^{[5]}) \leq f(x^{[6]})$, Bounded $x^{[a]} = x^{[\text{left}]} = x^{[4]}$ and $x^{[b]} = x^{[\text{right}]} = x^{[6]}$

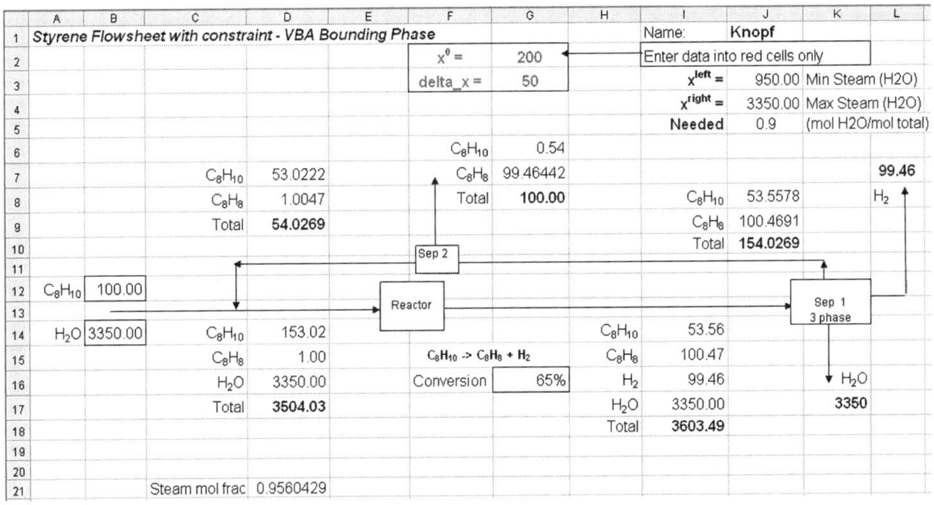

Figure 3.23 Bounding—styrene successive substitution flow sheet.

EXAMPLE 3.13 *VBA Code for Bounding Phase: Styrene Process with Alternative Specification*

We want to develop the VBA code needed for bounding the styrene process of Example 3.8. The solution is provided in **Example 3.13.xls**. Here, we start with the successive substitution flow sheet developed in Example 3.4, and on the sheet we add cells for the initial guess and the step size; these cells are highlighted in red in the provided Excel solution file and in cells F2 : G3 in Figure 3.23. We calculate the water (steam) mole fraction entering the reactor, and provisions are made on the sheet for the upper and lower bound water feed flow rates (to be determined by the VBA code).

VBA Bounding Code

Example 3.13.xls provides the VBA bounding code, which is also shown in Figure 3.24. If we take a quick look back at Table 3.1, we will notice that we do not need to keep all the values x and $f(x)$ as developed in Table 3.1. For any iteration, we actually only need to have values $(f(x^{[0]}), f(x^{[1]}), f(x^{[2]}))$, then $(f(x^{[1]}), f(x^{[2]}), f(x^{[3]}))$, . . . , $(f(x^{[4]}), f(x^{[5]}), f(x^{[6]}))$ to determine if the function was bounded. In the code mentioned next, we successively name these points—$x^{[\text{left}]}$, $x^{[1]}$, $x^{[\text{right}]}$ and $f(x^{[\text{left}]})$, $f(x^{[1]})$, $f(x^{[\text{right}]})$.

Examining the macro `Sub Bound()`, we read the initial guess for the water feed rate in `line 3` as `x1 = Range("G2").Value`. This line could also have been coded as `x1 = Sheet1.Cells(2,7)`. Here, `x1` = 200, and the Sub procedure `Sub Flowsheet(x, F)` is called from `line 5`, `Call Flowsheet(x1, F)`. In the Sub procedure, the value for $x^{[1]}$ is placed on the flow sheet by `line 44` and the flow sheet converges. The value for the objective function F is calculated in `line 45` and `F` = 0.1123 is returned to `line 5` and set equal to `f1` in `line 6`.; here, `f1` represents $f(x^{[1]})$.

In `line 8`, the step size `delta_x` is read from the spreadsheet. In `lines 8-14`, `x + delta_x` = 250 and `x - delta_x` = 150 mol/h are calculated along with their resulting objective function values.

```
Sub Bound()                                                             line 1
    ' We want to bound the water feed rate between [xa, xb] or [xleft, xright]
    ' We get the initial guess, x0 now called x1, from cell G2 on the flowsheet   line 2
    x1 = Range("G2").Value                                              line 3
    ' We next calculate our objective function
    ' This is done on the flowsheet in a separate Sub Procedure        line 4
    Call Flowsheet(x1, F)                                               line 5
    f1 = F                                                              line 6

    ' Next we determine the direction for the step delta_x             line 7

    delta_x = Range("G3").Value                                         line 8

    xright = x1 + delta_x                                               line 9
    Call Flowsheet(xright, F)                                           line 10
    fright = F                                                          line 11
    xleft = x1 - delta_x                                                line 12
    Call Flowsheet(xleft, F)                                            line 13
    fleft = F                                                           line 14

    Iteration_limit = 10000                                             line 15
    Iteration = 0                                                       line 16
    Do                                                                  line 17
        If fleft > f1 And fright > f1 Then Exit Do                      line 18

        If f1 > fright Then                                             line 19
            Iteration = Iteration + 1                                   line 20
            xleft = x1                                                  line 21
            fleft = f1                                                  line 22
            x1 = xright                                                 line 23
            f1 = fright                                                 line 24
            xright = x1 + (2 ^ Iteration) * delta_x                     line 25
            Call Flowsheet(xright, F)                                   line 26
            fright = F                                                  line 27

        Else                                                            line 28
            Iteration = Iteration + 1                                   line 29
            xright = x1                                                 line 30
            fright = f1                                                 line 31
            x1 = xleft                                                  line 32
            f1 = fleft                                                  line 33
            xleft = x1 - (2 ^ Iteration) * delta_x                      line 34
            ' Normally the next two statements (currenly comments) would not
            ' be used as the flow rate can not be negative
            '               If xleft < 0 Then xleft = 0
            '               If xleft = 0 Then Exit Do                   line 35
            Call Flowsheet(xleft, F)                                    line 36
            fleft = F                                                   line 37

        End If                                                          line 38
    Loop Until Iteration > Iteration_limit                              line 39
    ' Place bounded water flow on spreadsheet as [xa, xb] or [xleft, xright]
    Range("J3").Value = xleft                                           line 40
    Range("J4").Value = xright                                          line 41
End Sub                                                                 line 42
Sub Flowsheet(x, F)                                                     line 43
    ' Here we place the water flow rate on the flowsheet in cell B14
    ' the flowsheet will converge before it returns control to the VBA code.
    Range("B14").Value = x                                              line 44
    ' The flowsheet is now converged with the x value (water flow rate)
    ' Cell D21 contains the converged (moles H2O)/(Total moles in Reator)
    ' Cell J5 contains the desired (moles H2O)/(Total moles in Reator)
    ' We square the difference between converged and desired (mole fraction)
    F = (Range("D21").Value - Range("J5").Value) ^ 2                    line 45
    ' The value in F will be returned to the Bounding Sub Procedure     line 46
End Sub                                                                 line 47
```

Figure 3.24 Bounding VBA code for styrene process.

In the VBA bounding procedure, we do not need to store all the information in Table 3.1. We only need to keep track of three trial x values and the resulting $f(x)$ values. Here, we assign $x^{[left]} = 150$, $x^{[1]} = 200$, and $x^{[right]} = 250$. In `line 18`, we check if $f(x^{[left]}) > f(x^{[1]})$ and $f(x^{[right]}) > f(x^{[1]})$; both conditions must be true for x to be bounded. Otherwise, we use Equation (3.66) to calculate a new $x^{[k+1]}$. In our case, $f(x^{[1]}) > f(x^{[right]})$ (currently, $f(x^{[1]}) = 0.1123$ and $f(x^{[right]}) = 0.079$), so `lines 19-27` are used; here, we set $x^{[left]} = 200$, $x^{[1]} = 250$, and $x^{[right]} = x^{[1]} + 2^k(\text{delta\_x}) = 350$. This process continues until `line 18` is satisfied; then, we exit and write both $x^{[left]}$ and $x^{[right]}$ on the flow sheet `lines 40-41`.

There are two structured statements used in the code:

Do Loop Until. Do looping allows repeating a set of instructions. The general form of the structure appears as

```
Do
  [code]
  If {condition A is met} Then Exit Do
  [code]
Loop Until {condition B is met}
```

The Do loop consists of `lines 17-39`. The Loop Until {condition B is met} is set in `line 39`; here, condition B is `Iteration > Iteration_limit`, then exit do. The If {condition A is met} is set in `line 18`. There only needs to be one exit path from the Do loop.

If/Then/Else. The general form of the structure is

```
If {condition A is met} Then
  [code for condition A met]
Else
  [code for condition A not met]
End If
```

The If/Then/Else is set up between `lines 19-38`. Line 19 sets up the condition {`f1 > fright`} to move an $x^{[right]}$ trial point in the positive `delta_x` direction (`lines 20-27`), while the Else moves an $x^{[left]}$ trial point in the negative `delta_x` direction (`lines 28-37`). ■

3.5.3 Interval Refinement Phase: Interval Halving

Once the single-variable problem has been bounded within an interval, the interval is refined or reduced. There are several strategies to reduce the bounded interval including region elimination methods, curve fitting techniques, and methods requiring derivative information. Region elimination methods include interval halving, golden section, and interpolation–extrapolation methods such as the regula falsi method. Curve fitting techniques include quadratic search, cubic search, and continued fractions. Methods requiring derivatives include bisection, Newton's method, and the secant method. See Ravindran et al. (2006), Edgar et al. (2001), or Pike (1986) for details of these methods. The interval halving method is detailed in this section, and both the regula falsi and Newton's method are shown in Chapter 3 problems.

Perhaps the best known region elimination method is *interval halving* (or the three-point trial method). When using interval halving, three equally spaced trial points are placed between $x^{[left]}$ and $x^{[right]}$ or $x^{[a]}$ and $x^{[b]}$. Here, $x^{[1]} = x^{[a]} + (x^{[b]} - x^{[a]})/4$; $x^{[m]} = x^{[a]} + (x^{[b]} - x^{[a]})/2$; and $x^{[2]} = x^{[b]} - (x^{[b]} - x^{[a]})/4$. Based on the objective function values at these three trial points, one-half of the region between ($x^{[a]}$ and $x^{[b]}$) will be eliminated.

The three possibilities for region elimination are shown in Figure 3.25; note how the $f(x)$ values change with x values from $x^{[a]}$ to $x^{[b]}$.

In Figure 3.25, Case (i), $f(x^{[1]}) < f(x^{[m]}) < f(x^{[2]})$. Here, the optimum lies between $x^{[a]}$ and $x^{[m]}$. On Figure 3.25 the optimum is shown between $x^{[1]}$ and $x^{[m]}$; however, it could also have been shown between $x^{[a]}$ and $x^{[1]}$. The next iteration, as indicated on Figure 3.25 (i), is accomplished by setting (renaming) $x^{[a]} = x^{[a]}$ (no change), $x^{[b]}$ = the current $x^{[m]}$, and $x^{[m]}$ = the current $x^{[1]}$. Interval halving continues by solving for a new $x^{[1]}$ and $x^{[2]}$. Note that the order of this variable renaming is important when coding; $x^{[b]}$ should be renamed before $x^{[m]}$ or else both will actually equal the current $x^{[1]}$.

In Figure 3.25, Case (ii), $f(x^{[1]}) > f(x^{[m]}) > f(x^{[2]})$. Here, the optimum lies between $x^{[m]}$ and $x^{[b]}$. The next iteration is accomplished by setting $x^{[a]} = x^{[m]}$, $x^{[m]} = x^{[2]}$, and $x^{[b]} = x^{[b]}$ (no change), and then solving for a new $x^{[1]}$ and $x^{[2]}$.

In Figure 3.25, Case (iii), the conditions in Case (i) or (ii) are not met. In Figure 3.25 (iii), $f(x^{[1]}) \geq f(x^{[m]}) \leq f(x^{[2]})$. Here, the optimum lies between $x^{[1]}$ and $x^{[2]}$. The next iteration is accomplished by setting $x^{[a]} = x^{[1]}$, $x^{[m]} = x^{[m]}$ (no change), and $x^{[b]} = x^{[2]}$, and then solving for a new $x^{[1]}$ and $x^{[2]}$.

For example, applying interval halving to our bounded result in Table 3.1,

$f(x^{[a]})$ =	$f(x^{[1]})$ =	$f(x^{[m]})$ =	$f(x^{[2]})$ =	$f(x^{[b]})$ =
0.001561	0.000092	0.001099	0.002205	0.003141
$x^{[a]} = 950$	$x^{[1]} = 1550$	$x^{[m]} = 2150$	$x^{[2]} = 2750$	$x^{[b]} = 3350$

which is Case (i), as $f(x^{[1]}) < f(x^{[m]}) < f(x^{[2]})$, giving for the next iteration,

$f(x^{[a]})$ =	$f(x^{[1]})$ =	$f(x^{[m]})$ =	$f(x^{[2]})$ =	$f(x^{[b]})$ =
0.001561	0.000094	0.000092	0.000536	0.001099
$x^{[a]} = 950$	$x^{[1]} = 1250$	$x^{[m]} = 1550$	$x^{[2]} = 1850$	$x^{[b]} = 2150$

which is Case (iii), as $f(x^{[1]}) \geq f(x^{[m]}) \leq f(x^{[2]})$, the process continues with $x^{[a]} = 1250$ and $x^{[b]} = 1850$.

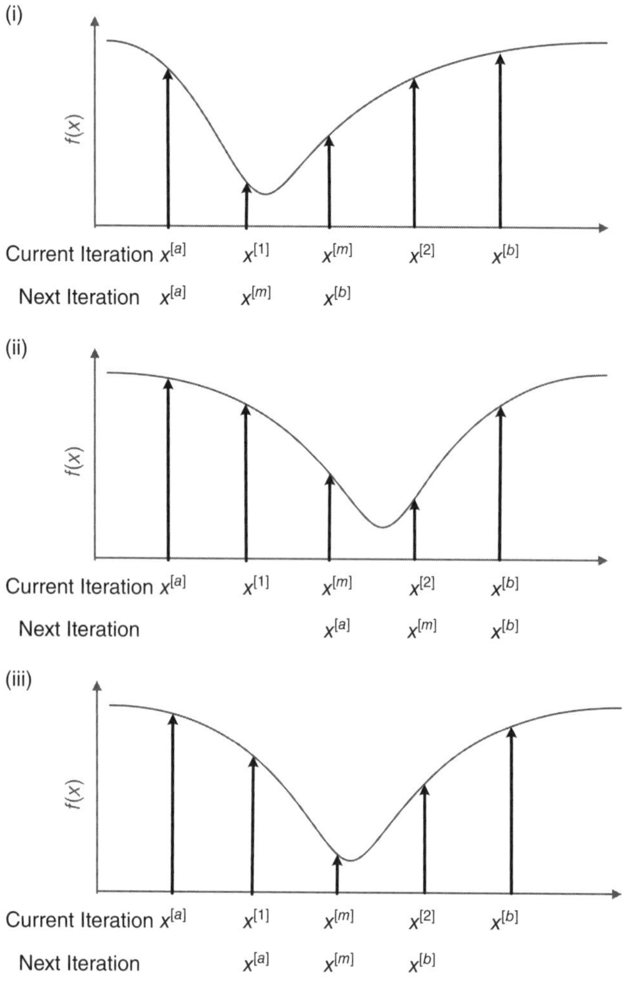

Figure 3.25 Interval halving—$f(x)$ versus x—the three possible cases are shown in the figure.

EXAMPLE 3.14 *VBA Code for Interval Halving: Styrene Process with Alternative Specification*

We want to develop the VBA code needed for interval refinement for the styrene process of Example 3.8; bounding of the water feed rate was accomplished in Example 3.13. The interval halving code to reduce the bound on the water feed rate for the styrene process is given in **Example 3.14.xls**. The final results are shown in Figure 3.26, and the VBA code is provided in Figure 3.27.

There are comments provided in the VBA code. Note that for each new iteration, we solve for two new trial points, $x^{[1]}$ and $x^{[2]}$, and we obtain $x^{[m]}$ by renaming a variable from the current iteration.

3.6 MATERIAL BALANCE PROBLEMS WITH LOCAL NONLINEAR SPECIFICATIONS

Section 3.5 showed that when natural unit operation specifications are replaced with alternative specifications, a search technique may be required to solve the material balance problem. So far in this chapter, both natural and alternative process specifications have been linear, but this is not always the case.

EXAMPLE 3.15 *Styrene Process with Local Nonlinear Specification*

Consider the following modification to Example 3.1. Specify that separator 2 is operated such that the molar concentration of styrene in the product stream and the molar concentration of ethylbenzene in the recycle stream are equal. This specification can be written as

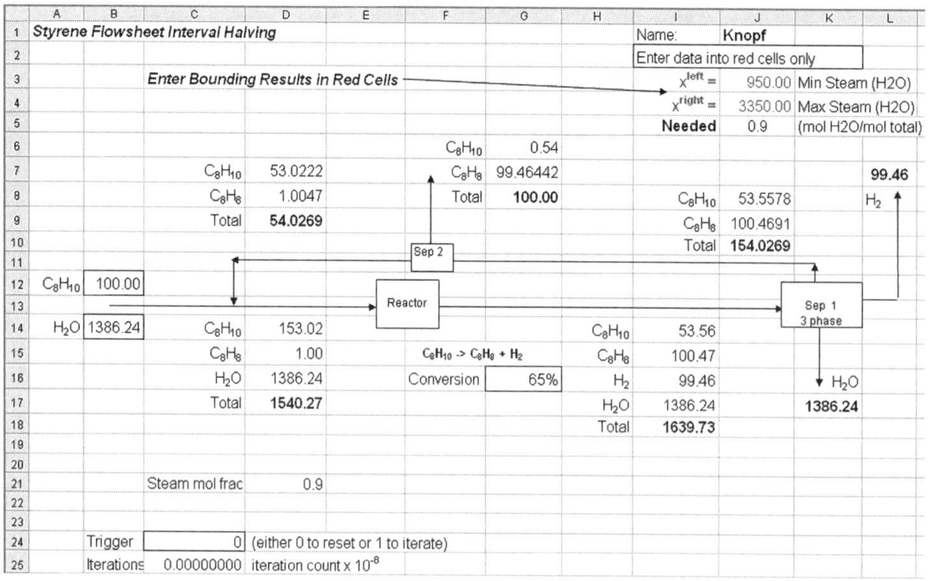

Figure 3.26 Interval halving solution—styrene process.

```
Sub Interval_Halving()
    ' We specify that our function has been bounded
    ' between [xa, xb] or [xleft, xright]
    ' and here xa and xb are given on the flowsheet (in cells J3 and J4,
respectively)
    '
    ' We get the lower bound, xa, from cell J3 on the flowsheet
    ' here xa is our lower bound on the water (steam) flow rate
    xa = Range("J3").Value
    ' Here we call our "function" - which is generally done in a separate Sub
Procedure
    Call Flowsheet(xa, F)
    fa = F
    ' Here we get the upper bound, xb, from cell J4 on the flowsheet
    xb = Range("J4").Value
    Call Flowsheet(xb, F)
    fb = F
    ' Here we determine the mid point
    xm = (xa + xb) / 2
    Call Flowsheet(xm, F)
    fm = F
    Max_Change = 0.000001
    Iteration_limit = 100000
    Iteration = 0
    Do
        Iteration = Iteration + 1
        delta_x = (xb - xa)
        If Abs(xb - xa) < Max_Change Then Exit Do
        ' Here we determine the points 1/4 and 3/4 of the total length
        x1 = xa + delta_x / 4
        Call Flowsheet(x1, F)
        f1 = F
        x2 = xb - delta_x / 4
        Call Flowsheet(x2, F)
        f2 = F
        If f1 < fm Then
            xa = xa
            xb = xm
            xm = x1
            fm = f1
        ElseIf f2 < fm Then
            xb = xb
            xa = xm
            xm = x2
            fm = f2
        Else
            xa = x1
            xb = x2
            xm = xm
            fm = fm
        End If
    Loop Until Iteration > Iteration_limit
    ' Here the optimum, xm, will be placed in the spreadsheet
    ' and one last convergence performed
    Range("B14").Value = xm
End Sub
Sub Flowsheet(x, F)
    ' Here we place the water flow rate on the flowsheet in cell B14
    ' the flowsheet will converge before it returns control to the VBA code.
    Range("B14").Value = x
    ' The flowsheet is now converged with the x value (water flow rate)
    ' Cell D21 contains the converged (moles H2O)/(Total moles in Reator)
    ' Cell J5 contains the desired (moles H2O)/(Total moles in Reator)
    ' We square the difference between converged and desired (mole fraction)
    F = (Range("D21").Value - Range("J5").Value) ^ 2
    ' F will be returned to the Interval Halving Sub Procedure
End Sub
```

Figure 3.27 Interval halving VBA code.

$$\frac{N_{C_8H_8}^{product}}{N_{C_8H_{10}}^{product}+N_{C_8H_8}^{product}}=\frac{N_{C_8H_{10}}^{recycle}}{N_{C_8H_{10}}^{recycle}+N_{C_8H_8}^{recycle}}. \tag{3.67}$$

The current solution to Example 3.1 does not satisfy Equation (3.67):

$$\frac{N_{C_8H_8}^{product}}{N_{C_8H_{10}}^{product}+N_{C_8H_8}^{product}}=\frac{99.4644}{0.5356+99.4644}=0.9946$$

$$\frac{N_{C_8H_{10}}^{recycle}}{N_{C_8H_{10}}^{recycle}+N_{C_8H_8}^{recycle}}=\frac{53.0222}{53.0222+1.0047}=0.9814.$$

Equation (3.67) can be used as an alternative to either Equation (3.44) or (3.45). The degree of freedom for the process does not change; we are simply substituting a nonlinear quadratic specification, Equation (3.67), for the linear specification given by Equation (3.44) or (3.45). Equation (3.67) can be rearranged and solved for $N_{C_8H_8}^{product}$:

$$N_{C_8H_8}^{product}=\frac{\left(N_{C_8H_{10}}^{recycle}\right)\left(N_{C_8H_{10}}^{product}\right)}{N_{C_8H_8}^{recycle}}. \tag{3.68}$$

Equation (3.68) can be used in place of Equation (3.45). For the sequential modular approach, the output stream is calculated based on feed streams to the unit. Using Equations (3.44), (3.46), and (3.47) in Equation (3.68) gives

$$\left(N_{C_8H_8}^{product}\right)^2-\left(N_{C_8H_8}^{organic\ out}\right)\left(N_{C_8H_8}^{product}\right)+0.0099\left(N_{C_8H_{10}}^{organic\ out}\right)^2=0,$$

which is of the form $ax^2+bx+c=0$. Solve using the quadratic formula

$$\left(N_{C_8H_8}^{product}\right)=\frac{\left(N_{C_8H_8}^{organic\ out}\right)+\sqrt{\left(-N_{C_8H_8}^{organic\ out}\right)^2-4\left(0.0099\left(N_{C_8H_{10}}^{organic\ out}\right)^2\right)}}{2}. \tag{3.69}$$

The Excel file with this modification to the styrene process is provided in **Example 3.15a.xls** and shown in Figure 3.28. Here, we use the Trace Precedents feature in Excel (Toolbars → Tools → Formula Auditing) to show that the styrene molar flow rate in the product stream is determined based on Organic Out stream information. Figure 3.29, from **Example 3.15b.xls**, shows the converged flow sheet after a successive substitution solution of the recycle loop. ■

Nonlinear process specifications that are local to a unit operation do not change the sequential modular solution approach, provided the nonlinear equation can be added to the material balance set for that unit operation.

3.7 CLOSING COMMENTS

Solving flow sheet material balance problems in Chapter 3 has involved assembling four elementary modules—there are additional modules that will be discussed in Chapter 5. We have used the sequential modular approach, which requires an iterative strategy if recycle loops are present in the flow sheet. Here, both successive substitution and the Wegstein method were used to converge the recycle stream. If a natural specification to a processing unit is missing, and

	A	B	C	D	E	F	G	H	I	J
1	*Styrene Production Nonlinear Specification - Problem Start*									
2										
3		Trigger	0	(either 0 to reset or 1 to iterate)						
4		Iterations	0.00000000	iteration count x 10^{-8}						
5										
6		Reaction	C8H10 ---> C8H8 + H2		Conversion	0.65				
7										
8	**Iteration #1**		(k=1)							
9			(x_n^1)							$g_n(x_n^1)$
10		**Feed**	**Recycle**	**Reactor_In**	**Reactor_Out**	**Water_Out**	**Vapor_Out**	**Organic_Out**	**Product**	**New Recycle**
11	N(C8H10)	100	0	100	35	0	0	35	0.35	34.65
12	N(C8H8)	0	0	0	65	0	0	65	64.812884	0.187115573
13	N(H2)	0	0	0	65	0	65	0	0	0
14	N(H2O)	3000	0	3000	3000	3000	0	0	0	0
15	N(total)	3100	0	3100	3165	3000	65	100	65.162884	34.83711557
16										
17										
18	**Iteration #2**		(k=2)							
19			(x_n^2)							$g_n(x_n^2)$
20		**Feed**	**Recycle**	**Reactor_In**	**Reactor_Out**	**Water_Out**	**Vapor_Out**	**Organic_Out**	**Product**	**New Recycle**
21	N(C8H10)	100	34.65	134.65	47.1275	0	0	47.1275	0.471275	46.656225
22	N(C8H8)	0	0.187115573	0.187115573	87.70961557	0	0	87.70961557	87.458205	0.251410516
23	N(H2)	0	0	0	87.5225	0	87.5225	0	0	0
24	N(H2O)	3000	0	3000	3000	3000	0	0	0	0
25	N(total)	3100	34.83711557	3134.837116	3222.359616	3000	87.5225	134.8371156	87.92948	46.90763552

Figure 3.28 Styrene process with nonlinear specification—problem start.

	A	B	C	D	E	F	G	H	I	J
1	*Styrene Production Nonlinear Specification*									
2										
3		Trigger	0	(either 0 to reset or 1 to iterate)						
4		Iterations	0.00000000	iteration count x 10^{-8}						
5										
6		Reaction	C8H10 ---> C8H8 + H2		Conversion	0.65	mol fraction C_8H_8 product		0.99464422	
7										
8	**Iteration #k**						mol fraction C_8H_{10} recycle		0.99464422	
9			(x_n^k)							$g_n(x_n^k)$
10		**Feed**	**Recycle**	**Reactor_In**	**Reactor_Out**	**Water_Out**	**Vapor_Out**	**Organic_Out**	**Product**	**New Recycle**
11	N(C8H10)	100	53.02218822	153.0221882	53.55776588	0	0	53.55776588	0.53557766	53.02218822
12	N(C8H8)	0	0.28550409	0.28550409	99.74992643	0	0	99.74992643	99.4644223	0.28550409
13	N(H2)	0	0	0	99.46442234	0	99.46442234	0	0	0
14	N(H2O)	3000	0	3000	3000	3000	0	0	0	0
15	N(total)	3100	53.30769231	3153.307692	3252.772115	3000	99.46442234	153.3076923	100	53.30769231
16										
17										
18	**Iteration #k+1**									
19			(x_n^{k+1})							$g_n(x_n^{k+1})$
20		**Feed**	**Recycle**	**Reactor_In**	**Reactor_Out**	**Water_Out**	**Vapor_Out**	**Organic_Out**	**Product**	**New Recycle**
21	N(C8H10)	100	53.02218822	153.0221882	53.55776588	0	0	53.55776588	0.53557766	53.02218822
22	N(C8H8)	0	0.28550409	0.28550409	99.74992643	0	0	99.74992643	99.4644223	0.28550409
23	N(H2)	0	0	0	99.46442234	0	99.46442234	0	0	0
24	N(H2O)	3000	0	3000	3000	3000	0	0	0	0
25	N(total)	3100	53.30769231	3153.307692	3252.772115	3000	99.46442234	153.3076923	100	53.30769231

Figure 3.29 Styrene process with nonlinear specification—converged solution.

if this is replaced by an alternative specification (not local to the unit operation), search techniques may be necessary to obtain a solution to the material balance problem. Nonlinear process specifications cause no additional difficulty in obtaining a sequential modular solution, provided the nonlinear material balance equation uses information local to that unit operation.

The solution of the tear stream in the recycle loop is of the form $f(x) = 0$. The rearrangement of $f(x) = 0$ to $x = g(x)$ allows convenient tear stream solution by successive substitution or by the Wegstein method. For some problems, $f(x) = 0$ cannot be rearranged to $x = g(x)$, and for some problems, it is simply more convenient to work directly with $f(x)$. We have used Goal Seek and Solver within Excel to solve problems of this form. An equivalent form of $f(x) = 0$ that is useful for some engineering problems can be formed by minimizing $f(x)$. We examined the use of bounding with interval halving to solve problems that involve minimizing $f(x)$. The choice of $f(x) = 0$, $x = g(x)$, or $\min f(x)$ is not meant to confuse; certain problems tend to lend themselves naturally to a specific choice of the form of $f(x)$. The Goal Seek algorithm uses a combination of bounding and interval refinement via regula falsi (see Problem 3.3) for unconstrained single-variable optimization problems. Excel Solver uses a generalized reduced gradient (GRG) algorithm for constrained multivariable optimization problems (Edgar et al., 2001; Lasdon et al., 1978).

Note that solving material balance problems has involved assembling n equations and n unknowns. This suggests the use of a simultaneous or an equation-based solution approach. We explore implementation of the simultaneous approach in Chapter 4. In Chapter 4 and in Appendix A, we develop an efficient method to solve a system of equations using a combination of the C language, VBA, and Excel.

The topics discussed in this chapter have been guided by two pioneering books: *Introduction to Material and Energy Balances* by Reklaitis (1983) and *Introduction to Chemical Engineering and Computer Calculations* by Myers and Seider (1976).

REFERENCES

DeLancey, G.B. 1999. Process analysis: An electronic version. *Chem. Eng. Ed.* (Winter): 40–45.

Edgar, T.F., D.M. Himmelblau, and L.S. Lasdon 2001. *Optimization of Chemical Processes* (2nd edition). McGraw-Hill Companies, New York.

Ferreira, E.C., R. Lima, and R. Salcedo 2004. Spreadsheets in chemical engineering education—A tool in process design and process integration. *Int. J. Eng. Ed.* 20(6): 928–938.

Lasdon, L.S., A.D. Warren, A. Jain, and M. Ratner 1978. Design and testing of a generalized reduced gradient code for nonlinear programming. *ACM Trans. Math Softw.* 1(4): 33–50.

Myers, A.L. and W.D. Seider 1976. *Introduction to Chemical Engineering and Computer Calculations*. Prentice Hall, Englewood Cliffs, NJ.

Pike, R.W. 1986. *Optimization for Engineering Systems*. Van Nostrand Reinhold, New York.

Ravindran, A., K.M. Ragsdell, and G.V. Reklaitis 2006. *Engineering Optimization: Methods and Applications* (2nd edition). John Wiley & Sons, Hoboken, NJ.

Redlich, O. and J.N.S. Kwong 1949. On the thermodynamics of solutions. V. An equation of state for fugacities of gaseous solutions. *Chem. Rev.* 44: 233–244.

REKLAITIS, G.V. 1983. *Introduction to Material and Energy Balances*. John Wiley & Sons, New York.
UPADHYE, R.S. 1983. Determining a set of independent chemical reactions. *Comp. Chem. Eng.* 7(2): 87–92.
WESTERBERG, A.W., H.P. HUTCHISON, R.L. MOTARD, and P. WINTER 1979. *Process Flowsheeting*. Cambridge University Press, London.

PROBLEMS

Problems 3.1–3.15 help explain bounding, interval refinement, and iteration strategies.

Table P3.1, for use with Problems 3.1–3.4—has been generated using a fifth-order polynomial, $f(x) = a_5(x)^5 + a_4(x)^4 + a_3(x)^3 + a_2(x)^2 + a_1(x) + a_0$, where a_i are known constants. All of the roots of the polynomial are real. For all $x \le -3.5$ and all $x \ge 4$, $f'(x)$ is positive.

3.1 ***Number of Roots*** $f(x) = 0$ How many roots are in the interval $-3.5 \le x \le 4.0$?

3.2 ***Location of Roots*** $f(x) = 0$ Consider the problem of locating the roots of the polynomial $f(x) = 0$. List each 0.5 interval in which one or more roots are located. For example, (3.5, 4.0) is a 0.5 interval, but no root exists in this interval.

3.3 ***Interpolation–Extrapolation/Regula Falsi Method/Method Of False Position*** Consider the problem of locating the roots of the polynomial $f(x) = 0$ with the trial points

$$x^{[1]} = 0 \quad f\left(x^{[1]}\right) = -0.39672$$

and

$$x^{[2]} = 0.5 \quad f\left(x^{[2]}\right) = 3.11808.$$

These two trial points bound $f(x) = 0$. Interpolation–extrapolation, also called the regula falsi method or the method of false position, is simply a linear interpolation between two bounding points to find the next trial point, $x^{[3]}$, where $f(x^{[3]}) = 0$ For subsequent iterations, two points are retained, which keep $f(x) = 0$ bounded.

Develop the needed regula falsi equation and solve for $x^{[3]}$; also, provide the regula falsi equation in general form. You cannot evaluate $f(x^{[3]})$ as the constants a_i in the polynomial are not provided.

Table P3.1 For Use with Problems 3.1–3.4

x	$f(x)$	$f'(x)$
−3.5	−437.944	806.1696
−3.0	−145.186	396.8211
−2.5	−13.5907	152.9176
−2.0	27.28908	26.6091
−1.5	25.85088	−22.4544
−1.0	12.19218	−27.1229
−0.5	1.86048	−12.7464
0	−0.39672	2.8251
0.5	3.11808	9.2416
1.0	6.80238	3.6531
1.5	5.50368	−9.2904
2.0	−1.73052	−17.4389
2.5	−7.90272	−1.1424
3.0	5.68458	66.7491
3.5	73.17888	220.8856
4.0	247.9277	503.4171

3.4 ***Newton's Method*** Consider the problem of locating the roots of the polynomial $f(x) = 0$. Newton's method begins with a linear approximation to $f(x)$ at $x^{[1]}$ as

$$f(x) \cong f\left(x^{[1]}\right) + \left.\frac{df}{dx}\right|_{x=x^{[1]}} \left(x - x^{[1]}\right) = f\left(x^{[1]}\right) + \left(f'\left(x^{[1]}\right)\right)\left(x - x^{[1]}\right),$$

with the equation depicted in Figure P3.4.

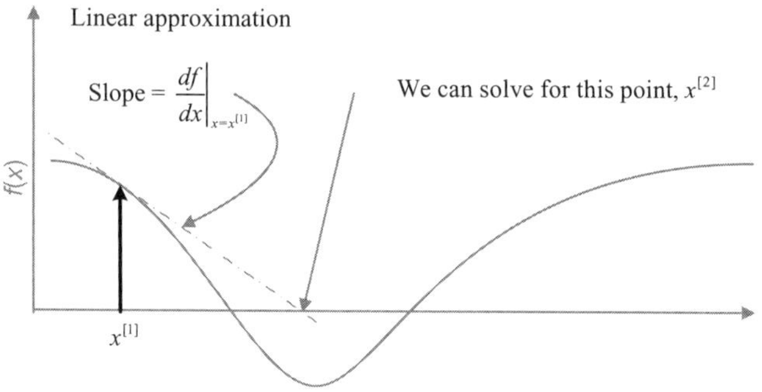

Figure P3.4 Linear approximation to $f(x)$ at $x = x^{[1]}$.

In the equation, we can use known values for $x^{[1]}$, $f(x^{[1]})$, and the slope $f'(x^{[1]})$ to determine point x, where $f(x) = 0$. As this is an iterative procedure, the second trial point is noted as $x^{[2]}$, where $f(x^{[2]}) = 0$. Show yourself Newton's method formula for $x^{[2]}$ is

$$x^{[2]} = x^{[1]} - \frac{f\left(x^{[1]}\right)}{f'\left(x^{[1]}\right)}.$$

Develop the general Newton method equation for $x^{[n+1]}$ where $f(x^{[n+1]}) = 0$. Also, solve for $x^{[2]}$ if $x^{[1]} = 0$ using the data provided in Table P3.1. You will not be able to evaluate $f(x^{[2]})$ as the constants a_i in the polynomial are not provided.

Table P3.2, for use with Problems 3.5 and 3.6—has been generated by rearranging a polynomial, $f(x) = 0$, to the form $x = g(x)$.

3.5 ***Successive Substitution (Table P3.2)*** Using the method of successive substitution, if $x^{[1]} = 0$, what would be the next trial point $x^{[2]}$?

3.6 ***Wegstein Method (Table P3.2)*** Using the bounded Wegstein method, if we have two trial points, $x^{[1]} = 1.5$ and $x^{[2]} = 2.5$, what would be the next trial point $x^{[3]}$?

Table P3.3, for use with Problems 3.7 and 3.8—assume provided data are from minimize f(x).

3.7 ***Bounding (Table P3.3)*** If an initial guess, $x^{[0]} = -2$, is given and the step size is 0.5, find the values for $x^{[1]}$, $x^{[2]}$, and $x^{[3]}$ using Swann's bounding method detailed in Section 3.5.2. Will the function be bounded with these points?

Table P3.2 For Use with Problems 3.5 and 3.6

x	$g(x)$
−3.5	27.14831
−3.0	7.305418
−2.5	−0.98277
−2.0	−2.9701
−1.5	−2.20778
−1.0	−0.84197
−0.5	0.088441
0	0.354259
0.5	0.238267
1.0	0.237573
1.5	0.765935
2.0	1.856097
2.5	2.862117
3.0	2.161703
3.5	−3.14146
4.0	−17.5154

Table P3.3 For Use with Problems 3.7 and 3.8

x	$f(x)$
−4.0	26.01
−3.5	21.16
−3.0	16.81
−2.5	12.96
−2.0	9.61
−1.5	6.76
−1.0	4.41
−0.5	2.56
0	1.21
0.5	0.36
1.0	0.01
1.5	0.16
2.0	0.81
2.5	1.96
3.0	3.61
3.5	5.76
4.0	8.41

3.8 ***Interval Halving (Table P3.3)*** Assume the function in Table P3.3 has been bounded between (−4, 4). With these values as ($x^{[\text{left}]}$ and $x^{[\text{right}]}$), determine the next trial points using interval halving. What region would be eliminated and what would be the new values for ($x^{[\text{left}]}$ and $x^{[\text{right}]}$)? This is just one complete iteration of interval halving.

3.9 ***Successive Substitution/Wegstein Method*** Recall Problem 2.3—a processing facility is considering shutting down or undertaking a major overhaul. The major overhaul will require \$2MM. If the overhaul is accomplished, it will extend the life of the plant for five more years with anticipated profits of \$500,000 per year.

(a) Solve for the rate of return of the possible overhaul investment using successive substitution with $i = 10\%$ per year as a starting guess. First, use Equation (3.50) to check if the successive substitution solution may diverge:

$$\left|\frac{dg\left(x^{[k]}\right)}{dx^{[k]}}\right| \geq 1.0.$$

(b) Solve using the bounded Wegstein method.

3.10 ***Successive Substitution/Wegstein Method*** Solve the equation $2x + 3\log(x) - 1 = 0$.

(a) Use the method of successive substitution starting from $x^{[1]} = 1.0$.

(b) Starting with results from the first two successive substitution iterations in (a), solve using the bounded Wegstein method. Comment on the required number of iterations for each method.

3.11 ***Newton's Method and the Redlich–Kwong EOS*** Solve Example 3.2 using Newton's method. For the first trial point, use the ideal gas result

$$v^{[1]} = \frac{RT}{P}$$

The reader can find background material for Newton's method in Problem 3.4.

3.12 ***Secant Method for* f(x) = 0** The secant method for finding the roots of a polynomial can be obtained by using a backward numerical derivative:

$$f'\left(x^{[n]}\right) \approx \frac{f\left(x^{[n]}\right) - f\left(x^{[n-1]}\right)}{\left(x^{[n]}\right) - \left(x^{[n-1]}\right)}$$

in Newton's method. Recall from Problem 3.4 that Newton's method provides

$$x^{[n+1]} = x^{[n]} - \frac{f\left(x^{[n]}\right)}{f'\left(x^{[n]}\right)}.$$

Derive the secant method. Does this method require that the root be bounded?

Problems 3.13–3.14 find the molar volume in the Redlich–Kwong EOS (Example 3.2) as a minimization problem.

3.13 ***Newton's Method and Numerical Derivatives to Find the Minimum of* f(x)** Solve Example 3.2 as a single-variable minimization problem using Newton's method. For the first trial point, use the ideal gas result,

$$v^{[1]} = \frac{RT}{P} = 6.39 \text{ ft}^3/\text{lb-mol}.$$

The reader can find background material for this application of Newton's method in Ravindran et al. (2006). Recall from Problem 3.4 that in Newton's method, we found the roots of the polynomial $f(x) = 0$ by using a linear approximation to $f(x)$ at $x^{[1]}$. In this application of Newton's method, we are again "looking for the roots of a polynomial," but here, the polynomial is $f'(x) = 0$. Recall from calculus that a necessary condition for the minimum of a function is $f'(x) = 0$. A linear approximation of $f'(x)$ at $x^{[1]}$ can be constructed as

$$f'(x) \cong f'(x^{[1]}) + \left.\frac{d^2 f}{dx^2}\right|_{x=x^{[1]}} (x - x^{[1]})$$
$$= f'(x^{[1]}) + (f''(x^{[1]}))(x - x^{[1]}).$$

For any given point $x^{[1]}$, we can use the known first derivative $f'(x^{[1]})$ and second derivative $f''(x^{[1]})$ information to determine point x, where $f'(x) = 0$. As this is an iterative procedure, the next trial point is noted as $x^{[2]}$, where $f'(x^{[2]}) = 0$ and, in general, $x^{[n+1]}$, where $f'(x^{[n+1]}) = 0$. Develop the general equation for this application of Newton's method. In terms of the molar volume, you should obtain

$$\upsilon^{[k+1]} = \upsilon^{[k]} - \frac{f'(\upsilon^{[k]})}{f''(\upsilon^{[k]})}$$

Here, the first and second derivative $f'(\upsilon^{[k]})$ and $f''(\upsilon^{[k]})$ evaluations are needed so the function $f(\upsilon)$ must be twice differentiable. We will find the analytic derivatives for the objective function (as a minimization problem) are cumbersome, and here it is easier to numerically determine values.

For the first derivative, we can use a forward difference approximation:

$$\left.\frac{df(\upsilon)}{d\upsilon}\right|_{\upsilon=\upsilon^{[k]}} = \frac{f(\upsilon^{[k]}+\varepsilon) - f(\upsilon^{[k]})}{\varepsilon},$$

and for the second derivative, we can use a central difference approximation:

$$\left.\frac{d^2 f(\upsilon)}{d\upsilon^2}\right|_{\upsilon=\upsilon^{[k]}} = \frac{d}{d\upsilon}\left(\left.\frac{df(\upsilon)}{d\upsilon}\right|_{\upsilon=\upsilon^{[k]}}\right)$$
$$= \frac{\left(\frac{f(\upsilon^{[k]}+\varepsilon) - f(\upsilon^{[k]})}{\varepsilon}\right) - \left(\frac{f(\upsilon^{[k]}) - f(\upsilon^{[k]}-\varepsilon)}{\varepsilon}\right)}{\varepsilon}$$

$$\left.\frac{d^2 f(\upsilon)}{d\upsilon^2}\right|_{\upsilon=\upsilon^{[k]}} = \frac{f(\upsilon^{[k]}-\varepsilon) - 2f(\upsilon^{[k]}) + f(\upsilon^{[k]}+\varepsilon)}{\varepsilon^2}.$$

In these equations, epsilon (ε) is a small increment; here, a value $\varepsilon = 0.00001$ was used.

3.14 ***Bounding with Interval Halving to Find the Minimum of* f(x)** Solve Example 3.2 as a single-variable minimization problem using the bounding and interval halving method developed in Section 3.5. For the first trial point, use the ideal gas result:

$$\upsilon^{[1]} = \frac{RT}{P} = 6.39 \text{ ft}^3\text{/lb-mol.}$$

3.15 ***Ammonia Process Material Balances: Excel Sheet Solution from Reklaitis (1983, p. 273)*** In an ammonia plant, a feed gas consisting of 24.5% N_2, 74% H_2, 1.2% CH_4, and 0.3% Ar is catalytically reacted to produce NH_3. The reaction is

$$N_2 + 3H_2 = 2NH_3,$$

with 65% of the N_2 entering the reactor converted per pass. The products of reaction are refrigerated to separate out 75% of the NH_3 product per pass. The remaining process stream is recycled back to the reactor. In order to stabilize the buildup of the inerts CH_4 and Ar in the process, part of the recycled gas is purged. The splitter purge rate (stream 6) is 5% of the stream into the splitter (stream 5); or from Equation (3.19), $N_6 = \alpha_6 N_5$ with $\alpha_6 = 0.05$. Calculate all flows in the process. A simplified flow sheet for the process is shown in Figure P3.15. Assume a basis of 1000 mol/h for the feed stream (stream 1). Solve using an Excel sheet with successive substitution. Be sure to name all cells.

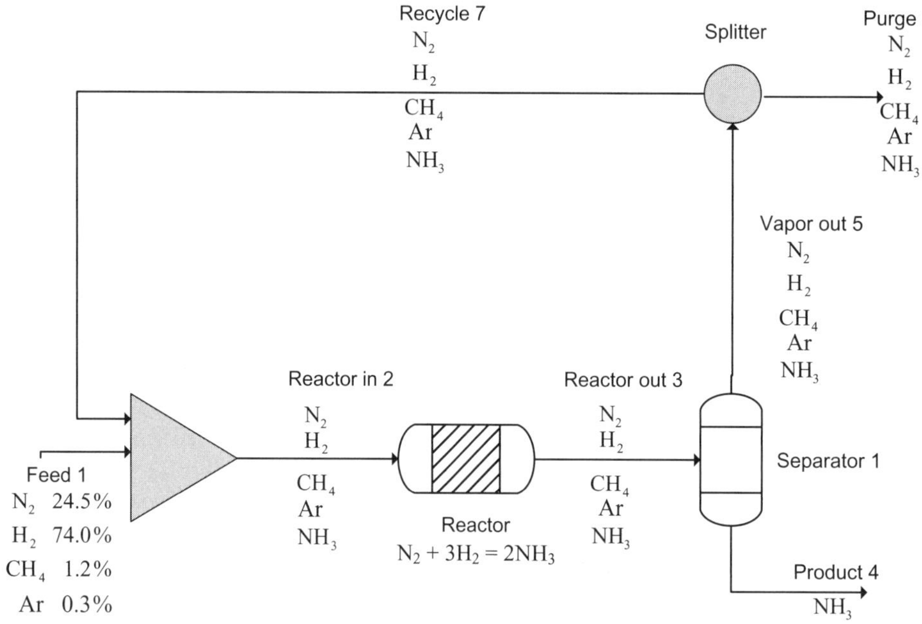

Figure P3.15 Ammonia plant flow sheet.

3.16 ***Ammonia Process Material Balances with Alternative Specification: Excel Sheet Solution Using Goal Seek and Solver*** Starting with Problem 3.15, suppose the splitter split fraction α_6 is adjusted so that the combined reactor feed gas (reactor in stream 2) contains 18 mol % CH_4. Calculate all flows in the process. Again, assume a basis of 1000 mol/h for the feed stream (stream 1).

(a) Use Goal Seek to vary the split fraction α_6, the goal being that the mole fraction of methane in stream 2 is 18%. Show yourself that Goal Seek will not work.

(b) Use Solver to vary the split fraction until the mole fraction of methane in stream 2 is 18%. Constraints on the split fraction may be needed to keep α_6 feasible.

3.17 ***Ammonia Process Material Balances with Alternative Specification: VBA Code*** Solve Problem 3.16 using a VBA code to first bound the splitter split fraction α^6 followed by interval refinement using interval halving. Write a single macro subroutine, which calls the bounding phase and then calls the interval refinement phase. In the bounding phase, you may need to add If statements to keep $0 \le \alpha_6 \le 1.0$.

3.18 ***Coding Improving: Wegstein Method*** In the Wegstein method developed in Example 3.7, we performed 2 iterations of successive substitution followed by 10 iterations of the Wegstein method. Improve the code by exiting the Wegstein method when the absolute value of the change in $|x_n^{[k]} - x_n^{[k-1]}|$ is less than the convergence criterion parameter (Max_Change) as $|x^{[k+1]} - x^{[k]}| < \varepsilon$. In some problems, a better selection may be

$$\left|\frac{x^{[k+1]} - x^{[k]}}{x^{[k]}}\right| < \varepsilon_1.$$

We can also include a convergence requirement on the objective function as

$$\left|\frac{f(x^{[k+1]}) - f(x^{[k]})}{f(x^{[k]})}\right| < \varepsilon_2.$$

Solution Hint

```
For i = 0 To n_components - 1
If abs (x_current(i) - x_old(i)) >=
Max_Change Then
Call Wegstein Method
Else
Next i
Stop
```

3.19 ***Coding Improvement: Bounding and Interval Halving*** Combine the bounding phase of Example 3.12 and the interval halving code of Example 3.13 into a single procedure as

```
Sub Bound_Then_Interval_Halving()
Call Bound
Call Interval_Halving
End Sub
```

3.20 ***Flow Sheet Decomposition: Vinyl Chloride Process Material Balance (Solution Provided)*** DeLancey (1999) presented a simplified process for vinyl chloride production from ethylene, solving the nonlinear material balance equation set using the software package Scientific Notebook. Ferreira et al. (2004) solved the vinyl chloride problem within Excel by using the material balance equations as constraints within Excel Solver. We will solve this problem in Chapter 5 using the Newton–Raphson method, but here we want to examine the solution using the sequential modular approach.

Figure P3.20a (DeLancey, 1999) provides a simplified flow sheet for vinyl chloride (C_2H_3Cl) production from ethylene (C_2H_4).

The reactions are the following:

Chlorination reactor: $C_2H_4 + Cl_2 = C_2H_4Cl_2$

Oxyhydrochlorination reactor:

$$C_2H_4 + 2HCl + \frac{1}{2}O_2 = C_2H_4Cl_2 + H_2O$$

Pyrolysis reactor: $C_2H_4Cl_2 = C_2H_4Cl + HCl$

Feed N_1 is 90 mol % ethylene and 10% inerts; N_2 is pure chlorine (Cl_2); and N_3 is pure oxygen (O_2). All ethylene, oxygen, chlorine, and hydrochloric acid (HCl) fed to the chlorination and oxyhydrochlorination units react completely—these species should not be present in N_6 or N_7.

In the pyrolysis reactor, 50% of the dichloroethane ($C_2H_4Cl_2$) fed to the unit is converted. The unreacted dichloroethane is separated and recycled with the inerts in stream N_{12}. The inert concentration in the recycle stream is 50 mol %. Pure hydrochloric acid is recycled in stream N_{13}. The final product stream, N_{12}, contains only vinyl chloride and water. We want to solve for all species flow rates when $N_1 = 100$.

Solution Hints and Solution Strategy Discussion

To obtain a sequential modular solution, we begin by redrawing Figure P3.20a in terms of our elementary modules as shown in Figure P3.20b.

The immediate problem with the sequential modular approach is that the split fractions from splitters 1 and 2 are not known. Only unique values for splitters 1 and 2 split fractions will produce an inert concentration in the recycle stream of 50 mol % and complete conversion of ethylene, oxygen, chlorine, and hydrochloric acid in the chlorination and oxyhrdrochlorination reactors.

Solution Approach 1

Let us assume the initial values for the split fractions (e.g., $\alpha_4 = 0.2$ with $N_4 = \alpha_4 N_1$ and $\alpha_{11} = 0.5$ with $N_{11} = \alpha_{11} N_{10}$) and assemble the needed material balances in an Excel sheet. We do need to consider the structure of the flow sheet in Figure P3.20b. There are two recycle loops as defined by mixer 1 (streams Recycle_1, N_7, Recycle_2, N_8, N_{13}) and mixer 2 (streams Recycle_2, N_8, N_{10}, N_{12}). Actually, only one of the two recycle loops is essential; the loop with mixer 2 is essential. For example, we should be able to guess or tear species flow rates in Recycle 2 and obtain a converged solution. For a discussion of flow sheet tear stream selection and updating, see Reklaitis (1983) or Westerberg et al. (1979).

Generally, flow rates in any stream in the essential recycle loop can be guessed and converged, and here it is convenient to start the solution by guessing values for N_8 and solving the material balance problem. An immediate problem

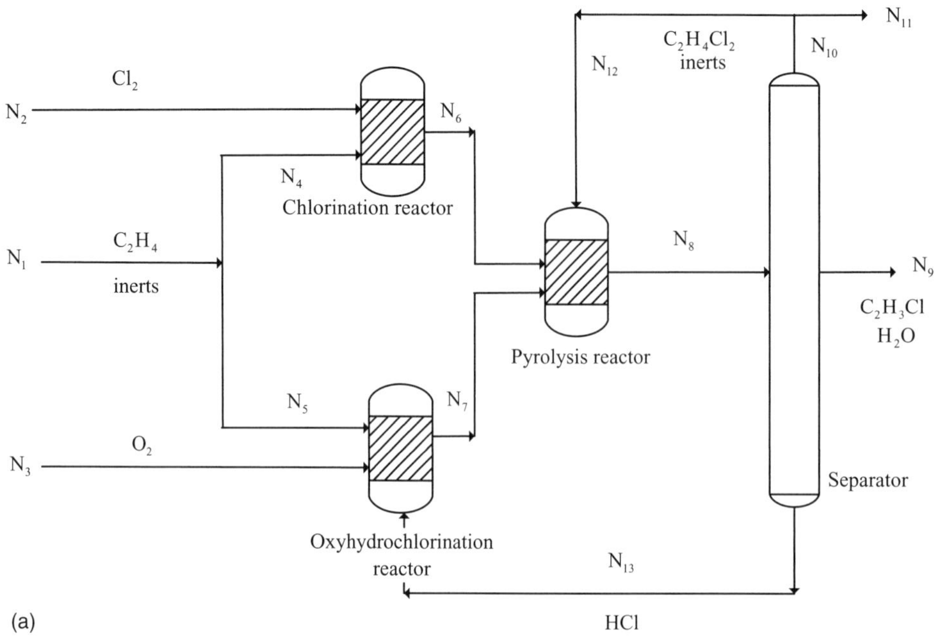

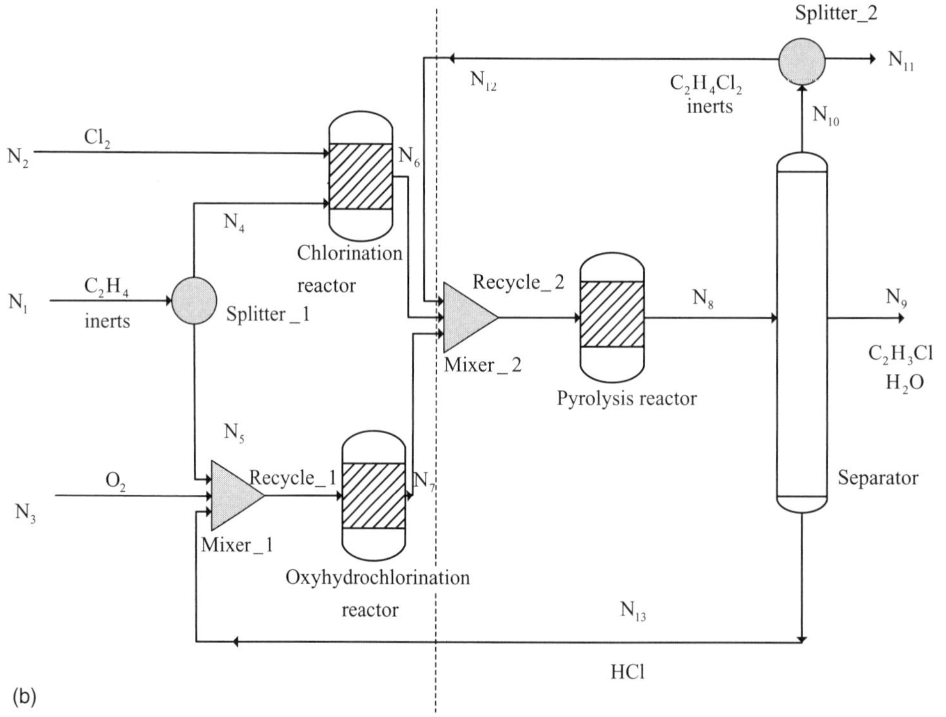

Figure P3.20 (a) Vinyl chloride flow sheet (DeLancey, 1999). (b) Vinyl chloride flow sheet in terms of needed elementary modules.

will surface. Regardless of the initial guess for stream N_8, there will be slight numerical inaccuracies in the HCl flow rate in stream 13, and these inaccuracies will accumulate causing the HCl flow rate to increase without bound.

Solution Approach 2

We must know the all split fractions to solve the original vinyl chloride flow sheeting problem with our general sequential modular approaches. To overcome this problem, we can decompose the flow sheet into two separate sequential modular flow sheeting problems as indicated by the dashed line in Figure P3.20b. We can fix N_{13} and solve the first part of the flow sheet. Specifically, we iterate for the split fraction, α_4 (and $\alpha_5 = 1 - \alpha_4$), which gives complete conversion of ethylene, oxygen, chlorine, and hydrochloric acid in the chlorination and oxyhydrochlorination reactors.

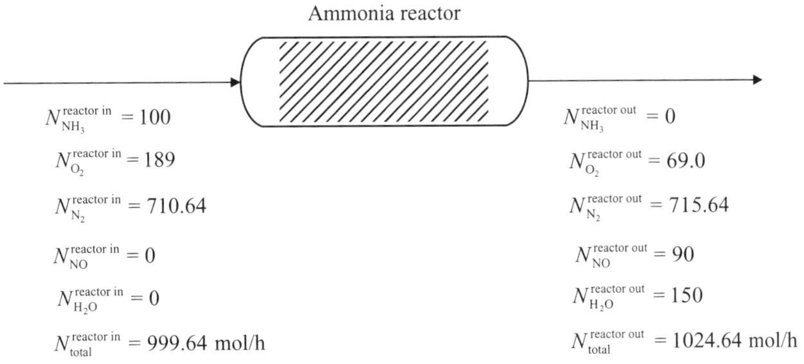

Figure P3.22 Ammonia reactor with species flow rates in and out (mole per hour).

We next solve the second part of the flow sheet for the split fraction, α_{11} (and $\alpha_{12} = 1 - \alpha_{11}$), which produces an inert concentration in the recycle stream of 50 mol %. In order to keep the two flow sheets separate, only flows for dichloroethane ($C_2H_4Cl_2$) and the inerts in N_6 and N_7 are allowed to pass to mixer 2 (flows of ethylene, oxygen, chlorine, and hydrochloric acid are set = 0). There is a recycle loop through splitter 2, which will need to be converged for each split fraction trial value. After α_{11} is converged, we will have an updated value for stream N_{13} and we can again solve for the split fraction α_4. This process is repeated until both $\alpha_4$and α_{11} are converged.

A straightforward approach here would be to use Excel Solver to alternately solve for α_4 and α_{11}. This will work for α_4 but not for α_{11}—there are numerical difficulties in determining α_{11}, and this is discussed in Problem 3.21.

Solution Approach 3

In Solution Approach 2, we are decomposing our initial flow sheeting problem into two smaller, but connected, problems in order to obtain a solution. Rather than using Excel Solver to alternately solve for $\alpha_4$and α_{11} here, use our bounding and interval halving approach—this will work and the solution is provided in **Problem 3.20.xls**; run the macro Run_Bound_and_Interval_Halving.

3.21 ***Numerical Accuracy: Vinyl Chloride Process Material Balance*** In Problem 3.20 Solution Approach 2, using Excel Solver should have worked to determine α_{11}. To understand the difficulty here, change the flow rate from $N_1 = 100$ to $N_1 = 1$ in the provided solution in **Problem 3.20.xls**; a value of $N_1 = 1$ was originally used in DeLancey (1999). Also change the values for α_4 and $\alpha_{11}$to their original starting guesses and run the solution macro Run_Bound_and_Interval_Halving. The value for α_{11} will not change—explain why.

Solution Hint

Look at the values for the objective function $f(x^{[0]})$ to $f(x^{[0]} + \Delta x)$ and $f(x^{[0]} - \Delta x)$. The objective function is bounded with the starting guess for α_{11} and the small step size. Vary the step size Δx for the α_{11} split fraction bounding phase until the correct solution is obtained.

3.22 ***Multiple Extents of Reaction: Ammonia Reaction*** Ammonia is converted in a reactor as shown in Figure P3.22. In this figure, the flow rates are in mole per hour. Here, we will first need to determine the number of reactions required to produce the observed products and then to determine the extent of each reaction.

In general (see Reklaitis, 1983; Upadhye, 1983), when given species in and out of a reactor, the maximum number of linearly independent reactions is found as the number of species minus the rank of the atomic matrix. The rank of the atomic matrix is generally the number of elements. For Figure P3.22, the number of expected reactions would be

$$n_{\text{species}} - n_{\text{elements}} = 5 \text{ species} - 3 \text{ elements (H, N, O)} = 2 \text{ reactions.}$$

One set of reactions (reaction set 1) could be

$$N_2 + O_2 \Leftrightarrow 2NO \qquad \text{(R1a)}$$

$$2NH_3 + \frac{3}{2}O_2 \Leftrightarrow N_2 + 3H_2O. \qquad \text{(R1b)}$$

A second set of reactions (reaction set 2) could be

$$4NH_3 + 5O_2 \Leftrightarrow 4NO + 6H_2O \qquad \text{(R2a)}$$

$$2NH_3 + \frac{3}{2}O_2 \Leftrightarrow N_2 + 3H_2O. \qquad \text{(R2b)}$$

A note of caution—there are systems where the rank of the atomic matrix ≠ the number of elements; here, for example, see Reklaitis (1983, p. 217).

Chapter 4

Computer-Aided Solutions of Process Material Balances: The Simultaneous Solution Approach

In Chapter 4, we explore the solution of material balance problems using the simultaneous or equation-based solution approach. To solve linear equation sets, we will use the Gauss–Jordan (G-J) elimination method, which is developed in Section 4.1. In Examples 4.1–4.4, the Gauss–Jordan method is used to solve linear equation sets including styrene material balance problems from Chapter 3. The Gauss–Jordan method is coded in Visual Basic for Applications (VBA) in Example 4.2. Also, in Appendix A, we show how the Gauss–Jordan algorithm can be written as a C program and linked to Excel as a dynamic link library (DLL). We briefly compare the C program with the VBA-based material (Example 4.2) presented here. Examples 4.5–4.9 address nonlinearities within the simultaneous approach, and here, the solution is obtained using the Newton–Raphson (NR) method.

After finishing Examples 4.1 and 4.2 in this chapter, you should read the provided Appendix A. An emphasis of this text is providing the tools necessary for Excel to be the pre- and postprocessor for programs written in low-level languages. In the Appendix, we set the foundation for the use of Excel as a pre- and postprocessor for C/C++ programs. Details are provided to allow single variables, vectors, and matrices to pass from Excel ↔ VBA ↔ C programs. The conduit for passing information between Excel and C programs is VBA. Passing single variables is straightforward. Passing vectors requires knowledge of pointers. Passing matrices requires understanding of pointers and row-major and column-major storage strategies. In the Appendix, the matrix transfer between Excel and C is demonstrated by solving n linear equations with n unknowns (Example 4.1) using the Gauss–Jordan reduction procedure written as a C program.

I do want to emphasize that it is not necessary to fully understand (or even read) the bridging techniques developed in the Appendix A in order to use the information and programs presented throughout this book. However, at some point if the reader wants to interface Excel to programs written in other languages, the material in the Appendix will be useful.

4.1 SOLUTION OF LINEAR EQUATION SETS: THE SIMULTANEOUS APPROACH

When we assemble independent material balances with natural specifications as detailed in Chapter 3, we actually create a system of n linear equations and n unknowns. The general form of this problem is $Ax = b$ where A is the matrix of coefficients in the material balances, x is the vector of species flow rates, and b is the vector of the right-hand side (RHS) for the material balance equations. Next, we developed the Gauss–Jordan solution method to solve a set of linear equations. We provided a simple "by-hand" example (Example 4.1) to show the method and the working equations.

4.1.1 The Gauss–Jordan Matrix Elimination Method

The Gauss–Jordan method begins by forming the augmented matrix. The coefficients in this matrix are a_{ij} (I rows, J columns). Step 1 is to select a_{11} as the pivot element and

Modeling, Analysis and Optimization of Process and Energy Systems, First Edition. F. Carl Knopf.

then divide all the elements of the first row by the pivot element, normalizing a_{11}. In Step 2, subtract multiples of the first row, termed the pivot row, from all remaining rows such that the first column, a_{i1}, of the resulting rows is zero. This process continues with the second row where a_{22} is selected as the pivot element. All elements of the second row are divided by a_{22}. Multiples of the second row, now the pivot row, are subtracted from all other rows such that the second column, a_{i2}, of the resulting rows is zero. The process continues until the last row is reached. These operations can be summarized as

Outline of the Gauss–Jordan Method

$k = 0$ to number of rows

Step 1: $k = k + 1$. Select a_{kk} as the pivot element. Divide all elements of row k by a_{kk}. Here,

$$a_{kj} = \frac{a_{kj}}{a_{kk}}, \quad j = 1, 2, \ldots J. \tag{4.1}$$

Step 2: For each row I, where $i \neq k$, calculate new elements as

$$a_{ij} = a_{ij} - a_{ik}a_{kj}, \quad i = 1, 2, \ldots, I \quad \text{for} \quad j = 1, 2, \ldots, J. \tag{4.2}$$

Next k

EXAMPLE 4.1 *The Gauss–Jordan Matrix Elimination Method*

Solve the following equation set using the Gauss–Jordan method:

$$\begin{aligned} 2x_1 \quad -2x_2 \quad +5x_3 \quad &= 13 \\ 2x_1 \quad +3x_2 \quad +4x_3 \quad &= 20 \\ 3x_1 \quad -x_2 \quad +3x_3 \quad &= 10. \end{aligned} \tag{4.3}$$

SOLUTION Here, we show the method and then comment on considerations when coding the method. The method begins by forming the augmented matrix

$$\begin{pmatrix} 2 & -2 & 5 & 13 \\ 2 & 3 & 4 & 20 \\ 3 & -1 & 3 & 10 \end{pmatrix}. \tag{4.4}$$

Step 1: Here, $k = 1$ and the pivot element $a_{11} = 2$. Dividing all elements of the first row by a_{11} or $a_{1j} = a_{1j}/a_{11}$ gives

$$\begin{pmatrix} 1 & -1 & 2.5 & 6.5 \\ 2 & 3 & 4 & 20 \\ 3 & -1 & 3 & 10 \end{pmatrix}. \tag{4.5}$$

Step 2 (for rows 2 and 3): Eliminate the a_{21} term from row 2 using $[a_{ij} = a_{ij} - a_{ik}a_{kj}]$ or $[a_{2j} = a_{2j} - a_{21}a_{1j}]$. Multiple the first row by 2 (a_{21}) and subtract the result from the second row. Eliminate the a_{31} term from the third row by $[a_{3j} = a_{3j} - a_{31}a_{1j}]$. Here, multiply the first row by 3 (a_{31}) and subtract the result from the third row:

$$\begin{pmatrix} 1 & -1 & 2.5 & 6.5 \\ 0 & 5 & -1 & 7 \\ 0 & 2 & -4.5 & -9.5 \end{pmatrix}. \tag{4.6}$$

Step 1 is repeated: Here, $k = 2$ and the pivot element $a_{22} = 5$. Dividing all elements of the first row by a_{22} or $a_{2j} = a_{2j} / a_{22}$ gives

$$\begin{pmatrix} 1 & -1 & 2.5 & 6.5 \\ 0 & 1 & -0.2 & 1.4 \\ 0 & 2 & -4.5 & -9.5 \end{pmatrix}. \tag{4.7}$$

Step 2 (for rows 1 and 3) is repeated: Eliminate the a_{12} term from the first row by $[a_{1j} = a_{1j} - a_{12}a_{2j}]$. Here, multiply the second row by −1 (a_{12}) and subtract the result from the first row. Eliminate the a_{32} term from the third row by $[a_{3j} = a_{3j} - a_{32}a_{2j}]$. Here, multiply the second row by 2 (a_{32}) and subtract the result from the third row:

$$\begin{pmatrix} 1 & 0 & 2.3 & 7.9 \\ 0 & 1 & -0.2 & 1.4 \\ 0 & 0 & -4.1 & -12.3 \end{pmatrix}. \tag{4.8}$$

Step 1 is repeated: Here, $k = 3$ and the pivot element $a_{33} = -4.1$. Dividing all elements of the third row by a_{33} or $a_{3j} = a_{3j} / a_{32}$ gives

$$\begin{pmatrix} 1 & 0 & 2.3 & 7.9 \\ 0 & 1 & -0.2 & 1.4 \\ 0 & 0 & 1 & 3 \end{pmatrix}. \tag{4.9}$$

Step 2 (for rows 1 and 2) is repeated: Eliminate the a_{13} term from the first row by $[a_{1j} = a_{1j} - a_{13}a_{3j}]$. Here, multiply the third row by 2.3 (a_{13}) and subtract the result from the first row. Eliminate the a_{23} term from the second row by $[a_{2j} = a_{2j} - a_{23}a_{3j}]$. Here, multiple the third row by −0.2 (a_{23}) and subtract the result from the second row:

$$\begin{pmatrix} 1 & 0 & 0 & 1 \\ 0 & 1 & 0 & 2 \\ 0 & 0 & 1 & 3 \end{pmatrix}. \tag{4.10}$$

The result, read from the diagonal, gives $x_1 = 1$, $x_2 = 2$, and $x_3 = 3$.

In the next example, we will develop a general Gauss–Jordan VBA program. We must ensure that the coefficient being normalized $a_{kk} \neq 0$; if $a_{kk} = 0$, we need to exchange rows with one of the following rows (one not yet normalized). We also will need to make adjustments with our indexing—in the VBA program, we will index from 0 to row −1 and 0 to column −1. ■

4.1.2 Gauss–Jordan Coding Strategy for Linear Equation Sets

EXAMPLE 4.2 *Gauss–Jordan Elimination Scheme Using VBA*

Develop a general Gauss–Jordan reduction algorithm using VBA. Solve the three-equation example, Example 4.1, using this code.

SOLUTION The solution is provided in **Example 4.2.xls**. The Gauss–Jordan algorithm VBA code is shown in Figure 4.1 (and it follows the C code of Appendix A). The final results from the Gauss–Jordan method are shown in Figure 4.2.

Here, we discuss key features of the code. `Line 2` defines `Sub Gauss_Jordan_Macro(),` a macro subroutine which is callable from the Excel sheet. `Line 7` defines the size of the augmented matrix named C, and this matrix is read from the sheet in `lines 8-19`. If no entry is found on the Excel sheet in `lines 13 and 14`, we set that element to zero. The Gauss–Jordan elimination subroutine is called in `line 21`.

In the Gauss–Jordan subroutine procedure (`line 29`), the augmented matrix is named A, and the number of rows is `row`. In `lines 34-37`, a check is made if the current diagonal element is zero. If true, this must be corrected, as a zero diagonal element cannot serve as a pivot element. Subsequent rows (`line 40`) are examined for the first nonzero element (`line 41`) in the pivot column. This row is marked as `nonZerIdx` in `line 42`. In `lines 47-54`, elements are exchanged, one at a time, between the current row (`i`) and the row with the nonzero pivot element (`nonZerIdx`). The process of checking the diagonal element and exchanging rows, if necessary, ends with the `End If` statement in `line 56`.

`Lines 57-63` divide each element of the pivot row by the pivot element as shown in Equation (4.1). The pivot column in all rows except the pivot row (`line 67`) is made zero by lines `lines 64-75`. Equation (4.2) is implemented in `line 71`. Indexing of the rows continues in `line 76`; `line 76` provides the `Next i` for `line 36`. The final diagonal matrix is written on the Excel sheet in `lines 22-28`.

Code Modifications Required for Each New Problem

We must specify the number of rows and columns in the augmented matrix (`lines 5-6`), the first cell of the input matrix on the Excel sheet (`lines 13 and 16 here row = 8, column = 2`), and the start cell on the Excel sheet where the results should be printed (`line 25 here row = 12, column = 2`).

Connecting Excel to VBA and C Programs: What Should the User Know?

At this point, you should read Appendix A—"Bridging Excel and C Codes." We now want to discuss what the user should know when solving a set of linear equations using Excel. The two options that we are developing are shown in Figure 4.3. As indicated in this figure, the user will always need to identify the location of the matrix coefficients for the solution engine (see the previous section, "Code Modifications Required for Each New Problem"). This part of the code (in some form) has to be under user control, and the user must make a few simple changes for each new problem.

If an all-VBA solution is used, it is usually convenient to also include the Gauss–Jordan solution engine as part of an Excel sheet (the VBA module attached to the sheet)—but no changes in the Gauss–Jordan solution method will be needed once it is developed. We did this in Example 4.2.

In some cases, we may have a high-level language program available to solve a problem, and we want to use Excel as our pre- and postprocessor. Let us assume that the Gauss–Jordan program is available as a C program. We can connect this C program (or any C program) to the Excel sheet through VBA as indicated in Figure 4.3 and detailed in the Appendix.

> *It is actually possible to hide all solution engines (here the Gauss–Jordan method in VBA or C) and all matrix transfer operations from the end user.*

C code, when connected to the Excel sheet as a DLL, will always be hidden. To see this, compare Figure A.22 (Excel file: Appendix Example A.5.xls) and Figure 4.1 (Excel file: **Example 4.2.xls**). The VBA subroutines shown in these figures are virtually identical; here, compare lines 3–26 of Figure A.22 to lines 2–28 of Figure 4.1. There is a small difference in how the Gauss–Jordan method is called. In Example 4.1, we call the Gauss–Jordan method using `GJ_Elimination(C, Nrows)`, while in Example A.5, we call using `GJ_Elimination_Main C(0, 0), Nrows, Ncolumns`. Here, `C(0, 0)` is the location of the first element in the C matrix. The obvious difference between the two examples is that in Example A.5, there was no direct inclusion of a C code for the Gauss–Jordan method, while in Example 4.2, the Gauss–Jordan method was provided as a VBA code (lines 29–77 of Figure 4.1). For Example A.5, the Gauss–Jordan method was written as a separate C program. This C program was compiled and then linked as a DLL to the Example by line 1 of Figure A.22. *Hopefully, you would agree that if we were provided an Excel sheet with the connection to the DLL already made (line 1 of Figure 4.11), and if the DLL exists, then all the developments of the Appendix would be transparent.* ■

4.1.3 Linear Material Balance Problems: Natural Specifications

As we will see in this section, the strength of the simultaneous solution approach is that naturally specified material balance problems with recycle loops can be solved directly without the need for an iterative solution to converge the recycle loop. Linear independent alternative specifications can also be used directly in the equation set. For linear alternative specifications, there will be no need for a single-variable search strategy as we developed in Chapter 3. However, if nonlinear specifications are present in the material balance problem, the simultaneous approach will require an iterative solution involving partial derivatives.

```
Option Explicit                                                   line 1

    Public Sub Gauss_Jordan_Macro()                               line 2
        Dim C() As Double                                         line 3
        Dim Nrows, Ncolumns As Long                               line 4

        Nrows = 3                                                 line 5
        Ncolumns = 4                                              line 6
        ReDim C(Nrows, Ncolumns)                                  line 7

'Read augmented matrix from Excel sheet                           line 8
        Dim i As Integer                                          line 9
        Dim j As Integer                                          line 10
        For i = 0 To Nrows - 1                                    line 11
            For j = 0 To Ncolumns - 1                             line 12
                If Sheet1.Cells(i + 8, j + 2) = " " Then          line 13
                    C(i, j) = 0                                   line 14
                Else                                              line 15
                    C(i, j) = Sheet1.Cells(i + 8, j + 2)          line 16
                End If                                            line 17
            Next j                                                line 18
        Next i                                                    line 19

    ' Call Gauss-Jordan matrix elimination method                line 20
        GJ_Elimination(C, Nrows)                                  line 21

    ' Place solution from Gauss-Jordan matrix elimination method on Excel Sheet line 22
        For i = 0 To Nrows - 1                                    line 23
            For j = 0 To Ncolumns - 1                             line 24
                Sheet1.Cells(i + 12, j + 2) = C(i, j)             line 25
            Next j                                                line 26
        Next i                                                    line 27
    End Sub                                                       line 28

    Public Sub GJ_Elimination(A, row)                             line 29

        Dim col As Long                                           line 30
        col = row + 1                                             line 31

        Dim nonZerIdx As Integer                                  line 32
        nonZerIdx = 0                                             line 33

' We first to check if the diagnol element is zero               line 34
        Dim i As Integer                                         line 35
        For i = 0 To row - 1                                     line 36
            If A(i, i) = 0 Then                                  line 37
' If the diagnol element = 0 we need to look at following rows
' to find a nonzero element in the i column                       line 38
                Dim i2 As Integer                                line 39
                For i2 = i + 1 To row - 1                         line 40

                    If A(i2, i) <> 0 Then                         line 41
                        nonZerIdx = i2                            line 42
                        Exit For                                  line 43
                    Else                                          line 44
                    End If                                        line 45
                Next i2                                           line 46

' Here we exchange the row with the diagnol element = 0 with the
' first following row in which the i column has a nonzero value   line 47
                Dim j1 As Integer                                line 48
                For j1 = 0 To col - 1                             line 49
                    Dim tmp As Double                            line 50
                    tmp = A(i, j1)                               line 51
                    A(i, j1) = A(nonZerIdx, j1)                  line 52
                    A(nonZerIdx, j1) = tmp                       line 53
                Next j1                                           line 54

            Else                                                 line 55
            End If                                               line 56
```

Figure 4.1   VBA code for the Gauss–Jordan matrix elimination method.

```
' Here we divide every element of the pivot row
' by the the diagnol element (the pivot element)              line 57
        Dim tmpAii As Double                                 line 58
        tmpAii = A(i, i)                                     line 59
        Dim j As Integer                                    line 60
        For j = 0 To (col - 1)                              line 61
            A(i, j) = A(i, j) / tmpAii                      line 62
        Next j                                              line 63

' Here we eliminate (make zero) all elements in the pivot column.  This is
' not done for the pivot row.  See equation (4.2) aij = aij - aik akj    line 64
        Dim i1 As Integer                                   line 65
        For i1 = 0 To (row - 1)                             line 66
            If i1 <> i Then                                 line 67
                Dim tmpAi1 As Double                        line 68
                tmpAi1 = A(i1, i)                           line 69
                For j = 0 To (col - 1)                      line 70
                    A(i1, j) = A(i1, j) - tmpAi1 * A(i, j)  line 71
                Next j                                      line 72
            Else                                            line 73

            End If                                          line 74
        Next i1                                             line 75
    Next i                                                  line 76
End Sub                                                     line 77
```

Figure 4.1   (*Continued*)

	A	B	C	D	E	F	G	H
1	*Gauss-Jordan Matrix Elimination VBA Code Only*							
2								
3			$2 x_1 - 2 x_2 + 5 x_3 = 13$					
4			$2 x_1 + 3 x_2 + 4 x_3 = 20$					
5			$3 x_1 - x_2 + 3 x_3 = 10$					
6	Augmented Matrix							
7		x1	x2	x3	RHS			Input Start
8	eq (1)	2	-2	5	13			Row 8 Column 2
9	eq (2)	2	3	4	20			
10	eq (3)	3	-1	3	10			
11								Answer Start
12		1	0	0	1			Row 12 Column 2
13		0	1	0	2			
14		0	0	1	3			

Figure 4.2   VBA code solution.

EXAMPLE 4.3   *Styrene Production (with Recycle)*

Solve Example 3.1 using the simultaneous solution approach.

If we examine Figure 4.4, there are 17 unknown flow rates and 1 unknown extent of reaction for a total of 18 unknowns. We previously assembled 13 independent material balances (Eqs. (3.32) and (3.33), (3.35)–(3.43), (3.46) and (3.47)). The problem description in Example 3.1 provides two feed flow rates (two independent specifications), a known extent of reaction, and two purge specifications (Eqs. (3.44) and (3.45)). In general terms then, we have 18 unknowns and 18 equations. However, the equation for the extent of reaction, $\xi = 0.65(N_{C_8H_{10}}^{\text{reactor in}})$, was directly incorporated in material balance equations for the reactor (Eqs. (3.36)–(3.38)). This eliminates ξ as a variable, leaving us with 17 equations and 17 unknowns. The 17 material balances are shown in the augmented matrix of Figure 4.5a. For example, row 8 reads, $N_{H_2O}^{\text{feed}} = 3000$; row 13, $(0.35) N_{C_8H_{10}}^{\text{reactor in}} - N_{C_8H_{10}}^{\text{reactor out}} = 0$, and row 24,

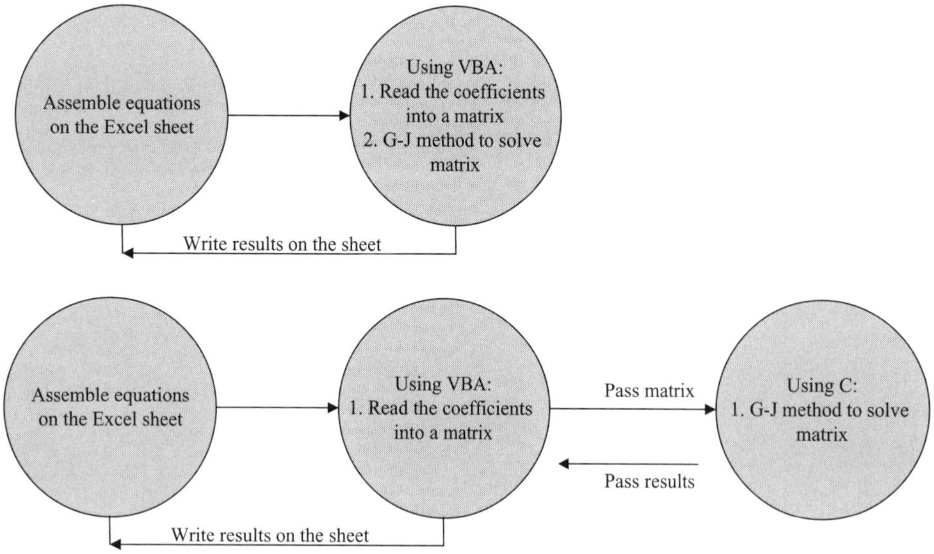

Figure 4.3   Overview of connecting Excel to the Gauss–Jordan program in VBA and the Gauss–Jordan program in C.

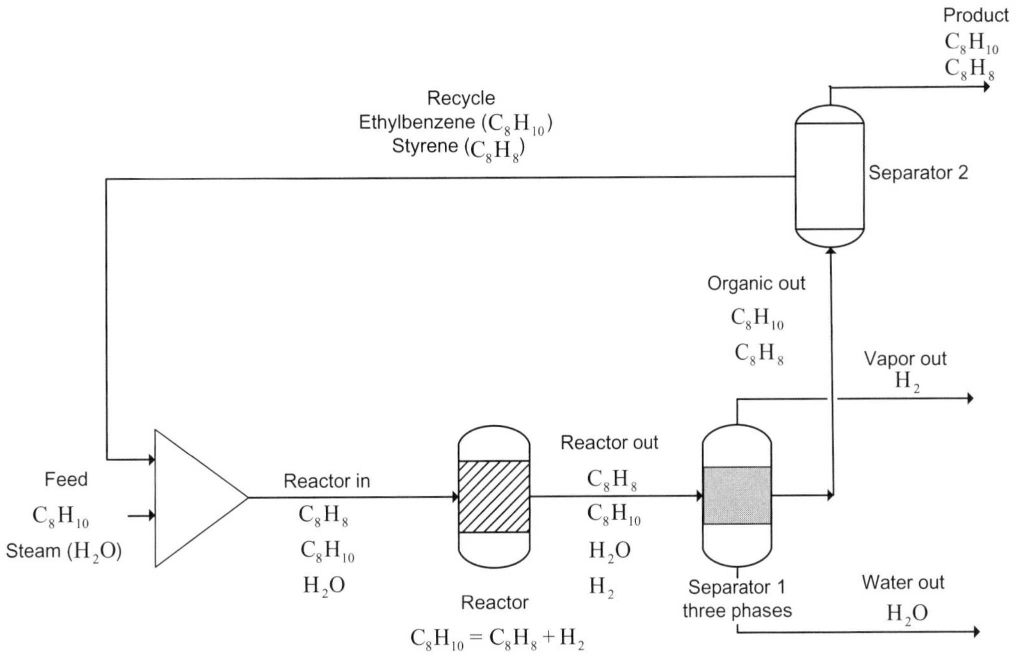

Figure 4.4 Flow sheet for styrene production (Figure 3.1 repeated).

Row 8 Column 2

	A	B	C	D	E	F	G	H	I	J	K	L	M	N	O	P	Q	R	S
1	*Styrene Material Balance using Gauss Jordan Elimination - VBA program*																		
2																			
3	Reactioi	C_8H_{10} ---> C_8H_8 + H_2			Conversion		0.65												
4																			
5																			
6		Feed			Reactor In			Reactor Out			Water Out	Organic Out		Vapor	Product		Recycle		
7		N(C8H10)	N(H2O)	N(C8H10)	N(H2O)	N(C8H8)	N(C8H10)	N(H2O)	N(H2)	N(C8H8)	N(H2O)	N(C8H10)	N(C8H8)	N(H2)	N(C8H10)	N(C8H8)	N(C8H10)	N(C8H8)	RHS
8	Feed		1																3000
9	Specs	1																	100
10	Mixer	1		-1													1		0
11	Material		1		-1														0
12	Balances					-1												1	0
13	Reactor			0.35			-1												0
14	Balances				1			-1											0
15				0.65					-1										0
16				0.65		1				-1									0
17	Separator 1						1					-1							0
18								1			-1								0
19									1					-1					0
20										1			-1						0
21	Separator 2											1			-1		-1		0
22													1			-1		-1	0
23												0.01			-1				0
24													0.99			-1			0

(a)

Row 30 Column 2

	A	B	C	D	E	F	G	H	I	J	K	L	M	N	O	P	Q	R	S	T	U
30		1	0	0	0	0	0	0	0	0	0	0	0	0	0	0	0	0	100	N(C8H10)	Feed
31		0	1	0	0	0	0	0	0	0	0	0	0	0	0	0	0	0	3000	N(H2O)	Feed
32		0	0	1	0	0	0	0	0	0	0	0	0	0	0	0	0	0	153.02	N(C8H10)	Reactor In
33		0	0	0	1	0	0	0	0	0	0	0	0	0	0	0	0	0	3000	N(H2O)	Reactor In
34		0	0	0	0	1	0	0	0	0	0	0	0	0	0	0	0	0	1.0047	N(C8H8)	Reactor In
35		0	0	0	0	0	1	0	0	0	0	0	0	0	0	0	0	0	53.558	N(C8H10)	Reactor Out
36		0	0	0	0	0	0	1	0	0	0	0	0	0	0	0	0	0	3000	N(H2O)	Reactor Out
37		0	0	0	0	0	0	0	1	0	0	0	0	0	0	0	0	0	99.464	N(H2)	Reactor Out
38		0	0	0	0	0	0	0	0	1	0	0	0	0	0	0	0	0	100.47	N(C8H8)	Reactor Out
39		0	0	0	0	0	0	0	0	0	1	0	0	0	0	0	0	0	3000	N(H2O)	Water Out
40		0	0	0	0	0	0	0	0	0	0	1	0	0	0	0	0	0	53.558	N(C8H10)	Organic Out
41		0	0	0	0	0	0	0	0	0	0	0	1	0	0	0	0	0	100.47	N(C8H8)	Organic Out
42		0	0	0	0	0	0	0	0	0	0	0	0	1	0	0	0	0	99.464	N(H2)	Vapor Out
43		0	0	0	0	0	0	0	0	0	0	0	0	0	1	0	0	0	0.5356	N(C8H10)	Product
44		0	0	0	0	0	0	0	0	0	0	0	0	0	0	1	0	0	99.464	N(C8H8)	Product
45		0	0	0	0	0	0	0	0	0	0	0	0	0	0	0	1	0	53.022	N(C8H10)	Recycle
46		0	0	0	0	0	0	0	0	0	0	0	0	0	0	0	0	1	1.0047	N(C8H8)	Recycle

(b)

Figure 4.5 (a) Styrene material balances augmented matrix. (b) Gauss–Jordan solution styrene process.

$(0.99)N_{C_8H_8}^{\text{organic out}} - N_{C_8H_8}^{\text{product}} = 0$. Do compare the equations in the augmented matrix with the equations developed in Chapter 3.

The solution is provided in **Example 4.3a.xls** or in **Example 4.3b.xls**. Regardless if the Gauss–Jordan algorithm is supplied as a VBA code (Example 4.3a.xls) or as a complied C program linked to the Excel sheet as a DLL (Example 4.3b.xls), the macro, `Public Sub Gauss_Jordan_Macro()`, must indicate the size of the augmented matrix, the start location for the augmented matrix, and the desired start location for the results matrix. The Excel input sheet of Figure 4.5a shows the 17 material balance equations starting in row = 8, column = 2.

The Gauss–Jordan solution starting in row = 30, column = 2 is shown in Figure 4.5b. Here for example: row 30, $N_{C_8H_{10}}^{\text{feed}} = 100$ (natural specification); row 38, $N_{C_8H_8}^{\text{reactor out}} = 100.47$; row 48, $N_{C_8H_8}^{\text{recycle}} = 1.0047$ mol/h. The results are identical to those found for Example 3.1, Figure 3.6b. ■

The simultaneous solution approach for a naturally specified material balance problem has the advantage that no iterative procedure is needed to converge the recycle loop. The sequential modular approach developed in Chapter 3 required tear stream selection and convergence for recycle loops. The simultaneous approach does require solution of n linear equations with n unknowns, which was accomplished using the Gauss Jordan algorithm.

4.1.4 Linear Material Balance Problems: Alternative Specifications

EXAMPLE 4.4 *Styrene Production (with Linear Alternative Specification)*

Solve Example 3.8 using the simultaneous approach. Here, the water feed rate to the styrene process, which is a natural specification, has been replaced by an alternative specification,

$$\frac{N_{H_2O}^{\text{reactor in}}}{N^{\text{reactor in}}} = 0.9.$$

Eliminating the natural specification on the water feed rate will reduce the overall degree of freedom by one. However, the system degree of freedom remains zero as the molar feed rate of water, $N_{H_2O}^{\text{feed}}$, is replaced with an alternative specification—water at 90% (molar basis) is maintained in the stream entering the reactor. This problem again requires solution of 17 independent material balances with 17 unknowns. After rearrangement, this alternative constraint would be written as

$$0.1N_{H_2O}^{\text{reactor in}} - 0.9N_{C_8H_{10}}^{\text{reactor in}} - 0.9N_{C_8H_8}^{\text{reactor in}} = 0.$$

The augmented matrix is shown in Figure 4.6a. The solution to this equation set is provided in **Example 4.4.xls** and is shown Figure 4.6b. In the Excel file, we are using the DLL link to the Gauss–Jordan method. ■

The simultaneous solution approach eliminates the need for a single-variable search technique as developed in Chapter 3 to satisfy the alternative specification. The results in Figure 4.6b are identical to those found in Figure 3.19.

At this point, it may seem the simultaneous solution approach is a much better approach for material balance problems. We have eliminated the need to converge recycle loops and we have eliminated single-variable search techniques to satisfy alternative specifications. However, the simultaneous approach cannot be used directly if nonlinear equations are present. Here, we must construct linear approximations to the nonlinear functions.

4.2 SOLUTION OF NONLINEAR EQUATION SETS: THE NEWTON–RAPHSON METHOD

In general, a set of nonlinear equations can be written as

$$\begin{aligned} f_1\{x_1, x_2, \dots, x_n\} &= 0 \\ f_2\{x_1, x_2, \dots, x_n\} &= 0 \\ f_n\{x_1, x_2, \dots, x_n\} &= 0 \end{aligned} \qquad (4.11)$$

The problem is how do we find the solution, $x = \{x_1, x_2, \dots, x_n\}$, such that $f_1 = 0, f_2 = 0, \dots, f_n = 0$. Here, we will use the Newton–Raphson method, which is the generalization of Newton's method to equation sets.

We begin with a discussion of constructing linear approximations to nonlinear functions. We next address the Newton–Raphson solution approach for solving a set of nonlinear equations. These results are then particularized to the solution of material balance problems involving both linear and nonlinear equations.

4.2.1 Equation Linearization via Taylor's Series Expansion

A Taylor series expansion, terminated after the first derivative, can be used to construct a linear approximation of a multivariable function, f_1, at a current point, x^*:

$$f_1\{x_1, x_2, \dots, x_n\} \cong f_1\{x_1^*, x_2^*, \dots, x_n^*\} + (x_1 - x_1^*)\left.\frac{\partial f_1}{\partial x_1}\right|_{x=x^*} + (x_2 - x_2^*)\left.\frac{\partial f_1}{\partial x_2}\right|_{x=x^*} + \dots + (x_n - x_n^*)\left.\frac{\partial f_1}{\partial x_n}\right|_{x=x^*}. \qquad (4.12)$$

EXAMPLE 4.5 *Equation Linearization*

Given $f_1 = (x_1)^{0.5} + (x_2)^2 + 10$ and a current point, $x^* = \{x_1^*, x_2^*\} = \{4, 3\}$, construct a linear approximation of f_1 at x^*.

SOLUTION Following the RHS of Equation (4.12), the value for f_1 at the current point x^* is

$$f_1\{x_1^*, x_2^*\} = (x_1)^{0.5} + (x_2)^2 + 10 = (4)^{0.5} + (3)^2 + 10 = 21,$$

and the first partial derivatives, evaluated at x^*, are

$$\left.\frac{\partial f_1}{\partial x_1}\right|_{x=x^*} = \frac{0.5}{x_1^{0.5}} = \frac{0.5}{(4)^{0.5}} = 0.25, \left.\frac{\partial f_1}{\partial x_2}\right|_{x=x^*} = 2x_2 = 2(3) = 6,$$

	A	B	C	D	E	F	G	H	I	J	K	L	M	N	O	P	Q	R	S
1	*Styrene Material Balance with Alternative Specification - Augmented Matrix*																		
2																			
3	Reactio	C_8H_{10} ---> C_8H_8 + H_2			Conversion		0.65												
4																			
5																			
6		Feed			Reactor In			Reactor Out			Water Out	Organic Out		Vapor	Product		Recycle		
7		N(C8H10)	N(H2O)	N(C8H10)	N(H2O)	N(C8H8)	N(C8H10)	N(H2O)	N(H2)	N(C8H8)	N(H2O)	N(C8H10)	N(C8H8)	N(H2)	N(C8H10)	N(C8H8)	N(C8H10)	N(C8H8)	RHS
8	Feed	1																	100
9	Mixer	1		-1													1		0
10	Material		1		-1														0
11	Balances					-1												1	0
12	Reactor			0.35			-1												0
13	Balances				1			-1											0
14				0.65					-1										0
15				0.65		1				-1									0
16	Separator 1						1					-1							0
17								1			-1								0
18									1					-1					0
19										1			-1						0
20	Separator 2											1			-1		-1		0
21													1			-1		-1	0
22												0.01			-1				0
23													0.99			-1			0
24	Alternative Spec			-0.9	0.1	-0.9													0

(a)

	A	B	C	D	E	F	G	H	I	J	K	L	M	N	O	P	Q	R	S	T	U
30		1	0	0	0	0	0	0	0	0	0	0	0	0	0	0	0	0	100	N(C8H10)	Feed
31		0	1	0	0	0	0	0	0	0	0	0	0	0	0	0	0	0	1386.24	N(H2O)	Feed
32		0	0	1	0	0	0	0	0	0	0	0	0	0	0	0	0	0	153.02	N(C8H10)	Reactor In
33		0	0	0	1	0	0	0	0	0	0	0	0	0	0	0	0	0	1386.24	N(H2O)	Reactor In
34		0	0	0	0	1	0	0	0	0	0	0	0	0	0	0	0	0	1.00469	N(C8H8)	Reactor In
35		0	0	0	0	0	1	0	0	0	0	0	0	0	0	0	0	0	53.5578	N(C8H10)	Reactor Out
36		0	0	0	0	0	0	1	0	0	0	0	0	0	0	0	0	0	1386.24	N(H2O)	Reactor Out
37		0	0	0	0	0	0	0	1	0	0	0	0	0	0	0	0	0	99.4644	N(H2)	Reactor Out
38		0	0	0	0	0	0	0	0	1	0	0	0	0	0	0	0	0	100.469	N(C8H8)	Reactor Out
39		0	0	0	0	0	0	0	0	0	1	0	0	0	0	0	0	0	1386.24	N(H2O)	Water Out
40		0	0	0	0	0	0	0	0	0	0	1	0	0	0	0	0	0	53.5578	N(C8H10)	Organic Out
41		0	0	0	0	0	0	0	0	0	0	0	1	0	0	0	0	0	100.469	N(C8H8)	Organic Out
42		0	0	0	0	0	0	0	0	0	0	0	0	1	0	0	0	0	99.4644	N(H2)	Vapor Out
43		0	0	0	0	0	0	0	0	0	0	0	0	0	1	0	0	0	0.53558	N(C8H10)	Product
44		0	0	0	0	0	0	0	0	0	0	0	0	0	0	1	0	0	99.4644	N(C8H8)	Product
45		0	0	0	0	0	0	0	0	0	0	0	0	0	0	0	1	0	53.0222	N(C8H10)	Recycle
46		0	0	0	0	0	0	0	0	0	0	0	0	0	0	0	0	1	1.00469	N(C8H8)	Recycle

(b)

Figure 4.6 (a) Styrene material balances augmented matrix—alternative specification. (b) Simultaneous solution styrene process—alternative specification.

giving

$$f_1\{x_1, x_2\} \cong 21 + (x_1 - 4)(0.25) + (x_2 - 3)(6) \cong 0.25x_1 + 6x_2 + 2.$$

Note that this approximation to f_1 at x^* is linear. We can use this linear approximation to estimate values for f_1 at any $\{x_1, x_2\}$. For example, at $x = \{x_1, x_2\} = \{9, 4\}$, the estimated value of f_1 would be $f_1\{9, 4\} \cong 0.25(9) + 6(4) + 2 \cong 28.25$. The true value for f_1 is $f_1\{9, 4\} \cong (9)^{0.5} + (4)^2 + 10 = 29$. Here, the error in the linear approximation is 2.6% ((29 − 28.25)/29). ■

4.2.2 Nonlinear Equation Set Solution via the Newton–Raphson Method

If we look back at Problem 3.4 (Newton's method), we constructed a linear approximation to $f(x)$ at a current point, x^1, allowing extrapolation to estimate the point x where $f(x) = 0$. This was Newton's method for a single-variable function. The Newton–Raphson method is the generalization of Newton's method to equation sets.

In Example 4.5, a linear approximation was used to provide an estimate of the value of a multivariable function at a point, x, near the current point x^*. This linear approximation can also be used to estimate the point x, which gives a desired function value; we want $f(x) = 0$. But to obtain a meaningful solution for an n-dimensional multivariable function, we will need to solve for the point where all $f(x)$ equations = 0 (there are n equations). The Newton–Raphson method for solving a set of n nonlinear equations with n unknowns is shown below.

From Equation (4.12), a linear approximation to each function in Equation (4.11), $f_1, f_2, \ldots, f_n$ can be constructed at a known point, $x^* = \{x_1^*, x_2^*, \ldots, x_n^*\}$, as

$$f_1\{x_1, x_2, \ldots, x_n\} \cong f_1\{x_1^*, x_2^*, \ldots, x_n^*\} + (x_1 - x_1^*)\left.\frac{\partial f_1}{\partial x_1}\right|_{x=x^*}$$
$$+ (x_2 - x_2^*)\left.\frac{\partial f_1}{\partial x_2}\right|_{x=x^*} + \ldots + (x_n - x_n^*)\left.\frac{\partial f_1}{\partial x_n}\right|_{x=x^*}$$
$$f_2\{x_1, x_2, \ldots, x_n\} \cong f_2\{x_1^*, x_2^*, \ldots, x_n^*\} + (x_1 - x_1^*)\left.\frac{\partial f_2}{\partial x_1}\right|_{x=x^*}$$
$$+ (x_2 - x_2^*)\left.\frac{\partial f_2}{\partial x_2}\right|_{x=x^*} + \ldots + (x_n - x_n^*)\left.\frac{\partial f_2}{\partial x_n}\right|_{x=x^*}$$
$$f_n\{x_1, x_2, \ldots, x_n\} \cong f_n\{x_1^*, x_2^*, \ldots, x_n^*\} + (x_1 - x_1^*)\left.\frac{\partial f_n}{\partial x_1}\right|_{x=x^*}$$
$$+ (x_2 - x_2^*)\left.\frac{\partial f_n}{\partial x_2}\right|_{x=x^*} + \ldots + (x_n - x_n^*)\left.\frac{\partial f_n}{\partial x_n}\right|_{x=x^*}. \quad (4.13)$$

We want the left-hand side (LHS) of Equation (4.13) to = 0, giving

$$0 \cong f_1\{x_1^*, x_2^*, \ldots, x_n^*\} + (x_1 - x_1^*)\left.\frac{\partial f_1}{\partial x_1}\right|_{x=x^*} + (x_2 - x_2^*)\left.\frac{\partial f_1}{\partial x_2}\right|_{x=x^*}$$
$$+ \ldots + (x_n - x_n^*)\left.\frac{\partial f_1}{\partial x_n}\right|_{x=x^*}$$
$$0 \cong f_2\{x_1^*, x_2^*, \ldots, x_n^*\} + (x_1 - x_1^*)\left.\frac{\partial f_2}{\partial x_1}\right|_{x=x^*} + (x_2 - x_2^*)\left.\frac{\partial f_2}{\partial x_2}\right|_{x=x^*}$$
$$+ \ldots + (x_n - x_n^*)\left.\frac{\partial f_2}{\partial x_n}\right|_{x=x^*}$$
$$0 \cong f_n\{x_1^*, x_2^*, \ldots, x_n^*\} + (x_1 - x_1^*)\left.\frac{\partial f_n}{\partial x_1}\right|_{x=x^*} + (x_2 - x_2^*)\left.\frac{\partial f_n}{\partial x_2}\right|_{x=x^*}$$
$$+ \ldots + (x_n - x_n^*)\left.\frac{\partial f_n}{\partial x_n}\right|_{x=x^*}. \quad (4.14)$$

Equation (4.14) provides n linear equations and n unknowns, $\{x_1, x_2, \ldots, x_n\}$, where $f_1 = f_2 =, \ldots f_n = 0$. In Equation (4.14), the current point $x^* = \{x_1^*, x_2^*, \ldots, x_n^*\}$ is known, and the partial derivative terms

$$\left.\frac{\partial f_i}{\partial x_i}\right|_{x=x^*}$$

at the current point are known. The unknowns are the x values, $\{x_1, x_2, \ldots, x_n\}$, which give function values (LHS) = 0. Equation (4.14) can be rearranged and written in augmented matrix form as

$$\begin{pmatrix} \left.\frac{\partial f_1}{\partial x_1}\right|_{x=x^*} & \left.\frac{\partial f_1}{\partial x_2}\right|_{x=x^*} & \cdots & \left.\frac{\partial f_1}{\partial x_n}\right|_{x=x^*} & \beta_1 \\ \left.\frac{\partial f_2}{\partial x_1}\right|_{x=x^*} & \left.\frac{\partial f_2}{\partial x_2}\right|_{x=x^*} & \cdots & \left.\frac{\partial f_2}{\partial x_n}\right|_{x=x^*} & \beta_2 \\ \cdots & \cdots & \cdots & \cdots & \cdots \\ \left.\frac{\partial f_n}{\partial x_1}\right|_{x=x^*} & \left.\frac{\partial f_n}{\partial x_2}\right|_{x=x^*} & \cdots & \left.\frac{\partial f_n}{\partial x_n}\right|_{x=x^*} & \beta_n \end{pmatrix}, \quad (4.15)$$

where

$$\beta_1 = -f_1\{x_1^*, x_2^*, \ldots, x_n^*\} + x_1^*\left.\frac{\partial f_1}{\partial x_1}\right|_{x=x^*} + x_2^*\left.\frac{\partial f_1}{\partial x_2}\right|_{x=x^*}$$
$$+ \ldots + x_n^*\left.\frac{\partial f_1}{\partial x_n}\right|_{x=x^*}$$
$$\beta_2 = -f_2\{x_1^*, x_2^*, \ldots, x_n^*\} + x_1^*\left.\frac{\partial f_2}{\partial x_1}\right|_{x=x^*} + x_2^*\left.\frac{\partial f_2}{\partial x_2}\right|_{x=x^*}$$
$$+ \ldots + x_n^*\left.\frac{\partial f_2}{\partial x_n}\right|_{x=x^*}$$
$$\beta_n = -f_n\{x_1^*, x_2^*, \ldots, x_n^*\} + x_1^*\left.\frac{\partial f_n}{\partial x_1}\right|_{x=x^*} + x_2^*\left.\frac{\partial f_n}{\partial x_2}\right|_{x=x^*}$$
$$+ \ldots + x_n^*\left.\frac{\partial f_n}{\partial x_n}\right|_{x=x^*}.$$

Equation (4.15) can be solved using the Gauss–Jordan algorithm we developed earlier in this chapter, giving

$$\begin{pmatrix} 1 & 0 & \ldots & 0 & x_1 \\ 0 & 1 & \ldots & 0 & x_2 \\ \ldots & \ldots & \ldots & \ldots & \ldots \\ 0 & 0 & \ldots & 1 & x_n \end{pmatrix}.$$

The solution is $x = \{x_1, x_2, \ldots, x_n\}$, and these values can be checked in Equation (4.11). If the RHS of Equation (4.11) is not sufficiently close to zero, x is set to x^* and the Newton–Raphson process continues.

EXAMPLE 4.6 *Newton–Raphson Method*

Solve the following equation set, which consists of one linear equation and two nonlinear equations, using the Newton–Raphson method:

$$3(x_1) + (x_2) + (x_3) = 8$$
$$(x_1)^{0.5} + (x_2)^2 + x_2 x_3 = 11 \quad (4.16)$$
$$(x_1)^2 + (x_2)^2 + (x_3)^2 = 14.$$

SOLUTION Following Equation (4.11), these equations can be written as

$$f_1\{x_1, x_2, x_3\} = 3(x_1) + (x_2) + (x_3) - 8 = 0,$$
$$f_2\{x_1, x_2, x_3\} = (x_1)^{0.5} + (x_2)^2 + x_2 x_3 - 11 = 0,$$

and

$$f_3\{x_1, x_2, x_3\} = (x_1)^2 + (x_2)^2 + (x_3)^2 - 14 = 0. \quad (4.17)$$

There is no restriction in the Newton–Raphson method that the equations all be nonlinear, and here f_1 is linear. The problem we are initially facing is determining our initial point, x^*. ■

*First Point: Starting Guess, x**

We must have an initial point, $x^* = \{x_1^*, x_2^*, \ldots, x_n^*\}$, for the linearization process. If all n equations in the equation set are nonlinear, then a *starting guess* for each of the n variables must be supplied. However, if there are L linear equations, then starting guesses for n-L variables can be made and values for the remaining L variables can be determined as part of a Gauss–Jordan elimination. This process helps find a feasible starting point, which can be especially useful in material balance problems.

In Example 4.6, we have two nonlinear equations and one linear equation. If we specify two values from $x = \{x_1, x_2, x_3\}$, then we can initially eliminate the two nonlinear equations from consideration and determine the third value of $x = \{x_1, x_2, x_3\}$ from the remaining linear equation. Here, for example, with an initial guess, $x = \{x_1, x_2, x_3\} = \{1, 1, \_\}$, x_3 from the first linear equation (Eq. (4.16)) would be = 4.

In material balance problems, the majority of the simultaneous equations will be linear. With initial guesses for the nonlinear equations, the Gauss–Jordan method can be used to solve the resulting set of linear equations. Using Example 4.6, this Gauss–Jordan matrix elimination would appear as

$$\begin{pmatrix} \text{Linear equations} \\ \text{Guesses for nonlinear equations} \end{pmatrix} \rightarrow \begin{pmatrix} x_1 & x_2 & x_3 & RHS \\ 3 & 1 & 1 & 8 \\ 1 & & & 1 \\ & 1 & & 1 \end{pmatrix}$$

$$\xrightarrow{\text{Gauss–Jordan}} \begin{pmatrix} x_1 & x_2 & x_3 & RHS \\ 1 & & & 1 \\ & 1 & & 1 \\ & & 1 & 4 \end{pmatrix}.$$

Our current point is now $x = \{x_1, x_2, x_3\} = \{1 \quad 1 \quad 4\}$. We can check if Equation (4.17) is satisfied (=0); here, $f_1 = 0$, $f_2 = -5$, and $f_3 = 4$. Equation (4.17) is not satisfied, and the Newton–Raphson method can begin by setting $x^* = x$ as the current point for linearization.

Before we leave our discussion on determining a starting guess, we do need to comment on the selection of the n-L variables used to initially replace the nonlinear equations. These n-L variables cannot be randomly selected; they must be independent of the remaining linear equation set.

The Newton–Raphson method requires partial derivatives at a current point. We have discussed obtaining a starting guess for linearization, and here $x^* = \{x_1^*, x_2^*, x_3^*\} = \{1 \quad 1 \quad 4\}$. The first partial derivatives for Example 4.6 are

$$\left.\frac{\partial f_1}{\partial x_1}\right|_* = 3 \qquad \left.\frac{\partial f_1}{\partial x_2}\right|_* = 1 \qquad \left.\frac{\partial f_1}{\partial x_3}\right|_* = 1$$

$$\left.\frac{\partial f_2}{\partial x_1}\right|_* = \frac{0.5}{x_1^{0.5}} = 0.5 \qquad \left.\frac{\partial f_2}{\partial x_2}\right|_* = 2x_2 + x_3 = 6 \qquad \left.\frac{\partial f_2}{\partial x_3}\right|_* = x_2 = 1$$

$$\left.\frac{\partial f_3}{\partial x_1}\right|_* = 2x_1 = 2 \qquad \left.\frac{\partial f_3}{\partial x_2}\right|_* = 2x_2 = 2 \qquad \left.\frac{\partial f_3}{\partial x_3}\right|_* = 2x_3 = 8,$$

and Equation (4.14) would be

$$0 \cong 0 + (x_1 - 1)(3) + (x_2 - 1)(1) + (x_3 - 4)(1)$$
$$0 \cong -5.0 + (x_1 - 1)(0.5) + (x_2 - 1)(6) + (x_3 - 4)(1)$$
$$0 \cong 4.0 + (x_1 - 1)(2) + (x_2 - 1)(2) + (x_3 - 4)(8),$$

and after rearrangement,

$$\begin{pmatrix} 3x_1 & +1x_2 & +1x_3 = & 8.0 \\ 0.5x_1 & +6x_2 & +1x_3 = & 15.5 \\ 2x_1 & +2x_2 & +8x_3 = & 32.0 \end{pmatrix},$$

giving the augmented matrix

$$\begin{pmatrix} 3 & 1 & 1 & 8 \\ 0.5 & 6 & 1 & 15.5 \\ 2 & 2 & 8 & 32.0 \end{pmatrix}. \quad (4.18)$$

Equation (4.18) can also be found directly from Equation (4.15). Do notice that there was actually no need to apply Equation (4.14) (or Eq. (4.15)) to the linear equation $f_1\{x_1, x_2, x_3\} = 3(x_1) + (x_2) + (x_3) - 8 = 0$. Application of Equation (4.14), to a linear equation, simply reproduces already available coefficients (here 3, 1, 1, 8).

These three linear equations (Eq. (4.18)) can be solved using the Gauss–Jordan algorithm:

$$\begin{pmatrix} 1 & 0 & 0 & 0.92 \\ 0 & 1 & 0 & 1.96 \\ 0 & 0 & 1 & 3.28 \end{pmatrix}. \quad (4.19)$$

At this value of $x = \{x_1, x_2, x_3\} = \{0.92, 1.96, 3.28\}$, we can evaluate our nonlinear equation set (Eq. (4.17)):

$$f_1\{x_1, x_2, x_3\} = 3(x_1) + (x_2) + (x_3) - 8 = 0$$
$$f_2\{x_1, x_2, x_3\} = (x_1)^{0.5} + (x_2)^2 + x_2x_3 - 11 = 0.23 \quad (4.20)$$
$$f_3\{x_1, x_2, x_3\} = (x_1)^2 + (x_2)^2 + (x_3)^2 - 14 = 1.446.$$

Here the RHS values all do not equal zero. We can use our current x values as new x^* values $x^* = \{x_1^*, x_2^*, x_3^*\} = \{0.92, 1.96, 3.28\}$, and the process continues. For the next iteration, Equation (4.14) after rearrangement would be

$$3x_1 + x_2 + x_3 = 8$$
$$0.521x_1 + 7.2x_2 + 1.96x_3 = 20.79 \quad (4.21)$$
$$1.84x_1 + 3.92x_2 + 6.56x_3 = 29.447,$$

and using Gauss–Jordan elimination,

$$\begin{pmatrix} 1 & 0 & 0 & 0.996 \\ 0 & 1 & 0 & 1.994 \\ 0 & 0 & 1 & 3.018 \end{pmatrix}. \tag{4.22}$$

The functional values at $x = \{x_1, x_2, x_3\} = \{0.996, 1.994, 3.018\}$ are

$$f_1\{x_1, x_2, x_3\} = 3(x_1) + (x_2) + (x_3) - 8 = 0,$$

$$f_2\{x_1, x_2, x_3\} = (x_1)^{0.5} + (x_2)^2 + x_2x_3 - 11 = -0.0085,$$

and

$$f_3\{x_1, x_2, x_3\} = (x_1)^2 + (x_2)^2 + (x_3)^2 - 14 = 0.0757. \tag{4.23}$$

These RHS values are close to zero, and continued iterations setting $x^* = x$ will improve the x values. ■

EXAMPLE 4.7 *Newton–Raphson Method Graphical Solution*

Use the Newton–Raphson method to solve the following equation set, which consists of one linear equation and one nonlinear equation:

$$(x_1) + (x_2) = 5 \quad \text{or} \quad f_1\{x_1, x_2\} = (x_1) + (x_2) - 5 = 0$$

and

$$(x_1)^2 + (x_2)^2 = 13 \quad \text{or} \quad f_2\{x_1, x_2\} = (x_1)^2 + (x_2)^2 - 13 = 0.$$

Here, let the starting point $x^0 = x^* = \{4, 1\}$; notice that the linear equation is satisfied. Make a plot as (x_1 vs. x_2) for the two functions as well as the linearized equation for f_2 at x^0. The linearized equation for f_2 at x^0 is $(8x_1) + (2x_2) = 30$.

In Figure 4.7, we have plotted the initial two equations (f_1, f_2) and the nonlinear equation linearized at x^0. The solution of the two linear equations is shown at x^1 ($x^1 = 3.33, 1.667$). At x^1, the nonlinear function f_2 would again be linearized and the process repeated until the final solution is obtained. In Figure 4.7, there are actually two possible final solutions (f_1 and f_2 share two points in common ($x^{\text{Final}} = 3, 2$) or ($x^{\text{Final}} = 2, 3$)). The final solution will depend on the starting guess, and this is explored in Problem 4.7. ■

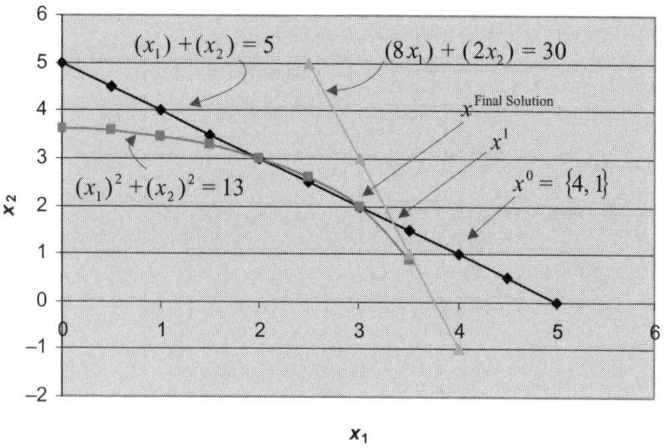

Figure 4.7 Graphical solution to the Newton–Raphson method—first iteration.

Our next task will be to write a general code to solve problems of n equations with n unknowns using the Newton–Raphson method. We can take advantage of our existing code for Gauss–Jordan elimination, but we will need to calculate partial derivatives.

4.2.3 Newton–Raphson Coding Strategy for Nonlinear Equation Sets

Figure 4.8 outlines the Newton–Raphson solution strategy for a set of equations containing both linear and nonlinear equations.

Let us assume that there are n equations with n unknowns and that L equations are linear. We need a good trial point to implement any solution strategy involving nonlinear equations. Here, a good initial point can by found by first "eliminating" the nonlinear equations. This can be done by suppling an initial guess for one independent variable appearing in each nonlinear equation. The result is a set of L linear equations with n-L initial guesses. It is straightfoward to solve this linear set of equations using the Gauss–Jordan elimination method, thereby providing the "good trial point." This result from the Gauss–Jordan elimination, x_i, can be checked for convergence using Equation (4.11). Also for material balance problems, we would like the initial solution from the Gauss–Jordan algorithm to be physically realistic. For example, flow rates should be positive. If the initial solution is unrealistic, it is straightfoward to simply supply new guesses for the nonlinear equations and to repeat the Gauss–Jordan matrix elimination.

These x_i values from the Gauss–Jordan matrix elimination can be used as the x_i^* values needed in the Newton–Raphson method to linearize the nonlinear equations. Here, we would only need to linearize the n-L nonlinear equations using Equation (4.14) or (4.15). If Equation (4.15) is applied to a linear equation, it will simply reproduce already known coefficients.

Numerical Partial Derivatives

Partial derivatives are required in Equation (4.14) or (4.15). Partial derivatives for a function can be numerically calculated using a forward difference approximation:

$$\left.\frac{\partial f_n}{\partial x_1}\right|_{x=x^*} \cong \frac{f_n\{(x_1^* + \varepsilon), x_2^*, \ldots, x_n^*\} - f_n\{x_1^*, x_2^*, \ldots, x_n^*\}}{\varepsilon}, \tag{4.24}$$

where ε is a small number determined in part by machine precision; a good initial value for ε is 1.0E − 04.

The generated equation set (linear equations and linearized nonlinear equations) can again be solved using the Gauss–Jordan method. This process would continue until the solution is obtained. The solution occurs when Equation (4.11) is satisfied or when the x_i values are not changing from iteration to iteration; the latter check is generally used.

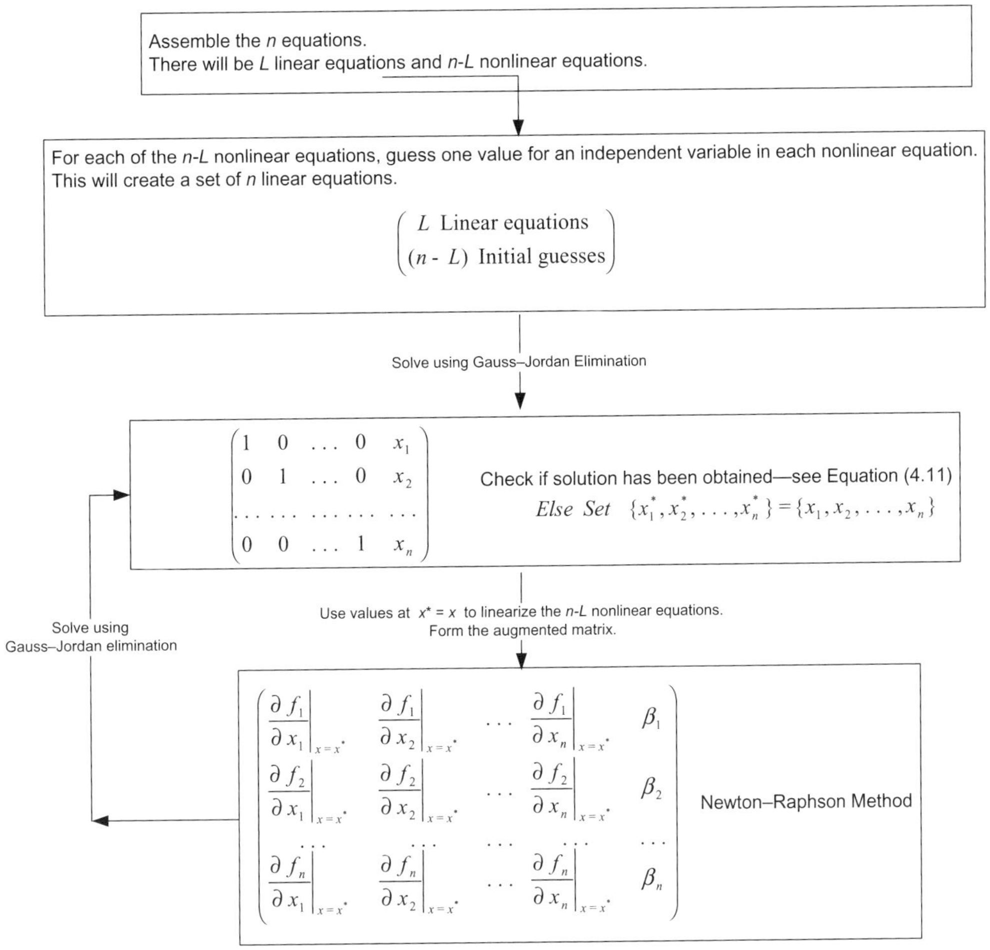

Figure 4.8 Solution strategy for a nonlinear equation set using the Newton–Raphson method.

EXAMPLE 4.8 *Newton–Raphson Method VBA Code*

Write a general VBA procedure that utilizes the Newton–Raphson solution strategy of Figure 4.8. Use this code to solve Example 4.6. The code provided in **Example 4.8.xls** is shown in Figure 4.9. Extensive comments are provided in the code.

SOLUTION The VBA code in Figure 4.9 contains two macro subroutine procedures, Public Sub Gauss_Jordan_Macro() and Public Sub NR_Gauss_Jordan_Macro(); one function procedure, Private Function NLFunc(ByVal Idx As Long, ByRef x() As Double) As Double, which holds the nonlinear equations; and one subroutine procedure, Private Sub Get_X_Values(ByVal N As Integer, ByRef x() As Double), which holds the current solution x.

The strategy used in the code follows Figure 4.8. The user first provides a matrix of the linear equation coefficients with intial guesses for the n-L nonlinear equations. In the VBA code, the user must specify the number of rows (the number of equations) and the start row on the Excel sheet for the first matrix. The Gauss–Jordan elimination method is called `Public Sub Gauss_Jordan_Macro()`, and a second matrix is generated, which contains a consistent initial guess, x; this can be important for material balance problems as will be shown in Example 4.9.

Next, the Newton–Raphson method is called `Public Sub NR_Gauss_Jordan_Macro()`. Here, we need an x^*, which is obtained from the second matrix ($x^* = x$). This x^* is used with Equation (4.24) to determine the partial derivative terms in each nonlinear function. In this example, we calculate the partial derivative terms for the two nonlinear functions (`cntNLeq = 2`). It is not necessary to calculate the partial derivatives for linear functions. The nonlinear equations are supplied in the VBA function `Function NLFunc`.

The Newton–Raphson method generates a third matrix consisting of the initial linear equations and the linearized nonlinear equations (see Eq. (4.15)). For this example, the last two equations (`cntNLeq = 2`) in the third matrix are the nonlinear equations, which have been linearized. This matrix is then solved using the Gauss–Jordan method and the results written to the current second matrix. The x values can be used in Equation (4.11) to check if all the RHS = 0. If convergence is not obtained, the NR/G-J iterations can continue by repeating the macro NR_Gauss_Jordan_Macro. We will automate this process, but in this example, the user must repeatedly call NR_Gauss_Jordan_Macro. This does allow the results shown in Equations (4.19)–(4.23) to be reproduced.

To run this example from the Excel sheet run, make sure the input matrix is correct, then run the macro Gauss_Jordan_Macro once followed by repeated NR_Gauss_Jordan_Macro until the solution (x values) in the second matrix remain unchanged.

The results from **Example 4.8.xls** are shown in Figure 4.10 and the key steps are summarized in Figure 4.11. ■

```
Option Explicit
'path to the Gauss-Jordan matrix elimination method
Public Declare Sub GJ_Elimination_Main Lib "C:\POEA\Bridging Excel and C
Codes\Examples\Simple_C_Matrix_dll\Debug\Simple_C_Matrix_dll.dll" (ByRef Matrix As
Double, ByVal Nrows As Long, ByVal Ncolumns As Long)

Dim Nrows, Ncolumns, cntNLeq, RowStartFirstMatrix, RowStartSecondMatrix,
RowStartThirdMatrix, RowStartNLEqsThirdMatrix As Long

Public Sub Gauss_Jordan_Macro()
    Dim C() As Double

    'The user must specify the number of rows (Nrows);
    'the number of nonlinear equations (cntNLeq);
    'and the row where the first matrix begins on the Excel sheet (RowStartFirstMatrix)
    'we assume all matrices will begin in column 2 - use column one for comments
    Nrows = 3
    Ncolumns = Nrows + 1
    cntNLeq = 2
    RowStartFirstMatrix = 8

    'Here we are keeping a gap of 4 rows between matrices on the Excel sheet
    RowStartSecondMatrix = RowStartFirstMatrix + Nrows + 4
    RowStartThirdMatrix = RowStartSecondMatrix + Nrows + 4

    'Here we identify the starting row in the thrid matrix for the nonlinear equations
    RowStartNLEqsThirdMatrix = RowStartThirdMatrix + Nrows - cntNLeq


    'Here we read the first matrix on the Excel sheet into the VBA matrix C
    ' if the entry on the Excel sheet is blank " " we set the entry to zero
    ReDim C(Nrows, Ncolumns)
    Dim i As Integer
    Dim j As Integer
    For i = 0 To Nrows - 1
        For j = 0 To Ncolumns - 1
            If Sheet1.Cells(i + RowStartFirstMatrix, j + 2) = " " Then
                C(i, j) = 0
            Else
                C(i, j) = Sheet1.Cells(i + RowStartFirstMatrix, j + 2)
            End If
        Next j
    Next i

    ' We call the Gauss-Jordan matrix elimination method which is a C program
    ' and we palce the results in the second matrix on the Excel sheet
    GJ_Elimination_Main C(0, 0), Nrows, Ncolumns

    For i = 0 To Nrows - 1
        For j = 0 To Ncolumns - 1
            Sheet1.Cells(i + RowStartSecondMatrix, j + 2) = C(i, j)
        Next j
    Next i
End Sub
Public Sub NR_Gauss_Jordan_Macro()
'Here we will use the Newton Raphson method to linearize the NL equations.

    Dim x() As Double
    Dim F() As Double
    Dim D() As Double

    Dim i, j As Integer

    Dim DELTA As Double
    DELTA = 0.0001

'First we copy all the current coeficients from the first matrix (the all linear matrix)
    'to the third matrix
    'Eventually we will then need to substitute in the linearized NL equations

    For i = 0 To Nrows - 1
        For j = 0 To Ncolumns - 1
            Sheet1.Cells(i + RowStartThirdMatrix, j + 2) = Sheet1.Cells(i +
RowStartFirstMatrix, j + 2)
        Next j
    Next i

 'Get the current solution x = x* from the second matrix
    ReDim x(Nrows)
    Get_X_Values(Nrows, x)

 'The Newton Raphson Method for the NL equations
    ReDim F(cntNLeq)
    ReDim D(cntNLeq, Nrows)
```

Figure 4.9 VBA code to solve a nonlinear equation set using the Newton–Raphson method.

```
 ' Evaluate each NL equation at x*
   For i = 0 To cntNLeq - 1
       F(i) = NLFunc(i, x)
   Next i

 ' Determine partial derivative for each NL equation wrt each variable at x*
   For i = 0 To cntNLeq - 1
       For j = 0 To Nrows - 1
           x(j) = x(j) + DELTA
           D(i, j) = (NLFunc(i, x) - F(i)) / DELTA
           x(j) = x(j) - DELTA
       Next j
   Next i
' Place the partial derivatives at x* for each NL equation on the Excel sheet - third
' matrix
   For i = 0 To cntNLeq - 1
       For j = 0 To Nrows - 1
           Sheet1.Cells(i + RowStartNLEqsThirdMatrix, j + 2) = D(i, j)
       Next j
   Next i

' Calculate Beta for each NL equation and place on Excel sheet - third matrix
   Dim TempSum As Double
   For i = 0 To cntNLeq - 1
       TempSum = 0
       For j = 0 To Nrows - 1
           TempSum = TempSum + x(j) * D(i, j)
       Next j
       Sheet1.Cells(i + RowStartNLEqsThirdMatrix, Ncolumns + 1) = TempSum - F(i)
   Next i

   ' Read coefficients from the third matrix into the VBA matrix C
   Dim C() As Double

   ReDim C(Nrows, Ncolumns)

   For i = 0 To Nrows - 1
       For j = 0 To Ncolumns - 1
           If Sheet1.Cells(i + RowStartThirdMatrix, j + 2) = " " Then
               C(i, j) = 0
           Else
               C(i, j) = Sheet1.Cells(i + RowStartThirdMatrix, j + 2)
           End If
       Next j
   Next i

  'Call the Gauss-Jordan matrix elimination mathod
  'and place the results on the Excel sheet in the second matrix
   GJ_Elimination_Main C(0, 0), Nrows, Ncolumns

   For i = 0 To Nrows - 1
       For j = 0 To Ncolumns - 1
           Sheet1.Cells(i + RowStartSecondMatrix, j + 2) = C(i, j)
       Next j
   Next i

End Sub

Private Sub Get_X_Values(ByVal N As Integer, ByRef x() As Double)
  ' Get the current solution x = x*
   Dim i As Integer

   For i = 0 To N - 1
       x(i) = Sheet1.Cells(i + RowStartSecondMatrix, Ncolumns + 1)
   Next i
End Sub

Private Function NLFunc(ByVal Idx As Long, ByRef x() As Double) As Double
  ' The NL equations
  ' --- Remember Index on Functions AND x(i) BOTH start at Zero
   If Idx = 0 Then
       NLFunc = x(0) ^ (1 / 2) + (x(1)) ^ 2 + x(1) * x(2) - 11
   End If
   If Idx = 1 Then
       NLFunc = x(0) * x(0) + x(1) * x(1) + x(2) * x(2) - 14
   End If

End Function
```

Figure 4.9 ***(Continued)***

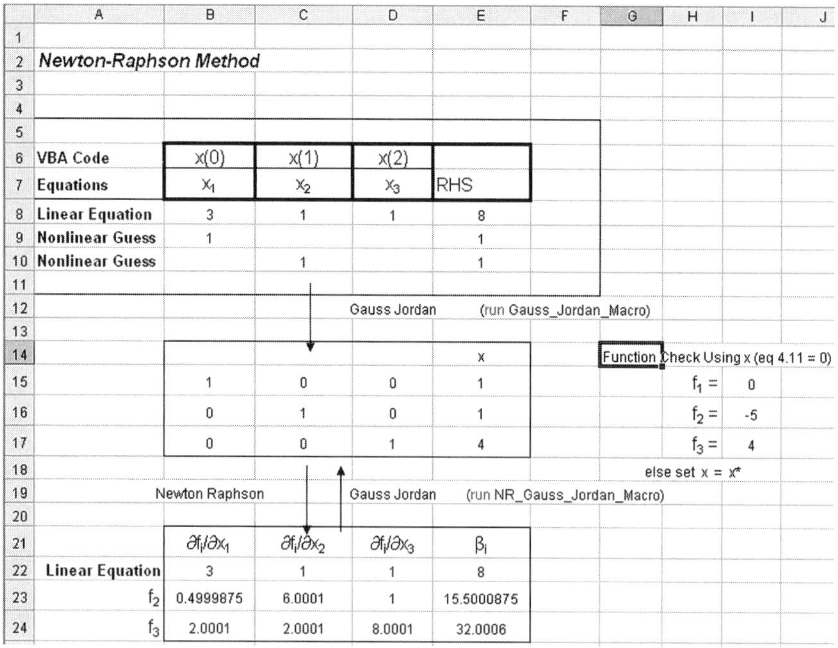

Figure 4.10 Solution of the nonlinear equation set using the Newton–Raphson method.

$$\begin{pmatrix} \text{Matrix 1 consisting of} \\ \text{Linear equations} \\ \text{Initial guess for independent variables in nonlinear equations} \end{pmatrix}$$

$$\downarrow \text{Gauss–Jordan}$$

$$\left\{\begin{pmatrix} 1 & 0 & \dots & 0 & x_1 \\ 0 & 1 & \dots & 0 & x_2 \\ \dots & \dots & \dots & \dots & \dots \\ 0 & 0 & \dots & 1 & x_n \end{pmatrix}\right\} \begin{array}{l} \text{Matrix 2} \\ \text{Equation, (4.11)} \cong 0 \\ \textit{else set } x^* = x \end{array}$$

$$\text{Newton–Raphson}\downarrow \quad \uparrow\text{Gauss–Jordan}$$

$$\begin{pmatrix} \left.\frac{\partial f_1}{\partial x_1}\right|_{x=x^*} & \left.\frac{\partial f_1}{\partial x_2}\right|_{x=x^*} & \dots & \left.\frac{\partial f_1}{\partial x_n}\right|_{x=x^*} & \beta_1 \\ \left.\frac{\partial f_2}{\partial x_1}\right|_{x=x^*} & \left.\frac{\partial f_2}{\partial x_2}\right|_{x=x^*} & \dots & \left.\frac{\partial f_2}{\partial x_n}\right|_{x=x^*} & \beta_2 \\ \dots & \dots & \dots & \dots & \dots \\ \left.\frac{\partial f_n}{\partial x_1}\right|_{x=x^*} & \left.\frac{\partial f_n}{\partial x_2}\right|_{x=x^*} & \dots & \left.\frac{\partial f_n}{\partial x_n}\right|_{x=x^*} & \beta_n \end{pmatrix} \text{Matrix 3}$$

Figure 4.11 Summary of the key steps—Newton–Raphson method.

The key steps of Figure 4.11 are

Step1: Guess the value for independent variable in each nonlinear equation (matrix 1).

Step 2: Solve the resulting equation set using the Gauss–Jordan algorithm (matrix 2). This is done by calling `Gauss_Jordan_Macro` from the Excel sheet, and this is done only *once*;

Step 3: If Equation (4.11) $\neq 0$, set $x^* = x$ and apply the NR method. This will result in a linear equation set (matrix 3), which can be soved as in Step 2. In matrix 3, the last `cntNLeq` equations are the nonlinear equations, which have been linearized. This is done by calling `NR_Gauss_Jordan_Macro` from the Excel sheet. Continue calling `NR_Gauss_Jordan_Macro` from the Excel sheet until the solution x in matrix 2 remains constant.

Code Modifications Required for Each New Problem

1. You will need to specify the total number of rows, `Nrows` (`Nrows` = number of equations).
2. You will need to specify the number of nonlinear equations, `cntNLeq`.
3. You will need to specify the start row for matrix 1, `RowStartFirstMatrix`. It is assumed that the second column will be the starting column.
4. You will need to supply the `cntNLeq` nonlinear equations to `Function NLFunc`.

4.2.4 Nonlinear Material Balance Problems: The Simultaneous Approach

An advantage of the simultaneous or equation-based solution approach is the elimination of iterations to converge recycle loops in material balance problems. The simultaneous approach also eliminates single-variable searches for linear alternative specifications. When using the simultaneous approach, nonlinear specifications must be linearized.

	A	B	C	D	E	F	G	H	I	J	K	L	M	N	O	P	Q	R	S
1	*Styrene Material Balance Nonlinear Specification*																		
2																			
3	Reactio	C_8H_{10} ---> C_8H_8 + H_2			Conversion		0.65												
4																			
5		x0	x1	x2	x3	x4	x5	x6	x7	x8	x9	x10	x11	x12	x13	x14	x15	x16	
6		Feed			Reactor In			Reactor Out			Water Out	Organic Out		Vapor	Product		Recycle		
7		N(C8H10)	N(H2O)	N(C8H10)	N(H2O)	N(C8H8)	N(C8H10)	N(H2O)	N(H2)	N(C8H8)	N(H2O)	N(C8H10)	N(C8H8)	N(H2)	N(C8H10)	N(C8H8)	N(C8H10)	N(C8H8)	RHS
8	Feed		1																3000
9	Specs	1																	100
10	Mixer	1		-1													1		0
11	Material		1		-1														0
12	Balances					-1												1	0
13	Reactor			0.35			-1												0
14	Balances				1			-1											0
15				0.65					-1										0
16				0.65		1				-1									0
17	Separator 1						1					-1							0
18								1			-1								0
19									1					-1					0
20										1			-1						0
21	Separator 2											1			-1		-1		0
22													1			-1		-1	0
23												0.01			-1				0
24													0.99			-1			0

(a)

30	1	0	0	0	0	0	0	0	0	0	0	0	0	0	0	0	0	100	N(C8H10)	Feed
31	0	1	0	0	0	0	0	0	0	0	0	0	0	0	0	0	0	1386.24	N(H2O)	Feed
32	0	0	1	0	0	0	0	0	0	0	0	0	0	0	0	0	0	153.02	N(C8H10)	Reactor In
33	0	0	0	1	0	0	0	0	0	0	0	0	0	0	0	0	0	1386.24	N(H2O)	Reactor In
34	0	0	0	0	1	0	0	0	0	0	0	0	0	0	0	0	0	1.00469	N(C8H8)	Reactor In
35	0	0	0	0	0	1	0	0	0	0	0	0	0	0	0	0	0	53.5578	N(C8H10)	Reactor Out
36	0	0	0	0	0	0	1	0	0	0	0	0	0	0	0	0	0	1386.24	N(H2O)	Reactor Out
37	0	0	0	0	0	0	0	1	0	0	0	0	0	0	0	0	0	99.4644	N(H2)	Reactor Out
38	0	0	0	0	0	0	0	0	1	0	0	0	0	0	0	0	0	100.469	N(C8H8)	Reactor Out
39	0	0	0	0	0	0	0	0	0	1	0	0	0	0	0	0	0	1386.24	N(H2O)	Water Out
40	0	0	0	0	0	0	0	0	0	0	1	0	0	0	0	0	0	53.5578	N(C8H10)	Organic Out
41	0	0	0	0	0	0	0	0	0	0	0	1	0	0	0	0	0	100.469	N(C8H8)	Organic Out
42	0	0	0	0	0	0	0	0	0	0	0	0	1	0	0	0	0	99.4644	N(H2)	Vapor Out
43	0	0	0	0	0	0	0	0	0	0	0	0	0	1	0	0	0	0.53558	N(C8H10)	Product
44	0	0	0	0	0	0	0	0	0	0	0	0	0	0	1	0	0	99.4644	N(C8H8)	Product
45	0	0	0	0	0	0	0	0	0	0	0	0	0	0	0	1	0	53.0222	N(C8H10)	Recycle
46	0	0	0	0	0	0	0	0	0	0	0	0	0	0	0	0	1	1.00469	N(C8H8)	Recycle

(b)

	A	B	C	D	E	F	G	H	I	J	K	L	M	N	O	P	Q	R	S
50			1																3000
51		1																	100
52		1		-1													1		0
53			1		-1														0
54						-1												1	0
55				0.35			-1												0
56					1			-1											0
57				0.65					-1										0
58				0.65		1				-1									0
59							1					-1							0
60								1			-1								0
61									1					-1					0
62										1			-1						0
63												1			-1		-1		0
64													1			-1		-1	0
65												0.01			-1				0
66		0	0	0	0	0	0	0	0	0	0	0	0	0	-0.0099	5E-05	-0.0001	0.0187	-1E-13

(c)

Figure 4.12 (a) Styrene material balances from Example 4.3—augmented matrix. (b) Simultaneous solution (same solution as Figure 4.6b). (c) Linearization via the Newton–Raphson method. (d) Solution of the styrene process with nonlinear specification.

	A	B	C	D	E	F	G	H	I	J	K	L	M	N	O	P	Q	R	S	T	U
29		1	0	0	0	0	0	0	0	0	0	0	0	0	0	0	0	0	100	N(C8H10)	Feed
30		0	1	0	0	0	0	0	0	0	0	0	0	0	0	0	0	0	3000	N(H2O)	Feed
31		0	0	1	0	0	0	0	0	0	0	0	0	0	0	0	0	0	153.02	N(C8H10)	Reactor In
32		0	0	0	1	0	0	0	0	0	0	0	0	0	0	0	0	0	3000	N(H2O)	Reactor In
33		0	0	0	0	1	0	0	0	0	0	0	0	0	0	0	0	0	0.2855	N(C8H8)	Reactor In
34		0	0	0	0	0	1	0	0	0	0	0	0	0	0	0	0	0	53.558	N(C8H10)	Reactor Out
35		0	0	0	0	0	0	1	0	0	0	0	0	0	0	0	0	0	3000	N(H2O)	Reactor Out
36		0	0	0	0	0	0	0	1	0	0	0	0	0	0	0	0	0	99.464	N(H2)	Reactor Out
37		0	0	0	0	0	0	0	0	1	0	0	0	0	0	0	0	0	99.75	N(C8H8)	Reactor Out
38		0	0	0	0	0	0	0	0	0	1	0	0	0	0	0	0	0	3000	N(H2O)	Water Out
39		0	0	0	0	0	0	0	0	0	0	1	0	0	0	0	0	0	53.558	N(C8H10)	Organic Out
40		0	0	0	0	0	0	0	0	0	0	0	1	0	0	0	0	0	99.75	N(C8H8)	Organic Out
41		0	0	0	0	0	0	0	0	0	0	0	0	1	0	0	0	0	99.464	N(H2)	Vapor Out
42		0	0	0	0	0	0	0	0	0	0	0	0	0	1	0	0	0	0.5356	N(C8H10)	Product
43		0	0	0	0	0	0	0	0	0	0	0	0	0	0	1	0	0	99.464	N(C8H8)	Product
44		0	0	0	0	0	0	0	0	0	0	0	0	0	0	0	1	0	53.022	N(C8H10)	Recycle
45		0	0	0	0	0	0	0	0	0	0	0	0	0	0	0	0	1	0.2855	N(C8H8)	Recycle

(d)

Figure 4.12 ***(Continued)***

EXAMPLE 4.9 *Styrene Process with Nonlinear Specification*

Here, solve Example 3.15, our styrene process with a nonlinear specification. The nonlinear specification came about from the requirement that separator 2 operate such that the molar concentration of styrene in the product stream and the molar concentration of ethylbenzene in the recycle stream are equal. This quadratic specification was

$$\frac{N_{C_8H_8}^{product}}{N_{C_8H_{10}}^{product}+N_{C_8H_8}^{product}}=\frac{N_{C_8H_{10}}^{recycle}}{N_{C_8H_{10}}^{recycle}+N_{C_8H_8}^{recycle}}, \qquad (3.65, \text{repeated})$$

and this specification can be rearranged to our needed form ($f(x) = 0$) as

$$\frac{N_{C_8H_8}^{product}}{N_{C_8H_{10}}^{product}+N_{C_8H_8}^{product}}-\frac{N_{C_8H_{10}}^{recycle}}{N_{C_8H_{10}}^{recycle}+N_{C_8H_8}^{recycle}}=0. \qquad (4.25)$$

The nonlinear Equation (4.25) will be used in place of the current linear specification given by Equation (3.45), $N_{C_8H_8}^{product}-0.99N_{C_8H_8}^{organic\ out}=0$.

Example 4.9 is somewhat different from a typical material balance problem with nonlinear specifications. For this example, we already have a good starting guess from Example 4.2—normally, this will not be the case and we solve the more typical problem a little later in this section.

The solution to Example 4.9 is provided in **Example 4.9a.xls** and the results are shown in Figure 4.12. Our existing solution from Example 4.3 is used to supply a good initial guess to the problem. Equation (4.25) is supplied to the VBA code in `Function NLFunc`.

In Figure 4.12a we have repeated the 17 equations developed in Example 4.3. The Gauss–Jordan solution is shown in Figure 4.12b, which is identical to Figure 4.6b.

We can check for convergence at our current solution, x (shown in Figure 4.12b), by using Equation (4.25):

$$\left(\frac{99.464}{0.5356+99.464}\right)-\left(\frac{53.022}{53.022+1.0047}\right)=(0.9946)-(0.9814)$$
$$=0.0132.$$

As Equation (4.25) ≠ 0 at x; we set our current solution, x, to x^* and use this point to linearize the nonlinear specification via the Newton–Raphson method. Equation (4.25) after linearization will replace $N_{C_8H_8}^{product}-0.99N_{C_8H_8}^{organic\ out}=0$. This creates a third matrix shown in Figure 4.12c.

Notice that row 66 in the Excel sheet (Figure 4.12c) is the linearized specification Equation (4.25) at the point found in Figure 4.12b. The Gauss–Jordan method can be used to solve Figure 4.12c providing a new x solution. This process continues until the x values converge, as shown in Figure 4.12d. Here, the converged values for molar flow rates are identical to those obtained in Example 3.15.

EXAMPLE 4.9 *(Repeated) Typical Problem*

The typical problem when solving Example 4.9 is how best to obtain a good initial solution that will be used in the subsequent linearization. Generally, we want the initial solution to come from a linear equation set that is solved using the Gauss–Jordan method. We then replace linear equations with nonlinear specifications and solve the resulting equation set using the Newton–Raphson method. For Example 4.9, we would have the first 16 linear equations from Example 4.3 and we would need to supply an estimate for an independent variable in the nonlinear concentration specification (Eq. (4.25)). Normally, we would not have $N_{C_8H_8}^{product}-0.99N_{C_8H_8}^{organic\ out}=0$ as provided from Example 4.3.

It would seem reasonable to simply set $N_{C_8H_8}^{product}$, the first term in Equation (4.25), to a value (say, $N_{C_8H_8}^{product}=100$) and to solve the linear equation set using Gauss–Jordan elimination. If we try this, the simultaneous solution will fail. Recall that Example 4.8 has 17 unknown flow rates, and when we assemble the 16 material balance equations, we will have specified 16 of the 17 unknowns. The only remaining independent variable available in Equation (4.25) is $N_{C_8H_8}^{recycle}$. We can set $N_{C_8H_8}^{recycle}$ to a value, say, $N_{C_8H_8}^{recycle}=0.3$, and solve the resulting equation set. This solution is provided in **Example 4.9b.xls**. The converged values for molar flow rates are identical to those obtained in Example 3.15 and those shown in Figure 4.12d.

REFERENCES

DeLancey, G.B. 1999. Process analysis: An electronic version. *Chem. Eng. Ed.* (Winter): 40–45.

Felder, R.M. and R.W. Rousseau 2005. *Elementary Principles of Chemical Processes* (3rd edition). John Wiley & Sons, Hoboken, NJ.

FERREIRA, E.C., R. LIMA, and R. SALCEDO 2004. Spreadsheets in chemical engineering education—A tool in process design and process integration. *Int. J. Eng. Ed.* 20(6): 928–938.

REKLAITIS, G.V. 1983. *Introduction to Material and Energy Balances*. John Wiley & Sons, New York.

PROBLEMS

4.1 ***Solutions to Linear Equation Sets*** Possible solutions to a set of linear equations are

(a) no solution,

(b) one unique solution, and

(c) infinite solutions.

Identify the solution to each of the following equation sets. Hint: Make a plot of x_2 versus x_1.

(I) $3x_1 + x_2 = 5$
$x_1 + x_2 = 3$

(II) $2x_1 + 4x_2 = 8$
$x_1 + 2x_2 = 2$

(III) $2x_1 + 4x_2 = 8$
$x_1 + 2x_2 = 4$

4.2 ***Gauss–Jordan Method*** During Gauss–Jordan elimination, the following augmented matrix is obtained:

$$\begin{pmatrix} 1 & 0 & 1 & 5 \\ 0 & -3 & 0 & -3 \\ 0 & 0 & -1 & 1 \end{pmatrix}.$$

What is the solution to the set of simultaneous equations?

4.3 ***Newton–Raphson Method "by Hand."*** Here, complete one iteration by hand of the Newton–Raphson method. The starting point is $x^0 = \{x_1^0, x_2^0\} = (2, 2)$. Solve for x^1:

$$f_1\{x_1, x_2\} = x_1 - 0.25(x_2)^2 + 3.25 = 0$$
$$f_2\{x_1, x_2\} = (x_1)^3 - (x_2)^2 - 2 = 0.$$

4.4 ***Newton–Raphson Method Applied to an Equation Set*** Solve the following equation set by hand using the Newton–Raphson method. Complete two full iterations of the method starting at the point (2, 2). Check your results using the Newton–Raphson computer code:

$$f_1\{x_1, x_2\} = (x_1)^3 + (x_2)^2 - 10 = 0$$
$$f_1\{x_1, x_2\} = \exp(x_1) + x_2 - 6 = 0.$$

4.5 ***VBA Matrix Operations*** What would be printed to the Excel sheet when the following VBA program segment is executed?

```
For I = 1 To 3

For J = 4 To 6
C(I, J) = I * J
' Place solution on Excel Sheet
Sheet1.Cells(I, J) = C(I, J)
Next J
Next I
```

Excel Sheet for Results

	A	B	C	D	E	F
1						
2						
3						
4						
5						
6						

4.6 ***VBA Matrix Operations*** An augmented array C is read from the Excel sheet as we have done in our chapter examples. If the elements of the first row of the array are 2.0, 2.0, 4.0, 8.0, what would be printed to the Excel sheet in row 12 when the following VBA program segment is executed for the first time?

```
For IP = 0 To 1
For IC = IP To 3
C(IP, IC) = C(IP, IC) / C(IP, IP)
Next IC
' Place solution on Excel Sheet
For j = 0 To Ncolumns - 1
Sheet1.Cells(12, j + 2) = C(0, j)
Next j
Next IP
```

4.7 ***Newton–Raphson Method as a VBA Code (Solution Provided)*** In Section 4.2.3, the Newton–Raphson method was developed as a VBA code. However, the Gauss–Jordan algorithm was provided by the complied C version. Replace the compiled C version of the Gauss–Jordan method with the VBA version of the Gauss–Jordan method. The code should now be all VBA. Solve Example 4.6 with this code. Also solve Example 4.7 from $x^0 = x^* = \{4, 1\}$ and $x^0 = x^* = \{1, 4\}$. Notice that two different answers are obtained for Example 4.7—see Figure 4.7.

Solution: **Problem 4.7a.xls**, **Problem 4.7b.xls**, **Problem 4.7c.xls**

4.8 ***Process A → B Simultaneous Solution***

(a) Consider the material balance problem for the mixer, reactor, and separator shown in Figure P4.8. In the reactor, 30% of the incoming A is converted to B on a molar basis. The recycle rate (stream 5) is set at 5% of the stream into the separator (stream 3); using Equation (3.19), $N_5 = \alpha_5 N_3$, with $\alpha_5 = 0.05$. Calculate all flows in the process using the simultaneous solution approach and using the Gauss–Jordan VBA code.

(b) An advantage of the simultaneous solution approach is the ease of solution when linear alternative specifications are utilized. Replace the natural specification $N_A^1 = 100$ mol/h with an alternative specification for the molar concentration of A entering the reactor, $N_A^2/(N_A^2 + N_B^2) = 0.895$. Compare the solution effort requiredhere (part b) with those in Section 3.5.

4.9 ***Dehydrogenation of Propane Simultaneous Solution (Felder and Rousseau, 2005, pp. 135–136)*** Propane is dehydrogenated to form propylene in a catalytic reactor:

$$C_3H_8 \rightarrow C_3H_6 + H_2.$$

The process, shown in Figure P4.9, is to be designed for a 95% overall conversion of propane. The reaction products

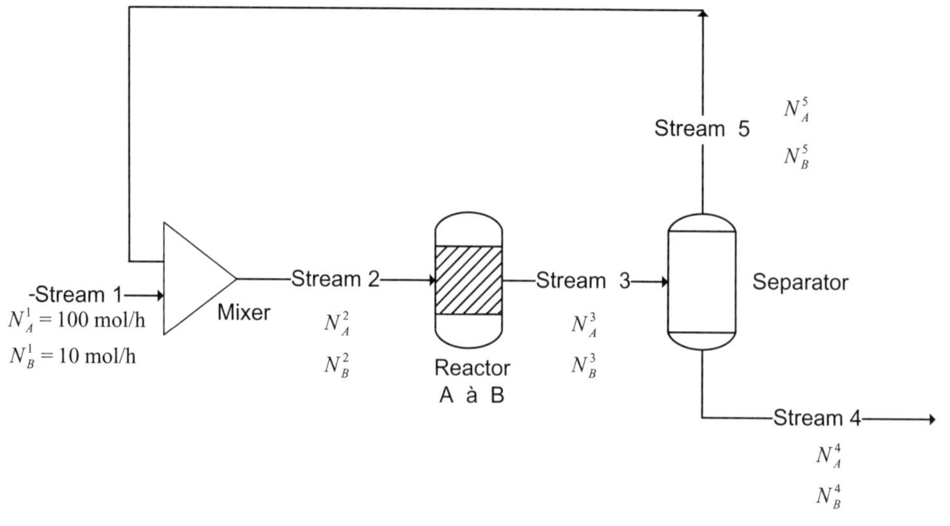

Figure P4.8 Simple process A → B.

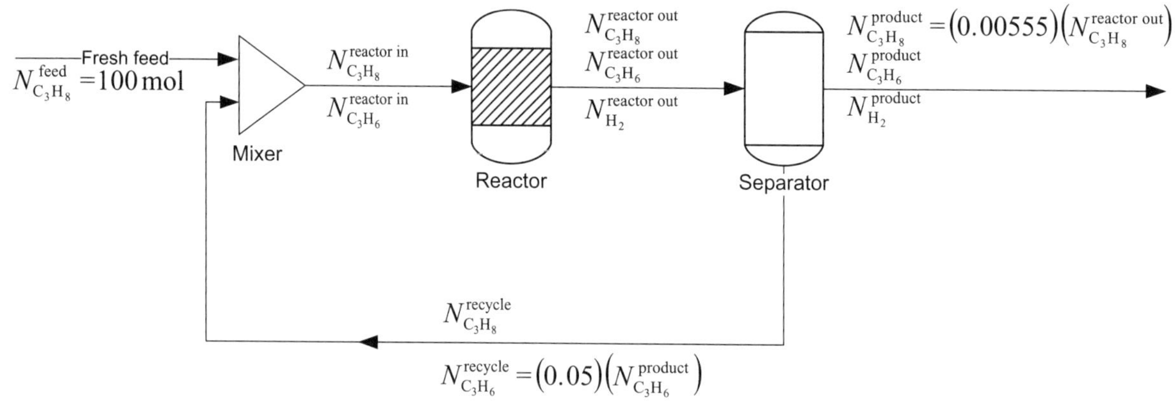

Figure P4.9 Propylene flow sheet.

are separated into two streams: The first, which contains H_2, C_3H_6, and 0.555% of the propane that leaves the reactor, is taken off as product; the second stream, which contains the balance of the unreacted propane and 5% of the propylene in the first stream, is recycled to the reactor. Calculate the composition of the product, the ratio (moles recycled)/(mole fresh feed), and the single-pass conversion using the simultaneous solution approach.

4.10 ***Ammonia Process: Simultaneous Solution (Reklaitis, 1983, p. 273)*** We have solved this problem using the sequential modular approach; see Chapter 3, Problem 3.15 (solution provided).

A simplified flow sheet for an ammonia process is shown in Figure P4.10. In the ammonia plant, a feed gas consisting of 24.5% N_2, 74% H_2, 1.2% CH_4, and 0.3% argon is catalytically reacted to produce NH_3. The reaction is

$$N_2 + 3H_2 = 2NH_3,$$

and 65% of the N_2 entering the reactor is converted per pass. The products of reaction are refrigerated to separate out 75% of the NH_3 product per pass. The remaining process stream is recycled back to the reactor. In order to stabilize the buildup of the inerts CH_4 and A in the process, part of the recycled gas is purged. Suppose the splitter purge rate (stream 6) is set at 5% of the stream into the splitter (stream 5) or from Equation (3.17), $N^6 = \alpha^6 N^5$, with $\alpha^6 = 0.05$. Calculate all flows in the process using the simultaneous solution approach. Assume a basis of 1000 mol/h for the feed stream (stream 1).

Solution: **Problem 4.10.xls**

4.11 ***Ammonia Process with Nonlinear Alternative Specifications: Simultaneous Solution (Reklaitis [1983], p. 273)*** We have solved this problem using the sequential modular approach with a single-variable search; see Chapter 3, Problem 3.17.

Starting with Problem 4.10, suppose the splitter split fraction α^6 is adjusted so that the combined reactor feed gas contains 18 mol % CH_4. Calculate all flows in the process. Again, assume a basis of 1000 mol/h for the feed stream (stream 1). Solve this problem using the simultaneous solution approach.

Solution Hint

This problem can be confusing to solve, especially as to the best way of incorporating the nonlinear equations into the

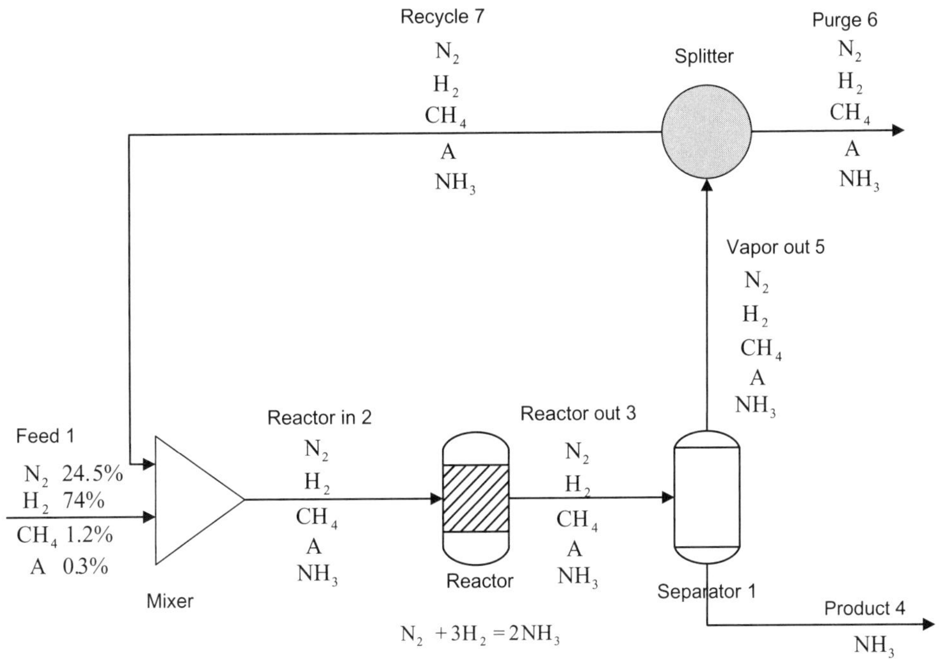

Figure P4.10 Ammonia flow sheet.

material balance set. We begin by taking advantage of the simultaneous solution we developed in Problem 4.10. Here, rearrange the equation set so that the last five equations solve for the purge flow rates for each species as

$$\alpha^6 N_{N_2}^{\text{vapor out}} - N_{N_2}^{\text{purge}} = 0 \qquad \text{(P4.11a)}$$

$$\alpha^6 N_{H_2}^{\text{vapor out}} - N_{H_2}^{\text{purge}} = 0 \qquad \text{(P4.11b)}$$

$$\alpha^6 N_{CH_4}^{\text{vapor out}} - N_{CH_4}^{\text{purge}} = 0 \qquad \text{(P4.11c)}$$

$$\alpha^6 N_{A}^{\text{vapor out}} - N_{A}^{\text{purge}} = 0 \qquad \text{(P4.11d)}$$

$$\alpha^6 N_{NH_3}^{\text{vapor out}} - N_{NH_3}^{\text{purge}} = 0. \qquad \text{(P4.11e)}$$

In Equation (P4.11a–e), when the splitter split fraction is fixed, $\alpha^6 \equiv 0.05$, these equations are linear. We do need to add an additional variable to the set, the purge split fraction. To do this, add the following equation as the last equation in the set (matrix 1):

$$\alpha^6 \equiv 0.05. \qquad \text{(P4.11f)}$$

Next, check that the same solution is obtained as in Problem 4.10 (matrix 2). There will be one additional variable, the split fraction, $\alpha^6 = 0.05$. It may appear we have not really accomplished much, but Problem 4.11 is almost done. Most important is that we have an initial solution, x, for every variable.

Equation (P4.11a–e) is a group of nonlinear species flow equations for the purge stream when the splitter split fraction, α^6, is a variable. These five equations are supplied to the subroutine `Function NLFunc` and these five equations will replace the linear versions (Eq. (P4.11a–e)), $\alpha^6 \equiv 0.05$) in matrix 3.

We also need to replace the linear purge split fraction equation, $\alpha^6 \equiv 0.05$ (Eq. (P4.11f)). We know the splitter split fraction α^6 must be adjusted so that the combined reactor feed gas contains 18 mol % CH_4:

$$\frac{N_{CH_4}^{\text{reactor in}}}{N_{N_2}^{\text{reactor in}} + N_{H_2}^{\text{reactor in}} + N_{CH_4}^{\text{reactor in}} + N_{A}^{\text{reactor in}} + N_{NH_3}^{\text{reactor in}}} = 0.18. \qquad \text{(P4.11g)}$$

Equation (P4.11g) can be rearranged to

$$(0.82)N_{CH_4}^{\text{reactor in}} - (0.18)N_{N_2}^{\text{reactor in}} - (0.18)N_{H_2}^{\text{reactor in}} - (0.18)N_{A}^{\text{reactor in}} - (0.18)N_{NH_3}^{\text{reactor in}} = 0. \qquad \text{(P4.11h)}$$

Equation (P4.11h) can be supplied to subroutine `Function NLFunc`, replacing (Eq. (P4.11f)) in matrix 3. Equation (P4.11h) deserves special comment. It is a linear equation; however, there is no requirement when applying the Newton–Raphson method that the equations be nonlinear. Adding this linear Equation (P4.11h) to matrix 3 via `Function NLFunc` is the easiest way to satisfy the requirement that that the combined reactor feed gas contains 18 mol % CH_4.

4.12 ***Production of Sodium Hydroxide (Reklaitis, 1983, pp. 338–344)*** A process for the production of sodium hydroxide contains an intermediate stream consisting of $CaCO_3$ precipitate slurried in a solution of NaOH and H_2O. This slurry is washed with water in three stages so as to reduce the NaOH concentration in the slurry to a sufficiently low level. The flow sheet for this countercurrent washing process is shown in Figure P4.12. The washing stages can be assumed to operate so that the slurry leaving each stage will contain 2 lb solution per 1 lb $CaCO_3$ solid. Also, the concentration of the solutions in the pair of streams leaving each stage can be assumed to be equal. If the slurry fed to the first stage contains 10% NaOH, 30% $CaCO_3$, and 60% H_2O, calculate

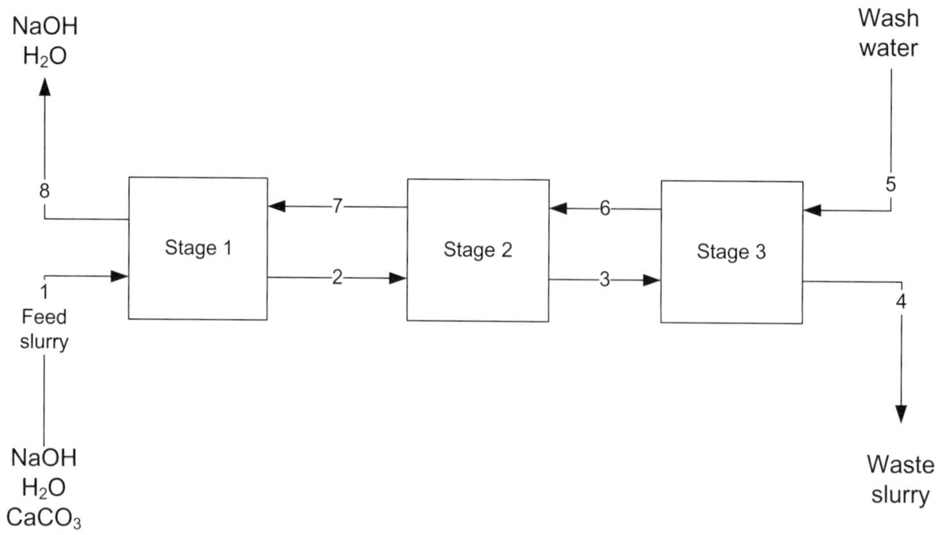

Figure P4.12 Sodium hydroxide flow sheet.

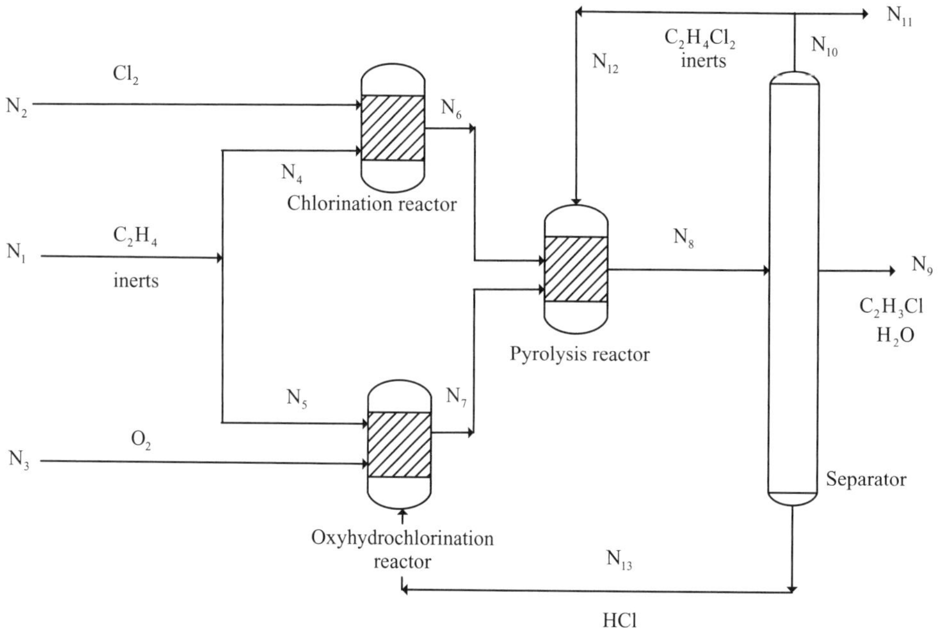

Figure P4.13 Vinyl chloride flow sheet (DeLancey, 1999).

the pounds of water wash required per pound of feed to achieve a 1% NaOH concentration in the solution leaving in stream 4.

There are 14 linear equations and two nonlinear equations needed to solve this material balance problem. The $CaCO_3$ balances are trivial and can be omitted.

Use the following species index: NaOH = 1 and H_2O = 2 so that, for example, F2(1) would be the flow rate of NaOH in stream 2 in pound per hour and F8(2) would be the flow rate of H_2O in stream 8.

The nonlinear specifications are

$$\frac{F^8_{\mathrm{NaOH}}}{F^8_{\mathrm{NaOH}}+F^8_{\mathrm{H_2O}}}=\frac{F^2_{\mathrm{NaOH}}}{F^2_{\mathrm{NaOH}}+F^2_{\mathrm{H_2O}}} \quad \text{or} \quad \frac{F8(1)}{F8(1)+F8(2)}=\frac{F2(1)}{F2(1)+F2(2)} \tag{P4.12a}$$

and

$$\frac{F^7_{\mathrm{NaOH}}}{F^7_{\mathrm{NaOH}}+F^7_{\mathrm{H_2O}}}=\frac{F^3_{\mathrm{NaOH}}}{F^3_{\mathrm{NaOH}}+F^3_{\mathrm{H_2O}}} \quad \text{or} \quad \frac{F7(1)}{F7(1)+F7(2)}=\frac{F3(1)}{F3(1)+F3(2)}. \tag{P4.12b}$$

Guesses are supplied to initially "eliminate" the two nonlinear equations.

4.13 ***Flow Sheet Decomposition: Vinyl Chloride Process Material Balance (Also See Problem 3.20)*** DeLancey (1999) presented a simplified process for vinyl chloride production from ethylene, solving the nonlinear material balance equation set using the software package Scientific Notebook. Ferreira et al. (2004) solved the vinyl chloride problem within Excel by using the material balance equations as constraints within Excel Solver. We solved this problem in Chapter 3 (Problem 3.20) using the sequential modular approach. Here, solve using the Newton–Raphson method.

Figure P4.13 (DeLancey, 1999) provides a simplified flow sheet for vinyl chloride (C_2H_3Cl) production from ethylene (C_2H_4).

The reactions are

Chlorination reactor: $C_2H_4 + Cl_2 = C_2H_4Cl_2$

Oxyhydrochlorination reactor:

$$C_2H_4 + 2HCl + \frac{1}{2}O_2 = C_2H_4Cl_2 + H_2O$$

Pyrolysis reactor: $C_2H_4Cl_2 = C_2H_4Cl + HCl$.

Feed N_1 is 90 mol % ethylene and 10% inerts; N_2 is pure chlorine (Cl_2); N_3 is pure oxygen (O_2). All ethylene, oxygen, chlorine, and hydrochloric acid (HCl) fed to the chlorination and oxyhydrochlorination units react completely—these species should not be present in N_6 or N_7.

In the pyrolysis reactor, 50% of the dichloroethane ($C_2H_4Cl_2$) fed to the unit is converted. The unreacted dichloroethane is separated and recycled with the inerts in stream N_{12}. The inert concentration in the recycle stream is 50 mol %. Pure hydrochloric acid is recycled in stream N_{13}. The final product stream, N_{12}, contains only vinyl chloride and water. We want to solve for all species flow rates when $N_1 = 100$.

Chapter 5

Process Energy Balances

5.1 INTRODUCTION

In Chapters 3 and 4, we assembled material balances (MBs) for styrene and ammonia (chapter problems) processes using elementary unit operations. We now want to examine energy balances for these processes. The energy balance for any unit operation requires knowledge of stream flow rates and stream enthalpies. Stream enthalpy depends on composition, temperature, and pressure. Energy balances can alter our sequential modular approach. This will be shown in the ammonia process when we replace the elementary component separator with a temperature-dependent flash unit. The ammonia process with a flash will also be solved using the simultaneous approach, which requires equation linearization. For both the styrene and ammonia processes, we will explore the need for energy balances to allow reactor design and sizing. Reactor design and sizing results in coupled ordinary differential equations (ODEs), and here we will develop both Euler's method and the Runge–Kutta (RK) fourth-order method as solution methods.

The energy balance (or first law of thermodynamics) can be written for an open flowing system as depicted in Figure 5.1 as

$$\begin{pmatrix}\text{Rate of energy}\\ \text{accumulation}\\ \text{in the system}\end{pmatrix} = \begin{pmatrix}\text{Rate of energy}\\ \text{entering the system}\\ \text{by the inlet stream}\end{pmatrix} - \begin{pmatrix}\text{Rate of energy}\\ \text{leaving the system}\\ \text{by the exit stream}\end{pmatrix} + \begin{pmatrix}\text{Rate of heat added}\\ \text{to the system}\\ \text{by the surroundings}\end{pmatrix} + \begin{pmatrix}\text{Rate of work}\\ \text{done on the system}\\ \text{by the surroundings}\end{pmatrix} \tag{5.1}$$

or

$$\frac{dU}{dt} = F_{\text{in}}(\hat{u}_{\text{in}}) - F_{\text{out}}(\hat{u}_{\text{out}}) + \dot{Q} + \dot{W}_T, \tag{5.2}$$

where U is the total internal energy (joule) of the system (the vessel contents), F_{out} is the flow rate out (kilogram per second), $\hat{u}_{\text{out}}$ is the specific internal energy at the vessel exit (joule per kilogram), and $\dot{Q}$ and $\dot{W}_T$ are the heat and work rate terms

$$\left(\text{here } \dot{Q} = \frac{dQ}{dt} \quad \text{and} \quad \dot{W}_T = \frac{dW_T}{dt}\right).$$

In writing Equation (5.2), the kinetic and potential energy terms, entering and leaving, have been neglected.

The total work rate can be written as

$$\dot{W}_T = \dot{W}_{\text{flow}} + \dot{W}_{\text{shaft}} + \dot{W}_{\text{boundary}}, \quad \frac{J}{s}, \tag{5.3}$$

where $\dot{W}_{\text{flow}}$ is the rate of flow work to move the fluid in and out of the vessel as $(F_{\text{in}} P_{\text{in}} \hat{\upsilon}_{\text{in}} - F_{\text{out}} P_{\text{out}} \hat{\upsilon}_{\text{out}})$ and where $\hat{\upsilon}_{\text{out}}$ is the exiting fluid specific volume; $\dot{W}_{\text{shaft}}$ is the rate of shaft work, and

$$\dot{W}_{\text{boundary}} = -P\frac{dV_r}{dt}$$

is the rate of work due to a change in volume or expansion of the system.

Recalling that $\hat{u} = \hat{h} - P\hat{\upsilon}$, we can write Equation (5.3) as

$$\frac{dU}{dt} = F_{\text{in}}\left(\hat{h}_{\text{in}} - P_{\text{in}}\hat{\upsilon}_{\text{in}}\right) - F\left(\hat{h}_{\text{out}} - P_{\text{out}}\hat{\upsilon}_{\text{out}}\right) + \dot{Q} + \dot{W}_T, \tag{5.4}$$

where $\hat{h}$ is the specific enthalpy in the exiting stream. Incorporating the definition for flow work in Equation (5.4) and defining

$$\dot{W} = \dot{W}_{\text{shaft}} - P\frac{dV_r}{dt}$$

Modeling, Analysis and Optimization of Process and Energy Systems, First Edition. F. Carl Knopf.

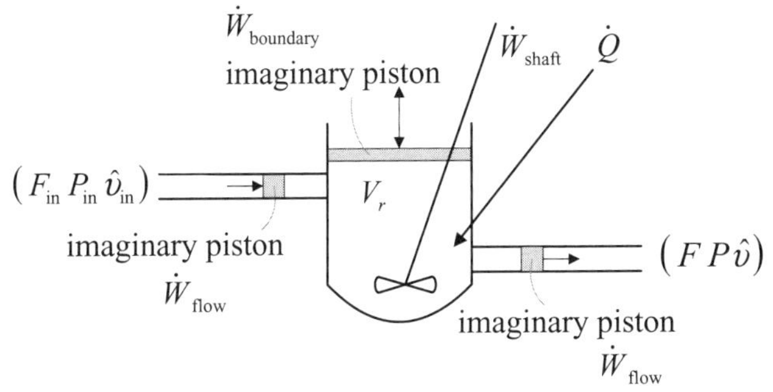

Figure 5.1 Open system for energy balance equation development.

as the combination of shaft and boundary work, we can write

$$\frac{dU}{dt} = F_{\text{in}}\left(\hat{h}_{\text{in}}\right) - F_{\text{out}}\left(\hat{h}_{\text{out}}\right) + \dot{Q} + \dot{W}. \tag{5.5}$$

A final simplification often used when applying Equation (5.5) is to neglect $P(dV_r/dt)$ and to simply set $\dot{W} = \dot{W}_{\text{shaft}}$. The development of Equation (5.5) has been completely general for any open system, including systems with reaction.

At steady state, the composition and temperature and flow rates in and out of the vessel will not change with time, and we can set the left-hand side (LHS) of Equation (5.5) to be zero, giving

$$0 = F_{\text{in}}\left(\hat{h}_{\text{in}}\right) - F_{\text{out}}\left(\hat{h}_{\text{out}}\right) + \dot{Q} + \dot{W}. \tag{5.6}$$

In forming Equation (5.1), the standard convention is to define dQ/dt as the heat added to the system by the surroundings. In Equation (5.1), it is common in many texts to define the work term to be the rate of work done *by* the system on the surroundings. Here then, the equation analogous to Equation (5.6) would be

$$0 = F_{\text{in}}\left(\hat{h}_{\text{in}}\right) - F_{\text{out}}\left(\hat{h}_{\text{out}}\right) + \dot{Q} - \dot{W}. \tag{5.7}$$

For computer applications, it is useful to continue the discussion of energy balances in the context of the process elementary modules introduced in Section 3.1. The elementary modules from Section 3.1 were a mixer, a separator, a splitter, and a reactor, and we will now add a heat–work module (Myers and Seider, 1976; Reklaitis, 1983). We will assume that all stream pressures are known, either specifically as stream pressure or as inlet stream pressure and known pressure drop (ΔP) across a given operation. Following our development of Section 3.1 and using Equation (5.6), the energy balance for each of the modules can be written:

Mixer—sums incoming streams:

$$\sum_{j=1}^{n_{\text{streams in}}} N_j h_j - N_{\text{out}} h_{\text{out}} = 0. \tag{5.8}$$

In the energy balance Equation (5.8), N_j is the total molar flow rate in each stream j, and h_j is the total enthalpy (joule per mole) in each stream j. Here, we have converted from a mass basis to a molar basis to match our development in Chapter 3. We have assumed stream mixing occurs adiabatically ($\dot{Q} = 0$, no energy exchange with the environment) and no work $(\dot{W} = 0)$ is done. The heat and work terms can be added (to any module) if appropriate. Also recall in the development of Equation (5.8) that potential and kinetic energy terms were neglected.

Separator—splits incoming stream into streams of different compositions:

$$N_{\text{in}} h_{\text{in}} - \sum_{j=1}^{n_{\text{streams out}}} N_j h_j + \frac{dQ}{dt} = 0. \tag{5.9}$$

Equation (5.9) accounts for energy exchange with the environment but assumes no work is done.

Splitter—splits incoming stream into streams of constant composition:

$$N_{\text{in}} h_{\text{in}} - \sum_{j=1}^{n_{\text{streams out}}} N_j h_j + \frac{dQ}{dt} = 0. \tag{5.10}$$

The enthalpy of the incoming stream, h_{in}, will be the same as the enthalpy in each splitter output streams, h_j. Equation (5.10) accounts for any energy exchange with the environment but assumes no work is done.

Reactors—feed stream is converted to a product stream with composition change and possible molar flow rate change. If the reactor operates adiabatically (no energy exchange with the surroundings) and no work is done, and there is single feed stream and output stream, we can write

$$N_{\text{out}} h_{\text{out}} - N_{\text{in}} h_{\text{in}} = 0. \tag{5.11}$$

This reactor energy balance (Eq. (5.11)) requires an explanation as you may be wondering about the heat of the reaction term. In Equation (5.11), we are implicitly writing the enthalpy balances in terms of the species enthalpy of formation from its elements. Using the species enthalpy of formation, and with all species having the same common reference state, eliminates the need for any heat of reaction information. Assuming the streams form ideal mixtures and the stream is a single phase, we can write Equation (5.11) as

$$\sum_{i=1}^{n_{\text{species}}} N_i^{\text{out}} h_i^{\text{out}} - \sum_{i=1}^{n_{\text{species}}} N_i^{\text{in}} h_i^{\text{in}} = 0. \tag{5.12}$$

Here, our enthalpy terms can be written as

$$h_i(T) = h_{f,i}^0 + \int_{T_{\text{ref}}}^{T} C_{p,i} dT, \quad \frac{\text{J}}{\text{mol}}, \tag{5.13}$$

where $h_{f,i}^0$ is the species standard molar enthalpy of formation from its elements at the standard reference conditions of $T_{\text{ref}}^0 = 298.15\text{ K}$ and $P_{\text{ref}}^0 = 1\text{ bar}$, and $C_{p,i}$ is the species molar heat capacity.

The reactor energy balance equation is also commonly written with the heat of reaction. To do this, we can rewrite Equation (5.12) with the reference state for each species in each stream both subtracted and added to each term:

$$\begin{aligned}&\sum_{i=1}^{n_{\text{species}}} N_i^{\text{out}}\left(h_i^{\text{out}} - h_{i,\text{out}}^{\text{ref}}\right) - \sum_{i=1}^{n_{species}} N_i^{\text{in}}\left(h_i^{\text{in}} - h_{i,\text{in}}^{\text{ref}}\right)\\ &+ \sum_{i=1}^{n_{\text{species}}} N_i^{\text{out}}\left(h_{i,\text{out}}^{\text{ref}}\right) - \sum_{i=1}^{n_{\text{species}}} N_i^{\text{in}}\left(h_{i,\text{in}}^{\text{ref}}\right) = 0,\end{aligned} \tag{5.14}$$

which can also be written as

$$\begin{aligned}&\sum_{i=1}^{n_{\text{species}}} N_i^{\text{out}}\left(h_i^{\text{out}} - h_i^{\text{ref}}\right) - \sum_{i=1}^{n_{\text{species}}} N_i^{\text{in}}\left(h_i^{\text{in}} - h_i^{\text{ref}}\right)\\ &+ \sum_{i=1}^{n_{\text{species}}} \left(N_i^{\text{out}} - N_i^{\text{in}}\right)\left(h_i^{\text{ref}}\right) = 0.\end{aligned} \tag{5.15}$$

Notice that if no reaction occurs, $N_i^{\text{in}} = N_i^{\text{out}}$ and the last term on the LHS is zero. Our reactor material balance equation from Chapter 3 (Eq. (3.26)) written for the case of a single reaction is

$$N_i^{\text{out}} = N_i^{\text{in}} + \xi \nu_i.$$

Using this equation, the last term on the LHS of Equation (5.15) can be written as

$$\sum_{i=1}^{n_{\text{species}}} \left(N_i^{\text{out}} - N_i^{\text{in}}\right)\left(h_i^{\text{ref}}\right) = \sum_{i=1}^{n_{\text{species}}} \xi \nu_i \left(h_i^{\text{ref}}\right). \tag{5.16}$$

Here, the grouping $\sum_{i=1}^{n_{\text{species}}} \nu_i\left(h_i^{\text{ref}}\right)$ is defined to be the heat of reaction (ΔH_r) at the reference temperature and pressure and with stoichiometric coefficients ν_i. So, for the case of a single reaction,

$$\sum_{i=1}^{n_{\text{species}}} \nu_i\left(h_i^{\text{ref}}\right) = \Delta H_r, \tag{5.17}$$

and using Equations (5.16) and (5.17), we can write our reactor energy balance as

$$\sum_{i=1}^{n_{\text{species}}} N_i^{\text{out}}\left(h_i^{\text{out}} - h_i^{\text{ref}}\right) + \xi \Delta H_r - \sum_{i=1}^{n_{\text{species}}} N_i^{\text{in}}\left(h_i^{\text{in}} - h_i^{\text{ref}}\right) = 0. \tag{5.18}$$

A convenient form accounting for multiple reaction can be written as

$$N_{\text{out}} h_{\text{out}} + \sum_{m=1}^{n_{\text{reactions}}} \xi_m \Delta H_{r,m} - N_{\text{in}} h_{\text{in}} = 0. \tag{5.19}$$

Enthalpy values are relative to the chosen reference state so there is an implicit reference state in Equation (5.19); the same reference state must be used for each term in the equation. In Equation (3.26), the choice of reaction stoichiometry determined the values for ξ_m, the extents of reaction for each reaction. The values for $\Delta H_{r,m}$, the heats of reaction at the system reference T and P, will be linked to the reaction stoichiometry. The choice of reference state is arbitrary, but we will need to form a thermodynamic path to adjust $\Delta H_{r,m}$ values if the reference state is changed from the standard reference state of $T_{\text{ref}}^0 = 298.15\text{ K}$ and $P_{\text{ref}}^0 = 1\text{ bar}$ (this is detailed in Example 5.5). The following simple example helps show the connection between ΔH_r and $h_{f,i}^0$.

EXAMPLE 5.1 *Determine the Outlet Temperature for a Hydrogen Combustor*

Consider the hydrogen + oxygen reaction shown in Figure 5.2. We want to determine the outlet temperature using Equations (5.12) and (5.19). Needed data are provided in the following table and for the reaction

$$H_2 + \frac{1}{2} O_2 \rightarrow H_2O,$$

$\Delta H_r = -57791.65$ cal/mol H_2 reacted at standard reference conditions, $T = 298.15$ K and $P = 1$ bar:

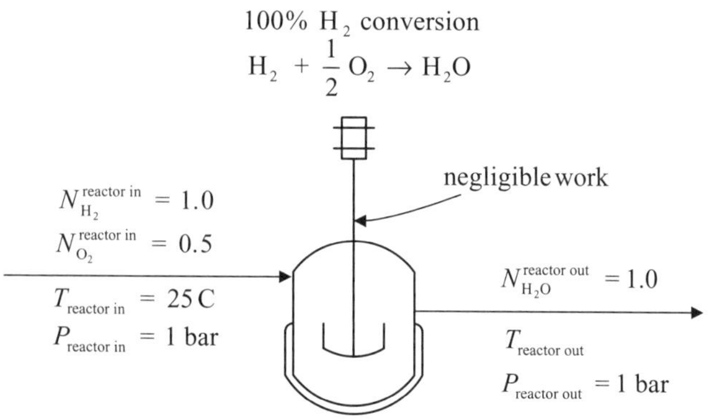

Figure 5.2 Flow sheet for the hydrogen + oxygen reaction.

Species	$C_{p,i}^{\text{average}}$ (cal/mol-K)	$h_{f,i}^0$ (cal/mol)
H_2	7.2	0
O_2	8.2	0
H_2O (vapor)	9.8	−57,791.65

In this problem, standard reference conditions are used for both ΔH_r and $h_{f,i}^0$ (25°C and 1 bar).

The energy balance for Equation (5.11), and where the streams form ideal mixtures, can be written as

$$N_{\mathrm{H_2O}}^{\mathrm{out}}\left(h_{f,\mathrm{H_2O}}^0+\int_{T_{\mathrm{ref}}}^{T_{\mathrm{out}}} C_{p,\mathrm{H_2O}}\,dT\right)-N_{\mathrm{H_2}}^{\mathrm{in}}\left(h_{f,\mathrm{H_2}}^0+\int_{T_{\mathrm{ref}}}^{T_{\mathrm{in}}} C_{p,\mathrm{H_2}}\,dT\right)$$
$$-N_{\mathrm{O_2}}^{\mathrm{in}}\left(h_{f,\mathrm{O_2}}^0+\int_{T_{\mathrm{ref}}}^{T_{\mathrm{in}}} C_{p,\mathrm{O_2}}\,dT\right)=0,$$

and using average heat capacity values,

$$(1)(-57791.65+9.8(T_{\mathrm{out}}-T_{\mathrm{ref}}))-(1)(0+7.2(T_{\mathrm{in}}-T_{\mathrm{ref}}))$$
$$-(0.5)(0+8.2(T_{\mathrm{in}}-T_{\mathrm{ref}}))=0.$$

For Equation (5.19),

$$N_{\mathrm{H_2O}}^{\mathrm{out}}\left(\int_{T_{\mathrm{ref}}}^{T_{\mathrm{out}}} C_{p,\mathrm{H_2O}}\,dT\right)+\xi\Delta H_r-N_{\mathrm{H_2}}^{\mathrm{in}}\left(\int_{T_{\mathrm{ref}}}^{T_{\mathrm{in}}} C_{p,\mathrm{H_2}}\,dT\right)$$
$$-N_{\mathrm{O_2}}^{\mathrm{in}}\left(\int_{T_{\mathrm{ref}}}^{T_{\mathrm{in}}} C_{p,\mathrm{O_2}}\,dT\right)=0$$

$$(1)(9.8(T_{\mathrm{out}}-T_{\mathrm{ref}}))+(1)(-57791.65)-(1)(7.2(T_{\mathrm{in}}-T_{\mathrm{ref}}))$$
$$-(0.5)(8.2(T_{\mathrm{in}}-T_{\mathrm{ref}}))=0.$$

You can see the calculated outlet temperature will be identical from both equations. It is also straightforward to see the connection between ΔH_r and $h_{f,i}^0$ (25°C and 1 bar). ■

Chemical engineers generally use the energy balance with heat of reactions. However, when there are more than three or four reactions, the energy balance utilizing species enthalpy of formation will prove more convenient. This is especially true in Chapter 15 when solving for species emissions and outlet temperature from combustion systems; here, there may be 50+ species and over 300 reactions.

The heat of reaction is defined to be negative for an exothermic reaction and positive for an endothermic reaction. Often when using the heat of reaction energy balance formulation, it is convenient to talk about the heat *generated* by the reaction. For an adiabatic reactor (and when using the heat of reaction formulation), the heat *generated* by reaction is the difference in the total enthalpy in the exiting stream minus the total enthalpy in the feed stream, which, for the case of a single reaction, can be written as

$$\text{Heat } generated \text{ by reaction} = N_{\mathrm{out}}h_{\mathrm{out}}-N_{\mathrm{in}}h_{\mathrm{in}}=-\xi\Delta H_r$$
$$=\xi(-\Delta H_r). \qquad (5.20)$$

Heat–Work

For processing units where work or heat is transferred and where $N_{\mathrm{out}} = N_{\mathrm{in}}$, the energy balance can be written as

$$N_{\mathrm{in}}h_{\mathrm{in}}+\left(\frac{dQ}{dt}+\frac{dW}{dt}\right)-N_{\mathrm{out}}h_{\mathrm{out}}=0. \qquad (5.21)$$

For chemical engineering, the two operations of perhaps greatest importance are reactors and species separation—react raw materials to product, separate out and sell the product, and recycle any unreacted materials back to the reactor. Next, we will show how equilibrium flash separation processes and reactors can be solved in computer-aided calculations. Do recall that our approach for chemical processes is somewhat general in nature due to the wide variety of chemical process species and operating conditions—we basically want to show the keys to creating your own simulation or for better understanding commercial flow sheet packages. Our approach in the heat and power chapters is more detailed as heat and power applications show much greater uniformity in almost all applications. But even with our more general approach to chemical processing computer-aided design, it is very instructive to see the complications that arise when we try to incorporate flash units within recycle loops.

5.2 SEPARATOR: EQUILIBRIUM FLASH

One example of a species or component separator is an equilibrium flash as shown in Figure 5.3. The equilibrium flash is not only an important individual operation but flash units can also be combined in series to represent distillation operations.

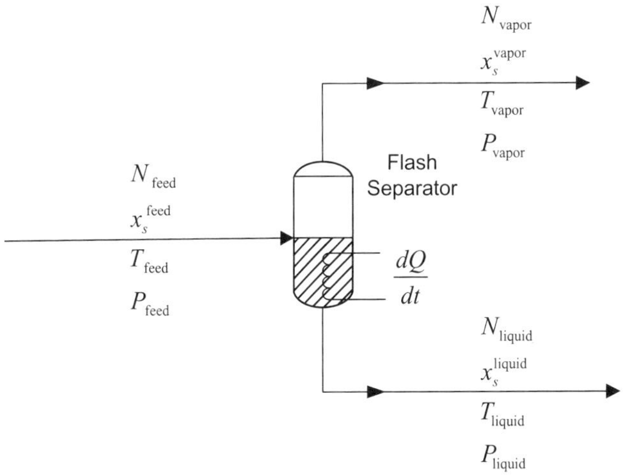

Figure 5.3 Vapor–liquid equilibrium flash.

In a multispecies flash operation, a single-phase feed stream (generally liquid) is flashed to a lower pressure in order to separate light species (predominately in the vapor product) from heavier species (predominately in the liquid product). The flash outlet streams are assumed to be in equilibrium. At equilibrium, these outlet streams will be the same temperature and pressure, which will also be the temperature and pressure in the flash vessel.

There are three main types of equilibrium flash units: an isothermal flash, an adiabatic flash, and a general flash.

Isothermal Flash. An isothermal flash shows $T_{\text{feed}} = T_{\text{vapor}} = T_{\text{liquid}}$ with both inlet pressure $P_{\text{in}} = P_{\text{feed}}$ and outlet pressure $P_{\text{out}} = P_{\text{vapor}} = P_{\text{liquid}}$ specified. An energy balance is required to determine the amount of energy (dQ/dt) that must be supplied to the system to maintain the outlet stream temperatures. Here, the material balance can be decoupled from the energy balance; in other words, first, the material balance can be solved (independent of the energy balance) and then the energy balance can be solved.

Adiabatic Flash. An adiabatic flash shows $T_{\text{feed}} \neq T_{\text{vapor}} = T_{\text{liquid}}$ with both inlet pressure $P_{\text{in}} = P_{\text{feed}}$ and outlet pressure $P_{\text{out}} = P_{\text{vapor}} = P_{\text{liquid}}$ specified. An energy balance with $dQ/dt = 0$ is required to determine the exiting stream temperature. Here, the energy balance cannot be decoupled from the material balance.

General Flash. A general flash shows $T_{\text{feed}} \neq T_{\text{vapor}} = T_{\text{liquid}}$, and again inlet pressure $P_{\text{in}} = P_{\text{feed}}$ and outlet pressure $P_{\text{out}} = P_{\text{vapor}} = P_{\text{liquid}}$ are specified. In the general flash, either the outlet temperature will be specified or a value for dQ/dt will be specified. If $dQ/dt =$ a specified value, the energy balance is required to determine the exiting streams temperature. If the outlet temperature is specified, the energy balance is required to determine the amount of energy (dQ/dt) that must be supplied to the system to obtain the outlet temperature.

For flash calculations, when $dQ/dt = 0$ or dQ/dt is specified, the energy balance cannot be decoupled from the material balance. For these flash units, an initial exit temperature is assumed and the flash solved as in the isothermal case. If (dQ/dt) does not equal the required value, temperature iteration must be performed.

All flash calculations require solution of material balances and energy balances. The solution of the material balance equation (Eq. (3.13)) and the energy balance equation (Eq. (5.9)) will require that the distribution of each species between the vapor and liquid phases be known. In Equation (3.13), we accounted for species distribution with $\alpha_{i,j}$ values. In equilibrium flash calculations, $\alpha_{i,j}$ can be determined using an equilibrium distribution factor for each species, K_i, which is defined as

$$K_i = \frac{x_i^{\text{vapor}}}{x_i^{\text{liquid}}} \quad \text{or} \quad K_i = \frac{y_i^{\text{vapor}}}{x_i^{\text{liquid}}} \quad i = 1, \ldots, n_{\text{species}}. \tag{5.22}$$

Generally y_i^{vapor} is used as the species mole fraction in the vapor phase, and x_i^{liquid} is the species mole fraction in the liquid phase. Thermodynamic equations of state (Chapter 8) can be used to determine K_i, as K_i is function of the flash temperature and pressure and the feed composition. The procedure for determining K_i values from an equation of state is detailed in Problem 8.5.

The Flash Material Balance Problem

By using Figure 5.3, we can write the material balance for each species, i, as

$$N_{\text{feed}} x_i^{\text{feed}} = N_{\text{vapor}} y_i^{\text{vapor}} + N_{\text{liquid}} x_i^{\text{liquid}} \quad i = 1, \ldots, n_{\text{species}}; \tag{5.23}$$

from Equation (5.22),

$$y_i^{\text{vapor}} = K_i x_i^{\text{liquid}} \quad i = 1, \ldots, n_{\text{species}}. \tag{5.24}$$

Substituting Equation (5.24) into Equation (5.23) and rearranging for x_i^{liquid} and y_i^{vapor} gives

$$x_i^{\text{liquid}} = \frac{N_{\text{feed}} x_i^{\text{feed}}}{(N_{\text{vapor}} K_i + N_{\text{liquid}})} \quad i = 1, \ldots, n_{\text{species}} \tag{5.25}$$

and

$$y_i^{\text{vapor}} = \frac{N_{\text{feed}} x_i^{\text{feed}} K_i}{(N_{\text{vapor}} K_i + N_{\text{liquid}})} \quad i = 1, \ldots, n_{\text{species}}. \tag{5.26}$$

Using $\sum_{i=1}^{n_{\text{species}}} x_i^{\text{liquid}} = 1$ and Equation (5.25) gives

$$\sum_{i=1}^{n_{\text{species}}} x_i^{\text{liquid}} = 1 = \sum_{i=1}^{n_{\text{species}}} \frac{N_{\text{feed}} x_i^{\text{feed}}}{(N_{\text{vapor}} K_i + N_{\text{liquid}})}, \tag{5.27}$$

and $\sum_{i=1}^{n_{\text{species}}} y_i^{\text{vapor}} = 1$ with Equation (5.26) gives

$$\sum_{i=1}^{n_{\text{species}}} y_i^{\text{vapor}} = 1 = \sum_{i=1}^{n_{\text{species}}} \frac{N_{\text{feed}} x_i^{\text{feed}} K_i}{(N_{\text{vapor}} K_i + N_{\text{liquid}})}. \tag{5.28}$$

Subtracting Equation (5.26) from Equation (5.25) and using from the overall balance $N_{\text{liquid}} = N_{\text{feed}} - N_{\text{vapor}}$

$$\sum_{i=1}^{n_{\text{species}}} \frac{x_i^{\text{feed}}(1 - K_i)}{1 + \dfrac{N_{\text{vapor}}}{N_{\text{feed}}}(K_i - 1)} = 0. \tag{5.29}$$

In Equation (5.29), there is only one unknown, N_{vapor}, the molar vapor flow rate; here, x_i^{feed}, N_{feed} and K_i are known. The quantity

$$\frac{N_{vapor}}{N_{feed}} \text{ (mole vapor per mole feed)}$$

is the total vapor flow ratio, and this quantity must fall between 0 and 1. It is convenient to define ν as the total vapor flow ratio:

$$\nu = \frac{N_{vapor}}{N_{feed}},$$

giving

$$\sum_{i=1}^{n_{species}} \frac{x_i^{feed}(1-K_i)}{1+\nu(K_i-1)} = 0 \quad \text{or} \quad \sum_{i=1}^{n_{species}} \frac{x_i^{feed}(K_i-1)}{1+\nu(K_i-1)} = 0. \qquad (5.30)$$

The solution of Equation (5.30) for ν requires an iterative procedure; here, Solver or Goal Seek can be used. With ν determined, $N_{vapor} = \nu N_{feed}$ and $N_{liquid} = N_{feed} - N_{vapor}$. The x_i^{liquid} values can be found using Equation (5.25) or

$$x_i^{liquid} = \frac{x_i^{feed}}{1+\nu(K_i-1)}, \qquad (5.31)$$

and the y_i^{vapor} values can be found from Equation (5.24). Once the flash equilibrium distribution problem has been solved, we can calculate $\alpha_{i,j}$ as

$$\alpha_i^{vapor} = \frac{N_{vapor}\, y_i^{vapor}}{N_{feed}\, x_i^{feed}} \quad \text{and} \quad \alpha_i^{liquid} = \frac{N_{liquid}\, x_i^{liquid}}{N_{feed}\, x_i^{feed}}.$$

5.2.1 Equilibrium Flash with Recycle: Sequential Modular Approach

EXAMPLE 5.2 *Ammonia Synthesis with General Flash*

In the chapter problems in Chapters 3 and 4, we solved material balance problems for an ammonia process with a small amount of methane and argon impurity. Now consider an ammonia plant with a feed stream containing 100 mol/h of nitrogen and 300 mol/h hydrogen (with no impurities). This feed is catalytically reacted to produce ammonia. The reactor operates at high pressure (≈2000 psia) with 25% conversion of nitrogen per pass in the reactor. The reaction is $N_2 + 3H_2 = 2NH_3$. To recover ammonia, reaction products are refrigerated and separated to −28°F in a general equilibrium flash. Products that are not condensed are recycled back to the reactor and part of the recycle gas is purged. This process is shown in Figure 5.4.

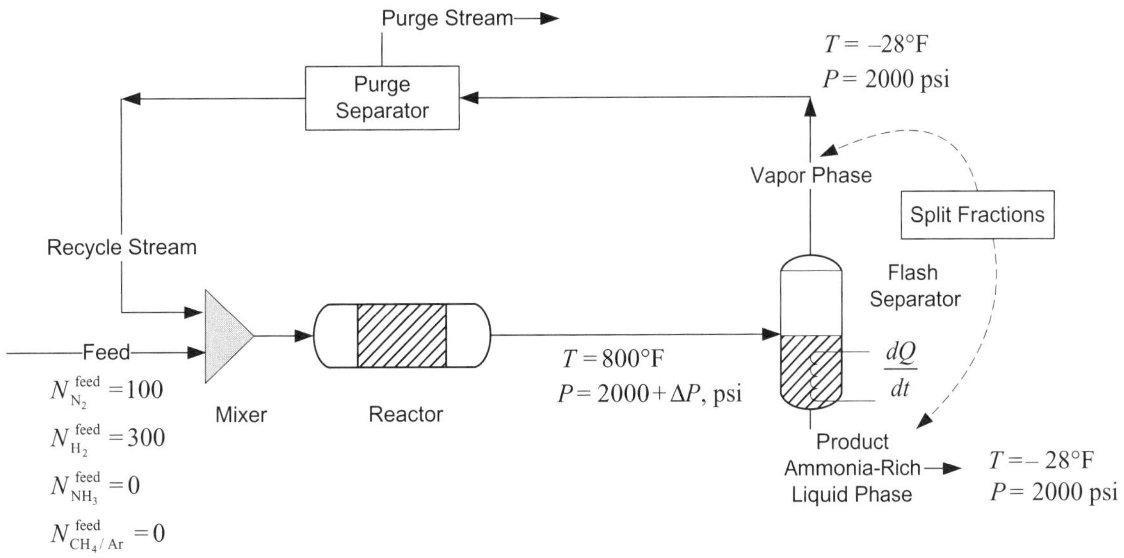

Figure 5.4 Ammonia process with equilibrium flash.

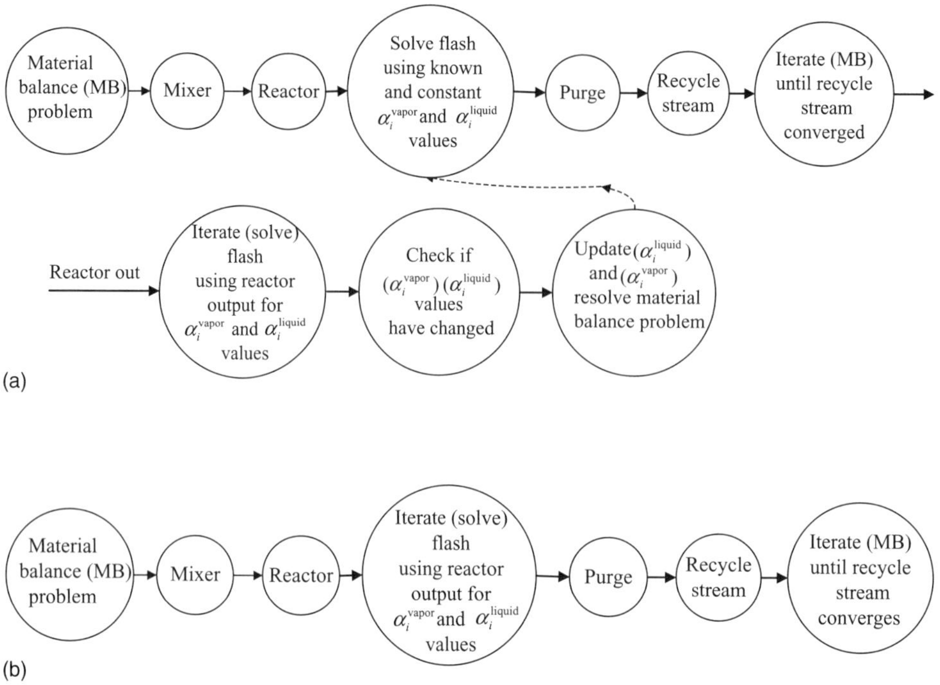

Figure 5.5 (a) Approach 1: solution strategy with flash updating calculations outside the recycle loop. (b) Approach 2: conceptual solution strategy with flash updating calculations inside the recycle loop.

In Example 5.2, the outlet temperature of the flash is known and we will solve the energy balance *after* the material balance problem is solved. Figure 5.5 shows two possible solution approaches to this problem. Both approaches follow our developments in Chapter 3 in which we want to iterate on the recycle loop until the species flow rates have converged. In *approach 1* (Figure 5.5a), we solve the mixer and reactor material balances, and for the flash unit, we assume values for α_i^{vapor} and α_i^{liquid}; for example, to start, we could assume the species split fractions for the flash separator are $\alpha_{N_2}^{liquid} = 0.005$, $\alpha_{H_2}^{liquid} = 0.005$, and $\alpha_{NH_3}^{liquid} = 0.98$. The purge is then solved and the recycle stream updated. We continue to iterate with these assumed split fractions until the material balance problem has converged. After the material balance has converged, we use the updated species flow rates into the flash (reactor out) and solve Equations (5.30) and (5.31) for new α_i^{vapor} and α_i^{liquid} values for the flash (solving the flash is an iterative process). These values are returned to the material balance problem, which is solved again. This process is continued until the α_i^{vapor} and α_i^{liquid} values for the flash have converged.

In the second solution approach, *approach 2* (Figure 5.5b), we solve the mixer and reactor material balances, but for the flash unit, we use the current species flow rates into the flash (reactor out) and solve Equations (5.30) and (5.31) for α_i^{vapor} and α_i^{liquid} values for the flash (solving the flash is an iterative process). The purge is then solved and the recycle stream updated. Each time the flash unit is encountered, new values for α_i^{vapor} and α_i^{liquid} are determined. This process continues until the material balance problem converges.

In approach 1, flash unit updating is "outside" the converged recycle loop and this procedure will *not* work. In approach 2, the flash unit updating is part of the recycle calculations (often termed "inside" the recycle loop) and, as we will see this procedure, will work. It is important to appreciate the difference in these calculations—inside versus outside the recycle loop. Recall in Section 3.4 that when we supplied alternative specifications to a flow sheeting problem (the molar feed rate of water to the process was replaced with a water molar composition entering the reactor), we solved for the missing natural specification, the water feed rate, outside the recycle loop.

Here, Example 5.2 will be solved in three steps. First, we will solve the problem using assumed species split fractions and the modular approach of Chapter 3. In step 2, we will solve the flash as a stand-alone unit with new, updated feed conditions. Finally, we will incorporate the flash into the flow sheet using the two approaches shown in Figure 5.5.

Step 1: Approach 1 Ammonia Process: Sequential Modular Solution with Species Split Fractions

Let us first solve this problem just as we did in Chapter 3. In order to construct a "Chapter 3" sequential modular solution using elementary units, we must have the species split fractions for the separators. Here, assume the species split fractions for the flash separator following the reactor are $\alpha_{N_2}^{liquid} = 0.005$, $\alpha_{H_2}^{liquid} = 0.005$, and $\alpha_{NH_3}^{liquid} = 0.98$, and the species split fractions for the purge separator are $\alpha_{N_2}^{purge} = 0.01$, $\alpha_{H_2}^{purge} = 0.01$, and $\alpha_{NH_3}^{purge} = 0.01$. For this example, assume the

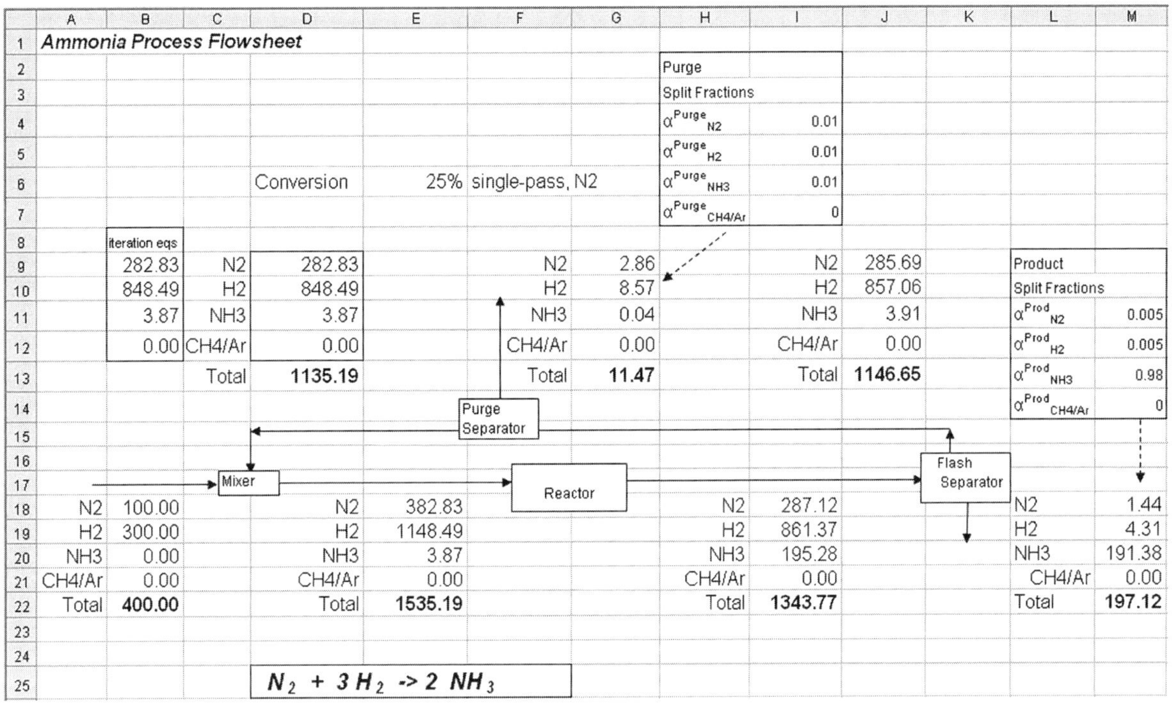

Figure 5.6 Solution of ammonia flow sheet with known species split fractions.

methane and argon impurities in the feed = 0 mol/h. Later, in Problem 5.5, we will explore the effect of feed impurities. The solution is provided in **Example 5.2a.xls** and the results are shown in Figure 5.6.

Step 2: Approach 1 Ammonia Process—Stand-Alone Flash Module

Next, we want to solve the flash as a stand-alone unit using Equations (5.22)–(5.31). These equations require that the feed condition is known and that the vapor–liquid equilibrium distribution factor K_i is known for each species. Generally, an equation of state is used to determine K_i (we detail this in Problem 8.5). In some cases, simplifications such as ideal solution (Raoult's law) or low concentration (Henry's law) can be used for the determination of K_i. Myers and Seider (1976) report vapor–liquid equilibrium factors for the ammonia process at −28°F and 2000 psia as $K_{N_2} = 66.67$, $K_{H_2} = 50.0$, and $K_{NH_3} = 0.015$.

With the K_i values from Myers and Seider (1976) and with the feed composition given by the converged stream out of the reactor determined in step 1, the flash can be solved. From Equation (5.30),

$$\sum_{i=1}^{n_{\text{species}}} \frac{x_i^{\text{feed}}(K_i-1)}{1+\nu(K_i-1)} = \frac{x_{\text{N}_2}^{\text{feed}}(K_{\text{N}_2}-1)}{1+\nu(K_{\text{N}_2}-1)} + \frac{x_{\text{H}_2}^{\text{feed}}(K_{\text{H}_2}-1)}{1+\nu(K_{\text{H}_2}-1)} + \frac{x_{\text{NH}_3}^{\text{feed}}(K_{\text{NH}_3}-1)}{1+\nu(K_{\text{NH}_3}-1)} = 0,$$

and from Figure 5.6, the flow rates out of the reactor are $N_{\text{N}_2}^{\text{reactor out}} = 287.12$ moles/h, $N_{\text{H}_2}^{\text{reactor out}} = 861.37$, $N_{\text{NH}_3}^{\text{reactor out}} = 195.28$, $N_{\text{CH}_4/\text{Ar}}^{\text{reactor out}} = 0$, and $N_{\text{rector out}} = 1343.772$, giving

$$\sum_{i=1}^{n_{\text{species}}} \frac{x_i^{\text{feed}}(K_i-1)}{1+\nu(K_i-1)} = \frac{\frac{287.12}{1343.77}(66.67-1)}{1+\nu(66.67-1)} + \frac{\frac{861.37}{1343.77}(50-1)}{1+\nu(50-1)} + \frac{\frac{195.28}{1343.77}(0.015-1)}{1+\nu(0.015-1)} = 0.$$

We use Excel Solver to solve for ν, the vapor flow ratio, then Equation (5.31) to solve for x_i^{liquid} and Equation (5.24) for y_i^{vapor}. The solution to the stand-alone flash problem is provided in **Example 5.2b.xls** and the results are shown in Figure 5.7.

In the Excel sheet, Equation (5.30) is solved for each species in cells J19:J22, and the results are summed in J24. Excel Solver is used to vary ν (I24), so the sum (J24) = 0. Equation (5.31) is used to determine x_i^{liquid} values in cells K19:K22, and Equation (5.24) is used to determine y_i^{vapor} values in cells L19:L22. Product and vapor species flow rates and split fractions can then be determined for the flash process.

	A	B	C	D	E	F	G	H	I	J	K	L	M	N
11								Vapor Out						
12								Split Fractions						
13					N2	286.45		α^{VO}_{N2}	0.99766287					
14					H2	858.69		α^{VO}_{H2}	0.9968861					
15					NH3	17.11		α^{VO}_{NH3}	0.08762636					
16					CH4/Ar	0.00		$\alpha^{VO}_{CH4/Ar}$	"0"					
17					Total	1162.25								
18										f(v)	x_s Liquid	y_s Vapor	V Flows	L Flows
19					Flash Separator			KN2	66.67	0.2427643	0.003696731	0.246461	286.448964	0.67103633
20		N2	287.12					KH2	50	0.7240397	0.014776321	0.738816	858.687779	2.68222072
21		H2	861.37					KNH3	0.015	-0.9668042	0.981527085	0.014723	17.1116749	178.168325
22		NH3	195.28		N2	0.67		KCH4/Ar	0	0	0	0	0	0
23		CH4/Ar	0		H2	2.68								
24								ν =	0.864916	-1.581E-07	1.000000137	1	1162.24842	181.521582
25		Total	1343.77		NH3	178.17		N^{vapor} =	1162.248					
26					CH4/Ar	0.00		N^{Liquid} =	181.5216	$\Sigma f(v)=0$	x_s Liquid			Σ L Flows
27					Total	181.52								
28								Product						
29								Split Fractions						
30								α^{Prod}_{N2}	0.002337					
31								α^{Prod}_{H2}	0.003114					
32								α^{Prod}_{NH3}	0.912374					
33								$\alpha^{Prod}_{CH4/Ar}$	"0"					

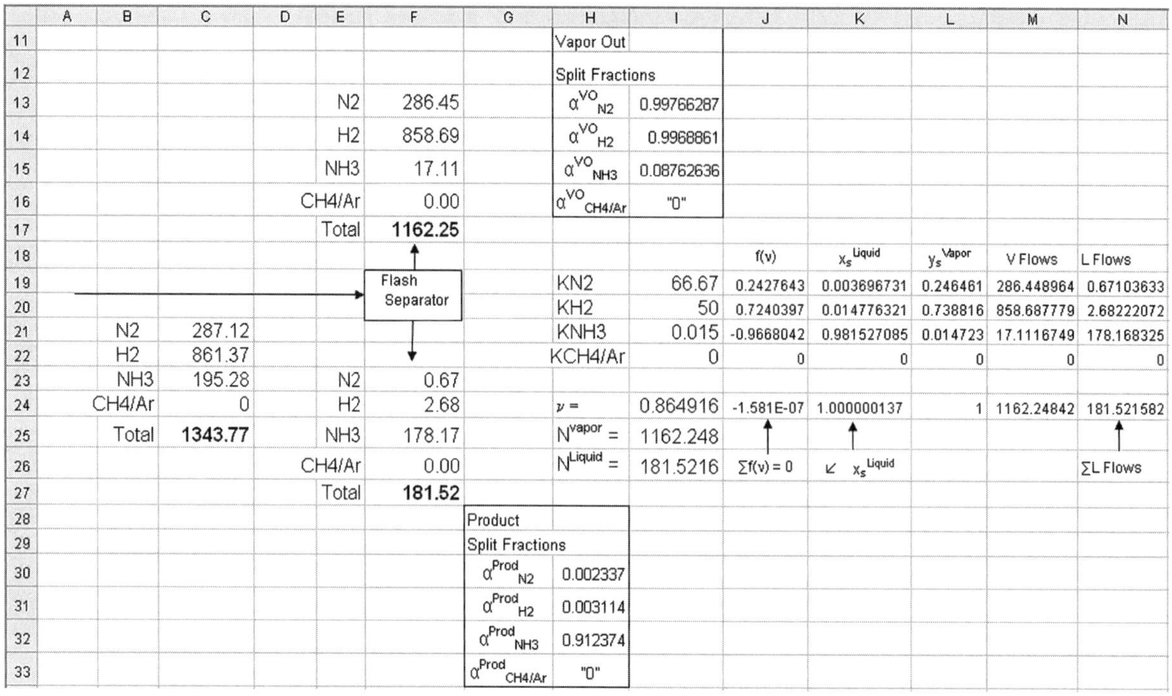

Figure 5.7   Solution of stand-alone flash problem—ammonia process.

Step 3: Ammonia Process: Sequential Modular with Flash

Finally, we want to incorporate the flash separator with the provided vapor–liquid equilibrium constant K_i into the flow sheet we developed in step 1. The conversion remains at 25% and the purge remains at 1%. This would seem to be a very minor modification, but there are immediate problems. We now have two iteration loops, one due to the recycle loop in the flow sheet and one due to the flash separator (Eq. (5.30)); these iteration loops have been addressed separately in steps 1 and 2, and here they must be combined.

Step 3: Approach 1 Flow Sheet and Solver Flash (Will *Not* Generate a Solution)

A straightforward approach as shown in Figure 5.5a is to directly combine our Solver flash solution developed in step 2 with our sequential modular solution developed in step 1. It is easy to show the problems with this idea by taking the split fractions determined from step 2 and using these as the known (fixed) values in our step 1 solution. From step 2, the split fractions from the flash would be $\alpha^{product}_{N_2} = 0.0023$, $\alpha^{product}_{H_2} = 0.0031$, and $\alpha^{product}_{NH_3} = 0.9124$. If you use these values in **Example 5.2a.xls**, we will quickly see that the values of the hydrogen flow rate in the recycle loop will become unstable. In the original problem (step 1), hydrogen and nitrogen are supplied in stoichiometric amounts and the problem specifies 25% conversion of nitrogen with $\alpha^{product}_{N_2} = 0.005$ and $\alpha^{product}_{H_2} = 0.005$. If equal amounts of hydrogen and nitrogen are

not separated in the separator following the reactor, one species will build in the recycle loop.

It may not be readily apparent, but in order to create successfully a solution to the recycle material balance problem containing any flash, we do need a stand-alone flash that can provide converged species split fractions (or species flow rates) leaving the flash for a given input. However, we do not want the material balance recycle loop to iterate using these split fractions. Each material balance loop (until the loop is converged) will change the input to the flash, and each new input to the flash requires that the flash distribution problem be solved again.

Step 3: Approach 2 Flow Sheet and Converged Flash (Will Generate a Solution)

Based on our discussion above, we want to construct a solution where the flash unit is converged for each new input to the flash. The converged species flow rates from the flash are then returned to the flow sheet and the material balance problem solved until the flash unit is encountered. Here then again, the species flow rates to the stand-alone flash are taken as fixed and this flash unit converged. This procedure is shown in Figure 5.5b.

The problem now is how to implement this solution strategy. Strategies that use Excel Solver to converge the flash are difficult, especially if flow sheet values are directly coupled to Solver. As Solver iterates on v for the flash, each

ν value will create an updated set of species split fractions (or species flows) leaving the flash. These updated species flows will actually be available to the material balance problem before the flash has converged. Recall from Example 3.10 that Solver will release and allow the flow sheet to converge for each trial point during the optimization. If these species flow values are used by the flow sheet before the flash has converged, Solver will be unable to find a solution. Each Solver iteration generates a new trial ν value, which in turn generates new species flows leaving the flash and in turn new inlet conditions to the flash. The inlet conditions to the flash (x_i^{feed} values) are used in the objective function (Eq. (5.30)) and no solution can be found. One easy solution is to use the Paste Values option in Excel to move needed reactor out species values to the flash, then converge the flash and again use the Paste Values option to move needed species flow rates to the Excel sheet material balances. This approach will work, but hundreds of Paste Values operations will be required. You can try different Solver-based solution methods using the provided file **Example 5.2c.xls**. ■

We need a better solution approach than using Solver with the Paste Values option in Excel. In commercial computer-aided design packages, each module in the sequential modular approach will be coded as a subprogram. These subprograms are then assembled to represent the process flow sheet. We can use a combination approach, where part of the flow sheet problem can be solved on the Excel sheet and part of the problem (the equilibrium flash) can be solved in a subprogram. Here, we will solve the equilibrium flash problem as a Visual Basic for Applications (VBA) subprogram with the remaining modules solved on the Excel sheet.

Equilibrium Flash Solution with User-Written VBA and Optimization Code

The conceptual process in Figure 5.5b is particularized in Figure 5.8. We can control the sequence of calculations and combine Excel material balance sheet calculations with a stand-alone flash (coded in VBA). Following Figure 5.8, we can write converged values for species flow rates leaving the flash to the Excel sheet where material balances can be performed on the purge (separator 2), the mixer, and the reactor. As shown in the figure, the flash is not part of these material balance calculations and no recycle loop would be present.

The flash calculation can be called as part of a VBA macro with the logic shown in Figure 5.8. The Excel sheet and VBA code are provided in **Example 5.2d.xls**. There are extensive comments provided in the VBA code and the problem solution is provided in Figure 5.9. One final comment here before we briefly look at the code—the material balance solution procedure is identical for an isothermal

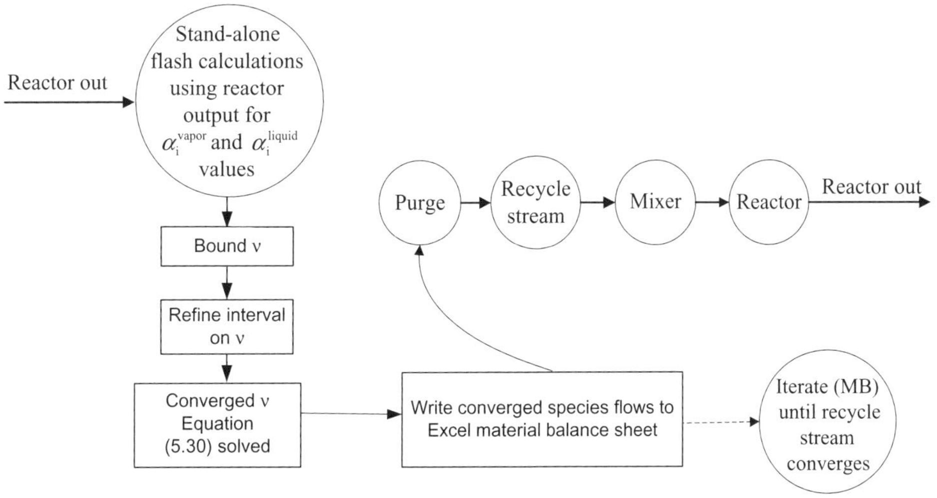

Figure 5.8 VBA calculations for stand-alone flash—inner loop.

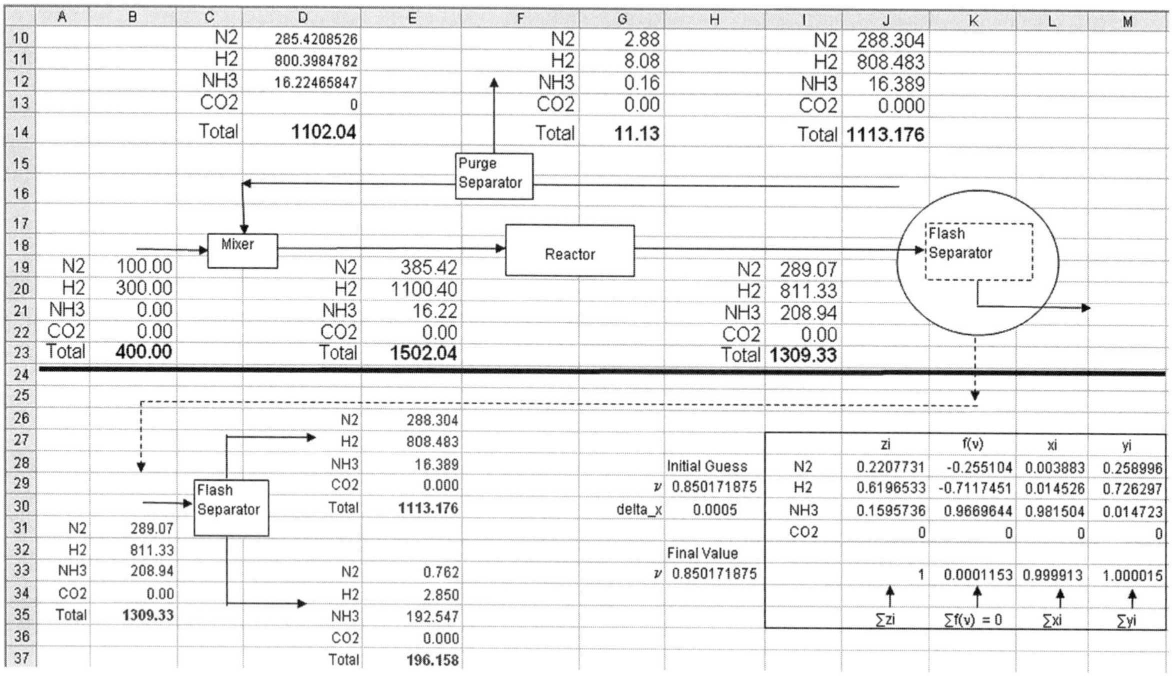

Figure 5.9 Ammonia flow sheet with VBA procedure for the general flash.

flash and a general flash with specified outlet temperatures. Actually, the material balance solution procedure is identical for all flash units (isothermal, adiabatic, and general)—all require that the outlet temperature be known or a guess for the outlet temperature must be supplied.

In the provided VBA–Excel solution, the macro Solve_Isothermal_Flash serves as the Main, which calls needed procedures. In the procedure Bound, an initial value for ν, the total vapor flow ratio, is read from the Excel sheet. Bound calls Stand_Alone_Flash, which reads the species flow rates to the flash (reactor out) from the Excel sheet and calculates Equation (5.30). The objective function for minimization is [Eq. (5.30)]^2. Once the value of ν has been bound, control is retuned to Solve_Isothermal_Flash. From Solve_Isothermal_Flash, the interval on ν is refined, here using interval halving in Interval_Halving. This bounding and interval refinement can be considered an inner iteration loop. Once ν has been determined, its value is used to determine new vapor species flow rates leaving the flash Update_Flash_Flows, and these values are returned to the Excel sheet for the solution of material balance problems of the purge, the mixer, and the reactor.

As part of the macro Solve_Isothermal_Flash, we keep track of the species vapor flow rates leaving the flash prior to the stand-alone flash being called and after the flash has converged. When these species flow rates have converged, the material balance problem has been solved—this can be considered an outer iteration loop. The macro will continue the inner and outer loops until the species flow rates leaving the flash have converged for two successive inner loop iterations.

Ammonia Synthesis with General Flash: The Energy Balance Example 5.2 started several pages ago and so far, all we have solved is the material balance problem. Once the material balance for the flash has been solved, it is straightforward to solve the energy balance problem.

The energy balance will actually be a combination of elementary modules including the heat—work and separator modules. We need the enthalpy of the incoming feed and exit vapor and liquid streams:

$$\frac{dQ}{dt} = \left(N_{\text{vapor}}h_{\text{vapor}} + N_{\text{liquid}}h_{\text{liquid}}\right) - N_{\text{feed}}h_{\text{feed}}. \quad (5.32)$$

In the flash in the ammonia process, the pressure drop is very small; the inlet stream to the flash is ~2000 psi and 800°F, and the outlet stream conditions are ~2000 psi and –28°F. The enthalpies in each stream can be determined from an equation of state as we detail in Chapter 8. In Table 5.1, we supply the average heat capacity and heat of vaporization information, which can also be used to determine the enthalpy in each stream (assuming ideal mixing).

Table 5.1 Species Physical Properties for Example 5.2

Species, i	$C_{p,i}^{\text{vapor}}\left(\frac{\text{Btu}}{\text{lb-mol°F}}\right)$	$\Delta H_{\text{vaporization}}\left(\frac{\text{Btu}}{\text{lb-mol}}\right)$
N_2	7.67	837.4
H_2	7.0	220.0
NH_3	16.08	4136

Here, $C_{p,i}^{\text{vapor}}$ are the average vapor heat capacity values for each species at 2000 psia and $\Delta H_{\text{vaporization}}$ is the heat of vaporization for each species at 2000 psia and −28°F. We can use the separator feed conditions as reference conditions so energy is removed to first take N_2 (298.07 mol/h), H_2 (811.33 mol/h), and NH_3 (208.94 mol/h) as vapor from 800 to −28°F and then energy as $\Delta H_{\text{vaporization}}$ is removed to condense N_2 (0.762 mol/h), H_2 (2.85 mol/h), and NH_3 (192.547) at −28°F. The energy balance using the feed temperature (T_{in}) as the reference state, and recalling the heat of condensation is opposite, is a sign to the heat of vaporization, will be

$$\frac{dQ}{dt} = \sum_{i}^{n_{\text{species}}} \left(N_i^{\text{feed}}\right)\left(C_{p,i}^{\text{vapor}}\right)\left(T_{\text{out}} - T_{\text{in}}\right) - \sum_{i}^{n_{\text{species}}} \left(N_i^{\text{liquid}}\right)\left(\Delta H_{\text{vaporization}}\right)$$

$$\frac{dQ}{dt} = (289.07)(7.67)(-28-800) + (811.33)(7.0)(-28-800) + (208.94)(16.08)(-28-800) - \{(0.762)(837.4) + (2.85)(220) + (192.547)(4136)\} = -10{,}131{,}114\frac{\text{Btu}}{\text{h}}. \tag{5.33}$$

The negative sign indicates heat is being removed from the system (the flash) by the surroundings.

We have solved the ammonia process with a flash separator using a sequential modular approach. Both an inner loop (for flash conversion) and an outer loop (for material balance conversion) were needed.

In the next section, we discuss the simultaneous solution approach to this problem. The strength of the simultaneous approach is the elimination of the two iteration loops as needed in the sequential modular approach with a flash, but successive linearization of the flash material balance equations will be required. The simultaneous approach is very attractive for flash and distillation column calculations.

5.3 EQUILIBRIUM FLASH WITH RECYCLE: SIMULTANEOUS APPROACH

You will recall from our developments of Chapter 4 that a linear equation set can be solved using the Gauss–Jordan method and a nonlinear equation set (or linear and nonlinear) can be solved using the Newton–Raphson method.

EXAMPLE 5.3 ***Ammonia Synthesis with General Flash***

Solve Example 5.2, the ammonia process, using the simultaneous methods we developed in Chapter 4.

In order to allow comparison with our solution of Example 5.2, Example 5.3 will also be solved in three steps. First, we will solve the problem using assumed species split fractions. Second, we will solve the flash as a stand-alone unit with known feed conditions. Finally, we will incorporate the flash into the equation-based flow sheet developed in step 1. The energy balance can be solved after the material balance problem is solved.

Step 1: Ammonia Process: Simultaneous Solution with Species Split Fractions (All Linear Equations)

Using the species split fractions for the flash separator following the reactor as $\alpha_{N_2}^{\text{liquid}} = 0.005$, $\alpha_{H_2}^{\text{liquid}} = 0.005$, and $\alpha_{NH_3}^{\text{liquid}} = 0.98$ and the species split fractions for the purge separator as $\alpha_{N_2}^{\text{purge}} = 0.01$, $\alpha_{H_2}^{\text{purge}} = 0.01$, and $\alpha_{NH_3}^{\text{purge}} = 0.01$, we can construct a linear material balance equation set for the ammonia process. The solution is provided in **Example 5.3a.xls** and the material balances are shown in Figure 5.10.

The solution obtained, after running the macro Gauss_Jordan_Macro, is shown in Figure 5.11. This solution is the same as that found in step 1 in Example 5.2.

Step 2: Ammonia Process: Stand-Alone Flash Simultaneous Solution (Linear and Nonlinear Equations)

You will recall from our developments in Chapter 4, which are summarized in Figure 5.12 that we can obtain solution to a set of linear and nonlinear equations using the Newton–Raphson method.

Here, a process consists of both linear and nonlinear equations. We first supply a starting guess for each variable as linear equations (n linear equations for n unknowns) in matrix 1. This equation set is solved using the Gauss–Jordan method and gives matrix 2 with solution x^*. In matrix 3, we linearize all process equations (linear and nonlinear equations) at x^* and we return this linearized set to matrix 2 for solution.

Now we want to solve the flash as a stand-alone unit using Equations (5.22)–(5.31). These equations require that that the feed condition is known and that the vapor–liquid equilibrium distribution factor, K_i is known for each species. There are nine stream variables associated with the flash; these variables are the nine species flow rates in and out of the flash.

	A	B	C	D	E	F	G	H	I	J	K	L	M	N	O	P	Q	R	S	T	U	V
5																						
6		Feed		Reactor In			Reactor Out			Product			Vapor Out			Purge			Recycle			
7		N(N2)	N(H2)	N(N2)	N(H2)	N(NH3)	N(N2)	N(H2)	N(NH3)	N(N2)	N(H2)	N(NH3)	N(N2)	N(H2)	N(NH3)	N(N2)	N(H2)	N(NH3)	N(N2)	N(H2)	N(NH3)	RHS
8	Feed	1																				100
9	Specs		1																			300
10	Mixer	1		-1															1			0
11	Material		1		-1															1		0
12	Balances					-1															1	0
13	Reactor			0.75			-1															0
14	Balances			-0.75	1			-1														0
15				0.5		1			-1													0
16	Separator 1						0.005			-1												0
17								0.005			-1											0
18									0.98			-1										0
19							1			-1			-1									0
20								1			-1			-1								0
21									1			-1			-1							0
22	Separator 2												0.01			-1						0
23														0.01			-1					0
24															0.01			-1				0
25													1			-1			-1			0
26														1			-1			-1		0
27															1			-1			-1	0

Figure 5.10 Simultaneous material balances for the ammonia process.

	A	B	C	D	E	F	G	H	I	J	K	L	M	N	O	P	Q	R	S	T	U	V	W	X
30		1	0	0	0	0	0	0	0	0	0	0	0	0	0	0	0	0	0	0	0	100	N(N2)	Feed
31		0	1	0	0	0	0	0	0	0	0	0	0	0	0	0	0	0	0	0	0	300	N(H2)	Feed
32		0	0	1	0	0	0	0	0	0	0	0	0	0	0	0	0	0	0	0	0	382.83	N(N2)	Reactor In
33		0	0	0	1	0	0	0	0	0	0	0	0	0	0	0	0	0	0	0	0	1148.5	N(H2)	Reactor In
34		0	0	0	0	1	0	0	0	0	0	0	0	0	0	0	0	0	0	0	0	3.8666	N(NH3)	Reactor In
35		0	0	0	0	0	1	0	0	0	0	0	0	0	0	0	0	0	0	0	0	287.12	N(N2)	Reactor Out
36		0	0	0	0	0	0	1	0	0	0	0	0	0	0	0	0	0	0	0	0	861.37	N(H2)	Reactor Out
37		0	0	0	0	0	0	0	1	0	0	0	0	0	0	0	0	0	0	0	0	195.28	N(NH3)	Reactor Out
38		0	0	0	0	0	0	0	0	1	0	0	0	0	0	0	0	0	0	0	0	1.4356	N(N2)	Product
39		0	0	0	0	0	0	0	0	0	1	0	0	0	0	0	0	0	0	0	0	4.3068	N(H2)	Product
40		0	0	0	0	0	0	0	0	0	0	1	0	0	0	0	0	0	0	0	0	191.38	N(NH3)	Product
41		0	0	0	0	0	0	0	0	0	0	0	1	0	0	0	0	0	0	0	0	285.69	N(N2)	Vapor Out
42		0	0	0	0	0	0	0	0	0	0	0	0	1	0	0	0	0	0	0	0	857.06	N(H2)	Vapor Out
43		0	0	0	0	0	0	0	0	0	0	0	0	0	1	0	0	0	0	0	0	3.9056	N(NH3)	Vapor Out
44		0	0	0	0	0	0	0	0	0	0	0	0	0	0	1	0	0	0	0	0	2.8569	N(N2)	Purge
45		0	0	0	0	0	0	0	0	0	0	0	0	0	0	0	1	0	0	0	0	8.5706	N(H2)	Purge
46		0	0	0	0	0	0	0	0	0	0	0	0	0	0	0	0	1	0	0	0	0.0391	N(NH3)	Purge
47		0	0	0	0	0	0	0	0	0	0	0	0	0	0	0	0	0	1	0	0	282.83	N(N2)	Recycle
48		0	0	0	0	0	0	0	0	0	0	0	0	0	0	0	0	0	0	1	0	848.49	N(H2)	Recycle
49		0	0	0	0	0	0	0	0	0	0	0	0	0	0	0	0	0	0	0	1	3.8666	N(NH3)	Recycle

Figure 5.11 Solution for the simultaneous material balances for the ammonia process.

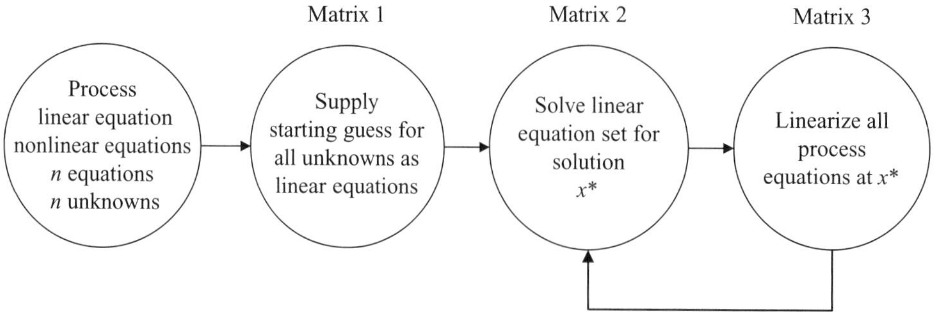

Figure 5.12 Simultaneous solution approach.

To account for these nine variables, three species flow rates are specified on the inlet, and there are three linear species material balances:

$$N_{N_2}^{\text{feed}} = N_{N_2}^{\text{vapor}} + N_{N_2}^{\text{liquid}},$$

$$N_{H_2}^{\text{feed}} = N_{H_2}^{\text{vapor}} + N_{H_2}^{\text{liquid}},$$

and

$$N_{NH_3}^{\text{feed}} = N_{NH_3}^{\text{vapor}} + N_{NH_3}^{\text{liquid}};$$

three nonlinear balance equations from Equation (5.30),

$$\frac{x_{N_2}^{\text{feed}}(1-K_{N_2})}{1+\nu(K_{N_2}-1)} + \frac{x_{H_2}^{\text{feed}}(1-K_{H_2})}{1+\nu(K_{H_2}-1)} + \frac{x_{NH_3}^{\text{feed}}(1-K_{NH_3})}{1+\nu(K_{NH_3}-1)} = 0;$$

and from Equation (5.22),

$$\frac{N_{N_2}^{\text{vapor}}}{N_{\text{vapor}}} - K_{N_2}\frac{N_{N_2}^{\text{liquid}}}{N_{\text{liquid}}} = 0$$

and

$$\frac{N_{H_2}^{\text{vapor}}}{N_{\text{vapor}}} - K_{H_2}\frac{N_{H_2}^{\text{liquid}}}{N_{\text{liquid}}} = 0.$$

The Newton–Raphson solution process is provided in **Example 5.3b.xls** and the results are shown in Figure 5.13. Here, the macro Gauss_Jordan_Macro is run once to obtain a good initial point for linearization, followed by the successive running of NR_Gauss_Jordan_Macro until the nonlinear equations converge. The final results are the same as found in step 2 of Example 5.2. In Figure 5.13, there are three detached coefficient matrices: The first provides an initial guess to the problem; the second matrix is the final solution; and the third matrix shows the linearization of the equations used in the solution of the flash.

Step 3: Flow Sheet and Flash Simultaneous Solution

In step 3 of Example 5.2, we faced the problem of sequencing two iteration loops, one for the flash and one for the recycle loop. Here we can assemble the equation-based material balances developed in step 1 with the flash balances of step 2. There is no concern with iteration loops; however, we will need to successively linearize the nonlinear flash equations. This computational difficulty was addressed in Example 5.3, step 2. The Newton–Raphson solution is provided in **Example 5.3c.xls** and the results shown in Figure 5.14. Again the macro Gauss_Jordan_Macro is run once to obtain a good initial point for linearization, followed by the successive running of

	A	B	C	D	E	F	G	H	I	J	K	L	M
5		x(0)	x(1)	x(2)	x(3)	x(4)	x(5)	x(6)	x(7)	x(8)			
6		Reactor Out				Product			Vapor Out				
7		N(N2)	N(H2)	N(NH3)	N(N2)	N(H2)	N(NH3)	N(N2)	N(H2)	N(NH3)	RHS		
8	Feed	1									287.12		
9			1								861.37		
10	Specs			1							195.28		
11	Separator 1	-1			1			1			0		
12			-1			1			1		0		
13	Flash			-1			1			1	0		
14	Separator	0.005			-1						0		
15			0.005			-1					0		
16				0.98			-1				0		
17													
18													
19													
20													
21		1	0	0	0	0	0	0	0	0	287.12	N(N2)	Reactor Out
22		0	1	0	0	0	0	0	0	0	861.37	N(H2)	Reactor Out
23		0	0	1	0	0	0	0	0	0	195.28	N(NH3)	Reactor Out
24		0	0	0	1	0	0	0	0	0	0.67104	N(N2)	Product
25		0	0	0	0	1	0	0	0	0	2.68222	N(H2)	Product
26		0	0	0	0	0	1	0	0	0	178.168	N(NH3)	Product
27		0	0	0	0	0	0	1	0	0	286.449	N(N2)	Vapor Out
28		0	0	0	0	0	0	0	1	0	858.688	N(H2)	Vapor Out
29		0	0	0	0	0	0	0	0	1	17.1117	N(NH3)	Vapor Out
30													
31													
32													
33													
34		1									287.12		
35			1								861.37		
36				1							195.28		
37		-1			1			1			0		
38			-1			1			1		0		
39				-1			1			1	0		
40		-8E-04	-0.0008	0.005	-0.005	-0.005	-0.0048	0	0.0008	0.0008	3.6E-07		
41		0	0	0	-0.366	0.001	0.0014	0	-2E-04	-2E-04	1.1E-10		
42		0	0	0	0.004	-0.271	0.0041	-0	0.0002	-6E-04	-5E-10		

Figure 5.13 Simultaneous material balances and solution for the flash.

NR_Gauss_Jordan_Macro until the nonlinear equations converge. The final results are the same as found in step 3 of Example 5.2. In Figure 5.14a–c, there are again three detached coefficient matrices: The first provides an initial guess to the problem; the second matrix is the final solution; and the third matrix shows the linearization of the equations used in solution of the material balances and the flash. ■

In Example 5.3, we solved the ammonia process with a general flash separator using an equation-based approach. The equation-based approach eliminated the inner loop (for flash conversion) and an outer loop (for material balance conversion) that were needed for the modular solution approach. The simultaneous approach does require successive linearization of the flash material balance equations as shown in the third matrix in Figures 5.13 and 5.14.

5.4 ADIABATIC PLUG FLOW REACTOR (PFR) MATERIAL AND ENERGY BALANCES INCLUDING RATE EXPRESSIONS: EULER'S FIRST-ORDER METHOD

In Chapters 3 and 4, we assembled material balances for styrene and ammonia (chapter problems) processes using elementary modules. In Section 5.2, we saw that when a flash unit was used as a component separator, the modular solution approach required that the flash unit had to be converged for each recycle loop calculation. This computational difficulty was avoided in Section 5.3, where a simultaneous approach was used, but here equations for the flash still required successive linearization. In this section, we discuss reactor energy balances and temperature dependency in reactor design and sizing. The material also serves as an important introduction to Chapter 15, where we use continuously stirred reactors (CSTR) and PFRs in series to model gas turbine systems.

5.4.1 Reactor Types

We are representing reactors using elementary material and energy balance modules. However, in order to properly use these modules, we must appreciate that there are many different reactor designs. The three most common reactor types are a PFR or tubular reactor, a perfectly stirred reactor (PSR) or a CSTR, and a batch reactor. In this section, we will focus on PFRs. In Chapter 15, we will need to utilize combinations of both PFR and PSR in order to predict the emissions from combustion systems. In this chapter, we will find numerical methods are needed to incorporate rate expressions into the reactor design and sizing problem; we will need numerical solution methods for ODEs. We will develop the Euler method and the fourth-order Runge–Kutta (RK4) solution methods. For the combustion reactor emissions problems in Chapter 15, these popular solution methods for ODEs will not work. Here, we utilize the commercial standard ODE solver from Lawrence Livermore National Laboratories—*CVODE*, which we will provide as a callable program from Excel. Even though we will use CVODE for emission calculations, it is important to understand how ODEs solvers work and we show this next.

In this section, it is convenient to particularize our initial discussion to a styrene PFR reactor, but the developed equations can be applied to general PFR systems. We will show examples for the styrene reactor and for the ammonia reactor (as Chapter 5 problems). In a PFR (see Figure 5.15), conditions change down the length, but it assumed there are no radial gradients in temperature, composition, or velocity.

The reaction ethylbenzene ↔ styrene + hydrogen or $C_8H_{10} = C_8H_8 + H_2$ is an endothermic reaction. Steam is supplied both to provide the heat necessary for the reaction to occur as well as to prevent coking. As each mole of ethylbenzene reacts, energy is removed from the steam and the temperature in the reactor decreases. As the temperature decreases, the rate at which ethylbenzene reacts will also decrease. At some temperature, the reaction to products will have slowed to a point where continued time in the reactor would not be useful/economical.

EXAMPLE 5.4 ***Determine the Outlet Temperature and Composition of a Styrene Reactor Using Elementary Reactor Modules***

The feed to an adiabatic styrene reactor is 18.4615 lb-mol/h ethylbenzene, 0.1212 lb-mol/h styrene, 0.0 lb-mol/h hydrogen, and 353.0745 lb-mol/h steam at 1616 R. Reactor ethylbenzene conversion is 65%. The heat of reaction at $T = 1616$ R $\Delta H_r^{T=1616\text{R}} = 60{,}000$ Btu/lb-mol ethylbenzene reacted, and because steam is in large excess, the heat capacity of the reacting fluid may be taken as a constant $\hat{C}_p = 0.52$ Btu/lb-R.

SOLUTION Using Equation (3.26) with 65% ethylbenzene conversion,

$$\xi = \frac{N_{C_8H_{10}}^{\text{reactor out}} - N_{C_8H_{10}}^{\text{reactor in}}}{\nu_{C_8H_{10}}} = \left(\frac{6.4615 - 18.4615}{-1}\right) = 12 \text{ lb-mol/h}.$$

The exiting molar flows are $N_{C_8H_{10}}^{\text{reactor out}} = 6.4615$ lb-mol/h, $N_{C_8H_8}^{\text{reactor out}} = 12.1212$ lb-mol/h, $N_{H_2O}^{\text{reactor out}} = 353.0745$ lb-mol/h, and $N_{H_2}^{\text{reactor out}} = 12.0$ lb-mol/h. Using Equation (5.19) with the feed temperature as reference, $\xi\Delta H_r = 720{,}000$ Btu/h = 200 Btu/s, and the water heat capacity expression for enthalpy change, we can write

$$N_{\text{out}}h_{\text{out}} = F_{\text{out}}\hat{C}_p(T_{\text{reactor out}} - T_{\text{reactor in}}) = -\xi\Delta H_r,$$

	A	B	C	D	E	F	G	H	I	J	K	L	M	N	O	P	Q	R	S	T	U	V
1	***Ammonia Material Balance using Gauss Jordan Elimination and Newton Raphson***																					
2																						
3	Reaction $N_2 + 3H_2 \rightarrow NH_3$					0.25	conversion															
4																						
5		x(0)	x(1)	x(2)	x(3)	x(4)	x(5)	x(6)	x(7)	x(8)	x(9)	x(10)	x(11)	x(12)	x(13)	x(14)	x(15)	x(16)	x(17)	x(18)	x(19)	
6		Feed		Reactor In			Reactor Out				Product			Vapor Out			Purge			Recycle		
7		N(N2)	N(H2)	N(N2)	N(H2)	N(NH3)	N(N2)	N(H2)	N(NH3)	N(N2)	N(H2)	N(NH3)	N(N2)	N(H2)	N(NH3)	N(N2)	N(H2)	N(NH3)	N(N2)	N(H2)	N(NH3)	RHS
8	Feed	1																				100
9	Specs		1																			300
10	Mixer	1		-1															1			0
11	Material		1		-1															1		0
12	Balances					-1															1	0
13	Reactor			0.75			-1															0
14	Balances			-0.75	1			-1														0
15				0.5		1			-1													0
16	Separator 2												0.01			-1						0
17														0.01			-1					0
18															0.01			-1				0
19													1			-1			-1			0
20														1			-1			-1		0
21															1			-1			-1	0
22	Separator 1						1			-1			-1									0
23								1			-1			-1								0
24	Flash								1			-1			-1							0
25	Separator						0.005			-1												0
26								0.005			-1											0
27									0.98			-1										0

(a)

	A	B	C	D	E	F	G	H	I	J	K	L	M	N	O	P	Q	R	S	T	U	V	W	X
32		1	0	0	0	0	0	0	0	0	0	0	0	0	0	0	0	0	0	0	0	100	N(N2)	Feed
33		0	1	0	0	0	0	0	0	0	0	0	0	0	0	0	0	0	0	0	0	300	N(H2)	Feed
34		0	0	1	0	0	0	0	0	0	0	0	0	0	0	0	0	0	0	0	0	385.42	N(N2)	Reactor In
35		0	0	0	1	0	0	0	0	0	0	0	0	0	0	0	0	0	0	0	0	1100.5	N(H2)	Reactor In
36		0	0	0	0	1	0	0	0	0	0	0	0	0	0	0	0	0	0	0	0	16.227	N(NH3)	Reactor In
37		0	0	0	0	0	1	0	0	0	0	0	0	0	0	0	0	0	0	0	0	289.07	N(N2)	Reactor Out
38		0	0	0	0	0	0	1	0	0	0	0	0	0	0	0	0	0	0	0	0	811.41	N(H2)	Reactor Out
39		0	0	0	0	0	0	0	1	0	0	0	0	0	0	0	0	0	0	0	0	208.94	N(NH3)	Reactor Out
40		0	0	0	0	0	0	0	0	1	0	0	0	0	0	0	0	0	0	0	0	0.762	N(N2)	Product
41		0	0	0	0	0	0	0	0	0	1	0	0	0	0	0	0	0	0	0	0	2.8494	N(H2)	Product
42		0	0	0	0	0	0	0	0	0	0	1	0	0	0	0	0	0	0	0	0	192.55	N(NH3)	Product
43		0	0	0	0	0	0	0	0	0	0	0	1	0	0	0	0	0	0	0	0	288.3	N(N2)	Vapor Out
44		0	0	0	0	0	0	0	0	0	0	0	0	1	0	0	0	0	0	0	0	808.56	N(H2)	Vapor Out
45		0	0	0	0	0	0	0	0	0	0	0	0	0	1	0	0	0	0	0	0	16.391	N(NH3)	Vapor Out
46		0	0	0	0	0	0	0	0	0	0	0	0	0	0	1	0	0	0	0	0	2.883	N(N2)	Purge
47		0	0	0	0	0	0	0	0	0	0	0	0	0	0	0	1	0	0	0	0	8.0856	N(H2)	Purge
48		0	0	0	0	0	0	0	0	0	0	0	0	0	0	0	0	1	0	0	0	0.1639	N(NH3)	Purge
49		0	0	0	0	0	0	0	0	0	0	0	0	0	0	0	0	0	1	0	0	285.42	N(N2)	Recycle
50		0	0	0	0	0	0	0	0	0	0	0	0	0	0	0	0	0	0	1	0	800.47	N(H2)	Recycle
51		0	0	0	0	0	0	0	0	0	0	0	0	0	0	0	0	0	0	0	1	16.227	N(NH3)	Recycle

(b)

	A	B	C	D	E	F	G	H	I	J	K	L	M	N	O	P	Q	R	S	T	U	V
56		1																				100
57			1																			300
58		1		-1															1			0
59			1		-1															1		0
60						-1															1	0
61				0.75			-1															0
62				-0.75	1			-1														0
63				0.5		1			-1													0
64													0.01			-1						0
65														0.01			-1					0
66															0.01			-1				0
67													1			-1			-1			0
68														1			-1			-1		0
69															1			-1			-1	0
70							1			-1			-1									0
71								1			-1			-1								0
72									1			-1			-1							0
73		0	0	0	0	0	-0	-0	0.0046	-0	-0	-0.0045	8E-04	0	0.0008	0	0	0	0	0	0	3E-07
74		0	0	0	0	0	0	0	0	-0.3	0.001	0.0013	7E-04	-0	-2E-04	0	0	0	0	0	0	-4E-10
75		0	0	0	0	0	0	0	0	0	-0.25	0.0037	-0	0	-7E-04	0	0	0	0	0	0	1E-09

(c)

Figure 5.14 (a) Simultaneous material balances for the ammonia process with flash—initial guess. (b) Simultaneous material balances for the ammonia process with flash—solution. (c) Simultaneous material balances for the ammonia process with flash—linearization.

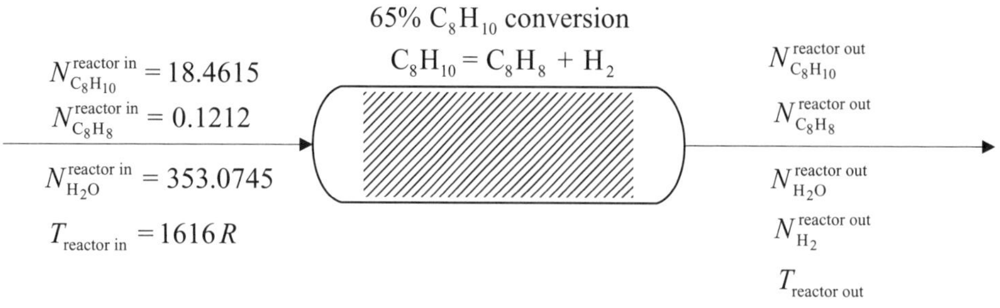

Figure 5.15 Plug flow reactor (PFR) for styrene production.

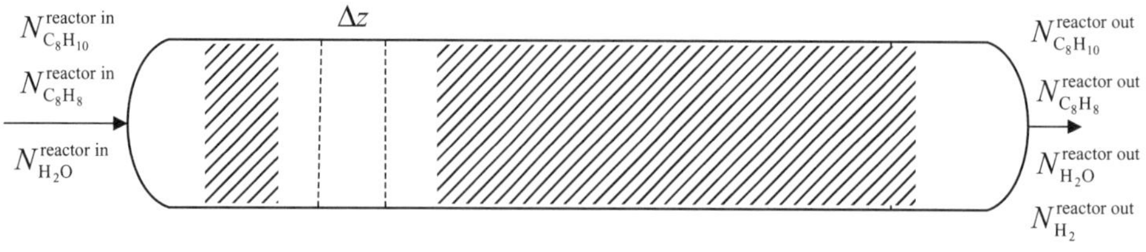

Figure 5.16 Plug flow reactor for styrene production—with Δz section indicated.

and using the species molecular weights, we find

$$F\hat{C}_p = \left(2.3148\frac{\text{lb}}{\text{s}}\right)\left(0.52\frac{\text{Btu}}{\text{lb-R}}\right) = 1.2037\frac{\text{Btu}}{\text{s-R}}.$$

Finally, solving for $T_{\text{reactor out}}$,

$$T_{\text{reactor out}} = T_{\text{reactor in}} + \frac{\xi(-\Delta H_r)}{F_{\text{out}}\hat{C}_p} = 1616 + \frac{(-200)}{1.2037} = 1449.84\text{R}.$$

■

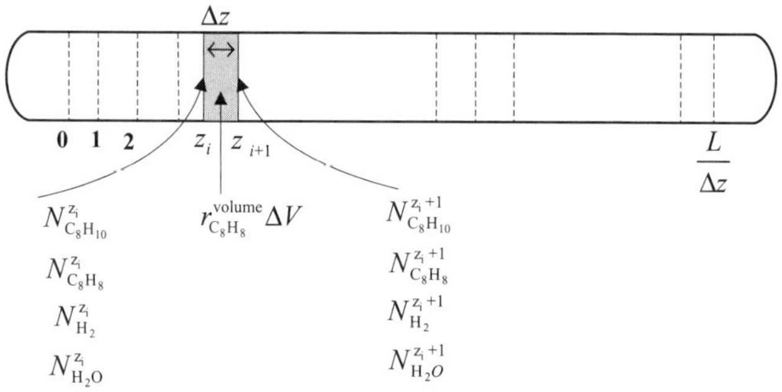

Figure 5.17 Differential volumes in a plug flow reactor.

The problem here is that this exit temperature from the adiabatic reactor is too low. Ethylbenzene conversion will virtually stop (as will be shown in Example 5.5) as temperatures near 1500 R—in other words, we may not reach 65% conversion regardless of the length of the adiabatic reactor. Often we cannot simply use the elementary modules, as developed in this chapter and in Chapter 3, independent of rate considerations.

To design (size) the needed reactor, we must solve the material and energy balance problems and account for the reaction kinetics all within the chosen reactor type. Styrene production is done on large volumes and, generally, PFRs are used. In a PFR, the ethylbenzene and steam are fed into what can be visualized as a large tube, and the reaction occurs down the length of the tube. We can design (size) this reactor using reaction rate equations that are available in the literature. Figure 5.16 shows the PFR, and we will discuss multiple tube PFRs later in this section.

From our material balance calculations in Chapters 3 and 4 (see Examples 3.1 and 4.3), at the entrance to the PFR we find ethylbenzene, steam, and a small amount of styrene. To solve the material and energy balance problems, we can divide the reactor into small Δz sections down the length of the reactor and account for the reactions and temperature change that occur over each of these volumes. Do note that each Δz actually represents a volume element or "plug" given by the tube cross-sectional area, $A_c = (\pi r^2)$, multiplied by the length, Δz.

A step from z_j to z_{j+1} is indicated in Figure 5.17. At z_j, ethylbenzene, $N^{z_j}_{C_8H_{10}}$, styrene, $N^{z_j}_{C_8H_8}$, hydrogen, $N^{z_j}_{H_2}$, and steam, $N^{z_j}_{H_2O}$, enter the volume element at temperature T_j. We can write material balances to determine species molar flows, which leave at z_{j+1} as

$$C_8H_{10} \text{ balance: } N^{z_{j+1}}_{C_8H_{10}} = N^{z_j}_{C_8H_{10}} + r^{\text{volume}}_{C_8H_{10}}\Delta V = N^{z_j}_{C_8H_{10}} - r^{\text{volume}}_{C_8H_8}\Delta V \tag{5.34}$$

$$\text{Styrene balance: } N^{z_{j+1}}_{C_8H_8} = N^{z_j}_{C_8H_8} + r^{\text{volume}}_{C_8H_8}\Delta V \tag{5.35}$$

$$H_2 \text{ balance: } N^{z_{j+1}}_{H_2} = N^{z_j}_{H_2} + r^{\text{volume}}_{H_2}\Delta V = N^{z_j}_{H_2} + r^{\text{volume}}_{C_8H_8}\Delta V \tag{5.36}$$

$$H_2O \text{ balance: } N^{z_{j+1}}_{H_2O} = N^{z_j}_{H_2O}. \tag{5.37}$$

$r^{volume}_{C_8H_8}$ is the molar rate of styrene produced per unit volume per unit time;

$$r^{\text{volume}}_{C_8H_8} [=] \frac{\text{lb-mol styrene produced}}{\text{s-ft}^3 \text{ reactor volume}},$$

which is multiplied by the volume ΔV to determine the change in moles of styrene over z_j to z_{j+1}. In these balances, water is not reacting. We have also taken advantage of the known reaction stoichiometric coefficients, allowing us to write

$$\left(-r_{C_8H_{10}}^{\text{volume}}\right) = r_{C_8H_8}^{\text{volume}} = r_{H_2}^{\text{volume}}. \tag{5.38}$$

In the PFR, the reacting fluid is assumed to be perfectly mixed in the radial direction and no mixing occurs between different Δz sections; this can be visualized as perfectly mixed ΔV_j volume elements (small batch reactors) moving down the length of the PFR.

As ΔV becomes small, we can write the differential material balance equation for the PFR reactor for any reacting species i:

$$\frac{dN_i}{dV} = r_i^{\text{volume}}. \tag{5.39}$$

The PFR tubes often contain catalysts to promote reaction. In this case, the weight of the catalyst, as opposed to the reactor volume, becomes important for the design. Here, the differential equation for the PFR (or packed bed reactor) is

$$\frac{dN_i}{dW} = r_i^{\text{global}}. \tag{5.40}$$

Here, W is the weight of the catalyst (lb); for example,

$$r_{C_8H_8}^{\text{global}} [=] \frac{\text{lb-mol styrene produced}}{\text{s-lb catalyst}}.$$

Examining Equation (5.40), if the density of the catalyst in the reactor bed, $\rho_{\text{catalyst bed}}\left(\text{lb catalyst/ft}^3\text{-reactor bed}\right)$, and cross-sectional area, A_c, of the PFR tube are known and constant, we can write

$$\frac{dN_i}{\left(\rho_{\text{catalyst bed}}\right)(A_c)(dz)} = r_i^{\text{global}}. \tag{5.41}$$

Equation (5.41) can be integrated between $z_j$and z_{j+1} to give

$$N_i^{z_{j+1}} = N_i^{z_j} + \left(\rho_{\text{catalyst}}\right)(A_c)\int_{z_j}^{z_{j+1}} r_i^{\text{global}} dz. \tag{5.42}$$

The problem is integrating r_i^{global}, which will be a function of the species N_i^z and T_z. We will explore the numerical integration of the ODE, Equation (5.42). The simplest numerical integration, Euler's first-order method, is to assume r_i^{global} is constant over $z_j$to z_{j+1} with all needed conditions determined at z_j. Here then,

$$N_i^{z_{j+1}} = N_i^{z_j} + \left(\rho_{\text{catalyst}}\right)(A_c)\left(r_i^{\text{global}}\Big|_{z_j}\right)\Delta z. \tag{5.43}$$

As all conditions (species molar feed rates and temperature) are known at $z_j = 0$, Equation (5.43) can be solved for species molar flow rates at $N_i^{z_{j+1}=1}$; this process continues down the length of the reactor.

The energy balance over each volume element allows the temperature at $T_{z_{j+1}}$ to be determined. This balance can be written for a steady-state, adiabatic reactor with a single reaction as

$$N_{z_{j+1}} h_{z_{j+1}} = N_{z_j} h_{z_j} - \left(\Delta H_{r,i}\right)\left(r_i^{\text{global}}\right)(\Delta W)$$

or, as often found,

$$N_{z_{j+1}} h_{z_{j+1}} = N_{z_j} h_{z_j} + \left(-\Delta H_{r,i}\right)\left(r_i^{\text{global}}\right)(\Delta W). \tag{5.44}$$

Here, (Nh) is the energy transfer rate (British thermal unit per second) and $\Delta H_{r,i}$ is the heat of reaction (British thermal unit per pound-mole) at z_i reaction conditions. The reaction stoichiometry allows r_i^{global} and $\Delta H_{r,i}$ to be based on the same species, and often the limiting reacting species is used. To emphasize the use of the limiting species, we can rewrite Equation (5.44) as

$$N_{z_{j+1}} h_{z_{j+1}} = N_{z_j} h_{z_j} + \left(-\Delta H_{r,i\,\text{limiting}}\right)\left(r_{i\,\text{limiting}}^{\text{global}}\right)(\Delta W). \tag{5.45}$$

For endothermic reactions, $\Delta H_{r,i}$ will be a positive quantity, and for exothermic reactions, $\Delta H_{r,i}$ will be a negative quantity. For the endothermic ethylbenzene reaction (styrene production), $\Delta H_{r,i}$ will be reported as a positive quantity (British thermal unit per pound-mole ethylbenzene reacted or British thermal unit per pound-mole styrene produced) and temperature down the reactor ($+\Delta z$) will decrease.

In Equation (5.45), as ΔW becomes small, we can write the differential equation for the PFR reactor energy balance as

$$\frac{d(Nh)}{dW} = \left(-\Delta H_{r,i\,\text{limiting}}\right)\left(r_{i\,\text{limiting}}^{\text{global}}\right). \tag{5.46}$$

If the catalyst bed density $\rho_{\text{catalyst bed}}$ and the cross-sectional area A_c of the PFR tube are known and constant, we can write

$$N_{z_{j+1}} h_{z_{j+1}} = N_{z_j} h_{z_j} + \left(\rho_{\text{catalyst bed}}\right)(A_c) \int_{z_j}^{z_{j+1}} \left(-\Delta H_{r,i\,\text{limiting}}\right)\left(r_{i\,\text{limiting}}^{\text{global}}\right) dz. \tag{5.47}$$

Here, $T_{z_{j+1}}$ is found from $\left(N_{z_{j+1}} h_{z_{j+1}}\right)$ using an iterative solution and an appropriate equation of state for h. A convenient means of determining $T_{z_{j+1}}$ is to calculate the enthalpy term as the product of flow rate, molar heat capacity, and temperature change (an ideal mixture):

$$\sum_i N_i^{z_{j+1}} \int_{T_{\text{ref}}}^{T_{z_{j+1}}} C_{p,i} dT = \sum_i N_i^{z_j} \int_{T_{\text{ref}}}^{T_{z_j}} C_{p,i} dT + \left(\rho_{\text{catalyst bed}}\right)(A_c) \int_{z_j}^{z_{j+1}} \left(-\Delta H_{r,i\,\text{limiting}}\right)\left(r_{i\,\text{limiting}}^{\text{global}}\right) dz. \tag{5.48}$$

A problem in Equations (5.47) and (5.48) will again be integrating $\left(r_{i\,\text{limiting}}^{\text{global}}\right)$, which is a function of the species N_i^z,

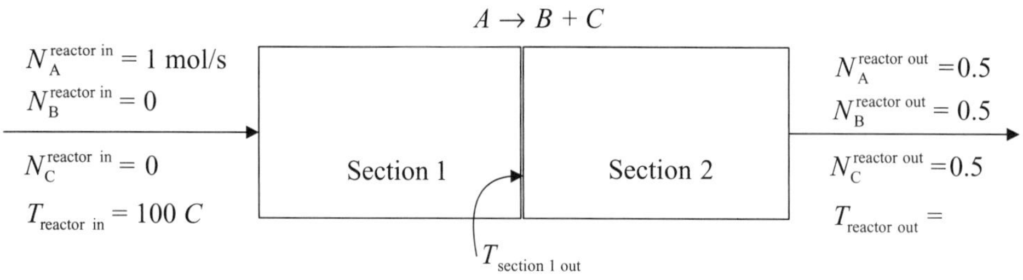

Figure 5.18 Simple $A \rightarrow B + C$ reaction system with ΔH_r^{zj} considerations.

T_z, and P_z, and integrating ($\Delta H_{r,i\ \mathrm{limiting}}$), which is a function of T_z and P_z. The simplest approximation, Euler's first-order method, assumes $\left(r_{i\ \mathrm{limiting}}^{\mathrm{global}}\right)$, ($\Delta H_{r,i\ \mathrm{limiting}}$), and N_i^z are constant over z_j to z_{j+1} with all needed conditions determined at z_j. For Equation (5.48),

$$\sum_i N_i^{zj}\int_{T_{zj}}^{T_{zj+1}} C_{p,i}dT$$

$$= \left(\rho_{\mathrm{catalyst\ bed}}\right)\left(A_c\right)\left(-\Delta H_{r,i\ \mathrm{limiting}}^{zj}\right)\left(r_{i\ \mathrm{limiting}}^{\mathrm{global}}\Big|_{zj}\right)\Delta z. \quad (5.49)$$

Equation (5.49) may still require an iterative solution to find T_{zj+1} depending on the form of the $C_{p,i}$ equations. To avoid an iterative procedure, the $\int C_{p,i}dT$ can be simplified to $\left(C_{p,i}^{\mathrm{avg}}\right)\left(T_{zj+1} - T_{zj}\right)$, which gives

$$T_{zj+1} = T_{zj} + \frac{\left(\rho_{\mathrm{catalyst}}\right)\left(A_c\right)\left(-\Delta H_{r,i\ \mathrm{limiting}}^{zj}\right)\left(r_{i\ \mathrm{limiting}}^{\mathrm{global}}\Big|_{zj}\right)\Delta z}{\sum_i N_i^{zj}\left(C_{p,i}^{\mathrm{avg}}\right)}. \quad (5.50)$$

In order to account for the change in ΔH_r with z_i (with T_{z_i} and P_{z_i}), we can write

$$\Delta H_r(T, P_{\mathrm{ref}}^0) = \Delta H_r(T_{\mathrm{ref}}^0, P_{\mathrm{ref}}^0) + \sum_{i=1}^{n_{\mathrm{species}}} \nu_i \int_{T_{\mathrm{ref}}^0}^{T} C_{p,i}dT, \quad (5.51)$$

where in Equation (5.51) we have assumed all reacting species remain in the same phase and the reacting streams form an ideal mixture. We would also need to correct ΔH_r for any species phase change and enthalpy effects due to pressure; generally, pressure effects are small. These topics are discussed in thermodynamic works (e.g., see Sandler, 1999).

If the effects of T and P on the heat of reaction are small, $\left(\Delta H_{r,i\ \mathrm{limiting}}^{z}\right)$ may be assumed constant, and here $\left(\Delta H_{r,i\ \mathrm{limiting}}^{zj=0}\right)$ can be used in Equation (5.50) giving

$$T_{zj+1} = T_{zj} + \frac{\left(\rho_{\mathrm{catalyst}}\right)\left(A_c\right)\left(-\Delta H_{r,i\ \mathrm{limiting}}^{zj=0}\right)\left(r_{i\ \mathrm{limiting}}^{\mathrm{global}}\Big|_{zj}\right)\Delta z}{\sum_i N_i^{zj}\left(C_{p,i}^{\mathrm{avg}}\right)}. \quad (5.52)$$

The following example will help clarify the use of ΔH_r^{zj}.

EXAMPLE 5.5 ***Calculating and Using ΔH_r^{zj}***

Calculate the temperature leaving the reactor system shown in Figure 5.18. The reaction $A \rightarrow B + C$ occurs in an adiabatic reactor with $T_{\mathrm{reactor\ in}}$ = 100°C and $P_{\mathrm{reactor\ in}}$ = 1 bar. The heat of reaction provided at the standard reference state ($T_{\mathrm{ref}}^0 = 25$°C, $P_{\mathrm{ref}}^0 = 1$ bar) is $\Delta H_r(T_{\mathrm{ref}}^0, P_{\mathrm{ref}}^0) = -1000$ cal/mol A reacted, and the species heat capacities are $C_{p,A} = 5$, $C_{p,B} = 2$, and $C_{p,C} = 1$ cal/mol-°C. Pressure drop in the reactor is negligible.

1. Use Equation (5.51) to calculate $\Delta H_r(T_{\mathrm{reactor\ in}}, P_{\mathrm{reactor\ in}})$ and calculate the exit temperature from the reactor; here, the overall conversion of A is 50%.
2. Determine the temperature leaving section 1 with 25% of the incoming A converted in this section of the reactor ($\xi = 0.25$ mol A/s). Then, using the temperature exiting section 1, and with $\xi = 0.25$ mol A/s in the second section of the reactor, determine the reactor outlet temperature. Here, assume the $\Delta H_r(T_{\mathrm{reactor\ in}}, P_{\mathrm{reactor\ in}})$ value determined in part 1 is constant for the entire reactor.
3. Solve part 2 but calculate and use $\Delta H_r(T_{\mathrm{section\ in}}, P_{\mathrm{section\ in}})$ leaving section 1.

SOLUTION

Part 1. The heat of reaction correction from Equation (5.51) is $\Delta H_r(T, P_{\mathrm{ref}}^0) = \Delta H_r(T_{\mathrm{ref}}^0, P_{\mathrm{ref}}^0) + \sum_{i=1}^{n_{\mathrm{species}}} \nu_i \int_{T_{\mathrm{ref}}^0}^{T} C_{p,i}dT$, and with ideal mixtures, we can write

$$\begin{aligned}\Delta H_r(T, P_{\mathrm{ref}}^0) = \Delta H_r(T_{\mathrm{ref}}^0, P_{\mathrm{ref}}^0) &+ \left(\nu_A\right)\left(C_{p,A}\right)\left(T_{\mathrm{reactor\ in}} - T_{\mathrm{ref}}\right)\\ &+ \left(\nu_B\right)\left(C_{p,B}\right)\left(T_{\mathrm{reactor\ in}} - T_{\mathrm{ref}}\right)\\ &+ \left(\nu_C\right)\left(C_{p,C}\right)\left(T_{\mathrm{reactor\ in}} - T_{\mathrm{ref}}\right).\end{aligned}$$

Here, then, at the reactor inlet $\Delta H_r(T = 100°\mathrm{C}, P_{\mathrm{ref}}^0)$, is

$$\begin{aligned}\Delta H_r(T = 100°\mathrm{C}, P_{\mathrm{ref}}^0) &= -1000 + (-1)(5)(100-25)\\ &\quad + (1)(2)(100-25) + (1)(1)(100-25)\\ &= -1150\ \mathrm{cal/mol}.\end{aligned}$$

The energy balance over the entire reactor is given by Equation (5.18):

$$\sum_{i=1}^{n_{\text{species}}} N_i^{\text{out}}\left(h_i^{\text{out}} - h_i^{\text{ref}}\right) + \xi \Delta H_r - \sum_{i=1}^{n_{\text{species}}} N_i^{\text{in}}\left(h_i^{\text{in}} - h_i^{\text{ref}}\right) = 0,$$

and using feed stream conditions as our reference,

$$\sum_{i=1}^{n_{\text{species}}} N_i^{\text{out}} \int_{T_{\text{reactor in}}}^{T_{\text{reactor out}}} C_{p,i} = \xi(-\Delta H_r),$$

we can solve for $T_{\text{reactor out}}$ as

$$\begin{aligned}(0.5)(5)(T_{\text{reactor out}} - 100) + (0.5)(2)(T_{\text{reactor out}} - 100)\\ + (0.5)(1)(T_{\text{reactor out}} - 100) = (0.5)(1150)\text{ cal}\end{aligned}$$

$$T_{\text{reactor out}} = 100 + \frac{(0.5)(1150)}{4} = 243.75°\text{C}.$$

Part 2. Out of section 1 with 25% conversion,

$$T_{\text{section 1 out}} = 100 + \frac{(0.25)(1150)}{4.5} = 163.888°\text{C}.$$

And out of section 2, which is out of the reactor, with constant $\Delta H_r(T_{\text{reactor in}}, P_{\text{reactor in}})$, we find

$$T_{\text{reactor out}} = 163.888 + \frac{(0.25)(1150)}{4} = 235.764°\text{C}.$$

This is not the same temperature as found in part 1.

Part 3. Here then, leaving reactor section 1 at 163.888°C, we calculate $\Delta H_r(T, P_{\text{ref}}^0)$as

$$\begin{aligned}\Delta H_r(T = 163.888°\text{C}, P_{\text{ref}}^0) &= -1000 + (-1)(5)(163.888 - 25)\\ &\quad + (1)(2)(163.888 - 25)\\ &\quad + (1)(1)(163.888 - 25)\\ &= -1277.777\text{ cal/mol}.\end{aligned}$$

And leaving the reactor,

$$T_{\text{reactor out}} = 163.888 + \frac{(0.25)(1277.777)}{4} = 243.75°\text{C}.$$

The part 3 exit temperature is the same temperature as found in part 1. This simple example shows that the heat of reaction may need to be calculated at each z_i of a PFR when numerical methods are used and if sensible heat contributions are significant. Heat of reaction considerations can be eliminated when using species enthalpy of formation as we do for the combustion reactors in Chapter 15. ■

A final simplification is to assume the product of flow rate and mass heat capacity is constant, which may be reasonable if an inert species dominates the flow rate. Here then, Equation (5.51) becomes

$$T_{z_{j+1}} = T_{z_j} + \frac{(\rho_{\text{catalyst}})(A_c)\left(-\Delta H_{r,i\text{ limiting}}^{z_j=0}\right)\left(r_{i\text{ limiting}}^{\text{global}}\Big|_{z_j}\right)\Delta z}{F\hat{C}_p}. \quad (5.53)$$

The accuracy of our energy balance calculation over each Δz step is being lessened as we move from Equation (5.47) to (5.53). In Example 5.6, we will write VBA code to numerically integrate the ODE system given by mass balance and energy balance equations using Euler's first-order method. In Example 5.7, will examine the use of the RK4 method to solve these equations.

5.5 STYRENE PROCESS: MATERIAL AND ENERGY BALANCES WITH REACTION RATE

EXAMPLE 5.6 ***Styrene Reactor Material and Energy Balances Using Euler's First-Order Method***

Example 5.6 is modified from Smith (1970). We want to determine the length of an adiabatic PFR needed to produce 30,000 lb of styrene per day. The styrene process flow sheet is shown in Figure 5.19. Steam is added to the fresh feed and recycle stream to provide heat (energy) for the endothermic reaction and to help prevent coke buildup in the reactor; here,

$$\frac{N_{H_2O}^{\text{reactor in}}}{N_{\text{reactor in}}} = 0.95.$$

Additional heat (energy) is supplied by a heat exchanger prior to the PFR. The product from the reactor bed is cooled, allowing separation into three phases. In this separator, 100% of the condensed steam (water) and 100% of the hydrogen (vapor stream) are removed. An organic liquid stream is sent to a distillation section where 99% (molar basis) of the incoming styrene is recovered to the styrene product and 99% (molar basis) of the incoming ethylbenzene is recycled.

The feed stream temperature to the reactor is 1616 R and the reactor pressure is 1.2 atm. The reactor feed rate and composition can be determined from the styrene material balance process flow sheet developed in Chapters 3 and 4. **Example 5.6a.xls** modifies Example 3.10 to provide for 30,000 lb of styrene per day, with

$$\frac{N_{H_2O}^{\text{reactor in}}}{N_{\text{reactor in}}} = 0.95,$$

and with 99% styrene recovery in the product stream and 99% ethylbeneze recovery in the recycle stream. The feed to the reactor is 18.4615 lb-mol/h ethylbenzene, 0.1212 lb-mol/h styrene, 0.0 lb-mol/h hydrogen, and 353.0745 lb-mol/h steam. The reactor ethylbenzene conversion is 65%. The heat of reaction (endothermic), $\Delta H_r^{T=1616\text{R}} = 60{,}000$ Btu/lb-mol ethylbenzene reacted or $\Delta H_r^{T=1616\text{R}} = 60{,}000$ Btu/lb-mol styrene produced as each mole of ethylbenzene reacted, produces 1 mol of styrene. We will assume the heat of reaction is constant, and because steam is in large excess, the heat capacity of the reacting fluid may also be taken as a constant, $\hat{C}_p = 0.52$ Btu/lb. These are the same reactor feed stream values as used in Example 5.4.

We want to determine the length of an adiabatic PFR needed to produce 30,000 lb of styrene per day. The PFR tube is 4.6 ft in diameter and packed with the catalyst studied by Wenner and Dybdal (1948); the rate expression and catalyst properties are discussed further.

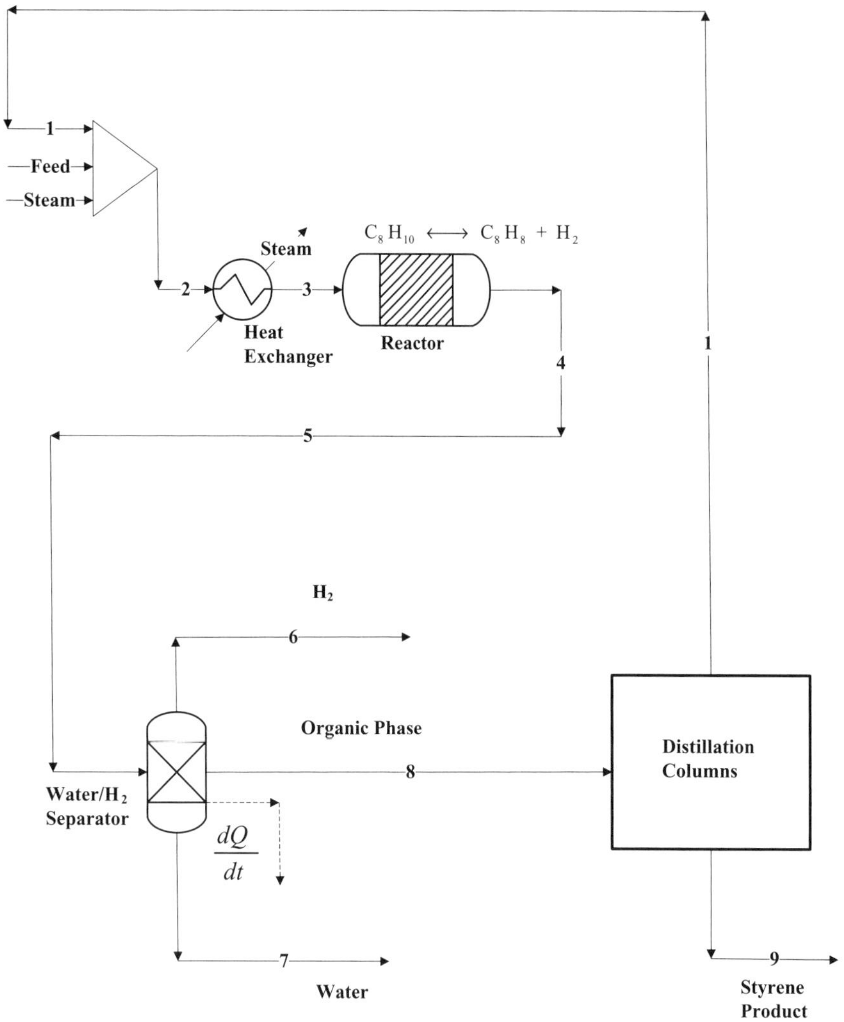

Figure 5.19 Flow sheet for the production of styrene from ethylbenzene.

Rate Expression for Ethylbenzene → Styrene + Hydrogen: $C_8H_{10} \leftrightarrow C_8H_8 + H_2$

Wenner and Dybdal (1948) provide the rate expression for a typical styrene catalyst as

$$\left(r_{C_8H_8}^{\text{global}}\right) = k\left(p_{C_8H_{10}} - \frac{1}{K_{eq}} p_{C_8H_8} p_{H_2}\right)$$
$$k = (3.5)\exp\left(\frac{-19800}{T}\right) \tag{5.54}$$
$$K_{eq} = \exp\left(15.596 - \frac{26,734.68}{T}\right),$$

with ρ_{catalyst} = 90-lb catalyst/ft^3 reactor volume, T [=] R, and reactor pressure = 1.2 atm.

Here,

k = the reaction rate constant $[=] \frac{\text{lb-mol styrene produced}}{\text{s-atm-lb catalyst}}$;

$p_{C_8H_{10}}$ = the partial pressure of ethylbenzene [=] atm;

$p_{C_8H_8}$ = the partial pressure of styrene [=] atm;

p_{H_2} = the partial pressure of hydrogen [=] atm; and

K_{eq} = the equilibrium constant, which is a function of the temperature [=] atm.

The partial pressure of each species, p_i, is straightforward to calculate as

$$p_i = \left(\frac{N_i}{N_{\text{total}}}\right) P_{\text{total}} = y_i P_{\text{total}}, \tag{5.55}$$

where N_i represents the moles of species i; N_{total}, the total number of moles; P_{total}, the system pressure; and y_i, the mole fraction of species i. This rate expression may seem unusual, but if we take a closer look at the ethylbenzene reaction, $C_8H_{10} \leftrightarrow C_8H_8 + H_2$, we see the reaction is actually reversible. In other words, there will be both forward and reverse reaction depending on the partial pressure of the reactants, the partial pressure of the products, and the reaction temperature. The forward reaction (to produce styrene and hydrogen) is governed by the rate constant and the ethylbenzene partial pressure, $kp_{C_8H_{10}}$, and the reverse reaction (to produce ethylbenzene) is governed by the rate constant, the equilibrium constant, and the styrene and hydrogen partial pressures,

$$\frac{k}{K_{eq}} p_{C_8H_8} p_{H_2}.$$

One final comment on rate expressions—some rate expressions appear with species concentrations (pound-mole per cubic foot), as opposed to partial pressures. In this case, if the reaction is gas phase and undergoes a change in the total number of moles, the volumetric flow rate will change. The change in the volumetric flow rate can be accounted for by using the ideal gas law or an appropriate equation of state. Here, the rate expression (Eq. (5.54)) is based on partial pressures, and we directly account for the change in total moles in Equation (5.55).

SOLUTION From Equations (5.34)–(5.37), the successive mass balances down the reactor length are

C_8H_{10} balance: $N_{C_8H_{10}}^{z_{j+1}} = N_{C_8H_{10}}^{z_j} - r_{C_8H_8}^{global}(\rho_{catalyst})(A_c)(\Delta z)$ (5.56)

Styrene balance: $N_{C_8H_8}^{z_{j+1}} = N_{C_8H_8}^{z_j} + r_{C_8H_8}^{global}(\rho_{catalyst})(A_c)(\Delta z)$ (5.57)

H_2 balance: $N_{H_2}^{z_{j+1}} = N_{H_2}^{z_j} + r_{C_8H_8}^{global}(\rho_{catalyst})(A_c)(\Delta z)$ (5.58)

H_2O balance: $N_{H_2O}^{z_{j+1}} = N_{H_2O}^{z_j}.$ (5.59)

The numerical integration is performed from known reactor feed conditions, ($N_{C_8H_{10}}^{z_j=0}$, $N_{C_8H_8}^{z_j=0}$, $N_{H_2}^{z_j=0}$, and $N_{H_2O}^{z_j=0}$, $T_{z_j=0}$). By selecting a value for Δz, and evaluating $r_{C_8H_8}^{global}$ at the $z_j = 0$ conditions, all of the conditions at $z_j = 1$ can be directly determined. This process continues until the end of the reactor is reached; if the reactor length is L, this would be $L/\Delta z$ steps. These iterative calculations are accomplished using a computer code, but here we perform the first Δz step by hand.

At the entrance to the PFR reactor $z_j = 0$, the known feed conditions are

Ethylbenzene $= 5.12821 \times 10^{-3}$ lb-mol/s

Styrene $= 3.36700 \times 10^{-5}$ lb-mol/s

Hydrogen $= 0.0$ lb-mol/s

Steam $= 9.80763 \times 10^{-2}$ lb-mol/s

Total moles $= 0.103238$ lb-mol/s

$$p_{C_8H_{10}} = \left(\frac{5.12821 \times 10^{-3}}{0.103238}\right) 1.2 \text{ atm} = 5.9608 \times 10^{-2} \text{ atm}$$

$$p_{C_8H_8} = \left(\frac{3.36700 \times 10^{-5}}{0.103238}\right) 1.2 \text{ atm} = 3.91368 \times 10^{-4} \text{ atm}$$

$$p_{H_2} = \left(\frac{0}{0.103238}\right) 1.2 \text{ atm} = 0.0 \text{ atm}$$

$$r_{C_8H_8}^{global}(\rho_{catalyst}) = (3.5)\left(\exp\left(\frac{-19,800}{1616}\right)\right)$$

$$\left(\left(5.9608 \times 10^{-2}\right) - \frac{(3.91368 \times 10^{-4})(0.0)}{(0.38762)}\right)(90)$$

$$= 8.96257 \times 10^{-5} \frac{\text{lb-mol } C_8H_8 \text{ produced}}{\text{ft}^3\text{-s}}.$$

Once we select our step size, Δz, we can calculate all the molar flow rates at $z_i = 1$ and the conversion over this Δz step. For $\Delta z = 0.1$ ft and $A_c = (\pi r^2) = \pi(2.3)^2$,

$$r_{C_8H_8}^{global}(\rho_{catalyst})(A_c)(\Delta z) = 1.48949 \times 10^{-4} \frac{\text{lb-mol } C_8H_8 \text{ produced}}{\text{s}}$$

$$= 1.48949 \times 10^{-4} \frac{\text{lb-mol } C_8H_{10} \text{ reacted}}{\text{s}}$$

C_8H_{10} balance:

$$N_{C_8H_{10}}^{z_{j+1}} = N_{C_8H_{10}}^{z_j} - r_{C_8H_8}^{global}(\rho_{catalyst})(A_c)(\Delta z) = 4.97926 \times 10^{-3} \text{ lb-mol/s}$$

Styrene balance:

$$N_{C_8H_8}^{z_{j+1}} = N_{C_8H_8}^{z_j} + r_{C_8H_8}^{global}(\rho_{catalyst})(A_c)(\Delta z) = 1.82619 \times 10^{-4} \text{ lb-mol/s}$$

H_2 balance:

$$N_{H_2}^{z_{j+1}} = N_{H_2}^{z_j} + r_{C_8H_8}^{global}(\rho_{catalyst})(A_c)(\Delta z) = 1.48949 \times 10^{-4} \text{ lb-mol/s}$$

H_2O balance: $N_{H_2O}^{z_{j+1}} = N_{H_2O}^{z_j} = 9.80763 \times 10^{-2}$ lb-mol/s

% conversion of ethylbenzene

$$= \frac{(5.12821 \times 10^{-3} - 4.97926 \times 10^{-3})}{5.12821 \times 10^{-3}} \times 100 = 2.905\%.$$

The heat that was generated by the reaction over this step (see Eq. (5.20)) is found as

$$\dot{Q}_0 = (-\Delta H_r)\left(\frac{\text{lb-mol } C_8H_8 \text{ produced}}{\text{s}}\right) = (-60,000)(1.48949 \times 10^{-4})$$

$$= -8.93694 \frac{\text{Btu}}{\text{s}}.$$

Finally, we calculate the temperature change from the energy balance, here using Equation (5.53),

$$F\hat{C}_p = \left(2.3148 \frac{\text{lb}}{\text{s}}\right)\left(0.52 \frac{\text{Btu}}{\text{lb-R}}\right) = 1.2037 \frac{\text{Btu}}{\text{s-R}}$$

$$T_1 = T_0 + \frac{\dot{Q}_0}{F\hat{C}_p} = 1616 + \frac{(-8.93694)}{1.2037} = 1608.58 \text{R}.$$

All conditions at $z_i = 1$ are known allowing conditions at $z_i = 2$ to be calculated. The VBA code and solution are provided in **Example 5.6b.xls**. The code with comments is shown in Figure 5.20 and the solution in Figure 5.21.

The output shows that even if the reactor length approached ∞, we will not reach the required 65% conversion. In an industrial process, the ethylbenzene reaction is performed in a series of two or three reactors. Between each reactor, the reacting fluid is heated. See Problem 5.2. ∎

```
Option Explicit
Public Sub Ethylbenzene_Kinetics()

    Dim N_EB(), P_EB(), Conversion_EB() As Double
    Dim N_S(), P_S(), delta_N_S() As Double
    Dim N_H2(), P_H2() As Double
    Dim N_H2O(), P_H2O() As Double

    Dim Temp(), N_Total(), Keq(), r_S(), Q() As Double

    Dim delta_z, Pi, radius, CSA As Double
    Dim rho_catalyst, delta_H_reaction, P_Total As Double
    Dim Cp, MW_EB, MW_S, MW_H2, MW_H2O, F_Total, FCp As Double

    Dim Number_zSteps, i As Integer

    'Set known reactor conditions and constants
    radius = 2.3 '  PFR tube radius
    Pi = 3.14159
    CSA = Pi * (radius) ^ 2 ' cross-sectional area ft^2

    rho_catalyst = 90    ' lbs/ft^3

    P_Total = 1.2    ' Reactor Pressure assumed constant
    delta_H_reaction = 60000.0#  ' Heat of Reaction (Btu/lb-mole)
    Cp = 0.52 ' Btu/(lb-R)

    MW_EB = 106.17
    MW_S = 104.15
    MW_H2 = 2.016
    MW_H2O = 18.015

    ' Here we set the step size down the reactor and the number of steps (=length)
    delta_z = 0.1         ' step size z direction [=] ft
    Number_zSteps = 90

    ReDim N_EB(Number_zSteps + 2), P_EB(Number_zSteps + 2), Conversion_EB(Number_zSteps
+ 2)
    ReDim N_S(Number_zSteps + 2), P_S(Number_zSteps + 2), delta_N_S(Number_zSteps + 2)
    ReDim N_H2(Number_zSteps + 2), P_H2(Number_zSteps + 2)
    ReDim N_H2O(Number_zSteps + 2), P_H2O(Number_zSteps + 2)

    ReDim Temp(Number_zSteps + 2), N_Total(Number_zSteps + 2), Keq(Number_zSteps + 2), _
                    r_S(Number_zSteps + 2), Q(Number_zSteps + 2)


    ' Establish PFR tube radius initial conditions: moles/s, temperature into PFR, z = 0
    Temp(0) = 1616  ' Temperature z = 0
    N_EB(0) = 18.4615 / 3600 ' Moles Ethylbenzene/hr, z = 0
    N_S(0) = 0.1212 / 3600
    N_H2(0) = 0.0 / 3600
    N_H2O(0) = 353.0745 / 3600
    N_Total(0) = N_EB(0) + N_S(0) + N_H2(0) + N_H2O(0)

    'Calculate the flow rate heat capacity, FCp, of the reacting fluid - assumed
constant
    F_Total = (N_EB(0) * MW_EB + N_S(0) * MW_S _
            + N_H2(0) * MW_H2 + N_H2O(0) * MW_H2O) ' Total Flow lbs/s
    FCp = F_Total * Cp ' Btu/lb-s

    ' We next set up the numerical integration to find the moles EB reacted
    ' here using a first-order Eulerian difference

    For i = 0 To Number_zSteps

        ' Monitor key values on the Excel sheet

        Sheet1.Cells(6 + i, 4) = delta_z * (i)
        Conversion_EB(i) = ((N_EB(0) - N_EB(i)) / N_EB(0)) * 100
        Sheet1.Cells(6 + i, 5) = Conversion_EB(i)
        Sheet1.Cells(6 + i, 6) = Temp(i)  ' R
        Sheet1.Cells(6 + i, 7) = Temp(i) / 1.8 - 273.15 ' C
        Sheet1.Cells(6 + i, 8) = Temp(i) - 459.67  ' F
        Sheet1.Cells(6 + i, 9) = N_EB(i)
        Sheet1.Cells(6 + i, 10) = N_S(i)
        Sheet1.Cells(6 + i, 11) = N_H2(i)
        Sheet1.Cells(6 + i, 12) = N_H2O(i)

        'Calculate species partial pressures

        P_EB(i) = (N_EB(i) / N_Total(i)) * P_Total
        P_S(i) = (N_S(i) / N_Total(i)) * P_Total
        P_H2(i) = (N_H2(i) / N_Total(i)) * P_Total
        P_H2O(i) = (N_H2O(i) / N_Total(i)) * P_Total
```

Figure 5.20 VBA code styrene reactor material balances and energy balance—Euler's method.

```
        ' Calculate Keq

        Keq(i) = Exp(15.596 - 26734.68 / Temp(i))

        '  Calcualte the reaction rate lb-moles Styrene/(ft^3-sec)

        r_S(i) = rho_catalyst * 3.5 * Exp(-19800 / Temp(i)) * (P_EB(i) - (P_S(i) *
P_H2(i) / Keq(i)))

        delta_N_S(i) = r_S(i) * CSA * delta_z

        Q(i) = delta_N_S(i) * (-delta_H_reaction)

        ' Calculate conditions at z + delta_z
        Temp(i + 1) = Temp(i) + Q(i) / (FCp)
        N_EB(i + 1) = N_EB(i) - delta_N_S(i)
        N_S(i + 1) = N_S(i) + delta_N_S(i)
        N_H2(i + 1) = N_H2(i) + delta_N_S(i)
        N_H2O(i + 1) = N_H2O(i)
        N_Total(i + 1) = N_EB(i + 1) + N_S(i + 1) + N_H2(i + 1) + N_H2O(i + 1)
    Next i

End Sub
```

Figure 5.20 (*Continued*)

	A	B	C	D	E	F	G	H	I	J	K	L
5				distance z (ft)	% conversion EB	Temperature (R)	Temperature (C)	Temperature (F)	EB mol/s	S mol/s	H2 mol/s	H2O mol/s
6				0	0.000	1616.00	624.63	1156.33	5.13E-03	3.37E-05	0.00E+00	9.81E-02
7				0.1	2.905	1608.58	620.50	1148.91	4.98E-03	1.83E-04	1.49E-04	9.81E-02
8				0.2	5.565	1601.77	616.72	1142.10	4.84E-03	3.19E-04	2.85E-04	9.81E-02
9				0.3	8.017	1595.51	613.24	1135.84	4.72E-03	4.45E-04	4.11E-04	9.81E-02
10				0.4	10.287	1589.70	610.02	1130.03	4.60E-03	5.61E-04	5.28E-04	9.81E-02
11				0.5	12.399	1584.31	607.02	1124.64	4.49E-03	6.70E-04	6.36E-04	9.81E-02
12				0.6	14.370	1579.27	604.22	1119.60	4.39E-03	7.71E-04	7.37E-04	9.81E-02
13				0.7	16.217	1574.55	601.60	1114.88	4.30E-03	8.65E-04	8.32E-04	9.81E-02
14				0.8	17.952	1570.11	599.13	1110.44	4.21E-03	9.54E-04	9.21E-04	9.81E-02
15				0.9	19.586	1565.93	596.81	1106.26	4.12E-03	1.04E-03	1.00E-03	9.81E-02
16				1	21.130	1561.99	594.62	1102.32	4.04E-03	1.12E-03	1.08E-03	9.81E-02
17				1.1	22.590	1558.25	592.55	1098.58	3.97E-03	1.19E-03	1.16E-03	9.81E-02
18				1.2	23.976	1554.71	590.58	1095.04	3.90E-03	1.26E-03	1.23E-03	9.81E-02
19				1.3	25.291	1551.35	588.71	1091.68	3.83E-03	1.33E-03	1.30E-03	9.81E-02
20				1.4	26.543	1548.15	586.93	1088.48	3.77E-03	1.39E-03	1.36E-03	9.81E-02
21				1.5	27.736	1545.10	585.24	1085.43	3.71E-03	1.46E-03	1.42E-03	9.81E-02
22				1.6	28.874	1542.19	583.62	1082.52	3.65E-03	1.51E-03	1.48E-03	9.81E-02
23				1.7	29.962	1539.41	582.08	1079.74	3.59E-03	1.57E-03	1.54E-03	9.81E-02
24				1.8	31.002	1536.75	580.60	1077.08	3.54E-03	1.62E-03	1.59E-03	9.81E-02
25				1.9	31.998	1534.21	579.19	1074.54	3.49E-03	1.67E-03	1.64E-03	9.81E-02
26				2	32.953	1531.76	577.83	1072.09	3.44E-03	1.72E-03	1.69E-03	9.81E-02
27				2.1	33.869	1529.42	576.53	1069.75	3.39E-03	1.77E-03	1.74E-03	9.81E-02
28				2.2	34.749	1527.17	575.28	1067.50	3.35E-03	1.82E-03	1.78E-03	9.81E-02
29				2.3	35.594	1525.01	574.08	1065.34	3.30E-03	1.86E-03	1.83E-03	9.81E-02
40				3.4	43.092	1505.85	563.43	1046.18	2.92E-03	2.24E-03	2.21E-03	9.81E-02
41				3.5	43.642	1504.44	562.65	1044.77	2.89E-03	2.27E-03	2.24E-03	9.81E-02
42				3.6	44.174	1503.08	561.90	1043.41	2.86E-03	2.30E-03	2.27E-03	9.81E-02
43				3.7	44.689	1501.77	561.16	1042.10	2.84E-03	2.33E-03	2.29E-03	9.81E-02
44				3.8	45.188	1500.49	560.46	1040.82	2.81E-03	2.35E-03	2.32E-03	9.81E-02
45				3.9	45.671	1499.26	559.77	1039.59	2.79E-03	2.38E-03	2.34E-03	9.81E-02
46				4	46.139	1498.06	559.10	1038.39	2.76E-03	2.40E-03	2.37E-03	9.81E-02
901				89.5	62.904376	1455.203045	535.2961363	995.5330453	0.001902	0.003259525	0.003226	0.098076
902				89.6	62.904376	1455.203045	535.2961363	995.5330453	0.001902	0.003259525	0.003226	0.098076
903				89.7	62.904376	1455.203045	535.2961363	995.5330453	0.001902	0.003259525	0.003226	0.098076
904				89.8	62.904376	1455.203045	535.2961363	995.5330453	0.001902	0.003259525	0.003226	0.098076
905				89.9	62.904376	1455.203045	535.2961363	995.5330453	0.001902	0.003259525	0.003226	0.098076
906				90	62.904376	1455.203045	535.2961363	995.5330453	0.001902	0.003259525	0.003226	0.098076

Figure 5.21 Results of styrene reactor material balances and energy balance—Euler's method.

5.6 EULER'S METHOD VERSUS FOURTH-ORDER RUNGE–KUTTA METHOD FOR NUMERICAL INTEGRATION

It is reasonable to ask how do we select the best step size, Δz, for use in Euler's method? The easy answer is to simply make Δz sufficiently small so that the use of properties at z_j are sensible over the step z_j to z_{j+1}; for example, an easy check is that the temperature at T_{z_j} should be close to $T_{z_{j+1}}$. However, if Δz is too small, numerical errors in computer calculations can accumulate.

5.6.1 The Euler Method: First-Order ODEs

A better answer is to understand the limitations of Euler's method and to consider the use of higher-order methods to solve our ODEs. Euler's method is intuitive, but the solution

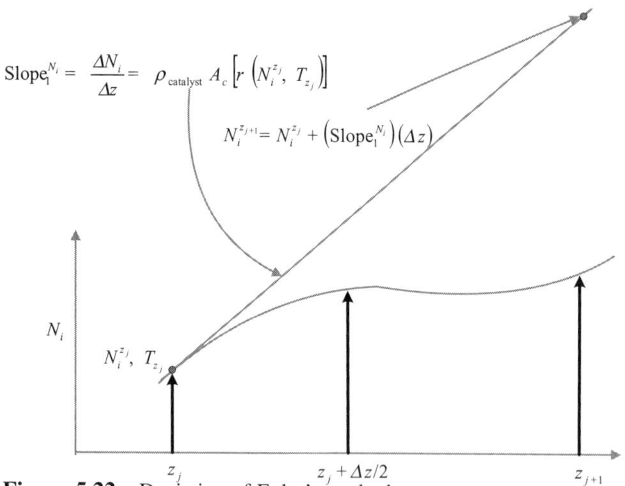

Figure 5.22 Depiction of Euler's method.

from z_j to z_{j+1} is based on derivative information evaluated at z_j only. Figure 5.22 depicts our PFR material balance solution using Euler's method over one step from z_j to z_{j+1}.

Allowing we are at N_i^{zj}, T_{zj}, we find the species molar flows at z_{j+1} from Equation (5.43),

$$N_i^{zj+1} = N_i^{zj} + \left(\rho_{\text{catalyst}}\right)\left(A_c\right)\left(r_i^{\text{global}}\Big|_{zj}\right)\Delta z,$$

where the slope of the line from N_i^{zj}, T_{zj} to N_i^{zj+1} is

$$\text{Slope}_1^{N_i} = \left(\rho_{\text{catalyst}}\right)\left(A_c\right)\left(r_i^{\text{global}}\Big|_{zj}\right) = \rho_{\text{catalyst}} A_c \left[r_{\text{global}}\left(N_i^{zj}, T_{zj}\right)\right].$$

Here, we have slightly altered our nomenclature to emphasize that the global reaction rate r_{global} is evaluated at z_j conditions N_i^{zj}, T_{zj}. The change in nomenclature will be helpful in our understanding the RK method. The figure shows that Equation (5.43) can be written as

$$N_i^{zj+1} = N_i^{zj} + \left(\text{Slope}_1^{N_i}\right)(\Delta z).$$

The slope of the line from N_i^{zj}, T_{zj} to T_{zj+1} is found via one of our energy balance equations (Eqs. (5.47)–(5.53)), here using Equation (5.52):

$$T_{zj+1} = T_{zj} + \frac{\left(\rho_{\text{catalyst}}\right)\left(A_c\right)\left(-\Delta H_{r,i\ \text{limiting}}^{T^{\text{ref}}}\right)\left(r_{i\ \text{limiting}}^{\text{global}}\Big|_{zj}\right)\Delta z}{\sum_i N_i^{zj}\left(C_{p,i}^{\text{avg}}\right)}$$

$$\text{Slope}_1^{T} = \frac{\left(\rho_{\text{catalyst}}\right)\left(A_c\right)\left(-\Delta H_{r,i\ \text{limiting}}^{T^{\text{ref}}}\right)\left(r_{i\ \text{limiting}}^{\text{global}}\Big|_{zj}\right)}{\sum_i N_i^{zj}\left(C_{p,i}^{\text{avg}}\right)}$$

$$= \frac{\left(\rho_{\text{catalyst}}\right)\left(A_c\right)\left(-\Delta H_{r,i\ \text{limiting}}^{T^{\text{ref}}}\right)\left[r^{\text{global}}\left(N_{i\ \text{limiting}}^{zj}, T_{zj}\right)\right]}{\sum_i N_i^{zj}\left(C_{p,i}^{\text{avg}}\right)}.$$

We can then write $T_{zj+1} = T_{zj} + \left(\text{Slope}_1^T\right)(\Delta z)$; the figure does not indicate the calculation of T_{zj+1}, but the process should be clear. Do note that in the calculation of Slope_1^T, the reaction rate and flow rate heat capacity terms are calculated using z_j conditions even though we have updated values for species flows at N_s^{zj+1}.

With the set of coupled reactor ODEs, there will be slopes associated with the change in species flows from z_j to z_{j+1} and the change in temperature from z_j to z_{j+1}. Examining Figure 5.22, it is clear that improvements in our determined species flows or temperature at z_{j+1} could be obtained if we had derivative information along the path z_j to z_{j+1}. The RK method utilizes additional slope information along the path z_j to z_{j+1}—the foundation for the RK method is the utilization of higher-order terms in a Taylor series expansion. Next, we develop the fourth-order Runge–Kutta (RK4) method, which is often used for solving ODEs.

5.6.2 RK4 Method: First-Order ODEs

It is easiest to show the RK4 method by a series of figures (see Figure 5.23a–c). As with Euler's method, we only indicate species flow rates on the figures. Also, for temperature calculations, we will use Equation (5.52). The RK4 method (Figure 5.23a) starts by determining $\text{Slope}_1^{N_i} = \rho_{\text{catalyst}} A_c \left[r_{\text{global}}\left(N_i^{zj}, T_{zj}\right)\right]$ and

$$\text{Slope}_1^{T} = \frac{\left(\rho_{\text{catalyst}}\right)\left(A_c\right)\left(-\Delta H_{r,i\ \text{limiting}}^{T^{\text{ref}}}\right)\left[r^{\text{global}}\left(N_{i\ \text{limiting}}^{zj}, T_{zj}\right)\right]}{\sum_i N_i^{zj}\left(C_{p,i}^{\text{avg}}\right)}.$$

This is identical to the Euler method. However, the species molar flow rates and the temperature are determined at the interval midpoint, not the end point, as in Euler's method:

$$N_i^2 = N_i^{zj} + \left(\text{Slope}_1^{N_i}\right)\left(\frac{\Delta z}{2}\right)$$

$$T_2 = T_{z_i} + \left(\text{Slope}_1^{T}\right)\left(\frac{\Delta z}{2}\right).$$

Here we have noted the midpoint conditions with a superscript 2 on the species molar flow rates and a subscript 2 on the temperature. We then calculate, using the midpoint conditions (N_s^2, T_2), the derivatives

$$\text{Slope}_2^{N_i} = \rho_{\text{catalyst}} A_c \left[r_{\text{global}}\left(N_i^2, T_2\right)\right]$$

and

$$\text{Slope}_2^{T} = \frac{\left(\rho_{\text{catalyst}}\right)\left(A_c\right)\left(-\Delta H_{r,i\ \text{limiting}}^{T^{\text{ref}}}\right)\left[r_{\text{global}}\left(N_{i\ \text{limiting}}^2, T_2\right)\right]}{\sum_i N_i^2\left(C_{p,i}^{\text{avg}}\right)}.$$

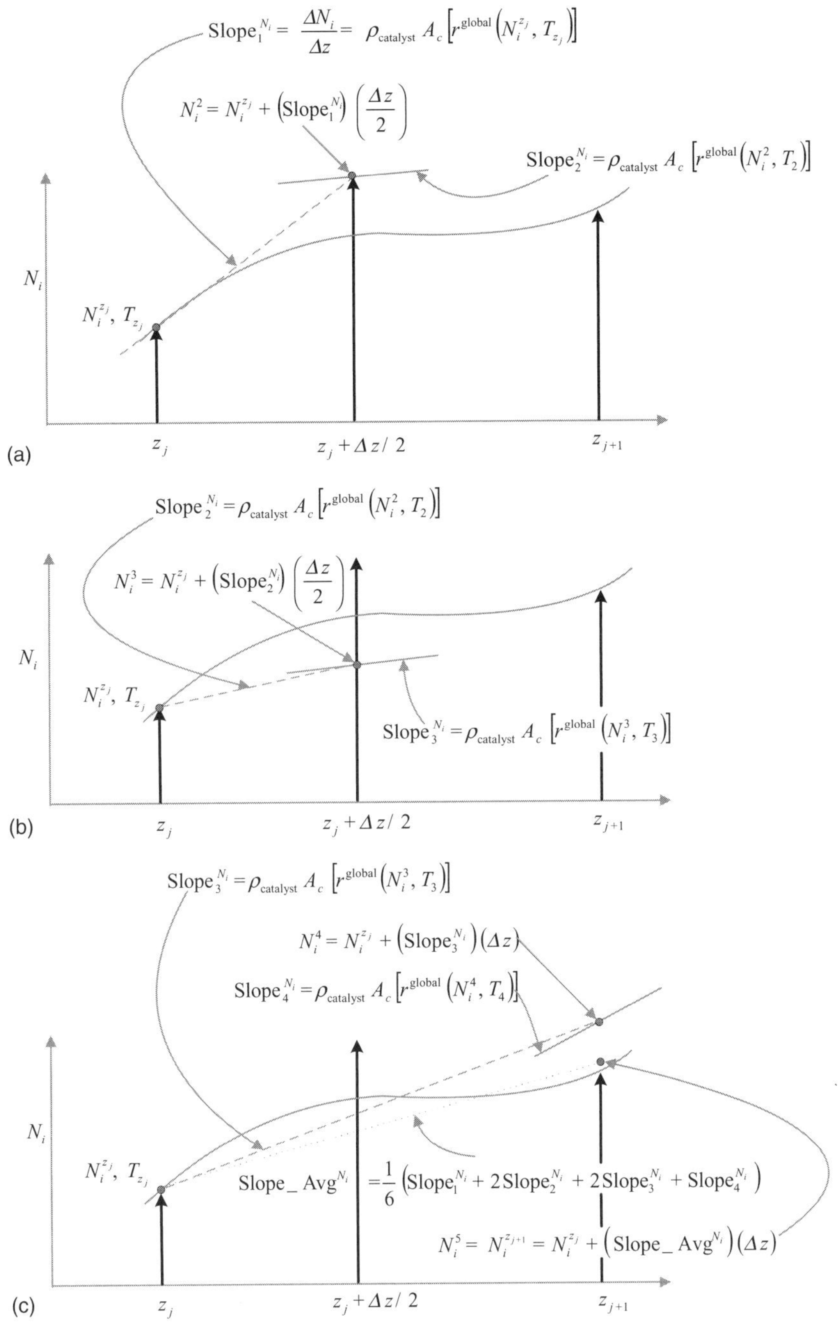

Figure 5.23 Runge–Kutta fourth-order method.

Next, as shown in Figure 5.23b, the determined slope$_2$ values are used from z_i to determine a new midpoint species molar flow rates and temperature:

$$N_i^3 = N_i^{zj} + \left(\text{Slope}_2^{N_i}\right)\left(\frac{\Delta z}{2}\right)$$

and

$$T_3 = T_{z_j} + \left(\text{Slope}_2^T\right)\left(\frac{\Delta z}{2}\right).$$

We then calculate Slope$_3$ using (N_i^3, T_3) conditions:

$$\text{Slope}_3^{N_i} = \rho_{\text{catalyst}} A_c \left[r_{\text{global}}\left(N_i^3, T_3\right)\right]$$

and

$$\text{Slope}_3^T = \frac{\left(\rho_{\text{catalyst}}\right)\left(A_c\right)\left(-\Delta H_{r,i\text{ limiting}}^{T^{\text{ref}}}\right)\left[r_{\text{global}}\left(N_{i\text{ limiting}}^3, T_3\right)\right]}{\sum_i N_i^3\left(C_{p,i}^{\text{avg}}\right)}.$$

Finally, as shown in Figure 5.23c, Slope$_3$ is used from z_j to determine new endpoint species molar flow rates and temperature: $N_i^4 = N_i^{z_j} + \left(\text{Slope}_3^{N_i}\right)(\Delta z)$ and $T_4 = T_{z_j} + \left(\text{Slope}_3^T\right)(\Delta z)$. We then calculate Slope_4:

$$\text{Slope}_4^{N_i} = \rho_{\text{catalyst}} A_c r\left(N_i^4, T_4\right)$$

$$\text{Slope}_4^T = \frac{\left(\rho_{\text{catalyst}}\right)\left(A_c\right)\left(-\Delta H_{r,i\,\text{limiting}}^{T^{\text{ref}}}\right)\left[r\left(N_{i\,\text{limiting}}^4, T_4\right)\right]}{\sum_i N_i^4\left(C_{p,i}^{\text{avg}}\right)}.$$

All intermediate slopes for the RK4 method have been determined, and a weighted average slope is calculated as

$$\text{Slope\_Avg}_{N_i} = \frac{1}{6}\left(\text{Slope}_1^{N_i} + 2\ \text{Slope}_2^{N_i} + 2\ \text{Slope}_3^{N_i} + \text{Slope}_4^{N_i}\right)$$

$$\text{Slope\_Avg}_T = \frac{1}{6}\left(\text{Slope}_1^T + 2\ \text{Slope}_2^T + 2\ \text{Slope}_3^T + \text{Slope}_4^T\right).$$

These slopes are used from z_j to determine the "true" endpoint conditions:

$$N_i^5 = N_i^{z_{j+1}} = N_i^{z_j} + \left(\text{Slope\_Avg}_{N_i}\right)(\Delta z)$$

$$T_5 = T_{z_{j+1}} = T_{z_j} + \left(\text{Slope\_Avg}_T\right)(\Delta z).$$

EXAMPLE 5.7 ***Styrene Reactor Material and Energy Balances Using RK Fourth-Order Method***

For the styrene PFR of Example 5.6 (modified from Smith, 1970), develop the numerical solution to the reactor material and energy balance problem using the RK4 method. Compare the solution from RK4 to those obtained using Euler's method (Example 5.6).

SOLUTION The RK4 solution is provided in **Example 5.7.xls**. Here we have modified our existing Euler method solution (from Example 5.6) and added the needed RK4 procedure following Figure 5.23a–c. Two VBA functions have been added to determine the needed slopes (derivatives) for the change in species molar flow rate with respect to z and the change in temperature with respect to z. The code is not "tight"; for example, the use of additional Sub procedures to determine species molar flow rates and species partial pressures would reduce the number of lines of code. Before "tightening" the code, be sure to confirm understanding of the RK4 method on Problem 5.3 (solution of two coupled ODEs by hand) and Problem 5.4 (computer solution of two coupled ODEs). The output for the catalyst bed using the RK4 method is shown in Figure 5.24. The code was run with a fixed step size of 0.1 ft to allow direct comparison with the results in Figure 5.21.

For the styrene material and energy balance problem, there is a small difference in the solution obtained from Euler's method and RK4. At 2.3 ft, Euler's method shows 35.59% conversion of ethylbenzene, while the RK4 method shows 35.06% ethylbenzene conversion. At 2.3 ft, there is and a ~1.36-R difference in temperature between the two solution methods. There will be greater differences in the solutions for strongly exothermic or endothermic reactions. ■

Improvements to the RK4 method can be made with an z adjustable—termed variable step size. We will see in Chapter 15 that in order to solve for reacting species concentrations from combustion systems, we will need to use commercial-grade codes. Here we provide CVODE from Lawrence Livermore National Laboratory, which we have modified to be callable from Excel.

5.7 CLOSING COMMENTS

In this chapter, we added energy balances to our elementary material balance modules. We saw for equilibrium flash

	A	B	C	D	E	F	G	H	I	J	K	L
5				distance z (ft)	% conversion EB	Temperature (R)	Temperature (C)	Temperature (F)	EB mol/s	S mol/s	H2 mol/s	H20 mol/s
6				0	0.00	1616.00	624.63	1156.33	5.13E-03	3.37E-05	0.00E+00	9.81E-02
7				0.1	2.78	1608.88	620.67	1149.21	4.99E-03	1.76E-04	1.43E-04	9.81E-02
8				0.2	5.35	1602.32	617.03	1142.65	4.85E-03	3.08E-04	2.74E-04	9.81E-02
9				0.3	7.73	1596.24	613.65	1136.57	4.73E-03	4.30E-04	3.96E-04	9.81E-02
10				0.4	9.94	1590.59	610.51	1130.92	4.62E-03	5.43E-04	5.10E-04	9.81E-02
11				0.5	12.01	1585.31	607.58	1125.64	4.51E-03	6.49E-04	6.16E-04	9.81E-02
12				0.6	13.94	1580.37	604.83	1120.70	4.41E-03	7.49E-04	7.15E-04	9.81E-02
13				0.7	15.76	1575.72	602.25	1116.05	4.32E-03	8.42E-04	8.08E-04	9.81E-02
14				0.8	17.47	1571.35	599.82	1111.68	4.23E-03	9.29E-04	8.96E-04	9.81E-02
15				0.9	19.08	1567.22	597.53	1107.55	4.15E-03	1.01E-03	9.79E-04	9.81E-02
16				1	20.61	1563.31	595.36	1103.64	4.07E-03	1.09E-03	1.06E-03	9.81E-02
17				1.1	22.06	1559.60	593.30	1099.93	4.00E-03	1.17E-03	1.13E-03	9.81E-02
18				1.2	23.44	1556.09	591.34	1096.42	3.93E-03	1.24E-03	1.20E-03	9.81E-02
19				1.3	24.75	1552.74	589.48	1093.07	3.86E-03	1.30E-03	1.27E-03	9.81E-02
20				1.4	26.00	1549.55	587.71	1089.88	3.80E-03	1.37E-03	1.33E-03	9.81E-02
21				1.5	27.19	1546.50	586.02	1086.83	3.73E-03	1.43E-03	1.39E-03	9.81E-02
22				1.6	28.32	1543.60	584.40	1083.93	3.68E-03	1.49E-03	1.45E-03	9.81E-02
23				1.7	29.41	1540.82	582.86	1081.15	3.62E-03	1.54E-03	1.51E-03	9.81E-02
24				1.8	30.45	1538.15	581.38	1078.48	3.57E-03	1.60E-03	1.56E-03	9.81E-02
25				1.9	31.45	1535.60	579.96	1075.93	3.52E-03	1.65E-03	1.61E-03	9.81E-02
26				2	32.41	1533.16	578.60	1073.49	3.47E-03	1.70E-03	1.66E-03	9.81E-02
27				2.1	33.33	1530.81	577.30	1071.14	3.42E-03	1.74E-03	1.71E-03	9.81E-02
28				2.2	34.21	1528.55	576.04	1068.88	3.37E-03	1.79E-03	1.75E-03	9.81E-02
29				2.3	35.06	1526.37	574.84	1066.70	3.33E-03	1.83E-03	1.80E-03	9.81E-02

Figure 5.24 Styrene production—the RK4 method solution.

modules that when iterative phase calculations are combined with recycle loops, we must be careful as to the solution order for each unit in the process. We also explored reactor energy balances using both a heat of reaction-based formulation and a formulation using species enthalpy of formation from its elements. Most commercial packages utilize species enthalpy of formation from its elements in energy balances. Understanding the heat of reaction formulation is often important for combustion processes as fuel is sold based on its higher heating value and energy balance calculation is performed using fuel lower heating value (*LHV*)—the *LHV* can be related to the heat of reaction.

The reactor ODE solution methods developed in this chapter (Euler's method and RK fourth-order method) are important tools for designing reactors. Chapter 15 combustion reactors will be modeled as combination PFR and PSRs in order to predict emissions. Here we will find that Euler's method and the RK fourth-order method will not be able to solve these reactors for temperature and emission concentrations. We will need to understand our solution formulation in this chapter to understand solution difficulties in Chapter 15. We will provide the commercial standard ODE solver CVODE from Lawrence Livermore National Laboratory to solve these combustion kinetics problems. CVODE is provided as a callable program from Excel, which was developed using the methods described in the Appendix.

REFERENCES

Chandra, P., P. Singh, and D.N. Saraf. 1979. Simulation of ammonia synthesis reactors. *Ind. Eng. Chem. Process Des. Dev.* 18(3): 364–370.

Dashti, A., K. Khorsand, M.A. Marvast, and M. Kakavand 2006. Modeling and simulation of ammonia synthesis reactor. *Petro. Coal* 48(2): 15–23.

Dyson D.C. and J.M. Simon 1968. A kinetic expression with diffusion correction for ammonia synthesis on industrial catalyst. *Ind. Eng. Chem. Fundam.* 7(4): 605–610.

Elnashaie, S.S., M.E. Abashar, and A.S. Al-Ubaid 1988a. Simulation and optimization of an industrial ammonia reactor. *Ind. Eng. Chem. Res.* 27: 2015–2022.

Elnashaie, S.S., A.T. Mahfouz, and S.S. Elshishini 1988b. Digital simulation of an industrial ammonia reactor. *Chem. Eng. Process.* 23(3): 165–177.

Gaines, L.D. 1977. Optimal temperatures for ammonia synthesis converters. *Ind. Eng. Chem. Process Des. Dev.* 16(3): 381–3389.

Gaines, L.D. 1979. Ammonia synthesis loop variable investigated by steady-state simulations. *Chem. Eng. Sci.* 34: 37–50.

Mansson, B. and B. Andresen 1986. Optimal temperature profile for an ammonia reactor. *Ind. Eng. Chem. Process Des. Dev.* 25: 59–65.

Myers, A.L. and W.D. Seider 1976. *Introduction to Chemical Engineering and Computer Calculations*. Prentice Hall, Englewood Cliffs, NJ.

Rase, H.F. 1977. *Chemical Reactor Design for Process Plants, Volume II Case Studies & Design Data*. John Wiley and Sons, New York.

Reid, R.C., J.M. Prausnitz, and T.K. Sherwood 1977. *The Properties of Gases and Liquids*. McGraw-Hill, New York.

Reklaitis, G.V. 1983. *Introduction to Material and Energy Balances*. John Wiley and Sons, New York.

Sandler, S.I. 1999. *Chemical and Engineering Thermodynamics*. John Wiley and Sons, New York.

Smith, J.M. 1970. *Chemical Engineering Kinetics*. McGraw-Hill, New York.

Walas, S.M. 1985. Chemical reactor data. *Chem. Eng.* 14: 79–83.

Wenner, R.R. and E.C. Dybdal 1948. Catalytic dehydrogenation of ethylbenzene. *Chem. Eng. Progr.* 44(4): 275–286.

PROBLEMS

5.1 Compare the K values used in Example 5.2 (Myers and Seider, 1976; $K_{N_2} = 66.67$, $K_{H_2} = 50.0$, and $K_{NH_3} = 0.015$) with those obtained using Raoult's law. For mixtures of similar molecular structure, Raoults's law may be applicable. Raoults's law can be written as

$$K_i = \frac{p_i}{P},$$

where p_i is the vapor pressure of species i at temperature T and P is the total pressure. Antoine's equation can be used for species vapor pressure:

$$\ln p_i = A - \frac{B}{T + C},$$

where the constants A, B, and C (T in kelvin and p_i in millimeter of mercury) are given in Reid et al. (1977) as

	A	B	C
N_2	14.9542	588.72	−6.6
H_2	13.6333	164.9	3.19
NH_3	16.9481	2132.5	−33

Determine the K_i values using Raoult's law and Antoine's equation. Compare the solution of the flash problem of Figure 5.7 with that obtained using the Raoult's law-based K_i values.

5.2 What ethlybenzene conversion would occur if two adiabatic PFR reactors in series, with heating between the PFR, were used? All properties and conditions remain as given in Example 5.6. The tube length in each reactor is now fixed at 4 ft. The stream leaving the first reactor is heated to 1650 R prior to entering the second reactor.

5.3 "By hand" numerically integrate the following set of ODEs:

$$\frac{dy_1}{dx} = -0.2y_1^3$$

$$\frac{dy_2}{dx} = 8 - 0.5y_1y_2 - 0.1y_1.$$

Integrate from $x = 0$ to $x = 0.5$ using a step size of 0.5. At $x = 0$, $y_1 = 2$, and $y_2 = 3$,

(a) perform one iteration by hand of Euler's method and

(b) perform one iteration by hand of the RK4 method.

5.4 Write a VBA code to numerically integrate the set of ODEs in Problem 5.3. Print intermediate results to allow comparison with Problem 5.3.

$$\frac{dy_1}{dx} = -0.2y_1^3$$

$$\frac{dy_2}{dx} = 8 - 0.5y_1y_2 - 0.1y_1.$$

Integrate from $x = 0$ to $x = 2.0$ using a step size of 0.5. At $x = 0$, $y_1 = 2$, and $y_2 = 3$,
(a) solve using Euler's method and
(b) solve using the RK4 method.

5.5 ***Ammonia Process Design: A Mini Project (Solution Provided)*** (modified from Chandra et al., 1979; Dashti et al., 2006; Elnashaie et al., 1988a,b; Gaines, 1977, 1979; Mansson and Andresen, 1986; Rase, 1977)

The ammonia process is exothermic—here, reactor design is necessary to prevent possible unsafe (high temperature) reactor operation. This problem may take several days to complete and should be considered a mini project best solved using student teams. Several mini-project extensions are given at the end of Problem 5.5 and in Problems 5.7–5.10. This problem is important as it includes safety considerations and many of the key topics detailed in this chapter.

The ammonia process flow sheet is shown in Figure P5.5a. We want to determine the size of the adiabatic PFR needed to

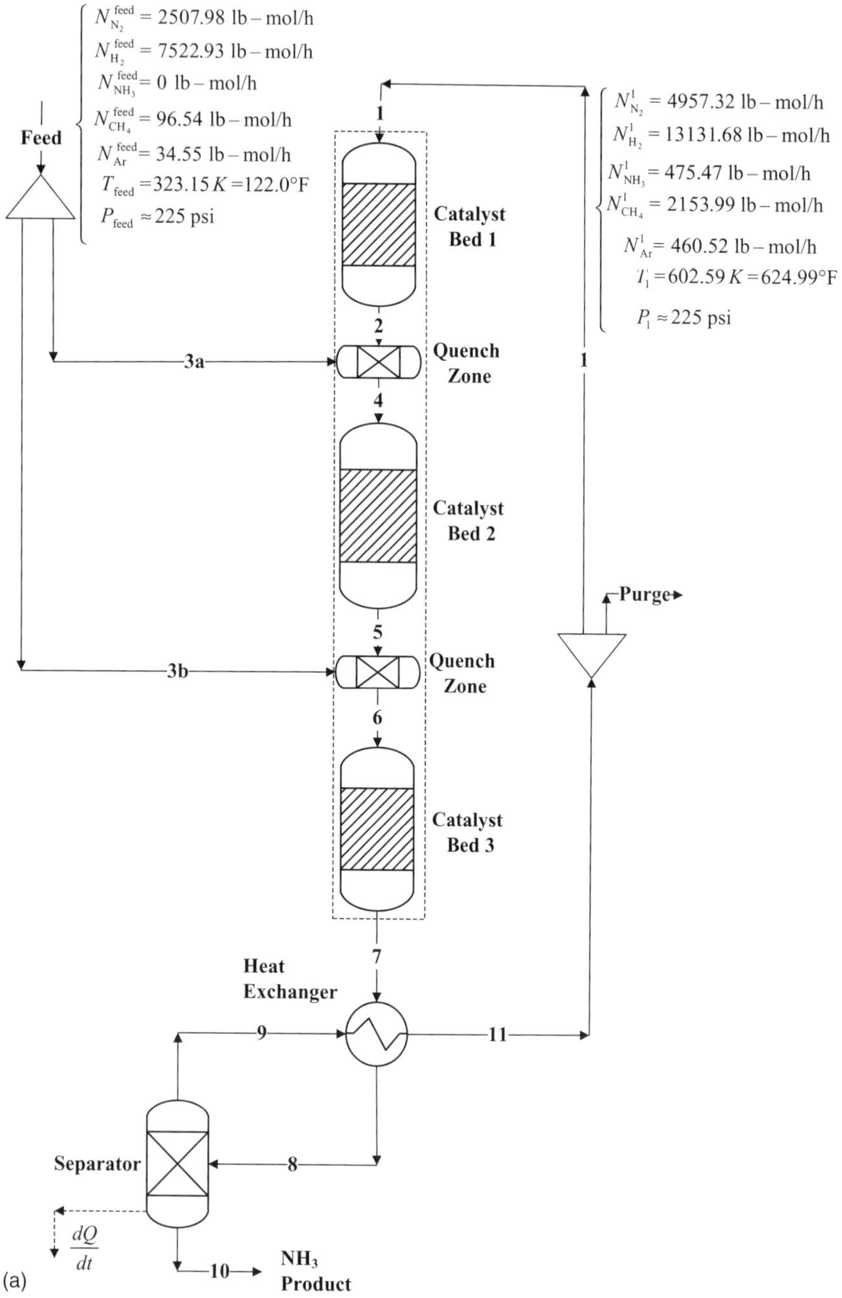

Figure P5.5 (a) Flow sheet for ammonia production. Results for stream 1 are from the Excel file **Problem 5.5a.xls**. (b) Solution procedure. (c) Ammonia production in the first catalytic bed—the Euler method solution.

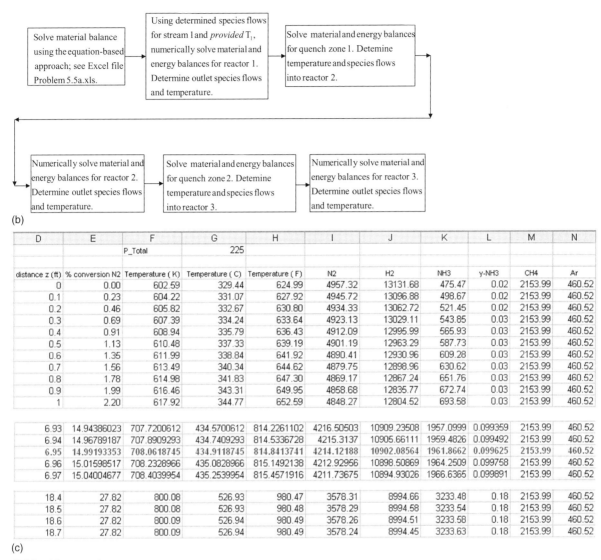

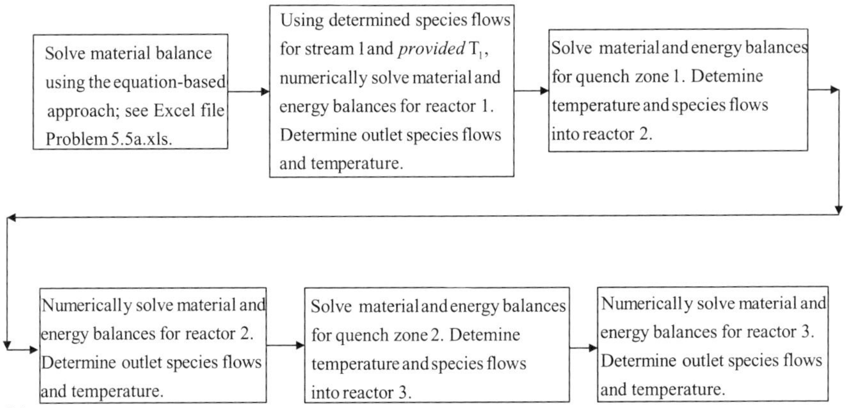

(b)

D	E	F	G	H	I	J	K	L	M	N
		P_Total	225							
distance z (ft)	% conversion N2	Temperature (K)	Temperature (C)	Temperature (F)	N2	H2	NH3	y-NH3	CH4	Ar
0	0.00	602.59	329.44	624.99	4957.32	13131.68	475.47	0.02	2153.99	460.52
0.1	0.23	604.22	331.07	627.92	4945.72	13096.88	498.67	0.02	2153.99	460.52
0.2	0.46	605.82	332.67	630.80	4934.33	13062.72	521.45	0.02	2153.99	460.52
0.3	0.69	607.39	334.24	633.64	4923.13	13029.11	543.85	0.03	2153.99	460.52
0.4	0.91	608.94	335.79	636.43	4912.09	12995.99	565.93	0.03	2153.99	460.52
0.5	1.13	610.48	337.33	639.19	4901.19	12963.29	587.73	0.03	2153.99	460.52
0.6	1.35	611.99	338.84	641.92	4890.41	12930.96	609.28	0.03	2153.99	460.52
0.7	1.56	613.49	340.34	644.62	4879.75	12898.96	630.62	0.03	2153.99	460.52
0.8	1.78	614.98	341.83	647.30	4869.17	12867.24	651.76	0.03	2153.99	460.52
0.9	1.99	616.46	343.31	649.95	4858.68	12835.77	672.74	0.03	2153.99	460.52
1	2.20	617.92	344.77	652.59	4848.27	12804.52	693.58	0.03	2153.99	460.52
6.93	14.94386023	707.7200612	434.5700612	814.2261102	4216.50503	10909.23508	1957.0999	0.099359	2153.99	460.52
6.94	14.96789187	707.8909293	434.7409293	814.5336728	4215.3137	10905.66111	1959.4826	0.099492	2153.99	460.52
6.95	14.99193353	708.0618745	434.9118745	814.8413741	4214.12188	10902.08564	1961.8662	0.099625	2153.99	460.52
6.96	15.01598517	708.2328966	435.0828966	815.1492138	4212.92956	10898.50869	1964.2509	0.099758	2153.99	460.52
6.97	15.04004677	708.4039954	435.2539954	815.4571916	4211.73675	10894.93026	1966.6365	0.099891	2153.99	460.52
18.4	27.82	800.08	526.93	980.47	3578.31	8994.66	3233.48	0.18	2153.99	460.52
18.5	27.82	800.08	526.93	980.48	3578.29	8994.58	3233.54	0.18	2153.99	460.52
18.6	27.82	800.09	526.94	980.49	3578.26	8994.51	3233.58	0.18	2153.99	460.52
18.7	27.82	800.09	526.94	980.49	3578.24	8994.45	3233.63	0.18	2153.99	460.52

(c)

Figure P5.5 ***(Continued)***

produce 2 million pounds of ammonia per day. The PFR, which is indicated by the dashed lines, consists of three adiabatic beds with fresh feed addition prior to beds 2 and 3. The product from reactor bed 1 is mixed with fresh feed prior to entering reactor bed 2. Feed addition serves to quench or reduce the temperature of the products leaving catalyst bed 1. The products from reactor bed 2 are also quenched with fresh feed prior to entering the final bed. The "cold shot" of feed is split 50/50 to each of the two quench zones. This configuration is quite interesting and raises several important questions we want to address after we discuss and examine the reaction kinetics.

The catalyst beds in the PFR can be considered three separate reactors. In each catalyst bed, the reaction is

$$N_2 + 3H_2 \longleftrightarrow 2NH_3,$$

and each bed has 15% conversion of the incoming nitrogen. Following the reactor, the products are cooled to −10°F in a flash and an ammonia-rich liquid stream is recovered. The vapor stream from the flash is sent to reactor 1 after heat exchange with the products from the reactor 3. To help prevent the buildup of inerts, 1% (molar flow basis) of the recycled vapor stream is purged.

The feed stream temperature is 323.15 K (122°F), the temperature of stream 1 into the first catalyst bed is 602.59 K (624.99°F), and the pressure throughout the process is taken as 225 atm. The mole fractions in the feed stream are $y_{N_2} = 0.2468$, $y_{H_2} = 0.7403$, $y_{NH_3} = 0.0$, $y_{CH_4} = 0.0095$, and $y_{Ar} = 0.0034$. The K values for the flash (−10°F, 225 atm) are $K_{N_2} = 132.52941$, $K_{H_2} = 72.2809$, $K_{NH_3} = 0.02317$, $K_{CH_4} = 6.86822$, and $K_{Ar} = 3.67273$.

The PFR tube in each reactor bed is 7.05 ft in diameter and packed with the catalyst studied by Dyson and Simon (1968); the rate expression and catalyst properties are discussed below. The heat of reaction, taken as a constant and determined at the inlet to the first catalyst bed, ΔH_r (602.59 K, 225 atm) = −23.63 kcal/g-mol nitrogen reacted. The average molar heat capacity of each component in the reacting fluid is $vC^{avg}_{p,N_2} = 7.42$ cal/g-mol-K, $C^{avg}_{p,H_2} = 7.02$ cal/g-mol-K, $C^{avg}_{p,NH_3} = 11.86$ cal/g-mol-K , $C^{avg}_{p,CH_4} = 14.57$ cal/g-mol-K , and $C^{avg}_{p,Ar} = 4.97$ cal/g-mol-K .

Rate Expression for Nitrogen + Hydrogen → Ammonia: $N_2 + 3H_2 \leftrightarrow 2NH_3$

Dyson and Simon (1968) provide the rate expression for an ammonia catalyst with $\rho_{catalyst}$ = 165-lb catalyst/ft^3 reactor volume. This rate expression appears more complex than the rate expression we used for the styrene process, but it is actually of the same form:

$$2\left(-r^{global}_{N_2}\right) = \left(r^{Global}_{NH_3}\right) = 2k\left(K^2_{eq}\left(\frac{a_{N_2}a^{3/2}_{H_2}}{a_{NH_3}}\right) - \left(\frac{a_{NH_3}}{a^{3/2}_{H_2}}\right)\right),$$

where

$$2k = \left(3.7836\times10^{-4}\right)\left(1.7689\times10^{15}\right)\exp\left(\frac{-40765}{RT}\right)$$

$$\log_{10} K_{eq} = -2.691122\log_{10} T - 5.519265\times10^{-5}T + 1.848863\times10^{-7}T^2 + \frac{2001.6}{T} + 2.6899.$$

Here,

$r^{global}_{NH_3}$ = is the reaction rate $[=]\dfrac{\text{lb-mol ammonia produced}}{\text{h-lb catalyst}}$;

K_{eq} = the equilibrium constant, which is a function of the temperature [=] K;

k = the reaction rate constant, which is a function of the temperature [=] K;

a_i = the activity of species i, which is given as $a_i = \varphi_i y_i P = \phi_i p_i$;

ϕ_i = is the fugacity coefficient of species i; and

R = gas law constant, 1.987

The fugacity coefficients account for vapor phase nonidealities. These nonidealities occur because of the elevated pressures and temperatures used in this reaction. For now, we do not need to be concerned with this detail and simply use the provided equations for the species fugacity coefficients:

$$\phi_{H_2} = \exp\left\{\begin{array}{l} P\left[\exp\left(-3.8402T^{0.125}+0.541\right)\right] \\ -P^2\left[\exp\left(-0.1263T^{0.5}-15.98\right)\right] \\ +300\left[\exp(-0.011901T-5.941)\right]\left(\exp\left(\frac{-P}{300}-1\right)\right) \end{array}\right\},$$

$$\phi_{N_2} = 0.93431737 + 0.3101804\times10^{-3}T + 0.295896\times10^{-3}P - 0.2707279\times10^{-6}T^2 + 0.4775207\times10^{-6}P^2,$$

$$\phi_{NH_3} = 0.1438996 + 0.2028538\times10^{-2}T - 0.4487672\times10^{-3}P - 0.1142945\times10^{-5}T^2 + 0.2761216\times10^{-6}P^2,$$

where temperature [=] K and pressure [=] atm.

Gathering terms, we find the rate expression as

$$2\left(-r^{global}_{N_2}\right) = \left(r^{global}_{NH_3}\right) = 2k\left(K^2_{eq}P^{3/2}\left(\frac{y_{N_2}\phi_{N_2}y^{3/2}_{H_2}\phi^{3/2}_{H_2}}{y_{NH_3}\phi_{NH_3}}\right) - \frac{1}{P}\left(\frac{y_{NH_3}\phi_{NH_3}}{y^{3/2}_{H_2}\phi^{3/2}_{H_2}}\right)\right).$$

Solution Hints

We want to determine the size of the adiabatic PFRs needed to produce 2 million pounds of ammonia per day.

Our solution procedure is shown in Figure 5.5b. Stream flow rates and composition can be determined for the ammonia process by using the techniques of Chapters 3 and 4; here we use the equation-based approach; see **Problem 5.5a.xls**. Results for key streams are summarized in Table P5.5a.

In Table P5.5a, we provide the temperatures entering reactors 2 and 3. In order to determine the temperature entering reactor 2, we first numerically determine the temperature and composition leaving reactor 1 (see solution below) and then solve the energy balance around the first quench zone. This same procedure is followed to determine the temperature entering bed 3. The quench unit calculations can be found in the end-of-the-chapter problem, Problem 5.6.

Detailed Solution

In order to determine the temperatures leaving reactors 1–3, we numerically solve the material and energy balances for

Table P5.5a Stream Data Summary for Ammonia Process

	Stream 3 (Feed)	Stream 1	Stream 4	Stream 6
T (F)	122.00	624.99	689.18	748.77
T (K)	323.15	602.59	638.25	671.36
P (atm)	225.00	225.00	225.00	225.00
N_{N_2} (lb-mol/h)	2507.98	4957.32	5467.72	5901.55
N_{H_2} (lb-mol/h)	7522.93	13,131.68	14,662.35	15,963.34
N_{NH_3} (lb-mol/h)	0	475.47	1962.67	3602.98
N_{CH_4} (lb-mol/h)	96.54	2153.99	2202.36	2250.53
N_{Ar} (lb-mol/h)	34.55	460.52	477.80	495.07
N_{total} (lb-mol/h)	10,162	21,178.98	24,772.80	28,213.48

each reactor. The process is shown by hand for the first step in R1 and computer solutions are provided for each of the three reactors.

The successive mass balances for any of the three sections of the PFR are

N_2 balance: $N_{N_2}^{z_{j+1}} = N_{N_2}^{z_j} - r_{N_2}^{\text{global}}(\rho_{\text{catalyst}})(A_c)(\Delta z)$

H_2 balance: $N_{H_2}^{z_{j+1}} = N_{H_2}^{z_j} - 3r_{N_2}^{\text{global}}(\rho_{\text{catalyst}})(A_c)(\Delta z)$

NH_3 balance: $N_{NH_3}^{z_{j+1}} = N_{NH_3}^{z_j} + 2r_{N_2}^{\text{global}}(\rho_{\text{catalyst}})(A_c)(\Delta z)$

CH_4 balance: $N_{CH_4}^{z_{j+1}} = N_{CH_4}^{z_j}$

Ar balance: $N_{Ar}^{z_{j+1}} = N_{Ar}^{z_j}$.

The solution procedure follows that developed for the styrene process and this procedure is identical for each of the three reactors. The numerical integration is performed from known reactor feed conditions, ($N_{N_2}^{z_j=0}$, $N_{H_2}^{z_j=0}$, $N_{NH_3}^{z_j=0}$, $N_{CH_4}^{z_j=0}$, $N_{Ar}^{z_j=0}$, and $T_{z_j=0}$). By selecting a value for Δz, and evaluating $r_{N_2}^{\text{global}}$ at the $z_j = 0$ conditions, all of the conditions at $z_j = 1$ can be directly determined. This process continues until the end of the reactor is reached; if the reactor length is L, this would be $L/\Delta z$ steps. These iterative calculations are accomplished using a computer code, but here we perform the first Δz step by hand.

At the entrance to any of the three sections of the PFR reactor $z_j = 0$, the feed conditions are known and provided in the table shown earlier. Using PFR catalyst bed 1 (S1, R1 in),

$$y_{N_2} = \left(\frac{4957.32}{21,178.98}\right) = 0.23407$$

$$y_{H_2} = 0.62003$$

$$y_{NH_3} = 0.02245$$

$$\phi_{N_2} = 1.11367$$

$$\phi_{H_2} = 1.07722$$

$$\phi_{NH_3} = 1.69430$$

$$k = \left(\frac{1}{2}\right)(3.7836 \times 10^{-4})(1.7689 \times 10^{15})\exp\left(\frac{-40,765}{(1.987)(602.59)}\right)$$

$$= 2.89508$$

$$K_{eq} = 3.66488 \times 10^{-2}$$

$$(-r_{N_2}^{\text{global}}) = k\left(K_{eq}^2 P^{3/2}\left(\frac{y_{N_2}\phi_{N_2}y_{H_2}^{3/2}\phi_{H_2}^{3/2}}{y_{NH_3}\phi_{NH_3}}\right) - \frac{1}{P}\left(\frac{y_{NH_3}\phi_{NH_3}}{y_{H_2}^{3/2}\phi_{H_2}^{3/2}}\right)\right)$$

$$(-r_{N_2}^{\text{global}}) = k\left(K_{eq}^2 P^{3/2}(3.74085) - \frac{1}{P}(6.96835 \times 10^{-2})\right)$$

$$= 2.97195 \frac{\text{lb-mol } N_2 \text{ reacted}}{\text{h-lb catalyst}}.$$

Once we select our step size, Δz, we can calculate the conversion over this Δz step. For $\Delta z = 0.1$ ft and $A_c = (\pi r^2) = \pi(3.525)^2$,

$$r_{N_2}^{\text{global}}(\rho_{\text{catalyst}})(A_c)(\Delta z) = 11.60135 \frac{\text{lb-mol } N_2 \text{ reacted}}{\text{h}}.$$

The heat that was generated by the reaction over this step (see Eq. (5.20)) is found as

$$\dot{Q}_j = (-\Delta H_r)\left(\frac{\text{g-mol } N_2 \text{ reacted}}{\text{h}}\right)$$

$$= \left(23.63\frac{\text{kcal}}{\text{g-mol } N_2 \text{ reacted}}\right)\left(11.60135\frac{\text{lb-mol } N_2 \text{ reacted}}{\text{h}}\right)$$

$$\left(\frac{454 \text{ g-mol}}{\text{lb-mol}}\right) = 124,459.5\frac{\text{kcal}}{\text{h}}.$$

We can calculate the temperature change from the energy balance (Eq. (5.52)) as

$$\left(\sum_i N_i^{z_j}(C_{p,i}^{\text{avg}})\right)\left(\frac{454 \text{ g-mol}}{\text{lb-mol}}\right)\left(\frac{\text{kcal}}{1000 \text{ cal}}\right) = 76,398.76\frac{\text{kcal}}{\text{h-K}}$$

$$T_1 = T_0 + \frac{Q_j}{\left(\sum_i N_i^{z_j}(C_{p,i}^{\text{avg}})\right)} = 602.59 + \frac{(124,459.5)}{(76,398.76)} = 604.22 \text{ K}.$$

Do note that the flow rate heat capacity term $\left(\sum_i N_i^{z_j}(C_{p,i}^{\text{avg}})\right)$ is evaluated at z_j conditions.

Finally, all the molar flow rates and conversion at $z_j = 1$ can be determined:

N_2 balance:

$$N_{N_2}^{z_{j+1}} = N_{N_2}^{z_j} - r_{N_2}^{\text{global}}(\rho_{\text{catalyst}})(A_c)(\Delta z) = 4945.72 \text{ lb-mol/h}$$

H_2 balance:

$$N_{H_2}^{z_{j+1}} = N_{N_2}^{z_j} - 3r_{N_2}^{\text{global}}(\rho_{\text{catalyst}})(A_c)(\Delta z) = 13,096.88 \text{ lb-mol/h}$$

NH_3 balance:

$$N_{NH_3}^{z_{j+1}} = N_{NH_3}^{z_j} + 2r_{N_2}^{\text{global}}(\rho_{\text{catalyst}})(A_c)(\Delta z) = 498.672 \text{ lb-mol/h}$$

CH_4 balance: $N_{CH_4}^{z_{j+1}} = N_{CH_4}^{z_j} = 2153.99 \text{ lb-mol/h}$

Ar balance: $N_{Ar}^{z_{j+1}} = N_{Ar}^{z_j} = 460.52 \text{ lb-mol/h}$

$$\% \text{ conversion of nitrogen} = \frac{(4957.32 - 4945.72)}{4957.32} \times 100 = 0.234\%.$$

All conditions at $z_j = 1$ are known, allowing conditions at $z = 2$ to be calculated.

The computer codes for each reactor are provided in **Problem 5.5b.xls**, **Problem 5.5c.xls**, and **Problem 5.5d.xls**. The output for catalyst bed 1 is shown in Figure P5.5c. The code was run with a fixed step size of 0.1 and 0.01 ft; the 0.01-ft step size allows a more accurate reactor length determination at 15% conversion.

We have combined a rate expression from the literature, with a constant, ΔH_r, and average species molar heat capacities to size the three PFR sections. Figure P5.5c shows that in the first catalyst, bed equilibrium is reached at about 27.8% conversion of the incoming nitrogen. We stop at 15% conversion of the incoming nitrogen and then add feed to lower the reaction temperature—this is termed a cold shot. The lower temperature mixture is sent to the second of three catalyst beds. After 15% conversion, cold feed is again added and the resulting mixture is sent to the third catalyst bed. Table P5.5b compares results from our three simulated catalyst beds with experimental values (plant data) reported in the literature (Chandra et al., 1979).

Although we have made several simplifying assumptions, there is an agreement between the plant data from Chandra and results from the simulation. The actual design of the PFR raises several interesting issues. For example, is there an optimal split of the incoming feed stream to the catalyst beds? Here we used a 50/50 feed split to catalyst beds 2 and 3. Would it be economical to add a fourth bed—increased ammonia concentration leaving the PFR will lower downstream separation costs, but will this savings justify the cost of an additional catalyst bed? Would it be economical for ammonia production to occur in a single bed—here we must consider equilibrium, the ultimate length of the bed, and the difficulty in maintaining a PFR center line temperature for the exothermic reaction? Would operating at a different pressure be economical—three ammonia process pressures are often cited in the literature: 150, 225, and 300 atm? What is the effect of the purge/recycle rates on process economics? Equations allowing a more rigorous simulation of the ammonia PFR are provided in Problem 5.8. Of course, we must appreciate the limitations of using a rate expression based on one catalyst. For example, a different rate expression would change the volume of catalyst needed for the same conversions and used in Problem 5.5 (see Walas, 1985).

5.6 ***(Needed for Solution of Problem 5.5) (Solution Provided)*** Use an energy balance to determine the temperatures of the stream (S4) into reactor 2 and the stream (S6) into reactor 3 (S6). Using results from the successive substitution solution of the ammonia process **Problem 5.5a.xls**, needed data for the mixer prior to reactor 2 are summarized in Figure P5.6a, and needed data for the mixer prior to reactor 3 are summarized in Figure P5.6b.

***Solution*: Problem 5.6a.xls, Problem 5.6b.xls**

5.7 ***Ammonia Process Design: Mini-Project Extension 1*** For the first bed of the ammonia PFR, compare the solution from Euler's method to that obtained from the RK4 method.

5.8 ***Ammonia Process Design: Mini-Project Extension 2*** Use the energy balance given by Equation (5.49) for the ammonia PFR. Here the heat of reaction $(\Delta H_r)_{T,P}$ cal/g-mol N_2 reacted is given by

$$(\Delta H_r)_{T,P} = (\Delta H_r^{z_j}) = 2\left\{\begin{array}{l} -9184.0 - 7.2949T + (0.34996\times10^{-2})T^2 \\ +(0.03356\times10^{-5})T^3 - (0.11625\times10^{-9})T^4 \\ -6329.3 + 3.1619P + (14.3595 + 4.4552\times10^{-3}P)T \\ -T^2(8.3395\times10^{-3} + 1.928\times10^{-6}P) - 51.21 \\ +0.14215P. \end{array}\right\}$$

For the vapor phase heat capacity (joule per gram-mole-kelvin), use the following data from Reklaitis (1983) and C_p expression:

	a	b	c	d	e
N_2	2.94119E + 01	−3.00681E − 03	5.45064E − 06	5.13186E − 09	−4.25308E − 12
H_2	1.76386E + 01	6.70050E − 02	−1.31485E − 04	1.05883E − 07	−2.91803E − 11
NH_3	2.75500E + 01	2.56278E − 02	9.90042E − 06	−6.68639E − 09	0.00000E + 00
CH_4	3.83870E + 01	−7.36639E − 02	2.90981E − 04	−2.63849E − 07	8.00679E − 11
Ar	2.07723E + 01	0.00000E + 00	0.00000E + 00	0.00000E + 00	0.00000E + 00

Table P5.5b Comparison of Problem 5.5 Solution with Literature Results (Chandra et al., 1979)

	Catalyst Bed 1 Out		Catalyst Bed 2 Out		Catalyst Bed 3 Out	
	S2	Chandra et al.	S5	Chandra et al.	S7	Chandra et al.
T (F)	814.84	940.60	864.70	935.60	911.09	851.00
T (K)	708.06	780.15	735.76	775.15	761.53	728.15
P (atm)	225	226	225	226	225	226
y_{N_2} (%)	21.40	20.1	20.09	18.2	18.97	17.8
y_{H_2} (%)	55.36	61.0	52.75	57.1	50.33	53.9
y_{NH_3} (%)	9.97	10.5	15.58	15.9	20.32	19.1
y_{CH_4} (%)	10.94	5.7	9.52	6.1	8.51	6.3
y_{Ar} (%)	2.34	2.7	2.07	2.7	1.87	2.9
Catalyst volume (ft^3)	271.3		437.2		554.3	

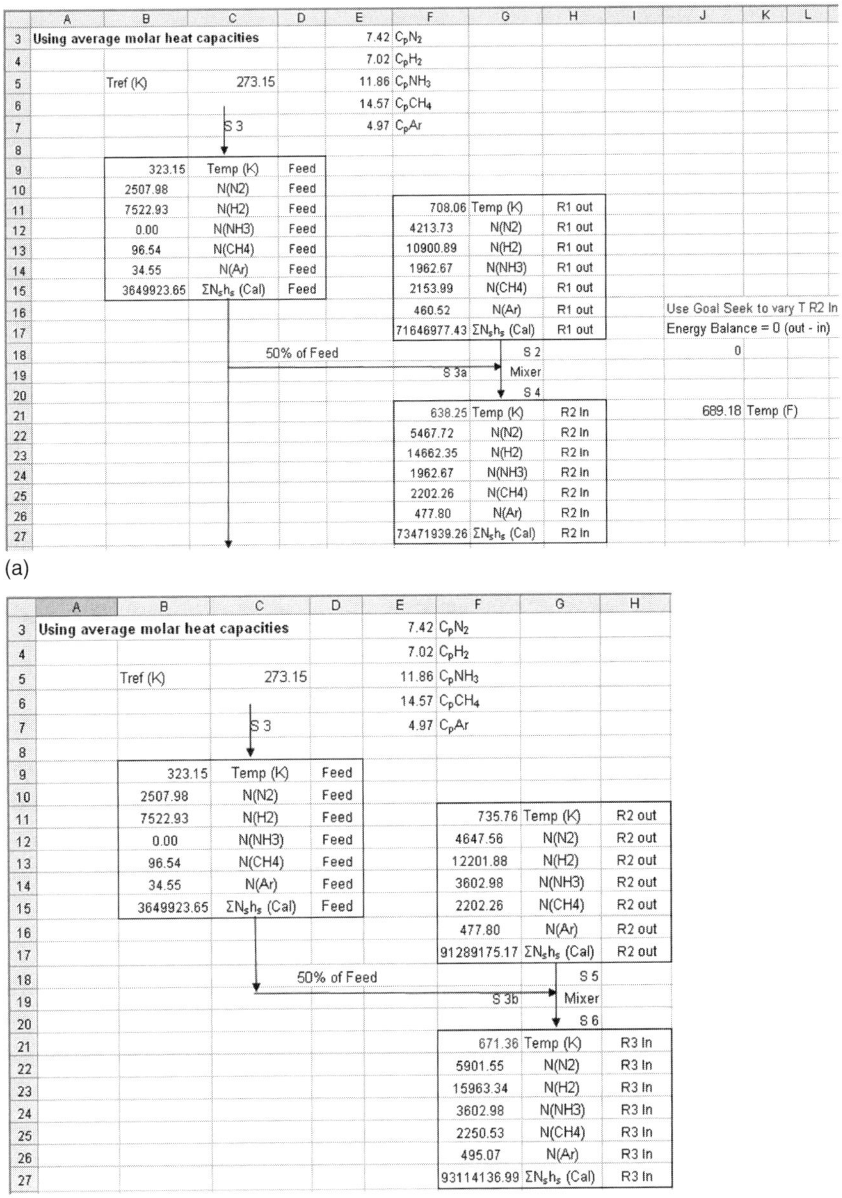

	A	B	C	D	E	F	G	H	I	J	K	L
3	Using average molar heat capacities				7.42	CpN₂						
4					7.02	CpH₂						
5		Tref (K)	273.15		11.86	CpNH₃						
6					14.57	CpCH₄						
7			S 3		4.97	CpAr						
8												
9		323.15	Temp (K)	Feed								
10		2507.98	N(N2)	Feed								
11		7522.93	N(H2)	Feed		708.06	Temp (K)	R1 out				
12		0.00	N(NH3)	Feed		4213.73	N(N2)	R1 out				
13		96.54	N(CH4)	Feed		10900.89	N(H2)	R1 out				
14		34.55	N(Ar)	Feed		1962.67	N(NH3)	R1 out				
15		3649923.65	ΣNₛhₛ (Cal)	Feed		2153.99	N(CH4)	R1 out				
16						460.52	N(Ar)	R1 out	Use Goal Seek to vary T R2 In			
17						71646977.43	ΣNₛhₛ (Cal)	R1 out	Energy Balance = 0 (out - in)			
18				50% of Feed			S 2			0		
19					S 3a		Mixer					
20							S 4					
21						638.25	Temp (K)	R2 In		689.18	Temp (F)	
22						5467.72	N(N2)	R2 In				
23						14662.35	N(H2)	R2 In				
24						1962.67	N(NH3)	R2 In				
25						2202.26	N(CH4)	R2 In				
26						477.80	N(Ar)	R2 In				
27						73471939.26	ΣNₛhₛ (Cal)	R2 In				

(a)

	A	B	C	D	E	F	G	H
3	Using average molar heat capacities				7.42	CpN₂		
4					7.02	CpH₂		
5		Tref (K)	273.15		11.86	CpNH₃		
6					14.57	CpCH₄		
7			S 3		4.97	CpAr		
8								
9		323.15	Temp (K)	Feed				
10		2507.98	N(N2)	Feed				
11		7522.93	N(H2)	Feed		735.76	Temp (K)	R2 out
12		0.00	N(NH3)	Feed		4647.56	N(N2)	R2 out
13		96.54	N(CH4)	Feed		12201.88	N(H2)	R2 out
14		34.55	N(Ar)	Feed		3602.98	N(NH3)	R2 out
15		3649923.65	ΣNₛhₛ (Cal)	Feed		2202.26	N(CH4)	R2 out
16						477.80	N(Ar)	R2 out
17						91289175.17	ΣNₛhₛ (Cal)	R2 out
18				50% of Feed			S 5	
19					S 3b		Mixer	
20							S 6	
21						671.36	Temp (K)	R3 In
22						5901.55	N(N2)	R3 In
23						15963.34	N(H2)	R3 In
24						3602.98	N(NH3)	R3 In
25						2250.53	N(CH4)	R3 In
26						495.07	N(Ar)	R3 In
27						93114136.99	ΣNₛhₛ (Cal)	R3 In

(b)

Figure P5.6 (a) Ammonia process stream composition and temperature entering reactor R2. (b) Ammonia process stream composition and temperature entering reactor R3.

$$\int_{T_1}^{T_2} C_p(T)dT = a(T_2 - T_1) + \frac{b}{2}\left(T_2^2 - T_1^2\right) + \frac{c}{3}\left(T_2^3 - T_1^3\right) + \frac{d}{4}\left(T_2^4 - T_1^4\right)$$
$$+ \frac{e}{5}\left(T_2^5 - T_1^5\right).$$

Be sure T is in kelvin for use in the heat capacity equation. Also, be sure to convert from joule to calories (1 cal = 4.1840 J) before using these heat capacities in the computer code.

5.9 *Ammonia Process Design: Mini-Project Extension 3* Evaluate the use of a single catalyst bed to produce ammonia. Here, combine the current feed and recycle streams into a single stream that will enter the bed. Determine the length necessary to obtain a mole fraction of ammonia at 19–20% leaving the bed. Discuss the temperature change $\Delta T/\Delta z$ that is occurring at the reactor outlet.

Chapter 6

Introduction to Data Reconciliation and Gross Error Detection

In order to improve the operation of an existing plant, we must first measure process variables including temperature, pressure, flow rates, and species compositions. Data should be reconciled to satisfy material and energy balances and to determine system parameters. We can then improve efficiency and profitability by utilizing computer-based techniques such as process economic optimization and energy management strategies to move appropriate variable set-point values. This overall strategy, shown in Figure 6.1, is termed online optimization. In this chapter, we introduce data reconciliation and gross error detection and gross error identification. In later chapters (Chapters 11 and 12), we apply data reconciliation, gross error detection, parameter estimation, and energy management optimization to a cogeneration system.

In Chapters 3–5, we explored computer-aided solution strategies for material and energy balance problems. The solution of the material balance problem, that is, the flow rates of species in each stream, is part of the "design case" as these flow rates can be used to size and cost equipment. Energy balances help determine needed utilities. In Figure 6.2, the converged solution of the styrene material balance problem in Example 3.1 (also Example 4.3) is repeated. Molar flow rates (pound-mole per hour) and mole fractions have been replaced by equivalent mass flow rates (pound per hour) and weight fractions for each component in each stream. Figure 6.2 can be considered the design case for material flows.

Actual measured process variables will not show the "design case values" of Figure 6.2 as changing process demands will dictate various production levels. Process measurements will also fluctuate with time due to random errors process measurements. Allow that the data in the boxes in Figure 6.3 are measured flow rates and compositions for the styrene process. Flow rate data may come from online flow meters and composition data may come from a combination of online analyzers including gas chromatographs as well as off-line laboratory analysis.

The data above the boxes in Figure 6.3 are calculated species flow rates (pound per hour) based on measurements. What is important to note is that the measured flow rates in and out of the unit operations are not consistent, and measured species mass fractions in process streams do not necessarily sum to 1.0. Table 6.1 summarizes residual values for the material balance for each unit operation and residual values for the weight fraction.

Before we can implement process improvement strategies, we must know species flows, temperatures, and pressures. The problem we address here is "How can we determine the most accurate values—termed reconciled values—for the flows and compositions in Figure 6.3 or for any process variable in a processing plant?" A statistically consistent approach (see Problem 6.10) is to minimize a weighted least squares for all of the i measured variables:

$$\text{Minimize} \sum_{i,\text{measured}} \left(\frac{\text{Measured value}_i - \text{Reconciled value}_i}{\text{Weighting factor}_i} \right)^2. \tag{6.1}$$

The weighting factor is chosen as the instrument standard deviation σ_i, and as will we see, σ_i is determined by the instrument accuracy. Data reconciliation is the process of solving Equation (6.1) subject to the condition that material and energy balances are satisfied. The material and energy balances then serve as constraints in the optimization problem.

Modeling, Analysis and Optimization of Process and Energy Systems, First Edition. F. Carl Knopf.

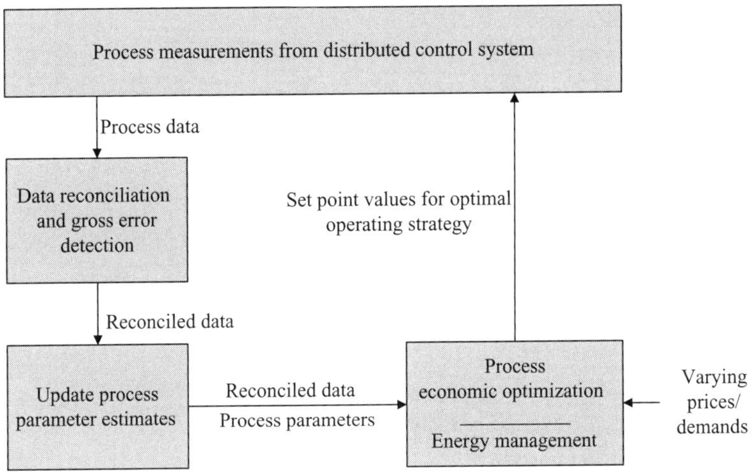

Figure 6.1 Overview of an online (real-time) optimization strategy.

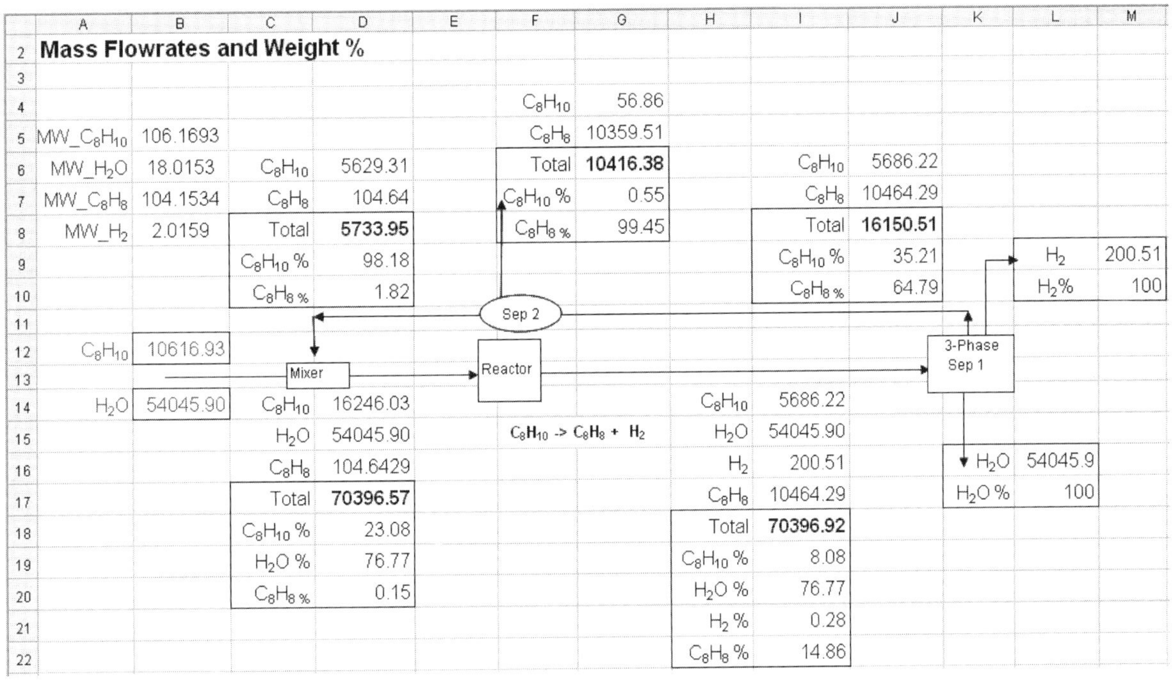

Figure 6.2 Styrene flow sheet with design flow rates and compositions; see **Figure 6.2.xls**.

6.1 STANDARD DEVIATION AND PROBABILITY DENSITY FUNCTIONS

Measurements taken by instruments, for example, mass flow meters, thermocouples, pressure gauges, and so on, are expected to have random errors in the reported values. It is important to understand the relation between the standard deviation, random errors, and instrument accuracy. Instrument σ_i impacts the solution to the data reconciliation problem.

EXAMPLE 6.1 *Standard Deviation*

A single thermocouple is placed in boiling water and the following seven temperatures (°C) are recorded at different times: 100.9, 99.2, 100.5, 100.6, 99.7, 99.8, and 99.3. What is the standard deviation of the thermocouple?

SOLUTION Here, the sum of these temperatures = 700.0 and the average or mean temperature is (700.0/7) = 100.0°C. An estimate of the standard deviation, s, is

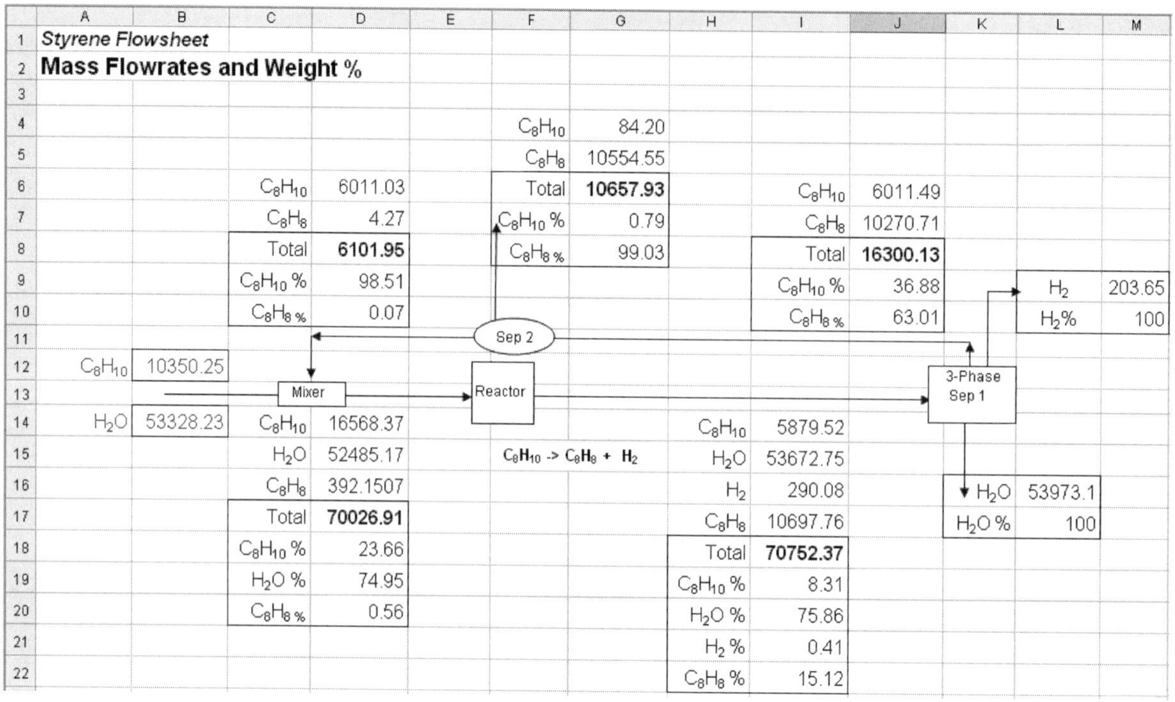

Figure 6.3 Styrene flow sheet with flow rates and compositions from plant data; see **Figure 6.3.xls**.

Table 6.1 Balance Residuals before Data Reconciliation—Values from Figure 6.3

	Mixer	Reactor	Separator 1	Separator 2
Out–in, total flow (lb/h)	246.48	725.46	−275.48	459.75
Out–in, C_8H_{10} (lb/h)	207.09		131.97	83.74
Out–in, H_2O (lb/h)	−843.06		300.36	
Out–in, H_2 (lb/h)			−86.43	
Out–in C_8H_8 (lb/h)	387.88		−427.05	288.11
Residual, $\sum w_i - 1.0$ (%)	−0.83 mixer out	−0.30 reactor out	−0.11 organic stream	−0.18 product stream
Residual, $\sum w_i - 1.0$ (%)				−1.42 recycle stream

$$s = \sqrt{\frac{\left(\sum_{j=1}^{N \text{ data points}} (x_j^+ - \bar{x})^2\right)}{(N-1)}}. \tag{6.2}$$

Here, x_j^+ is the measured variable (the seven temperatures in this example), $\bar{x}$ is the mean of the measured variable (100.0°C), and N is the number of measurements (here $N = 7$). We will use a + superscript to indicate a measured variable. Solving,

$$s = \sqrt{\frac{(100.9-100)^2 + (99.2-100)^2 + (100.5-100)^2 + \cdots}{6}}$$

$$= \sqrt{\frac{2.68}{6}} = 0.668°\text{C}.$$

A general Excel/Visual Basic for Applications (VBA) program to calculate an estimate of the standard deviation, s, for any data set is given in **Example 6.1.xls**. ■

Generally, we collect a limited number of data points to determine an estimate of the standard deviation and we set s to be the standard deviation σ:

$$\sigma \cong s. \tag{6.3}$$

This standard deviation is used in probability density functions, and a widely used probability density function to represent measured variables in engineering and the physical sciences is the Gaussian or normal distribution function, given as

$$F(x^+) = \frac{1}{\sigma\sqrt{2\pi}} \exp\left(-\frac{(x^+ - \bar{x})^2}{2\sigma^2}\right). \tag{6.4}$$

Here, $F(x^+)$ represents the expected density of observing a measurement, x^+, and again $\bar{x}$ is the mean of the measured variable, σ is the standard deviation, and σ^2 is the variance.

Equation (6.4) represents a family of functions, all of which have a bell-shaped symmetric density curve with a single peak at $\bar{x}$. The dispersion or width of the bell-shaped curve is controlled by σ. Each different value of σ produces a different density curve.

EXAMPLE 6.2 *Density Plot for Normal Distribution Function*

Generate the plot for the density of observing various thermocouple values for a thermocouple with $\bar{x} = 100°C$ and $\sigma = 1°C$ (this is a different σ than that found in our previous example; however, for the discussion below, it is convenient to use $\sigma = 1°C$). Using Equation (6.4),

$$F(x^+) = \frac{1}{\sigma\sqrt{2\pi}} \exp\left(-\frac{(x^+ - \bar{x})^2}{2\sigma^2}\right)$$
$$= \frac{1}{\sqrt{2\pi}} \exp\left(-\frac{(x^+ - 100)^2}{2}\right).$$

For $x^+ = 100°C$,

$$F(100) = \frac{1}{\sqrt{2\pi}} \exp(0) = 0.4.$$

For $x^+ = 101$ or $99°C$,

$$F(101) = F(99) = \frac{1}{\sqrt{2\pi}} \exp\left(-\frac{1}{2}\right) = 0.242.$$

For $x^+ = 102$ or $98°C$,

$$F(102) = F(98) = \frac{1}{\sqrt{2\pi}} \exp\left(-\frac{(2)^2}{2}\right) = 0.054.$$

The Excel File **Example 6.2.xls** generates the density plot versus thermocouple values shown in Figure 6.4. Do note that in the Excel file, the term

$$\exp\left(-\frac{(x^+ - \bar{x})^2}{2\sigma^2}\right)$$

should be input as

$$\exp\left(-1 \times \frac{(x^+ - \bar{x})^2}{2\sigma^2}\right). \quad \blacksquare$$

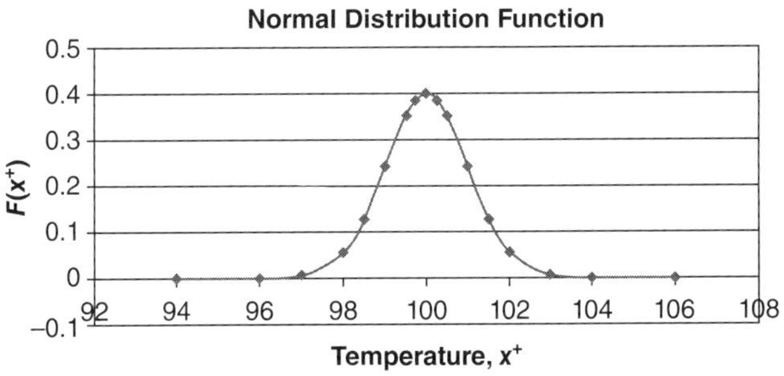

Figure 6.4 Density plot for normal distribution function with $\bar{x} = 100°C$ and $\sigma = 1°C$.

There are important features for the density curve. For example, use the Excel file **Example 6.2.xls** to reduce σ (say, $\sigma = 0.5$) and notice how the density distribution plot becomes a sharp peak centered at the mean. In contrast, increase σ and notice how the distribution plot flattens out.

We are interested in the probability that any measured variable x^+ (e.g., the temperature in Example 6.2) falls within a range from x_1 to x_2. This is indicated as $P(x_1 \le x^+ \le x_2)$, and mathematically, this would be expressed as

$$P(x_1 \le x^+ \le x_2) = \frac{1}{\sigma\sqrt{2\pi}} \int_{x_1}^{x_2} \exp\left(-\frac{(x^+ - \bar{x})^2}{2\sigma^2}\right) dx^+. \quad (6.5)$$

In Equation (6.5), the factor $\left(1/\sigma\sqrt{2\pi}\right)$ actually serves to normalize the probability density function. The integral of $P(x^+)$, from $x_1 = -\infty$ to $x_2 = \infty$, equals 1.0 = 100%.

EXAMPLE 6.3 *Probability of One Standard Deviation from the Mean*

What is the probability that the measured value from the thermocouple in Example 6.2 will fall between 99 and 101°C, which is $\pm 1\sigma$ from the expected mean?

$$P(99 \le x^+ \le 101) = \frac{1}{(1)\sqrt{2\pi}} \int_{99}^{101} \exp\left(-\frac{(x^+ - 100)^2}{2(1)^2}\right) dx^+$$
$$= 0.6827$$

This integration involves use of the error function. The important finding is that 68.27% of the measured thermocouple values should fall between 99 and 101°C. In other words, 68.27% of all measured thermocouple values are expected to fall within $\pm 1\sigma$ (± one standard deviation) of the mean.

EXAMPLE 6.4 *Probability of Two Standard Deviations from the Mean*

What is the probability that the measured value from the thermocouple in Example 6.2 will fall between 98 and 102°C, which is $\pm 2\sigma$ from the expected mean?

$$P(98 \le x^+ \le 102) = \frac{1}{(1)\sqrt{2\pi}} \int_{98}^{102} \exp\left(-\frac{(x^+ - 100)^2}{2(1)^2}\right) dx^+$$
$$= 0.9545 = 95.45\%.$$

Here, 95.45% of all measured thermocouple values are expected to fall within ±2σ (±two standard deviations) of the mean. Continuing, 99.73% of all measured thermocouple values are expected to fall within ±3σ of the mean. Solutions for Examples 6.3 and 6.4 are found using the Maple software program; see file **Probability Integration.mw**. ■

With this background, we can now discuss what is important for our data reconciliation problem. We accept that all plant measurements will have some error. If the errors are random, then the effect of the errors on a measured variable may be assumed to create a Gaussian or normal distribution about the mean (expected) value of the measured variable. The random error is directly tied to the standard deviation.

For any instrument, we could directly measure the standard deviation as we did for the thermocouple in Example 6.1. Often the vendor provides accuracy information (or maximum error information) on the measurement device, typically expressed as ± some percentage of full scale or ± some percentage of the actual measurement. For example, a mass flow meter reports an accuracy of ±2% of full scale, and the full scale is 5000 kg/h. Then any measured value (say, 3325 kg/h) could actually be that value (3325 kg/h) ± 100 kg/h. Here then, the expected range would be 3225–3425 kg/h even if the flow rate was constant with a "true" value of 3325 kg/h.

We are now in the position to ask: What does this instrument accuracy actually mean? It is generally accepted that a reported accuracy (or maximum error) implies 95% of measured values will fall within the expected range (a 95% confidence level). For our mass flow meter, we would expect 95% of all measurements for the mean flow rate of 3325 kg/h would fall between 3225 and 3425 kg/h.

Our story is finally done. A 95% confidence level is a 95% probability, which for a Gaussian distribution is

$$P(x_1 \le x^+ \le x_2) = \frac{1}{\sigma\sqrt{2\pi}} \int_{x_1}^{x_2} \exp\left(-\frac{\left(x^+ - \bar{x}\right)^2}{2\sigma^2}\right) dx^+ = 0.95.$$

Using symmetry and solving for x_1 and x_2, we find $x_1 = \bar{x} - 1.96\,\sigma$ and $x_2 = \bar{x} + 1.96\,\sigma$.

So, for our mass flow meter,

$$x_1 = \bar{x} - 1.96\,\sigma = 3325 - 1.96\,\sigma = 3325 - 100 \text{ lb/h}$$

and

$$x_2 = \bar{x} + 1.96\,\sigma = 3325 + 1.96\,\sigma = 3325 + 100 \text{ lb/h},$$

giving σ = (100/1.96) = 51.02 kg/h.

For most measurement devices,

$$\sigma = \frac{|\text{error}|_{\max}}{1.96} = \frac{|\text{accuracy}|}{1.96}. \tag{6.6}$$

In many applications, for convenience, the value of 1.96 in the denominators in Equation (6.6) is replaced by 2. Here then, for our mass flow meter, σ = (100/2) = 50 kg/h.

6.2 DATA RECONCILIATION: EXCEL SOLVER

6.2.1 Single-Unit Material Balance: Excel Solver

Data reconciliation is a constrained optimization problem in which the objective function given by Equation (6.1) is solved subject to the constraints that material and energy balances are satisfied. The least squares weighting factors in Equation (6.1) are given by the standard deviation for each measurement device, which is either measured or determined from vendor information using Equation (6.6). The data reconciliation problem is

$$\text{Minimize } f = \sum_{\substack{i,\text{measured}\\ \text{variables}}} \left(\frac{\left(x_i^+ - x_i\right)}{\sigma_i}\right)^2 \tag{6.7}$$

$$\text{Subject to } h_k(x_i, u_j) = 0 \quad k = 1, \ldots, K, \tag{6.8}$$

where x_i^+ is the measured process variable, x_i is the reconciled value, $h_k(x_i, u_j)$ are the material and energy balances, and u_j is the estimate of nonmeasured process variables; nonmeasured process variables will be discussed later in this chapter.

EXAMPLE 6.5 ***Mixer Data Reconciliation***

Let us start with a simple mass balance problem around the mixer in Figure 6.5. Measured quantities are indicated by adding a (+) symbol as a superscript.

The linear mass balance (in = out) is $F_1 + F_2 = F_3$, but here using the measured values, 245 + 240 ≠ 500 kg/h. Assuming the standard deviation for each measurement device is σ = 10 kg/h, we can determine the reconciled flow rates as

$$\text{Minimize}\left(\left(\frac{245 - F_1}{10}\right)^2 + \left(\frac{240 - F_2}{10}\right)^2 + \left(\frac{500 - F_3}{10}\right)^2\right)$$

$$\text{Subject to } F_1 + F_2 - F_3 = 0.$$

The optimization variables are the reconcilable flow rates F_1, F_2, and F_3. We can use Excel Solver to solve this problem and the solution is provided in **Example 6.5.xls** and shown in Figure 6.6. To start the optimization procedure, the measured flow rates should be used as the initial guess for the reconciled flow rates.

In case the reader is not familiar with Excel Solver, details on setting this problem up in Solver are provided here. Each term in the objective function is formed in an Excel cell—cells G15, G16, and G17. For example, cell G15 contains

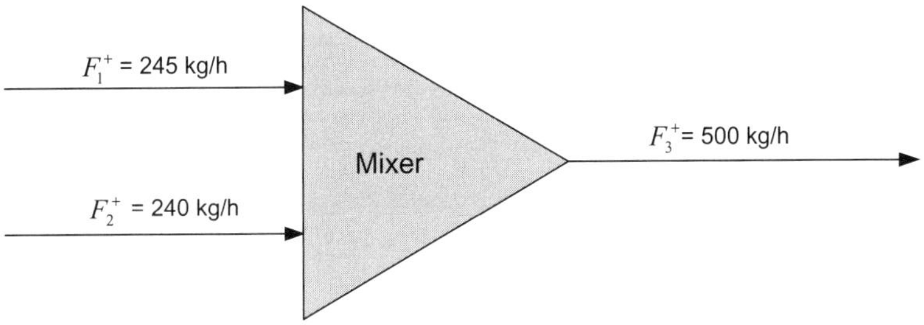

Figure 6.5 Measured flow rates in and out of a mixer.

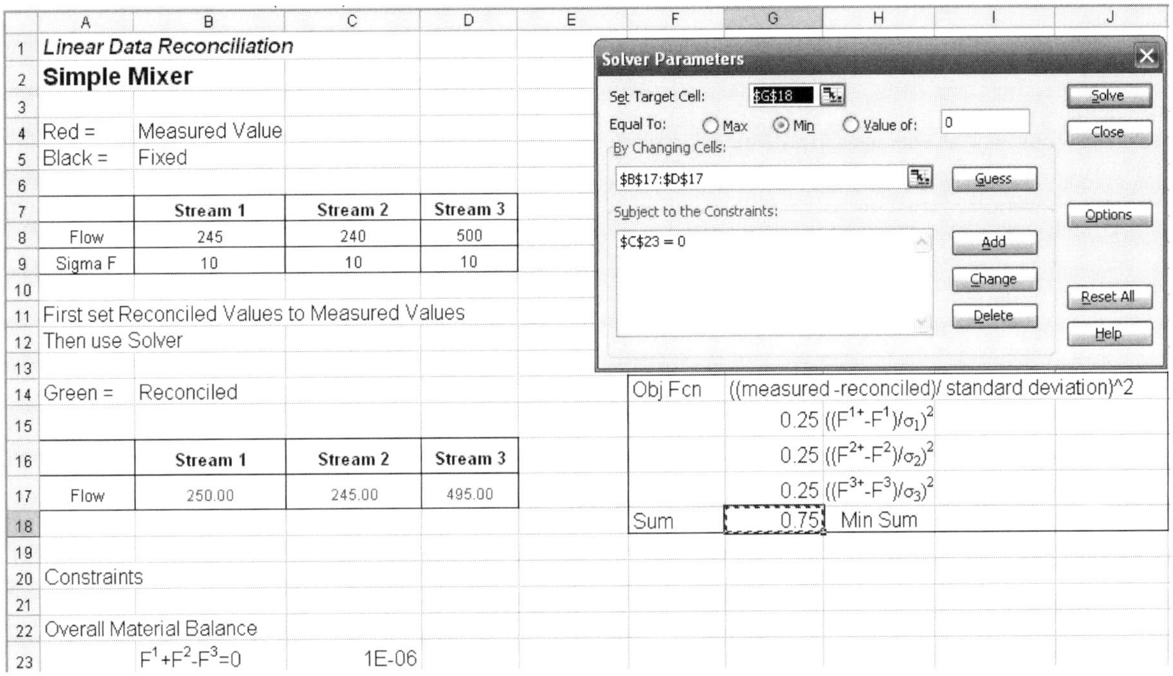

Figure 6.6 Excel Solver solution to Example 6.5—mixer data reconciliation.

$$\left(\frac{245-F_1}{10}\right)^2,$$

which is formed as

$$\left(\frac{\text{cell (B8)} - \text{cell (B17)}}{\text{cell (B9)}}\right)^2.$$

The three terms in the objective function are summed in cell G18. The measured flow rates in cells B8, C8, and D8 remain fixed in value. Cells B17, C17, and D17 contain the adjustable variables F_1, F_2, and F_3. The values in these cells are varied by Solver to minimize the objective function, subject to the material balance constraint, which is supplied in cell C23. The optimum "reconciled" flow rate values are found in cells B17, C17, and D17 and are listed in Table 6.2.

The reconciled flow rates are found as $F_1 = 250$ kg/h, $F_2 = 245$ kg/h, and $F_3 = 495$ kg/h, and with these reconciled values, the material balance closes. ■

Table 6.2 Measured and Reconciled Data for Example 6.5

	Stream 1	Stream 2	Stream 3
Measured flow	245	240	500
σ Flow	10	10	10
Reconciled flow	250	245	495

In Excel Solver, once the problem has been formulated, *do not* push the Guess button to highlight the Solver parameters (variables). The Guess button will make everything in the current objective function Solver optimization variables including measured variables, standard deviations, and reconciled variables. You can push the icon to the left of the Guess button to highlight problem variables.

Simple data reconciliation problems, such as Example 6.5, can also be solved by using the equality constraint(s) to eliminate variable(s) in the objective function and then by employing the necessary conditions for an optimal solution—here see Problems 6.1 and 6.2.

6.2.2 Multiple-Unit Material Balance: Excel Solver

The same process we have used for single-unit data reconciliation can be extended to processes with multiple-unit operations.

EXAMPLE 6.6 *Multiple-Unit Data Reconciliation*

Consider the three-unit operation flow sheet shown in Figure 6.7. The measured values for F_1^+, F_2^+, F_3^+, F_4^+, and F_5^+ are 99.3, 123.2, 125.1, 100.5, and 23.9. The standard deviation for each measurement is 2.0. Determine the reconciled flow rates using Excel Solver. It is actually not necessary to indicate flow rate units; we just need to work with a consistent set of units. To start the optimization procedure, the measured flow rates should be used as the initial guess for the reconciled flow rates.

The data reconciliation problem would be

$$\text{Minimize}\left(\left(\frac{99.3-F_1}{2.0}\right)^2+\left(\frac{123.2-F_2}{2.0}\right)^2+\left(\frac{125.1-F_3}{2.0}\right)^2+\left(\frac{100.5-F_4}{2.0}\right)^2+\left(\frac{23.9-F_5}{2.0}\right)^2\right)$$

$$\text{Subject to } F_1+F_5-F_2=0$$
$$F_2-F_3=0$$
$$F_3-F_4-F_5=0.$$

The equality constraints are the linear mass flow balances around the mixer, reactor, and separator. The Excel Solver solution is provided in **Example 6.6.xls**. The optimization variables are the reconcilable flow rates F_1, F_2, F_3, F_4, and F_5, and the reconciled values are listed in Table 6.3.

The three material balances close with the reconciled values. We do want to comment on the overall material balance, $F_1 = F_4$, which is not included in the constraint set. The overall balance is simply the sum of the three material balances (constraints) currently used in the data reconciliation formulation, and it would not provide any independent information.

6.3 DATA RECONCILIATION: REDUNDANCY AND VARIABLE TYPES

A requirement for data reconciliation is that redundant measurements must be available. Redundant measurements occur when the reconciled values for the measured variables and any fixed variables (discussed later in this chapter) complete a material or energy balance. For example, in Example 6.5, the reconciled values for the three measured values complete the mass balance, $F_1 + F_2 - F_3 = 0$; here, F_1^+, F_2^+, and F_3^+ are considered redundant or correctable measurements. If in Example 6.5 there were only two measured flow rates ($F_1^+ = 245$ kg/h and $F_2^+ = 240$ kg/h), we could not determine a value for the material balance $F_1 + F_2 - F_3$. Here we could only solve for the F_3 value making the material balance = 0, giving $F_3 = 485$ kg/h. In this case, F_1^+ and F_2^+ would be measured but not redundant or correctable. There is no possibility for data reconciliation

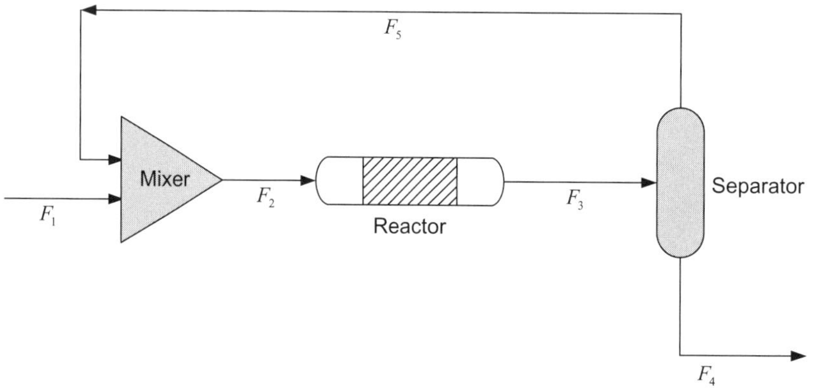

Figure 6.7 Three-unit operation material balance data reconciliation.

Table 6.3 Measured and Reconciled Data for Example 6.6

	Stream 1	Stream 2	Stream 3	Stream 4	Stream 5
Measured flow	99.3	123.2	125.1	100.5	23.9
Σ Flow	2.0	2.0	2.0	2.0	2.0
Reconciled flow	99.99	124.06	124.06	99.99	24.07

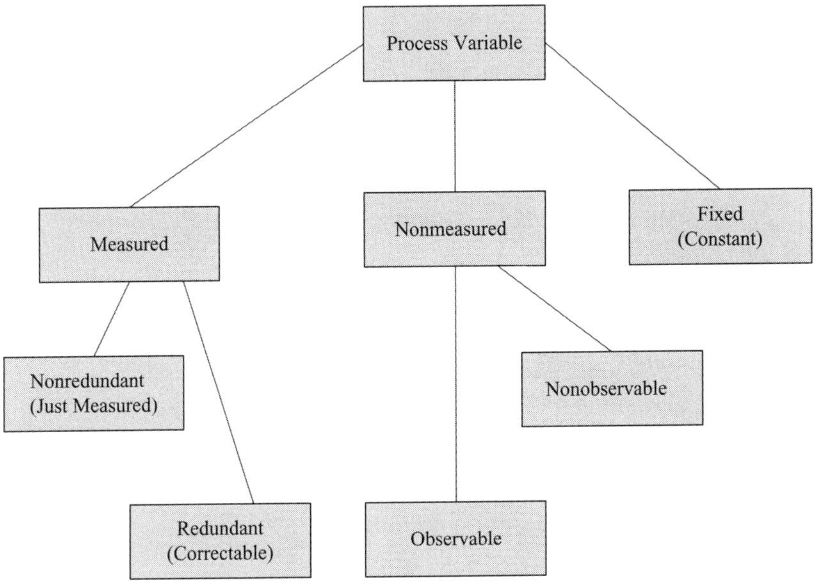

Figure 6.8 Variable classifications, Veverka (2004).

because there is no redundancy or "check" available on the flow rates.

In a processing plant, process variables include temperatures, pressures, stream compositions, species flow rates, and stream flow rates. In order for a process variable to be reconciled, it must be measured and its adjusted or reconciled value must appear in a usable material or energy balance constraint in the data reconciliation problem. Simply measuring a variable does not guarantee that it can be used in the data reconciliation problem. Reconciled values must provide unique values in the process constraints. In other words, all the variables in a given material or energy balance must be measured or fixed (constant value) in order for the measured variables to be reconciled. Process variables can be defined as measured redundant (reconcilable or correctable), measured nonredundant, nonmeasured observable, nonmeasured nonobservable, or fixed. Figure 6.8 provides an overview of variable classifications from Veverka (2004).

We do need to understand the effect these process variable types have on data reconciliation problems. In Examples 6.7–6.10, we apply data reconciliation to the mass flows in and out of the three units (e.g., distillation column sequence) shown in Figure 6.9. We begin with the case where all flow rates are measured and then we will examine the effect of reducing the measurement set.

Using Equations (6.7) and (6.8), we can immediately establish the data reconciliation problem for the mass flow problem of Figure 6.9 as

Formulation 1

$$\text{Minimize } f = \sum_{\substack{i,\text{measured}\\ \text{variables}}} \left(\frac{(x_i^+ - x_i)}{\sigma_i} \right)^2$$

$$\text{Subject to } F_1 - F_2 - F_3 = 0$$
$$F_2 - F_4 - F_5 = 0$$
$$F_3 + F_4 - F_6 - F_7 = 0.$$

The constraints are the material balances over the three units: $F_1 - F_2 - F_3 = 0$ (unit I), $F_2 - F_4 - F_5 = 0$ (unit II), and $F_3 + F_4 - F_6 - F_7 = 0$ (unit III). The overall balance would simply be the sum of these three balances and it does not supply an additional constraint.

In order to better understand our discussion of redundancy as well as Figure 6.8, we will also form the data reconciliation problem using an equivalent formulation of

Formulation 2

$$\text{Minimize } f = \sum_{\substack{i,\text{measured}\\ \text{variables}}} \left(\frac{(x_i^+ - x_i)}{\sigma_i} \right)^2 \tag{6.9}$$

$$\text{Subject to } h_k(x_i) = 0 \quad k = 1, \ldots, K. \tag{6.10}$$

Again x_i^+ is the measured process variable, x_i is the reconciled value, and $h_k(x_i)$ are the material and energy balances. There are differences between Equations (6.8) and (6.10). When using Equation (6.10), the material and energy balance constraints can be formed using *only* measured process variables and fixed variables—this is the definition of redundancy. Here then, nonmeasured process variables (u_j in Eq. (6.8)) are not used in the constraints. After the data reconciliation problem is solved with formulation 2, fixed variables and reconciled estimates can be used to determine values for nonmeasured observable variables. This formulation (only fixed and reconcilable variables in the constraints) is also a first step in using Lagrange multipliers (Section 6.5) for the solution of data reconciliation problems.

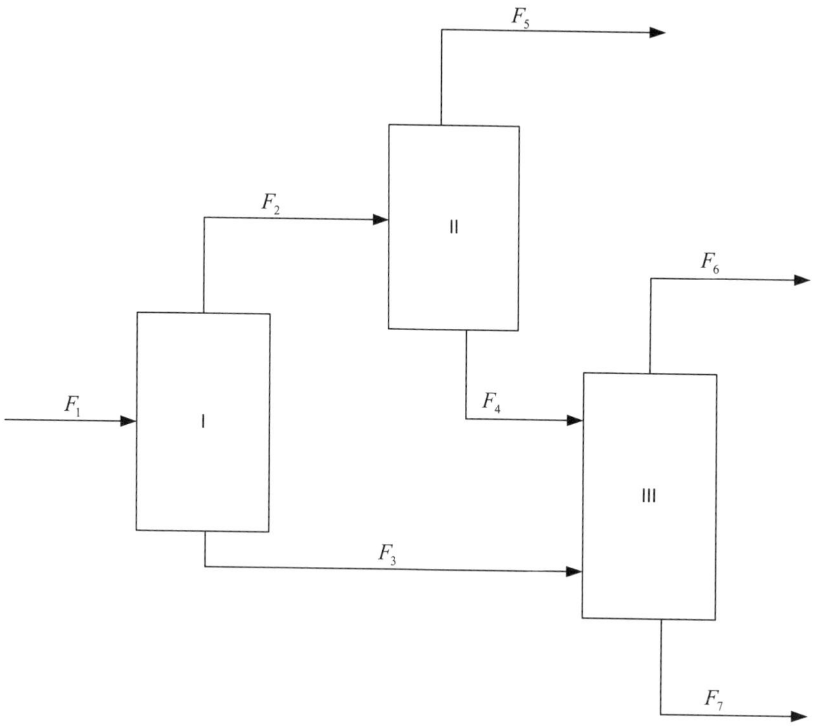

Figure 6.9 Three-unit operations in sequence for Examples 6.7–6.10.

Table 6.4 Measured and Reconciled Data for Example 6.7

	Stream 1	Stream 2	Stream 3	Stream 4	Stream 5	Stream 6	Stream 7
Measured flow	102.2	50.2	52.4	25.1	27.8	50.3	25.1
σ Flow	1.02	1.02	1.02	1.02	1.02	1.02	1.02
Reconciled flow	102.56	50.77	51.82	23.95	26.83	50.48	25.28

Based on formulation 2, we can also define a degree of redundancy (DoR), which will be needed for gross error detection (Section 6.6). For linear data reconciliation balance problems, the DoR is

$$\begin{aligned} \text{DoR} &= (\text{row rank of Eq. (6.10)}) \\ &= \text{the number of independent balances in Eq. (6.10).} \end{aligned} \tag{6.11}$$

In Equation (6.11), we are saying as all the system variables are known (measured), the number of independent balances provides the measure of system redundancy. The partitioning of variables into measured and nonmeasured is instructive, but it can quickly become cumbersome for large problems.

EXAMPLE 6.7 *Base Case: All Measured and Redundant Variables*

In Figure 6.9, let all flow rates be measured (kilogram per hour) with $(F_1^+, F_2^+, F_3^+, F_4^+, F_5^+, F_6^+, F_7^+) = (102.2, 50.2, 52.4, 25.1, 27.8, 50.3, 25.1)$ and the maximum instrument error is ±2 kg/h. From Equation (6.6), $\sigma_i = 2/1.96 = 1.02$ kg/h. Form and solve the data reconciliation problem. As all variables are measured, there is no difference in the data reconciliation problem given by Equations (6.7), (6.8), (6.10), and (6.11). The data reconciliation optimization problem for formulation 1 or 2 is

$$\text{Minimize} \left(\begin{aligned} &\left(\frac{102.2-F_1}{1.02}\right)^2 + \left(\frac{50.2-F_2}{1.02}\right)^2 + \left(\frac{52.4-F_3}{1.02}\right)^2 \\ &+ \left(\frac{25.1-F_4}{1.02}\right)^2 + \left(\frac{27.8-F_5}{1.02}\right)^2 \\ &+ \left(\frac{50.3-F_6}{1.02}\right)^2 + \left(\frac{25.1-F_7}{1.02}\right)^2 \end{aligned} \right)$$

$$\begin{aligned} \text{Subject to } & F_1 - F_2 - F_3 = 0 \\ & F_2 - F_4 - F_5 = 0 \\ & F_3 + F_4 - F_6 - F_7 = 0. \end{aligned}$$

The optimization variables are the reconcilable flow rates F_1, F_2, F_3, F_4, F_5, F_6, and F_7, and the Excel solution is given in **Example 6.7.xls**. Key results are summarized in Table 6.4 and show the measured values $(F_1^+, F_2^+, F_3^+, F_4^+, F_5^+, F_6^+, F_7^+) = (102.2, 50.2, 52.4, 25.1, 27.8, 50.3, 25.1)$ are reconciled to (102.56, 50.77,

Table 6.5 Measured and Reconciled Data for Example 6.8

	Stream 1	Stream 2	Stream 3	Stream 4	Stream 5	Stream 6	Stream 7
Measured flow	102.2	0	52.4	25.1	27.8	50.3	25.1
σ Flow	1.02	Not measured	1.02	1.02	1.02	1.02	1.02
Reconciled flow	102.88	51.35	51.53	24.23	27.11	50.48	25.28

Table 6.6 Measured and Reconciled Data for Example 6.9

	Stream 1	Stream 2	Stream 3	Stream 4	Stream 5	Stream 6	Stream 7
Measured flow	102.2	0	0	0	27.8	50.3	25.1
σ Flow	1.02	Not measured	Not measured	Not measured	1.02	1.02	1.02
Reconciled flow	102.45	40.33	62.12	12.77	27.55	50.05	24.85

51.82, 23.95, 26.83, 50.48, 25.28). The three linear material balances are satisfied (constraints = 0).

Here, all process variables, the flow rates, have been measured. Each constraint is completed by the measured variables (their adjusted value) and all variables appear in the constraints. The process variables F_1, F_2, F_3, F_4, F_5, F_6, and F_7 would be considered *measured and redundant* (or correctable). From Equation (6.11), the DoR = 3.

EXAMPLE 6.8 *Nonmeasured Observable Variables*

For Figure 6.9 let assume that the measured flow rates are $(F_1^+, F_3^+, F_4^+, F_5^+, F_6^+, F_7^+) = (102.2, 52.4, 25.1, 27.8, 50.3, 25.1)$ and that F_2^+ is not measured. As F_2^+ is not measured, it will not appear in the objective function of either formulation. In Equation (6.8), $u_j = F_2$. The objective function and constraints in the data reconciliation problem using formulation 1 (Eqs. (6.7) and (6.8)) are

$$\text{Minimize}\left(\begin{array}{l}\left(\frac{102.2-F_1}{1.02}\right)^2+\left(\frac{52.4-F_3}{1.02}\right)^2+\left(\frac{25.1-F_4}{1.02}\right)^2\\+\left(\frac{27.8-F_5}{1.02}\right)^2+\left(\frac{50.3-F_6}{1.02}\right)^2+\left(\frac{25.1-F_7}{1.02}\right)^2\end{array}\right)$$

$$\begin{aligned}\text{Subject to } & F_1 - F_2 - F_3 = 0\\ & F_2 - F_4 - F_5 = 0\\ & F_3 + F_4 - F_6 - F_7 = 0.\end{aligned}$$

For formulation 2 (Eqs. (6.10) and (6.11)), the objective function and constraints in the data reconciliation problem can only involve measured process variables giving

$$\text{Minimize}\left(\begin{array}{l}\left(\frac{102.2-F_1}{1.02}\right)^2+\left(\frac{52.4-F_3}{1.02}\right)^2+\left(\frac{25.1-F_4}{1.02}\right)^2\\+\left(\frac{27.8-F_5}{1.02}\right)^2+\left(\frac{50.3-F_6}{1.02}\right)^2+\left(\frac{25.1-F_7}{1.02}\right)^2\end{array}\right)$$

$$\begin{aligned}\text{Subject to } & F_1 - F_3 - F_4 - F_5 = 0\\ & F_3 + F_4 - F_6 - F_7 = 0.\end{aligned}$$

The material balance constraints are $F_1 - F_3 - F_4 - F_5 = 0$ (for units I and II) and $F_3 + F_4 - F_6 - F_7 = 0$ (for unit III).

Key results from either formulation in **Example 6.8a.xls** or **Example 6.8b.xls** are provided in Table 6.5.

For formulation 2, the optimization variables are F_1, F_3, F_4, F_5, F_6, and F_7, and the reconciled values are $(F_1, F_3, F_4, F_5, F_6, F_7) = (102.88, 51.53, 24.23, 27.11, 50.48, 25.28)$. The process variables F_1, F_3, F_4, F_5, F_6, and F_7 would be measured and redundant (or correctable). After the optimization is complete, the reconciled values can be used to determine a value for F_2 as $F_2 = F_1 - F_3 = 51.35$. The variable F_2 is termed *nonmeasured observable* as it is not measured; however, an equation (here a mass balance) involving only measured redundant variables can be formed to determine its value (hence observable).

For formulation 1, the optimization variables are F_1, F_2, F_3, F_4, F_5, F_6, and F_7. F_2^+ is not measured; however, a constraint equation (here the mass balance on unit I) allows the value for F_2 to be determined as part of the optimization process. For either formulation, the DoR = 2.

EXAMPLE 6.9 *Nonmeasured Nonobservable Variables*

For Figure 6.9, assume the measured values are $(F_1^+, F_5^+, F_6^+, F_7^+) = (102.2, 27.8, 50.3, 25.1)$ and that F_2^+, F_3^+, and F_4^+ are not measured.

For formulation 1, terms involving F_2, F_3, and F_4 cannot appear in the objective function, but these terms can appear in the constraints. In Equation (6.8), $u_j = F_2$, F_3, and F_4. For formulation 2, the objective function and constraints can only involve measured variables so the data reconciliation problem is

$$\text{Minimize}\left(\begin{array}{l}\left(\frac{102.2-F_1}{1.02}\right)^2+\left(\frac{27.8-F_5}{1.02}\right)^2\\+\left(\frac{50.3-F_6}{1.02}\right)^2+\left(\frac{25.1-F_7}{1.02}\right)^2\end{array}\right)$$

$$\text{Subject to } F_1 - F_5 - F_6 - F_7 = 0.$$

The constraint is the overall balance.

Key results from either formulation in **Example 6.9a.xls** or **Example 6.9b.xls** are summarized in Table 6.6.

For formulation 2, the optimization variables are F_1, F_5, F_6, and F_7, and reconciled values are $(F_1, F_5, F_6, F_7) = (102.45, 27.55,$

50.05, 24.85). The process variables F_1, F_5, F_6, and F_7 would be measured and redundant (or correctable). The reconciled values cannot be used to determine values for F_2, F_3, and F_4 unless one of these flow rates is specified; F_2, F_3, and F_4 are termed *nonmeasured nonobservable*.

For formulation 1, the constraints are the material balances over the three units and the optimization variables are F_1, F_2, F_3, F_4, F_5, and F_6, and F_7, F_2, F_3, and F_4 do not appear in the objective function but values for F_2, F_3, and F_4 are determined as part of the optimization process. There is an important point here: The values for F_2, F_3, and F_4 in Table 6.6 are not unique. If the starting guess for these variables is changed from (0, 0, 0) to (10, 10, 10), the final values from the optimization problem become $F_2 = 43.66$, $F_3 = 58.79$, and $F_4 = 16.11$. You may be asking, "Which constraint equation was used to reconcile the flow rates (F_1, F_5, F_6, F_7) in formulation 1?" The answer is the three balances sum to give an overall balance, which involves only F_1, F_5, F_6, and F_7, allowing solution of the reconciliation problem. The DoR = 1.

EXAMPLE 6.10 *Measured Nonredundant Variable*

As a final example using Figure 6.9, assume the measured values are $(F_1^+, F_2^+, F_5^+, F_6^+, F_7^+) = (102.2, 50.2, 27.8, 50.3, 25.1)$ and that F_3^+ and F_4^+ are not measured.

For formulation 1, terms involving F_3 and F_4 cannot appear in the objective function, but these terms can appear in the constraints. In Equation (6.8), $u_j = F_3$ and F_4. For formulation 2, the objective function and constraints can only involve measured variables so the data reconciliation problem is

$$\text{Minimize} \left(\left(\frac{102.2 - F_1}{1.02}\right)^2 + \left(\frac{50.2 - F_2}{1.02}\right)^2 + \left(\frac{27.8 - F_5}{1.02}\right)^2 + \left(\frac{50.3 - F_6}{1.02}\right)^2 + \left(\frac{25.1 - F_7}{1.02}\right)^2 \right)$$

$$\text{Subject to } F_1 - F_5 - F_6 - F_7 = 0.$$

Only the overall material balance constraint, $F_1 - F_5 - F_6 - F_7 = 0$, is available.

Key results from either formulation in **Example 6.10a.xls** or **Example 6.10b.xls** are summarized in Table 6.7.

For formulation 2, the optimization variables are F_1, F_5, F_6, and F_7. Here, F_2 is in the objective function (with F_1, F_5, F_6, and F_7), but we find that it will not move from its measured value even if F_2 is declared an optimization variable. This occurs because F_2 is not involved in the constraint. The reconciled values are (F_1, F_5, F_6, F_7) = (102.45, 27.55, 50.05, 24.85). The process variables F_1, F_5, F_6, and F_7 would be measured and redundant (or correctable). We can use the reconciled values and the *measured nonredundant* value of $F_2 = 50.2$ to determine values for F_3 and F_4. Here, $F_3 = F_1 - F_2 = 52.25$ and $F_4 = F_2 - F_5 = 22.65$, and F_3 and F_4 are termed *nonmeasured and observable*.

For formulation 1, the constraints are again the material balances over the three units and the optimization variables are again F_1, F_2, F_3, F_4, F_5, F_6, and F_7. Here, we do have F_2 in the objective function (with F_1, F_5, F_6, and F_7), but we find that its value will not move from its measured value. This occurs because F_2 is not uniquely determined in any of the constraints. As part of the optimization process, values for F_3 and F_4 are determined. The DoR = 1. ■

In each example problem (Example 6.7–6.10), formulations 1 and 2 gave identical results. Using only adjustable and fixed variables in the formulation (formulation 2) emphasizes the fact that nonmeasured variables are not actually part of the data reconciliation problem. The values for nonmeasured observable variables can be determined after the data reconciliation problem is solved using reconciled and fixed variable values. Values for nonmeasured observable variables can be determined as part of the data reconciliation problem as we did in formulation 1, but here some care as to result interpretation (see Example 6.9, nonmeasured nonobserved) is required.

Finally, we want to discuss fixed variables. A *fixed* variable is generally measured; however, its value is not allowed to change. Fixed variables are generally used for the amount of product sold or the amount of materials purchased. They may be used in constraints, but not in the objective function.

So far in this chapter, we have provided a statistical basis for the weighting factor in the data reconciliation problem. We have shown how Excel Solver can be used to solve simple data reconciliation problems and we have shown the effect of process variable types on the data reconciliation problem. The examples have involved linear material balance constraints. There is actually no additional difficulty when using Excel Solver to solve data reconciliation problems with nonlinear constraints. Data reconciliation problems are classified as linear or nonlinear based on the constraints, and problems with linear and nonlinear constraints are examined in the next section.

Table 6.7 Measured and Reconciled Data for Example 6.10

	Stream 1	Stream 2	Stream 3	Stream 4	Stream 5	Stream 6	Stream 7
Measured flow	102.2	50.2	0	0	27.8	50.3	25.1
σ Flow	1.02	1.02	Not measured	Not measured	1.02	1.02	1.02
Reconciled flow	102.45	50.2	52.25	22.65	27.55	50.05	24.85

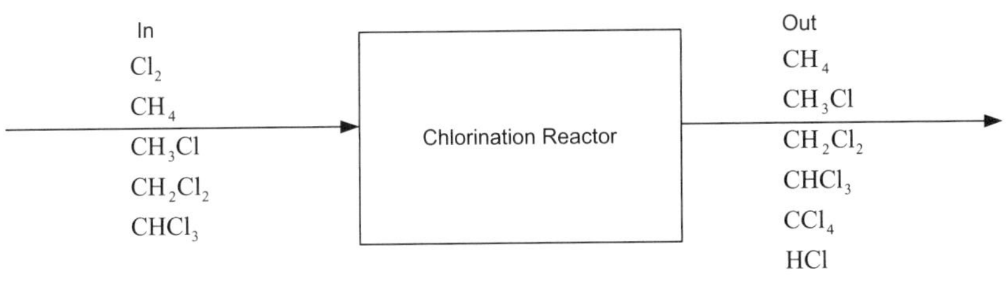

Figure 6.10 Observed species in a chlorination reactor.

Table 6.8 Measured Molar Flow Rates and Instrument Standard Deviations Example 6.11

	Inflow Measured	σ	Outflow Measured	σ
Flow	100	Fixed		
Cl_2	62.1	3.105	0	Fixed
CH_4	26.9	1.345	1.57	0.0785
CH_3Cl	10.8	0.54	6.83	0.3415
CH_2Cl_2	0.78	0.039	24.8	1.24
$CHCl_3$	0.38	0.019	4.82	0.241
CCl_4	0	Fixed	0.39	0.0195
HCl	0	Fixed	59.3	2.965

6.4 DATA RECONCILIATION: LINEAR AND NONLINEAR MATERIAL AND ENERGY BALANCES

Because of the importance of chemical reactions, it is instructive to examine a reaction data reconciliation problem.

EXAMPLE 6.11 *Chlorination of Methane: Linear Constraints*

The following problem is from Madron et al. (2007). We are asked to determine the reconciled species flow rates in a chlorination reactor. Observed species are indicated in Figure 6.10 and Table 6.8 provides measured molar species flows (mole per hour) and instrument standard deviations.

This data reconciliation problem will require species mass balances as constraints. As discussed in Chapter 3 (Section 3.1.4), we can supply needed species material balances using either extents of reaction or elemental balances. The extent of reaction formulation is often preferred because it allows straightforward incorporation of the energy balance into the constraint set.

SOLUTION ***Extent of Reaction Formulation***

Species material balances in reacting systems can be formulated using extents of reaction. A problem is determining how many independent reactions and therefore extents of reaction (one for each independent reaction) are required. The simple answer is that the number of independent reactions = the number of species − the rank of the atomic matrix; see Problem 3.22. Generally, the rank of the atomic matrix is the number of available elemental balances. So, for this example, we would have 7 species—3 elemental balances (C, H, Cl) = 4 independent reactions to account for species reacting and being formed. There are many sets of four reactions, with one possible set:

$$Cl_2 + CH_4 = CH_3Cl + HCl$$

$$Cl_2 + CH_3Cl = CH_2Cl_2 + HCl$$

$$Cl_2 + CH_2Cl_2 = CHCl_3 + HCl$$

$$Cl_2 + CHCl_3 = CCl_4 + HCl.$$

Defining an extent of reaction for each of these four reactions (ξ_1, ξ_2, ξ_3, ξ_4) based on the moles of chlorine reacting, the species mass balances become

$$Cl_2\text{: } N_{Cl_2}^{\text{reactor in}} - (\xi_1) - (\xi_2) - (\xi_3) - (\xi_4) - N_{Cl_2}^{\text{reactor out}} = 0$$

$$CH_4\text{: } N_{CH_4}^{\text{reactor in}} - (\xi_1) - N_{CH_4}^{\text{reactor out}} = 0$$

$$CH_3Cl\text{: } N_{CH_3Cl}^{\text{reactor in}} + (\xi_1) - (\xi_2) - N_{CH_3Cl}^{\text{reactor out}} = 0$$

$$CH_2Cl_2\text{: } N_{CH_2Cl_2}^{\text{reactor in}} + (\xi_2) - (\xi_3) - N_{CH_2Cl_2}^{\text{reactor out}} = 0$$

$$CHCl_3\text{: } N_{CHCl_3}^{\text{reactor in}} + (\xi_3) - (\xi_4) - N_{CHCl_3}^{\text{reactor out}} = 0$$

$$CCl_4\text{: } N_{CCl_4}^{\text{reactor in}} + (\xi_4) - N_{CCl_4}^{\text{reactor out}} = 0$$

$$HCl\text{: } N_{HCl}^{\text{reactor in}} + (\xi_1) + (\xi_2) + (\xi_3) + (\xi_4) - N_{HCl}^{\text{reactor out}} = 0.$$

The data reconciliation problem can be formulated as

$$\text{Minimize} \left(\begin{aligned} &\left(\frac{62.1 - N_{Cl_2}^{\text{reactor in}}}{3.105}\right)^2 + \left(\frac{26.9 - N_{CH_4}^{\text{reactor in}}}{1.345}\right)^2 \\ &+ \left(\frac{10.8 - N_{CH_3Cl}^{\text{reactor in}}}{0.54}\right)^2 + \left(\frac{0.78 - N_{CH_2Cl_2}^{\text{reactor in}}}{0.039}\right)^2 \\ &+ \left(\frac{0.38 - N_{CHCl_3}^{\text{reactor in}}}{0.019}\right)^2 + \left(\frac{1.57 - N_{CH_4}^{\text{reactor out}}}{0.0785}\right)^2 \\ &+ \left(\frac{6.83 - N_{CH_3Cl}^{\text{reactor out}}}{0.3415}\right)^2 + \left(\frac{24.8 - N_{CH_2Cl_2}^{\text{reactor out}}}{1.24}\right)^2 \\ &+ \left(\frac{4.82 - N_{CHCl_3}^{\text{reactor out}}}{0.241}\right)^2 + \left(\frac{0.39 - N_{CCl_4}^{\text{reactor out}}}{0.0195}\right)^2 \\ &+ \left(\frac{59.3 - N_{HCl}^{\text{reactor out}}}{2.965}\right)^2 \end{aligned} \right)$$

Table 6.9 Measured and Reconciled Data and Extents of Reaction for Example 6.11

	Inflow Measured	σ	Inflow Reconciled	Outflow Measured	σ	Outflow Reconciled
Flow	100		Fixed			
Cl_2	62.1	3.105	60.7157	0		Fixed
CH_4	26.9	1.345	27.3276	1.57	0.0785	1.5702
CH_3Cl	10.8	0.54	10.7972	6.83	0.3415	6.8618
CH_2Cl_2	0.78	0.039	0.77961	24.8	1.24	25.5975
$CHCl_3$	0.38	0.019	0.37982	4.82	0.241	4.8644
CCl_4	0		Fixed	0.39	0.0195	0.3904
HCl	0		Fixed	59.3	2.965	60.7157
Sum			100			100

Reaction	Extent
$Cl_2 + CH_4 - CH_3Cl - HCl = 0$	25.75747
$Cl_2 + CH_3Cl - CH_2Cl_2 - HCl = 0$	29.69288
$Cl_2 + CH_2Cl_2 - CHCl_3 - HCl = 0$	4.874971
$Cl_2 + CHCl_3 - CCl_4 - HCl = 0$	0.390384

Subject to

$$N_{Cl_2}^{\text{reactor in}} - (\xi_1) - (\xi_2) - (\xi_3) - (\xi_4) - N_{Cl_2}^{\text{reactor out}} = 0$$

$$N_{CH_4}^{\text{reactor in}} - (\xi_1) - N_{CH_4}^{\text{reactor out}} = 0$$

$$N_{CH_3Cl}^{\text{reactor in}} + (\xi_1) - (\xi_2) - N_{CH_3Cl}^{\text{reactor out}} = 0$$

$$N_{CH_2Cl_2}^{\text{reactor in}} + (\xi_2) - (\xi_3) - N_{CH_2Cl_2}^{\text{reactor out}} = 0$$

$$N_{CHCl_3}^{\text{reactor in}} + (\xi_3) - (\xi_4) - N_{CHCl_3}^{\text{reactor out}} = 0$$

$$N_{CCl_4}^{\text{reactor in}} + (\xi_4) - N_{CCl_4}^{\text{reactor out}} = 0$$

$$N_{HCl}^{\text{reactor in}} + (\xi_1) + (\xi_2) + (\xi_3) + (\xi_4) - N_{HCl}^{\text{reactor out}} = 0$$

$$100 - N_{Cl_2}^{\text{reactor in}} - N_{CH_4}^{\text{reactor in}} - N_{CH_3Cl}^{\text{reactor in}} - N_{CH_2Cl_2}^{\text{reactor in}} - N_{CHCl_3}^{\text{reactor in}} = 0.$$

There are eight constraints, the seven species mass balances and the requirement that the total moles into the reactor, $N_{total}^{\text{reactor in}} = 100$. In the material balance constraints, we define *fixed variables*, $N_{CCl_4}^{\text{reactor in}} = 0$, $N_{HCl}^{\text{reactor in}} = 0$, and $N_{Cl_2}^{\text{reactor out}} = 0$, and the total molar flow in, $N_{total}^{\text{reactor in}} = 100$. The use of fixed variables $N_{CCl_4}^{\text{reactor in}} = 0$, $N_{HCl}^{\text{reactor in}} = 0$, and $N_{Cl_2}^{\text{reactor out}} = 0$ allows their inclusion in the material balance constraints, but these three values are not allowed to change and they are not included in the objective function. The total flow rate in is also a fixed variable, and this will force adjustments to the species molar flows in (Σ species molar flows in = 100). There is no change in moles with any reaction, so $N_{total}^{\text{reactor in}} = N_{total}^{\text{reactor out}}$.

The variables in Excel Solver are the species molar flow rates in and out of the chlorination reactor ($N_{Cl_2}^{\text{reactor in}}$, $N_{CH_4}^{\text{reactor in}}$, $N_{CH_3Cl}^{\text{reactor in}}$, $N_{CH_2Cl_2}^{\text{reactor in}}$, $N_{CHCl_3}^{\text{reactor in}}$, $N_{CH_4}^{\text{reactor out}}$, $N_{CH_3Cl}^{\text{reactor out}}$, $N_{CH_2Cl_2}^{\text{reactor out}}$, $N_{CHCl_3}^{\text{reactor out}}$, $N_{CCl_4}^{\text{reactor out}}$, and $N_{HCl}^{\text{reactor out}}$) and the four extents of reaction (ξ_1, ξ_2, ξ_3, and ξ_4). In actuality, the extents of reaction are not variables, but they are uniquely determined parameters once the data reconciliation problem with species molar flow rates is solved.

To see this, we will need to formulate the constraints without extents of reaction. This is done next by using elemental balances for the species material balances. For now accept that the extents of reaction must be included as variables in the current formulation of the optimization problem. The next table summarizes the variable types.

Variable Classification	Process Variable
Measured and redundant	$N_{Cl_2}^{\text{reactor in}}$, $N_{CH_4}^{\text{reactor in}}$, $N_{CH_3Cl}^{\text{reactor in}}$, $N_{CH_2Cl_2}^{\text{reactor in}}$, $N_{CHCl_3}^{\text{reactor in}}$, $N_{CH_4}^{\text{reactor out}}$, $N_{CH_3Cl}^{\text{reactor out}}$, $N_{CH_2Cl_2}^{\text{reactor out}}$, $N_{CHCl_3}^{\text{reactor out}}$, $N_{CCl_4}^{\text{reactor out}}$, $N_{HCl}^{\text{reactor out}}$
Fixed	$N_{total}^{\text{reactor in}}$, $N_{CCl_4}^{\text{reactor in}}$, $N_{HCl}^{\text{reactor in}}$, $N_{Cl_2}^{\text{reactor out}}$
Nonmeasured observable	$N_{total}^{\text{reactor out}}$

The Excel solution is provided in **Example 6.11a.xls** and key results are summarized in Table 6.9.

In Example 6.11, if the total flow rate in is not fixed (currently set at 100 mol/h), the molar flow in and out of the reactor will balance after reconciliation at a value between the current measured molar flow rates in and out. This solution is provided in **Example 6.11b.xls**, and $N_{total}^{\text{reactor in}} = N_{total}^{\text{reactor out}} = 98.8067$.

Elemental Balance Formulation

As we noted, the use of extents of reaction in the above formulation is convenient especially when an energy balance is included in the reactor constraints—heat of reaction data are often available as kilocalorie per mole of reacting species. We can avoid using extents of reaction by formulating the problem using elemental balances. Elemental balances can always be written for any system

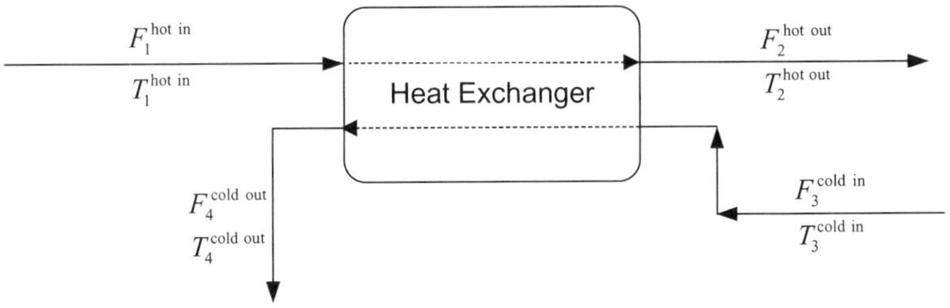

Figure 6.11 Heat exchanger process variables.

regardless if reactions are occurring. For Example 6.11, the elemental balances on carbon, hydrogen, and chlorine are

$$\text{C: } N_{CH_4}^{\text{reactor in}} + N_{CH_3CL}^{\text{reactor in}} + N_{CH_2CL_2}^{\text{reactor in}} + N_{CHCL_3}^{\text{reactor in}} - N_{CH_4}^{\text{reactor out}} - N_{CH_3Cl}^{\text{reactor out}} - N_{CH_2Cl_2}^{\text{reactor out}} - N_{CHCl_3}^{\text{reactor out}} - N_{CCl_4}^{\text{reactor out}} = 0$$

$$\text{H: } 4N_{CH_4}^{\text{reactor in}} + 3N_{CH_3Cl}^{\text{reactor in}} + 2N_{CH_2Cl_2}^{\text{reactor in}} + N_{CHCl_3}^{\text{reactor in}} - 4N_{CH_4}^{\text{reactor out}} - 3N_{CH_3Cl}^{\text{reactor out}} - 2N_{CH_2Cl_2}^{\text{reactor out}} - N_{CHCl_3}^{\text{reactor out}} - N_{HCl}^{\text{reactor out}} = 0$$

$$\text{Cl: } 2N_{Cl_2}^{\text{reactor in}} + N_{CH_3Cl}^{\text{reactor in}} + 2N_{CH_2Cl_2}^{\text{reactor in}} + 3N_{CHCl_3}^{\text{reactor in}} - 2N_{Cl_2}^{\text{reactor out}} - N_{CH_3Cl}^{\text{reactor out}} - 2N_{CH_2Cl_2}^{\text{reactor out}} - 3N_{CHCl_3}^{\text{reactor out}} - 4N_{CCl_4}^{\text{reactor out}} - N_{HCl}^{\text{reactor out}} = 0.$$

These three element balances can replace the seven species mass balance constraints used in the formulation above. This replacement will eliminate the use of extents of reaction, and the optimization variables are the species molar flow rates in and out of the chlorination reactor ($N_{Cl_2}^{\text{reactor in}}$, $N_{CH_4}^{\text{reactor in}}$, $N_{CH_3Cl}^{\text{reactor in}}$, $N_{CH_2Cl_2}^{\text{reactor in}}$, $N_{CHCl_3}^{\text{reactor in}}$, $N_{CH_4}^{\text{reactor out}}$, $N_{CH_3Cl}^{\text{reactor out}}$, $N_{CH_2Cl_2}^{\text{reactor out}}$, $N_{CHCl_3}^{\text{reactor out}}$, $N_{CCl_4}^{\text{reactor out}}$, and $N_{HCl}^{\text{reactor out}}$). The solution to Example 6.11 using elemental balances is provided in **Example 6.11c.xls**. Results for the reconciled species flows from **Example 6.11c.xls** are identical to those reported in Table 6.9. The extents of reaction, if needed, can be uniquely determined from these reconciled species flows. For example, using the reconciled species flows in **Example 6.11c.xls** and $N_{CH_4}^{\text{reactor in}} - (\xi_1) - N_{CH_4}^{\text{reactor out}} = 0$, we find $\xi_1 = 25.75747$. ■

We next examine nonlinear data reconciliation where at least one of the constraints is nonlinear. Most energy balances are nonlinear; for example, they often involve the product of the flow rate and enthalpy. Material balance problems can also be nonlinear; for example the product of the total flow rate and composition (mole fraction or weight fraction) is nonlinear.

EXAMPLE 6.12 *Adiabatic Heat Exchanger with Nonlinear Energy Balance and the "No-Leak" Assumption*

Consider the adiabatic heat exchanger problem shown in Figure 6.11 with measured data in Table 6.10. The stream heat capacities are $\hat{C}_{p,h}$ (hot streams) = 5 + (0.1)T, and $\hat{C}_{p,c}$ (cold streams) = 3 + (0.2)T. Recall that when using heat capacities, we assume the streams are single phase and constant pressure. Here the heat capacity is given by $\hat{C}_p = a + bT$, where $\hat{C}_p$ is in kilojoule per kilogram-Celsius, T is the temperature in Celsius, and a and b are stream (species) dependent constants. We can estimate the change in enthalpy as

$$\Delta\hat{h} = \hat{h}_{(T_2,P)} - \hat{h}_{(T_1,P)} = \int_{T_1}^{T_2} \hat{C}_p dT = a(T_2 - T_1) + \frac{b}{2}(T_2^2 - T_1^2).$$

Table 6.10 Measured Data and Instrument Standard Deviations for Example 6.12

	Hot In	Hot Out	Cold In	Cold Out
Flow (kg/s)	10		20	
Temperature (°C)	90	51	20	43
σ Flow	0.2		0.4	
σ Temperature	1	1	1	1

The measured flow rates and temperatures are shown in Table 6.10, and the change in enthalpy (kilojoule per kilogram) from the inlet to the outlet conditions in each stream is given by

$$\Delta\hat{h}_{\text{hot}} = \int_{T_1}^{T_2} \hat{C}_{p,h} dT = \int_{T\ \text{Hot\_In}}^{T\ \text{Hot\_Out}} (5 + 0.1T) = \left[5T + \frac{0.1}{2}T^2\right]_{T\ \text{Hot\_In}}^{T\ \text{Hot\_Out}}$$

$$\Delta\hat{h}_{\text{cold}} = \int_{T_1}^{T_2} \hat{C}_{p,c} dT = \int_{T\ \text{Cold\_In}}^{T\ \text{Cold\_Out}} (3 + 0.2T) = \left[3T + \frac{0.2}{2}T^2\right]_{T\ \text{Cold\_In}}^{T\ \text{Cold\_Out}}.$$

The rate of heat transfer $\dot{Q}$ (kilojoule per second) for each stream is given as $\dot{Q}_{\text{hot}} = F_{\text{hot}}\Delta\hat{h}_{\text{hot}}$ and $\dot{Q}_{\text{cold}} = F_{\text{cold}}\Delta\hat{h}_{\text{cold}}$. Here, $F_1 = F_2 = F_{\text{hot}}$ and $F_3 = F_4 = F_{\text{cold}}$ as only inlet flow rates are measured; this is making the assumption that *there are no losses or leaks in the system.* Unless there are expected "leaks" in process streams, the no-leak assumption is standard in data reconciliation problem formulation. The overall energy balance constraint for the adiabatic process is formed as the following: The rate of heat given up by the hot stream (a negative quantity) is equal to the rate of heat absorbed by the cold stream (a positive quantity), $\dot{Q}_{\text{hot}} + \dot{Q}_{\text{cold}} = 0$ or $|\dot{Q}_{\text{hot}}| - |\dot{Q}_{\text{cold}}| = 0$. If the outlet flow rates are measured, then two additional material balance constraints would be possible: $F_1 - F_2 = 0$ and $F_3 - F_4 = 0$.

Table 6.11 Measured and Reconciled Data for Example 6.12—Heat Exchanger

	Hot In	Hot Out	Cold In	Cold Out
Measured flow	10		20	
Measured temperature	90	51	20	43
σ Flow	0.2		0.4	
σ Temperature	1	1	1	1
Reconciled flow	9.9359586		20.1261888	
Reconciled temperature	89.515928	51.35156873	19.515903	43.82560344

The data reconciliation optimization problem is

$$\text{Minimize}\left(\begin{array}{l}\left(\dfrac{10-F_{\text{hot}}}{0.2}\right)^2+\left(\dfrac{20-F_{\text{cold}}}{0.4}\right)^2\\+\left(\dfrac{90-T_{\text{Hot\_In}}}{1.0}\right)^2+\left(\dfrac{51-T_{\text{Hot\_Out}}}{1.0}\right)^2\\+\left(\dfrac{20-T_{\text{Cold\_In}}}{1.0}\right)^2+\left(\dfrac{43-T_{\text{Cold\_Out}}}{1.0}\right)^2\end{array}\right)$$

$$\text{Subject to } F_{\text{hot}}\left[\begin{array}{l}\left(5T_{\text{Hot\_Out}}+\dfrac{0.1}{2}(T_{\text{Hot\_Out}})^2\right)\\-\left(5T_{\text{Hot\_In}}+\dfrac{0.1}{2}(T_{\text{Hot\_In}})^2\right)\end{array}\right]$$

$$+F_{\text{cold}}\left[\begin{array}{l}\left(3T_{\text{Cold\_Out}}+\dfrac{0.2}{2}(T_{\text{Cold\_Out}})^2\right)\\-\left(3T_{\text{Cold\_In}}+\dfrac{0.2}{2}(T_{\text{Cold\_In}})^2\right)\end{array}\right]=0.$$

The optimization variables are F_{hot}, F_{cold}, $T_{\text{Hot\_In}}$, $T_{\text{Hot\_Out}}$, $T_{\text{Cold\_In}}$, and $T_{\text{Cold\_Out}}$. The data reconciliation problem is solved in **Example 6.12.xls**, and the final reconciled values are summarized in Table 6.11.

Using the reconciled values, we can calculated the rate of heat transfer between the hot and cold streams as

$$\begin{aligned}\dot{Q}_{\text{hot}}&=F_{\text{hot}}\Delta\hat{h}_{\text{hot}}\\&=F^{\text{hot}}\left[\begin{array}{l}\left(5T_{\text{Hot\_Out}}+\dfrac{0.1}{2}(T_{\text{Hot\_Out}})^2\right)\\-\left(5T_{\text{Hot\_In}}+\dfrac{0.1}{2}(T_{\text{Hot\_In}})^2\right)\end{array}\right]\\&=-4566.84167\text{ kJ/s}\end{aligned}$$

and

$$\begin{aligned}\dot{Q}_{\text{cold}}&=F_{\text{cold}}\Delta\hat{h}_{\text{cold}}\\&=F_{\text{cold}}\left[\begin{array}{l}\left(3T_{\text{Cold\_Out}}+\dfrac{0.2}{2}(T_{\text{Cold\_Out}})^2\right)\\-\left(3T_{\text{Cold\_In}}+\dfrac{0.2}{2}(T_{\text{Cold\_In}})^2\right)\end{array}\right]\\&=4566.84167\text{ kJ/s.}\end{aligned}$$

The $\dot{Q}$ values would be classified as nonmeasured observable. Following Figure 6.1, one use of these results would be to monitor the performance of the heat exchanger via the overall heat transfer coefficient U (kilojoule per second-Celsius-square meter). The overall heat transfer coefficient is considered a process parameter. Here we can write $|\dot{Q}_{\text{hot}}|=|\dot{Q}_{\text{cold}}|=\dot{Q}_{\text{Transferred}}=UA\Delta T_{\text{lm}}$. With $\dot{Q}_{\text{Transferred}}$ and a known heat exchanger area, U, can be calculated from the heat exchanger log mean temperature difference ΔT_{lm} using reconciled temperatures. For countercurrent heat exchanger operation,

$$\Delta T_{\text{lm}}=\frac{(T_{\text{Hot\_Out}}-T_{\text{Cold\_In}})-(T_{\text{Hot\_In}}-T_{\text{Cold\_Out}})}{\ln\left(\dfrac{(T_{\text{Hot\_Out}}-T_{\text{Cold\_In}})}{(T_{\text{Hot\_In}}-T_{\text{Cold\_Out}})}\right)}.$$

For example, using the reconciled values from Example 6.12, $\Delta T_{\text{lm}} = 38.35$, and if A = 10 m^2, $U = 11.9W/°\text{C m}^2$. The trend in U can be monitored and at some decline in U value, the heat exchanger may need to be cleaned. This cleaning could be determined as part of the economic optimization/energy management strategy (see Figure 6.1).

EXAMPLE 6.13 *Nonlinear Data Reconciliation: Styrene Process*

We now solve the styrene material balance problem, which started the discussion in this chapter. Examining Figure 6.3, we identify streams as shown in Figure 6.12 and process measurements are provided in Table 6.12.

As shown in Tables 6.1 and 6.12, the sum of the weight fractions in any stream (with more than one component) does not equal 1.0 and the flows (out–in) for any unit do not sum to zero. For flow rate measurements, we assume the maximum instrument error is 10% of the measured flow rate value, giving σ_i = (Measured value$_i$)(0.10)/(1.96), and for weight fraction measurements all $\sigma_i = 0.05$.

The objective function for the data reconciliation problem is given below—the objective function will involve every measured variable.

$$
\text{Minimize}\left(\begin{array}{l}
\left(\frac{1035.25-F_{\text{ETB}}}{528.07}\right)^2+\left(\frac{53328.23-F_{\text{H}_2\text{O}}}{2720.83}\right)^2+\left(\frac{70026.91-F_2}{3572.80}\right)^2+\left(\frac{70752.37-F_3}{3609.81}\right)^2\\
+\left(\frac{53973.1-F_4}{2753.73}\right)^2+\left(\frac{203.65-F_5}{10.39}\right)^2+\left(\frac{16300.13-F_6}{831.64}\right)^2+\left(\frac{10657.93-F_7}{543.77}\right)^2\\
+\left(\frac{6101.95-F_8}{311.32}\right)^2+\left(\frac{0.2366-w^2_{\text{C}_8\text{H}_{10}}}{0.05}\right)^2+\left(\frac{0.7495-w^2_{\text{H}_2\text{O}}}{0.05}\right)^2+\left(\frac{0.0056-w^2_{\text{C}_8\text{H}_8}}{0.05}\right)^2\\
+\left(\frac{0.0831-w^3_{\text{C}_8\text{H}_{10}}}{0.05}\right)^2+\left(\frac{0.7586-w^3_{\text{H}_2\text{O}}}{0.05}\right)^2+\left(\frac{0.0041-w^3_{\text{H}_2}}{0.05}\right)^2+\left(\frac{0.1512-w^3_{\text{C}_8\text{H}_8}}{0.05}\right)^2\\
+\left(\frac{0.3688-w^6_{\text{C}_8\text{H}_{10}}}{0.05}\right)^2+\left(\frac{0.6301-w^6_{\text{C}_8\text{H}_8}}{0.05}\right)^2+\left(\frac{0.0079-w^7_{\text{C}_8\text{H}_{10}}}{0.05}\right)^2+\left(\frac{0.9903-w^7_{\text{C}_8\text{H}_8}}{0.05}\right)^2\\
+\left(\frac{0.9851-w^8_{\text{C}_8\text{H}_{10}}}{0.05}\right)^2+\left(\frac{0.0007-w^8_{\text{C}_8\text{H}_8}}{0.05}\right)^2
\end{array}\right).
$$

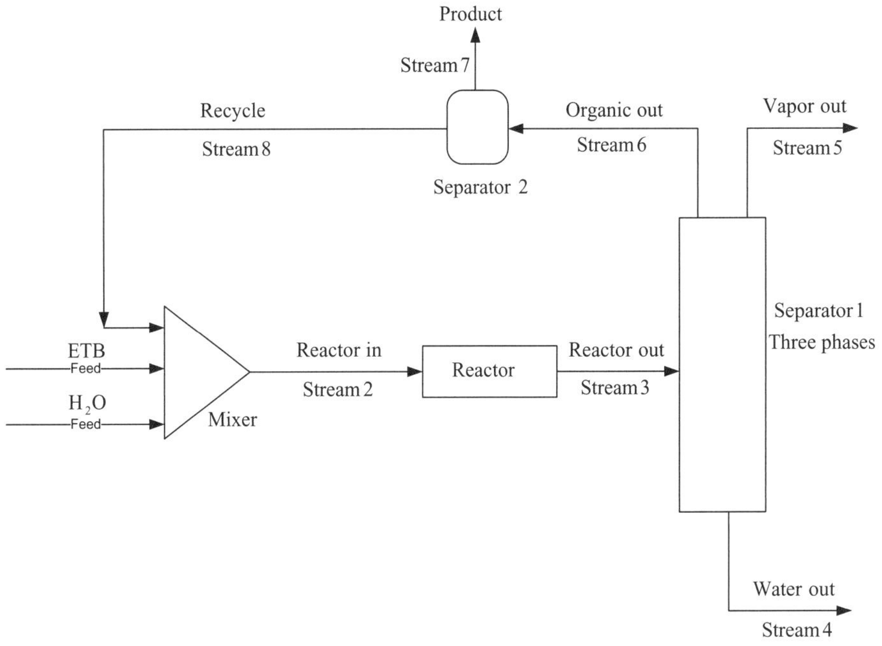

Figure 6.12 Stream identification—styrene process.

Table 6.12 Measured Data and Instrument Standard Deviations for the Styrene Process

	Feed ETB	Feed H_2O	Stream 2 Reactor In	Stream 3 Reactor Out	Stream 4 Water Out	Stream 5 H_2 Vapor	Stream 6 Organic Out	Stream 7 Product	Stream 8 Recycle
Total mass flow	10,350.25	53,328.23	70,026.9100	70,752.37	53,973.1	203.65	16,300.13	10,657.93	6101.95
$w_{C_8H_{10}}$	1	0	0.2366	0.0831	0	0	0.3688	0.0079	0.9851
w_{H_2O}	0	1	0.7495	0.7586	1	0	0	0	0
w_{H_2}	0	0	0	0.0041	0	1	0	0	0
$w_{C_8H_8}$	0	0	0.0056	0.1512	0	0	0.6301	0.9903	0.0007
σ Flow	528.07	2720.83	3572.80	3609.81	2753.73	10.39	831.64	543.77	311.32
σ w			0.05	0.05			0.05	0.05	0.05
$\sum w_i$	1	1	0.9917	0.997	1	1	0.9989	0.9982	0.9858

ETB, ethylbenzene.

Possible constraints for this problem are

1. Total flow mass balance in and out of each unit operation
2. Mass flow for each species in and out of each unit operation
3. Normalization equations for the weight fractions of each stream.

We also can write overall mass balances (total and component) around the entire plant; however, these would not be independent balances; they are just the sum of (1) and (2). In fact, the possible constraints (1)–(3) are not all independent. The sum of the mass flows of each species in and out of a unit operation (2) will give (1) the total mass flows in and out of that unit operation. A common mistake is to specify the species mass balances around each unit operation and the total mass balance around each unit operation and to assume the weight fractions will sum to one in each stream. This will *not* occur as will be seen in Problem 6.4.

For this example, the constraints are the species mass balances (2) around each unit operation and the normalization of the weight fractions (3). Below when a species weight fraction is fixed = 0, we do not include that species in the unit balance. Do note that product of weight fraction and flow rate is nonlinear.

C_8H_{10} Balances

Mixer: $(w_{C_8H_{10}}^{\text{feed ETB}})F_{\text{feed ETB}} + (w_{C_8H_{10}}^{\text{recycle}})F_{\text{recycle}} = (w_{C_8H_{10}}^{\text{reactor in}})F_{\text{reactor in}}$

Reactor: $(w_{C_8H_{10}}^{\text{reactor in}})F_{\text{reactor in}} - (\xi)MW_{C_8H_{10}} = (w_{C_8H_{10}}^{\text{reactor out}})F_{\text{reactor out}}$

Separator 1: $(w_{C_8H_{10}}^{\text{reactor out}})F_{\text{reactor out}} = (w_{C_8H_{10}}^{\text{organic out}})F_{\text{organic out}}$

Separator 2: $(w_{C_8H_{10}}^{\text{organic out}})F_{\text{organic out}} = (w_{C_8H_{10}}^{\text{product}})F_{\text{product}} + (w_{C_8H_{10}}^{\text{recycle}})F_{\text{recycle}}$

H_2O Balances

Mixer: $(w_{H_2O}^{\text{feed H}_2\text{O}})F_{\text{feed H}_2\text{O}} = (w_{H_2O}^{\text{reactor in}})F_{\text{reactor in}}$

Reactor: $(w_{H_2O}^{\text{reactor in}})F_{\text{reactor in}} = (w_{H_2O}^{\text{reactor out}})F_{\text{reactor out}}$

Separator 1: $(w_{H_2O}^{\text{reactor out}})F_{\text{reactor out}} = (w_{H_2O}^{\text{water out}})F_{\text{water out}}$

Separator 2:

H_2 Balances

Mixer:

Reactor: $(\xi)MW_{H_2} = (w_{H_2}^{\text{reactor out}})F_{\text{reactor out}}$

Separator 1: $(w_{H_2}^{\text{reactor out}})F_{\text{reactor out}} = (w_{H_2}^{\text{vapor out}})F_{\text{vapor out}}$

Separator 2:

C_8H_8 Balances

Mixer: $(w_{C_8H_8}^{\text{recycle}})F_{\text{recycle}} = (w_{C_8H_8}^{\text{reactor in}})F_{\text{reactor in}}$

Reactor: $(w_{C_8H_8}^{\text{reactor in}})F_{\text{reactor in}} + (\xi)MW_{C_8H_8} = (w_{C_8H_8}^{\text{reactor out}})F_{\text{reactor out}}$

Separator 1: $(w_{C_8H_8}^{\text{reactor out}})F_{\text{reactor out}} = (w_{C_8H_8}^{\text{organic out}})F_{\text{organic out}}$

Separator 2: $(w_{C_8H_8}^{\text{organic out}})F_{\text{organic out}} = (w_{C_8H_8}^{\text{product}})F_{\text{product}} + (w_{C_8H_8}^{\text{recycle}})F_{\text{recycle}}$

Weight Fraction Normalization

Stream 2: $(w_{C_8H_{10}}^{\text{reactor in}}) + (w_{H_2O}^{\text{reactor in}}) + (w_{C_8H_8}^{\text{reactor in}}) = 1$

Stream 3: $(w_{C_8H_{10}}^{\text{reactor out}}) + (w_{H_2O}^{\text{reactor out}}) + (w_{H_2}^{\text{reactor out}}) + (w_{C_8H_8}^{\text{reactor out}}) = 1$

Stream 6: $(w_{C_8H_{10}}^{\text{organic out}}) + (w_{C_8H_8}^{\text{organic out}}) = 1$

Stream 7: $(w_{C_8H_{10}}^{\text{product}}) + (w_{C_8H_8}^{\text{product}}) = 1$

Stream 8: $(w_{C_8H_{10}}^{\text{recycle}}) + (w_{C_8H_8}^{\text{recycle}}) = 1$

For the reactor, there is one reaction ($C_8H_{10} = C_8H_8 + H_2$) with known stoichiometric coefficients. The extent of reaction ξ is the total number moles of C_8H_{10} reacting, and ξ also accounts for the increase in the number of moles of C_8H_8 and H_2 leaving the reactor.

The optimization variables are all the reconcilable variables in the objective function and the extent of reaction ξ. For the optimization, it is *important* to initially set the reconciled values equal to the measured values, and for this problem, initially estimate the extent of reaction at 50 mol/h. Within Solver → Options, select assume nonnegative to keep the variables positive and then select automatic scaling. The Excel optimization solution is found in **Example 6.13.xls** and the reconciled values are given in Table 6.13.

Table 6.13 Reconciled Data for Example 6.13—Styrene Process

	Feed ETB	Feed H_20	Stream 2 Reactor In	Stream 3 Reactor Out	Stream 4 Water Out	Stream 5 H_2 Vapor	Stream 6 Organic Out	Stream 7 Product	Stream 8 Recycle
Total mass flow	10,643.96	53,628.08	70,245.2356	70,245.23564	53,628.0807	201.20302	16,415.952	10,442.75968	5973.19223
$w_{C_8H_{10}}$	1	0	0.23655917	0.08570838	0	0	0.3667533	0.00454029	1
w_{H_2O}	0	1	0.76344083	0.763440826	1	0	0	0	0
w_{H_2}	0	0	0	0.002864294	0	1	0	0	0
$w_{C_8H_8}$	0	0	6.8475E − 12	0.147986499	0	0	0.6332467	0.99545971	8.0527E − 11
$\sum w_i$	1	1	1	1	1	1	1	1	1

Examining the Excel solution in **Example 6.13.xls**, all the constraints are satisfied and the extent of reaction is 99.808 mol/h. The extent of reaction is being used as an optimization variable, but, as we discussed in Example 6.11, it is actually a uniquely determined parameter. If elemental balances were used in the reaction balances (e.g., carbon in = carbon out), the extent of reaction would not be required; see Problem 6.5. ■

Next, we examine a general analytic approach to solving data reconciliation problems—the use of Lagrange multipliers. This approach is often used as the solution method in the data reconciliation literature.

6.5 DATA RECONCILIATION: LAGRANGE MULTIPLIERS

The Lagrange multiplier method is a technique allowing the conversion of an equality constrained optimization problem to an equivalent unconstrained problem. The Lagrange multiplier technique is a specialized topic but one that is used extensively in the data reconciliation literature.

The Lagrangian function is formed as

$$\text{Minimize } L(x,\lambda) = f(x_i) - \sum_k \lambda_k h_k(x_i), \tag{6.12}$$

where $f(x_i)$ is the objective function as given in Equation (6.10) and λ_k are the Lagrange multipliers, one for each equality constraint $h_k(x_i)$. The equality constraints are provided by the independent material and energy balances using adjustable and fixed variables; see Equation (6.10). The necessary and sufficient conditions for the minimization are found by setting the partial derivatives with respect to each variable, x_i and λ_k, to zero.

EXAMPLE 6.14 ***Mixer Data Reconciliation by Lagrange Multipliers***

Solve Example 6.5, the simple mixer, using Lagrange multipliers. Here, Equation (6.12) would be

$$\text{Min } L(F,\lambda) = \left(\left(\frac{245-F_1}{10}\right)^2 + \left(\frac{240-F_2}{10}\right)^2 + \left(\frac{500-F_3}{10}\right)^2\right) - \lambda(F_1+F_2-F_3). \tag{6.13a}$$

Taking the partial derivatives with respect to each variable, x_i and λ_k, and setting the results to zero,

$$\frac{\partial L}{\partial F_1} = -2\left(\frac{245-F_1}{100}\right) - \lambda = 0 \tag{6.13b}$$

$$\frac{\partial L}{\partial F_2} = -2\left(\frac{240-F_2}{100}\right) - \lambda = 0 \tag{6.13c}$$

$$\frac{\partial L}{\partial F_3} = -2\left(\frac{500-F_3}{100}\right) + \lambda = 0 \tag{6.13d}$$

$$\frac{\partial L}{\partial \lambda} = (F_1+F_2-F_3) = 0. \tag{6.13e}$$

Lagrange multipliers do convert the constrained optimization problem (Example 6.5) to an unconstrained optimization problem termed the Lagrangian function (Eq. (6.13a)). The solution of the Lagrangian function $L(F, \lambda)$ requires solution of the Lagrange multiplier problem, which is given by Equation (6.13b–e).

Equation (6.13b–e) provides four equations with four unknowns, (F_1, F_2, F_3, λ). One solution approach is variable elimination. Here, Equation (6.13b–d) is solved for the variables in the original objective function (F_1, F_2, F_3) in terms of λ. These expressions are then substituted into the remaining equation, Equation (6.13e). This process results in $F_1 = (245 + 50\lambda)$, $F_2 = (240 + 50\lambda)$, and $F_3 = (500 - 50\lambda)$, which, when substituted into Equation (6.13e), gives $\lambda = 0.1$. Using this value for λ in Equation (6.13b–d) gives $(F_1, F_2, F_3) = (250, 245, 495 \text{ kg/h})$, which is the same result we found in Example 6.5.

A second, more general, solution approach is to use the Gauss–Jordan matrix elimination method we developed in Chapter 4 to directly solve the Lagrange multiplier problem, Equation (6.13b–e). Rearranging Equation (6.13b–e) gives

$$\begin{pmatrix} 1 & 0 & 0 & -50 \\ 0 & 1 & 0 & -50 \\ 0 & 0 & 1 & 50 \\ 1 & 1 & -1 & 0 \end{pmatrix} \begin{pmatrix} F^1 \\ F^2 \\ F^3 \\ \lambda \end{pmatrix} = \begin{pmatrix} 245 \\ 240 \\ 500 \\ 0 \end{pmatrix} \text{ with the augmented matrix}$$

$$\begin{pmatrix} 1 & 0 & 0 & -50 & 245 \\ 0 & 1 & 0 & -50 & 240 \\ 0 & 0 & 1 & 50 & 500 \\ 1 & 1 & -1 & 0 & 0 \end{pmatrix}.$$

The Gauss–Jordan solution is given in **Example 6.14.xls** and the results are shown in Figure 6.13.

The results from the Gauss–Jordan elimination show $(F_1, F_2, F_3) = (250, 245, 495 \text{ kg/h})$, as expected.

EXAMPLE 6.15 ***Multiple-Unit Data Reconciliation by Lagrange Multipliers***

Solve Example 6.6, the three-unit operation process, using Lagrange multipliers. Here, Equation (6.12) would be

$$\text{Min } L = \left(\left(\frac{99.3-F_1}{2}\right)^2 + \left(\frac{123.2-F_2}{2}\right)^2 + \left(\frac{125.1-F_3}{2}\right)^2 + \left(\frac{100.5-F_4}{2}\right)^2 + \left(\frac{23.9-F_5}{2}\right)^2\right) - \lambda_1(F_1+F_5-F_2) - \lambda_2(F_2-F_3) - \lambda_3(F_3-F_4-F_5). \tag{6.14a}$$

	A	B	C	D	E	F	G
1	Lagrange Multiplier using Gauss Jordan Elimination						
2	Mixer Example						
3							
4							
5							
6							
7		F1	F2	F3	λ	RHS	
8	eq 6.13b	1	0	0	-50	245	
9	eq 6.13c	0	1	0	-50	240	
10	eq 6.13d	0	0	1	50	500	
11	eq 6.13e	1	1	-1	0	0	
12							
13	Solution from Gauss Jordan Elimination						
14							
15		1	0	0	0	250	F1
16		0	1	0	0	245	F2
17		0	0	1	0	495	F3
18		0	0	0	1	0.1	λ

Figure 6.13 Gauss–Jordan solution of the Lagrange multiplier problem—Example 6.14.

Taking the partial derivatives, the Lagrangian function with respect to the flow rates gives

$$\frac{\partial L}{\partial F_1} = -2\left(\frac{99.3 - F_1}{4}\right) - \lambda_1 = 0 \tag{6.14b}$$

$$\frac{\partial L}{\partial F_2} = -2\left(\frac{123.2 - F_2}{4}\right) + \lambda_1 - \lambda_2 = 0 \tag{6.14c}$$

$$\frac{\partial L}{\partial F_3} = -2\left(\frac{125.1 - F_3}{4}\right) + \lambda_2 - \lambda_3 = 0 \tag{6.14d}$$

$$\frac{\partial L}{\partial F_4} = -2\left(\frac{100.5 - F_4}{4}\right) + \lambda_3 = 0 \tag{6.14e}$$

$$\frac{\partial L}{\partial F_5} = -2\left(\frac{23.9 - F_5}{4}\right) - \lambda_1 + \lambda_3 = 0, \tag{6.14f}$$

and, taking the partial derivatives with respect to λ_1, λ_2, and λ_3, gives the material balance constraints:

$$F_1 + F_5 - F_2 = 0 \tag{6.14g}$$

$$F_2 - F_3 = 0 \tag{6.14h}$$

$$F_3 - F_4 - F_5 = 0. \tag{6.14i}$$

Equation (6.14b–i) provides eight equations with eight unknowns (F_1, F_2, F_3, F_4, F_5, λ_1, λ_2, and λ_3). One solution approach is again to use variable elimination, but this process quickly becomes cumbersome; see Problem 6.3. The more general approach is to directly solve the Lagrange multiplier problem using the Gauss–Jordan algorithm as done in **Example 6.15.xls** with the results shown in Figure 6.14.

The flow rate values are $F_1 = 99.988$, $F_2 = 124.06$, $F_3 = 124.06$, $F_4 = 99.988$, and $F_5 = 24.075$ kg/h, which is the same result obtained in Example 6.6. ■

These two Lagrange multiplier examples (Examples 6.14 and 6.15) involved linear constraints—the linear material balances. We show in the next example that the Lagrange multiplier method can be applied to problems with nonlinear constraints. Example 6.16 utilizes the Newton–Raphson method (which we developed in Chapter 4) for the solution of nonlinear simultaneous equations.

EXAMPLE 6.16 *Data Reconciliation: Nonlinear Lagrange Multiplier*

Solve Example 6.12, the adiabatic heat exchanger, using Lagrange multipliers. Although this problem involves nonlinear constraints, it can be solved using Lagrange multipliers.

Taking the partial derivatives the Lagrangian function (Eq. (6.15a)) with respect to the flow rate, temperatures, and λ gives

$$\text{Minimize } L = \left(\begin{array}{l} \left(\dfrac{10 - F_{\text{hot}}}{0.2}\right)^2 + \left(\dfrac{20 - F_{\text{cold}}}{0.4}\right)^2 \\ + \left(\dfrac{90 - T_{\text{Hot\_In}}}{1.0}\right)^2 + \left(\dfrac{51 - T_{Hot\_Out}}{1.0}\right)^2 \\ + \left(\dfrac{20 - T_{\text{Cold\_In}}}{1.0}\right)^2 + \left(\dfrac{43 - T_{\text{Cold\_Out}}}{1.0}\right)^2 \end{array}\right) - \lambda \left\{\begin{array}{l} F_{\text{hot}}\left[\left(5T_{\text{Hot\_In}} + \dfrac{0.1}{2}(T_{\text{Hot\_In}})^2\right) \right. \\ \left. -\left(5T_{\text{Hot\_Out}} + \dfrac{0.1}{2}(T_{\text{Hot\_Out}})^2\right)\right] \\ - F_{\text{cold}}\left[\left(3T_{\text{Cold\_Out}} + \dfrac{0.2}{2}(T_{\text{Cold\_Out}})^2\right) \right. \\ \left. -\left(3T_{\text{Cold\_In}} + \dfrac{0.2}{2}(T_{\text{Cold\_In}})^2\right)\right] \end{array}\right\} \tag{6.15a}$$

$$\frac{\partial L}{\partial F_{\text{hot}}} = -2\left(\frac{10 - F_{\text{hot}}}{0.04}\right) - \lambda\left[\left(5T_{\text{Hot\_In}} + \frac{0.1}{2}(T_{\text{Hot\_In}})^2\right) - \left(5T_{\text{Hot\_Out}} + \frac{0.1}{2}(T_{\text{Hot\_Out}})^2\right)\right] = 0 \tag{6.15b}$$

$$\frac{\partial L}{\partial F_{\text{cold}}} = -2\left(\frac{20 - F_{\text{cold}}}{0.16}\right) + \lambda\left[\left(3T_{\text{Cold\_Out}} + \frac{0.2}{2}(T_{\text{Cold\_Out}})^2\right) - \left(3T_{\text{Cold\_In}} + \frac{0.2}{2}(T_{\text{Cold\_In}})^2\right)\right] = 0 \tag{6.15c}$$

$$\frac{\partial L}{\partial T_{\text{Hot\_In}}} = -2\left(\frac{90 - T_{\text{Hot\_In}}}{1.0}\right) - \lambda F_{\text{hot}}(5 + 0.1(T_{\text{Hot\_In}})) = 0 \tag{6.15d}$$

$$\frac{\partial L}{\partial T_{\text{Hot\_Out}}} = -2\left(\frac{51 - T_{\text{Hot\_Out}}}{1.0}\right) + \lambda F_{\text{hot}}(5 + 0.1(T_{\text{Hot\_Out}})) = 0 \tag{6.15e}$$

$$\frac{\partial L}{\partial T_{\text{Cold\_In}}} = -2\left(\frac{20 - T_{\text{Cold\_In}}}{1.0}\right) - \lambda F_{\text{cold}}(3 + 0.2(T_{\text{Cold\_In}})) = 0 \tag{6.15f}$$

	A	B	C	D	E	F	G	H	I	J	K
1	*Lagrange Multiplier using Gauss Jordan Elimination*										
2	*Multiple Unit Example*										
3											
4											
5											
6											
7		F1	F2	F3	F4	F5	λ1	λ2	λ3	RHS	
8	eq 6.14b	1	0	0	0	0	-2	0	0	99.3	
9	eq6.14c	0	1	0	0	0	2	-2	0	123.2	
10	eq 6.14d	0	0	1	0	0	0	2	-2	125.1	
11	eq 6.14e	0	0	0	1	0	0	0	2	100.5	
12	eq 6.14f	0	0	0	0	1	-2	0	2	23.9	
13	eq 6.14g	1	-1	0	0	1	0	0	0	0	
14	eq 6.14h	0	1	-1	0	0	0	0	0	0	
15	eq 6.14i	0	0	1	-1	-1	0	0	0	0	
16											
17	Solution from Gauss-Jordan Matrix Elimination										
18											
19		1	0	0	0	0	0	0	0	99.988	F1
20		0	1	0	0	0	0	0	0	124.06	F2
21		0	0	1	0	0	0	0	0	124.06	F3
22		0	0	0	1	0	0	0	0	99.988	F4
23		0	0	0	0	1	0	0	0	24.075	F5
24		0	0	0	0	0	1	0	0	0.3438	λ1
25		0	0	0	0	0	0	1	0	0.775	λ2
26		0	0	0	0	0	0	0	1	0.2563	λ3

Figure 6.14 Gauss–Jordan solution of the Lagrange multiplier problem—Example 6.15.

$$\frac{\partial L}{\partial T_{\text{Cold\_Out}}} = -2\left(\frac{43 - T_{\text{Cold\_Out}}}{1.0}\right) + \lambda F_{\text{cold}}\left(3 + 0.2(T_{\text{Cold\_Out}})\right) = 0 \tag{6.15g}$$

$$\begin{aligned}\frac{\partial L}{\partial \lambda} = F_{\text{hot}}&\left[\left(5T_{\text{Hot\_In}} + \frac{0.1}{2}(T_{\text{Hot\_In}})^2\right) - \left(5T_{\text{Hot\_Out}} + \frac{0.1}{2}(T_{\text{Hot\_Out}})^2\right)\right] \\ - F_{\text{cold}}&\left[\left(3T_{\text{Cold\_Out}} + \frac{0.2}{2}(T_{\text{Cold\_Out}})^2\right)\right. \\ &\left.- \left(3T_{\text{Cold\_In}} + \frac{0.2}{2}(T_{\text{Cold\_In}})^2\right)\right] = 0\end{aligned} \tag{6.1.5h}$$

The Lagrange multiplier problem shows seven nonlinear equations with seven unknowns (Eq. (6.15b–h)). The solution to this set of nonlinear equations can be found in **Example 6.16a.xls** and in Figure 6.15. The solution in the Excel file uses the Newton–Raphson method we developed in Chapter 4. Our Newton–Raphson solution method requires an initial guess for a variable associated with each nonlinear equation. As shown in Figure 6.15, we initially set $F_{\text{hot}} = 10$ kg/s, $F_{\text{cold}} = 20$ kg/s, $T_{\text{Hot\_In}} = 90°\text{C}$, $T_{\text{Hot\_Out}} = 51°\text{C}$, $T_{\text{Cold\_In}} = 20°\text{C}$, $T_{\text{Cold\_Out}} = 43°\text{C}$, and $\lambda = 1$. The seven nonlinear equations (Eq. (6.13b–h)) are then supplied to VBA code in `Function NLFunc`. In **Example 6.16a.xls**, the Gauss–Jordan algorithm is called once (Gauss_Jordan_Macro) to obtain a good initial point for linearization, followed by successive calls to the Newton–Raphson method (NR_Gauss_Jordan_Macro) until desired accuracy is met. In **Example 6.16a.xls**, a loop is established in the NR_Gauss_Jordan_Macro so once the macro is called, the variables will converge. The code controlling the loop, shown next, compares variable values in the second matrix, before (`CTemp(i)`) and after the Newton–Raphson (`C(i, Ncolumns - 1)`) iteration. When the normalized difference in these values is <= `Epsilon` for every variable (row), the Newton–Raphson iteration process is complete:

```
 ' Check if values have converged - part of
the outer loop k
   Dim TmpCounter As Integer
   Dim Epsilon As Double
   Epsilon = 0.000001
   TmpCounter = 0
   For i = 0 To Nrows-1
      If Abs(CTemp(i)-C(i, Ncolumns-1)) / C(i,
Ncolumns-1) <= Epsilon Then
         TmpCounter = TmpCounter + 1
      Else
         Exit For
      End If
   Next i
   If TmpCounter = Nrows Then
      Exit For
   End If
```

	A	B	C	D	E	F	G	H	I	J	K	L	M	N
1	*Newton Raphson Method*													
2	Lagrange Multiplier													
3														
4														
5														
6	**VBA Code**	x(0)	x(1)	x(2)	x(3)	x(4)	x(5)	x(6)						
7	**Equations**	FH	FC	TH_In	TH_Out	TC_In	TC_Out	λ	RHS					
8	**Nonlinear**	1							10					
9	**Nonlinear**		1						20					
10	**Nonlinear**			1					90					
11	**Nonlinear**				1				51					
12	**Nonlinear**					1			20					
13	**Nonlinear**						1		43					
14	**Nonlinear**							1	1					
15														
16														
17		x(0)	x(1)	x(2)	x(3)	x(4)	x(5)	x(6)						
18		FH	FC	TH_In	TH_Out	TC_In	TC_Out	λ	RHS					
19		1	0	0	0	0	0	0	9.9359	FH		-9.6153	eq(6.15b)	
20		0	1	0	0	0	0	0	20.127	FC		-8.8261	eq(6.15c)	
21		0	0	1	0	0	0	0	89.517	TH_In		0	eq(6.15d)	
22		0	0	0	1	0	0	0	51.351	TH_Out		3.4E-15	eq(6.15e)	
23		0	0	0	0	1	0	0	19.516	TC_In		1E-15	eq(6.15f)	
24		0	0	0	0	0	1	0	43.826	TC_Out		-2E-14	eq(6.15g)	
25		0	0	0	0	0	0	1	-0.007	λ		Q hot	Q cold	
26												4566.97	4566.97	eq(6.15h)
27														
28												5	NL Iterations	
29														
30		50	0	0.0972854	-0.070672452	0	0	-459.64	505.08					
31		0	12.5	0	0	0.048136	-0.08204	226.91	247.34					
32		0.0972854	0	2.0069283	0	0	0	-138.62	181.59					
33		-0.07067242	0	0	1.993071669	0	0	100.7	100.94					
34		0	0.048135654	0	0	2.028069	0	-138.94	41.517					
35		0	-0.082038494	0	0	0	1.97193	236.79	83.119					
36		459.643112	-226.9122103	138.62241	-100.7014649	138.9364	-236.792	0	-428.22					

Figure 6.15 Newton–Raphson solution of the nonlinear Lagrange multiplier problem—Example 6.16.

If desired, the user can follow the iteration process using **Example 6.16b.xls**.

The reconciled values for flows and temperatures from the nonlinear Lagrange multiplier problem provided in Figure 6.15 are in agreement with the Excel Solver solution in Example 6.13. ■

Special note: Care must be taken when coding equations in VBA that the number of mathematical operations concatenated on a single line of code is not excessive. In **Example 6.16b.xls**, the nonlinear Equation (6.15b,c,h) needs to be broken into two lines in the function holding the nonlinear equations, Function NLFunc.

6.5.1 Data Reconciliation: Lagrange Multiplier Compact Matrix Notation

We have shown in Examples 6.5–6.16 that data reconciliation problems with linear or nonlinear constraints can be a solved using Excel Solver or as a Lagrange multiplier problem. For the Lagrange multiplier problem, the Gauss–Jordan algorithm provided a convenient solution platform for problems with linear constraints, and the Newton–Raphson method was needed for problems with nonlinear constraints.

It is possible to formalize the Lagrange multiplier procedure in a compact matrix notation, and this compact matrix notation is the starting point for many data reconciliation problems in the literature. Here we follow the development provided in Mah (1990) for the data reconciliation problem in matrix form:

$$\text{Objective function} \quad \text{Min } (x^+ - x)^T Q^{-1} (x^+ - x) \qquad (6.16)$$

$$\text{Subject to } Ax = 0, \qquad (6.17)$$

where x^+ = measured values and x = reconciled values. Equations (6.16) and (6.17) are actually the same data reconciliation formulation we have used previously; see Equations (6.9) and (6.10). If the measurement error has a normal distribution with zero mean, a variance–covariance matrix Q can be defined as

$$Q = \begin{bmatrix} \sigma_1^2 & & & \\ & \sigma_2^2 & & \\ & & \sigma_3^2 & \\ & & & \sigma_n^2 \end{bmatrix} \text{ or } Q^{-1} = \begin{bmatrix} \frac{1}{\sigma_1^2} & & & 0 \\ & \frac{1}{\sigma_2^2} & & \\ & & \frac{1}{\sigma_3^2} & \\ 0 & & & \frac{1}{\sigma_n^2} \end{bmatrix}. \quad (6.18)$$

A is the incidence matrix for the linear constraints. The row rank of A is also the DoR. For example, using Example 6.6, the multiple-unit data reconciliation problem, $Ax = 0$ (Eq. (6.17)), would be

$$\begin{bmatrix} 1 & -1 & 0 & 0 & 1 \\ 0 & 1 & -1 & 0 & 0 \\ 0 & 0 & 1 & -1 & -1 \end{bmatrix} \begin{bmatrix} F_1 \\ F_2 \\ F_3 \\ F_4 \\ F_5 \end{bmatrix} = \begin{bmatrix} 0 \\ 0 \\ 0 \end{bmatrix}.$$

It is convenient to define an adjustment vector, a, which is the difference between the reconciled and measured value ($a = x - x^+$). The data reconciliation problem can then be expressed as

$$\text{Minimize} \quad (a)^T Q^{-1}(a) \quad (6.19)$$

$$\text{Subject to} \quad Ax^+ + Aa = 0, \quad (6.20)$$

which, for Example 6.6, is

$$\text{Min}\begin{bmatrix} F_1 - 99.3 & F_2 - 123.2 & F_3 - 125.1 & F_4 - 100.5 & F_5 - 23.9 \end{bmatrix} \begin{bmatrix} \frac{1}{2^2} & 0 & 0 & 0 & 0 \\ 0 & \frac{1}{2^2} & 0 & 0 & 0 \\ 0 & 0 & \frac{1}{2^2} & 0 & 0 \\ 0 & 0 & 0 & \frac{1}{2^2} & 0 \\ 0 & 0 & 0 & 0 & \frac{1}{2^2} \end{bmatrix} \begin{bmatrix} F_1 - 99.3 \\ F_2 - 123.2 \\ F_3 - 125.1 \\ F_4 - 100.5 \\ F_5 - 23.9 \end{bmatrix}$$

$$\text{Subject to} \begin{bmatrix} 1 & -1 & 0 & 0 & 1 \\ 0 & 1 & -1 & 0 & 0 \\ 0 & 0 & 1 & -1 & -1 \end{bmatrix} \begin{bmatrix} 99.3 \\ 123.2 \\ 125.1 \\ 100.5 \\ 23.9 \end{bmatrix}$$

$$+ \begin{bmatrix} 1 & -1 & 0 & 0 & 1 \\ 0 & 1 & -1 & 0 & 0 \\ 0 & 0 & 1 & -1 & -1 \end{bmatrix} \begin{bmatrix} F^1 - 99.3 \\ F^2 - 123.2 \\ F^3 - 125.1 \\ F^4 - 100.5 \\ F^5 - 23.9 \end{bmatrix} = \begin{bmatrix} 0 \\ 0 \\ 0 \end{bmatrix}.$$

Next, we can form the Lagrangian function as

$$\underset{a,\lambda}{\text{Min}}\ L = a^T Q^{-1} a - 2\lambda^T (Ax^+ + Aa). \quad (6.21)$$

Here, the constant 2 before the λ has no impact on the final solution—it is just convenient for the solution derivation used here. We can write Equation (6.21) as

$$\underset{a,\lambda}{\text{Min}}\ L = a^T Q^{-1} a - 2\lambda^T Ax^+ - 2\lambda^T Aa,$$

and using

$$\lambda^T Aa = (Aa)^T \lambda = A^T a^T \lambda$$

gives

$$\underset{a,\lambda}{\text{Min}}\ L = a^T Q^{-1} a - 2\lambda^T Ax^+ - 2A^T a^T \lambda. \quad (6.22)$$

To solve Equation (6.21) or the equivalent (Eq. (6.22)), the necessary conditions are

$$\frac{\partial L}{\partial \lambda} = 0 \quad \text{and} \quad \frac{\partial L}{\partial a} = 0 \quad (6.23)$$

For any vectors p and q, and any matrix M, the product of $p^T Mq$ is a scalar. Therefore, differentiation follows the product rule:

$$\frac{\partial (p^T Mq)}{\partial q} = \frac{\partial p^T}{\partial q} Mq + \frac{\partial q^T}{\partial q} M^T p.$$

For the special case where M is symmetric and $p = q$,

$$\frac{\partial (q^T Mq)}{\partial q} = 2Mq. \quad (6.24)$$

Using Equations (6.23b) and (6.24), Equation (6.22) becomes

$$\frac{\partial L}{\partial a} = 2Q^{-1} a - 2A^T \lambda = 0. \quad (6.25)$$

Rearranging Equation (6.25),

$$a = QA^T \lambda. \quad (6.26)$$

Now using Equation (6.23a), Equation (6.21) gives back the constraint equation, Equation (6.20), or

$$\frac{\partial L}{\partial \lambda} = Ax^+ + Aa = 0$$

$$Ax^+ = -Aa; \quad (6.27)$$

substituting Equation (6.26) into Equation (6.27),

$$Ax^+ = -AQA^T \lambda. \quad (6.28)$$

Solving Equation (6.28) for λ,

$$\lambda = -(AQA^T)^{-1} Ax^+. \quad (6.29)$$

Finally, substitution of Equation (6.29) into Equation (6.26) gives

$$a = -QA^T(AQA^T)^{-1}Ax^+,$$

or using the definition of a, $a = x - x^+$,

$$x = x^+ - QA^T(AQA^T)^{-1}Ax^+. \qquad (6.30)$$

EXAMPLE 6.17 ***Multiple-Unit Data Reconciliation by Lagrange Multiplier Compact Matrix Notation***

Solve Example 6.6 using the Lagrange multiplier compact matrix notation—Equation (6.30). Here, it is straightforward to evaluate each term in Equation (6.30):

$$QA^T = \begin{bmatrix} 2^2 & 0 & 0 & 0 & 0 \\ 0 & 2^2 & 0 & 0 & 0 \\ 0 & 0 & 2^2 & 0 & 0 \\ 0 & 0 & 0 & 2^2 & 0 \\ 0 & 0 & 0 & 0 & 2^2 \end{bmatrix} \begin{bmatrix} 1 & 0 & 0 \\ -1 & 1 & 0 \\ 0 & -1 & 1 \\ 0 & 0 & -1 \\ 1 & 0 & -1 \end{bmatrix} = \begin{bmatrix} 4 & 0 & 0 \\ -4 & 4 & 0 \\ 0 & -4 & 4 \\ 0 & 0 & -4 \\ 4 & 0 & -4 \end{bmatrix},$$

$$AQA^T = \begin{bmatrix} 1 & -1 & 0 & 0 & 1 \\ 0 & 1 & -1 & 0 & 0 \\ 0 & 0 & 1 & -1 & -1 \end{bmatrix} \begin{bmatrix} 4 & 0 & 0 \\ -4 & 4 & 0 \\ 0 & -4 & 4 \\ 0 & 0 & -4 \\ 4 & 0 & -4 \end{bmatrix} = \begin{bmatrix} 12 & -4 & -4 \\ -4 & 8 & -4 \\ -4 & -4 & 12 \end{bmatrix},$$

$$(AQA^T)^{-1} = \begin{bmatrix} \left(\frac{1}{6.4}\right) & \left(\frac{1}{8}\right) & \left(\frac{1}{10.6667}\right) \\ \left(\frac{1}{8}\right) & \left(\frac{1}{4}\right) & \left(\frac{1}{8}\right) \\ \left(\frac{1}{10.6667}\right) & \left(\frac{1}{8}\right) & \left(\frac{1}{6.4}\right) \end{bmatrix},$$

$$Ax^+ = \begin{bmatrix} 1 & -1 & 0 & 0 & 1 \\ 0 & 1 & -1 & 0 & 0 \\ 0 & 0 & 1 & -1 & -1 \end{bmatrix} \begin{bmatrix} 99.3 \\ 123.2 \\ 125.1 \\ 100.5 \\ 23.9 \end{bmatrix} = \begin{bmatrix} 0 \\ -1.9 \\ 0.7 \end{bmatrix},$$

$$a = -QA^T(AQA^T)^{-1}Ax^+ = \begin{bmatrix} 0.6875 \\ 0.8625 \\ -1.0375 \\ -0.5125 \\ 0.175 \end{bmatrix},$$

and

$$x = x^+ + a = (x^+ - QA^T(AQA^T)^{-1}Ax^+) = \begin{bmatrix} F_1 \\ F_2 \\ F_3 \\ F_4 \\ F_5 \end{bmatrix} = \begin{bmatrix} 99.9875 \\ 124.0625 \\ 124.0625 \\ 99.9875 \\ 24.075 \end{bmatrix},$$

giving us the same results we obtained from Example 6.6, where Excel Solver was used to solve the data reconciliation problem. Example 6.17 also gives the same results as Example 6.15, where we used the Lagrange multiplier method with the Gauss–Jordan algorithm to obtain a solution. The matrix operations in Example 6.17 are provided in **Example 6.17.xls**. Matrix operations in Excel are somewhat cumbersome. For example, to multiply two matrices (say, A and B), each matrix must be named and the location for the product matrix (rows and columns) must be highlighted on the Excel sheet. The Excel sheet operation = MMULT(A,B) requires keystrokes Crtl + Shift + Enter to complete the matrix multiplication. This same procedure is required for all matrix operations. One advantage of Excel is that any matrix operation will be automatically updated with changes in matrix A or B. ■

You can show yourself if the right-hand side of Equation (6.17), the material and energy balance constraint set, is not identically zero, but rather a known constant vector c; the reconciled values are obtained as

$$x = x^+ - QA^T(AQA^T)^{-1}(Ax^+ - c). \qquad (6.31)$$

The Lagrange multiplier compact matrix notation approach developed in this section is often the basis for more extensive developments found in the literature and data reconciliation texts. The matrix A must involve linear constraints. If nonlinear constraints are present, they must be linearized as we did in Example 6.16.

6.6 GROSS ERROR DETECTION AND IDENTIFICATION

We have used data reconciliation to resolve random errors in measured data. These errors are assumed to be normally distributed around the true measurement value. Here we discuss gross errors, which are considered to be nonrandom events, such as instrument bias or drift, malfunctioning sensors, or even process leaks. We first introduce the global test (GT) method for the detection of gross errors in our measurement system, and we then show the measurement test (MT) method for the identification of suspect measurements. The discussion provided here is an overview, and the reader is referred to Mah (1990) and the references therein for the development of the statistics and hypothesis testing used in these methods.

6.6.1 Gross Error Detection: The Global Test (GT) Method

Again consider our data reconciliation problem as given by

$$\text{Minimize } f = \sum_{\substack{i,\text{measured} \\ \text{variables}}} \left(\frac{(x_i^+ - x_i)}{\sigma_i} \right)^2 \qquad (6.32)$$

$$\text{Subject to } h_k(x_i) = 0 \quad k = 1, \ldots, K. \qquad (6.33)$$

The objective function is the sum of the terms

$$\left(\frac{(x_i^+ - x_i)}{\sigma_i}\right)^2,$$

which is also a chi-square probability distribution, $\chi^2(\nu)$, with degree of freedom ((ν). The degree of freedom (ν) equals the DoR in the data reconciliation problem, which is given by the number of constraints in Equation (6.33); we have previously discussed the DoR in Equation (6.11). This brief discussion allows the presentation of the system global test for gross errors as

$$GT = \sum_{\substack{i,\text{measured}\\ \text{variables}}} \left(\frac{(x_i^+ - x_i)}{\sigma_i}\right)^2 < \chi^2_{(1-\alpha)}(\nu), \qquad (6.34)$$

where $\chi^2_{(1-\alpha)}(\nu)$ is the upper limit value of the chi-square distribution where gross errors are not expected and α is the level of significance, which is generally taken as $\alpha = 5\%$ (0.05). Full development of Equation (6.34) requires the aid of hypothesis testing and a discussion of errors of the first and second types (see Mah, 1990). For our purposes, we can simply say if GT is less than $\chi^2_{(1-0.05)}(\nu)$, gross errors are not expected in the system. In Excel, the value for $\chi^2_{(1-\alpha)}(\nu) = \text{CHIINV(level of significance, DoR)} = \text{CHIINV}$ (0.05, DoR).

EXAMPLE 6.18 *Gross Error Detection: Global Test*

Use the global test to determine if the three-unit operation flow sheet in Figure 6.7 (used in Examples 6.6 and 6.17) contains gross errors.

SOLUTION From the Excel solution file for Example 6.6, the value for

$$GT = \sum_{\substack{i,\text{measured}\\ \text{variables}}} \left(\frac{(x_i^+ - x_i)}{\sigma_i}\right)^2 = 0.64656.$$

We calculate

$$\sum_{\substack{i,\text{measured}\\ \text{variables}}} \left(\frac{(x_i^+ - x_i)}{\sigma_i}\right)^2$$

as the objective function of our Excel minimization, or this value can be calculated as $(a)^T Q^{-1}(a)$. Using Example 6.17, with all the flow rates measured, the DoR = the row rank of the A matrix = 3. From Excel $\chi^2_{(1-\alpha)}(\nu) = \text{CHIINV}$ (0.05, 3) = 7.815, so no gross errors are expected. Calculations are shown in **Example 6.18.xls**. ■

6.6.2 Gross Error (Suspect Measurement) Identification: The Measurement Test (MT) Method: Linear Constraints

We can use the global test (Eq. (6.34)) to determine if gross errors are present. The measurement test can be used to locate the most probable suspect measurement. This is done by listing MT_i values from highest to lowest with the higher values the more suspect measurements; the highest MT_i value is the most likely suspect measurement. Here, utilizing our definitions from Section 6.5.1, Mah (1990) provides the following MT for each variable:

$$MT_i = \frac{|a_i|}{\sqrt{V_{ii}}}, \qquad (6.35)$$

where a_i are the elements of the adjustment vector a, and V_{ii} are the diagonals of the covariance of a:

$$V = \text{cov}(a) = QA^T(AQA^T)^{-1}AQ. \qquad (6.36)$$

In Equation (6.35) MT_i is a standardized adjustment, which follows a standard normal distribution with mean = 0 and variance = 1. Again, higher values of MT_i point to more suspect measurements. In order to use Equation (6.36), the matrix A must involve linear constraints and, generally, Q, the variance–covariance matrix, will be diagonal. The next example shows the use of these Equations (6.35) and (6.36).

EXAMPLE 6.19 *Gross Error Detection and Identification: Linear System*

Narasimhan and Jordache (2000) provide a heat exchanger bypass problem as shown in Figure 6.16. The "true" and measured flow rates are shown in Table 6.14; the measured flow rate of stream 2 contains a positive 4 bias. Use the global test to determine if gross errors are present, and if gross errors are present, use the measurement test to determine the most likely measurement location.

SOLUTION The reconciled flow rates and the global test can be determined using the Excel-based minimization procedure in **Example 6.19a.xls**. The reconciled flow rates are $F_1 = 100.887$,

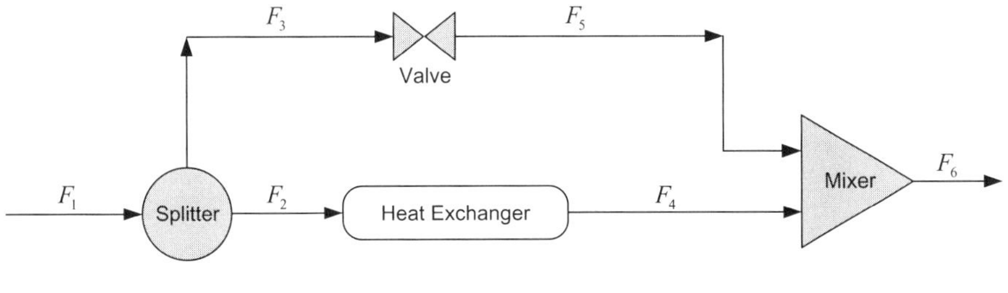

Figure 6.16 Heat exchanger with bypass valve.

Table 6.14 Data for Example 6.19—Heat Exchanger with Bypass Valve

Stream Number	True Flow Rate Values	Measured Flow Rates	σ Flow
1	100	101.91	1
2	64	68.45	1
3	36	34.65	1
4	64	64.20	1
5	36	36.44	1
6	100	98.88	1

$F_2 = 65.833$, $F_3 = 35.053$, $F_4 = 65.833$, $F_5 = 35.053$, and $F_6 = 100.887$. The objective function value from the Excel minimization,

$$\sum_{i,\text{measured variables}} \left(\frac{(x_i^+ - x_i)}{\sigma_i} \right)^2 = 16.674,$$

is larger than the $\chi^2_{(1-\alpha)}(\nu)$ value = CHIINV (0.05, 4) = 9.488, so gross errors are detected.

The Lagrange multiplier compact matrix technique in **Example 6.19b.xls** can be used to determine the reconciled flow rates, the *GT* and the *MT*. The matrix calculations produce the same reconciled flow rates; the same objective function value, as $(a)^T Q^{-1}(a) = 16.674$; and for the *MT* (Eq. (6.35)),

$$V = \begin{bmatrix} 0.66667 & -0.16667 & -0.16667 & -0.16667 & -0.16667 & -0.33333 \\ -0.16667 & 0.66667 & 0.16667 & -0.33333 & 0.16667 & -0.16667 \\ -0.16667 & 0.16667 & 0.66667 & 0.16667 & -0.33333 & -0.16667 \\ -0.16667 & -0.33333 & 0.16667 & 0.66667 & 0.16667 & -0.16667 \\ -0.16667 & 0.16667 & -0.33333 & 0.16667 & 0.66667 & -0.16667 \\ -0.33333 & -0.16667 & -0.16667 & -0.16667 & -0.16667 & 0.66667 \end{bmatrix}. \quad (6.37)$$

The variable order is important and is in the adjustment vector a, and we have used $a_1 = F_1 - F_1^+$, $a_2 = F_2 - F_2^+$, and so on.

We find

$$MT_i = \frac{|a_i|}{\sqrt{V_{ii}}} = \left\{ \frac{|-1.02333|}{\sqrt{0.66667}}, \frac{|-2.61667|}{\sqrt{0.66667}}, \frac{|0.40333|}{\sqrt{0.66667}}, \frac{|1.63333|}{\sqrt{0.66667}}, \frac{|-1.38667|}{\sqrt{0.66667}}, \frac{|2.00667|}{\sqrt{0.66667}} \right\}$$
$$= \{1.25332, 3.20475, 0.49398, 2.00042, 1.69831, 2.45765\},$$

and here the largest MT_i is $MT_2 = 3.20475$, which is the flow rate measurement, F_2. The gross error is most likely associated with F_2. In the Excel-based matrix solution, any value in Q, the variance–covariance matrix, or the measured values vector, x_i^+, can be changed and all other sheet values will be updated. This is useful to explore gross error detection and identification (see Problems 6.11 and 6.12). ■

6.6.3 Gross Error (Suspect Measurement) Identification: The Measurement Test Method: Nonlinear Constraints

Equations (6.35) and (6.36) can be used to determine the most likely location of suspect measurements. However, matrix A in Equation (6.36) must involve linear constraints. If some constraints are nonlinear, as, for example, energy balances, we can use Equation (4.12), the Taylor series expansion, to construct a linear approximation. This suggests solving the data reconciliation problem with nonlinear constraints as an Excel minimization. The presence of gross errors can be determined from Equation (6.35). If gross errors are present, nonlinear constraints can be linearized at the solution using the Taylor expansion, and Equation (6.36) can be used for the measurement test. The next example shows this application.

EXAMPLE 6.20 ***Gross Error Detection and Identification: Nonlinear System***

Consider the heat exchanger bypass problem of Example 6.19 with additional measurements allowing a heat exchanger energy balance; see Figure 6.17. The true and measured flow rates and temperatures are provided in Table 6.15; the bias in the flow rate of stream 2 (as in Example 6.19) has been removed, but now the temperature of stream 7 contains a positive 5 bias. Use the global test to determine if gross errors are present, and if gross errors are present, use the measurement test to determine the most likely location. Stream heat capacities are taken from Example 6.12 with $\hat{C}_{p,h}$ (hot streams, F_2, F_4) = 5 + (0.1)T, and $\hat{C}_{p,c}$ (cold streams, F_7, F_8) = 3 + (0.2)T.

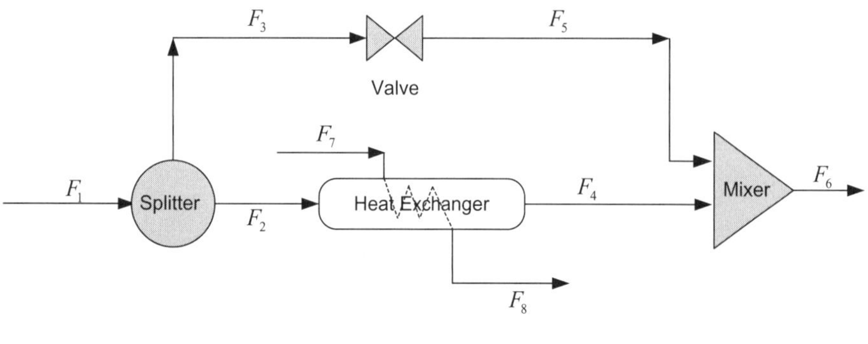

Figure 6.17 Heat exchanger with bypass valve and energy balance.

Table 6.15 Data for Example 6.20—Heat Exchanger with Bypass Valve

Stream Number	True Flow Rate Values	True Temperature Values	Measured Values	σ Flow or σ Temperature
1	100		101.91	1
2	64		64.45	1
3	36		34.65	1
4	64		64.20	1
5	36		36.44	1
6	100		98.88	1
7	140.6		140.0	1
2		90	90	1
4		51	51	1
7		20	25	1
8		43	43	1

SOLUTION The reconciled flow rates and the *GT* can be determined using the Excel-based minimization procedure in **Example 6.20a.xls**. Here we assume no leaks in the heat exchanger giving reconciled flow rates and temperatures: $F_1 = 100.141, F_2 = 64.341, F_3 = 35.799, F_4 = 64.341, F_5 = 35.799, F_6 = 100.141, F_7 = 140.218, T_2 = 89.054,\ T_4 = 51.691,\ T_7 = 23.849$, and $T_8 = 44.774$. The objective function value from the Excel minimization,

$$\sum_{\substack{i,\text{measured}\\ \text{variables}}} \left(\frac{\left(x_i^+ - x_i\right)}{\sigma_i}\right)^2 = 12.376,$$

is larger than the $\chi^2_{(1-\alpha)}(\nu)$ value = CHIINV (0.05, 5) = 11.070, so gross errors are detected.

Equations (6.35) and (6.36) can be used to determine the most likely location of the suspect measurement. However, the immediate problem is that matrix *A* in Equation (6.36) must involve linear constraints and the heat exchanger energy balance is nonlinear. Here, we can use Equation (4.12), the Taylor series expansion, to construct a linear approximation to the heat exchanger energy balance at the data reconciliation solution given by the Excel minimization. The result of the linearization is

$$\begin{aligned} &-841.8F_2 + 392.06F_4 + 206.37F_7 - 894.69T_2 \\ &\quad + 654.3T_4 - 1089.45T_7 + 1676.28T_8 = 3217.44. \end{aligned} \tag{6.38}$$

Linearization can be accomplished "by hand"; however, this quickly becomes cumbersome. We recommend using the Newton–Raphson method we developed in Chapter 4 for linearization. In **Example 6.20b.xls**, we provide the data reconciliation solution as the starting guess and we supply the nonlinear equation to the VBA code in `Function NLFunc`. The Gauss–Jordan macro is run one time to establish the point for linearization and then the Newton–Raphson macro is run one time to obtain the needed coefficients. Any number of nonlinear functions can be supplied to `Function NLFunc`.

The Newton–Raphson macro will give an error as it tries to generate a new trial point. This is not of concern. The nonlinear equations will have been linearized at data reconciliation solution, and we can obtain the linearized coefficients from matrix 3 (see Figure 4.11).

We will also need to modify our constraint equation from the current Equation (6.17), $Ax = 0$, to

$$Ax - c = 0, \tag{6.39}$$

where here,

$$A = \begin{pmatrix} F_1 & F_2 & F_3 & F_4 & F_5 & F_6 & F_7 & T_2 & T_4 & T_7 & T_8 \\ -1 & 1 & 1 & 0 & 0 & 0 & 0 & 0 & 0 & 0 & 0 \\ 0 & 1 & 0 & -1 & 0 & 0 & 0 & 0 & 0 & 0 & 0 \\ 0 & -1 & 0 & 0 & 1 & 0 & 0 & 0 & 0 & 0 & 0 \\ 0 & 0 & 0 & 1 & 1 & -1 & 0 & 0 & 0 & 0 & 0 \\ 0 & -841.8 & 0 & 392.06 & 0 & 0 & 206.37 & -894.69 & 654.3 & -1098.45 & 1676.28 \end{pmatrix} \tag{6.40}$$

and

$$c = \begin{pmatrix} 0 & 0 & 0 & 0 & 3217.44 \end{pmatrix}^T. \tag{6.41}$$

Using Equation (6.40) for *A* allows us to determine *V* from Equation (6.36) and MT_i from Equation (6.35); calculations are provided in **Example 6.20c.xls** with key results:

$$V = \begin{pmatrix} 0.668 & -0.165 & -0.168 & -0.165 & -0.168 & -0.332 & -0.003 & 0.013 & -0.009 & 0.015 & -0.024 \\ -0.165 & 0.671 & 0.165 & -0.329 & 0.165 & -0.165 & -0.006 & 0.025 & -0.018 & 0.031 & -0.047 \\ -0.168 & 0.165 & 0.668 & 0.165 & -0.332 & -0.168 & 0.003 & -0.013 & 0.009 & -0.015 & 0.024 \\ -0.165 & -0.329 & 0.165 & 0.671 & 0.165 & -0.165 & -0.006 & 0.025 & -0.018 & 0.031 & -0.047 \\ -0.168 & 0.165 & -0.332 & 0.165 & 0.668 & -0.168 & 0.003 & -0.013 & 0.009 & -0.015 & 0.024 \\ -0.332 & -0.165 & -0.168 & -0.165 & -0.168 & 0.668 & -0.003 & 0.013 & -0.009 & 0.015 & -0.024 \\ -0.003 & -0.006 & 0.003 & -0.006 & 0.003 & -0.003 & 0.008 & -0.035 & 0.025 & -0.042 & 0.065 \\ 0.013 & 0.025 & -0.013 & 0.025 & -0.013 & 0.013 & -0.035 & 0.150 & -0.110 & 0.183 & -0.281 \\ -0.009 & -0.018 & 0.009 & -0.018 & 0.009 & -0.009 & 0.025 & -0.110 & 0.080 & -0.134 & 0.206 \\ 0.015 & 0.031 & -0.015 & 0.031 & -0.015 & 0.015 & -0.042 & 0.183 & -0.134 & 0.222 & -0.342 \\ -0.024 & -0.047 & 0.024 & -0.047 & 0.024 & -0.024 & 0.065 & -0.281 & 0.206 & -0.342 & 0.527 \end{pmatrix} \tag{6.42}$$

and

$$MT_i = \frac{|a_i|}{\sqrt{V_{ii}}} = \left\{ \begin{array}{l} \frac{|-1.769|}{\sqrt{0.668}}, \frac{|-0.109|}{\sqrt{0.671}}, \frac{|1.149|}{\sqrt{0.668}}, \frac{|0.141|}{\sqrt{0.671}}, \\ \frac{|-0.641|}{\sqrt{0.668}}, \frac{|1.261|}{\sqrt{0.668}}, \frac{|0.218|}{\sqrt{0.008}}, \frac{|-0.946|}{\sqrt{0.150}}, \\ \frac{|0.691|}{\sqrt{0.080}}, \frac{|-1.151|}{\sqrt{0.222}}, \frac{|1.774|}{\sqrt{0.527}} \end{array} \right\}$$
$$= \left\{ \begin{array}{l} 2.165, 0.133, 1.406, 0.173, 0.784, \\ 1.543, 2.438, 2.442, 2.441, 2.441, 2.445 \end{array} \right\},$$

with the variable order F_1, F_2, F_3, F_4, F_5, F_6, F_7, T_2, T_4, T_7, T_8.

Here it is not possible to identify a single suspect measurement as F_7, T_2, T_4, T_7, T_8 all show the same single measurement error $MT_i \cong 2.44$. This result can be explained; the error in T_7 is spread among the unique measurements forming the heat exchanger energy balance. See Problem 6.14 where energy balances can be used to identify the suspect measurement. ■

6.7 CLOSING REMARKS

In this chapter, we have shown how linear and nonlinear data reconciliation problems can be solved as constrained optimization problems using Excel. We have also shown how Lagrange multipliers can be used to solve data reconciliation problems. For both linear and nonlinear data reconciliation problems, we introduced gross error detection and identification.

Data reconciliation and gross error detection is a mature topic and there are excellent works available that greatly extend the topics introduced in this chapter. The interested reader is encouraged to read Narasimhan and Jordache (2000), Madron et al. (2007), Veverka (2004), and Romagnoli and Sanchez (2000). Then, depending on the area of interest, read, for material and energy balances, Veverka and Madron (1997); for plant monitoring, Madron (1992); for gross error detection and identification, Mah (1990); and for plant instrumentation and optimization, Bagajewicz (2001).

There are commercial and academic software packages for data reconciliation and gross error detection. A code I find very useful is Recon developed by Dr. Frantisek (Frank) Madron and ChemPlant Technologies. The Recon program is available for download at www.chemplant.cz; there is a no cost-reduced version (in terms of allowable number of streams, components, etc.) of the commercial code. Dr. Ralph W. Pike at Louisiana State University provides a downloadable online optimization program that includes gross error detection and data reconciliation, parameter estimation, and economic optimization (see Figure 6.1).

REFERENCES

BAGAJEWICZ, M.J. 2001. *Process Plant Instrumentation—Design and Upgrade*. Technomic, Lancaster, CA.

COHEN, E.R. 1953. The basis for the criterion of least squares. *Rev. Mod. Phys.* 28(3): 709–713.

MADRON, F. 1992. *Process Plant Performance—Measurement and data Processing for Optimization and Retrofits*. Ellis Horwood, New York.

MADRON, F., V. VEVERKA, and M. HOSTALEK 2007. *Process Data Validation in Practice: Applications from Chemical, Oil, Mineral and Power Industries*, Report CPT-229-07. ChemPlant Technology, Czech Republic.

MAH, R.S.H. 1990. *Chemical Process Structure and Information Flows*. Butterworths, Stoneham, MA.

NARASIMHAN, S. and C. JORDACHE 2000. *Data Reconciliation and Gross Error Detection: An Intelligent Use of Process Data*. Gulf Publishing Company, Houston, TX.

NARASIMHAN, S., R.S.H. MAH, and G. LIKEHOOD 1987. Ratio method for gross error identification. *AIChE J.* 33(9): 1514–1521.

OZYURT, D.B. and R.W. PIKE 2004. Theory and practice of simulation data reconciliation and gross error detection for chemical processes. *Comput. Chem. Eng.* 28: 381–402.

ROMAGNOLI, J.A. and M.C. SANCHEZ 2000. *Data Processing and Reconciliation for Chemical Process Operations*. Academic Press, London.

VEVERKA, V.V. 2004. *Balancing and Data Reconciliation Minibook*, Report CPT-189-04. ChemPlant Technology, Czech Republic. http://www.chemplant.cz/dwnld.htm

VEVERKA, V.V. and F. MADRON 1997. *Material and Energy Balancing in the Process Industries—From Microscopic Balances to Large Plants*. Elsevier, Amsterdam.

PROBLEMS

6.1 ***Data Reconciliation—Variable Elimination Calculus Solution*** A simple solution approach to equality constrained optimization problems is to use the equality constraints to eliminate variables from the problem. If the equality constraints are complex, this approach is often not possible. However, if all the equality constraints can be expressed in terms of individual variables, then a constrained optimization problem becomes an unconstrained optimization problem after variable substitution. The optimum to an unconstrained optimization problem can be found by setting the partial derivative with respect to each variable equal to zero—this is a necessary condition for multivariable optimization.

Use variable elimination to solve Example 6.5 (mixer data reconciliation):

$$\text{Minimize}\left(\left(\frac{245-F_1}{10}\right)^2+\left(\frac{240-F_2}{10}\right)^2+\left(\frac{500-F_3}{10}\right)^2\right)$$

$$\text{Subject to } F_1 + F_2 - F_3 = 0.$$

6.2 ***Data Reconciliation—Variable Elimination Calculus Solution*** Consider the process flow sheet shown in Figure P6.2. Measured flow rate values are provided in the Table P6.2. Do note that the instrument standard deviations are not the same ($\sigma_1 \neq \sigma_2 \neq \sigma_3$). Provide the numeric solution to this flow rate data reconciliation problem using variable elimination and the necessary conditions for an optimal solution.

6.3 ***Data Reconciliation—Variable Elimination Lagrange Multiplier*** In Example 6.15, we used the Gauss–Jordan algorithm to solve the set of eight equations and eight unknowns

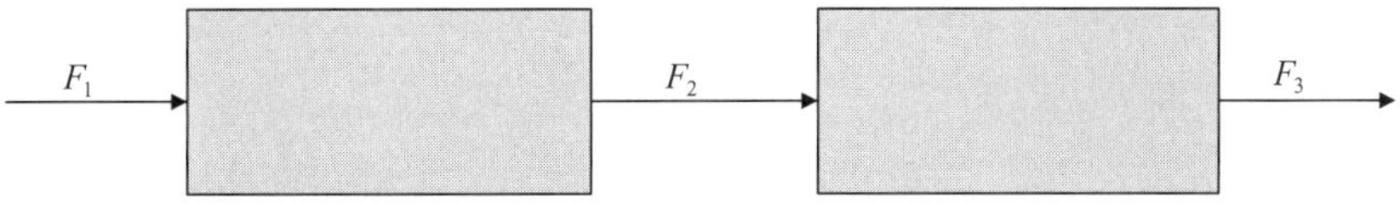

Figure P6.2 Two single-input single-output units in series.

Table P6.2 Measurement Data for Problem 6.2

	F_1^+	F_2^+	F_3^+
Measured flow rate (kg/h)	530.0	518.0	536.0
Standard deviation σ (kg/h)	2	4	6

in the Lagrange multiplier problem. Here, use variable elimination to solve this equation set.

6.4 ***Data Reconciliation—Styrene Process Total Flow Balances*** For Example 6.13, eliminate the use of the $\sum w_i = 1.0$ for streams 2, 3, 6, 7, and 8, and add as constraints the total flow balances for the mixer, reactor, separator, and purge.

Total Flow Balances

Mixer: $F_{\text{feed ETB}} + F_{\text{feed H}_2\text{O}} + F_{\text{recycle}} = F_{\text{reactor in}}$

Reactor: $F_{\text{reactor in}} = F_{\text{reactor out}}$

Separator 1: $F_{\text{reactor out}} = F_{\text{vapor out}} + F_{\text{water out}} + F_{\text{organic out}}$

Separator 2: $F_{\text{organic out}} = F_{\text{product}} + F_{\text{recycle}}$.

6.5 ***Data Reconciliation—Styrene Process Elemental Balances*** For Example 6.13, eliminate the use of the extent of reaction. Here elemental balances, carbon in = carbon out and hydrogen in = hydrogen out, must be added as constraints. The three reactor balance equations involving the extent of reaction (for C_8H_{10}, H_2, and C_8H_8) are removed from the constraint set.

6.6 ***Data Reconciliation—Chlorination of Methane Reactor*** This is a modification of Example 6.11, the chlorination of methane reactor (Madron et al., 2007). Here, some chlorine is present in the reactor outlet. We are asked to determine the reconciled species flow rates in the chlorination reactor. Observed species are indicated in Figure P6.6, and Table P6.6 provides measured molar species flows (mole per hour) and instrument standard deviations.

This data reconciliation problem will require species mass balances as constraints. Supply needed species material balances using extents of reaction and then elemental balances.

6.7 ***Data Reconciliation—Ammonia Reactor System*** The following problem is from Madron et al. (2007). Figure P6.7 shows a four-stage ammonia reactor system. Here the feed stream is split to allow cold shots into all four stages to control temperature. Data for this process is provided in Table P6.7.

The flows in streams 3, 5, 7, and 9 are not measured. The values in parentheses are estimated molar flow rates (mole per second). Measured molar flow rates and measured mol % are provided without parentheses. The composition of streams 2, 4, 6, 8, and 10 are the same as the composition of stream 1. Measurement precision (maximum errors): maximum flow error = 5% and maximum mol % error = 2%.

The reaction in each reactor is

$$N_2 + 3H_2 = 2NH_3.$$

We are provided with estimates of the extent of reaction for each reactor as ξ_1 = 50 mol of N_2/s, ξ_2 = 40 mol of N_2/s, ξ_3 = 40 mol of N_2/s, and ξ_4 = 25 mol of N_2/s.

First, solve the data reconciliation problem for the first reactor—reactor 1 only. Here use the extent of reaction formulation. Reconciled variables will involve measured molar flow rates and measured mol %—these variables will appear in the objective function. For this problem it is convenient to make the molar flow rate leaving each reactor an optimization variable (notice that estimates have been provided). These molar flow rates leaving each reactor will not appear in the objective function, but they will appear in the constraints. The molar flow rates leaving each reactor are nonmeasured observable variables. Also note that the composition of streams 1, 2, 4, 6, 8, and 10 has only been measured once—do not double count this in the objective function. Be sure to use Options in Solver—keep variables positive (assume nonnegative) and select automatic scaling.

Continue to add reactors and solve the data reconciliation problem.

6.8 ***Data Reconciliation—Compact Matrix Notation*** Recall our data reconciliation problem for the simple mixer in Example 6.5. The data reconciliation problem was formulated as

$$\text{Minimize } f = \left(\left(\frac{245 - F_1}{10} \right)^2 + \left(\frac{240 - F_2}{10} \right)^2 + \left(\frac{500 - F_3}{10} \right)^2 \right)$$

Subject to $F_1 + F_2 - F_3 = 0$.

Solve this problem using Equation (6.30)—compact matrix notation.

6.9 ***Data Reconciliation—Elementary Gasification Reaction*** In a gasification process, carbon is reacted with oxygen to form carbon dioxide. Here the known reaction is

$$C + O_2 \rightarrow CO_2.$$

Observed species are indicated in Figure P6.9 and Table P6.9 provides measured molar species flows (mole per hour) and instrument standard deviations. No CO_2 enters the process and no C leaves. There is an excess of O_2. Provide the numeric solution to this data reconciliation problem by combining elemental balances with (1) variable substitution; (2) Equation (6.30)—compact matrix notation; and (3) Excel Solver.

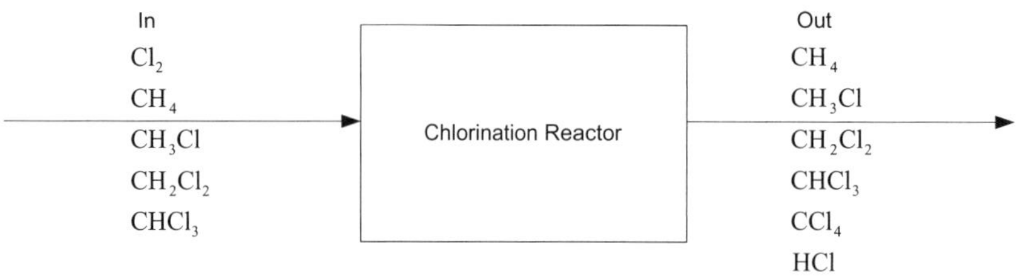

Figure P6.6 Chlorination reactor.

Table P6.6 Measured Molar Flow Rates and Instrument Standard Deviations

	Inflow Measured	σ	Outflow Measured	σ
Flow	100	Fixed		
Cl_2	62.1	3.105	1.55	0.0775
CH_4	26.9	1.345	1.57	0.0785
CH_3Cl	10.8	0.54	6.83	0.3415
CH_2Cl_2	0.78	0.039	24.8	1.24
$CHCl_3$	0.38	0.019	4.82	0.241
CCl_4	0	Fixed	0.39	0.0195
HCl	0	Fixed	59.3	2.965

Table P6.7 Data for the Four-Stage Ammonia System (Madron et al., 2007)

Stream	Flow (mol/s)	% H_2	% N_2	% NH_3	% CH_4 + Ar
1	655	63.8	21.2	3.0	12.0
2	655				
3	(1200)	57.2	19.0	10.9	12.9
4	330				
5	(1400)	53.4	17.7	15.4	13.5
6	360				
7	(1800)	51.7	17.1	17.5	13.7
8	300				
9	(2000)	50.7	16.8	18.7	13.8
10	2300				

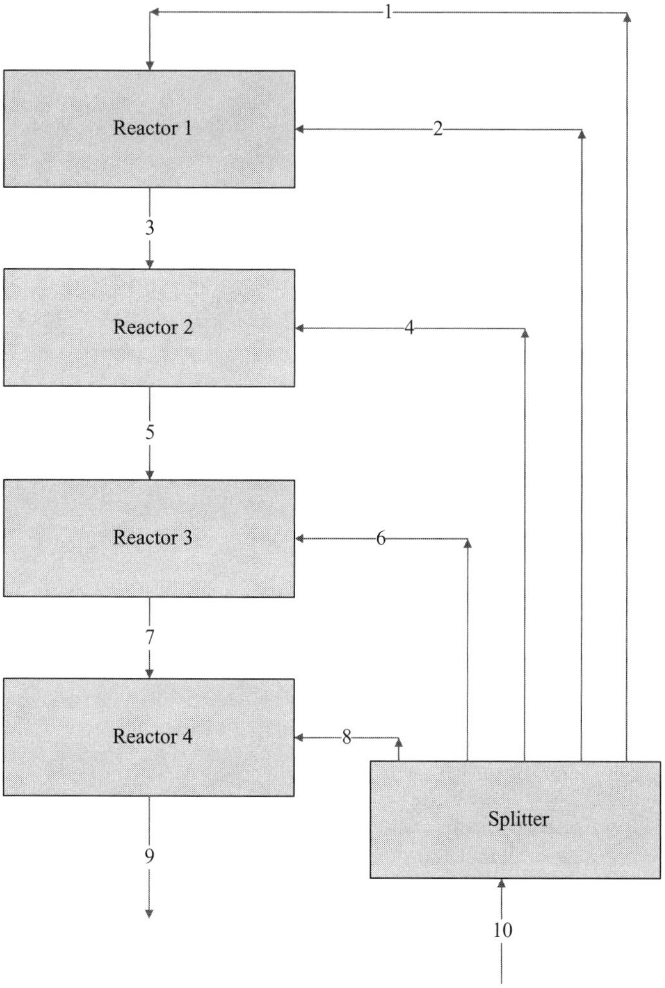

Figure P6.7 Four-stage ammonia process.

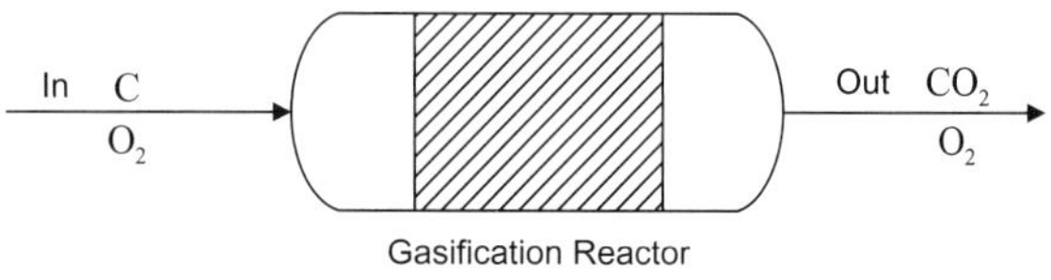

Figure P6.9 Gasification reactor.

Table P6.9 Measured Molar Flow Rates and Instrument Standard Deviations

	Inflow Measured	σ	Outflow Measured	σ
C	62.1	4	0	Fixed
O_2	65.3	2	1.37	2
CO_2	0	Fixed	63.5	2

Note that all the σ values are *not* the same.

6.10 ***Data Reconciliation—Statistical Basis*** Following Cohen (1953) and Ozyurt and Pike (2004), the maximum likelihood function is formed by maximizing the product of the probability distribution function for each measured variable, x_i^+. Assuming normal distributions for each x_i^+, this product can be written as

$$\max P = \max \prod_i P_i = \max \prod_i \left\{ \frac{1}{\sigma_i \sqrt{2\pi}} \exp\left(-\frac{\left(x_i^+ - \overline{x}_i\right)^2}{2\sigma_i^2} \right) \right\}.$$

Reform P to the equivalent a natural log function and covert the objective function from a maximization to minimization problem (max ln(P) = min − (ln(P)). Compare the final result to Equation (6.1).

6.11 ***Gross Error Detection and Identification—Single Unit*** Consider a single mixer as in Figure P6.11. Show yourself that regardless of the values chosen for instrument accuracy and flow rates, we can only identify if a gross error may exist. The error will be distributed over all the measurements and we cannot determine which instrument (MT_i) is causing the gross error.

To get started, assume $F_1^+ = 245$ and $\sigma_1 = 2$; $F_2^+ = 240$ and $\sigma_2 = 2$; and $F_3^+ = 530$ and $\sigma_3 = 10$. But be sure to solve this problem with other values.

6.12 ***Gross Error Detection and Identification—Multiple Units*** Consider a simple three-unit process in Figure P6.12. Show yourself we can identify if a gross error may exist (GT) and we can determine which instrument (MT_i) is most likely causing the gross error.

To get started, assume $F_1^+ = 99.3$ and $\sigma_1 = 1$; $F_2^+ = 104.1$ and $\sigma_2 = 1$; $F_3^+ = 100.1$ and $\sigma_3 = 1$; and $F_4^+ = 100.5$ and $\sigma_4 = 1$ (which are values from Example 6.19). But be sure to solve this problem with other values.

6.13 ***Gross Error Detection and Identification—Recycle Process*** A problem often found in the literature to test gross error algorithms is the recycle system shown in Figure P6.13 (Narasimhan et al., 1987). Using the flow rate data in Table P6.13, determine if gross errors may exist (GT) and if true, determine which instrument (MT_i) is most likely causing the gross error.

6.14 ***Gross Error Detection and Identification—Nonlinear System (Solution Provided)*** Consider the heat exchanger bypass problem of Example 6.20 (also shown in Figure P6.14a). Here, we allow for energy balances for both the heat exchanger and the mixer. The true and measured flow rates and temperatures are provided in Table P6.14. The temperature of stream 4 contains a positive 5 bias. Use the global test to determine if gross errors are present, and if gross errors are present, use the measurement test to determine the most likely suspect measurement.

Solution:

The reconciled flow rates and the GT can be determined using the Excel-based minimization procedure in **Problem 6.14a.xls**. The reconciled flow rates and temperatures are $F_1 = 100.254$, $F_2 = 64.706$, $F_3 = 35.548$, $F_4 = 64.706$, $F_5 = 35.548$, $F_6 = 100.254$, $F_7 = 139.847$, $T_2 = 90.873$, $T_4 = 54.528$, $T_5 = 79.836$, $T_6 = 64.146$, $T_7 = 20.432$, and $T_8 = 42.610$. The objective function value from the Excel minimization,

$$\sum_{i,\text{measured variables}} \left(\frac{\left(x_i^+ - x_i\right)}{\sigma_i} \right)^2 = 14.3,$$

is larger than the $\chi^2_{(1-\alpha)}(\nu)$ value = CHIINV (0.05, 6) = 12.592, so gross errors are detected.

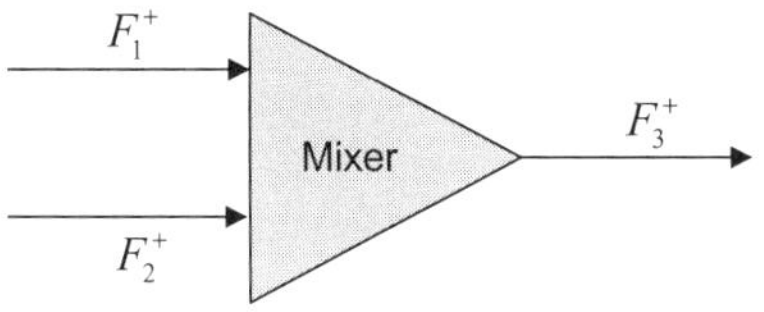

Figure P6.11 Simple mixer.

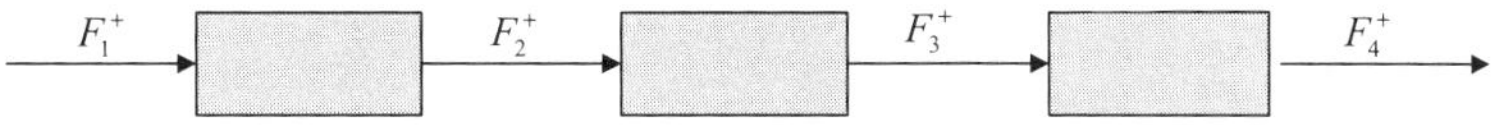

Figure P6.12 Three units in series.

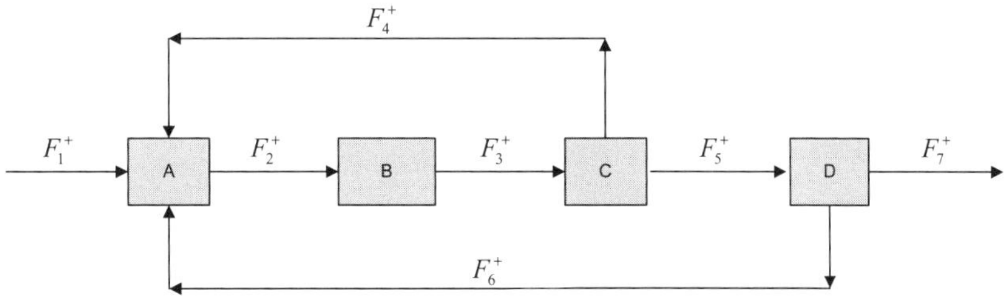

Figure P6.13 Four units with recycle.

In order to linearize the energy balances, the data reconciliation solution is supplied to our Newton–Raphson code as a starting point and the energy balances are supplied to the VBA code in `Function NLFunc`. Within **Problem 6.14b.xls** the Gauss–Jordan and Newton–Raphson macros are each run one time. Linearization of the heat exchanger energy balance gives

$$-867.27F_2 + 421.3F_4 + 206.34F_7 - 911.54T_2 + 676.36T_4 - 991.02T_7 + 1611.31T_8 = 2454.56,$$

and linearization of the mixer energy balance gives

$$-421.3F_4 - 717.87F_5 + 526.46F_6 - 676.36T_4 - 461.54T_5 + 1144.35T_6 = -322.77.$$

The solution for the measurement test to determine the most likely suspect measurement is provided in **Problem 6.14c.xls**; key results are shown below.

The A matrix is shown in Figure P6.14b, and the V matrix is shown in Figure P6.14c,

and

$$MT_i = \frac{|a_i|}{\sqrt{V_{ii}}} = \left\{ \begin{array}{l} \frac{|-1.656|}{\sqrt{0.668}}, \frac{|0.256|}{\sqrt{0.673}}, \frac{|0.898|}{\sqrt{0.671}}, \frac{|0.506|}{\sqrt{0.673}}, \frac{|-0.892|}{\sqrt{0.671}}, \\ \frac{|1.374|}{\sqrt{0.668}}, \frac{|-0.153|}{\sqrt{0.009}}, \frac{|0.673|}{\sqrt{0.171}}, \frac{|-1.472|}{\sqrt{0.282}}, \\ \frac{|-0.664|}{\sqrt{0.109}}, \frac{|1.646|}{\sqrt{0.670}}, \frac{|0.732|}{\sqrt{0.202}}, \frac{|-1.190|}{\sqrt{0.535}} \end{array} \right\}$$
$$= \left\{ \begin{array}{l} 2.026, 0.312, 1.096, 0.617, 1.089, 1.681, 1.631, \\ 1.628, 2.771, 2.011, 2.011, 1.628, 1.628 \end{array} \right\}.$$

Table P6.13 Data for Problem 6.13—Four Units with Recycle

Stream Number	True Flow Rate Values	Measured Flow Rates	σ Flow
1	100	100.9	1
2	150	145.1	1
3	150	150.5	1
4	30	30.3	1
5	120	122.9	1
6	20	18.7	1
7	100	99.1	1

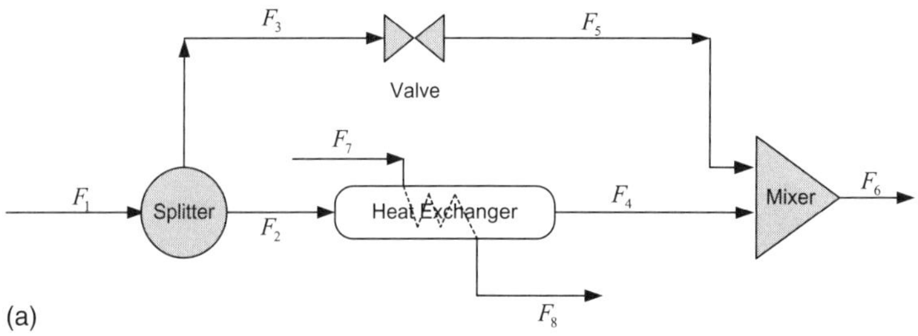

(a)

					A							
F1	F2	F3	F4	F5	F6	F7	T2	T4	T5	T6	T7	T8
-1	1	1	0	0	0	0	0	0	0	0	0	0
0	1	0	-1	0	0	0	0	0	0	0	0	0
0	0	-1	0	1	0	0	0	0	0	0	0	0
0	0	0	1	1	-1	0	0	0	0	0	0	0
0.00	-867.27	0.00	421.30	0.00	0.00	206.34	-911.54	676.36	0.00	0.00	-991.02	1611.31
0.00	0.00	0.00	-421.30	-717.87	526.46	0.00	0.00	-676.36	-461.54	1144.35	0.00	0.00

(b)

					QATIAQATAQ or V							
0.668	−0.165	−0.167	−0.165	−0.167	−0.332	−0.003	0.015	−0.003	0.005	−0.013	0.016	−0.026
−0.165	0.673	0.162	−0.327	0.162	−0.165	−0.006	0.025	−0.037	−0.012	0.031	0.027	−0.044
−0.167	0.162	0.671	0.162	−0.329	−0.167	0.002	−0.010	0.033	0.017	−0.043	−0.011	0.018
−0.165	−0.327	0.162	0.673	0.162	−0.165	−0.006	0.025	−0.037	−0.012	0.031	0.027	−0.044
−0.167	0.162	−0.329	0.162	0.671	−0.167	0.002	−0.010	0.033	0.017	−0.043	−0.011	0.018
−0.332	−0.165	−0.167	−0.165	−0.167	0.668	−0.003	0.015	−0.003	0.005	−0.013	0.016	−0.026
-0.003	−0.006	0.002	−0.006	0.002	−0.003	0.009	−0.039	0.022	−0.005	0.012	−0.042	0.068
0.015	0.025	−0.010	0.025	−0.010	0.015	−0.039	0.171	−0.096	0.021	−0.052	0.186	−0.302
−0.003	−0.037	0.033	−0.037	0.033	−0.003	0.022	−0.096	0.282	0.144	−0.357	−0.104	0.170
0.005	−0.012	0.017	−0.012	0.017	0.005	−0.005	0.021	0.144	0.109	−0.270	0.023	−0.037
−0.013	0.031	−0.043	0.031	−0.043	−0.013	0.012	−0.052	−0.357	−0.270	0.670	−0.057	0.092
0.016	0.027	−0.011	0.027	−0.011	0.016	−0.042	0.186	−0.104	0.023	−0.057	0.202	−0.329
−0.026	−0.044	0.018	−0.044	0.018	−0.026	0.068	−0.302	0.170	−0.037	0.092	−0.329	0.535

(c)

Figure P6.14 (a) Heat exchanger with bypass valve and energy balance. (b) A matrix. (c) V matrix.

Table P6.14 Data for Problem 6.14—Heat Exchanger with Bypass Valve

Stream Number	True Flow Rate Values	True Temperature Values	Measured Values	σ Flow or σ Temperature
1	100		101.91	1
2	64		64.45	1
3	36		34.65	1
4	64		64.20	1
5	36		36.44	1
6	100		98.88	1
7	140.6		140.0	1
2		90	90.2	1
4		51	56	1
5		80	80.5	1
6		62.3	62.5	1
7		20	19.7	1
8		43	43.8	1

Here, as expected, T_4 with $MT_i = 2.771$ can be identified as the most likely suspect measurement.

6.15 ***Gross Error Detection and Identification—Matrix Operations in Excel*** Matrix operations in Excel are somewhat cumbersome. For example, to multiple two matrices (say, A and B), each matrix must be named and the location for the product matrix (rows and columns) must be highlighted on the Excel sheet. The Excel sheet operation = MMULT(A,B) requires keystrokes Crtl + Shift + Enter to complete the matrix multiplication. This same procedure is required for all matrix operations. Develop VBA procedures to allow matrix calculations for $(a)^T Q^{-1}(a)$, GT (Eq. (6.32)), $\chi^2_{(1-\alpha)}(\nu)$, and MT (Eq. (6.35)). Here, the user will supply Q, A, and x^+.

Chapter 7

Gas Turbine Cogeneration System Performance, Design, and Off-Design Calculations: Ideal Gas Fluid Properties

In this chapter, we introduce material and energy balance equations for cogeneration systems. These material and energy balances or performance calculations allow determination of system temperatures, pressures, and flow rates, which in turn can be used for design, sizing, and costing. These performance calculations also help identify any discrepancies that might indicate problems or degradation in system operation. The material developed in this chapter provides the basis for heat and power calculations throughout the remainder of the text.

To understand what we need to accomplish, it is instructive to reexamine the cogeneration system introduced in Chapter 1. The system is shown in Figure 7.1.

For *cogeneration turbine system performance calculations*, given air feed conditions (T_1, P_1, F_{air}), air compressor efficiency, and known P_2, we will want to determine the conditions at 2 (T_2, P_2, F_{air}) in Figure 7.1. With known fuel flow rate and combustion efficiency, we want to determine the conditions at 3. From 3 and with known turbine efficiency and known P_4, we want to determine the conditions at 4 and the net power generated from the turbine.

We will utilize these performance calculations to determine *turbine system performance in design and off-design operation*. Design or benchmark turbine performance is established using standard design conditions of air feed at 59°F/15°C, 14.7 psia/1.013 bar, and 60% relative humidity—these conditions are commonly referred to as the International Standards Organization (ISO) conditions. Turbine performance at nondesign conditions (e.g., air compressor efficiency at $T_1 > 59°F$) will often be given as a percent of

design values in manufacturer-provided performance curves. Here, we develop general equations that can be used to predict turbine performance in off-design cases when the design case data are available. In order to use these equations, we will need to understand how the turbine system control system operates. We will also need a more detailed understanding of the components and flows within the turbine system.

Using the heat recovery steam generator *(HRSG) design and off-design performance calculations*, we want to determine the steam that could be raised from the waste heat boiler or HRSG. The HRSG design performance is based on expected conditions at 4 $(T_4, P_4, F_{exhaust})$, values selected for the pinch and approach temperatures (defined below) and the condensate return/steam conditions. Off-design HRSG performance (e.g., when T_4 is increased by supplemental firing of the exhaust gas) requires an understanding of overall heat transfer coefficients.

In this chapter, performance, design, and off-design calculations are developed in general form to allow application to either ideal gas (constant heat capacity) or real fluid cogeneration systems.

In this chapter, the performance, design, and off-design equations are particularized to an ideal gas working fluid (as $h_{out} - h_{in} = C_p \Delta T$) for the power side of the combined heat and power cycle. We also introduce the use of callable Excel functions from our rigorous thermodynamics program (Chapter 8) to allow accurate steam property calculations in the HRSG. In Chapter 9, the developed equations are particularized to real fluid cogeneration systems; h_{out} and h_{in} are

Modeling, Analysis and Optimization of Process and Energy Systems, First Edition. F. Carl Knopf.
© 2012 John Wiley & Sons, Inc. Published 2012 by John Wiley & Sons, Inc.

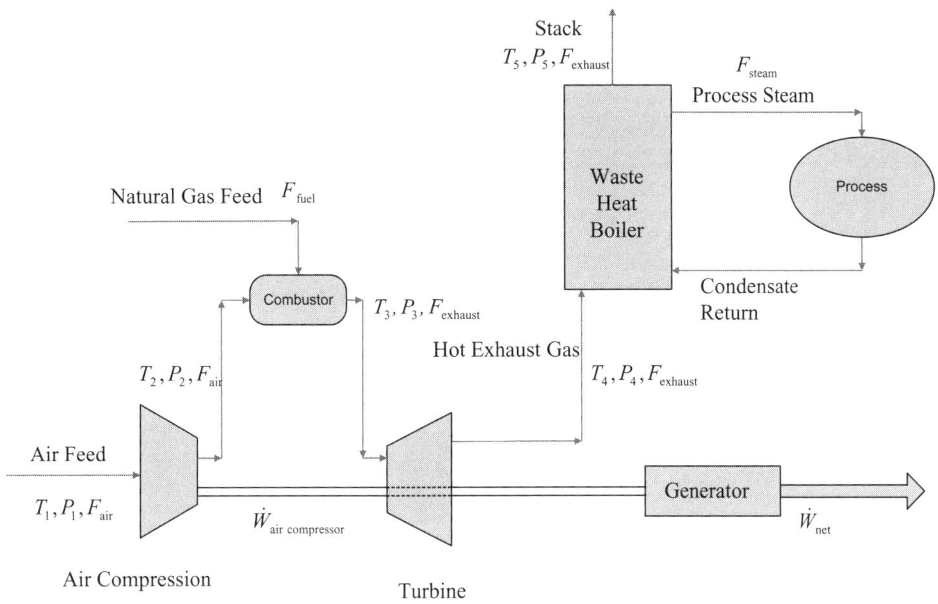

Figure 7.1 Typical gas turbine engine-based cogeneration system—the gas turbine includes the air compressor, combustor, and turbine.

calculated from rigorous equations of state (EOSs) with ideal mixing rules (Chapter 8).

We begin this chapter with a general discussion of thermodynamics and the equilibrium state of a simple compressible fluid. The equilibrium state of a simple compressible fluid or system is completely specified by fixing two independent properties. This equilibrium state is independent of either the path used to obtain the state or the system (open or closed). The assumption of a simple compressible system allows development of "*T ds* equations," which ultimately allow entropy calculations in terms of easily determined variables. The *Tds* equations are applicable to systems of ideal gas, real fluid, or mixtures of real fluids.

Power generating systems are generally open systems where mass flows in and out of the system. Once the general energy balance for an open system is developed, it can be applied to systems of real gases or liquids, simple compressible systems, or ideal gas systems. The general energy balance equation is coupled with thermodynamic equilibrium equations to allow gas turbine-based cogeneration system material and energy balance calculations.

7.1 EQUILIBRIUM STATE OF A SIMPLE COMPRESSIBLE FLUID: DEVELOPMENT OF THE *T dS* EQUATIONS

The equilibrium state of a simple compressible fluid can be specified by two independent properties (often the intensive properties *T*, *P*, or ρ are used). For properties of ultimate interest to heat and power system, $\Delta\hat{h}$ and $\Delta\hat{s}$, we begin with

$$\hat{u} = \hat{u}(\hat{s}, \hat{\upsilon}), \tag{7.1}$$

$$\hat{h} = \hat{h}(\hat{s}, P), \tag{7.2}$$

and

$$\hat{s} = \hat{s}(\hat{u}, \hat{\upsilon}), \tag{7.3}$$

where $\hat{u}$, $\hat{h}$, and $\hat{s}$ are the specific internal energy (joule per kilogram), specific enthalpy (joule per kilogram), and specific entropy (joule per kilogram-kelvin) of the fluid; $\hat{\upsilon}$ is the specific volume (cubic meter per kilogram). Using the fundamental theorem of partial differentiation, we can write, for Equation (7.1),

$$d\hat{u} = \left(\frac{\partial \hat{u}}{\partial \hat{s}}\right)_{\hat{\upsilon}} d\hat{s} + \left(\frac{\partial \hat{u}}{\partial \hat{\upsilon}}\right)_{\hat{s}} d\hat{\upsilon}. \tag{7.4}$$

Since the thermodynamic temperature of a substance can be defined as

$$T \equiv \left(\frac{\partial \hat{u}}{\partial \hat{s}}\right)_{\hat{\upsilon}} \tag{7.5}$$

and the thermodynamic pressure can be defined as

$$P \equiv -\left(\frac{\partial \hat{u}}{\partial \hat{\upsilon}}\right)_{\hat{s}}, \tag{7.6}$$

we find that when these definitions are used in Equation (7.4),

$$T\,d\hat{s} = d\hat{u} + P\,d\hat{\upsilon}. \tag{7.7}$$

Similarly, from Equation (7.2),

$$d\hat{h} = \left(\frac{\partial \hat{h}}{\partial \hat{s}}\right)_P d\hat{s} + \left(\frac{\partial \hat{h}}{\partial P}\right)_{\hat{s}} dP, \tag{7.8}$$

and with the identities

$$T \equiv \left(\frac{\partial \hat{h}}{\partial \hat{s}}\right)_P \tag{7.9}$$

and

$$\hat{\upsilon} \equiv \left(\frac{\partial \hat{h}}{\partial P}\right)_{\hat{s}}, \tag{7.10}$$

we obtain

$$T\,d\hat{s} = d\hat{h} - \hat{\upsilon}\,dP. \tag{7.11}$$

Equations (7.7) and (7.11) are commonly referred to as the *Tds* equations. They allow the calculation of entropy change in terms of easily determined quantities. Equations (7.7) and (7.11) describe changes in equilibrium properties; here there is no concern if the system is open or closed or about the path taken from state 1 to state 2.

A simple compressible system can consist of a single component or a mixture. For example, air, which is primarily a mixture of oxygen and nitrogen, may be considered a simple compressible system as long as it remains in the vapor state. However, this assumption is invalid if the temperature is lowered and/or the pressure increased to where liquid air is in coexistence with vapor air in the system. Here the composition of the liquid phase air will not be the same as the vapor phase. For a single-phase simple system, once two intensive properties (e.g., T, P) have been fixed, the value of a state property such as entropy is determined (relative to a reference state).

7.1.1 Application of the *T ds* Equations to an Ideal Gas

The ideal gas EOS is typically written as $P\upsilon = RT$, where υ is the molar volume and R is the universal gas constant. The ideal gas EOS can also be written for any species on a unit mass basis as $P\hat{\upsilon} = \hat{R}T$, where $\hat{R}$ is the gas constant for the species i, which is equal to the universal gas constant R divided by the molecular weight of the species, $\hat{R} = R/MW_i$. For an ideal gas, both enthalpy and internal energy are a function of temperature only and with constant heat capacity, $d\hat{h} = \hat{C}_P^{ig}dT$. Applying Equation (7.11) to a species that behaves as an ideal gas with constant heat capacity and using $\hat{\upsilon} = \hat{R}T/P$ gives

$$d\hat{s} = \hat{C}_P^{ig}\frac{dT}{T} - \hat{R}\frac{dP}{P}. \tag{7.12}$$

Dividing by $\hat{R}$ and integrating, term by term, between states 1 and 2, we obtain

$$\frac{1}{\hat{R}}\int_{\hat{s}_1}^{\hat{s}_2} d\hat{s} = \frac{\hat{C}_P^{ig}}{\hat{R}}\int_{T_1}^{T_2}\frac{dT}{T} - \int_{P_1}^{P_2}\frac{dP}{P} \tag{7.13}$$

and

$$\frac{\hat{s}_2 - \hat{s}_1}{\hat{R}} = \frac{\hat{C}_P^{ig}}{\hat{R}}\ln\frac{T_2}{T_1} - \ln\frac{P_2}{P_1}. \tag{7.14}$$

Equation (7.14) gives us the change in entropy (from state 1 to state 2) for an ideal gas. Here state 1 is defined by the measurable quantities (T_1, P_1) and state 2 by (T_2, P_2). Again there are no requirements on the path taken or if the system is open or closed—Equation (7.14) gives the entropy change between two equilibrium states for an ideal gas. It is possible that Equation (7.14) may be used to define the entropy of the system at a given T and P. Here (T_1, P_1, s_1) would be taken as $(T_{\text{ref}}, P_{\text{ref}}, s_{\text{ref}})$.

7.1.2 Application of the *T ds* Equations to an Ideal Gas: Isentropic Process

At this point, it is now possible to impose certain process conditions. Often used is the assumption that the process is isentropic, or a constant entropy process, $(\hat{s}_2 - \hat{s}_1) = 0$. An isentropic process is also reversible, but this fact is not important to our current discussion. For an ideal gas of constant heat capacity,

$$\frac{\hat{C}_P^{ig}}{\hat{C}_V^{ig}} = \frac{C_P^{ig}}{C_V^{ig}} = \gamma,$$

where γ is the specific heat ratio and $\hat{C}_P^{ig} - \hat{C}_V^{ig} = \hat{R}$. Using this equation for $\hat{R}$ and setting the right-hand side of Equation (7.14) to zero for an isentropic process,

$$\hat{C}_P^{ig}\ln\frac{T_2^{\text{isentropic}}}{T_1}=\hat{C}_P^{ig}\ln\frac{T_2^{\text{isen}}}{T_1}=\left(\hat{C}_P^{ig}-\hat{C}_V^{ig}\right)\ln\frac{P_2}{P_1}. \quad (7.15)$$

Here we have indicated that T_2^{isen} is the result of an isentropic process by the addition of the additional subscript on T_2. Dividing by $\hat{C}_V^{ig}$,

$$\frac{\hat{C}_P^{ig}}{\hat{C}_V^{ig}}\ln\left(\frac{T_2^{\text{isen}}}{T_1}\right)=\left(\frac{\hat{C}_P^{ig}}{\hat{C}_V^{ig}}-1\right)\ln\frac{P_2}{P_1} \quad (7.16)$$

or

$$\gamma\ln\left(\frac{T_2^{\text{isen}}}{T_1}\right)=(\gamma-1)\ln\frac{P_2}{P_1}. \quad (7.17)$$

Taking the exponential of both sides of Equation (7.17),

$$\left(\frac{T_2^{\text{isen}}}{T_1}\right)^{\gamma}=\left(\frac{P_2}{P_1}\right)^{(\gamma-1)}, \quad (7.18)$$

and solving for T_2^{isen},

$$T_2^{\text{isen}}=T_1\left(\frac{P_2}{P_1}\right)^{\frac{(\gamma-1)}{\gamma}}. \quad (7.19)$$

Here T_2^{isen} is the temperature that results from the isentropic compression or isentropic expansion from (P_1, T_1) to P_2 of an ideal gas with constant heat capacity. Here we have specified the path (isentropic path); however, there is no requirement on the system, it may be either open or closed.

7.2 GENERAL ENERGY BALANCE EQUATION FOR AN OPEN SYSTEM

The mass balance for a steady-state open system (an open system or process is one in which mass both enters and exits) simply states

$$\sum_{\substack{\text{Outlet}\\ \text{Streams},k}} F_k=\sum_{\substack{\text{Inlet}\\ \text{Streams},j}} F_j. \quad (7.20)$$

Here the mass flow rates of the streams into the system are F_j (kilogram per second) and the outlet streams are F_k. The energy balance for this steady-state open system is from Chapter 5:

$$\frac{dQ}{dt}-\frac{dW_s}{dt}=\sum_{\substack{\text{Outlet}\\ \text{Streams},k}} F_k\left(\hat{h}+gz+\frac{1}{2}\vartheta^2\right)_k-\sum_{\substack{\text{Inlet}\\ \text{Streams},j}} F_j\left(\hat{h}+gz+\frac{1}{2}\vartheta^2\right)_j, \quad (7.21)$$

where dW_s/dt is the rate of shaft work (joule per second) done *by* the system on the surroundings. We will define the rate of shaft work done by the system on the surroundings as a positive quantity; shaft work done by the surroundings on the system would be a negative quantity. The rate heat being added to the system is dQ/dt (J / s), and this would be a positive quantity. If heat is removed from the system, dQ/dt would be a negative quantity. The enthalpy per unit mass is $\hat{h}$ (joule per kilogram), and the potential and kinetic energies (joule per kilogram) are given by gz and $\frac{1}{2}\vartheta^2$, respectively.

In most power devices, the potential energy terms are generally small. The kinetic energy terms (in and out) generally cancel in most power devices, as the inlet and outlet pipes are sized to make the fluid velocities nearly equal. With these assumptions, the steady-state energy balance reduces to

$$\frac{dQ}{dt}-\frac{dW_s}{dt}=\sum_{\substack{\text{Outlet}\\ \text{Streams},k}} F_k\left(\hat{h}_k\right)-\sum_{\substack{\text{Inlet}\\ \text{Streams},j}} F_j\left(\hat{h}_j\right), \quad (7.22)$$

and for a single-inlet single-outlet process (also see Eq. (5.7)),

$$\frac{dQ}{dt}-\frac{dW_s}{dt}=F\left(\hat{h}_{\text{out}}-\hat{h}_{\text{in}}\right)=F\left(\hat{h}_2-\hat{h}_1\right)=F\left(\Delta\hat{h}\right). \quad (7.23)$$

Equation (7.22) is the energy balance for an open, steady-state process. It is the first law of thermodynamics for an open process and states that energy changes in the process occur as a result of the rate of work done *by* the system and the rate of heat transferred *to* the system. Do note that we have made no requirements on the type of fluid in the open system. Equations (7.20)–(7.23) are appropriate for real gases or liquids, simple compressible systems, or ideal gas systems.

7.3 COGENERATION TURBINE SYSTEM PERFORMANCE CALCULATIONS: IDEAL GAS WORKING FLUID

Here we will develop the working equations for an open power system that utilizes an ideal gas as the working fluid. Most important will be the equations for air compressors (compression) and turbines (expansion).

7.3.1 Compressor Performance Calculations

Following Figure 7.2, we want to compress the ideal gas from P_1 to P_2. This will result in a temperature increase from T_1 to T_2. The actual temperature increase will depend

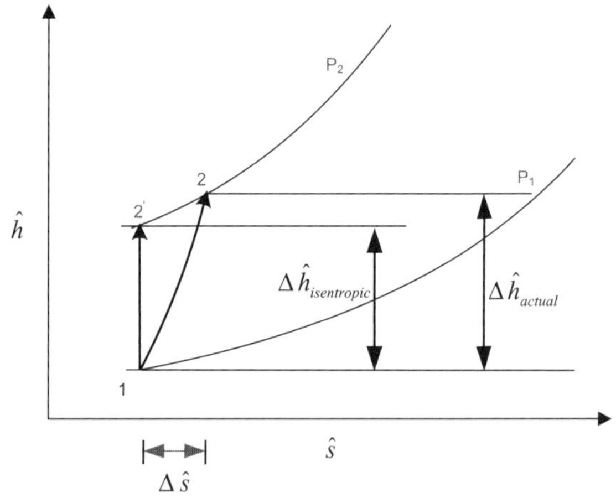

Figure 7.2 Enthalpy–entropy diagram for the compression process.

on the assumptions we make about the actual compression process.

To begin, we recall that for an ideal gas system or a system with constant heat capacity,

$$d\hat{h} = \hat{C}_P^{ig} dT \tag{7.24}$$

or

$$\int_{\hat{h}_1}^{\hat{h}_2} d\hat{h} = \hat{C}_P^{ig} \int_{T_1}^{T_2} dT, \tag{7.25}$$

which gives

$$\hat{h}_2 - \hat{h}_1 = \Delta\hat{h} = \hat{C}_P^{ig}(T_2 - T_1). \tag{7.26}$$

If we allow that the power device (the compressor) is well designed, heat transfer should be negligible; this is equivalent to assuming the process is adiabatic or $(dQ/dt) = 0$.

Combining Equation (7.26) with Equation (7.23) gives

$$-\frac{dW_s}{dt} = F\Delta\hat{h} = F\hat{C}_P^{ig}(T_2 - T_1). \tag{7.27}$$

For compressors, shaft work will be required from the surroundings, so dW_s/dt will be a negative quantity. This is easy to see as $T_2 > T_1$ for compression processes.

If we further assume that the process is isentropic, then we can take advantage of Equation (7.19), giving

$$\frac{-dW_s}{dt}(\text{isentropic}) = F\Delta\hat{h} = F\hat{C}_P^{ig}\left(T_1\left(\frac{P_2}{P_1}\right)^{\frac{(\gamma-1)}{\gamma}} - T_1\right). \tag{7.28}$$

It is instructive to examine the difference between an isentropic and a real compression process on an enthalpy versus entropy plot.

The process from $1 \rightarrow 2'$ is isentropic ($\Delta\hat{s} = 0$), and this will result in an energy change given by $\Delta\hat{h}_{\text{isentropic}}$. The actual system path will be $1 \rightarrow 2$, and here the actual energy change will be $\Delta\hat{h}_{\text{actual}}$, giving

$$-\frac{dW_s}{dt}(\text{actual}) = F\Delta\hat{h} = F\hat{C}_P^{ig}(T_2 - T_1). \tag{7.29}$$

The rate of shaft work, dW_s/dt (actual) and dW_s/dt (isentropic), is related by the isentropic compressor efficiency $\eta_{\substack{\text{isentropic}\\\text{compression}}}$:

$$\frac{dW_s}{dt}(\text{actual}) = \frac{\frac{dW_s}{dt}(\text{isentropic})}{\eta_{\substack{\text{isentropic}\\\text{compression}}}}. \tag{7.30}$$

Now, using the definitions from Equations (7.29) and (7.28), we can write

$$\hat{C}_P^{ig}(T_2 - T_1) = \frac{\hat{C}_P^{ig}\left(T_1\left(\frac{P_2}{P_1}\right)^{\frac{(\gamma-1)}{\gamma}} - T_1\right)}{\eta_{\substack{\text{isentropic}\\\text{compression}}}}. \tag{7.31}$$

After rearrangement we find

$$T_2 \equiv T_2(\text{actual}) = T_1\left\{1 + \frac{1}{\eta_{\substack{\text{isentropic}\\\text{compression}}}}\left[\left(\frac{P_2}{P_1}\right)^{\frac{(\gamma-1)}{\gamma}} - 1\right]\right\}. \tag{7.32}$$

With T_2(actual) determined from Equation (7.32), we can calculate the actual rate of shaft work from Equation (7.29).

It is also common to report shaft work or work as the *rate* of shaft work divided by the mass flow rate of the working fluid through the process:

$$\left(\frac{-dW_s}{Fdt}\right);$$

the units are joule per kilogram or British thermal unit per pound of working fluid. Work is often reported as a positive quantity. We must determine whether this work is being done by the process or on the process to ensure the proper sign.

7.3.2 Turbine Performance Calculations

We have actually developed all the equations we need for turbine (expansion) calculation in the previous section

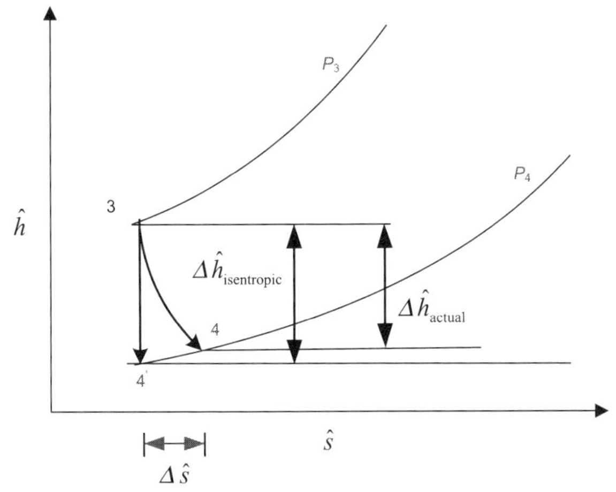

Figure 7.3 Enthalpy–entropy diagram for the expansion process.

(compressor calculations). Because of the importance of these expansion calculations to cogeneration (turbine) processes, we will repeat the equations with the discussion particularized to the expansion process. For power systems, after the gas compression step, heat (energy) will be added to the working fluid by a combustion process. This high-temperature and high-energy gas will expanded through a turbine and power will be extracted. Basically, the heat added from the combustion process will be converted to shaft work.

Following Figure 7.3, the exhaust gas from the combustion chamber is expanded from P_3 to P_4 in the turbine. This will result in a temperature decrease from T_3 to T_4. The actual temperature decrease will depend on the assumptions we make about the actual expansion process. For an adiabatic $(dQ/dt = 0)$ system or a system with constant heat capacity, Equation (7.23) gives

$$-\frac{dW_s}{dt} = F\Delta\hat{h} = F\hat{C}_P^{ig}(T_4 - T_3). \tag{7.33}$$

For turbines, shaft work will be generated by the process, so dW_s/dt will be a positive quantity. This is easy to see as $T_3 > T_4$ for expansion processes.

If we further assume that the process is isentropic, then we can take advantage of Equation (7.19):

$$-\frac{dW_s}{dt}(\text{isentropic}) = F\Delta\hat{h} = F\hat{C}_P^{ig}\left(T_3\left(\frac{P_4}{P_3}\right)^{\frac{(\gamma-1)}{\gamma}} - T_3\right). \tag{7.34}$$

It is instructive to examine the difference between an isentropic and a real expansion process on an enthalpy versus entropy plot (Mollier diagram).

The process from $3 \to 4'$ is isentropic $(\Delta\hat{s} = 0)$ and this will result in an energy change given by $\Delta\hat{h}_{\text{isentropic}}$. The actual system path will be $3 \to 4$ and here the actual energy change will be $\Delta\hat{h}_{\text{actual}}$. Here,

$$-\frac{dW_s}{dt}(\text{actual}) = F\Delta\hat{h} = F\hat{C}_P^{ig}(T_4 - T_3). \tag{7.35}$$

The rate of shaft work, dW_s/dt (actual) and dW_s/dt (isentropic), are related by the isentropic turbine efficiency $\eta_{\substack{\text{isentropic}\\\text{expansion}}}$:

$$\frac{dW_s}{dt}(\text{actual}) = (\eta_{\substack{\text{isentropic}\\\text{expansion}}})\frac{dW_s}{dt}(\text{isentropic}). \tag{7.36}$$

Now using the definitions from Equations (7.35) and (7.34), we can write

$$\hat{C}_P^{ig}(T_4 - T_3) = (\eta_{\substack{\text{isentropic}\\\text{expansion}}})\hat{C}_P^{ig}\left(T_3\left(\frac{P_4}{P_3}\right)^{\frac{(\gamma-1)}{\gamma}} - T_3\right). \tag{7.37}$$

After rearrangement, we find

$$T_4 \equiv T_4(\text{actual}) = T_3\left\{1 + (\eta_{\substack{\text{isentropic}\\\text{expansion}}})\left[\left(\frac{P_4}{P_3}\right)^{\frac{(\gamma-1)}{\gamma}} - 1\right]\right\} \tag{7.38}$$

or

$$T_4 \equiv T_4(\text{actual}) = T_3\left\{1 - (\eta_{\substack{\text{isentropic}\\\text{expansion}}})\left[1 - \left(\frac{P_3}{P_4}\right)^{\frac{(1-\gamma)}{\gamma}}\right]\right\}. \tag{7.39}$$

With T_4(actual) determined from Equation (7.38) or (7.39), we can calculate the actual rate of shaft work from Equation (7.35).

7.4 AIR BASIC GAS TURBINE PERFORMANCE CALCULATIONS

The air basic power cycle assumes that air behaves as an ideal gas with known and constant heat capacity and specific heat ratio. The cycle that is shown in Figure 7.4 begins by air compression, which is taken as an adiabatic and reversible (isentropic) process. The compressed air is mixed with fuel and combustion occurs, which raises the temperature of the mixture. The combustion gas is then expanded through two turbines. The first turbine is termed a gas generating turbine and here the power generated is used to compress the incoming air; this is an adiabatic and reversible (isentropic) process. The second turbine generates the net power from the system. The net power is always the difference between the total power generated by the turbine(s) and that used for air compression.

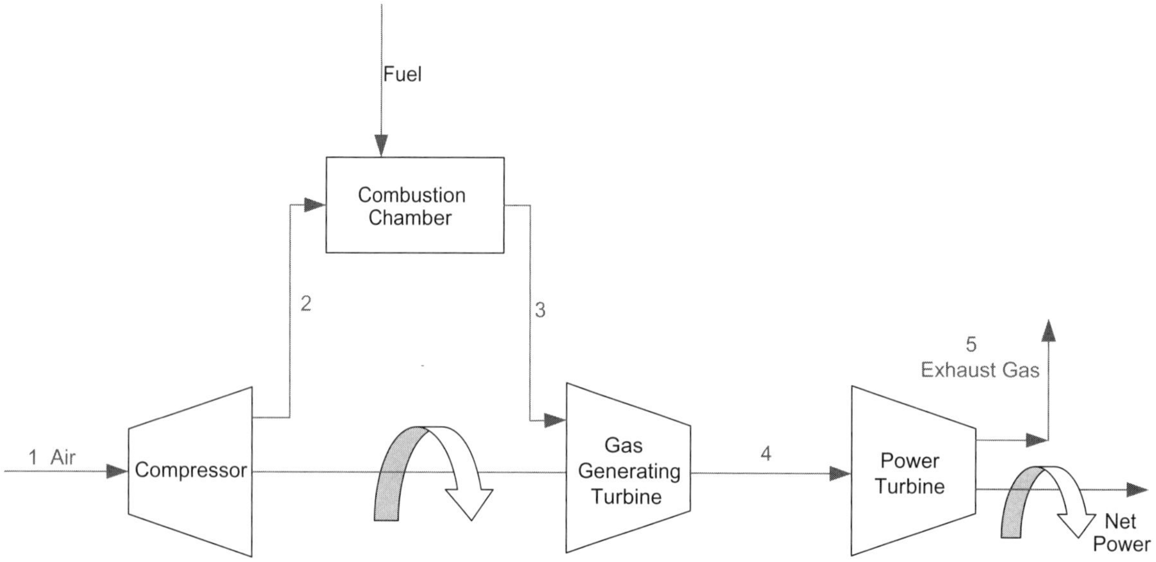

Figure 7.4 Air basic gas turbine compression and expansion process.

EXAMPLE 7.1 ***Air Basic Power Cycle Gas Turbine Compression and Expansion: Ideal Gas Working Fluid***

We want to solve for temperatures, flow rates, and power production for the air basic gas turbine engine shown in Figure 7.4. We first generate a solution with the initial values provided below, and then generate a general solution to this problem in Excel. In the Excel solution, we will want to be sure that the conditions, marked as Variable below, are allowed to accept any value—in the sheet we use named cells. The values in these cells can then be varied to solve any ideal air cycle. This type of general solution will be needed for a cogeneration system economic optimization, as performed in Chapter 10.

For an ideal air cycle, air can be taken as an ideal gas with the following properties: heat capacity $\hat{C}_{P,\text{air}}^{ig} = 0.24$ Btu/lb-R (Variable) and specific heat ratio

$$\frac{\hat{C}_P}{\hat{C}_V} = \gamma = 1.4 \text{ (Variable)}.$$

We also know the following: The air inlet temperature $T_1 = 519$ R (Variable), $P_1 = 1$ atm, and air enters the compressor at 1 lb/s (Variable). The compression ratio (P_2/P_1) = 15.9 (Variable). There is no pressure drop in the combustion chamber ($P_3 = P_2$) and the turbine inlet temperature $T_3 = 2520$ R (Variable). The compressor efficiency (isentropic) = 87% (Variable %). The gas generating turbine efficiency (isentropic) = 89% (Variable %). The power turbine efficiency (isentropic) = 89% (Variable %) and P_5 is atmospheric. For Example 7.1, the flow rate of gas through the system will be taken as constant at 1 lb/s. We will account for the mass of fuel added in the next example. When the Excel solution is complete, you should be able to change any "variable" and see the effect. For example, if you double the air flow rate from 1 to 2 lb/s, the net power produced should double.

SOLUTION To solve for the needed temperatures, flow rates, and power production, we will use the equations developed in Section 7.3. We will systematically use these equations (steps a–o below) and then generalize these calculations in an Excel sheet.

(a) Determine T_2 actual, the actual temperature out of the compressor (Eq. (7.32)):

$$T_2 \equiv T_2(\text{actual}) = T_1\left\{1 + \frac{1}{\eta_{\substack{\text{isentropic}\\ \text{compression}}}}\left[\left(\frac{P_2}{P_1}\right)^{\frac{(\gamma-1)}{\gamma}} - 1\right]\right\}$$

$$T_2 \equiv T_2(\text{actual}) = 519\left\{1 + \frac{1}{0.87}\left[\left(\frac{15.9}{1.0}\right)^{\frac{(0.4)}{1.4}} - 1\right]\right\} = 1237.38 \text{ R}.$$

(b) Determine the actual rate of work in the compressor (Eq. (7.29)):

$$-\frac{dW_C}{dt} = F\Delta\hat{h} = F\hat{C}_{P,\text{air}}^{ig}(T_2 - T_1)$$

$$\frac{dW_C}{dt}(\text{actual}) = -\left(1\frac{\text{lb}}{\text{s}}\right)\left(0.24\frac{\text{Btu}}{\text{lb-R}}\right)(1237.38 - 519) = -172.41\frac{\text{Btu}}{\text{s}}.$$

(c) Determine the actual work done on the compressor per pound of working fluid:

$$W_C = \left|\frac{\frac{dW}{dt}}{F}\right| = \left|\frac{172.4\frac{\text{Btu}}{\text{s}}}{1\frac{\text{lb}}{\text{s}}}\right| = 172.41\frac{\text{Btu}}{\text{lb}}.$$

(d) Determine the rate of heat addition to the combustion chamber. The gas flow rate through the system is assumed constant, so we can write

$$\frac{dQ}{dt} = F\Delta\hat{h} = F\hat{C}^{ig}_{P,\text{air}}(T_3 - T_2)$$

$$\frac{dQ}{dt} = \left(\frac{1\ \text{lb}}{\text{s}}\right)\left(0.24\frac{\text{Btu}}{\text{lb-R}}\right)(2520 - 1237.38) = 307.82\frac{\text{Btu}}{\text{s}}.$$

(e) Determine the heat addition in the combustion chamber per pound of working fluid:

$$\hat{q} = \frac{\left(\frac{dQ}{dt}\right)}{F} = \frac{307.82\frac{\text{Btu}}{\text{s}}}{1\frac{\text{lb}}{\text{s}}} = 307.82\frac{\text{Btu}}{\text{lb of air}}.$$

(f) Determine the actual rate of work of the gas generating turbine:

$$\frac{dW_{GT}}{dt}(\text{actual}) = -\frac{dW_C}{dt}(\text{actual}) = 172.41\frac{\text{Btu}}{\text{s}}$$
$$= -\text{rate of the compressor work.}$$

(g) Determine the actual temperature out of the gas generating turbine (Eq. (7.33)):

$$-\frac{dW_{GT}}{dt} = F\Delta\hat{h} = F\hat{C}^{ig}_{P,\text{air}}(T_4 - T_3)$$

$$-\left(172.41\frac{\text{Btu}}{\text{s}}\right) = \left(1\frac{\text{lb}}{\text{s}}\right)\left(0.24\frac{\text{Btu}}{\text{lb-R}}\right)(T_4 - 2520)$$

$$T_4 \equiv T_4(\text{actual}) = 1801.62\,\text{R}.$$

(h) Determine the pressure out of the gas generating turbine (rearranging Eq. (7.38)):

$$P_4 = P_3\left[\frac{1}{(\eta_{\substack{\text{isentropic}\\ \text{expansion}}})}\left\{\frac{T_{4,\text{actual}}}{T_3} - 1\right\} + 1\right]^{\frac{\gamma}{(\gamma-1)}}$$

$$P_4 = P_3\left[\frac{1}{(\eta_{\substack{\text{isentropic}\\ \text{expansion}}})}\left\{\frac{T_{4,\text{actual}}}{T_3} - 1\right\} + 1\right]^{\frac{\gamma}{(\gamma-1)}}$$

$$= 15.9\left[\frac{1}{(0.89)}\left\{\frac{1801.62}{2520} - 1\right\} + 1\right]^{\frac{1.4}{0.4}} = 4.116\ \text{atm}.$$

(i) Determine the actual temperature (exhaust gas temperature) from the power turbine (from Eq. (7.38)):

$$T_5 \equiv T_5(\text{actual}) = T_4\left\{1 + (\eta_{\substack{\text{isentropic}\\ \text{expansion}}})\left[\left(\frac{P_5}{P_4}\right)^{\frac{(\gamma-1)}{\gamma}} - 1\right]\right\}$$

$$T_5 = 1801.62\left\{1 + (0.89)\left[\left(\frac{1}{4.116}\right)^{\frac{0.4}{1.4}} - 1\right]\right\} = 1268.45\ \text{R}.$$

(j) Determine the actual rate of work of the power turbine:

$$-\frac{dW_{PT}}{dt}(\text{actual}) = FC^{ig}_{P,\text{air}}(T_{5,\text{actual}} - T_4)$$
$$= \left(1\frac{\text{lb}}{\text{s}}\right)\left(0.24\frac{\text{Btu}}{\text{lb-R}}\right)(1268.45 - 1801.62)$$
$$\frac{dW_{PT}}{dt}(\text{actual}) = 127.96\frac{\text{Btu}}{\text{s}}.$$

(k) Determine the net work rate:

$$\frac{dW_{PT}}{dt}(\text{actual}) = 127.96\frac{\text{Btu}}{\text{s}}.$$

(l) Determine the net work of the engine per pound entering the compressor:

$$\frac{127.96\frac{\text{Btu}}{\text{s}}}{1\frac{\text{lb}}{\text{s}}} = 127.96\frac{\text{Btu}}{\text{lb entering air}}.$$

(m) Determine the net power (in kilowatts) generated by the engine:

$$\frac{\left(127.96\frac{\text{Btu}}{\text{s}}\right)\left(3600\frac{\text{s}}{\text{h}}\right)}{\left(3.413\frac{\text{Btu}}{\text{h-W}}\right)\left(1000\frac{\text{W}}{\text{kW}}\right)} = 134.97\ \text{kW}.$$

(n) Determine the thermal efficiency:

$$\eta = \frac{\text{Net work}}{\text{Heat addition}} = \frac{127.96\frac{\text{Btu}}{\text{lb-air}}}{307.82\frac{\text{Btu}}{\text{lb-air}}} = 0.4157 = 41.57\%.$$

(o) Finally, determine the *heat rate*. The heat rate is a term that is often used in cogeneration and power systems. The heat rate is the rate of heat addition to the combustion chamber (d above) divided by the net power generated by the engine (m above). Our units would be British thermal unit per kilowatt-second. The preferred units are British thermal unit per kilowatt-hour, so multiply by 3600 s/h. This is the British thermal unit of heat added to produce a kilowatt-hour of electricity:

$$\frac{\left(307.82\frac{\text{Btu}}{\text{s}}\right)\left(3600\frac{\text{s}}{\text{h}}\right)}{(134.97\,\text{kW})} = 8210.36\frac{\text{Btu}}{\text{kW-h}}.$$

The Excel solution is provided in **Example 7.1a.xls** and shown in Figure 7.5. You may want to develop your solution independent of the Excel provided solution. Here you should take advantage of an Excel sheet template with named cells and the necessary conversion factors provided in **Example 7.1b.xls**. ■

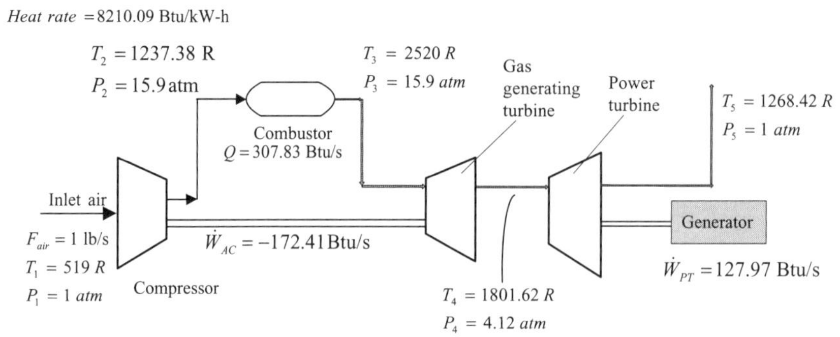

Figure 7.5 Solution of the air basic gas turbine engine in Example 7.1.

Before we leave this example, we want to comment on calculating T_5 directly from Equation (7.38) using P_5, T_3, and P_3 with a single $\eta = \eta_{GT} = \eta_{PT}$. This would result in $T_5 = 1294.70$ R. This is not correct and would actually produce less net power. Multistage processes will produce more power than a single-stage process if the efficiencies are all equal. Also note that in the Excel sheet, we calculated the isentropic work and isentropic temperatures; these values can be compared to the actual work and actual temperatures in the cogeneration system.

7.5 ENERGY BALANCE FOR THE COMBUSTION CHAMBER

In Example 7.1, the simplifying assumption of a constant flow rate of gas through the system was made. To account for the mass of fuel added, we will need to perform an energy balance around the combustion chamber. Following Figure 7.4, the mass and energy balances are

$$F_{\text{gas}} = F_{\text{air}} + F_{\text{fuel}}, \tag{7.40}$$

$$F_{\text{air}}\left(\hat{h}_2 - \hat{h}_{\text{air},(77°\text{F},1\text{atm})}\right) + F_{\text{fuel}}(LHV) = F_{\text{gas}}\left(\hat{h}_3 - \hat{h}_{\text{gas},(77°\text{F},1\text{atm})}\right), \tag{7.41}$$

or

$$F_{\text{air}}\left(\hat{h}_2 - \hat{h}_{\text{air,ref}}\right) + F_{\text{fuel}}(LHV) = F_{\text{gas}}\left(\hat{h}_3 - \hat{h}_{\text{gas,ref}}\right). \tag{7.42}$$

Equations (7.41) and (7.42) require explanation. Recall that there is no absolute value for enthalpy; enthalpy values are always relative to a reference state. In typical "in–out" energy balance calculations involving an inert stream, the reference state "subtracts out." For example, for an inert nitrogen stream in states 1 and 2, $\hat{h}_{N_2,1} \equiv \hat{h}_{N_2,1} - \hat{h}_{N_2,\text{ref}}$, $\hat{h}_{N_2,2} \equiv \hat{h}_{N_2,2} - \hat{h}_{N_2,\text{ref}}$, and $\Delta\hat{h}_{N_2} = \hat{h}_{N_2,2} - \hat{h}_{N_2,1}$; $\Delta\hat{h}_{N_2}$ is independent of any chosen reference state.

In Equations (7.41) and (7.42), the enthalpy of each stream is calculated with respect to a specific reference state, ref of 77°F and 1 atm. Here, in order to remain consistent in our energy balance, we cannot pick an arbitrary reference state, but rather we must use the reference state associated with the fuel lower heating value (*LHV*). The *LHV* for fuels is provided at 77°F and 1 atm. The *LHV* is defined as the heat liberated when 1 lb of fuel is mixed with stoichiometric oxygen at 77°F and is completely burned at constant pressure. The products are then cooled to 77°F with the water formed remaining in the vapor state. The process in Equation (7.41) is shown for methane fuel in Figure 7.6.

Equation (7.41) or (7.42) provides the needed energy balance to determine the outlet temperature of the combustion chamber given a fuel flow rate and fuel *LHV*. Alternatively, Equation (7.41) or (7.42) can be used to determine the fuel flow rate needed to produce a given outlet temperature from the combustion chamber. If needed, we can also account for energy loss, $\dot{Q}_{\text{loss}}$, from the combustion chamber as

$$\begin{aligned} F_{\text{air}}\left(\hat{h}_2 - \hat{h}_{\text{air},(77°\text{F},1\text{atm})}\right) + F_{\text{fuel}}(LHV) \\ = F_{\text{gas}}\left(\hat{h}_3 - \hat{h}_{\text{gas},(77°\text{F},1\text{atm})}\right) + \dot{Q}_{\text{loss}}. \end{aligned} \tag{7.43}$$

7.5.1 Energy Balance for the Combustion Chamber: Ideal Gas Working Fluid

Equation (7.41) provides the needed energy balance to determine the outlet temperature of the combustion chamber given a fuel flow rate. Equation (7.41) can also be used to determine the fuel flow rate given an outlet temperature from the combustion chamber. For a system where the air into and the product gas from the combustion chamber are taken as ideal gases, we can write

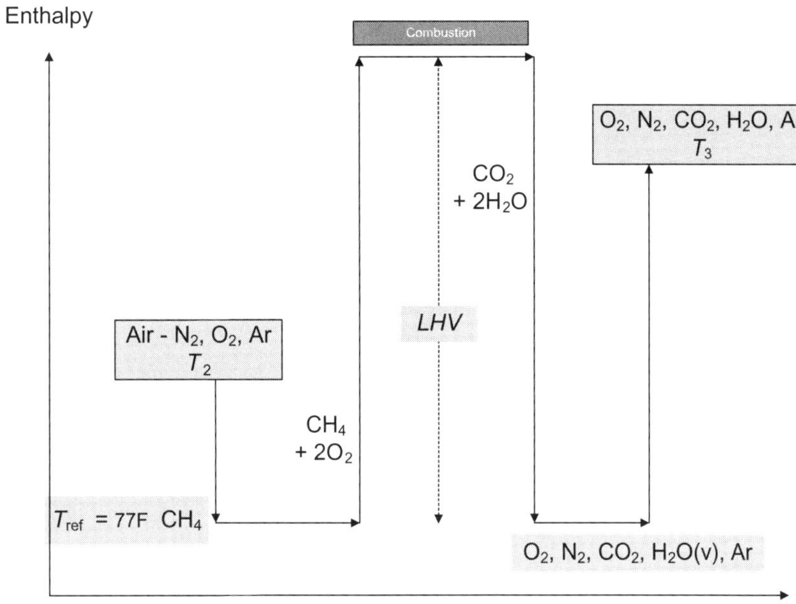

Figure 7.6 Air and fuel combustion process with reference state determined by fuel *LHV*.

$$F_{\text{air}}\hat{C}^{ig}_{P,\text{air}}(T_2 - T_{\text{ref}}) + F_{\text{fuel}}(LHV) = (F_{\text{air}} + F_{\text{fuel}})\hat{C}^{ig}_{P,\text{gas}}(T_3 - T_{\text{ref}}), \tag{7.44}$$

where $T_{\text{ref}} = 77°\text{F} = 536.67$ R. We can also account for combustion chamber losses using a combustion chamber efficiency, η_{CC}, as

$$F_{\text{air}}\hat{C}^{ig}_{P,\text{air}}(T_2 - T_{\text{ref}}) + F_{\text{fuel}}(LHV) = (F_{\text{air}} + F_{\text{fuel}})\hat{C}^{ig}_{P,\text{gas}}(T_3 - T_{\text{ref}}) + F_{\text{fuel}}(LHV)(1 - \eta_{\text{CC}}). \tag{7.45}$$

Solving Equation (7.45) for the fuel flow rate to obtain a known T_3 from the combustion chamber gives

$$F_{\text{fuel}} = \frac{F_{\text{air}}\hat{C}^{ig}_{P,\text{gas}}(T_3 - T_{\text{ref}}) - F_{\text{air}}\hat{C}^{ig}_{P,\text{air}}(T_2 - T_{\text{ref}})}{\left((LHV)(\eta_{\text{CC}}) - \hat{C}^{ig}_{P,\text{gas}}(T_3 - T_{\text{ref}})\right)}. \tag{7.46}$$

7.6 THE HRSG: DESIGN PERFORMANCE CALCULATIONS

In a cogeneration system, both steam and electricity are produced. This combination of electricity generation and steam production provides the economic justification for the installation and continued operation of cogeneration systems. Steam is generated in the HRSG, also called the waste heat boiler. Figure 7.7 shows a gas turbine system with the addition of an HRSG. Here, recycled water from the process is exchanged with the exhaust gas from the turbine to create steam. The exhaust gas, after losing energy to the water, exits in the stack. Figure 7.7 indicates steam production at a single pressure level. Condensate (hot water) from the steam system is returned to the HRSG along with makeup water. This hot water is first heated to near boiling in a section of the HRSG termed the economizer or preheater and then converted to saturated steam in a section termed the evaporator. Figure 7.7 also indicates the possibility that additional fuel (supplemental fuel) may be added to the exhaust gas and fired to increase the exhaust stream temperature and raise additional steam when compared to the unfired case.

In this section, we want to develop the equations that determine the design performance of the HRSG. First, we establish the normal expected exhaust gas conditions entering the evaporator section of the HRSG. For these conditions, the HRSG is taken to operate in the unfired mode (no supplemental firing) so $(T_6, P_6, F_6) = (T_5, P_5, F_5)$. These HRSG standard or design entering conditions are dependent on the turbine system configuration; for example, compare Figure 7.7 with Figure 9.1 (with air preheat); assuming similar turbine system performance, the temperature of the exhaust gas entering the evaporator will be lower with air preheat compared to the configuration without air preheat. All nonstandard conditions (e.g., supplemental firing) are considered the off-design operation of the HRSG. Later in this chapter we examine off-design HRSG performance.

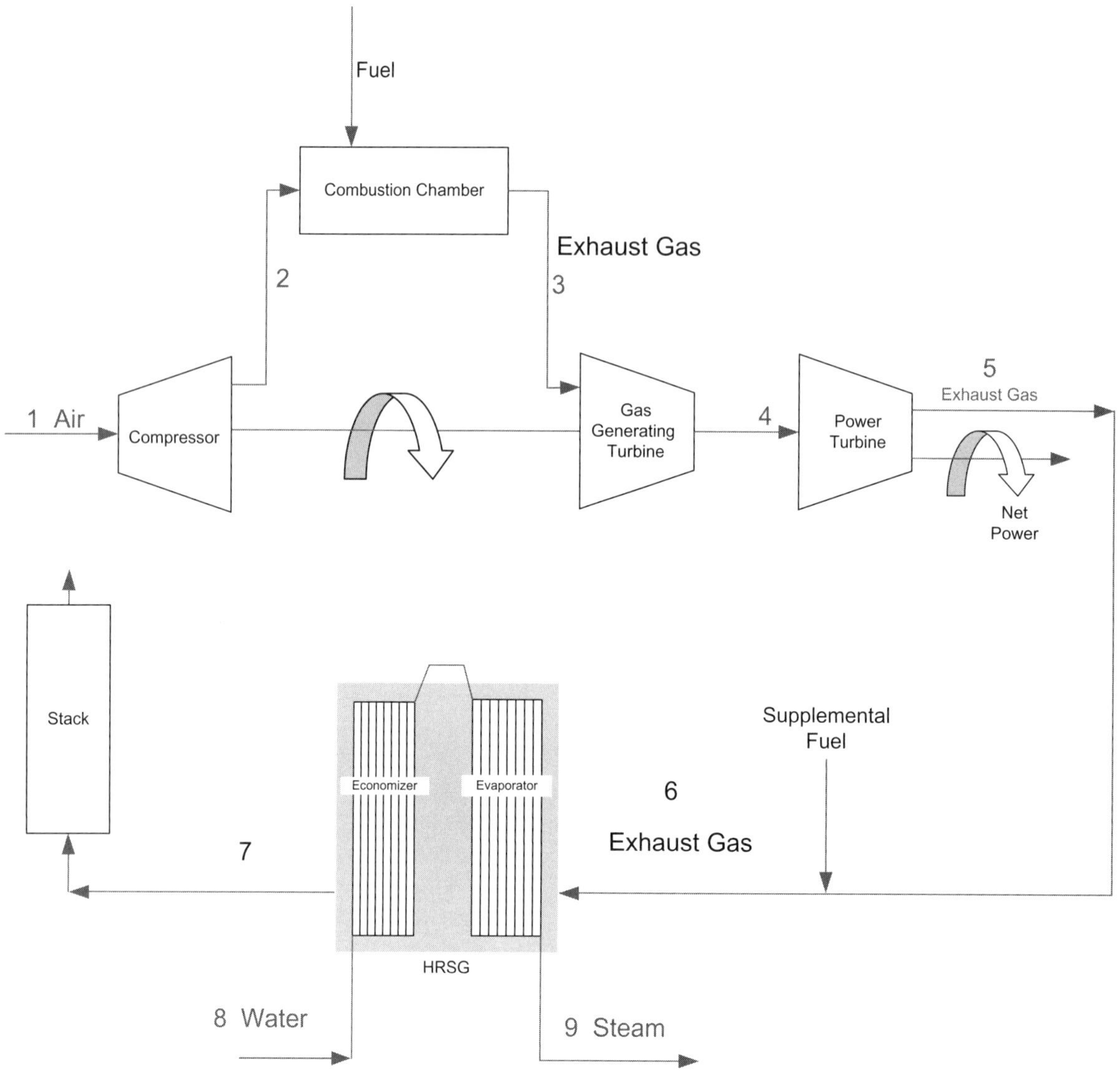

Figure 7.7 Cogeneration system—gas turbine with heat recovery steam generator.

The HRSG Design Performance

Two key parameters in the HRSG design performance are the pinch temperature difference, ΔT_{Pinch}, and the approach temperature difference, $\Delta T_{\text{Approach}}$. As indicated in Figure 7.8, in the economizer section of the HRSG, hot water at T_8, P_8 is heated to $\Delta T_{\text{Approach}}$ degrees below the saturation temperature of the steam. In the evaporator section, the water is converted to saturated steam at T_9, P_9.

Water is converted to steam by heat exchange with the exhaust gas from the turbine. The exhaust gas enters the HRSG at T_6, P_6. Here (the HRSG design case), T_6, P_6 are equal to T_5, P_5 from the turbine; we develop the material and energy balance equations using T_6, P_6, which will prove convenient when we discuss supplemental firing later in this chapter. The exhaust gas leaves the evaporator section and enters the economizer section of the HRSG at ΔT_{Pinch} degrees above the saturation temperature of the steam. The exhaust gas leaving the economizer section at T_7, P_7, as well as the condensate entering at T_8, P_8, must be of sufficient temperature to ensure that acid condensation (generally sulfuric) does not occur in the stack or in the economizer.

The recommended approach (Ganapathy, 1991) for the "HRSG design case" is to first fix both the pinch temperature difference ΔT_{Pinch} and the approach temperature difference $\Delta T_{\text{Approach}}$. The ΔT_{Pinch} and $\Delta T_{\text{Approach}}$ temperature differences are indicated in Figure 7.8 and are shown in an expanded view in Figure 7.9. With these values fixed, the amount of steam generated can be determined by material and energy balances around the economizer and evaporator sections of the HRSG, as we do below.

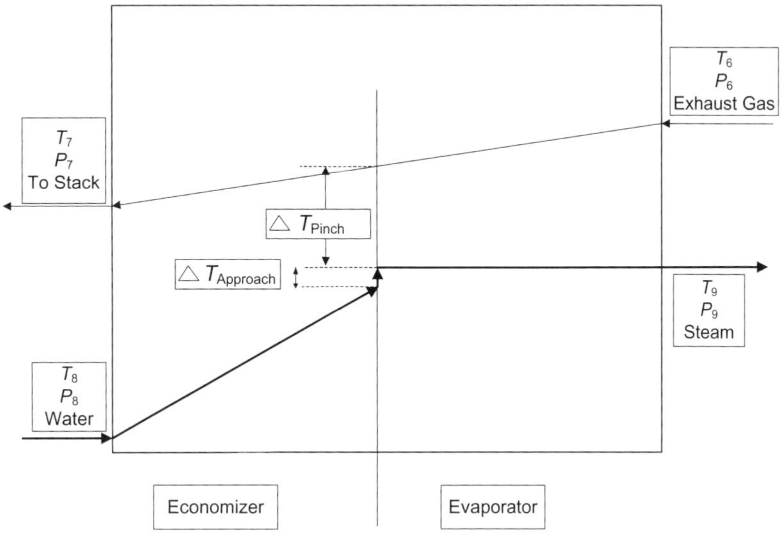

Figure 7.8 Single pressure-level HRSG sections with temperature and pressure notations.

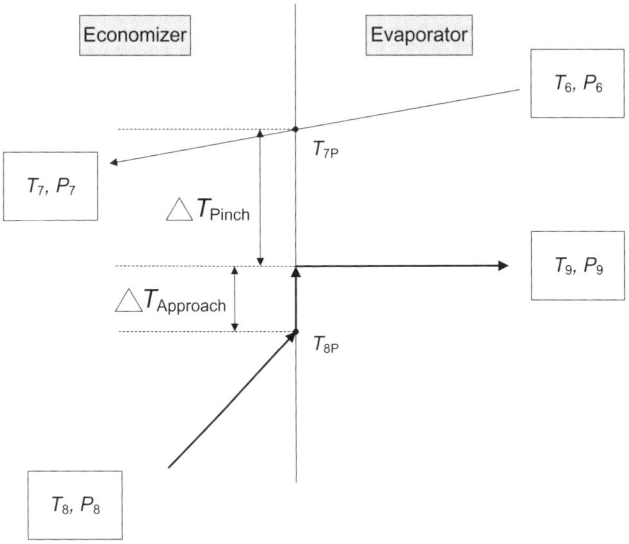

Figure 7.9 Single pressure-level HRSG sections with expanded pinch point.

As ΔT_{Pinch} increases in value, less energy will be recovered from the exhaust gas. However, too low of a value of ΔT_{Pinch} may lead to a temperature crossover in an HRSG off-design operation. Too low of a value of $\Delta T_{\text{Approach}}$ may allow steaming to occur in the economizer in off-design conditions. We will discuss off-design HRSG performance later in this chapter and in Chapters 9 and 10.

Recommended values for ΔT_{Pinch} and $\Delta T_{\text{Approach}}$ are based on the evaporator inlet exhaust gas temperature and the type of evaporator tube construction. Table 7.1 provides typical values.

The range in pinch and approach temperature values in the table can be attributed to the "cleanliness" of the fuel being used. For "clean burning" natural gas-fired turbines, ΔT_{Pinch} is generally between 15 and 20°F (or 15–20 R), and $\Delta T_{\text{Approach}}$ also between 15 and 20°F. Values chosen within these ranges will produce an HRSG design with good energy recovery, and practical size (area) and cost.

Figure 7.9 provides an expanded view of the transition between the economizer and evaporator sections of the HRSG, which will be useful in the development of material and energy balances. Often ΔT_{Pinch} is referred to as the *pinch point* and $\Delta T_{\text{Approach}}$ is termed the *approach point*. In this section, we will assume both the gas-side pressure drop and the water-side pressure drop in the HRSG are negligible. We will address pressure drop in the HRSG in Chapter 9.

Beginning with an energy balance on the gas side of the evaporator,

$$\frac{dQ_{\text{Evaporator, gas}}}{dt} = \dot{Q}_{\text{Evaporator, gas}} = F_{\text{gas}}\left(\hat{h}_{(T_{7P},P_6)} - \hat{h}_{(T_6,P_6)}\right) = F_{\text{gas}}\left(\hat{h}_{7P} - \hat{h}_6\right). \tag{7.47}$$

Here we assume $P_6 \approx P_{7\text{P}} \approx P_7$. Furthermore, P_7 is often taken as the atmospheric pressure, although it would be slightly above atmospheric. The exhaust temperature entering the evaporator will be calculated as part of the turbine system calculations. Using the definition of the pinch point,

$$T_{7P} = T_9 + \Delta T_{\text{Pinch}}, \tag{7.48}$$

where T_9 is the saturation temperature of the generated steam. The gas enthalpies at $\hat{h}_6$ and $\hat{h}_{7P}$ are known, as both are functions of known pressure and temperatures.

Table 7.1 Typical Values for ΔT_{Pinch} and $\Delta T_{\text{Approach}}$ for HRSG Design Unfired Operation (Ganapathy, 1991)

	ΔT_{Pinch} (°F)	ΔT_{Pinch} (°F)	$\Delta T_{\text{Approach}}$ (°F)
Evaporator type	Bare tube	Finned tube	
Evaporator inlet temperature (°F)			
750–1200	80–130	10–30	10–40
1200–1800	130–150	30–60	40–70

The flow rate of steam can be calculated from an overall energy balance around the evaporator. The evaporator section is assumed adiabatic and the overall balance sums the gas-side and water/steam-side balances:

$$\frac{dQ_{\text{Evaporator}}}{dt} = \frac{dQ_{\text{Evaporator,gas}}}{dt} + \frac{dQ_{\text{Evaporator,steam}}}{dt} = 0,$$

giving

$$F_{\text{steam}} = \frac{-\dot{Q}_{\text{Evaporator,gas}}}{\hat{h}^{\text{sat vapor}}_{(T_9,P_9)} - \hat{h}^{\text{liquid}}_{(T_{8P},P_9)}} = \frac{-\dot{Q}_{\text{Evaporator,gas}}}{\hat{h}_9 - \hat{h}_{8P}}. \quad (7.49)$$

Here we have added superscripts on the enthalpy terms to emphasize the water phase as either saturated vapor or liquid:

$$T_{8P} = T_9 - \Delta T_{\text{Approach}}. \quad (7.50)$$

For now, we will set the flow rate of water equal to the flow rate of steam, but later, we will modify the water flow rate to account for blowdown:

$$F_{\text{water}} = F_{\text{steam}}. \quad (7.51)$$

An energy balance on the water side of the economizer or preheat section gives

$$\frac{dQ_{\text{Economizer, water}}}{dt} = \dot{Q}_{\text{Economizer, water}} = F_{\text{water}}\left(\hat{h}^{\text{liquid}}_{(T_{8P},P_9)} - \hat{h}^{\text{liquid}}_{(T_8,P_8)}\right) = F_{\text{water}}\left(\hat{h}_{8P} - \hat{h}_8\right). \quad (7.52)$$

Here, we assume $P_8 \approx P_{8P} \approx P_9$, with P_9 being the saturation pressure of the steam. Finally, we can calculate the exhaust gas temperature T_7 leaving the HRSG from an overall energy balance on economizer. The economizer section is assumed adiabatic, and the overall balance sums the gas-side and water/steam-side balances,

$$\frac{dQ_{\text{Economizer}}}{dt} = \frac{dQ_{\text{Economizer,gas}}}{dt} + \frac{dQ_{\text{Economizer,water}}}{dt} = 0,$$

giving

$$F_{\text{gas}}\left(\hat{h}_{7P} - \hat{h}_7\right) = \dot{Q}_{\text{Economizer,water}} \quad (7.53)$$

and

$$\hat{h}_7 = \hat{h}_{7P} - \frac{\dot{Q}_{\text{Economizer,water}}}{F_{\text{gas}}}, \quad (7.54)$$

where h_7 is a function of T_7 and P_7; P_7 will be known ($P_7 \approx P_6 \approx$ atmospheric pressure), allowing T_7 to be determined. Before leaving this section, we note that all the enthalpy terms in the HRSG calculations are much stronger functions of temperature than pressure.

7.6.1 HRSG Design Calculations: Exhaust Gas Ideal and Water-Side Real Properties

HRSG design calculations can be performed using an ideal gas assumption for the exhaust gas. However, calculations involving water or steam must be performed using real water properties.

Following our developments of the previous section, the energy balance on the gas side of the evaporator becomes

$$\frac{dQ_{\text{Evaporator,gas}}}{dt} = \dot{Q}_{\text{Evaporator,gas}} = F_{\text{gas}}\hat{C}_p^{ig}(T_{7P} - T_6). \quad (7.55)$$

The flow rate of steam can be calculated from Equation (7.49) with $\dot{Q}_{\text{Evaporator,gas}}$ determined from Equation (7.55):

$$F_{\text{steam}} = \frac{-\dot{Q}_{\text{Evaporator,gas}}}{\hat{h}^{\text{sat vapor}}_{(T_9,P_9)} - \hat{h}^{\text{liquid}}_{(T_{8P},P_9)}} = \frac{F_{\text{gas}}\hat{C}_p^{ig}(T_6 - T_{7P})}{\hat{h}_9 - \hat{h}_{8P}}. \quad (7.56)$$

Equation (7.56) provides the flow rate of water as the $F_{\text{water}} = F_{\text{steam}}$. The energy balance on the water side of the economizer or preheat section is given by Equation (7.52) and provides $\dot{Q}_{\text{Economizer,water}} = F_{\text{water}}\left(\hat{h}_{8P} - \hat{h}_8\right)$.

Finally, we can calculate the exhaust gas temperature T_7 leaving the HRSG from an overall energy balance on the economizer:

$$F_{\text{gas}}\hat{C}_p^{ig}(T_{7P} - T_7) = \dot{Q}_{\text{Economizer,water}} \quad (7.57)$$

$$T_7 = T_{7P} - \frac{\dot{Q}_{\text{Economizer,water}}}{F_{\text{gas}}\hat{C}_P^{ig}}. \quad (7.58)$$

7.7 GAS TURBINE COGENERATION SYSTEM PERFORMANCE WITH DESIGN HRSG

At this point, let us add an HRSG to Example 7.1 and solve the HRSG design case.

EXAMPLE 7.2 ***Gas Turbine Cogeneration System Using $\hat{C}_{P,\text{air}}^{ig}$ and $\hat{C}_{P,\text{gas}}^{ig}$ with Design HRSG***

Starting with the basic processing conditions established in Example 7.1, we want to add an HRSG for energy recovery (steam production) as well as account for heat losses from the combustion process. Assume a realistic air feed of $F_{\text{air}} = 200$ lb/s (Variable) and assume the cogeneration system has the configuration in Figure 7.7.

We will also want to account for the mass of fuel added and to allow for different heat capacities for the air feed, $\hat{C}_{P,\text{air}}^{ig}$, and the combustion exhaust gas, $\hat{C}_{P,\text{gas}}^{ig}$. The air feed has heat capacity, $\hat{C}_{P,\text{air}}^{ig} = 0.24$ Btu/lb-R (Variable) and a specific heat ratio,

$$\frac{\hat{C}_{P,\text{air}}^{ig}}{\hat{C}_{V,\text{air}}^{ig}} = \gamma_{\text{air}} = 1.4\,(\text{Variable}).$$

The combustion products (gas) have a heat capacity, $\hat{C}_{P,\text{gas}}^{ig} = 0.28$ Btu/lb-R (Variable), and a specific heat ratio,

$$\frac{\hat{C}_{P,\text{gas}}^{ig}}{\hat{C}_{V,\text{gas}}^{ig}} = \gamma_{\text{gas}} = 1.33\,(\text{Variable}).$$

Examining Figure 7.7, the fuel and air streams are mixed in the combustion chamber. Here, we assume the fuel is natural gas (methane) with an *LHV* of 21,500 Btu/lb-methane at 77°F. The outlet temperature from the combustor remains at 2520 R. However, for material and energy balances starting with the combustion chamber, we will want to account for both air and fuel in the combustion exhaust gas; for example, for the combustion chamber mass balance, $F_{\text{air}} + F_{\text{fuel}} = F_{\text{gas}}$.

We also want to account for heat loss from the combustion chamber. The rate of heat loss from the combustion chamber dQ_{loss}/dt is taken as a percentage of the fuel *LHV*. This can be given as $dQ_{\text{loss}}/dt = (F_{\text{fuel}})\,(LHV)\,(1 - \eta_{\text{cc}})$; with the efficiency of the combustion chamber $\eta_{\text{cc}} = 0.98$, this is equivalent to a 2% loss.

As we discussed earlier, the HRSG design case requires specification of pinch and approach temperature differences. We set the pinch temperature difference at $\Delta T_{\text{Pinch}} = 15$ R and the approach temperature difference at $\Delta T_{\text{Approach}} = 15$ R. But we still allow both ΔT_{Pinch} and $\Delta T_{\text{Approach}}$ to be variable; there is no requirement that ΔT_{Pinch} and $\Delta T_{\text{Approach}}$ have the same value. The water feed (condensate return) conditions are 700 R and 290 psia (20 bar). Saturated steam is needed by the process at 874 R and 290 psia.

For Example 7.2, we will follow the same solution steps as Example 7.1 (steps a–o) and determine the rate of heat loss from the combustion chamber. For the HRSG, we will determine all the temperatures both in and out of the HRSG economizer and evaporator and the quantity of steam raised.

SOLUTION We will provide details of the calculations and assemble all calculations in an Excel sheet; both the complete solution and a start template are provided in **Example 7.2.a.xls** and in **Example 7.2.b.xls**, respectively.

The calculations from Example 7.1 are repeated, but now with $F_{\text{air}} = 200$ lb/s until we reach the combustion chamber. This gives

$$\frac{dW_C}{dt}(\text{actual}) = -34{,}482\frac{\text{Btu}}{\text{s}} \text{ and } T_2(\text{actual}) = 1237.38 \text{ R}.$$

In the combustion chamber, we must determine the mass of fuel added to raise the temperature of the 200 lb/s entering air from 1237.38 to 2520 R. We can solve Equation (7.46) for the flow rate of fuel needed to raise the air (200 lb/s) + fuel to $T_3 = 2520$ R, including heat losses from the combustion chamber:

$$F_{\text{fuel}} = \frac{F_{\text{air}}\hat{C}_{P,\text{gas}}^{ig}(T_3 - T_{\text{ref}}) - F_{\text{air}}\hat{C}_{P,\text{air}}^{ig}(T_2 - T_{\text{ref}})}{\left((LHV)(\eta_{CC}) - \hat{C}_{P,\text{gas}}^{ig}(T_3 - T_{\text{ref}})\right)}$$

$$F_{\text{fuel}} = \frac{\left(200\frac{\text{lb}}{\text{s}}\right)\left(0.28\frac{\text{Btu}}{\text{lb-R}}\right)(2520 - 536.67 \text{ R}) - \left(200\frac{\text{lb}}{\text{s}}\right)\left(0.24\frac{\text{Btu}}{\text{lb-R}}\right)(1237.38 - 536.67 \text{ R})}{\left(\left(21500\frac{\text{Btu}}{\text{lb-methane}}\right)(0.98) - \left(0.28\frac{\text{Btu}}{\text{lb-R}}\right)(2520 - 536.67 \text{ R})\right)}$$

$$F_{\text{fuel}} = 3.7745\frac{\text{lb}}{\text{s}},$$

and where $T_{\text{ref}} = 77 + 459.67 = 536.67$ R.

We have performed the needed gas generating and power turbine calculation in Example 7.1. The only changes are the flow rate of gas, F_{gas}, is now 203.7745 lb/s and $\hat{C}_{P,\text{gas}}^{ig} = 0.28$ Btu/lb-R, and the specific heat ratio

$$\frac{\hat{C}_{P,gas}^{ig}}{\hat{C}_{V,gas}^{ig}} = \gamma_{gas} = 1.33.$$

Following the same steps as Example 7.1:

Determine the rate of total heat addition to the combustion chamber (step d):

$$\frac{dQ}{dt} = F_{\text{fuel}}(LHV) = \left(3.7745\frac{\text{lb-methane}}{\text{s}}\right)\left(21500\frac{\text{Btu}}{\text{lb-methane}}\right) = 81151\frac{\text{Btu}}{\text{s}}.$$

Determine the rate of heat of heat loss from the combustion chamber:

$$\frac{dQ_{\text{loss}}}{dt} = F_{\text{fuel}}(LHV)(1 - \eta_{CC}) = \left(3.7745\frac{\text{lb-methane}}{\text{s}}\right)\left(21{,}500\frac{\text{Btu}}{\text{lb-methane}}\right)(0.02) = 1623\frac{\text{Btu}}{\text{s}}.$$

Determine the actual temperature out of the gas generating turbine (step g):

$$-\frac{dW_T}{dt} = F_{\text{gas}}\Delta\hat{h} = F\hat{C}^{ig}_{P,\text{gas}}(T_4 - T_3)$$

$$-\left(34{,}482\frac{\text{Btu}}{\text{s}}\right) = \left(203.7745\frac{\text{lb}}{\text{s}}\right)\left(0.28\frac{\text{Btu}}{\text{lb-R}}\right)(T_4 - 2520)$$

$$T_4 = 1915.65\text{ R}.$$

Determine the pressure out of the gas generating turbine (step h):

$$P_4 = P_3\left[\frac{1}{(\eta_{\substack{\text{isentropic}\\ \text{expansion}}})}\left\{\frac{\text{T}_{4,\text{actual}}}{\text{T}_3} - 1\right\} + 1\right]^{\frac{\gamma}{(\gamma-1)}}$$

$$= 15.9\left[\frac{1}{(0.89)}\left\{\frac{1915.65}{2520} - 1\right\} + 1\right]^{\frac{1.33}{0.33}} = 4.486\text{ atm}.$$

Determine the actual exhaust gas temperature from the power turbine (step j):

$$T_5 \equiv T_5(\text{actual}) = T_4\left\{1 + (\eta_{\substack{\text{isentropic}\\ \text{expansion}}})\left[\left(\frac{P_5}{P_4}\right)^{\frac{(\gamma-1)}{\gamma}} - 1\right]\right\}$$

$$T_5 = 1915.65\left\{1 + (0.89)\left[\left(\frac{1}{4.486}\right)^{\frac{0.33}{1.33}} - 1\right]\right\} = 1385.53\text{ R}.$$

Determine the net work rate from the power turbine (step k):

$$-\frac{dW_{PT}}{dt}(\text{actual}) = F_{\text{gas}}C^{ig}_{P,\text{gas}}(T_5 - T_4)$$

$$= 203.7745\frac{\text{lb-gas}}{\text{s}}\left(0.28\frac{\text{Btu}}{\text{lb-R}}\right)(1385.53 - 1915.65\text{ R})$$

$$\frac{dW_{PT}}{dt}(\text{actual}) = 30{,}246\frac{\text{Btu}}{\text{s}}.$$

We now solve the HRSG design calculation. First, we establish known temperatures, pressures, and enthalpies for the ideal gas side and the real properties of the water side of the HRSG.

> At this point, steam tables are needed. Later in this chapter, we utilize the physical properties package we have developed, which includes steam and water properties.

For the water side of the HRSG, utilizing the steam tables,

$$T_9 = T_9^{\text{sat}} = 874\text{ R};\ P_9^{\text{sat}} = P_8 = 290\text{ psia} = 20\text{ bar};$$

$$\hat{h}_9 = \hat{h}^{\text{sat vapor}}_{874\text{R},290\text{psia}} = \hat{h}^{\text{sat vapor}}_{874\text{R}} = 1203.09\frac{\text{Btu}}{\text{lb}}$$

$$T_{8P} = T_9 - \Delta T_{\text{Approach}} = 874 - 15 = 859\text{ R};$$

$$\hat{h}^{\text{liquid}}_{8P} = \hat{h}^{\text{liquid}}_{859\text{R},290\text{psia}} = 374.42\frac{\text{Btu}}{\text{lb}};$$

$$T_8 = 700\ R;\quad \hat{h}_8 = \hat{h}^{\text{liquid}}_{700\text{R},\,290\text{psia}} = 209.37\frac{\text{Btu}}{\text{lb}}.$$

For the gas side of the HRSG, we have design conditions of

$$F_{\text{gas}} = 203.7745\frac{\text{lb}}{\text{s}};$$

$$T_6 = T_5 = 1385.53\text{ R (as there is no supplemental firing)};$$

$$P_6 = P_7 = 1\text{ atm}$$

$$T_{7P} = T_9 + \Delta T_{\text{Pinch}} = 874 + 15 = 889\text{ R}.$$

Next, determine the quantity of steam raised using Equations (7.55) and (7.56). The energy balance on the gas side of the evaporator (Eq. (7.55)) gives

$$\frac{dQ_{\text{Evaporator,gas}}}{dt} = \dot{Q}_{\text{Evaporator,gas}} = F_{\text{gas}}\hat{C}^{ig}_{P,\text{gas}}(T_{7P} - T_6)$$

$$= \left(203.7745\frac{\text{lb}}{\text{s}}\right)\left(0.28\frac{\text{Btu}}{\text{lb-R}}\right)(889 - 1385.53)$$

$$= -28331\frac{\text{Btu}}{\text{s}}.$$

The flow rate of steam (Eq. (7.56)) is

$$F_{\text{steam}} = \frac{-\dot{Q}_{\text{Evaporator}}}{\hat{h}^{\text{sat vapor}}_{(T_9,P_9)} - \hat{h}^{\text{liquid}}_{(T_{8P},P_9)}} = \frac{\left(28{,}331\frac{\text{Btu}}{\text{s}}\right)}{\hat{h}_9 - \hat{h}_{8P}}$$

$$= \frac{\left(28{,}331\frac{\text{Btu}}{\text{s}}\right)}{1203.09 - 374.42\frac{\text{Btu}}{\text{lb}}} = 34.188\frac{\text{lb}}{\text{s}},$$

and the flow rate of water = the flow rate of steam:

$$F_{\text{water}} = F_{\text{steam}} = 34.188\frac{\text{lb-steam}}{\text{s}}\ \text{or}\ \frac{\text{lb-water}}{\text{s}}.$$

Finally, the energy balance on the water side of the economizer or preheat section (Eq. (7.57)) gives

$$\dot{Q}_{\text{Economizer,water}} = F_{\text{water}}\left(\hat{h}_{8P} - \hat{h}_8\right)$$

$$= \left(34.188\frac{\text{lb}}{\text{s}}\right)\left(374.42 - 209.37\frac{\text{Btu}}{\text{lb}}\right)$$

$$= 5643\frac{\text{Btu}}{\text{s}}.$$

This allows the last missing temperature, T_7, to be solved (Eq. (7.58)) as

$$T_7 = T_{7P} - \frac{\dot{Q}_{\text{Economizer,water}}}{F_{\text{gas}}\hat{C}^{ig}_{P,\text{gas}}}$$

$$= 889\text{ R} - \frac{\left(5643\dfrac{\text{Btu}}{\text{s}}\right)}{\left(203.7745\dfrac{\text{lb}}{\text{s}}\right)\left(0.28\dfrac{\text{Btu}}{\text{lb-R}}\right)} = 790.10\text{ R}.$$

The Excel solution provided in **Example 7.2a.xls** is shown in Figure 7.10. ■

7.7.1 HRSG Material and Energy Balance Calculations Using Excel Callable Sheet Functions

As will be developed in Chapter 9 and utilized in the remaining chapters, our main goal is to seamlessly incorporate real physical properties in cogeneration calculations. As a preview to this process, we want to show how steam properties from our library can be used in solving the HRSG in Example 7.2.

EXAMPLE 7.3 ***Basic Gas Turbine Engine with HRSG Sheet Functions***

The thermodynamic library is accessed through a macro that links the Excel sheet and the Combustion Library Field Units. If you go to **Example 7.3.xls** and examine the macro (Tools → Macro → Visual Basic Editor), you will see links to steam properties need to solve Example 7.2:

```
Public Declare Function H_Steam Lib "C:\POEA\
Combustion Library Field Units\Combustion_
Library.dll" (ByVal Pressure As Double, ByVal
Temperature As Double) As Double

Public Declare Function H_Water Lib "C:\POEA\
Combustion Library Field Units\Combustion_
Library.dll" (ByVal Pressure As Double, ByVal
Temperature As Double) As Double
```

These links establish the following functions:

Function	Input Parameters	Input Units	Output Units
H_Steam	(P, T)	(psia, R)	Btu/lb-mol
H_Water	(P, T)	(psia, R)	Btu/lb-mol

These functions can be accessed from any cell in the Excel sheet in **Example 7.3.xls**. For example, on the sheet in cell G40 type = H_Steam (290, 874) or H_Steam (P_9, T_9) and divide by the molecular weight of water, $MW_{\text{water}} = 18.0153$; you should find the value in the cell = 1203.43 Btu/lb. Similarly in cell G41 type = H_Water(P_{8P}, T_{8P})/MW_{water} and in cell G42 type = H_Water(P_8, T_8)/MW_{water}; here the values should be 374.36 and 209.25 Btu/lb. Compare these three values with those found using the steam tables at these same conditions—the values should be extremely close. The values found from the sheet functions could replace those values currently in cells D40:D42.

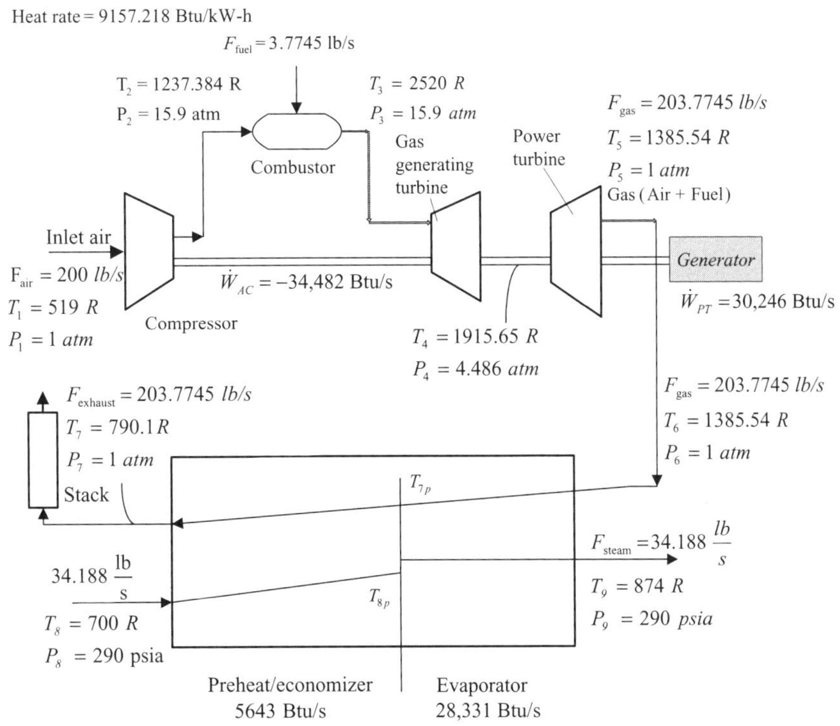

Figure 7.10 Solution basic gas turbine engine with design HRSG.

What is most important to see right now is that the use of the sheet functions coupled to the thermodynamic library allows us to change (P, T) values in (P_9, T_9), (P_{8P}, T_{8P}), and (P_8, T_8) and to obtain new, accurate values for enthalpies. These (P, T) values can now become as part of a design optimization process. For example, as (P, T) values are varied; new values for enthalpy are determined, which in turn would result in new steam flow rates. ■

7.8 HRSG OFF-DESIGN CALCULATIONS: SUPPLEMENTAL FIRING

It is important to predict the performance of the HRSG in an off-design operation, especially the determination of the steam production rate. An off-design operation includes all cases where the HRSG is not receiving design values for F_{gas}, F_{water}, and T_6. For example, F_{gas} may be below the design flow rate when the gas turbine is operating below the design capacity. Generally, the most important off-design operation of the HRSG occurs when supplemental firing is used to increase steam production (above design). Supplemental firing adds fuel to the hot exhaust gas from the gas turbine and this fuel is combusted. This combustion increases the gas inlet temperature to the HRSG and there is also an increase in the gas flow rate. In order to predict the effect of supplemental firing, two approaches are commonly used (Bustami, 2001). In the first approach, an overall energy balance on the HRSG is used with the assumptions that $\Delta T_{\text{Approach}}$ and T_7, the gas exit temperature to the stack, remain fixed at design conditions. In the second *more accurate* approach (Ganapathy, 1991), the overall heat transfer coefficient for the HRSG is determined for the design case. This coefficient is then modified to account for off-design conditions and the HRSG performance determined. The equations developed for HRSG off-design performance under supplemental firing are not limited to just the case of supplemental firing—these equations can be used to predict any off-design performance.

7.8.1 HRSG Off-Design Performance: Overall Energy Balance Approach

To solve for the performance of the HRSG under supplemental firing, an overall energy balance can be used with the assumptions that $\Delta T_{\text{Approach\_SF}}$ and T_{7_SF} remain fixed at design conditions ($\Delta T_{\text{Approach\_SF}} = \Delta T_{\text{Approach}}$ and $T_{7\_\text{SF}} = T_7$). Here we will add the subscript SF to indicate supplemental firing. The temperature of the gas resulting from supplemental firing, T_{6_SF}, can be assumed and the supplemental fuel flow rate determined. An equivalent formulation would start with an assumed supplemental fuel flow rate and determine the gas temperature. An energy balance on the supplemental fuel firing point gives

$$\begin{aligned} &F_{\text{gas}}\hat{C}^{ig}_{P,\text{gas}}(T_5 - T_{\text{ref}}) + F_{\text{supp fuel}}(LHV) \\ &\quad = (F_{\text{gas}} + F_{\text{supp fuel}})\hat{C}^{ig}_{P,\text{gas}}(T_{6\_\text{SF}} - T_{\text{ref}}). \end{aligned} \tag{7.59}$$

Recall that in Figure 7.7 we allowed for the possibility of supplemental firing to raise the exhaust temperature entering the HRSG from T_5 to T_{6_SF} (in the design case $T_5 = T_6$). With T_{6_SF} assumed, we can solve for the needed flow rate of supplemental fuel, $F_{\text{supp fuel}}$, as

$$F_{\text{supp fuel}} = \frac{(F_{\text{gas}})\hat{C}^{ig}_{P,\text{gas}}(T_{6\_SF} - T_5)}{LHV - \hat{C}^{ig}_{P,\text{gas}}(T_{6\_\text{SF}} - T_{\text{ref}})}. \tag{7.60}$$

The mass flow rate of steam under supplemental firing, $F_{\text{steam\_SF}}$, can be determined from an overall energy balance on the HRSG if we assume that the gas exit temperature from the HRSG, T_{7_SF}, is unchanged from the design case ($T_{7\_\text{SF}} = T_7$):

$$\begin{aligned} F_{\text{steam\_SF}} &= \frac{(F_{\text{gas}} + F_{\text{supp fuel}})\hat{C}^{ig}_{P,\text{gas}}(T_{6\_\text{SF}} - T_{7\_\text{SF}})}{\hat{h}^{\text{sat vap}}_{(T_9,P_9)} - \hat{h}^{\text{liquid}}_{(T_8,P_8)}} \\ &= \frac{(F_{\text{gas\_SF}})\hat{C}^{ig}_{P,\text{gas}}(T_{6\_\text{SF}} - T_{7\_\text{SF}})}{\hat{h}^{\text{sat vap}}_{(T_9,P_9)} - \hat{h}^{\text{liquid}}_{(T_8,P_8)}}. \end{aligned} \tag{7.61}$$

Here $F_{\text{gas\_SF}}$ is the gas flow rate after supplemental firing: $F_{\text{gas\_SF}} = (F_{\text{gas}} + F_{\text{supp fuel}})$. The pinch temperature difference under supplemental firing, $\Delta T_{\text{Pinch\_SF}}$, can be found if we assume that the approach temperature difference under supplemental firing, $\Delta T_{\text{Approach\_SF}}$, remains unchanged from the design case. This assumption allows (T_{8P_SF}, P_{8P_SF}) to remain unchanged from the design case (T_{8P_SF}, P_{8P_SF}) = (T_{8P}, P_{8P}). Here, an overall energy balance on the evaporator section of the HRSG yields

$$\begin{aligned} &F_{\text{steam\_SF}}\left(\hat{h}^{\text{sat vap}}_{(T_9,P_9)} - \hat{h}^{\text{liquid}}_{(T_{8P\_\text{SF}},P_{8P\_\text{SF}})}\right) = \\ &\quad (F_{\text{gas\_SF}})\hat{C}^{ig}_{P,\text{gas}}(T_{6\_\text{SF}} - T_{7\text{P\_SF}}). \end{aligned} \tag{7.62}$$

Solving for the gas-side pinch temperature under supplemental firing, $T_{7\text{P\_SF}}$, gives

$$T_{7P\_\text{SF}} = \frac{-\left\{F_{\text{steam\_SF}}\left(\hat{h}^{\text{sat vap}}_{(T_9,P_9)} - \hat{h}^{\text{liquid}}_{(T_{8P\_\text{SF}},P_{8P\_\text{SF}})}\right) - (F_{\text{gas\_SF}})\hat{C}^{ig}_{P,\text{gas}}(T_{6\_\text{SF}})\right\}}{(F_{\text{gas\_SF}})\hat{C}^{ig}_{P,\text{gas}}}, \tag{7.63}$$

and $\Delta T_{\text{Pinch\_SF}}$ can be found from

$$\Delta T_{\text{Pinch\_SF}} = T_{7\text{P\_SF}} - T_9. \tag{7.64}$$

EXAMPLE 7.4 ***HRSG Off-Design Performance Based on Overall Energy Balance: Supplemental Firing***

Starting with Example 7.2, we want to see the impact on steam production if supplemental firing is used to increase the temperature entering the HRSG; here assume $T_{6\_SF} = 1870$ R. Conditions in Example 7.2 give our design HRSG with $\Delta T_{\text{Pinch}} = 15$ R and $\Delta T_{\text{Approach}} = 15$ R. In Example 7.4, use water and steam properties as determined using the Excel sheet functions introduced in Example 7.3. Also, be sure to define new variables on the Excel sheet to account for supplemental firing.

SOLUTION The solution to Example 7.4 provided in **Example 7.4.xls** first shows that the system produces $F_{\text{gas}} = 203.7745$ lb/s and 31.9 MW electricity. The temperature entering the HRSG is 1385.5 R, $F_{\text{steam}} = 34.17$ lb/s, and the transferred Q's are $\dot{Q}_{\text{Evaporator}} = 28{,}331.3$ Btu/s and $\dot{Q}_{\text{Economizer}} = 5642.3$ Btu/s. These values for the HRSG represent the HRSG *design* case—remember here we are using water and steam properties as determined using the provided Excel sheet functions.

We can now use Equations (7.59)–(7.64) to solve the off-design HRSG where $T_{6\_SF} = 1870$ R. These equations are also used in the provided Excel solution. The flow of supplemental fuel from Equation (7.60) is

$$F_{\text{supp fuel}} = \frac{(F_{\text{gas}})\hat{C}^{ig}_{P,\text{gas}}(T_{6\_SF} - T_5)}{LHV - \hat{C}^{ig}_{P,\text{gas}}(T_{6\_SF} - T_{\text{ref}})} = \frac{\left(203.77\frac{\text{lb}}{\text{s}}\right)\left(0.28\frac{\text{Btu}}{\text{lb-R}}\right)(1870 - 1385.5\text{ R})}{\left(21{,}500\frac{\text{Btu}}{\text{lb-methane}}\right) - \left(0.28\frac{\text{Btu}}{\text{lb-R}}\right)(1870 - 536.67\text{ R})} = 1.3084\frac{\text{lb}}{\text{s}}.$$

The new steam production rate from Equation (7.61) is

$$F_{\text{steam\_SF}} = \frac{(F_{\text{gas\_SF}})\hat{C}^{ig}_{P,\text{gas}}(T_{6\_SF} - T_{7\_SF})}{\hat{h}^{\text{sat vap}}_{(T_9,P_9)} - \hat{h}^{\text{liquid}}_{(T_8,P_8)}} = \frac{\left(205.078\frac{\text{lb}}{\text{s}}\right)\left(0.28\frac{\text{Btu}}{\text{lb-R}}\right)(1870 - 790.11\text{ R})}{\left(1203.427\frac{\text{Btu}}{\text{lb}}\right) - \left(209.249\frac{\text{Btu}}{\text{lb}}\right)} = 62.37\frac{\text{lb}}{\text{s}}.$$

Supplemental firing is increasing steam production by 82.5%. The new gas-side pinch temperature T_{7P_SF} is found from Equation (7.63) as

$$T_{7P\_SF} = \frac{-\left\{F_{\text{steam\_SF}}\left(\hat{h}^{\text{sat vap}}_{(T_9,P_9)} - \hat{h}^{\text{liquid}}_{(T_{8P\_SF},P_{8P\_SF})}\right) - (F_{\text{gas\_SF}})\hat{C}^{ig}_{P,\text{gas}}(T_{6\_SF})\right\}}{(F_{\text{gas\_SF}})\hat{C}^{ig}_{P,\text{gas}}} = \frac{-\left\{62.37\frac{\text{lb}}{\text{s}}\left(1203.427 - 374.36\frac{\text{Btu}}{\text{lb}}\right) - \left(205.078\frac{\text{lb}}{\text{s}}\right)\left(0.28\frac{\text{Btu}}{\text{lb-R}}\right)(1870\text{ R})\right\}}{\left(205.078\frac{\text{lb}}{\text{s}}\right)\left(0.28\frac{\text{Btu}}{\text{lb-R}}\right)} = 969.49\text{ R}.$$

The new pinch temperature difference under supplemental firing, $\Delta T_{\text{Pinch\_SF}}$, is found from Equation (7.64): $\Delta T_{\text{Pinch_SF}} = \Delta T_{7P_SF} - T_9 = 969.49 - 874 = 95.49$R. On the Excel sheet, we have also calculated the transferred dQ/dt under supplemental firing: $\dot{Q}_{\text{Evaporator\_SF}} = 51{,}711.98$ Btu/s and $\dot{Q}_{\text{Economizer\_SF}} = 10{,}298.74$ Btu/s. ■

7.8.2 HRSG Off-Design Performance: Overall Heat Transfer Coefficient Approach

In Section 7.8.1, we solved for the performance of the HRSG under supplemental firing using overall energy balances with the assumptions that $\Delta T_{\text{Approach\_SF}}$ and ΔT_{7_SF} remain fixed at design conditions. These assumptions are not needed if we employ the overall heat transfer coefficient method as recommended by Ganapathy (1991). In the following discussion, some knowledge of heat transfer is helpful but not required. In the HRSG, we have seen that $\dot{Q}$ transferred can be given by $\dot{Q} = FC_p\Delta T$ or $\dot{Q} = F\Delta\hat{h}$; this calculation can be applied to the evaporator section, the economizer section, or over the entire HRSG. Equivalently the $\dot{Q}$ transferred can be given by

$$\dot{Q} = UA\Delta T_{\text{LMTD}}, \tag{7.65}$$

where U is the overall heat transfer coefficient Btu/s-ft^2-R; A is the area available for heat transfer in the HRSG; and ΔT_{LMTD} is log mean temperature difference LMTD in the HRSG. The log mean temperature difference represents a temperature driving force at each end of the HRSG heat exchanger sections (evaporator section, economizer section, or overall) given in the design case by

$$\Delta T_{\text{LMTD,Evaporator}} = \left(\frac{(T_6 - T_9) - (T_{7P} - T_{8P})}{\ln\left(\frac{(T_6 - T_9)}{(T_{7P} - T_{8P})}\right)} \right),$$

$$\Delta T_{\text{LMTD,Economizer}} = \left(\frac{(T_{7P} - T_{8P}) - (T_7 - T_8)}{\ln\left(\frac{(T_{7P} - T_{8P})}{(T_7 - T_8)}\right)} \right),$$

and

$$\Delta T_{\text{LMTD,HRSG}} = \left(\frac{(T_6 - T_9) - (T_7 - T_8)}{\ln\left(\frac{(T_6 - T_9)}{(T_7 - T_8)}\right)} \right).$$

If we group the terms UA, we can use Equation (7.65) and the definition for the log mean temperature difference to solve for the design case $UA_{\text{Evaporator\_D}}$ and $UA_{\text{Economizer\_D}}$, and for the *supplemental firing* case $UA_{\text{Evaporator\_SF}}$ and $UA_{\text{Economizer\_SF}}$; here, using our results from Example 7.4,

$$UA_{\text{Evaporator\_D}} = \frac{\dot{Q}_{\text{Evaporator\_D}}}{\Delta T_{\text{LMTD,Evaporator\_D}}} = \frac{\dot{Q}_{\text{Evaporator\_D}}}{\left(\frac{(T_6 - T_9) - (T_{7P} - T_{8P})}{\ln\left(\frac{(T_6 - T_9)}{(T_{7P} - T_{8P})}\right)} \right)}$$

$$= \frac{28{,}331.27}{\left(\frac{(1385.54 - 874) - (889 - 859)}{\ln\left(\frac{(1385.54 - 874)}{(889 - 859)}\right)} \right)}$$

$$= \frac{28{,}331.27}{169.78} = 166.87 \frac{\text{Btu}}{\text{s-R}},$$

$$UA_{\text{Economizer\_D}} = \frac{\dot{Q}_{\text{Economizer\_D}}}{\Delta T_{\text{LMTD,Economizer\_D}}} = \frac{\dot{Q}_{\text{Economizer\_D}}}{\left(\frac{(T_{7P} - T_{8P}) - (T_7 - T_8)}{\ln\left(\frac{(T_{7P} - T_{8P})}{(T_7 - T_8)}\right)} \right)}$$

$$= \frac{5642.3381}{\left(\frac{(889 - 859) - (790.11 - 700)}{\ln\left(\frac{(889 - 859)}{(790.11 - 700)}\right)} \right)}$$

$$= \frac{5642.3381}{54.6537} = 103.238 \frac{\text{Btu}}{\text{s-R}},$$

$$UA_{\text{Evaporator\_SF}} = \frac{\dot{Q}_{\text{Evaporator\_SF}}}{\Delta T_{\text{LMTD,Evaporator\_SF}}}$$

$$= \frac{\dot{Q}_{\text{Evaporator\_SF}}}{\left(\frac{(T_{6\_\text{SF}} - T_9) - (T_{7P\_\text{SF}} - T_{8P\_SF})}{\ln\left(\frac{(T_{6\_\text{SF}} - T_9)}{(T_{7P\_\text{SF}} - T_{8P\_\text{SF}})}\right)} \right)}$$

$$= \frac{51{,}711.98}{\left(\frac{(1870 - 874) - (969.46 - 859)}{\ln\left(\frac{(1870 - 874)}{(969.46 - 859)}\right)} \right)}$$

$$= \frac{51{,}711.98}{402.68} = 128.419 \frac{\text{Btu}}{\text{s-R}},$$

and

$$UA_{\text{Economizer\_SF}} = \frac{\dot{Q}_{\text{Economizer\_SF}}}{\Delta T_{\text{LMTD,Economizer\_SF}}}$$

$$= \frac{\dot{Q}_{\text{Economizer\_SF}}}{\left(\frac{(T_{7P\_\text{SF}} - T_{8P\_\text{SF}}) - (T_{7\_\text{SF}} - T_8)}{\ln\left(\frac{(T_{7P\_\text{SF}} - T_{8P\_\text{SF}})}{(T_{7\_\text{SF}} - T_8)}\right)} \right)}$$

$$= \frac{10{,}298.74}{\left(\frac{(969.46 - 859) - (790.11 - 700)}{\ln\left(\frac{(969.46 - 859)}{(790.11 - 700)}\right)} \right)}$$

$$= \frac{10{,}298.74}{99.94} = 103.05 \frac{\text{Btu}}{\text{s-R}}.$$

In an HRSG that does not have a radiant section, the overall heat transfer coefficient U accounts for three resistances to heat transfer: first the convective heat transfer resistance from the exhaust gas to the outside of the HRSG tubes, then the conductive resistance through the boiler tubes, and finally the convective resistance to the water inside the tubes. For the HRSG, the outside or gas-side resistance (gas-side heat transfer coefficient), $h_{\text{gas side}}$, is the rate-limiting step, allowing us to write

$$U = h_{\text{gas side}}. \tag{7.66}$$

The gas-side heat transfer coefficient can be estimated as (Ganapathy, 1991, 2003)

$$h_{\text{gas side}} \cong \frac{(F_{\text{gas}})^{0.6}(k)^{0.7}(\hat{C}_{P,\text{gas}})^{0.3}}{(d)^{0.4}(\mu)^{0.3}}, \tag{7.67}$$

where k is the thermal conductivity of the exhaust gas, μ is the viscosity of the exhaust gas, and d is the tube diameter in the HRSG. Equation (7.67) can be applied to both the HRSG design case and off-design case, allowing us to write the following ratio for the design and off-design case with supplemental firing:

$$\frac{U_D}{U_{SF}} = \frac{h_{\text{out\_}D}}{h_{\text{out\_SF}}} = \frac{\left(F_{\text{gas\_}D}\right)^{0.6}\left(\dfrac{(k)^{0.7}\left(\hat{C}_{P,\text{ gas}}\right)^{0.3}}{(\mu)^{0.3}}\right)_D}{\left(F_{\text{gas\_SF}}\right)^{0.6}\left(\dfrac{(k)^{0.7}\left(\hat{C}_{P,\text{ gas}}\right)^{0.3}}{(\mu)^{0.3}}\right)_{SF}}. \quad (7.68)$$

The tube diameter cancels and the second grouping of terms involves only gas physical properties which are evaluated at the exhaust gas average temperature, for example, the average gas temperature of the evaporator, or the average gas temperature of the economizer, evaluated at design or supplemental firing temperatures. Rearranging Equation (7.68) gives

$$U_{SF} = U_D\frac{\left(F_{\text{gas\_SF}}\right)^{0.6}}{\left(F_{\text{gas\_}D}\right)^{0.6}}\frac{\left(\dfrac{(k)^{0.7}\left(\hat{C}_{P,\text{ gas}}\right)^{0.3}}{(\mu)^{0.3}}\right)_{SF}}{\left(\dfrac{(k)^{0.7}\left(\hat{C}_{P,\text{ gas}}\right)^{0.3}}{(\mu)^{0.3}}\right)_D}. \quad (7.69)$$

Multiplying both sides of Equation (7.69) by the area available for heat transfer (which of course is the same for both design and supplemental firing) gives

$$\text{Predicted }(UA)_{SF} = (UA)_D\frac{\left(F_{\text{gas\_SF}}\right)^{0.6}}{\left(F_{\text{gas\_}D}\right)^{0.6}}\frac{\left(\dfrac{(k)^{0.7}\left(\hat{C}_{P,\text{ gas}}\right)^{0.3}}{(\mu)^{0.3}}\right)_{SF}}{\left(\dfrac{(k)^{0.7}\left(\hat{C}_{P,\text{ gas}}\right)^{0.3}}{(\mu)^{0.3}}\right)_D}. \quad (7.70)$$

Equation (7.70) shows that $(UA)_{SF}$ can be predicted from $(UA)_D$ by accounting for the changes in exhaust gas flow rate and changes in physical properties between the design case and the supplemental firing case.

To see how to use Equation (7.70), let us simplify the equation by making the ratio of the physical properties = 1; we do expect the contribution from the physical properties ratio to be *significant* and we *will* include the physical properties ratio in all chapter problems and in all future examples. Allowing the simplification here gives

$$\text{Predicted }(UA)_{SF} = (UA)_D\frac{\left(F_{\text{gas\_SF}}\right)^{0.6}}{\left(F_{\text{gas\_}D}\right)^{0.6}} \quad \text{Simplified version.} \quad (7.71)$$

We can use our $(UA)_D$ and $(F_{\text{gas\_}D})$ design results from Example 7.4, as well as our computed $(F_{\text{gas\_SF}})$ for supplemental firing in Example 7.4, to predict values for $(UA)_{SF}$ from Equation (7.71):

$$\begin{aligned}\text{Predicted } UA_{\text{Evaporator\_SF}} &= UA_{\text{Evaporator\_}D}\frac{\left(F_{\text{gas\_SF}}\right)^{0.6}}{\left(F_{\text{gas\_}D}\right)^{0.6}}\\ &= (166.87)\frac{(205.0839)^{0.6}}{(203.7745)^{0.6}}\\ &= 167.51\frac{\text{Btu}}{\text{s-R}}\end{aligned}$$

$$\begin{aligned}\text{Predicted } UA_{\text{Economizer\_SF}} &= UA_{\text{Economizer\_}D}\frac{\left(F_{\text{gas\_SF}}\right)^{0.6}}{\left(F_{\text{gas\_}D}\right)^{0.6}}\\ &= (103.238)\frac{(205.0839)^{0.6}}{(203.7745)^{0.6}}\\ &= 103.636\frac{\text{Btu}}{\text{s-R}}.\end{aligned}$$

We can compare these predicted values for $(UA)_{SF}$ with the values we determined from Example 7.4, where $UA_{\text{Evaporator\_SF}} = 128.419\frac{\text{Btu}}{\text{s-R}}$ and $UA_{\text{Economizer\_SF}} = 103.05\frac{\text{Btu}}{\text{s-R}}$. The predicted value for the economizer is close (0.57%) to the value found from Example 7.4, but the predicted $(UA)_{SF}$ value for the evaporator is really not close (30.4%) to the value found from Example 7.4.

For the overall heat transfer coefficient method, we want the predicted values for $(UA)_{SF}$ to equal the calculated values for $(UA)_{SF}$. The differences in $(UA)_{SF}$ values can be traced to the assumptions made when we solved the supplemental firing case in Example 7.4. Recall in Example 7.4 we assumed that under supplemental firing, T_{7_SF}, the HRSG exit temperature, and $\Delta T_{\text{Approach\_SF}}$ remain fixed at design conditions. These assumptions are questionable and can now be addressed.

Let us assume that the Predicted $UA_{\text{Evaporator\_SF}}$ and Predicted $UA_{\text{Economizer\_SF}}$ values are correct. Note that once we pick T_{6_SF}, this fixes Predicted $UA_{\text{Evaporator\_SF}}$ and Predicted $UA_{\text{Economizer\_SF}}$ values when using Equation (7.71), as $F_{\text{gas\_SF}}$ is uniquely determined from T_{6_SF}, and the other values needed in Equation (7.71) are determined from the HRSG design case (which does not change).

When we calculate $UA_{\text{Evaporator\_SF}}$ and $UA_{\text{Economizer\_SF}}$ as in Example 7.4, values on the gas side depend on T_{6_SF}, T_{7P_SF}, and T_{7_SF}, and, on the water side, T_9, T_{8P_SF}, and T_8. Here, both T_{7_SF} and T_{8P_SF} ($T_{8P\_SF} = T_9 - \Delta T_{\text{Approach\_SF}}$) have been fixed at design case values. If we do allow T_{7_SF} and $\Delta T_{\text{Approach\_SF}}$ (or T_{8P_SF}) to vary, we can still solve the HRSG supplemental firing case where calculated and predicted $(UA)_{SF}$ values are equal.

Summary HRSG Off-Design Procedure Using Overall Heat Transfer Coefficient Approach

We can detail the needed steps here:

Step 1: Solve the design and supplemental firing problems using the overall energy balance approach as we did in Example 7.4. Then determine $(\Delta T_{LMTD})_D$, $(\Delta T_{LMTD})_{SF}$, $(UA)_D$, and $(UA)_{SF}$ for the evaporator and economizer sections of the HRSG. Here, $T_{7\_SF} = T_7$ and $\Delta T_{Approach\_SF} = \Delta T_{Approach}$; the exit temperature from the gas side of the HRSG and the approach temperature difference remain at the same values for both the design and supplemental firing cases.

Step 2: Use values from step 1 for $(UA)_D$, (F_{gas_D}), (F_{gas_SF}) and Equation (7.70) to solve for Predicted $UA_{Evaporator\_SF}$ and Predicted $UA_{Economizer\_SF}$.

Step 3: For the *evaporator section*, pick a new value for $\Delta T_{Approach\_SF}$ (with $T_{8P\_SF} = T_9 - \Delta T_{Approach\_SF}$) and solve for new calculated values for $\Delta T_{LMTD,Evaporator\_SF}$, $\dot{Q}_{Evaporator\_SF}$, and $UA_{Evaporator\_SF}$. Vary $\Delta T_{Approach\_SF}$ (or T_{8P_SF}) until the calculated and predicted values for $UA_{Evaporator\_SF}$ are equal; here, Excel Solver can be used to minimized the absolute value difference in these values by varying $\Delta T_{Approach\_SF}$.

Step 4: For the *economizer section*, pick a new value T_{7_SF} and solve for new calculated values for $\Delta T_{LMTD,Economizer\_SF}$, $\dot{Q}_{Economizer\_SF}$, and $UA_{Economizer\_SF}$. Vary T_{7_SF} until the calculated and predicted values for $UA_{Economizer\_SF}$ are equal; here, Excel Solver can be used to minimized the absolute value difference in these values by varying T_{7_SF}.

The new value for T_{7_SF} will produce a new F_{steam_SF}, as an overall energy balance on the HRSG is used to determine F_{steam_SF}. The new flow rate of steam will change $\dot{Q}_{Evaporator\_SF}$ and $\dot{Q}_{Economizer\_SF}$ values. Return to step 3 and repeat steps 3 and 4 if the calculated and predicted $(UA)_{SF}$ values are not close (one or two iteration loops is generally sufficient).

Step 5: After iterations of steps 3 and 4, the calculated and predicted $(UA)_{SF}$ values should be close. But there will still be some differences due to the changing F_{steam_SF}. A new objective function can be formed as the absolute value of the difference between the calculated and predicted values for $UA_{Evaporator\_SF}$ + the absolute value of the difference between the calculated and predicted values for $UA_{Economizer\_SF}$; here, both $\Delta T_{Approach\_SF}$ and T_{7_SF} are allowed to vary using Solver. Step 5 should not be used until the difference values from steps 3 and 4 are close (see Example 7.5).

Step 6: Use the HRSG temperatures determined in step 5 to update exhaust gas physical properties in Equation (7.70) and obtain updated values for Predicted $UA_{Evaporator\_SF}$ and Predicted $UA_{Economizer\_SF}$. Repeat steps 3–6 until the HRSG temperatures (in step 5) remain unchanged.

> In the overall heat transfer coefficient approach developed here, we have used the exponent on the gas flow rate of 0.6 as $(F_{gas})^{0.6}$. In the commercial code GateCycle™ for HRSG off-design performance, the default exponent on the gas flow rate is 0.8, but the user can change this value. In practice, this exponent should be varied to best match existing HRSG performance.

We can now apply the HRSG off-design procedure using the overall heat transfer coefficient to Example 7.4.

EXAMPLE 7.5 ***HRSG Off-Design Performance Based on Overall Heat Transfer Coefficient: Supplemental Firing***

Here we want to resolve Example 7.4 using the overall heat transfer coefficient method.

SOLUTION The solution is provided in **Example 7.5.xls**. As detailed earlier, step 1 solves the design and supplemental firing problems using the overall energy balance approach as we did in Example 7.4. We determine $(\Delta T_{LMTD})_D$, $(\Delta T_{LMTD})_{SF}$, $(UA)_D$, and $(UA)_{SF}$ for the evaporator and economizer sections of the HRSG. From Example 7.4 we determined $F_{gas\_D} = 203.77$ lb/s, and supplemental firing raised the temperature entering the HRSG to $T_{6\_SF} = 1870$ R and $F_{gas\_SF} = 205.08$ lb/s. Results from Example 7.4 (**step 1**) are repeated here:

		ΔT_a	ΔT_p	T_{7P}	T_7	T_{8P}	$\dot{Q}_{Evap}$
Step 1	D	15	15	889	790.1	859	28,331.3

$\dot{Q}_{Econ}$	$\Delta T_{LM,Evap}$	$\Delta T_{LM,Econ}$	UA_{Evap}	UA_{Econ}	F_{steam}
5642.3	169.78	54.65	166.87	103.24	34.17

and

		ΔT_{a_SF}	ΔT_{p_SF}	T_{7P_SF}	T_{7_SF}	T_{8P_SF}	$\dot{Q}_{Evap\_SF}$
Step 1	SF	15	95.46	969.5	790.1	859	51,712

$\dot{Q}_{Econ\_SF}$	$\Delta T_{LM,Evap\_SF}$	$\Delta T_{LM,Econ\_SF}$	UA_{Evap_SF}	UA_{Econ_SF}	F_{steam_SF}
10,298.7	402.68	99.94	128.42	103.05	62.37

Step 2: From the overall energy balance approach solution and using Equation (7.70) or (7.71) as we do here, solve for the predicted $(UA)_{SF}$. Here,

$$\text{Predicted } UA_{Evaporator\_SF} = 167.51 \frac{\text{Btu}}{\text{s-R}}$$

and

$$\text{Predicted } UA_{Economizer\_SF} = 103.64 \frac{\text{Btu}}{\text{s-R}}.$$

Step 3: Vary $\Delta T_{\text{Approach\_SF}}$ until the calculated value for $UA_{\text{Evaporator\_SF}}$ (as determine from an energy balance on either the gas or water side of the HRSG evaporator section) equals the Predicted $UA_{\text{Evaporator\_SF}}$. This (step 3) gives

		ΔT_{a_SF}	ΔT_{p_SF}	T_{7P_SF}	T_{7_SF}	T_{8P_SF}	$\dot{Q}_{\text{Evap\_SF}}$
Step3	SF	61.81	41.40	915.40	790.1	812.18	54,816

$\dot{Q}_{\text{Econ\_SF}}$	$\Delta T_{\text{LM,Evap\_SF}}$	$\Delta T_{\text{LM,Econ\_SF}}$	$UA_{\text{Evap\_SF}}$	$UA_{\text{Econ\_SF}}$	$F_{\text{steam\_SF}}$
7194.6	327.24	71.94	167.51	100.0	62.37

Step 4: Vary T_{7_SF} until the calculated value for $UA_{\text{Economizer\_SF}}$ (as determined from an energy balance on either the gas or water side of the HRSG economizer section) equals the Predicted $UA_{\text{Economizer\_SF}}$. This (step 4) gives

		ΔT_{a_SF}	ΔT_{p_SF}	T_{7P_SF}	T_{7_SF}	T_{8P_SF}	$\dot{Q}_{\text{Evap\_SF}}$
Step 4	SF	61.81	39.2	913.2	787.6	812.18	54,942

$\dot{Q}_{\text{Econ\_SF}}$	$\Delta T_{\text{LM,Evap\_SF}}$	$\Delta T_{\text{LM,Econ\_SF}}$	$UA_{\text{Evap\_SF}}$	$UA_{\text{Econ\_SF}}$	$F_{\text{steam\_SF}}$
7211.1	323.53	69.58	169.82	103.64	62.51

Step 5: Because the predicted and calculated values of UA_{Evap} and UA_{Econ} determined in step 4 are close, we can use step 5 without iteration on steps 3 and 4. Vary both $\Delta T_{\text{Approach\_SF}}$ and T_{7_SF} until the calculated value for $UA_{\text{Evaporator\_SF}} + UA_{\text{Economizer\_SF}}$ (as determined from an energy balance on either the gas or water side of the appropriate HRSG section) equals the Predicted $UA_{\text{Evaporator\_SF}}$ + Predicted $UA_{\text{Economizer\_SF}}$. This (step 5) gives

		ΔT_{a_SF}	ΔT_{p_SF}	T_{7P_SF}	T_{7_SF}	T_{8P_SF}	$\dot{Q}_{\text{Evap\_SF}}$
Step5	SF	59.81	41.4	915.4	787.5	814.19	54,816

$\dot{Q}_{\text{Econ\_SF}}$	$\Delta T_{\text{LM,Evap\_SF}}$	$\Delta T_{\text{LM,Econ\_SF}}$	$UA_{\text{Evap\_SF}}$	$UA_{\text{Econ\_SF}}$	$F_{\text{steam\_SF}}$
7343.4	327.24	70.82	167.51	103.69	62.52

Step 6 is required when exhaust gas physical properties are included; see Problem 7.3. The results in step 5 show that both $\Delta T_{\text{Approach\_SF}}$ (59.81 R) and $\Delta T_{\text{Pinch\_SF}}$ (41.4 R) increase in value compared to the design case (both were set at 15 R). Also, the exhaust gas temperature from the HRSG decreases from 790.1 to 787.5 R. When the gas-side temperature entering the HRSG increases, the approach and pinch temperature differences increase and the stack gas temperature decreases (compared to the design case). It is instructive to compare the supplemental firing results from Examples 7.4 and 7.5 (step 1 SF and step 5 SF). The differences in steam produced (62.3738 vs. 62.5235 lb/s) is about 540 lb/h and the difference in the exhaust gas temperature (790.1 vs. 787.5 R) is about 2.5 R; these are not large differences. However, there are significant differences in the predicted temperatures at the pinch point of the HRSG. ■

In general, when solving supplemental firing problems, T_{6_SF} will not be fixed; instead, the $F_{\text{steam\_SF}}$ will be set by process demands. We do not need to change our solution procedure, but rather we simply add an iteration loop in which T_{6_SF} is varied until the required steam production is met. The equations developed in Section 7.8 focused on the HRSG off-design performance case of supplemental firing. The same procedures developed in this section can be applied to other off-design cases; for example, see Problem 7.4.

7.9 GAS TURBINE DESIGN AND OFF-DESIGN PERFORMANCE

We have determined gas turbine performance based on thermodynamic changes across each component in the turbine. In general, there are two approaches in predicting gas turbine off-design performance: The first involves the use of manufacturer-provided performance curves, and the second uses a general approach in which off-design performance equations are developed based on thermodynamic considerations as well as an understanding of turbine internal fluid flows and the turbine control algorithm. Software programs including GateCycle, IPSEpro, and GasTurb provide for off-design turbine performance using these type general equations and/or the use of manufacturer-provided correction (or performance) curves, which are specific for each turbine. These correction curves can actually be used independently of any software to predict turbine performance at off-design conditions. For example, typical engine correction curves provide the change in turbine exhaust temperature, exhaust flow rate, and fuel flow as the system is moved from design power to reduced-load operations.

7.9.1 Gas Turbines Types and Gas Turbine Design Conditions

There are two basic gas turbine types for heat and power generation: a heavy-duty gas turbine (also called an industrial or frame turbine) and an aeroderivative gas turbine, which is based on aircraft turbine technology. In general, at design conditions (discussed next), industrial turbines show a lower overall efficiency when compared to aeroderivative gas turbines. However, industrial gas turbine off-design performance is less sensitive to ambient and operating conditions when compared to aeroderivative gas turbines, and generally, industrial gas turbines require less maintenance.

Within these gas turbine types, several shaft configurations are possible. In a single-shaft gas turbine configuration (see Figure 7.14), a common shaft from the turbine drives both the air compressor and the generator; these units must operate at the same rotational speed. We have explored a two-shaft configuration (Figure 7.5) where the high-pressure gas turbine (gas generating turbine) drives the air compressor and the low-pressure power turbine drives the generator; here, the gas generating turbine and the power turbine can operate at different speeds. Other configurations with two and three shafts are available.

We have also seen that different gas turbine system configurations are available. Figure 7.14 shows a simple-cycle single-shaft gas turbine and, if you look ahead, Figure 9.1 shows a gas turbine system with air preheat. Any gas turbine can be operated in a combined cycle if heat is recovered from the turbine exhaust. The article by Poullikkas (2005) reviews current and emerging gas turbine system configurations.

Gas turbine performance at standard design conditions establishes a benchmark that must be known before off-design calculations can be performed. Standard design conditions require air feed to the gas turbine at 59°F/15°C, 14.7 psia/1.013 bar, and 60% relative humidity—these conditions are commonly referred to as the ISO conditions. Using these air feed conditions, reported gas turbine design performance from the manufacture will include net power, exhaust flow rate, heat rate, and heat consumption (discussed further).

7.9.2 Gas Turbine Design and Off-Design Using Performance Curves

Gas turbine performance at off-design conditions will often be given as a percent change from the design performance in manufacturer-provided performance curves (often called correction curves). For example, changes in gas turbine performance with ambient temperature can be determined using the performance curve shown in Figure 7.11.

EXAMPLE 7.6 ***Gas Turbine Off-Design Performance Using Performance Curves***

Allow that the gas turbine used to generate Figure 7.11 shows design performance as indicated in Table 7.2 (column 2). Using Figure 7.11, estimate gas turbine performance if the ambient air temperature is increased from 59°F (design) to 80°F. Also at design conditions and off-design conditions (80°F), how much natural gas (LHV = 21,500 Btu/lb) is being used by the gas turbine?

SOLUTION Off-design performance is shown in the third column of the table.

Using a natural gas LHV of 21,500 Btu/lb: For design conditions, fuel consumption is (728,041,600)/(21,500) = 33,862.4 lb/h (~9.4 lb/s) and for off-design conditions, 32,169.28 lb/h.

Inlet humidity, altitude (ambient pressure), and steam or water addition in the combustion chamber all effect gas turbine performance, and each can be given by a correction curve similar to that shown in Figure 7.11. These charts are machine specific and require design performance values. ■

7.9.3 Gas Turbine Internal Mass Flow Patterns

Developing general equations for off-design gas turbine performance calculations begins with the understanding that there are two basic gas turbine types: an industrial gas turbine (sometimes called a heavy-duty frame turbine) and an aeroderivative gas turbine. Gas turbines, for example, the GE LMS 100 shown in Figure 7.12, can even be combination frame and aeroderivative technologies. Manufacturers have developed sophisticated control algorithms for each gas turbine type to maximize turbine efficiency as operation

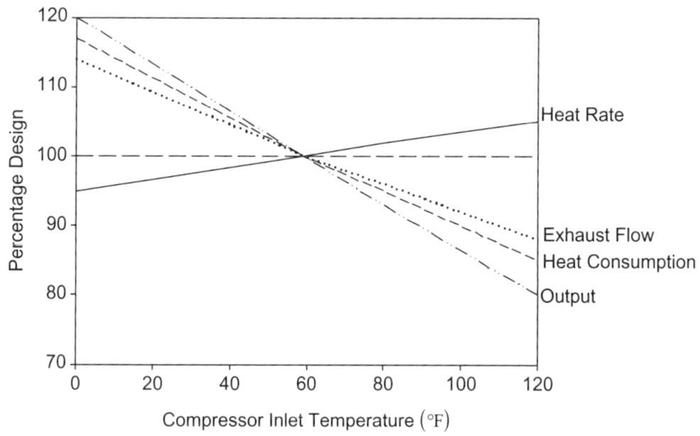

Figure 7.11 Typical effect of ambient temperature on a single-shaft industrial gas turbine.

Table 7.2 Turbine Design and Off-Design Performance Based on Figure 7.11

Turbine Performance	Design Conditions	Off-Design—80°F Compressor Inlet T
Output (kW)	65,120	≈ (65,120)(0.93) = 60,562
Heat rate (Btu/kW-h)	11,180	≈ (11,180)(1.02) = 11,404
Heat consumption (Btu/h)	728,041,600	≈ (728,041,600)(0.95) = 691,639,520
Exhaust flow (lb/h)	2,024,000	≈ (2,024,000)(0.96) = 1,943,040

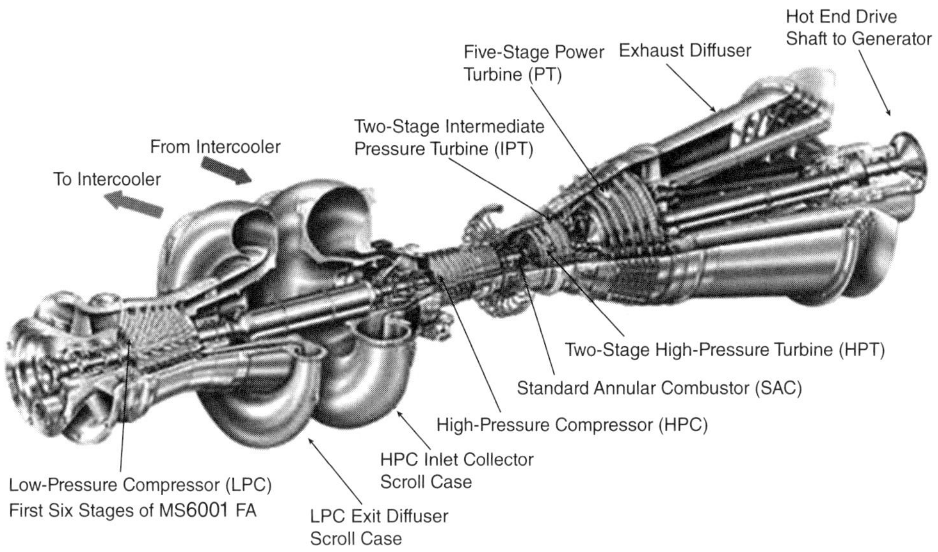

Figure 7.12 GE LMS 100 gas turbine (Reale and Prochaska, 2005; used with permission).

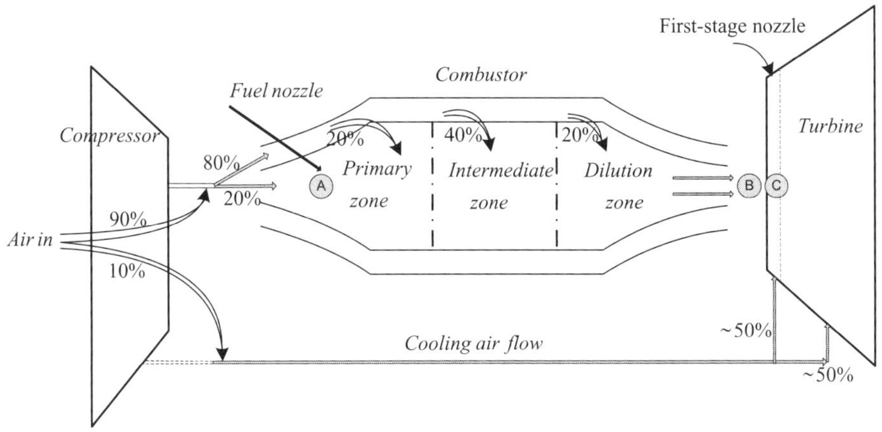

Figure 7.13 Annular combustor fuel and air flow.

moves away from design conditions. We must understand these control algorithms as well as mass flow patterns and temperatures within the gas turbine to predict off-design performance.

Consider a standard annular combustor, which is depicted in both Figures 7.13 and 7.14. In our previous gas turbine calculations, we used (or determined) the temperature leaving the combustion chamber (T_3 in Figure 7.14) with an energy balance (e.g., see Eq. (7.45)), which involved the total mass flow rate of air entering the compressor and fuel flow rate. A more accurate depiction of flows within the gas turbine is provided in Figure 7.13. Here, ~90% of the air from the compressor at (T_2, P_2) is actually sent to the combustor with the remaining ~10% used as cooling flow for the turbine. Within this ~10% cooling flow, ~½ is immediately mixed with the exit gas from the combustor at the turbine first-stage nozzle inlet with the remaining ~½ distributed to cool the turbine blades. For the actual combustor, a large portion of the incoming air ~80% is used in an annular passage to help cool the combustor. This cooling air is gradually introduced in different zones (primary, intermediate, and dilution) to help lower the temperature of the gas exiting the combustor. The zones will be important when we use a combination of elementary reactions, perfectly stirred reactors, and plug flow reactors to predict emissions from gas turbines in Chapter 15.

There is no uniform agreement in the terms used to define the different temperatures in the combustor. At point A in Figure 7.13, the temperature would be near the adiabatic flame temperature of the fuel. Point B represents the combustor exit temperature; point C, which immediately follows the first-stage nozzle, is the turbine rotor inlet temperature or the firing temperature. This turbine inlet

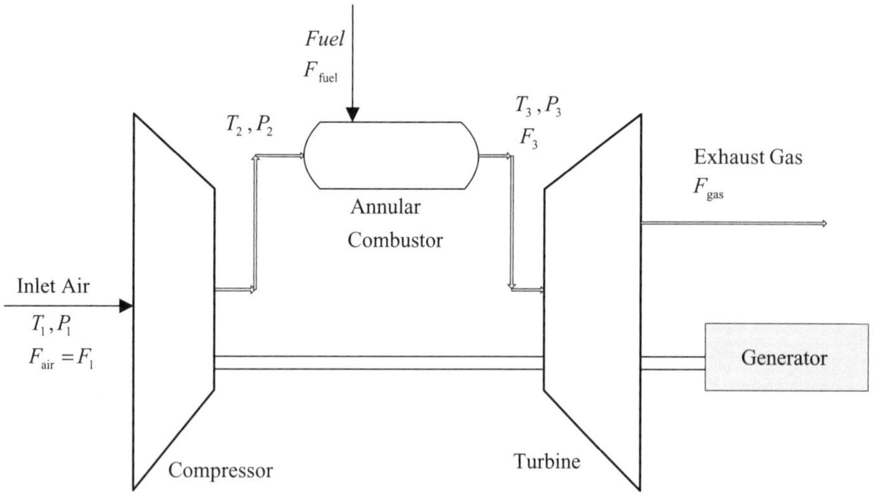

Figure 7.14 Simple-cycle single-shaft gas turbine.

temperature (point C) or ISO firing temperature is often considered the temperature that would result if the combustor exit was immediately combined with all the cooling air or, in other words, if no cooling air was extracted and all incoming air to the combustor was used in the combustion process.

We will see these temperature details are important for gas turbine design and off-design performance calculations. We will also need to account for the actual temperature profile in the combustor in order to predict emissions in Chapter 15. For our current performance discussion, we will use T_3 (see Figure 7.14) as the turbine inlet temperature. This T_3 will also be loosely defined as the combustor exit temperature as well as the rotor inlet temperature, $T_3 \approx T_B \approx T_C$. The distinction between T_3, T_B, and T_C will not be important for the simple off-design algorithm presented next, but in more rigorous calculations, the temperature and flow rate entering the first-stage nozzle of the turbine must be known (this is discussed further).

7.9.4 Industrial Gas Turbine Off-Design (Part Load) Control Algorithm

We have indicated that the gas turbine type will dictate the off-design control strategy and this control strategy impacts the general equations we will use to predict off-design performance. An excellent discussion of off-design control for industrial turbines is provided in the book by Gay et al. (2004), which is repeated here (with permission):

Industrial gas turbines and combined cycles often need to operate at below their base-loaded conditions. There are two distinct methods for part-loading engines. The first is called fuel control. In this method, the fuel flow is reduced below the base-load value (as determined by the control system) until the desired load is met. The second method is called guide-vane control. In this method the airflow into the gas turbine is reduced by closing the first set of non-rotating vanes (inlet guide vanes) at the intake of the compressor. Turning these inlet guide vanes allows the gas turbine to reduce its power while maintaining a high firing temperature. Guide-vane control also maintains a high exhaust temperature, which is necessary for HRSG operation. Because of this, guide-vane control is the preferred method for part-loading cogeneration and combined cycle plants. Part-loading an industrial engine on guide vane control occurs in a number of steps. A generic sequence of these is shown in Figure 7.15, and described below.

Starting at the base load condition (denoted A in Figure 7.15), the engine will slightly decrease its firing temperature as soon as control is turned off of the base-load control curve. This initial temperature drop is implemented for engine protection, and not efficiency purposes. The initial drop in firing temperature causes an exhaust temperature drop of a few (approximately 5°F to 15°F) degrees (B on Figure 7.15). When further decreases in load are requested, the engine starts to close its inlet guide vanes. As the vanes turn, airflow decreases, causing engine pressure ratio to decrease. As the control algorithm tries to maintain a constant firing temperature, the exhaust temperature increases with the decreasing pressure ratio. At point C, the exhaust temperature has reached its maximum allowed value. Further reductions in power require simultaneous reduction in firing temperature and airflow (through turning the guide vanes) so that the exhaust temperature does not exceed its limit. At point D, the guide vanes have reached their 'fully closed' position. This occurs somewhere between 60–70% of the base-load power. Reducing power below this point can only be accomplished by reducing firing temperature. Some engine types may not reach the maximum exhaust temperature limit, particularly at colder ambient temperatures. In this situation guide vane control can be used until the guide vanes are fully closed.

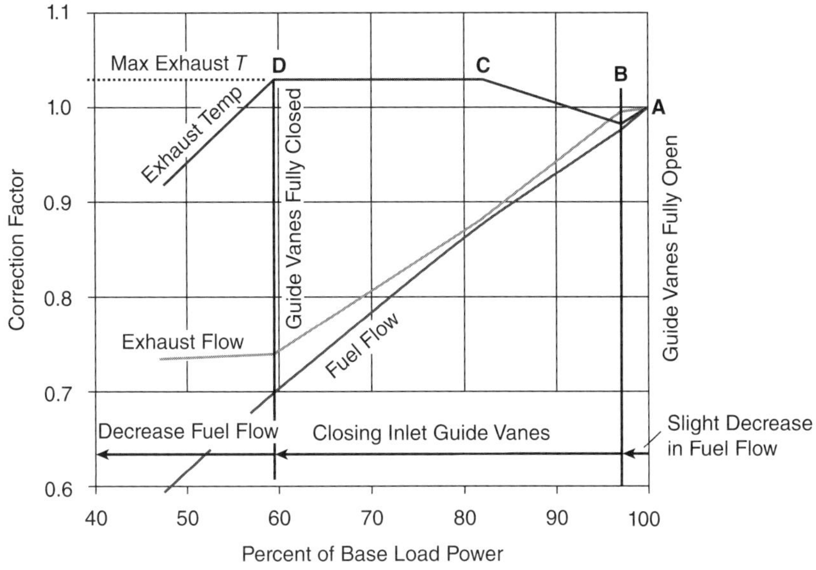

Figure 7.15 Typical part load performance using guide vane control (Gay et al., 2004).

Most industrial engines generally follow most of the part-load sequence detailed above. However, some do not have the initial temperature drop (A-B). Others maintain the value of the exhaust temperature at base load by not allowing the exhaust temperature to rise (B-C) through simultaneously reducing firing temperature and turning the guide vanes (as in segment C-D).

7.9.5 Aeroderivative Gas Turbine Off-Design (Part Load) Control Algorithm

An aeroderivative gas turbine adds several layers of control to the control algorithm described for the industrial gas turbine. In an aeroderivative turbine; for example, some or all sections of the compressor are free spinning and change revolutions per minute with operating conditions. Here then, limits on compressor discharge temperature and pressure become important in the control strategy.

7.9.6 Off-Design Performance Algorithm for Gas Turbines

Accepting the complexities of gas turbine off-design performance, it is possible to assemble equations that show the factors affecting performance. These equations, in combination with the turbine control strategy, allow the development of a simple off-design performance algorithm for both industrial and aeroderivative gas turbines. Gay and Erbes (M.R. Erbes, pers. comm.) used such an approach within GateCycle for user-defined engines. In this strategy, the turning of the inlet guide vanes (IGVs) affects the mass flow to the compressor as well as the compressor efficiency. The compressor discharge pressure is determined by assuming the turbine-swallowing capacity (defined next) remains constant (choked flow) between the design and off-design operation—this will require a simple iterative calculation based on the temperature in the combustion chamber. Remaining properties are determined using standard thermodynamic relations we have previously developed:

1. First, we account for the change in mass flow to the compressor with the change in IGV angle. The compressor flow correction is

$$F_{\text{air}}^{off\_d} = F_{\text{air}}^{d}\left(\frac{P_1^{off\_d}}{P_1^{d}}\right)\left(\frac{T_1^{d}}{T_1^{off\_d}}\right)(1-(\Delta\text{angle})(VFC))\left[1+TFC\left(\frac{T_1^{off\_d}-T_1^{d}}{T_1^{d}}\right)\right]. \tag{7.72}$$

Here using Figure 7.14,

off_d	= indicates off-design value,
d	= indicates design value,
F_{air}	= F_1 = mass flow rate into the compressor,
P_1	= absolute pressure into the compressor,
T_1	= absolute temperature into the compressor,
Δangle	= change in IGV angle from design conditions,
VFC	= vane flow correction factor, and
TFC	= temperature flow correction factor.

2. The air compressor efficiency will change with the inlet mass flow rate as well as the compressor speed. The compressor efficiency correction can be given by,

$$\eta_{AC}^{off\_d} = \eta_{AC}^{\max}\left(1-\left|\frac{F_{air}^{d}-F_{air}^{off\_d}}{F_{air}^{d}}\right|(FC)\right)\left[1+SEC\left|\frac{CS_{off\_d}-CS_{\max\eta}}{CS_{\max\eta}}\right|\right]. \quad (7.73)$$

Here,

η^{AC} = air compressor efficiency,
max = denotes values at maximum,
off_d = indicates off-design value,
d = indicates design value,
F_{air} = F_1 = mass flow rate into the compressor,
FC = mass flow correction factor,
SEC = compressor speed efficiency correction factor, and
CS = compressor speed.

The maximum air compressor efficiency, $\eta_{AC}^{\max}$, may not be at the design point, but we will assume $\eta_{AC}^{\max} = \eta_{AC}^{d}$.

3. The first-stage nozzle of the turbine is generally choked; in other words, the swallowing capacity of the turbine is constant for both design and off-design cases. The relation for the flow, temperature, and pressure into the turbine can then be written,

$$\left(\frac{F_3^d\sqrt{T_3^d}}{P_3^d}\right)_{\text{turbine nozzel in}} = \text{constant} = \left(\frac{F_3^{off\_d}\sqrt{T_3^{off\_d}}}{P_3^{off\_d}}\right)_{\text{turbine nozzel in}}. \quad (7.74)$$

Here using Figure 7.14,

off_d = indicates off-design value,
d = indicates design value,
F_3 = mass flow rate into the turbine,
P_3 = absolute pressure into the turbine, and
T_3 = absolute temperature into the turbine.

For our purposes, and following our discussion of Section 7.9.3, we will assume that $F_3 = F_{gas}$ and that $T_3 = T_B = T_C$ (in Figure 7.13). For a complete development of Equation (7.74), including the case where the turbine is not choked, see Streeter and Wylie (1979).

You will recall that our control strategy may change the IGV angle, which will change the mass flow rate into the compressor, and that the control strategy may also change the temperature in the combustor to prevent the turbine exhaust temperature limit from being exceeded. Here then, we will know $T_3^{off\_d}$ and an estimate for $F_3^{off\_d}$ allowing us to solve for $P_3^{off\_d}$ in Equation (7.74), which can be taken as the pressure out of the compressor (or adjusted if there is a pressure drop in the combustion chamber). We can then calculate the temperature leaving the compressor using Equation (7.32) with the compressor off-design efficiency as given by Equation (7.73). The compressor outlet temperature will affect the fuel required to reach the combustor temperature, and this will impact $F_3^{off\_d}$, so an iterative calculation will be required.

4. Finally, a reasonable assumption is that the turbine efficiency will not change in an off-design operation, allowing us to calculate the turbine exhaust temperature (Eq. (7.39)) as well as the power generated.

We next examine the use of Equations (7.72)–(7.74) to predict gas turbine performance for two off-design cases: first where the inlet air to the compressor is increased from design conditions and a second case where the turbine required power generation is reduced. In Example 9.7, we solve the same problem using real gas properties (Chapter 8) and performance equation as developed in Chapter 9.

Design Case Example 7.7a.xls

EXAMPLE 7.7 *Gas Turbine Design and Off-Design Performance Ideal Gas Physical Properties*

In Table 7.3 and in the column 2, we provide design performance data for a 21.8-MW gas turbine as well as physical properties for the system. The gas turbine is shown in Figure 7.14. We will want to modify our solution in Example 7.2 (with water/steam functions added) in order to also examine HRSG off-design performance. The needed modification to Example 7.2 (Figure 7.10) is setting $T_3 = T_4$ and $P_3 = P_4$ in order to reduce the system to a single turbine.

Off-Design Case 1

We want to predict the performance of the system using Equations (7.72)–(7.74) if the ambient air temperature is raised from design conditions 15–30°C.

Off-Design Case 2

We also want to predict system performance if the turbine power requirements are reduced 25% from 21,807.7 kJ/s (design) to 16,355.78 kJ/s when the ambient air temperature is 15°C.

For the off-design cases, the turbine exhaust temperature should not exceed 110% of the design value.

In order to use Equations (7.72)–(7.74), additional information, including various system specific correction factors, is needed. For Equation (7.72), the IGV angle can change from 85° (design) to a minimum of 55° (this would be a maximum Δangle = 30°). The change in flow rate with change in IGV angle is accounted for by the vane flow correction factor, and here, assume *VFC* = 0.0135 per degree change. In this example, there is no change in ambient pressure and we will ignore the temperature flow correction factor (*TFC*), allowing Equation (7.72) to be written,

$$F_{air}^{off\_d} = 50\left(\frac{T_1^d}{T_1^{off\_d}}\right)(1-(\Delta\text{angle})(0.0135)), \frac{\text{kg}}{\text{s}}.$$

Table 7.3 Design and Off-Design Turbine Performance Calculations Using Ideal Gas Properties

Variable/Parameter	Design	Off-Design 1	Off-Design 2
Air gamma, γ_a	1.38	1.38	1.38
Air specific heat capacity, $C_{p,\text{air}}$ (kJ/kg-K)	1.046	1.046	1.046
Compressor inlet guide vane angle change	0	0	9.1425
Compressor inlet flow, F_1 (kg/s)	50	47.526	43.829
Compressor inlet temperature, T_1 (K)	288.15	303.15	288.15
Compressor inlet pressure, P_1 (bar)	1.013	1.013	1.013
Compressor isentropic efficiency, η_{AC}	0.89	0.8397	0.7644
Compressor outlet temperature, T_2 (K)	659.09	705.833	691.054
Compressor outlet pressure, P_2 (bar)	16.208	15.391	14.198
Compressor work (kJ/s)	−19,400	−20,018.26	−18,471.10
Fuel lower heating value, *LHV* (kJ/kg)	50,010	50,010	50,010
Fuel flow, F_{fuel} (kg/s)	1.1731	1.0671	0.9981
Exhaust gas gamma, γ_g	1.31	1.31	1.31
Exhaust heat capacity, $C_{p,g}$ (kJ/kg-K)	1.237	1.237	1.237
Turbine inlet flow, F_g (kg/s)	51.1731	48.5931	44.8269
Turbine inlet temperature, T_3 (K)	1523.15	1523.15	1523.15
Turbine inlet pressure, P_3 (bar)	15.560	14.775	13.630
Turbine isentropic efficiency, η_{PT}	0.90	0.90	0.90
Turbine outlet temperature, T_5 (K)	872.171	881.038	895.083
Turbine outlet pressure, P_5 (bar)	1.023	1.023	1.023
Turbine work (kJ/s)	41,207.701	38,597.167	34,826.878
Turbine net work (kJ/s)	21,807.701	18,578.908	16,355.78
Heat rate (kJ/kW-h)	9684.688	10,340.93	10,986.51
Fuel flow correction, $F_{\text{fuel}}/F_{\text{fuel,design}}$	1.0	0.90967	0.85082
Exhaust flow correction, $F_g/F_{g,\text{design}}$	1.0	0.94958	0.87599
Exhaust temp correction, $T_5/T_{5,\text{design}}$	1.0	1.0102	1.02627

A change in the mass flow rate to the air compressor as well as the compressor speed will affect the compressor efficiency as given by Equation (7.73). For this example, we will ignore the compressor speed efficiency factor (*SEC*), set the maximum efficiency to be design efficiency ($\eta_{\max} = \eta_d$), and for the mass correction factor set as $FC = 1.143$. These assumptions allow Equation (7.73) to be written,

$$\eta_{AC}^{off\_d} = \eta_{AC}^{d}\left(1 - \left|\frac{F_{\text{air}}^{d} - F_{\text{air}}^{off\_d}}{F_{\text{air}}^{d}}\right|(FC)\right)$$
$$= 0.89\left(1 - \left|\frac{50 - F_{\text{air}}^{off\_d}}{50}\right|(1.143)\right).$$

Equation (7.74) utilizes the temperature and flow into the first-stage nozzle of the turbine. As we have discussed in Section 7.9.3, the actual flow and temperature into the first-stage nozzle of the turbine are not known. For our purposes, we will simply use T_3 in Equation (7.74).

SOLUTION Performance results for the design case and the two off-design cases are shown in Table 7.3. The solution for the design case performance is provided in **Example 7.7a.xls**; off-design case 1 is provided in **Example 7.7b.xls**, and off-design case 2 is provided in **Example 7.7c.xls**.

All solutions begin with the Excel file **Example 7.2a.xls** with sheet functions added for steam and water properties. In this file, we simply set (T_4, P_4) = (T_3, P_3) as we have eliminated the gas generating turbine.

For the design case, key variables include $F_{\text{air}} = F_1 = 50$ kg/s; $\eta_{AC} = 0.89$; $P_2/P_1 = 16$; 4% pressure drop in the combustion chamber; $\eta_{CC} = 100\%$ combustion chamber efficiency; $T_3 = 1523.15$ K (1250°C); and $\eta_{PT} = 0.90$. The main performance equations include the following:

For the air compressor,

$$T_2 = T_1\left\{1 + \frac{1}{\eta_{AC}}\left[\left(\frac{P_2}{P_1}\right)^{\frac{(\gamma_a - 1)}{\gamma_a}} - 1\right]\right\}$$
$$= 288.15\left\{1 + \frac{1}{0.89}\left[\left(\frac{16.208}{1.013}\right)^{\frac{(1.38-1)}{1.38}} - 1\right]\right\} = 659.09 \text{ K}.$$

For the combustion chamber,

$$P_3 = (16.208)(0.96) = 15.56 \text{ bar}$$

$$F_{\text{fuel}} = \frac{F_{\text{air}}\hat{C}^{ig}_{P,\text{gas}}(T_3 - T_{\text{ref}}) - F_{\text{air}}\hat{C}^{ig}_{P,\text{air}}(T_2 - T_{\text{ref}})}{\left((LHV)(\eta_{CC}) - \hat{C}^{ig}_{P,\text{gas}}(T_3 - T_{\text{ref}})\right)}$$
$$= \frac{(50)(1.237)(1523.15 - 298.15) - (50)(1.046)(659.09 - 298.15)}{((50010)(1) - (1.237)(1523.15 - 298.15))}$$
$$= 1.1731 \text{ kg/s}.$$

And for the power turbine,

$$T_5 = T_4\left\{1 + (\eta_{PT})\left[\left(\frac{P_5}{P_4}\right)^{\frac{(\gamma_g - 1)}{\gamma_g}} - 1\right]\right\}$$
$$= 1523.15\left\{1 + (0.90)\left[\left(\frac{1.023}{15.56}\right)^{\frac{(1.31-1)}{1.31}} - 1\right]\right\} = 872.17 \text{ K}.$$

With this design case solved, we can determine the relation for the flow, temperature, and pressure into the turbine Equation (7.74) as

$$\left(\frac{F_3^d\sqrt{T_3^d}}{P_3^d}\right)_{\text{turbine nozzel in}} = \left(\frac{51.1731\sqrt{1523.15}}{15.56}\right)_{\text{turbine in}}$$
$$= 128.355.$$

Off-Design Case 1 Example 7.7b.xls

For off-design case 1, the ambient air temperature changes from design conditions 15 to 30°C. For Equation (7.72), the only change will be in temperature:

$$F_{\text{air}}^{off\_d} = 50\left(\frac{T_1^d}{T_1^{off\_d}}\right)(1 - (\Delta\text{angle})(0.0135))$$
$$= 50\left(\frac{288.15}{303.15}\right) = 47.526\frac{\text{kg}}{\text{s}}.$$

The air compressor isentropic efficiency can be found as

$$\eta_{AC}^{off\_d} = \eta_{AC}^d\left(1 - \left|\frac{F_{\text{air}}^d - F_{\text{air}}^{off\_d}}{F_{\text{air}}^d}\right|(FC)\right)$$
$$= 0.89\left(1 - \left|\frac{50 - 47.526}{50}\right|(1.143)\right) = 0.8397.$$

Next, for the air compressor, we can estimate $T_2^{off\_d}$ as

$$T_2^{off\_d} = T_1^{off\_d}\left\{1 + \frac{1}{\eta_{AC}^{off\_d}}\left[\left(\frac{P_2^{off\_d}}{P_1^{off\_d}}\right)^{\frac{(\gamma_a - 1)}{\gamma_a}} - 1\right]\right\}$$
$$= 303.15\left\{1 + \frac{1}{0.8397}\left[\left(\frac{16.208}{1.013}\right)^{\frac{(1.38-1)}{1.38}} - 1\right]\right\}$$
$$= 716.79 \text{ K}.$$

For the combustion chamber,

$$P_3^{off\_d} = (16.208)(0.96) = 15.56 \text{ bar}$$

$$F_{\text{fuel}}^{off\_d} = \frac{F_{\text{air}}^{off\_d}C^{ig}_{P,\text{gas}}(T_3^{off\_d} - T_{\text{ref}}) - F_{\text{air}}^{off\_d}C^{ig}_{P,\text{air}}(T_2^{off\_d} - T_{\text{ref}})}{\left((LHV)(\eta_{CC}) - C^{ig}_{P,\text{gas}}(T_3 - T_{\text{ref}})\right)}$$
$$= \frac{(47.526)(1.237)(1523.15 - 298.15) - (47.526)(1.046)(716.79 - 298.15)}{((50010)(1) - (1.237)(1523.15 - 298.15))}$$
$$= 1.056 \text{ kg/s}.$$

Here we check if Equation (7.74),

$$\left(\frac{F_3^{off\_d}\sqrt{T_3^{off\_d}}}{P_3^{off\_d}}\right)_{\text{turbine nozzel in}} = 128.355$$

as determined from our design case:

$$\left(\frac{F_3^{off\_d}\sqrt{T_3^{off\_d}}}{P_3^{off\_d}}\right)_{\text{turbine nozzel in}}$$
$$= \left(\frac{(48.582)\sqrt{1523.15}}{(15.56)}\right)_{\text{turbine nozzel in}}$$
$$= 121.855 \neq 128.355.$$

In the provided Excel solution, we use the optimization routine in Solver to iterate on $P_3^{off\_d}$ until Equation (7.74) is solved. Each new value of $P_2^{off\_d}$, will require that a new temperature, $T_2^{off\_d}$, be determined, which will in turn affect the fuel needed to reach 1250°C, and each new value of $P_2^{off\_d}$ will affect the value of $P_3^{off\_d}$ entering the turbine. The objective function in Solver is to minimize:

$$\min\left[\left(\left(\frac{F_3^d\sqrt{T_3^d}}{P_3^d}\right) - \left(\frac{F_3^{off\_d}\sqrt{T_3^{off\_d}}}{P_3^{off\_d}}\right)_{\text{turbine nozzel in}}\right)^2\right].$$

After Equation (7.74) is converged, the power turbine exhaust temperature is determined for as

$$T_5^{off\_d} = T_4^{off\_d}\left\{1+(\eta_{PT})\left[\left(\frac{P_5}{P_4^{off\_d}}\right)^{\frac{(\gamma_g-1)}{\gamma_g}}-1\right]\right\}$$

$$=1523.15\left\{1+(0.90)\left[\left(\frac{1.023}{14.775}\right)^{\frac{(1.31-1)}{1.31}}-1\right]\right\}$$

$$=881.038\text{ K}.$$

A check is made to make sure the turbine exhaust temperature is not violated.

Off-Design Case 2 Example 7.7c.xls

For off-design case 2, the net power is reduced 25% from the design case with ISO ambient conditions. In order to determine the system performance, we must utilize the off-design part load control algorithm (Section 7.9.4). For reduced power, the control algorithm increases the Δangle for the IGVs (Eq. (7.72)), which reduces air flow to the compressor. Reduced air flow calculations then follow those as outlined in off-design case 1. We must modify our Excel-based solution, as developed in off-design case 1, to now include both Δangle and $P_2^{off\_d}$ as variables.

In the provided Excel solution, we use the optimization routine in Solver to iterate on $P_2^{off\_d}$ and Δangle until both Equation (7.74) and the required net power are met. Each new value of Δangle will change the air flow rate (Eq. (7.72)) and air compressor efficiency (Eq. (7.73)). Each new value of $P_2^{off\_d}$ will require a new temperature $T_2^{off\_d}$ be determined. The air flow rate and $T_2^{off\_d}$ will in turn affect the fuel needed to reach 1250°C. Each new value of $P_2^{off\_d}$ will affect the value of $P_3^{off\_d}$ entering the turbine. The objective function in Solver is to minimize:

$$\min\left[\left(\left(\frac{F_3^d\sqrt{T_3^d}}{P_3^d}\right)-\left(\frac{F_3^{off\_d}\sqrt{T_3^{off\_d}}}{P_3^{off\_d}}\right)_{\substack{\text{turbine}\\\text{nozzel in}}}\right)^2+\left(\begin{matrix}(\text{net kW}_d\times 0.75)-\\(\text{net kW}_{off\_d})\end{matrix}\right)^2\right].$$

It is often better in optimization problems to first solve the problem without constraints or bounds and then to check if violations occur—in other words, first solve the problem as unconstrained. There is a bound on the maximum allowable exhaust temperature in this problem, which is not violated here. If the exhaust temperature was violated (following the control algorithm requirements), we would need to include the fuel flow rate as a variable in order to allow a lower exhaust temperature. In this problem, it will be necessary to add a constraint to keep $P_2^{off\_d} \geq P_1$.

In the two off-design cases of Example 7.7, we did not address changes in steam flow and temperatures in the HRSG. Using the HRSG overall energy balance approach with Equation (7.61),

$$F_{\text{steam\_SF}} = \frac{(F_{\text{gas\_SF}})\hat{C}_{P,\text{gas}}^{ig}(T_{6\_\text{SF}} - T_{7\_\text{SF}})}{\hat{h}_{(T_9,P_9)}^{\text{sat vap}} - \hat{h}_{(T_8,P_8)}^{\text{liquid}}},$$

and without supplemental firing, $F_{\text{gas\_SF}}$ is simply the F_{gas} in the off-design case. For off-design case 1,

$$F_{\text{steam}}^{off\_d} = \frac{(48.5931)(1.237)(881.038-422.301)}{(2797.25077-486.391892)}$$

$$=11.9326\text{ kg/s},$$

and for off-design case 2,

$$F_{\text{steam}}^{off\_d} = \frac{(44.8269)(1.237)(895.083-422.301)}{(2797.25077-486.391892)}$$

$$=11.3448\text{ kg/s}.$$

Remaining off-design HRSG calculations including the HRSG overall heat transfer coefficient approach are left for the reader. ■

We do want to comment on the turbine exhaust temperature. Recall that in the off-design cases, we did not change the efficiency of the power turbine. Lowering the efficiency of the power turbine from the design value will increase the turbine exhaust temperature.

We also want to emphasize that in Example 7.7, we used a generic control algorithm with Equations (7.72)–(7.74), where many of the key parameters to obtain realistic results from a specific engine were ignored. Manufacturer-provided correction curves can be used to determine these parameters or plant data at off-design conditions can be used.

With manufacturer performance curves, we can predict off-design performance by simple multiplication. These curves allow system performance to be determined without the details of the control algorithm. The advantage of using equations (Eq. (7.72) and (7.73)) is that it allows for a straightforward computer implementation of the off-design performance, which may be valuable in an optimization strategy including online performance monitoring or optimal energy dispatch as we develop in Chapter 12. In Equations (7.72) and (7.73), nonlinear performance fits can be used to increase accuracy—with the optimization approach used here for off-design performance, there is no increased difficulty in using nonlinear performance fits.

7.10 CLOSING REMARKS

Gas turbine cogeneration system design is a mature topic and there are excellent books that greatly expand the topics introduced in this chapter. For readers interested in gas turbine design and applications, see *Gas Turbine Theory*, Sixth Edition (Saravanamuttoo et al., 2009); *Fundamentals of Gas Turbines*, Second Edition (Bathie, 1996); and *Gas Turbine Engineering Handbook*, Third Edition (Boyce, 2006). For readers interested in details of HRSG design and operation, I suggest first reading *Waste Heat Boiler Deskbook* (Ganapathy, 1991) followed by *Industrial Boilers and Heat Recovery Steam Generators: Design, Applications, and Calculations* (Ganapathy, 2003). Additional details of both gas and steam turbine design and off-design operations can be found in the book by Gay et al. (2004).

In this chapter, we have introduced cogeneration system material and energy balance calculations. Here we developed the general thermodynamic and energy balance equations to allow the design or analysis of both the power section (turbine) and energy recovery section (HRSG) of the cogeneration system. These general thermodynamic and energy balance performance equations were particularized to ideal gas power-side working fluid.

There are several important features in this chapter. Here the construction of general Excel sheets for the solution of mass and energy balances in gas turbine-based cogeneration systems is developed. We also introduced use of Excel callable sheet functions for steam properties. The use of Excel sheets allows complete cogeneration system performance updating as the processing conditions are changed. We used named cells in the Excel sheet for variables—values we wanted complete access to change. It is a straightforward to extend the current Excel formulation to an optimal design problem where process design variables are varied to minimize an objective function that combines capital and fuel and maintenance costs (Chapter 10). It is also straightforward to replace the ideal gas working fluid with real physical properties all callable as Excel sheet functions (Chapters 9 and 10). Several of the variables in Chapter 7 will be true *design or optimization decision variables* in Chapter 10. The design variables used in Chapter 10 include F_{air}, P_2, the temperature leaving the combustion chamber T_3, the compression and turbine efficiencies $\eta_{\text{compression}}$ and $\eta_{\text{expansion}}$, and for the HRSG, ΔT_{Pinch} and $\Delta T_{\text{Approach}}$. In Chapter 7, we also let $\hat{C}^{ig}_{P,\text{air}}$ and $\hat{C}^{ig}_{P,\text{gas}}$ be taken as variables, but these are not true decision variables; generally, $\hat{C}^{ig}_{P,\text{air}}$ and $\hat{C}^{ig}_{P,\text{gas}}$ values are known, or the needed enthalpies will be determined from the physical properties package; here, we simply wanted to be able to change enthalpy values to reflect an air basic system or a system of both air and product gases.

Energy recovery in the HRSG is a key component in the economic justification for cogeneration. The HRSG should be designed using values for ΔT_{Pinch} and $\Delta T_{\text{Approach}}$ based on HRSG gas-side inlet temperature and HRSG tube design. The performance of the HRSG in the off-design mode, for example, under supplemental firing, may best be determined using a method based on heat transfer coefficients. Here, Excel provides a powerful tool for the iteration and convergence of off-design HRSG variables.

The careful reader will note that the modular approach was used to solve the cogeneration unit operations. Here, the simultaneous approach offers few advantages. The steam recycle loop does not require iterative calculations as the steam flow rate can be uniquely determined from HRSG energy balances.

REFERENCES

Bathie, W.W. 1996. *Fundamentals of Gas Turbines* (2nd edition). John Wiley & Sons, New York.

Bolland, O. 2008. Thermal Power Generation (Compendium). Norwegian University of Science and Technology, September.

Boyce, M.P. 2006. *Gas Turbine Engineering Handbook* (3rd edition). Gulf Professional Publishing, Burlington, MA.

Bustami, L.M. 2001. Design of heat recovery steam generators. MS Thesis, Louisiana State University.

Ganapathy, V. 1991. *Waste Heat Boiler Deskbook*. The Fairmont Press, Inc., Liburn, GA.

Ganapathy, V. 2003. *Industrial Boilers and Heat Recovery Steam Generators: Design, Applications, and Calculations*. Marcel Dekker, New York.

Gay, R.R., C.A. Palmer, and M.R. Erbes. 2004. *Power Plant Performance Monitoring*. R-Squared Publishing, Woodland, CA.

Poullikkas, A. 2005. An overview of current and future sustainable gas turbine technologies. *Renew. Sustain. Energ. Rev.* 9: 409–443.

Reale, M.J. and J.K. Prochaska. 2005. New high efficiency simple cycle gas turbine—GE's LMS 100™. Paper No. 05-IAGT-1.2, Presented at the 16th Symposium on Industrial Applications of Gas Turbines (IAGT), Banff, Alberta, Canada, October 12–14, 2005.

Saravanamuttoo, H.I.H., G.F.C. Rogers, H. Cohen, and P.V. Straznicky. 2009. *Gas Turbine Theory* (6th edition). Prentice Education Limited, Essex, England.

Streeter, V.L. and E.B. Wylie. 1979. *Fluid Mechanics*. McGraw-Hill, New York.

PROBLEMS

As discussed in the closing remarks, the strength of the developments of this chapter is that we have actually assembled a general solution to gas turbine-based cogeneration problems. In this chapter, our limitation is that the working fluid is an ideal gas or a system with constant heat capacity. We will solve the cogeneration system design with real fluids in Chapter 9. *Using the results of Chapter 7, the user can vary any of the inputs (variables) on the Excel performance and design sheet and immediately see the effect.*

7.1 ***Solve Example 7.2 Using SI Units (Solution Provided)*** Here we want to solve the basic gas turbine with HRSG of Example 7.2, and shown in Figure 7.10, but now in SI units. The air feed and combustion products will be taken as ideal gases but with different heat capacities. The air feed can be taken as an ideal gas with heat capacity $\hat{C}^{ig}_{P,\text{air}} = 1.004$ kJ/kg-K (Variable) and specific heat ratio

$$\frac{\hat{C}^{ig}_{P,\text{air}}}{\hat{C}^{ig}_{V,\text{air}}} = \gamma_{\text{air}} = 1.4 \text{ (Variable)}.$$

The combustion products (gas) have a heat capacity, $\hat{C}^{ig}_{P,gas} = 1.17$ kJ/kg-K (Variable), and specific heat ratio

$$\frac{\hat{C}^{ig}_{P,\text{gas}}}{\hat{C}^{ig}_{V,\text{gas}}} = \gamma_{\text{gas}} = 1.33\,(\text{Variable}).$$

We also know the following: The air inlet temperature $T_1 = 298.15$ K (Variable), $P_1 = 1.013$ bar, and air enters the compressor at 1 kg/s (Variable). The compression ratio $(P_2/P_1) = 15.9$ (Variable). There is no pressure drop in the combustion chamber ($P_3 = P_2$) and the turbine inlet temperature $T_3 = 1400$ K (Variable). The compressor efficiency (isentropic) = 87% (Variable%). The gas generating turbine efficiency (isentropic) = 89% (Variable%). The power turbine efficiency (isentropic) = 89% (Variable %) and P_5 is at atmospheric conditions.

Here assume the fuel is natural gas (methane) with an *LHV* of 50,000 kJ/kg-methane at 298.15 K. The rate of heat loss from the combustion chamber, dQ_{loss}/dt, is taken as a percentage of the fuel *LHV*. This can be given as $dQ_{\text{loss}}/dt = (F_{\text{fuel}})(LHV)(1 - \eta_{cc})$; with the efficiency of the combustion chamber, $\eta_{cc} = 0.98$, this is equivalent to a 2% loss.

As we have previously discussed, the HRSG design case requires specification of pinch and approach temperature differences; use $\Delta T_{\text{Pinch}} = 8.4$ K and the approach temperature difference $\Delta T_{\text{Approach}} = 8.4$ K. But do allow both ΔT_{Pinch} and $\Delta T_{\text{Approach}}$ to be variable. It may seem unusual that we are solving this problem with $F_{\text{air}} = 1$ kg/s, but our results here are "scalable" as we will see in Problem 7.2.

The water feed conditions are 388.89 K and 20 bar. Saturated steam is needed by the process at 485.55 K and 20 bar. Use our Excel sheet function for steam properties:

$$\hat{h}^{\text{vapor}}_{\text{saturated}}(485.55\text{ K}) = \text{H\_Steam\_SI(Pressure\_9/10, Temperature\_9)/MW\_Water}$$

$$\hat{h}^{\text{liquid}}(470.55\text{ K, 20 bar}) = \text{H\_Water\_SI(Pressure\_8\_p/10, Temperature\_8\_p)/MW\_Water}$$

$$\hat{h}^{\text{liquid}}(388.89\text{ K, 20 bar}) = \text{H\_Water\_SI(Pressure\_8\_p/10, Temperature\_8)/MW\_Water.}$$

The factor of 10 in the pressure is needed to convert from bar (Excel sheet) to megapascal (Excel sheet functions).

SOLUTION The solution can be found in the Excel file **Problem 7.1.xls**. ■

7.2 ***Effect of*** $\left(\frac{P_2}{P_1}\right)$ ***and*** F_{air} ***Using SI Units*** Here take the solution obtained in Problem 7.1 and change

$$\left(\frac{P_2}{P_1}\right) = 8.5232$$

and $F_{\text{air}} = 99.4559$ kg/s. Set the pinch temperature difference at $\Delta T_{\text{Pinch}} = 1.64$ K and the approach temperature difference $\Delta T_{\text{Approach}} = 15$ K. These are the optimal values that will be found in our economic optimization of Chapter 10 (the original CGAM problem).

7.3 ***HRSG Off-Design Supplemental Firing—Equation (7.70)*** Here we want to resolve Example 7.5 using Equation (7.70) as opposed to the simplified Equation (7.71). Equation (7.70) requires addition of exhaust gas physical properties including heat capacity, viscosity, and thermal conductivity as a function of temperature. We could develop correlations for these properties, but here, to keep the problem straightforward, let us simply use tabulated air properties to represent the exhaust gas. We will also keep $\hat{C}^{ig}_{P,\text{gas}} = 0.28$ Btu/lb-R as a constant for the HRSG material and energy balance calculations. The reader is encouraged to improve on these assumptions.

Using the temperatures found for the design and off-design cases in Example 7.4, we can estimate exhaust gas physical properties in the HRSG. Recall we will need average physical properties for the evaporator and economizer sections of the HRSG for use in Equation (7.70). I found $\hat{C}_{p,\text{air}}$ using our Excel sheet function {*H_Air*} with a 1° temperature change; μ_{air} from a Web search for "gas viscosity calculator"; and k_{air} from a Web search for "air property calculator."

Air Properties

Using results from Example 7.4	T (R)	$\hat{C}_{p,\text{air}}\left(\frac{\text{Btu}}{\text{lb-R}}\right)$	μ_{air} (mPa-s)	$k_{\text{air}}\left(\frac{\text{W}}{\text{m-K}}\right)$
Design, T_6	1385	0.261	0.0363	0.0561
Design, T_{7P}	889	0.246	0.0270	0.0399
Design, T_7	790.1	0.243	0.0249	0.0362
Design, average evaporator		0.254	0.0317	0.048
Design, average economizer		0.245	0.026	0.0381
Off-design, T_{6_SF}	1870	0.274	0.0437	0.0700
Off-design, T_{7P_SF}	969.5	0.248	0.0287	0.0427
Off-design, T_{7_SF}	790.1	0.243	0.0249	0.0362
Off-design, average evaporator		0.261	0.0362	0.0564
Off-design, average economizer		0.246	0.0268	0.0395

7.4 ***HRSG Steaming—Off-Design Operation with Decreased Ambient Temperature Using Equation (7.70)*** Determine the HRSG off-design performance using Equation (7.70) if due to off-design gas turbine performance, $T_{6\_Off\text{-}D} = 1350$ R and $F_{gas\_Off\text{-}D} = 214$ lb/s. Needed HRSG design data are found in Example 7.5. HRSG temperatures and physical properties, needed for the first iteration of Equation (7.70), are provided in the next table. We will also keep

$$\hat{C}^{ig}_{P,\text{gas}} = 0.28 \frac{\text{Btu}}{\text{lb-R}}$$

as a constant for the HRSG material and energy balance calculations. Determine if steaming may be a concern in the economizer section of the HRSG. Determine if the design steam production rate be met or exceeded.

Air Properties

	T (R)	$\hat{C}_{p,\text{air}}\left(\frac{\text{Btu}}{\text{lb-R}}\right)$	μ_{air} (mPa-s)	$k_{\text{air}}\left(\frac{\text{W}}{\text{m-K}}\right)$
Design, T_6	1385	0.261	0.0363	0.0561
Design, T_{7P}	889	0.246	0.0270	0.0399
Design, T_7	790.1	0.243	0.0249	0.0362
Design, average evaporator		0.254	0.0317	0.048
Design, average economizer		0.245	0.026	0.0381
Off-design, $T_{6\_Off\text{-}D}$	1350	0.260	0.0357	0.0551
Off-design, $T_{7P\_Off\text{-}D}$	889	0.246	0.0270	0.0399
Off-design, $T_{7\_Off\text{-}D}$	790.1	0.243	0.0249	0.0362
Off-design, average evaporator		0.253	0.0314	0.0475
Off-design, average economizer		0.245	0.026	0.0381

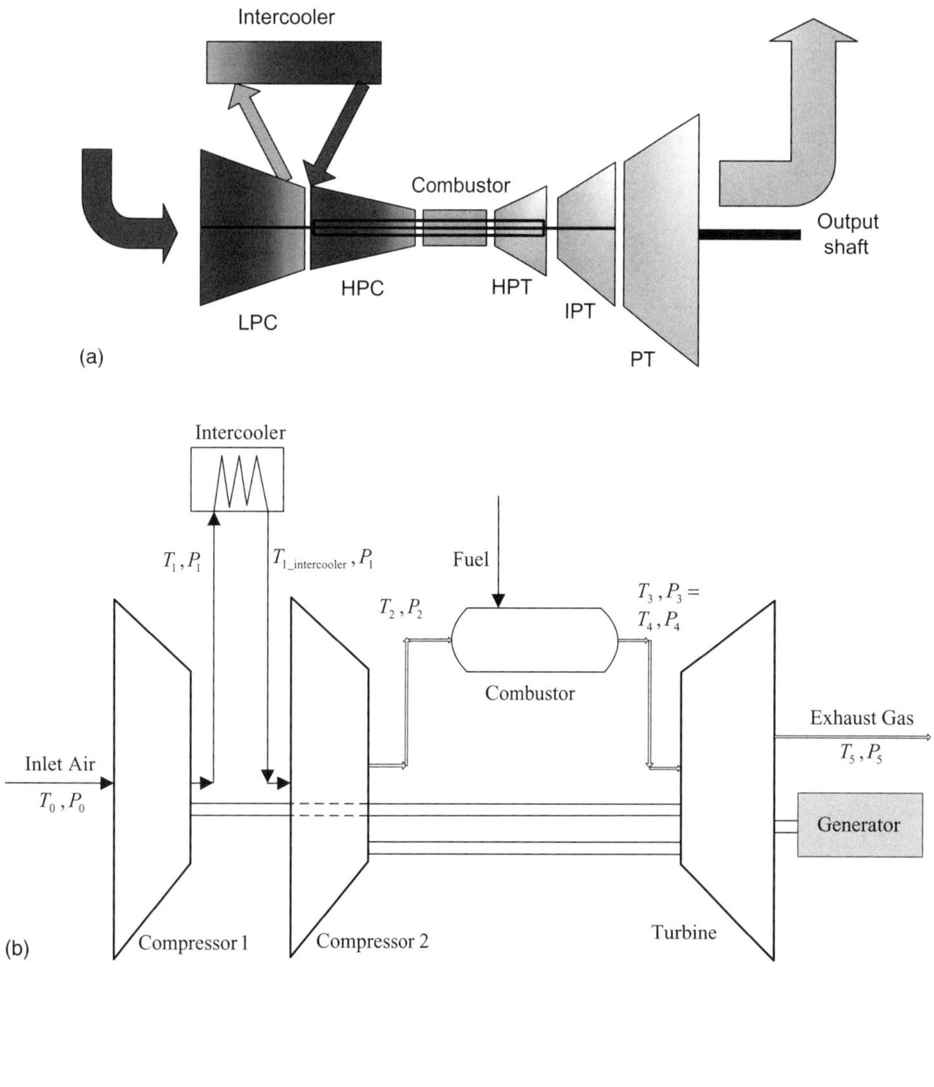

Figure P7.5 (a) Overview schematic of the GE LMS 100 (Reale and Prochaska, 2005; used with permission). (b) Schematic for GE LMS 100 performance calculations (nomenclature used in provided Excel solutions). LPC, low pressure compressor; HPC, high pressure compressor; HPT, high pressure turbine; IPT, intermediate pressure turbine; PT, power turbine.

Table P7.5 Design Turbine Performance Calculations Using Ideal Gas and Real Gas Properties (see also Bolland, 2008)

Variable/Parameter	GE Reported Design	Design Ideal Gas	Design Real Gas
Air gamma, γ_a		1.38	
Air specific heat capacity, $\hat{C}_{p,\text{air}}$(kJ/kg-K)		1.046	
Compressor inlet flow, F_0 (kg/s)	209	209	209
Compressor inlet temperature, T_0 (K)	288.15	288.15	288.15
Compressor inlet pressure, P_0 (bar)	1.013	1.013	1.013
Compressor isentropic efficiency, η_{AC}		0.89	0.89
Compressor outlet temperature, T_2 (K)			
Compressor outlet pressure, P_2 (bar)	42.546	42.546	42.546
Compressor work (kJ/s)			
Fuel lower heating value, *LHV* (kJ/kg)		50,010	50,010
Fuel flow, F_{fuel} (kg/s)			
Exhaust gas gamma, γ_g		1.31	
Exhaust heat capacity, $\hat{C}_{p,g}$(kJ/kg-K)		1.237	
Turbine inlet flow, F_g (kg/s)			
Turbine inlet temperature, T_3 (K)		1523.15	1523.15
Turbine inlet pressure, P_3 (bar)		40.844	40.844
Turbine isentropic efficiency, η_{PT}		0.90	0.90
Turbine outlet temperature, T_5 (K)	683.15		
Turbine outlet pressure, P_5 (bar)		1.023	1.023
Turbine work (kJ/s)			
Turbine net work (kJ/s)	98,700		
Heat rate (kJ/kW-h)	7922.4		

Here you may need to add constraints to help keep the solution feasible. These constraints include $T_{7P\_Off\text{-}D} \geq T_9$, $T_{7\_Off\text{-}D} \geq 770$ R, $\Delta T_{Approach\_Off\text{-}D} \geq 1$ R, and $\Delta T_{Pinch\_Off\text{-}D} \geq 1$ R.

7.5 The GE LMS100 shown in Figure P7.5a uses an air cooler (intercooler) between low- and high-pressure air compressors; these are compressors 1 and 2 in Figure P7.5b. General Electric reports the following benchmarks for the LMS100 when using an air flow rate of 209 kg/s at the following design conditions: net power 98.7 MW; a heat rate of 7509 Btu/kW-h; thermal efficiency of 46%; and an exhaust temperature of 410°C. The pressure ratio in the LMS is reported as 42:1.

Using the component efficiencies and ideal gas physical properties in Table P7.5 (which are those used in Table 7.3), determine the performance of the LMS 100 and compare these values to those reported by GE. Use $F_{\text{air}} = 209$ kg/s; a compression ratio, $P_2/P_0 = 42$; $T_{1\_intercooler} = 15°C$; a 4% pressure drop in the combustion chamber; $T_3 = 1250°C$; and $P_5 = 1.023$ bar. Use the optimum pressure ratio for P_1/P_0 and P_2/P_1; show yourself this will occur when $P_1/P_0 = P_2/P_1$. In Table P7.5, complete the ideal gas solution; we will determine the real gas solution in Chapter 9, Problem 9.5.

Chapter 8

Development of a Physical Properties Program for Cogeneration Calculations

Selecting appropriate thermodynamic models is essential for the accurate design and analysis of process and energy transformation systems. Combined heat and power (CHP) systems require the solution of mass, energy, and entropy balances and specific performance equations. Realize that because of the large flow rates of materials through many of these systems, inaccuracies in thermodynamic properties will be magnified. This chapter develops and illustrates the approaches that we have found useful for the calculation of thermodynamic properties.

CHP systems present an interesting challenge—pure components, such as steam and refrigerants, cycle from low pressures and low temperature to high pressures and high temperatures. Mixtures, for example, combustion products, may be found at high temperatures and over a range of pressures.

In this chapter, we start with a somewhat "reverse" order in developing thermodynamic properties. In Section 8.1, we demonstrate the use of our Excel function calls for cogeneration calculations. We next show how pure species properties for internal energy ($\hat{u}_s$ or u_s), enthalpy ($\hat{h}_s$ or h_s), and entropy ($\hat{s}_s$ or s_s) can be determined from an equation of state (EOS). In Section 8.3, the derivations to allow calculation of internal energy, enthalpy, and entropy from an EOS are developed. We appreciate that some readers may have a strong background in thermodynamics, and these readers will only need to quickly review Sections 8.2 and 8.3.

In Section 8.4, ideal mixtures of real fluids are introduced and Excel functions specifically developed for combustion calculations are discussed. Typical problems when using ideal mixtures of real fluids are detailed in Section 8.5. One solution to the problems of Section 8.5, mixing rules for EOSs, is developed in Section 8.6. In the chapter problems, we also use EOSs with mixing rules to solve for equilibrium distribution factors (introduced in Chapter 5) and how to determine if a stream is a vapor-phase, a liquid-phase, or a two-phase mixture (bubble point and dew point temperature calculations).

8.1 AVAILABLE FUNCTION CALLS FOR COGENERATION CALCULATIONS

Thermodynamic properties of pure species including internal energy, enthalpy, and entropy are not measured, but rather they are calculated from EOSs. EOSs are fit to pure component *P*-*V*-*T* data. In this chapter, vapor-phase properties for species of importance in CHP calculations are provided as Excel callable functions in Tables 8.1 and 8.2. Vapor-phase and compressed liquid-state properties are also provided for water. These specific Excel callable functions are based on the work of Reynolds (1979), *Thermodynamic Properties in SI: Graphs, Tables, and Computational Equations for Forty Substances* (*TPSI*). Reynolds's work is still considered an industry standard for pure species properties. Vapor-phase and saturated liquid-state properties are provided for refrigerant R134a (Dupont (1993)). The user is given full access to change or modify the thermodynamic programs. In addition to the thermodynamic programs for these species property determinations, the user has access to programs and Excel function calls specifically developed for

Modeling, Analysis and Optimization of Process and Energy Systems, First Edition. F. Carl Knopf.

Table 8.1a Pure Species Properties in Field Units, $P - \upsilon - T$ EOS from Reynolds (1979)

Species	$h(P, T)$	$P(\upsilon, T)$	$P(h, T)$	$P(s, T)$	$s(P, T)$	$T(h, P)$	$T(s, P)$	$\upsilon(P, T)$
Air	H_Air	P_Air	P_HT_Air	P_ST_Air	S_Air	T_HP_Air	T_SP_Air	V_Air
Ammonia	H_Ammonia	P_Ammonia	P_HT_Ammonia	P_ST_Ammonia	S_Ammonia	T_HP_Ammonia	T_SP_Ammonia	V_Ammonia
Argon	H_Argon	P_Argon	P_HT_Argon	P_ST_Argon	S_Argon	T_HP_Argon	T_SP_Argon	V_Argon
Butane	H_Butane	P_Butane	P_HT_Butane	P_ST_Butane	S_Butane	T_HP_Butane	T_SP_Butane	V_Butane
CarbDiox	H_CarbDiox	P_CarbDiox	P_HT_CarbDiox	P_ST_CarbDiox	S_CarbDiox	T_HP_CarbDiox	T_SP_CarbDiox	V_CarbDiox
Ethane	H_Ethane	P_Ethane	P_HT_Ethane	P_ST_Ethane	S_Ethane	T_HP_Ethane	T_SP_Ethane	V_Ethane
Ethylene	H_Ethylene	P_Ethylene	P_HT_Ethylene	P_ST_Ethylene	S_Ethylene	T_HP_Ethylene	T_SP_Ethylene	V_Ethylene
Hydrogen	H_Hydrogen	P_Hydrogen	P_HT_Hydrogen	P_ST_Hydrogen	S_Hydrogen	T_HP_Hydrogen	T_SP_Hydrogen	V_Hydrogen
Methane	H_Methane	P_Methane	P_HT_Methane	P_ST_Methane	S_Methane	T_HP_Methane	T_SP_Methane	V_Methane
Nitrogen	H_Nitrogen	P_Nitrogen	P_HT_Nitrogen	P_ST_Nitrogen	S_Nitrogen	T_HP_Nitrogen	T_SP_Nitrogen	V_Nitrogen
Octane	H_Octane	P_Octane	P_HT_Octane	P_ST_Octane	S_Octane	T_HP_Octane	T_SP_Octane	V_Octane
Oxygen	H_Oxygen	P_Oxygen	P_HT_Oxygen	P_ST_Oxygen	S_Oxygen	T_HP_Oxygen	T_SP_Oxygen	V_Oxygen
Propane	H_Propane	P_Propane	P_HT_Propane	P_ST_Propane	S_Propane	T_HP_Propane	T_SP_Propane	V_Propane
Steam	H_Steam	P_Water	P_HT_Steam	P_ST_Steam	S_Steam	T_HP_Steam	T_SP_Steam	V_Steam
Water	H_Water	P_Water	P_HT_Water	P_ST_Water	S_Water	T_HP_Water	T_SP_Water	V_Water

Units T (R), P (psia), h (Btu/lb-mol), s (Btu/lb-mol-R), υ (ft^3/lb-mol), $\rho = \frac{1}{\upsilon}$ (lb-mol/ft^3).

Table 8.1b Saturated Water Properties, Saturated Density Equation from Reynolds (1979)

Species	$T_{sat}(P)$	$P_{sat}(T)$	$\rho_{sat}(T)$
Water	Tsat_Water	Psat_Water	rhosat_Water

Units T (R), P (psia), $\rho = \frac{1}{\upsilon}$ (lb-mol/ft^3).

Table 8.1c Combustion Product Properties, Ideal Mixing of Real Gases

Species	$\hat{h}\left(P, T, \frac{H}{C}, DAR\right)$	$\hat{s}\left(P, T, \frac{H}{C}, DAR\right)$	$T\left(P, \hat{h}, \frac{H}{C}, DAR\right)$	$T\left(P, \hat{s}, \frac{H}{C}, DAR\right)$
Products	H_Products	S_Products	TfromH_Products	TfromS_Products

Units T (R), P (psia), $\hat{h}$ (Btu/lb), $\hat{s}$ (Btu/lb-R), $\frac{H}{C}\left(\frac{\text{fuel hydrogen}}{\text{fuel carbon}}\right)$, $DAR\left(\frac{\text{dry air used}}{\text{theoretical air}}\right)$.

Table 8.1d Refrigerant R134a Properties from Dupont (1993)

Species	$\hat{h}(P,T)$	$P(\hat{\upsilon},T)$	$P(\hat{h},T)$	$P(\hat{s},T)$	$\hat{s}(P,T)$	$T(\hat{h},P)$	$T(\hat{s},P)$	$\hat{\upsilon}(P,T)$
R134a	H_R134a	P_R134a	P_HT_R134a	P_ST_R134a	S_R134a	T_HP_R134a	T_SP_R134a	V_R134a

Species	$T_{sat}(P)$	$P_{sat}(T)$	$\hat{\rho}_{sat}(T)$
R134a	Tsat_ R134a _SI	Psat_ R134a _SI	RHOsat_L_R134a

Species	$\hat{\upsilon}_{sat}(T)$	$\hat{h}_{sat}(T)$	$\hat{s}_{sat}(T)$	$T_{sat}(\hat{h})$	$T_{sat}(\hat{s})$
R134a	Vsat_L_R134a	Hsat_L_R134a	Ssat_L_R134a	Tsat_Hsat_L_R134a	Tsat_Ssat_L_R134a

Units T (R), P (psia), $\hat{h}$ (Btu/lb), $\hat{s}$ (Btu/lb-R), $\hat{\upsilon}$ (ft^3/lb), $\hat{\rho} = \frac{1}{\hat{\upsilon}}$ (lb/ft^3).

Table 8.2a Pure Species Properties in SI Units, $P - \upsilon - T$ EOS from Reynolds (1979)

Species	$h(P, T)$	$P(\upsilon, T)$	$P(h, T)$	$P(s, T)$	$s(P, T)$	$T(h, P)$	$T(s, P)$	$\upsilon(P, T)$
Air_SI	H_Air_SI	P_Air_SI	P_HT_Air_SI	P_ST_Air_SI	S_Air_SI	T_HP_Air_SI	T_SP_Air_SI	V_Air_SI
Ammonia_SI	H_Ammonia_SI	P_Ammonia_SI	P_HT_Ammonia_SI	P_ST_Ammonia_SI	S_Ammonia_SI	T_HP_Ammonia_SI	T_SP_Ammonia_SI	V_Ammonia_SI
Argon_SI	H_Argon_SI	P_Argon_SI	P_HT_Argon_SI	P_ST_Argon_SI	S_Argon_SI	T_HP_Argon_SI	T_SP_Argon_SI	V_Argon_SI
Butane_SI	H_Butane_SI	P_Butane_SI	P_HT_Butane_SI	P_ST_Butane_SI	S_Butane_SI	T_HP_Butane_SI	T_SP_Butane_SI	V_Butane_SI
CarbDiox_SI	H_CarbDiox_SI	P_CarbDiox_SI	P_HT_CarbDiox_SI	P_ST_CarbDiox_SI	S_CarbDiox_SI	T_HP_CarbDiox_SI	T_SP_CarbDiox_SI	V_CarbDiox_SI
Ethane_SI	H_Ethane_SI	P_Ethane_SI	P_HT_Ethane_SI	P_ST_Ethane_SI	S_Ethane_SI	T_HP_Ethane_SI	T_SP_Ethane_SI	V_Ethane_SI
Ethylene_SI	H_Ethylene_SI	P_Ethylene_SI	P_HT_Ethylene_SI	P_ST_Ethylene_SI	S_Ethylene_SI	T_HP_Ethylene_SI	T_SP_Ethylene_SI	V_Ethylene_SI
Hydrogen_SI	H_Hydrogen_SI	P_Hydrogen_SI	P_HT_Hydrogen_SI	P_ST_Hydrogen_SI	S_Hydrogen_SI	T_HP_Hydrogen_SI	T_SP_Hydrogen_SI	V_Hydrogen_SI
Methane_SI	H_Methane_SI	P_Methane_SI	P_HT_Methane_SI	P_ST_Methane_SI	S_Methane_SI	T_HP_Methane_SI	T_SP_Methane_SI	V_Methane_SI
Nitrogen_SI	H_Nitrogen_SI	P_Nitrogen_SI	P_HT_Nitrogen_SI	P_ST_Nitrogen_SI	S_Nitrogen_SI	T_HP_Nitrogen_SI	T_SP_Nitrogen_SI	V_Nitrogen_SI
Octane_SI	H_Octane_SI	P_Octane_SI	P_HT_Octane_SI	P_ST_Octane_SI	S_Octane_SI	T_HP_Octane_SI	T_SP_Octane_SI	V_Octane_SI
Oxygen_SI	H_Oxygen_SI	P_Oxygen_SI	P_HT_Oxygen_SI	P_ST_Oxygen_SI	S_Oxygen_SI	T_HP_Oxygen_SI	T_SP_Oxygen_SI	V_Oxygen_SI
Propane_SI	H_Propane_SI	P_Propane_SI	P_HT_Propane_SI	P_ST_Propane_SI	S_Propane_SI	T_HP_Propane_SI	T_SP_Propane_SI	V_Propane_SI
Steam_SI	H_Steam_SI	P_ Water_SI	P_HT_Steam_SI	P_ST_Steam_SI	S_Steam_SI	T_HP_Steam_SI	T_SP_Steam_SI	V_Steam_SI
Water_SI	H_Water_SI	P_Water_SI	P_HT_Water_SI	P_ST_Water_SI	S_Water_SI	T_HP_Water_SI	T_SP_Water_SI	V_Water_SI

Units T (K), P (MPa), h (kJ/kg-mol), s (kJ/kg-mol-K), υ (m^3/kg-mol), $\rho = \frac{1}{\upsilon}$ (kg-mol/m^3).

Table 8.2b Saturated Water Properties in SI Units, Saturated Density Equation from Reynolds (1979)

Species	$T_{sat}(P)$	$P_{sat}(T)$	$\rho_{sat}(T)$
Water_SI	Tsat_Water_SI	Psat_Water_SI	rhosat_Water_SI

Units T (K), P (MPa), $\rho = \frac{1}{\upsilon}$ (kg-mol/m^3).

Table 8.2c Combustion Product Properties; Ideal Mixing of Real Gases

Species	$\hat{h}\left(P, T, \frac{H}{C}, DAR\right)$	$\hat{s}\left(P, T, \frac{H}{C}, DAR\right)$	$T\left(P, \hat{h}, \frac{H}{C}, DAR\right)$	$T\left(P, \hat{s}, \frac{H}{C}, DAR\right)$
Products_SI	H_Products_SI	S_Products_SI	TfromH_Products_SI	TfromS_Products_SI

Units T (K), P (MPa), $\hat{h}$ (kJ/kg), $\hat{s}$ (kJ/kg-K), $\frac{H}{C}\left(\frac{\text{fuel hydrogen}}{\text{fuel carbon}}\right)$, $DAR\left(\frac{\text{dry air used}}{\text{theoretical air}}\right)$.

Table 8.2d Refrigerant R134a Properties in SI Units from Dupont (1993)

Species	$\hat{h}(P,T)$	$P(\hat{\upsilon},T)$	$P(\hat{h},T)$	$P(\hat{s},T)$	$\hat{s}(P,T)$	$T(\hat{h},P)$	$T(\hat{s},P)$	$\hat{\upsilon}(P,T)$
R134a_SI	H_R134a_ SI	P_R134a_ SI	P_HT_ R134a_SI	P_ST_ R134a_SI	S_R134a_ SI	T_HP_ R134a_SI	T_SP_ R134a_SI	V_R134a_ SI

Species	$T_{sat}(P)$	$P_{sat}(T)$	$\hat{\rho}_{sat}(T)$
R134a_SI	Tsat_ R134a _SI	Psat_ R134a _SI	RHOsat_L_R134a_SI

Species	$\hat{\upsilon}_{sat}(T)$	$\hat{h}_{sat}(T)$	$\hat{s}_{sat}(T)$	$T_{sat}(\hat{h})$	$T_{sat}(\hat{s})$
R134a_SI	Vsat_L_R134a_SI	Hsat_L_R134a_SI	Ssat_L_R134a_SI	Tsat_Hsat_L_R134a_SI	Tsat_Ssat_L_R134a_SI

Units T (K), P (MPa), $\hat{h}$ (kJ/kg), $\hat{s}$ (kJ/kg-K), $\hat{\upsilon}$ (m^3/kg), $\hat{\rho} = \frac{1}{\hat{\upsilon}}$ (kg/m^3).

CHP mixture calculations. The collection of Excel functions in Tables 8.1 and 8.2 is called *TPSI+* throughout the remainder of the text.

Tables 8.1 and 8.2 provide Excel function calls to the available physical properties. These same function calls can be used in Visual Basic for Applications (VBA) code. I do want to emphasize that the user is provided with the source code for all programs in this text. The source code for the functions in Table 8.1 can be found in C:\POEA\Combustion Library Field Units\Source Code\Combustion_Library.dsw and it consists of almost 7000 lines of code. The source code for the functions in Table 8.2 can be found in C:\POEA\Combustion Library SI Units\Source Code\Combustion_Library_SI.dsw.

The first row of any table provides the calculated physical property based on values supplied in the function parameter list. For example, from Table 8.1a, h (P, T) determines species enthalpy (British thermal unit per pound-mole) at the specified pressure (pound force per square inch absolute) and temperature (Rankine). The actual species calling function is identified in each appropriate row. For steam, H_Steam (P, T) determines the enthalpy of steam at the user-specified pressure and temperature. As a second example, P (h, T) P_HT_Steam (h, T) determines the steam pressure (pound force per square inch absolute) at the user-specified enthalpy (British thermal unit per pound-mole) and temperature (R). Example 8.1 shows the use of Tables 8.1a,b,d. The use of Tables 8.1c and 8.2c, for estimation of the physical properties of combustion products, is explained in Section 8.5.

EXAMPLE 8.1 ***Thermodynamic Properties from Excel Function Calls***

Use the function calls in Table 8.1 to determine the properties of steam, saturated water, and refrigerant R-134a and compare these values to those provided in Table 8.3. Here, the Excel sheet is provided in **Example 8.1a.xls**. The Excel sheet will appear empty, but the user should go to Tools → Macro → Visual Basic Editor to see the VBA calls to the C functions of Table 8.1. Be sure that in the VBA project, the dynamic link library (DLL) function is selected.

Table 8.3 Selected Steam, Saturated Water, and R-134a Properties

	Steam (Keenan et al., 1969)	Saturated Water (Keenan et al., 1969)
T (R)	1059.67	859.67
P (psia)	260	247.1
υ (ft^3/lb-mol)	41.959	0.3358
h (Btu/lb-mol)	23737.88	6758.16
s (Btu/lb-mol-R)	29.6291	10.21

	R-134a (Vapor) (Dupont, 1993)
T (R)	581.67
P (psia)	29
$\hat{\upsilon}$ (ft^3/lb)	2.045
$\hat{h}$ (Btu/lb)	191.486
$\hat{s}$ (Btu/lb-R)	0.4572

SOLUTION The solution is provided in **Example 8.1b.xls** and shown in Figure 8.1. From the Excel sheet, all the available function calls can be found using Inset → function (or f_x) → User Defined. In Figure 8.1, we show → V_Steam→ OK and the cells with the pressure (B5) and the temperature (B4). All the function calls used to generate the results in Figure 8.1 can be seen by using reveal equations (Ctrl + tilde).

Do note the strong agreement between literature values and physical property prediction using the EOS package developed here. The user can confirm SI properties using the template in **Example 8.1c.xls**. In Tables 8.1 and 8.2, the same P-ρ-T (or P-υ-T) EOS is used for both water and steam calculations. ■

8.2 PURE SPECIES THERMODYNAMIC PROPERTIES

Experimental data can be fit to EOSs, which are typically of the form $P = f(\upsilon, T)$ or $P = f(\rho, T)$. These equations can account for both the compressed liquid state and vapor states of pure species. The details of how the curve fitting is performed can be found in Reynolds (1979). Reynolds (1979) actually fit species P-$\hat{\rho}$-T data to 11 different EOSs of the

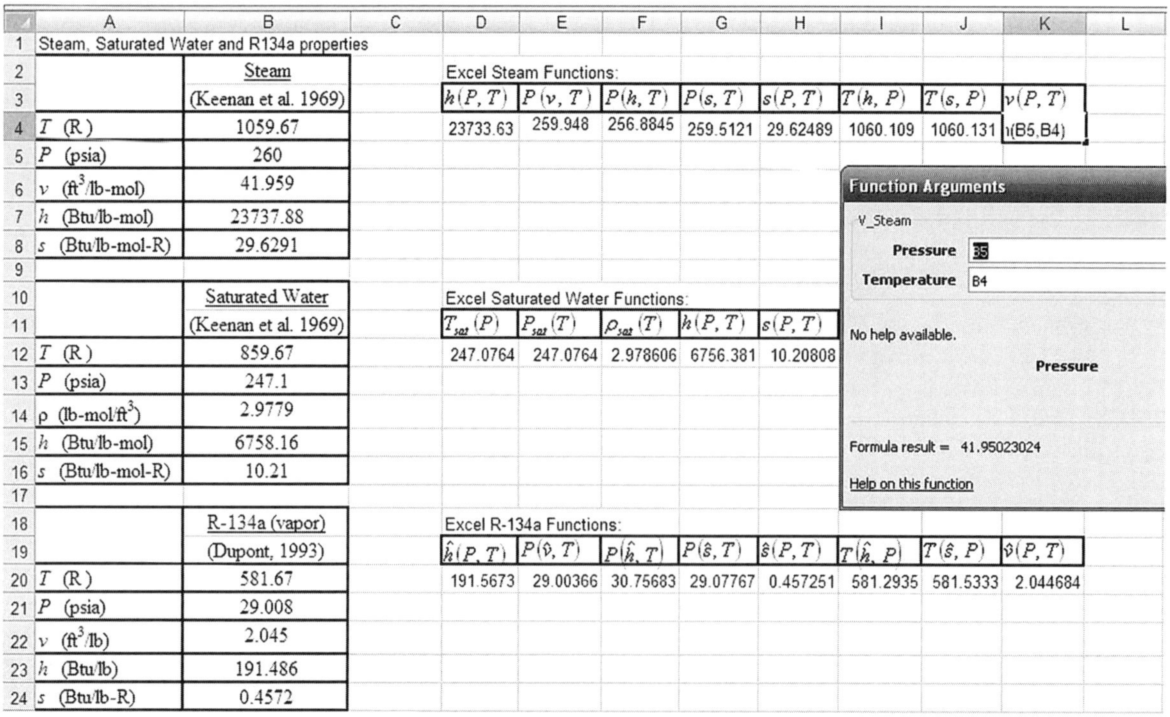

Figure 8.1 Comparison between data and EOS predictions.

form $P = \hat{\rho}RT + f(\hat{\rho}, T)$ where $f(\hat{\rho}, T)$ represents a function of $\hat{\rho}$ and T.

Here we provide the equations relating species specific internal energy $\hat{u}_s$ (joule per kilogram), specific enthalpy $\hat{h}_s$ (joule per kilogram), and specific entropy $\hat{s}_s$ (joule per kilogram-kelvin) to the system temperature and specific density $\hat{\rho}$ (kilogram per cubic meter) or specific volume $\hat{\upsilon}$ (cubic meter per kilogram). In Section 8.3, we will show the derivation for these equations. Without loss of generality, and to reduce the number of subscripts, in Sections 8.2 and 8.3 we will let $\hat{u}$, $\hat{h}$, and $\hat{s}$ represent the pure species properties. The basic working equations can be provided with either specific density integrals or specific volume integrals. Depending on the form of the EOS, the density or volume integral equation may be more convenient to use.

Specific Internal Energy

$$\hat{u} = \int_{T_0}^{T} \hat{C}_\upsilon^o dT + \int_0^{\hat{\rho}} \frac{1}{\hat{\rho}^2}\left[P - T\left(\frac{\partial P}{\partial T}\right)_{\hat{\rho}}\right] d\hat{\rho} + \hat{u}_0 \quad (8.1)$$

$$\hat{u} = \int_{T_0}^{T} \hat{C}_\upsilon^o dT + \int_{\hat{\upsilon}=\infty}^{\hat{\upsilon}} \left[T\left(\frac{\partial P}{\partial T}\right)_{\hat{\upsilon}} - P\right] d\hat{\upsilon} + \hat{u}_0 \quad (8.2)$$

Specific Enthalpy

$$\hat{h} = \int_{T_0}^{T} \hat{C}_\upsilon^o dT + \int_0^{\hat{\rho}} \frac{1}{\hat{\rho}^2}\left[P - T\left(\frac{\partial P}{\partial T}\right)_{\hat{\rho}}\right] d\hat{\rho} + \frac{P}{\hat{\rho}} + \hat{h}_0 \quad (8.3)$$

$$\hat{h} = \int_{T_0}^{T} \hat{C}_\upsilon^o dT + \int_{\hat{\upsilon}=\infty}^{\hat{\upsilon}} \left[T\left(\frac{\partial P}{\partial T}\right)_{\hat{\upsilon}} - P\right] d\hat{\upsilon} + P\hat{\upsilon} + \hat{h}_0 \quad (8.4)$$

Specific Entropy

$$\hat{s} = \int_{T_0}^{T} \frac{\hat{C}_\upsilon^o}{T} dT + \int_0^{\hat{\rho}} \frac{1}{\hat{\rho}^2}\left[\hat{\rho}R - \left(\frac{\partial P}{\partial T}\right)_{\hat{\rho}}\right] d\hat{\rho} - R\ln\hat{\rho} + \hat{s}_0 \quad (8.5)$$

$$\hat{s} = \int_{T_0}^{T} \frac{\hat{C}_\upsilon^o}{T} dT + \int_{\hat{\upsilon}=\infty}^{\hat{\upsilon}} \left[\left(\frac{\partial P}{\partial T}\right)_{\hat{\upsilon}} - \frac{R}{\hat{\upsilon}}\right] d\hat{\upsilon} + R\ln(\hat{\upsilon}) + \hat{s}_0 \quad (8.6)$$

Here $\hat{C}_\upsilon^o$ is the species specific heat capacity (joule per kilogram-kelvin) at constant volume and at zero density, and R is the appropriate gas constant. If we examine any of the density integral forms, there are two integrals, with the first integral at zero density from the reference temperature T_0 to the system T and the second integral from zero density to the system $\hat{\rho}$. This integration process is shown in Figure 8.2 (Reynolds, 1979).

Molar Basis for Pure Species Properties

Thermodynamic property equations equivalent to Equations (8.1)–(8.6) can also be developed on a molar basis. Here, for example, the volume-based integral equations would be

$$u = \int_{T_0}^{T} C_\upsilon^o dT + \int_{\upsilon=\infty}^{\upsilon} \left[T\left(\frac{\partial P}{\partial T}\right)_{\upsilon} - P\right] d\upsilon + u_0, \quad (8.7)$$

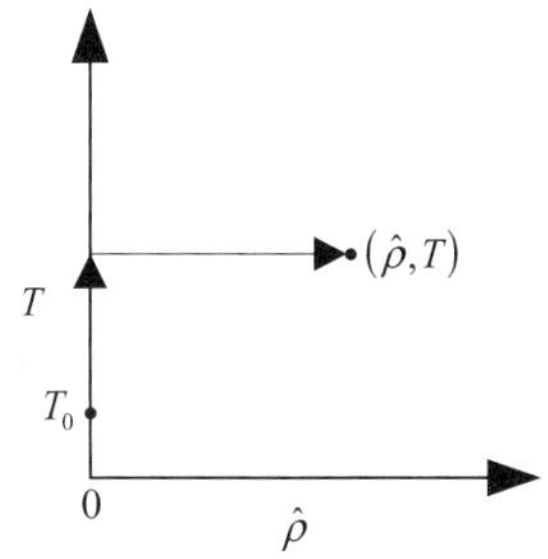

Figure 8.2 Integration at zero density from $(T_0, \hat{\rho} = 0)$ to $(T, \hat{\rho} = 0)$ and at constant temperature from $(T, \hat{\rho} = 0)$ to $(T, \hat{\rho})$.

$$h = \int_{T_0}^{T} C_\upsilon^o dT + \int_{\upsilon=\infty}^{\upsilon} \left[T\left(\frac{\partial P}{\partial T}\right)_{\upsilon} - P\right] d\upsilon + P\upsilon + h_0, \quad (8.8)$$

and

$$s = \int_{T_0}^{T} \frac{C_\upsilon^o}{T} dT + \int_{\upsilon=\infty}^{\upsilon} \left[\left(\frac{\partial P}{\partial T}\right)_{\upsilon} - \frac{R}{\upsilon}\right] d\upsilon + R\ln(\upsilon) + s_0, \quad (8.9)$$

where u, h, s, C_υ^o, υ, u_0, h_0, and s_0 are all molar properties. The extension to the density-based integral equations would follow the same pattern.

In Equations (8.1)–(8.9), the constants $\hat{u}_0$, u_0, $\hat{h}_0$, h_0, $\hat{s}_0$, and s_0 set property values at the species reference state. The main reason to allow these constants to take on a value other than zero is to allow comparison of results from Equations (8.1)–(8.9) with standard tabulated data. In Example 8.2, we show the h and s equations (Eq. (8.8) and (8.9)) applied to the Redlich–Kwong equation of state (RK-EOS); we have introduced the RK-EOS in Chapters 2 and 3. These equations are then used in Examples 8.3 and 8.4 for h and s predictions for vapor- and liquid-phase carbon dioxide. We will explore reference states in more detail later in this section, after Example 8.4.

EXAMPLE 8.2 ***Thermodynamic Properties from the Redlich–Kwong PVT EOS***

Solve for h and s for the RK-EOS (Redlich and Kwong, 1949). The RK-EOS is provided in Equation (8.10). The constants a and b depend on the critical temperature and critical pressure of the species:

$$P = \frac{RT}{\upsilon - b} - \frac{a}{\upsilon(\upsilon + b)T^{1/2}} \quad (8.10a)$$

$$a = \frac{0.4278\, R^2 T_{\text{critical}}^{2.5}}{P_{\text{critical}}} \quad (8.10b)$$

$$b = \frac{0.0867\, RT_{\text{critical}}}{P_{\text{critical}}}. \quad (8.10c)$$

SOLUTION Depending on the form of the EOS, the integrals over $d\rho$ or over $d\upsilon$ may be more convenient. Here use the integrals over $d\upsilon$. For the molar enthalpy using Equation (8.8),

$$h=\int_{T_0}^{T} C_\upsilon^o dT+\int_{\upsilon=\infty}^{\upsilon}\left[T\left(\frac{\partial P}{\partial T}\right)_\upsilon - P\right]d\upsilon + P\upsilon + h_0.$$

Solving for needed terms in the enthalpy equation,

$$\left(\frac{\partial P}{\partial T}\right)_\upsilon=\frac{R}{\upsilon-b}+\frac{a}{2\upsilon(\upsilon+b)T^{1.5}};$$

$$T\left(\frac{\partial P}{\partial T}\right)_\upsilon=\frac{RT}{\upsilon-b}+\frac{a}{2\upsilon(\upsilon+b)T^{0.5}};$$

$$\left[T\left(\frac{\partial P}{\partial T}\right)_\upsilon - P\right]=\frac{3a}{2T^{0.5}\upsilon(\upsilon+b)}=\frac{3a}{2T^{0.5}}\left[\frac{1}{\upsilon b}-\frac{1}{b(\upsilon+b)}\right];$$

$$\int_{\upsilon=\infty}^{\upsilon}\left[T\left(\frac{\partial P}{\partial T}\right)_\upsilon - P\right]d\upsilon=\frac{3a}{2T^{0.5}b}(\ln(\upsilon)-\ln(\upsilon+b))\Big]_{\upsilon=\infty}^{\upsilon=\upsilon}$$

$$=\frac{-3a}{2T^{0.5}b}\ln\left(\frac{\upsilon+b}{\upsilon}\right),$$

giving us

$$h=\int_{T_0}^{T} C_\upsilon^o dT-\frac{3a}{2T^{0.5}b}\ln\left(\frac{\upsilon+b}{\upsilon}\right)+P\upsilon+h_0. \tag{8.11}$$

For the molar entropy using Equation (8.9),

$$s=\int_{T_0}^{T}\frac{C_\upsilon^o}{T}dT+\int_{\upsilon=\infty}^{\upsilon}\left[\left(\frac{\partial P}{\partial T}\right)_\upsilon-\frac{R}{\upsilon}\right]d\upsilon+R\ln(\upsilon)+s_0.$$

Solving for needed terms in the entropy equation,

$$\left[\left(\frac{\partial P}{\partial T}\right)_\upsilon-\frac{R}{\upsilon}\right]=\frac{R}{(\upsilon-b)}+\frac{a}{2(\upsilon)(\upsilon+b)T^{1.5}}-\frac{R}{\upsilon}$$

$$=\frac{R}{(\upsilon-b)}+\frac{a}{2T^{1.5}}\left[\frac{1}{\upsilon b}-\frac{1}{b(\upsilon+b)}\right]-\frac{R}{\upsilon};$$

$$\int_{\upsilon=\infty}^{\upsilon}\left[\left(\frac{\partial P}{\partial T}\right)_\upsilon-\frac{R}{\upsilon}\right]d\upsilon$$

$$=R\ln(\upsilon-b)+\frac{a}{2T^{1.5}b}(\ln(\upsilon)-\ln(\upsilon+b))-R\ln(\upsilon)\Big]_{\upsilon=\infty}^{\upsilon=\upsilon};$$

$$=R\ln\left(\frac{\upsilon-b}{\upsilon}\right)-\frac{a}{2T^{1.5}b}\left(\ln\left(\frac{\upsilon+b}{\upsilon}\right)\right)\Big]_{\upsilon=\infty}^{\upsilon=\upsilon},$$

giving us

$$s=\int_{T_0}^{T}\frac{C_\upsilon^o}{T}dT+R\ln\left(\frac{\upsilon-b}{\upsilon}\right)-\frac{a}{2T^{1.5}b}\left(\ln\left(\frac{\upsilon+b}{\upsilon}\right)\right)+R\ln(\upsilon)+s_0. \tag{8.12}$$

In order to solve Equations (8.11) and (8.12) we will need the species-dependent equation for the molar heat capacity at zero density, C_υ^o. C_υ^o will be a function of temperature only and for an ideal gas, $C_\upsilon = C_P - R$; for liquids and solids, $C_\upsilon \approx C_P$. ■

Before we move on, let us examine how we would use Equation (8.11) to calculate the species enthalpy. Generally, we know the system T and P. An immediate problem is that to evaluate Equation (8.11), we need the system υ at T and P. Here we use the secant search method to vary υ in the EOS (Eq. (8.10a)) until the known system P is obtained. The RK-EOS (Eq. (8.10a)) is a cubic equation EOS with three roots. If the three roots are real, the largest root is the vapor volume and the smallest is the liquid (the intermediate root has no significance); there can also be one real and two complex roots. For the initial guess in the secant method, when the vapor volume is needed, the ideal gas is used as the estimate, and for the liquid volume, we can bound the solution between b and $\upsilon_{\text{critical}}$:

$$\upsilon_{\text{vapor}}^{\text{initial}}=\frac{P}{RT} \tag{8.13a}$$

$$b<\upsilon_{\text{liquid}}<\upsilon_{\text{critical}}. \tag{8.13b}$$

With known T and υ (with known P), and the equation for C_υ^o, Equation (8.11) can be evaluated.

EXAMPLE 8.3 *Solve for h and s Values for Vapor-Phase CO_2 Using the RK-EOS*

Compare h and s values from the RK-EOS for vapor-phase carbon dioxide to those provided in Reynolds (1979); selected Reynolds vapor-phase values for carbon dioxide are given in Table 8.4. In Example 8.2, we obtained the general equations for species h and s when using the RK-EOS. In order to use Equations (8.11) and (8.12), we will need the species heat capacity at constant volume, C_υ^o, and the species-dependent constants a and b for the RK-EOS.

The following data are adapted from Myers and Seider (1976) for carbon dioxide with $T_{\text{critical}} = 304.2$ K and $P_{\text{critical}} = 72.9$ atm:

$$a=6.377\times10^7\ \text{atm K}^{0.5}\left(\frac{\text{cm}^3}{\text{g-mol}}\right)^2,$$

$$b=29.7\left(\frac{\text{cm}^3}{\text{g-mol}}\right),$$

Table 8.4 Carbon Dioxide Vapor-Phase Data from Reynolds (1979)

T(K)	P(atm)	$\upsilon\left(\frac{\text{cm}^3}{\text{g-mol}}\right)$	$h\left(\frac{\text{cal}}{\text{g-mol}}\right)$	$s\left(\frac{\text{cal}}{\text{g-mol-K}}\right)$
500	9.869233	4126.378	6450.242	23.8237
700	9.869233	5818.122	8729.421	27.6472
900	9.869233	7494.903	11,196.67	30.7439
500	49.34616	802.7424	6315.708	20.4135
700	49.34616	1163.184	8666.625	24.3612
900	49.34616	1509.103	11,165.01	27.4968
500	98.69233	389.4885	6150.566	18.7716
700	98.69233	583.1325	8593.415	22.8791
900	98.69233	761.8131	11,128.93	26.0620

and

$$C_v^o = 16.0488 - 0.00004474T - \frac{158.08}{T^{0.5}}\left(\frac{\text{cal}}{\text{g-mol-K}}\right).$$

For Equations (8.11) and (8.12) with carbon dioxide, use

$$h_0 = 3101.2708\left(\frac{\text{cal}}{\text{g-mol}}\right), \quad s_0 = 0.7569\left(\frac{\text{cal}}{\text{g-mol-K}}\right),$$

and $T_0 = 216.54$ K. The choice of these reference state values is explained immediately after Example 8.4. An important aspect of solving this problem is keeping track of units. You will need several conversions including 1 L-atm = 24.217 cal, 1 L = 1000 cm^3, and R = 1.987 cal/g-mol-K. Every summation term of Equation (8.11) should have the units $(\text{cal}/\text{g-mol})$ and every summation term of Equation (8.12) should have units $(\text{cal}/\text{g-mol-K})$.

SOLUTION The solution is provided in **Example 8.3.xls** and shown in Figure 8.3.

As shown in Table 8.5, there is reasonable agreement between Reynolds (1979) h and s values and those obtained from the RK-EOS. The EOS equation in Reynolds for carbon dioxide uses more adjustable constants compared with the RK-EOS. ■

EXAMPLE 8.4 ***Solve for h and s Values for Liquid-Phase CO_2 Using the RK-EOS***

Compare h and s values from the RK-EOS for liquid-phase carbon dioxide to those provided by Reynolds (1979); selected Reynolds liquid-phase values for carbon dioxide are given in Table 8.6.

SOLUTION The solution is provided in **Example 8.4.xls** and shown in Figure 8.4.

Table 8.7 shows the agreement between Reynolds h and s values and those obtained from the RK-EOS.

An examination of the values in Tables 8.5 and 8.7 shows that the RK-EOS is more accurate at predicting vapor-phase properties (as predicted by Reynolds) than liquid-phase properties. ■

Reference Values and Reference States and $\hat{u}_0$, $\hat{h}_0$, $\hat{s}_0$

The values for $\hat{u}$, $\hat{h}$, $\hat{s}$ are not measured, and as shown in this section, they can be calculated from an EOS. In order to better understand reference values and reference states, let us reexamine, for example, Equation (8.1). Here we could have written

	A	B	C	D	E	F	G	H	I	J	K	L	M	N
1	Solution Example 8.3													
2					Cv_a	Cv_b	Cv_c			a	b		h_0	s_0
3	T_0	216.54		Cv	16.0488	-0.00004474	-158.08		RK-EOS	63770000	29.7		3101.2708	0.7569
4	Rgas	1.9872												
5				h cal/g-mol		h cal/g-mol		h total		s cal/g-mol-K	s cal/g-mol-K			s total
6	T (K)	P (atm)		∫Cv dT		RK-EOS				∫Cv/T dT		RK-EOS		
7	500	9.8692		2127.4854		961.2016		6189.9578		6.0715		16.5127		23.3411
8	700	9.8692		4036.6218		1375.5395		8513.4321		9.2731		17.2092		27.2392
9	900	9.8692		6119.2307		1781.0238		11001.5253		11.8864		17.7181		30.3614
10	500	49.3462		2127.4854		832.5675		6061.3237		6.0715		13.1311		19.9594
11	700	49.3462		4036.6218		1315.6973		8453.5899		9.2731		13.9407		23.9707
12	900	49.3462		6119.2307		1752.7324		10973.2339		11.8864		14.4866		27.1299
13	500	98.6923		2127.4854		674.5636		5903.3198		6.0715		11.5248		18.3532
14	700	98.6923		4036.6218		1247.2586		8385.1512		9.2731		12.4817		22.5117
15	900	98.6923		6119.2307		1721.3255		10941.8271		11.8864		13.0706		25.7139

Figure 8.3 Excel sheet screen shot solution for the Redlich–Kwong EOS (Example 8.3).

Table 8.5 Carbon Dioxide Vapor-Phase Values from Reynolds (1979) Compared with Values Predicted from the Redlich–Kwong EOS

T(K)	P(atm)	$v\left(\frac{\text{cm}^3}{\text{g-mol}}\right)$	$h\left(\frac{\text{cal}}{\text{g-mol}}\right)$	h from RK-EOS	$s\left(\frac{\text{cal}}{\text{g-mol-K}}\right)$	s from RK-EOS
500	9.869233	4126.378	6450.242	6189.9578	23.8237	23.3411
700	9.869233	5818.122	8729.421	8513.4321	27.6472	27.2392
900	9.869233	7494.903	11,196.67	11,001.5253	30.7439	30.3614
500	49.34616	802.7424	6315.708	6061.3237	20.4135	19.9594
700	49.34616	1163.184	8666.625	8453.5899	24.3612	23.9707
900	49.34616	1509.103	11,165.01	10,973.2339	27.4968	27.1299
500	98.69233	389.4885	6150.566	5903.3198	18.7716	18.3532
700	98.69233	583.1325	8593.415	8385.1512	22.8791	22.5117
900	98.69233	761.8131	11,128.93	10,941.8271	26.0620	25.7139

$$\int_{\hat{u}_0}^{\hat{u}} d\hat{u} = (\hat{u} - \hat{u}_0) = \int_{T_0}^{T} \hat{C}_v^o dT + \int_0^{\hat{\rho}} \frac{1}{\hat{\rho}^2}\left[P - T\left(\frac{\partial P}{\partial T}\right)_{\hat{\rho}}\right] d\hat{\rho}.$$

Rigorously, the reference state for $\hat{u}_0$ (or $\hat{h}_0$ from Eq. (8.3) or $\hat{s}_0$ from Eq. (8.5)) for any pure species would be $T = T_0$ and $\hat{\rho} = 0$, and the reference value for $\hat{u}_0$ should be given at these conditions. However, for energy balance calculation, we are interested in differences in internal energy, or enthalpy or entropy at two separate states. It is the difference in the values at the two states, not the actual values at a given state, that is important. The reference value will cancel in the difference calculation. This will be discussed again in Section 8.4 for mixtures without reaction and mixtures with reaction; for mixtures with reaction, we will find care must be taken in the selection of the reference state.

The numerical values for $\hat{u}_0$, $\hat{h}_0$, and $\hat{s}_0$ can be chosen to make the saturated liquid $\hat{h} = 0$ and the saturated liquid $\hat{s} = 0$ at a convenient temperature, T_0. In Examples 8.3 and 8.4, we found $h_0 = 3101.2708(\text{cal/g-mol})$ and $s_0 = 0.7569(\text{cal/g-mol-K})$, with the RK-EOS to match saturated liquid data from Reynolds, $h = 0$ and $s = 0$ at T_0 = 216.54 K (and P = 5.1054 atm). The values for h_0 and s_0 are uniquely determined to match T_0 saturated liquid data from Reynolds, $h = 0$ and $s = 0$ at T_0 = 216.54 K and P = 5.1054 atm.

For water, $\hat{h}_0$ and $\hat{s}_0$ are chosen so that at T_0 = 273.16 K and saturated liquid conditions P_0 = 0.0006113 MPa, $\hat{h} = 0$

Table 8.6 Carbon Dioxide Liquid-Phase Data from Reynolds (1979)

T(K)	P(atm)	$\upsilon\left(\frac{\text{cm}^3}{\text{g-mol}}\right)$	$h\left(\frac{\text{cal}}{\text{g-mol}}\right)$	$s\left(\frac{\text{cal}}{\text{g-mol-K}}\right)$
216.54	5.1054	37.27647	0.0	0.0
240	19.7385	40.3170	514.6772	2.1963
250	19.7385	42.0304	734.5168	3.0935
230	49.3462	38.6019	303.0421	1.1760
250	49.3462	41.5541	732.5183	2.9663
270	49.3462	45.8342	1187.0288	4.7134
230	98.6923	38.2031	314.9282	1.0277
250	98.6923	40.8749	733.6753	2.7738
280	98.6923	46.8690	1392.2475	5.2583

	A	B	C	D	E	F	G	H	I	J	K	L	M	N
1	Solution Example 8.4													
2					Cv_a	Cv_b	Cv_c			a	b		h_0	s_0
3	T_0	216.54		Cv	16.0488	-0.00004474	-158.08		RK-EOS	63770000	29.7		3101.2708	0.7569
4	Rgas	1.9872												
5				h cal/g-mol		h cal/g-mol		h total		s cal/g-mol-K		s cal/g-mol-K		s total
6	T (K)	P (atm)		∫Cv dT		RK-EOS				∫Cv/T dT		RK-EOS		
7	216.54	5.1054		0.0000		-3101.2708		0.0000		0.0000		-0.7569		0.0000
8	240	19.7385		130.7243		-2759.6585		472.3365		0.5727		0.8350		2.1646
9	250	19.7385		190.1038		-2616.6472		674.7274		0.8151		1.4763		3.0483
10	230	49.3462		73.4673		-2888.5929		286.1452		0.3291		0.0913		1.1773
11	250	49.3462		190.1038		-2610.4318		680.9427		0.8151		1.3669		2.9389
12	270	49.3462		314.7367		-2316.4536		1099.5540		1.2945		2.5988		4.6502
13	230	98.6923		73.4673		-2866.7051		308.0329		0.3291		-0.0335		1.0525
14	250	98.6923		190.1038		-2596.4479		694.9267		0.8151		1.2043		2.7762
15	280	98.6923		379.7719		-2175.8378		1305.2049		1.5310		2.9262		5.2141

Figure 8.4 Excel sheet screen shot solution for the Redlich–Kwong EOS (Example 8.4).

Table 8.7 Carbon Dioxide Liquid-Phase Data from Reynolds (1979) Compared with Values Predicted from the Redlich–Kwong EOS

T(K)	P(atm)	$\upsilon\left(\frac{\text{cm}^3}{\text{g-mol}}\right)$	$h\left(\frac{\text{cal}}{\text{g-mol}}\right)$	h from RK-EOS	$s\left(\frac{\text{cal}}{\text{g-mol-K}}\right)$	s from RK-EOS
216.54	5.1054	37.27647	0.0	0.0000	0.0	0.0000
240	19.7385	40.3170	514.6772	472.3365	2.1963	2.1646
250	19.7385	42.0304	734.5168	674.7274	3.0935	3.0483
230	49.3462	38.6019	303.0421	286.1452	1.1760	1.1773
250	49.3462	41.5541	732.5183	680.9427	2.9663	2.9389
270	49.3462	45.8342	1187.0288	1099.5540	4.7134	4.6502
230	98.6923	38.2031	314.9282	308.0329	1.0277	1.0525
250	98.6923	40.8749	733.6753	694.9267	2.7738	2.7762
280	98.6923	46.8690	1392.2475	1305.2049	5.2583	5.2141

and $\hat{s}=0$. The reference state for refrigerants is often set at saturated liquid conditions at −40°C (−40°F) with $\hat{h}$ and $\hat{s}$ equal to zero. For any species, it is not a requirement that $\hat{h}$ and $\hat{s}$ take zero values at the reference state, although this is generally the case. For example, to generate the National Institute of Standards and Technology (NIST) table of thermodynamic properties for R-134a, $\hat{s}_0$ and $\hat{h}_0$ must be chosen so that at −40°C and saturated liquid conditions, $\hat{h}=148.4$ kJ/kg and $\hat{s}=0.7967$ kJ/kg-K. Finally, note that $\hat{h}_0$ and $\hat{u}_0$ are not independent as $\hat{h}=\hat{u}+P\hat{\upsilon}$.

We next examine how Equations (8.1)–(8.6) are obtained. We then address how pure species properties are combined to obtain mixture properties.

8.3 DERIVATION OF WORKING EQUATIONS FOR PURE SPECIES THERMODYNAMIC PROPERTIES

The specific entropy $\hat{s}$ of a compressible species can be expressed as a function of temperature T and specific volume $\hat{\upsilon}$:

$$\hat{s}=\hat{s}(T,\hat{\upsilon}), \tag{8.14}$$

taking the total derivative,

$$d\hat{s}=\left(\frac{\partial \hat{s}}{\partial T}\right)_{\hat{\upsilon}} dT+\left(\frac{\partial \hat{s}}{\partial \hat{\upsilon}}\right)_T d\hat{\upsilon}, \tag{8.15}$$

and using a Maxwell relation, $\left(\frac{\partial P}{\partial T}\right)_{\hat{\upsilon}}=\left(\frac{\partial \hat{s}}{\partial \hat{\upsilon}}\right)_T$ (see Problem 8.1),

$$d\hat{s}=\left(\frac{\partial \hat{s}}{\partial T}\right)_{\hat{\upsilon}} dT+\left(\frac{\partial P}{\partial T}\right)_{\hat{\upsilon}} d\hat{\upsilon}. \tag{8.16}$$

The specific internal energy $\hat{u}$ of a compressible species can also be expressed as a function of temperature T and specific volume $\hat{\upsilon}$:

$$\hat{u}=\hat{u}(T,\hat{\upsilon}), \tag{8.17}$$

taking the total derivative,

$$d\hat{u}=\left(\frac{\partial \hat{u}}{\partial T}\right)_{\hat{\upsilon}} dT+\left(\frac{\partial \hat{u}}{\partial \hat{\upsilon}}\right)_T d\hat{\upsilon}, \tag{8.18}$$

and using the definition for the heat capacity at constant volume, $\hat{C}_\upsilon$,

$$\hat{C}_\upsilon=\left(\frac{\partial \hat{u}}{\partial T}\right)_{\hat{\upsilon}}, \tag{8.19}$$

we find that

$$d\hat{u}=\hat{C}_\upsilon dT+\left(\frac{\partial \hat{u}}{\partial \hat{\upsilon}}\right)_T d\hat{\upsilon}. \tag{8.20}$$

The energy balance in differential form, $d\hat{u}$, which accounts for the internal energy change between two equilibrium states, independent of the path, is

$$d\hat{u}=Td\hat{s}-Pd\hat{\upsilon}. \tag{8.21}$$

Substituting Equations (8.20) and (8.16) into Equation (8.21),

$$\hat{C}_\upsilon dT+\left(\frac{\partial \hat{u}}{\partial \hat{\upsilon}}\right)_T d\hat{\upsilon}=T\left[\left(\frac{\partial \hat{s}}{\partial T}\right)_{\hat{\upsilon}} dT+\left(\frac{\partial P}{\partial T}\right)_{\hat{\upsilon}} d\hat{\upsilon}\right]-Pd\hat{\upsilon}, \tag{8.22}$$

and collecting on dT and $d\hat{\upsilon}$,

$$\left[\hat{C}_\upsilon-T\left(\frac{\partial \hat{s}}{\partial T}\right)_{\hat{\upsilon}}\right]dT=\left[T\left(\frac{\partial P}{\partial T}\right)_{\hat{\upsilon}}-P-\left(\frac{\partial \hat{u}}{\partial \hat{\upsilon}}\right)_T\right]d\hat{\upsilon}. \tag{8.23}$$

The volume and temperature can be varied independently to move between two equilibrium states. Setting $d\hat{\upsilon}=0$ and $dT\neq 0$, we find

$$\frac{\hat{C}_\upsilon}{T}=\left(\frac{\partial \hat{s}}{\partial T}\right)_{\hat{\upsilon}}, \tag{8.24}$$

and setting $dT=0$ and $d\hat{\upsilon}\neq 0$,

$$\left(\frac{\partial \hat{u}}{\partial \hat{\upsilon}}\right)_T=T\left(\frac{\partial P}{\partial T}\right)_{\hat{\upsilon}}-P. \tag{8.25}$$

Internal Energy

Now, substituting Equation (8.25) into Equation (8.20),

$$d\hat{u}=\hat{C}_\upsilon dT+\left[T\left(\frac{\partial P}{\partial T}\right)_{\hat{\upsilon}}-P\right]d\hat{\upsilon} \tag{8.26}$$

and $\hat{\upsilon}=\hat{\rho}^{-1}$ so

$$d\hat{\upsilon}=-\frac{d\hat{\rho}}{\hat{\rho}^2}.$$

Substituting for $d\hat{\upsilon}$ we can integrate over temperature and density. As shown in Figure 8.2, this is done by integrating from a reference state along a line of constant density, followed by integration along a line of constant temperature until we reach the conditions $(T,\hat{\rho})$. With the reference state chosen as $(T_0,\hat{\rho}=0)$, the first integration is carried out at zero density, which allows $\hat{C}_\upsilon$ to be replaced by $\hat{C}_\upsilon^o$. $\hat{C}_\upsilon^o$ is the ideal gas specific heat capacity at constant specific volume and $\hat{C}_v^0$ is a function of temperature only. These steps produce Equation (8.1), which is repeated here:

$$\hat{u} = \int_{T_0}^{T} \hat{C}_\upsilon^o dT + \int_0^{\hat{\rho}} \frac{1}{\hat{\rho}^2}\left[P - T\left(\frac{\partial P}{\partial T}\right)_{\hat{\rho}}\right] d\hat{\rho} + \hat{u}_0. \quad (8.27)$$

Enthalpy

The specific enthalpy $\hat{h}$ is defined as $\hat{h} = \hat{u} + P\hat{\upsilon}$, so we simply add $P\hat{\upsilon}$ (as $P/\hat{\rho}$) to Equation (8.27) and we obtain Equation (8.3), which is repeated here:

$$\hat{h} = \int_{T_0}^{T} \hat{C}_\upsilon^o dT + \int_0^{\hat{\rho}} \frac{1}{\hat{\rho}^2}\left[P - T\left(\frac{\partial P}{\partial T}\right)_{\hat{\rho}}\right] d\hat{\rho} + \frac{P}{\hat{\rho}} + \hat{h}_0. \quad (8.28)$$

Entropy

For the specific entropy $\hat{s}$, substituting Equation (8.24) into Equation (8.16),

$$d\hat{s} = \frac{\hat{C}_\upsilon}{T} dT + \left(\frac{\partial P}{\partial T}\right)_{\hat{\upsilon}} d\hat{\upsilon}, \quad (8.29)$$

and again using $d\hat{\upsilon} = -d\hat{\rho}/\hat{\rho}^2$, we have

$$d\hat{s} = \frac{\hat{C}_\upsilon}{T} dT - \frac{1}{\hat{\rho}^2}\left(\frac{\partial P}{\partial T}\right)_{\hat{\rho}} d\hat{\rho}, \quad (8.30)$$

giving

$$\int_{\hat{s}_0}^{\hat{s}} d\hat{s} = \int_{T_0}^{T} \frac{\hat{C}_\upsilon^o}{T} dT - \int_0^{\hat{\rho}} \frac{1}{\hat{\rho}^2}\left[\left(\frac{\partial P}{\partial T}\right)_{\hat{\rho}}\right] d\hat{\rho},$$
$$\text{but not the same as Equation (8.5),} \quad (8.31)$$

or

$$\int_{\hat{s}_0}^{\hat{s}} d\hat{s} = \int_{T_0}^{T} \frac{\hat{C}_\upsilon^o}{T} dT + \int_{\hat{\upsilon}=\infty}^{\hat{\upsilon}} \left[\left(\frac{\partial P}{\partial T}\right)_{\hat{\upsilon}}\right] d\hat{\upsilon},$$
$$\text{but not the same as Equation (8.6).} \quad (8.32)$$

Note that Equation (8.31) does not match Equation (8.5) and Equation (8.32) does not match Equation (8.6). Often, depending on the form of the EOS, a direct integration to the lower limit of $\hat{\rho} = 0$ in Equation (8.31) or $\hat{\upsilon} = \infty$ in Equation (8.32) is not possible.

The solution to this problem comes from examining the virial EOS which is shown in Equation (8.33). Here B, C, and so on, are the virial coefficients, which are functions of temperature, for a given species. Most modern EOS are modifications of the viral EOS:

$$P = RT\hat{\rho} + RTB\hat{\rho}^2 + RTC\hat{\rho}^3 + \cdots. \quad (8.33)$$

Consider the $d\hat{\rho}$ integration in Equation (8.31) with the virial EOS:

$$\int_0^{\hat{\rho}} \frac{1}{\hat{\rho}^2}\left[\left(\frac{\partial P}{\partial T}\right)_{\hat{\rho}}\right] d\hat{\rho} = \left[R\ln(\hat{\rho}) + RB\hat{\rho} + \frac{RC\hat{\rho}^2}{2} + \cdots\right]_0^{\hat{\rho}}$$
$$= R\ln(\hat{\rho}) + RB\hat{\rho} + \frac{RC\hat{\rho}^2}{2} + \cdots - R\ln(\hat{\rho})_{\hat{\rho}\to 0}. \quad (8.34)$$

The integral in Equation (8.34) does not have a finite lower limit as $\hat{\rho} \to 0$ because the $\left(\lim_{\hat{\rho}\to 0} \ln(\hat{\rho})\right) = \infty$. To find a solution to the lower limit integration problem, we can add a term,

$$\left(\frac{R}{\hat{\rho}} d\hat{\rho}\right),$$

to both sides of Equation (8.31), giving

$$\int_{\hat{s}_0}^{\hat{s}} d\hat{s} + \int_0^{\hat{\rho}} \frac{R}{\hat{\rho}} d\hat{\rho} = \int_{T_0}^{T} \frac{\hat{C}_\upsilon^o}{T} dT + \int_0^{\hat{\rho}} \frac{1}{\hat{\rho}^2}\left[\hat{\rho}R - \left(\frac{\partial P}{\partial T}\right)_{\hat{\rho}}\right] d\hat{\rho}. \quad (8.35)$$

Integration of the left-hand side of Equation (8.35) gives

$$\hat{s} - \left(\lim_{\hat{s}\to\hat{s}_0} \hat{s}\right) + R\ln(\hat{\rho}) - \left(\lim_{\hat{\rho}\to 0} R\ln(\hat{\rho})\right) = \int_{T_0}^{T} \frac{\hat{C}_\upsilon^o}{T} dT + \int_0^{\hat{\rho}} \frac{1}{\hat{\rho}^2}\left[\hat{\rho}R - \left(\frac{\partial P}{\partial T}\right)_{\hat{\rho}}\right] d\hat{\rho}, \quad (8.36)$$

or rearranging,

$$\hat{s} - \left(\lim_{\hat{s}\to\hat{s}_0} \hat{s}\right) - \left(\lim_{\hat{\rho}\to 0} R\ln(\hat{\rho})\right) = \int_{T_0}^{T} \frac{\hat{C}_\upsilon^o}{T} dT + \int_0^{\hat{\rho}} \frac{1}{\hat{\rho}^2}\left[\hat{\rho}R - \left(\frac{\partial P}{\partial T}\right)_{\hat{\rho}}\right] d\hat{\rho} - R\ln(\hat{\rho}). \quad (8.37)$$

Every term on the right-hand side (RHS) of Equation (8.37) is finite. Therefore, the singularity in the $\ln(\hat{\rho})$ as $\hat{\rho} \to 0$ on the left-hand side of the equation must cancel with the limit on $\hat{s}$ as $\hat{s} \to \hat{s}_0$. This allows us to write

$$\hat{s} = \int_{T_0}^{T} \frac{\hat{C}_\upsilon^o}{T} dT + \int_0^{\hat{\rho}} \frac{1}{\hat{\rho}^2}\left[\hat{\rho}R - \left(\frac{\partial P}{\partial T}\right)_{\hat{\rho}}\right] d\hat{\rho} - R\ln(\hat{\rho}),$$

and, recalling our discussion of reference values, allows the final result to be expressed as

$$\hat{s} = \int_{T_0}^{T} \frac{\hat{C}_{\upsilon}^{o}}{T} dT + \int_{0}^{\hat{\rho}} \frac{1}{\hat{\rho}^2}\left[\hat{\rho} R - \left(\frac{\partial P}{\partial T}\right)_{\hat{\rho}}\right] d\hat{\rho} - R\ln(\hat{\rho}) + \hat{s}_0, \tag{8.38}$$

where here $\hat{s}_0$ is simply a constant used to set the datum for $\hat{s}$. Equation (8.38) is our working equation for specific entropy equation; Equation (8.38) is Equation (8.5). Following our development of Equation (8.38), we can make changes to Equation (8.32) to allow direct integration from $\hat{\upsilon} = \infty$ to $\hat{\upsilon}$; Equation (8.39) is Equation (8.6):

$$\hat{s} = \int_{T_0}^{T} \frac{\hat{C}_{\upsilon}^{o}}{T} dT + \int_{\hat{\upsilon}=\infty}^{\hat{\upsilon}} \left[\left(\frac{\partial P}{\partial T}\right)_{\hat{\upsilon}} - \frac{R}{\hat{\upsilon}}\right] d\hat{\upsilon} + R\ln(\hat{\upsilon}) + \hat{s}_0. \tag{8.39}$$

8.4 IDEAL MIXTURE THERMODYNAMIC PROPERTIES: GENERAL DEVELOPMENT AND COMBUSTION REACTION CONSIDERATIONS

Most thermodynamic problems require calculation of mixture properties. For cogeneration calculations involving work, energy transfer, and system optimization, ideal mixture properties can be used. The ideal mixture assumption is reasonable as mixtures in cogeneration processes are generally at low pressure. An ideal mixture can apply to either the vapor or liquid phase, and here real species properties are used with the assumption that mixing properties, Δu_{mix}, $\Delta \upsilon_{\text{mix}}$, and Δh_{mix}, are each equal to 0.

8.4.1 Ideal Mixture

For an ideal mixture, which may be either vapor or liquid (Sandler, 1999), the mixture properties are

$$u = u^{IM}(P, T, y_i) = \sum y_i u_i(P, T), \tag{8.40}$$

$$\upsilon = \upsilon^{IM}(P, T, y_i) = \sum y_i \upsilon_i(P, T), \tag{8.41}$$

$$h = h^{IM}(P, T, y_i) = \sum y_i h_i(P, T), \tag{8.42}$$

and

$$s = s^{IM}(P, T, y_i) = \sum y_i s_i(P, T) - R\sum y_i \ln y_i. \tag{8.43}$$

For example, the estimation of a binary mixture enthalpy at a given temperature and pressure, $h^{IM}(P, T)$, would be

$$h^{IM}(P, T, y_i) = y_1 h_1(P, T) + y_2 h_2(P, T),$$

where the pure species enthalpies h_1 and h_2 are taken at the mixture P and T.

8.4.2 Changes in Enthalpy and Entropy

Pure Species

As we have discussed, the absolute value for the enthalpy or entropy of a species is reported relative to a reference state. For streams containing a single species, changes in enthalpy or entropy are independent of the reference state, as it cancels out. For example, we can view Equation (8.8) as $h = f(h(T, \upsilon)) + h_0$. The change in enthalpy between states 1 (in) and 2 (out) would be $N_{\text{out}}(f(h(T_{\text{out}}, \upsilon_{\text{out}})) + h_0) - N_{\text{in}}(f(h(T_{\text{in}}, \upsilon_{\text{in}})) + h_0)$, and with constant molar flow rates $N_{\text{out}} = N_{\text{in}} = N_{\text{species}}$, the h_0 terms cancel.

Mixtures: Without Reaction

Consider that several pure species streams are mixed to create a single outlet stream with ideal mixture properties. The terms h_0 and s_0 will again cancel as there will be no change in the molar flow rates of any species in and out. However, we will need to account for a change of entropy on mixing given by $-R\sum y_i \ln y_i$.

Mixtures: With Reaction, Excel Functions for Combustion Products Enthalpy (Tables 8.1c and 8.2c)

Generally, cogeneration process calculations involve pure species streams (here air is treated as a single species stream) until the combustion reaction process. Streams downstream of the combustion process are mixtures. Consider the general combustion process shown in Figure 8.5.

The enthalpy balance around the combustion chamber must account for the energy of the combustion reaction. As the molar flow rates for individual species in and out are not all equal, the h_0 terms will not cancel. The s_0 terms will not cancel in the entropy balance. This problem can be solved by utilizing the fact that the enthalpy of the combustion reaction (the fuel lower heating value [*LHV*]) is defined as the energy transferred when the reaction products and reactants are at the same temperature and pressure. This defined reaction temperature and pressure can be used to eliminate h_0 in the enthalpy.

As discussed in Chapter 7, the enthalpy contribution from the reaction of the fuel with air can be given by the fuel *LHV*. Here, in order to remain consistent in our energy balance, we should use the reference state associated with the fuel *LHV*. The *LHV* for fuels is provided at 77°F and 1 atm. The *LHV* is defined as the heat liberated when 1 lb of fuel is mixed with stoichiometric oxygen at 77°F and completely burned at constant pressure. The products are then cooled to 77°F with the water formed remaining in the vapor state.

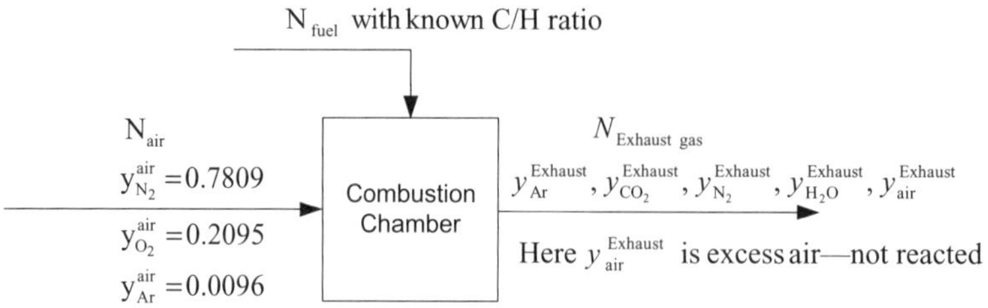

Figure 8.5 General combustion process with air, fuel, and a mixture of exhaust gas products.

For now, allow that we know both the exit composition from the combustor and the molecular weight of these products. These calculations are detailed in Chapter 9 and require knowledge of the excess air used in the combustion process and the fuel carbon to hydrogen ratio. The assumption of complete fuel combustion combined with the fuel *LHV* reference conditions allows us to create the Excel functions provided in combustion products (Tables 8.1c and 8.2c) to simplify cogeneration calculations. The use of combustion products is explained in detail in Chapter 9. With the exit composition and molecular weight known, we can write the enthalpy of the combustion products (the exhaust gas) as

$$\hat{h}_{\text{Products}} = \frac{\left\{\begin{array}{l} y_{\text{H}_2\text{O}}\, H\_Steam(P,T) + y_{\text{CO}_2}\, H\_CarbDiox(P,T) \\ \quad + y_{\text{Air}} H\_Air(P,T) + y_{\text{Ar}}\, H\_Argon(P,T) \\ \quad + y_{\text{N}_2}\, H\_Nitrogen(P,T) - y_{\text{H}_2\text{O}}\,(19{,}595.414) \\ \quad - y_{\text{CO}_2} H\_CarbDiox(14.696, 536.67) \\ \quad - y_{\text{Air}} H\_Air(14.696, 536.67) \\ \quad - y_{\text{Ar}} H\_Argon(14.696, 536.67) \\ \quad - y_{\text{N}_2} H\_Nitrogen(14.696, 536.67) \end{array}\right\}}{MW\_Products}.$$

Here the enthalpy (British thermal unit per pound) of the combustion products is determined as

$$\sum \frac{y_i h_i(P,T) - y_i h_i(P_{\text{ref}}, T_{\text{ref}})}{MW\_Products};$$

this eliminates species h_0. However, at the conditions 14.696 psia and 536.67 R, steam will not exist. Here we curve fit steam enthalpy at 14.696 psia over a wide temperature range and then extended the resulting polynomial to 536.67 R; the value obtained for the pseudoenthalpy of steam (14.696 psia, 536.67 R) is 19,505.414 Btu/lb-mol. An enthalpy balance on the combustion chamber would then be

$$N_{\text{air}}\left(H\_Air(P,T) - H\_Air(14.696, 536.67)\right) + F_{\text{fuel}}(LHV) = (F_{\text{air}} + F_{\text{fuel}})\left(\hat{h}_{\text{Products}}\right) + \dot{Q}_{\text{loss}}. \tag{8.44}$$

Enthalpy balances on downstream single-stream input, single-stream output units would then be

$$(F_{\text{air}} + F_{\text{fuel}})\Delta\hat{h}_{\text{Products}}, \text{ where } \Delta\hat{h}_{\text{Products}} = \hat{h}_{\text{Products},(P_{\text{out}},T_{\text{out}})} - \hat{h}_{\text{Products},(P_{\text{in}},T_{\text{in}})}. \tag{8.45}$$

Mixtures: Excel Functions for Combustion Products Entropy (Tables 8.1c and 8.2c)

An entropy balance on the combustion chamber will generally not be needed. For energy transformation systems, calculation of entropy is needed for power devices including air compressors, gas turbines, and power turbines. We can write the entropy of the products (the exhaust gas from the combustor) as

$$\hat{s}_{Products} = \frac{\left\{\begin{array}{l} y_{\text{H}_2\text{O}} S\_Steam(P,T) + y_{\text{CO}_2}\, S\_CarbDiox(P,T) \\ \quad + y_{\text{Air}} S\_Air(P,T) + y_{\text{Ar}} S\_Argon(P,T) \\ \quad + y_{\text{N2}} S\_Nitrogen(P,T) - y_{\text{H}_2\text{O}}(29.6476) \\ \quad - y_{\text{CO}_2} S\_CarbDiox(14.696, 536.67) \\ \quad - y_{\text{Air}} S\_Air(14.696, 536.67) \\ \quad - y_{\text{Ar}} S\_Argon(14.696, 536.67) \\ \quad - y_{\text{N}_2} S\_Nitrogen(14.696, 536.67) \end{array}\right\}}{MW\_Products}.$$

The entropy (British thermal unit per pound-Rankine) of the combustion products is

$$\sum \frac{y_i s_i(P,T) - y_i s_i(P_{\text{ref}}, T_{\text{ref}})}{MW\_Products};$$

this eliminates species s_0. At reference conditions 14.696 psia and 536.67 R, steam will not exist. Here, we curve fit steam entropy at 14.696 psia over a wide temperature range and then extended the resulting polynomial to 536.67 R; the value obtained for the pseudoentropy of steam (14.696 psia, 536.67 R) was 29.6476 Btu/lb-mol-R. Do note in the calculation of entropy we have used ($y_i s_i(P, T) - y_i s_i(P_{\text{ref}}, T_{\text{ref}})$). For a power device (gas turbine or power turbine) with a single input and single output stream and with no change in composition, the term $-R\sum(y_i \ln y_i)$ will cancel. The entropy balance for units downstream of the combustor with a single-stream input and a single-stream output (and no reaction) will be

$$(F_{\text{air}} + F_{\text{fuel}})\Delta\hat{s}_{\text{Products}},$$

where $\Delta\hat{s}_{\text{Products}} = \hat{s}_{\text{Products},(P_{\text{out}},T_{\text{out}})} - \hat{s}_{\text{Products},(P_{\text{in}},T_{\text{in}})}$. (8.46)

8.5 IDEAL MIXTURE THERMODYNAMIC PROPERTIES: APPARENT DIFFICULTIES

The ideal mixture assumption will prove useful for thermodynamic properties of the combustion systems studied in this text. However, there can be difficulties in applying the ideal mixture assumption. For example, consider the C2/C3 distillation system shown in Figure 8.6. Distillation is a key unit operation in chemical processing plants where a feed mixture is generally separated into two products. Distillation columns operate by preferentially boiling the more volatile components from the feed mixture to the overhead product. Each stage in the column (except the condenser and reboiler) can be considered an adiabatic flash unit (Chapter 5.2). Column simulations (e.g., HYSYS or Aspen Plus simulations) are generally used to determine the number of stages or trays needed to obtain desired product purities as well as condenser and reboiler duties. Columns (condenser and reboiler duties) are often the most energy-intensive operations in a processing plant.

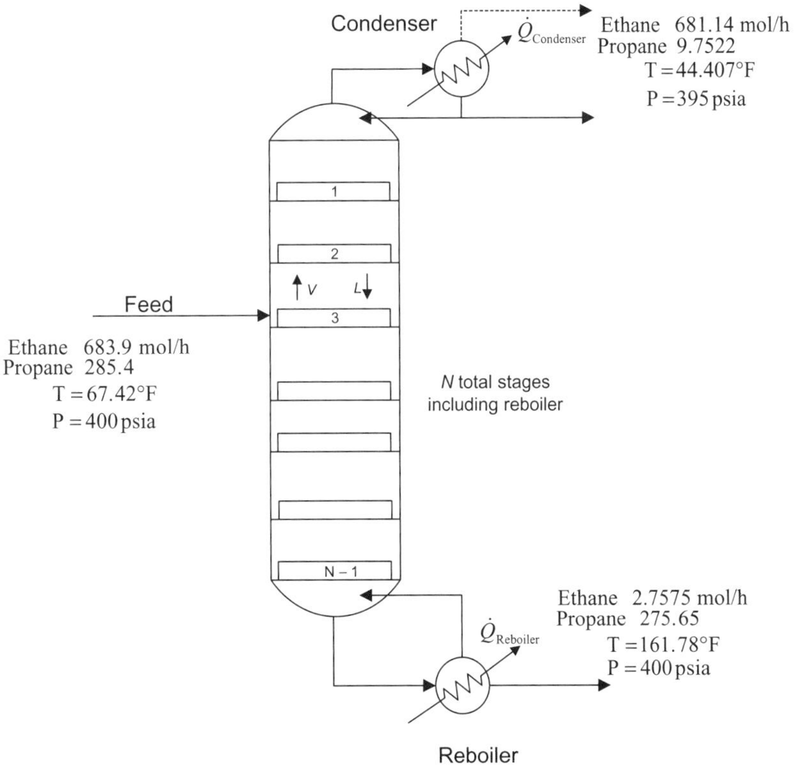

Figure 8.6 Ethane/propane (C2/C3) distillation column.

EXAMPLE 8.5 ***Solve for the Reboiler and Condenser Duties in a C2/C3 Distillation Column Assuming Ideal Mixture and Using the RK-EOS***

Figure 8.7 provides detailed results from a HYSYS simulation (with Peng–Robinson EOS, 15 trays, reflux ratio = 2.5, and C2/C3 ratio in reboiler = 0.01) showing species flows and duties at both the partial condenser and the reboiler for the C2/C3 column in Figure 8.6.

SOLUTION We want to use the RK-EOS to estimate reboiler and condenser duties and compare values with those reported in Figure 8.7. For the RK-EOS, we first determine parameters for ethane and propane.

Ethane

For ethane C_v^o,

$$C_v^o = -0.3392 + 0.04124T - (1.53\times10^{-5})T^2 + (1.74\times10^{-9})T^3\left(\frac{\text{cal}}{\text{gmol-K}}\right), \tag{8.47}$$

and $T_{\text{critical}} = 305.5$ K and $P_{\text{critical}} = 48.2$ atm (adapted from Myers and Seider, 1976). For the RK-EOS,

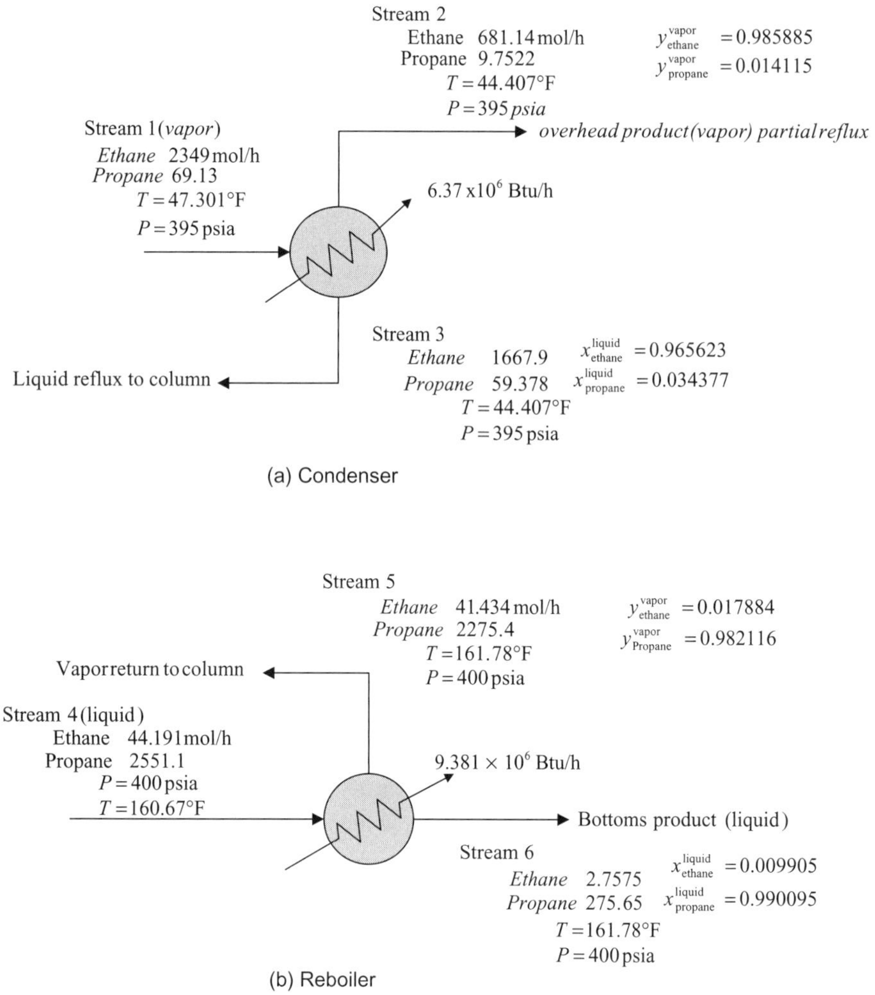

Figure 8.7 Condenser and reboiler duties and species flows (C2/C3) distillation column.

$$a = 9.7496\times10^{7}\ \text{atm K}^{0.5}\left(\frac{\text{cm}^3}{\text{g-mol}}\right)^2$$

$$\text{and}\quad b = 45.09\left(\frac{\text{cm}^3}{\text{g-mol}}\right).$$

When using Equations (8.11) and (8.12) and the pure component ethane data provided by Reynolds (1979), T_0 = 150 K and h_0 = 4067.67 cal/gmol and s_0 = 5.5412 cal/gmol-K. These calculations can be found in the Excel files **Example 8.5a (vapor ethane).xls** or **Example 8.5b (liquid ethane).xls**.

Propane

For propane C_v^o,

$$C_v^o = -2.9532 + 0.07279T - (3.755\times10^{-5})T^2 + (7.58\times10^{-9})T^3\left(\frac{\text{cal}}{\text{gmol-K}}\right), \tag{8.48}$$

and T_{critical} = 370 K and P_{critical} = 42 atm (adapted from Myers and Seider, 1976). For the RK-EOS,

$$a = 1.806\times10^{8}\ \text{atm K}^{0.5}\left(\frac{\text{cm}^3}{\text{g-mol}}\right)^2$$

$$\text{and}\quad b = 62.68\left(\frac{\text{cm}^3}{\text{g-mol}}\right).$$

When using Equations (8.11) and (8.12) and the pure component propane data provided by Reynolds (1979), T_0 = 200 K and h_0 = 4576.62 cal/gmol and s_0 = 2.9448 cal/gmol-K. These calculations can be found in the Excel files **Example 8.5c (vapor propane).xls** or **Example 8.5d (liquid propane).xls**.

There are problems when using these pure component files to determine enthalpy in and out of the condenser and reboiler. For the *condenser*,

Ethane Entering the Condenser (in Stream 1, Figure 8.7). The mixture entering the condenser is vapor phase (T_{in} = 47.301°F,

P_{in} = 395 psia), and we can solve for the needed molar volume υ using Equation (8.10). In the Excel file **Example 8.5a (vapor ethane).xls**, Equation (8.10) is provided in cell 6J and Excel Solver is used to vary υ in cell 3L; here we find υ = 591.283 cm^3/g-mol and h_{ethane} = 9254.57 Btu/lb-mol.

Propane Entering the Condenser (in Stream 1, Figure 8.7). In the Excel file **Example 8.5c (vapor propane).xls**, Equation (8.10) is provided in cell 6J and Solver is used to vary υ in cell 3L, but here we find Solver cannot find a vapor-phase value for υ to solve Equation (8.10). The problem here is that at 395 psia, the bubble/dew point for propane is ~162.2°F; in other words, at 395 psia and 47°F, pure propane is a liquid.

A similar problem occurs in the reboiler, but here problems occur with ethane. For the *reboiler*,

Propane Entering the Reboiler (in Stream 4, Figure 8.7). The mixture entering the reboiler is liquid phase (T_{in} = 160.67°F, P_{in} = 400 psia), and we can solve for the needed molar volume υ using Equation (8.11). In the Excel file **Example 8.5d (liquid propane).xls**, Equation (8.10) is provided in cell 6J and Solver is used to vary υ in cell 3L. Here with υ bounded, $105 \leq \upsilon \leq 150$ cm^3/g-mol (for the liquid molar volume root), we find υ = 134.14 cm^3/g-mol and $h_{propane}$ = 8197.88 Btu/lb-mol.

Ethane Entering the Reboiler (in Stream 4, Figure 8.7). In the Excel file **Example 8.5b (liquid ethane).xls**, Equation (8.10) is provided in cell 6J and Solver is used to vary υ in cell 3L. Here even with υ bounded, $50 \leq \upsilon \leq 150$ cm^3/g-mol (to find the liquid molar volume root), we cannot find a υ value to satisfy Equation (8.10). The problem here is that at 400 psia, the bubble/dew point for ethane is ~42.5°F; in other words, at 400 psia and 160°F, pure ethane is a vapor.

The problems encountered in this example are not unique to our using the RK-EOS; the same problems would occur if we were using pure component data or EOS fits from Reynolds (1979). One solution to this problem is to use mixing rules with the EOS to obtain thermodynamic properties for the mixture. ■

8.6 MIXING RULES FOR EOS

Mixing rules allow the formation of a single EOS that can then be used to determine thermodynamic properties for mixtures. For example, the RK-EOS for a mixture can be written as

$$P = \frac{RT}{\upsilon_{mixture} - b_{mixture}} - \frac{a_{mixture}}{\upsilon_{mixture}\left(\upsilon_{mixture} + b_{mixture}\right)T^{1/2}}. \quad (8.49)$$

Mixing rules for various EOSs can have a theoretical basis; however, for the Redlich–Kwong, the following "heuristic mixing rules" have been found to give reasonable results:

$$a_{mixture} = \sum_{i=1}^{n_{species}} \sum_{k=1}^{n_{species}} x_i x_k a_{ik}, \quad (8.50)$$

$$b_{mixture} = \sum_{i=1}^{n_{species}} x_i b_i, \quad (8.51)$$

and

$$a_{ik} = \sqrt{a_i a_k}. \quad (8.52)$$

For our binary distillation example, which consists of ethane (C2) and propane (C3), Equations (8.50)–(8.52) would be written:

$$a_{ik} = \sqrt{a_i a_k} = a_{C2C3} = \sqrt{(a_{C2})(a_{C3})}$$

$$\begin{aligned} a_{mixture} &= \sum_{i=1}^{n_{species}} \sum_{k=1}^{n_{species}} x_i x_k a_{ik} \\ &= \left(x_{C2}^2\right)(a_{C2}) + 2(x_{C2})(x_{C3})(a_{C2C3}) + \left(x_{C3}^2\right)(a_{C3}) \end{aligned}$$

$$b_{mixture} = x_{C2} b_{C2} + x_{C3} b_{C3}.$$

EXAMPLE 8.6 ***Solve for the Reboiler and Condenser Duties in a C2/C3 Distillation Column Using the RK-EOS with Mixing Rules***

Here we estimate thermodynamic properties using the RK-EOS and mixing rules provided by Equations (8.49)–(8.52). We want to determine $\dot{Q}_{reboiler}$ and $\dot{Q}_{condenser}$. It is straightforward to determine these quantities by constructing thermodynamic paths around the reboiler and condenser, which is discussed next.

SOLUTION For the Reboiler

To determine the reboiler duty, we can construct two thermodynamic paths, which are shown in Table 8.8. In the first path, the column vapor return stream (stream 5) is taken from the reboiler inlet conditions (T = 160.7°F, P = 400 psia, liquid phase) to stream 5 conditions (T = 161.78°F, P = 400 psia, vapor phase). In the second path, the column bottom product stream (stream 6) is taken from the reboiler inlet conditions (T = 160.7°F, P = 400 psia, liquid phase) to stream 6 conditions (T = 161.78°F, P = 400 psia, liquid phase). The combined streams into the reboiler (streams 4a + 4b in Table 8.8) do account for stream 4 as found in Figure 8.7.

The column reboiler duty is found as the sum of these two paths, $\Delta h = (h_{out} - h_{in})$, for the column vapor return ($h_{stream\ 5} - h_{stream\ 4a}$) and for the bottom product stream ($h_{stream\ 6} - h_{stream\ 4b}$). This is analogous to making the conditions of stream 4 the reference conditions.

Using Equations (8.49) and (8.11), we can then write for the column vapor return,

Table 8.8 Reboiler Duty Calculations

	In	Out	x_i or y_i
Column vapor return (path 1)			
Stream (see Figure 8.7)	4a	5	
Ethane (lb-mol/h)	41.434	41.434	0.017884
Propane (lb-mol/h)	2275.4	2275.4	0.982116
Total (lb-mol/h)	2316.834	2316.834	
T (F)	160.7	161.78	
P (psia)	400	400	
Phase	liquid	vapor	
Product stream (path 2)			
Stream (see Figure 8.7)	4b	6	
Ethane (lb-mol/h)	2.7575	2.7575	0.009905
Propane (lb-mol/h)	275.65	275.65	0.990095
Total (lb-mol/h)	278.4075	278.4075	
T (F)	160.7	161.78	
P (psia)	400	400	
Phase	liquid	liquid	

$$\left(h_{\text{stream 5}} - h_{\text{stream 4a}}\right) = \sum_{i=1}^{n_{\text{species}}} x_i \int_{T_4}^{T_5} C^o_{\upsilon,i} dT$$
$$+\left[-\frac{3a_{\text{mixture}}}{2T_5^{0.5} b_{\text{mixture}}} \ln\left(\frac{\upsilon_{\text{mixture},5} + b_{\text{mixture}}}{\upsilon_{\text{mixture},5}}\right) + P_5 \upsilon_{\text{mixture},5}\right]$$
$$-\left[-\frac{3a_{\text{mixture}}}{2T_4^{0.5} b_{\text{mixture}}} \ln\left(\frac{\upsilon_{\text{mixture},4} + b_{\text{mixture}}}{\upsilon_{\text{mixture},4}}\right) + P_4 \upsilon_{\text{mixture},4}\right]. \tag{8.53}$$

Note that the values for the $x_i \int_{T_4}^{T_5} C^o_{\upsilon,i} dT$, a_{mixture} and b_{mixture} terms, will depend on the species mole fractions (x_i or y_i), which is here simply denoted as x_i. In developing Equation (8.53), the h_0 terms cancel, and by setting $T_0 = T_4$, the lower limit on the species $C^o_{\upsilon,i}$ integral becomes T_4.

The calculation for

$$\sum_{i=1}^{n_{\text{species}}} x_i \int_{T_4}^{T_5} C^o_{\upsilon,i} dT + \left[-\frac{3a_{\text{mixture}}}{2T_5^{0.5} b_{\text{mixture}}} \ln\left(\frac{\upsilon_{\text{mixture},5} + b_{\text{mixture}}}{\upsilon_{\text{mixture},5}}\right) + P_5 \upsilon_{\text{mixture},5}\right]$$
$$= -109.4711 \text{ Btu/lb-mol}$$

can be found in the Excel file **Example 8.6a (vapor reboiler).xls**; here,

$$a_{\text{mixture}} = 1.789 \times 10^8 \text{ atm K}^{0.5}\left(\frac{\text{cm}^3}{\text{g-mol}}\right)^2,$$
$$b_{\text{mixture}} = 62.36\left(\frac{\text{cm}^3}{\text{g-mol}}\right),$$

and using Solver (as described earlier), we find

$$\upsilon_{\text{mixture},5} = 659.65 \frac{\text{cm}^3}{\text{g-mol}}.$$

The calculation for

$$\left[-\frac{3a_{\text{mixture}}}{2T_4^{0.5} b_{\text{mixture}}} \ln\left(\frac{\upsilon_{\text{mixture},4} + b_{\text{mixture}}}{\upsilon_{\text{mixture},4}}\right) + P_4 \upsilon_{\text{mixture},4}\right]$$
$$= -3664.7768 \text{ Btu/lb-mol}$$

can be found in the Excel file **Example 8.6b (liquid reboiler).xls**; when using Solver to determine $\upsilon_{\text{mixture}}$, be sure υ is bounded, $50 \le \upsilon \le 150$ cm^3/g-mol (for the liquid molar volume root). Here then, $(h_{\text{stream 5}} - h_{\text{stream 4a}}) = (-109.4711 - (-3664.7768)) =$ 3555.3057 Btu/lb-mol. The energy rate needed to create stream 5 is (2316.834 lb-mol/h) (3555.3057 Btu/lb-mol) = 8,237,053 Btu/h; it is positive quantity as heat is being added to the reboiler to vaporize stream 4a to stream 5.

For the bottom product stream, $(h_{\text{stream 6}} - h_{\text{stream 4b}}) =$ 53.6095 Btu/lb-mol, which can be determined by appropriate changes to the Excel file **Example 8.6b (liquid reboiler).xls**. The energy rate needed to create stream 6 is (278.4075 lb-mol/h) (53.6095 Btu/lb-mol) = 14,925 Btu/h; it is positive quantity as heat is being added to the reboiler to heat stream 4b to stream 6.

Finally, we can determine $\dot{Q}_{\text{rebolier}} = 8{,}251{,}978$ Btu/h. When compared to the value determined from the HYSYS simulation, there is a ~12% error in the reboiler predicted heat duty when using the RK-EOS as compared to the Peng–Robinson EOS.

For the Condenser

Determining the condenser duty is left as an exercise for the reader. But remember to again construct two thermodynamic paths. In the first path, the column liquid reflux stream (stream 3) is taken from the condenser inlet conditions (T = 47.301°F, P = 395 psia, liquid phase) to stream 3 conditions (T = 44.407°F, P = 395 psia, liquid phase). In the second path, the column overhead product stream (stream 2) is taken from the condenser inlet conditions (T = 47.301°F, P = 395 psia, liquid phase) to stream 2 conditions (T = 44.407, P = 395 psia, vapor phase). The files **Example 8.6c (vapor condenser).xls** and **Example 8.6d (liquid condenser).xls** are provided to check your calculations. You should find that $\dot{Q}_{\text{condenser}} = -5{,}956{,}225$ Btu/h, which has a ~6.5% error when compared to the value determined from the HYSYS simulation.

There are many available EOS for thermodynamic property calculations. In the HYSYS simulation of the C2/C3 column, we utilized the Peng–Robinson EOS (Peng and Robinson, 1976). The Peng–Robinson EOS utilizes three parameters (a, b, and one additional), and mixing rules for the Peng–Robinson equation generally include an adjustable interaction parameter, which is obtained from experimental equilibrium data. Regardless of the EOS, the calculation procedure for changes in enthalpy follow Example 8.6.

For some combustion systems, you may want to check the results obtained from the provided *TPSI+* (which uses the EOS from Reynolds, 1979). We have provided species-dependent T_{critical}, P_{critical}, and C^o_υ data for use with the RK-EOS in Table 8.9. ■

Table 8.9 Species Critical Properties (K, atm) and Molar Heat Capacity at Zero Density, $C_v^o = a + bT + cT^2 + dT^3$, $\left(\frac{\text{cal}}{\text{gmol-K}}\right)$ (Adapted from Myers and Seider, 1976)

Species	$T_{critical}$	$P_{critical}$	a	$b \times 10^2$	$c \times 10^5$	$d \times 10^9$
Air			4.7258	0.04697	0.1147	−0.4696
Ammonia	405.5	111.3	4.5974	0.61251	0.23663	−1.5981
Methane	191.1	45.8	2.7628	1.200	0.3030	−2.630
Ethane	305.5	48.2	−0.3392	4.124	−1.530	1.740
Propane	370.0	42.0	−2.9532	7.279	−3.755	7.580
n-butane	425.2	37.5	−1.0422	8.873	−4.380	8.360
i-butane	408.1	36.0	−3.8772	9.936	−5.495	11.92
n-pentane	469.8	33.3	−0.3692	10.85	−5.365	10.10
n-hexane	507.9	29.9	−0.3302	13.19	−6.844	13.78
Hydrogen	33.3	12.8	4.4368	0.1039	−0.007804	–
Carbon dioxide*	304.2	72.9				
Nitrogen	126.2	33.5	4.9158	−0.03753	0.1930	−0.6861
Oxygen	154.8	50.1	4.0978	0.3631	−0.1709	0.3133
Water vapor	647.4	218.3	5.7128	0.04594	0.2521	−0.8587

*For carbon dioxide, $C_v^o = 16.0488 - 0.00004474T - \frac{158.08}{T^{0.5}}\left(\frac{\text{cal}}{\text{g-mol-K}}\right)$.

8.7 CLOSING REMARKS

It is important for students working in the energy area to have an appreciation and understanding of how thermodynamic databases are constructed and how thermodynamic properties are determined. The work of Reynolds (1979) remains an industry standard for pure components as well as for the determination of steam and water properties. We have shown how pure species information can be used to construct mixture properties—with the assumption of ideal mixtures. We have also shown how any EOSs can be used to determine thermodynamic properties including enthalpy and entropy. The RK-EOS was used to develop enthalpy and entropy terms primarily because mathematical operations are straightforward with this EOS. The RK-EOS generally provides reasonable results for vapor-phase thermodynamic properties (at low to medium pressure), but liquid-phase properties may not be as accurate. We also showed the use of EOS mixing rules, which may be necessary to predict mixture properties. There are many EOSs; for example, Nasri and Binous (2009) show mixture property development using the Peng–Robinson EOS, and Nasri and Binous (2007) show the use of the Soave–RK-EOS. In the chapter problems (Problem 8.5), we provide discussion on how equilibrium distribution factors (coefficients), which were first introduced in Chapter 5, can be determined from EOS. We also show (in Problem 8.6) how EOS can be used to determine if a stream is a vapor-phase, a liquid-phase, or a two-phase mixture (bubble point and dew point temperature calculations).

In this chapter, pure species vapor- and liquid-phase enthalpies were calculated from an EOS with the inclusion of a reference state to allow matching with tabulated values. A more modern approach is to use the ideal gas as the reference state (at T and P reference) for all species. Here then, for example, pure species liquid enthalpy is calculated from the ideal gas enthalpy and a liquid enthalpy departure function. The liquid enthalpy departure function includes the heat of vaporization, the vapor enthalpy departure from ideal pressure to saturation pressure, and the liquid-phase pressure correction from saturation pressure to real pressure (Carlson, 1996).

Finally, for those who may need steam tables with expanded physical properties, Magnus Holmgren at http://www.x-eng.com provides a downloadable International Association for Properties of Water and Steam Industrial Formulation 1997 (IAPWS IF-97) Excel-based steam table add-in.

REFERENCES

DUPONT, 1993. Thermodynamic Properties of HFC-134a, Dupont Technical Information T-134a-SI.

BARRIE, P.J. 2005. JavaScript programs to calculate thermodynamic properties using cubic equations of state. *J. Chem. Educ.* 82(6): 958–960.

BAZMI, M., K. GHANBARI, and J. KLOMFAR. 2004. Liquid-vapor equilibrium calculations for MTBE/MEOH binary system using Patel & Teja equation of state. *Petroleum Coal* 46(2): 34–40.

BINOUS, H. 2008. Applications of the Peng–Robinson equation of state using Mathematica. *Chem. Eng. Educ.* 42(1): 47–51.

CARLSON, E.C. 1996. Don't gamble with physical properties for simulations. *Chem. Eng. Prog.* 35–46.

ELLIOTT, F.G., R. KURZ, C. ETHERIDGE, and J.P. O'CONNELL. 2004. Fuel system suitability considerations for industrial gas turbines. *J. Eng. Gas Turb. Power* 126: 119–126.

KEENAN, J.H., F.G. KEYES, P.G. HILL, and J.G. MOORE. 1969. *Steam Tables: Thermodynamic Properties of Water Including Vapor, Liquid and Solid Phases*. John Wiley & Sons, New York.

LAWAL, A.S. 1987. A consistent rule for selecting roots in cubic equations of state. *Ind. Eng. Chem. Res.* 26: 857–859.

MYERS, A.L. and W.D. SEIDER. 1976. *Introduction to Chemical Engineering and Computer Calculations*. Prentice-Hall, Englewood Cliffs, NJ.

NASRI, Z. and H. BINOUS. 2007. Applications of the Soave–Redlich–Kwong equation of state using mathematica. *J. Chem. Eng. Jpn.* 40(6): 534–538.

NASRI, Z. and H. BINOUS 2009. Applications of the Peng–Robinson equation of state using MATLAB. *Chem. Eng. Educ.* 43(2): 1–10.

PATEK, J. and J. KLOMFAR. 2009. A simple formulation for thermodynamic properties of steam from 273 to 523 K, explicit in temperature and pressure. *Int. J. Refrig.* 32: 1123–1125.

PENG, D.-Y. and D.B. ROBINSON. 1976. A new two-constant equation of state. *Ind. Eng. Chem. Fund.* 15: 59–64.

REDLICH, O. and J.N.S. KWONG. 1949. On the thermodynamics of solutions: V: An equation of state. Fugacities of gaseous solutions. *Chem. Rev.* 44: 233–244.

REYNOLDS, W.C. 1979. *Thermodynamic Properties in SI: Graphs, Tables, and Computational Equations for Forty Substances*. Department of Mechanical Engineering, Stanford University, Stanford, CA.

SANDLER, S.I. 1999. *Chemical and Engineering Thermodynamics*. John Wiley & Sons, New York.

SMITH, J.M., H.C. VAN NESS, and M.M. ABBOTT. 2001. *Introduction to Chemical Engineering Thermodynamics* (6th edition). McGraw-Hill, New York.

SPIEGEL, M.R. and J. LIU. 1999. *Mathematical Handbook of Formulas and Tables* (2nd edition). Schaum's Outlines McGraw Hill, New York.

WALAS, S.M. 1985. *Phase Equilibria in Chemical Engineering*. Butterworth Publishers, Stoneham, MA.

PROBLEMS

8.1 ***Maxwell Relation Showing*** $\left(\frac{\partial P}{\partial T}\right)_\upsilon = \left(\frac{\partial s}{\partial \upsilon}\right)_T$ Problem 8.1 will require outside reading. Starting with the Helmholtz energy in differential form, obtain the relation

$$\left(\frac{\partial P}{\partial T}\right)_\upsilon = \left(\frac{\partial s}{\partial \upsilon}\right)_T.$$

Maxwell relations are provided in standard engineering thermodynamics texts; see for example, Smith et al. (2001).

8.2 ***Thermodynamic Properties from Reynolds EOS*** We mentioned in Section 8.2 that Reynolds (1979) used 11 different EOSs to model pure species properties. One of the EOS, [EOS P-2], is given in Equation (P8.2a). Derive the expressions for $\hat{u}$, $\hat{h}$, and $\hat{s}$. Show that the lower limit of the density integral is finite. [EOS P-2] is used in the computer program provided in this text (TPSI+) for butane, ethane, octane, and propane vapor state properties. [EOS P-2] from Reynolds is

$$P = \hat{\rho}RT + \left(B_oRT - A_o - \frac{C_o}{T^2} + \frac{D_o}{T^3} - \frac{E_o}{T^4}\right)\hat{\rho}^2 + \left(bRT - a - \frac{d}{T}\right)\hat{\rho}^3 + \alpha\left(a + \frac{d}{T}\right)\hat{\rho}^6 + c\frac{\hat{\rho}^3}{T^2}(1+\gamma\hat{\rho}^2)e^{-\gamma\hat{\rho}^2}, \qquad \text{(P8.2a)}$$

and the $\hat{C}_\upsilon^0$ equation used with [EOS P-2] is

$$\hat{C}_\upsilon^0 = \sum_{i=1}^{6} G_i T^{i-2}. \qquad \text{(P8.2b)}$$

The species specific constants are provided in Reynolds (1979) as well as the computer code supplied here. Values for the constants are not needed to solve this example.

8.3 ***Development of a Simple Thermodynamic Property Package for Steam and the Use of the Secant Method*** Patek and Klomfar (2009) provide molar thermodynamic properties for steam, including g, h, s, υ and C_p, which are explicit in temperature (kelvin) and pressure (pascal).

The provided equations are

$$P_r = \frac{P}{P_c}, \quad T_r = \frac{T}{T_c}, \quad \Pi_i = a_i\left(\frac{1}{T_r}\right)^{n_i}(P_r)^{m_i};$$

$$g(T,P) = RT\left[\ln P_r + a_1 \ln\left(\frac{1}{T_r}\right) + \sum_{i=2}^{N}\Pi_i\right], \quad \frac{\text{J}}{\text{mol}};$$

$$h(T,P) = RT\left[a_1 + \sum_{i=2}^{N} n_i\Pi_i\right], \quad \frac{\text{J}}{\text{mol}};$$

$$s(T,P) = RT\left[-\ln P_r + a_1\left(1 - \ln\left(\frac{1}{T_r}\right)\right) + \sum_{i=2}^{N}(n_i - 1)\Pi_i\right], \quad \frac{\text{J}}{\text{mol-K}};$$

$$\upsilon(T,P) = \frac{RT}{P}\left[1 + \sum_{i=2}^{N} m_i\Pi_i\right], \quad \frac{\text{m}^3}{\text{mol}};$$

and

$$C_p(T,P) = R\left[a_1 - \sum_{i=2}^{N} n_i(n_i - 1)\Pi_i\right], \quad \frac{\text{J}}{\text{mol-K}}.$$

Coefficients are provided in Table P8.3.

Table P8.3 Coefficients and Exponents for Use with g, h, s, υ and C_p equations in Problem 8.3

i	n_i	m_i	a_i
1	0	−1	4.0374000
2	1	0	6.6686000
3	0	0	−8.7847400
4	−4	0	−0.0267859
5	4	1	−0.8240160
6	5	1	0.7398480
7	6	1	−0.2735580
8	2	2	0.2513250
9	12	2	−0.0864644
10	13	2	0.0404713
11	12	3	−2.0771400
12	15	3	2.4894100
13	16	3	−1.2638200
14	15	4	6.4675500
15	16	4	−4.6340400
16	15	5	−3.5951800
17	16	5	1.7165300

Known physical constants are

$$T_c = 647.096 \text{ K}, P_c = 22.064 \times 10^6 \text{ Pa},$$
$$R = 8.314371 \frac{\text{J}}{\text{mol-K}}.$$

(a) Develop VBA functions to calculate g, h, s, υ and C_p when provided (T, P) for vapor-phase steam. Determine g, h, s, υ and C_p values for (T [kelvin], P (pascal)) of (275, 698.451); (523, 698.451); (400, 245,769); (523, 245,769); and (523, 39,966,110).

(b) Determine the temperature for each steam property when provided a value of g, h, s, υ or C_p and a known pressure. The secant method is commonly used for this problem; we use the secant method in the provided thermodynamic package (*TPSI+*) to solve "inverse property" problems.

The secant method can be used to solve single-variable problems of the form $f(x) = 0$. The secant method requires no bounding. To start the secant method, two points, $(x^{[1]} f(x^{[1]}))$ and $(x^{[2]} f(x^{[2]}))$, are required. The secant method places a straight line between these points and then solves for the $x^{[3]}$ value, where $f(x^{[3]}) = 0$. This equation would be

$$x^{[3]} = x^{[2]} - f\left(x^{[2]}\right)\left(\frac{x^{[2]} - x^{[1]}}{f\left(x^{[2]}\right) - f\left(x^{[1]}\right)}\right).$$

The process can continue, retaining the two most recent trial points and $f(x)$ values as

$$x^{[k]} = x^{[k-1]} - f\left(x^{[k-1]}\right)\left(\frac{x^{[k-1]} - x^{[k-2]}}{f\left(x^{[k-1]}\right) - f\left(x^{[k-2]}\right)}\right).$$

8.4 ***Solve Example 8.6 for the Reflux Liquid Molar Volume Using the Bounding and Interval Halving Methods Developed in Chapter 3*** In Example 8.6, we used the Excel file **Example 8.6d (liquid condenser).xls** and Solver to determine $\upsilon_{\text{mixture}}$ for the liquid reflux stream (stream 3). For stream 3 with $T_3 = 44.407$°F and $P_3 = 395$ psia, we determined $\upsilon_{\text{mixture},3} = 91.1442$ cm³/g-mol; here when using Solver, we bounded $\upsilon_{\text{mixture}}$ as $50 \le \upsilon \le 100$ cm³/g-mol (for the liquid molar volume root).

Modify the Excel file **Example 8.6d (liquid condenser).xls** so that the bounding and interval halving methods developed in Chapter 3 are used to determine $\upsilon_{\text{mixture}}$; for the initial guess, use $\upsilon_{\text{mixture}} = 75$ cm³/g-mol.

8.5 ***Determination of Equilibrium Distribution Factor (Coefficient) Values, K_i from EOS (Solution Provided)*** In Chapter 5, we used equilibrium distribution factors (Eq. (5.22)) to determine species mole fractions in vapor and liquid phases for flash operations. Distribution factors are also used in the tray-to-tray calculations found in distillation columns and in other equilibrium stage separation devices. In Chapter 5, we indicated that generally, an EOS is used to determine K_i values. Here we want to provide the equations used to determine K_i values from an EOS and then solve for the species distribution coefficient in our C2/C3 column of Example 8.6. We will again utilize the RK-EOS.

We can write the equilibrium distribution factor (see Eq. (5.22)) as

$$K_i = \frac{x_i^{\text{vapor}}}{x_i^{\text{liquid}}} = \frac{y_i^{\text{vapor}}}{x_i^{\text{liquid}}}. \quad \text{(P8.5a)}$$

At equilibrium, $f_i^{\text{vapor}} = f_i^{\text{liquid}}$, where f_i is the species fugacity. We can also write $f_i^{\text{vapor}} = \phi_i^{\text{vapor}} y_i^{\text{vapor}} P$ and $f_i^{\text{liquid}} = \phi_i^{\text{liquid}} x_i^{\text{liquid}} P$, where ϕ_i is the species fugacity coefficient. So in addition to Equation (P8.5a), we can write the equilibrium distribution factor as

$$K_i = \frac{\phi_i^{\text{liquid}}}{\phi_i^{\text{vapor}}}. \quad \text{(P8.5b)}$$

The fugacity coefficient can be determined from any EOS as

$$RT \ln \phi_i = \int_{\infty}^{\upsilon} \left[\frac{RT}{\upsilon} - \left(\frac{\partial P}{\partial n_i} \right)_{T,\upsilon,n_j} \right] d\upsilon - RT \ln\left(\frac{P\upsilon}{RT} \right). \quad \text{(P8.5c)}$$

When using the RK-EOS with mixing rules as defined by Equations (8.49)–(8.52), we find (Walas, 1985)

$$\ln \phi_i = -\ln\left[\left(\frac{P\upsilon}{RT} \right)\left(1 - \frac{b_{\text{mixture}}}{\upsilon} \right) \right] + \left(\frac{b_i}{b_{\text{mixture}}} \right)\left(\frac{P\upsilon}{RT} - 1 \right) + \left(\frac{1}{b_{\text{mixture}} RT^{1.5}} \right)\left(\frac{a_{\text{mixture}} b_i}{b_{\text{mixture}}} - 2\left(a_{\text{mixture}} a_i\right)^{0.5} \right) \ln\left(1 + \frac{b_{\text{mixture}}}{\upsilon} \right). \quad \text{(P8.5d)}$$

For Problem 8.5, determine K_{ethane} and K_{propane} for the vapor (stream 5) and liquid stream (stream 6) leaving the reboiler of Example 8.6—use Equation (P8.5b) with values for ϕ_i^{vapor} and ϕ_i^{liquid} determined from Equation (P8.5d). We actually have determined all the terms on the RHS of Equation (P8.5d) in Example 8.6; be sure to use the known species mole fractions.

Compare the results obtained from Equation (P8.5b) to results from Equation (P8.5a), where

$$K_{\text{ethane}}^{\text{reboiler}} = \frac{y_{\text{ethane}}^{\text{stream 5}}}{x_{\text{etahne}}^{\text{stream 6}}} = \frac{0.017884}{0.009905} = 1.80555$$

and

$$K_{\text{propane}}^{\text{reboiler}} = \frac{y_{\text{propane}}^{\text{stream 5}}}{x_{\text{propane}}^{\text{stream 6}}} = \frac{0.982116}{0.990095} = 0.99194.$$

Note that normally, we do *not* know the equilibrium species mole fractions in the vapor and liquid phases. Normally, a "guess" must be made for each y_i and x_i value. With these guesses, we can immediately solve for K_i values from Equation (P8.5a). With these guesses for y_i and x_i values, we can also determine K_i values from Equation (P8.5b) and our selected EOS. Values for y_i and x_i are varied until the determined K_i values (Eq. (P8.5a,b)) match.

SOLUTION The solution for Equation (P8.5b) can be found in the Excel files **Problem 8.5 V.xls** and **Problem 8.5 L.xls**; here we find

$$K_{\text{ethane}}^{\text{reboiler}} = \frac{\phi_{\text{ethane,stream 6}}^{\text{liquid}}}{\phi_{\text{ethane,stream 5}}^{\text{vapor}}} = \frac{1.601093}{0.922903} = 1.7348$$

and

$$K_{\text{propane}}^{\text{reboiler}} = \frac{\phi_{\text{propane,stream 6}}^{\text{liquid}}}{\phi_{\text{propane,stream 5}}^{\text{vapor}}} = \frac{0.747374}{0.735536} = 1.0161.$$

These K_i values are close (but not exactly the same) to those found using Equation (P8.5a) or Equation (5.22). The discrepancy occurs because equilibrium values in Figure 5.5a where determined using the Peng–Robinson EOS and in this problem we used the RK-EOS.

8.6 ***Determination of Mixture Dew Point and Bubble Point Temperatures (*****Solution Provided*****)*** We have actually been examining the flash/vapor-liquid equilibrium problems in a reverse order. In Chapter 5, we determined species vapor and liquid flows from equilibrium flash operations when the species distribution factors (coefficients) were known. In Problem 8.5, we showed how these distribution coefficients can be determined from EOSs and with known T and P. A final consideration (Problem 8.6) is how do we know if a stream is a vapor-phase, a liquid-phase, or a two-phase mixture? A two-phase mixture can be flashed into vapor and liquid streams, which are in equilibrium. For problems in this chapter, the pressure is known—recall (starting in Chapter 3) we always know the pressure into a unit and the outlet pressure is known by the unit pressure drop; even in distillation columns, we can specify a tray-to-tray pressure drop. A stream with known species composition and pressure contains two phases if the stream temperature falls between the bubble point temperature and the dew point temperature. The mixture bubble point is defined as the temperature where the first infinitesimal amount of vapor appears from a liquid phase, and the mixture dew point temperature occurs when the first infinitesimal amount of liquid appears from a vapor phase.

We want to determine the bubble point and dew point temperatures for the mixture entering the reboiler in Example 8.6; this is stream 4 in Figure 8.7. For this stream and using the simulation program HYSYS (with the Peng–Robinson EOS), the bubble point temperature is 160.67°F and the dew point temperature is 161.85°F. The streams leaving the reboiler (streams 5 and 6) have been flashed to 161.78°F, which is between these bubble point and dew point temperatures. The feed to the reboiler is at the bubble point temperature.

Bubble Point Temperature Algorithm

For the bubble point, a liquid stream with known species composition and pressure will be brought to a temperature where an infinitesimal amount of vapor will appear in the liquid phase. The liquid-phase composition (x_i values) can be taken as constant. The algorithm in Figure P8.6a (modified from Bazmi et al., 2004) can be used to determine the bubble point temperature.

(a) Solve the bubble point temperature algorithm (Figure P8.6a) for the reboiler stream in Figure 8.7 (stream 4) with T_4 = 160.67°F. For comparison, also solve the bubble point algorithm for stream 4 with the temperature changed from T_4 = 160.67°F to T_4 = 155°F. Use the Excel files developed in Problem 8.5. The mole fractions for stream 4 are $x_{\text{ethane}}^{\text{stream 4}} = 0.017028$ and $x_{\text{propane}}^{\text{stream 4}} = 0.982972$.

(b) After you solve Problem 8.6a, develop an algorithm similar to Figure P8.6a for the dew point calculation. As a check, a dew point algorithm is provided in Figure P8.6b. Solve for the dew point temperature of stream 4 in Figure 8.7 using the RK-EOS.

SOLUTION Solution to Problem 8.6a: The solution using Figure P8.6a at T = 160.67°F can be found in the Excel files **Problem 8.6 L (160.67F).xls** and **Problem 8.6 V (160.67F).xls**. In Table P8.6a, we summarize the iteration results from these Excel files. Note in Table P8.6a that sum 2 (or $\sum y_i^{\text{vapor}}$ in Figure P8.6a) iteration 5 is $\sum y_i = 1.0223$. As this $\sum y_i \neq 1.0$ (using the RK-EOS), T = 160.67°F is not the bubble point temperature for the given species composition in stream 4 (at 400 psia). Here then, as shown in Figure P8.6a, a temperature adjustment is needed.

To help understand the needed temperature adjustment, next we try T = 155°F.

The solution at T = 155°F can be found in the Excel files **Problem 8.6 L (155F).xls** and **Problem 8.6 V (155F).xls**. In Table P8.6b, we summarize the iteration results. Note that sum 2 or $\sum y_i = 0.9963$, so from the RK-EOS, T = 155°F at P = 400 psia is not the bubble point; again, a temperature adjustment is needed.

With these two trial temperatures, T = 160.67°F and T = 155°F, we have bounded the $\sum y_i$ values between $\sum y_i = 1.0223$ and $\sum y_i = 0.9963$ (the desired value is $\sum y_i = 1.0$). We can search

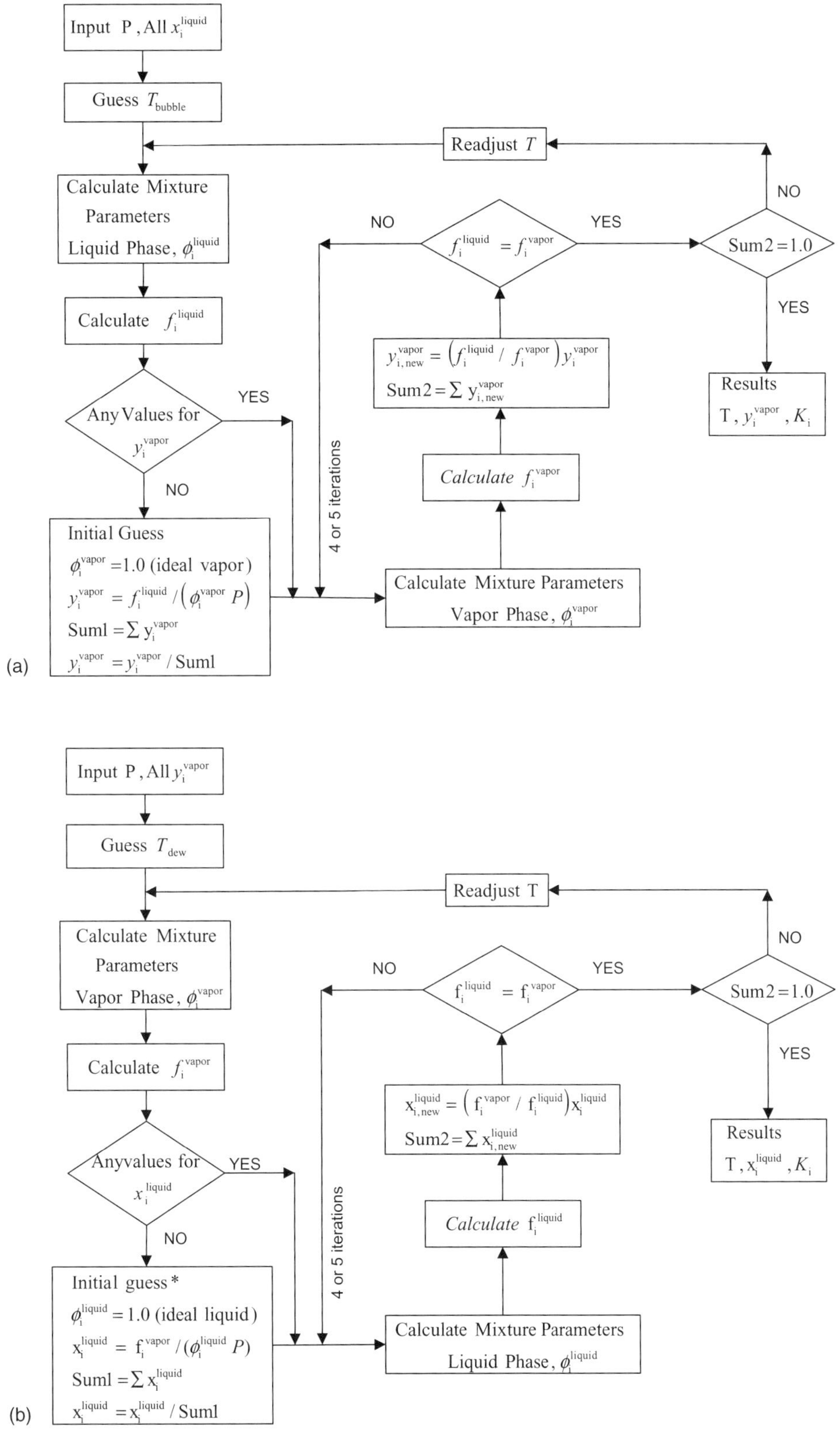

Figure P8.6 (a) Bubble point algorithm (modified from Bazmi et al., 2004). (b) Dew point algorithm.
*A new initial guess (x_i^{liquid} values) will be required if you cannot find an associated liquid volume root. For example, a problem with the initial guess may occur if a species is above its critical point (or is noncondensable); here the liquid-phase mole fraction for that species should be small.

Table P8.6a Iteration Results for Bubble Point Algorithm (T = 160.67°F, P = 400 psia)

Problem 8.6 L (160.67°F).xls		Problem 8.6 V (160.67°F).xls				
Iteration		1	2	3	4	5
$x_{\text{ethane}}^{\text{liquid}}$	0.017028					
$x_{\text{propane}}^{\text{liquid}}$	0.982972					
$f_{\text{ethane}}^{\text{liquid}}$ (atm)	0.739116					
$f_{\text{propane}}^{\text{liquid}}$(atm)	19.85234					
$y_{\text{ethane}}^{\text{vapor}}$	0.035894	0.035894	0.0295	0.0295	0.0289	0.0289
$y_{\text{propane}}^{\text{vapor}}$	0.964106	0.964106	0.9935	0.9938	0.9933	0.9934
$\sum y_i$ (Sum 2)	1.0	1.0	1.0231	1.0233	1.0222	1.0223
$f_{\text{ethane}}^{\text{vapor}}$ (atm)		0.8987	0.7393	0.7552	0.7383	0.7392
$f_{\text{propane}}^{\text{vapor}}$(atm)		19.2630	19.8479	19.8628	19.8512	19.8524

Table P8.6b Iteration Results for Bubble Point Algorithm (T = 155°F, P = 400 psia)

Problem 8.6 L (155°F).xls		Problem 8.6 V (155°F).xls			
Iteration		1	2	3	4
$x_{\text{ethane}}^{\text{liquid}}$	0.017028				
$x_{\text{propane}}^{\text{liquid}}$	0.982972				
$f_{\text{ethane}}^{\text{liquid}}$ (atm)	0.731236				
$f_{\text{propane}}^{\text{liquid}}$ (atm)	19.11776				
$y_{\text{ethane}}^{\text{vapor}}$	0.03684	0.03684	0.0292	0.0292	0.0292
$y_{\text{propane}}^{\text{vapor}}$	0.96316	0.96316	0.9672	0.9672	0.9672
$\sum y_i$ (Sum 2)	1.0	1.0	0.9963	0.9963	0.9963
$f_{\text{ethane}}^{\text{vapor}}$ (atm)		0.9234	0.7302	0.7312	0.7312
$f_{\text{propane}}^{\text{vapor}}$(atm)		19.0389	19.1188	19.1178	19.1178

temperature between these bounds to find the temperature where $\sum y_i = 1.0$; this would be the bubble point temperature.

SOLUTION Solution to Problem 8.6b: There is no need to keep the liquid-phase and vapor-phase portions of Figure 8.6a (bubble point) or Figure 8.6b (dew point) in separate files (as we did in "Solution to Problem 8.6a"). We have combined the bubble point files in **Problem 8.6 bubble.xls** and the dew point calculations are provided in **Problem 8.6 dew.xls**. For the ethane–propane feed stream in Problem 8.6, and when using the Redlich–Kwong, the bubble point is ~156.4°F and the dew point is ~156.8°F. For comparison, from the Peng–Robinson EOS in HYSYS, the bubble point is ~160.67°F and the dew point is ~161.85.

For Problems 8.7 and 8.8, use the Peng–Robinson EOS.

8.7 ***Determination of Equilibrium Distribution Factor (Coefficient) Values,*** K$_i$ ***for the Ammonia System*** Recall in Chapter 5 we simply provided K$_i$values from Myers and Seider (1976) for the ammonia system. With the experience gained after solving Problems 8.5 and 8.6, the reader may want to calculate these κ_i values using the Peng–Robinson EOS. The article by Nasri and Binous (2009) details how this can be done for the ammonia reaction system when using the Peng–Robinson EOS.

8.8 ***Determination of Fuel Suitability Mixture Dew Point Temperature*** A feed stream is proposed for an industrial turbine that contains 98% methane and 2% n-hexane (molar basis). The feed pressure to the turbine is 10 bar. There is concern that if any liquid is present in the feed, it could impact turbine blade life. Use the Peng–Robinson EOS to calculate the mixture dew point (needed data are provided in Table 8.9). Suggest a means to ensure the feed will always be vapor phase. For an advanced version of this problem (more species) and with the solution using the Peng–Robinson EOS, see Elliott et al. (2004) and Binous (2008).

8.9 ***Thermodynamic Properties Using Cubic EOS—a Mini Project*** An easy-to-use Web-based program for thermodynamic properties of pure species and binary mixtures has been developed by Barrie (2005). As a mini project, develop similar programs. For pure species, the EOS Barrie (2005) includes van der Waals, Redlich–Kwong, Soave–Redlich–Kwong, Peng–Robinson, and Peng–Robinson–Gasem. Calculations provide υ and ϕ values. EOS for binary mixtures include Soave–Redlich–Kwong, Peng–Robinson, and Peng–Robinson–Gasem. Calculations are available for bubble point, dew point, and flash calculations including υ, ϕ_i, y_i, x_i and enthalpy and entropy departure functions values. Here, rather than using Excel Solver or the secant method, to find υ values as we did in Chapter 8, you should use an analytic scheme to find υ (see Spiegel and Liu, 1999; Lawal, 1987).

Chapter 9

Gas Turbine Cogeneration System Performance, Design, and Off-Design Calculations: Real Fluid Properties

In Chapter 7, we developed performance, design, and off-design calculations in general form for cogeneration systems. These calculations allow determination of system temperatures, pressures, and flow rates, which, as we will see in Chapter 10, can be used for design, sizing, and costing. These calculations can also help identify operations that are showing degradation in performance. In Chapter 7, our developed equations assumed an ideal gas (or constant heat capacity) working fluid (as $h_{out} - h_{in} = C_p\Delta T$) for the power side of the combined heat and power cycle. In this chapter, the developed equations are particularized to real fluid cogeneration systems; h_{out} and h_{in} are calculated from rigorous equations of state (EOS) with ideal mixing rules as we developed in Chapter 8.

We do assume you have read Chapters 7 and 8, so here we can focus on applying real gas mixtures to the solution of cogeneration problems. Our efforts in this chapter will focus on the cogeneration system shown in Figure 9.1. The gas turbine or turbine system includes the air compressor (AC), air preheater (APH), combustor and a single turbine (gas and power turbine (G&PT) for air compression and power generation). A heat recovery steam generator (HRSG) is used to recover energy from the exhaust gas and to generate steam. This cogeneration system configuration is different from those discussed in Chapter 7 as here an air preheater is used to raise the air temperature from the compressor prior to the air entering the combustion chamber (CC). The tradeoff here is that the energy extracted from the exhaust stream will reduce the amount of steam that could have been raised, but the fuel need to produce each kilowatt of electricity will be reduced. The configuration in Figure 9.1 is important as it will be used in our economic optimization studies in Chapter 10.

You will recall from Chapter 7 that for *cogeneration gas turbine system performance calculations*, given any air feed conditions (T_1, P_1, F_{air}), AC efficiency, and known P_2, we will want to determine the conditions at 2 (T_2, P_2, F_{air}) in Figure 9.1. For the air preheater, we will utilize known T information (as either known T_3, or known $\Delta T_{cold\_APH} = T_6 - T_2$, or known $\Delta T_{hot\_APH} = T_5 - T_3$) to determine conditions at 3. With known fuel flow rate (or known T_4) and combustion efficiency, we want to determine the conditions at 4. From 4 and with known turbine efficiency and known P_5, we want to determine the conditions at 5 and the net power generated from the turbine. The conditions at 6 will be coupled to our air preheater calculations for (T_2, P_2, F_{air}) to (T_3, P_3, F_{air}).

We will utilize these performance calculations to determine *gas turbine system performance in design and off-design operation*. Design or benchmark gas turbine performance is established using standard design conditions of air feed at 59°F/15°C, 14.7 psia/1.013 bar, and 60% relative humidity—International Standards Organization (ISO) conditions. Gas turbine performance at off-design conditions can be determined using manufacturer-provided performance curves or general equations. But in order to use the general equations, we will need to understand how the gas turbine system control system operates (recall our discussion of Chapter 7, Section 7.9).

The *HRSG design and off-design performance calculations* allow determination of the amount of steam that can be raised from the waste heat boiler or HRSG. The HRSG design performance is based on expected conditions at 6 (T_6, P_6, F_{gas}), values selected for the pinch and approach temperatures, and the condensate return/steam conditions. Off-design HRSG performance requires an understanding of overall heat transfer coefficients.

Modeling, Analysis and Optimization of Process and Energy Systems, First Edition. F. Carl Knopf.

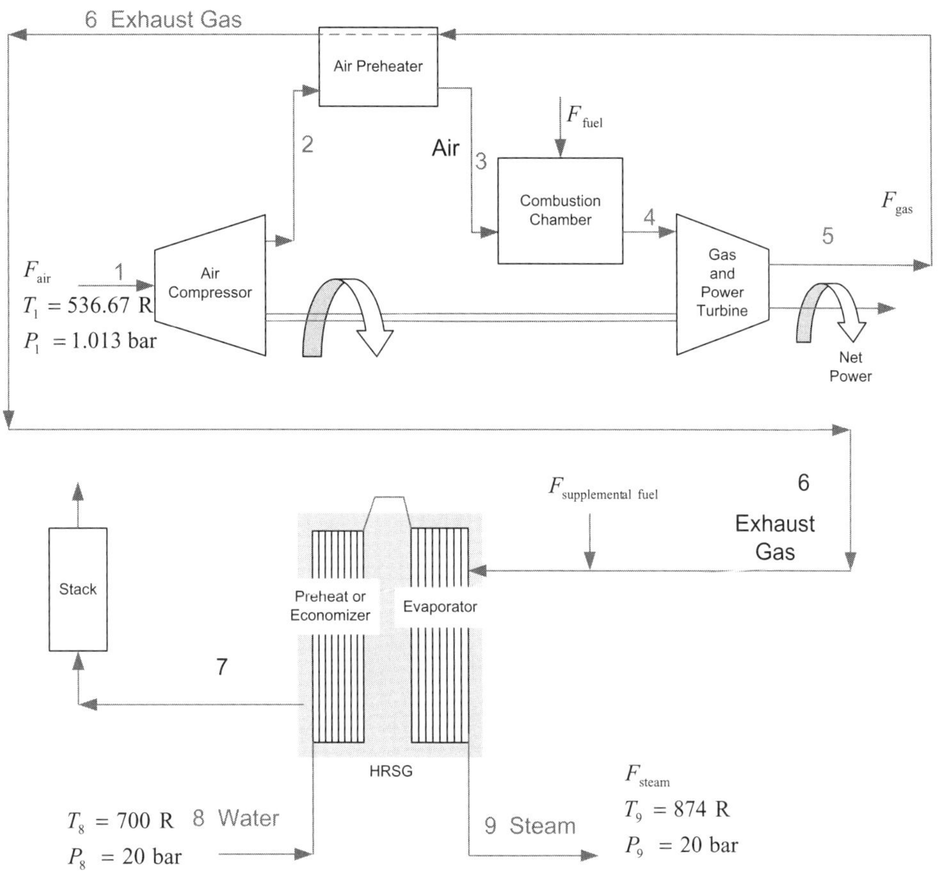

Figure 9.1 Gas turbine-based cogeneration system with air preheat.

9.1 COGENERATION GAS TURBINE SYSTEM PERFORMANCE CALCULATIONS: REAL PHYSICAL PROPERTIES

In Chapter 7, we developed general material and energy balances for the operations in Figure 9.1 and these equations will be assembled here (with some additions for the air preheater). The air feed and the combustion products are taken as real gases using our thermodynamic properties library (*TPSI+*) developed in Chapter 8. The library also allows for variable steam properties within the HRSG. In Chapter 8, thermodynamic property determinations were based in part on the pure-component EOS work of Reynolds (1979) and the assumption of ideal mixtures of real gases. The properties of air, combustion products, and steam, over the temperature and pressure ranges for cogeneration projects, were assembled as Excel callable dynamic link libraries (DLLs). The original C source code is provided in the Appendix.

Formulating the cogeneration performance problem using callable functions enables the user to change any independent variable in the Excel sheet and to find new values for all dependent variables. For example, the user can changeT_1 and obtain updated values for all other system dependent variables. Any set of operating conditions (any set of independent variables, e.g., T_1, P_1, F_{air}) for the cogeneration system can be evaluated. The solution procedure for performance calculations using real physical properties is outlined next; the Excel function calls to DLLs are indicated by = {*Excel function call*}. Do recall that the pressure at each point of the cogeneration system is known—either as a system parameter, an independent variable, or a dependent variable calculated with known unit ΔP.

For cogeneration problems, it is generally convenient to use pound-mass as the flow basis. To aid in this process, we developed functions within the cogeneration library, which calculate physical properties of combustion products on a pound-mass flow basis. These functions all contain the key word *Products* and for field units use T (Rankine), P (pound per square inch absolute), $\hat{h}$ (British thermal unit per pound), and $\hat{s}$ (British thermal unit per pound-Rankine), while for SI units use T (kelvin), P (megapascal), $\hat{h}$ (kilojoule per kilogram), $\hat{s}$ (kilojoule per kilogram-kelvin). The development of these *Products* functions is detailed in Section 9.1.3.

9.1.1 Air Compressor (AC) Performance Calculation

The minimum or ideal work required for gas compression is found by using an adiabatic and reversible process—this would be an isentropic process. Here, following Figure 9.1,

$$\hat{s}_1(T_1, P_1) = \hat{s}_2^{\text{isen}}(T_2^{\text{isen}}, P_2). \tag{9.1}$$

For the cogeneration process, $\hat{s}_1$ and $\hat{s}_2^{\text{isen}}$ are the entropies of air (British thermal unit per pound -Rankine) at the inlet and exit of the compressor, and T_2^{isen} is the isentropic outlet temperature of the air. Since P_1 and T_1 are known, $\hat{s}_1$ can be found using the entropy function for real air, $\hat{s}_1 = \{S\_Air(P_1, T_1)/MW_{\text{air}}\}$. P_2 is known (alternatively P_2 can be set equal to P_1 multiplied by the compressor compression ratio. T_2^{isen} can be found from $\hat{s}_2^{\text{isen}}$ and P_2 as $T_2^{\text{isen}} = \{T\_SP\_Air(\hat{s}_2^{\text{isen}} \times MW_{\text{air}}, P_2)\}$; here within *TPSI+*, the C code uses the function {*S_Air*(*P*, *T*)} and the secant method to vary *T*, with fixed *P*, until the needed $\hat{s}$ value is found.

$\hat{h}_2^{\text{isen}}$ is the enthalpy at T_2^{isen} and P_2, found using $\hat{h}_2^{\text{isen}} = \{H\_Air(P_2, T_2^{\text{isen}})/MW_{\text{air}}\}$. With η_{comp} known, $\hat{h}_2^{\text{actual}}$ can be found as

$$\hat{h}_2^{\text{actual}} = \hat{h}_1^{\text{actual}} + \frac{\hat{h}_2^{\text{isen}} - \hat{h}_1^{\text{actual}}}{\eta_{\text{comp}}}, \tag{9.2}$$

where $\hat{h}_2^{\text{actual}}$ is the actual air enthalpy at P_2 and T_2; here, of course, T_2 would be the actual air temperature leaving the compressor. T_2 can be found from $T_2 = \{T\_HP\_Air(\hat{h}_2^{\text{actual}} \times MW_{\text{air}}, P_2)\}$; the C code uses the function {*H_Air*(*P*, *T*)} and the secant method to vary *T*, with fixed *P*, until the needed $\hat{h}$ value is found. The work (British thermal unit per second) in the air compressor is

$$-\frac{W_{\text{air compressor}}}{dt} = -\dot{W}_{\text{air compressor}} = F_{\text{air}}(\hat{h}_2^{\text{actual}} - \hat{h}_1^{\text{actual}}) = F_{\text{air}}(\hat{h}_2 - \hat{h}_1), \tag{9.3}$$

where the air flow rate F_{air} (pound per second) is a variable. The symbols $\hat{h}_1^{\text{actual}}$ and $\hat{h}_1$ are used interchangeably in this chapter. We use $\hat{h}_1^{\text{actual}}$ when it is important to emphasize that the actual enthalpy is being used or calculated. We next discuss the combustion chamber and we will examine the air preheater later in this section.

9.1.2 Energy Balance for the Combustion Chamber (CC)

In Chapter 7, we allowed the temperature entering (T_3) and leaving (T_4) the combustion chamber to be variables. The flow rate of fuel F_{fuel} was then determined using Equation (7.46). For the economic optimization of Chapter 10, it will be more convenient to choose the temperature entering, T_3, and the flow rate of fuel, F_{fuel}, as variables and to solve for the temperature leaving the combustion chamber, T_4.

The combustion chamber mass and energy balances can be written as

$$F_{\text{gas}} = F_{\text{air}} + F_{\text{fuel}} \tag{9.4}$$

and

$$F_{\text{air}}(\hat{h}_3 - \hat{h}_{\text{air,ref}}) + F_{\text{fuel}}(LHV) = (F_{\text{air}} + F_{\text{fuel}})(\hat{h}_4 - \hat{h}_{\text{gas,ref}}) + \dot{Q}_{\text{loss}}, \tag{9.5}$$

where F_{gas} is the combustion products gas mass flow rate (pound per second); $\hat{h}_3$ is the enthalpy of the inlet air to the combustion chamber; $\hat{h}_4$ is the enthalpy of the outlet stream from the combustion chamber; *LHV* is the lower heating value of fuel (for natural gas *LHV* = 21,500 Btu/lbm); and $\dot{Q}_{\text{loss}}$ is the rate of heat loss from the combustion chamber, $\dot{Q}_{\text{loss}} = F_{\text{fuel}}(LHV)(1 - \eta_{\text{CC}})$. Do recall from our discussion in Chapter 7 (Section 7.5) that we must subtract the reference enthalpy for air, $\hat{h}_{\text{air,ref}}$, and the reference enthalpy of the gas products of combustion, $\hat{h}_{\text{gas,ref}}$, at the *LHV* reference conditions—see Figure 7.6.

We can directly solve the energy balance (Eq. (9.5)) for $\hat{h}_4$,

$$(\hat{h}_4 - \hat{h}_{\text{gas,ref}}) = \frac{F_{\text{air}}(\hat{h}_3 - \hat{h}_{\text{air,ref}}) + F_{fuel}(LHV)(\eta_{\text{cc}})}{(F_{\text{air}} + F_{\text{fuel}})}. \tag{9.6}$$

In Equation (9.6) F_{air} and F_{fuel} are known values (known variables); T_3 is also known (known variable), which allows determination of $\hat{h}_3$. The reference conditions $\hat{h}_{\text{air,ref}}$ and $\hat{h}_{\text{gas,ref}}$ are known. The problem then is to use the calculated values of $\hat{h}_4$ and P_4 to find the combustion product temperature, T_4. This is a common cogeneration calculation and here we developed a C function, callable from the Excel sheet to solve this problem—this and other *Products* functions are discussed in the next section.

9.1.3 C Functions for Combustion Temperature and Exhaust Gas Physical Properties

The problem we are now facing is that the "pure" components, air and fuel, have been combusted to give CO_2, H_2O, N_2, and Ar and some unreacted air. We need to determine the combustion chamber temperature and the physical properties of the exhaust gas. It quickly becomes cumbersome to calculate the amounts of each combustion species and determine the physical properties for the mixture in each stream after the combustion chamber. However, we can use our existing library (*TPSI+*) to create C functions, which will provide these calculations.

Conceptually determining T_4 is a straightforward process. If a stoichiometric amount of air is used to combust the fuel, the products will be F_{air}, H_2O, and the unreacted N_2 and Ar from the air. If F_{air} contains excess air, then the product stream will also contain unreacted excess air. As F_{air} and F_{fuel} are known values (known variables), the amount of each species in the product stream is known. We can assume a value for T_4 and determine the enthalpy for each of these species at the guessed T_4 and known P_4 values. For an ideal mixture real gas, the enthalpy of the combustion gas is found as the sum of each species enthalpy contribution divided by the total flow. The secant method can be used to vary the value of T_4 until the needed $\hat{h}_4$ value is found.

In order to find the combustion chamber outlet conditions (temperature, enthalpy, etc.), we must first determine the composition of the exhaust gas. The procedure for finding the composition of the exhaust gas begins with the consideration of a generalized combustion reaction as shown in Figure 9.2. Here an air stream (N_{air}) is reacted with a fuel stream (N_{fuel}), giving a product or exhaust gas stream (N_{gas}); do recall N represents molar flow rate.

The complete combustion of 1 mol of any C/H-based fuel is given by

$$C_xH_y + \left(x + \frac{y}{4}\right)O_2 = xCO_2 + \frac{y}{2}H_2O. \quad (9.7)$$

For example, for the complete combustion of methane ($x = 1$, $y = 4$),

$$CH_4 + 2O_2 = CO_2 + 2H_2O.$$

Examining Figure 9.2, the mole fraction of O_2in dry air, $y_{O_2}^{air}$, is 0.2095 so the moles of dry air required for complete combustion would be

Moles dry air required for complete combustion

$$= \left(\frac{x + \frac{y}{4}}{y_{O_2}^{air}}\right) = \left(\frac{x + \frac{y}{4}}{0.2095}\right). \quad (9.8)$$

The term

$$\left(\frac{\left(x + \frac{y}{4}\right)}{y_{O_2}^{air}}\right),$$

the moles of dry air required, deserves comment. The composition of dry air to the combustion process is taken as $y_{N_2}^{air} = 0.7809$, $y_{O_2}^{air} = 0.2095$, and $y_{Ar}^{air} = 0.0096$. The moles of dry air required will account for nitrogen, oxygen, and argon being fed to the combustion process.

We also know that excess air will be supplied to the combustion process. Allow that the moles of dry air actually used in the combustion process is

$$\text{Moles dry air used} = DAR\left(\frac{x + \frac{y}{4}}{y_{O_2}^{air}}\right), \quad (9.9)$$

where the dry air ratio (DAR)

$$DAR = \left(\frac{\text{Mole dry air used}}{\text{Mole dry air required}}\right).$$

The general combustion reaction for 1 mol of C/H-based fuel, accounting for excess air, can then be written as

$$[C_xH_y] + (DAR)\left(\frac{\left(x + \frac{y}{4}\right)}{y_{O_2}^{air}}\right)[\text{Air}] = x[CO_2] + \frac{y}{2}[H_2O]$$
$$+ (DAR)\left(\frac{\left(x + \frac{y}{4}\right)}{y_{O_2}^{air}}\right)[\text{Air}] - \left(x + \frac{y}{4}\right)[O_2]. \quad (9.10)$$

EXAMPLE 9.1 ***Methane Combustion Products***

Use Equation (9.10) to determine the moles of the product (CO_2, H_2O, O_2, N_2, and Ar) that result from the complete combustion of 1 mol of methane when using a $DAR = 1$ and $DAR = 2$.

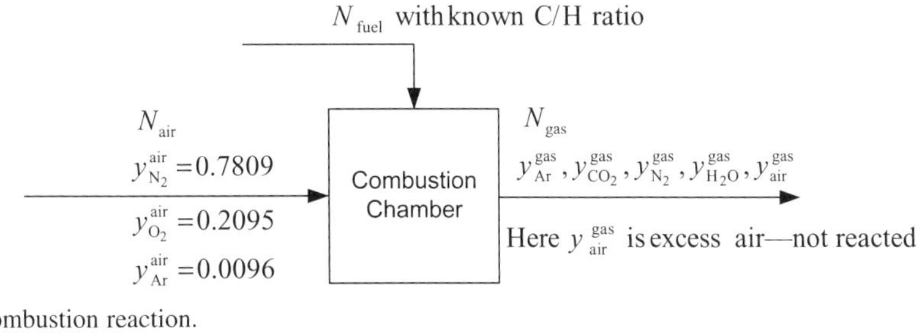

Figure 9.2 Generalized combustion reaction.

SOLUTION A *DAR* = 1 means the exact amount of air for combustion will be used. There should be no O_2 in the product gas. Equation (9.10) for methane ($x = 1$ and $y = 4$),

$$[CH_4]+(1)\left(\frac{(2)}{0.2095}\right)[Air]=1[CO_2]+2[H_2O]$$
$$+(1)\left(\frac{(2)}{0.2095}\right)[Air]-(2)[O_2]$$
$$[CH_4]+(9.5465)[Air]=[CO_2]+2[H_2O]+(9.5465)[Air]-(2)[O_2]$$

shows that 9.5465 mol of air will be needed to completely combust 1 mol of methane. The product stream (right-hand side [RHS] of the equation) will contain 1 mol CO_2, 2 mol H_2O, $((9.5465\times y_{O_2}^{air})-2)=0$ mol O_2, $(9.5465\times y_{N_2}^{air})=7.455$ mol N_2, and $(9.5465\times y_{Ar}^{air})=0.09164$ mol Ar.

For *DAR* = 2, Equation (9.10),

$$[CH_4]+(2)\left(\frac{(2)}{0.2095}\right)[Air]=1[CO_2]+2[H_2O]$$
$$+(2)\left(\frac{(2)}{0.2095}\right)[Air]-(2)[O_2]$$
$$[CH_4]+(19.093)[Air]=[CO_2]+2[H_2O]+(19.093)[Air]-(2)[O_2]$$

shows that 19.093 mol of air will be used in the combustion of 1 mol of methane. The product stream (RHS of the equation) will contain 1 mol CO_2, 2 mol H_2O, $((19.093\times y_{O_2}^{air})-2)=2$ mol O_2, $(19.093\times y_{N_2}^{air})=14.91$ mol N_2, and $(19.093\times y_{Ar}^{air})=0.183$ mol Ar; here then, $y_{CO_2}^{exhaust}=0.0498$, $y_{H_2O}^{exhaust}=0.0995$, $y_{O_2}^{exhaust}=0.0995$, $y_{N_2}^{exhaust}=0.7420$, and $y_{Ar}^{exhaust}=0.0091$. ■

Equation (9.10) provides a convenient means of determining the mole fraction for the combustion products in the exhaust gas stream. The mole fraction of Co_2 in the combustion products, $y_{CO_2}^{exhaust}$, would be

$$\left(\frac{x}{\text{Total moles out}}\right),$$

$$y_{CO_2}^{exhaust}=\frac{x}{x+\frac{y}{2}+(DAR)\left(\frac{\left(x+\frac{y}{4}\right)}{y_{O_2}^{Air}}\right)-\left(x+\frac{y}{4}\right)}. \quad (9.11)$$

If we define

$$HC\_ratio=\frac{y}{x} \quad (9.12)$$

and

$$SpentO_2=\frac{(HC\_ratio+4)}{4}, \quad (9.13)$$

where Spent O_2 is the oxygen used in combustion weighted by the carbon atoms as

$$\frac{\left(x+\frac{y}{4}\right)}{x}.$$

We can rearrange Equation (9.11) to

$$y_{CO_2}^{exhaust}=\frac{1}{\left(\frac{HC\_ratio+2}{2}\right)+SpentO_2\left(\frac{(DAR)}{y_{O_2}^{Air}}-1\right)}. \quad (9.14)$$

The mole fraction of H_2O in the combustion products can be found by recognizing that Total moles out $= x / y_{CO_2}^{exhaust}$, giving

$$y_{H_2O}^{exhaust}=\frac{\left(\frac{y}{2}\right)}{\text{Total moles out}}=\frac{\left(y_{CO_2}^{exhaust}\right)\left(\frac{y}{x}\right)}{2}$$
$$=\frac{\left(y_{CO_2}^{exhaust}\right)(HC\_ratio)}{2}. \quad (9.15)$$

We have accounted for the mole fraction of CO_2 and H_2O in the exhaust gas—as products from the reaction of the C/H and O_2 from air. We next need to account for the remaining components, which are (1) the argon and nitrogen in the reacted air and (2) any excess air. We have developed a physical properties library that includes all these species; recall air is available as a species in our library. We can also break the excess air into its individual components of O_2, N_2, and Ar:

Fuel	Air	Exhaust
C/H fuel +	Stoichiometric air →	CO_2, H_2O, N_2, Ar – Reacted air: $y_{CO_2}^{exhaust}$, $y_{H_2O}^{exhaust}$, $y_{Ar}^{exhaust}$, $y_{N_2}^{exhaust}$
	Excess air →	Excess air: $y_{air}^{exhaust}$.

Next, we will account for the mole fraction of the nitrogen and argon in reacted air and we will account for the mole fraction of excess air. This approach is correct, but do note that if we specifically need to account for the argon or nitrogen or oxygen mole fraction in the exhaust gas, we will need to account for these species in the excess air.

The mole fraction of argon in the combustion gas from the *reacted air* is

$$y_{Ar}^{exhaust}=\frac{y_{Ar}^{air}(\text{moles dry air required})}{\text{Total moles out}}=\frac{y_{Ar}^{air}\left(\frac{x+\frac{y}{4}}{y_{O_2}^{air}}\right)\left(y_{CO_2}^{exhaust}\right)}{x}$$
$$=\left(y_{CO_2}^{exhaust}\right)\left(\frac{y_{Ar}^{air}}{y_{O_2}^{air}}\right)(SpentO_2). \quad (9.16)$$

The mole fraction of nitrogen in the combustion gas from the reacted air is

$$y_{N_2}^{\text{exhaust}} = \frac{y_{N_2}^{\text{air}}(\text{moles dry air required})}{\text{Total moles out}}$$
$$= \left(y_{CO_2}^{\text{exhaust}}\right)\left(\frac{y_{N_2}^{\text{air}}}{y_{O_2}^{\text{air}}}\right)(\text{SpentO}_2). \quad (9.17)$$

The mole fraction of excess air in the exhaust is

$$y_{\text{excess air}}^{\text{exhaust}} = \frac{((DAR)-1)(\text{Mole of dry air required})}{\text{Total moles out}}$$
$$= \frac{((DAR)-1)\left(\dfrac{x+\frac{y}{4}}{y_{O_2}^{\text{air}}}\right)}{\text{Total moles out}}$$
$$= \frac{((DAR)-1)\left(y_{CO_2}^{\text{exhaust}}\right)(\text{SpentO}_2)}{y_{O_2}^{\text{air}}}. \quad (9.18)$$

The following example will help clarify the use of Equations (9.14)–(9.18).

EXAMPLE 9.2 ***Use of Combustion Exhaust Mole Fraction Equations (9.14)–(9.18)***

Solve for the exhaust composition using Equations (9.14)–(9.18) with methane fuel and a *DAR* of 2. From Example 9.1 for methane combustion with $DAR = 2$, $y_{CO_2}^{\text{exhaust}} = 0.0498$, $y_{H_2O}^{\text{exhaust}} = 0.0995$, $y_{O_2}^{\text{exhaust}} = 0.0995$, $y_{N_2}^{\text{exhaust}} = 0.7420$, and $y_{Ar}^{\text{exhaust}} = 0.0091$.

SOLUTION Results are provided in the Excel file **Example 9.2.xls**, and using Equations (9.14)–(9.18),

$$y_{CO_2}^{\text{exhaust}} = \frac{1}{\left(\dfrac{HC\_ratio+2}{2}\right)+\text{SpentO}_2\left(\dfrac{(DAR)}{y_{O_2}^{\text{air}}}-1\right)}$$
$$= \frac{1}{\left(\dfrac{4+2}{2}\right)+2\left(\dfrac{2}{0.2095}-1\right)} = 0.0498,$$

$$y_{H_2O}^{\text{exhaust}} = \frac{\left(y_{CO_2}^{\text{exhaust}}\right)(HC\_ratio)}{2} = \frac{(0.0498)(4)}{2} = 0.0996,$$

$$y_{Ar}^{\text{exhaust}} = \left(y_{CO_2}^{\text{exhaust}}\right)\left(\frac{y_{Ar}^{\text{air}}}{y_{O_2}^{\text{air}}}\right)(\text{SpentO}_2)$$
$$= (0.0498)\left(\frac{0.0096}{0.2095}\right)(2) = 0.00456,$$

$$y_{N_2}^{\text{exhaust}} = \left(y_{CO_2}^{\text{exhaust}}\right)\left(\frac{y_{N_2}^{\text{air}}}{y_{O_2}^{\text{air}}}\right)(\text{SpentO}_2)$$
$$= (0.0498)\left(\frac{0.7809}{0.2095}\right)(2) = 0.3712,$$

and

$$y_{\text{excess air}}^{\text{exhaust}} = \frac{((DAR)-1)\left(y_{CO_2}^{\text{exhaust}}\right)(\text{SpentO}_2)}{y_{O_2}^{\text{air}}}$$
$$= \frac{(2-1)(0.0498)(2)}{0.2095} = 0.475.$$

■

The difference in Examples 9.1 and 9.2 is that in Example 9.2, we account for excess air (which includes oxygen, argon, and nitrogen). It is actually beneficial because we will need to account for excess air in supplemental firing of the HRSG for us to retain excess air as a species.

It is now instructive to examine the C code from the combustion library, which is used to calculate the molecular weight of the exhaust gas and the enthalpy of the exhaust gas. In the code, we use the variable PercentAir in place of DAR and we calculate the exhaust gas mole fractions as xCO2, xH2O, and so on.

Molecular Weight of Products

```
MwAir = 28.96
MwAr = 39.948
MwN2 = 28.0134
MwCO2 = 44.01
MwO2 = 31.9994
MwH2O = 18.016
// mole fractions in air
yN2 = 0.7809
yO2 = 0.2095
yAr = 0.0096
    xCO2 = 1/((HC_ratio + 2)/2
           + SpentO2*(PercentAir/yO2
- 1.0));
    xH2O = (xCO2)*HC_ratio/2;
    xAr = (xCO2)*SpentO2*yAr/yO2;
    xN2 = (xCO2)*SpentO2*yN2/yO2;
    xAir = (xCO2)*SpentO2*(PercentAir
- 1.0)/yO2;
Mw_Products =
        {(xCO2)*MwCO2 + (xH2O)*MwH2O +
(xAir)*MwAir
           + (xAr)*MwAr + (xN2)*MwN2}
```

Enthalpy of Products

```
H_Products =
        {xH2O*H_Steam(P,T) +
xCO2*H_CarbDiox(P,T)
        + xAir*H_Air(P,T) +
xAr*H_Argon(P,T)
        + xN2*H_Nitrogen(P,T)
        - xH2O*(19505.414)
- xCO2*H_CarbDiox(14.696,536.67)
        - xAir*H_Air(14.696,536.67)
- xAr*H_Argon(14.696,536.67)
        - xN2*H_Nitrogen(14.696,536.67)
        }/Mw_Products;
```

The code to calculate the molecular weight of the combustion products directly follows the equations we have developed (Eqs. (9.11)–(9.18)). The enthalpy (British thermal unit per pound) of the combustion products is determined as

$$\sum \frac{y_i h_i(P,T) - y_i h_i(P_{\text{ref}}, T_{\text{ref}})}{MW\_Products}.$$

At the reference conditions (14.696 psia and 536.67 R), steam will not exist; here we curve fit steam enthalpy at 14.696 psia over a wide temperature range and then extended the resulting polynomial to 536.67 R; the value obtained for the pseudoenthalpy of steam (14.696 psia, 536.67 R) was 19,505.414 Btu/lb-mol.

Entropy of Products

```
S_Products =
        {xH2O*S_Steam(P,T) +
xCO2*S_CarbDiox(P,T)
        + xAir*S_Air(P,T) +
xAr*S_Argon(P,T)
        + xN2*S_Nitrogen(P,T)
        - xH2O*(29.6476)- xCO2*S_
CarbDiox(14.696,536.67)
        - xAir*S_Air(14.696,536.67) -
xAr*S_Argon(14.696,536.67)
        - xN2*S_Nitrogen(14.696,536.67)
        }/MwProducts;
```

The entropy (Brtitish thermal unit per pound-Rankine) of the combustion products is

$$\sum \frac{y_i s_i(P,T) - y_i s_i(P_{\text{ref}}, T_{\text{ref}})}{MW\_Products}.$$

At the reference conditions (14.696 psia and 536.67 R), steam will not exist; here we curve fit steam entropy at 14.696 psia over a wide temperature range and then extend the resulting polynomial to 536.67 R; the value obtained for the pseudoentropy of steam (14.696 psia, 536.67 R) was 29.6476 Btu/lb-mol-R. Do note that in the calculation of entropy, we have used $(s_i - s_{i,\text{ref}})$. S_Products is intended for use with single-input, single-output cogeneration system power devices with no change in composition (including gas and power turbines); here the term $-R\sum(y_i \ln y_i)$ cancels.

Temperature of Products

The combustion chamber outlet temperature is determined by trial and error and here we describe the overall process. Following Figure 9.2, the inlet air stream total enthalpy H_{air} is known, $H_{\text{air}} = N_{\text{air}}\left(h_{\text{air}(P_{\text{in}},T_{\text{in}})} - h_{\text{air,ref}}\right)$; here the inlet air temperature and pressure are known and the air flow rate would be a known variable. The inlet fuel enthalpy H_{fuel} is known, $H_{\text{fuel}} = N_{\text{fuel}}(LHV)$; here the fuel flow rate would be a known variable. The enthalpy of the exhaust gas is $H_{\text{exhaust gas}} = H_{\text{air}} + H_{\text{fuel}}$. If the combustion chamber is not adiabatic, heat losses can be accounted for by setting $H_{\text{fuel}} = N_{\text{fuel}}(LHV)(\eta_{CC})$, where η_{CC} is the combustion chamber efficiency. We can take advantage of the C function H_Products, described earlier, and vary T_{out} using the secant method until H_{gas} is obtained. A C function, {*T* from *H_Products*}, has been written for this task (see Table 9.1).

Table 9.1 Exhaust Gas "*Products*" Properties Available in Either Field or SI Units

= {*H_Products* (*Pressure, Temperature, H/C Ratio, DAR*)}
This function calculates the enthalpy of the exhaust gas in Btu/lb at *Pressure* in psia, *Temperature* in R, and subtracts the enthalpy of the exhaust gas at the reference state (14.696 psia, 536.67 R).

= {*TfromG_Products* (*Pressure, Enthalpy, H/C Ratio, DAR*)}
This function calculates the temperature of the exhaust gas in R using the enthalpy of the exhaust gas *Enthalpy* in Btu/lb at *Pressure* in psia. The *Enthalpy* parameter accounts for the reference state (14.696 psia, 536.67 R).

= {*S_Products* (*Pressure, Temperature, H/C Ratio, DAR*)}
This function calculates the entropy of the exhaust gas in Btulb-R at *Pressure* in psia, *Temperature* in R, and subtracts the entropy of the exhaust gas at the reference state (14.696 psia, 536.67 R).

= {*TfromS_Products* (*Pressure, Entropy, H/C Ratio, DAR*)}
This function calculates the temperature in of the exhaust gas in R using the entropy of the products *Entropy* in Btu/lb-R at *Pressure* in psia. The *Entropy* parameter accounts for the reference state (14.696 psia, 536.67 R).

= {*H_Products_SI* (*Pressure, Temperature, H/C Ratio, DAR*)}
This function calculates the enthalpy of the exhaust gas in kJ/kg at *Pressure* in MPa, *Temperature* in K, and subtracts the enthalpy of the exhaust gas at the reference state (0.101326 MPa, 298.15 K).

= {*TfromH_Products_SI* (*Pressure, Enthalpy, H/C Ratio, DAR*)}
This function calculates the temperature of the exhaust gas in K using the enthalpy of the exhaust gas *Enthalpy* in kJ/kg at *Pressure* in MPa. The *Enthalpy* parameter accounts for the reference state (0.101326 MPa, 298.15 K).

= {*S_Products_SI* (*Pressure, Temperature, H/C Ratio, DAR*)}
This function calculates the entropy of the exhaust gas in kJ/kg-K at *Pressure* in MPa, *Temperature* in K, and subtracts the entropy of the exhaust gas at the reference state (0.101326 MPa, 298.15 K).

= {*TfromS_Products_SI* (*Pressure, Enthalpy, HC Ratio, DAR*)}
This function calculates the temperature in of the exhaust gas in K using the entropy of the products *Entropy* in kJ/kg-K at *Pressure* in MPa. The *Entropy* parameter accounts for the reference state (0.101326 MPa, 298.15 K).

= {*MW_Products*}
This function calculates the molecular weight of the exhaust gas. This function is used internally by the C program combustion library.

Excel Callable Products Functions to Calculate Exhaust Gas Properties

We have developed a series of C functions for physical property calculations involving combustion exhaust gases. These functions are summarized in Table 9.1. Notice that the key word *Products* is in each of these functions. Use of these functions is *not* restricted to just the combustion chamber, but they can also be used throughout the cogeneration process (where exhaust gas is present). Do note the Field units here are enthalpy (British thermal unit per pound), entropy (British thermal unit per pound-Rankine), pressure (pound per square inch absolute), and temperature (R). The SI units are enthalpy (kilojoule per kilogram), entropy (kilojoule per kilogram-kelvin), pressure (megapascal), and temperature (kelvin).

EXAMPLE 9.3 ***(a) Methane Adiabatic Flame Temperature and (b) Combustion Temperature***

(a) Determine the adiabatic flame temperature for methane combustion with air using the functions in Table 9.1. The methane LHV = 21,501 Btu/lb; methane molecular weight = 16.043 lb/lb-mol; and stoichiometric air (from Eq. (9.8)) = 9.5465 mol air/mol CH_4 = 276.5155 lb air/mol CH_4. The process is shown here:

Fuel		Air		Products
CH_4 fuel	+	Stoichiometric air	$\longrightarrow$	CO_2, H_2O, N_2, Ar.
77°F, 1 atm		77°F, 1 atm		$T_{\text{adiabatic}}$, 1 atm

(b) Determine the combustion temperature if methane at 77°F and 1 atm is combusted with two times the stoichiometric air and if the air stream is at 100°F and 1 atm.

SOLUTION Part (a) solution—we pick a basis of 1 mol of methane. The incoming air will not contribute to the product enthalpy as the air is at the reference condition $H_{\text{air}} = 0$. $H_{\text{fuel}} = N_{\text{fuel}}(LHV)$ = (16.043 lb/mol) (21,501 Btu/lb) = 344,940.54 Btu. In the provided Excel sheet in **Example 9.3.xls**, we use the function = {*TfromH_Products* (*Pressure*, *Enthalpy*, *H/C ratio*, *DAR*)} to solve for the adiabatic flame temperature. Here, *Pressure* = 14.696 psia, *H_Products* = 344,940.54 Btu/(16.043 + 276.5155 lb) = 1179.048 Btu/lb, *H/C ratio* = 4, and *DAR* = 1.0; $T_{\text{adiabatic}}$ = 4210.327 R. This value can be compared to a measured value of 3866.4 R and a computed value of 4190.4 R (Myers and Seider, 1976). The measured adiabatic flame temperature falls 8.89% below the calculated value because of energy lost in dissociation reactions—these dissociation reactions are discussed in Chapter 15.

Part (b) solution—we again pick a basis of 1 mol of methane. The solution is also provided on the Excel sheet in **Example 9.3.xls**. For Part (b), the incoming air will contribute to the exit enthalpy. H_{air} is found from (H_air (14.696 psia, 559.67 R) − H_air (14.696 psia, 536.67 R)) = 160.05 Btu/lb-mol air; here, we do need to account for the reference state, $H_{\text{fuel}} = N_{\text{fuel}}(LHV)$ = (16.043) (21,501) = 344,940.54 Btu. We again use the function = {*TfromH_Products* (*Pressure*, *Enthalpy*, *H/C ratio*, *DAR*)} to solve for the combustion temperature. Here, *Pressure* = 14.696 psia, *H_Products* = 347,996.4 Btu/(16.043 + 2 × 276.5155) = 611.51 Btu/lb, *H/C ratio* = 4, and *DAR* = 2.0; $T_{\text{combustion}}$ = 2682.9 R. ■

Fuel Lower Heating Value (LHV) and Higher Heating Value (HHV)

We discussed combustion and fuel *LHV*s in Chapters 1 and 7. The fuel *LHV* is defined as the heat liberated when 1 lb of fuel is mixed with stoichiometric oxygen at 77°F and is completely burned at constant pressure. The products are then cooled to 77°F with the water formed remaining in the vapor state. The higher heating value of the fuel may also be provided. The higher heating value (HHV) is defined as the heat liberated when 1 lb of fuel is mixed with stoichiometric oxygen at 77°F and is completely burned at constant pressure. The products are then cooled to 77°F with the water formed from combustion condensed to the liquid state. Fuel LHV and higher heating value are provided in Table 9.2.

It is important to remember that fuel is generally priced based on the higher heating value. Manufacturers generally report turbine performance data based on the fuel *LHV*. This can cause problems in an economic analysis where costs are based on gas flow rates.

9.1.4 Gas and Power Turbine (G&PT) Performance Calculations

The gas and power turbine (Figure 9.1) material and energy balance procedure follows that used for the AC. Isentropic work provides

$$\hat{s}_4(T_4, P_4) = \hat{s}_5^{\text{isen}}(T_5^{\text{isen}}, P_5). \tag{9.19}$$

Table 9.2 Fuel Lower and Higher Heating Values at 77°F (25°C), from Bathie (1996)

Compound	Formula	State	*LHV* (Btu/lb)	*LHV* (kJ/kg)	*HHV* (Btu/lb)	*HHV* (kJ/kg)
Methane	CH_4	Gas	21,501	50,012	23,860	55,499
Ethane	C_2H_6	Gas	19,141	44,521	20,911	48,638
Propane	C_3H_8	Liquid	19,927	46,351	21,644	50,343
n-Butane	C_4H_{10}	Liquid	19,493	45,342	21,121	49,128
n-Heptane	C_7H_{12}	Liquid	19,155	44,555	20,666	48,069
n-Octane	C_8H_{18}	Liquid	19,098	44,422	20,589	47,890

For the gas and power turbine, $\hat{s}_4$ and $\hat{s}_5^{\text{isen}}$ are the entropies of the exhaust gas (British thermal unit per pound-Rankine) at the inlet and exit of the turbine and T_5^{isen} is the isentropic outlet temperature of the exhaust gas. Since P_4 and T_4 are known following our combustion chamber calculations, $\hat{s}_4$ can be found using the entropy function for the exhaust gas, $\hat{s}_4 = \{S\_Products(P_4, T_4, H/C\ ratio, DAR)\}$. For methane fuel, the $H/C\ ratio = 4$ and DAR is

$$\frac{F_{\text{air}}}{\left(17.2359 \dfrac{\text{lb-air required}}{\text{lb-methane}}\right)}.$$

Do recall when using the *Products* functions the reference state is included.

T_5^{isen} can be found from $\hat{s}_5^{\text{isen}}$ and P_5 as $T_5^{\text{isen}} = \{TfromS\_Products(P_5, \hat{s}_5^{\text{isen}}, H/C\ ratio, DAR)\}$; the C code uses $\{TfromS\_Products(P, \hat{s}, H/C\ ratio, DAR)\}$ and the secant method to vary T, with fixed P, $H/C\ ratio$, and DAR until the needed entropy value is found.

$\hat{h}_5^{\text{isen}}$ is the enthalpy at T_5^{isen} and P_5, found using $\hat{h}_5^{\text{isen}} = \{H\_Products(P_5, T_5^{\text{isen}}, H/C\ ratio, DAR)\}$. With $\eta_{\text{G\&PT}}$ known (known variable), $\hat{h}_5^{\text{actual}}$ can be found as

$$\hat{h}_5^{\text{actual}} = \hat{h}_4^{\text{actual}} + \left(\hat{h}_5^{\text{isen}} - \hat{h}_4^{\text{actual}}\right)\left(\eta_{\text{G\&PT}}\right), \tag{9.20}$$

where $\hat{h}_5^{\text{actual}}$ is the actual exhaust enthalpy at P_5 and T_5. T_5 can be found from $T_5 = \{TfromH\_Products(P_5, \hat{h}_5^{\text{actual}} H/C\ ratio, DAR)\}$; here the C code uses $\{TfromH\_Products(P, \hat{h}, H/C\ ratio, DAR)\}$ and the secant method to vary T, with fixed P, $H/C\ ratio$, and DAR until the needed enthalpy value is found. The work (British thermal unit per second) produced by the power turbine is

$$\begin{aligned} -\frac{W_{\text{power turbine}}}{dt} &= -\dot{W}_{\text{power turbine}} = F_{\text{gas}}(\hat{h}_5^{\text{actual}} - \hat{h}_4^{\text{actual}}) \\ &= F_{\text{gas}}(\hat{h}_5 - \hat{h}_4). \end{aligned} \tag{9.21}$$

The net work available from the gas and power turbine must account for the work needed for the gas compression:

$$\dot{W}_{\text{Net\_power turbine}} = \dot{W}_{\text{power turbine}} + \dot{W}_{\text{air compressor}}. \tag{9.22}$$

With our sign convention, $dW_{\text{air compressor}}/dt$ will be a negative quantity.

9.1.5 Air Preheater (APH)

For the adiabatic air preheater, an overall energy balance provides

$$\frac{dQ_{air\ preheater}}{dt} = \frac{dQ_{air\ preheater,air}}{dt} + \frac{dQ_{air\ preheater,gas}}{dt} = 0$$

giving

$$F_{\text{gas}}(\hat{h}_5 - \hat{h}_6) = F_{\text{air}}(\hat{h}_3 - \hat{h}_2), \tag{9.23}$$

where $\hat{h}_2$ and $\hat{h}_3$ are air enthalpies evaluated at (P_2,T_2) and (P_3,T_3) using $\{H\_Air(P, T)/MW_{\text{air}}\}$, and $\hat{h}_5$ is the exhaust enthalpy evaluated at (P_5,T_5) using $\{H\_Products\ (P_5, T_5, H/C\ ratio, DAR)\}$. Equation (9.23) can be solved for $\hat{h}_6$:

$$\hat{h}_6 = \hat{h}_5 - \frac{F_{\text{air}}(\hat{h}_3 - \hat{h}_2)}{F_{\text{gas}}}. \tag{9.24}$$

T_6 can be found from $T_6 = \{TfromH\_Products(P_6, \hat{h}_6, H/C\ ratio, DAR)\}$. The rate that heat is transferred from the gas side to the air side in the air preheater is

$$\begin{aligned} \frac{dQ_{\text{air preheater,air}}}{dt} &= -\frac{dQ_{\text{air preheater,gas}}}{dt} = F_{\text{air}}(\hat{h}_3 - \hat{h}_2) \\ &= -F_{\text{gas}}(\hat{h}_6 - \hat{h}_5). \end{aligned} \tag{9.25}$$

9.2 HRSG: DESIGN PERFORMANCE CALCULATIONS

The HRSG consists of an economizer or preheat section and an evaporator section as shown in Figure 9.3. A recommended approach to HRSG design (Ganapathy, 1991, 2003) is to utilize expected values (design operation values) for T_6, P_6, F_{gas}^6 and to fix both the pinch and approach temperatures. This allows the amount of steam generated to be determined by material and energy balance around the evaporator section of the HRSG. Values for pinch temperature difference ΔT_{Pinch} and the approach temperature difference $\Delta T_{\text{Approach}}$ are provided in Table 7.1. Table 7.1 values are chosen to produce an HRSG of reasonable size (area) and cost as well as to allow acceptable HRSG performance in off-design cases. HRSG performance in off-design is discussed in Section 9.3 (and previously in Section 7.8).

From Figure 9.3, the temperature equations that govern the HRSG design are

$$T_{8P} = T_9 - \Delta T_{\text{Approach}} \tag{9.26}$$

and

$$T_{7P} = T_9 + \Delta T_{\text{Pinch}}. \tag{9.27}$$

From Figure 9.1 and using Figure 9.3, the overall material and energy balance on the evaporator section allows the steam flow rate to be determined:

$$F_{\text{steam}} = \frac{F_{\text{gas}}(\hat{h}_6 - \hat{h}_{7P})}{\hat{h}_9^{\text{sat vapor}} - \hat{h}_{8P}^{\text{liquid}}}. \tag{9.28}$$

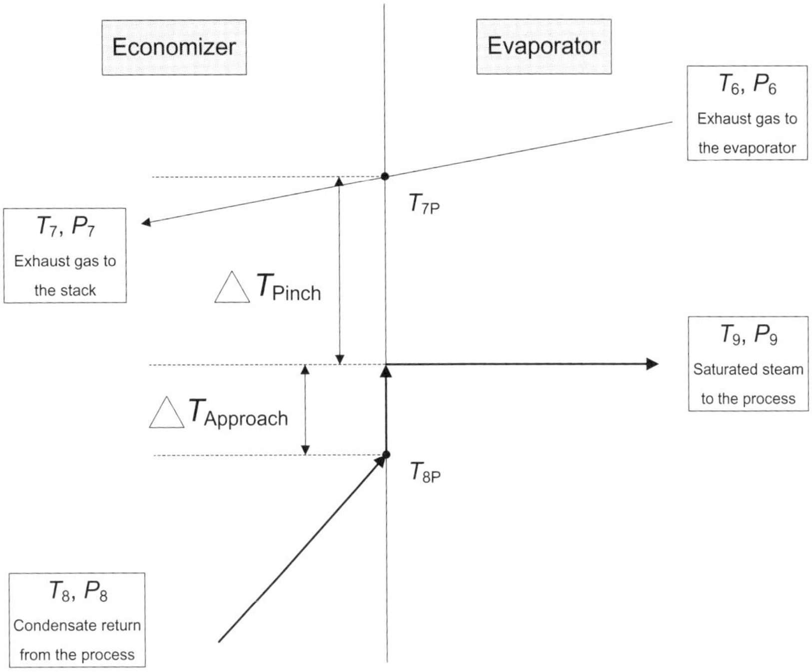

Figure 9.3 Heat recovery steam generator with expanded pinch point.

From Figure 9.1, the exhaust gas enthalpy $\hat{h}_6$ would be available from air preheater calculations. The exhaust gas enthalpy at (P_{7P}, T_{7P}) can be evaluated using $\hat{h}_{7P} = \{H\_Products(P_{7P}, T_{7P}, H/C\ ratio, DAR)\}$ here as the HRSG gas-side pressure drop is known and P_{7P} can be taken as the average of P_7 and P_6. The steam-side enthalpies, $\hat{h}_9^{\text{sat vapor}}$ and $\hat{h}_{8P}^{\text{liquid}}$, can be evaluated using $\hat{h}_9^{\text{sat vapor}} = \{H\_Steam(P_9, T_9)/MW_{\text{water}}\}$ and $\hat{h}_{8P}^{\text{liquid}} = \{H\_Water(P_8, T_{8P})/MW_{\text{water}}\}$; to properly use the steam functions from the *TPSI*+ library, the state of the steam (vapor or liquid) must be known. Here we ignore pressure losses on the water/steam side, so$P_8 \approx P_{8P} \approx P_9$ and $P_9$is the saturation pressure of the steam.

The rate that heat is transferred from the gas side to the steam side in the evaporator section is

$$\frac{dQ_{\text{Evaporator,steam}}}{dt} = -\frac{dQ_{\text{Evaporator,gas}}}{dt} = -F_{\text{gas}}(\hat{h}_{7P} - \hat{h}_6). \tag{9.29}$$

If the flow rate of water is equal to the flow rate of steam (no blowdown), then

$$F_{\text{water}} = F_{\text{steam}}. \tag{9.30}$$

The rate that heat is transferred from the gas side to the water side in the economizer section is

$$\frac{dQ_{\text{Economizer,water}}}{dt} = -\frac{dQ_{\text{Economizer,gas}}}{dt} = F_{\text{water}}(\hat{h}_{8P}^{\text{liquid}} - \hat{h}_8^{\text{liquid}}), \tag{9.31}$$

where h_8^{liquid} can be evaluated using$h_8^{\text{liquid}} = \{H_Water(P_8, T_8)/MW_{\text{water}}\}$.

Finally, we can calculate the exhaust gas temperature T_7 leaving the HRSG from an overall energy balance on the gas side of the HRSG as

$$\begin{aligned}\hat{h}_7 &= \hat{h}_6 + \frac{(\dot{Q}_{\text{Evaporator,gas}} + \dot{Q}_{\text{Economizer,gas}})}{F_{\text{gas}}}\\ &= \hat{h}_6 - \frac{(\dot{Q}_{\text{Evaporator,steam}} + \dot{Q}_{\text{Economizer,water}})}{F_{\text{gas}}}.\end{aligned} \tag{9.32}$$

T_7 can be found from $T_7 = \{TfromH\_Products(P_7, \hat{h}_7, H/C\ ratio, DAR)\}$.

Equations (9.26)–(9.32) provide the "HRSG design case." As discussed in Chapter 7, Table 7.1 provides pinch and approach temperature values in order to produce HRSG units of reasonable size and cost. When using methane as fuel, $\Delta T_{\text{Approach}}$ should fall between 15 and 20 R and ΔT_{Pinch} should fall between 15 and 20 R.

EXAMPLE 9.4 ***Cogeneration System with HRSG Design***

We want to solve the cogeneration system of Figure 9.1 using real physical properties and using conditions that will be similar to those found in the economic optimization problems of Chapter 10. In setting up the Excel-based solution, assume air enters the compressor at 230 lb/s (variable); the efficiency of the air compressor = 0.84 (variable); the air pressure at the exit of the air compressor $P_2 = 9.0$ (variable); the combustion chamber inlet temperature

$T_3 = 1625$ R (variable); the mass flow of fuel to the combustion chamber is 3.6 lb/s (variable); and the efficiency of the power turbine = 0.87 (variable). With our *Products* functions, it is more convenient to use the fuel mass flow rate as an independent variable; recall in Chapter 7 we used known T_4 to solve for the fuel flow rate.

We also set the pinch temperature difference at $\Delta T_{\text{Pinch}} = 15$ R and the approach temperature difference $\Delta T_{\text{Approach}} = 15$ R. We do want to allow for the possibility that both ΔT_{Pinch} and $\Delta T_{\text{Approach}}$ may be variable; we will solve an economic optimization problem with variable ΔT_{Pinch} and $\Delta T_{\text{Approach}}$ in Example 10.6.

Solution Hint: When assembling the material and energy balance equations in Excel, avoid using the flow rate of the product gas F_{gas} in the balance equations; instead, use $F_{\text{air}} + F_{\text{fuel}}$, which will help avoid a circular reference error from Excel.

SOLUTION The solution is provided in **Example 9.4.xls** and shown in Figure 9.4. In the solution file, we have also included the equations to determine the fuel and equipment costs, which we will utilize in Chapter 10—these equations are not important for present discussion, but if you change any variable in the problem, for example, the air flow rate, you can see how the system costs change in cell F96. ■

9.3 HRSG OFF-DESIGN CALCULATIONS: SUPPLEMENTAL FIRING

Supplemental firing of the HRSG is critical to meet increased (above design) steam requirements in a process. In Section 7.8, we detailed the effects of supplemental firing on HRSG performance with an ideal exhaust gas. Here we want to examine the effects of supplemental firing on HRSG performance when real exhaust gas properties are used. We follow the developments of Section 7.8 again using two common approaches to solving the HRSG off-design case. In the first approach, an overall energy balance on the HRSG is used with the assumptions that $\Delta T_{\text{Approach}}$ and T_7, the gas exit temperature to the stack, remain fixed at design conditions. In the second approach, the overall heat transfer coefficient for the HRSG is determined for the design case. This coefficient is then modified to account for off-design conditions and the HRSG performance is determined. The reader may want to reread Section 7.8 before continuing Section 9.3.

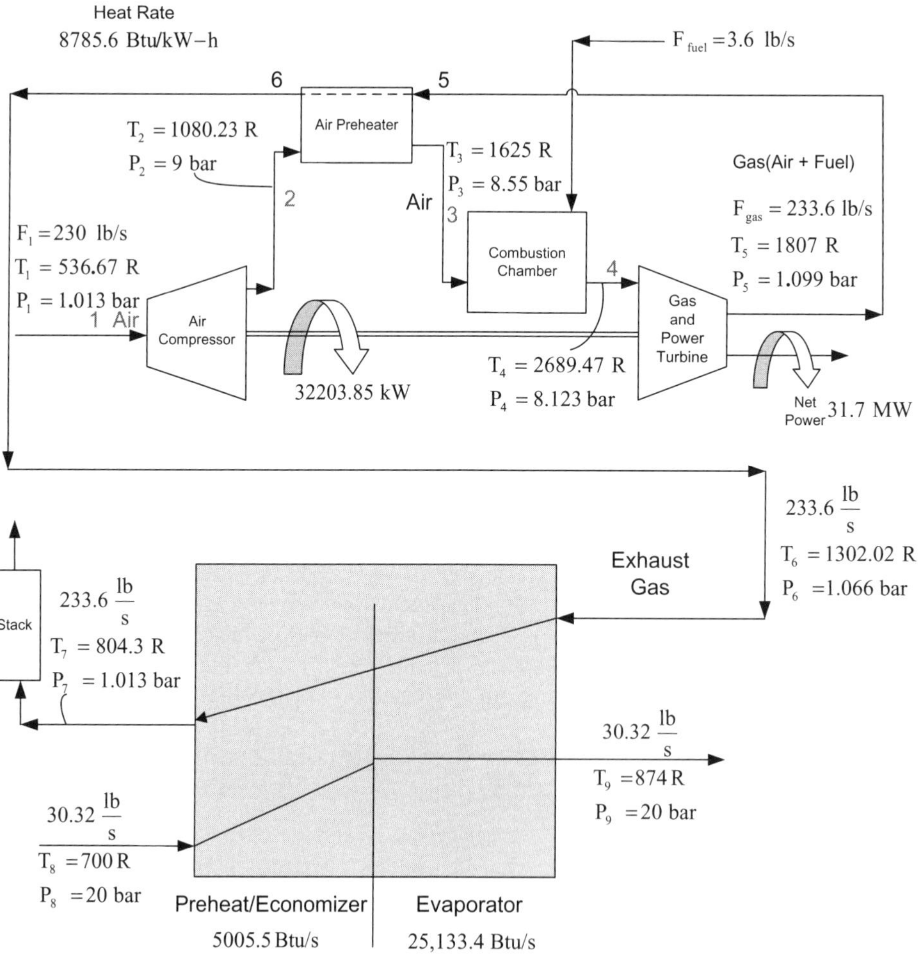

Figure 9.4 Cogeneration solution for real fluid properties, $\Delta T_{\text{Pinch}} = 15$ R and $\Delta T_{\text{Approach}} = 15$ R.

9.3.1 HRSG Off-Design Performance: Overall Energy Balance Approach

To solve for the performance of the HRSG under supplemental firing, an overall energy balance can be used with the assumptions that $\Delta T_{approach\_SF}$ and T_{7_SF} remain fixed at design conditions ($\Delta T_{approach\_SF} = \Delta T_{Approach}$ and $T_{7\_SF} = T_7$). Again we use the subscript SF to indicate supplemental firing. Here, a supplemental fuel flow rate, $F_{supp\ fuel}$, is assumed and the gas temperature after supplemental firing, T_{6_SF}, is determined. An energy balance on the supplemental fuel firing point gives (see Figure 9.5):

$$F_{gas\_SF} = F_{gas} + F_{supp\ fuel} \tag{9.33}$$

and

$$\begin{aligned} &F_{gas}\left(\hat{h}_6 - \hat{h}_{gas,ref}\right) + F_{supp\ fuel}(LHV) \\ &= (F_{gas} + F_{supp\ fuel})\left(\hat{h}_{6\_SF} - \hat{h}_{gas\_SF,ref}\right), \end{aligned} \tag{9.34}$$

where F_{gas_SF} is the combustion products gas mass flow rate (pound per second) after supplemental firing; $\hat{h}_6$ is the enthalpy of the inlet exhaust gas; $\hat{h}_{6\_SF}$ is the enthalpy of the outlet exhaust gas; *LHV* is the *LHV* of fuel (for natural gas *LHV* = 21,500 Btu/lb). Do recall from our discussion in Chapter 7, Section 7.5, that we must subtract the reference enthalpy of the exhaust gas before supplemental firing, $\hat{h}_{gas,ref}$, and the reference enthalpy of the exhaust gas after supplemental firing, $\hat{h}_{gas\_SF,ref}$, at the *LHV* reference conditions—see Figure 7.6. Also note that the compositions of the two streams (stream 6 and stream 6_SF) are not the same.

We can directly solve the energy balance (Eq. (9.34)) for $\hat{h}_{6\_SF}$:

$$\left(\hat{h}_{6\_SF} - \hat{h}_{gas\_SF,ref}\right) = \frac{F_{gas}\left(\hat{h}_6 - \hat{h}_{gas,ref}\right) + F_{supp\ fuel}(LHV)}{(F_{gas} + F_{supp\ fuel})}. \tag{9.35}$$

With an assumed value for $F_{supp\ fuel}$, all the terms on the RHS of Equation (9.35) are known. Here, T_{6_SF} can be found from $T_{6\_SF} = \{TfromH\_Products(P_6, \hat{h}_{6\_SF}, H/C\ ratio, DAR)\}$; recall for $\hat{h}_{6\_SF}$ in {*TfromH_Pro ducts*} we are actually using $\left(\hat{h}_{6\_SF} - \hat{h}_{gas\_SF,ref}\right)$. The fuel used in supplemental firing *will* change the *DAR* value, which at (T_{6_SF}, P_{6_SF}, F_{gas_SF}) is

$$DAR = \frac{F_{air}}{(F_{fuel} + F_{supp\ fuel})(\text{Theoretical air})},$$

and for methane combustion, Theoretical air = 17.2359 lb air/lb CH_4.

With T_{6_SF} and $\hat{h}_{6\_SF}$ determined, the mass flow rate of steam under supplemental firing, F_{steam_SF}, can be determined from an overall energy balance on the HRSG if we assume that the gas exit temperature from the HRSG, T_{7_SF}, is unchanged from the design case ($T_{7\_SF} = T_7$):

$$\begin{aligned} F_{steam\_SF} &= \frac{(F_{gas} + F_{supp\ fuel})\left(\hat{h}_{6\_SF} - \hat{h}_{7\_SF}\right)}{\hat{h}^{sat\ vap}_{(T_9,P_9)} - \hat{h}^{liquid}_{(T_8,P_8)}} \\ &= \frac{(F_{gas\_SF})\left(\hat{h}_{6\_SF} - \hat{h}_{7\_SF}\right)}{\hat{h}^{sat\ vap}_{(T_9,P_9)} - \hat{h}^{liquid}_{(T_8,P_8)}}. \end{aligned} \tag{9.36}$$

The exhaust gas enthalpy at (P_7, T_{7_SF}) can be evaluated using {*H_Products* (*H_Products* (P_7, T_7, *H/C ratio*, *DAR*)}.

The pinch temperature difference under supplemental firing, ΔT_{Pinch_SF}, can be found if we assume that the approach temperature difference under supplemental firing, $\Delta T_{Approach\_SF}$, remains unchanged from the design case. This assumption allows (T_{8P_SF}, P_{8P_SF}) to remain unchanged from the design case (T_{8P_SF}, P_{8P_SF}) = (T_{8P}, P_{8P}). Here, an overall energy balance on the evaporator section of the HRSG yields

$$F_{steam\_SF}\left(\hat{h}^{sat\ vap}_{(T_9,P_9)} - \hat{h}^{liquid}_{(T_{8P\_SF},P_{8P\_SF})}\right) = (F_{gas\_SF})\left(\hat{h}_{6\_SF} - \hat{h}_{7P\_SF}\right). \tag{9.37}$$

Solving for the gas-side pinch enthalpy under supplemental firing, $\hat{h}_{7P\_SF}$, gives

$$\hat{h}_{7P\_SF} = \left(\hat{h}_{6\_SF}\right) - \frac{\left\{F_{steam\_SF}\left(\hat{h}^{sat\ vap}_{(T_9,P_9)} - \hat{h}^{liquid}_{(T_{8P\_SF},P_{8P\_SF})}\right)\right\}}{(F_{gas\_SF})}. \tag{9.38}$$

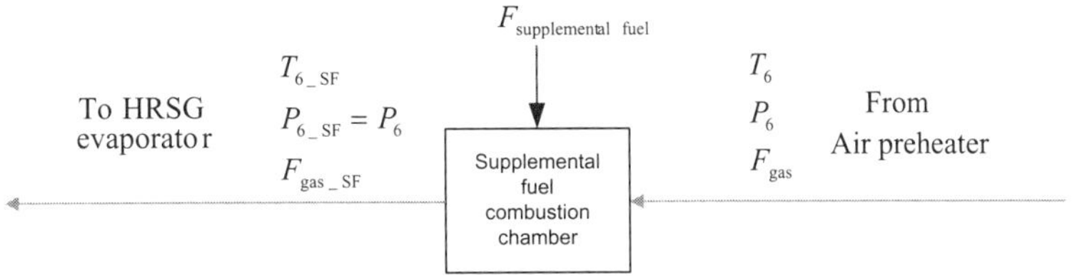

Figure 9.5 Supplemental firing with variables/nomenclature indicated.

Here, T_{7P_SF} can be found from $T_{7P\_SF} = \{TfromH\_Products(P_{7P}, \hat{h}_{7P\_SF}, H/C\ ratio, DAR)\}$ and $\Delta T_{\text{Pinch\_SF}}$ is

$$\Delta T_{\text{Pinch\_SF}} = T_{7P\_SF} - T_9. \tag{9.39}$$

EXAMPLE 9.5 ***HRSG Performance Based on Overall Energy Balance***

Starting with Example 9.4, we want to see the impact on steam production if supplemental firing is used to increase the temperature entering the HRSG; here assume $F_{\text{supp fuel}} = 1.4$ lb/s. From Example 9.4, $F_{\text{air}} = 230$ lb/s, $F_{\text{fuel}} = 3.6$ lb/s, $F_{\text{gas}} = 233.6$ lb/s, $T_6 = 1302$ R, $P_6 = 1.066$ bar, $T_7 = 804.3$ R, $P_7 = 1.013$ bar, $\Delta T_{\text{Pinch}} = 15$ R and $\Delta T_{\text{Approach}} = 15$ R, $F_{\text{steam}} = 30.32$ lb/s; and the transferred Q's are $\dot{Q}_{\text{Evaporator}} = 25{,}133.4$ Btu/s and $\dot{Q}_{\text{Economizer}} = 5005.5$ Btu/s. These values for the HRSG represent the HRSG *design* case.

SOLUTION The solution to Example 9.5 is provided in **Example 9.5.xls**. We use Equations (9.33)–(9.39) to solve the off-design HRSG where $F_{\text{supp fuel}} = 1.4$ lb/s. These equations are also used in the provided Excel solution. Finding $\hat{h}_{6\_SF}$ from Equation (9.35),

$$\left(\hat{h}_{6\_SF} - \hat{h}_{\text{gas\_SF,ref}}\right) = \frac{F_{\text{gas}}\left(\hat{h}_6 - \hat{h}_{\text{gas,ref}}\right) + F_{\text{supp fuel}}(LHV)}{(F_{\text{gas}} + F_{\text{supp fuel}})}$$

$$= \frac{\left(233.6\frac{\text{lb}}{\text{s}}\right)\left(195.99\frac{\text{Btu}}{\text{lb}}\right) + \left(1.4\frac{\text{lb}}{\text{s}}\right)\left(21{,}500\frac{\text{Btu}}{\text{lb}}\right)}{\left(233.6 + 1.4\frac{\text{lb}}{\text{s}}\right)} = 322.91\frac{\text{Btu}}{\text{lb}}.$$

The $DAR = 2.67$:

$$DAR = \frac{F_{\text{air}}}{(F_{\text{fuel}} + F_{\text{supp fuel}})(\text{Theoretical air})}$$

$$= \frac{230\frac{\text{lb}}{\text{s}}}{\left(3.6 + 1.4\frac{\text{lb}}{\text{s}}\right)\left(17.2359\frac{\text{lb}}{\text{lb}}\right)} = 2.67,$$

and from {*TfromH_Products* ((1.066 × 14.5 psia), 322.91 Btu/lb, 4, 2.67)} we find $T_{6\_SF} = 1748.03$ R. The exhaust gas enthalpy at (P_7, T_{7_SF}) can be evaluated using {*H_Products* ((1.013 × 14.5 psia), 804.29 R, 4, 2.67)} giving $\hat{h}_{7\_SF} = 67.86$ Btu/lb. The new steam production rate from Equation (9.36) is

$$F_{\text{steam\_SF}} = \frac{(F_{\text{gas}} + F_{\text{supp fuel}})\left(\hat{h}_{6\_SF} - \hat{h}_{7\_SF}\right)}{\hat{h}^{\text{sat vap}}_{(T_9,P_9)} - \hat{h}^{\text{liquid}}_{(T_8,P_8)}} = \frac{(F_{\text{gas\_SF}})\left(\hat{h}_{6\_SF} - \hat{h}_{7\_SF}\right)}{\hat{h}^{\text{sat vap}}_{(T_9,P_9)} - \hat{h}^{\text{liquid}}_{(T_8,P_8)}}$$

$$= \frac{\left(233.6 + 1.4\frac{\text{lb}}{\text{s}}\right)\left(322.91 - 67.86\frac{\text{Btu}}{\text{lb}}\right)}{\left(1203.41 - 209.25\frac{\text{Btu}}{\text{lb}}\right)} = 60.289\frac{\text{lb}}{\text{s}}.$$

Supplemental firing is nearly doubling steam production. The new gas-side pinch temperature T_{7P_SF} is found via Equation (9.38) as

$$\hat{h}_{7P\_SF} = \left(\hat{h}_{6\_SF}\right) - \frac{\left\{F_{\text{steam\_SF}}\left(\hat{h}^{\text{sat vap}}_{(T_9,P_9)} - \hat{h}^{\text{liquid}}_{(T_{8P\_SF},P_{8P\_SF})}\right)\right\}}{(F_{\text{gas\_SF}})}$$

$$= \left(322.91\frac{\text{Btu}}{\text{lb}}\right) - \frac{\left\{60.289\frac{\text{lb}}{\text{s}}\left(1203.41 - 374.36\frac{\text{Btu}}{\text{lb}}\right)\right\}}{\left(233.6 + 1.4\frac{\text{lb}}{\text{s}}\right)}$$

$$= 110.22\frac{\text{Btu}}{\text{lb}}.$$

From {*TfromH_Products* ((1.04 × 14.5 psia), 110.22 Btu/lb, 4, 2.67)} we find $T_{7P\_SF} = 968.93$ R.

The new pinch temperature difference under supplemental firing, $\Delta T_{\text{Pinch\_SF}}$, is found from Equation (9.39), $\Delta T_{\text{Pinch\_SF}} = T_{7P\_SF} - T_9 = 968.93 - 874 = 94.93$ R. On the Excel sheet, we have also calculated the transferred dQ/dt's under supplemental firing, $\dot{Q}_{\text{Evaporator\_SF}} = 49{,}982.26$ Btu/s and $\dot{Q}_{\text{Economizer\_SF}} = 9954.43$ Btu/s. ■

9.3.2 HRSG Off-Design Performance: Overall Heat Transfer Coefficient Approach

We solved for the performance of the HRSG under supplemental firing using the overall energy balance approach with the assumptions that $\Delta T_{\text{Approach\_SF}}$ and T_{7_SF} remain fixed at design conditions. These assumptions are not needed if we employ the overall heat transfer coefficient method as recommended by Ganapathy (1991, 2003). This method is detailed in Section 7.8. The idea here is that from $\dot{Q}_D = (UA)_D(\Delta T_{\text{LMTD}})_D$ we can solve for

$$(UA)_D = \frac{(\Delta T_{\text{LMTD}})_D}{\dot{Q}_D},$$

where (ΔT_{LMTD}) and $\dot{Q}_D$ are known from the design solution to the HRSG. Under the assumption of controlling gas-side resistance, $(UA)_D$ and $(UA)_{\text{SF}}$ can be predicted from exhaust gas properties, $f\left(F_{\text{gas\_}D}, F_{\text{gas\_SF}}, \hat{C}_{P,\text{gas}}, k, \mu, d_{\text{tube}}\right)$. A ratio of the function for $(UA)_D$ and $(UA)_{\text{SF}}$ allows $(UA)_{\text{SF}}$ to be predicted as

$$\text{Predicted }(UA)_{\text{SF}} = (UA)_D \frac{(F_{\text{gas\_SF}})^{0.6}}{(F_{\text{gas\_}D})^{0.6}} \frac{\left(\dfrac{(k)^{0.7}\left(\hat{C}_{P,\text{gas}}\right)^{0.3}}{(\mu)^{0.3}}\right)_{\text{SF}}}{\left(\dfrac{(k)^{0.7}\left(\hat{C}_{P,\text{gas}}\right)^{0.3}}{(\mu)^{0.3}}\right)_D}. \tag{9.40}$$

The $(UA)_{\text{SF}}$ values from Equation (9.40) are matched to $(UA)_{\text{SF}}$ values from our material and energy balances by varying values for $\Delta T_{\text{Approach\_SF}}$ and T_{7_SF}.

EXAMPLE 9.6 ***HRSG Performance Based on Overall Heat Transfer Coefficient***

Here we want to resolve Example 9.5 using the overall heat transfer coefficient method. Using the temperatures found for the design and off-design cases in Example 9.5, we can estimate exhaust gas (air) physical properties in the HRSG. We will need average physical properties for the evaporator and economizer sections of the HRSG for use in Equation (9.40). We found $\hat{C}_{p,\text{air}}$ using our Excel sheet function {*H_Air*} with a 1° temperature change, μ_{air} from a Web search for "gas viscosity calculator," and k_{air} from a Web search for "air property calculator."

Air Properties

Using results from Example 9.5	$T(R)$	$\hat{C}_{p,\text{air}}\left(\frac{\text{Btu}}{\text{lb-R}}\right)$	μ_{air} (mPa-s)	$k_{\text{air}}\left(\frac{\text{W}}{\text{m-K}}\right)$
Design, T_6	1302	0.258	0.0349	0.0536
Design, T_{7P}	889	0.246	0.0270	0.0399
Design, T_7	804.3	0.244	0.0252	0.0368
Design, average evaporator		0.252	0.031	0.0468
Design, average economizer		0.245	0.0261	0.0384
Off-design, T_{6_SF}	1748	0.271	0.0419	0.0666
Off-design, T_{7P_SF}	968.9	0.246	0.0287	0.0427
Off-design, T_{7_SF}	804.3	0.244	0.0252	0.0368
Off-design, average evaporator		0.259	0.0353	0.0547
Off-design, average economizer		0.245	0.0270	0.0398

SOLUTION The Excel solution to Example 9.5 (following the six steps in Section 7.8) can be found in **Example 9.6.xls**. Using Equation (9.40), Predicted $(UA)_{\text{Evaporator\_SF}} = 182.19$ Btu/s-R; Predicted $(UA)_{\text{Economizer\_SF}} = 85.52$ Btu/s-R. These values for predicted $(UA)_{\text{SF}}$ can be used in a solution technique detailed in Example 8.5. Here you may need to add constraints to help keep the solution feasible. These constraints include $T_{7P\_SF} \geq T_9$, $\Delta T_{\text{Approach\_SF}} \geq 1$ R, $\Delta T_{\text{Approach\_SF}} \geq 150$ R, and $\Delta T_{\text{Pinch\_SF}} \geq 1$ R. After reaching step 5 with these $(UA)_{\text{SF}}$ values, new temperatures will be found for the HRSG. The results from step 5 are summarized here:

		ΔT_{a_SF}	ΔT_{p_SF}	ΔT_{7P_SF}	T_{7_SF}	T_{8P_SF}	$\dot{Q}_{\text{Evap\_SF}}$
Step 5	SF	69	38.2	912.2	804.4	751.2	53,431

$\dot{Q}_{\text{Econ\_SF}}$	$\Delta T_{\text{LM,Evap\_SF}}$	$\Delta T_{\text{LM,Econ\_SF}}$	$UA_{\text{Evap\_SF}}$	$UA_{\text{Econ\_SF}}$	$F_{\text{steam\_SF}}$
6497	293.3	76	182.18	85.52	60.28

In step 6, these new temperatures can be used to update the exhaust gas physical properties and to obtain new values for predicted $(UA)_{\text{SF}}$. This process is repeated until temperatures in the HRSG remain unchanged.

The results from step 5 show that both $\Delta T_{\text{Approach\_SF}}$ (69 R) and $\Delta T_{\text{Pinch\_SF}}$ (38.2 R) increase in value compared to the design case (both were set at 15 R). The exhaust gas temperature from the HRSG is basically unchanged from the design case. The overall energy balance approach, as used in Example 9.5, cannot account for changes in the approach temperature or the exhaust gas temperature as we are able to do here. ■

In general, when solving supplemental firing problems, Fsupp fuel will not be set; instead, the Fsteam_SF will be set by process demands. We do not need to change our solution procedure, but Fsupp fuel is varied until the required steam production is met.

9.4 GAS TURBINE DESIGN AND OFF-DESIGN PERFORMANCE

In Chapter 7, Section 7.9 we discussed the factors that control gas turbine performance in both design and off-design operation—the reader should review this section. Gas turbine performance at design conditions establishes a performance benchmark. Here, ISO air feed conditions are used and key components of the gas turbine system performance including heat rate, power output, exhaust flow, and exhaust temperature are reported. Gas turbine off-design can be determined using manufacturer-provided performance curves which provide for changes in system performance as percent of design values. Off-design conditions included, for example, ambient air conditions different from ISO conditions or nondesign system power requirements. We also established in Chapter 7 that gas turbine off-design performance could be determined using general equations—provided the operation of the gas turbine control system was understood and information was available to determine the needed constants in the general equations (Eqs. (7.72)–(7.74)). In Chapter 7, Example 7.7, we solved a gas turbine design and off-design problem, which we now want to solve using real gas properties.

EXAMPLE 9.7 ***Solve Example 7.7 Gas Turbine Design Off-Design Performance Using Real Gas Properties***

In Table 9.3, column 2, we provide design performance data for a 21.8-MW gas turbine as well as needed properties for the system. The gas turbine is shown in Figure 7.13. We want to modify our solution in Example 7.7 in order to account for real gas properties.

Table 9.3 Design and Off-Design Gas Turbine Performance Calculations Using Real Gas Properties

Variable/Parameter	Design	Off-Design 1	Off-Design 2
Compressor inlet guide vane angle change	0	0	9.044
Compressor inlet flow, F_1 (kg/s)	50	47.526	43.895
Compressor inlet temperature, T_1 (K)	288.15	303.15	288.15
Compressor inlet pressure, P_1 (bar)	1.013	1.013	1.013
Compressor isentropic efficiency, η_{AC}	0.89	0.8397	0.7658
Compressor outlet temperature, T_2 (K)	669.52	714.313	700.397
Compressor outlet pressure, P_2 (bar)	16.208	15.390	14.216
Compressor work (kJ/s)	−19,611	−20,210.44	−18,668.77
Fuel lower heating value *LHV* (kJ/kg)	50,010	50,010	50,010
Fuel flow, F_{fuel} (kg/s)	1.0897	0.9850	0.9242
Turbine inlet flow, F_g (kg/s)	51.0897	48.5110	44.8193
Turbine inlet temperature, T_3 (K)	1523.15	1523.15	1523.15
Turbine inlet pressure, P_3 (bar)	15.560	14.774	13.648
Turbine isentropic efficiency, η_{PT}	0.90	0.90	0.90
Turbine outlet temperature, T_5 (K)	872.57	880.922	895.578
Turbine outlet pressure, P_5 (bar)	1.023	1.023	1.023
Turbine work (kJ/s)	41,222.035	38,557.007	34,877.208
Turbine net work (kJ/s)	21,611.269	18,346.57	16,208.44
Heat rate (kJ/kW-h)	9078.236	9666.09	10,265.67
Fuel flow correction, $F_{\text{fuel}}/F_{\text{fuel, design}}$	1.0	0.9039	0.8481
Exhaust flow correction, $F_g/F_{g,\text{design}}$	1.0	0.9495	0.8773
Exhaust temp correction, $T_5/T_{5,\text{design}}$	1.0	1.0096	1.026

We begin this process by taking the solution file from Example 9.4, and we modify the sheet functions for SI properties. The new file is named **Example 9.7a.xls**. In this file, we have eliminated the air preheater and we do set $(T_4, P_4) = (T_3, P_3)$ to allow easy comparison with the Chapter 7 results.

Off-Design Case 1: We want to predict the performance of the system using Equations (7.72)–(7.74) if the ambient air temperature is raised from design conditions 15–30°C.

Off-Design Case 2: We also want to predict system performance if the turbine power requirements are reduced 25% from 21,611.3 kJ/s (design) to 16,355.78 kJ/s when the ambient air temperature is 15°C.

For the off-design cases, the turbine exhaust temperature should not exceed 110% of the design value.

The same general equations (Eqs. (7.72)–(7.74)) as utilized in Example 7.7 will be valid here.

For this gas turbine system, Equation (7.72) can be written as

$$F_{\text{air}}^{off\_d} = 50\left(\frac{T_1^d}{T_1^{off\_d}}\right)(1-(\Delta\text{angle})(0.0135)), \frac{\text{kg}}{\text{s}};$$

Equation (7.73) can be written as

$$\eta_{\text{AC}}^{off\_d} = \eta_{\text{AC}}^d\left(1-\left|\frac{F_{\text{air}}^d - F_{\text{air}}^{off\_d}}{F_{\text{air}}^d}\right|(FC)\right)$$
$$= 0.89\left(1-\left|\frac{50 - F_{\text{air}}^{off\_d}}{50}\right|(1.143)\right),$$

and for our purposes, we will use T_3 in Equation (7.74) as the temperature into the turbine.

SOLUTION **Design Case: Example 9.7a.xls**

For the design case, key variables include $F_{\text{air}} = F_1 = 50$ kg/s, $\eta_{\text{AC}} = 0.89$, $P_2/P_1 = 16$, 4% pressure drop in the combustion chamber, $\eta_{\text{CC}} = 100\%$ combustion chamber efficiency, $T_3 = 1523.15$ K (1250°C), and $\eta_{PT} = 0.90$. All remaining design performance values are calculated and reported in Table 9.3. You will need to use Excel Goal Seek to vary F_{fuel} so $T_3 = T_4 = 1523.15$ K. We also retain $\Delta T_{\text{Pinch}} = 10$ K and $\Delta T_{\text{Approach}} = 10$ K as used in Example 7.7.

With this design case solved, we can determine the relation for the flow, temperature, and pressure into the turbine Equation (7.74) as

$$\left(\frac{F_3^d\sqrt{T_3^d}}{P_3^d}\right)_{\text{turbine nozzel in}} = \left(\frac{51.0897\sqrt{1523.15}}{15.56}\right)_{\text{turbine in}} = 128.146.$$

Off-Design Case 1: Example 9.7b.xls

For off-design case 1, the ambient air temperature changes from design conditions (15°C) to 30°C. For Equation (7.72), the only change will be in temperature:

$$F_{\text{air}}^{off\_d} = 50\left(\frac{T_1^d}{T_1^{off\_d}}\right)(1-(\Delta\text{angle})(0.0135)) = 50\left(\frac{288.15}{303.15}\right)$$
$$= 47.526\frac{\text{kg}}{\text{s}}.$$

The air compressor isentropic efficiency can be found as

$$\eta_{\text{AC}}^{off\_d} = \eta_{\text{AC}}^{d}\left(1-\left|\frac{F_{\text{air}}^{d} - F_{\text{air}}^{off\_d}}{F_{\text{air}}^{d}}\right|(FC)\right)$$
$$= 0.89\left(1-\left|\frac{50-47.526}{50}\right|(1.143)\right) = 0.8397.$$

In Solver, you will need to vary both $P_2^{off\_d}$ and $F_{\text{fuel}}^{off\_d}$. Changing $P_2^{off\_d}$ allows Equation (7.74) to be satisfied, and $F_{\text{fuel}}^{off\_d}$ must be varied, so $T_{3,off\_d} = T_{4,off\_d} = 1523.15$ K. The objective function in Solver is

$$\min\left[\left(\left(\frac{F_d\sqrt{T_d}}{P_d}\right)-\left(\frac{F_{off\_d}\sqrt{T_{off\_d}}}{P_{off\_d}}\right)_{\text{turbine nozzel in}}\right)^2 + \left(T_{3,d} - T_{3,off\_d}\right)^2\right].$$

A check is made to make sure the turbine exhaust temperature is not violated and key results are provided in Table 9.3.

Off-Design Case 2: Example 9.7c.xls

For off-design case 2, the net power is reduced 25% from the design case with ISO ambient conditions. In order to determine system performance, we must utilize the off-design part load control algorithm (Section 7.9.4). For reduced power, the control algorithm increases the Δ angle for the inlet guide vanes (Eq. (7.72)), which reduces air flow to the compressor. Reduced air-flow calculations then follow those as outlined in off-design case 1. We must modify our Excel-based solution, as developed in off-design case 1, to now include Δ angle as well as $P_2^{off\_d}$ and $F_{\text{fuel}}^{off\_d}$ as variables. In Solver, you will need to vary P_2, F_{fuel}, and Δ angle. The objective function in Solver is

$$\min\left[\left(\left(\frac{F_d\sqrt{T_d}}{P_d}\right)-\left(\frac{F_{off\_d}\sqrt{T_{off\_d}}}{P_{off\_d}}\right)_{\text{turbine nozzel in}}\right)^2 + \left(T_{3,d} - T_{3,off\_d}\right)^2 + \left((\text{Net kW}_d \times 0.75) - (\text{Net kW}_{off\_d})\right)^2\right],$$

and key results are provided in Table 9.3.

In the two off-design cases of Example 7.7, we did not address changes in steam flow and temperatures in the HRSG. Using the HRSG overall energy balance approach with Equation (9.36),

$$F_{\text{steam\_SF}} = \frac{\left(F_{\text{gas\_SF}}\right)\left(\hat{h}_{6\_\text{SF}} - \hat{h}_{7\_\text{SF}}\right)}{\hat{h}^{\text{sat vap}}_{(T_9,P_9)} - \hat{h}^{\text{liquid}}_{(T_8,P_8)}},$$

and without supplemental firing, $F_{\text{gas\_SF}}$ is simply the F_{gas} in the off design case. For off-design case 1,

$$F_{\text{steam}}^{off\_d} = \frac{(48.5110)(642.1897-127.5596)}{(2797.24456-486.390812)} = 10.8035 \text{ kg/s},$$

and for off-design case 2,

$$F_{\text{steam}}^{off\_d} = \frac{(44.8193)(659.8718-127.5596)}{(2797.24456-486.390812)} = 10.3243 \text{ kg/s}.$$

Remaining off-design HRSG calculations including the HRSG overall heat transfer coefficient approach are left for the reader. ■

9.5 CLOSING REMARKS

As noted in the closing comments of Chapter 7, gas turbine cogeneration system design is a mature topic. Gas turbine research and development remains extremely active as small improvements in component efficiency or reliability will translate into large dollars. For readers interested in gas turbine design and application details, I recommend *Gas Turbine Theory*, Sixth Edition (Saravanamuttoo et al., 2009), *Fundamentals of Gas Turbines*, Second Edition (Bathie, 1996), and *Gas Turbine Engineering Handbook*, Third Edition (Boyce, 2006). For readers interested in details of HRSG design and operation, I suggest first reading *Waste Heat Boiler Deskbook* (Ganapathy, 1991) followed by *Industrial Boilers and Heat Recovery Steam Generators: Design, Applications, and Calculations* (Ganapathy, 2003). Additional details of gas turbine design and off-design operation are provided in Kurz (2005), and for both gas and steam turbine design and off-design operation, see the book by Gay et al. (2004).

Once we have selected the cogeneration system configuration, the use of real physical properties allows us to vary any of the independent variables and to determine system performance based on the solution of material and energy balances. These performance calculations can be modified to evaluate potential system performance enhancements including turbine steam injection (see Problem 9.3) and HRSG off-design operation. Our HRSG discussions were limited to single pressure units. The methods developed in this chapter can be applied to multipressure HRSGs and HRSGs with superheating (see Ganapathy, 1991, 2003, and Bustami, 2001).

REFERENCES

BATHIE, W.W. 1996. *Fundamentals of Gas Turbines* (2nd edition). John Wiley & Sons, New York.

BOLLAND, O. 2008. Thermal Power Generation (Compendium). Norwegian University of Science and Technology.

BOYCE, M.P. 2006. *Gas Turbine Engineering Handbook* (3rd edition). Gulf Professional Publishing, Burlington, MA.

BUSTAMI, L.M. 2001. Design of Heat Recovery Steam Generators. MS Thesis, Louisiana State University.

GANAPATHY, V. 1991. *Waste Heat Boiler Deskbook*. The Fairmont Press, Inc., Liburn, GA.

GANAPATHY, V. 2003. *Industrial Boilers and Heat Recovery Steam Generators: Design, Applications, and Calculations*. Marcel Dekker, New York.

GAY, R.R., C.A. PALMER, and M.R. ERBES. 2004. *Power Plant Performance Monitoring*. R-Squared Publishing, Woodland, CA.

KURZ, R. 2005. Gas turbine performance. *Proceedings of the Thirty-Forth Turbomachinery Symposium*, College Station, TX, pp. 131–146.

MYERS, A.L. and W.D. SEIDER. 1976. *Introduction to Chemical Engineering and Computer Calculations*. Prentice-Hall, Englewood Cliffs, NJ.

REALE, M.J. and J.K. PROCHASKA. 2005. New high efficiency simple cycle gas turbine—GE's LMS 100™. *Paper No. 05-IAGT 1.2, Presented at the 16th Symposium on Industrial Applications of Gas Turbines (IAGT)*, Banff, Alberta, Canada, October 12–14, 2005.

REYNOLDS, W.C. 1979. *Thermodynamic Properties in SI: Graphs, Tables, and Computational Equations for Forty Substances*. Department of Mechanical Engineering, Stanford University, Stanford, CA.

SARAVANAMUTTOO, H.I.H., G.F.C. ROGERS, H. COHEN, and P.V. STRAZNICKY. 2009. *Gas Turbine Theory* (6th edition). Prentice Education, Essex, England.

PROBLEMS

9.1 ***Adiabatic Flame Temperature C_8H_{18}*** Determine the adiabatic flame temperature for n-octane combustion with air using the *Products* functions in Table 9.1. For liquid octane, the *LHV* = 19,098 Btu/lb, octane molecular weight = 114.2336 lb/lb-mol, and stoichiometric air (from Eq. (9.8)) = 59.6658 mol air/mol C_8H_{18} = 1728.2199 lb air/mol C_8H_{18}.

9.2 ***(a) Methane Adiabatic Flame Temperature and (b) Combustion Temperature in SI Units*** Solve Example 9.3 in SI units.

(a) Determine the adiabatic flame temperature for methane combustion with air using the SI *Products* functions in Table 9.1. The methane LHV = 50,012 kJ/kg, methane molecular weight = 16.043 kg/kg-mol, and stoichiometric air (from Eq. (9.8)) = 9.5465 mol air/mol CH_4 = 276.5155 kg air/kg-mol CH_4. The process is shown here:

Fuel		Air		Products
CH_4 fuel	+	Stoichiometric air	$\longrightarrow$	CO_2, H_2O, N_2, Ar
298.15°K,		298.15°K,		$T_{adiabatic}$,
0.101326 MPa		0.101326 MPa		0.101326 MPa.

(b) Determine the combustion temperature if methane at 298.15 K and 0.101326 MPa is combusted with two times the stoichiometric air and if the air stream is at 298.15 K and 0.101326 MPa.

9.3 ***Gas Turbine Performance Enhancement—Steam Injection (Solution Provided)*** In a gas turbine cogeneration system, steam or water injection after the air compressor can used to increase turbine power output and to help reduce NO_x emissions. Generally, steam injection is used at 1–5% of the air flow rate. Steam injection will increase the mass flow rate through the power turbine. It will also lower the temperature in the combustion chamber allowing the use of additional fuel while maintaining the same (nonsteam injected) outlet temperature of the combustion chamber.

Gas Turbine Steam Injection—Using Species Functions

Determine the performance improvement for the gas turbine system provided in the Excel file **Problem 9.3a.xls** if steam addition at 3% of the air flow rate from the compressor is utilized. Steam from the HRSG is available at 874 R. The starting gas turbine system from the Excel file, without steam injection, is shown in Figure P9.3a.

If the outlet temperature from the combustion chamber remains constant, T_4 = 2674.56 R, determine how much additional fuel will be required and how much additional power will be generated if steam injection is used. The gas turbine system with steam injection is depicted in Figure P9.3b.

To solve for the dependent variables, we can use our thermodynamics library, taking advantage of individual species functions. We will find we need to sequentially iterate on the fuel flow rate, then T_5^{isen}, and finally T_5.

SOLUTION The solution is provided in **Problem 9.3b.xls** and shown in Figure P9.3c. Solution steps are detailed next. The reader may be curious as to why we do not use our existing *Products* functions, which already incorporate needed iterations for fuel combustion. The problem here is that the inclusion of $F\hat{h}$ and $F\hat{s}$ contributions from steam injection ultimately necessitate the same iterations as the individual species functions. This is shown in the next problem, Problem 9.4.

The key steps (see Figure P9.3c) when using individual species real properties include

Step 1: Determine the fuel flow rate.

(a) For stream 3, see the In Combustion Chamber section in Figure P9.3c. Guess a fuel flow rate, say, $F_{\text{fuel}} = 5.0$ lb/s. Determine the species flow rates in stream 3. Then, using known T_3 and P_3, determine *Nh* for each species and $\hat{h}_3^{\text{actual}}$. For example, $h_{N_2,3} = \{H\_Nitrogen(P_3, T_3)\} =$ 13,400.84 Btu/lb-mol; $h_{N_2,ref} = \{H\_Nitrogen(14.696$ psia, 536.67 R$)\} = 5530.52$ Btu/lb-mol; $H_{N_2}^3 = N_{N_2}^3\left(h_{3,N_2} - h_{N_2,\text{ref}}\right) = 48{,}779.83$ Btu; and

$$\hat{h}_3^{\text{actual}} = \frac{\sum_{\text{species}} H_{\text{species}}^3}{F_{\text{Total}}^3} = 700.75 \text{ Btu/lb}.$$

The fuel *LHV* and combustion chamber efficiency, η_{CC}, are accounted for in $\hat{h}_3^{\text{actual}}$.

(b) For stream 4, see the Out Combustion Chamber section in Figure P9.3c. Using known T_4 and P_4, determine Nh for each species, including the products from combustion, and determine $\hat{h}_4^{\text{actual}}$ ($\hat{h}_4^{\text{actual}} = 614.37$ Btu/lb). Here we want $\hat{h}_4^{\text{actual}} = \hat{h}_3^{\text{actual}}$. We can accomplish this by using Solver to vary F_{fuel} until $\left(\hat{h}_4^{\text{actual}} - \hat{h}_3^{\text{actual}}\right)^2$ is minimized; this gives $F_{\text{fuel}} = 3.902$ lb/s and $\hat{h}_4^{\text{actual}} = \hat{h}_3^{\text{actual}} = 607.81$ Btu/lb.

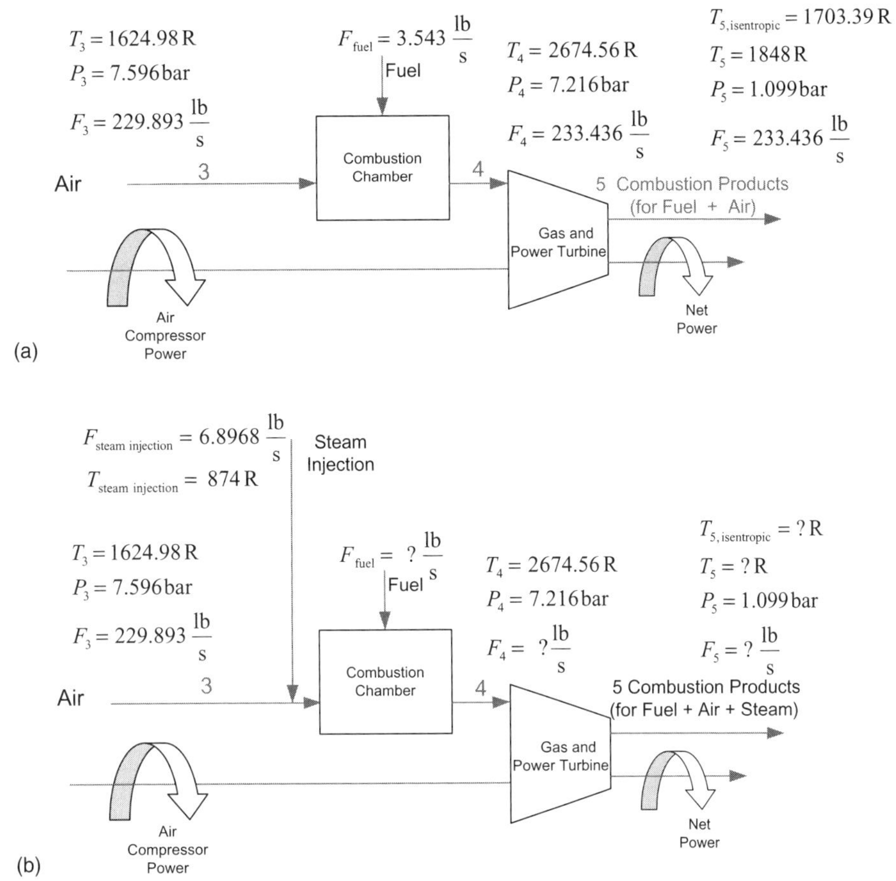

Figure P9.3 (a) Start to Problem 9.3 before steam injection provided in Excel file **Problem 9.3a.xls**. (b) Cogeneration system with steam injection—dependent variables indicated.

(c) Using known T_4 and P_4, determine Ns for each species and determine the entropy, $\hat{s}_4^{\text{actual}}$. For example, $s_{N_2,4} = \{S\_Nitrogen(P_4, T_4)\} = 37.42$ Btu/lb-mol-R; $s_{N_2,\text{ref}} = \{S\_Nitrogen(14.696 \text{ psia}, 536.67 \text{ R})\} = 29.43$ Btu/lb-mol-R; $S_{N_2}^4 = N_{N_2}^4(s_{N_2,4} - s_{N_2,\text{ref}}) = 49.52$ Btu/R; and

$$\hat{s}_4^{\text{actual}} = \frac{\sum_{\text{species}} S_{\text{species}}^4}{F_{\text{Total}}^4} = 0.304 \text{ Btu/lb-R}.$$

Do note that in the calculation of entropy we have used $(s - s_{\text{ref}})$. Here, for the mixture, there is no need for the term $-R\sum(y_i \ln y_i)$; it will cancel.

Step 2: Perform an isentropic expansion through the gas and power turbine.

(d) For stream 5i, see the Out Gas and Power Turbine—Isentropic section in Figure P9.3c. Guess a value for T_5^{isen}; the value for P_5 is known. Solve for $\hat{h}_5^{\text{isen}}$ and $\hat{s}_5^{\text{isen}}$ as detailed in (b) and (c). Here we want $\hat{s}_5^{\text{isen}} = \hat{s}_4^{\text{actual}}$. We can accomplish this by using Solver to vary T_5^{isen} until $(\hat{s}_5^{\text{isen}} - \hat{s}_4^{\text{actual}})^2$ is minimized; this gives $T_5^{\text{isen}} = 1719.3$ R and $\hat{s}_5^{\text{isen}} = \hat{s}_4^{\text{actual}} = 0.304$ Btu/lb-R.

Step 3: Perform the real expansion through the gas and power turbine.

(e) For stream 5a, the real or actual stream 5, see the Out gas and power turbine—Isentropic section in Figure P9.3c. We can now use Equation (9.20) to directly calculate the actual enthalpy leaving the $\hat{h}_5^{\text{actual}}(\text{calculated}) = \hat{h}_4^{\text{actual}} + (\hat{h}_5^{\text{isen}} - \hat{h}_4^{\text{actual}})(\eta_{\text{G\&PT}})$. To determine T_5^{actual}, guess a value for T_5^{actual}; the value for P_5 is known. Solve for $\hat{h}_5^{\text{actual}}$ and $\hat{s}_5^{\text{actual}}$ as detailed in (b) and (c). We want $\hat{h}_5^{\text{actual}}(\text{calculated}) = \hat{h}_5^{\text{actual}}$. We can accomplish this by using Solver to vary T_5^{actual} until $(\hat{h}_5^{\text{actual}} - \hat{h}_5^{\text{actual}}(\text{calculated}))^2$ is minimized; this gives $T_5^{\text{actual}} = 1857.5$ R and $\hat{h}_5^{\text{actual}}(\text{calculated}) = \hat{h}_5^{\text{actual}} = 359.6$ Btu/lb .

The power has been increased from 30,000 to 33,158 kW, which represents a 10.5% increase in power output. The additional costs for the increased power output come from the increased fuel flow rate and the increased makeup water to the HRSG; injected steam, supplied from the HRSG, is lost to the atmosphere.

The sequential iteration on the F_{fuel}, then T_5^{isen}, and finally T_5 does not lend itself to our desired Excel sheet solution. We want to be able to change any design variable and immediately to see the effect on the entire system. There are several alternatives to reach this goal. We can make use of Visual Basic for Applications (VBA) and the single-variable search techniques

(c)

	A	B	C	D	E	F	G	H	I	J	K	L
21		IN Combustion Chamber										
22												
23		F (lb/s)	N (mol/s)									
24	Air	229.8931156	7.936927865									
25	CH4	3.901999928	0.243221338			combustion efficiency						
26	Steam/H2O	6.896793468	0.382814913			η_{cc}	0.98					
27												
28	P_3	111.630816	psia									
29	T_3	1624.98	R	T_{steam} INJ=	874	R						
30						Reference						
31			N_{in}	F_{in}	H_{in} Btu/mol	H_{ref} Btu/mol	H_{in} Total Btu					
32		N2	6.19794697	173.62557	13400.839	5530.51877	48779.82759					
33		O2	1.662786388	53.208167	14708.766	6387.00653	13837.3085					
34		Ar	0.076194508	3.0438182	10184.39	4776.80469	412.028273					
35		H20	0.382814913	6.8967935	22217.494	19505.414	1038.224583					
36		CO2	0	0	20206.884	8079.54664	0					
37		CH4	0.243221338	3.9019999	338041.73	0	82218.96244					
38		Sum		240.67635			146286.3514	Btu				
39												
40						h_{3a}	607.8135797	Btu/lb				
41												
42		Out Combustion Chamber						vary F_{CH4} and min $(h_{4a}-h_{3a})^2$				
43						$(h_{4a}-h_{3a})^2$	6.47033E-22	← using Solver				
44	P_4	106.046336	psia									
45	T_4	2674.614001	R									
46						Reference				Reference		
47			N_{out}	F_{out}	H_{out} Btu/mol	H_{ref} Btu/mol	H_{out} Total Btu		S_{out} Btu/mol-R	S_{ref} Btu/mol-R	S_{out} Total Btu/R	
48		N2	6.19794697	173.62557	21811.149	5530.51877	100906.4856		37.42010858	29.43081159	49.51723907	
49		O2	1.176343711	37.642293	23635.466	6387.00653	20290.11672		41.57380119	32.9296038	10.16854724	
50		Ar	0.076194508	3.0438182	15399.571	4776.80469	809.3964169		28.51994472	24.46692565	0.308817792	
51		H20	0.869257589	15.660545	40111.308	19505.414	17911.82944		40.65804148	29.6476	9.570909815	
52		CO2	0.243221338	10.704171	34263.611	8079.54664	6368.523224		38.0341501	23.34909172	3.571719551	
53		CH4										
54		Sum	8.562964116	240.6764			146286.3514	Btu			73.13723346	Btu/R
55												
56						h_{4a}	607.8134569	Btu/lb		s_{4a}	0.303882039	Btu/lb-R
57												
58		Out Gas and Power Turbine - Isentropic										
59												
60	P_5	16.150904	psia					vary $T_{5,isen}$ and min $(s_{5i}-s_{4a})^2$				
61	$T_{5,\ isentropic}$	1719.303146	R			$(s_{5i}-s_{4a})^2$	4.81959E-17	← using Solver				
62												
63						Reference				Reference		
64			N_{out}	F_{out}	H_{out} Btu/mol	H_{ref} Btu/mol	H_{out} Total Btu		S_{out} Btu/mol-R	S_{ref} Btu/mol-R	S_{out} Total Btu/R	
65		N2	6.19794697	173.62557	14121.677	5530.51877	53247.54169		37.6132336	29.43081159	50.71421775	
66		O2	1.176343711	37.642293	15482.495	6387.00653	10699.42041		41.54858664	32.9296038	10.13888626	
67		Ar	0.076194508	3.0438182	10651.686	4776.80469	447.6337109		30.06411497	24.46692565	0.426475083	
68		H20	0.869257589	15.660545	30106.654	19505.414	9215.207918		39.79063731	29.6476	8.816912156	
69		CO2	0.243221338	10.704171	21419.867	8079.54664	3244.650462		35.85105355	23.34909172	3.040743887	
70		CH4										
71		Sum	8.562964116	240.6764			76854.4542	Btu			73.13723513	Btu/R
72												
73						h_{5i}	319.3269299	Btu/lb		s_{5i}	0.303882046	Btu/lb-R
74												
75												
76							$\eta_{G\&P\ Turbine}$	0.8604				
77		Out Gas and Power Turbine - Real				$h_{5a,calc} = h_{4a}+(h_{5i}-h_{4a})(\eta_{G\&P\ Turbine}) =$			359.5996491	Btu/lb		
78												
79	P_5	16.150904	psia					vary $T_{5,a}$ and min $(h_{5a} - h_{5calculated})^2$				
80	$T_{5,a\ (real)}$	1857.546824	R		$(h_{5a} - h_{5calculated})^2$		2.84136E-18	← using Solver				
81												
82						Reference				Reference		
83			N_{out}	F_{out}	H_{out} Btu/mol	H_{ref} Btu/mol	H_{out} Total Btu		S_{out} Btu/mol-R	S_{ref} Btu/mol-R	S_{out} Total Btu/R	
84		N2	6.19794697	173.62557	15197.902	5530.51877	59917.92906		38.21525935	29.43081159	54.44554142	
85		O2	1.176343711	37.642293	16633.158	6387.00653	12052.99605		42.19229454	32.9296038	10.896108	
86		Ar	0.076194508	3.0438182	11338.129	4776.80469	499.9368496		30.44813159	24.46692565	0.455735041	
87		H20	0.869257589	15.660545	31465.527	19505.414	10396.41887		40.55071692	29.6476	9.477617123	
88		CO2	0.243221338	10.704171	23209.248	8079.54664	3679.866223		36.85198185	23.34909172	3.284191007	
89		CH4										
90		Sum	8.562964116	240.6764			86547.14704	Btu			78.55919259	Btu/R
91												
92						h_{5a}	359.5996491	Btu/lb		s_{5a}	0.326410044	Btu/lb-R
93												
94			Power Air Comp		-28303.558	Btu/s						
95			Power G&P Turbine		59739.2043	Btu/s						
96			Net Power		33158.0217	kW						

Figure P9.3 (*Continued*)

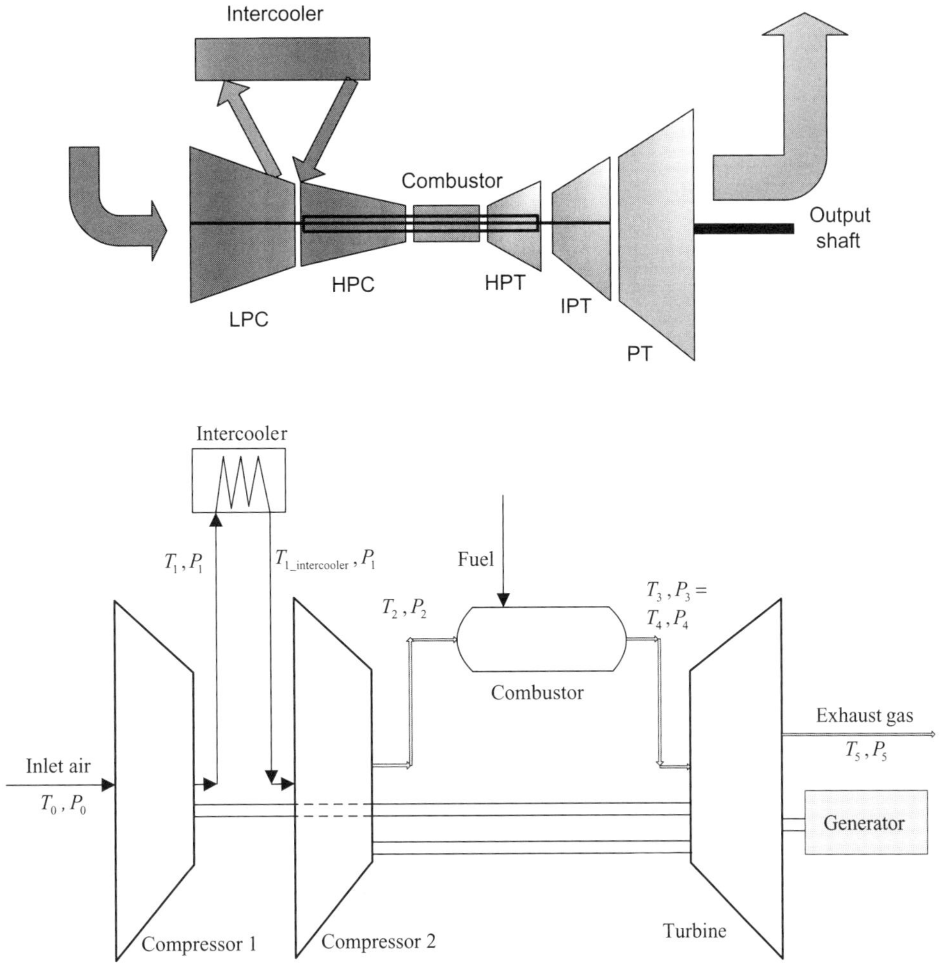

Figure P9.5 Overview schematic of the GE LMS 100 (Reale and Prochaska, 2005; used with permission). Schematic for GE LMS 100 performance calculations (nomenclature used in provided Excel solutions).

we developed in Chapter 3. A second alternative is to write additional C functions, which account for steam injection, and to calculate needed properties.

9.4 ***Steam Injection—Using Products Functions*** Solve Problem 9.3 using the exhaust gas *Products* functions provided in Table 9.1.

9.5 The GE LMS100 shown in Figure P9.5 uses an air cooler (intercooler) between low- and high-pressure air compressor; these are compressors 1 and 2 in Figure P9.5. General Electric (GE) reports the following benchmarks for the LMS100 when using an air flow rate of 209 kg/s at design conditions: net power of 98.7 MW, a heat rate of 7509 Btu/kW-h, thermal efficiency of 46%, and an exhaust temperature 410°C. The pressure ratio in the LMS is reported as 42 : 1.

Using the component efficiencies and properties in Table P9.5 (which are those used in Table 9.3), determine the performance of the LMS 100 and compare these values to those reported by GE. Use $F_{air} = 209$ kg/s; a compression ratio, $P_2/P_0 = 42$; $T_{1\_intercooler} = 15$°C; a 4% pressure drop in the combustion chamber; $T_3 = 1250$°C; and $P_5 = 1.023$ bar. Use the optimum pressure ratio for P_1/P_0 and P_2/P_1; show yourself this will occur when $P_1/P_0 = P_2/P_1$. Remember you will have to use Goal Seek in Excel to vary the fuel flow until $T_3 = 1250$°C.

In Table P9.5, complete the real gas solution; we have determined the ideal gas solution in Chapter 7, Problem 7.5.

Table P9.5 Design Turbine Performance Calculations Using Ideal Gas and Real Gas Properties (See Also Bolland, 2008)

Variable/Parameter	GE Reported Design	Design Ideal Gas	Design Real Gas
Air gamma, γ_a		1.38	
Air specific heat capacity, $\hat{C}_{p,\text{air}}$ (kJ/kg-K)		1.046	
Compressor inlet flow, F_0 (kg/s)	209	209	209
Compressor inlet temperature, T_0 (K)	288.15	288.15	288.15
Compressor inlet pressure, P_0 (bar)	1.013	1.013	1.013
Compressor isentropic efficiency, η_{AC}		0.89	0.89
Compressor outlet temperature, T_2 (K)			
Compressor outlet pressure, P_2 (bar)	42.546	42.546	42.546
Compressor work (kJ/s)			
Fuel lower heating value *LHV* (kJ/kg)		50,010	50,010
Fuel flow, F_{fuel} (kg/s)			
Exhaust gas gamma, γ_g		1.31	
Exhaust heat capacity, $\hat{C}_{p,g}$ (kJ/kg-K)		1.237	
Turbine inlet flow, F_g (kg/s)			
Turbine inlet temperature, T_3 (K)		1523.15	1523.15
Turbine inlet pressure, P_3 (bar)		40.844	40.844
Turbine isentropic efficiency, η_{PT}		0.90	0.90
Turbine outlet temperature, T_5 (K)	683.15		
Turbine outlet pressure, P_5 (bar)		1.023	1.023
Turbine work (kJ/s)			
Turbine net work (kJ/s)	98,700		
Heat rate (kJ/kW-h)	7922.4		

Chapter 10

Gas Turbine Cogeneration System Economic Design Optimization and Heat Recovery Steam Generator Numerical Analysis

Cogeneration systems can take a variety of configurations ranging from a simple diesel engine with a water jacket for heat recovery to combined cycle systems with gas turbines, heat recovery steam generators (HRSGs), and steam turbines with multiple extraction points. Cogeneration system design needs to account for both daily and seasonal variations in the electrical and steam demands of a process or plant or site. Within established daily or seasonal demands, short-term steam or electrical demand may often exceed (or fall below) expected levels. A qualifying cogeneration facility can purchase or sell electricity, in either an open or a regulated electric market. The potential profit from the sale of electricity can be a factor in the construction and sizing of a cogeneration facility. The sale of surplus steam can also be a factor in the sizing of the cogeneration facility. Generally, the sale of surplus steam is not regulated.

Before we examine cogeneration system design, we do need to caution that even in the most straightforward case, where electrical and steam demands are known and relatively constant, fluctuation in cogeneration fuel costs (e.g., the cost of natural gas) can quickly move a cogeneration project from being economically viable to nonviable. For example, many industrial natural gas fuel-based cogeneration projects are economically viable with natural gas prices in the \$4–\$6 per MMBtu range. However, as natural gas prices rise above \$6, the fuel mix for utility companies, which generally includes coal, natural gas, hydroelectric, and perhaps nuclear energy, allows the sale of electricity at a price, which eliminates many of the advantages for cogeneration. This \$6 benchmark will certainly move upward if CO_2 sequestration from coal-fired power plants becomes a reality, as discussed in Chapter 15.

The design of an optimal cogeneration system requires minimizing total costs. Total costs include capital equipment, installation costs, and operating costs including fuel purchase and maintenance costs, and the cost of (or profit from) supplemental electricity. The design requires selection of cogeneration configuration and equipment as well as the determination of processing conditions. Equipment selection must address which type and size of units to use, for example, a gas turbine or steam turbine or both. Equipment must be configured; for example, in some cases, parallel units may be better at matching processing needs than a single larger unit; parallel units allow increased reliability or the ability to cycle units for widely varying loads.

Often the best cogeneration system design is one that targets the base power and base heat requirements of the site. Additional electricity can be purchased and additional steam production can be realized by supplemental firing of the HRSG.

Modeling, Analysis and Optimization of Process and Energy Systems, First Edition. F. Carl Knopf.

In this chapter, we explore the optimal economic design of cogeneration systems. The optimal design problem begins by first selecting the cogeneration configuration that matches the site utility needs and then assembling material and energy balances for the chosen cogeneration configuration. These material and energy balance performance equations provide needed flow rates, temperatures, and pressures that are used to size and then the cost equipment. Often fuel cost, as determined from the fuel flow rate, is the dominant cost component for the cogeneration system.

In a final section of this chapter, Section 10.7, we examine heat transfer in the HRSG in greater detail. Here we use numerical methods to determine expected temperature profiles in the HRSG tubes and at the HRSG walls.

10.1 COGENERATION SYSTEM: ECONOMY OF SCALE

We start our cogeneration system equipment cost discussion by examining economy of scale and efficiency in cogeneration systems. To simplify the discussions, we consider simple-cycle plants, which consist of just the gas turbine system. Figure 10.1 shows simple-cycle total capital costs (Figure 10.1a) and heat rates (Figure 10.1b) as a function of power output over a 0- to 50-MW range (*Gas Turbine World Handbook*, 2010, with permission).

There is an economy of scale for gas turbine systems as indicated in Figure 10.1a. The polynomial curve fit in Figure 10.1a is,

$$\text{Total plant price (\$)} = 431{,}900 + (471{,}600)(\text{MW}) - (6057)(\text{MW}^2) + (58.69)(\text{MW}^3).$$

Using this polynomial, a 10-MW system will have a capital cost of \$4.601 MM; a 20-MW system will cost \$7.911 MM; and a 50-MW system will cost \$16.206 MM. The heat rate plots in Figure 10.1b indicate gas turbine system and therefore component efficiencies; recall the heat rate is defined as British thermal unit per kilowatt-hour, which is the fuel rate (British thermal unit per hour) needed to produce each kilowatt of electricity. At a delivered power level, there is generally a range of heat rates, which indicates a range of gas turbine component efficiencies. The idea from Figure 10.1a,b is that equations can be developed, which account for the gas turbine cost as a function of both size and efficiency—these types of equations for the cogeneration system in Figure 10.2 are provided in Section 10.3.

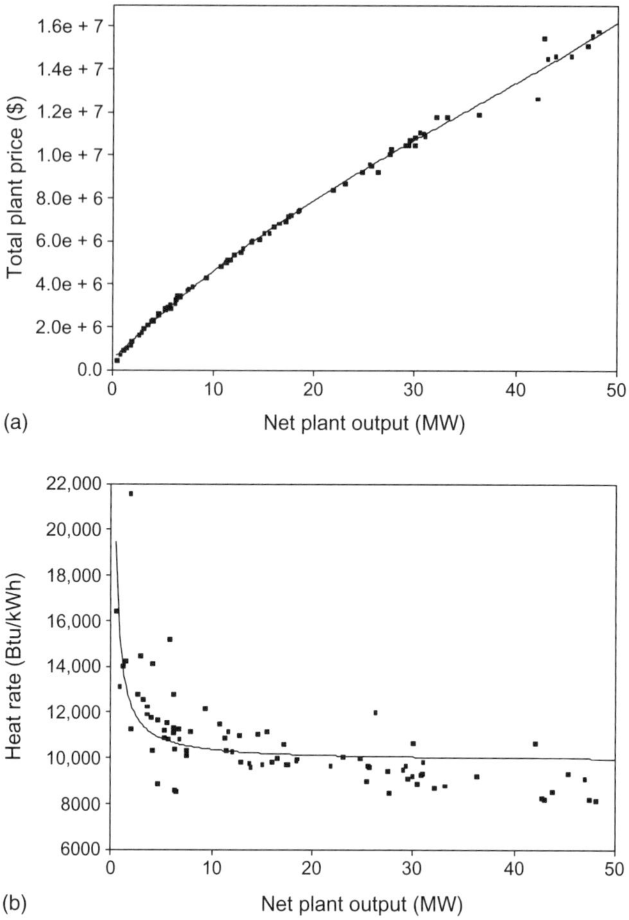

Figure 10.1 (a) Simple-cycle plant total capital cost. (b) Simple-cycle plant heat rate.

10.2 COGENERATION SYSTEM CONFIGURATION: SITE POWER-TO-HEAT RATIO

As stated in the 4th paragraph of this chapter, "Often the best cogeneration design is one that targets the base power and base heat requirements of the site. Additional electricity can be purchased and additional steam production can be realized by supplemental firing of the HRSG."

In many cogeneration designs, the cogeneration system is "overbuilt" with emphasis placed on meeting peak electrical demands or future anticipated electrical demands. Here, a justification often cited is that as the site increases capacity, the overbuilt cogeneration facility will be viewed as a better and better design. This justification is often quickly followed by the economy of scale discussions in Section 10.1. Unfortunately, in many cases, expansion at the processing site is slower than expected. Here then, when the cogeneration facility is meeting typical baseline loads, the efficiency loss in the oversized gas turbine can be quite costly. To avoid

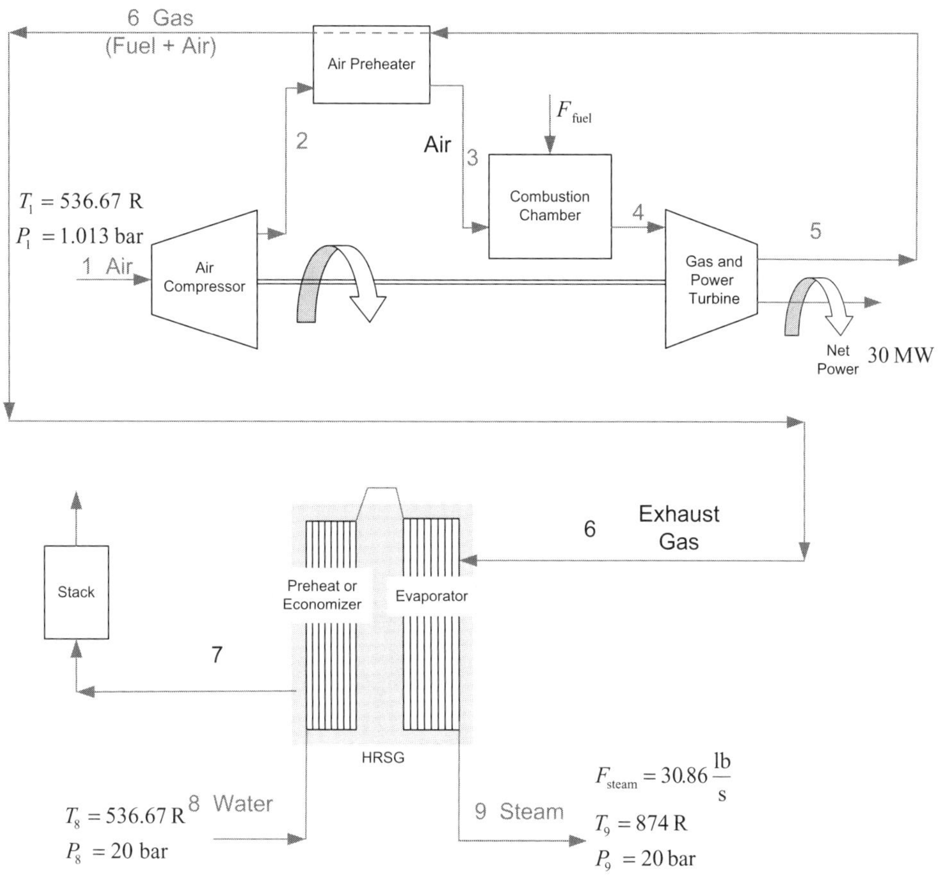

Figure 10.2 Cogeneration system with air preheat and HRSG.

potential problems with overbuilding, a first step toward the design of an optimal cogeneration system is to determine a site's base power-to-base heat ratio, $\alpha_{\text{site\_PHR}}$:

$$\alpha_{\text{site\_PHR}} = \frac{\text{Base power need (MW)}}{\text{Base thermal load (MW)}}. \tag{10.1}$$

This value can then be matched to typical power-to-heat ratios for cogeneration systems, $\alpha_{\text{cogen\_PHR}}$:

$$\alpha_{\text{cogen\_PHR}} = \frac{\begin{array}{c}\text{Net power available from}\\ \text{cogeneration system (MW)}\end{array}}{\text{Available heat from cogeneration system (MW)}}. \tag{10.2}$$

Some cogeneration configurations and $\alpha_{\text{cogen\_PHR}}$ values are indicated in Table10.1 (Rossiter, 1990; Spiewak and Weiss, 1997). We suggest matching the site's base power-to-heat ratio, $\alpha_{\text{site\_PHR}}$, to the most appropriate cogeneration system configuration and then designing this cogeneration system.

Table 10.1 Cogeneration Configuration and Power-to-heat Ratio (Spiewak and Weiss, 1997)

Cogeneration System Configuration	Typical $\alpha_{\text{cogen\_PHR}}$ Range
Gas turbine with HRSG	0.6–0.9
Combined cycle—gas turbine, HRSG, back-pressure steam turbine	0.7–1.3
Diesel engine	1.0–2.0

The article by Poullikkas (2005) reviews current and emerging cogeneration configurations and can be used to expand the available systems in Table 10.1.

10.3 ECONOMIC OPTIMIZATION OF A COGENERATION SYSTEM: THE CGAM PROBLEM

A problem to allow comparison of different methods of thermoeconomic optimization and analysis for cogeneration system design was developed by Valero et al. (1994a). The CGAM cogeneration flow sheet is shown in Figure 10.2. We are asked to determine the optimal equipment sizing and design parameters (T, P, flow rates, etc.) which minimize total system costs for fixed baseline loads of 30 MW of electricity and 30.86 lbm/s of saturated steam at 290 psia.

Checking the site power-to-heat ratio,

$$\alpha_{\text{site\_PHR}} = \frac{\text{Base power needed (MW)}}{\text{Base thermal load (MW)}} = \frac{30\ \text{MW}}{\left[\left(30.86\frac{\text{lbm}}{\text{s}}\right)\left(1203\frac{\text{Btu}}{\text{lbm steam}}\right)\left(3600\frac{\text{s}}{\text{h}}\right)\left(\frac{1\ \text{W}}{3.413\ \text{Btu}}\right)\left(\frac{1\ \text{MW}}{10^6\ \text{W}}\right)\right]} = \frac{30\ \text{MW}}{39.16\ \text{MW}} = 0.76.$$

The value of 0.76 falls near within the expected range for a gas turbine cogeneration configuration as indicated in Table 10.1.

Here we show the process for the economic optimization of the cogeneration system in Figure 10.2. This configuration is *exactly* the same as the system we utilized in Chapter 9, so we have developed real gas performance equations for the gas turbine system and HRSG. We will need to make a small modification to our work in Chapter 7, to account for the air preheater (APH), in order to also have ideal gas performance equations for the system. All of these performance calculations are independent of the cost equations that are used to determine the optimal design. For the nonlinear optimization employed here, equipment costs are approximated as continuous functions and are expressed in terms of the design variables. Recall that any independent variable can be changed in our Excel-based performance calculations, and values will be determined for dependent variables—we will utilize this for system costing and optimization. In fact, you may recall we have already utilized this in Example 9.3 when we changed the air flow rate and saw a change in system costs (although in Example 9.3 we did not explain how the costs were obtained). The independent variables in our cogeneration system formulation of Chapters 7 and 9, which impact the economic optimization, will be referred to as "design" variables in our economic optimization—as these design variables are changed, the system cost changes.

10.3.1 The Objective Function: Cogeneration System Capital and Operating Costs

The objective function accounts for capital and operating costs, which are minimized by varying the design variables (Valero et al., 1994a). The total cost rate including fuel, equipment, and maintenance, C_{Total} (dollar per second), can be found as

$$C_{\text{Total}} = (c_f)(F_{\text{fuel}})(LHV\ \text{CH}_4) + \sum_{i=1}^{5}\frac{(Z_i)(\text{CRF})(\phi)}{(3600)(N)}, \qquad (10.3)$$

where c_f is the fuel cost per energy unit (based on the fuel lower heating value [*LHV*]); F_{fuel} is the fuel flow rate (pound per second); i indexes the five equipment items: air compressor, combustion chamber, gas and power turbine, air preheater, and HRSG; Z_i is the purchase cost of the ith component (dollar) (see Tables 10.2 and 10.3); CRF is the annual capital recovery factor (CRF = 18.2%); N is the number of hours of plant operation per year (N = 8000 hours and 3600 s/h); and ϕ is the maintenance factor (ϕ = 1.06).

For the cogeneration system, the fuel and equipment costs are being brought to an equivalent dollar-per-second basis through the capital recovery factor (see Chapter 2). For cogeneration systems, fuel costs often dominate. Here, c_f = \$4.2204 per MMBtu based on the *LHV* of the methane fuel. Do recall that natural gas is actually priced on the higher heating value of the fuel, which is ~110% the *LHV*. Therefore, fuel costs in this problem would be ~\$4.60 natural gas, which would be ~\$4.60 per MMBtu or ~\$4.60 per mSCF (1000 standard cubic feet).

In the costing equation, the use of the *LMTD* in the air preheater and HRSG represents a temperature driving force at each end of the heat exchanger given by

$$\Delta T_{\text{LMTD,Air preheater}} = \left(\frac{(T_5 - T_3) - (T_6 - T_2)}{\ln\left(\frac{(T_5 - T_3)}{(T_6 - T_2)}\right)}\right),$$

$$\Delta T_{\text{LMTD,Economizer}} = \left(\frac{(T_{7P} - T_{8P}) - (T_7 - T_8)}{\ln\left(\frac{(T_{7P} - T_{8P})}{(T_7 - T_8)}\right)}\right),$$

and

$$\Delta T_{\text{LMTD,Evaporator}} = \left(\frac{(T_6 - T_9) - (T_{7P} - T_{8P})}{\ln\left(\frac{(T_6 - T_9)}{(T_{7P} - T_{8P})}\right)}\right).$$

For the HRSG economizer and evaporator, temperatures have been defined in Figures 7.8 and 7.9.

Table 10.2 Cost Equations—Purchase Cost (Valero et al., 1994a)

Component	Capital Investment Costs ($)
Air compressor	$Z_{\text{Comp}} = \left(\dfrac{C_{11}F_{\text{air}}}{C_{12} - \eta_{\text{comp}}}\right)\left(\dfrac{P_2}{P_1}\right)\ln\left(\dfrac{P_2}{P_1}\right)$
Combustion chamber	$Z_{\text{CC}} = \left(\dfrac{C_{21}F_{\text{air}}}{C_{22} - \dfrac{P_4}{P_3}}\right)\left(1+\exp\left(C_{23}T_4 - C_{24}\right)\right)$
Gas and power turbine	$Z_{\text{Turb}} = \left(\dfrac{C_{31}F_{\text{gas}}}{C_{32} - \eta_{\text{G\&P turbine}}}\right)\ln\left(\dfrac{P_4}{P_5}\right)\left(1+\exp(C_{33}T_4 - C_{34})\right)$
Air preheater	$Z_{\text{air PreH}} = C_{41}\left(\dfrac{F_{\text{gas}}\left(\hat{h}_5 - \hat{h}_6\right)}{\left(U_{\text{air preheater}}\right)\ \left(\text{LMTD}_{\text{air prheaeater}}\right)}\right)^{0.6}$
HRSG	$Z_{\text{HRSG}} = C_{51}\left[\left(\dfrac{\dot{Q}_{\text{Ecomizer}}}{\text{LMTD}_{\text{Economizer}}}\right)^{0.8} + \left(\dfrac{\dot{Q}_{\text{Evaporator}}}{\text{LMTD}_{\text{Evaporator}}}\right)^{0.8}\right] + C_{52}F_{\text{steam}} + C_{53}F_{\text{gas}}^{1.2}$

C_{ij} are cost coefficients (see Table 10.3); $U_{\text{air preheater}}$ is the overall heat transfer coefficient in the air preheater; and *LMTD* is the log mean temperature difference in the specified heat recovery unit.

Table 10.3 Cost Coefficients for the Cogeneration Components (Valero et al., 1994a)

	Field Units	SI Units
Air compressor	$C_{11} = 17.95$ \$/(lb/s) $C_{12} = 0.9$	$C_{11} = 39.5$ \$/(kg/s) $C_{12} = 0.9$
Combustion chamber	$C_{21} = 11.64$ \$/(lb/s) $C_{22} = 0.995$ $C_{23} = 0.01$ (1/R) $C_{24} = 26.4$	$C_{21} = 25.6$ \$/(kg/s) $C_{22} = 0.995$ $C_{23} = 0.018$(1/K) $C_{24} = 26.4$
Gas and power turbine	$C_{31} = 120.79$ \$/(lb/s) $C_{32} = 0.92$ $C_{33} = 0.02$ (1/R) $C_{34} = 54.4$	$C_{31} = 266.3$ \$/(kg/s) $C_{32} = 0.92$ $C_{33} = 0.036$(1/K) $C_{34} = 54.4$
Air preheater	$C_{41} = 2290$ \$/(m$^{1.2}$) $U_{\text{air preheater}} = 0.009478\left(\dfrac{\text{Btu}}{\text{s-m}^2\text{-R}}\right)$	$C_{41} = 2290$ \$/(m$^{1.2}$) $U_{\text{air preheater}} = 0.018\left(\dfrac{\text{kW}}{\text{m}^2\text{-K}}\right)$
HRSG	$C_{51} = 6097.2$ \$/(Btu/[s R])$^{0.8}$ $C_{52} = 5361.5$\$/(lb/s) $C_{53} = 254.8$\$/(lb/s)$^{1.2}$	$C_{51} = 3650$ \$/(kW/K)$^{0.8}$ $C_{52} = 11{,}820$ \$/(kg/s) $C_{53} = 658$ \$/(kg/s)$^{1.2}$

EXAMPLE 10.1 ***Design Problem Component Costing***

In Example 9.4, the cogeneration system component costs as given in Tables 10.3 and 10.4 were actually included in the solution file **Example 9.4.xls**. Using the results from Example 9.4 (Figure 9.4), determine the purchase cost of the air compressor component of the gas turbine system.

SOLUTION

$$Z_{\text{Comp}} = \left(\frac{C_{11}F_{\text{air}}}{C_{12} - \eta_{\text{comp}}}\right)\left(\frac{P_2}{P_1}\right)\ln\left(\frac{P_2}{P_1}\right) = \left(\frac{17.95F_{\text{air}}}{0.9 - \eta_{\text{comp}}}\right)\left(\frac{P_2}{P_1}\right)\ln\left(\frac{P_2}{P_1}\right)$$
$$= \left(\frac{17.95(230)}{0.9-(0.84)}\right)\left(\frac{9}{1.013}\right)\ln\left(\frac{9}{1.013}\right) = \$1{,}335{,}328$$

Table 10.4 Possible Constraints for the CGAM Cogeneration Design Problem

$4 < P_2/P_1$
$75\% < \eta_{\text{air compressor}} < 90\%$
$75\% < \eta_{\text{G\&P turbine}} < 92\%$
$T_2 + 10 \le T_6$
$T_3 + 10 \le T_5$
$T_7 > 709.67$ R (250°F)

Table 10.5 Physical Properties, Design Parameters, and Constants

	Field Units	SI Units
T_1, P_1	536.67 R, 1.013 bar	298.15 K, 1.013 bar
γ_{air}	1.4	1.4
$\hat{C}^{ig}_{P,\text{air}}$	$0.24\left(\frac{\text{Btu}}{\text{lb-R}}\right)$	$1.004\left(\frac{\text{kJ}}{\text{kg-K}}\right)$
T_{ref}	536.67 (R)	298.15 (K)
LHV CH_4	$21,500\left(\frac{\text{Btu}}{\text{lb}}\right)$	$50,000\left(\frac{\text{kJ}}{\text{kg}}\right)$
γ_{gas}	1.33	1.33
$\hat{C}^{ig}_{P,\text{gas}}$	$0.28\left(\frac{\text{Btu}}{\text{lb-R}}\right)$	$1.17\left(\frac{\text{kJ}}{\text{kg-K}}\right)$
η_{CC}	0.98	0.98
ΔP_{CC}, $\Delta P_{\text{APH, air}}$, $\Delta P_{\text{APH, gas}}$	0.05, 0.05, 0.03	0.05, 0.05, 0.03
ΔT_{Pinch}	3 R	1.64 K
$\Delta T_{\text{Approach}}$	27 R	15 K
$\hat{h}_9^{\text{sat vapor}}$(874 R, 20 bar)	$1203.6\left(\frac{\text{Btu}}{\text{lb}}\right)$	$2797.2\left(\frac{\text{kJ}}{\text{kg}}\right)$
$\hat{h}_{8P}^{\text{liquid}}$(847 R, 20 bar)	$361.4\left(\frac{\text{Btu}}{\text{lb}}\right)$	$841.2\left(\frac{\text{kJ}}{\text{kg}}\right)$
$\hat{h}_8^{\text{liquid}}$(536.7 R, 20 bar)	$45.09\left(\frac{\text{Btu}}{\text{lb}}\right)$	$110.9\left(\frac{\text{kJ}}{\text{kg}}\right)$

You should determine the cost of each cogeneration system component using the results from Example 9.4 and compare your results with those provided in the Excel file, cells H98–H102. Equipment purchase costs will be multiplied by the capital recovery factor to bring costs to a yearly basis and then converted to a per-second basis. Fuel and maintenance costs are included to form the system costs on a dollar-per-second basis. ■

10.3.2 Optimization: Variable Selection and Solution Strategy

The nonlinear programming solution approach to the design problem consists of minimizing the capital and operational costs (given by Eq. (10.3)) while delivering 30 MW of electricity and 30.865 lbm/s of saturated steam at 290 psia. The objective function is minimized by varying the independent design variables. We can walk through the optimization process by recalling our performance equation developments of Chapters 7 and 9.

For the *ideal gas* optimization formulation, we use the following for the independent design variables: the air flow rate, F_{air}; P_2; the air compressor efficiency, η_{comp}; T_3; T_4; the gas and power turbine efficiency, $\eta_{\text{G\&P turbine}}$; and for the HRSG values for ΔT_{Pinch} and $\Delta T_{\text{Approach}}$. The system dependent design variables (P_4, P_5, $\dot{Q}_{\text{Economizer}}$, $\dot{Q}_{\text{Evaporator}}$, etc.) are calculated as functions of these independent design variables and known system parameters including pressure drops, fuel LHV, and so on (see Table 10.5). The *real gas* optimization formulation follows the ideal gas formulation with one change; the fuel flow rate F_{fuel} rather than T_4 will be used as an independent variable.

The inlet air temperature and pressure to the compressor are fixed at ambient conditions. The optimization program will select values for P_2 and $\eta_{\text{air compressor}}$, and as we have seen in Example 10.1, a cost for the air compressor can be determined. A value for T_3 will be selected and from P_2 and the known pressure drop in the air preheater, all conditions at point 3 will be known. As T_3 is increased in value, the cost of the air preheater will increase, but the fuel flow rate (and fuel cost) needed to bring the combustion products to T_4 will decrease. The power turbine efficiency $\eta_{\text{G\&P turbine}}$ will impact the delivered power. Values for ΔT_{Pinch} and $\Delta T_{\text{Approach}}$ will control the amount of steam raised and the cost of the HRSG.

This economic design optimization problem does require comment. We have discussed in Chapters 7 and 9 that values for ΔT_{Pinch} and $\Delta T_{\text{Approach}}$ should be selected based on fuel type and expected exhaust gas inlet temperature to the evaporator. Values for ΔT_{Pinch} and $\Delta T_{\text{Approach}}$ are not generally considered design optimization variables. We will solve the optimization problem with constraints added to keep ΔT_{Pinch} and $\Delta T_{\text{Approach}}$ at appropriate values or simply fix values for ΔT_{Pinch} and $\Delta T_{\text{Approach}}$ and solve the remaining optimization problem. A larger problem is that this optimal system design assumes continuous variables, for example, the efficiency of the air compressor can take on any value. We know from our discussion of gas turbine design calculations that there is just one single design efficiency for the air compressor as provided from the manufacturer. *The optimization problem here should be viewed as a framework for helping to select among several alternative cogeneration systems, or possible system configurations, or help identify needed ranges in operating conditions.*

10.3.3 Process Constraints

The inlet air temperature and pressure to the compressor are fixed at ambient conditions. And the cogeneration system must generate 30 MW of electricity and 30.86 lb/s of saturated steam at 290 psia. The power and steam requirements will necessitate the use of at least two constraints in the formulation

$$F_{\text{steam}} \geq 30.86\frac{\text{lb}}{\text{s}}$$

$$\dot{W}_{\text{net}} \geq 30{,}000\ \text{kW}.$$

There are other processing conditions or limitations that *may* need to be imposed on the design solution depending on the problem formulation and the initial guess used. For example, materials of construction impose certain practical limits on the inlet temperature to the gas and power turbine. Similarly, concerns exist for the compression ratio in the air compressor. The form of the cost expression for the air compressor and the gas and power turbine require that $\eta_{\text{air compressor}} < 0.90$ and that $\eta_{\text{G\&P turbine}} < 0.92$; a practical lower limit on efficiency for each of these units is 75%. During the optimization process, temperature crossover in the air preheater heat exchanger may occur, but this can be avoided by temperature constraints $T_2 + 10 \leq T_6$ and $T_3 + 10 \leq T_5$; here a minimum 10 R approach temperature is being used. The exhaust gas temperature at the stack, T_7, should be >~759.67 R when using methane or natural gas fuel in order to avoid acid condensation. Allowable T_7 values will increase as the sulfur content of the fuel increases. Table 10.4 summarizes possible constraints that *may* need to be added to the optimization model to obtain a reasonable solution.

10.4 ECONOMIC DESIGN OPTIMIZATION OF THE CGAM PROBLEM: IDEAL GAS

The cogeneration system in Figure 10.2 consists of an air compressor, an air preheater, a combustion chamber, gas and power generating turbine and HRSG. We developed material and energy balances for these operations in Chapter 7 (we will need to add equations for the air preheater) for use in the Excel-based optimal design solution. In this section, both the air feed and the combustion products are taken as ideal gas with known heat capacity. In addition, both ΔT_{Pinch} and $\Delta T_{\text{Approach}}$ will be taken as known values. With $\Delta T_{\text{Approach}}$ fixed in value, water and steam properties are fixed in the HRSG. In Section 10.5, real gas and water/steam properties from our combustion library (*TPSI+*) will be used allowing for variable steam properties within the HRSG.

10.4.1 Air Preheater Equations

We will need to add the air preheater from Figure 10.2 to our ideal gas formulation of Chapter 7. Energy from the exhaust gas leaving the power turbine is used to heat the compressed air that will enter the combustion chamber. With adiabatic heat exchange,

$$\frac{dQ_{\text{air}}}{dt} = -\frac{dQ_{\text{gas}}}{dt}, \tag{10.4}$$

$$F_{\text{air}}\hat{C}^{ig}_{P,\text{air}}(T_3 - T_2) = F_{\text{gas}}\hat{C}^{ig}_{P,\text{gas}}(T_5 - T_6), \tag{10.5}$$

and

$$T_6 = T_5 - \frac{F_{\text{air}}\hat{C}^{ig}_{P,\text{air}}(T_3 - T_2)}{F_{\text{gas}}\hat{C}^{ig}_{P,\text{gas}}}, \tag{10.6}$$

where the $F_{\text{gas}} = F_{\text{air}} + F_{\text{fuel}}$. The pressure drop on each side of the air preheater can be found as

$$P_3 = P_2(1 - \Delta P_{\text{APH,air}}) \tag{10.7}$$

and

$$P_6 = P_5(1 - \Delta P_{\text{APH,gas}}). \tag{10.8}$$

10.4.2 CGAM Problem Physical Properties

Table 10.5 provides physical properties for the air, exhaust gas, and fuel and known system parameters. Steam physical properties are found from steam tables.

The pinch and approach temperature difference and the condensate return temperature deserve comment. As we discussed in Chapter 7, the "HRSG design case" requires specification of pinch and approach temperature differences. Here we have set the pinch temperature difference at $\Delta T_{\text{Pinch}} = 3$ R and the approach temperature difference $\Delta T_{\text{Approach}} = 27$ R; both values are taken from the CGAM problem optimal solution (Valero et al., 1994a). The $\Delta T_{\text{Pinch}} = 3$ R value is questionable and this value will be changed in Example 10.2. Recall from our discussion in Chapters 7 and 9 that typical values for ΔT_{Pinch} and $\Delta T_{\text{Approach}}$ with methane fuel in gas turbines are both ~15 R. Also in Figure 10.2, the condensate return temperature $T_8 = 536.7$ R used in this problem is very low. T_8 should be kept as high as possible. Too low of a temperature for T_8 may necessitate the use of special materials of construction in the economizer.

EXAMPLE 10.2 *Original CGAM Design Problem: Ideal Gas Working Fluid*

Solve the original CGAM problem as shown in Figure 10.2 with system costs provided in Tables 10.2 and 10.3 and with physical properties in Table 10.5. Here we provide a solution template to this problem—**Example 10.2a.xls**. The reader is encouraged to use this template as the starting point for obtaining the solution to this problem. In this template, we have provided named variables and the design variables are indicated on the Excel sheet. In this template, we have also provided the objective function—the total cost equation, which combines fuel and all equipment costs.

SOLUTION A good initial guess is an important step in any optimization process. This is especially true for this problem, which shows several local minima. The independent design variables, which will be varied during the optimization process, are shown on the Excel sheet in the column named Variables. To start, assume air enters the compressor at 230 lb/s; the efficiency of the air compressor = 0.84; the air pressure at the exit of the air compressor P_2 = 9.0; the combustion chamber inlet temperature T_3 = 1625 R; the turbine inlet temperature T_4 = 2650 R; and the efficiency of the power turbine = 0.87.

With the provided initial guess, the remaining dependent variables in the solution template can be determined using the equations we have developed in Chapter 7 and in Section 10.4.1. When assembling the material and energy balance equations, avoid using the flow rate of the product gas F_{gas} in the balance equations; instead, use $F_{air} + F_{fuel}$, which will help avoid a circular reference error from Excel. We also caution that depending on the initial guess used, constraints as provided in Table 10.4 may be required. Two constraints, $F_{steam} \geq 30.86$ lbm/s and $\dot{W}_{net} \geq 30{,}000$ kW, will be required in all formulations.

The independent design variables can be varied by Solver and the optimal solution for the cogeneration design can be determined. The solution is provided in **Example 10.2b.xls** and shown in Figure 10.3.

The optimal cogeneration system cost, while meeting the requirements of 30-MW net power and 30.86 lb/s of saturated steam at 20 bar, is found as \$0.3697 per second. The results found in the Excel solution file are also summarized in Table 10.6.

Before we leave Example 10.2, it is instructive to see how sensitive the solution is to the starting guess. From the solution in the Excel file (also Figure 10.3), set the air mass flow rate to 226 lb/s, P_2 = 8.4, T_3 = 1650 R, and T_4 = 2670 R, and run Excel Solver. A new solution with a lower optimal cost of \$0.3690 per second should be found. The multiple optimums in this problem are, in part, due to the form of the equipment costing equations. ■

EXAMPLE 10.3 *Original CGAM Design Problem with ΔT_{Pinch} = 15 R and $\Delta T_{Approach}$ = 15 R*

In Example 10.2, we solved the original CGAM problem. Here we used ΔT_{Pinch} = 3 R and $\Delta T_{\text{Approach}}$ = 27 R; both these values are taken from the CGAM problem optimal solution (Valero et al., 1994a). For Example 10.3, let us solve the CGAM problem with the pinch temperature difference fixed at ΔT_{Pinch} = 15 R and the approach temperature difference also fixed at $\Delta T_{\text{Approach}}$ = 15 R; both these values are typical of methane-fired gas turbine/HRSG design calculations.

SOLUTION We allowed for the possibility of variable ΔT_{Pinch} and $\Delta T_{\text{Approach}}$ values in our formulation of Example 10.2. We will need to change the enthalpy value for water to reflect $\Delta T_{\text{Approach}}$ = 15 R. Here, $\hat{h}^{\text{liquid}}_{(T8P,P9)} = \hat{h}^{\text{liquid}}_{(859\,R,20\,bar)} = 374.42$ Btu/lb. With the *same* starting guess as used in Example 10.2, the solution to Example 10.3 is provided in **Example 10.3.xls**. Here the optimal cogeneration system cost, while again meeting the requirements of 30-MW net power and 30.86 lb/s of saturated steam at 20 bar, is found as \$0.3709 per second. The key results are summarized in Table 10.6.

In this design problem, the impact of ΔT_{Pinch} and $\Delta T_{\text{Approach}}$ selection on the total cost rate is relatively small; the difference in the total cost rate for the two examples is ~\$34,650 per year. For both examples, fuel costs dominate the total cost rate. The importance of ΔT_{Pinch} and $\Delta T_{\text{Approach}}$ selection occurs in the off-design operation of the HRSG; see Problems 10.3 and 10.4. ■

10.5 THE CGAM COGENERATION DESIGN PROBLEM: REAL PHYSICAL PROPERTIES

We want to reexamine the CGAM problem to see the impact of rigorous physical properties on the optimal system design. In Section 10.4, we used ideal gas properties, for the "power-side" and real steam properties in the HRSG when solving the optimal design problem. However, to find steam properties, we had to specify $\Delta T_{\text{Approach}}$ prior to solving the performance equations for the HRSG. With *TPSI+* providing rigorous gas and steam properties, we will no longer need to fix $\Delta T_{\text{Approach}}$ or ΔT_{Pinch} prior to design optimization; however, values for these independent design variables may need to be constrained.

EXAMPLE 10.4 *CGAM Design Problem: Real Fluid Solution*

Solve the original CGAM problem as developed in Example 10.2 using real gas properties.

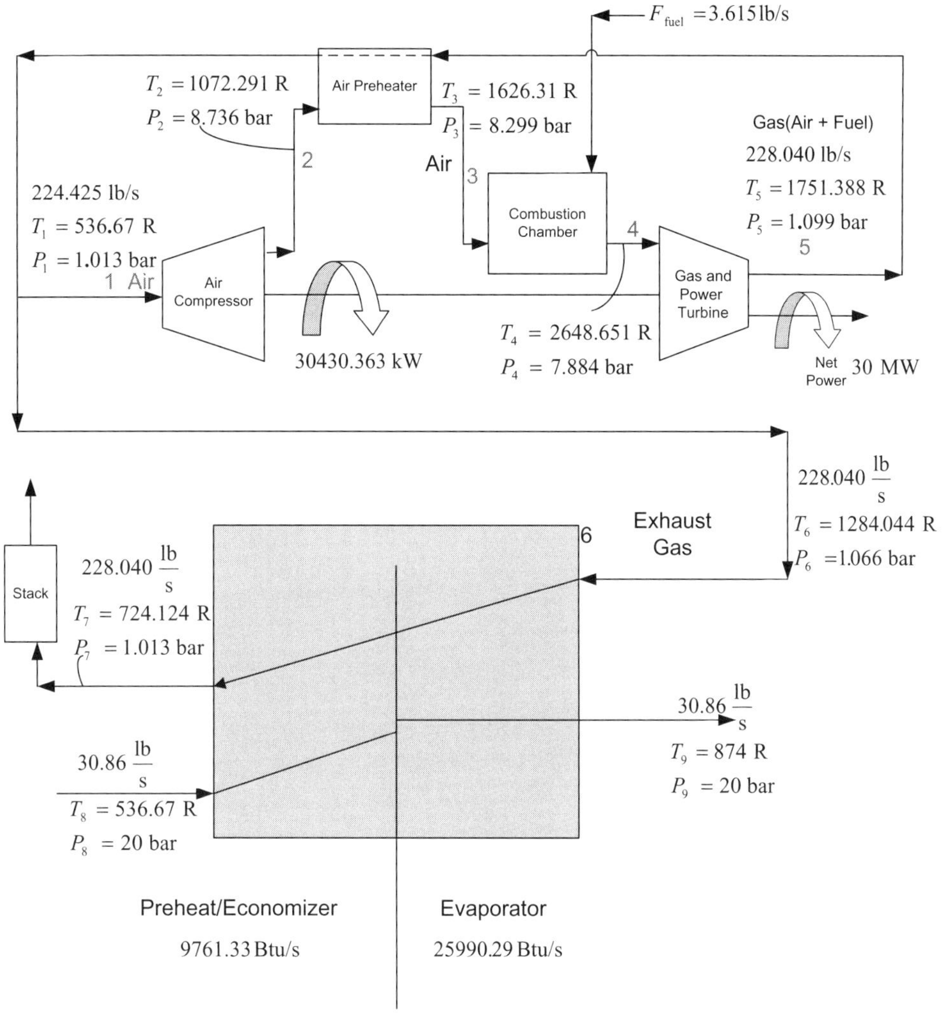

Figure 10.3 Solution CGAM problem, ideal gas, $\Delta T_{\text{Pinch}} = 3$ R and $\Delta T_{\text{Approach}} = 27$ R.

SOLUTION The optimal solution is provided **Example 10.4.xls** and shown in Figure 10.4. As we discussed in Example 10.2, a good initial guess is an important step in the optimization process. The independent design variables, which will be varied during the optimization process, are shown on the Excel sheet in the column named Variables. To start, assume air enters the compressor at 230 lb/s (design variable); the efficiency of the air compressor = 0.84 (design variable); the air pressure at the exit of the air compressor $P_2 = 9.0$ (design variable); the combustion chamber inlet temperature $T_3 = 1625$ R (design variable); the mass flow of fuel to the combustion chamber is 3.6 lb/s (design variable); and the efficiency of the air compressor = 0.87 (design variable). With our combustion library, it is more convenient to the fuel mass flow rate, as opposed to T_4, as an independent design variable. We also set the pinch temperature difference at $\Delta T_{\text{Pinch}} = 3$ R and the approach temperature difference $\Delta T_{\text{Approach}} = 27$ R. We will solve the design problem with variable ΔT_{Pinch} and $\Delta T_{\text{Approach}}$ in Example 10.6. With the provided initial guess, the remaining dependent variables can be determined using the equations developed in Chapter 9 and information from Tables 10.2, 10.3 and 10.5. When assembling the material and energy balance equations, again avoid using the flow rate of the product gas F_{gas} in the balance equations; instead, use $F_{\text{air}} + F_{\text{fuel}}$, which will help avoid a circular reference error from Excel.

Here the optimal cogeneration system cost (fuel + equipment), while meeting the requirements of 30-MW net power and 30.86 lb/s of saturated steam at 20 bar, is found as \$0.355 per second. The results found in Figure 10.4 are summarized in Table 10.6. ■

EXAMPLE 10.5 *CGAM Design Problem: Real Fluid Solution $\Delta T_{Pinch} = 15$ R and $\Delta T_{Approach} = 15$*

In Example 10.4, we solved the original CGAM problem using real fluid properties and using $\Delta T_{\text{Pinch}} = 3$ R and $\Delta T_{\text{Approach}} = 27$ R. For Example 10.5, solve the CGAM problem using real fluid properties and with more realistic values $\Delta T_{\text{Pinch}} = 15$ R and $\Delta T_{\text{Approach}} = 15$ R.

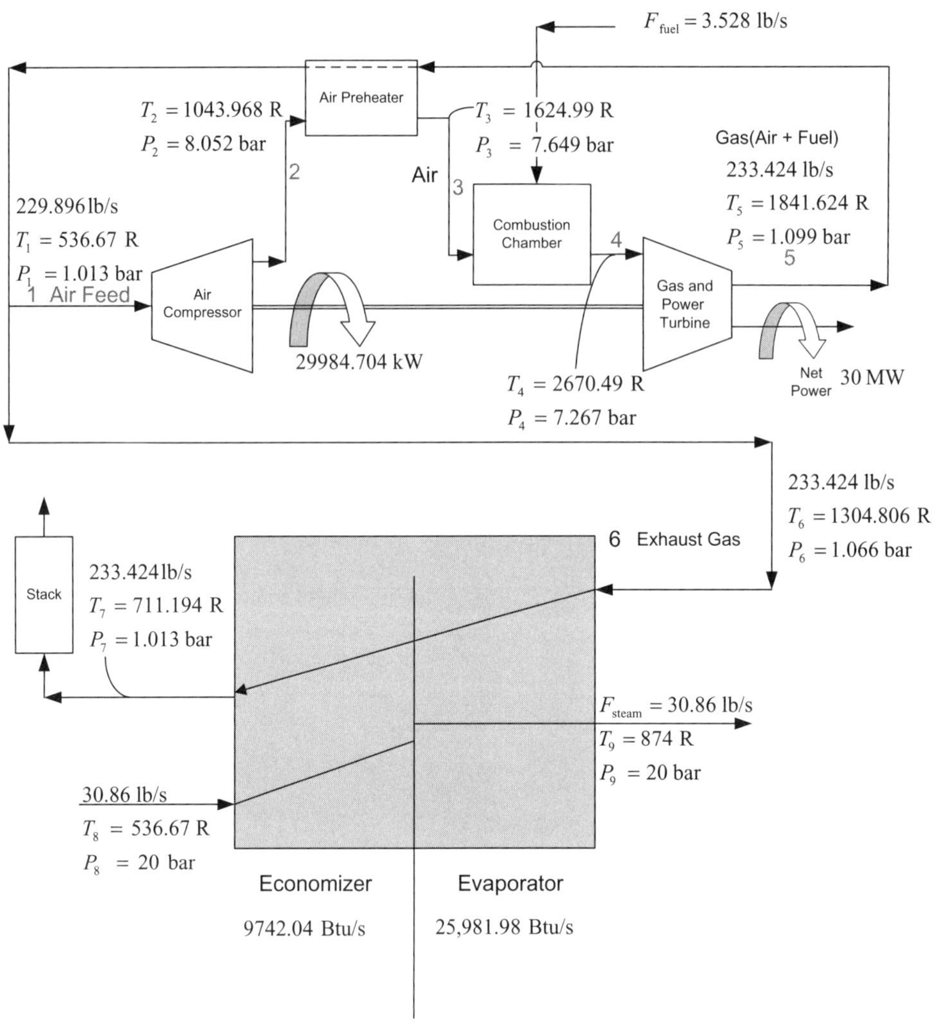

Figure 10.4 Solution to the CGAM problem, real fluid properties, ΔT_{Pinch} = 3 R and $\Delta T_{\text{Approach}}$ = 27 R.

SOLUTION We allowed for the possibility of varying ΔT_{Pinch} and $\Delta T_{\text{Approach}}$ values in our formulation of Example 10.4. With the *same* starting guess as used in Example 10.4, the solution to Example 10.5 is provided in **Example 10.5.xls**. Here the optimal cogeneration system cost (fuel + equipment), while again meeting the requirements of 30-MW net power and 30.86 lb/s of saturated steam at 20 bar, is found as \$0.3563 per second. The key results are again summarized in Table 10.6. ■

EXAMPLE 10.6 *CGAM Design Problem: Real Fluid Solution Variable* ΔT_{Pinch} *and* $\Delta T_{Approach}$

Finally, we solve the CGAM problem using real fluid properties and allowing ΔT_{Pinch} and $\Delta T_{\text{Approach}}$ to both be independent design variables in the optimization process. Here we can use the same starting guess as our two previous examples, plus the starting guesses for ΔT_{Pinch} = 15 R and $\Delta T_{\text{Approach}}$ = 15 R. We do need to provide lower bound constraints on both ΔT_{Pinch} and $\Delta T_{\text{Approach}}$; here we use $\Delta T_{\text{Pinch}} \geq 10$ R and $\Delta T_{\text{Approach}} \geq 10$ R.

SOLUTION The solution to Example 10.6 is provided in **Example 10.6.xls**. You will find that at the optimal solution, ΔT_{Pinch} and $\Delta T_{\text{Approach}}$ both reach their lower bound of 10.0 R. Here, the optimal cogeneration system cost, while again meeting the requirements of 30-MW net power and 30.86 lb/s of saturated steam at 20 bar, is found as \$0.3549 per second. The key results are summarized in Table 10.6, but this design would not be recommended as the exhaust gas temperature from the HRSG is 707.7 R—this temperature is of concern as acid condensation may occur at T_7 values below ~710 R with methane fuel. ■

In Table 10.6, for all examples, fuel costs dominate the total cost rate. It is interesting to note that within ideal gas solutions (Examples 10.2 and 10.3) and real gas solutions (Examples 10.4–10.6), the selection of ΔT_{Pinch} and $\Delta T_{\text{Approach}}$ did not greatly impact the total cost rate. However, there is significant difference in the total cost rate when ΔT_{Pinch} and $\Delta T_{\text{Approach}}$ are common and ideal gas and real gas solutions are compared (Examples 10.2 and 10.4 and Examples 10.3 and 10.5). For example, when comparing the total cost rate of Example 10.2 (ideal gas formulation) and Example 10.4

Table 10.6 Optimal Solutions to the GCAM Design Problem

Variable/Cost	Example 10.2 Ideal Gas	Example 10.3 Ideal Gas	Example 10.4 Real Fluid	Example 10.5 Real Fluid	Example 10.6 Real Fluid
F_{air} (lb/s)	224.4254	225.7829	229.8965	229.8931	228.6751
$\eta_{air\ compressor}$	0.8524	0.8512	0.8411	0.8410	0.8418
P_2 (bar)	8.7356	8.6320	8.0516	8.0000	8.1067
T_3 (R)	1626.31	1625.98	1624.99	1624.98	1624.68
F_{fuel} (lb/s)	3.6147	3.6385	3.5280	3.5430	3.5158
$\eta_{G\&P\ turbine}$	0.8761	0.8748	0.8612	0.8604	0.8620
ΔT_{Pinch} (R)	3.0	15.0	3.0	15.0	10.0
$\Delta T_{Approach}$ (R)	27.0	15.0	27.0	15.0	10.0
T_4 (R)	2648.65	2648.98	2670.49	2674.57	2672.01
$\dot{W}_{air\ compressor}$ (kW)	−30,430.36	−30,428.10	−29,984.70	−29,854.32	−29,927.41
$\dot{W}_{G\ \&P\ turbine}$ (kW)	60,430.36	60,428.10	59,984.70	59,854.32	59,927.41
$\dot{W}^{net}_{G\ \&P\ turbine}$ (kW)	3000	3000	3000	3000	3000
F_{steam} (lb/s)	30.86	30.86	30.86	30.86	30.86
Transferred $\dot{Q}_{Evaporator}$ (Btu/s)	25,990.29	25,588.49	25,981.98	25,584.45	25,417.91
Transferred $\dot{Q}_{Economizer}$ (Btu/s)	9761.33	10,163.12	9742.04	10,138.58	10,306.11
Cost air compressor ($)	1,571,619	1,571,689	1,153,588	1,139,670	1,174,642
Cost comb chamber ($)	121,348	122,295	140,132	143,485	140,617
Cost G&P turbine ($)	2,691,912	2,617,611	2,181,144	2,192,872	2,224,209
Cost air preheater ($)	848,751	835,130	743,419	732,636	743,006
Cost of the HRSG ($)	992,817	991,430	996,707	994,351	1,076,375
Equipment cost rate ($/s)	0.0417086	0.0407554	0.0349332	0.034853	0.0358969
Fuel cost rate ($/s)	0.3279959	0.3301521	0.3201226	0.3214886	0.3190214
Total cost rate ($/s)	0.3697045	0.3709075	0.3550558	0.3563415	0.3549182

(real gas formulation), a difference of $421,881 per year occurs. This cost savings is almost equally split between fuel costs and equipment costs. Before we leave our discussion of the results in Table 10.6, we want to again emphasize that *this optimization problem and our developed solutions should be viewed as a framework for helping to select among several alternative cogeneration systems, or possible system configurations, or to help identify needed ranges in operating conditions.*

10.6 COMPARING COGEND AND GENERAL ELECTRIC'S GATECYCLE™

It was expected that the use of real properties from our combustion library (*TPSI+*) would produce results different from the original CGAM optimal solution where ideal gas properties were used. It is reasonable to ask, "Are the values from *TPSI+* correct?" Here we can compare results obtained from our real gas cogeneration software with results obtained from the commercial code from General Electric—GateCycle—when solving the same problem.

The CGAM cogeneration system solved using our software can be constructed and solved in GateCycle. Here we will not perform a design optimization, but rather we will use the same values from our Excel sheet solution as independent variables in GateCycle and compare values for resulting dependent variables. For the CGAM problem using our software, allowable independent variables include F_{air}, $\eta_{air\ compressor}$, P_2, T_3, F_{fuel}, $\eta_{G\&P\ turbine}$, ΔT_{Pinch}, and $\Delta T_{Approach}$, and the solution is found for T_2, T_4, T_5, T_6, T_7, $\dot{W}_{air\ compressor}$, $\dot{W}^{net}_{G\ \&P\ turbine}$, F_{steam}, $\dot{Q}_{Evaporator}$, and $\dot{Q}_{Economizer}$. For the CGAM problem using GateCycle, F_{air}, $\eta_{air\ compressor}$, P_2, T_3, T_4, $\eta_{G\&P\ turbine}$, ΔT_{Pinch}, and $\Delta T_{Approach}$ are specified and the solution is found for T_2, F_{fuel}, T_5, T_6, T_7, $\dot{W}_{air\ compressor}$, $\dot{W}^{net}_{G\ \&P\ turbine}$, F_{steam}, $\dot{Q}_{Evaporator}$, and $\dot{Q}_{Economizer}$. In Table 10.7, solutions for the CGAM problem using GateCycle and our Excel-based formulation are presented.

In Table 10.7, the variables in italics were specified both in GateCycle and in our formulation. We do not specify T_4, but rather we specify the equivalent F_{fuel}, which was set to 3.62 lb/s (matching the GateCycle calculated value). There is good agreement between all calculated values from GateCycle and our *TPSI+*/Excel-based solution.

Table 10.7 Solution of the CGAM System—*TPSI+* with Excel and GateCycle™

Variable	*TPSI+*/Excel Real Fluid	GateCycle Real Fluid
F_{air} (lb/s)	*203.48*	*203.48*
$\eta_{air\ compressor}$	*0.86*	*0.86*
P_2	*10.13 bar*	*10.13 bar*
T_3 (R)	*1530*	*1530*
F_{fuel} (lb/s)	*3.62*	3.62
$\eta_{G\&P\ turbine}$	*0.86*	*0.86*
ΔT_{Pinch} (R)	*20.0*	*20.0*
$\Delta T_{Approach}$ (R)	*10.0*	*10.0*
T_4 (R)	2739	*2736*
T_2 (R)	1106	1104.9
T_5 (R)	1814.3	1811.2
T_6 (R)	1428.6	1424.9
T_7 (R)	673.6	668
$\dot{W}^{net}_{G\&P\ turbine}$ (kW)	2995	3000
F_{steam} (lb/s)	35.48	35.34

T_1 = 536.67 R, P_1 = 1.013 bar, *LHV* = 21,500 Btu/lbm, HHV = 23,861 Btu/lbm, carbon content = 74.8%, % humidity = 0%, $\Delta P^{gas}_{APH} = 0.03$, $\Delta P^{air}_{APH} = 0.05$, $\Delta P_{CC} = 0.05$, $\Delta P^{gas}_{HRSG} = 0.05$, combustion efficiency = 98%.

10.7 NUMERICAL SOLUTION OF HRSG HEAT TRANSFER PROBLEMS

For the cogeneration system, changes in process steam demands must be met by the HRSG. It can be important to understand the transient response expected from the HRSG. We will introduce this topic by using finite differences to solve the unsteady-state heat transfer problem in the HRSG wall (an initial value partial differential equation). We will also examine heat transfer in the HRSG tubes at steady state using Euler's method to solve for the temperature profile (an initial value ordinary differential equation [ODE]). The methods developed in this section serve as a basis for determining the transient response of the boiler tubes.

10.7.1 Steady-State Heat Conduction in a One-Dimensional Wall

We have discussed that energy from the hot exhaust gas exiting the turbine will be recovered in the HRSG—here steam will be generated and cooled combustion products will be vented to the atmosphere from the stack. Energy loss from the walls of the HRSG should be minimal and maximum steam generation within the HRSG should occur. We know from Figure 10.4 that the exhaust gas leaving the gas turbine is about 1390°F (1850 R). Both heat recovery and employee protection dictate that the walls of the transition region between the turbine to the HRSG should be a reasonable temperature. Figure 10.5 shows the steady-state heat transfer problem. There would be convective heat transfer from the exhaust gas to the wall, conduction through the metal wall, conduction through the insulation, and convection from the insulated wall to the room; here we are ignoring any radiation heat transfer considerations. We are considering this problem to be heat transfer in one direction—the temperature is changing in the x-direction and uniform in the y- and z-directions.

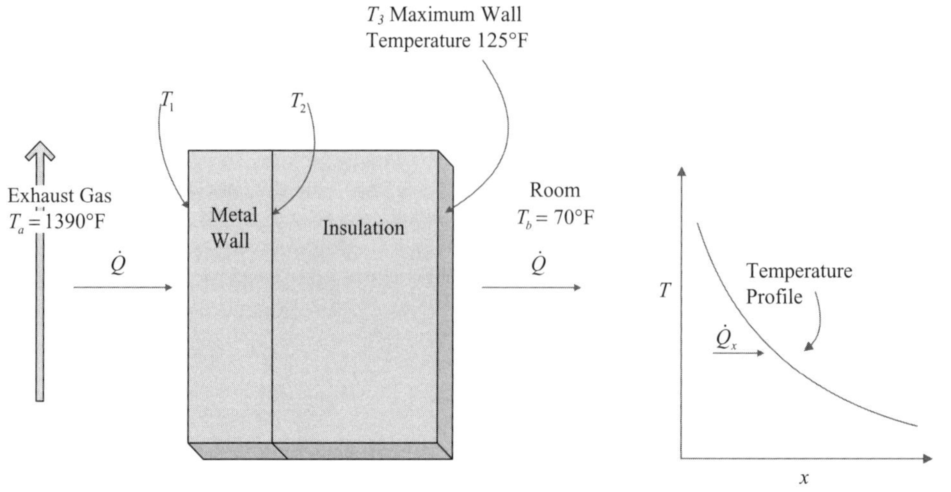

Figure 10.5 One-dimension heat transfer with convection and conduction.

The convection process can be represented by Newton's law of heating:

$$\dot{Q}_{a\to 1} = h_i A(T_a - T_1). \tag{10.9}$$

Here, the heat transfer rate $\dot{Q}$ is given by the temperature difference between the exhaust gas and the inside wall and the surface area A, which is taken perpendicular to the direction of heat flow. Similarly from the insulated wall to the room, Newton's law of cooling provides

$$\dot{Q}_{3\to b} = h_o A(T_3 - T_b). \tag{10.10}$$

Here, h_i and h_o are the convection heat transfer coefficients (inside and outside), which are generally determined from available correlations.

The heat transfer rate by conduction can be represented by Fourier's law of heat conduction, which for the one dimensional heat flow in Figure 10.5 is

$$\dot{Q}_x = -kA\frac{\partial T}{\partial x}, \tag{10.11}$$

where k is the thermal conductivity of the material. Using Equation (10.11) and applying the forward difference numerical definition for $\partial T/\partial x$, we can write for the metal wall and the insulation:

$$\dot{Q}_{1\to 2} = -\frac{k_{\text{metal}}A}{\Delta x_{\text{metal}}}(T_2 - T_1) = \frac{k_{\text{metal}}A}{\Delta x_{\text{metal}}}(T_1 - T_2) \tag{10.12}$$

and

$$\dot{Q}_{2\to 3} = \frac{k_{\text{insulation}}A}{\Delta x_{\text{insulation}}}(T_2 - T_3). \tag{10.13}$$

At steady state, the heat transfer rate through each section must be equal: $\dot{Q}_{a\to 1} = \dot{Q}_{1\to 2} = \dot{Q}_{2\to 3} = \dot{Q}_{3\to b}$. Simultaneous solution of these four equations gives

$$\dot{Q}_{a\to b} = \frac{(T_a - T_b)}{\frac{1}{h_i A} + \frac{\Delta x_{\text{metal}}}{k_{\text{metal}}A} + \frac{\Delta x_{\text{insulation}}}{k_{\text{insulation}}A} + \frac{1}{h_o A}}, \tag{10.14}$$

or in general, for one-dimensional heat transfer,

$$\dot{Q}_{\text{overall}} = \frac{\Delta T_{\text{overall}}}{\sum R_{\text{thermal}}}, \tag{10.15}$$

where $\sum R_{\text{thermal}}$ is the sum of the thermal resistances.

EXAMPLE 10.7 ***Determining Insulation Thickness***

The HRSG (or HRSG transition) in Figure 10.5 has a 0.25-in. carbon steel metal wall with a thermal conductivity, k_{metal}, of 25 Btu/h-ft-°F. A glass fiber insulation with a thermal conductivity, $k_{\text{insulation}}$, of 0.025 Btu/h-ft-°F is available. If the exhaust gas is 1390°F, determine the insulation thickness needed to maintain a room-side wall temperature of 125°F. The inside (exhaust gas to steel wall) convection heat transfer coefficient, h_i, is 100 Btu/h-ft^2-°F and the outside (insulation to room) convection heat transfer coefficient, h_o, is 1 Btu/h-ft^2-°F.

SOLUTION The solution to Example 10.7 is provided in **Example 10.7.xls**. We use Equation (10.10) to determine the heat flux at steady state. This heat flux is then used in Equation (10.14) to determine the needed insulation thickness, $\Delta x_{\text{insulation}} = 6.9$ in. ■

10.7.2 Unsteady-State Heat Conduction in a One-Dimensional Wall

Unsteady-state heat transfer is needed to predict system response to temperature transients. Unsteady-state heat transfer can be especially important when system physical properties (fluid side or metallurgical) depend on temperature. Consider the one-dimensional heat transfer problem of Section 10.7.1. To account for transients, we can write the energy balance as

$$\text{Rate of heat input} - \text{Rate of heat output} = \text{Rate of heat accumulation}. \tag{10.16}$$

Unsteady-State Conduction in a Homogeneous Medium

Consider heat conduction in one direction, for example, though the metal wall of the HRSG, a small section of which is depicted in Figure 10.6.

The rate of heat input into and output from the block shown in Figure 10.6 is given by Equation (10.11):

$$\text{Rate of heat input} = \dot{Q}_x = -k(\Delta y \Delta z)\left.\frac{\partial T}{\partial x}\right|_x \tag{10.17}$$

and

$$\text{Rate of heat output} = \dot{Q}_{x+\Delta x} = -k(\Delta y \Delta z)\left.\frac{\partial T}{\partial x}\right|_{x+\Delta x}. \tag{10.18}$$

The rate of heat accumulation in the block ($\Delta x \Delta y \Delta z$) is

$$\text{Rate of heat accumulation} = \rho \hat{C}_p(\Delta x \Delta y \Delta z)\frac{\partial T}{\partial t}, \tag{10.19}$$

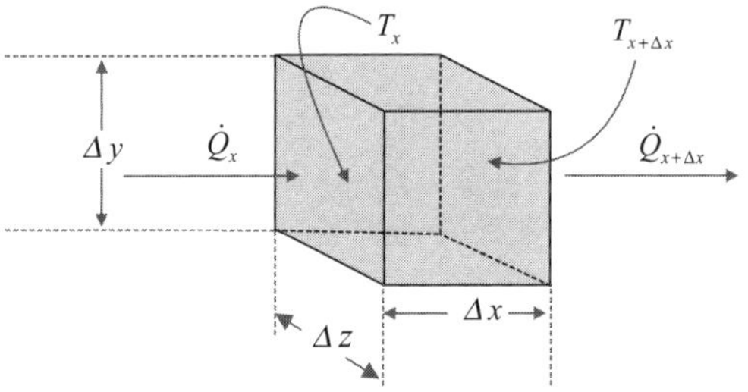

Figure 10.6 Elemental volume for unsteady-state heat conduction in the x-direction.

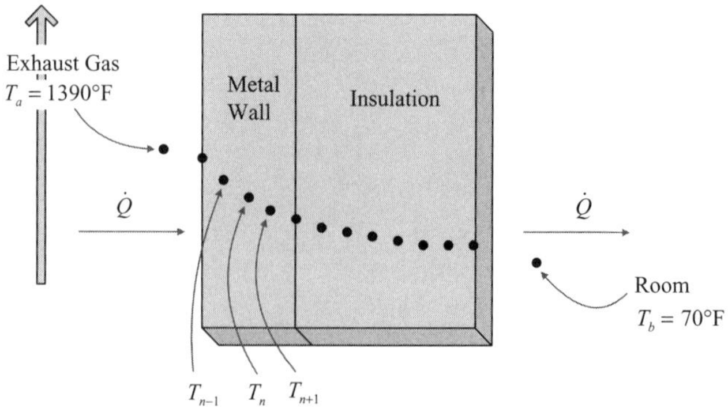

Figure 10.7 Unsteady-state heat conduction at time t.

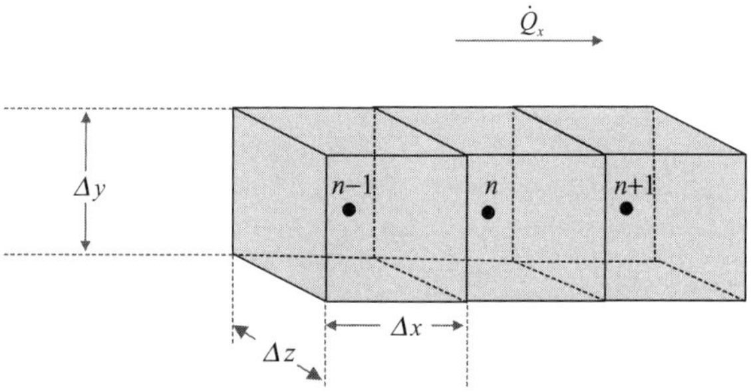

Figure 10.8 Heat conduction through three adjacent blocks (elemental volumes).

where ρ is the density and $\hat{C}_p$ is the heat capacity. Substituting Equations (10.17)–(10.19) into Equation (10.16) and dividing by ($\Delta x \Delta y \Delta z$) gives

$$\left(\frac{-k}{\rho \hat{C}_p}\right) \frac{\left(\left.\frac{\partial T}{\partial x}\right|_x - \left.\frac{\partial T}{\partial x}\right|_{x+\Delta x}\right)}{\Delta x} = \frac{\partial T}{\partial t}. \quad (10.20)$$

Taking the limit as $\Delta x \to 0$,

$$\left(\frac{k}{\rho \hat{C}_p}\right) \frac{\partial^2 T}{\partial x^2} = \frac{\partial T}{\partial t} \quad (10.21)$$

or

$$(\alpha) \frac{\partial^2 T}{\partial x^2} = \frac{\partial T}{\partial t}, \quad (10.22)$$

where $\alpha = k / \rho C_p$ and α is the thermal diffusivity. Equation (10.21) (or Eq. (10.22)) relates temperature with time and position, $T(x, t)$, during heat conduction. The analytic solution of Equation (10.21), an initial value, partial differential equation (PDE), is known for many fixed geometries provided physical properties and boundary conditions are constant with time. The initial temperature profile at time = 0 must be known. For example, Figure 10.5 boundary conditions would include the convection equations with fixed T_a and $T_b$values for $t > 0$, and the initial temperature profile could be $T(x, 0) = T_b$.

Unsteady-State Heat Conduction in a Homogeneous Medium: Finite Difference Formulation

In many situations, it is convenient to use numerical methods to solve Equation (10.21). To develop the needed equations, we can use temperature nodes as indicated in Figure 10.7.

The conducting body can be divided into blocks of size ($\Delta x \Delta y \Delta z$) as shown in Figure 10.8, with nodes 1, . . . , $n - 1$, n, $n + 1$ located at the center of each block.

Using Equations (10.16)–(10.19) and the forward difference numerical definition for $\partial T / \partial x$ and $\partial T / \partial t$, we can write the energy balance for block n as

$$-k(\Delta y \Delta z)\frac{({}_tT_n - {}_tT_{n-1})}{\Delta x} + k(\Delta y \Delta z)\frac{({}_tT_{n+1} - {}_tT_n)}{\Delta x} = \rho \hat{C}_p (\Delta x \Delta y \Delta z)\frac{({}_{t+\Delta t}T_n - {}_tT_n)}{\Delta t}. \quad (10.23)$$

Combining terms and assuming constant k, ρ, $\hat{C}_p$ (or thermal diffusivity) give

$$\left(\frac{k}{\rho \hat{C}_p}\right) \frac{({}_tT_{n+1} + {}_tT_{n-1} - 2\,{}_tT_n)}{(\Delta x)^2} = \left\{\frac{({}_{t+\Delta t}T_n - {}_tT_n)}{\Delta t}\right\}; \quad (10.24)$$

solving for ${}_{t+\Delta t}T_n$,

$$ {}_{t+\Delta t}T_n = \left(\frac{k\Delta t}{\rho\hat{C}_p(\Delta x)^2}\right)({}_tT_{n+1} + {}_tT_{n-1}) + \left(1 - \frac{2k\Delta t}{\rho\hat{C}_p(\Delta x)^2}\right)({}_tT_n). \quad (10.25) $$

Defining

$$ M = \frac{\rho\hat{C}_p(\Delta x)^2}{k\Delta t} = \frac{(\Delta x)^2}{\alpha\Delta t}, $$

Equation (10.25) can be written as

$$ {}_{t+\Delta t}T_n = \left(\frac{1}{M}\right)({}_tT_{n+1} + {}_tT_{n-1}) + \left(1 - \frac{2}{M}\right)({}_tT_n). \quad (10.26) $$

M must be ≥2. Do note that if $M < 2$, the coefficient of ${}_tT_n$ in Equation (10.26) becomes negative and values for ${}_{t+\Delta t}T_n$ are obtained, which violate the second law of thermodynamics. You can choose any combination of Δx and Δt so $M \geq 2$.

Special Cases

$M = 2$, the Schmidt method. When the time and distance increments can be chosen such that

$$ M = \frac{\rho\hat{C}_p(\Delta x)^2}{k\Delta t} = \frac{(\Delta x)^2}{\alpha\Delta t} = 2, \quad (10.27) $$

the temperature at ${}_{t+\Delta t}T_n$ becomes

$$ {}_{t+\Delta t}T_n = \left(\frac{1}{2}\right)({}_tT_{n+1} + {}_tT_{n-1}), \quad (10.28) $$

which is simply the arithmetic average of the two adjacent nodes.

$M = 3$ gives

$$ {}_{t+\Delta t}T_n = \left(\frac{1}{3}\right)({}_tT_{n+1} + {}_tT_n + {}_tT_{n-1}). \quad (10.29) $$

Unsteady-State Heat Conduction at a Nonhomogenous Boundary: Finite Difference Formulation

Figure 10.7 shows the plane wall is a composite wall consisting of metal and glass fiber insulation. The thermal diffusivity may be taken as constant in both sections, but we need to account for the interface between the two layers. Using Figure 10.7, we can imagine that block n represents the perfect contact between the two materials; A = metal wall and B = glass fiber insulation. Equation (10.23) can be written as

$$ -k_A(\Delta y\Delta z)\frac{({}_tT_n - {}_tT_{n-1})}{\Delta x} + k_B(\Delta y\Delta z)\frac{({}_tT_{n+1} - {}_tT_n)}{\Delta x} = \left(\rho_A\hat{C}_{p,A}\frac{\Delta x}{2} + \rho_B\hat{C}_{p,B}\frac{\Delta x}{2}\right)(\Delta y\Delta z)\frac{({}_{t+\Delta t}T_n - {}_tT_n)}{\Delta t}. \quad (10.30) $$

Notice that the accumulation term consists of the average density and heat capacity of the two materials. In the accumulation term, we are approximating that ${}_tT_{n-0.5} \cong {}_tT_n \cong {}_tT_{n+0.5}$. Solving for ${}_{t+\Delta t}T_n$ gives

$$ {}_{t+\Delta t}T_n = \left(\frac{2k_A\Delta t}{(\rho_A\hat{C}_{p,A} + \rho_B\hat{C}_{p,B})(\Delta x)^2}\right)({}_tT_{n-1}) + \left(\frac{2k_B\Delta t}{(\rho_A\hat{C}_{p,A} + \rho_B\hat{C}_{p,B})(\Delta x)^2}\right)({}_tT_{n+1}) + \left(1 - \left[\frac{2(k_A + k_B)\Delta t}{(\rho_A\hat{C}_{p,A} + \rho_B\hat{C}_{p,B})(\Delta x)^2}\right]\right)({}_tT_n). \quad (10.31) $$

Unsteady-State Heat Convection Boundary: Finite Difference Formulation

For the one-dimensional system in Figure 10.5, we still need to account for the convection boundaries at the exhaust gas metal wall interface and the insulation room interface. For the convection boundary shown in Figure 10.9, we can write the energy balance on the one-half block as

$$ h(\Delta y\Delta z)({}_tT_a - {}_tT_1) + k(\Delta y\Delta z)\frac{({}_tT_2 - {}_tT_1)}{\Delta x} = \rho\hat{C}_p\left(\frac{\Delta x\Delta y\Delta z}{2}\right)\frac{({}_{t+\Delta t}T_1 - {}_tT_1)}{\Delta t}. \quad (10.32) $$

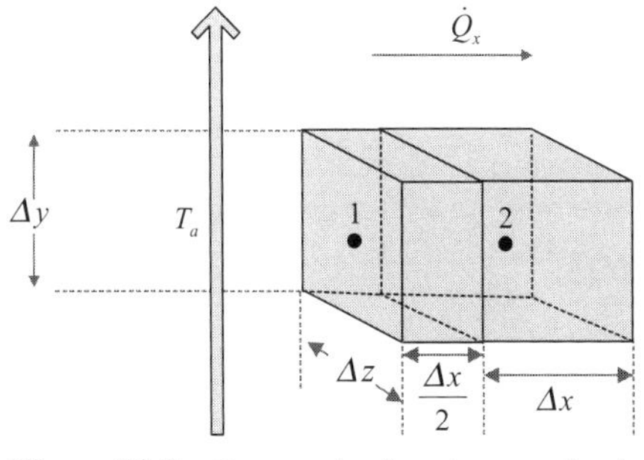

Figure 10.9 Heat conduction at convection boundary.

Solving for ${}_{t+\Delta t}T_1$ gives

$$
{}_{t+\Delta t}T_1 = \left(\frac{2h\Delta t}{\rho\hat{C}_p(\Delta x)}\right){}_tT_a + \left(1-\left[\frac{2h\Delta t}{\rho\hat{C}_p(\Delta x)}+\frac{2k\Delta t}{\rho\hat{C}_p(\Delta x)^2}\right]\right)({}_tT_1) + \frac{2k\Delta t}{\rho\hat{C}_p(\Delta x)^2}({}_tT_2). \quad (10.33)
$$

Here the environment is at a temperature T_a and we are neglecting heat accumulation in the exhaust gas (a front one-half block, which is not shown). If T_a varies with time, a new value for ${}_tT_a$ can be used at each time step:

With

$$
M = \frac{\rho\hat{C}_p(\Delta x)^2}{k\Delta t}
$$

and

$$
N = \frac{h(\Delta x)}{k},
$$

$$
{}_{t+\Delta t}T_1 = \left(\frac{2N}{M}\right){}_tT_a + \left(1-\left[\frac{2N}{M}+\frac{2}{M}\right]\right)({}_tT_1)+\frac{2}{M}({}_tT_2). \quad (10.34)
$$

Here we require that $M \geq 2N + 2$, where N is the Nusselt number for this problem and M is the inverse Fourier number.

In a series of example problems, we want to first examine heating a metal wall with convection on both surfaces. We then examine the case for the metal wall with a convective surface and a perfectly insulated surface. Then, we examine the HRSG composite wall problem of Figure 10.7.

EXAMPLE 10.8 *Heating a Metal Wall with Two Convective Surfaces*

Consider the heat conduction problem shown in Figure 10.10. The 0.25-in. carbon steel metal wall has a thermal conductivity of $k_{\text{metal}} = 25$ Btu/h-ft-°F, $\rho_{\text{metal}} = 500$ lb/ft^3, and $\hat{C}_{p,\text{metal}} = 0.12$ Btu/lb-°F. Solve for the time-dependent metal wall temperature if the wall is initially at 70°F and then exposed to exhaust gas at 1390°F at both exterior surfaces. The convective heat transfer coefficient on both exterior surfaces, h, is 100 Btu/h-ft^2-°F.

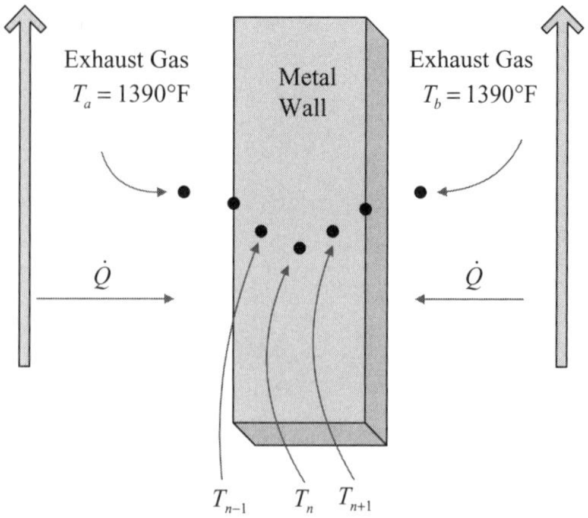

Figure 10.10 Sketch of heat conduction problem for Example 10.8.

SOLUTION The solution to Examplc 10.8 is provided in **Example 10.8.xls**. We use Equation (10.34) for the convective surfaces and Equation (10.29) for the conduction in the metal. As $M \geq 2N + 2$, it is convenient to set

$$
M = \frac{\rho\hat{C}_p(\Delta x)^2}{k\Delta t} = 3
$$

and to use $\Delta x = 0.05$ in. to create six temperature nodes. Here then,

$$
\Delta t = \frac{\left(500\frac{\text{lb}}{\text{ft}^3}\right)\left(0.12\frac{\text{Btu}}{\text{lb-°F}}\right)\left(\frac{0.05}{12}\right)^2}{\left(25\frac{\text{Btu}}{\text{hr-ft-°F}}\right)(3)} = 1.3889\text{ E-05 hours (0.05 second)}.
$$

At 20 seconds, the temperature 0.05 in. from the surface is 841.5°F. ■

EXAMPLE 10.9 *Heating a Metal Wall with One Convective Surface and a Perfectly Insulated Surface*

The 0.25-in. carbon steel metal wall of the HRSG has a thermal conductivity of $k_{\text{metal}} = 25$ Btu/h-ft-°F, $\rho_{\text{metal}} = 500$ lb/ft^3, and $\hat{C}_{p,\text{metal}} = 0.12$ Btu/lb-°F. Solve for the time-dependent metal wall temperature if the wall is initially at 70°F and then exposed to exhaust gas at 1390°F at one surface and the other surface is perfectly insulated. The convection heat transfer coefficient h is 100 Btu/h-ft^2-°F. The simplifying assumption of perfect insulation or zero heat flux at a metal wall surface is often made. Zero heat flux is equivalent to setting $h = 0$.

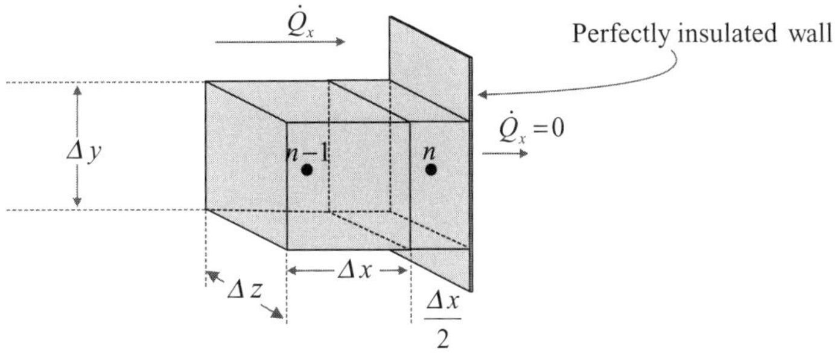

Figure 10.11 Heat conduction at a perfectly insulated wall.

SOLUTION Accounting for the perfectly insulated wall as shown in Figure 10.11.

Here the energy balance on the one-half block would be

$$-k(\Delta y \Delta z)\frac{({}_tT_n - {}_tT_{n-1})}{\Delta x} = \rho\hat{C}_p\left(\frac{\Delta x \Delta y \Delta z}{2}\right)\frac{({}_{t+\Delta t}T_n - {}_tT_n)}{\Delta t}, \quad (10.35)$$

and solving for ${}_{t+\Delta t}T_n$,

$${}_{t+\Delta t}T_n = \frac{2k\Delta t}{\rho\hat{C}_p(\Delta x)^2}({}_tT_{n-1}) + \left(1 - \frac{2k\Delta t}{\rho\hat{C}_p(\Delta x)^2}\right)({}_tT_n) \quad (10.36)$$

or

$${}_{t+\Delta t}T_n = \frac{2}{M}({}_tT_{n-1}) + \left(1 - \frac{2}{M}\right)({}_tT_n). \quad (10.37)$$

The solution to Example 10.9 is provided in **Example 10.9.xls**. We use Equation (10.34) for the convective surface, Equation (10.29) for the conduction in the metal, and Equation (10.37) for the insulated surface. In the Excel solution, we set $M = 3$ and $\Delta x = 0.05$ in., giving $\Delta t = 1.3889$ E-05 hours (0.05 second). At 20 seconds, the temperature 0.05 in. from the convective surface is 544.3°F and the temperature 0.05 in. from the insulated surface is 523.3°F. ■

EXAMPLE 10.10 ***Heating a Composite Wall with Two Convective Surfaces***

The HRSG wall shown in Figure 10.5 is a composite wall consisting of carbon steel and a perfectly contacted insulating glass fiber. The 0.25-in. carbon steel metal wall of the HRSG has a thermal conductivity of $k_{\text{metal}} = 25$ Btu/h-ft-°F, $\rho_{\text{metal}} = 500$ lb/ft^3, and $\hat{C}_{p,\text{metal}} = 0.12$ Btu/lb-°F. The 6.9-in. glass fiber wall (see Example 10.7 for thickness) has $k_{\text{insulation}} = 0.025$ Btu/h-ft-°F, $\rho_{\text{insulation}} = 5$ lb/ft^3, and $\hat{C}_{p,\text{insulation}} = 0.2$ Btu/lb-°F. The inside (exhaust gas to steel wall) convection heat transfer coefficient h_i is 100 Btu/h-ft^2-°F and the outside (insulation to room) convection heat transfer coefficient h_o is 1 Btu/h-ft^2-°F. Solve for the time-dependent wall temperature if the wall is initially at 70°F and then exposed to exhaust gas at 1390°F. At exposure time = 60 seconds, determine the temperature at the metal/insulation interface and 0.15 in. from the interface in both the metal and insulation.

SOLUTION The solution to Example 10.10 is provided in **Example 10.10.xls**. We use Equation (10.34) for the convective surfaces, Equation (10.29) for the conduction in the metal and insulating fiber, and Equation (10.31) for the interface between the metal wall and the fiber. We again use a value of $M = 3$ and $\Delta x = 0.05$ in. and $\Delta t = 1.3889$ E-05 hours (0.05 second). At 60 seconds, the temperature at the interface is 1017.2°F, and 0.15 in. from the interface, the metal-side temperature is 1026.5°F and the insulation-side temperature is 699.5°F. ■

In Section 10.7.2, we have examined the solution of initial value PDEs using the explicit finite difference method. The method is explicit in that nodes at the current time can be determined using information from nodes at the previous time interval. To keep M at allowed values, the explicit method often necessitates the use of a small time step. This problem can be overcome using implicit methods with the Crank–Nicholson method being widely used (Chapra and Canale, 2010).

10.7.3 Steady-State Heat Conduction in the HRSG

In Sections 10.7.1 and 10.7.2, we discussed heat transfer considerations in walls typical of the transition from the gas turbine to the HRSG. We now want to examine heat transfer in the boiler tubes. In the HRSG, condensate returning from the process enters the HRSG and is heated to steam by exchanging energy with the hot exhaust gas leaving the turbine. As the water to steam phase change occurs, physical properties, which control the effectiveness of the heat transfer, will also change. Here we want to show how numerical methods can be used to account for changing system properties and the effects on the heat transfer.

The HRSG water/steam tubes may have a horizontal or, more typically, a vertical tube arrangement. In an idealized case, as shown in Figure 10.12, we can consider the HRSG as a cross-flow heat exchanger with water/steam on the tube side and exhaust gas flowing outside the tubes.

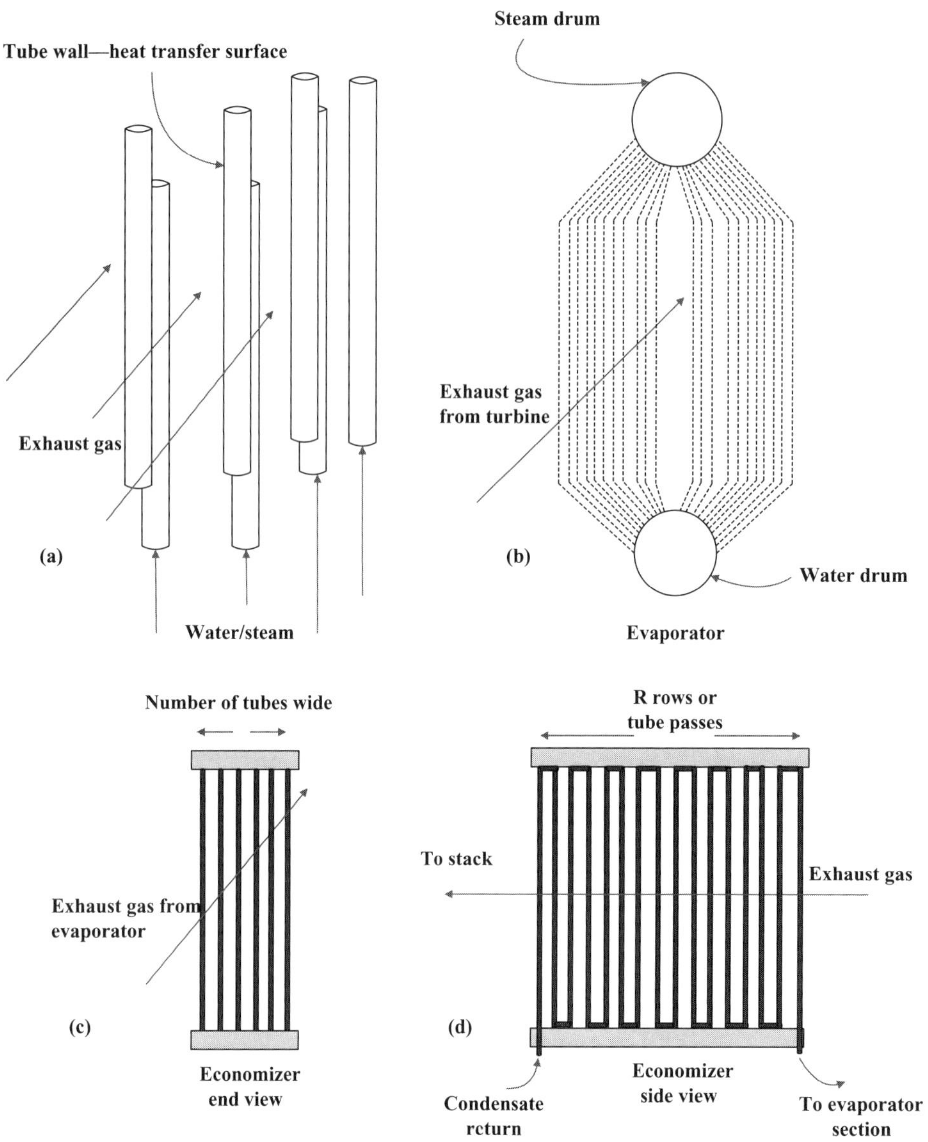

Figure 10.12 (a) HRSG as a cross-flow heat exchanger. (b) End view of evaporator. (c) HRSG economizer vertical configuration end view. (d) Side view.

Figure 10.13a shows the representation of the evaporator section for use in LMTD calculations. The evaporator (as well as the economizer) is considered a countercurrent heat exchanger. For example, comparing Figures 9.3 and 10.13a, $T_{h,2} = T_6$, $T_{h,1} = T_{7P}$, $T_{c,2} = T_9$, and $T_{c,1} = T_{8P}$. Here we are following the standard convention that the exhaust gas is a hot (h) stream and the water/steam stream is a cold (c) stream. The use of h and c, as opposed to *exhaust_gas* and *water/steam*, helps keep the equations less cumbersome.

For model development, a single heat exchanger tube can be considered with Figure 10.13b serving as a basis for our approach to account for changing physical properties. Figure 10.13b represents the view normal to the flowing exhaust gas. The temperature of the gas around any section of a vertical tube is uniform with $T_{h(r,n-1)} = T_{h(r,n)} = T_{h(r,n+1)} = \ldots$, where r is the row number (or tube pass number) in the tube bank and n is the node indicator in each tube.

The Cold-Side Equations (c = Cold = Water/Steam)

At node n and assuming steady state, we can write the incremental energy transfer rate through dA on the cold (water/steam) side as

$$\dot{Q}_{(r,n)} = F_c d\hat{h}_{c(r,n)} = F_c\left(\hat{h}_{c(r,n+1)} - \hat{h}_{c(r,n)}\right). \qquad (10.38)$$

Here the cold fluid flow direction is taken as positive and a forward difference is used for $d\hat{h}_{c(r,n)}$.

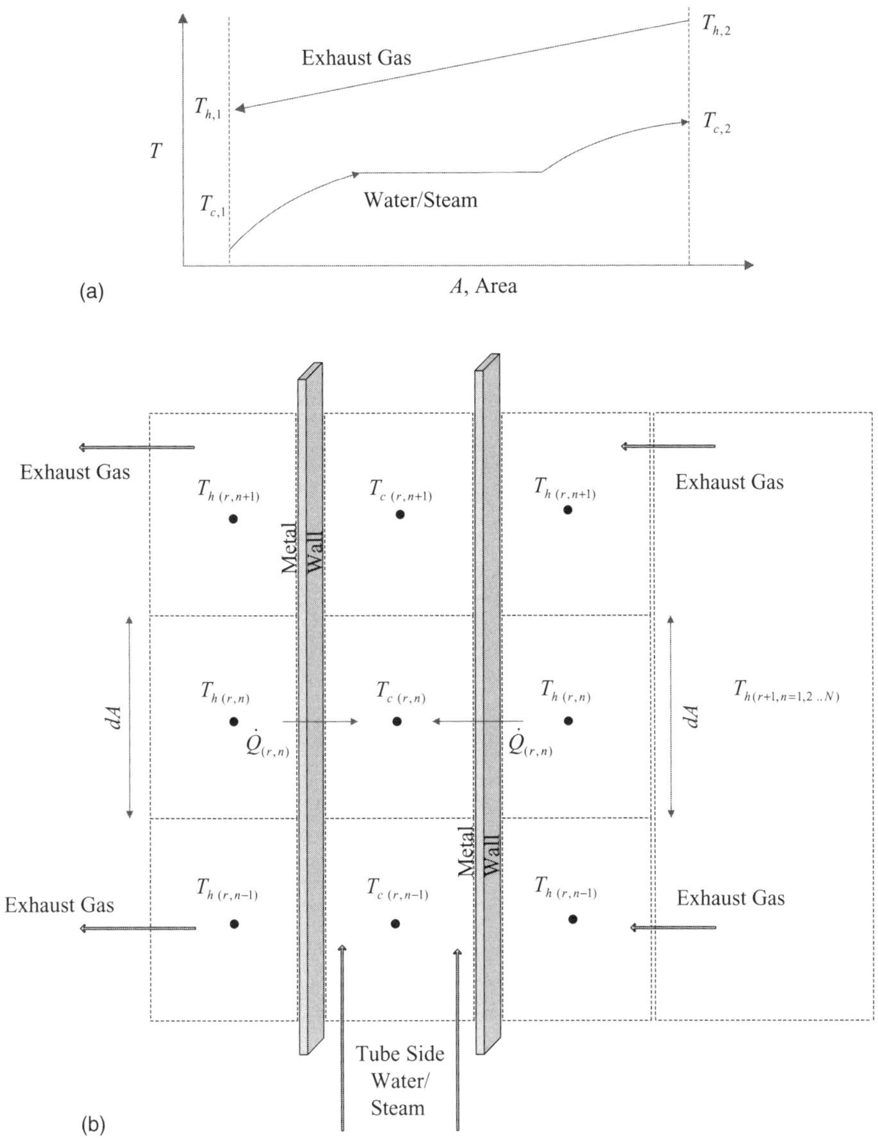

Figure 10.13 (a) Temperature profile in the HRSG evaporator section. (b) Nomenclature for numerical analysis of a single tube in the HRSG (evaporator or economizer section).

The Hot-Side Equations (h = Hot = Exhaust Gas)

The temperature of the exhaust gas surrounding any tube in a given row is assumed uniform with $T_{h(r,n-1)} = T_{h(r,n)} = T_{h(r,n+1)} = \ldots$ ($n = 1, 2, \ldots N$).

Heat Transfer between Hot and Cold Sides

The incremental heat transfer rate can be expressed as

$$\dot{Q}_{(r,n)} = U_{(r,n)}(dA)\left(T_{h(r,n)} - T_{c(r,n)}\right). \qquad (10.39)$$

In Equation (10.39), dA must be chosen small enough to allow accurate system and physical property determination in the determination of $U_{(r,n)}$.

If the cold fluid heat capacities are constant through the volume of area dA, we can substitute for Equation (10.38):

$$\begin{aligned}\dot{Q}_{(r,n)} &= F_c d\hat{h}_{c(r,n)} = F_c\left(\hat{C}_{Pc,(r,n)}\right)\left(dT_{c(r,n)}\right) \\ &= F_c\left(\hat{C}_{Pc,(r,n)}\right)\left(T_{c(r,n+1)} - T_{c(r,n)}\right).\end{aligned} \qquad (10.40)$$

Numerical Solution Strategy: Euler's Method (for Initial Value ODE)

Assume all physical and system properties at $r = 1$ and $n = 1$ are known (initial values). The numerical solution of Figure 10.13b can be constructed as follows:

1. Fix an appropriate length, dL, in the heat exchanger, which in turn fixes dA as $dA = A_o dL$; A_o is the external surface area per foot length of tube.

Table 10.8 Evaporator and Economizer Tube Data

Outside Diameter, d_o (in.)	Inside Diameter, d_i (in.)	Wall Thickness (in.)	Internal Cross-Sectional Area for Flow (in.^{2})	External Surface Area per Foot Length, A_o (ft^2)	Internal Surface Area per Foot Length, A_i (ft^2)
2	1.81	0.095	2.573	0.5236	0.4739

2. Calculate $U_{(r,n)}$, which, based on the outside surface area of the water/steam tube, is

$$U_{(r,n)} = \frac{1}{\dfrac{A_o}{A_i h_{i,(r,n)}} + \dfrac{A_o \ln\left(\dfrac{d_o}{d_i}\right)}{2\pi k_{wall}(dL)} + \dfrac{1}{h_{o,(r,n)}}}. \quad (10.41)$$

In Equation (10.41), i = inside and o = outside.

3. Calculate $\dot{Q}_{(r,n)}$ using Equation (10.39).
4. Calculate $T_{c(r,n+1)}$ using Equation (10.38) or (10.40). Equation (10.38) should be used when a phase change is occurring.
5. After all cold-side tube calculations are completed ($n = N$), calculate $\dot{Q}_r = \sum_{n=1}^{N} \dot{Q}_{(r,n)}$.
6. Using $\dot{Q}_r$, calculate the hot stream temperature for the next row of tubes, $T_{h(r=r+1,n=1,\ldots,N)}$.

This numerical solution strategy will be shown in Example 10.13.

The numerical solution requires accurate calculation of the inside and outside heat transfer coefficients, $h_{i,(r,n)}$ and $h_{o,(r,n)}$, for use in Equation (10.41). Here we must first have knowledge of the HRSG actual physical design in order to determine fluid velocities in the HRSG. We show how the physical design calculations can be performed in the next example.

Before we leave this numerical solution strategy, do show yourself that Equations (10.40) and (10.39) can be combined to give the explicit equation

$$T_{c(r,n+1)} = T_{c(r,n)} + \frac{U_{(r,n)}}{F_c\left(\hat{C}_{Pc,(r,n)}\right)}\left(T_{h(r,n)} - T_{c(r,n)}\right)(A_0\, dL)$$

and $T_{h(r,n+1)} = T_{h(r,n)}$.

This is Euler's method for $dT_{c(r,n)}/dL \neq 0$ and $dT_{h(r,n)}/dL = 0$.

In the numerical solution strategy, we have broken Euler's method for $T_{c(r,n+1)}$ into two steps, which is easier to code (Example 10.13); see also Problem 10.7.

EXAMPLE 10.11 *Determination of the HRSG Physical Design*

For the HRSG of Example 10.4, assume the evaporator is 30 tubes wide and the economizer is 20 tubes wide. The tubes are O.D. = 2 in., wall thickness = 0.095 in., length = 15 ft, and tube spacing = 4 in.^{2}. Tube data are also provided in Table 10.8. Using a typical value for overall heat transfer coefficient based on the tube outside area of 12.3 Btu/h-ft^2-°F, determine the equivalent number of rows needed in the evaporator and economizer sections.

***SOLUTION** Evaporator Section*

Using data from Example 10.4,

$$\frac{dQ_{\text{Evaporator,steam}}}{dt} = \dot{Q}_{\text{Evaporator,steam}} = F_{\text{steam}}\left(\hat{h}_9 - \hat{h}_{8P}\right) =$$

$$\dot{Q}_{\text{Evaporator,steam}} = 30.86\frac{\text{lb}}{\text{s}}\left(1203.41 - 361.48\frac{\text{Btu}}{\text{lb}}\right)\left(\frac{3600\ \text{s}}{\text{h}}\right)$$

$$= 93{,}535{,}055\frac{\text{Btu}}{\text{h}}$$

$$\Delta T_{\text{LMTD}} = \frac{(T_6 - T_9) - (T_{7P} - T_{8P})}{\ln\left(\dfrac{(T_6 - T_9)}{(T_{7P} - T_{8P})}\right)} = \frac{430.8 - 30}{\ln\left(\dfrac{430.8}{30}\right)} = 150.425°\text{F}.$$

Next, calculate the total area needed for the evaporator section using $\dot{Q} = UA\Delta T_{\text{LMTD}}$:

$$A = \frac{\dot{Q}}{U\Delta T_{\text{LMTD}}} = \frac{93{,}535{,}055\dfrac{\text{Btu}}{\text{h}}}{\left(12.3\dfrac{\text{Btu}}{\text{h-ft}^2\text{- F}}\right)(150.425\ \text{F})} = 50{,}553\ \text{ft}^2.$$

Determine the total number of tubes needed for the evaporator section:

$$N_{\text{tubes\_total}} = \frac{A}{\left(\dfrac{\text{External surface area}}{\text{Foot of tube}}\right)(\text{Tube length})} = \frac{A}{(A_o)(L_{\text{tube}})} = \frac{50{,}553\ \text{ft}^2}{\left(0.5236\dfrac{\text{ft}^2}{\text{ft}}\right)(15\ \text{ft})} = 6435\ \text{tubes}.$$

Determine the number of rows needed:

$$N_{\text{rows\_Evaporator}} = \frac{\text{Number tubes}}{\text{Number of tubes wide}} = \frac{6435\ \text{tubes}}{30\ \text{tubes wide}} = 214.46\ \text{rows}.$$

The equivalent of about 215 rows is needed in the economizer. The number of rows in the evaporator would be substantially reduced (generally by a factor of 5 or more) if finned tubes as opposed to bare tubes are used.

Economizer Section

Again using data from Example 10.4,

$$\dot{Q}_{\text{Economizer,water}} = F_{\text{water}}\left(\hat{h}_{8P} - \hat{h}_8\right) = 30.86\frac{\text{lb}}{\text{s}}\left(361.48 - 45.79\frac{\text{Btu}}{\text{lb}}\right)\left(\frac{3600\ \text{s}}{\text{h}}\right) = 35{,}071{,}896\frac{\text{Btu}}{\text{h}}$$

$$\Delta T_{\text{LMTD}} = \frac{(T_8 - T_7) - (T_{7P} - T_{8P})}{\ln\left(\dfrac{(T_8 - T_7)}{(T_{7P} - T_{8P})}\right)} = \frac{174.524 - 30}{\ln\left(\dfrac{174.524}{30}\right)} = 82.075°\text{F}.$$

Next, calculate the total area needed for the economizer section using $\dot{Q} = UA\Delta T_{\text{LMTD}}$:

$$A = \frac{\dot{Q}}{U\Delta T_{\text{LMTD}}} = \frac{35{,}071{,}896\dfrac{\text{Btu}}{\text{h}}}{\left(12.3\dfrac{\text{Btu}}{\text{h-ft}^2\text{-°F}}\right)(82.075°\text{F})} = 34{,}741\ \text{ft}^2.$$

Determine the total number of tubes needed for the economizer section:

$$N_{\text{tubes\_total}} = \frac{A}{\left(\dfrac{\text{External surface area}}{\text{Foot of tube}}\right)(\text{Tube length})} = \frac{A}{(A_o)(L_{\text{tube}})} = \frac{34{,}741\ \text{ft}^2}{\left(0.5236\dfrac{\text{ft}^2}{\text{ft}}\right)(15\ \text{ft})} = 4423.4\ \text{tubes}.$$

Determine the number of rows needed:

$$N_{\text{rows\_Economizer}} = \frac{\text{Number tubes}}{\text{Number of tubes wide}} = \frac{4423.4\ \text{tubes}}{20\ \text{tubes wide}} = 221.2\ \text{rows}.$$

The equivalent of about 222 rows is needed in the economizer. This economizer is large (large number or rows) as the condensate return temperature of 77°F from Example 10.4 is low. Generally, the condensate is returned to the HRSG at temperatures >~200°F to help prevent acid condensation.

Using the economizer configuration shown in Figure 10.12, the water to the economizer will be fed to 20 tubes. Each of the 20 tubes will then have ~222 passes through the exhaust gas. The nominal length of each tube (222 × 15 ft) = 3330 ft. However, accounting for all pipe bends, the equivalent length of each tube is about 7400 ft and the pressure drop would be about 12 psi (Geankoplis, 2003). ■

Determination of the Inside and Outside Heat Transfer Coefficients

We can now determine the inside and outside heat transfer coefficients for the cross-flow HRSG.

Outside Heat Transfer Coefficient, h_o (Btu/h-ft^2-°F)

The outside heat transfer coefficient is a function of the exhaust gas properties, outside tube diameter, and velocity often expressed in terms of the Nusselt (Nu), Prandtl (Pr), and Reynolds (Re) numbers. Ganapathy (1991, 2003) provides a conservative estimate for the outside heat transfer coefficient h_o in HRSG applications (exhaust gas external to tubes) as

$$h_o = 0.9G_h^{0.6}\frac{P_h}{d_o^{0.4}}, \tag{10.42}$$

where

P_h = a function of exhaust gas physical properties,

$$P_h = \frac{\left(k_h^{0.6}\right)\left(\hat{C}_{P,h}^{0.33}\right)}{\mu_h^{0.27}};$$

G_h = exhaust gas mass velocity (pound per hour-square foot);

d_o = outside tube diameter (inch);

k_h = exhaust gas thermal conductivity (British thermal unit per hour-foot-Fahrenheit);

$\hat{C}_{P,h}$ = exhaust gas specific heat (British thermal unit per pound-Fahrenheit); and

μ_h = exhaust gas viscosity (pound per hour-foot).

We can fit P_h to typical exhaust gas data (see Example 9.6) over the temperature range typically found in HRSG (200–1200°F), giving

$$P_h = 3.512 \times 10^{-5}\, T_{\text{film}} + 0.088, \qquad (10.43)$$

where T_{film} is the film temperature (Fahrenheit) and T_{film} is the average temperature of the exhaust gas and the tube wall. For example, from Figure 10.13b,

$$T_{\text{film}(r,n)} = \frac{T_{h(r,n)} + T_{\text{wall}(r,n)}}{2}.$$

$T_{\text{wall}(r,n)}$ can be estimated as

$$T_{\text{wall}(r,n)} = T_{h(r,n)} - \frac{h_{i(r,n)}}{h_{o(r,n)} + h_{i(r,n)}}\left(T_{h(r,n)} - T_{c(r,n)}\right). \qquad (10.44)$$

Generally, it is sufficient to estimate $T_{\text{film}(r,n)}$ as

$$T_{\text{film}(r,n)} = \frac{T_{h(r,n)} + T_{c(r,n)}}{2}. \qquad (10.45)$$

We can determine h_o from Equation (10.42) using Equations (10.43) and (10.45) once a value for the exhaust gas mass flow velocity G_h (pound per hour-square foot) has been determined. Here,

$$G_h = \frac{F_h}{(N_{\text{tubes\_wide}})(L_{\text{tube}})(S_{\text{tube}} - d_o)\left(\frac{1\text{ ft}}{12\text{ in.}}\right)}, \qquad (10.46)$$

where

F_h = the exhaust gas flow rate (pound per hour),

$N_{\text{tubes\_wide}}$ = the number of tubes wide,

L_{tube} = tube length (foot),

S_{tube} = tube spacing (inch), and

d_o = outside tube diameter (inch).

For h_o, we are only considering convective heat transfer. If thermal radiation is significant, the total outside heat transfer coefficient would be the sum of $h_{o,\text{convection}} + h_{o,\text{radiation}}$. Generally, $h_{o,\text{radiation}}$ can be neglected for gas temperatures below 800–900°F.

Inside Heat Transfer Coefficient, h_i (Btu/h-ft²-°F)

The inside heat transfer coefficient is a function of the water/steam properties, inside tube diameter, and velocity often expressed in terms of the Nusselt (*Nu*), Prandtl (*Pr*), and Reynolds (*Re*) numbers. Generally, for turbulent flow in smooth pipes, the Dittus–Boelter equation (Holman, 1972) is used:

$$Nu = 0.23 Re^{0.8} Pr^{0.4} \Rightarrow h_o = 2.44 \frac{w_c^{0.8} k_c^{0.6} \hat{C}_{P,c}^{0.4}}{d_i^{1.8} \mu_c^{0.4}}, \qquad (10.47)$$

where

w_c = water/steam flow rate in the single tube (pound per hour),

d_i = inside tube diameter (inch),

k_c = water/steam thermal conductivity (British thermal unit per hour-foot-Fahrenheit),

$\hat{C}_{P,c}$ = water/steam specific heat (British thermal unit per pound-Fahrenheit), and

μ_c = water/steam viscosity (pound per hour-foot).

The inside heat transfer coefficient is determined using water or steam or two-phase properties at the bulk fluid temperature ($T_{c(r,n)}$).

For hot water flowing inside tubes ($T_{c(r,n)} < 300$°F), Ganapathy (1991, 2003) provides

$$h_i = \left(150 + 1.55 T_{c(r,n)}\right) \frac{v_c^{0.8}}{d_i^{0.2}}, \qquad (10.48)$$

where

$T_{c(r,n)}$ = the water temperature (Fahrenheit)

v_c = the water velocity in the tube (foot per second)

d_i = the inside tube diameter (inch).

EXAMPLE 10.12 *Determination of* h$_o$, h$_i$, *and* U *for* (r = 1, n = 1) *in the Economizer*

Using data from the HRSG economizer of Example 10.4, determine $h_{o(r=1,n=1)}$ and $h_{i(r=1,n=1)}$. Also, determine $U_{(r=1,n=1)}$ based on the tube outside the surface area if $k_{\text{wall}} = 25$ Btu/h-ft-°F. Here, r and n follow the nomenclature of Figure 10.13b and $r = 1$, $n = 1$ would be the first node in the economizer row before the exhaust gas exits the stack. Do note that any consistent numbering of the rows (here tube passes) can be used, and here $r = 1, \ldots, N_{\text{rows\_economizer}}$. Allow that $T_{h(r=1,n=1)} = T_{h(r=1,n=2)} = T_{h(r=1,n=3)} \ldots = 711.194$ R (251.524°F); this would also be true for any of the first 20 tubes ($N_{\text{tubes\_wide}} = 20$) in the economizer. The cold stream enters at $T_{c(r=1,n=1)} = 536.67$ R (77°F).

SOLUTION ***Outside Heat Transfer Coefficient,* h$_o$ *(Btu/h-ft^2-°F)***

From Equation (10.45),

$$T_{\text{film}(r=1,n=1)} = \frac{T_{h(r,n)} + T_{c(r,n)}}{2} = \frac{251.524 + 77}{2} = 164.262°\text{F}.$$

From Equation (10.46),

$$G_h = \frac{F_h}{(N_{\text{tubes\_wide}})(L_{\text{tube}})(S_{\text{tube}} - d_o)\left(\frac{1\text{ ft}}{12\text{ in.}}\right)}$$

$$= \frac{840{,}328\frac{\text{lb}}{\text{h}}}{(20\text{ tubes wide})(15\text{ ft})(4\text{ in.} - 2\text{ in.})\left(\frac{1\text{ ft}}{12\text{ in.}}\right)}$$

$$= 16{,}806.56\frac{\text{lb}}{\text{h-ft}^2}.$$

From Equation (10.43),

$$P_h = 3.512 \times 10^{-5}\, T_{\text{film}} + 0.088 = 3.512 \times 10^{-5}(164.262) + 0.088$$
$$= 0.0938.$$

From Equation (10.42),

$$h_{o(r=1,n=1)} = 0.9G_h^{0.6}\frac{P_h}{d_o^{0.4}} = 0.9(16806.56)^{0.6}\frac{0.0938}{2^{0.4}}$$

$$= 21.944\frac{\text{Btu}}{\text{h-ft}^2\text{-°F}}.$$

Inside Heat Transfer Coefficient,* h$_i$ *(Btu/h-ft^2-°F)

Here we need the velocity of water in the tube at conditions of 536.67 R (77°F) and 20 bar (290.08 psia):

$$F_{c/tube} = \frac{F_c}{N_{tubes\_wide}} = \frac{111{,}096\frac{lb}{h}}{20\ tubes\ wide} = 5554.8\frac{lb}{h - tube}$$

$$v_c = \frac{F_{c/\text{tube}}}{\pi\left(\frac{d_i}{2}\frac{1\text{ ft}}{12\text{ in.}}\right)^2 \rho_c}\left(\frac{1\text{ h}}{3600\text{ s}}\right)$$

$$= 0.05093\frac{F_{c/\text{tube}}}{(d_i)^2\rho_c} = 1.386\text{ ft/s},$$

where

v_c = the water velocity in the tube (foot per second),

d_i = the inside tube diameter (inch), and

ρ_c = water density 77°F, 20 bar (for $r = 1$, $n = 1$) = 62.3 lb/ft^3.

Using Equation (10.48),

$$h_{i(r=1,n=1)} = (150 + 1.55T_{c(r,n)})\frac{v_c^{0.8}}{d_i^{0.2}}$$

$$= (150 + 1.55(77))\frac{(1.386)^{0.8}}{(1.81)^{0.2}} = 310.214\frac{\text{Btu}}{\text{h-ft}^2\text{-°F}}.$$

From Equation (10.41),

$$U_{(r=1,n=1)} = \frac{1}{\frac{A_o}{A_i h_{i(r,n)}} + \frac{A_o \ln\left(\frac{d_o}{d_i}\right)}{2\pi k_{\text{wall}}(dL)} + \frac{1}{h_{o(r,n)}}}$$

$$= \frac{1}{\frac{0.5236}{(0.4739)(310.214)} + \frac{0.5236\ln\left(\frac{2}{1.81}\right)}{2\pi(25)(1\text{ ft})} + \frac{1}{21.944}}$$

$$= 20.216\frac{\text{Btu}}{\text{h-ft}^2\text{-°F}}.$$

Do note that $U_{(r=1,n=1)} \sim (0.95)(h_{o(r=1,n=1)})$. ■

EXAMPLE 10.13 ***Steady-State Temperature Profile for the Economizer Tube Passes***

Using data from the HRSG economizer of Example 10.4, develop the numerical solution for the temperature profile in the first two tube passes of the economizer. Use the thermodynamic library for physical properties for the exhaust gas and water. Then, solve for the temperature profile in the entire economizer and determine the needed number of tube passes. From Example 10.12, the number of passes for each tube was estimated at 222, but here the overall U was assumed to be 12.3 Btu/h-ft^2-°F.

SOLUTION The solution to Example 10.13 is provided in **Example 10.13.xls** and the results are shown in Figure 10.14. The solution in the Excel file can be outlined as follows:

1. Pick an appropriate dL; in the provided Excel code, $dL = 1$ ft, but this value can be changed.
2. Calculate h_o, h_i, and U as in Example 10.12 and for the water density use ρ_c = `(1 / (V_Water(Pressure_c, T_c(r, n) + 459.67))) * MW_water`.
3. Use Equation (10.39), $\dot{Q}_{(r,n)} = U_{(r,n)}(A_o dL)(T_{h(r,n)} - T_{c(r,n)})$ to calculate the heat transfer rate into the node.
4. Using Equation (10.38), calculate the total enthalpy into the node as `H_c_in=(H_Water(Pressure_c, T_c(r, n)+459.67)) / MW_water *` $F_{c/tube}$. Calculate the total enthalpy out of the node as `H_c_out = H_c_in +` $\dot{Q}_{(r,n)}$ and determine the temperature of the next node as `T_c(r, n + 1) = T_HP_Water(H_c_out / ` $F_{c/tube}$ ` * MW_water, Pressure_c) - 459.67`.

	A	B	C	D	E	F	G	H	I	J	K	L
1	Numerical Solution Economizer Tubes Temperature Profile											
2					**To Stack**					**Exhaust Gas**		
3	**Tube Data**	data input in green cells			←							
4	Outside Diameter	2	in			Tube Pass =	1			Tube Pass =	2	
5	Inside Diameter	1.81	in		Position	$U_{(r=1,\ n=1..N)}$	$T_{h(r=1,\ n=1..N)}$	$T_{c(r=1,\ n=1..N)}$	Position	$U_{(r=2,\ n=1..N)}$	$T_{h(r=2,\ n=1..N)}$	$T_{c(r=2,\ n=1..N)}$
6	Number of Tubes Wide	20			1	20.2121146	251.524	77	1	20.239291	254.12729	81.941269
7	Tube Spacing	4	in		2	20.2161156	251.524	77.33331824	2	20.24329	254.12729	82.270724
8	Tube Length	15	ft		3	20.2201009	251.524	77.66607783	3	20.247272	254.12729	82.599623
9	Number Tube Passes	200			4	20.2240706	251.524	77.99827911	4	20.251239	254.12729	82.927966
10	k_{wall}	25	btu/hr-ft-°F		5	20.2280248	251.524	78.32992239	5	20.255191	254.12729	83.255754
11					6	20.2319636	251.524	78.66100802	6	20.259127	254.12729	83.582987
12	**Water Flow Data**				7	20.235887	251.524	78.99153632	7	20.263047	254.12729	83.909665
13	$F_c = F_{water}$	111096	lb/hr		8	20.2397952	251.524	79.32150764	8	20.266952	254.12729	84.235788
14	$T_{c(r=R,\ n=1)}$	77	°F		9	20.2436883	251.524	79.65092233	9	20.270842	254.12729	84.561358
15	P_{water}	290.08	psia		10	20.2475663	251.524	79.97978073	10	20.274717	254.12729	84.886374
16					11	20.2514294	251.524	80.3080832	11	20.278577	254.12729	85.210837
17					12	20.2552776	251.524	80.6358301	12	20.282421	254.12729	85.534747
18	**Exhaust Flow Data**				13	20.259111	251.524	80.9630218	13	20.286251	254.12729	85.858104
19	$F_h = F_{exhaust}$	840327.8	lb/hr		14	20.2629297	251.524	81.28965866	14	20.290066	254.12729	86.18091
20	$G_h = G_{exhaust}$	16806.56	lb/ft^2-hr		15	20.2667338	251.524	81.61574106	15	20.293867	254.12729	86.503164
21	$T_{h(r=R,\ n=1)} \ldots T_{h(r=R,\ n=N)}$	251.524	°F									
22												
23												
24	**Finite Difference Data**											
25	dL	1	ft									
26	number nodes / tube	15										

Figure 10.14 Temperature profile in the first two tube passes of the economizer.

5. Continue until the all the nodes in the current tube pass are solved. Determine the total $\dot{Q}_r$ for the row as $\dot{Q}_r = \sum_{n=1}^{N} \dot{Q}_{(r,n)}$.
6. Solve for the temperature of the exhaust gas surrounding the $r + 1$ tube pass. First, determine the enthalpy of the current exhaust gas in row r using `H_Products` $*F_h$ / N_{tubes_wide}. Add $\dot{Q}_r$ to this value. The temperature of the exhaust in row $r + 1$ can then be found using `TfromH_Products`.

A simple modification of the Excel file indicates that ~132 tube passes would be necessary to bring T_c to ~387.33°F and here T_h ~ 417.33°F; the results are provided in the Excel file **Example 10.13.xls**. ■

10.8 CLOSING REMARKS

As noted in the closing comments of Chapters 7 and 9, gas turbine cogeneration system design is a mature topic. For readers interested in gas turbine design and application details, I recommend *Gas Turbine Theory*, Sixth Edition (Saravanamuttoo et al., 2009), *Fundamentals of Gas Turbines*, Second Edition (Bathie, 1996), and *Gas Turbine Engineering Handbook*, Third Edition (Boyce, 2006). For readers interested in details of HRSG design and operation, I suggest first reading *Waste Heat Boiler Deskbook* (Ganapathy, 1991) followed by *Industrial Boilers and Heat Recovery Steam Generators: Design, Applications, and Calculations* (Ganapathy, 2003).

Cogeneration system design includes the consideration of the system configuration (equipment type and number) and the determination of the optimal size and operating conditions in order to minimize costs. In Chapter 10, equipment costs were approximated as continuous functions and costs (capital + operating) were minimized for a given cogeneration configuration with fixed process utility demands. Excel was shown as the ideal solution platform for this cogeneration design problem as Excel provides both optimization routines and sheet function capabilities. Sheet functions allow inclusion of rigorous physical properties providing more accurate system design and performance calculations. Sheet functions also allow calculation of all system variables when one or more design variables are changed—this is ideal for design optimization, evaluation of control strategies, or improved operability of cogeneration facilities. We also used sheet functions, combined with numerical methods, to determine temperature profiles in the HRSG tubes and at the HRSG walls.

The equipment costs provided in the CGAM problem provide a starting point for cost estimates (1994 dollars) for many significant cogeneration equipment items. These included individual costs for the air compressor, combustion chamber, gas and power turbine, the air preheater heat exchanger, and both the economizer and evaporator sections of the HRSG. Cogeneration systems often include additional equipment items including a main boiler, auxiliary boilers, a feedwater heat exchanger, a back-pressure steam turbine, an absorptive chiller, a compression chiller, as well as a capital investment to allow connection to the grid for the sale of excess power or the import of power. Additional costing expressions for utility systems can be found in Woods et al. (1979), Manninen and Zhu (1999), and Rodriguez (2003).

The approach developed in this chapter, which optimizes an economic model of capital and operating costs, is a powerful approach to determining needed operating conditions and equipment sizes. There are different solution approaches to the optimization of the CGAM problem or any energy system. For example, another commonly used method is exergoeconomic evaluation and optimization, and here see Valero et al. (1994b), Tsatsaronis and Pisa (1994), Hua et al. (1997), and the text by Bejan et al. (1996).

REFERENCES

BATHIE, W.W. 1996. *Fundamentals of Gas Turbines* (2nd edition). John Wiley & Sons, New York.

BEJAN, A., G. TSATSARONIS, and M. MORAN. 1996. *Thermal Design and Optimization*. John Wiley & Sons, New York.

BOYCE, M.P. 2006. *Gas Turbine Engineering Handbook* (3rd edition). Gulf Professional Publishing, Burlington, MA.

CHAPRA, S.C. and R.P. CANALE. 2010. *Numerical Methods for Engineers* (6th edition). McGraw Hill, Boston.

GANAPATHY, V. 1991. *Waste Heat Boiler Deskbook*. The Fairmont Press, Liburn GA.

GANAPATHY, V. 2003. *Industrial Boilers and Heat Recovery Generators: Design, Applications and Calculations*. Marcel Dekker, New York.

GEANKOPLIS, C.J. 2003. *Transport Processes and Separation Processes (Includes Unit Operations)* (4th edition). Prentice Hall, Upper Saddle River, NJ.

HOLMAN, J.P. 1972. *Heat Transfer* (3rd edition). McGraw-Hill, New York.

HUA, B., Q.L. CHEN, and P. WANG. 1997. A new exergoeconomic approach for analysis and optimization of energy systems. *Energy* 23(11): 1071–1078.

MANNINEN, J. and X.X. ZHU. 1999. Optimal gas turbine integration to the process industries. *Ind. Eng. Chem. Res.* 38: 4317–4329.

POULLIKKAS, A. 2005. An overview of current and future sustainable gas turbine technologies. *Renewable Sustainable Energy Rev.* 9: 409–443.

RODRIGUEZ, R. 2003. Combined heat and power technologies applied studies of options including microturbines. MS Thesis, Technische Universitat Wien.

ROSSITER, A.P. 1990. Criteria for the integration of combined cycle cogeneration systems in the process industries. *Heat Recov. Syst. CHP* 10(1): 37–48.

SARAVANAMUTTOO, H.I.H., G.F.C. ROGERS, H. COHEN, and P.V. STRAZNICKY. 2009. *Gas Turbine Theory* (6th edition). Prentice Education, Essex, England.

SPIEWAK, S.A. and L. WEISS. 1997. *Cogeneration & Small Power Production Manual*. The Fairmont Press, Liburn, GA.

TSATSARONIS, G. and J. PISA. 1994. Exergoeconomic evaluation and optimization of energy systems—Application to the CGAM problem. *Energy* 19(3): 287–321.

VALERO, A., M.A. LOZANO, L. SERRA, G. TSATSARONIS, J. PISA, C. FRANGOPOULOS, and M. von SPAKOVSKY. 1994a. CGAM problem: Definition and conventional solution. *Energy* 20(3): 279–286.

VALERO, A., M.A. LOZANO, L. SERRA, and C. TORRES. 1994b. Application of the exergetic cost theory to the CGAM problem. *Energy* 19(3): 365–381.

WOODS, D.R., S.J. ANDERSON, and S.L. NORMAN. 1979. Evaluation of capital cost data: Offsite utilities (supply). *Can. J. Chem. Eng.* 57(5): 533–565.

PROBLEMS

10.1 ***Solve Example 10.2, the Original CGAM Design Problem—Ideal Gas Solution in SI Units*** Use the costing data provided in Tables 10.2 and 10.3 and the physical property data in Table 10.5. Here the power requirement is 30 MW and the required steam flow is 14 kg/s.

10.2 ***Solve Example 10.5, CGAM Design Problem—Real Fluid Solution in SI Units*** Here the power requirement is 30 MW and the required steam flow is 14 kg/s; use fixed values of $\Delta T_{\text{Pinch}} = 15$ R and $\Delta T_{\text{Approach}} = 15$ R.

10.3 ***HRSG Off-Design Performance Based on Overall Energy Balance*** Starting with Example 10.5 and recalling our HRSG off-design calculations of Chapter 9, we want to see the impact on steam production if supplemental firing is used to increase the temperature entering the HRSG; here assume $F_{\text{supp fuel}} = 1.4$ lb/s. From Example 10.5, $F_{\text{air}} = 229.89$ lb/s, $F_{\text{fuel}} = 3.543$ lb/s, $F_{\text{gas}} = 233.436$ lb/s, $T_6 = 1309.74$ R, $P_6 = 1.066$ bar, $T_7 = 716.59$ R, $P_7 = 1.013$ bar, $\Delta T_{\text{Pinch}} = 15$ R and $\Delta T_{\text{Approach}} = 15$ R, $F_{\text{steam}} = 30.86$ lb/s; and the transferred Q's are $\dot{Q}_{\text{Evaporator}} = 25{,}584.5$ Btu/s and $\dot{Q}_{\text{Economizer}} = 10{,}138.6$ Btu/s. These values for the HRSG represent the HRSG *design* case.

10.4 ***HRSG Off-Design Performance Based on Overall Heat Transfer Coefficient*** Here we want to resolve Problem 10.3 using the overall heat transfer coefficient method. Using the temperatures found for the design and off-design cases in Problem 10.3, we can estimate exhaust gas (air) physical properties in the HRSG. We will need average physical properties for the evaporator and economizer sections of the HRSG for use in Equation (9.40).

Air Properties

Using Results from Problem 10.3	T(R)	$C_{p,\text{air}}\left(\frac{\text{Btu}}{\text{lb-R}}\right)$	μ_{air} (mPa-s)	$k_{\text{air}}\left(\frac{\text{W}}{\text{m-K}}\right)$
Design, T_6	1309.7	0.259	0.0350	0.0538
Design, T_{7P}	889	0.246	0.0270	0.0399
Design, T_7	716.6	0.242	0.0232	0.0334
Design, average evaporator		0.253	0.031	0.0469
Design, average economizer		0.244	0.0251	0.0367
Off-design, T_{6_SF}	1756	0.272	0.0421	0.0668
Off-design, T_{7P_SF}	1025	0.250	0.0298	0.0447
Off-design, T_{7_SF}	716.6	0.242	0.0232	0.0334
Off-design, average evaporator		0.261	0.0360	0.0558
Off-design, average economizer		0.246	0.0265	0.0391

10.5 ***Unsteady-State Heat Radiation Boundary—Finite Difference Formulation*** Develop the unsteady-state energy balance in finite difference form for a planar wall with a radiative

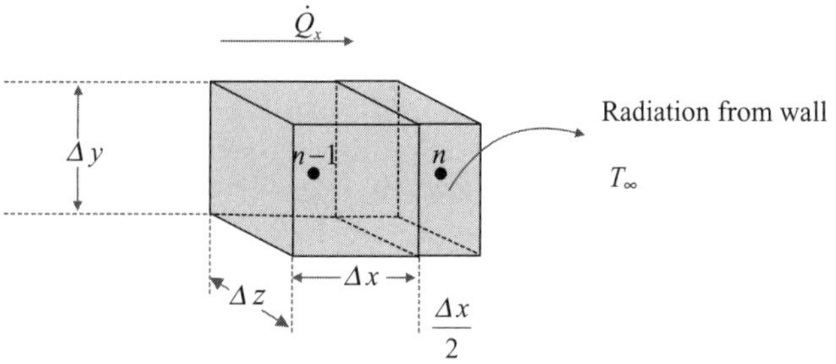

Figure P10.5 Elemental volumes and nomenclature for radiation boundary.

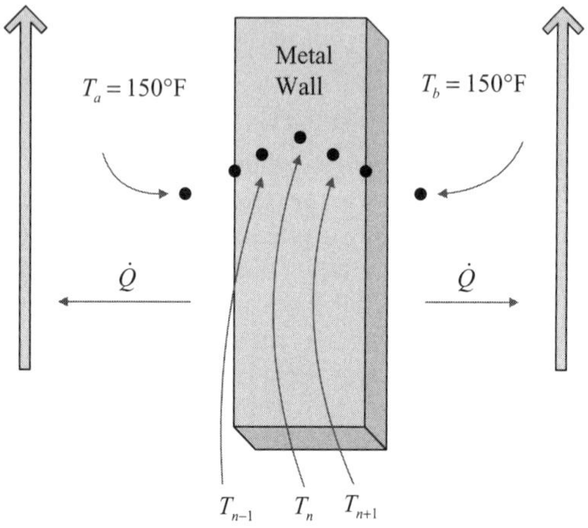

Figure P10.6 Sketch of the heat conduction problem.

boundary shown in Figure P10.5. The solution to this problem will require outside reading of a standard engineering heat transfer text. Remember that for radiation, temperature must be in absolute units (Kelvin or Rankine).

10.6 ***Heating a Metal Wall with Two Convection Surfaces—Finite Difference Solution (from Holman, 1972)*** A 2-in. thick aluminum metal wall, shown in Figure P10.6, has a thermal conductivity of $k_{metal} = 120$ Btu/h-ft-°F, $\hat{\rho}_{metal} = 169$ lb/ft^3, and $\hat{C}_{p,metal} = 0.214$ Btu/lb-°F. Solve for the time-dependent metal wall temperature if the wall is initially at 450°F and then exposed to the convection environment of 150°F at both exterior surfaces. The convection heat transfer coefficient on both exterior surfaces, h, is 100 Btu/h-ft^2-°F. Although not detailed here, the Heisler charts (see Holman, 1972) can be used to solve for temperature in the metal slab as a function of time. Using the Heisler charts, the temperature 0.5 in. from the face after 1 minute of exposure to the 150°F environment should be near 326°F. Compare this value to that found using the finite difference approach.

10.7 ***Steady-State Temperature Profile Single-Pass Countercurrent Double Pipe Heat Exchanger (Solution Provided)*** Develop a numerical strategy for the hot and cold temperature profiles in a countercurrent double pipe heat exchanger, which accounts for changing physical properties. A single countercurrent double pipe heat exchanger and a temperature profile are indicated in Figure P10.7a,b. For model development purposes, the single heat exchanger can be considered a series of connected heat exchangers nodes each of area dA as shown in Figure P10.7c. Figure P10.7c serves as a basis for our numerical approach to account for changing physical properties in each of the connected heat exchangers. Here, fluid 1 is taken as the hot stream or the hot side (the stream giving up energy) and fluid 2 is assumed as the cold stream or the cold side (the stream accepting energy).

Solution:

The Cold-Side Equations (*c* = Cold)

At node n and assuming steady state, we can write the incremental energy transfer rate through dA on the cold side as

$$\dot{Q}_n = F_c d\hat{h}_{c,n} = F_c\left(\hat{h}_{c,n+1} - \hat{h}_{c,n}\right). \qquad \text{(P10.7a)}$$

Here the cold fluid flow direction is taken as positive and a forward difference is used for $d\hat{h}_{c,n}$.

The Hot-Side Equations (*h* = Hot)

At node n and assuming steady state, we can write the energy transfer rate through dA on the hot side as

$$\dot{Q}_n = F_h d\hat{h}_{h,n} = F_h\left(\hat{h}_{h,n+1} - \hat{h}_{h,n}\right). \qquad \text{(P10.7b)}$$

Our normal convention follows heat given up by the hot side is a negative quantity as in $\dot{Q} = F_h\left(\hat{h}_{out} - \hat{h}_{in}\right)$, where out and in are given by the hot fluid flow direction. In Equation (P10.7b), $\dot{Q}_n$ is a positive quantity because we are indexing by the nodes.

Heat Transfer between Hot and Cold Sides

The heat transfer rate can also be expressed as

$$\dot{Q}_n = U_n(dA)(T_{h,n} - T_{c,n}). \qquad \text{(P10.7c)}$$

In Equation (P10.7c) dA must be chosen small enough to allow accurate system and physical property determination in the determination of U_n.

If the hot and cold fluid heat capacities are constant through the volume of area dA, we can substitute for Equation (P10.7a,b):

$$\dot{Q}_n = F_c d\hat{h}_{c,n} = F_c\left(\hat{C}_{Pc,n}\right)(dT_{c,n}) = F_c\left(\hat{C}_{Pc,n}\right)(T_{c,n+1} - T_{c,n}) \qquad \text{(P10.7d)}$$

and

$$\dot{Q}_n = F_h d\hat{h}_{h,n} = F_h\left(\hat{C}_{Ph,n}\right)(dT_{h,n}) = F_h\left(\hat{C}_{Ph,n}\right)(T_{h,n+1} - T_{h,n}). \qquad \text{(P10.7e)}$$

Numerical Solution Strategy

Assume all physical and system properties at $n = 1$ are known; the numerical solution for the temperature profiles (Figure P10.7) can be constructed as follows:

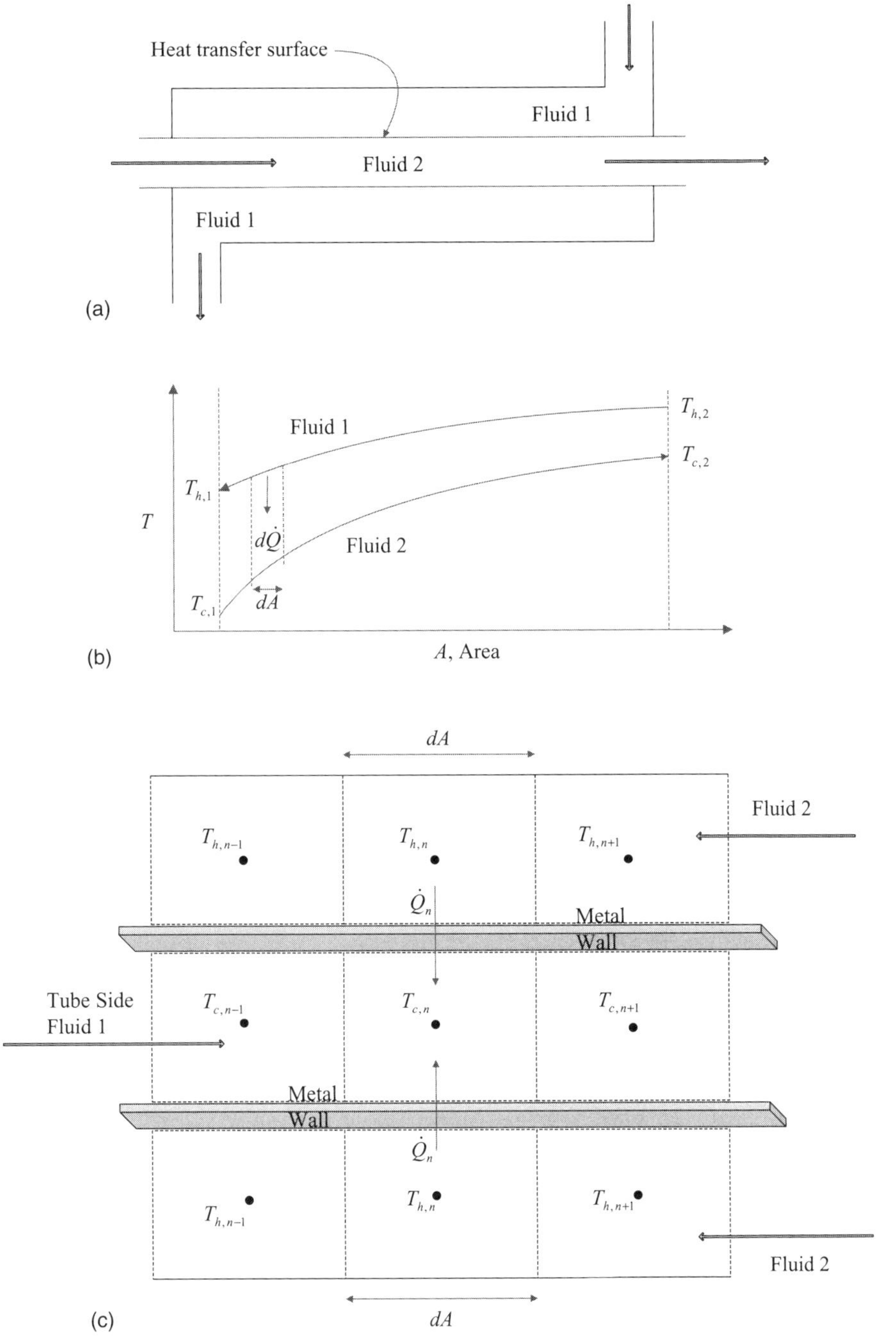

Figure P10.7 (a) a Single-pass double pipe heat exchanger. (b) Single-pass double pipe heat exchanger—temperature profile. (c) Single-pass double pipe heat exchanger—nomenclature for numerical analysis of the double pipe heat exchanger.

1. Fix an appropriate length, dL, in the heat exchanger, which in turn fixes dA as $dA = A_o dL$; A_o is the external surface area per foot length of tube.
2. Calculate U_n, which for the double pipe heat exchanger of Figure P10.7 based on the inside surface area of the fluid 2 tube is

$$U_n = \frac{1}{\dfrac{1}{h_{i,n}} + \dfrac{A_i \ln\left(\dfrac{d_o}{d_i}\right)}{2\pi k_n (dL)} + \dfrac{A_i}{A_o h_{o,n}}}. \qquad \text{(P10.7f)}$$

The value for U_n based on the outside area of the fluid 2 tube can be found using Equation (10.41).

3. Calculate $\dot{Q}_n$ using Equation (P10.7c).
4. Calculate $T_{c,n+1}$ using Equation (P10.7d) and calculate $T_{h,n+1}$ using Equation (P10.7e).

Here we have a system of two ODEs, which are solved using Euler's method:

$$T_{c,n+1} = T_{c,n} + \frac{U_n}{F_c\left(\hat{C}_{Pc,n}\right)}(T_{h,n} - T_{c,n})(A_0\, dL)$$

$$T_{h,n+1} = T_{h,n} + \frac{U_n}{F_h\left(\hat{C}_{Ph,n}\right)}(T_{h,n} - T_{c,n})(A_0\, dL).$$

Two initial conditions at the starting dL value ($n = 1$) are required. If, for example, the hot stream properties are only known at the hot inlet, $n = N$, guess $T_{h,1}$ and solve until the cold stream outlet temperature is reached. $T_{h,1}$ can be varied until the correct hot stream inlet temperature is obtained.

10.8 Log Mean Temperature Difference. Using the results from Problem 10.7, show how the LMTD arises from the integration of the differential temperature change equation for the countercurrent double pipe heat exchanger.

10.9 *Solve Example 10.13 Using $C_p = a + bT$ for Both the Water-Side and Exhaust Gas-Side Enthalpy (Solution Provided)* This problem requires the user to gather data and to determine C_p for both water and exhaust gas. The enthalpy term will be a quadratic in T (when using $C_p = a + bT$). T from $\hat{h}$ can be found using the quadratic equation.

Solution:

The solution for the needed physical properties is given in **Problem 10.9a.xls**. The solution for the temperature profile is given in **Problem 10.9b.xls** and shown in Figure P10.9.

The solution in the Excel file **Problem 10.9b.xls** can be outlined as follows. The thermodynamic properties are found in **Problem 10.9a.xls**.

1. Pick an appropriate dL; in the provided Excel code, $dL = 1$ ft, but this value can be changed.
2. Calculate h_o, h_i, and U as in Example 10.12 and for the water density, use ρ_c = `(63.74450083 - 0.01779699 * T_c(r, n))`. Here, ρ_c is in pound per cubic foot and T is in Fahrenheit.
3. Use Equation (10.39), $\dot{Q}_{(r,n)} = U_{(r,n)}(A_o\, dL)\left(T_{h(r,n)} - T_{c(r,n)}\right)$, to calculate the heat transfer rate into the node.
4. Using Equation (10.38), calculate the total enthalpy into the node as `H_c_in = (0.0000689 * T_c(r, n) ^ 2 + 0.9829722 * T_c(r, n) - 76.097377) *` $F_{c/tube}$.

   Here,
   $\hat{h}_{\text{water},T} = (0.0000689)T^2 + (0.9829722)T - 76.097377$ is an enthalpy fit for water in British thermal unit per pound with T in Fahrenheit. In this equation, we have accounted for the reference state as $\hat{h}_{\text{water},T} = \hat{h}_{\text{water},T} - \hat{h}_{\text{water},T=T_{\text{ref}}}$.

   Calculate the total enthalpy out of the node as `H_c_out = H_c_in +` $\dot{Q}_{(r,n)}$ and determine the temperature of the next node as `T_c(r, n + 1) = (-0.9829722 + (0.966234346 + 0.0002756 * (76.097377 + H_c_out /` $F_{c/tube}$`)) ^ 0.5) / 0.0001378`.

   Here,

$$T_{\text{water}} = (-0.9829722 + (0.966234346 + 0.0002756 \times (76.097377 + \hat{h}_{\text{water},T}))\wedge 0.5)/0.0001378$$

   is the quadratic equation used to determine T in Fahrenheit (from the enthalpy fit we are using for water).
5. Continue until the all the nodes in the current tube pass are solved. Determine the total $\dot{Q}_r$ for the row as $\dot{Q}_r = \sum_{n=1}^{N} \dot{Q}_{(r,n)}$.

	A	B	C	D	E	F	G	H	I	J	K	L
1	Numerical Solution Ecnnomizer Tubes Temperature Profile											
2					**To Stack**					**Exhaust Gas**		
3	**Tube Data**	data input in green cells										
4	Outside Diameter	2	in			Tube Pass =	1			Tube Pass =	2	
5	Inside Diameter	1.81	in		Position	$U_{(r=1,\ n=1..N)}$	$T_{h(r=1,\ n=1..N)}$	$T_{c(r=1,\ n=1..N)}$	Position	$U_{(r=2,\ n=1..N)}$	$T_{h(r=2,\ n=1..N)}$	$T_{c(r=2,\ n=1..N)}$
6	Number of Tubes Wide	20			1	20.2107242	251.524	77	1	20.238597	254.08703	81.957699
7	Tube Spacing	4	in		2	20.2147981	251.524	77.33462074	2	20.242659	254.08703	82.287956
8	Tube Length	15	ft		3	20.2188551	251.524	77.66865172	3	20.246704	254.08703	82.617631
9	Number Tube Passes	200			4	20.2228954	251.524	78.00209364	4	20.250732	254.08703	82.946724
10	k_{wall}	25	btu/hr-ft-°F		5	20.2269191	251.524	78.33494717	5	20.254744	254.08703	83.275236
11					6	20.2309264	251.524	78.66721298	6	20.25874	254.08703	83.603168
12	**Water Flow Data**				7	20.2349172	251.524	78.99889178	7	20.262719	254.08703	83.930519
13	$F_c = F_{water}$	111096	lb/hr		8	20.2388918	251.524	79.32998423	8	20.266681	254.08703	84.257291
14	$T_{c(r=R,\ n=1)}$	77	°F		9	20.2428502	251.524	79.66049104	9	20.270628	254.08703	84.583485
15	P_{water}	290.08	psia		10	20.2467926	251.524	79.99041289	10	20.274558	254.08703	84.909101
16					11	20.250719	251.524	80.31975048	11	20.278473	254.08703	85.234139
17					12	20.2546295	251.524	80.64850451	12	20.282372	254.08703	85.558602
18	**Exhaust Flow Data**				13	20.2585243	251.524	80.97667566	13	20.286254	254.08703	85.882488
19	$F_h = F_{exhaust}$	840327.8	lb/hr		14	20.2624034	251.524	81.30426465	14	20.290122	254.08703	86.2058
20	$G_h = G_{exhaust}$	16806.56	lb/ft^2-hr		15	20.266267	251.524	81.63127217	15	20.293973	254.08703	86.528537
21	$T_{h(r=R,\ n=1)} \ldots T_{h(r=R,\ n=N)}$	251.524	°F									
22												
23												
24	**Finite Difference Data**											
25	dL	1	ft									
26	number nodes / tube	15										

Figure P10.9 Temperature profile in the first two tube passes of the economizer.

6. Solve for the temperature of the exhaust gas surrounding the $r+1$ tube pass. First, determine the enthalpy of the current exhaust gas in row r using

```
H_exhaust_in = 0.0000149363 * T_h(r)
^ 2 + 0.246619961 * T_h(r)
- 19.07829448.
```

Here,

$$\hat{h}_{\text{gas},T} = (0.0000149363)T^2 + (0.246619961)T - 19.07829448$$

is the enthalpy fit for the exhaust gas in British thermal unit per pound with T in Fahrenheit. In this equation, we have again accounted for the reference state as

$$\hat{h}_{\text{gas},T} = \hat{h}_{\text{gas},T} - \hat{h}_{\text{gas},T=T_{\text{ref}}}.$$

The total enthalpy in would be, H_exhaust_in $*F_h$/Ntubes_wide. Add $\dot{Q}_r$ to this value. The temperature of the exhaust in row $r+1$ can then be found using a quadratic equation for T:

```
T_h(r + 1) = (-0.246619961 + (0.060821405
+ 0.0000597452 * (19.07829448 + H_exhaust_
out)) ^ 0.5) / 0.0000298726.
```

Here,

$$T_{\text{gas}} = (-0.246619961 + (00.060821405 + 0.0000597452 \times (19.07829448 + \hat{h}_{\text{gas},T}))\wedge 0.5)/0.0000298726$$

is the quadratic equation used to determine T in Fahrenheit (from the enthalpy fit we are using for the exhaust gas).

Be sure H_exhaust_out has been divided by $F_h/N_{\text{tubes\_wide}}$.

Chapter 11

Data Reconciliation and Gross Error Detection in a Cogeneration System

An industrial power plant may produce steam, electricity, refrigeration, and chilled water for process use. This text focuses on power plants utilizing cogeneration systems for combined heat and power generation. A cogeneration facility may have gas turbines, heat recovery steam generators (HRSGs), stand-alone boilers, steam-driven turbines, steam-driven chillers, and electric chillers. Steam must be raised for heating purposes and steam may be used to produce electricity from steam turbines and to drive chillers to produce chilled water. There may also be electric chillers available to produce chilled water. Supplemental firing of the HRSGs can be used to increase steam generation when needed.

One way to improve the operation of the cogeneration system is to use an online optimization strategy as introduced in Chapter 6 (see Figure 6.1). Here data are used to determine process parameters including heat transfer coefficients and equipment efficiencies. These parameters can be compared to test data from the manufacturer or values when the equipment was recently cleaned and trends can be monitored to determine when a unit should be taken off-line for maintenance or cleaning. As will be described in Chapter 12, we can use efficiency data from the cogeneration system and process energy demand data in an energy management optimization strategy to determine the amounts of self-generated electricity, purchased electricity, and purchased natural gas needed to minimize utility costs. A key in any online optimization strategy is gathering needed data and making sure the data are accurate and do not contain errors. In Chapter 6, we introduced data reconciliation and gross error detection to ensure that plant data are accurate—satisfy known material and energy balance and do not contain gross errors. In this chapter, we specifically look at data from key components in a gas turbine cogeneration system and perform data reconciliation and gross error detection. We will integrate our developed thermodynamics package (Chapter 8, *TPSI+*) into the data reconciliation problem.

11.1 COGENERATION SYSTEM DATA RECONCILIATION

Figure 11.1 shows the main components of an aeroderivative-type gas turbine cogeneration system for electricity and steam generation. These components include an air compressor, combustion chamber, gas generating turbine (GGT) for air compression, power turbine for electricity generation, and an HRSG for steam production. The air cooler is used to set $T_{\text{air}}^{\text{compressor in}}$ ~60°F in order to help maintain the performance of the aeroderivative gas turbine.

There are seven key steps in the process:

1. Ambient air (shown as state 0 in Figure 11.1) is sent through a heat exchanger (the air cooler) to adjust its temperature to a nominal 60°F (state 1). Chilled water is used as the cold fluid in the air cooler. The incoming air is adjusted to 60°F prior to entering the air compressor in order to help maintain gas turbine efficiency. Generally, chilled water is needed as the design condition for the system calls for ambient air at 87.5°F.
2. The cooled air is then sent to the compressor to increase pressure (state 2).

Modeling, Analysis and Optimization of Process and Energy Systems, First Edition. F. Carl Knopf.

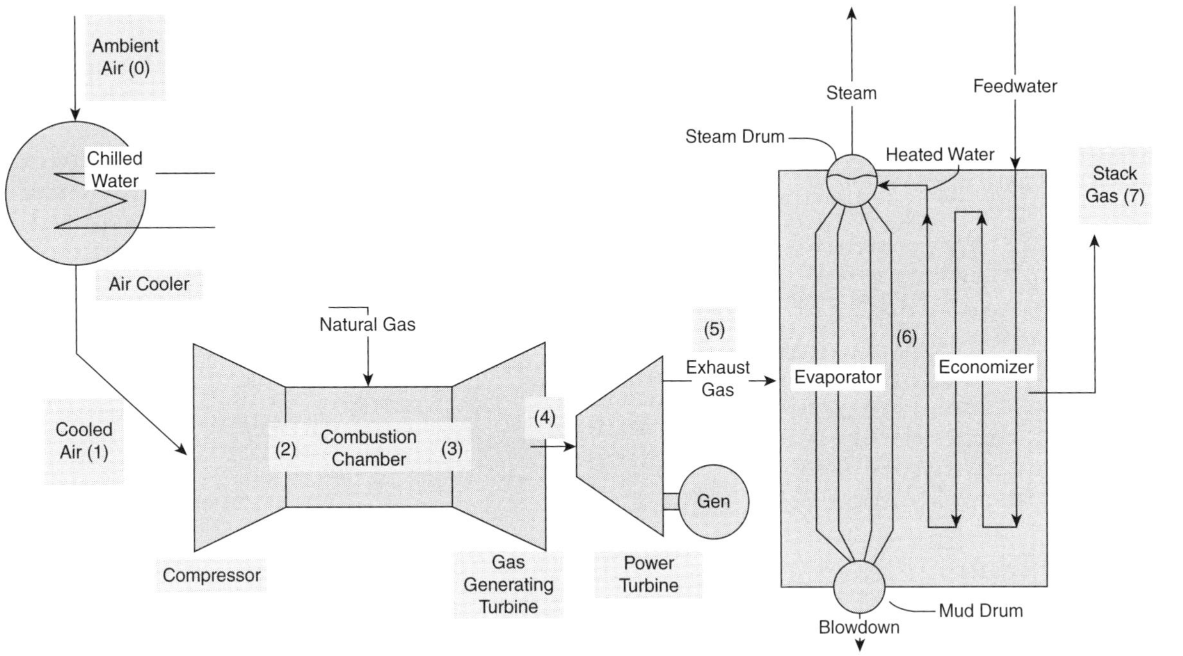

Figure 11.1 Gas turbine cogeneration system—gas turbine and HRSG.

3. Natural gas and compressed air are burned in the combustion chamber (state 3).
4. The combustion products are sent through the gas generating turbine (state 4). The shaft of this turbine is directly connected to the compressor. All work done by the gas generating turbine is used to power the compressor.
5. The combustion products then expand to nearly atmospheric pressure in the power turbine (state 5). The shaft of this turbine is directly connected to a generator to produce electricity for the process.
6. The combustion products are sent through the HRSG, consisting of two heat exchangers to recover heat before venting to the atmosphere. In the evaporator, the combustion products transfer heat to vaporize heated water into steam. Some of the heated water from the economizer is not vaporized in the evaporator and exits as blowdown. In the evaporator, there is a natural circulation between two drums—one drum providing saturated steam to the process and the other drum (the lower or mud drum) saturated liquid. Blowdown occurs from the saturated liquid drum and is used to control suspended and dissolved solid concentration in the steam system. Blowdown P and T are not generally measured, but we know $P_{\text{Steam}} = P_{\text{Blowdown}}$ and $T_{\text{Steam}} = T_{\text{Blowdown}}$.
7. In the economizer, the combustion products heat the feedwater before being sent to the evaporator.

The data in Table 11.1 are available from the distributed control system.

In order to help keep the data reconciliation problem more tractable, we will assume the incoming air is dry (no humidity) and there is no pressure drop in the air cooler. Heat loss from the combustion chamber is estimated at ~2%(F_{NG})(LHV) and accounted for by a combustion chamber heat loss term, $\dot{Q}_{\text{CC,Loss}}$. For safety considerations, the gas turbine components themselves (compressor, combustion chamber, gas generating turbine, and power turbine) are enclosed and the enclosure is maintained under a slight vacuum. $\dot{Q}_{\text{CC,Loss}}$ could be replaced with the flow rate and temperature change for the air passing through the enclosure. We will initially ignore possible inaccuracies in pressure measurements. It is straightforward to add pressure considerations to the final formulation, but generally, pressure effects on enthalpy are small. Pressure considerations are added to the final formulation in Problem 11.2.

EXAMPLE 11.1 *Mass and Energy Balances for the Data Reconciliation Problem*

In order to solve the data reconciliation problem given by Equations (6.7) and (6.8), material and energy balance are needed. Write the mass and energy balances for the processes shown in Figure 11.1. Use the variables in Table 11.1.

Table 11.1 Measure Values, Instrument Standard Deviations, and System Nomenclature (Buckley, 2006)

Name	Description	Value	Units	Standard Deviation
P_0	Ambient pressure	14.696	psia	1
T_0	Ambient temperature	547.17	R	2
P_1	Air P leaving air cooler	14.696	psia	1
T_1	Air T leaving air cooler	519.94	R	5
P_2	Air P leaving compressor	243.7	psia	1
T_2	Air T leaving Compressor	1260.48	R	10
P_3	Product gas P leaving combustion chamber	243.7	psia	1
T_3	Product gas T leaving combustion chamber	2400	R	150
P_4	Product gas P leaving gas generating turbine	56.9	psia	1
T_4	Product gas T leaving gas generating turbine	1836.34	R	30
P_5	Product gas P leaving power turbine	14.82	psia	1
T_5	Product gas T leaving power turbine	1386.67	R	60
P_6	Product gas P leaving evaporator	14.82	psia	1
T_6	Product gas T leaving evaporator	914.67	R	50
P_7	Product gas P leaving economizer	14.82	psia	1
T_7	Product gas T leaving economizer	787.67	R	20
F_{air}	Air flow rate	145	lb/s	20
$\hat{h}_{\text{air},0} - \hat{h}_{\text{air},2}$	Air enthalpy at states 0–2		Btu/lb	
$\hat{h}_{\text{Prod},3} - \hat{h}_{\text{Prod},7}$	Combustion product enthalpy states 3–7		Btu/lb	
F_{Prod}	Combustion products flow rate	147.6208	lb/s	20
F_{CW}	Chilled water flow rate	200	lb/s	44
$F_{\text{CW},a}$	Chilled water P in	60	psia	1
$T_{\text{CW},a}$	Chilled water T in	504.57	R	5
$F_{\text{CW},b}$	Chilled water P out	60	psia	1
$T_{\text{CW},b}$	Chilled water T out	515.07	R	5
$\hat{h}_{\text{CW},a}$	Chilled water enthalpy in		Btu/lb	
$\hat{h}_{\text{CW},b}$	Chilled water enthalpy out		Btu/lb	
F_{NG}	Natural gas flow	2.6208	lb/s	0.07
LHV	Natural gas lower heating value	21,501	Btu/lb	200
$F_{\text{Water,Econ}}$	Feedwater flow entering economizer	24.4444	lb/s	1.2222
$P_{\text{Water,Econ}}$	Feedwater P entering economizer	239	psia	1
$T_{\text{Water,Econ}}$	Feedwater T entering economizer	667.67	R	30
$\hat{h}_{\text{Water,Econ}}$	Feedwater enthalpy entering economizer		Btu/lb	
$P_{\text{Water,Evap}}$	Heated water P entering evaporator	198	psia	1
$T_{\text{Water,Evap}}$	Heated water T entering evaporator	780.67	R	30
$\hat{h}_{\text{Water,Evap}}$	Heated water enthalpy entering evaporator		Btu/lb	
F_{Steam}	Steam flow	24.17	lb/s	0.725
P_{Steam}	Steam P	140.9	psia	NA
T_{Steam}	Steam T	821.67	R	1
$\hat{h}_{\text{Steam}}$	Steam enthalpy		Btu/lb	
$F_{\text{Water,BlowD}}$	Blowdown flow—saturated water	0.28	lb/s	0.1
$P_{\text{Water,BlowD}}$	Blowdown P, $P_{\text{Steam}} = P_{\text{Blowdown}}$			
$T_{\text{Water,BlowD}}$	Blowdown T, $T_{\text{Steam}} = T_{\text{Blowdown}}$			
$\hat{h}_{\text{Water,BlowD}}$	Blowdown enthalpy—saturated water		Btu/lb	
$\dot{W}_{\text{Net}}$	Net power produced by power turbine	19.27	MW	
$\dot{W}_{\text{Net}}$	Net power produced by power turbine	18,264.44	Btu/s	0.01
$\dot{Q}_{\text{Air Cooler}}$	Heat transfer in air cooler		Btu/s	
$\dot{W}_{\text{GG Turbine}}$	Work done by gas generating turbine (GGT)/used by compressor		Btu/s	
$\dot{Q}_{\text{CC,Loss}}$	Heat loss in combustion chamber	1125	Btu/s	1000
$\dot{Q}_{\text{Evaporator}}$	Heat transfer in evaporator		Btu/s	
$\dot{Q}_{\text{Economizer}}$	Heat transfer in economizer		Btu/s	

SOLUTION Mass balances are needed to account for fuel addition in the combustion chamber and for blowdown in the HRSG as

Combustion Chamber

$$F_{\text{Air}} + F_{\text{NG}} = F_{\text{Prod}} \tag{11.1}$$

HRSG

$$F_{\text{Water,Econ}} = F_{\text{Blowdown}} + F_{\text{Steam}} \tag{11.2}$$

Energy balances are needed for the air cooler, compressor, combustion chamber, gas generating turbine, power turbine, evaporator, and economizer as

Air Cooler

$$F_{\text{air}}\left(\hat{h}_{\text{air},0} - \hat{h}_{\text{air},1}\right) = F_{\text{CW}}\left(\hat{h}_{\text{CW},b} - \hat{h}_{\text{CW},a}\right) = \dot{Q}_{\text{Air Cooler}} \tag{11.3}$$

Compressor

$$F_{\text{air}}\left(\hat{h}_{\text{air},2} - \hat{h}_{\text{air},1}\right) = \dot{W}_{\text{GG Turbine}} \tag{11.4}$$

Combustion Chamber

$$F_{\text{air}}\left(\hat{h}_{\text{air},2}\right) + F_{\text{NG}}\left(LHV\right) = F_{\text{Prod}}\left(\hat{h}_{\text{Prod},3}\right) + \dot{Q}_{\text{CC,Loss}} \tag{11.5}$$

Gas Generating Turbine

$$F_{\text{Prod}}\left(\hat{h}_{\text{Prod},3} - \hat{h}_{\text{Prod},4}\right) = \dot{W}_{\text{GG Turbine}} \tag{11.6}$$

Power Turbine

$$F_{\text{Prod}}\left(\hat{h}_{\text{Prod},4} - \hat{h}_{\text{Prod},5}\right) = \dot{W}_{\text{Net}} \tag{11.7}$$

Evaporator

$$\begin{aligned} F_{\text{Prod}}\left(\hat{h}_{\text{Prod},5} - \hat{h}_{\text{Prod},6}\right) &= F_{\text{Steam}}\left(\hat{h}_{\text{Steam}} - \hat{h}_{\text{Water,Evap}}\right) \\ &\quad + F_{\text{Water,BlowD}}\left(\hat{h}_{\text{Water,BlowD}} - \hat{h}_{\text{Water,Evap}}\right) \\ &= \dot{Q}_{\text{Evaporator}} \end{aligned} \tag{11.8}$$

Economizer

$$F_{\text{Prod}}\left(\hat{h}_{\text{Prod},6} - \hat{h}_{\text{Prod},7}\right) = F_{\text{Water,Econ}}\left(\hat{h}_{\text{Water,Evap}} - \hat{h}_{\text{Water,Econ}}\right) = \dot{Q}_{\text{Economizer}} \tag{11.9}$$

In the energy balance for the compressor and gas generating turbine, we are following the sign convention that work done by the system is a positive quantity and work done on the system is a negative quantity. Here, $\dot{W}_{\text{GG Turbine}}$ is a positive quantity and $\dot{W}_{\text{Compressor}}$ would be a negative quantity. The net work from the cogeneration system is the work done by the power turbine. ■

EXAMPLE 11.2 *Thermodynamic Database for the Data Reconciliation Problem*

Here replace the enthalpy values in Example 11.1 ($\hat{h}_{\text{air},0}$, $\hat{h}_{\text{Prod},3}$, etc.) with the appropriate functions from the thermodynamic database (Table 9.1). We will have to address the calculation of the dry air ratio.

SOLUTION These substitutions for the air stream would be

$$\begin{aligned} \hat{h}_{\text{air},0} &= \frac{(H\_Air(P_0, T_0) - H\_Air(P_{\text{Ref}}, T_{\text{Ref}}))}{MW_{\text{air}}} \\ &= \frac{(H\_Air(14.696, T_0) - H\_Air(14.696, 536.67))}{28.965} \\ \hat{h}_{\text{air},1} &= \frac{(H\_Air(P_1, T_1) - H\_Air(P_{\text{Ref}}, T_{\text{Ref}}))}{MW_{\text{air}}} \\ &= \frac{(H\_Air(14.696, T_1) - H\_Air(14.696, 536.67))}{28.965} \\ \hat{h}_{\text{air},2} &= \frac{(H\_Air(P_2, T_2) - H\_Air(P_{\text{Ref}}, T_{\text{Ref}}))}{MW_{\text{air}}} \\ &= \frac{(H\_Air(243.7, T_2) - H\_Air(14.696, 536.67))}{28.965}. \end{aligned}$$

For the air stream, we will subtract the reference condition to emphasize this condition must match the reference condition for the fuel lower heating value (*LHV*).

The substitutions for the chilled water stream would be

$$\begin{aligned} \hat{h}_{\text{CW},a} &= \frac{H\_Water(P_{\text{CW},a}, T_{\text{CW},a})}{MW_{\text{water}}} = \frac{H\_Water(60.0, T_{\text{CW},a})}{18.01534} \\ \hat{h}_{\text{CW},b} &= \frac{H\_Water(P_{\text{CW},b}, T_{\text{CW},b})}{MW_{\text{water}}} = \frac{H\_Water(60.0, T_{\text{CW},b})}{18.01534}. \end{aligned}$$

For the chilled water, there is no need for the inclusion of a reference condition as this term will cancel out in the enthalpy balance.

For the combustion products, we first need to calculate the dry air ratio (DAR), which is defined as

$$\begin{aligned} DAR &= \frac{\left(\dfrac{\text{Mole air used}}{\text{Mole fuel used}}\right)}{\left(\dfrac{\text{Theoretical mole air}}{\text{Mole fuel}}\right)} = \frac{\left(\dfrac{\text{Mole air used}}{\text{Mole fuel used}}\right)}{(9.54654)} \\ &= \frac{\left(\dfrac{F_{\text{air}}/MW_{\text{air}}}{F_{\text{NG}}/MW_{\text{NG}}}\right)}{(9.54654)} = \frac{\left(\dfrac{F_{\text{air}}/28.965}{F_{\text{NG}}/16.043}\right)}{(9.54654)}, \end{aligned} \tag{11.10}$$

the actual air-to-fuel ratio used divided by the theoretical minimum air-to-fuel ratio for combustion. For methane combustion, the minimum air-to-fuel ratio for combustion is 9.54654. The enthalpy of the exhaust gas can be found as

$$\begin{aligned} \hat{h}_{\text{Prod},3} &= H\_Products\left(P_3, T_3, \frac{H}{C}, DAR\right) \\ &= H\_Products(243.7, T_3, 4, DAR). \end{aligned}$$

The hydrogen to carbon ratio (H/C) for methane is 4. All Product functions (Table 9.1) are on a per-pound basis and the reference state has been subtracted (during coding) from the product functions:

$$\hat{h}_{\text{Prod},4} = H\_Products\left(P_4, T_4, \frac{H}{C}, DAR\right)$$
$$= H\_Products(56.9, T_4, 4, DAR)$$
$$\hat{h}_{\text{Prod},5} = H\_Products\left(P_5, T_5, \frac{H}{C}, DAR\right)$$
$$= H\_Products(14.82, T_5, 4, DAR)$$
$$\hat{h}_{\text{Prod},6} = H\_Products\left(P_6, T_6, \frac{H}{C}, DAR\right)$$
$$= H\_Products(14.82, T_6, 4, DAR)$$
$$\hat{h}_{\text{Prod},7} = H\_Products\left(P_7, T_7, \frac{H}{C}, DAR\right)$$
$$= H\_Products(14.82, T_7, 4, DAR).$$

And finally, we determine the enthalpy for the water streams in the HRSG as

$$\hat{h}_{\text{Water,Econ}} = \frac{H\_Water(P_{\text{Water,Econ}}, T_{\text{Water,Econ}})}{MW_{\text{water}}} = \frac{H\_Water(239, T_{\text{Water,Econ}})}{18.01534}$$

$$\hat{h}_{\text{Water,Evap}} = \frac{H\_Water(P_{\text{Water,Evap}}, T_{\text{Water,Evap}})}{MW_{\text{water}}} = \frac{H\_Water(198, T_{\text{Water,Evap}})}{18.01534}$$

$$\hat{h}_{\text{Steam}} = \frac{H\_Steam(P_{\text{Steam}}, T_{\text{Steam}})}{MW_{\text{water}}} = \frac{H\_Steam(140.9, T_{\text{Steam}})}{18.01534}$$

The blowdown stream is saturated water at steam (P_{Steam}, T_{Steam}) conditions. In order to ensure water properties are used, we can set the blowdown pressure to the saturation pressure at the steam temperature:

$$\hat{h}_{\text{Blowdown}} = \frac{H\_Water(Psat\_Water(T_{\text{Steam}}), T_{\text{Steam}})}{MW_{\text{water}}} = \frac{H\_Water(Psat\_Water(T_{\text{Steam}}), T_{\text{Steam}})}{18.01534}.$$ ■

The data reconciliation problem, as developed in Chapter 6, is

$$\text{Minimize } f = \sum_{i,\text{measured variables}} \left(\frac{(x_i^+ - x_i)}{\sigma_i}\right)^2 \qquad (11.11)$$

$$\text{Subject to } h_k(x_i, u_j) = 0 \qquad k = 1, \ldots, K, \qquad (11.12)$$

where x_i^+ are the measured process variables, x_i are the reconciled values, h_k (x_i, u_j) are the material and energy balances, and u_j are estimates of nonmeasured process variables. For the cogeneration data reconciliation problem, the process variables (x_i) and system parameters are summarized in Table 11.2. There are eight material and energy balances for the cogeneration system: the two linear mass balances for the combustion chamber and HRSG, Equations (11.1) and (11.2), and six nonlinear energy balances for the air cooler, compressor/gas generating turbine, combustion chamber, power turbine, evaporator, and economizer, Equations (11.3)–(11.9) with Equation 11.4 set equal to Equation 11.6.

Table 11.2 Cogeneration Data Reconciliation—Process Variables and Process Parameters

Variables	P_{air}, P_{NG}, P_{Prod}, $F_{\text{Water,Econ}}$, F_{Steam}, F_{Blowdown}, F_{CW}, T_0, T_1, T_2, T_3, T_4, T_5, T_6, T_7, $T_{\text{CW},a}$, $T_{\text{CW},b}$, LHV, $T_{\text{Water,Econ}}$, $T_{\text{Water,Evap}}$, T_{Steam}, $\dot{W}_{\text{Net}}$, $\dot{Q}_{\text{CC,Loss}}$
Parameters	$\dot{Q}_{\text{Air Cooler}}$, $\dot{W}_{\text{GG Turbine}}$, $\dot{Q}_{\text{Evaporator}}$, $U_{\text{Evaporator}}$, $\dot{Q}_{\text{Economizer}}$, $U_{\text{Economizer}}$, Heat Rate, $\eta_{\text{Compressor}}$, $\eta_{\text{GGTurbine}}$, $\eta_{\text{Power Turbine}}$

EXAMPLE 11.3 *Data Reconciliation for the Cogeneration System*

Solve the cogeneration data reconciliation problem using data from Table 11.1, combined with the material and energy balances developed in Example 11.1 and the enthalpy functions developed in Example 11.2. When using Excel Solver for this data reconciliation problem, it will be important to utilize the automatic scaling feature: Tools → Solver → Options → Use Automatic Scaling. Ideally, optimization variables should fall within three orders of magnitude of each other and in the range 0.1–10 (Ravindran et al., 2006, p. 555). If variables fall outside this range, scaling (either manual or here automatic) can be helpful in obtaining a good solution. For the cogeneration data reconciliation problem, variables span the range from ~0.28 lb/s for the blowdown flow rate to ~2400 R for the combustion chamber temperature.

SOLUTION The solution to the data reconciliation problem is provided in **Example 11.3.xls** and shown in Figure 11.2.

From Figure 11.2, the material and energy balances close (column N) with the reconciled values. We test for the presence of gross errors using the global test (*GT*) method (Eq. (6.34)):

$$GT = \sum_{i,\text{measured variables}} \left(\frac{(x_i^+ - x_i)}{\sigma_i}\right)^2 < \chi^2_{(1-\alpha)}(\nu).$$

Here $GT = 2.753$ and $\chi^2_{(1-\alpha)}(\nu) = 15.507$ so no gross errors are detected; in the chi-square distribution, we are using $\alpha = 0.05$ and $\nu = 8$ (see Chapter 6, Section 6.6.1). ■

EXAMPLE 11.4 *Updating Process Parameters*

Using the reconciled values provided in Figure 11.2, update the process parameters $\dot{Q}_{\text{Air Cooler}}$, $\dot{W}_{\text{GG Turbine}}$, $\dot{Q}_{\text{Evaporator}}$, $U_{\text{Evaporator}}$, $\dot{Q}_{\text{Economizer}}$, and $U_{\text{Economizer}}$. Data from the manufacturer show $A_{\text{Evaporator}} = 56{,}248$ ft^2 and $A_{\text{Economizer}} = 10{,}565$ ft^2. Also calculate the apparent pinch and approach temperatures, Heat Rate, $\eta_{\text{Compressor}}$, $\eta_{\text{GGTurbine}}$, and $\eta_{\text{Fower Turbine}}$.

SOLUTION For the HRSG, air cooler, or any heat exchanger, $\dot{Q}$ transferred can be given by $\dot{Q} = F\Delta\hat{h}$.

Air Cooler

From Equation (11.3), $\dot{Q}_{\text{Air Cooler}} = F_{\text{air}}\left(\hat{h}_{\text{air},0} - \hat{h}_{\text{air},1}\right) = 152.7237$ $(2.541 - (-4.222)) =$ 1032.82 Btu/s.

	A	B	C	D	E	F	G	H	I	J	K	L	M	N
60	**Variables**	**Measured**	**Reconciled**	**St. Dev**	**OBJ**		$h_{Reconciled}$			**Material Balances (lb/s)**				
61	F_{Air}	145	152.7237	20	0.149137							IN	OUT	DIFF
62	F_{NG}	2.6208	2.581389	0.07	0.316977					Combustion Chamber		155.3050	155.3050	0
63	F_{Prod}	147.6208	155.305	20	0.147619					HRSG		24.5659	24.5659	0
64	$F_{Water, Econ}$	24.4444	24.56587	1.2222	0.009878									
65	F_{Steam}	24.17	24.28586	0.725	0.025538									
66	$F_{Blowdown}$	0.28	0.280013	0.1	1.65E-08					**Energy Balances (Btu/s)**				
67	F_{CW}	200	193.9783	44	0.01873							IN	OUT	DIFF
68	T_0	547.17	547.2477	2	0.001509		2.541	$h_{Air,0}$		Air Cooler		3421.3824	3421.382	-2.9E-05
69	T_1	519.94	519.0908	5	0.028846		-4.222	$h_{Air,1}$		Compressor/GGT		27910.108	27910.11	4.9E-09
70	T_2	1260.48	1260.165	10	0.00099		178.527	$h_{Air,2}$		Combustion		82656.543	82656.54	-3.7E-09
71	T_3	2400	2446.649	150	0.096715		523.564	$h_{Prod,3}$		Power Turbine		53401.993	53401.99	2.2E-08
72	T_4	1836.34	1833.771	30	0.007332		343.852	$h_{Prod,4}$		Evap		42522.272	42522.27	1.3E-08
73	T_5	1386.67	1412.059	60	0.179049		226.249	$h_{Prod,5}$		Eco		17553.348	17553.35	-2.7E-09
74	T_6	914.67	877.5671	50	0.55065		85.732	$h_{Prod,6}$						
75	T_7	787.67	797.6416	20	0.248581		65.475	$h_{Prod,7}$						
76	T_{CWa}	504.57	507.1745	5	0.271345		15.637	$h_{CW,a}$						
77	T_{CWb}	515.07	512.4828	5	0.267752		20.962	$h_{CW,b}$						
78	LHV	21501	21457.9	200	0.046441									
79	$T_{Water, Econ}$	677.67	663.6964	30	0.216959		172.547	$h_{Water, Econ}$						
80	$T_{Water, Evap}$	780.67	789.3733	30	0.084165		300.609	$h_{Water, Evap}$						
81	T_{Steam}	821.67	821.6748	1	2.29E-05		334.390	$h_{Water,BlowD}$						
82	W_{Net}	18264.44	18264.44	0.01	0.036918		1198.807	h_{Steam}						
83	$Q_{CC,loss}$	1125	1344.441	1000	0.048154					Chiinv (0.05,8)				
84					2.753307	ΣOBJ				15.50731				

Figure 11.2 Data reconciliation solution using Excel Solver.

Evaporator

From Equation (11.8),

$$\dot{Q}_{\text{Evaporator}} = F_{\text{Steam}}\left(\hat{h}_{\text{Steam}} - \hat{h}_{\text{Water,Evap}}\right) + F_{\text{Water,BlowD}}\left(\hat{h}_{\text{Water,BlowD}} - \hat{h}_{\text{Water,Evap}}\right)$$
$$= 24.2859(1198.807 - 300.609) + 0.28(334.39 - 300.609)$$
$$= 21{,}822.98 \text{ Btu/s.}$$

Economizer

From Equation (11.9),

$$\dot{Q}_{\text{Economizer}} = F_{\text{Water,Econ}}\left(h_{\text{Water,Evap}} - h_{\text{Water,Econ}}\right)$$
$$= 24.5659(300.609 - 172.547) = 3145.95 \text{ Btu/s.}$$

From our discussions in the previous chapter, $\dot{Q}$ transferred can be also given by

$$\dot{Q} = UA\Delta T_{\text{LMTD}}, \tag{11.13}$$

where U is the overall heat transfer coefficient $\frac{\text{Btu}}{\text{h-ft}^2\text{-R}}$; A is the area available for heat transfer; and ΔT_{LMTD} is the log mean temperature difference (LMTD). In the HRSG, the LMTD (Chapter 7) represents a temperature driving force at each end of the HRSG heat exchanger sections (evaporator section and economizer section) given by

$$\Delta T_{\text{LMTD,Evaporator}} = \left(\frac{(T_5 - T_{\text{Steam}}) - (T_6 - T_{\text{Water,Evap}})}{ln\left(\dfrac{(T_5 - T_{\text{Steam}})}{(T_6 - T_{\text{Water,Evap}})}\right)}\right) \tag{11.14}$$

and

$$\Delta T_{\text{LMTD,Economizer}} = \left(\frac{(T_7 - T_{\text{Water,Econ}}) - (T_6 - T_{\text{Water,Evap}})}{\ln\left(\dfrac{(T_7 - T_{\text{Water,Econ}})}{(T_6 - T_{\text{Water,Evap}})}\right)}\right). \tag{11.15}$$

Using the provided $A_{\text{Evaporator}}$ and $A_{\text{Economizer}}$ values, $U_{\text{Evaporator}} = 5.29$ Btu/h-ft²-R and $U_{\text{Economizer}} = 9.79$ Btu/h-ft²-R. We can determine the pinch and approach temperature (Chapter 7) as

$$\Delta T_{\text{Pinch}} = T_6 - T_{\text{Steam}} = 55.89 \text{ R} \tag{11.16}$$

and

$$\Delta T_{\text{Approach}} = T_{\text{Steam}} - T_{\text{Water,Evap}} = 32.13 \text{ R}, \tag{11.17}$$

and the heat rate (energy rate in from natural gas/net electric power generated) as

$$\text{Heat Rate} = \frac{(F_{\text{NG}})(LHV)\left(3600\dfrac{\text{s}}{\text{h}}\right)}{(\dot{W}_{\text{Net}})\left(3600\dfrac{\text{s}}{\text{h}}\right)\left(0.293\,07\dfrac{\text{W-h}}{\text{Btu}}\right)\left(\dfrac{1\text{ kW}}{1000\text{ W}}\right)} \tag{11.18}$$
$$= 10{,}348.16\frac{\text{Btu}}{\text{kW-h}}.$$

In Chapter 9, we developed the efficiency equations for the gas turbine components using our physical property programs. For the compressor, we determined the entropy (British thermal unit per pound-Rankine) at the compressor inlet. We then assume isentropic compression from the inlet to the outlet. With known entropy and known P_2, the isentropic temperature at the compressor outlet is determined. This allows the isentropic enthalpy to be calculated. Finally, the efficiency of the compression process is determined as

$$\frac{\hat{h}_{2,\text{isen}} - \hat{h}_{\text{air},1}}{\hat{h}_{\text{air},2} - \hat{h}_{\text{air},1}}.$$

For the gas generating turbine, we determine the entropy (British thermal unit per pound-Rankine) at the turbine inlet. We then assume isentropic expansion from the inlet to the outlet. With known entropy and known P_4, the isentropic temperature at the turbine outlet is determined. This allows the isentropic enthalpy to be calculated. Finally, the efficiency of the expansion process is determined as

$$\frac{\hat{h}_{\text{Prod},4} - \hat{h}_{\text{Prod},3}}{\hat{h}_{4,\text{isen}} - \hat{h}_{\text{Prod},3}}.$$

The same procedure is used for the power turbine. The calculations are shown next.

Air Compressor

$$\hat{s}_{\text{air},1} = \frac{S\_Air(P_1, T_1)}{MW_{\text{air}}} = 1.07318\frac{\text{Btu}}{\text{lb-R}} \tag{11.19}$$

$$\hat{s}_{2,\text{isen}} = \hat{s}_{\text{air},1} \tag{11.20}$$

$$T_{2,\text{isen}} = T\_SP\_Air(\hat{s}_{2,\text{isen}} \times MW_{\text{air}}, P_2) = 1144.15\ \text{R} \tag{11.21}$$

$$\hat{h}_{2,\text{isen}} = \frac{(H\_Air(P_2, T_{2,\text{isen}}) - H\_Air(P_{\text{Ref}}, T_{\text{Ref}}))}{MW_{\text{air}}} = 148.88\frac{\text{Btu}}{\text{lb}} \tag{11.22}$$

$$\eta_{\text{Comp}} = \frac{\hat{h}_{2,\text{isen}} - \hat{h}_{\text{Air},1}}{\hat{h}_{\text{air},2} - \hat{h}_{\text{air},1}} = 83.78\% \tag{11.23}$$

Gas Generating Turbine

$$\hat{s}_{\text{Prod},3} = S\_Products(P_3, T_3, \frac{\text{H}}{\text{C}}, DAR) = 0.21008\frac{\text{Btu}}{\text{lb-R}} \tag{11.24}$$

$$\hat{s}_{4,\text{isen}} = \hat{s}_{\text{Prod},3} \tag{11.25}$$

$$T_{4,\text{isen}} = TfromS\_Products\left(P_4, \hat{s}_{4,\text{isen}}, \frac{\text{H}}{\text{C}}, DAR\right) = 1727.9\ \text{R} \tag{11.26}$$

$$\hat{h}_{4,\text{isen}} = H\_Products\left(P_4, T_{4,\text{isen}}, \frac{\text{H}}{\text{C}}, DAR\right) = 313.83\frac{\text{Btu}}{\text{lb}} \tag{11.27}$$

$$\eta_{\text{GG Turbine}} = \frac{\hat{h}_{\text{Prod},4} - \hat{h}_{\text{Prod},3}}{\hat{h}_{4,\text{isen}} - \hat{h}_{\text{Prod},3}} = 85.69\% \tag{11.28}$$

Power Turbine

$$\hat{s}_{\text{Prod},4} = S\_Products(P_4, T_4, \frac{\text{H}}{\text{C}}, DAR) = 0.22694\frac{\text{Btu}}{\text{lb-R}} \tag{11.29}$$

$$\hat{s}_{5,\text{isen}} = \hat{s}_{\text{Prod},4} \tag{11.30}$$

$$T_{5,\text{isen}} = TfromS\_Products\left(P_5, \hat{s}_{5,\text{isen}}, \frac{\text{H}}{\text{C}}, DAR\right) = 1307.72\ \text{R} \tag{11.31}$$

$$\hat{h}_{5,\text{isen}} = H\_Products\left(P_5, T_{5,\text{isen}}, \frac{\text{H}}{\text{C}}, DAR\right) = 198.05\frac{\text{Btu}}{\text{lb}} \tag{11.32}$$

$$\eta_{\text{Power Turbine}} = \frac{\hat{h}_{\text{Prod},5} - \hat{h}_{\text{Prod},4}}{\hat{h}_{5,\text{isen}} - \hat{h}_{\text{Prod},4}} = 80.66\% \tag{11.33}$$

Parameter calculations are provided on the Excel sheet in **Example 11.3.xls** and they will be automatically updated with any change on the sheet. ■

11.2 COGENERATION SYSTEM GROSS ERROR DETECTION AND IDENTIFICATION

For systems with a large number of nonlinear balances, it becomes increasingly difficult to apply and interpret results from gross error detection and to suspect measurement identification.

EXAMPLE 11.5 *Data Reconciliation for the Cogeneration System*

In Example 11.3, we solved the cogeneration data reconciliation problem using the measured data provided in Figure 11.3 (column B). Now we want to change the measured data from Example 11.3 so that F_{NG}, the measured flow rate of natural gas, is given a positive bias of 0.4 lb/s. Again use the automatic scaling option in Excel Solver.

SOLUTION The data reconciliation solution is also shown in Figure 11.3 and provided in **Example 11.5.xls**. We test for the presence of gross errors using the global test (*GT*) method (Eq. (6.34)):

$$GT = \sum_{\substack{i,\text{measured} \\ \text{variables}}} \left(\frac{(x_i^+ - x_i)}{\sigma_i}\right)^2 < \chi^2_{(1-\alpha)}(\nu).$$

Here, $GT = 22.009$ and $\chi^2_{(1-\alpha)}(\nu) = 15.507$, so gross errors are detected.

Using the procedure of Section 6.6.3, we linearize the six nonlinear energy balances at the data reconciliation solution (column C) of the provided Excel solution file. Linearization of the energy balances on the air cooler, compressor/turbine, evaporator, and economizer is straightforward. These balances only involve $F\hat{h}$ terms and here we will assume that $\hat{h} = f(T)$ and that P is known. The remaining two balances, in addition to $F\hat{h}$ terms, involve $\dot{Q}_{\text{CC,Loss}}$ for the combustion chamber and $\dot{W}_{\text{Net}}$ for the power turbine. In order to linearize these two balances, we must address how to handle these terms ($\dot{Q}_{\text{CC,Loss}}$ and $\dot{W}_{\text{Net}}$). If we take a step back from the problem, we can actually consider $\dot{Q}_{\text{CC,Loss}}$ to be an energy leak from the combustion process. Similarly, we can view $\dot{W}_{\text{Net}}$ as a leak from the power turbine. Values for these "leaks" can be taken as known and fixed at the data reconciliation solution. This allows linearization of the combustion chamber and power turbine balances involving only $F\hat{h}$ terms with known and constant values for $\dot{Q}_{\text{CC,Loss}}$ and $\dot{W}_{\text{Net}}$.

In order to linearize the energy balances, the data reconciliation solution is supplied to our Newton–Raphson code as a starting

	A	B	C	D	E	F	G	H
60	**Variables**	**Measured**	**Reconciled**	**St. Dev**	**OBJ**	**MT Error**	$h_{Reconciled}$	
61	F_{Air}	145	163.40328	20	0.846701647	0.95		
62	F_{NG}	3.0208	2.824449	0.07	7.868106285	4.14		
63	F_{Prod}	147.6208	166.22773	20	0.865544375	0.96		
64	$F_{Water, Econ}$	24.4444	25.265102	1.2222	0.450906654	0.76		
65	F_{Steam}	24.17	24.984429	0.725	1.261914706	1.86		
66	$F_{Blowdown}$	0.28	0.2806736	0.1	4.5377E-05	0.10		
67	F_{CW}	200	199.30044	44	0.000252778	0.09		
68	T_0	547.17	547.25088	2	0.001635355	0.75	2.542	$h_{Air,0}$
69	T_1	519.94	517.37196	5	0.26379318	3.17	-4.634	$h_{Air,1}$
70	T_2	1260.48	1260.2859	10	0.00037659	0.35	178.558	$h_{Air,2}$
71	T_3	2400	2436.5984	150	0.0595307	0.25	521.028	$h_{Prod,3}$
72	T_4	1836.34	1822.5512	30	0.211255338	0.96	340.948	$h_{Prod,4}$
73	T_5	1386.67	1429.0617	60	0.499182057	0.75	231.072	$h_{Prod,5}$
74	T_6	914.67	910.83319	50	0.005888456	0.09	94.288	$h_{Prod,6}$
75	T_7	787.67	822.33253	20	3.003727505	3.40	71.777	$h_{Prod,7}$
76	T_{CWa}	504.57	506.84308	5	0.206675565	0.66	15.305	$h_{CW,a}$
77	T_{CWb}	515.07	512.70879	5	0.223013224	0.69	21.188	$h_{CW,b}$
78	LHV	21501	21258.17	200	1.474163347	4.71		
79	$T_{Water, Econ}$	677.67	632.39662	30	2.277420842	3.28	141.162	$h_{Water, Econ}$
80	$T_{Water, Evap}$	780.67	778.44941	30	0.005478928	0.23	289.270	$h_{Water, Evap}$
81	T_{Steam}	821.67	821.69665	1	0.000710286	2.96	334.413	$h_{Water,BlowD}$
82	W_{Net}	18264.44	18264.435	0.01	0.276566115		1198.820	h_{Steam}
83	$Q_{CC,loss}$	1125	2610.3969	1000	2.206404001			
84					22.00929331	ΣOBJ		

	I	J	K	L	M	N
60		**Material Balances (lb/s)**				
61				IN	OUT	DIFF
62		Combustion Chamber		166.2277	166.2277	-2.7E-11
63		HRSG		25.2651	25.2651	-4.6E-14
66		**Energy Balances (Btu/s)**				
67				IN	OUT	DIFF
68		Air Cooler		3465.58052	3465.5804	0.00014
69		Compressor/GGT		29934.3127	29934.313	-2.1E-06
70		Combustion		89219.6385	89219.638	2.3E-05
71		Power Turbine		56674.9288	56674.929	-1.7E-05
72		Evap		45718.9373	45718.937	2.9E-05
73		Eco		19239.7224	19239.722	-3E-05
83		Chiinv (0.05,8)				
84		15.5073131				

Figure 11.3 Data reconciliation solution using Excel Solver.

point and the nonlinear energy balances are supplied to the Visual Basic for Applications (VBA) code in `Function NLFunc`. Within **Example 11.5.xls**, the Gauss–Jordan and Newton–Raphson macros are each run one time.

> The Newton–Raphson macro will give an error as it tries to generate a new trial point. This is not of concern. The nonlinear equations will have been linearized at data reconciliation solution and we can obtain the linearized coefficients from matrix 3 (see Figure 4.11).

The material and linearization of the energy balances gives

Combustion Chamber

$$F_{\text{air}} + F_{\text{NG}} - F_{\text{Prod}} = 0 \tag{11.34}$$

HRSG

$$F_{\text{Water,Econ}} - F_{\text{Steam}} - F_{\text{Blowdown}} = 0 \tag{11.35}$$

Air Cooler

$$\begin{aligned} &7.176F_{\text{air}} - 5.884F_{\text{CW}} + 39.256T_0 - 39.236T_1 \\ &+ 199.993T_{CW,a} - 199.817T_{CW,b} = 100.097 \end{aligned} \tag{11.36}$$

Compressor/Gas Generating Turbine

$$\begin{aligned} &183.193F_{\text{air}} - 179.692F_{\text{Prod}} - 39.236T_1 + 42.036T_2 \\ &- 49.657T_3 + 47.297T_4 = -2114.033 \end{aligned} \tag{11.37}$$

Combustion Chamber

$$\begin{aligned} &178.558F_{\text{air}} + 21258.170F_{\text{NG}} - 519.999F_{\text{Prod}} + 42.036T_2 \\ &- 49.657T_3 + 2.824LHV = -5363.239 \end{aligned} \tag{11.38}$$

Power Turbine

$$109.658F_{\text{Prod}} + 47.297T_4 - 45.261T_5 = 3256.582 \tag{11.39}$$

Evaporator

$$\begin{aligned} &136.531F_{\text{Prod}} - 909.550F_{\text{Steam}} - 45.143F_{\text{Blowdown}} + 45.261T_5 \\ &- 42.400T_6 + 26.168T_{\text{Water,Evap}} - 15.244T_{\text{Steam}} = 33905.872 \end{aligned} \tag{11.40}$$

Economizer

$$\begin{aligned} &22.472F_{\text{Prod}} - 148.108F_{\text{Water,Econ}} + 42.400T_6 - 42.028T_7 \\ &+ 25.286T_{\text{Water,Econ}} - 26.168T_{\text{Water,Evap}} = -320.557 \end{aligned} \tag{11.41}$$

We next determine V and the measurement test (MT), MT_i, using Equations (6.36), (6.35), and (6.39). The variable order is important and for this problem, we used

$$\left\{ \begin{array}{l} F_{\text{air}}, F_{\text{NG}}, F_{\text{Prod}}, F_{\text{Water,Econ}}, F_{\text{Steam}}, F_{\text{Blowdown}}, F_{\text{CW}}, \\ T_0, T_1, T_2, T_3, T_4, T_5, T_6, T_7, \\ T_{CWa}, T_{CWb}, LHV, T_{\text{Water,Econ}}, T_{\text{Water,Evap}}, T_{\text{Steam}} \end{array} \right\},$$

giving V (which is a 21×21 matrix), which is shown in Figure 11.4, and

373.76481	-0.162232	-26.397418	-0.52799025	-0.52926015	0.001269896	-0.70259486	0.00968549	-1.5421679	-0.3838434	102.021547
-0.1622317	0.0022521	-0.1648796	-0.009813706	-0.00983873	2.50253E-05	-0.010808608	0.000149	0.0204207	-0.0005577	0.14823458
-26.397418	-0.16488	373.4377	-0.537803956	-0.53909888	0.001294922	-0.713403468	0.00983449	-1.5217472	-0.3844011	102.169781
-0.5279902	-0.009814	-0.537804	1.155322734	-0.33095203	-0.007498075	-0.000699936	9.6488E-06	-0.093324	0.0040237	-1.0694684
-0.5292601	-0.009839	-0.5390989	-0.330952031	0.1922209	0.002452071	-0.000698658	9.6312E-06	-0.0935622	0.004036	-1.07272
0.0012699	2.503E-05	0.0012949	-0.007498075	0.00245207	4.98536E-05	-1.27763E-06	1.7613E-08	0.00023818	-1.223E-05	0.00325156
-0.7025949	-0.010809	-0.7134035	-0.000699936	-0.00069866	-1.27763E-06	61.49357362	-0.8477078	5.19169896	0.0013531	-0.3596268
0.0096855	0.000149	0.0098345	9.64883E-06	9.6312E-06	1.76125E-08	-0.847707766	0.01168591	-0.0715692	-1.865E-05	0.00495757
-1.5421679	0.0204207	-1.5217472	-0.093324015	-0.0935622	0.000238181	5.191698963	-0.0715692	0.65705432	-0.0764745	20.3261277
-0.3838434	-0.000558	-0.3844011	0.004023742	0.00403598	-1.22336E-05	0.001353051	-1.865E-05	-0.0764745	0.3085746	-82.015897
102.02155	0.1482346	102.16978	-1.069468399	-1.07271996	0.003251557	-0.359626761	0.00495757	20.3261277	-82.015897	21798.9653
107.49876	0.1750155	107.67378	-1.202014622	-1.20563792	0.003623298	3.939018062	-0.0543006	1.40585934	2.4577426	-653.2422
48.379903	-0.216578	48.163324	-2.559075692	-2.56599932	0.006923625	2.387812189	-0.0329167	-2.2177551	1.6369957	-435.09628
-19.34514	-0.484514	-19.829655	2.324428394	2.3512017	-0.026773311	0.234558912	-0.0032335	-4.6252306	0.372023	-98.879799
-18.984268	-0.335861	-19.320128	1.968479944	1.97046304	-0.001983096	-0.060565243	0.00083491	-3.1914775	0.1141221	-30.332456
0.3083956	0.0047443	0.3131399	0.000307228	0.00030667	5.608E-07	-26.9918659	0.37209115	-2.2788339	-0.0005939	0.15785385
-0.3081251	-0.00474	-0.3128652	-0.000306959	-0.0003064	-5.60308E-07	26.96819177	-0.3717648	2.2768352	0.0005934	-0.1577154
-180.97052	2.444218	-178.52631	-10.63373115	-10.6608482	0.027117	-11.91672081	0.16427565	22.1289726	-0.5282213	140.395676
25.699312	0.4546601	26.153972	-2.664763298	-2.66744785	0.002684549	0.081988154	-0.0011302	4.32035494	-0.1544889	41.0615388
4.2980361	0.1076477	4.4056838	-0.516433428	-0.52238183	0.005948401	-0.052113485	0.0007184	1.02761767	-0.0826548	21.9687703
-0.0199962	-0.000374	-0.0203704	0.002119214	0.00212486	-5.64839E-06	-2.1187E-05	2.9207E-07	-0.0035591	0.000157	-0.041724

107.49876	48.3799	-19.34514	-18.98426768	0.308395579	-0.308125	-180.97052	25.69931174	4.29803613	-0.019996194
0.1750155	-0.21658	-0.484514	-0.335860721	0.004744309	-0.00474	2.444218	0.454660117	0.10764768	-0.000374223
107.67378	48.1633	-19.82965	-19.3201284	0.313139888	-0.312865	-178.52631	26.15397186	4.40568381	-0.020370417
-1.202015	-2.55908	2.324428	1.968479944	0.000307228	-0.000307	-10.633731	-2.664763298	-0.51643343	0.002119214
-1.205638	-2.566	2.351202	1.97046304	0.000306668	-0.000306	-10.660848	-2.667447847	-0.52238183	0.002124862
0.0036233	0.00692	-0.026773	-0.001983096	5.608E-07	-5.6E-07	0.027117	0.002684549	0.0059484	-5.64839E-06
3.9390181	2.38781	0.234559	-0.060565243	-26.9918659	26.968192	-11.916721	0.081988154	-0.05211348	-2.1187E-05
-0.054301	-0.03292	-0.003233	0.00083491	0.372091147	-0.371765	0.1642756	-0.001130232	0.0007184	2.92069E-07
1.4058593	-2.21776	-4.625231	-3.191477478	-2.278833932	2.2768352	22.128973	4.320354939	1.02761767	-0.003559054
2.4577426	1.637	0.372023	0.114122095	-0.000593905	0.0005934	-0.5282213	-0.154488936	-0.08265477	0.000156981
-653.2422	-435.096	-98.8798	-30.33245606	0.157853849	-0.157715	140.39568	41.06153883	21.9687703	-0.04172397
205.85797	-464.501	-108.3889	-34.46528132	-1.728984689	1.7274682	188.90931	46.65621155	24.0814804	-0.046838946
-464.5012	3231.29	-161.3083	-82.82435886	-1.048101494	1.0471822	-235.12224	112.1206809	35.8389225	-0.098299335
-108.3889	-161.308	1946.589	-169.6545389	-0.102956819	0.1028665	-525.04986	229.6641071	-432.486418	0.126097362
-34.46528	-82.8244	-169.6545	103.6699734	0.026584386	-0.026561	-363.91816	-140.3397281	37.6932563	0.06960648
-1.728985	-1.0481	-0.102957	0.026584386	11.84775549	-11.83736	5.2307015	-0.035987716	0.02287459	9.29978E-06
1.7274682	1.04718	0.102867	-0.02656107	-11.83736402	11.826982	-5.2261138	0.035956152	-0.02285453	-9.29163E-06
188.90931	-235.122	-525.0499	-363.9181621	5.230701537	-5.226114	2652.7558	492.6419314	116.653755	-0.405493337
46.656212	112.121	229.6641	-140.3397281	-0.035987716	0.0359562	492.64193	189.9801712	-51.0259737	-0.094227424
24.08148	35.8389	-432.4864	37.69325627	0.022874588	-0.022855	116.65375	-51.02597371	96.0883304	-0.028015874
-0.046839	-0.0983	0.126097	0.06960648	9.29978E-06	-9.29E-06	-0.4054933	-0.094227424	-0.02801587	8.12503E-05

Figure 11.4 *V* (21 × 21 matrix) Excel screen shot.

$$MT_i = \frac{|a_i|}{\sqrt{V_{ii}}} \quad \text{(Eq. (6.35))},$$

gives

$$\left\{\begin{array}{l} \frac{|-18.403|}{\sqrt{373.765}}, \frac{|0.196|}{\sqrt{0.002}}, \frac{|-18.607|}{\sqrt{373.438}}, \frac{|-0.821|}{\sqrt{1.155}}, \frac{|-0.814|}{\sqrt{0.192}}, \\ \frac{|-6.736\text{E-}04|}{\sqrt{4.985\text{E}-05}}, \frac{|0.70|}{\sqrt{61.494}}, \frac{|-0.081|}{\sqrt{0.012}}, \frac{|2.568|}{\sqrt{0.657}}, \frac{|0.194|}{\sqrt{0.309}}, \\ \frac{|-36.598|}{\sqrt{21799}}, \frac{|13.789|}{\sqrt{205.858}}, \frac{|-42.392|}{\sqrt{3231}}, \frac{|3.837|}{\sqrt{1946.6}}, \frac{|-34.662|}{\sqrt{103.67}}, \\ \frac{|-2.273|}{\sqrt{11.848}}, \frac{|2.361|}{\sqrt{11.827}}, \frac{|242.830|}{\sqrt{2652.756}}, \frac{|45.273|}{\sqrt{189.980}}, \\ \frac{|2.22|}{\sqrt{96.088}}, \frac{|-0.026|}{\sqrt{8.125\text{E}-05}} \end{array}\right\}$$

$$= \left\{\begin{array}{l} 0.95, 4.14, 0.96, 0.76, 1.86, 0.10, 0.09, 0.75, 3.17, 0.35, 0.25, \\ 0.96, 0.75, 0.09, 3.40, 0.66, 0.69, 4.71, 3.28, 0.23, 2.96 \end{array}\right\}. \tag{11.42}$$

All the matrix calculations to solve for V and MT_i can be found in the file **Example 11.5.xls**. The suspect measurement order is LHV first with $MT_{LHV} = 4.71$ followed by F_{NG} with $MT_{F_{NG}} = 4.14$ and next, T_7 with $MT_{T_7} = 3.40$. Recall at the start of this problem that the bias was placed with F_{NG} and that F_{NG} was expected to show the largest MT_i. However, the most likely suspect measurement appears as the fuel LHV. This shows the difficulties with error identification in larger nonlinear problems. Here, the extra energy input to the system by the bias in F_{NG} can also appear as increased energy availability in the fuel LHV. This occurs even with redundant equations involving F_{NG}, in Equations (11.34) and (11.38). An additional difficulty is that the fuel LHV is generally not measured. Typically, the natural gas supplier takes hourly samples and determines the fuel LHVs. These grab sample values are then averaged over a 1-month period, and customers are billed based on the natural gas used and the average fuel energy content. Generally, changes in fuel energy content are gradual, but this value can change depending on components that are removed during upstream processing. The third suspect measurement (T_7), shows that the extra energy input to the system by the bias in F_{NG}

can also appear as an increased energy loss from the stack with an increase in T_7.

A final consideration is system sensitivity to instrument bias. In this gross error example, if the bias in F_{NG} is 0.3 (as opposed to 0.4), the objective function value for the data reconciliation problem becomes 14.813. This value for the global test falls below $\chi^2_{(1-\alpha)}(v) = 15.507$, so gross errors are not expected. ■

11.3 VISUAL DISPLAY OF RESULTS

EXAMPLE 11.6 ***Visual Display of the Data Reconciliation Results***

A final consideration is presentation of the data reconciliation results (here also see Problem 11.1). A convenient means of presenting the reconciled data is directly on a process flow diagram. However, the actual drawing capabilities in Excel are limited. A detailed process flow diagram can be created and pasted in the Excel sheet. A problem then is transferring the data to the picture and we show how that is accomplished using a VBA code with the results from Example 11.3 presented as in Figure 11.5. The measured and reconciled values are moved to the picture by running the macro `Write_Data_Recon_Values` found in **Example 11.3.xls**.

A representative section of the macro `Write_Data_Recon_Values()`, which is found in module 1 of the Excel file, is provided as in Figure 11.6. Here, a Visio-generated diagram of a cogeneration system (picture 8) has been placed on the Excel sheet. Excel will sequentially number any diagram or picture placed on the sheet. This numbering also includes any data in cells on the Excel sheet, which may be pasted onto the picture. The picture number can be found in the Name Box on the Excel sheet (upper left corner).

In the code, we first clear any existing pictures except the cogeneration system (here picture 8). We then select a cell and copy its value. The cogeneration system (picture 8) is made active and the copied value from the cell is pasted onto the active picture. In order to place the data, a cell width and a cell height for the cells in the Excel sheet are defined. Picture placement is adjusted by using multiples of the cell width and cell height. For example, the measured value for the air flow rate is placed using, `2 * cell_height` and `2 * cell_width`.

11.4 CLOSING COMMENTS

We examined steady-state data reconciliation and gross error detection and identification in a cogeneration system. This is an industrial application of the data reconciliation materials we developed in Chapter 6. In Example 11.5, we combined data reconciliation, gross error detection, and suspect measurement identification in a single Excel sheet. For large industrial problems, it will be time-consuming to create this sheet, but once the sheet is completed, a single macro can be developed to:

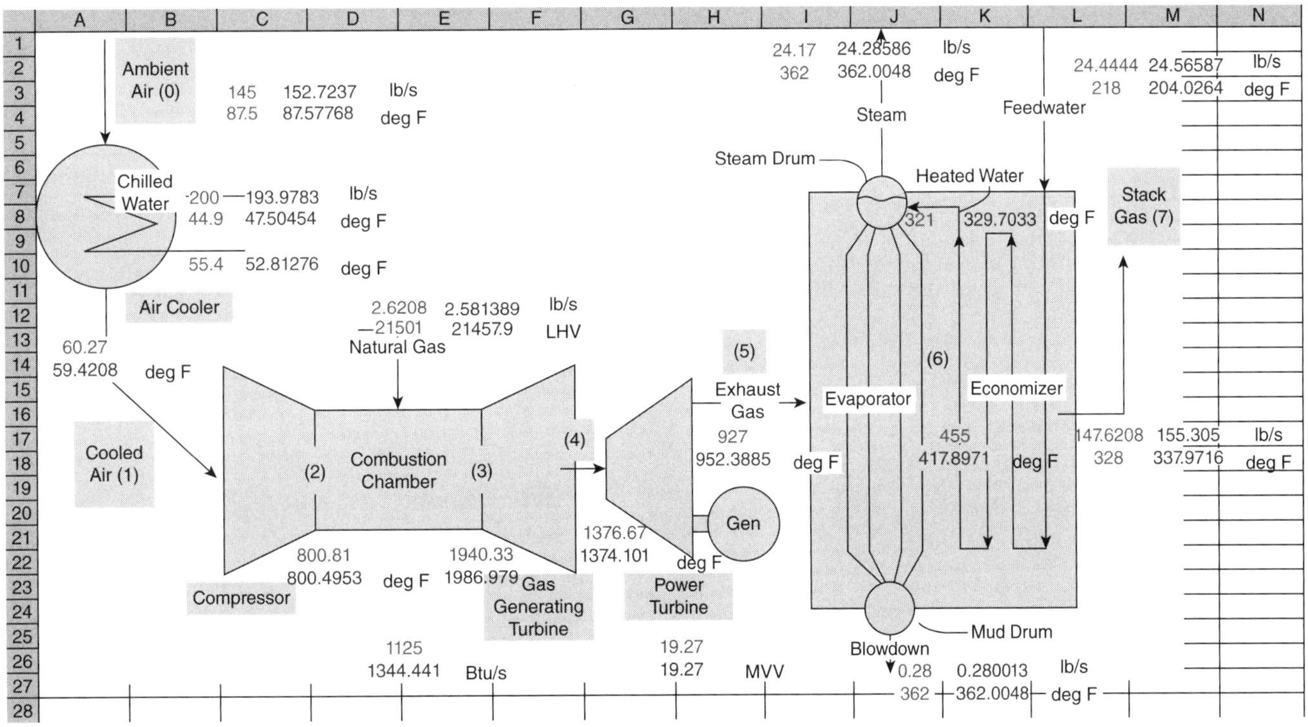

Figure 11.5 Data reconciliation solution transferred to a process flow diagram.

```
Public Sub Write_Data_Recon_Values()
'
' First clear the figure of previous calculations
' But here do not use the following command as it will clear the sheet of ALL
pictures
' ActiveSheet.Pictures.Delete
'
' Loop and eliminate all pictures EXCEPT "Picture 8" - the cogen Picture Name
' You can see the cogen Picture Name in the name box
 Dim myPict As Picture
 For Each myPict In ActiveSheet.Pictures
 If myPict.Name <> "Picture 8" Then
 myPict.Delete
 End If
 Next myPict

' We need the height and width of the sheet cells
' Set cell width and height
    cell_width = 42#
    cell_height = 12#
' AF - Air Flow - Measured
    Range("B61").Select
    Selection.Copy
    ActiveSheet.Shapes("Picture 8").Select
    ActiveSheet.Paste
    column_number = 2
    row_number = 2
    column_distance = cell_width * column_number
    row_distance = cell_height * row_number
    Selection.ShapeRange.IncrementLeft column_distance
    Selection.ShapeRange.IncrementTop row_distance
' AF - Air Flow - Reconciled
    Range("C61").Select
    Selection.Copy
    ActiveSheet.Shapes("Picture 8").Select
    ActiveSheet.Paste
    column_distance = cell_width * (column_number + 1)
    row_distance = cell_height * (row_number)
    Selection.ShapeRange.IncrementLeft column_distance
    Selection.ShapeRange.IncrementTop row_distance
```

Figure 11.6 Part of the VBA code allowing results to be written to an Excel picture.

1. Move measured values from the distributed control system to the Excel sheet.
2. Solve the data reconciliation problem as a constrained nonlinear minimization problem.
3. Determine if gross errors are present using the global test method.
4. Linearize the nonlinear constraints at the data reconciliation solution using the Newton–Raphson method (Taylor series expansions).
5. Copy the linearized coefficients to the *A* matrix.
6. The Excel sheet will automatically update values for MT_i.

Data reconciliation is a mature field within the chemical processing industry. In addition to the software mentioned in Chapter 6 (Section 6.7), commercial flow sheeting packages generally provide data reconciliation options. After completing Chapters 6 and 11, you should be able to use data reconciliation in any appropriate system, determine if gross errors are present, and address suspect measurement identification. Note that it is not necessary that rigorous thermodynamic properties be available in order to apply data reconciliation. In Problem 11.3, we show that thermodynamic properties can often be obtained by linearizing the enthalpy of each stream over typical operating conditions. This same linearization process can generally be applied to any industrial system. As you further explore data reconciliation, please see the texts cited in the references of Chapter 6.

REFERENCES

BUCKLEY, R.A. 2006. Overview of Cogeneration at LSU. MS Thesis, Louisiana State University.

RAVINDRAN, A., K.M. RAGSDELL, and G.V. REKLAITIS. 2006. *Engineering Optimization: Methods and Applications* (2nd edition). John Wiley and Sons, New York.

REYNOLDS, W.C. 1979. *Thermodynamic Properties in SI: Graphs, Tables, and Computational Equations for Forty Substances*. Department of Mechanical Engineering, Stanford University, Stanford, CA.

PROBLEMS

11.1 ***Visual Alarm of Suspect Measurement*** In Example 11.6, we showed how results from the data reconciliation solution can be placed on a picture on the Excel sheet. Enhance the Example 11.6 macro by adding a section of the code, which checks if gross errors are present, and if true, turn the largest MT_i reconciled value to red. Also, modify the macro so that a red value is displayed on the picture. Use data from Example 11.5 as a gross error is present.

11.2 ***Complete Cogeneration Data Reconciliation (Including Measured Pressures)*** Solve the complete data reconciliation problem for the cogeneration system of Chapter 11. Recall we solved the cogeneration system assuming all pressure measurements were fixed. For Problem 11.2, include all measured pressure given in Table 11.1 in the formulation and assume the standard deviation for each pressure measurement is 1 psia. Be sure the natural gas flow rate is returned to the measured value, $F_{NG} = 2.6208$ lb/s.

11.3 ***Cogeneration Data Reconciliation Linearized Enthalpies*** The problems in Chapter 11 have utilized the Excel callable physical properties we have developed (Chapter 8, *TPSI+*, based on Reynolds, 1979). Data reconciliation can, in some cases, be performed using linearized physical properties. Linearized stream enthalpies, as $\hat{h} = a + bT$, for the cogeneration system operating near design conditions are provided in Table P11.3.

Table P11.3 requires explanation. The air and exhaust gas stream enthalpies have been calculated using a reference state of 1 atm and 77°F. For example, the enthalpy of the chilled air stream (stream 1), $\hat{h}_{\text{air},1}$, relative to the reference state, would be (using $T_1 = 519.94$ R from Table 11.1) $\hat{h} = a + bT = -128.879 + 0.240148(519.94) = -4.016$ Btu/lb. We have discussed in previous chapters the inclusion of the reference state in combustion calculations. Recall that in order to account for the energy input from the fuel in the combustion chamber (Eq. (11.5)), we use the *LHV* of the fuel. This *LHV* assumes a reference state of 1 atm and 77°F with the water in the combustion products to be in the vapor state. This same reference state actually needs to be common to all streams included in the combustion chamber energy balance. It is therefore convenient to simply use 1 atm and 77°F as the reference state for all gas-side cogeneration enthalpies as done in Table P11.3.

Use linearized stream enthalpies to solve the cogeneration problem of Example 11.3. Compare the results of Problem 11.3 (linearized enthalpies) with the results of Examples 11.3 and 11.4 (rigorous properties).

Table P11.3 Enthalpy Coefficients—Linear Fit of Enthalpy Data (Buckley, 2006)

$\hat{h}\left(\frac{\text{Btu}}{\text{lb}}\right)$	$a\left(\frac{\text{Btu}}{\text{lb}}\right)$	$b\left(\frac{\text{Btu}}{\text{lb-R}}\right)$
$\hat{h}_{\text{air},0}$	−128.879	0.240148
$\hat{h}_{\text{air},1}$	−128.879	0.240148
$\hat{h}_{\text{air},2}$	−146.446	0.257997
$\hat{h}_{\text{Prod},3}$	−199.574	0.295265
$\hat{h}_{\text{Prod},4}$	−174.799	0.282716
$\hat{h}_{\text{Prod},5}$	−153.664	0.269029
$\hat{h}_{\text{Prod},6}$	−138.187	0.255162
$\hat{h}_{\text{Prod},7}$	−135.778	0.252288
$\hat{h}_{\text{CW},a}$	−235.649	0.439476
$\hat{h}_{\text{CW},b}$	−235.649	0.439476
$\hat{h}_{\text{Water,Econ}}$	−1579.62	1.009221
$\hat{h}_{\text{Water,Evap}}$	−1596.55	1.032152
$\hat{h}_{\text{Water,BlowD}}$	−1596.55	1.032152
$\hat{h}_{\text{Steam}}$	−398.553	0.617606

11.4 ***Cogeneration Data Reconciliation Linearized Enthalpies with Gross Error*** Here again, solve the cogeneration data reconciliation problem using the measured data provided in Table 11.1 except F_{NG}, the measured flow rate of natural gas, is given a positive bias of 0.4 lb/s. Use the linearized enthalpies as provided in Table P11.3. Use the global test method to determine if gross errors are present. If gross errors are present, use the measurement test method to locate suspect measurements. Compare the results of Example 11.5 with those of Problem 11.4.

Chapter 12

Optimal Power Dispatch in a Cogeneration Facility

Determining the optimal power dispatch or energy management strategy in a cogeneration facility is conceptually a challenging task. Consider an industrial processing plant, university, or hospital that needs electricity, steam, and chilled water and which is a qualifying cogeneration facility (QF). The facility may produce all its electricity needs, and in some cases produce additional electricity for open market sale. The facility may also produce part of its electrical needs and purchase the remaining load.

The cogeneration facility may have gas turbines, heat recovery steam generators (HRSGs), stand-alone boilers, steam-driven turbines, steam-driven chillers, and electric chillers. Steam must be raised for heating purposes and steam may be used to produce electricity from steam turbines and may be used to drive chillers to produce chilled water. There may also be electric chillers available to produce chilled water. The ability to provide supplemental firing to the HRSGs for additional steam generation further complicates the optimal operational strategy. Equipment efficiencies vary with ambient temperature, humidity, and operating load. The loads themselves (electricity, steam, and chilled water) may vary with the day, the time of the day, and the season. Clearly, operating complex cogeneration utility systems based on heuristic rules may not be optimal. A more systematic approach is needed.

Given that the utility requirements (steam, electricity, and cooling) for a production period are known, we want to determine the "optimal energy position" or energy dispatch. The optimal energy dispatch determines the amounts of self-generated electricity, purchased electricity, and purchased natural gas needed to minimize costs for each period. It determines which units in the cogeneration facility should be operational and at what levels. As we allow the utility requirements to change from period to period, this can also be considered a multiperiod optimal power dispatch procedure. Once the optimal energy dispatch has been determined, optimal control strategies may be used to further minimize the cost of operating the utility system under the targets set by the optimal energy position. As part of a real-time (online) optimization strategy, the processing plant itself may follow a similar path of determining a product slate and equipment utilization for a production period and then utilize control strategies to minimize production costs (see Figure 6.1).

The goals of minimizing the needed energy in a process, minimizing the cost of purchased and generated utilities, and optimally operating the cogeneration system and the process are inextricably linked. However, they can be treated as separate problems, with information exchange between solutions as indicated in Figure 12.1. An additional consideration must be the match between cogeneration system design and the energy needs of the process—a cogeneration system with overcapacity seldom shows favorable economics when compared to electricity purchase and stand-alone boilers.

12.1 DEVELOPING THE OPTIMAL DISPATCH MODEL

In order to develop an optimal power dispatch strategy, we suggest a systematic approach that begins by looking at the energy efficiency of each significant energy consumer or generator in the cogeneration system. We recommend evaluating each significant piece of equipment in the cogeneration energy suite (generating or using electricity, steam, or cooling water) as a stand-alone unit. Each unit will have a power stream in and a power stream out. Initially, we take

Modeling, Analysis and Optimization of Process and Energy Systems, First Edition. F. Carl Knopf.

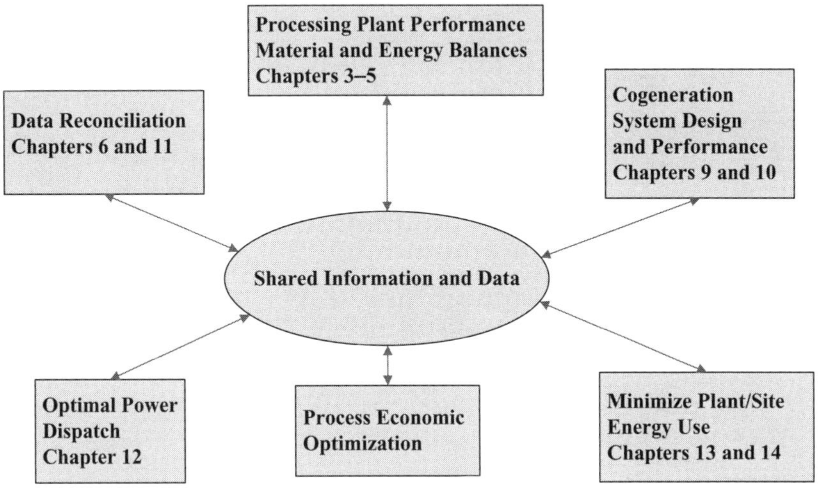

Figure 12.1 Overview of process and cogeneration system analysis.

a very simplistic approach—the power out of a unit operation can be modeled as a simple linear function of the power in. A first improvement on this model will be to include other factors such as the ambient conditions.

A nominal efficiency model of the process can be developed using both manufacturer's test data and operational data. Here, best available operational data should be used—the data should not include periods just prior to either scheduled or unplanned maintenance. For some units, for example, steam or electric driven chillers, very simple energy efficiency equations will prove accurate. However, for some units, for example, an air cooler prior to a gas turbine, this simple approach will not work. Here, more detailed material and energy balances will be required to properly determine energy efficiency. Once the optimal power dispatch model has been developed, a near-continuous performance evaluation and cost minimization is possible as part of an online optimization strategy.

Allowing that we can account for the operating energy efficiency of each utility unit operation, we must assemble this information in a fashion that allows us to meet target electrical, steam, and cooling water demands at a minimum cost. Here, by the construction of a mixed-integer linear programming (MILP) problem, we can determine an optimal operational strategy, all within Excel Solver or an equivalent optimization program. For MILP, we use "What's Best" from LINDO Systems; a student version of What's Best, which can solve all the problems in this chapter, has been supplied by LINDO Systems (see Chapter 1, Section 1.7). The need for mixed-integer variables occurs because at the optimal solution, some available energy unit operations may be on = 1 or off = 0.

We can provide an overview of the procedures we use in this chapter by considering the gas turbine and HRSG system that we have examined in Chapters 7 and 9–11. Our new view of the gas turbine and HRSG system is shown in Figure 12.2. Here, energy enters as natural gas, which fires the gas turbine to produce electricity. The energy in (energy rate) is British thermal unit per hour for the natural gas. The energy out (energy rate or power) is kilowatt of electricity. We will develop a linear function, which relates the energy rate out to the energy rate in for the gas turbine. The same natural gas feed firing the gas turbine also produces steam from the HRSG as pound per hour or British thermal unit per hour. The HRSG only produces steam if the gas turbine is in operation. A function will be developed, which relates the natural gas energy rate in to the steam energy production rate out. Finally, if the gas turbine is operational, there exists the possibility of supplemental firing of the HRSG, which will produce additional steam. Here, a function will be developed that relates the additional steam production from supplemental firing to the additional energy input from the natural gas. The steam (energy rate out) from the HRSG will be the sum of both of these HRSG operational modes.

Before we move into the implementation details, we comment on the importance of careful accounting of the cost of electricity. As detailed later in this chapter, for a qualified cogeneration facility, purchased electricity costs include a base rate, a fuel adjustment cost, and a demand charge. The demand charge includes both a firm component and a standby component. Fuel adjustment charges have perhaps the largest impact in a power dispatch model. However, fuel adjustment charges are not known until the bill for the month is received. We develop a simple estimation strategy for fuel adjustment charges in this chapter.

Starting in Section 12.2, we present a detailed example involving a 25-MW university system with two separate gas turbine-based cogeneration systems. Power dispatch in a university system is a challenging task as electricity, steam, and chilled water demands fluctuate with season, day, and time of day. A university cogeneration system may be closer in operation to a cogeneration system for a batch processing plant where process equipment start-up and shutdown is common. In a continuous processing plant, the goal is

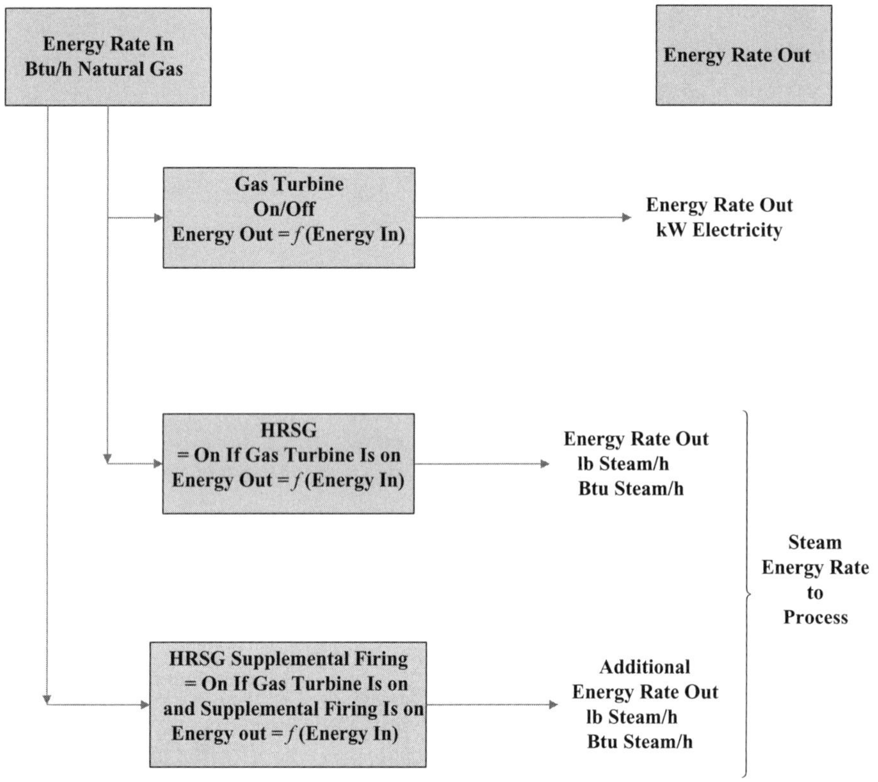

Figure 12.2 Conceptual overview of the dispatch model formulation for the gas turbine system.

typically to maintain steady-state operation for long time periods. We emphasize that regardless of the "process" (university, hospital, batch, or continuous plant), the techniques developed here are applicable to any cogeneration-based utility system. As a background, in Problems 12.1 and 12.2, we provide an extended discussion of data regression for both linear and nonlinear functions. Recall from Figure 12.2 that we want to develop functions that relate energy rate in to energy rate out for each significant operation in the cogeneration system; these functions will be first assumed linear and may incorporate material and energy balance information. If linear efficiency equations are not sufficiently accurate, then nonlinear equations such as polynomial approximations may be used. We will need nonlinear data regression to predict the cost of purchased electricity (the fuel adjustment charge). Once determined, this fuel adjustment cost will become a constant (dollar per kilowatt-hour) in the dispatch formulation.

12.2 OVERVIEW OF THE COGENERATION SYSTEM

Here we develop the optimal power dispatch for a cogeneration system using as an example a university system which has the complexity of units that would also be anticipated at an industrial site. An overview of the cogeneration system is given in Figure 12.3.

Electricity Generation

The cogeneration facility is a qualifying facility that allows electricity purchase from the local utility. The details of purchased electricity cost are discussed in Section 12.5. For electricity generation, there is one gas turbine, a General Electric (GE) LM-2000 aeroderivative gas turbine, capable of producing 20 MW of electricity. The exhaust gas of this gas turbine drives an HRSG unit (boiler 8).

Chilled Water Production

There are 10 chillers available to produce chilled water. One chiller is directly driven by a second aeroderivative gas turbine, an Allison 501-KC; six chillers are driven by electricity (chillers 1–5 and 7); and three chillers are driven by steam (chillers 8–10).

The Allison gas turbine is rated at 5000 hp (3.728 MW). The shaft of the Allison gas turbine is directly coupled to a chiller that can produce 6400 tons of refrigeration. The Allison gas turbine technically does not produce electricity. The use of a gas turbine to drive a refrigeration unit to produce chilled water, as opposed to electricity, is interesting. This may be a method of avoiding conflicts with the local utility concerning cogeneration qualification while still obtaining energy recovery benefits. The Allison gas turbine also has an HRSG (boiler 7) unit for steam generation.

The six electrically driven centrifugal chillers (chillers 1–5 and 7) have a combined capability of 9400 tons. The

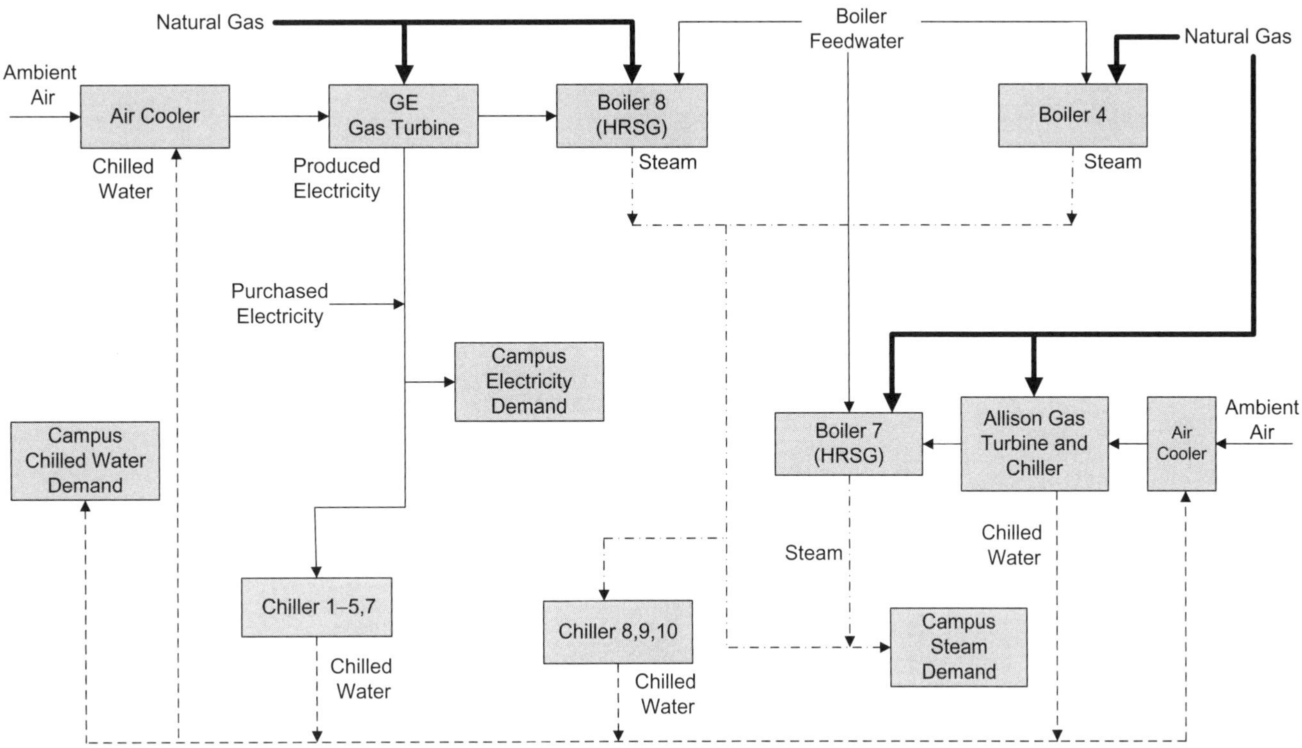

Figure 12.3 Cogeneration system overview (Buckley, 2006).

three York centrifugal steam-driven chillers (chillers 8–10) each have a capacity of 2060 tons.

Steam Production

There are three steam generation units. There is a stand-alone boiler (boiler 4) capable of producing 100,000 lb/h of steam at 115 psig. There are two HRSGs (boiler 7 and boiler 8), both with supplemental firing. Boiler 7 from the Allison gas turbine can produce 22,000 lb/h of steam in the design case and with supplemental firing can produce an additional 75,000 lb/h; the total with supplemental firing is ≈100,000 lb/h at 115 psig. Boiler 8 from the GE gas turbine can produce 90,000 lb/h of steam in the design case and with supplemental firing can produce an additional 62,000 lb/h of steam; the total with supplemental firing is ≈150,000 lb/h at 140 psig.

The system we are modeling has no steam turbines for electricity generation and no provision to sell electricity to the grid. These considerations would be straightforward to add to the problem formulation.

12.3 GENERAL OPERATING STRATEGY CONSIDERATIONS

Given the cogeneration system capabilities described earlier, we want to develop a power dispatch management strategy or program that will minimize total utility cost, including purchased electricity and purchased natural gas costs. The program can be run at any time, but we must have an appreciation for the time needed for equipment start-up or shut down. For a university setting, we would anticipate running this optimization program at least three times a day—in order to determine the best positions for the system during the 7 AM–5 PM, 5 PM–11 PM, and 11 PM–7 AM time periods. Each of these time periods show substantial differences in utility demands. The program developed here could also be used for long-range planning studies. For example, it could be used to determine if long-term natural gas contracts are advantageous. It can also be used to determine the profitability of the cogeneration system.

12.4 EQUIPMENT ENERGY EFFICIENCY

The first step in developing an optimal operating strategy is to determine the energy efficiency of each piece of equipment of importance in the cogeneration facility. Initially, linear energy efficiency expressions will be evaluated, and if necessary, expressions utilizing material and energy balances will be developed. These equations will be assembled in an MILP problem to determine the most cost-effective and efficient equipment utilization to meet known electricity, steam, and cooling water requirements.

The situation we face in developing energy efficiency expressions is typical of most industrial settings. For some units, the needed information is readily available from manufacturer's design and test data, literature data, and most important, existing instrumentation. However, for some operations, key data will be missing. In this later case, we must use any available data combined with material and energy balances to solve for the efficiencies. In some cases, we may need to take additional measurements. Clearly, if the power dispatch management model proves useful, then improvements to the model, including additional instrumentation, should be easy to justify.

12.4.1 Stand-Alone Boiler (Boiler 4) Performance (Based on Fuel Higher Heating Value (*HHV*))

Boiler efficiency accounts for the energy available in the fuel, which is transferred to the steam:

$$\text{Energy rate steam out} = (\text{Energy rate fuel in})(\text{Boiler efficiency}).$$

In general, boiler efficiency can be represented by (Pattison and Sharma, 1980)

$$\text{Boiler efficiency} = \frac{F_{\text{steam}}^{\text{boiler\_out}}}{\left[(1+B)F_{\text{steam}}^{\text{boiler\_out}} + (A)F_{\text{steam,Max}}^{\text{boiler\_out}}\right]}, \quad (12.1)$$

where, $F_{\text{steam}}^{\text{boiler\_out}}$ = mass flow of steam (lb/h), $F_{\text{steam,max}}^{\text{boiler\_out}}$ = maximum steam flow , and A and B are efficiency correlation coefficients with $A = 0.0126$, $B = 0.2156$ (Shang and Kokossis, 2004).

If $F_{\text{steam,max}}^{\text{boiler\_out}} = 100{,}000$ lb/h, a plot of efficiency versus steam flow would appear as shown in Figure 12.4.

What is apparent from this generalized plot is that boiler efficiency is expected to be near 80% over a wide flow range of steam flow—from 100% capacity to about 20% capacity. This 80% efficiency is based on the *HHV* British thermal unit content of the natural gas feed.

Boiler 4 efficiency data were collected over a 1-year period. Table 12.1 reports the average boiler efficiency was 86% based on natural gas *HHV*. The fuel *HHV* averaged 1054.8677 Btu/SCF of natural gas feed. The efficiency for boiler 4 is slightly above the value reported by Pattison and Sharma (1980), but within the expected efficiency range (80–86%) at the rated steam flow. The steam from the boiler is saturated at 115 psig (347.2°F), and the feedwater temperature is 218.4°F at 239 psig. The boiler provides some sensible heat to the water, but the primary energy input is latent heat. The energy needed per pound of steam produced is ($\hat{h}_{\text{steam\_out}} - \hat{h}_{\text{water\_in}} = 1192.3434 - 187.0427 = 1005.3008$ Btu/lb), 1005.3008 Btu/lb.

Over an output steam range 20,000–100,000 lb/h, we can estimate the needed energy rate in as linear function (with intercept = 0):

$$\frac{\text{mlb steam out}}{\text{h}} = \left(0.9015067 \frac{\frac{\text{mlb steam}}{\text{h}}}{\frac{\text{mSCF natural gas}}{\text{h}}}\right)\left(\frac{\text{mSCF natural gas}}{\text{h}}\right) \quad (12.2)$$

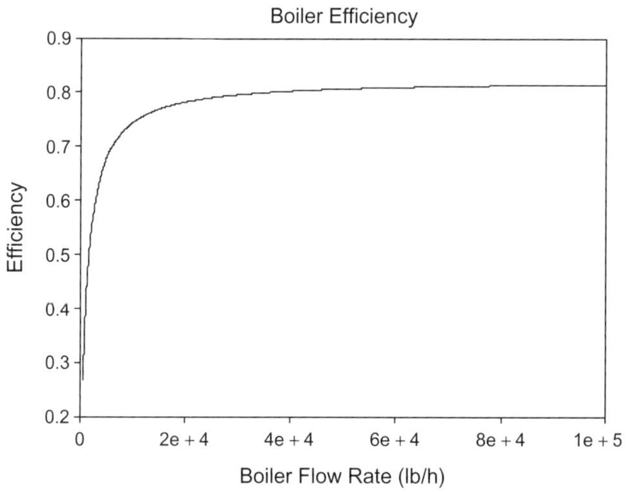

Figure 12.4 Stand-alone boiler efficiency versus steam flow rate.

Table 12.1 Averaged Data—Efficiency of Stand-Alone Boiler (Buckley, 2006)

Natural Gas (SCF/h)	Natural Gas (*HHV*) (Btu/h)	Rated Capacity Steam (lb/h)	Rated Capacity Steam (Btu/h)	$\frac{\text{mlb Steam}}{\text{m SCF}}$	$\frac{\text{mlb Steam}}{\text{MMBtu } (HHV)}$	Efficiency (*HHV*) (Btu out/Btu in), %
110,925.406	117,011,628	100,000	100,530,080	0.9015067	0.85461592	85.915

Prefix MM = 10^6 and m = 10^3.

or

$$\frac{\text{mlb steam out}}{\text{h}} = \left(0.85461589 \frac{\frac{\text{mlb steam}}{\text{h}}}{\frac{\text{MMBtu}(HHV)\text{natural gas}}{\text{h}}} \right) \left(\frac{\text{MMBtu}(HHV)\text{natural gas}}{\text{h}} \right). \quad (12.3)$$

So at this gas *HHV*, 50,000 lb/h (50 mlb/h) of steam would require 58.5058 MMBtu/h natural gas or 55.4627 mSCF/h natural gas.

It is implied by Equations (12.2) and (12.3) that we could consider formulating the energy rate in using either the SCF or the *HHV* British thermal unit content of the natural gas; the *HHV* British thermal unit content of the fuel is the *better* choice. Natural gas fuel is priced on its *HHV* and the *HHV* varies (see Problem 12.5). It is straightforward to account for this variation by using the *HHV* British thermal unit content as the basis in the energy rate in term. Plant data generally provide natural gas SCF (Eq. (12.2)), which can then be converted to an *HHV* basis (Eq. (12.3)).

12.4.2 Electric Chiller Performance

The electric chillers (chillers 1–5 and 7) use electricity to drive compressors in a refrigeration cycle with R-134a to produce chilled water. The chilled water is a closed cycle, here used in the heating, ventilation, and air conditioning (HVAC) system. Chilled water enters campus at 45°F and returns at a nominal 55°F. The chillers also use cooling water or condenser water in a closed cycle to remove heat from the refrigerant. Heat from the condenser water is removed in cooling towers.

Determining the energy efficiency of the chillers presents a typical problem. The total flow rate of chilled water to all the electric chillers is known. The flow rate of chilled water to each chiller is not known, and each chiller has a different refrigeration capacity. The temperature of the chilled water in and out of each chiller, as well as the amps drawn by each chiller, is known. The manufacturer's test data in Table 12.2 provide the kilowatt of electricity needed to produce 1 ton of refrigeration at full load for each chiller. We need to determine the efficiency when chiller operation is not at full load.

In order to determine chiller performance when operating at off-design capacity, we can use Figure 12.5. Figure 12.5 provides performance data from three manufacturers showing how the required kilowatt per ton of refrigeration changes as a function of load and the entering condenser water temperature (ECWT) from the cooling towers. To see how Figure 12.5 is used, assume that the chiller "design case" (100% design load and 100% design kilowatt) is 2000 tons at 1300 kW with 85°F ECWT. If the ECWT is lowered to 75°F, with the load remaining at 2000 tons, the needed kilowatt is reduced to ~88% of the design case, or about 1144 kW. With 75°F ECWT, if the refrigeration load is reduced 50% to 1000 tons, the required kilowatt is reduced to ~44% of the design case, or about 572 kW. Most important is to see that at 75°F ECWT, a 50% reduction in

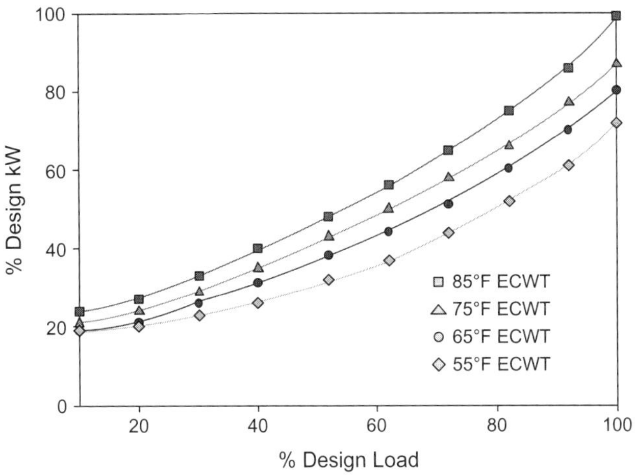

Figure 12.5 Typical chiller off-design efficiency as a function of design load—figure adapted from York International Corporation, HVAC&R Engineering Update, "New NPLV Rating Works 'Weather' Your Plant Has a Single or Multiple Chillers," 1999.

Table 12.2 Electric Chiller Efficiency Data (Buckley, 2006)

Chiller	Voltage In (kV)	Amps In	kW In	Max Tons Refrigeration Out	kW/ton Test Data from Manufacturer	Efficiency (Btu out/Btu in), %
Chiller 1	6.42	161.0	1033.62	1600	0.646	544.4
Chiller 2	6.42	161.0	1033.62	2000	0.690	509.7
Chiller 3	6.42	161.0	1033.62	1000	0.604	582.3
Chiller 4	6.42	161.0	1033.62	1700	0.635	553.8
Chiller 5	6.42	161.0	1033.62	2000	0.655	536.9
Chiller 7	6.42	161.0	1033.62	1100	0.630	558.2

refrigeration load also results in a 50% reduction in power required (compared to that producing 2000 tons with 75°F ECWT). As a first approximation, we can assume a linear relation between refrigeration out and power in, provided the ECWT remains constant. Generally, a chiller is not operated below 10–20% capacity as there can be a sharp drop-off in efficiency.

Plant operational data confirms that the ECWT is relatively constant between 79 and 82°F. For chiller 1, using Table 12.2 and Figure 12.5 for a refrigeration range of 320–1600 tons, we can estimate the needed energy rate in as a linear function (with intercept = 0):

$$\text{tons refrigeration out} = \left(\frac{1}{0.646}\ \frac{\text{tons refrigeration}}{\text{kW electricity}}\right)(\text{kW electricity}). \tag{12.4}$$

For example, 800 tons of refrigeration from chiller 1 would require 516.8 kW of electricity. This same procedure can be used for the remaining chillers. The last column in Table 12.2 is the efficiency of the chiller. This can be found by converting column 6 (kilowatt per ton) to a common basis; for example, the tons of refrigeration to British thermal unit per hour (1 ton = 12,000 Btu/h) and kilowatt to British thermal unit per hour (1 W = 3.4121 Btu/h). Recall the efficiency is (energy rate out/energy rate in) and for chiller 1,

$$\left(\frac{1\ \text{ton}}{0.646\ \text{kW}}\right)\left(\frac{1\ \text{W}}{3.4121\ \text{Btu/h}}\right)\left(\frac{1\ \text{kW}}{1000\ \text{W}}\right)\left(\frac{12{,}000\ \text{Btu/h}}{1\ \text{ton}}\right) = 5.444 = 544.4\%.$$

12.4.3 Steam-Driven Chiller Performance

There are three York steam-driven chillers (chillers 8–10). Here, steam is used to drive centrifugal compressors in a refrigeration cycle with R-134a to produce chilled water. The closed cycle chilled water exit temperature is 45°F and the return is at a nominal 55°F. The chillers also use cooling water in a closed cycle to remove heat from the refrigerant. Heat is removed from the cooling water in cooling towers.

Available plant data include the total flow rate of chilled water to all the steam turbine-driven chillers, the chilled water temperatures in and out of each chiller, and the steam flow to each chiller. The flow rate of chilled water to each chiller is not known. However, just as with the electric chillers, the effect of the actual refrigeration load on steam-driven chiller efficiency is minor. York (Millennium YIA Single-Effect Absorption Chillers, Steam and Hot Water Chillers; Form 155, 16-EG1 (604), 1997) presented performance data showing that at constant inlet steam pressure, the required (Btu steam/h)/(ton of refrigeration) varied about 5% over loads ranging from 10–100% of design. As a first approximation, we can assume a linear relation between refrigeration out and power in, provided the inlet steam pressure remains constant. Generally, a steam-driven chiller is not operated below 10% capacity as there can be a sharp drop-off in efficiency.

As with the electrically driven chillers, the ECWT to the steam-driven chillers was maintained at 79–82°F. The steam chillers use steam from boiler 8, which is delivered at 140 psig. Steam from boiler 4 can also be used to drive the steam turbines, although this option was never employed during our 1-year study. If boiler 4 were utilized, the inlet pressure to the steam-driven chiller turbines would be 115 psig, which would lower the chiller efficiency.

Data for the steam-driven chillers were collected for a 1 year + period in 15-minute intervals (27,000+ data points). At each time period, the tons of refrigeration generated could be determined from the total water flow rate to the chillers and its temperature change. The refrigeration duty of each chiller was estimated by weighting the total refrigeration duty by the known steam flow to each chiller. The results from this procedure for chiller 8 gave a linear fit of

$$\text{tons refrigeration out} = \left(87.7438\frac{\text{tons refrigeration}}{\left(\frac{\text{m lb of steam at 140 psig}}{\text{h}}\right)}\right)\left(\frac{\text{m lb of steam at 140 psig}}{\text{h}}\right) \tag{12.5}$$

with a coefficient of determination, $R^2 = 0.9636$ (see Problem 12.1). To determine efficiency, tons of refrigeration are converted to British thermal unit per hour (1 ton = 12,000 Btu/h) and the pound per hour of steam to British thermal unit per hour. The condition of the steam to the chiller is set at 140 psig (361°F). The steam driving the centrifugal chiller is partially condensed and exits at vacuum conditions of 1 psia (101.7°F, 95% steam quality). Therefore, the available steam energy $\left(\left(\hat{h}_{\text{steam\_in}} - \hat{h}_{\text{steam\_out}}\right) = 1195.4\ \text{Btu/lb} - 1054\ \text{Bu/lb}\right)$ is 141.4 Btu/lb. The efficiency is (energy out/energy in)

$$\left(\frac{87.7438\ \text{ton refrigeration}}{1000\ \text{lb steam/h}}\right)\left(\frac{1\ \text{lb steam}}{141.4\ \text{Btu}}\right)\left(\frac{12000\ \text{Btu/h}}{1\ \text{ton}}\right) = 7.446 = 744.6\%.$$

Table 12.3 provides the efficiency for each of the three steam chillers based on data collection for over a 1-year period. The third column is the rated capacity (tons of refrigeration) for each chiller.

Before we leave the steam chillers, additional discussion of efficiency is needed. There is no need in the process for the low-pressure steam (1 psia) from the chillers.

Table 12.3 Steam Chiller Efficiency Data

Chiller	$\frac{\text{ton refrigeration}}{\text{lb steam/h}}$	Maximum Tons of Refrigeration Out	Reported Efficiency, %	Actual Efficiency, %
Chiller 8	87.7438	2060	744.6	93.54
Chiller 9	88.2369	2060	748.8	94.06
Chiller 10	89.1679	2060	756.7	95.06

This steam is condensed to water by heat exchange with condenser cooling water. Therefore, the actual steam energy used by the chillers may be taken as $(\hat{h}_{\text{steam\_in}} - \hat{h}_{\text{steam\_out}} = 1195.4\ \text{Btu/lb} - 69.74\ \text{Btu/lb})$, is 1125.66 Btu/lb. The actual efficiency is (energy out/energy in)

$$\left(\frac{87.7438\ \text{ton refrigeration}}{1000\ \text{lb steam/h}}\right)\left(\frac{1\ \text{lb steam}}{1125.66\ \text{Btu}}\right)\left(\frac{12000\ \text{Btu/h}}{1\ \text{ton}}\right)$$
$$= 0.9354 = 93.54\%.$$

12.4.4 GE Air Cooler Chiller Performance

The GE gas turbine is an aeroderivative unit which necessitates that the inlet air temperature be maintained at 60°F for optimum gas turbine performance. When the ambient air temperature is above 60°F, the inlet air is cooled using chilled water as shown in Figure 12.6.

In relating the energy transferred in the air cooler to a function of the ambient air temperature and gas turbine loading (or incoming air flow rate), neither a linear fit nor a polynomial fit proved accurate. Here an additional consideration must be the air humidity. If 60°F is below the dew point of the ambient air then water will be condensed during the air cooling process. Condensing water has a significant impact on the required refrigeration load. The energy required to condense water can be 50% or more of the energy needed to cool the inlet air to 60°F. Material and energy balances must be used to determine the needed refrigeration load in the air cooler. Matters are further complicated if the ambient air is cooler than 60°F; here, the air cooler can actually generate chilled water and this scenario is explored in Example 12.3 and in Problem 12.3.

EXAMPLE 12.1 *Air Cooler Water Condensation Rate*

Figure 12.6 shows the GE gas turbine delivers 19,270 kW of power with 145.5 lb/s of air flow at design conditions of 60°F and 100% saturated air. If the ambient air temperature is 76.7°F with a relative humidity of 73.3%, determine the water condensation rate needed to bring this air stream to design conditions. The incoming air can be taken as a mixture of dry air and saturated water vapor. If the ambient air contains more water vapor than design, this extra water will be condensed during the air cooling process.

SOLUTION The problem solution is provided in the Excel file **Example 12.1.xls** and shown in Figure 12.7. Here, a basis of 1 second of flow to the compressor at design conditions is chosen.

We need to determine the amount of water vapor in the 100% humidity 60°F stream leaving the air cooler. This value can be compared with the amount of water vapor in the ambient air stream. This allows determination of the amount of water (if any) that needs to be condensed from the ambient air stream.

The process begins by determining the water vapor saturation pressure as a function of temperature, at 1-atm pressure. Here,

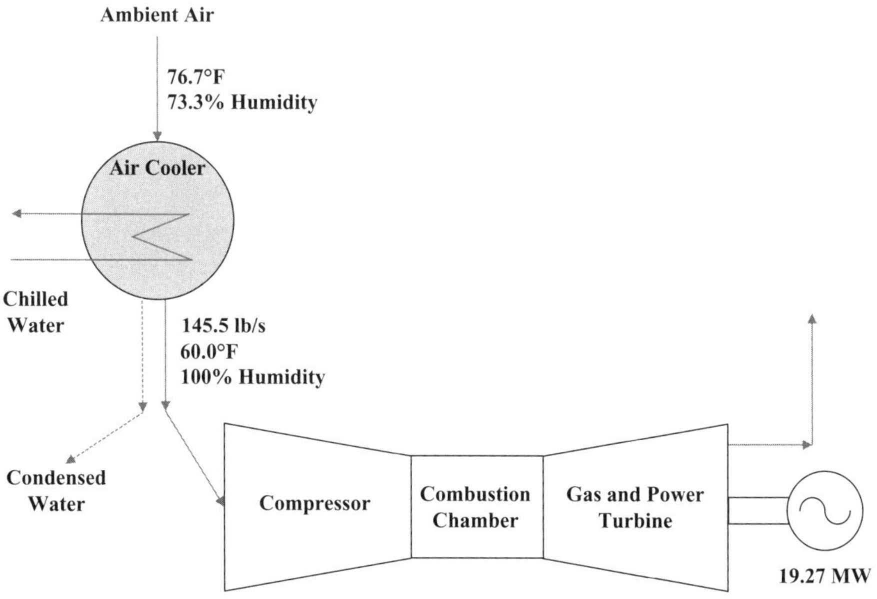

Figure 12.6 Gas turbine with air cooler to precondition incoming air.

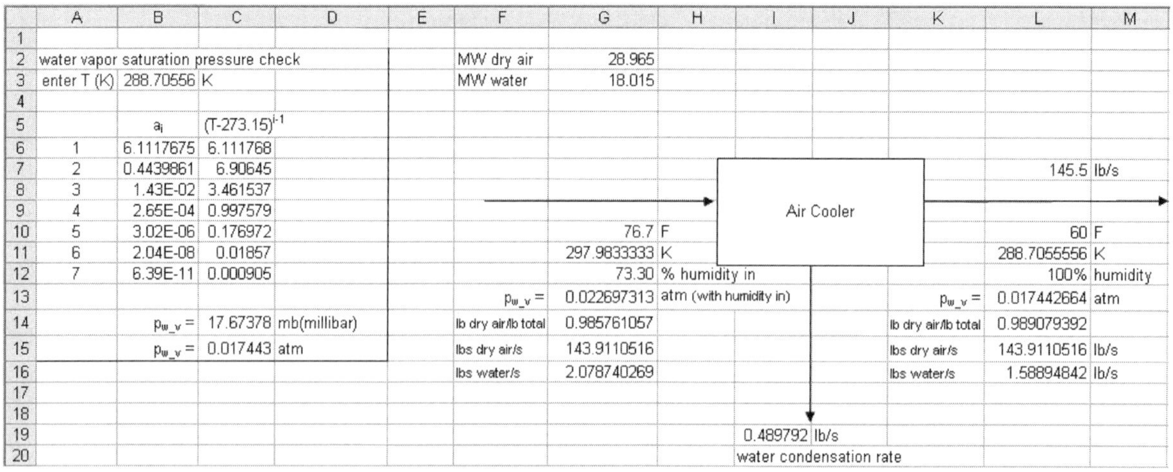

Figure 12.7 Solution to Example 12.1—water condensation rate.

Flatau et al. (1992) provide the following polynomial (–60 to 125°F):

$$p_{\text{water\_sat}}(\text{millibar}) = a_1 + a_2(T - 273.15) + a_3(T - 273.15)^2 + \cdots + a_7(T - 273.15)^6, \tag{12.6}$$

where T is the temperature in kelvin and the coefficients are given as

a_1	a_2	a_3	a_4
6.1117675	0.443986062	0.0143053301	0.265027242E-03
a_5	a_6	a_7	
0.302246994E-05	0.203886313E-07	0.638780966E-10	

For the stream leaving the air cooler, at 60°F (288.7 K), $p_{\text{water\_sat}}$ = 17.673 millibar = 0.01744 atm. Recall that the saturation pressure is the partial pressure exerted by water vapor in 100% saturated air. Condensation will occur with any increase in the water partial pressure or decrease in temperature. Taking the air as an ideal gas mixture,

$$y_{\text{water\_vapor}}^{\text{Air Cooler\_Out}} = \frac{N_{\text{water\_vapor}}^{\text{Air Cooler\_Out}}}{N_{\text{Total}}^{\text{Air Cooler\_Out}}} = \frac{p_{\text{water\_sat}}}{P_{\text{Total}}} = \frac{0.01744\text{ atm}}{1\text{ atm}} = 0.01744. \tag{12.7}$$

As the stream leaving the air cooler contains only dry air and water vapor, we can write

$$\frac{\text{lb dry air}}{\text{lb total}} = \frac{\left(1 - y_{\text{water\_vapor}}^{\text{Air Cooler\_Out}}\right)\left(MW_{\text{dry\_air}}\right)}{\left(1 - y_{\text{water\_vapor}}^{\text{Air Cooler\_Out}}\right)\left(MW_{\text{dry\_air}}\right) + \left(y_{\text{water\_vapor}}^{\text{Air Cooler\_Out}}\right)\left(MW_{\text{water}}\right)} = 0.989, \tag{12.8}$$

giving us, for the outlet stream,

$$\frac{\text{lb dry air}}{\text{s}} = (0.989)\left(145.5\frac{\text{lb}}{\text{s}}\right) = 143.91\frac{\text{lb}}{\text{s}}$$

and

$$\frac{\text{lb water vapor}}{\text{s}} = 145.5 - 143.91 = 1.59\frac{\text{lb}}{\text{s}}.$$

We can follow the same procedure for the inlet stream to the air cooler, at 76.7°F (297.98 K), $p_{\text{water\_sat}}$ = 0.031 atm (100% humidity air). Using the definition of relative humidity,

$$\text{Relative humidity}(\%) = \left(\frac{\text{Actual water vapor partial pressure}}{\text{Saturated water vapor partial pressure}}\right)(100), \tag{12.9}$$

the actual water vapor partial pressure at 73.3% relative humidity is

$$p_{\text{water\_actual}} = (0.031)(0.733) = 0.0227\text{ atm}, \tag{12.10}$$

and for ideal air,

$$y_{\text{water\_vapor}}^{\text{Air Cooler\_In}} = \frac{N_{\text{water\_vapor}}^{\text{Air Cooler\_In}}}{N_{\text{Total}}^{\text{Air Cooler\_In}}} = \frac{p_{\text{water\_actual}}}{P_{\text{Total}}} = \frac{0.0227\text{ atm}}{1\text{ atm}} = 0.0227. \tag{12.11}$$

For the inlet stream,

$$\frac{\text{lb dry air}}{\text{lb total}} = \frac{\left(1 - y_{\text{water\_vapor}}^{\text{Air Cooler\_In}}\right)\left(MW_{\text{dry\_air}}\right)}{\left(1 - y_{\text{water\_vapor}}^{\text{Air Cooler\_In}}\right)\left(MW_{\text{dry\_air}}\right) + \left(y_{\text{water\_vapor}}^{\text{Air Cooler\_In}}\right)\left(MW_{\text{water}}\right)} = 0.98576. \tag{12.12}$$

The pound of dry air leaving the air chiller is the same as the pound of dry air entering the air chiller = 143.91 lb/s. This combined with Equation (12.12) provides

$$\frac{\text{lb water vapor}}{\text{s}} = 2.079\frac{\text{lb}}{\text{s}}.$$

The air cooler will condense 0.49 lb/s. ■

EXAMPLE 12.2 *Air Cooler Refrigeration Duty*

Figure 12.6 shows the GE gas turbine delivers 19,270 kW of power with 145.5 lb/s of air flow at design conditions of 60°F and 100% saturated air. If the ambient air temperature is 76.7°F with a relative humidity of 73.3%, determine the refrigeration rate needed to bring the air stream to design conditions. The heat capacity C_p of water vapor can be taken as 0.445 Btu/lb-F over the temperature range of interest for the air cooler. The enthalpy of dry air can be found using our H_Air(P,T) function and for water condensing at 60°F $(h_v - h_l) = 1059.36$ Btu/lb.

SOLUTION The problem solution is provided in the Excel file **Example 12.2.xls**. The energy transfer rate is the sum of the energy rate needed to reduce the incoming dry air from 76.7°F to 60°F, to reduce the water vapor from 76.7 to 60.0°F, and to condense 0.49 lb/s of water at 60°F (water condensation rate from Example 12.1).

For dry air,

$$\frac{F_{\text{dry\_air}}^{\text{Air Cooler\_In}}\begin{pmatrix} H\_Air(14.696\ psia, 519.67\ R) \\ -H\_Air(14.696\ \text{psia}, 536.37\ \text{R}) \end{pmatrix}}{(MW_{\text{dry\_air}})} = -577.17\ \text{Btu/s};$$

for water vapor,

$$F_{\text{water\_vapor}}^{\text{Air Cooler\_In}}\left(0.445\frac{\text{Btu}}{\text{lb-°F}}\right)(76.7-60.0°\text{F}) = -15.45\ \text{Btu/s};$$

and for condensation,

$$0.49(h_l - h_v) = 518.87\ \text{Btu/s}.$$

The energy transfer rate sum is −1111.486 Btu/s or −333.446 tons of refrigeration; here, the negative sign indicates that heat is removed in this air cooling process. ■

EXAMPLE 12.3 *General (Actual) Air Cooler Refrigeration Duty*

In Examples 12.1 and 12.2, the air flow rate to the compressor was fixed at 145.5 lb/s and at design conditions of 60°F and 100% saturated air. The ambient air was supplied at 76.7°F and 73.3% relative humidity. Here we want to solve for the air cooler duty at any ambient air conditions (temperature and relative humidity). The air flow rate to the compressor will remain at 145.5 lb/s and 60°F, but the humidity will not necessarily be 100%. We want to determine the actual relative humidity of the air stream leaving the air cooler. Also, if the ambient air is below 60°F, the air cooler can generate refrigeration duty.

SOLUTION The problem solution is provided in the Excel file **Example 12.3.xls**. This is an excellent exercise for the reader to try, and here we provide some solution hints. To start, solve the problem as we did in Examples 12.1 and 12.2 with the assumption that the air leaving the air cooler is at 60°F and 100% humidity. Compare the (lb dry air)/(lb total) in the ambient air to the (lb dry air)/(lb total) leaving the air cooler when assuming saturation. Use an Excel IF statement to select the larger of the two values as the actual (lb dry air)/(lb total). Based on this actual (lb dry air)/(lb total), determine the conditions leaving the air cooler and the air cooler duty. ■

Air Cooler Efficiency

Data from the GE gas turbine were collected for a 1 year + period (27,000+ data points) including ambient air temperature and relative humidity, air temperature to the gas turbine (generally 60°F), air cooler duty, and power generated. The air flow rate is assumed to be directly proportional to the power generated as

$$\text{Air flow rate}\frac{\text{lb}}{\text{s}} = \left(\frac{\text{Power generated}}{19{,}270\ \text{kW}}\right)\left(145.5\frac{\text{lb}}{\text{s}}\right). \tag{12.13}$$

The equations to predict the needed air refrigeration (or generated refrigeration if ambient air $T < 60°\text{F}$) were assembled in the Excel spreadsheet as in Example 12.3. The predicted refrigeration was compared to the actual refrigeration. The fit of the actual plant data required a premultiplier of 71.4% in order to obtain the best fit between measured and predicted refrigeration duties:

$$\begin{aligned} &\text{Actual refrigeration load} = \\ &\left(\frac{1}{71.4\%}\right)(\text{Predicted refrigeration load}); \\ &T_{\text{air\_Temp(°F)}} > 60°\text{F} \end{aligned} \tag{12.14}$$

and

$$\begin{aligned} &\text{Actual refrigeration generation} = \\ &(71.4\%)(\text{Predicted refrigeration generation}); \\ &T_{\text{air\_Temp(°F)}} < 60°\text{F}. \end{aligned} \tag{12.15}$$

There are several possible interpretations of the 71.4% value. One interpretation is that the coil effectiveness or efficiency is 71.4%; here, the "extra" refrigeration is simply lost to the environment.

EXAMPLE 12.4 *Air Cooler Refrigeration Duty as a Function of GE Kilowatt*

If air is at 91°F and 42% relative humidity, develop the relation between the air cooler duty and the power generated by the GE gas turbine.

SOLUTION Using the Excel solution for Example 12.3, with air at 91°F and 42% relative humidity and with 100% coil efficiency, 422.69593 tons of refrigeration are needed to cool 145.5 lb/s of air to 60°F. Using Equation (12.14), with the measured 71.4% coil efficiency, 592.0111 tons of refrigeration are actually required to cool the air and to generate 19,270 kW of power. The relation between the air cooler duty and the power generated by the GE gas turbine would be

$$\text{Air cooler duty (ton)} = 592.0111\left(\frac{\text{kW actually generated}}{19{,}270\text{ kW}}\right)$$

for ambient air at 91°F and 42% relative humidity. If, for example, the gas turbine generated 11,500 kW, the air cooler would require 353.302 tons of refrigeration to cool the incoming air. In our formulation, the 350.302 tons would be in addition to the refrigeration load needed by the campus (or the process in an industrial setting). ■

12.4.5 GE Gas Turbine Performance (Based on Fuel *HHV*)

As noted earlier, the GE gas turbine is operated with a constant inlet air temperature of 60°F. Therefore, the ambient air temperature is not a consideration in accounting for the gas turbine energy efficiency. The gas turbine can produce a maximum of 20 MW of electric power. The gas turbine uses water injection to reduce NO_x, but if the power output drops below 12 MW, the water injection is turned off and some NO_x control is lost.

Data were again collected at 15-minute intervals over a 1 year + period; this represented 27,000+ data points. The gas turbine did not operate below 4 MW and the maximum power output was 20.1 MW of generated power. A very small sampling of data is given in Table 12.4.

The natural gas fed to the gas turbine combustion chamber is continuously monitored as thousand standard cubic foot (mSCF) per hour. The energy content of the natural gas is 1.0548677 MMBtu/mSCF based on the average natural gas *HHV*. Column 3 is obtained by multiplying column 2 by the natural gas *HHV*. Column 4 is the net (usable) electrical power delivered to the grid. The fifth column is simply column 3 divided by column 4, the natural gas energy input per hour per net kilowatt of electricity produced. The final column is the stand-alone gas turbine efficiency based on the *HHV*. Utilizing 1 W = 3.4121 Btu/h, we have for (energy out/energy in)

$$\left(\frac{18{,}985\text{ kW}}{(206.5\text{ mSCF/h})(1.0548677\text{ MMBtu/mSCF})}\right)\left(\frac{1000\text{ W}}{\text{kW}}\right)\left(\frac{\text{MMBtu}}{10^6\text{Btu}}\right)\left(\frac{3.4121\text{ Btu/h}}{1\text{ W}}\right).$$
$$= .29738 = 29.74\%.$$

The data collected for the 1 year + period is plotted in Figure 12.8 and shows a strong correlation ($R^2 = 0.99+$) to a linear fit of kilowatt of electricity produced to the natural gas energy rate:

Table 12.4 GE Gas Turbine Efficiency Data

Date	mSCF/h Natural Gas Feed	Natural Gas in Energy Content *HHV* (MMBtu/h)	kW of Net Power Produced	$\left(\frac{\text{MMBtu}}{\text{kW-h}}\right)$	Efficiency, % (*HHV*)
June 1, 9:00 AM	206.5	217.8302	18,985	0.011474	29.74
June 1, 9:15 AM	207.1	218.4631	19,006	0.011494	29.68
	206.7	218.0412	19,025	0.011461	29.77
	206.7	218.0412	19,026	0.011460	29.77
	206.6	217.9357	19,015	0.011461	29.77
	206.7	218.0412	18,996	0.011478	29.73
	207.1	218.4631	19,027	0.011482	29.72
	206.8	218.1466	19,034	0.011461	29.77
	207.1	218.4631	19,027	0.011482	29.72
	207.7	219.096	19,041	0.011507	29.65
	207.4	218.7796	19,039	0.011491	29.69
	206.3	217.6192	19,015	0.011445	29.81
	205.5	216.7753	19,020	0.011397	29.94
	205.8	217.0918	18,972	0.011443	29.82
	206.1	217.4082	18,942	0.011478	29.73
	206.0	217.3027	18,950	0.011467	29.76
	205.8	217.0918	18,931	0.011468	29.75
	207.0	218.3576	19,029	0.011475	29.74
	207.4	218.7796	19,043	0.011489	29.70
June 1, 1:45 PM	206.7	218.0412	19,042	0.011451	29.80

$$\text{kW} = 113.66533 \times (\text{Natural gas feed in mSCF/h}) - 4078.15 \quad (12.16)$$

or

$$\text{kW} = 107.7532 \times (\text{Natural gas feed in MMBtu/h}) - 4078.15 \quad (12.17)$$

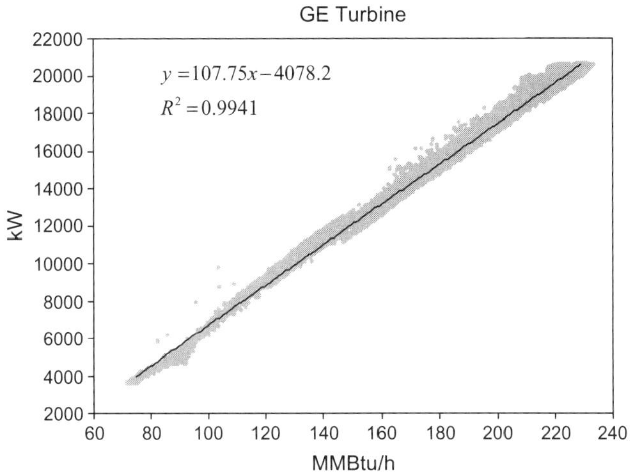

Figure 12.8 Linear fit of GE gas turbine net power to energy rate of natural gas feed.

over the net generated power range of 4.0–20.1 MW. The negative intercept occurs because some natural gas is required to power the air compressor.

12.4.6 GE Gas Turbine HRSG Boiler 8 Performance (Based on Fuel *HHV*)

Energy is recovered from the exhaust gases leaving the GE gas turbine to produce steam. Data were collected over a 1 year + period, with a small sample shown in Table 12.5 (same period as the GE gas turbine is shown).

The natural gas fed to the GE gas turbine is continuously monitored as mSCF per hour. The energy content of the natural gas is taken as the average *HHV* of 1.0548677 MMBtu/mSCF. Column 3 is obtained by multiplying column 2 by the *HHV*. Column 4 is the thousand pound per hour of steam generated from the HRSG. The operating pressure of the HRSG is saturation conditions at 140 psig (361°F), and the feedwater temperature is 218°F at 239 psig. The HRSG transfers some sensible heat to the water, but the primary energy input is latent heat. The energy needed per pound of steam produced is $(\hat{h}_{\text{steam\_out}} - \hat{h}_{\text{water\_in}}) = 1195.338 - 186.6398 = 1008.698$ Btu/lb. The fifth column is simply column 4 multiplied by the energy content of the steam. The final column is the steam generation efficiency, (column 5)/(column 3).

Table 12.5 GE Gas Turbine HRSG (Boiler 8) Efficiency Data

Date	mSCF/h Natural Gas Feed to Gas Turbine	Natural Gas to Gas Turbine Energy Content *HHV* (MMBtu/h)	mlb/h Steam Out	MMBtu/h Steam Out	Efficiency, % (*HHV*)
June 1, 9:00 AM	206.5	217.8302	87.3	88.05934	40.43
June 1, 9:15 AM	207.1	218.4631	87.4	88.16021	40.35
	206.7	218.0412	87.2	87.95847	40.34
	206.7	218.0412	87.1	87.8576	40.29
	206.6	217.9357	86.9	87.65586	40.22
	206.7	218.0412	87.0	87.75673	40.25
	207.1	218.4631	86.9	87.65586	40.12
	206.8	218.1466	87.2	87.95847	40.32
	207.1	218.4631	87.2	87.95847	40.26
	207.7	219.096	86.9	87.65586	40.01
	207.4	218.7796	87.2	87.95847	40.20
	206.3	217.6192	86.7	87.45412	40.19
	205.5	216.7753	86.8	87.55499	40.39
	205.8	217.0918	86.9	87.65586	40.38
	206.1	217.4082	86.7	87.45412	40.23
	206.0	217.3027	87.0	87.75673	40.38
	205.8	217.0918	86.8	87.55499	40.33
	207.0	218.3576	86.6	87.35325	40.00
	207.4	218.7796	87.0	87.75673	40.11
June 1, 1:45 PM	206.7	218.0412	87.0	87.75673	40.25

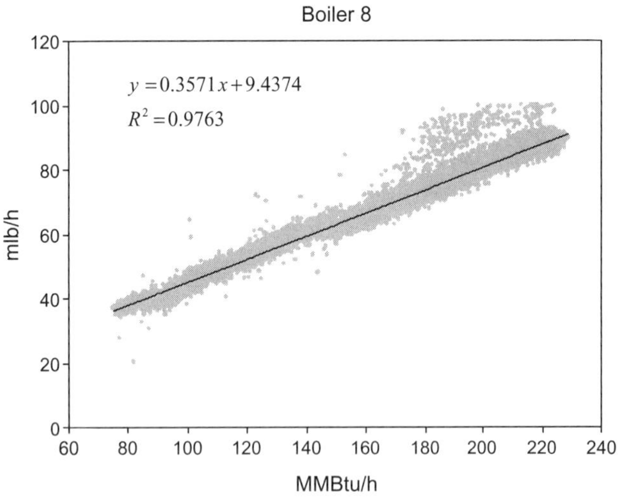

Figure 12.9 Linear fit of HRSG (boiler 8) steam production to *HHV* energy rate of natural gas feed to GE gas turbine.

The data collected for the 1 year + period is plotted in Figure 12.9 and shows a good correlation (R^2 = 0.97+) to a linear fit of steam production rate to energy rate of natural gas feed to the GE gas turbine:

$$\frac{\text{mlb steam}}{\text{h}} = 0.37664 \times \left(\text{Natural gas feed to GE gas turbine in}\frac{\text{mSCF}}{\text{h}}\right) + 9.4374 \tag{12.18}$$

or

$$\frac{\text{mlb steam}}{\text{h}} = 0.35705 \times \left(\text{Natural gas feed to GE turbine in } \frac{\text{MMBtu}}{\text{h}}\right) + 9.4374 \tag{12.19}$$

over the net generated power range of 4.0–20.1 MW.

12.4.7 GE Gas Turbine HRSG Boiler 8 Performance Supplemental Firing (Based on Fuel *HHV*)

Steam generation from the HRSG under supplemental firing requires complex calculations, which were detailed in Chapters 7 and 9. Here we take advantage of available data giving the increased steam flow with supplemental fuel flow to the HRSG. The best fit of the data gave

$$\frac{\text{mlb steam out (increased)}}{\text{h}} = \left(1.023839 \frac{\frac{\text{mlb steam (increased)}}{\text{h}}}{\frac{\text{mSCF natural gas}}{\text{h}}}\right) \left(\frac{\text{mSCF natural gas (supplemental)}}{\text{h}}\right) \tag{12.20}$$

and

$$\frac{\text{mlb steam out (increased)}}{\text{h}} = \left(0.9705852 \frac{\frac{\text{mlb steam (increased)}}{\text{h}}}{\frac{\text{MMBtu natural gas}}{\text{h}}}\right) \left(\frac{\text{MMBtu natural gas (supplemental)}}{\text{h}}\right) \tag{12.21}$$

The efficiency based on the fuel *HHV* would be

$$\frac{\left(\frac{\text{mlb steam (increased)}}{\text{h}}\right)\left(1008.698\frac{\text{Btu}}{\text{lb}}\right)}{\left(\frac{1}{1.023839}\frac{\text{mSCF natural gas}}{\text{h}}\right)\left(1054.8677\frac{\text{Btu}}{\text{SCF natural gas}}\right)} = 97.9\%.$$

12.4.8 Allison Gas Turbine Performance (Based on Fuel *HHV*)

An Allison aeroderivative gas turbine rated at 5000 hp (3.727 MW) is available to produce chilled water. The shaft of the Allison gas turbine is directly coupled to a chiller that can produce 6400 tons of refrigeration. An air cooler reduces the air temperature entering the gas turbine; however, neither the air flow rate nor the exiting air temperature from the cooler is known. The air cooler duty could not be calculated. On a limited number of days when only the Allison gas turbine was operating (the GE gas turbine and chillers 1–5 and 7 were not operating), the actual tons of refrigeration delivered to campus could be determined, and these data were regressed:

$$\text{tons of refrigeration out} = (0.9828)\left(\frac{5594.1 \text{ tons refrigeration}}{40.6151894\frac{\text{mSCF natural gas}}{\text{h}}}\right)\left(1 + 0.002\left(84.9 - T_{\text{air\_Temp(°F)}}\right)\right)\left(\frac{\text{mSCF natural gas}}{\text{h}}\right) \tag{12.22}$$

tons of refrigeration out

$$= (0.9828)\left(\frac{5594.1 \text{ tons refrigeration}}{42.8436514 \dfrac{\text{MMBtu natural gas}}{\text{h}}}\right)$$

$$\left(1 + 0.002\left(84.9 - T_{\text{air\_Temp}(^\circ\text{F})}\right)\right)\left(\frac{\text{MMBtu natural gas}}{\text{h}}\right). \tag{12.23}$$

The form of this energy efficiency equation requires some explanation. For the Allison gas turbine, we picked a base ambient temperature of 84.9°F and here, data showed 5594.1 tons of refrigeration was delivered at full gas firing (40,615.1894 SCF/h). We regressed available data (when the GE gas turbine and electric chillers were not operating) to the form of Equation (12.22) ($R^2 = 0.88+$). Equation (12.23) provides the tons of refrigeration delivered to the process; this would be the total tons of refrigeration generated minus the air cooler load. Equation (12.23) allows us to eliminate the calculation of the air cooler load.

12.4.9 Allison Gas Turbine HRSG Boiler 7 Performance (Based on Fuel *HHV*)

The exhaust gas from the Allison gas turbine is sent to an HRSG (boiler 7). The steam side of the HRSG operates at saturation conditions at 115 psig (347.2°F), and the boiler feedwater temperature is 218.4°F at 239 psig. The HRSG transfers some sensible heat to the water, but the primary energy input is latent heat.

A linear fit to estimate steam flow as a function of natural gas loading to the gas turbine was performed using

lb steam/h = a + b (mSCF natural gas/h).

The best fit of the data ($R^2 = 0.93+$) gave

$$\text{lb steam/h} = 13.1594 + 0.176183 \times (\text{mSCF natural gas/h}) \tag{12.24}$$

and

$$\text{lb steam/h} = 13.1594 + 0.167019 \times (\text{MMBtu natural gas/h}). \tag{12.25}$$

12.4.10 Allison Gas Turbine HRSG Boiler 7 Performance Supplemental Firing (Based on Fuel *HHV*)

Steam generation from the HRSG under supplemental firing requires complex calculations, which were detailed in Chapters 7 and 9. Supplemental firing data for the Allison HRSG were limited, and here we use the efficiency determined for supplemental firing of the GE HRSG (97.9% for boiler 8) and adjust for the $\left(\hat{h}_{\text{steam\_out}} - \hat{h}_{\text{water\_in}}\right) = 1005.3$ Btu/lb in boiler 7, giving

$$\frac{\text{lb steam out (increased)}}{\text{h}} = \left(1.02729 \frac{\dfrac{\text{lb steam (increased)}}{\text{h}}}{\dfrac{\text{SCF natural gas}}{\text{h}}}\right)\left(\frac{\text{SCF natural gas (supplemental)}}{\text{h}}\right) \tag{12.26}$$

and

$$\frac{\text{lb steam out (increased)}}{\text{h}} = \left(0.9738567 \frac{\dfrac{\text{lb steam (increased)}}{\text{h}}}{\dfrac{\text{MMBtu natural gas}}{\text{h}}}\right)\left(\frac{\text{MMBtu natural gas (supplemental)}}{\text{h}}\right). \tag{12.27}$$

Table 12.6 provides a summary of the energy rate in and energy rate out for each unit operation in the cogeneration system of Figure 12.3. The four columns (after the unit operation column) on the left-hand side are the inputs to each unit operation, and the three columns on the right-hand side are the power generated for each unit of power input. Table 12.6 provides cogeneration unit operation efficiencies for all major energy units. For example, from Table 12.6 for boiler 4, 1 MMBtu/h (1 MMBtu/h of natural gas) will produce 0.85461589 mlb steam/h or 8546.16 lb steam/h. Table 12.7 provides minimum and maximum capacities (energy rate out) for each operation in the cogeneration system.

The energy rate for the air cooler prior to the GE gas turbine is not included in this table, but this rate will be included in the final formulation. From Example 12.4, once the ambient air temperature and humidity have been fixed, we can predict the air cooler duty as a function of the power generated by the GE gas turbine. You should show yourself that the duty of the air cooler before the GE gas turbine is linear with respect to the natural gas flow (million British thermal unit per hour [MMBtu]) to the gas turbine once the ambient air temperature and humidity are fixed.

We can now provide an overview of our solution approach to the optimal dispatch problem. As indicated in Figure 12.10, we specify the electrical, steam, and cooling demands (energy rate out) of the process. These demands are independent of how the cogeneration system generates the energy streams. For example, if chilled water is

Table 12.6 Summary of Cogeneration Unit Operation Efficiencies

Unit Operation	In				Out		
	kW	Steam (mlb/h)	CW (ton)	Natural Gas (MMBtu/h)	kW	Steam (mlb/h)	Chilled Water (ton)
Boiler 4				1		0.85461589	
Chiller 1	1						$\frac{1}{0.646}$
Chiller 2	1						$\frac{1}{0.690}$
Chiller 3	1						$\frac{1}{0.604}$
Chiller 4	1						$\frac{1}{0.635}$
Chiller 5	1						$\frac{1}{0.655}$
Chiller 7	1						$\frac{1}{0.630}$
Chiller 8		1					87.7438
Chiller 9		1					88.2369
Chiller 10		1					89.1679
GE gas turbine			See below	1	107.7532(MMBtu/h) − 4078.15		See below
Boiler 8				1		0.35705(MMBtu/h) + 9.437	
Boiler 8 supplemental				1		0.9705852	
Allison gas turbine (air cooler duty included)				1			0.9828 × (5594.1/ 42.8436514) × (1 + 0.002 × (84.9 − Air Temp)) × (MMBtu/h)
Allison boiler 7				1		0.167019(MMBtu/h) + 13.1594	
Boiler 7 supplemental				1		0.9738567	

generated from steam (chillers 8–10), the additional steam load (above the process needs) is internal to the cogeneration system and will be accounted for by the equations in Table 12.6 and the system constraints, as we will see later in this chapter. The optimal solution to the energy dispatch is found by varying which units are operational (on/off), at what input energy rates, so as to meet the required electrical, steam, and chilled water demands of the process while minimizing the cost of purchased natural gas and purchased electricity.

For each of the operations in the cogeneration system, once the ambient air temperature and humidity are fixed, we have related the energy rate out to the energy rate in using linear equations. This is easily seen for the units in Table 12.6, and in the previous paragraph, we discussed the linear energy relation for the air cooler. There is one more important detail, the pricing of electricity and natural gas, that we need to consider before presenting the details of the dispatch formulation and solution.

12.5 PREDICTING THE COST OF NATURAL GAS AND PURCHASED ELECTRICITY

We want to determine the best operating strategy for an existing cogeneration system. As indicated in Figure 12.10, for the cogeneration system studied here (see Figure 12.3), the problem is to determine the best mix of purchased natural gas and purchased electricity to meet the campus (process)

Table 12.7 Summary of Cogeneration Unit Operation Capacities

Unit Operation	Minimum kW	Minimum Steam (mlb/h)	Minimum CW (ton)	Maximum kW	Maximum Steam (mlb/h)	Maximum CW (ton)
Boiler 4		20			100	
Chiller 1			320			1600
Chiller 2			400			2000
Chiller 3			200			1000
Chiller 4			340			1700
Chiller 5			400			2000
Chiller 7			220			1100
Chiller 8			500			2060
Chiller 9			500			2060
Chiller 10			500			2060
GE gas turbine	4000			20,100		
Boiler 8		36.2			89.6	
Boiler 8 supplemental		0			60.4	
Allison gas turbine (air cooler duty included)			2500			6400
Allison boiler 7		16.6			22.0	
Boiler 7 supplemental		0			71.5	

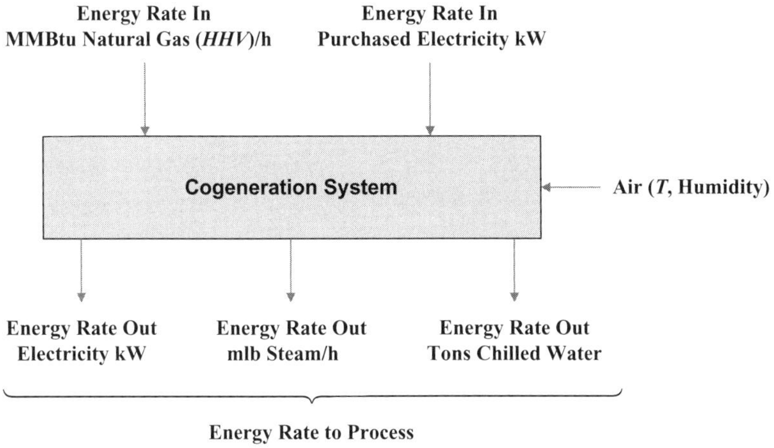

Figure 12.10 Overview of the energy dispatch solution process.

needs of electricity, steam, and chilled water. The optimal operating strategy can be developed without consideration of cogeneration capital equipment costs, or operation and maintenance (O&M) costs, although these costs certainly impact the ultimate profitability of the system. The cogeneration system requires the purchase of natural gas and may require electricity purchase. The cost of natural gas is straightforward. Determining the cost of purchased electricity is more complex.

12.5.1 Natural Gas Cost

The cost of natural gas is based on the Henry Hub (NYMEX) cost plus additional charges for transportation and distribution. The Henry Hub price is quoted as dollar per MMBtu. For the cogeneration system, we will use the Henry Hub price + 0.18 in dollar per MMBtu for the cost of natural gas. Here, the $0.18 per MMBtu is the pipeline transportation and distribution cost. The MMBtu value is based on the *HHV* of natural gas, and do recall there is roughly a 10% difference between the *HHV* and lower heating value (*LHV*) for natural gas.

12.5.2 Purchased Electricity Cost

There are regional and local differences in how electricity is priced. Typically, the purchased cost of electricity includes at least three components. There is a base rate, which is generally set by the state public utilities commission. Second, there is a fuel adjustment charge that allows utilities to recover increased costs for both fuel and electricity purchase. Finally, for QFs needing standby electricity, there is a demand charge. In order to explain these charges, it is easiest to use 2006 cost data for purchased electricity at the cogeneration facility we are considering now.

Base Rate

The base rate must be paid by the cogeneration system for any purchased electricity. Currently, this rate is set at \$0.01034 per kilowatt-hour.

Fuel Adjustment Charge

The fuel adjustment charge allows utilities to recover increased costs for both fuel and electricity purchase above the base rate. Because of the reliance of many utilities on natural gas for power generation, the fuel adjustment charge often tracks linearly with increased natural gas cost, with some leveling off generally observed at high natural gas prices.

The observed leveling off of the fuel adjustment charges at high natural gas prices can be attributed to a number of factors. First, most utilities have a mix of power generation capabilities including natural gas-fired units, nuclear power, coal-fired units, hydropower, and purchased power. This mix generally helps cap or level off the fuel adjustment charge even with increasing natural gas prices. Leveling off also occurs as a substantial percentage of the natural gas used by utility companies comes from long-term natural gas contacts. These long-term contracts do, however, have a mixed benefit, as they shelter against fuel adjustment spikes, but they also tend to slow fuel adjustment declines as natural gas prices drop. Finally, leveling off also occurs because the utilities can carry forward (to future billing months) unrecovered fuel adjustment charges.

There is one final consideration to the fuel adjustment charge. The fuel adjustment charge on the electric utility bill is for fuel and electricity purchases actually made 2 months earlier. For example, the March electric bill (say, received sometime in early April) will show a March fuel adjustment charge, which is actually for fuel and electricity purchases made in February. Here see Problem 12.2.

In Figure 12.11, we have plotted the actual fuel adjustment charge against the monthly average Henry Hub price and the monthly average electricity spot market price. Both the Henry Hub price and the spot market price have been shifted forward 2 months to reflect the actual billing cycle.

Using this shifted data, we minimized the absolute value of the difference between the actual fuel adjustment cost and a predicted fuel adjustment cost:

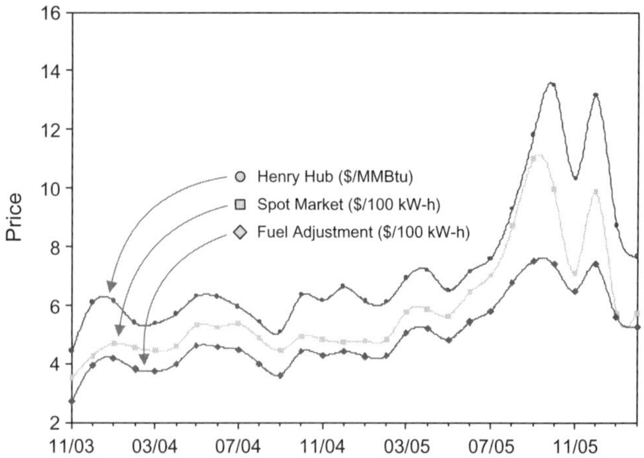

Figure 12.11 Henry hub, spot market, and actual fuel adjustment costs.

$$\text{Minimize}\left|\text{Actual fuel adjustment} - \text{Predicted fuel adjustment}\right|, \tag{12.28}$$

where the predicted fuel adjustment is given by the polynomial expression

$$\text{Predicted fuel adjustment} = a + b\times(HH) + c\times(HH)^2 + d\times(SM) + e\times(SM)^2, \tag{12.29}$$

where (HH) is the shifted Henry Hub fuel price (dollar per MMBtu) and (SM) is the shifted spot market electricity price (dollar per 100 kW-h). The best fit for the coefficients in the predicted fuel adjustment equation was found as

a	*b*	*c*	*d*	*e*
−2.8272051	0.787515392	−0.03083611	0.889166	−0.036752

Demand Charge

Qualifying industrials often want to purchase part of their electrical load from the local utility or have the ability to obtain backup power when needed. Industrials must pay a fee, termed a demand charge, for this standby power regardless if the power is actually used. For a qualified cogeneration facility, the demand charge generally consists of a billing load charge, which for the billing months of May–October was \$6.49 per kilowatt, and for November–April was \$5.67 per kilowatt.

There are certain stipulations applied to the demand load charge in order to determine the applicable kilowatt. In general, the applicable kilowatt is the larger of

(a) 75% of the current contract power or

(b) The maximum measured 30-minute demand.

This requires explanation. Assume that the current contract power is 17.5 MW. That means that during the peak periods of 8 AM–11 PM, Monday–Friday, up to 17.5 MW of electricity can be used without changing the current contract. During off-peak periods, 11 PM–8 AM, Monday–Friday and all day Saturday and Sunday, up to 17.5 (3/2) = 26.25 MW of electricity can be used without changing the current contract. Under the conditions that neither 17.5 MW (peak) nor 26.25 MW (off-peak) are violated, the demand load charge would be billed at 75% of the current contract price for 17.5 MW. For May, this cost would be

$$(17.5\text{ MW})\left(1000\frac{\text{kW}}{\text{MW}}\right)(75\%)\left(6.49\frac{\$}{\text{kW}}\right)$$
$$= \$85{,}181.25\text{per month}$$

or

$$\frac{\$85{,}181.25}{\left(24\dfrac{\text{h}}{\text{day}}\right)\left(31\dfrac{\text{days}}{\text{May}}\right)} = \$114.49/\text{h}.$$

However, if the maximum measured 30-minute demand exceeds 17.5-MW peak period contract power, then 100% of this demand must be paid. For example, assume that peak demand in May is 20 MW (for some time exceeding 30 minutes). Then, the bill for May would show a demand charge,

$$(20\text{ MW})\left(1000\frac{\text{kW}}{\text{MW}}\right)(100\%)\left(6.49\frac{\$}{\text{kW}}\right)$$
$$= \$129{,}800\text{ per month}$$

or

$$\frac{\$129{,}800}{\left(24\dfrac{\text{h}}{\text{day}}\right)\left(31\dfrac{\text{days}}{\text{May}}\right)} = \$174.462\text{ per hour.}$$

Most important, this 20 MW would become the new contract power for the next 12-month period. In June, the contract power would be 20 MW and normal billing would be 75% of this 20 MW subject to conditions a and b mentioned earlier.

The demand charge actually supplies a constraint to the optimal power dispatch problem. For example, with contract power set at 17.5 MW, purchased power during peak periods would be limited to 17.5 MW and purchase power during nonpeak periods would be limited to 26.25 MW (17.5 MW × 3/2).

Additional Charges: Facility Charges

As discussed, even though a facility generates electricity, they also often want to purchase electricity, on a daily basis or on a standby basis from the local utility company. Demand, base, and fuel adjustment costs have been detailed earlier. However, in order to purchase electricity, the utility companies typically require some facility upgrades, and these costs must also be paid by the cogeneration facility. These costs are termed facility charges. For the cogeneration system described earlier, the facility charges are \$25,524 per month. Facility charges appear on the monthly bill, along with demand, base, and fuel adjustment charges.

> Monthly utility charges include fixed costs (regardless if any electricity is purchased) consisting of facility charges, and demand charges + variable costs (based on the amount of electricity purchased) consisting of base rate charges and fuel adjustment charges.

Electricity Purchased

Purchased electricity is billed as a running total of all the electricity used during each month. Electricity usage is recorded on an instantaneous basis with both the running total (kilowatt-hour) and average kilowatt determined at the end of each month. This billing average can be viewed as

$$\frac{\sum^{\text{month}}(\text{kW used over period}_i)(\text{period}_i)}{\sum^{\text{month}}(\text{period}_i)}.$$

For example, if over a 2-hour period we used (17,000 kW for 15 minutes) + (18,000 kW for 10 minutes) + (24,000 kW for 5 minutes) + (16,000 kW for 90 minutes), the running total for this 2-hour period would be

$$\begin{matrix}(4250\text{ kW-h})+(3000\text{ kW-h})\\+(2000\text{ kW-h})+(24{,}000\text{ kW-h})\end{matrix} = 33{,}250\text{ kW-h},$$

and the 2 hour period average would be

$$\frac{33{,}250\text{ kW-h}}{2\text{ h}} = 16{,}625\text{ kW}.$$

EXAMPLE 12.5 ***Cost of Purchased Electricity***

The average kilowatt purchased by the cogeneration facility in May was 15,225 kW, and the cogeneration facility has a contract power of 17,500 kW, which was never exceeded (for a 30-minute period). The fuel adjustment charge is estimated at \$0.05085 per kilowatt-hour, the base rate is \$0.01034 per kilowatt-hour, and the facility charge is \$25,524 per month. Determine the hourly and monthly purchased utility charge.

Variable Costs

Base rate	15,225 kW ($0.01034 per kilowatt-hour) = $157.4265 per hour
Fuel adjustment	15,225 kW ($0.05085 per kilowatt-hour) = $774.19125 per hour

Fixed Costs

Facility charges	$25,524 per month (month May/31 day) (1 day/24 hour) = $34.30645/h
Demand billing	(0.75) (17,500 kW) = 13,125 kW, as the contract power is not exceeded
Demand charge	13,125 kW (6.49 $/kW) = $85,181.25 per month = $114.49093/h

Total cost = Variable + Fixed charges

= $931.61775 + $148.79738

= $1080.41513 per hour or $803,828.86 per month or 7.096¢ per kilowatt-hour. ■

12.6 DEVELOPMENT OF A MULTIPERIOD DISPATCH MODEL FOR THE COGENERATION FACILITY

The power dispatch model for the cogeneration system can be formulated as an MILP optimization problem:

Minimize Costs = Natural gas cost + Fixed electricity cost + Variable electricity cost

Subject to
- Material and energy balances—Table 12.6 equipment efficiencies
- Table 12.7 equipment capacities
- Logical constraints
- Known needs of electricity, steam, and chilled water for the period

The easiest approach to understanding the formulation of the binary integer power dispatching problem is to first examine a simplified version of the problem. The complete the problem, involving all units, all needed utilities (steam, electricity, and chilled water), peak and nonpeak demand, and variable and fixed costs, is solved in Example 12.7.

EXAMPLE 12.6 ***Simplified Dispatch Problem***

Let us consider a simplified version of the power dispatch problem for the cogeneration system of Figure 12.3. Here we will limit our discussion to only three units: the GE gas turbine, its HRSG (boiler 8), and the stand-alone boiler (boiler 4). We will assume only steam and electricity are needed in the amounts of 25,000 kW and 100 mlb/h. The costs for the problem are $7.781 per MMBtu (*HHV*) delivered natural gas and purchased electricity at $0.06119 per kilowatt-hour. Cogeneration unit operation efficiency and capacity data are provided in Tables 12.6 and 12.7, respectively.

SOLUTION An overview of this dispatch problem is provided in Figure 12.12. The solution begins with variable definitions and then the formulation of process constraints.

Variables

We define eight decision or adjustable variables to solve this problem. Let the binary integer variable δ_i denote whether equipment i is operating:

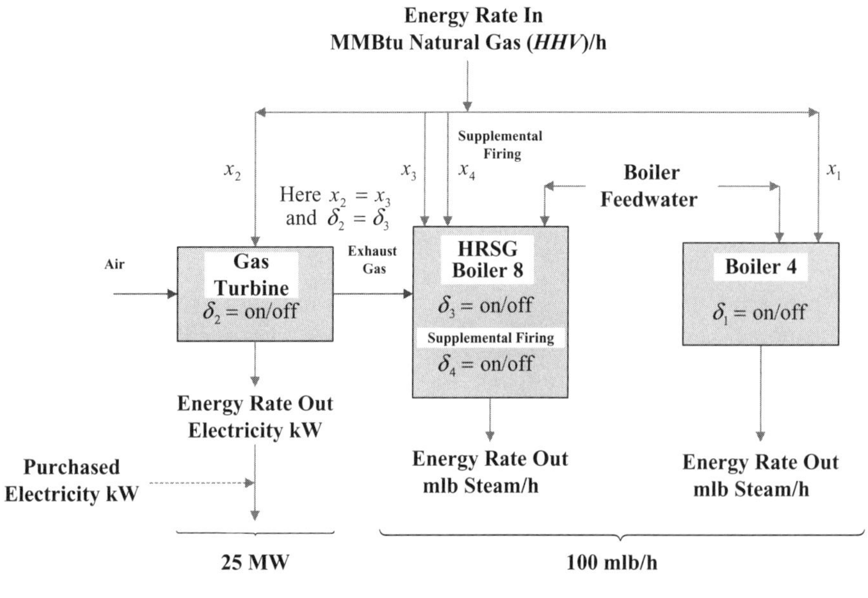

Figure 12.12 Simplified cogeneration system dispatch problem.

$$\delta_1 = \begin{cases} 1 & \text{if boiler 4 is on} \\ 0 & \text{if boiler 4 is off} \end{cases}$$

$$\delta_2 = \begin{cases} 1 & \text{if GE gas turbine is on} \\ 0 & \text{if GE gas turbine is off} \end{cases}$$

$$\delta_3 = \begin{cases} 1 & \text{if HRSG (boiler 8) is on} \\ 0 & \text{if HRSG (boiler 8) is off} \end{cases}$$

$$\delta_4 = \begin{cases} 1 & \text{if boiler 8 supplemental firing is on} \\ 0 & \text{if boiler 8 supplemental firing is off} \end{cases}.$$

And we define the continuous variables, x_i, as

x_1 = the amount of natural gas (MMBtu per hour) used by boiler 4

x_2 = the amount of natural gas (MMBtu per hour) used by GE gas turbine

x_3 = the amount of natural gas (MMBtu per hour) used by boiler 8 for steam production

x_4 = the amount of natural gas (MMBtu per hour) used by boiler 8 supplemental firing

The eight variables, δ_i and x_i, will be varied by the optimization code to solve for the minimum cost (defined next) subject to the natural processing constraints of the problem. The need for binary variables will be apparent as we form the constraints to the problem.

Equipment Constraints

$\delta_2 = \delta_3$ — If the gas turbine is operating, boiler 8 must be on, or both units must be off.

$x_2 = x_3$ — The fuel to the gas turbine is also used in boiler 8.

$\delta_4 \le \delta_3$ — If there is no supplemental firing of boiler 8, then $\delta_4 = 0$. Supplemental firing of boiler 8 occurs with $\delta_4 = 1$, and this is only allowed if both the gas turbine and boiler 8 are operating $\delta_3 = 1$.

Electricity Constraint (25,000-kW Demand)

$$\text{Imported electricity (a calculated value)} \ge 0$$

$$\text{Imported electricity} + \text{Self-generated electricity} = \text{Process electricity demand}$$

Using Table 12.6 and the problem statement,

$$\text{Imported electricity} + \delta_2(107.7532 \times x_2 - 4078.15) = 25{,}000\ \text{kW}.$$

Steam Constraint (100-mlb/h Demand)

$$\text{Boiler 8} + \text{Boiler 8 supplemental steam} + \text{Boiler 4} = \text{Process steam demand}$$

Using Table 12.6 and the problem statement,

$$\begin{matrix} \delta_3(0.35705 \times x_3 + 9.437) + \delta_4(0.9705852 \times x_4) \\ +\delta_1(0.85461589 \times x_1) \end{matrix} = 100\ \text{mlb/h}.$$

Natural Gas Constraint

GE gas turbine + Boiler 8 supplemental natural gas + Boiler 4 = Natural gas purchased

$$(\delta_2 \times x_2) + (\delta_4 \times x_4) + (\delta_1 \times x_1) = \text{Natural gas purchased (a calculated value)}$$

Equipment Capacity Constraints (Tables 12.6 and 12.7):

Boiler 4:	20 mlb/h ≤ $(0.85461589 \times x_1)$ ≤ 100 mlb/h
GE gas turbine:	4000 kW ≤ $(107.7532 \times x_2 - 4078.15)$ ≤ 20,100 kW
Boiler 8:	36.2 mlb/h ≤ $(0.35705 \times x_3 + 9.437)$ ≤ 89. 6 mlb/h
Boiler 8 supplemental:	0 mlb/h ≤ $(0.9705852 \times x_4)$ ≤ 60.4 mlb/h

For those not familiar with integer programming, the equipment capacity constraints require explanation. Equipment capacity constraints represent physical bounds on equipment operation that must be satisfied. The energy rate *actually produced* from each unit is a calculated output multiplied by the appropriate on/off binary variable. For example, boiler 4 equipment capacity constraint requires that

$$x_1 \ge \frac{20\dfrac{\text{mlb}}{\text{h}}}{0.85461589\dfrac{\text{mlb/h}}{\text{MMBtu/h}}} = 23.402326\frac{\text{MMBtu}}{\text{h}}\ \text{of natural gas.}$$

However, the natural gas actually used by Boiler 4 = $(\delta_1 \times x_1)$. If $\delta_1 = 0$, then no natural gas will be purchased for boiler 4 regardless of the value for x_1 because boiler 4 is off. Minimum and maximum capacities of real equipment always exist; however, the equipment may be turned off. Unless the equipment minimum capacity is zero, the only way to handle this mathematically is to add a binary variable for each piece of equipment.

Objective Function

The objective function is to minimize the cost of purchased natural gas and purchased (imported) electricity. We can pick any time basis and here we use 1 hour of operation:

$$\begin{aligned} &\{(\delta_1 \times x_1) + (\delta_2 \times x_2) + (\delta_4 \times x_4)\} \times (\$7.781/\text{MMBtu}) \\ &+ \{25{,}000\ \text{kW} - \delta_2 \times (107.7532 \times x_2 - 4078.15)\} \\ &\times (\$0.06119/\text{kW-h}) \times (1\ \text{h}). \end{aligned}$$

Solution Technique

This problem is an MILP problem. The solution can be found in the Excel file **Example 12.6.xls** and in Figure 12.13. In the Excel file, the solution using either Excel Solver or the Excel add-in What's Best is provided. Variables and constraints are summarized in Table 12.8.

Using Solver

Example 12.6 can be directly solved using the standard Excel Solver routine. Note that in addition to the 15 process constraints, the four binary variables are declared in the Excel constraint box.

Using What's Best (See Installation, Chapter 1, Section 1.7)

What's Best is an add-in to Excel to allow solutions of MILP and MINLP and NLP problems. The constraints for the What's Best solution are provided below the equipment selection table in Figure 12.13.

When using What's Best, you will need to highlight the integer variables, select WB! → Integer . . . → Integer Type • Binary-WBBIN. You can supply a name for the selected binary variables in the workbook. For example, supply the name Binary; What's Best will actually name the selected binary variables WBBINBinary.

Also, when using What's Best for all the MILP and MINLP problems in this chapter, you will need to select WB! → Options → General → Linearization → Degree . . . Math → OK. See the What's Best Manual for other options.

Results

The spreadsheet solution consists of three basic parts. The conditions for the power dispatch problem are provided in the two tables at the top of the spreadsheet in Figure 12.13. The third table checks that the kW produced = kW used; the steam produced = steam used; and the natural gas purchased = natural gas used. The fourth table calculates the energy rate out based on the energy rate in (optimization variables). The minimum and maximum equipment capacities are provided and will be used in the constraints.

The optimization process minimizes the objective function by varying the natural gas feed in MMBtu/h to boiler 4, to the GE gas turbine and HRSG, and for supplemental firing of the HRSG. The optimum variable values are shown in cells B39:C42 in Figure 12.13. At the optimal solution, the GE gas turbine receives (IN) 224.385 MMBtu/h and the energy rate out (cell E40) is OUT = 107.7532 × 224.385 MMBtu/h − 4078.15 = 20,100 kW. As the GE gas turbine (ON/OFF) = 1, this 20,100 kW contributes to the amount of electricity produced as shown in the third table of Figure 12.13. Notice that boiler 4 receives (IN) 34.585 MMBtu/h and the energy rate out is OUT = 0.85461589 × 34.585 MMBtu/h = 29.6 mlb/h steam (cell E39). However, as boiler 4 (ON/OFF) = 0, this 29.6 mlb/h does not contribute to the amount of steam produced. The process variables and 15 process constraints are summarized in Table 12.8.

The important aspect of this problem is developing the formulation. The solution indicates that the gas turbine is producing maximum electricity and that the steam demand is met by using both the HRSG and supplemental firing of the HRSG. Boiler 4 is not operating.

It is also instructive to explore the cost of purchased (imported) electricity, which shuts down the GE gas turbine and heat recovery. Here, steam will be produced by the boiler (boiler 4) and all electricity will be purchased at *\$0.045 per kilowatt*. You can also check the impact on the original problem, of reducing the natural gas *HHV* British thermal unit content from 1054.8677 to 1000 Btu/SCF. The only change in the solution is the natural flow rate to the process will be increased from 222.92 to 235.15 mSCF/h. ■

EXAMPLE 12.7 *Complete Dispatch Problem*

We next examine the Excel solution to the complete dispatch problem for one set of peak summer utility needs. Open the Excel file **Example 12.7.xls**.

The spreadsheet solution again consists of three basic parts. The known conditions and requirements for the power dispatch problem are given in the table on the right-hand side of the spreadsheet shown in Figure 12.14. The utility requirements are specified as if no utility system were actually present. The utilities actually delivered may be increased by the optimization code. For example, the steam requirement is 38 mlb/h, but if the optimization procedure determines that the steam chillers are the most economical means of producing chilled water, then the steam actually delivered will be 38 mlb/h plus the steam needed for the steam chillers.

The Henry Hub natural gas price is input (\$7.601 per MMBtu) and \$0.18 will be added for delivery (\$7.781 per MMBtu). The ambient air conditions, electricity base rate, contract power, and the period (peak vs. off-peak, month for demand rate) are input. The 2-month prior average NYMEX Henry Hub natural gas (\$7.061 per MMBtu) and average spot market electricity (\$0.05731 per kilowatt-hour) costs are also input. With these values, the predicted fuel adjustment is calculated using Equation (12.29). Any value in the Input table (except the predicted fuel adjustment) can be changed by the user.

The next table, shown in Figure 12.15, checks that the kW produced = kW used; the steam produced = steam used; the natural gas purchased = natural gas used; and the chilled water produced = chilled water used. For the air cooler, we utilize Example 12.4, which shows that with ambient air to the GE gas turbine at 91°F and 42% relative humidity, the air cooler duty (ton) will be

$$\text{Air cooler duty (ton)} = 592.0111\left(\frac{\text{kW actually generated}}{19{,}270\ \text{kW}}\right).$$

Process Energy Dispatch Optimization

	Predicted	
NATURAL GAS		
LSU NG Price Delivered (HHV)	7.781	$/MMBTU
NG Purchased (HHV)	235.15	MMBTU/hr
Natural Gas (HHV)	1054.8677	BTU/SCF
NG flow rate (based on NG_HHV)	222.92	mSCF/hr
NG Cost	1829.68	$/hr
VARIABLE ELECTRICITY		
Rate	0.06119	$/kW-hr
Electricity Purchased	4900.0	kW
Electricity Cost	299.831	
TOTAL COST	2129.51	$/hr

INPUTS

Campus Electricity Demand	25000	kW
Campus Steam Demand	100	mlb/hr

ELECTRICITY BALANCE			
PRODUCED (kW)		USED (kW)	
GE Gas Turbine	20100	Campus	25000
Imported	4900		
SUM	25000	SUM	25000

STEAM BALANCE			
PRODUCED (mlb/hr)		USED (mlb/hr)	
Boiler 8	89.6	Campus	100.0
Boiler 8 Supp.	10.4		
Boiler 4	0.0	SUM	100.0
SUM	100.0		

NATURAL GAS			
PURCHASED (MMBTU/hr)		USED (MMBTU/hr)	
SUM	235.1	GE Gas Turbine	224.4
		Boiler 8 Supp.	10.8
		Boiler 4	0.0
		SUM	235.1

EQUIPMENT SELECTION

NAME	ON/OFF	IN	OUT	EFFICIENCY	MAX CAPACITY	MIN CAPACITY
Boiler 4	0	34.6 MMBTU/hr	29.6 mlb/hr	85.9 %	100 mlb/hr	20 mlb/hr
GE GasTurbine	1	224.4 MMBTU/hr	20100 kW	30.6 %	20100 kW	4000 kW
Boiler 8	1	224.4 MMBTU/hr	89.6 mlb/hr	40.3 %	89.6 mlb/hr	36.2 mlb/hr
Boiler 8 Supp.	1	10.8 MMBTU/hr	10.4 mlb/hr	97.9 %	60.4 mlb/hr	0.0 mlb/hr

CONSTRAINTS for use in What's Best Solution

NAME	CONSTRAINT
GE Gas Turbine = Boiler 8 binary	=
GE Gas Turbine = Boiler 8 mcf/hr	=
Boiler 8 Supp <= Boiler 8 binary	=<=
Electricity Imported >= 0	>=
Electricity Produced = Consumed	=
Steam Produced = Consumed	=
Natural Gas Purchased = Used	=

Equipment Capacity	MAX	MIN
Boiler 4	<=	>=
GE Gas Turbine	=<=	>=
Boiler 8	<=	>=
Boiler 8 Supp.	<=	>=

Figure 12.13 Solution to the simplified cogeneration system dispatch problem.

The value 592.011 is provided in cell F38. The actual air cooler duty depends on the kilowatt generated by the GE gas turbine, with the final value in cell E38. Notice that the air cooler duty adds to the total demand.

The next table, shown in Figure 12.16, calculates the energy rate out based on the energy rate in (optimization variables). Minimum and maximum equipment capacities are provided and are used in the constraints.

The equipment selection table as shown in Figure 12.16 provides the optimum variable values (cells B51:C66); for example, at the optimal solution, chiller 1 is on and it receives (IN) 1033.6 kW of electricity. The energy rate out of chiller 1 is OUT = (1033.6 kW)

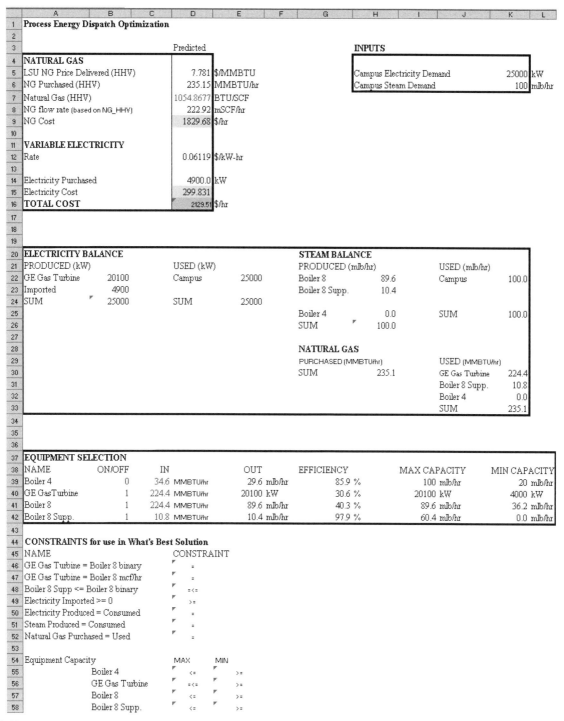

Table 12.8 Descriptive Form of Problem Variables and Constraints for Example 12.6

$\delta_{Boiler4}$, $\delta_{GE\ Turbine}$, $\delta_{HRSG\ (8)}$, $\delta_{HRSG\ Sup\ Fire\ (8)}$	4 binary variables
$x_{Boiler4}$, $x_{GE\ Turbine}$, $x_{HRSG\ (8)}$, $x_{HRSG\ Sup\ Fire\ (8)}$	4 continuous variables (MMBtu/h) IN
20 mlb/h $\leq$ Boiler 4 OUT = $(0.85461589\ x_{Boiler4}) \leq$ 100 mlb/h	2 equipment capacity constraints, boiler 4
4000kw $\leq$ GE Gas Turbine OUT = $(107.7532\ x_{GE\ Turbine} - 4078.15) \leq$ 20100 kW	2 equipment capacity constraints, GE gas turbine
36.2 mlb/h $\leq$ HRSG (8) OUT = $(0.35705\ x_{HRSG\ (8)} + 9.437) \leq$ 89.6 mlb/h	2 equipment capacity constraints, HRSG (8)
0 mlb/h $\leq$ HRSG (8) Sup Fire OUT = $(0.9705852\ x_{HRSG\ Sup\ Fire\ (8)}) \leq$ 60.4 mlb/h	2 equipment capacity constraints, HRSG (8) SF
Imported electricity ≥ 0	1 electricity constraint
Elec produced = Imported elec + $\delta_{GE\ Turbine}\ (107.7532\ x_{GE\ Turbine} - 4078.15) \leq$ 25,000 Kw	1 electricity constraint
Steam produced $= \delta_{Boiler\ 4}(0.85461589 x_{Boiler\ 4}) + \delta_{HRSG\ (8)}(0.35705 x_{HRSG\ (8)} + 9.437) + \delta_{HRSG\ Sup\ Fire\ (8)}(0.9705852 x_{HRSG\ Sup\ Fire\ (8)}) = 100$ mlb/h	1 steam constraint
NG Purchased = NG Used $= \delta_{Boiler\ 4}(x_{Boiler\ 4}) + \delta_{GE\ Turbine}(x_{GE\ Turbine}) + \delta_{HRSG\ Sup\ Fire\ (8)}(x_{HRSG\ Sup\ Fire\ (8)})$	1 NG constraint
$\delta_{GE\ Turbine} = \delta_{HRSG\ (8)}$	1 equipment constraint
$x_{GE\ Turbine} = x_{HRSG\ (8)}$	1 equipment constraint
$\delta_{HRSG\ Sup\ Fire\ (8)} \leq \delta_{HRSG\ (8)}$	1 equipment constraint

Process Energy Dispatch Optimization

	Predicted	
NATURAL GAS		
LSU NG Price (HHV)	7.781	\$/MMBTU
NG Purchased (HHV)	167.36	MMBTU/hr
NG flow rate (based on NG_HHV)	158.65	mSCF/hr
NG Cost	1302.20	\$/hr
VARIABLE ELECTRICITY		
Base Rate	0.01034	\$/kW-hr
Fuel Adjustment	0.05085	\$/kW-hr
Electricity Purchased	17500.0	kW
Variable Electricity Cost	1070.79	\$/hr
FIXED ELECTRICITY		
Facilities Charge	35.45	\$/hr
Demand Rate	6.49	\$/kW-month
Demand Billing Load	13125	kW
Demand Charge	118.31	\$/hr
Fixed Electricity Cost	153.76	\$/hr
TOTAL COST	2526.75	\$/hr

INPUTS

Henry Hub (NYMEX) Price	7.601	\$/MMBTU
Natural Gas (HHV)	1054.8677	BTU/SCF
Campus Electricity Demand	25000	kW
Campus Steam Demand	38	mlb/hr
Campus CHW Demand	12500	tons
Ambient Air Temperature	91	deg F
Relative Humidity	42	%
Electricity Base Rate	0.01034	\$/kW-hr
May-Oct Demand Rate	6.49	\$/kW
Nov-Apr Demand Rate	5.67	\$/kW
Contract Power	17500	kW
Prior Monthly Average NYMEX	7.061	\$/MMBTU
Prior Monthly Average Spot Market	0.05731	\$/kW-hr
Predicted Fuel Adjustment	0.05085	\$/kW-hr
Peak Period	1	0 or 1
Off Peak Period	0	0 or 1
May-Oct	1	0 or 1
Nov-Apr	0	0 or 1

Figure 12.14 Costing data cogeneration system dispatch problem.

(1 ton/0.646 kW) = 1600 tons of refrigeration (cell E51). As chiller 1 (ON/OFF) = 1, this 1600 tons contributes to the amount of chilled water produced as shown in Figure 12.15. For chiller 2, OUT = (276 W) (1 ton/0.69 kW) = 400 tons, but ON/OFF = 0, so no refrigeration will actually be available from chiller 2 as shown in Figure 12.15. For the GE gas turbine, the input is 146.756 MMBtu/h and the output (cell E61) is OUT = 107.7532 × 146.756 MMBtu/h − 4078.15 = 11,735 kW; here ON/OFF = 1. The process variables and constraints are summarized in Table 12.9.

The formulation of Example 12.7 parallels Example 12.6, but there are several differences. For example, in Example 12.6, the electricity and steam usage were fixed at 25,000 kW and 100 mlb, respectively. For Example 12.7, the amount of electricity, steam, and chilled water actually used is not known until the final equipment mix and energy inputs to the equipment are determined. However, the electricity, steam, and chilled water demands of the plant (campus) are known for the given period. This is reflected in the constraints in Table 12.9 and is described next.

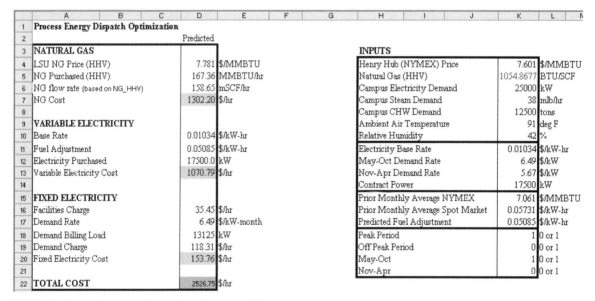

	A	B	C	D	E	F	G	H	I	J	K
24	**ELECTRICITY BALANCE**						**STEAM BALANCE**				
25	PRODUCED (kW)			USED (kW)			PRODUCED (mlb/hr)			USED (mlb/hr)	
26	GE Gas Turbine	11735		Campus	25000		Boiler 8	61.8		Campus	38.0
27	Imported	17500		Chiller 1	1034		Boiler 8 Supp.	0.0		Chiller 8	0.0
28	SUM	29235		Chiller 2	0		Boiler 7	16.6		Chiller 9	17.3
29				Chiller 3	604		Boiler 7 Supp.	0.0		Chiller 10	23.1
30	MAX IMPORTED (kW)			Chiller 4	1080		Boiler 4	0.0		SUM	78.4
31		17500		Chiller 5	825		SUM	78.4			
32				Chiller 7	693						
33				SUM	29235						
34											
35	**CHILLED WATER BALANCE**						**NATURAL GAS**				
36	PRODUCED (tons)			USED (tons)			PURCHASED (MMBTU/hr)			USED (MMBTU/hr)	
37	Chiller 1	1600		Campus	12500		SUM	167.4		GE G T	146.8
38	Chiller 2	0		Air Cooler	361	592.01				Boiler 8 Supp.	0.0
39	Chiller 3	1000		SUM	12861					Allison G T	20.6
40	Chiller 4	1700								Boiler 7 Supp.	0.0
41	Chiller 5	1260								Boiler 4	0.0
42	Allison Chiller	2611								SUM	167.4
43	Chiller 7	1100									
44	Chiller 8	0									
45	Chiller 9	1529									
46	Chiller 10	2060									
47	SUM	12861									

Figure 12.15 Equipment levels and constraints for the cogeneration system dispatch problem.

	A	B	C–D	E–F	G	I–J	K–L
49	**EQUIPMENT SELECTION**						
50	NAME	ON/OFF	IN	OUT	EFFICIENCY	MAX CAPACITY	MIN CAPACITY
51	Chiller 1	1	1033.6 kW	1600.0 tons	544.4 %	1600 tons	320 tons
52	Chiller 2	0	276.2 kW	400.3 tons	509.7 %	2000 tons	400 tons
53	Chiller 3	1	604.0 kW	1000.0 tons	582.3 %	1000 tons	200 tons
54	Chiller 4	1	1079.5 kW	1700.0 tons	553.8 %	1700 tons	340 tons
55	Chiller 5	1	825.2 kW	1259.8 tons	536.9 %	2000 tons	400 tons
56	Chiller 7	1	693.0 kW	1100.0 tons	558.2 %	1100 tons	220 tons
57	Chiller 8	0	5.7 mlb/hr	500.0 tons	93.5 %	2060 tons	500 tons
58	Chiller 9	1	17.3 mlb/hr	1529.5 tons	94.1 %	2060 tons	500 tons
59	Chiller 10	1	23.1 mlb/hr	2060.0 tons	95.1 %	2060 tons	500 tons
60	Boiler 4	0	23.7 MMBTU/hr	20.3 mlb/hr	85.9 %	100 mlb/hr	20 mlb/hr
61	GE Gas Turbine	1	146.8 MMBTU/hr	11735 kW	27.3 %	20100 kW	4000 kW
62	Boiler 8	1	146.8 MMBTU/hr	61.8 mlb/hr	42.5 %	89.6 mlb/hr	36.2 mlb/hr
63	Boiler 8 Supp.	0	0.3 MMBTU/hr	0.3 mlb/hr	97.9 %	60.4 mlb/hr	0.0 mlb/hr
64	Allison G T/Chill	1	20.6 MMBTU/hr	2611 tons	152.1 %	6400 tons	2500 tons
65	Boiler 7	1	20.6 MMBTU/hr	16.6 mlb/hr	81.0 %	22 mlb/hr	16.6 mlb/hr
66	Boiler 7 Supp.	0	0.2 MMBTU/hr	0.2 mlb/hr	97.9 %	71.5 mlb/hr	0 mlb/hr

Figure 12.16 Optimal equipment levels and capacities for the cogeneration system.

Electricity Constraints

$$\text{Electric power imported} \geq 0$$

Electric power imported ≤ Maximum allowed input without contract power violation

The sum of the electric power generated
+ Electricity purchased
= Electric power used
(Campus electric + Chillers 1, 2, 3, 4, 5, 7)

Steam Constraint

Steam used (Campus steam + Chillers 8, 9, 10)
= Steam produced (Boiler 8 + Boiler 8 supplemental
+ Boiler 7 + Boiler 7 supplemental + Boiler 4)

Chiller Water Constraint

Chilled water used (Campus + GE air cooler)
= Chilled water produced
(Chillers 1, 2, 3, 4, 5, 7, 8, 9, 10, Allison)

Table 12.9 Descriptive Form of Problem Variables and Constraints for Example 12.7

$\delta_{\text{Chiller 1}}, \delta_{\text{Chiller 2}}, \delta_{\text{Chiller 3}}, \delta_{\text{Chiller 4}}, \delta_{\text{Chiller 5}}, \delta_{\text{Chiller 7}}, \delta_{\text{Chiller 8}}, \delta_{\text{Chiller 9}}, \delta_{\text{Chiller 10}}, \delta_{\text{Boiler4}}, \delta_{\text{GE Turbine}}, \delta_{\text{HRSG (8)}}, \delta_{\text{HRSG Sup Fire (8)}}, \delta_{\text{Allison Turbine}}, \delta_{\text{HRSG (7)}}, \delta_{\text{HRSG Sup Fire (7)}}$	16 binary variables
$x_{\text{Chiller 1}}, x_{\text{Chiller 2}}, x_{\text{Chiller 3}}, x_{\text{Chiller 4}}, x_{\text{Chiller 5}}, x_{\text{Chiller 7}}, x_{\text{Chiller 8}}, x_{\text{Chiller 9}}, x_{\text{Chiller 10}}, x_{\text{Boiler4}}, x_{\text{GE Turbine}}, x_{\text{HRSG (8)}}, x_{\text{HRSG Sup Fire (8)}}, x_{\text{Allison Turbine}}, x_{\text{HRSG (7)}}, x_{\text{HRSG Sup Fire (7)}}$	16 continuous variables (kW, mlb/h [steam], MMBtu/h)
$320 \text{ tons} \le \text{Chiller 1 OUT} = \left(\frac{1}{0.646} x_{\text{Chiller 1}}\right) \le 1600 \text{ tons}$	2 equipment capacity constraints, chiller 1
$400 \text{ tons} \le \text{Chiller 2 OUT} = \left(\frac{1}{0.690} x_{\text{Chiller 2}}\right) \le 2000 \text{ tons}$	2 equipment capacity constraints, chiller 2
$200 \text{ tons} \le \text{Chiller 3 OUT} = \left(\frac{1}{0.604} x_{\text{Chiller 3}}\right) \le 1000 \text{ tons}$	2 equipment capacity constraints, chiller 3
$340 \text{ tons} \le \text{Chiller 4 OUT} = \left(\frac{1}{0.635} x_{\text{Chiller 4}}\right) \le 1700 \text{ tons}$	2 equipment capacity constraints, chiller 4
$400 \text{ tons} \le \text{Chiller 5 OUT} = \left(\frac{1}{0.655} x_{\text{Chiller 5}}\right) \le 2000 \text{ tons}$	2 equipment capacity constraints, chiller 5
$220 \text{ tons} \le \text{Chiller 7 OUT} = \left(\frac{1}{0.630} x_{\text{Chiller 7}}\right) \le 1100 \text{ tons}$	2 equipment capacity constraints, chiller 7
500 tons ≤ Chiller 8 OUT = (87.7438 $x_{\text{Chiller 8}}$) ≤ 2060 tons	2 equipment capacity constraints, chiller 8
500 tons ≤ Chiller 9 OUT = (87.7438 $x_{\text{Chiller 9}}$) ≤ 2060 tons	2 equipment capacity constraints, chiller 9
500 tons ≤ Chiller 10 OUT = (87.1679 $x_{\text{Chiller 10}}$) ≤ 2060 tons	2 equipment capacity constraints, chiller 10
20 mlb/h ≤ Boiler 4 OUT = (0.85461589 x_{Boiler4}) ≤ 100 mlb/h	2 equipment capacity constraints, boiler 4
4000kw ≤ GE gas turbine OUT = (107.7532 $x_{\text{GE Turbine}}$ − 4078.15) ≤ 20,100 kW	2 equipment capacity constraints, GE gas turbine
36.2 mlb/h ≤ HRSG (8) OUT = (0.35705 $x_{\text{HRSG (8)}}$ + 9.437) ≤ 89.6 mlb/h	2 equipment capacity constraints, HRSG (8)
0 mlb/h ≤ HRSG (8) Sup Fire OUT = (0.9705852 $x_{\text{HRSG Sup Fire (8)}}$) ≤ 60.4 mlb/h	2 equipment capacity constraints, HRSG(8) SF
$2500 \text{ tons} \le \text{Allison gas turbine OUT} = \left\{(0.9828)\left(\frac{5594.1}{42.8436514}\right)(1 + 0.002(84.9 - \text{Air Temp}))\right\}(x_{\text{Allison Turbine}}) \le 6400 \text{ tons}$	2 equipment capacity constraints, Allison gas turbine
16.6 mlb/h ≤ HRSG (7) OUT = (0.167019 $x_{\text{HRSG (7)}}$ + 13.1594) ≤ 22.0 mlb/h	2 equipment capacity constraints, HRSG (7)
0 mlb/h ≤ HRSG (7) Sup Fire OUT = (0.9738567 $x_{\text{HRSG Sup Fire (7)}}$) ≤ 71.5 mlb/h	2 equipment capacity constraints, HRSG(7) SF
Imported electricity ≥ 0	1 electricity constraint
Imported electricity ≤ Contract power = 17500 kW (for Example 12.7)	1 electricity constraint
$\text{Elec produced} = \text{Imported elec} + \delta_{\text{GE Turbine}}(107.7532 x_{\text{GE Turbine}} - 4078.15) = \text{Elec used} = \text{Elec demand} + \delta_{\text{Chiller 1}}(x_{\text{Chiller 1}}) + \delta_{\text{Chiller 2}}(x_{\text{Chiller 2}}) + \delta_{\text{Chiller 3}}(x_{\text{Chiller 3}}) + \delta_{\text{Chiller 4}}(x_{\text{Chiller 4}}) + \delta_{\text{Chiller 5}}(x_{\text{Chiller 5}}) + \delta_{\text{Chiller 7}}(x_{\text{Chiller 7}})$	1 electricity constraint
$\text{Steam produced} = \delta_{\text{Boiler 4}}(0.85461589 x_{\text{Boiler 4}}) + \delta_{\text{HRSG (8)}}(0.35705 x_{\text{HRSG (8)}} + 9.437) + \delta_{\text{HRSG Sup Fire (8)}}(0.9705852 x_{\text{HRSG Sup Fire (8)}}) + \delta_{\text{HRSG (7)}}(0.167019 x_{\text{HRSG (7)}} + 13.1594) + \delta_{\text{HRSG Sup Fire (7)}}(0.9738567 x_{\text{HRSG Sup Fire (7)}}) = \text{Steam used} = \text{Steam demand} + \delta_{\text{Chiller 8}}(x_{\text{Chiller 8}}) + \delta_{\text{Chiller 9}}(x_{\text{Chiller 9}}) + \delta_{\text{Chiller 10}}(x_{\text{Chiller 10}})$	1 steam constraint

Table 12.9 (*Continued*)

NG used $= \delta_{\text{Boiler 4}}(x_{\text{Boiler 4}}) + \delta_{\text{GE Turbine}}(x_{\text{GE Turbine}}) + \delta_{\text{HRSG Sup Fire (8)}}(x_{\text{HRSG Sup Fire (8)}}) + \delta_{\text{Allison Turbine}}(x_{\text{Allison Turbine}}) + \delta_{\text{HRSG Sup Fire (7)}}(x_{\text{HRSG Sup Fire (7)}})$ = NG purchased	1 NG constraint
Chilled water produced $= \delta_{\text{Chiller 1}}\left(\frac{1}{0.646}x_{\text{Chiller 1}}\right) + \delta_{\text{Chiller 2}}\left(\frac{1}{0.690}x_{\text{Chiller 2}}\right) + \delta_{\text{Chiller 3}}\left(\frac{1}{0.604}x_{\text{Chiller 3}}\right) + \delta_{\text{Chiller 4}}\left(\frac{1}{0.635}x_{\text{Chiller 4}}\right) + \delta_{\text{Chiller 5}}\left(\frac{1}{0.655}x_{\text{Chiller 5}}\right) + \delta_{\text{Chiller 7}}\left(\frac{1}{0.630}x_{\text{Chiller 7}}\right) + \delta_{\text{Chiller 8}}(87.7438x_{\text{Chiller 8}}) + \delta_{\text{Chiller 9}}(88.2369x_{\text{Chiller 9}}) + \delta_{\text{Chiller 10}}(89.1679x_{\text{Chiller 10}}) + \delta_{\text{Allison Turbine}} \left\{(0.9828)\left(\frac{5594.1}{42.8436514}\right)(1+0.002(84.9-\text{Air Temp}))\right\}(x_{\text{Allison Turbine}})$ = Chilled water used = Chilled water demand + Air cooler demand	1 chilled water constraint
$\delta_{\text{GE Turbine}} = \delta_{\text{HRSG (8)}}$	1 equipment constraint
$x_{\text{GE Turbine}} = x_{\text{HRSG (8)}}$	1 equipment constraint
$\delta_{\text{HRSG Sup Fire (8)}} \le \delta_{\text{HRSG (8)}}$	1 equipment constraint
$\delta_{\text{Allison Turbine}} = \delta_{\text{HRSG (7)}}$	1 equipment constraint
$x_{\text{Allison Turbine}} = x_{\text{HRSG (7)}}$	1 equipment constraint
$\delta_{\text{HRSG Sup Fire (7)}} \le \delta_{\text{HRSG (7)}}$	1 equipment constraint

Natural Gas Constraint

Natural gas used (GE gas turbine + Boiler 8 supplemental
+ Allison gas turbine + Boiler 7 supplemental + Boiler 4)
= Natural gas purchased

Problem Size
This MILP problem involves 16 binary variables representing equipment on/off positions and 16 continuous variables representing the power input to the unit operations of the cogeneration system. There are 44 process constraints.

Solution Technique: Use What's Best (See Installation, Chapter 1, Section 1.7)
What's Best is an add-in to Excel to allow solutions of MILP and MINLP and NLP problems. What's Best provides a convenient and robust solution platform for MILP problems. The reduced size version (Academic Version) will directly solve Example 12.7 and all problems in this text.

Results
The MILP solution indicates that to meet this summer day utility load, maximum electricity should be purchased and the cost of operation will be \$2526.76 per hour. Cogeneration equipment position (on/off) and energy rates (in and out) are provided in Figure 12.16. ■

12.7 CLOSING COMMENTS

In this chapter, cogeneration equipment size and configuration were known and equipment was modeled assuming linear power efficiencies over appropriate ranges. The linear assumption allowed formulation of a multiperiod MILP power dispatching problem. Solution was possible once the utility loads and ambient conditions were fixed for a given time period. For some equipment, and all equipment over extended power ranges, accuracy may be improved by using nonlinear efficiencies. The reader interested in multiperiod formulations should also see Arivalagan et al. (1995), Iyer and Grossman (1997), Marechal and Kalitventzeff (2003), and Ashok and Banerjee (2003).

We developed equipment efficiency based on the fuel *HHV*—this is the natural choice as the energy rate in is based on the British thermal unit content of the fuel, and we pay for the fuel based on the *HHV*. In the previous chapters, where we developed cogeneration system material and energy balances, we used the fuel *LHV*—this is the natural choice in these chapters as the fuel *LHV* reflects the actual energy used because water is in the vapor state in the exhaust gas leaving the cogeneration system. In general, equipment manufacturers use natural gas *LHV* when reporting efficiencies, while *HHV* is used for pricing natural gas.

The formulation in this chapter made several simplifying assumptions. To simplify discussions, we use an average value for the fuel *HHV*. We ignored the energy requirements of the cooling towers as these loads are considered small in comparison with the rest of the cogeneration system. Capital costs for the cogeneration system have not been included in the formulation. These costs are important for the design phase but are not needed in the dispatch model. O&M costs could be included in the dispatch formulation. In general, O&M costs, including gas turbine repair/replacement insurance from the manufacturer, would range from \$30 to \$50 per kilowatt of generation capacity; for the 25-MW two-gas turbine cogeneration system studied here, the expected range would be \$750K to \$1.25 MM per year.

The energy dispatch model can help identify areas where increased instrumentation would be beneficial. The saving possible from the dispatch model may help justify the additional instrumentation. In the dispatch problem of Chapter 12, the Allison gas turbine requires additional instrumentation to allow the air inlet conditions to the gas turbine to be determined.

Not specifically shown in the examples of Chapter 12 is that the dispatch formulation may recommend that the 20-MW GE gas turbine be shut down in low electrical usage periods, especially 11 PM–7 AM and weekend periods. Although the expected savings when purchasing electricity as opposed to running the gas turbine approaches \$200 per hour over these time periods, in reality, this type of operational mode is questionable. Gas turbine systems cannot be cycled on and off over short time periods. The gas turbine can be forced to remain in operation by requiring $\delta_{\text{GE Turbine}} = 1.0$. Continued cycling of any unit may indicate that the unit is oversized.

We are now in a position to better understand the complex interaction between optimal cogeneration system design and optimal cogeneration system operation. In Chapter 10, we developed an optimal cogeneration system design based on a single set of process utility demands—a known full-load steam demand and a known full-load electricity demand. In Chapter 12, we developed optimal multiperiod power dispatch for a cogeneration facility. Here, the cogeneration system design was fixed and utility loads for the process were known, but these loads varied over time of day, day, and season. Clearly, an iterative or simultaneous solution of the combined cogeneration dispatch and cogeneration design problem should improve the overall economics when compared to solutions based solely on the optimal design for a single set of operating conditions or solely on optimal dispatching. Here readers should see Aguilar et al. (2007a,b, 2008).

In Chapters 12 and 13, we will address how the optimal full-load steam and electrical demands for the process can be determined. We will also address energy interactions between processes within a site and the site utility system.

Finally and perhaps most important, the efficiencies and costing provided in this chapter are not intended as baselines for unit operation performance or system costs. My hope is that this chapter will provide a blueprint you can follow and modify for optimal energy dispatch problems.

REFERENCES

Aguilar, O., S.J. Perry, J.-K. Kim, and R. Smith. 2007a. Design and optimization of flexible utility systems subject to variable conditions Part 1: Modelling framework. *Trans. IChemE A Chem. Eng. Res. Des.* 85(A8): 1136–1148.

Aguilar, O., S.J. Perry, J.-K. Kim, and R. Smith. 2007b. Design and optimization of flexible utility systems subject to variable conditions Part 2: Methodology and applications. *Trans. IChemE A Chem. Eng. Res. Des.* 85(A8): 1149–1168.

Aguilar, O., J.-K. Kim, S. Perry, and R. Smith. 2008. Availability and reliability considerations in the design and optimization of flexible utility systems. *Chem. Eng. Sci.* 63(14): 3569–3584.

Arivalagan, A., B.G. Raghavendra, and A.R.K. Rao. 1995. Integrated energy optimization model for a cogeneration based energy supply system in the process industry. *Electr. Power Energy Syst.* 17(4): 227–233.

Ashok, S. and R. Banerjee. 2003. Optimal operation of industrial cogeneration for load management. *IEEE Trans. Power Syst.* 18(2): 931–937.

Buckley, R.A. 2006. Overview of Cogeneration at LSU. MS Thesis, Louisiana State University.

Flatau, P.J., R.L. Walko, and W.R. Cotton. 1992. Polynomial fits to saturation vapor pressure. *J. Appl. Meteor.* 31: 1507–1513.

Iyer, R.R. and I.E. Grossmann. 1997. Optimal multiperiod operational planning for utility systems. *Comput. Chem. Eng.* 21(8): 787–800.

Marechal, F. and B. Kalitventzeff. 2003. Targeting the integration of multi-period utility systems for site scale process integration. *Appl. Therm. Eng.* 23: 1763–1784.

Navidi, W. 2008. *Statistics for Engineers and Scientists*. McGraw Hill, New York.

Pattison, J.R. and V. Sharma. 1980. Selection of boiler plant and overall system efficiency. Studies in Energy Efficiency in Buildings. British Gas, London.

Shang, Z. and A. Kokossis. 2004. A transhipment model for the optimization of steam levels of total site utility system for multiperiod operation. *Comput. Chem. Eng.* 28: 1673–1688.

PROBLEMS

12.1 ***Least Squares Linear and Nonlinear Data Regression—Single Independent Variable*** In data reconciliation, we determine the best values for measured variables such that material and energy balances are conserved. The same form of the objective function, as used in data reconciliation, can be used to fit measured data to linear and nonlinear process performance models. Here we can define the measured dependent variable as y^+ and the measured independent variable as x^+. A predictive model for the dependent variable, y^p, is developed using the independent variable, $y^p = f(x^+)$. Typical examples would include a linear model, $y^p = a_0 + a_1 x^+$; a polynomial model, $y^p = a_0 + a_1 x^+ + a_2 \left(x^+\right)^2$; or an exponential model $y^p = a_0 \exp(a_1 x^+)$. Using provided data, the least squares fit objective function (or the sum of the squares of the residuals between measured and predicted dependent variable) can be formed as

$$\text{Minimize} \quad \sum_{j=1}^{\substack{N \\ \text{data points}}} \left(y_j^+ - y_j^p\right)^2, \qquad \text{(P12.1a)}$$

where the coefficients a_0, a_1, a_2, and so on, in the predictive model will be varied to minimize the objective function. Once the optimization is complete (and the values for the coefficients determined), the standard error of the estimate is found as

$$\text{Standard error} = S_{y/x} = \sqrt{\frac{\sum_{j=1}^{\substack{N \\ \text{data points}}} \left(y_j^+ - y_j^p\right)^2}{N - \xi}}, \qquad \text{(P12.1b)}$$

where ξ is the degree of freedom lost from the determined coefficients. For example, when using $y^p = a_0 + a_1 x^+ + a_2 \left(x^+\right)^2$, $\xi = 3$ to account for a_0, a_1, and a_2; for linear predictive models, $\xi = 2$.

In the same fashion as the standard deviation, the standard error provides the "spread" of the data about the chosen performance model. Do recall that the standard deviation provides the spread of the data around the mean or average of the measured data as

$$\text{Standard deviation} = \sigma = \sqrt{\frac{\sum_{j=1}^{\substack{N \\ \text{data points}}} \left(y_j^+ - \bar{y}\right)^2}{N - 1}}, \qquad \text{(P12.1c)}$$

where the average

$$\bar{y} = \frac{\sum_{j=1}^{N} y_j^+}{N}.$$

Comparing the standard error and the standard deviation, y_j^p is replaced with $\bar{y}$, the average (or mean) of the measured data. We can also define a coefficient of determination, R^2:

$$R^2 = \frac{\left(\sum_{j=1}^{\substack{N \\ \text{data points}}} \left(y_j^+ - \bar{y}\right)^2\right) - \left(\sum_{j=1}^{\substack{N \\ \text{data points}}} \left(y_j^+ - y_j^p\right)^2\right)}{\left(\sum_{j=1}^{\substack{N \\ \text{data points}}} \left(y_j^+ - \bar{y}\right)^2\right)}, \qquad \text{(P12.1d)}$$

which provides indication of the improvement in the overall fit of the data as we move from the simplest model (the mean of available data) to a more complex model. The correlation coefficient r is defined as $r = \sqrt{R^2}$. For additional discussion, see Navidi (2008).

Consider that we want to fit steam saturation temperature to its corresponding enthalpy of vaporization; data are provided in Table P12.1.

Table P12.1 Saturated Steam Properties

x^+, Saturation Temperature, T (°F)	y^+, Enthalpy of Vaporization $\hat{h}_{\text{vap}}$ (Btu/lb)
200	977.9
240	952.3
280	924.9
320	895.3
360	862.9
400	826.8

Develop a general Excel sheet solution for a linear fit, a polynomial fit, and an exponential fit of the data. Also calculate the standard error, standard deviation, coefficient of determination, and correlation coefficient.

12.2 ***Least Squares Linear and Nonlinear Data Regression—Multiple Independent Variables*** Our discussions in Problem 12.1 hold true for data regression with either single or multiple independent variables. Here we examine data regression for a multivariable problem—developing a model to predict utility company fuel adjustment charges.

It is important to understand the general billing process for fuel adjustment as the fuel adjustment charge is often the major component of the electric utility bill. Consider electricity used in December. We pay for the December electricity usage in January. The utility company that produced the electricity in December will actually want to recover its costs for fuel (here primarily natural gas) and spot market electricity purchases made in November. So the electric bill paid in January has a fuel adjustment component based on November costs. For example, using Table P12.2, the utility bill received in January will have a fuel adjustment charge of $3.126 per 100 kW-h used. This fuel adjustment charge will be based in large part on fuel actually used in November (at $4.420 per MMBtu) and on spot market electricity purchases in November (at $3.500 per 100 kW-h). From the utility company perspective, all these costs average out. From the industrial consumer perspective, there may be opportunities to take advantage of the timing of the charges.

Develop a general polynomial fit for the fuel adjustment charge. The dependent variable y^+ is the actual fuel adjustment charge for a given month. There are two independent variables associated with y^+: the 1-month shifted average natural gas cost and the shifted average electricity spot market cost.

12.3 ***Nonlinear Equipment Efficiency*** Here modify the simple cogeneration dispatch formulation of Example 12.6. Assume that the needed steam rate is 50 mlb/h and account for a nonlinear boiler efficiency for boiler 8 as

Table P12.2 Key Components in the Electricity Fuel Adjustment Charge

Electricity Usage Month	Actual Fuel Adjustment Charge (\$/100 kW-h), y^+	Utility Cost Recovery Month	Average Natural Gas Cost for Recovery Month (\$/MMBtu)	Average Spot Market Electricity Cost for Recovery Month (\$/100 kW-h)	Predicted Fuel Adjustment Charge, y^p (\$/100 kW-h) See Solution Least Squares	Predicted Fuel Adjustment Charge, y^p (\$/100 kW-h) See Solution Absolute Value
December	3.126	November	4.420	3.500	3.135	3.130
January	3.566	December	6.085	4.268	3.419	3.342
February	3.780	January	6.137	4.684	3.818	3.780
March	4.092	February	5.398	4.555	4.021	4.043
April	3.859	March	5.376	4.434	3.905	3.917
May	3.652	April	5.700	4.598	3.944	3.942
June	3.965	May	6.300	5.334	4.354	4.375
July	4.643	June	6.292	5.260	4.290	4.303
August	4.649	July	5.937	5.373	4.582	4.649
September	4.464	August	5.423	4.874	4.330	4.386

Boiler 8: $36.2 \text{ mlb/h} \le (0.003254 \times x_3 + 0.001817 \times x_3^2)$
$\le 89.6 \text{ mlb/h},$

where x_3 = the amount of natural gas (MMBtu per hour) used by boiler 8. Please note there is no physical significance to this energy equation; it is simply a nonlinear equation that can be substituted in the problem formulation.

12.4 ***Cost of Purchased Electricity—Off-Peak*** Determine the hourly charge if for Example 12.5 we are at off-peak conditions and 20,000 kW is purchased from the local utility.

12.5 ***Optimal Dispatch Winter Day (Solution Provided)*** Determine the optimal dispatch for a winter day with conditions as shown in Figure P12.5.

INPUTS

Henry Hub (NYMEX) Price	9.41	\$/MMBtu
Natural Gas (HHV)	1054.8677	Btu/SCF
Campus Electricity Demand	23735	kW
Campus Steam Demand	40.731	mlb/h
Campus CHW Demand	3526	tons
Ambient Air Temperature	33	°F
Relative Humidity	33	%
Electricity Base Rate	0.01034	\$/kW-h
May-October Demand Rate	6.49	\$/kW
November-April Demand Rate	5.67	\$/kW
Contract Power	17500	kW
Prior Monthly Average NYMEX	13.157	\$/MMBtu
Prior Monthly Average Spot Market	0.09897	\$/kW-h
Predicted Fuel Adjustment	0.07396	\$/kW-h
Peak Period	0	0 or 1
Off-Peak Period	1	0 or 1
May-October	0	0 or 1
November-April	1	0 or 1

Figure P12.5 Conditions for winter day.

Solution:
The solution is provided in **Problem 12.2.xls**.

12.6 ***Formulation Using Natural Gas SCF*** Show the difficulties of formulating the optimal dispatch problem when using the natural gas mSCF per hour rather than MMBtu per hour as the energy rate in term. Here modify Example 12.6—the simplified dispatch problem.

From Equations (12.2), (12.16), (12.18), (12.20), and Table 12.7,

Boiler 4:	$20 \text{ mlb/h} \le (0.9015067 \times x_1) \le 100 \text{ mlb/h}$
GE gas turbine:	$4000 \text{ kW} \le (113.66533 \times x_2 - 4078.15) \le 20{,}100 \text{ kW}$
Boiler 8:	$36.2 \text{ mlb/h} \le (0.37664 \times x_3 + 9.437) \le 89.6 \text{ mlb/h}$
Boiler 8 supplemental:	$0 \text{ mlb/h} \le (1.023839 \times x_4) \le 60.4 \text{ mlb/h},$

where

x_1 = the amount of natural gas (mSCF per hour) used by boiler 4;
x_2 = the amount of natural gas (mSCF per hour) used by GE Gas Turbine;
x_3 = the amount of natural gas (mSCF per hour) used by boiler 8; and
x_4 = the amount of natural gas (mSCF per hour) used by boiler 8 supplemental firing.

These equations are only valid for a natural gas feed with an *HHV* of 1054.8677 Btu/SCF. The problem is trying to modify these equations to account for natural gas feed of any *HHV*.

12.7 ***HHV Content of Natural Gas*** Natural gas is priced on the NYMEX as dollar per MMBtu based on the *HHV*. Natural gas is delivered to the end user as SCF, which is totalized for each month. The British thermal unit per SCF content of the

Table P12.7a Molar Composition (mol %) of Five Natural Gas Stream and Species *HHV* Values

	Molecular Weight	*HHV** (Btu/mol)	*HHV** (Btu/lb)	mol % NG-1	mol % NG-2	mol % NG-3	mol % NG-4	mol % NG-5
Methane	16.0429	380,362.41	23,709.08	85.4000	94.7500	94.75	85.2500	70.0000
Ethane	30.0699	667,042.11	22,183.05	5.0000	2.0000	2.0262	5.0621	9.0995
Propane	44.0970	94,9703.32	21,536.69	3.0000	0.7500	0.75	3.0000	6.0000
Isobutane	58.1240	1,228,675.80	21,138.87	1.0000	0.3100	0.3	1.0000	3.0000
n-Butane	58.1240	1,231,577.78	21,188.80	1.0000	0.3100	0.3	1.0000	3.0000
Isopentane	72.1510	1,509,729.10	20,924.58	0.5000	0.1500	0.15	0.5000	1.0000
n-Pentane	72.1510	1,509,729.10	20,924.58	0.5000	0.1500	0.15	0.5000	1.0000
n-Hexane	86.1779	1,795,171.91	20,831.00	0.1000	0.0400	0.0324	0.0810	0.1295
n-Heptane	100.2050	2,077,205.44	20,729.56	0.0000	0.0300	0.0302	0.0754	0.1207
n-Octane	114.2320	2,359,238.96	20,653.05	0.0000	0.0090	0.0089	0.0222	0.0355
n-Nonane	128.2590	2,641,293.98	20,593.44	0.0000	0.0010	0.0013	0.0033	0.0053
n-Decane	142.2850	2,923,349.00	20,545.73	0.0000	0.0000	0.001	0.0026	0.0041
Methylcyclopentane	84.1619	1,699,552.83	20,193.85	0.0000	0.0000	0.0000	0.0034	0.0054
Benzene	78.1100	1,416,165.04	18,130.39	0.0000	0.0000	0.0000	0.0000	0.0500
Toluene	92.1408	1,692,901.92	18,372.99	0.0000	0.0000	0.0000	0.0000	0.0500
CO_2	44.0097	0.00	0.00	1.0000	0.5000	0.5	1.0000	1.5000
Nitrogen	28.0130	0.00	0.00	2.5000	1.0000	1	2.5000	5.0000

**HHV* values from HYSYS.

Table P12.7b Reported Values for the Natural Gas Streams

	NG-1	NG-2	NG-3	NG-4	NG-5	Units
HHV (Btu/lb-mol)	428,184.75	394,599.10	394,429.32	429,998.28	495,545.29	Btu/lb-mol
MW	19.6376	17.2900	17.2775	19.7276	23.9787	lb/lb-mol
v (ft^3/lb-mol)	377.3011	377.5488	377.5497	377.2897	376.7640	ft^3/lb-mol
Btu/SCF	1134.8621	1045.1604	1044.7082	1139.7032	1315.2672	

fuel is not known when it is delivered. To determine the British thermal unit content of the delivered fuel, the pipeline company takes hourly samples of the natural gas. These samples are analyzed for species content, generally by gas chromatograph and according to American Gas Association (AGA) standards. The hourly British thermal unit per SCF values are averaged and the monthly fuel bill is determined.

Estimate the *HHV* British thermal unit per SCF content for the five natural gas samples (suggested by M. Firmin, personal communication) shown in Table P12.7a (NG-1, NG-2, NG-3, NG-4, and NG-5). Compare calculated values to the values reported in Table P12.7b. In Table P12.7a, we have provided species *HHV* values as reported in HYSYS. Here use 60°F and 14.73 psia as reference conditions for natural gas SCF.

Chapter 13

Process Energy Integration

Energy integration explores the interconnection between the utility heating and cooling requirements of a process. These heating and cooling requirements may be determined as part of the design phase of the plant where a heat exchanger network (HEN) is used to exchange energy between hot and cold process streams and unmet demand is met by the utility plant as depicted in Figure 13.1.

In the grassroots design phase, process integration often targets two distinct heat exchanger networks: one that uses minimum hot and cold utilities as delivered from the central utility plant and a second network that has the fewest heat exchangers and possibly the lowest capital cost. For an existing plant, with an existing heat exchanger network, process energy integration establishes minimum heating and cooling requirements, which can be compared to actual usage. The cost savings realized by moving closer to the minimum utilities may provide justification for modification of the existing heat exchanger network. In this chapter, we are primarily concerned with the mechanics of hot and cold stream identification and the design of the two networks—one using minimum utilities and one using the fewest heat exchangers. We will explore an important industrial application—options for energy integration of distillation columns and process streams.

In Chapter 14, we extend our energy analysis to include the design of the utility system, which will supply the heating and cooling for the processing streams and meet the process electric power requirements. A processing site may consist of a number of separate production facilities that share a central utility plant. In Chapter 14 we also examine the opportunity for energy exchange between these individual facilities to reduce the overall utility needs of the site.

13.1 INTRODUCTION TO PROCESS ENERGY INTEGRATION/ MINIMUM UTILITIES

An early recognition of the connection between utility heating and cooling loads in a processing plant was provided by Hohmann (1971), and later, Linnhoff and Flower (1978a,b) presented the basis of what is commonly referred to today as pinch analysis. Linnhoff et al. (1982) made the concepts of pinch analysis and process energy integration very accessible in their book, *A User Guide on Process Integration for the Efficient Use of Energy*.

The process energy integration problem can be explained as follows. Given a set of process streams needing heating and a set of process streams needing cooling, we want to determine the minimum required utilities. For each stream needing heating or cooling, the inlet and outlet temperatures are known and the product of flow rate and heat capacity is known.

To determine the minimum required utilities, we begin with energy balances on the individual hot stream (HS) and cold stream (CS):

$$\frac{dQ_{\text{HS}}}{dt} = \dot{Q}_{\text{HS}} = F_{\text{HS}}\hat{C}_{p,HS}(T_{\text{HS,out}} - T_{\text{HS,in}}) \quad (13.1)$$

$$\frac{dQ_{\text{CS}}}{dt} = \dot{Q}_{\text{CS}} = F_{\text{CS}}\hat{C}_{p,\text{CS}}(T_{\text{CS,out}} - T_{\text{CS,in}}). \quad (13.2)$$

Here, F_{HS} is the flow rate (pound per hour) of a hot stream (a stream needing cooling); F_{CS} is the flow rate (pound per second or kilogram per second) of a cold stream (a stream needing heating); $\hat{C}_p$ is the stream heat capacity (British thermal unit per pound-Fahrenheit or kilojoule per kilogram-Celsius); T is the temperature (°F or °C) and $\dot{Q}$ is the energy

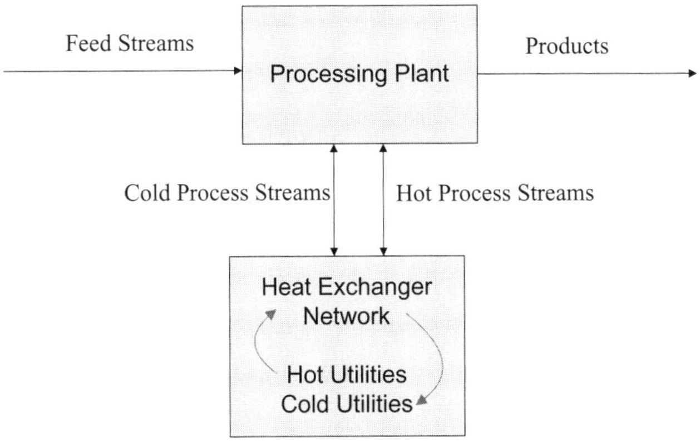

Figure 13.1 Process heat recovery network with hot and cold utilities.

Table 13.1 Stream Data Summary

Stream	Type	$F\hat{C}_p\left(\frac{\text{kW}}{°\text{C}}\right)$	T_{in} (°C)	T_{out} (°C)
H1	Hot	1	100	20
H2	Hot	1	80	50
C3	Cold	3	30	90

flow (British thermal unit per hour or kilowatt). Often the $F\hat{C}_p$ term is simply represented as C, but we will retain $F\hat{C}_p$. Any set of consistent units can be used, and in many problems, this fact is emphasized with no actual units being provided.

EXAMPLE 13.1 *Heating and Cooling Utility Duties*

Consider the three-stream problem of Table 13.1 in which two streams need to be cooled and one stream needs to be heated. There are no phase changes in these streams. What are the minimum heating and cooling loads from utilities?

SOLUTION Let us first ignore the possibility of energy exchange between the hot and cold streams and simply use utilities to bring each stream to its needed outlet condition. Hot stream 1 would require $\dot{Q}_{\text{H1}} = F_{\text{H1}}\hat{C}_{p,H1}(T_{\text{H1,out}} - T_{\text{H1,in}}) = 1(20-100) = -80$ kW to lower the temperature from 100 to 20°C. Hot stream 2 would require −30 kW to lower its temperature from 80 to 50°C and cold stream 3 would require $\dot{Q}_{\text{C3}} = F_{\text{C3}}\hat{C}_{p,\text{C3}}(T_{\text{C3,out}} - T_{\text{C3,in}}) =$ to $3(90-30) = 180$ kW raise its temperature from 30 to 90°C. The 180 kW of needed heating could be supplied by exchange with the steam system and the 110 kW (total) of needed cooling could be supplied by exchange with a cold utility, for example, cooling water or chilled water.

The heat exchanger network would appear as shown in Figure 13.2.

If we allow energy exchange between hot and cold streams, we must also fix an approach temperature at each end of the heat exchanger. For now, let us assume a 0° approach temperature, which we recognize is *not physically possible*. However, it is actually easier to use a 0° approach for this problem and see how the pinch method is implemented. Future problems will use appropriate approach temperatures.

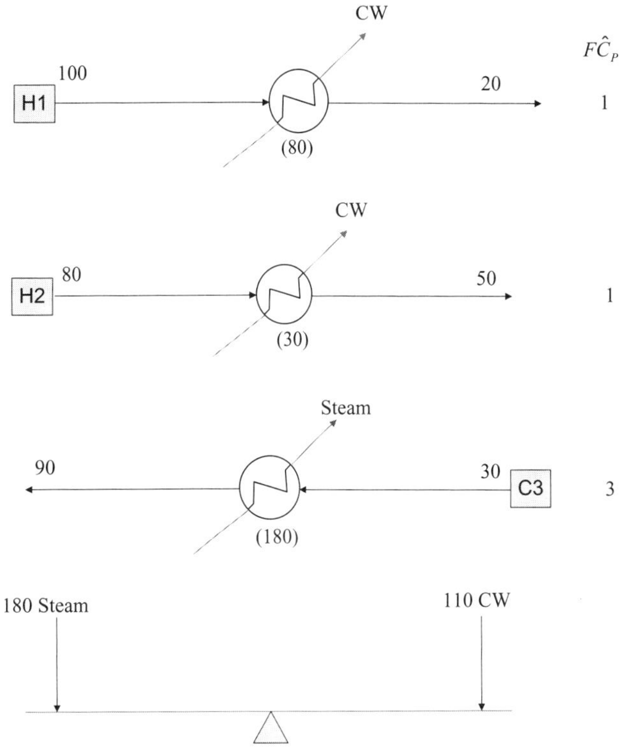

Figure 13.2 Heat exchanger network (HEN) using only utilities.

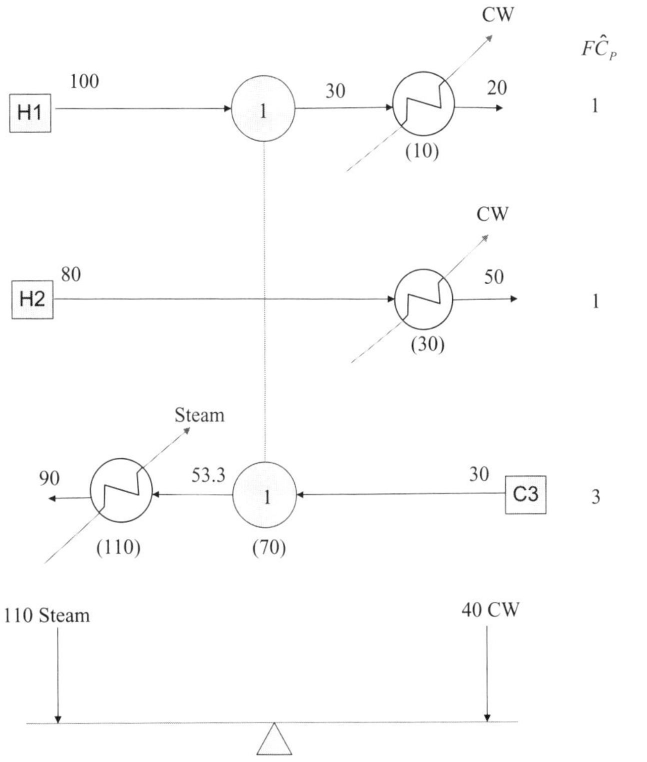

Figure 13.3 HEN with energy exchange between H1 and C3 and utilities (0° approach).

Allowing energy exchange between streams H1 and C3 and using a 0° approach, we would obtain the network as shown in Figure 13.3.

For heat exchanger 1, the temperature on the hot side of the exchanger is controlled or limited by the hot stream inlet temperature (100°C). The temperature on the cold side of the heat exchanger is controlled by the cold stream inlet temperature (30°C). The actual energy transfer rate is determined by the smaller of the energy available in the hot stream and needed by the cold stream. For the hot stream of heat exchanger 1, $\dot{Q}_{\mathrm{H1,max}} = F_{\mathrm{H1}}\hat{C}_{p,\mathrm{H1}}(T_{\mathrm{H1,out}} - T_{\mathrm{H1,in}}) = 1(30-100) = -70$ kW; here the minimum hot stream outlet temperature is set by the cold stream inlet temperature. For the cold stream, $\dot{Q}_{\mathrm{C3,max}} = F_3\hat{C}_{p,\mathrm{C3}}(T_{\mathrm{C3,out}} - T_{\mathrm{C3,in}}) = 3(90-30) = 180$ kW; here only a maximum temperature of 90°C is needed. The smaller of the two $|\dot{Q}s|$ is the maximum allowable energy transfer rate, and in this case, it is the 70 kW from the hot stream. Following the transfer of 70 kW, the hot stream will be reduced from 100 to 30°C, and the cold stream in heat exchanger 1 will be heated from 30°C to

$$T_{\mathrm{C3,out}} = T_{\mathrm{C3,in}} + \frac{\dot{Q}_{\mathrm{H1,max}}}{F_{\mathrm{C3}}\hat{C}_{p,\mathrm{C3}}} = 30 + \frac{70}{3} = 53.3°\mathrm{C}.$$

The remainder of the network is satisfied with utilities. Comparing Figure 13.2 with Figure 13.3, the amount of hot utility has been reduced from 180 to 110 kW and the amount of utility cooling needed has been reduced from 110 to 40 kW. Both utilities (steam and cooling) have been reduced by 70 kW when moving from the design in Figure 13.2 to the design in Figure 13.3.

Continuing adding heat exchangers, we can exchange energy from stream H2 to the remainder of stream C3. This heat exchanger network would appear as shown in Figure 13.4.

Comparing Figure 13.4 with Figure 13.3, the amount of hot utility needed has been reduced from 110.0 to 83.3 kW and the amount of utility cooling needed has been reduced from 40.0 to 13.3 kW. Both utilities (steam and cooling) have been reduced by 26.7 kW when moving from the design in Figure 13.3 to the design in Figure 13.4. ■

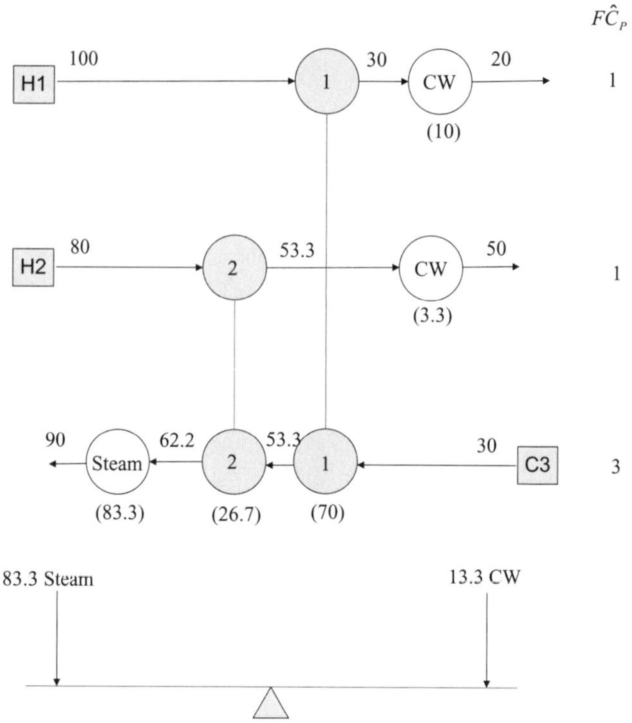

Figure 13.4 HEN with energy exchange between H1 and C3, H2 and C3, and utilities (0° approach).

13.2 TEMPERATURE INTERVAL/PROBLEM TABLE ANALYSIS WITH 0° APPROACH TEMPERATURE

There are other possible designs; for example, show yourself the design that would be obtained if the first match (heat exchanger 1) was made between H2 and C3. However, at this point, an important question is "Can the minimum utilities be determined before I consider any possible networks?" The answer is yes. Following Linnhoff et al. (1982), an easy approach is to formulate a temperature interval analysis or problem table as shown in Figure 13.5.

Here the inlet and outlet temperatures for each stream are used to establish temperature intervals. Over each temperature interval i, energy balances on all streams present are performed and the resulting duties are summed:

$$\dot{Q}_i = \left(\sum_{\mathrm{HS}} F_{\mathrm{HS}}\hat{C}_{p,\mathrm{HS}}\right)_i (T_{\mathrm{HS,out}_i} - T_{\mathrm{HS,in}_i}) + \left(\sum_{\mathrm{CS}} F_{\mathrm{CS}}\hat{C}_{p,\mathrm{CS}}\right)_i (T_{\mathrm{CS,out}_i} - T_{\mathrm{CS,in}_i}). \tag{13.3}$$

For example, in the first temperature interval, from 100 to 90°C, only hot stream 1 is present:

$$\dot{Q}_1 = (F_{\mathrm{H1}}\hat{C}_{p,\mathrm{H1}})(T_{\mathrm{HS,out}_1} - T_{\mathrm{HS,in}_1}) = (1)(90-100) = -10\ \mathrm{kW}.$$

Here the −10 kW of energy flow is a surplus of energy. For the second interval, 90–80°C,

$$\begin{aligned}\dot{Q}_2 &= \left(\sum_{\mathrm{HS}} F_{\mathrm{HS}}\hat{C}_{p,\mathrm{HS}}\right)_2 (T_{\mathrm{HS,out}_2} - T_{\mathrm{HS,in}_2}) + \left(\sum_{\mathrm{CS}} F_{\mathrm{CS}}\hat{C}_{p,\mathrm{CS}}\right)_2 (T_{\mathrm{CS,out}_2} - T_{\mathrm{CS,in}_2}) = \dot{Q}_2 \\ &= (F_{\mathrm{H1}}\hat{C}_{p,\mathrm{H1}})(T_{\mathrm{H1,out}_2} - T_{\mathrm{H1,in}_2}) + (F_{\mathrm{C3}}\hat{C}_{p,\mathrm{C3}})(T_{\mathrm{C3,out}_2} - T_{\mathrm{C3,in}_2}) \\ &= (1)(80-90) + (3)(90-80) = -10 + 30 = 20\ \mathrm{kW}.\end{aligned}$$

Here the 20 kW of energy flow is a net deficit of energy. The complete problem table is shown in Figure 13.5.

Heat can flow from a high-temperature interval to a lower-temperature interval (the second law of thermodynamics). The surplus of 10 kW in the first interval can cascade down to help satisfy the deficit of 20 kW in the second interval. The fifth interval (30–20°C) has a surplus

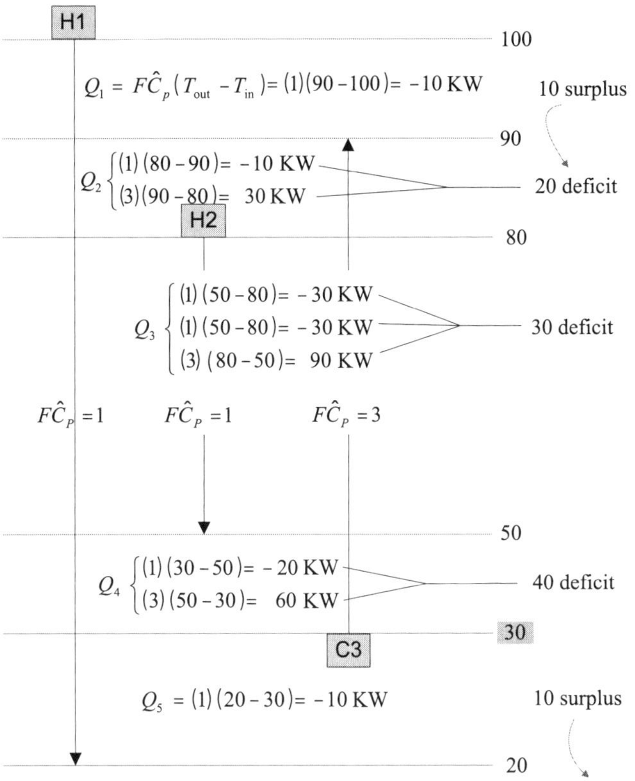

Figure 13.5 Problem table or temperature interval analysis (0° approach) with energy cascade.

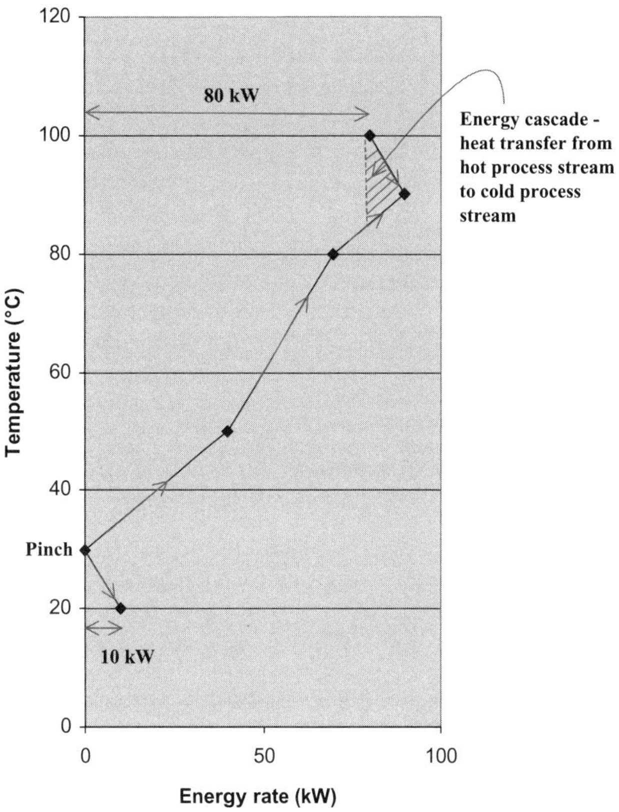

Figure 13.6 Grand composite curve from temperature interval analysis data of Figure 13.5.

of 10 kW, but there are no deficits below this temperature level, so no cascade occurs. After energy cascading, the temperature interval analysis depicted in Figure 13.5 shows a net deficit of 80 kW above 30°C and a surplus of 10 kW below 30°C. The temperature 30°C is termed the pinch temperature or the process pinch temperature. In order to satisfy the energy needs of the network, 80 kW of a hot utility and 10 kW of a cold utility must be supplied to the network—a heat exchanger network using these levels of utilities would be termed a maximum energy recovery (MER) or minimum utility network. Ultimately, to develop a MER heat exchanger network, two separate networks must be developed: one network above the pinch and a second network below the pinch. Above the pinch, 80 kW of energy from a hot utility will be added to heat exchanger network. Below the pinch, 10 kW of energy will be removed from the heat exchanger network. For MER, we require that no energy flows across the pinch temperature.

Our initial heat exchanger networks in Figures 13.2–13.4 were developed without consideration of minimum heating and cooling requirements, the pinch temperature, and energy transfer across the pinch. If we look at Figures 13.2–13.4, we can see that our designs are approaching 80 kW of hot utility and 10 kW of cooling, but we cannot actually reach these minimum utilities by continued evolving of these networks. Right now, we want to appreciate that the MER utility levels may not have been found without our discussions of Section 13.2.

13.3 THE GRAND COMPOSITE CURVE (GCC)

A practical representation of the information in Figure 13.5 is the GCC, which plots residual enthalpies (without energy cascade) versus temperature as shown in Figure 13.6. The GCC will prove especially useful in Chapter 14 for matching the utility system with the needs of the process.

The GCC is straightforward to construct using the data of Figure 13.5. Starting at the pinch of 30°C, the energy rate is assigned a 0-kW value. From 30 to 20°C, the surplus of 10 kW is plotted with a negative slope. The interval from 30 to 50°C has a 40-kW deficit enthalpy residual, which is plotted with a positive slope. The interval from 50 to 80°C has a 30-kW deficit residual, which is added to (40 kW, 50°C). The completed curve in Figure 13.6 also shows a "knee" or pocket where energy cascade is possible. In the pocket of the GCC process, heating and cooling requirements can be satisfied by hot process stream to cold process stream heat exchange.

13.4 TEMPERATURE INTERVAL/PROBLEM TABLE ANALYSIS WITH "REAL" APPROACH TEMPERATURE

We have used a 0° approach temperature to introduce the concept of a pinch temperature and to demonstrate the procedure to determine the minimum utilities required in the heat exchange network using the temperature interval table. To account for a realistic minimum approach temperature, $\Delta T_{\min}$, we can shift the hot stream temperatures down and the cold stream temperatures up in the temperature interval table.

EXAMPLE 13.2 *Temperature Interval Analysis with Real Approach Temperature*

Solve the three-stream problem of Table 13.1 for the pinch location and minimum utility consumptions if the minimum approach temperature $\Delta T_{\min} = 10$.

SOLUTION In Figure 13.7, we solve our three-stream problem for $\Delta T_{\min} = 10$. Figure 13.7a shows the temperature interval table with the hot stream temperatures shifted down $\Delta T_{\min}/2$ and the cold stream temperatures shifted up $\Delta T_{\min}/2$.

In Figure 13.7b we have shifted the hot stream temperatures down $\Delta T_{\min}$ and left the cold stream temperatures unchanged. In Figure 13.7c we have left the hot stream temperatures unchanged and shifted the cold stream temperatures up $\Delta T_{\min}$.

In Figure 13.7a, stream H1 with actual T_{in} (°C) = 100 and T_{out} (°C) = 20 is plotted with shifted T_{in} (°C) = 95 and T_{out} (°C) = 15. Stream C3 with actual T_{in} (°C) = 30 and T_{out} (°C) = 90 is plotted with shifted T_{in} (°C) = 35 and T_{out} (°C) = 95. Similarly in Figure 13.7b, stream H1 is plotted with shifted T_{in} (°C) = 90 and T_{out} (°C) = 10. Stream C3 is plotted with its actual T_{in} (°C) = 30 and T_{out} (°C) = 90. In Figure 13.7c, stream H1 is plotted with its actual T_{in} (°C) = 100 and T_{out} (°C) = 20. Stream C3 is plotted with shifted T_{in} (°C) = 40 and T_{out} (°C) = 100.

We determine the energy deficit or surplus for each temperature interval using the same procedure as in Figure 13.5 with Equation (13.3). For the first interval of each figure in Figure 13.7,

$$\dot{Q}_i = \left(\sum_{\text{HS}} F_{\text{HS}}\hat{C}_{p,\text{HS}}\right)_i (T_{\text{HS,out}_i} - T_{\text{HS,in}_i}) + \left(\sum_{\text{CS}} F_{\text{CS}}\hat{C}_{p,\text{CS}}\right)_i (T_{\text{CS,out}_i} - T_{\text{CS,in}_i})$$

$$\text{Figure 13.7a: } \dot{Q}_1 = (1)(75-95) + (3)(95-75) = -20 + 60 = 40 \text{ kW}$$

$$\text{Figure 13.7b: } \dot{Q}_1 = (1)(70-90) + (3)(90-70) = -20 + 60 = 40 \text{ kW}$$

$$\text{Figure 13.7c: } \dot{Q}_1 = (1)(80-100) + (3)(100-80) = -20 + 60 = 40 \text{ kW.}$$

After the surplus/deficit for each interval is found, surplus energy cascading allows the minimum utilities and pinch temperature to be determined. Here there is no residual surplus energy cascading.

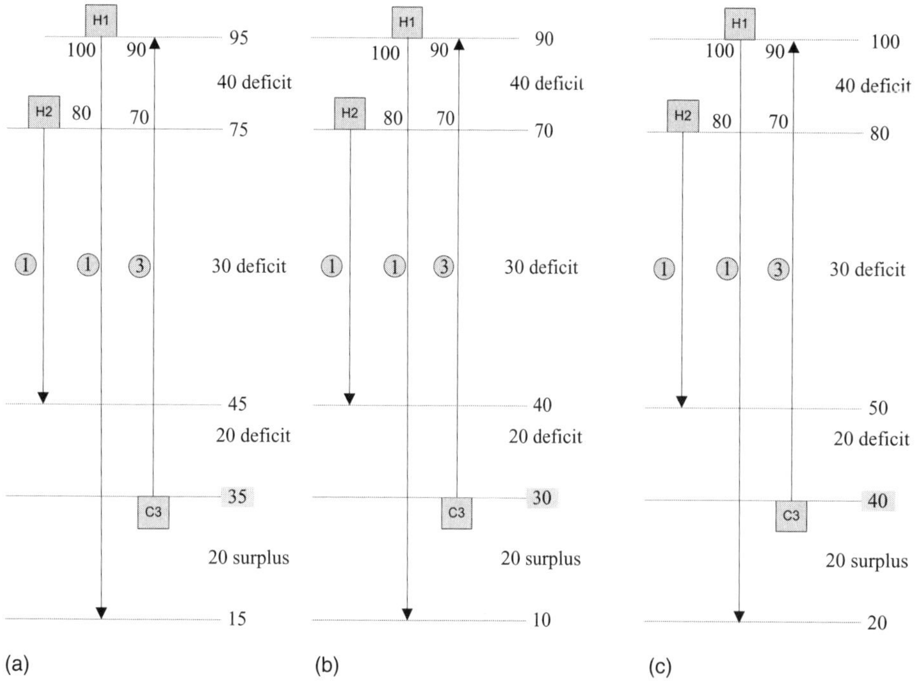

Figure 13.7 Temperature interval analysis (10° approach).

To account for the 10° minimum approach temperature, we shifted hot and cold stream interval temperature scales. Regardless of how the 10° total shifting is accomplished, the energy exchange in the *i* temperature interval in each figure is identical. To understand this, in each of the first intervals in Figure 13.7, we show the actual temperatures next to the streams. Even though shifting of the interval temperatures has occurred, the actual stream temperatures are identical in each figure. The shifting insures there is a 10° approach at the beginning and at the end of each *i* interval. The pinch temperature is 40° hot/30° cold, and the minimum utilities are 90-kW hot utility and 20-kW cold utility. ■

13.5 DETERMINING HOT AND COLD STREAM FROM THE PROCESS FLOW SHEET

We are often facing two difficulties in starting the pinch methodology. We must determine which streams in a plant, or subsection of the plant or process, should be included in the analysis. Second, we must determine $F\hat{C}_p$, T_{in}, and T_{out} for each of these streams. This later information would be available from material and energy balances. However, for a "first-pass" energy targeting analysis, we often will only have a process flow sheet with design temperatures and design duties indicated.

EXAMPLE 13.3 ***Data Extraction***

Determine the hot and cold streams for the flow sheet in Figure 13.8. Temperatures are in Celsius and the heat exchanger duties are in kilowatt.

SOLUTION Here, we examine stream data available from Figure 13.8 and then we discuss possible limitations. Starting with stream feed 1, it is a cold stream being heated from 15° to 98.33° and then from 98.33° to 220° prior to entering reactor 1. The heat exchanger duties, $\dot{Q}$, are known, and using

$$F\hat{C}_p = \frac{\dot{Q}}{(T_{out} - T_{in})},$$

we find

$$F\hat{C}_p = \frac{500}{(98.33 - 15)} = 6\frac{\text{kW}}{°\text{C}} \quad \text{and} \quad F\hat{C}_p = \frac{730}{(200 - 98.33)} = 6\frac{\text{kW}}{°\text{C}}$$

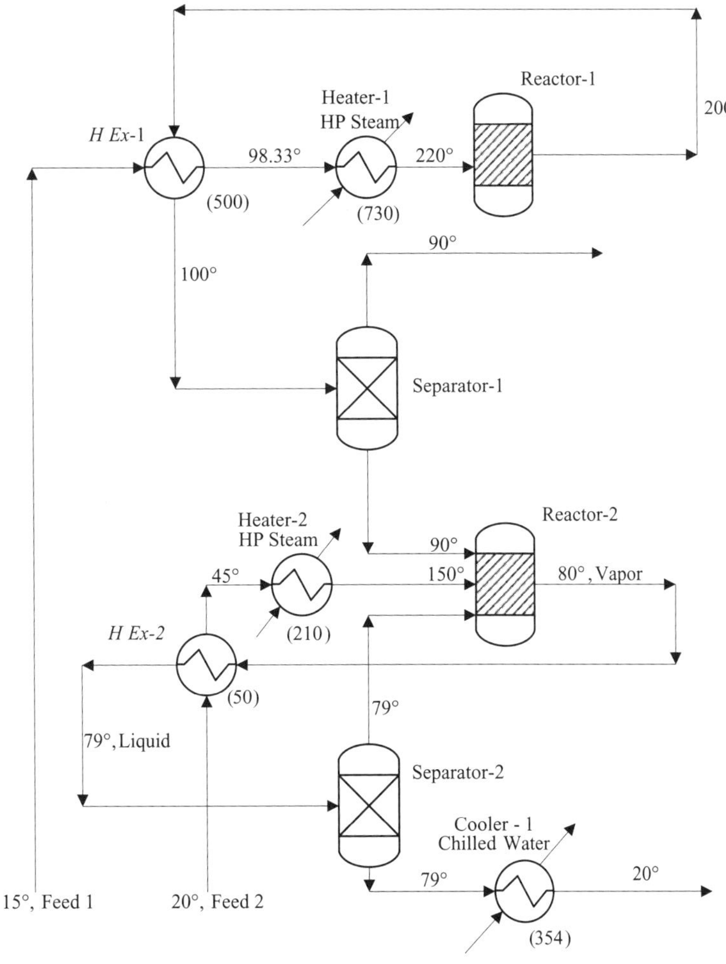

Figure 13.8 Process and heat exchanger network.

for each heating segment of the feed stream. Unless there is a reason (termed a process constraint) for first heating feed 1 to 98.33° and then to 220°, we can consider this to be a single cold stream with $F\hat{C}_p = 6$ kW/°C, $T_{in} = 15$°C, and $T_{out} = 220$°C. Remember we want to identify hot and cold streams $F\hat{C}_p$, T_{in}, and T_{out} and then to determine the minimum utilities. We are not actually concerned with how these temperature changes are accomplished. For example, the fact that high-pressure (HP) steam was used in heater −1 is not needed for our analysis. The existing design of the heat exchanger network will be important if changes are considered (retrofitting) to meet energy targets.

The stream leaving the reactor, R1-out, is cooled from 200° to 100° before entering separator 1. Here,

$$F\hat{C}_p = \frac{-500}{(100-200)} = 5\frac{\text{kW}}{\text{C}}.$$

R1-out is a hot stream with $F\hat{C}_p = 5$ kW/°C, $T_{in} = 200°$ and $T_{out} = 100°$. For stream feed 2, there are two heating segments with

$$F\hat{C}_p = \frac{50}{(45-20)} = 2\frac{\text{kW}}{°\text{C}} \quad \text{and} \quad F\hat{C}_p = \frac{210}{(150-45)} = 2\frac{\text{kW}}{°\text{C}}.$$

Following our discussion for feed 1, we can take feed 2 as a single cold stream with $F\hat{C}_p = 2$ kW/°C, $T_{in} = 20°$and $T_{out} = 150°$.

The inlet to reactor 2 requires additional engineering detail. Stream mixing should occur isothermally for MER. If, however, the three streams are being sent to specific reactor locations, for example, different catalyst beds, then energy recovery consideration for these streams may not be needed.

Figure 13.9 provides a second possibility where stream mixing is occurring before the reactor. Energy recovery may be possible if the pinch temperature falls between the maximum and minimum inlet temperatures to the mixer (150° and 79°). Here then, we may want to account for three streams: 79° → $T_{R2,in}$, 90° → $T_{R2,in}$, and 150° → $T_{R2,in}$.

The outlet of reactor 2, R2-out, condenses in heat exchanger 2. The phase change occurs at constant temperature, but it is convenient in our problem table to account for the heat of vaporization/condensation as the product of $F\hat{C}_p$ and a 1° temperature change. For this condensation process, $F\left(\hat{h}^{liq}_{80°} - \hat{h}^{vap}_{80°}\right) \Leftrightarrow F\hat{C}_p\left(T_{out,79°} - T_{in,80°}\right)$. The outlet of separator 2, Sep2-out, is cooled from 79° to 20°; this is a hot stream with $F\hat{C}_p = 6$ kW/°C, $T_{in} = 79°$, and $T_{out} = 20°$. We summarize the stream results from Figure 13.8 in Table 13.2. Here we are not considering possible energy recovery from the streams entering reactor 2. ■

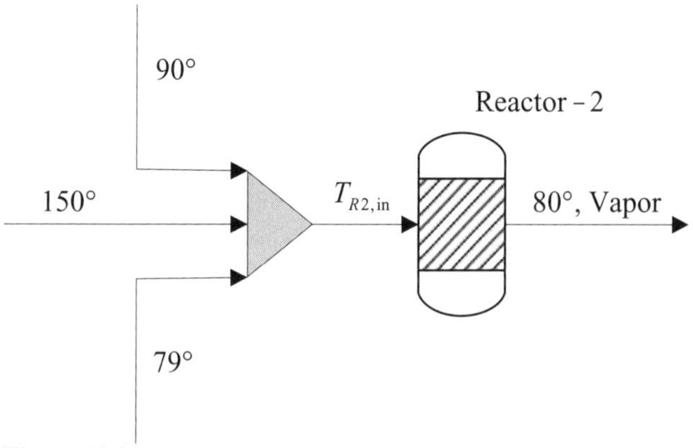

Figure 13.9 Stream analysis at a mixing point.

Before we solve the energy targeting problem, a few comments are needed. If a vapor stream is cooled, condensed to liquid, and the resulting liquid further cooled in a single heat exchanger, three streams would be required in our problem table: one for the vapor, one for the condensation process, and one for the liquid cooling. We have also taken the heat capacity of a single-phase stream to be constant, which may not be accurate. A plot of temperature versus enthalpy for a stream may be nonlinear. Here, it is generally it is possible to linearize the plot with two or three linear segments. Each linearized segment must be treated as a separate stream in the problem table. The conservative linearization of a hot stream should fall below the stream's actual temperature versus enthalpy curve. The conservative linearization of a cold stream should fall above the stream's actual temperature versus enthalpy curve.

EXAMPLE 13.4 *Process Minimum Heating and Cooling Requirements and Pinch Location*

Using the stream data from Table 13.2 for the process shown in Figure 13.8, determine the pinch location and minimum heating and cooling requirements when using a minimum approach temperature of 10°. Also draw the GCC.

SOLUTION The temperature interval table with the hot stream temperatures shifted down $\Delta T_{min}/2$ and the cold stream temperatures shifted up $\Delta T_{min}/2$ is shown in Figure 13.10a.

Table 13.2 Stream Data Summary for the Process of Figure 13.8

Stream	Type/#	$F\hat{C}_p\left(\frac{\text{kW}}{°\text{C}}\right)$	T_{in} (°C)	T_{out} (°C)
Feed 1	Cold 1	6	15	220
Feed 2	Cold 2	2	20	150
R1-out	Hot 3	5	200	100
R2-out	Hot 4	50	80	79
Sep2-out	Hot 5	6	79	20

The residual surplus available in temperature interval 5, $\dot{Q}_5 = 42$ kW, can cascade down to satisfy part of the deficit in temperature interval 6. The resulting network minimum heating requirement is 616 kW and the minimum cooling requirement is 30 kW. The pinch temperature occurs over the range 25°/20°. The current network is using 940-kW heating and 354-kW cooling. There is the possible savings of 324 kW of heating and 324 kW of cooling. In other words, there is a possible savings of 324 kW, which are currently being transferred across the pinch. A range for the pinch is a little different, but it is straightforward to interpret. The pinch range is occurring at a low temperature so we really want to determine at what hot stream temperature will we allow cooling utility. Examining the limits of the pinch range, the lower-limit pinch temperature of 20° means utility cooling is limited to actual hot stream temperatures of 25° (pinch temperature + ½ $\Delta T_{\min}$) and below.

Using the temperature interval analysis of Figure 13.10a, we can generate the GCC as shown in Figure 13.10b.

The GCC is constructed using the data of Figure 13.10a. Staring at the lower pinch temperature of 20°C, the energy rate is assigned a 0-kW value. From 20 to 15°C, the surplus of 30 kW is plotted with a negative slope. The pinch range of 20–25°C is assigned a 0-kW value. Starting at the upper pinch temperature of 25°C, the interval from 25 to 74°C has a 98-kW deficit enthalpy residual, which is plotted with a positive slope. The interval from 74 to 75°C has a 42-kW surplus enthalpy, which is subtracted from (98 kW, 74°C). This process continues until the GCC is completed. ■

EXAMPLE 13.5 *Costing Heat Exchanger Networks (with one being a Maximum Energy Recovery Network)*

We saw in Example 13.4 that energy savings were possible in the heat exchanger network depicted in Figure 13.8. For the moment, allow that Figure 13.8 represents a proposed grassroots heat exchanger network design. We want to determine the capital cost and utility cost of this proposed network and compare these costs to a MER network. We will then discuss the situation where the network in Figure 13.8 is the existing operational network. Here we will want to see if energy savings may justify a retrofit to the existing network. Table 13.3a provides cost data for steam and chilled water (also see Problem 13.1), and for heat exchanger costs, use (Biegler et al., 1997)

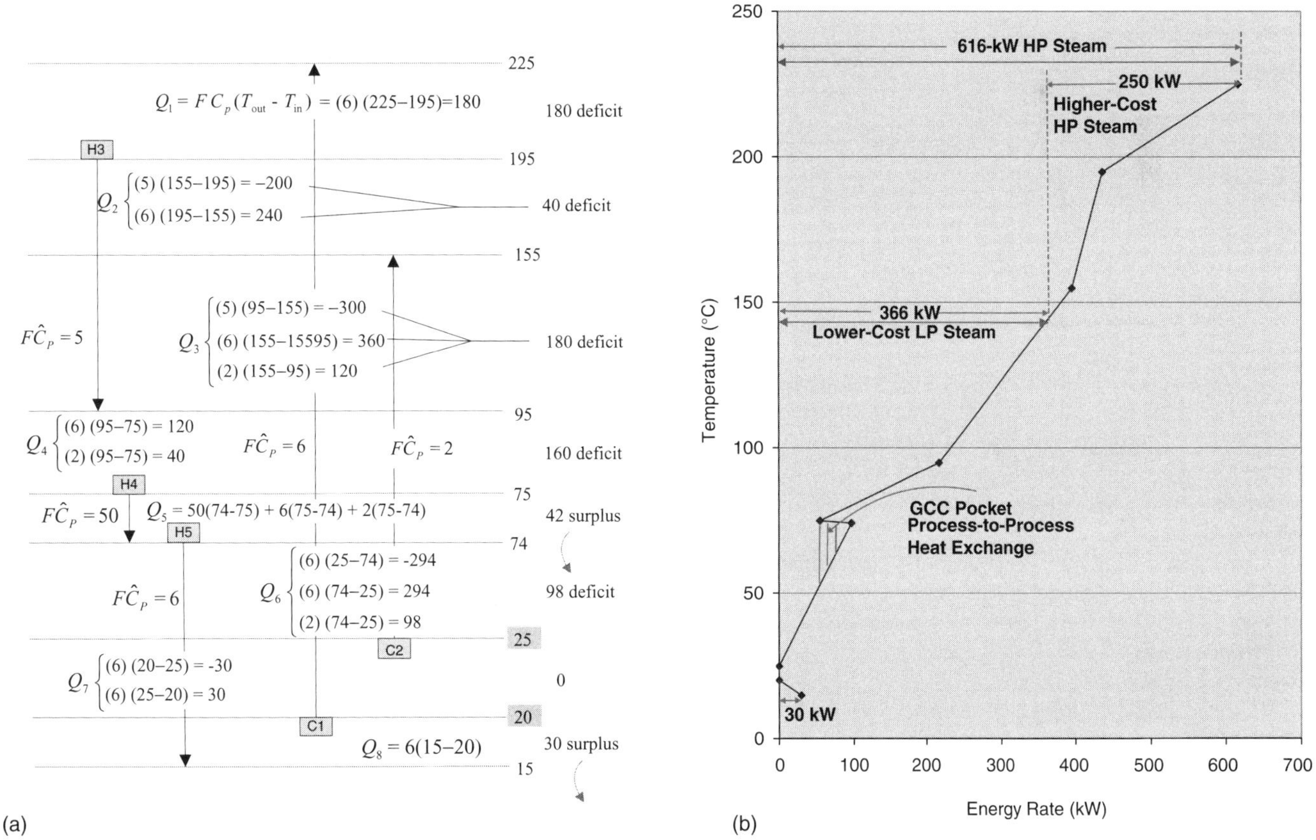

Figure 13.10 (a) Temperature interval analysis (10° approach) for Example 13.4. (b) Grand composite curve (GCC) from temperature interval analysis data of Figure 13.10a. Note that GCC shows the use of 616 kW of HP steam as the hot utility. Later, in Problem 14.1, we will explore the use of LP steam (366 kW) + HP steam (250 kW) as the hot utility.

Table 13.3a Stream Data Summary for the Process of Figure 13.8

Stream	Type/#	$h\left(\frac{\text{kW}}{\text{m}^2\text{-K}}\right)$	$\text{Cost}\left(\frac{\$}{\text{kW-year}}\right)$	T_{in} (°C)	T_{out} (°C)
Feed 1	Cold 1	1		15	220
Feed 2	Cold 2	1		20	150
R1-out	Hot 3	1		200	100
R2-out	Hot 4	3		80	79
Sep2-out	Hot 5	1		79	20
HP steam		5	80	240	240
Chilled water		1	30	7	15

Table 13.3b Heat Exchanger Summary for the Process of Figure 13.8

Heat Exchanger	Streams	$\lvert\dot{Q}(\text{kW})\rvert$	ΔT_1 (°C)	ΔT_2 (°C)	$\sim\Delta T_{LMTD}$
H Ex-1	H3/C1	500	101.67	85.00	93.09
Heater-1	Steam/C1	730	20.00	141.67	61.18
H Ex-2	H4/C2	50	35.00	59.00	45.96
Heater-2	Steam/C2	210	90.00	195.00	135.74
Cooler-1	Water/H5	354	64.00	13.00	31.76

Table 13.3c Costing of Heat Exchangers for the Process of Figure 13.8

Heat Exchanger	Streams	$U\left(\frac{\text{kW}}{\text{m}^2\text{-K}}\right)$	Area (m^2)	$\text{Capital}\left(\frac{\$}{\text{year}}\right)$	$\text{Utility}\left(\frac{\$}{\text{year}}\right)$	$\text{Total}\left(\frac{\$}{\text{year}}\right)$
H Ex-1	H3/C1	0.50	10.74	7,111.34		
Heater-1	Steam/C1	0.83	14.32	7,647.69	58,401.60	
H Ex-2	H4/C2	0.75	1.45	5,717.60		
Heater-2	Steam/C2	0.83	1.86	5,778.48	16,800.00	
Cooler-1	Water/H5	0.50	22.29	8,843.98	10,620.00	
Total cost				35,099.09	85,821.60	**120,920.69**

Heat exchanger annualized capital costs:

$$\$5500 + \$150 \times \text{Area}\left(\text{m}^2\right), \$/\text{year}.$$

Values for stream heat transfer coefficients, h, are also provided in Table 13.3a.

Saturated steam is available at 240°C and chilled water is delivered at 7°C and returned at 15°C. In Table 13.3b, we calculate the duty of each heat exchanger as well as the approach temperature (ΔT_1 and ΔT_2) at each end of each heat exchanger based on Figure 13.8.

For the log mean temperature difference, ΔT_{LMTD} (see Chapter 7), we now use the approximation by Chen (1987) to allow the possibility that the temperature difference at each end of the heat exchanger is the same:

$$\Delta T_{LMTD} = \frac{\Delta T_2 - \Delta T_1}{\ln\left(\frac{\Delta T_2}{\Delta T_1}\right)} \cong \left(\Delta T_1 \Delta T_2 \frac{(\Delta T_2 + \Delta T_1)}{2}\right)^{1/3}.$$

SOLUTION The heat exchanger network costing is provided in the Excel file **Example 13.5a.xls** and key results are shown in the Table 13.3c. The heat exchanger area A is calculated from the design equation $\dot{Q} = UA\Delta T_{LMTD}$, and the overall heat transfer coefficient U is given by

$$\frac{1}{U} = \frac{1}{h_{\text{hot stream}}} + \frac{1}{h_{\text{cold stream}}}$$

or

$$= \frac{1}{h_{\text{hot stream}}} + \frac{1}{h_{\text{cold utility}}}$$

or

$$= \frac{1}{h_{\text{hot utility}}} + \frac{1}{h_{\text{cold stream}}}.$$

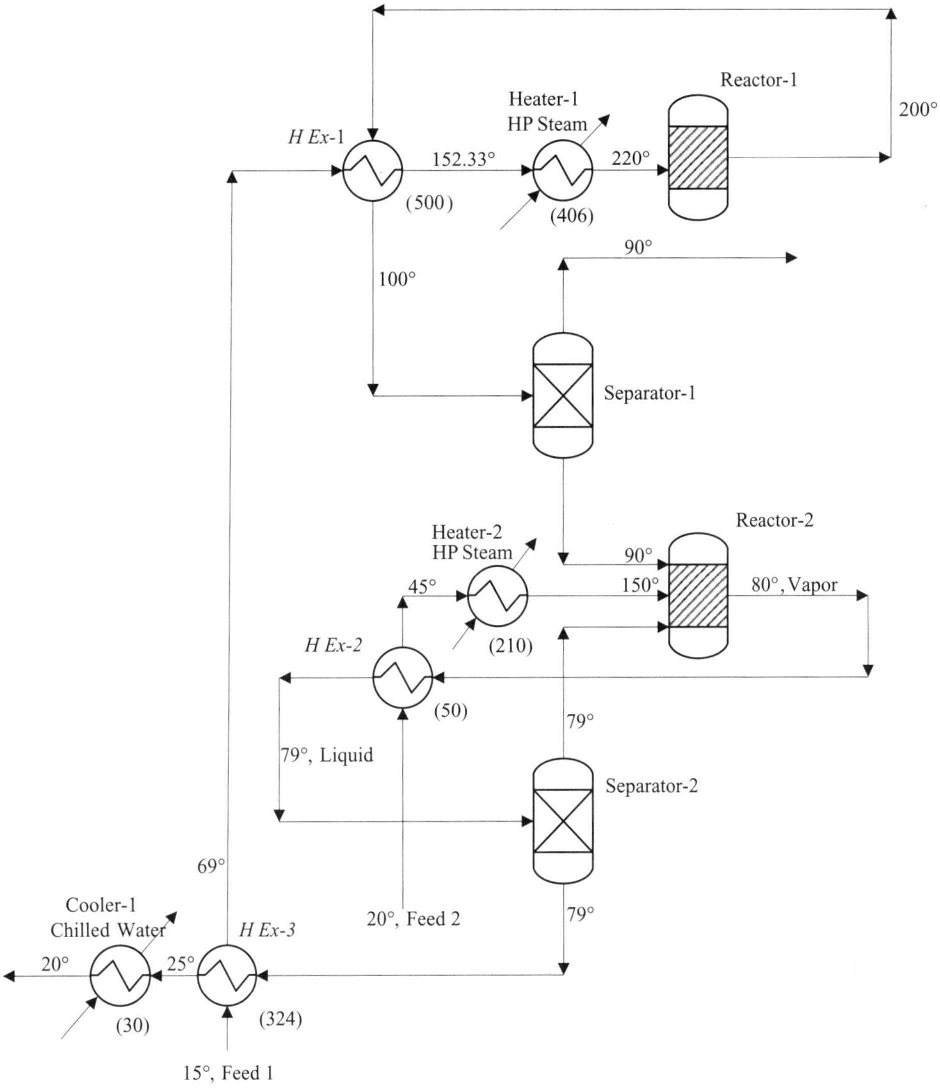

Figure 13.11 Modified HEN meeting maximum energy recovery (MER) requirements.

The annualized capital cost for the heat exchanger network is \$35,099.09 and the annual utility cost is \$85,821.60, giving a total network cost of \$120,920.69 per year. We know from Example 13.4 that a savings of 324 kW of both heating and cooling is possible if we use a MER network. The utility savings for the MER network would be \$35,640 per year, but these savings may be offset by the cost of additional heat exchangers.

We will discuss heat exchanger network design in Section 13.6. For now, starting with Figure 13.8, see if you can add a new heat exchanger, H Ex-3, between two process streams, which will ultimately result in utility cooling only for a hot stream temperature of 25° and below; recall this 25° comes from our discussion at the end of Example 13.4. The modified heat exchanger network is shown in Figure 13.11.

It is straightforward to see that the addition of a new heat exchanger, H Ex-3, allows stream feed 1 to cool the product stream from Separator-2 (Sep 2-out) to 25°. Cooling Sep 2_Out to 20° is then accomplished with chilled water, requiring 30 kW of the cold utility. The increase in the feed 1 temperature lowers the steam requirement in Heater-1 to 406 kW. As calculated in the pinch analysis of Example 13.4, both the hot and cold utilities have been reduced by 324 kW.

Now using the same data and procedure as mentioned earlier, we can cost this MER network. The heat exchanger network costing is provided in the Excel file **Example 13.5b.xls** and key results are shown in the Table 13.4a,b.

The annualized capital cost for the MER heat exchanger network is \$49,480.38 and the annual utility cost is \$50,480.60, giving a total network cost of \$99,661.98 per year. Comparing the proposed (Table 13.3c) and the MER (Table 13.4b) heat exchanger network, the MER network does show the expected \$35,640 savings in yearly utility costs. But there are additional heat exchanger capital costs that reduce the overall savings to \$21,259 when the two networks are compared. For a grassroots design, the MER network of Figure 13.11 may be preferred over Figure 13.8 (proposed) as it has a lower annual cost.

The MER design of Figure 13.11 used HP steam at 240°C as the hot utility. One of the strengths of the GCC of Figure 13.10b

Table 13.4a MER Heat Exchanger Summary for the Process of Figure 13.11

Heat Exchanger	Streams	$\lvert\dot{Q}(\text{kW})\rvert$	ΔT_1 (°C)	ΔT_2 (°C)	~ΔT_{LMTD}
H Ex-1	H3/C1	500.0	47.67	31.00	38.74
Heater-1	Steam/C1	406.0	20.00	87.67	45.53
H Ex-2	H4/C2	50.0	35.00	59.00	45.96
Heater-2	Steam/C2	210.0	90.00	195.00	135.74
H Ex-3	H5/C1	324.0	10.00	10.00	10.00
Cooler-1	Water/H5	30.0	10.00	13.00	11.43

Table 13.4b Costing of Heat Exchangers for Process of Figure 13.11

Heat Exchanger	Streams	$U\left(\frac{\text{kW}}{\text{m}^2\text{-K}}\right)$	Area (m^2)	Capital$\left(\frac{\$}{\text{year}}\right)$	Utility$\left(\frac{\$}{\text{year}}\right)$	Total$\left(\frac{\$}{\text{year}}\right)$
H Ex-1	H3/C1	0.50	25.81	9,372.09		
Heater-1	Steam/C1	0.83	10.70	7,105.11	32,481.60	
H Ex-2	H4/C2	0.75	1.45	5,717.60		
Heater-2	Steam/C2	0.83	1.86	5,778.48	16,800.00	
H Ex-3	H5/C1	0.50	64.80	15,220.00		
Cooler-1	Water/H5	0.50	5.25	6,287.10	900.00	
Total cost				49,480.38	50,181.60	**99,661.98**

is it allows for the evaluation of multiple utility levels and duties—this will be discussed in more detail in Chapter 14. As indicated in the GCC (Figure 13.10b), there also exists the possibility of creating a MER heat exchanger network using both HP steam at 240°C (plotted at 235°C on the GCC) and lower-cost low-pressure (LP) steam at 150°C (plotted at 145°C). The utility duties in a MER network could be 366 kW of LP steam and 250 kW of HP steam. The maximum LP steam that can be used is 366 kW. ■

Now let us discuss the situation where the network in Figure 13.8 exists and is operational. Here, we want to consider the possibility that energy savings as determined by the pinch analysis can justify retrofitting the existing network. Retrofitting an existing network is complicated and here we briefly introduce some of the difficulties. Moving from the heat exchanger network in Figures 13.8–13.11 initially appears to be the simple addition of one heat exchanger. However, when we compare the heat exchanger areas in Table 13.3c and Table 13.4b, problems arise. For example, heat exchanger 1 (Table 13.3c) shows an available area of 10.74 m^2, while in the MER network, heat exchanger 1 (Table 13.4b) requires 25.81 m^2—this is nearly 2.5 times the available area of the existing heat exchanger. Retrofitting requires experience, innovation, and careful consideration of operability; the reader is referred to Tjoe and Linnhoff (1986) and Nordman (2008). What is important is that the tools we are discussing for heat exchanger design can be utilized in grassroots designs, and these tools often serve as a basis for the design evolution in retrofit situations.

13.6 HEAT EXCHANGER NETWORK DESIGN WITH MAXIMUM ENERGY RECOVERY (MER)

We next want to explore the procedure to match hot and cold streams with the resulting heat network having no $\Delta T_{\min}$ violations and having MER. We show the procedure using the process of Figure 13.12, which also introduces column condensers and reboilers.

Step 1: Data extraction—identify hot and cold streams in the process. Here there are two cold streams and two hot streams as summarized in Table 13.5.

Do note that we did not include the column condenser and reboiler duties in the stream table. The condensing of the light product will be a hot stream and the reboiling of the heavy product will be a cold stream. Generally, columns (condensing and reboiling) are treated separately from the background process and appropriate inclusion of these column duties is discussed in Section 13.10.

Step 2: With know $\Delta T_{\min} = 10$°C, determine the pinch location and utility requirements. This is left as an exercise for the reader. For $\Delta T_{\min} = 10$°C, the pinch temperature is 180°C hot/170°C cold, with hot utility = 56 kW and cold utility = 155 kW.

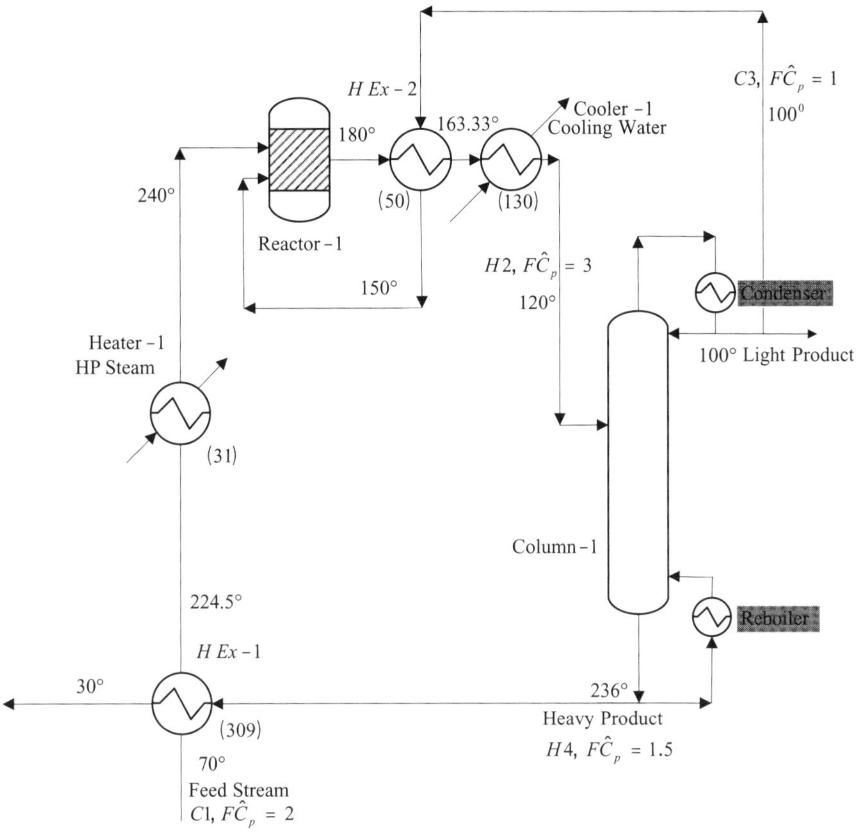

Figure 13.12 Process and heat exchanger network.

Table 13.5 Stream Data Summary for the Process of Figure 13.12

Stream	Type	$F\hat{C}_p\left(\frac{\text{kW}}{°\text{C}}\right)$	T_{in} (°C)	T_{out} (°C)
C1	Cold	2.0	70	240
C3	Cold	1.0	100	150
H2	Hot	3.0	180	120
H4	Hot	1.5	236	30

Step 3: Design the MER network. This process begins with a grid diagram for heat exchanger placement as shown in Figure 13.13.

In the grid diagram, the actual stream temperatures, without adjustment, are plotted. Hot streams above the pinch are ≥180°C and cold streams above the pinch are ≥170°C. Hot streams below the pinch are ≤180°C and cold streams below the pinch are ≤170°C. To place heat exchangers on the grid diagram, we use the pinch design method ("heuristic rules") developed by Linnhoff et al. (1982) and Linnhoff and Hindmarsh (1983). The following two figures present the design rules for the MER heat exchanger network above and below the pinch. We will next see how to apply these heuristic rules.

13.6.1 Design above the Pinch

We start our design at the pinch and move away from the pinch. The logic here is that the matches at the pinch are the most constrained and these matches should be made first following the rules of Figure 13.14. As shown in Figure 13.16, there are two streams above the pinch and both touch the pinch, $N_{\text{Hot}} = N_{\text{Cold}} = 1$. We do not want to use a cold utility above the pinch and we require that for each hot stream touching the pinch there be one unique cold stream such that $F\hat{C}_{p,\text{Hot}} \le F\hat{C}_{p,\text{Cold}}$. A heat exchanger can placed between H4 ($F\hat{C}_{p,\text{Hot}} = 1.5$) and C1 ($F\hat{C}_{p,\text{Cold}} = 2$) as shown in Figure 13.16.

To determine the energy transferred in this exchanger (between H4 and C1), we determine the maximum energy transfer rate possible from H4, which is $\dot{Q}_{\text{H4,max}} = 1.5(236-180) = -84$ kW and the maximum energy rate to C1, which is $\dot{Q}_{\text{C1,max}} = 2(240-170) = 140$ kW. The actual energy rate transferred is limited by the smaller $\left|\dot{Q}_{\max}\right|$ of these two values, which, in this case, is $\dot{Q}_{\text{H4,max}}$. The energy available in H4 will be exhausted and the outlet side of cold stream for heat exchanger 1 will be

$$T_{\text{C1,out}} = T_{\text{C1,in}} + \frac{\dot{Q}_{\text{Transferred}}}{F\hat{C}_{p,\text{C1}}} = 170 + \frac{84\ \text{kW}}{2\dfrac{\text{kW}}{°\text{C}}} = 212°\text{C}.$$

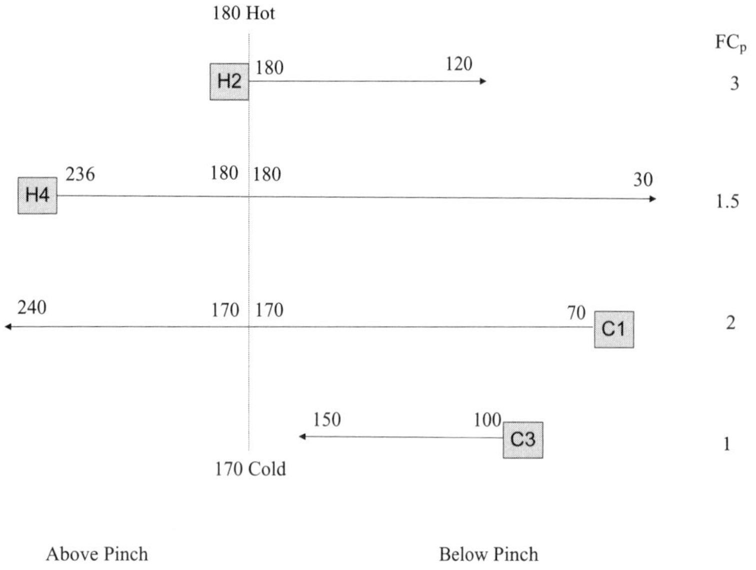

Figure 13.13 Grid diagram for heat exchanger placement.

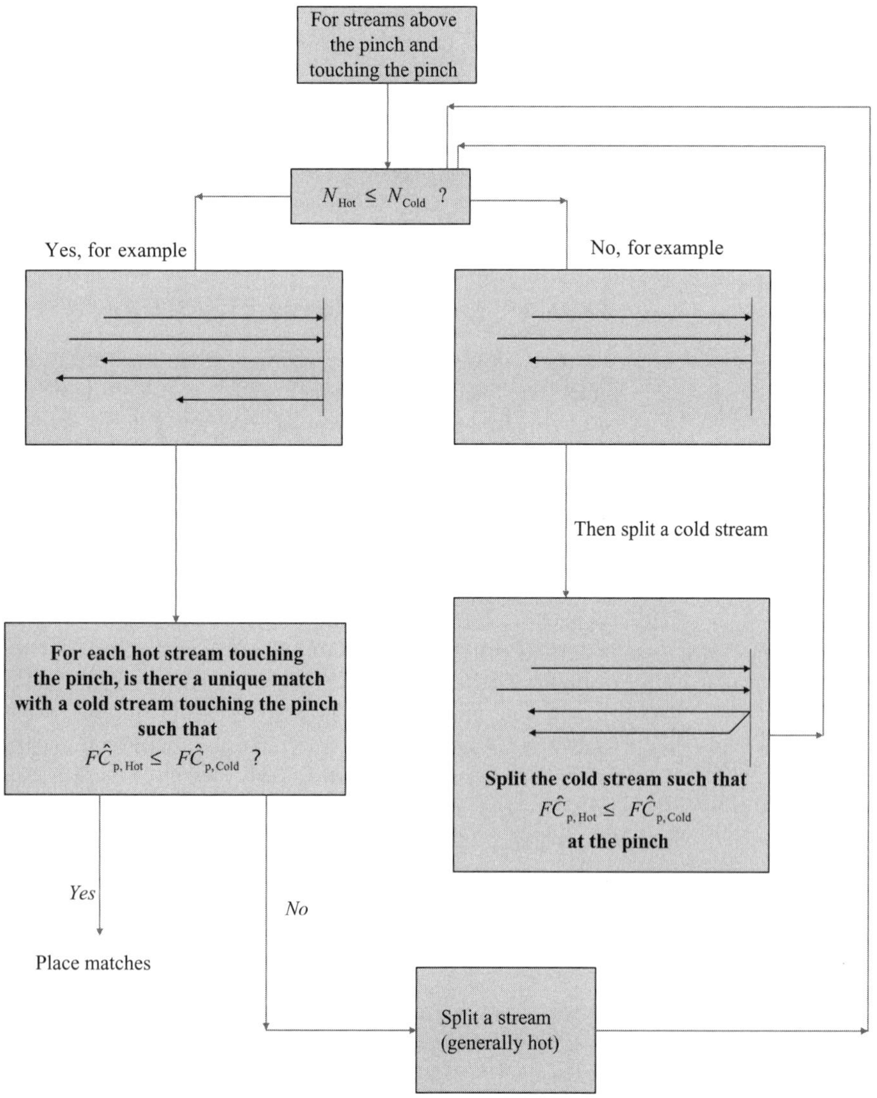

Figure 13.14 Heat exchanger placement heuristic rules for MER design above the pinch.

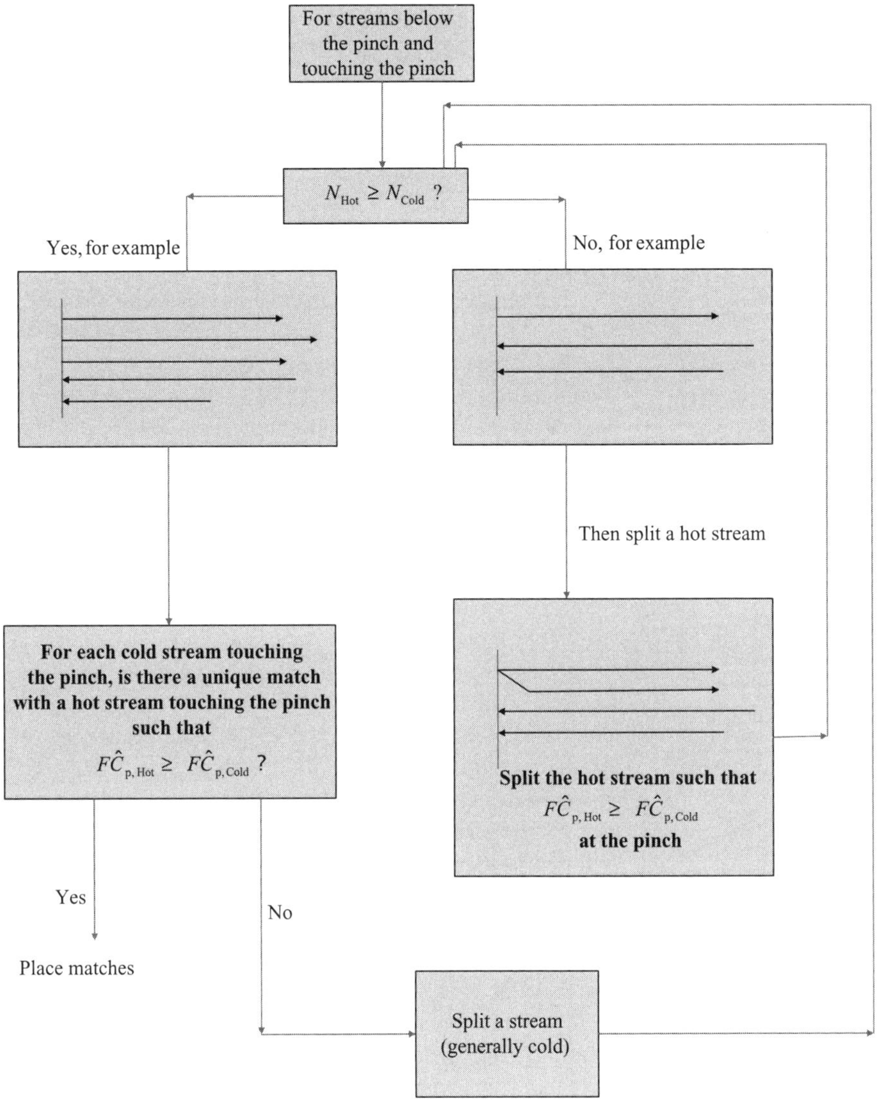

Figure 13.15 Heat exchanger placement heuristic rules for MER design below the pinch.

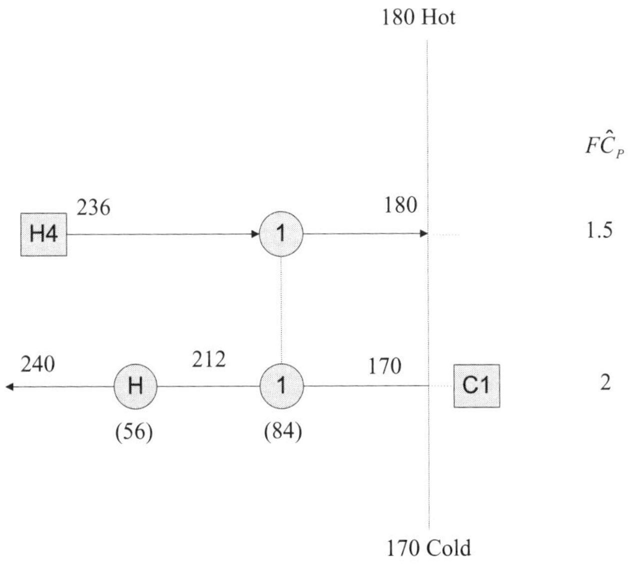

Figure 13.16 Heat exchanger placement above the pinch.

There is no remaining hot stream and the remaining cold stream is heated by a hot utility with a requirement, $\dot{Q}_{C1,Utility} = 2(240 - 212) = 56$ kW.

13.6.2 Design below the Pinch

We again start our design at the pinch and move away from the pinch following the rules of Figure 13.15. As shown in Figure 13.17, there are four streams below the pinch, but only three of these streams touch the pinch, giving $N_{Hot} = 2$ and $N_{Cold} = 1$. We do not want to use a hot utility below the pinch and we require that for each cold stream touching the pinch there be one hot stream such that $F\hat{C}_{p,Hot} \geq F\hat{C}_{p,Cold}$. Heat exchanger 2, matching H2 and C1, is shown in Figure 13.17.

To determine the energy transferred in exchanger 2 (between H2 and C1), we determine the maximum energy transfer rate possible from H2, which is $\dot{Q}_{H2,max} =$

$3(120-180)=-180$ kW, and the maximum energy rate to C1, which is $\dot{Q}_{\text{C1,max}}=2(170-70)=200$ kW. The actual energy rate transferred is limited by the smaller $|\dot{Q}_{\max}|$ of these two values, which, in this case, is $\dot{Q}_{\text{H2,max}}$. The energy available in H2 will be exhausted, and the inlet side of cold stream in heat exchanger 2 will be

$$T_{\text{C1,in}}=T_{\text{C1,out}}-\frac{\dot{Q}_{\text{Transferred}}}{F\hat{C}_{p,\text{C1}}}=170-\frac{180\text{ kW}}{2\dfrac{\text{kW}}{°\text{C}}}=80°\text{C}.$$

At this point, it is instructive for the reader to see the result if instead of matching H2 and C1 we tried to match H4 to C1—here a temperature violation would occur explaining the need for the $F\hat{C}_p$ inequality.

After matching H2 and C1, there are no remaining cold streams touching the pinch, and we now make practical matches between H4 and C3 and H4 and the remaining cold stream C1. In general, these matches are guided by the need to maintain $\Delta T_{\min} \geq 10°\text{C}$ at both ends of the countercurrent heat exchanger. Matching H4 and C3, we determine the maximum energy transfer rate possible from H4, which is $\dot{Q}_{\text{H4,max}}=1.5(30-180)=-225$ kW, and the maximum energy rate to C3, which is $\dot{Q}_{\text{C3,max}}=1(150-100)=50$ kW.

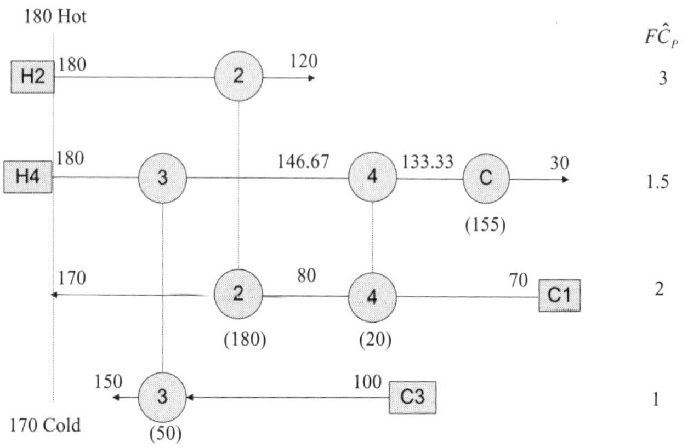

Figure 13.17 Heat exchanger placement below the pinch.

Here C3 will be exhausted and the outlet side of the hot stream in heat exchanger 3 will be

$$T_{\text{H4,out}}=T_{\text{H4,in}}+\frac{\dot{Q}_{\text{Transferred}}}{F\hat{C}_{p,\text{H4}}}=180+\frac{-50\text{ kW}}{1.5\dfrac{\text{kW}}{°\text{C}}}=146.67°\text{C}.$$

We do need to check that there are no $\Delta T_{\min}$ violations at the ends of the heat exchanger.

The remainder of H4 is matched to the remainder of C1 with the hot-side outlet temperature determined as 133.33°C. The final unmet requirement of H4 (133.33 → 30°C) uses a cold utility, $\dot{Q}_{\text{H4,Utility}}=1.5(133.33-30)=-155$ kW. The final network, both above and below the pinch, is shown in Figure 13.18.

Before we leave this problem, we can comment on the minimum number of heat exchangers expected above and below the pinch. The minimum number of expected heat exchangers is provided by Hohmann (1971) as

$$\mu_{\text{Expected}}=N-1, \tag{13.4}$$

where N is the total number of hot and cold process streams plus utilities. For example, above the pinch, there are two process streams and a hot utility, with $\mu_{\text{Expected}} = 3 - 1 = 2$, as found in Figure 13.16. Below the pinch, there are four process streams and a cold utility with $\mu_{\text{Expected}} = 5 - 1 = 4$, as found in Figure 13.17.

13.7 HEAT EXCHANGER NETWORK DESIGN WITH STREAM SPLITTING

As indicated in the heuristic design rules (Figures 13.14 and 13.15), in some cases, during the heat exchanger design, we may need to split a stream in order to obtain a MER design. From Figure 13.14, for the MER design above the pinch, if the number of hot streams (N_{Hot}) touching the pinch is greater than the number of cold streams (N_{Cold}) touching the pinch, then a cold stream must be split. From Figure 13.15, for the MER design below the pinch, if N_{Hot} touching the pinch is

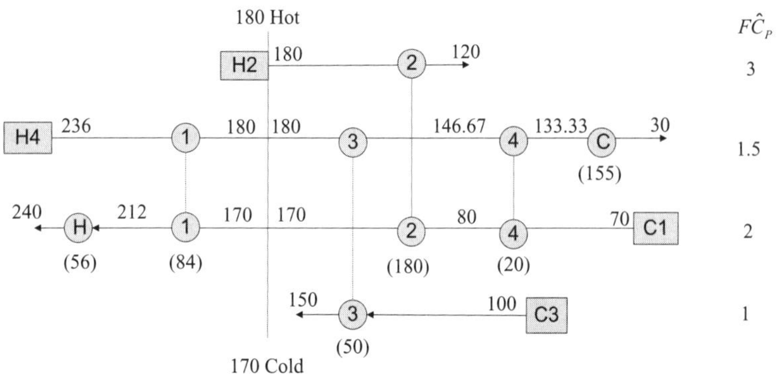

Figure 13.18 Heat exchanger placement for MER network.

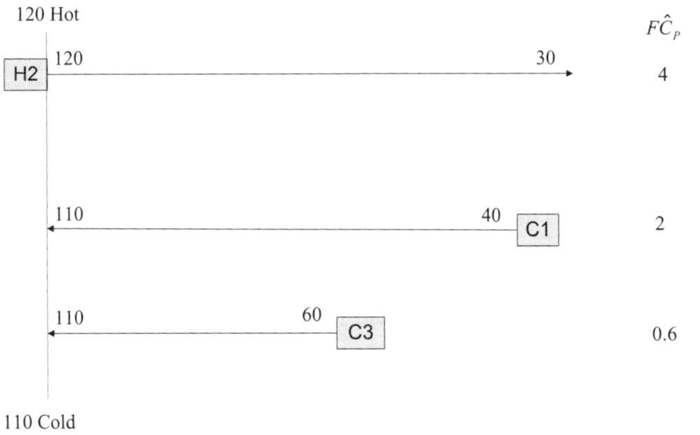

Figure 13.19 Heat exchanger design problem below the pinch.

less than N_{Cold} touching the pinch, then a hot stream must be split. The following example will help clarify why stream splitting is necessary

EXAMPLE 13.6 *Stream Splitting*

Obtain a MER design for the streams below the pinch shown in Figure 13.19; here, $\dot{Q}_{\text{cold utility}} = 190$ kW. Here, $N_{\text{Hot}} < N_{\text{Cold}}$. We can match H2 *without* splitting to either C1 or C3. However, if we match H2 to C1, we will not be able to bring C3 to 110° without the use of a hot utility below the pinch. Similarly, if we match H2 to C3, we will not be able to bring C1 to 110° without the use of a hot utility. Recall that a hot utility cannot be used below the pinch for a MER design. The solution is to split H2 into two streams that allow matching with both C1 and C3.

SOLUTION Here, $N_{\text{Hot}} = 1$ and $N_{\text{Cold}} = 2$. Following Figure 13.15, we want to split the hot stream so that after splitting, $F\hat{C}_{p,\text{Hot}} \geq F\hat{C}_{p,\text{Cold}}$. In other words, we want one split of the hot stream to have an $F\hat{C}'_{p,\text{Hot}} \geq F\hat{C}_{p,\text{C1}}$ and the other split of the hot stream to have an $F\hat{C}''_{p,\text{Hot}} \geq F\hat{C}_{p,\text{C2}}$. This split can be somewhat arbitrary as shown in Figure 13.20, where we simply set one of the splits of H2 such that $F\hat{C}'_{p,\text{Hot}} = F\hat{C}_{p,\text{C1}}$ and then solve the remaining MER network.

The network of Figure 13.20 shows the H2 split through heat exchanger 1 exits at 50°C and the H2 split through heat exchanger 2 exists at 105°C. These streams are then mixed to give a stream at 77.5°C, which is then cooled with a cold utility to 30°C.

From second law considerations, we would prefer mixing to be isothermal. Here we can solve for the $FC_{p,\text{Hot}}$ H2 splits, which will give the same $T_{\text{H2,out}}$ for exchangers 1 and 2:

$$x(120 - T_{\text{H2,out}}) = 140 \text{ kW}$$

$$(4 - x)(120 - T_{\text{H2,out}}) = 30 \text{ kW}.$$

Here $x = F\hat{C}'_{p,\text{H2}}$ split value for the match with C1 and $(4 - x) = F\hat{C}''_{p,\text{H2}}$ split value for the match with C3. Solving the two equations, we find $x = 3.294$ kW/C and $T_{\text{H2,out}} = 77.5$°C. The network is shown in Figure 13.21. ■

When isothermal mixing is used after stream splitting, be sure to check that an approach temperature is not violated; here, for example, see Problem 13.5. Stream splitting

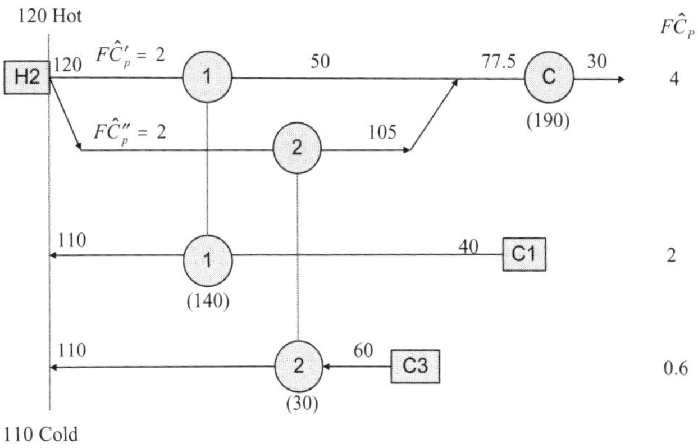

Figure 13.20 Heat exchanger design with "arbitrary" stream splitting.

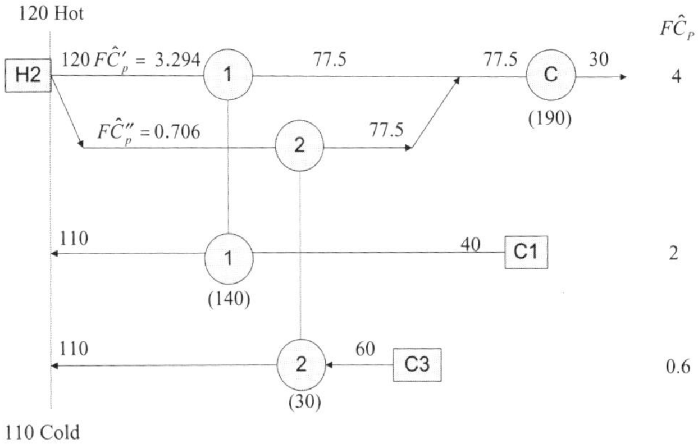

Figure 13.21 Heat exchanger design with isothermal mixing after stream splitting.

does allow for a MER design, but careful consideration must be given to the operability of the network.

13.8 HEAT EXCHANGER NETWORK DESIGN WITH MINIMUM NUMBER OF UNITS (MNU)

We have established the procedures to develop a MER heat exchanger network. It is common to also consider a second design, one that uses the minimum number of heat exchanger units, termed an MNU design. Here the idea is that the MER design uses minimum utilities and the MNU design will have the minimum capital cost. This allows a final design selection based on whichever cost dominates—utility or capital.

The MNU design can be evolved from the MER design, which hopefully allows retention of some aspects of the MER design. Let us reexamine the MER design of Figure 13.18 but eliminate any consideration of the pinch. Equation (13.4) can be applied to the entire network, which now consists of two hot and two cold streams as well as a hot and cold utility, $\mu_{\text{Expected}} = 6 - 1 = 5$. Figure 13.18 shows that six heat exchangers are being used and the minimum expected number of heat exchangers is five. Here the "extra"

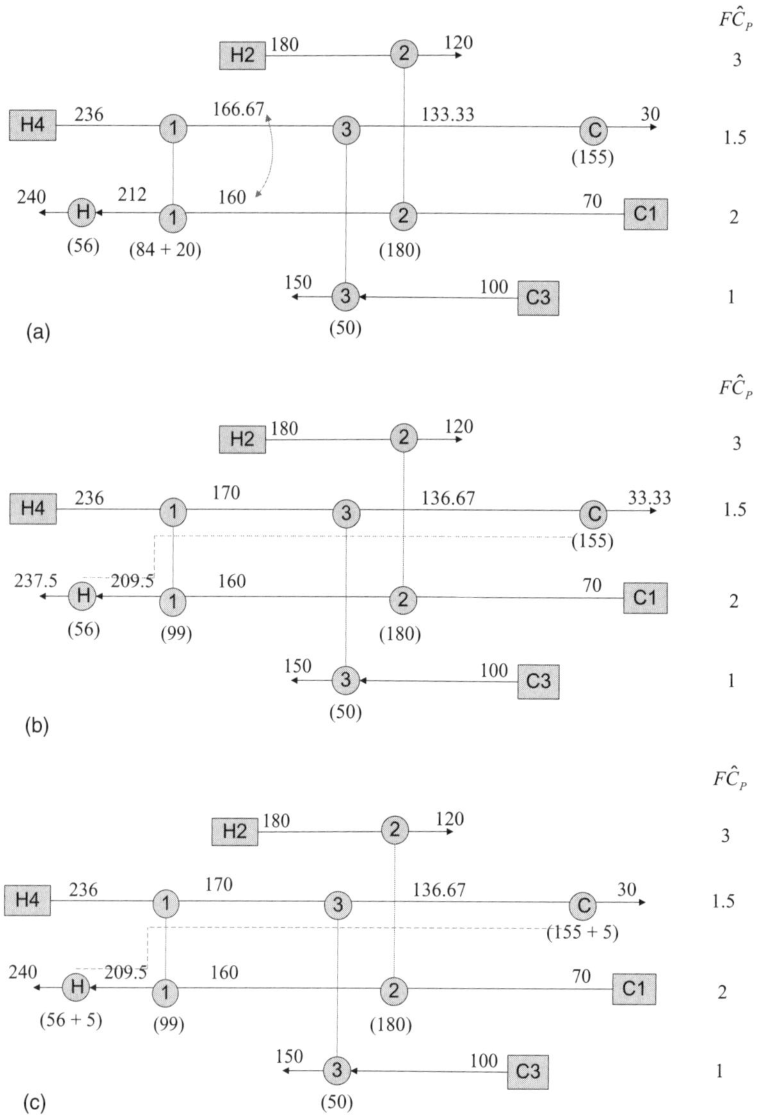

Figure 13.22 (a) Loop breaking with smallest $\dot{Q}$ transfer in the MER design of Figure 13.18. (b) MNU heat exchanger network after loop breaking and heat load adjustment. (c) MNU heat exchanger network after loop breaking and utility load adjustment.

heat exchanger occurs because energy is exchanged between H4 and C1 above the pinch (heat exchanger 1) and H4 and C1 below the pinch (heat exchanger 4). Without consideration of the pinch, we may only need to match these streams one time and we can eliminate the "loop" that exists between the two heat exchangers. Notice in Figure 13.18 that you can start at HEx-1, 180°, and follow H4 to 146.67°, then drop through HEx-4 to 80° and trace back to HEx-1 following C1—this is the loop in Figure 13.18.

A general procedure to eliminate heat exchanger loops is to remove the heat exchanger with the smallest $\dot{Q}$ and to transfer its energy to the "next" heat exchanger in the loop. The idea here is that eliminating the heat exchanger with the smallest $\dot{Q}$ may help preserve much of the MER design.

EXAMPLE 13.7 ***Evolving from a MER Network to an MNU Network***

Evolve the MER network of Figure 13.18 to a network with MNU.

SOLUTION Eliminate HEx-4 by transferring its duty to the next heat exchanger in the loop, which is HEx-1. Here the duty of HEx-1 will be increased from 84 to 104 kW, but HEx-4 can be totally removed from the network. The temperatures in the network must be recalculated, resulting in Figure 13.22a.

The immediate problem in Figure 13.22a is that the approach temperature on the cold side of heat exchanger 1 is not allowed. We need to raise the hot stream outlet temperature of heat exchanger 1 from 166.67 → 170 to avoid a $\Delta T_{\min}$ violation. We also want to preserve the existing energy transfer in HEx-3. We can raise the hot stream outlet temperature of heat exchanger 1 by reducing the

	A	B	C	D	E	F	G	H	I	J	K
1	Name	F	Cp	Tin	Tout		DeltaT min	10.00		Scale Factor	5
2	c1	1	2	70	240						
3	h2	1	3	180	120						
4	c3	1	1	100	150						
5	h4	1	1.5	236	30		Verify Input Data				
6											
7											
8											
9	Delta T min	10		#Hot Streams	2		#Cold Streams	2			
10	Streams Data										
11	h2	1	3	180	120						
12	h4	1	1.5	236	30						
13	c1	1	2	70	240		Run HEN				
14	c3	1	1	100	150						
15											

Figure 13.23 Data input to program THEN for Example 13.8.

energy transfer rate in the heat exchanger 1 by an amount, $(170 - 166.67) \times 1.5 = 5$ kW. This would lower the duty on HEx-1 from its current value of 104 to 99 kW. We again need to recalculate all the temperatures in the network, giving the results shown in Figure 13.22b.

The problem in Figure 13.22b is that C1 is no longer heated to the required 240° and H4 is no longer cooled to the required 30°. We can reestablish required temperatures by increasing the utilities used. We are actually creating a path (shown as the dotted line in Figure 13.22b) between the hot and cold utility through heat exchanger 1. After increasing the utilities so $T_{H4,out} = 30$ and $T_{C1,out} = 210$, we find the results shown in Figure 13.22c.

In Figure 13.22c, we have broken the heat loop by the removal of heat exchanger 4. However, both the hot and cold utilities have increased by 5 kW. This 5 kW can be traced to the reduced heat transfer in HEx-1 as we evolved from Figure 13.22a (with ΔT_{min} violation) to Figure 13.22b. We should appreciate that any heuristic rule (here we eliminated the heat exchanger with the smallest $\dot{Q}$ in the loop) may not always be the best approach. In Problem 13.2, we cost the two networks—the MER network (Figure 13.18) and the MNU network (Figure 13.22b). In Problem 13.3, we explore elimination of heat exchanger 1 from the loop. ■

13.9 SOFTWARE FOR TEACHING THE BASICS OF HEAT EXCHANGER NETWORK DESIGN (TEACHING HEAT EXCHANGER NETWORKS (THEN))

Several years ago, we (Pethe et al., 1989) developed a Fortran-based program, THEN, for reinforcing the basics of heat exchanger network design. The program calculated the temperature interval analysis and drew the GCC; constructed and drew the MER network following the rules of Figures 13.14 and 13.15; and, located loops in the MER network. Fortran no longer supports the drawing routines used and the program was retired. However, Excel provides a powerful pre- and postprocessing platform for legacy codes such as THEN.

Here we describe the use of THEN now called from Excel. The development of the Excel-based preprocessor for data input and postprocessor for results display is an advanced Excel application. The reader is encouraged to examine the VBA macros for these tasks as well as the source code for THEN.

EXAMPLE 13.8 *Heat Exchanger Design Using THEN*

Solve the stream data in Table 13.5 with $\Delta T_{min} = 10$ for the MER network using THEN.

SOLUTION Open the file **Example13.8.xls**. There are four Excel sheets: Input, Output, GCC, and Grid Diagram (GDD). Stream data are input on the Input sheet as shown in Figure 13.23 and ΔT_{min} is set = 10. The scale factor controls the size of the grid diagram for the MER heat exchanger network. The input data are verified (Verify Input Data) to establish the hot and cold streams and to create the needed data file for the Fortran program. THEN is executed from (Run THEN).

The GCC sheet will become active and you will need to delete (Delete) any existing sheets. The GDD sheet will then become active, showing the MER heat exchanger design (Figure 13.24), which here is identical to the results we obtained in Figure 13.18.

The Output sheet provides details of the heat exchanger matches and indicates loops that may exist in the system. The Output sheet (an Excel sheet) is convenient for heat exchanger costing calculations. ■

As practice, a series of heat exchanger design problems have been assembled in Problem 13.5; the user should develop "by hand" the MER network and GCC for each problem and compare these results with those obtained using THEN. When using THEN, be sure the executable file, Then.exe, is located in the same folder as the Excel file containing the pre- and postprocessing VBA modules. We also note that THEN may fail on some problems, especially those involving streams with large latent heat changes or if ΔTmin is set = 0.

13.10 HEAT EXCHANGER NETWORK DESIGN: DISTILLATION COLUMNS

One important application of pinch technology is guiding the appropriate placement of distillation columns. In many processes, the largest energy demand is distillation for product separation. Townsend and Linnhoff (1983a,b), Linnhoff et al. (1983), and Dhole and Linnhoff (1993) show that the appropriate placement of a distillation column

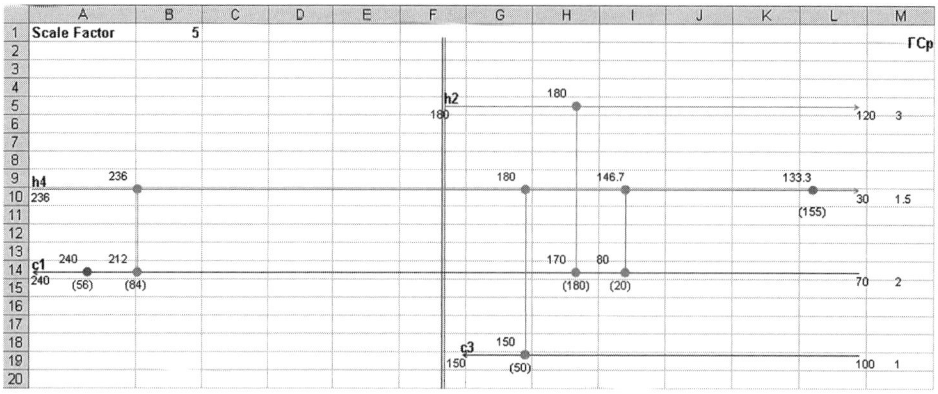

Figure 13.24 MER heat exchanger design for Example 13.8.

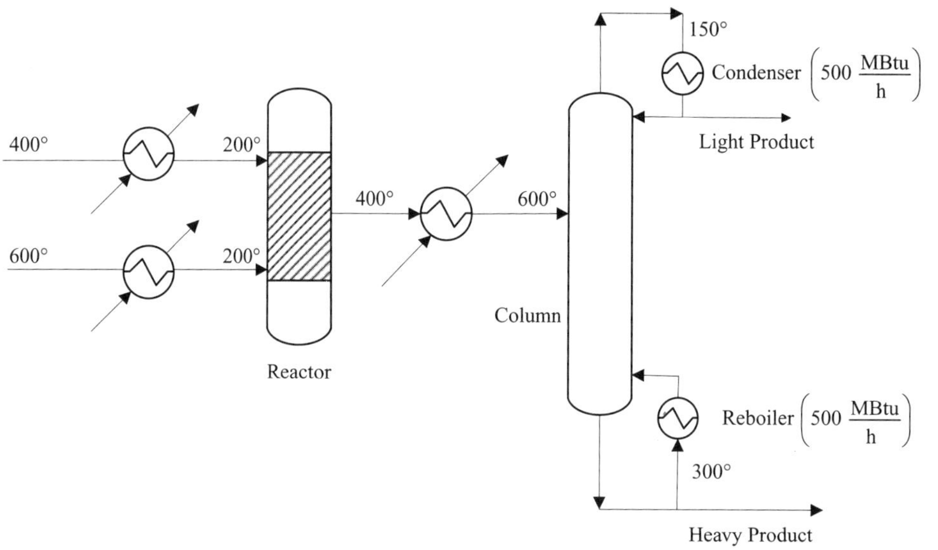

Figure 13.25 Simplified process for discussion of distillation column placement.

occurs when the column reboiler and condenser are both located either entirely above or entirely below the background pinch. This concept is first explained using simple examples. A more realistic problem with detailed solution is provided in Problem 13.7.

The distillation column heat integration process can be outlined as follows:

1. Construct the GCC for the background section (sometimes also called the reaction section). The background section includes all streams needing heating and cooling *except* the column condenser and reboiler duties. The background will also include any column product or feed streams needing either heating or cooling.
2. As needed, adjust the column temperature to allow distillation columns to fit entirely above or entirely below the background pinch—this is the appropriate placement of the column. Here, changing the column pressure may allow needed changes in the column operating temperature. Advanced options for column temperature adjustment are discussed in Smith (2005).

The column appropriate placement principle does require that the background be able to supply the needed heating or cooling duty for the columns. The case where the columns require more energy than the background can supply is detailed in Problem 13.7.

EXAMPLE 13.9 *Distillation Column Placement*

For the simplified process of Figure 13.25, examine distillation column placement in relation to the background pinch. Stream data are provided in Table 13.6. Allow a 1° temperature change for the vaporization and condensation processes. To simplify calculations and discussion, let us again use $\Delta T_{\min} = 0$ as the approach temperature. In this example, we will evaluate three column placements (labeled as column placements A, B, and C).

SOLUTION **Solution Column Placement A: Column Operates below the Background Pinch**

First consider the background process, which consists of the process streams *without* column condensers and reboilers. The temperature interval analysis for the three process streams (Figure 13.26a) with $\Delta T_{\min} = 0$ shows $\dot{Q}_{\text{hot,utility}} = 1800$ MBtu/h, $\dot{Q}_{\text{cold,utility}} = 1200$ MBtu/h, and the pinch temperature = 400°F. Next, add the column condenser and reboiler to the three background process streams. From the temperature interval analysis (Figure 13.26b) with $\Delta T_{\min} = 0$, we obtain the somewhat surprising result that $\dot{Q}_{\text{hot,utility}} = 1800$ MBtu/h, $\dot{Q}_{\text{cold,utility}} = 1200$ MBtu/h, and the pinch temperature remains at 400°F. Before we discuss this result, we also show in Figure 13.26c that the same result is obtained if the condensation and evaporation processes occur at constant temperature.

The values for the pinch and utility duties are the same for the analysis with just the three background streams and the analysis with the three background streams and the column condenser and reboiler. This result can be explained by comparing the GCC for the background process in Figure 13.27a with the GCC for the entire process (background plus columns) in Figure 13.27b.

Figure 13.27a shows the background process above the pinch is a heat sink requiring $\dot{Q}_{\text{hot,utility}} = 1800$ MBtu/h and the background process below the pinch is a heat source rejecting $\dot{Q}_{\text{cold,utility}} = 1200$ MBtu/h.

Column integration in Figure 13.27b shows that the reboiler accepts $\dot{Q}_{\text{reboiler}}$ from the background process streams and rejects $\dot{Q}_{\text{condenser}}$. As the condenser and reboiler duties are the same in this example (and they are typically similar in value), there is no effect on the overall utility requirements. To emphasize the appropriate placement of distillation columns, Figure 13.27b is typically drawn as shown in Figure 13.27c. In Figure 13.27c, the background GCC is drawn and the column placed with the appropriate condenser and reboiler temperatures and duties ($\dot{Q}_{\text{reboiler}} = 500$ MBtu/h and $\dot{Q}_{\text{condenser}} = 500$ MBtu/h). The column can clearly "fit" under the background pinch as shown in Figure 13.27c.

Table 13.6 Stream Data Summary for Process of Figure 13.25 with Column Placement A

Stream	Type	$F\hat{C}_p\left(\frac{\text{MBtu}}{\text{h-°F}}\right)$	T_{in} (°F)	T_{out} (°F)
H1	Hot	5	400	200
H2	Hot	1	600	200
C3	Cold	10	400	600
Condenser	Hot	500	150	149
Reboiler	Cold	500	300	301

Column Placement B: Column Operates across the Background Pinch

Next, consider the problem as provided in Table 13.6 except the reboiler temperature is changed from 300 to 500°F.

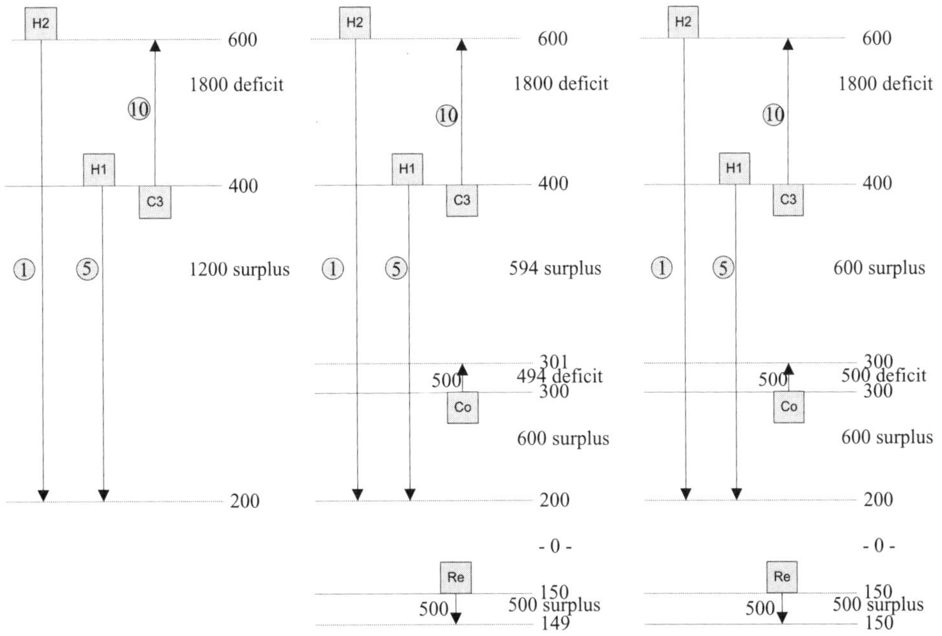

Figure 13.26 Problem table for discussion of distillation column placement (a) background, (b) background plus column, and (c) background plus column (constant T evaporation and condensation).

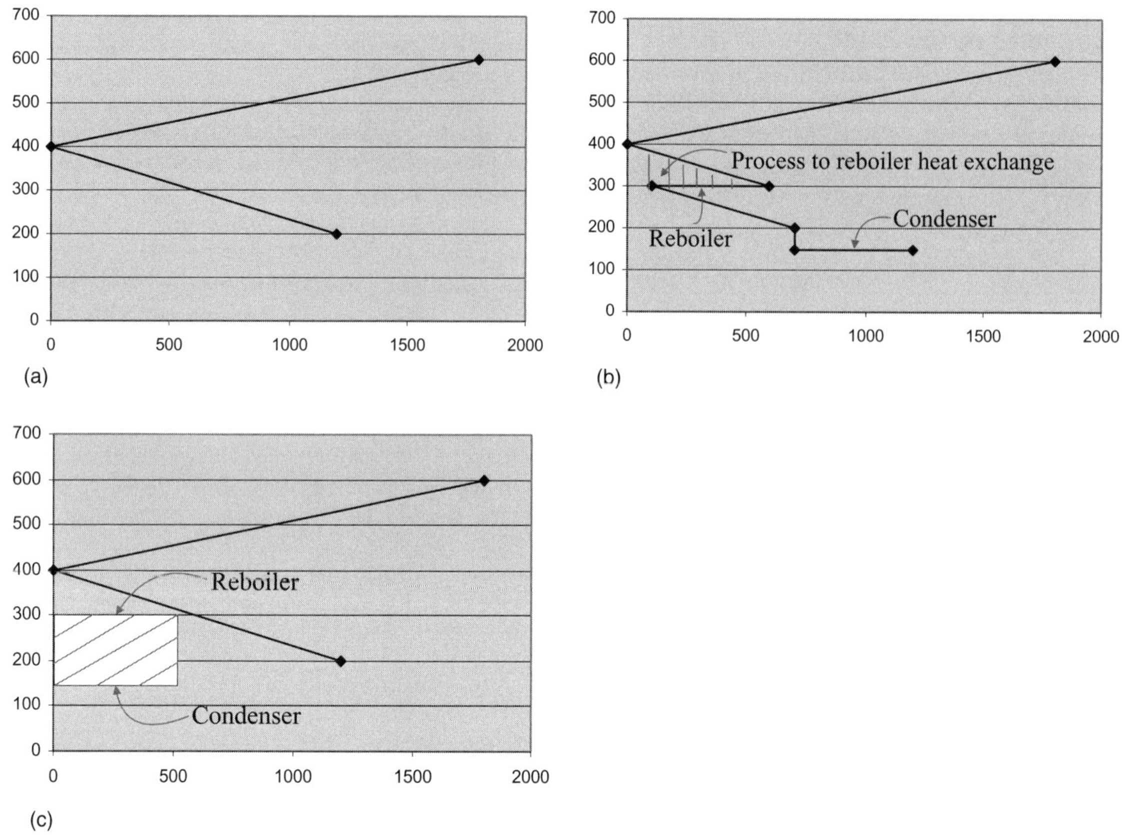

Figure 13.27 (a) GCC for the background and (b) GCC for the entire process with column below the background pinch—column placement A. (c) GCC for the background with column fitting against the background GCC.

Stream	Type	$F\hat{C}_p\left(\frac{\text{MBtu}}{\text{h-°F}}\right)$	T_{in} (°F)	T_{out} (°F)
H1	Hot	5	400	200
H2	Hot	1	600	200
C3	Cold	10	400	600
Condenser	Hot	500	150	149
Reboiler	Cold	500	500	501

Solution Column Placement B

The temperature interval analysis for the background process remains unchanged, but for the entire process (background and column), we find with $\Delta T_{min} = 0$, $\dot{Q}_{hot,utility} = 2300$ MBtu / h, $\dot{Q}_{cold,utility} = 1700$ MBtu / h, and the pinch temperature = 400°F. The GCC is shown in Figure 13.28 and can be used to explain this result.

Figure 13.28a shows that the reboiler is entirely above the background pinch and the condenser is entirely below the background pinch. Figure 13.28b highlights that column integration is not possible. Here the reboiler adds an additional requirement, $\dot{Q}_{reboiler}$, to the background heat sink and the condenser adds an additional load of $\dot{Q}_{condenser}$ to the background heat source. This is indicated by the $\dot{Q}_{reboiler} = 500$ MBtu/h and the $\dot{Q}_{condenser} = 500$ MBtu/h to the right of the y-axis in Figure 13.28b. There are no energy savings from operating the column across the pinch. In column placement B, the energy requirements are the same regardless if the column is integrated with the background process (Figure 13.28a) or if the column is operated stand-alone. This (placement B) is not considered an appropriate/optimal placement of the column.

Column Placement C: Column with Partial Energy Integration

Next, consider the problem again as provided in Table 13.6 except the reboiler temperature is 350°F.

Stream	Type	$F\hat{C}_p\left(\frac{\text{MBtu}}{\text{h-°F}}\right)$	T_{in} (°F)	T_{out} (°F)
H1	Hot	5	400	200
H2	Hot	1	600	200
C3	Cold	10	400	600
Condenser	Hot	500	150	149
Reboiler	Cold	500	350	351

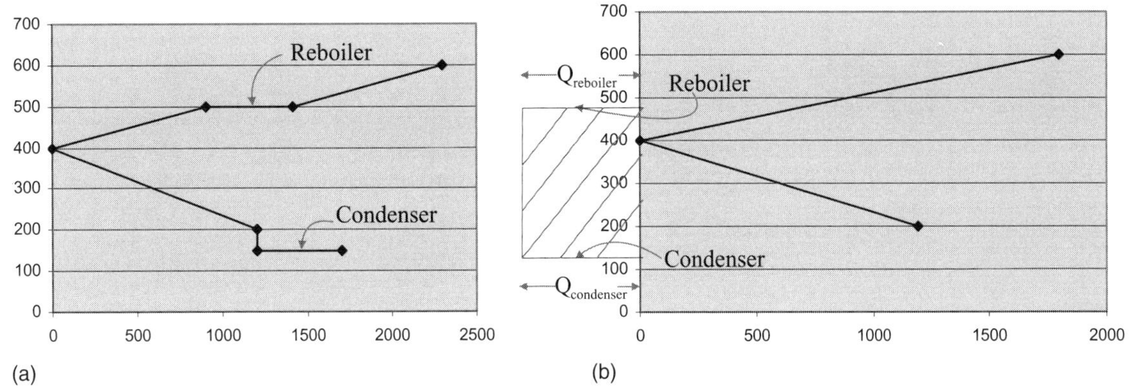

Figure 13.28 GCC for the entire process with column operating across the background pinch—column placement B; (a) entire process, (b) background plus column.

Solution Column Placement C

The temperature interval analysis for the background process remains unchanged, but for the entire process (background and column), we find with $\Delta T_{min} = 0$, $\dot{Q}_{hot,utility} = 2000$ MBtu / h, $\dot{Q}_{cold,utility} = 1400$ MBtu / h, and the entire process pinch temperature is 350°F. The GCC shown in Figure 13.29 can be used to explain this result.

Figure 13.29b is again the best way to understand possible column integration. Figure 13.29b shows that the column reboiler duty is actually too large to be entirely satisfied by the background process. Part of the reboiler duty can be satisfied by the process streams, and if the column is integrated with these process streams, there will be an energy savings compared to operating the column with stand-alone utilities. Ultimately, part of the reboiler duty $\dot{Q}_{reboiler}$ and part of the condenser duty $\dot{Q}_{condenser}$, the portion to the left of the y-axis (each 200 MBtu / h) in Figure 13.29b, will require external utilities. ■

An important aspect of column energy integration is using column pressure/temperature change to allow better column heat integration. As an exercise, show yourself that if in column placement C the column temperature can be lowered 34°F by lowering the pressure (here change the reboiler temperature from 350 to 316°F and the condenser from 150 to 116°F) and the reboiler and condenser duties remain constant, the column can be integrated below the pinch.

We can now identify two limiting cases. In limiting case 1, the column can be integrated entirely below the background pinch. The column can be operated at no additional energy cost compared to the background process; this is column placement A. In limiting case 2, the column operates across the pinch (column placement B) and there are no energy savings. Here the energy cost for the entire process is the sum of the background process and the column, which is the same as if each where independent and operating stand-alone. This is not considered an appropriate/optimal placement of the column.

In column placement C, the background cannot supply enough energy to satisfy the column loads. Some energy savings, compared to stand-alone column operation, are possible by integration with process streams. Changing the column pressure/temperature can increase integration. For column placement C (as discussed earlier), lowering the pressure in the column would reduce the reboiler temperature and condenser temperature and may allow the column to operate below the background pinch. Of course the impact of any column pressure change must be evaluated with rigorous simulation programs such as Aspen or HYSYS. Careful consideration of the economics of a pressure change must also be made. For example, it is seldom economical to lower the column temperature such that chilled water as opposed to cooling water is needed in the condenser.

Parallel to our discussions in cases A–C, if the background pinch temperature is low, process stream to the distillation column condenser heat exchange may be possible. Here, for example, in the equivalent to case A, the column would operate entirely above the pinch.

We can summarize this section by saying the appropriate placement of distillation columns is entirely below or entirely above the background pinch, provided the background process has the capacity to accept or supply the required duties. The reader is encouraged to see Problem

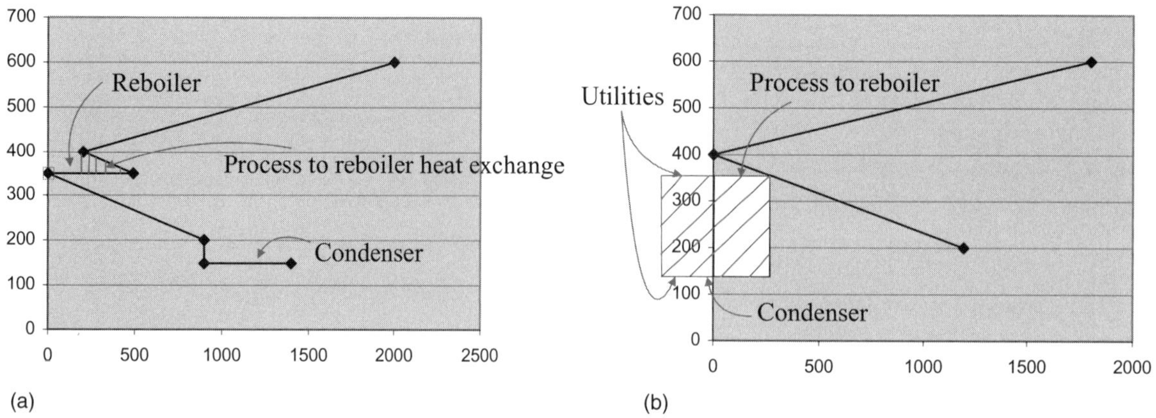

Figure 13.29 GCC for the entire process with column determining pinch location—column placement C: (a) entire process and (b) background plus column.

13.7, which provides a more complex application—column integration in the hydrodealkylation (HDA) process.

13.11 CLOSING REMARKS

Process energy integration is a mature topic and there are excellent books available, which greatly expand the topics introduced in this chapter. The interested reader is encouraged to first read, *A User Guide on Process Integration for the Efficient Use of Energy* (Linnhoff et al., 1982), followed by *Heat Exchanger Network Synthesis: Process Optimization by Energy and Resource Analysis* (Shenoy, 1995) and finally *Chemical Process Design and Integration* (Smith, 2005). In this chapter, we detailed the mechanics of hot and cold stream identification, the location of the pinch temperature, the formation of the GCC, and the design of MER and MNU networks. We also explored one application, the energy integration of distillation columns and process streams. Many related topics including hot and cold composite curves and the energy integration of heat pumps were not discussed, and here the reader is referred to the three books cited above. Also, the principles of energy integration are being extended to a number of important processing problems, for example, waste minimization (El-Halwagi, 1997) and water and hydrogen management (Agrawal and Shenoy, 2006).

The temperature interval analysis developed in this chapter is rigorous for energy targeting. Heuristic rules were used for stream matching, leading to two heat exchanger network designs, one utilizing minimum utilities and a second network utilizing the fewest total heat exchangers. Optimization-based solution approaches, as discussed in the works of Floudas (1995) and Biegler et al. (1997) can be used for both energy targeting and stream matching. Energy targeting can be formulated as a linear programming transshipment model (Papoulias and Grossmann, 1983). Stream matching with the goal of minimizing total cost (total cost = heat exchanger capital cost + utilities cost) can be accomplished with a heat exchanger superstructure—see Yee et al. (1990a,b) and Yee and Grossmann (1990).

There are software packages for energy integration and heat exchanger design calculations. A code I find very useful is Hint, developed by Professor Angel Martin (Martin and Mato, 2008); the Hint program is available for download at http://www.iq.uva.es/integ/Hint.zip. Available academic and commercial codes are listed in Biegler et al. (1997).

REFERENCES

AGRAWAL, V. and U.V. SHENOY. 2006. Unified conceptual approach to targeting and design of water and hydrogen networks. *AIChE J.* 52(3): 1071–1082.

BIEGLER, L.T., I.E. GROSSMANN, and A.W. WESTERBERG. 1997. *Systematic Methods of Chemical Process Design*. Prentice Hall, Upper Saddle River, NJ.

CHEN, J.J.J. 1987. Letter to the Editor: Comments on improvement on a replacement for the logarithmic mean. *Chem. Eng. Sci.* 42: 2488.

Dhole, V.R. and B. Linnhoff. 1993. Distillation column targets. *Comput. Chem. Eng.* 17(5–6): 549–560.

Douglas, J.M. 1988. *Conceptual Design of Chemical Processes.* McGraw Hill, New York.

El-Halwagi, M.M. 1997. *Pollution Prevention through Process Integration: Systematic Design Tool.* Academic Press, San Diego, CA.

Floudas, C.A. 1995. *Nonlinear and Mixed-integer Optimization: Fundamentals and Applications.* Oxford University Press, Oxford.

Hohmann, E.C. 1971. Optimum networks for heat exchange. PhD Dissertation, University of Southern California, Los Angeles.

Lee, K.F., A.H. Masso, and D.F. Rudd. 1970. Branch and bound synthesis of integrated process design. *Ind. Eng. Chem. Fundam.* 9: 48–58.

Linnhoff, B., H. Dunford, and R. Smith. 1983. Heat integration of distillation columns into overall processes. *Chem. Eng. Sci.* 38(8): 1175–1188.

Linnhoff, B. and J.R. Flower. 1978a. Synthesis of heat exchanger networks: I. Systematic generation of energy optimal networks. *AIChE J.* 24: 633–642.

Linnhoff, B. and J.R. Flower. 1978b. Synthesis of heat exchanger networks: II. Evolutionary generation of networks with various criteria of optimality. *AIChE J.* 24: 642–653.

Linnhoff, B. and E. Hindmarsh. 1983. The pinch design method for heat exchanger networks. *Chem. Eng. Sci.* 38: 745–763.

Linnhoff, B., D.W. Townsend, D. Boland, G.E. Hewitt, B.E.A. Thomas, A.R. Guy, and R.H. Marsland. 1982. *A User Guide on Process Integration for the Efficient Use of Energy.* The Institution of Chemical Engineers, Rugby, England.

Martin, A. and F.A. Mato. 2008. Hint: An educational software for heat exchanger network design with the pinch method. *Educ. Chem. Eng.* 3: e6–e14.

Masso, A.H. and D.F. Rudd. 1969. The synthesis of system designs: II. Heuristic structuring. *AIChE J.* 15: 10–17.

Nishida, N., Y.A. Liu, and L. Lapidus. 1977. Studies in chemical process design and synthesis: III. A simple and practical approach to optimal synthesis of heat exchanger networks. *AIChE J.* 23: 77–93.

Nordman, R. 2008. *New Methods for Process Integration: New Process Integration Methods for Heat-saving Retrofit Projects in Industrial Systems.* VDM Verlag Dr. Müller, Germany. December 10.

Papoulias, S.A. and I.E. Grossmann. 1983. A structural optimization approach to process synthesis. II. Heat recovery networks. *Comput. Chem. Eng.* 7(6): 707–721.

Pethe, S., R. Singh, and F.C. Knopf. 1989. A simple technique for locating loops in heat exchanger networks. *Comput. Chem. Eng.* 13(7): 859–860.

Shenoy, U.V. 1995. *Heat Exchanger Network Synthesis: Process Optimization by Energy and Resource Analysis.* Gulf Publishing, Houston, TX.

Smith, R. 2005. *Chemical Process Design and Integration.* John Wiley & Sons, West Sussex, England.

Tensa Services and F.C. Knopf, 1990. *An Introduction to Pinch Technology and Industrial Energy Optimization.* Workbook for industrial short course. Tensa Services, Houston, TX.

Terrill, D.L. and J.M. Douglas. 1987. Heat-exchanger network analysis 1. Optimization. *Ind. Eng. Chem. Res.* 26(4): 685–691.

Tjoe, T.N. and B. Linnhoff 1986. Using pinch technology for process retrofit. *Chem. Eng.* April: 47–60.

Townsend, D.W. and B. Linnhoff. 1983a. Heat and power networks

Yee, T.F. and I.E. Grossmann. 1990. Simultaneous optimization models for heat integration. II. Heat exchanger network synthesis. *Comput. Chem. Eng.* 14(10): 1165–1184.

Yee, T.F., I.E. Grossmann, and Z. Kravanja. 1990a. Simultaneous optimization models for heat integration. I. Area and energy targeting and modeling of multi-stream exchangers. *Comput. Chem. Eng.* 14(10): 1151–1164.

Yee, T.F., I.E. Grossmann, and Z. Kravanja. 1990b. Simultaneous optimization models for heat integration. III. Process and heat exchanger network optimization. *Comput. Chem. Eng.* 14(10): 1185–1200.

Zamora, J.M. and I.E. Grossmann. 1997. A comprehensive global optimization approach for the synthesis of heat exchanger networks with no stream splits. *Comput. Chem. Eng.* 21: S65–S70.

PROBLEMS

13.1 ***Steam and Chilled Water Costs in Example 13.5*** In Example 13.5, HP steam cost is given as \$80 / kW-year.

Determine steam cost as dollar per 1000 lb. Chilled water cost is given as \$30 / kW-year.

Determine chilled water cost as dollar per 1000 gal. For comparison, typical 2009 utility costs are provided in Table P13.1.

13.2 ***Costing MER and MNU Heat Exchanger Networks*** In Figure 13.18, we developed the MER network for the process in Figure 13.12; this is a two-hot stream, two-cold stream process as summarized in Table P13.2. In Example 13.7, starting with this MER design, we developed the MNU network shown in Figure 13.22c. Cost these two networks using the following data (Biegler et al., 1997):

Heat exchanger annualized capital costs:

$$\$5500 + \$150 \times \text{Area}\left(\text{m}^2\right).$$

Saturated steam is available at 250°C and cooling water is delivered at 20°C and returned at 40°C.

13.3 ***Loop Breaking and MNU Heat Exchanger Networks Design*** In Example 13.7, we identified the heat exchanger loop consisting of HEx-1 and HEx-4. In the example, we eliminated HEx-4 by transferring energy to HEx-1 and solving the resulting network. For Problem 13.3, eliminate HEx-1 by transferring energy to HEx-4 and solve the resulting network. Cost this network using the data provided in Problem 13.2. In this problem, we are transferring the largest $\dot{Q}$ in the loop.

13.4 ***Heat Exchanger Network Design*** The following six problems in Table P13.4a–f are taken from the literature. The user should develop by hand the MER network and GCC for each problem and compare results with those obtained using the

Table P13.1 2009 Utility Costs

Utility	Typical Conditions	Typical Conditions	~2009 Cost
Very high-pressure (VHP) steam	600 psig	488.9°F (253.8°C)	
High-pressure (HP) steam	450 psig	459.6°F (237.5°C)	\$7.00/1000 lb
Medium-pressure (MP) steam	150 psig	365.9°F (185.5°C)	\$5.00/1000 lb
Low-pressure (LP) steam	50 psig	297.7°F (147.6°C)	\$4.00/1000 lb
Chilled water	Out 45°F return 60°F	Out 7.2°C return 15.6°C	\$0.60/1000 gal
Cooling water	Out 70°F return 105°F	Out 21.1°C return 40.6°C	\$0.10/1000 gal
Purchased electricity			\$0.075/kW-h
Natural gas			\$8.00/1000 SCF

Table P13.2 Stream Data Summary for the Process of Figure 13.12

Stream	Type/#	$h\left(\frac{\text{kW}}{\text{m}^2\text{-K}}\right)$	$\text{Cost}\left(\frac{\$}{\text{kW-year}}\right)$	T_{in} (°C)	T_{out} (°C)
Cold 1	Cold	1		70	240
Cold 3	Cold	1		100	150
Hot 2	Hot	1		180	120
Hot 4	Hot	1		236	30
HP steam		5	80	250	250
Cooling water		1	5	20	40

Table P13.4a Stream Data Summary $\Delta T_{min} = 10$ (Linnhoff et al., 1982)

Stream	Type	$F\hat{C}_p$	T_{in}	T_{out}
H1	Hot	3.0	170	60
H2	Hot	1.5	150	30
C1	Cold	2.0	20	135
C2	Cold	4.0	80	140

Table P13.4b Stream Data Summary $\Delta T_{min} = 20$ (Shenoy, 1995)

Stream	Type	$F\hat{C}_p$	T_{in}	T_{out}
H1	Hot	10	175	45
H2	Hot	40	125	65
C1	Cold	20	20	155
C2	Cold	15	40	112

provided program THEN and Hint (see Martin and Mato, 2008 for download instructions). Also, identify any loops that may exist in each network. For some problems, we will also develop an MNU network. Also, for the problem in Table P13.4a, develop the MNU network.

13.5 ***Isothermal Mixing after Stream Splitting*** Design a MER network for the one-hot stream and two-cold stream problem shown in Figure P13.5. Is isothermal mixing possible after stream splitting?

13.6 ***Threshold Problem*** A process that only requires a hot utility or a cold utility is termed a threshold problem—here no pinch exists. Often processes showing a pinch temperature (and a hot and cold utility) at a given minimum approach temperature, ΔT_{min}, can be converted to a threshold problem by reducing ΔT_{min}. The minimum approach temperature below which no pinch exists is termed the threshold approach temperature, $\Delta T_{threshold}$.

From an economic point of view, as ΔT_{min} is decreased, the cost of utilities decrease, but capital costs increase.

Table P13.4c Stream Data Summary $\Delta T_{min} = 10$ (Modified from Linnhoff and Flower, 1978a,b)

Stream	Type	$F\hat{C}_p$	T_{in}	T_{out}
H1	Hot	2.0	180	40
H2	Hot	4.0	150	40
C1	Cold	3.0	60	180
C2	Cold	2.6	30	105

Table P13.4d Stream Data Summary $\Delta T_{min} = 10$ (Lee et al., 1970)

Stream	Type	$F\hat{C}_p$	T_{in}	T_{out}
H1	Hot	8.79	160	93
H2	Hot	10.55	249	138
C1	Cold	7.62	60	160
C2	Cold	6.08	116	260

At $\Delta T_{threshold}$ and below, the cost of the single required utility will remain constant but capital costs will continue to increase.

For the stream data in Table P13.6, start with ΔT_{min} = 10°C and use the program THEN (provided) or Hint (see Martin and Mato, 2008 for download instructions) to locate $\Delta T_{threshold}$. Cost the MER network at ΔT_{min} = 10°C, $\Delta T_{min} = \Delta T_{threshold}$, and $\Delta T_{min} = (\Delta T_{threshold}/2)$; again, use for heat exchanger annualized capital costs \$5500 + \$150 × Area (m^2). Finally, sketch a plot of cost versus ΔT_{min}; include both capital cost and utility cost.

Saturated steam is available at 240°C and cooling water is delivered at 20°C and returned at 40°C.

Table P13.4e Stream Data Summary ΔT_{min} = 20 (Masso and Rudd, 1969)

Stream	Type	$F\hat{C}_p$	T_{in}	T_{out}
H1	Hot	2.37	590	400
H2	Hot	1.58	471	200
H3	Hot	1.32	533	150
C1	Cold	1.60	200	400
C2	Cold	1.60	100	430
C3	Cold	4.13	300	400
C4	Cold	2.62	150	280

Table P13.4f Stream Data Summary ΔT_{min} = 30 (Nishida et al., 1977)

Stream	Type	$F\hat{C}_p$	T_{in}	T_{out}
H1	Hot	10.55	533.2	316.5
H2	Hot	26.38	494.3	383.2
H3	Hot	15.83	477.5	316.5
C1	Cold	36.93	269.3	488.7

13.7 ***Distillation Column Train Energy Integration in the HDA Process (Adapted from Tensa Services and Knopf, 1990; Douglas, 1988; Terrill and Douglas, 1987)*** Figure P13.7 shows the reaction and fractionation sections for the HDA process. The numbers inside the circles are stream numbers. Heat exchanger duties are provided in Table P13.7. The required approach temperature is ΔT_{min} = 40. Develop possible energy integration strategies for the distillation columns.

The following *steps* and comments will help guide you through the problem:

1. Identify the hot and cold streams in the background (or reaction) section.
2. Determine the pinch temperature and hot and cold utilities for the background section. Here, remember to use an approach temperature of 40°F.
3. Plot the GCC for the background section.

   Comments: For the background, you should find three hot and three cold streams with $\dot{Q}_{hot,utility}$ = 6.43 MMBtu/h, $\dot{Q}_{cold,utility}$ = 12.75 MMBtu/h, and the pinch temperature = 655°F.

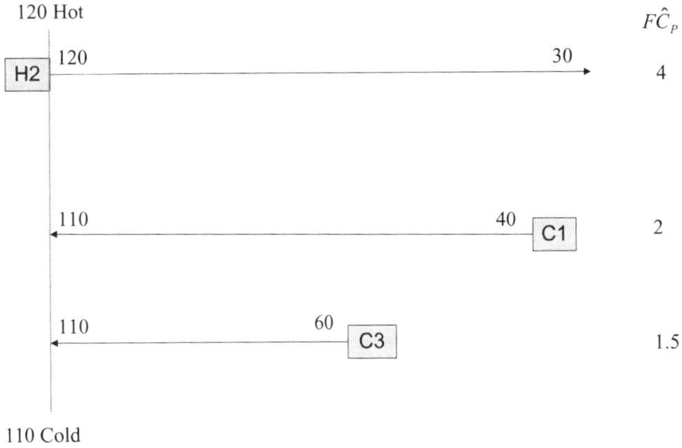

Figure P13.5 Problem 13.5 grid diagram for heat exchanger placement.

Table P13.6 Stream Data Summary

Stream	$F\hat{C}_p\left(\frac{kW}{°C}\right)$	$h\left(\frac{kW}{m^2\text{-}K}\right)$	$\text{Cost}\left(\frac{\$}{kW\text{-year}}\right)$	T_{in} (°C)	T_{out} (°C)
Cold 1	2.0	1		20	195
Cold 3	1.0	1		100	150
Hot 2	3.0	1		180	120
Hot 4	1.5	1		206	30
HP steam		5	80	240	240
Cooling water		1	5	20	40

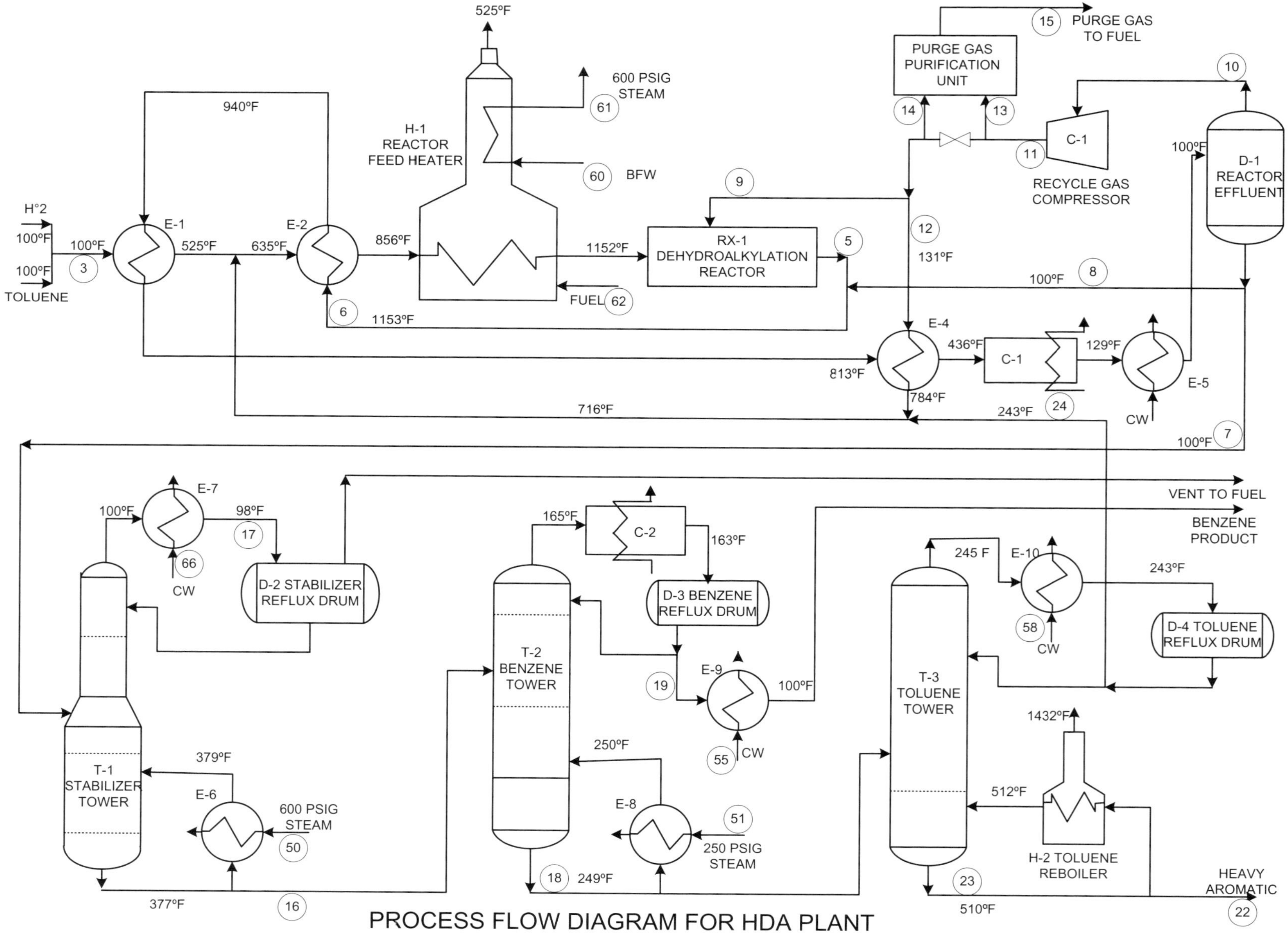

Figure P13.7 Reaction and fractionation sections for the HDA process.

4. Identify the hot and cold streams (condensers and reboilers) in the distillation/fractionation section.
5. Determine the pinch temperature and hot and cold utilities for just the distillation section. Here, use an approach temperature of 40°F.
6. Plot the GCC for just the distillation section.

   Comments: We did not plot the GCC for the distillation section in our chapter example. The reason we are making this plot now is to show that with the current column temperatures (and with a 40°F approach), none of the column condensers can supply heat for any of the reboilers. For the distillation section, you should find three hot and three cold streams with $\dot{Q}_{\text{hot,utility}} = 9.56\ \text{MMBtu/h}$, $\dot{Q}_{\text{cold,utility}} = 9.09\ \text{MMBtu/h}$, and the pinch temperature over a range from 269 to 225°F.
7. Determine the pinch temperature and hot and cold utilities for the entire process (background plus columns). The approach temperature is 40°F.
8. Plot the GCC for the entire process. Here, be sure to make both GCC plots as in Figures 13.28a,b and 13.29a,b.

   Comments: For the entire process (background and distillation), you should find $\dot{Q}_{\text{hot,utility}} = 11.0\ \text{MMBtu/h}$ and $\dot{Q}_{\text{cold,utility}} = 16.86\ \text{MMBtu/h}$.

   We make the GCC plot, which isolates the columns in relation to the background process and background pinch to help evaluate column placement strategies. It is often useful to think of a column in this plot as a "block." The block width is determined by the reboiler/condenser duty, and the block length is determined by the column reboiler and condenser temperatures. The column blocks can be moved up and down by adjusting the pressure/temperature.
9. Determine the optimum column placement.

   Comments: As noted, the column blocks can be moved by adjusting the pressure/temperature. See the effect of raising the toluene column reboiler and condenser temperatures by 60°F. Keep the reboiler and condenser duties at their current levels. To help with plotting the GCC, assume that the stabilizer column will be operated stand-alone; therefore, it can be removed from this placement analysis.

Table P13.7 Exchanger Duties and Stream Names (Tensa Services and Knopf, 1990)

Exchanger	Service	Duty (MMBtu/h)
E-1	Reactor effluent/total reactor feed	7.79
E-2	Reactor effluent/H2 + toluene feed	13.08
E-3	Out of service/removed	—
E-4	Reactor effluent/recycle gas	23.17
E-5	Reactor effluent/cooling water	4.32
E-6	Stabilizer reboiler	0.372
E-7	Stabilizer condenser	0.083
E-8	Benzene reboiler	7.99
E-9	Benzene product cooler	0.52
E-10	Toluene condenser	1.15
C-1	Reactor effluent cooler	18.95
C-2	Benzene condenser	7.86
H-1	Total reactor feed heater (plant section)	17.47
H-2	Toluene reboiler heater	1.198

Be sure you download the program Hint at http://www.iq.uva.es/integ/Hint.zip to help with these calculations.

13.8 ***MER Design and Costing (Zamora and Grossmann, 1997)*** Develop a MER network for a two-hot stream, two-cold stream process with data provided in Table P13.8. Cost the network (operating + capital) using the following annualized (\$/year) capital costs:

Heat exchangers and coolers (\$/year):

$$15000 + 30 \times \text{Area}\,(\text{m}^2)^{0.8}$$

Table P13.8 Stream Data Summary

Stream	Type/#	$h\left(\frac{\text{kW}}{\text{m}^2\text{-K}}\right)$	$\text{Cost}\left(\frac{\$}{\text{kW-year}}\right)$	T_{in} (°C)	T_{out} (°C)
Hot 1	Hot	0.15		180	75
Hot 2	Hot	0.10		240	60
Cold 1	Cold	0.20		40	230
Cold 2	Cold	0.10		120	300
Steam		0.5	110	325	325
Cooling water		2.0	10	25	40

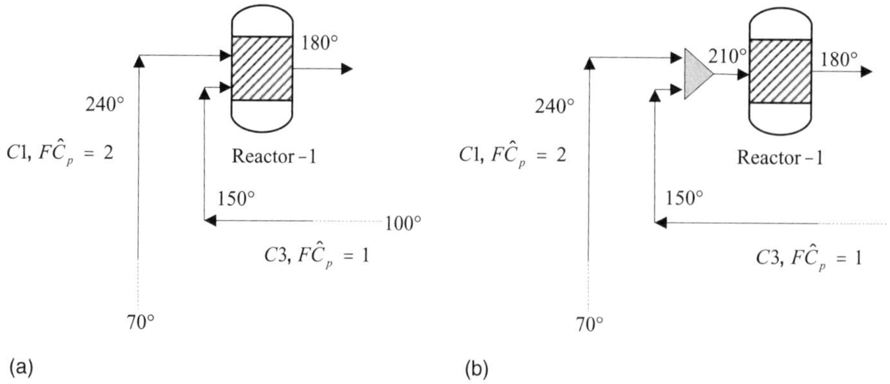

Figure P13.9 Feed stream configuration to the reactor (a) current; (b) with mixer.

and

$$\text{Heaters(\$/year)}: \quad 15000 + 60 \times \text{Area}\left(\text{m}^2\right)^{0.8}.$$

Saturated steam is available at 325°C and cooling water is delivered at 25°C and returned at 40°C.

13.9 ***Isothermal Mixing of Process Streams*** In Figure 13.12, we assumed the two feed streams to the reactor were separately fed to the reactor; we highlight this in Figure P13.9a. For the MER design associated with Figure 13.12 (found in Figure 13.18), we found heating requirements of 56 kW, cooling requirements of 155 kW, and a pinch temperature of 175°.

For Problem 13.9, allow that the two feed streams entering the reactor are mixed as shown in Figure P13.9b. Determine the needed MER utilities if these feed streams are isothermally mixed—both brought to 210°—prior to entering the reactor. Recall in Section 13.5 we discussed that energy recovery may be possible if the pinch temperature falls between the maximum and minimum inlet temperatures to the mixer; in this process, the pinch temperature is 175°.

For isothermal mixing (Figure P13.9b plus data from Figure 13.12), the stream summary should be

Stream	Type	$F\hat{C}_p\left(\frac{\text{kW}}{°\text{C}}\right)$	T_{in} (°C)	T_{out} (°C)
C1	Cold	2	70	210
C3	Cold	1	100	210
H2	Hot	3	180	120
H4	Hot	1.5	236	30

Chapter 14

Process and Site Utility Integration

In Chapter 13, we explored energy integration for a processing plant, or distinct processing area, with clearly defined hot and cold streams. In Chapter 14, we want to extend our energy analysis to include the design of the utility system, which will supply the heating and cooling for the processing streams and will meet the process electricity and power requirements. This overall concept is depicted in Figure 14.1.

In order to meet the electrical, heating, and cooling demands of a process, two distinctly different cogeneration designs—first, a gas turbine-based cogeneration system and, second, a steam turbine-based cogeneration system—will be explored. Our cogeneration design work of Chapter 10 will be directly applicable to the gas turbine system. Our rigorous thermodynamics package will be used to design the steam turbine-based system.

We also address the situation where there are several processing plants/areas all located at the same corporate site. At chemical and petroleum processing sites, different products are often made in separate processing areas. For example, at a single chemical processing site, ethylene and oxygen may be reacted to ethylene oxide. The ethylene oxide may be combined with ammonia to make ethanolamines, with water to make ethylene glycols, with acetic acid to make glycol ether acetates, and with alcohols to make glycol ethers. Each of these products may be produced in a separate process/area/plant (e.g., the ethylene oxide plant or the ethylene glycol plant). Each of these plants may share a common utility system or the situation may be more complex where part of the utility system may be distributed. Our energy analysis can be extended from a single processing unit to a processing plant to a site-wide analysis—here, the possibility may exist for separate plants to exchange energy from existing hot and cold streams, thereby reducing the overall utility load on the central utility system. Energy integration, which reduces utilities, will reduce operating costs and emissions.

14.1 GAS TURBINE-BASED COGENERATION UTILITY SYSTEM FOR A PROCESSING PLANT

Here we detail a gas turbine-based cogeneration system design for a process with known fixed electrical demand and known hot and cold process streams. From Chapter 13 the temperature interval analysis can be used to identify utility heating and cooling requirements for the process. A modified version of the grand composite curve (GCC) can be used to guide cogeneration system configuration and design.

EXAMPLE 14.1 ***The Modified GCC for Utility System Design***

Let us consider the design of a cogeneration-based utility system to produce 4500 kW of electric power and to meet the heating requirements of a six-stream plant with data provided in Table 14.1. For process-to-process heat exchange, $\Delta T_{\min} = 36°F$, and for cogeneration exhaust-to-process heat exchange, $\Delta T_{\min} = 54°F$. Develop the GCC and modified GCC for utility system design (adapted from Shenoy, 1995).

SOLUTION In order to determine the heating and cooling requirements for the process streams and to draw the GCC, we first construct the problem table or temperature interval analysis as shown in Figure 14.2.

Here, the interval analysis shows a construction with 0^0 temperature change (311 → 311°F and 419 → 419°F, adjusted temperatures) for the two streams undergoing liquid to vapor phase change. The problem table shows the process requires only heating (no cooling) with a total heating requirement of 5288.83 Btu/s. A process requiring only a hot or only a cold utility is termed a threshold problem—see Problem 13.6.

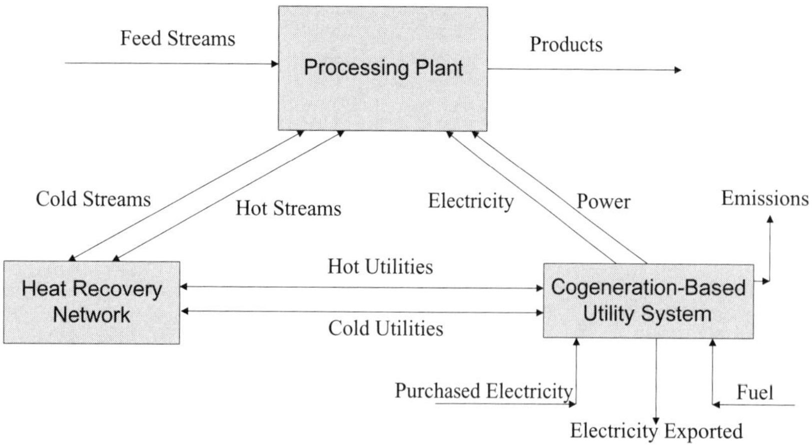

Figure 14.1 Overview of a processing plant, process heat recovery, and utilities.

Table 14.1 Stream Data Summary

Stream	$F\hat{C}_p\left(\frac{\text{Btu}}{\text{s-°F}}\right)$	$\Delta\dot{H}_{L\to V}\left(\frac{\text{Btu}}{\text{s}}\right)$	T_{in}(°F)	T_{out}(°F)
H1	5.26566		347	113.0
H2	7.89852		257	149.0
C3	10.5313		68	311.0
C4	7.8985		104	233.6
C5		1137.384	401	401.0
C6		2653.896	293	293.

We next construct the GCC for the process that is shown in the Figure 14.3. On the GCC we indicate points that will be important in our future discussions; note $a(4151.45\ \text{Btu/s}, 419)$ and $b(1402.77\ \text{Btu/s}, 311)$, where 419 and 311°F are adjusted temperatures.

We will find it useful in utility design problems to prepare a modified GCC by the following steps:

Step 1: Modify the GCC by removing any pockets in the process GCC. In a GCC pocket, hot process streams can exchange heat with cold process streams to satisfy heating and cooling requirements.

Step 2: Redraw the GCC with the interval temperatures shifted back to actual temperatures. For the utility design above the pinch, or a threshold problem requiring only hot utilities, shift interval temperatures by $-\frac{1}{2}\Delta T_{\min}$; for Figure 14.2, $-\frac{1}{2}\Delta T_{\min} = -18$°F.

In step 2, we are taking advantage of the fact that the net stream above the pinch, after combining hot and cold streams, in each temperature interval is a cold stream (this is also true for the threshold problem of Figure 14.2). Remember that GCC pockets (also called "knees") are removed in step 1. The residual enthalpy in each temperature interval above the process pinch is a deficit enthalpy (also true for Figure 14.2). This allows us to convert the interval temperatures to actual temperatures—here also see Problem 14.1. In the literature, this conversion to actual temperatures (step 2) is often not done and problems are solved with adjusted temperatures. For us, the advantage of using actual temperatures will become apparent as we connect results from the GCC to the actual optimization of the utility system design.

In Figure 14.4, the modified process GCC is now plotted with actual temperatures; notice that $a(4151.45\ \text{Btu/s}, 401°\text{F})$ and $b(1402.77\ \text{Btu/s}, 293°\text{F})$.

In the figure we have also plotted a possible cogeneration exhaust profile (in light gray), which will be discussed in later examples.

In order to meet the electric power and heating needs of the process, we have many choices including gas turbine-based cogeneration systems, steam turbine-based cogeneration systems, or combined steam and gas turbine cogeneration systems. ■

EXAMPLE 14.2 ***Gas Turbine Cogeneration System Exhaust: Targeting the Modified Process GCC Profile***

Here we match the exhaust of a regenerative cogeneration system to the process heating needs of Example 14.1. Figure 14.5 shows

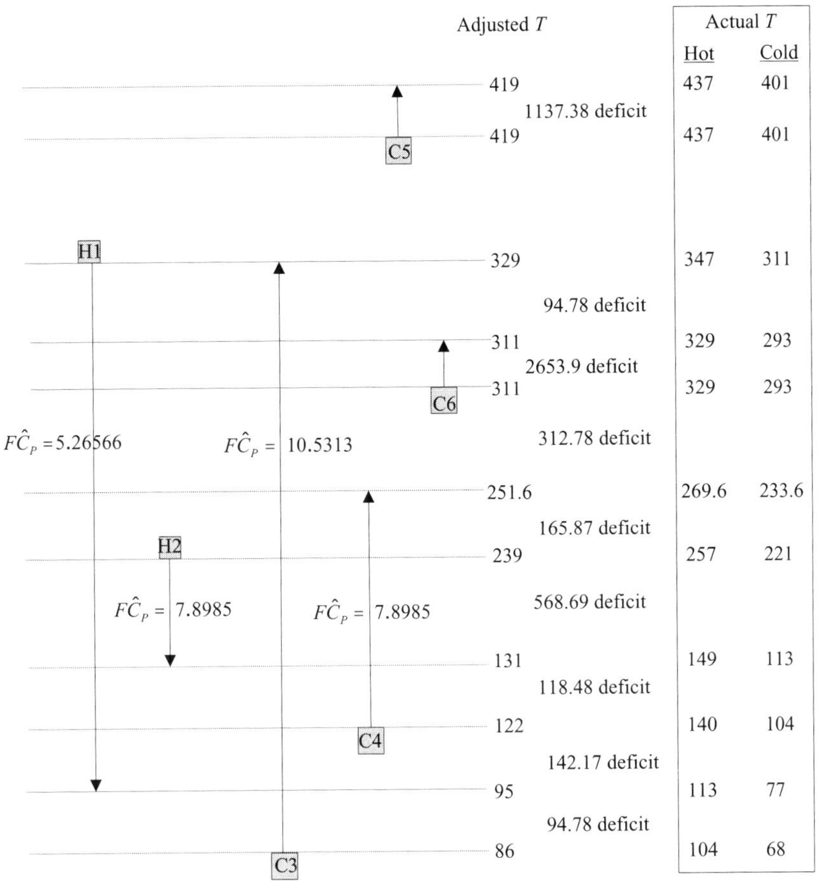

Figure 14.2 Problem table or temperature interval analysis (36° approach) for Example 14.1.

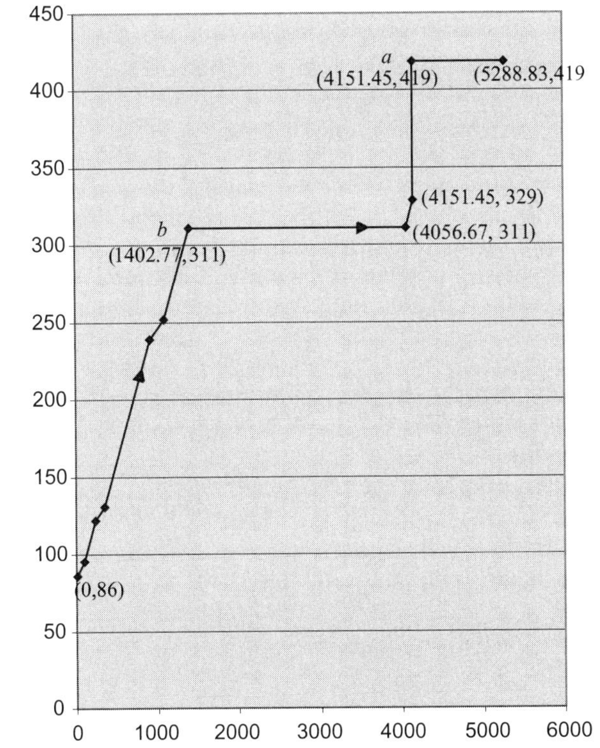

Figure 14.3 Grand composite curve with adjusted temperatures for Example 14.1 (energy rate and temperature).

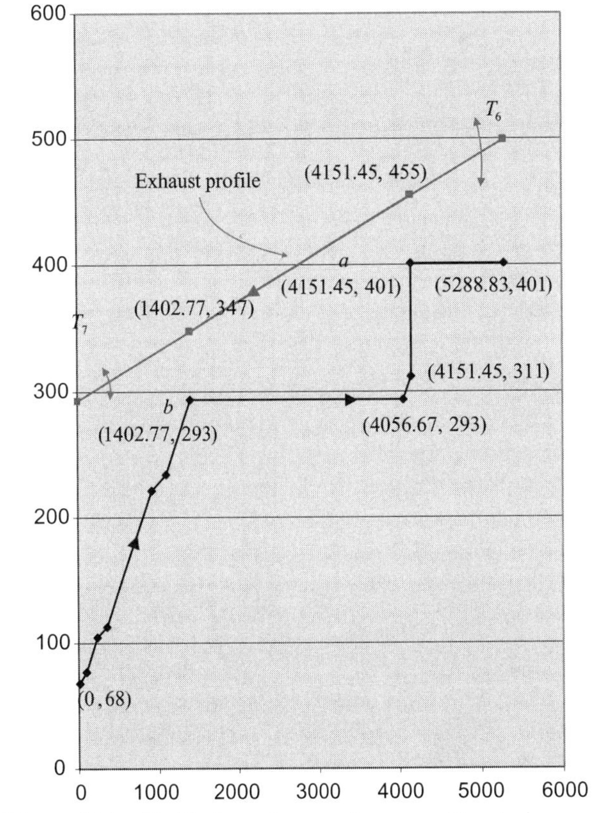

Figure 14.4 Modified grand composite curve with actual temperatures for Example 14.1 (energy rate and temperature).

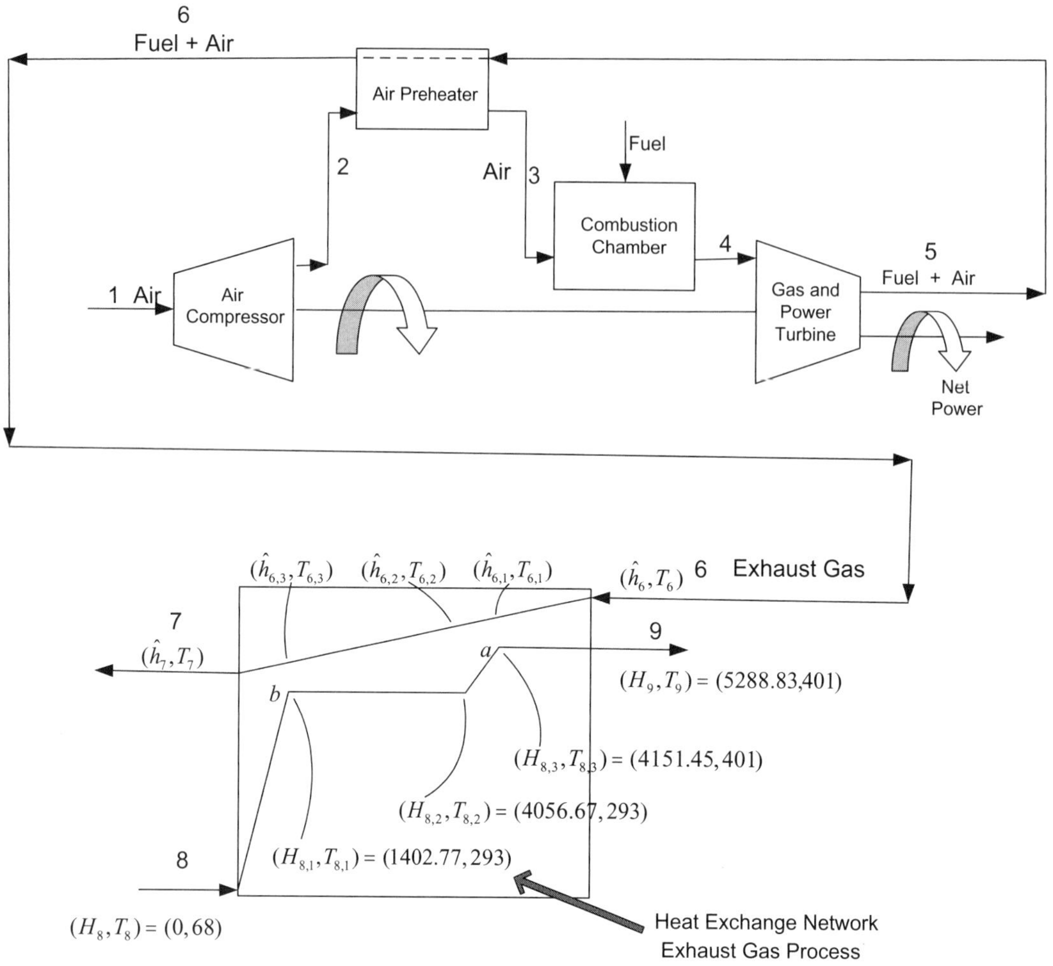

Figure 14.5 Gas turbine cogeneration system cycle with exhaust gas recovery.

the regenerative cogeneration system developed in Chapter 10 (the CGAM problem), where we have replaced the heat recovery steam generator (HRSG) with energy exchange between the gas turbine exhaust gas and the process.

SOLUTION For this cogeneration system design, we initially want to match the exhaust profile from the gas turbine as closely as possible to the modified GCC for the process. For cogeneration exhaust-to-process heat exchange, $\Delta T_{\min} = 54°\text{F}$. Therefore, the cogeneration exhaust profile, which would most closely match the process GCC, would pass through

$$\left(4151.45\frac{\text{Btu}}{\text{s}}, 401 + 54 = 455°\text{F}\right)$$

and

$$\left(1402.77\frac{\text{Btu}}{\text{s}}, 293 + 54 = 347°\text{F}\right).$$

This exhaust profile is shown in light gray in Figure 14.4, and the equation for this line is

$$T_6 = \left(\frac{1}{25.45}\right)(\dot{Q}) + 291.88°\text{F}. \tag{14.1}$$

Here T_6 or $T_{\text{Exhaust\_gas}}$ is the temperature of the exhaust gas from the cogeneration system and $\dot{Q}$ is the available British thermal unit per second in the exhaust stream. As the exhaust needs to supply 5288.83 Btu/s to the process, we find $T_6 = 499.69°\text{F}$. The use of the nomenclature T_6 follows our development of the CGAM problem from Chapter 10. The intercept, $T_7 = 291.88°\text{F}$, is above the methane fuel acid condensation temperature $\approx 250°\text{F}$. However, because the exhaust gas is matching a process stream entering at 68°F, condensation of the exhaust gas on the tube wall may occur and this possibility should be checked (see Chapter 10). For heavier fuels, the acid condensation temperature will increase and $T_7 = 291.88°\text{F}$ may require special materials in the heat exchangers. For example, the acid condensation temperature for low sulfur content fuel oil is ≈300°F.

We can estimate the exhaust gas flow rate needed for the process heating requirements. From the CGAM ideal gas solution we can estimate the heat capacity of the exhaust gas as

$$\hat{C}_{p,\text{Exhaust\_gas}} = 0.28\frac{\text{Btu}}{\text{lb-°F}},$$

and the estimated exhaust gas flow rate would be

$$F_{\text{Exhaust\_gas}} \equiv F_{\text{gas}} = \left(\frac{\dot{Q}}{\hat{C}_{p,\text{Exhaust\_gas}}(\Delta T)}\right)$$
$$= \frac{5288.83}{0.28(499.69 - 291.88)} = 90.894 \frac{\text{lb}}{\text{s}}. \quad (14.2)$$

We now have reasonable values for the exhaust from the cogeneration system—here, we expect $F_{\text{gas}} \approx 90.89$ lb/s, the hot end temperature of the exhaust matching the process $T_6 = 499.69$°F, and the exiting temperature of the exhaust after matching with the process $T_7 = 291.88$°F.

At this point, we can design a heat exchanger network (HEN), which will satisfy the process heating requirements using the exhaust from the cogeneration system. There will be some hot process-to-cold process stream heat exchange as indicated by the temperature interval analysis of Figure 14.2. ■

EXAMPLE 14.3 *Design the Heat Exchanger Network Using Estimated Cogeneration Exhaust Stream Properties*

This design is unlike our past heat exchanger network designs as here there is no pinch, and we have two $\Delta T_{\min}$: $\Delta T_{\min} = 36$°F for process stream-to-process stream heat exchange and $\Delta T_{\min} = 54$°F for cogeneration exhaust-to-process stream heat exchange. This heat exchanger network design is guided by Figures 14.5 and 14.4 and the problem table in Figure 14.2. The final heat exchanger network design is provided in Figure 14.6 and the development of the heat exchanger matches is discussed next.

Heat Exchanger (HEX) 1

If we examine Figures 14.4 and 14.5, we see that exhaust gas will be used to heat the process from $(\dot{H}_{8,3}, T_{8,3}) \rightarrow (\dot{H}_9, T_9)$:

$$\left(4151.45\frac{\text{Btu}}{\text{s}}, 401°\text{F}\right) \rightarrow \left(5288.83\frac{\text{Btu}}{\text{s}}, 401°\text{F}\right).$$

From Figure 14.2 we can identify this as cold stream C5, which needs 1137.38 Btu/s for the heat of vaporization at 401°F; do recall that hot and cold temperatures must be shifted from those in the problem table by $\pm\frac{1}{2}\Delta T_{\min}$ to obtain actual temperatures. C5 heating is accomplished using the exhaust gas from the cogeneration system with the exhaust stream outlet temperature from HEX 1:

$$T_{\text{Exhaust\_gas,Out HEX\_1}} = T_{6,1}$$
$$= 499.69°\text{F} - \left(\frac{1137.38\frac{\text{Btu}}{\text{s}}}{\left(0.28\frac{\text{Btu}}{\text{lb-°F}}\right)\left(90.89\frac{\text{lb}}{\text{s}}\right)}\right)$$
$$= 455°\text{F}.$$

HEXs 2 and 3

Examining Figures 14.4 and 14.5, we next see that the exhaust gas will be used to heat the process from $(\dot{H}_{8,2}, T_{8,2}) \rightarrow (\dot{H}_{8,3}, T_{8,3})$:

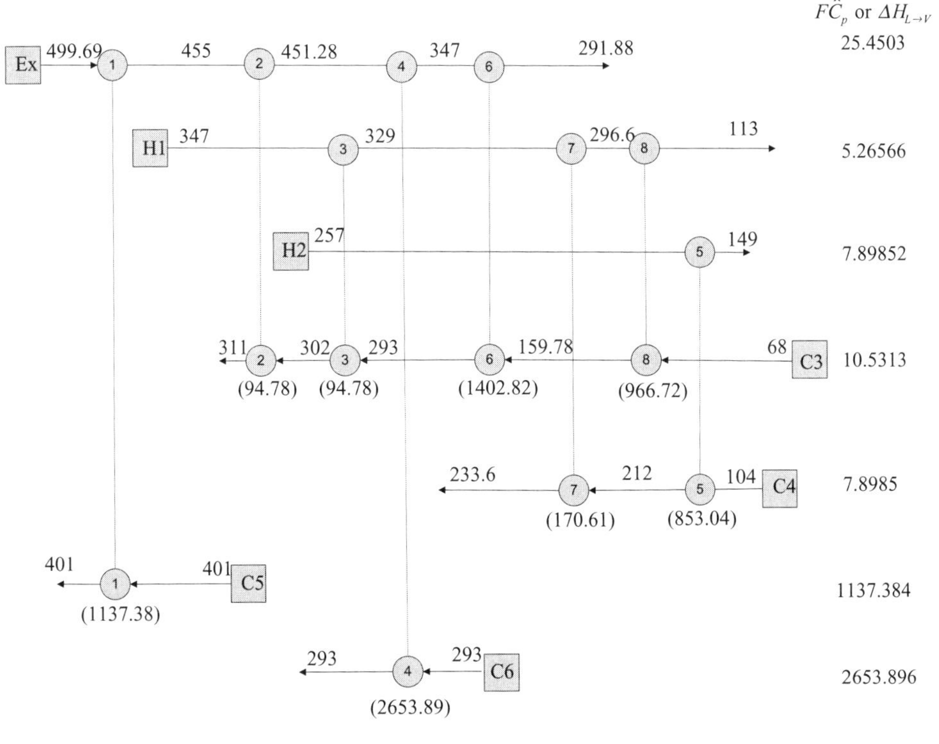

Figure 14.6 Heat exchanger matches for Example 14.1 with estimated cogeneration exhaust stream properties.

$$\left(4056.67\frac{\text{Btu}}{\text{s}}, 293°\text{F}\right) \rightarrow \left(4151.45\frac{\text{Btu}}{\text{s}}, 401°\text{F}\right).$$

A close examination of Figure 14.2 shows there are actually no process streams needing heating from 311 → 401°F; so, this energy exchange could be viewed as (4056.67, 293) → (4151.45, 311). Also, from Figure 14.2, there will also be heat exchange between H1 and C3. The exhaust gas stream will provide 4151.45 − 4056.67 = 94.78 Btu/s, and the heat exchange between H1 and C3 is ((10.5313 − 5.26566) × (329 − 311)) = 94.78 Btu/s:

$$
\begin{aligned}
T_{\text{Exhaust\_gas,Out HEX\_2}} &= T_{6,2} \\
&= 455°\text{F} - \left(\frac{94.78\frac{\text{Btu}}{\text{s}}}{\left(0.28\frac{\text{Btu}}{\text{lb-°F}}\right)\left(90.89\frac{\text{lb}}{\text{s}}\right)}\right) \\
&= 451.28°\text{F},
\end{aligned}
$$

$$T_{\text{C3,In HEX\_2}} = 311°\text{F} - \left(\frac{94.78\frac{\text{Btu}}{\text{s}}}{10.5313}\right) = 302°\text{F},$$

$$T_{\text{H1,Out HEX\_3}} = 347°\text{F} - \left(\frac{94.78\frac{\text{Btu}}{\text{s}}}{5.26566}\right) = 329°\text{F},$$

$$T_{\text{C3,In HEX\_2}} = 302°\text{F} - \left(\frac{94.78\frac{\text{Btu}}{\text{s}}}{10.5313}\right) = 293°\text{F}.$$

HEX 4

Examining Figures 14.4 and 14.5,we see that exhaust gas will next be used to heat the process from $(\dot{H}_{8,1}, T_{8,1}) \rightarrow (\dot{H}_{8,2}, T_{8,2})$:

$$\left(1402.77\frac{\text{Btu}}{\text{s}}, 293°\text{F}\right) \rightarrow \left(4056.67\frac{\text{Btu}}{\text{s}}, 293°\text{F}\right).$$

This is the heat of vaporization for C6 requiring 2653.89 Btu/s:

$$
\begin{aligned}
T_{\text{Exhaust\_gas,Out HEX\_4}} &= T_{6,1} \\
&= 451.28°\text{F} - \left(\frac{2653.89\frac{\text{Btu}}{\text{s}}}{\left(0.28\frac{\text{Btu}}{\text{lb-°F}}\right)\left(90.89\frac{\text{lb}}{\text{s}}\right)}\right) \\
&= 347°\text{F}.
\end{aligned}
$$

HEXs 5–8

Here H2 is used heat C4; the remaining exhaust gas is matched against C3; heat exchangers 7 and 8 are placed to complete the network. Heat exchangers 1, 2, 4, and 6 will be heat exchanger banks (coils) within the heat recovery unit. ■

At this point, there are many options to allow the design of the complete cogeneration system to meet the heating and power requirements of Example 14.1 including

Ideal gas approximations can be combined with typical gas turbine operating conditions to design the cogeneration gas turbine section (see Shenoy, 1995; Townsend and Linnhoff, 1983a,b).

You will recall that in Chapter 10 we developed the gas turbine design equations using both ideal gas assumptions as well as real gas rigorous thermodynamics. In the next example, Example 14.4, we take advantage of our rigorous thermodynamics work of Chapter 10 to design the cogeneration system based on typical cogeneration operational conditions.

In Example 14.5, we optimize the cogeneration design to meet the requirement that all process–exhaust and exhaust–exhaust heat exchanges (the gas turbine air preheater) maintain ΔT_{min} = 54°F.

In Example 14.6, we optimize the cogeneration design to meet the power requirement of 4.5 MW and the requirement that all process–exhaust and exhaust–exhaust heat exchange maintain ΔT_{min} = 54°F.

In Example 14.7, we optimize the cogeneration system design utilizing the cost equations developed in Chapter 10.

For comparison, the results of all these different options are summarized in Table 14.5

Finally, in Example 14.8, we design a cogeneration system using steam and steam turbines to meet the heating and power requirements of Example 14.1.

EXAMPLE 14.4 ***Gas Turbine Cogeneration System Design: Real Gas with Fixed Operating Parameters***

Using real gas properties and typical gas turbine operating conditions will yield the design of the gas turbine cogeneration system, which will supply the heating needs for the process streams of Example 14.1. Here our goal is not to optimize the gas turbine system but rather to design and evaluate system performance based on typical gas turbine system operating conditions. The motivation here is that we can compare these rigorous results with more approximate methods (Shenoy, 1995; Townsend and Linnhoff, 1983a,b) if desired.

Table 14.2 Typical Gas Turbine Operating Conditions

	Industrial Gas Turbine	Aeroderivative Gas Turbine
Pressure ratio, P_2/P_1	$4 \le P_2/P_1 \le 15$	$4 \le P_2/P_1 \le 30$
Overall efficiency	80–85%	82–89%
Combustion temperature (°F), T_4	$T_4 \le 2700°F$ (3160 R)	$T_4 \le 3000°F$ (3460 R)

Table 14.3 Fixed Design Variables for Example 14.4

F_{air}	90.89 lb/s
P_2	4.65 bar
η_{Air_Comp}	0.83
$\eta_{Power\_Turbine}$	0.86
T_4	2471 R

T_1 = 536.67 R; P_1 = 1.013 bar; percent humidity = 0%; lower heating value (*LHV*) = 21,500 Btu/lb; higher heating value (*HHV*) = 23,861 Btu/lb; ΔT_{min} = 54°F; ΔP_{APH}^{gas} = 3%; ΔP_{APH}^{air} = 5%; ΔP_{CC} = 5%; η_{CC} = 98%.

SOLUTION In Chapter 10, the design equations to allow sizing, performance determination, and costing of a regenerative cogeneration system were developed using real gas properties. The cogeneration system in Chapter 10 used an HRSG to generate steam, and this steam was available for use by the process. In Example 14.4, we want the exhaust gas from the gas turbine (after the regenerator) to be used directly to heat process streams.

For Example 14.4, we are not trying to optimize the design, but rather, we want to pick typical operating conditions and to evaluate system performance. We first set the air flow rate to 90.89 lb/s which is the estimated exhaust flow rate found in Example 14.3; the fuel flow rate will be small compared to the air flow rate. We next need to fix typical operating conditions to allow design of the gas turbine system. Table 14.2 indicates typical operating conditions for industrial and aeroderivative gas turbines.

With the use of our Excel sheet solutions of Chapter 10 (Section 10.3), any values from the ranges given in Table 14.2 can be selected and the cogeneration system performance determined. In the Excel solution in **Example 14.4.xls**, which is shown in Figure 14.7, we used the values in Table 14.3 as "typical" to allow the Excel sheet to be solved—again no optimization procedure is being used and no choice of specific gas turbine type is being made. In the Excel sheet, we do have to vary the fuel flow rate to a value of 0.83 lb/s to obtain T_4 near 2471 R.

We also fix the pressure drop in the combustion chamber, ΔP_{CC} = 5%; the pressure drop on the air side of the air preheater, ΔP_{APH}^{air} = 5%; and the pressure drop on the gas side of the air preheater, ΔP_{APH}^{gas} = 3%.

Using the values in Table 14.3, the resulting real gas solution shows a minimum approach temperature in the air preheater, ΔT_{APH} = 43.1 R, T_6 = 964.94 R (505.27°F), P_6 = 1.066 bar, and the exhaust gas flow rate, F_{gas} = 91.72 lb/s.

In the Excel sheet, we must replace the HRSG of Chapter 10 with a heat recovery unit matching the process and the gas turbine exhaust. This can be accomplished by using linear segments for the process T versus $\dot{Q}$.

Gas Turbine Exhaust/Process Heat Exchange: Use Linearized Process Segments

As indicated in Figure 14.5, the process streams needing heating can be represented by linear segments.

$(\dot{H}_8, T_8) \rightarrow (\dot{H}_{8,1}, T_{8,1})$ represents process heating from

$$\left(0\frac{\text{Btu}}{\text{s}}, 68°\text{F}\right) \rightarrow \left(1402.77\frac{\text{Btu}}{\text{s}}, 293°\text{F}\right);$$

$(\dot{H}_{8,1}, T_{8,1}) \rightarrow (\dot{H}_{8,2}, T_{8,2})$ represents process heating from

$$\left(1402.77\frac{\text{Btu}}{\text{s}}, 293°\text{F}\right) \rightarrow \left(4056.67\frac{\text{Btu}}{\text{s}}, 293°\text{F}\right);$$

$(\dot{H}_{8,2}, T_{8,2}) \rightarrow (\dot{H}_{8,3}, T_{8,3})$ represents process heating from

$$\left(4056.67\frac{\text{Btu}}{\text{s}}, 293°\text{F}\right) \rightarrow \left(4151.45\frac{\text{Btu}}{\text{s}}, 401°\text{F}\right);$$

and $(\dot{H}_{8,3}, T_{8,3}) \rightarrow (\dot{H}_9, T_9)$ represents process heating from

$$\left(4151.45\frac{\text{Btu}}{\text{s}}, 401°\text{F}\right) \rightarrow \left(5288.83\frac{\text{Btu}}{\text{s}}, 401°\text{F}\right).$$

An energy balance between the gas turbine exhaust gas and each linear process stream segment can be performed, and beginning with the hot turbine exhaust,

$$F_{gas}\left(\hat{h}_6 - \hat{h}_{6,1}\right) = \left(\dot{H}_9 - \dot{H}_{8,3}\right), \qquad (14.3)$$

where $\hat{h}_6$ (British thermal unit per pound) is the exhaust enthalpy evaluated at (P_6, T_6) using H_Products (P_6, T_6, H/C ratio, dry air ratio [*DAR*]). The argument list contains all known quantities; recall H_Products (Table 8.1) is in British thermal unit per pound of combustion products, P is in pound per square inch absolute, and T must be in R. Equation (14.3) can be solved for $\hat{h}_{6,1}$,

$$\hat{h}_{6,1} = \hat{h}_6 - \frac{\left(\dot{H}_9 - \dot{H}_{8,3}\right)}{F_{gas}}, \qquad (14.4)$$

and $T_{6,1}$ can be found from TfromH_Products ($P_{6,1}$, $h_{6,1}$, H/C ratio, *DAR*).

Successive energy balances give

$$\hat{h}_{6,2} = \hat{h}_{6,1} - \frac{\left(\dot{H}_{8,3} - \dot{H}_{8,2}\right)}{F_{gas}}, \qquad (14.5)$$

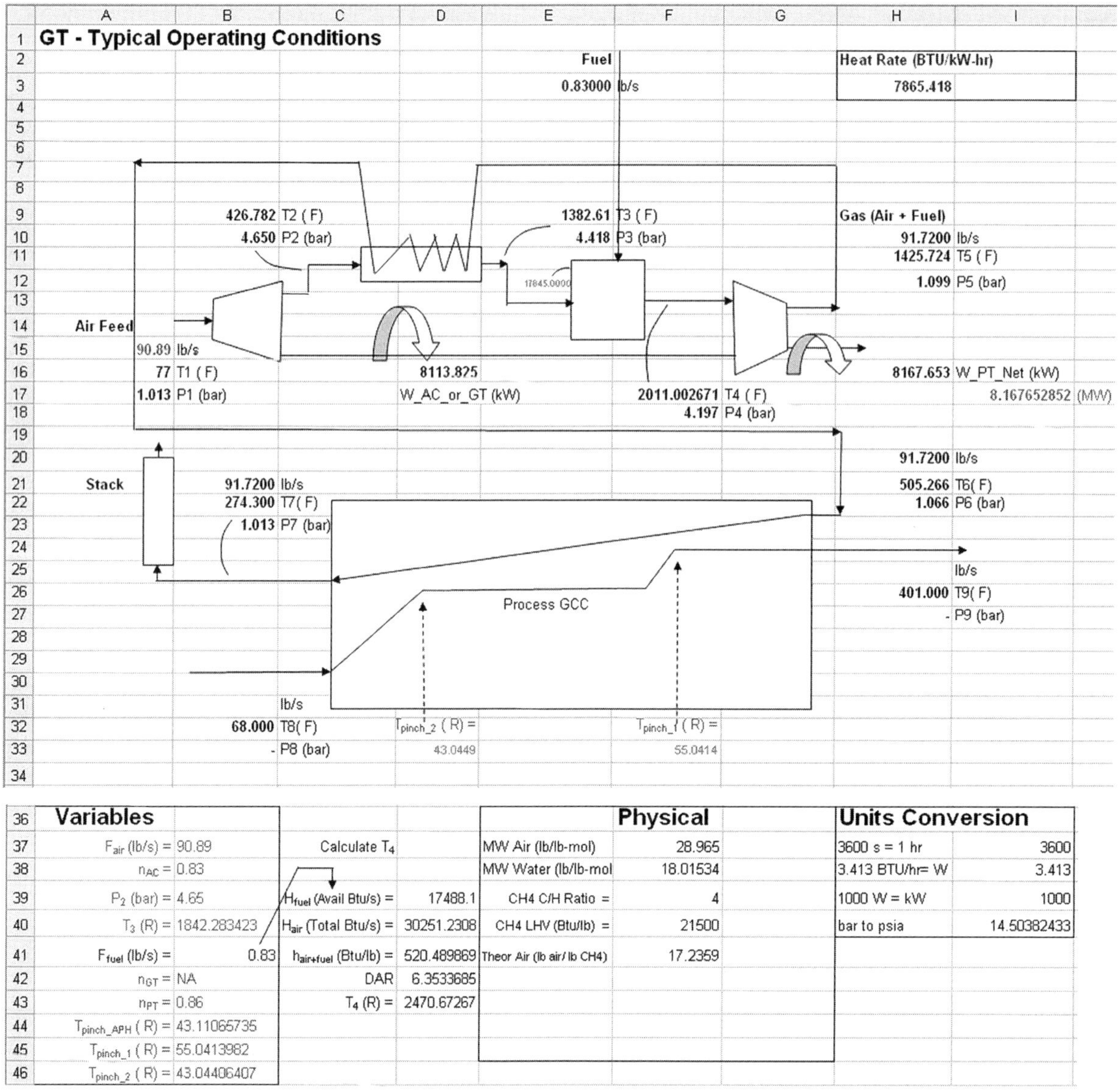

Figure 14.7 Example 14.4 solution: cogeneration design with typical operating conditions—real gas properties.

and $T_{6,2}$ can be found from TfromH_Products ($P_{6,2}$, $h_{6,2}$, H/C ratio, *DAR*):

$$\hat{h}_{6,3} = \hat{h}_{6,2} - \frac{\left(\dot{H}_{8,2} - \dot{H}_{8,1}\right)}{F_{\text{gas}}}; \tag{14.6}$$

$T_{6,3}$ can be found from TfromH_Products ($P_{6,3}$, $h_{6,3}$, H/C ratio, *DAR*):

$$\hat{h}_7 = \hat{h}_{6,3} - \frac{\left(\dot{H}_{8,1} - \dot{H}_8\right)}{F_{\text{gas}}}; \tag{14.7}$$

and T_7 can be found from TfromH_Products (P_7, h_7, H/C ratio, *DAR*).

Equations (14.3)–(14.7) replace the HRSG equations in the CGAM formulation we developed in Chapter 10.

For the heat recovery unit, the real gas solution shows $\Delta T_{6,1-8,3} = 55$ R and $\Delta T_{6,3-8,1} = 43$ R. These values for pinch 1 ($\Delta T_{6,1-8,3}$) and pinch 2 ($\Delta T_{6,3-8,1}$) are reasonably close to 54°F value used in Example 14.2 to obtain an exhaust flow rate estimate of 90.89 lb/s.

The net power produced is 8.17 MW and this exceeds the 4.5 MW required in the problem statement. The value of $T_6 = 964.94$ R is close to the value found in Example 14.2 where $T_6 = 959.36$ R. We do caution that different values in Table 14.3 will produce much different results—for example, use the Excel sheet to examine the effect of changing P_2 to10 bar; the monitored approach temperatures become negative, which is physically impossible. The strength of Example 14.3 is that any combination of operating conditions can be quickly evaluated using real gas properties. ■

In the next three examples, we use the optimization features in Excel to ensure that design conditions (approach temperatures and electricity generated) are met.

EXAMPLE 14.5 ***GT Cogeneration System Design: Targeting with ΔT_{min} = 54°F***

Here let us explore the effect of requiring that the process–exhaust and air preheater matches have a minimum approach temperature of 54°F (ΔT_{min} = 54°F). We also want to keep the combustion chamber temperature, T_4, at or near 2471°F.

SOLUTION We can use optimization procedures to bring our design more in line with any set of fixed design conditions. For Example 14.5, the objective function to be minimized can be formulated as

$$\text{Obj Fcn: } (T_{\text{Pinch\_APH}} - 54)^2 + (T_{\text{Pinch\_1}} - 54)^2 + (T_{\text{Pinch\_2}} - 54)^2 + (T_4 - 2471)^2,$$

where

$$T_{\text{Pinch\_APH}} = T_5 - T_3,$$

$$T_{\text{Pinch\_1}} = T_{6,1} - T_{8,3}, \text{ and}$$

$$T_{\text{Pinch\_2}} = T_{6,3} - T_{8,1}.$$

Here we vary the air flow rate, F_{air}; the pressure out of the compressor, P_2; the air temperature leaving the air preheater, T_3; and the flow rate of fuel, F_{fuel}. The efficiency air compressor, $\eta_{\text{Air\_Comp}}$, and the efficiency of the power turbine, $\eta_{\text{Power\_Turbine}}$, are fixed at the values used in Example 14.4. Table 14.4 summarizes variables and fixed values.

The Excel solution is provided in **Example 14.5.xls**. In the formulation, constraints may be added to ensure that approach temperatures are ≥54°F and that T_7 remains realistic (see Table 10.4).

The results from the Excel sheet are summarized in Table 14.5 and show ΔT_{APH} = 54 R, T_6 = 964.94 R, $\Delta T_{6,1\text{–}8,3}$ = 54 R, $\Delta T_{6,3\text{–}8,1}$ = 54 R, and T_4 = 2471 R, which are all desired values. The net power produced is 8.78 MW, which again exceeds the 4.5 MW required in the problem statement. Compared to the solution of Example 14.4, the exhaust flow rate is increased from 91.72 to 101.91 lb/s and the pressure leaving the compressor, P_2, is lowered from 4.65 to 4.35 bar, and the value of T_6 = 959 R remains fairly constant. ■

Table 14.4 Variables and Fixed Operating Conditions for Example 14.5

Variables	Fixed Values
$F_{\text{air}}, P_2, T_3, F_{\text{fuel}}$	$\eta_{\text{Air\_Comp}} = 0.83$, $\eta_{\text{Power\_Turbine}} = 0.86$

T_1 = 536.67 R; P_1 = 1.013 bar; percent humidity = 0%;
LHV = 21,500 Btu/lb; HHV = 23,861 Btu/lb; ΔT_{min} = 54°F; $\Delta P_{\text{APH}}^{\text{gas}}$ = 3%;
$\Delta P_{\text{APH}}^{\text{air}}$ = 5%; ΔP_{CC} = 5%; η_{CC} = 98%.

In both Examples 14.4 and 14.5, we provided the required heating for the process, but we exceeded the amount of power needed. We want to examine the design next where both the process heating requirements and the power requirements are targeted. It is instructive to reexamine Figure 14.4. To change the power generated, while still providing the needed $\dot{Q}$ for process heating, the exhaust line in Figure 14.4 will need to be moved as indicated by the light gray arrow on the exhaust line. Consider that the exhaust line is moved counterclockwise. This will increase temperature T_6 while still maintaining the needed $\dot{Q}$ from the exhaust stream. We also want to maintain a 54°F approach temperature—this will now occur at just one point, b, on the figure. So, moving the exhaust line counterclockwise will increase the total ΔT (T_6–T_7); T_6 will be increased and T_7 will be lowered. The larger ΔT on the exhaust stream will result in a lower F_{gas} needed to deliver the same $\dot{Q}$ to the process streams. The lower exhaust gas flow rate will result in less power being generated from the gas turbine section. But a concern here may be that the temperature T_7 becomes too low.

If the exhaust line is moved in the clockwise direction, the amount of power generated will be increased. Moving the exhaust line clockwise will decrease temperature T_6 while still maintaining the needed $\dot{Q}$ from the exhaust stream. We also want to maintain a 54°F approach temperature—this will now occur at just one point, a, on the figure. Moving the exhaust line clockwise will decrease the total ΔT (T_6–T_7); T_6 will be decreased and T_7 will be increased. The smaller ΔT on the exhaust stream will result in a higher F_{gas} needed to deliver the same $\dot{Q}$ to the process streams. The higher exhaust flow rate will result in increased power being generated from the gas turbine section. The relation between the exhaust line and the electric power generation may be clearer after the following example.

EXAMPLE 14.6 ***Gas Turbine Cogeneration System Design: Targeting Heat and Electricity Requirements***

Here we want to meet the heating and power requirements of Example 14.1 without producing excess electricity or violating the minimum approach temperature. We want the process–exhaust and air preheater matches to have a minimum approach temperature of 54°F (ΔT_{min} = 54°F) and we want to keep the combustion chamber temperature T_4 at or near 2471°F.

SOLUTION Based on our discussion of the exhaust line and Figure 14.4, we require that the process–exhaust match at point b (pinch 2) and the air preheater matches have a minimum approach temperature of 54°F (ΔT_{min} = 54°F). We also want the combustion chamber temperature T_4 to be 2471°F.

Table 14.5 Summary of Cogeneration System Designs

Variable/Cost	Example 14.2	Example 14.4	Example 14.5	Example 14.6	Example 14.7
F_{air} (lb/s)		**90.89**	101.01	41.53	36.81
$\eta_{Air\ Compressor}$		**0.83**	**0.83**	**0.83**	0.841
T_2 (R)		886.45	867.76	1,090.06	994.18
P_2 (bar)		**4.65**	4.35	9.10	6.857
T_3 (R)		1,842.28	1,857.20	1,599.83	1,662.67
F_{fuel} (lb/s)		0.83	0.90	0.52	0.53
$F_{Exhaust\_gas}$ (lb/s)	90.89	91.72	101.91	42.05	37.34
T_4 (R)		**2,470.67**	**2,471.00**	**2,471.19**	2,644.33
$\eta_{G\&P\ Turbine}$		**0.86**	**0.86**	**0.86**	0.86
T_5 (R)		1,885.39	1,911.20	1,653.84	1,881.01
ΔT_{APH} (R)		43.1	**54**	**54**	218.34
T_6 (R)	959.36	964.94	959.01	1,168.60	1,261.73
ΔT_{Pinch_1} (R)	54	55.0	**54**	203.60	285.38
ΔT_{Pinch_2} (R)	54	43.0	**54**	**54**	106.69
T_7 (R)	751.55	733.97	751.13	672.78	709.67
$\dot{W}_{Air\ Compressor}$ (kW)		8,113.83	8,528.64	5,922.56	4,318.77
$\dot{W}_{G\&P\ Turbine}$ (kW)		16,281.48	17,309.73	10,422.60	8,818.77
$\dot{W}^{Net}_{G\&P\ Turbine}$ (kW)		8,167.65	8,781.09	**4,500.05**	4,500.00
Cost air compressor ($)		163,041	161,838	209,851	145,507
Cost comb chamber ($)		27,834	30,948	12,727	19,463
Cost G&P turbine ($)		437,103	461,175	300,753	275,634
Cost air preheater,($)		1,327,701	1,292,578	534,358	266,541
Equipment cost rate ($/s)		0.0131	0.01304	0.00708	0.00474
Fuel cost rate ($/s)		0.0753	0.08180	0.04726	0.04808
Total cost rate ($/s)		0.0884	0.09484	0.05434	0.05282

T_1 = 536.67 R; P_1 = 1.013 bar; percent humidity = 0%; LHV = 21,500 Btu/lb; HHV = 23,861 Btu/lb; ΔT_{min} = 54°F; ΔP^{gas}_{APH} = 3%; ΔP^{air}_{APH} = 5%; ΔP_{CC} = 5%; η_{CC} = 98%; CRF = 18.2%; N = 8000 h; ϕ = 1.06; cf = \$0.004 per MJ.

The objective function to be minimized can be formulated as

$$\text{Obj Fcn:} (T_{\text{Pinch\_APH}} - 54)^2 + (T_{\text{Pinch\_2}} - 54)^2 + (T_4 - 2471)^2 + (\text{Net\_Power(kW)} - 4500)^2,$$

where

$$T_{\text{Pinch\_APH}} = T_5 - T_3 \text{ and}$$

$$T_{\text{Pinch\_2}} = T_{6,3} - T_{8,1}.$$

Here we again vary the air flow rate, F_{air}; the pressure out of the compressor, P_2; the air temperature leaving the air preheater, T_3; and the flow rate of fuel, F_{fuel}. The efficiency air compressor, η_{Air_Comp}, and the efficiency of the power turbine, $\eta_{Power\_Turbine}$, are fixed at the values used in Example 14.4 (see Table 14.4).

The Excel sheet is provided in **Example 14.6.xls**. In the formulation, constraints may be added to ensure that all approach temperatures are ≥54°F. For the initial optimization allow T_7 to take on any value—this is for discussion purposes.

The results from the Excel sheet are summarized in Table 14.5 and show ΔT_{APH} = 54 R, T_6 = 1168.60 R, $\Delta T_{6,1-8,3}$ = 203.60 R, $\Delta T_{6,3-8,1}$ = 54 R, and T_4 = 2471 R. The net power produced is 4.5 MW, which is the required power in the problem statement. Compared to the solution of Example 14.4, the exhaust flow rate is decreased from 91.72 to 42.05 lb/s and the pressure leaving the compressor, P_2, increases from 4.65 to 9.10, and the value of T_6 increases from 959 to 1169 R. If we examine this solution, compared to Figure 14.4, we see that the increase in T_6 to 1169 R eliminates the possible pinch at point a, which is now $\Delta T_{6,1-8,3}$ = 203.60 R. The exhaust line has moved in a counterclockwise direction with T_6 increased to 1169 R and T_7 decreased to 672.78 R (213°F). The exhaust flow rate has been reduced, which lowers the net power that can be produced from the gas turbine system. However, the exhaust gas temperature from the heat recovery unit is too low at (213°F). In Problem 14.3, Example 14.6 is resolved with the constraint added requiring $T_7 \geq$ 709.67 R; here T_6 increases from 1168.6 to 1204.8 R. ■

EXAMPLE 14.7 *Gas Turbine Cogeneration System Design: Minimize Total Costs*

In Examples 14.3–14.6, we did not actually consider the cost of the cogeneration system (capital cost + fuel costs). Rather, the cogeneration system design was varied by forming an objective function to meet specific design requirements. We have developed needed costing equations in Chapter 10 and it is straightforward to utilize these cost equations in the optimization process. Here we include costs for fuel and maintenance, as well as capital costs for air compression, the combustion chamber, the power turbine, and the air preheater. Costing of the heat recovery unit is not considered. Constraints, from Chapter 10, which may be used in the optimization process, are provided in Table 10.4

The Excel optimization is provided in **Example 14.7.xls**. In the formulation, constraints are used to ensure that all approach temperatures are ≥54°F; that T_7, $\eta_{\text{Air Compressor}}$, and $\eta_{\text{G\&P Turbine}}$ remain realistic; and that 4.5 MW of net power is produced. The optimization solution is summarized in Table 14.5. ■

The intent of the examples explored thus far in this chapter is to show how any cogeneration system, using real physical properties, can be constructed and evaluated to meet targeted process heating requirements. We next examine the design of a cogeneration system using steam turbines to meet the heating and power requirements of Example 14.1.

14.2 STEAM TURBINE-BASED UTILITY SYSTEM FOR A PROCESSING PLANT

In Chapter 9, we developed the equations to determine the power generated by a gas turbine with known isentropic efficiency and using real gas properties. These same working equations are applicable to a steam turbine with the real gas properties replaced by steam properties. We will need to check the steam quality after expansion. If the quality q is ≥1, then the turbine exhaust is all vapor. If the quality q is <1, we will need to modify exiting steam properties based on the vapor and liquid fractions present in the exiting stream.

EXAMPLE 14.8 *Steam Turbine Cogeneration System Design: Targeting GCC and Power Requirements*

Design a cogeneration system using steam and steam turbines to meet the heating and power requirements of Example 14.1. For steam to process heat exchange, ΔT_{min} = 36°F. Steam is available from a boiler at 1414.14 psia and 887°F and steam turbine isentropic efficiency is 85%.

SOLUTION A general design for a steam cogeneration system is presented in Figure 14.8. The boiler, makeup water, and condensate return system are not shown in Figure 14.8. Figure 14.8 shows high-pressure (HP) steam from the boiler is passed through an HP turbine and electricity (power 1) is generated. The exhaust from the HP turbine is medium-pressure (MP) steam. Part of the MP steam will be sent to the process to satisfy heating requirements and part will be sent to an MP turbine to generate additional electricity (power 2). Low-pressure (LP) steam will exhaust the MP turbine. Part of this LP steam will be sent to the process to satisfy heating requirements and part will be sent to an LP turbine to generate additional electricity (power 3). The steam output of the LP turbine is generally condensed against cooling water.

The heating requirements of the process can be used to guide the pressure levels of the turbines. As indicated in Figure 14.8, the process can use MP steam at 401°F + ΔT_{min} = 437°F to satisfy part of the heating requirements. This heating load would be 5288.83 – 4056.67 = 1232.16 Btu/s. Using our C-based thermodynamic functions (*TPSI+*), saturated steam at 437°F would have pressure of Psat_Water (R) = Psat_Water (437 + 459.67 R) = 369.51 psia. We can initially specify the MP steam level from the HP turbine at P_2 = 369.51 psia and T_2 = 437°F.

Similarly, the process can use LP steam at 293°F + ΔT_{min} = 329°F to satisfy part of the heating requirements. This heating load would be 4056.67 – 0 = 4056.67 Btu/s. Again, using our C-based thermodynamic functions, saturated steam at 329°F would have pressure of Psat_Water (R) = Psat_Water (329 + 459.67 R) = 101.59 psia. We can initially specify the exit from the LP stream level from the MP turbine at P_3 = 101.59 psia and T_3 = 329°F.

LP steam, not sent to the process, can be exhausted through an LP turbine to generate additional electricity (power 3). Here typical exhaust conditions from the LP turbine are P_4 = 1.45 psia and T_4 = 114.45°F.

With these set conditions (fixed steam inlet conditions to the turbine network and pressure levels for the exhaust of the turbines) and known heating and power requirements, we have a deterministic system. The Excel sheet to directly solve for needed steam flow rates is given in **Example 14.8.xls** and is shown in Figure 14.9.

There are really two separate problems being addressed on the Excel sheet. The first is determining the enthalpy, entropy, and steam quality in and out of each turbine. This information allows the second problem of determining steam flows to meet process heating and power requirements to be solved.

Our thermodynamics database (*TPSI+*) provides the needed functions to determine steam properties, and we are familiar (Chapters 7–9) with the use of Excel sheet functions to call these C functions. Here we detail the process for turbine 1; the calculation procedures for turbines 2 and 3 are identical. Values from the calculations and thermodynamic functions are shown for turbine 1; recall from Tables 8.1 and 8.2 that our *TPSI+* thermodynamic Excel sheet functions all contain an underscore.

Step 1: Determine the turbine inlet enthalpy and entropy conditions:

P_1 = 1414.14 psia

T_1 = 887°F = 1346.67 R

$\hat{h}_1$ = H_Steam (P_1, T_1)/MW_H_2O = H_Steam(1414.14, 1346.67)/18.01534 = 1423.62 Btu/lb

$\hat{s}_1$ = S_Steam (P_1, T_1)/MW_H_2O = S_Steam(1414.14, 1346.67)/18.01534 = 1.5585 Btu/lb-R

Step 2: Determine the turbine outlet saturation conditions (vapor and liquid states). These values will be needed to determine the steam quality.

Saturated Vapor

P_2 = 369.51 psia

$T_2^{\text{Sat Vapor}} = T_2^{\text{Sat Liquid}}$ = Tsat_Water (P2) = 896.67 R

$\hat{h}_2^{\text{Sat Vapor}}$ = H_Steam(P_2, $T_2^{\text{Sat Vapor}}$)/MW_H_2O = 1205.04 Btu/lb

$\hat{s}_2^{\text{Sat Vapor}}$ = S_Steam(P_2, $T_2^{\text{Sat Vapor}}$)/MW_H_2O = 1.4927 Btu/lb-R

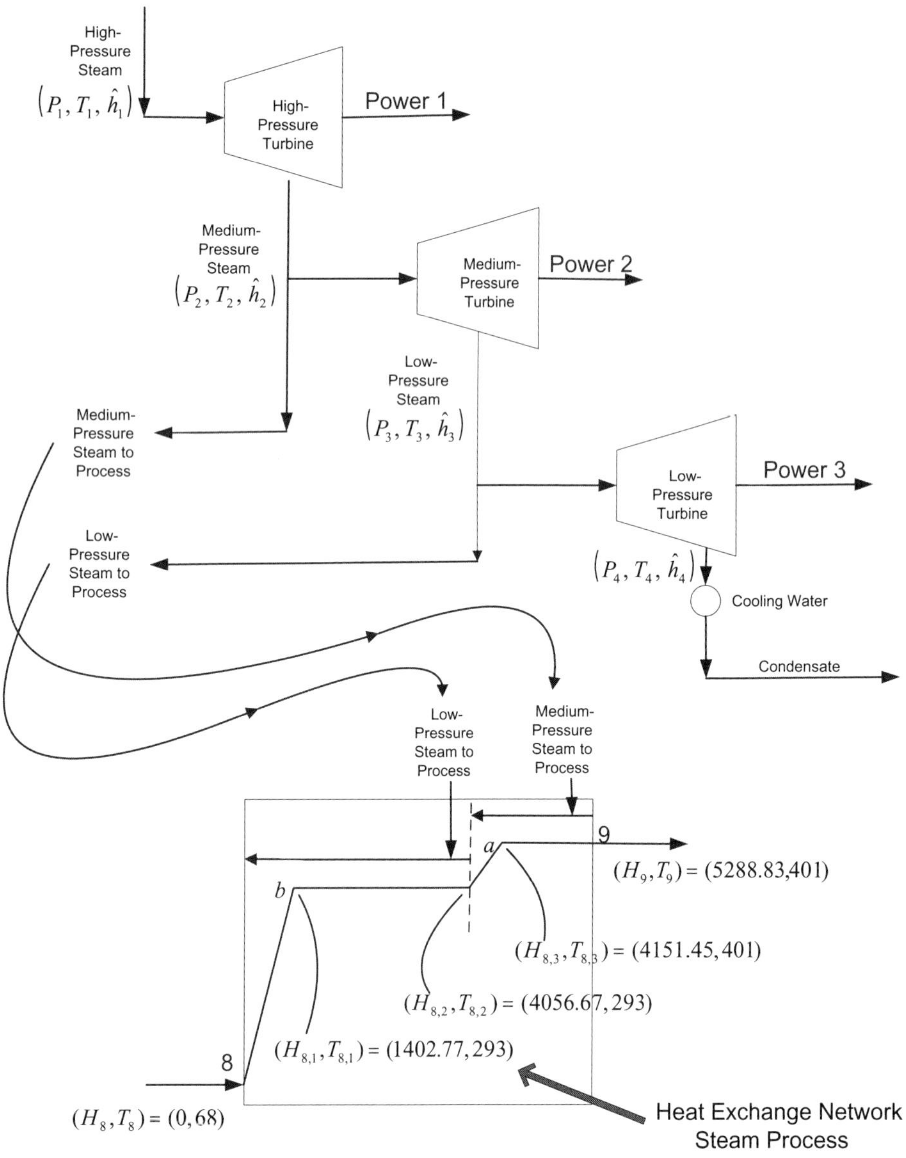

Figure 14.8 Steam power cycle with heat recovery.

Saturated Liquid

$P_2 = 369.51$ psia

$T_2^{\text{Sat Liquid}} = \text{TSat\_Water (P2)} = 896.67$ R

$\hat{h}_2^{\text{Sat Liquid}} = \text{H\_Water}(P_2, T_2^{\text{Sat Liquid}})/\text{MW\_H}_2\text{O} = 415.56$ Btu/lb

$\hat{s}_2^{\text{Sat Liquid}} = \text{S\_Water}(P_2, T_2^{\text{Sat Liquid}})/\text{MW\_H}_2\text{O} = 0.6123$ Btu/lb-R

Step 3: Determine the turbine outlet conditions (entropy, $\hat{s}_2^{\text{isentropic}}$; quality, $q_1^{\text{isentropic}}$; temperature, $T_2^{\text{isentropic}}$; and enthalpy, $\hat{h}_2^{\text{isentropic}}$) assuming a 100% efficient isentropic expansion.

$P_2 = 369.51$ psia

$\hat{s}_2^{\text{isentropic}} = \hat{s}_1 = 1.5585$ Btu/lb-R

$$q_1^{\text{isentropic}} = \frac{\hat{s}_2^{\text{isentropic}} - \hat{s}_2^{\text{Sat Liquid}}}{\hat{s}_2^{\text{Sat Vapor}} - \hat{s}_2^{\text{Sat Liquid}}} = 1.075 \rightarrow \text{steam leaving the}$$

HP turbine is all vapor as $q_1^{\text{isentropic}} \geq 1$.

$T_2^{\text{isentropic}} = \text{If } (q_1^{\text{isentropic}} < 1.0,\ T_2^{\text{Sat Vapor}},\ \text{T\_SP\_Steam}(\hat{s}_2^{\text{isentropic}} \times \text{MW\_H}_2\text{O}, P_2)) = 988.31$ R

$\hat{h}_2^{\text{isentropic}} = \text{If}(q_1^{\text{isentropic}} < 1.0, (1 - q_1^{\text{isentropic}}) \times \hat{h}_2^{\text{Sat Liquid}} + q_1^{\text{isentropic}} \times \hat{h}_2^{\text{Sat Vapor}},\ \text{H\_Steam}(P_2,\ T_2^{\text{isentropic}})/\text{MW\_H}_2\text{O}) = 1266.91$ Btu/lb

In step 3, we determine the steam quality leaving the HP turbine as

$$q_1^{\text{isentropic}} = \frac{\hat{s}_2^{\text{isentropic}} - \hat{s}_2^{\text{Sat Liquid}}}{\hat{s}_2^{\text{Sat Vapor}} - \hat{s}_2^{\text{Sat Liquid}}}.$$

Excel If statements (If (calculation, calculation true, calculation false)) are used to determine temperature and enthalpy values. If $q_1^{\text{isentropic}}$ is ≥ 1, then the MP stream leaving the HP turbine is all vapor. Here, $T_2^{\text{isentropic}}$ can be found from the known steam entropy and pressure as T_SP_Steam ($\hat{s}_2^{\text{isentropic}} \times \text{MW\_H}_2\text{O}, P_2$). The steam enthalpy can be found

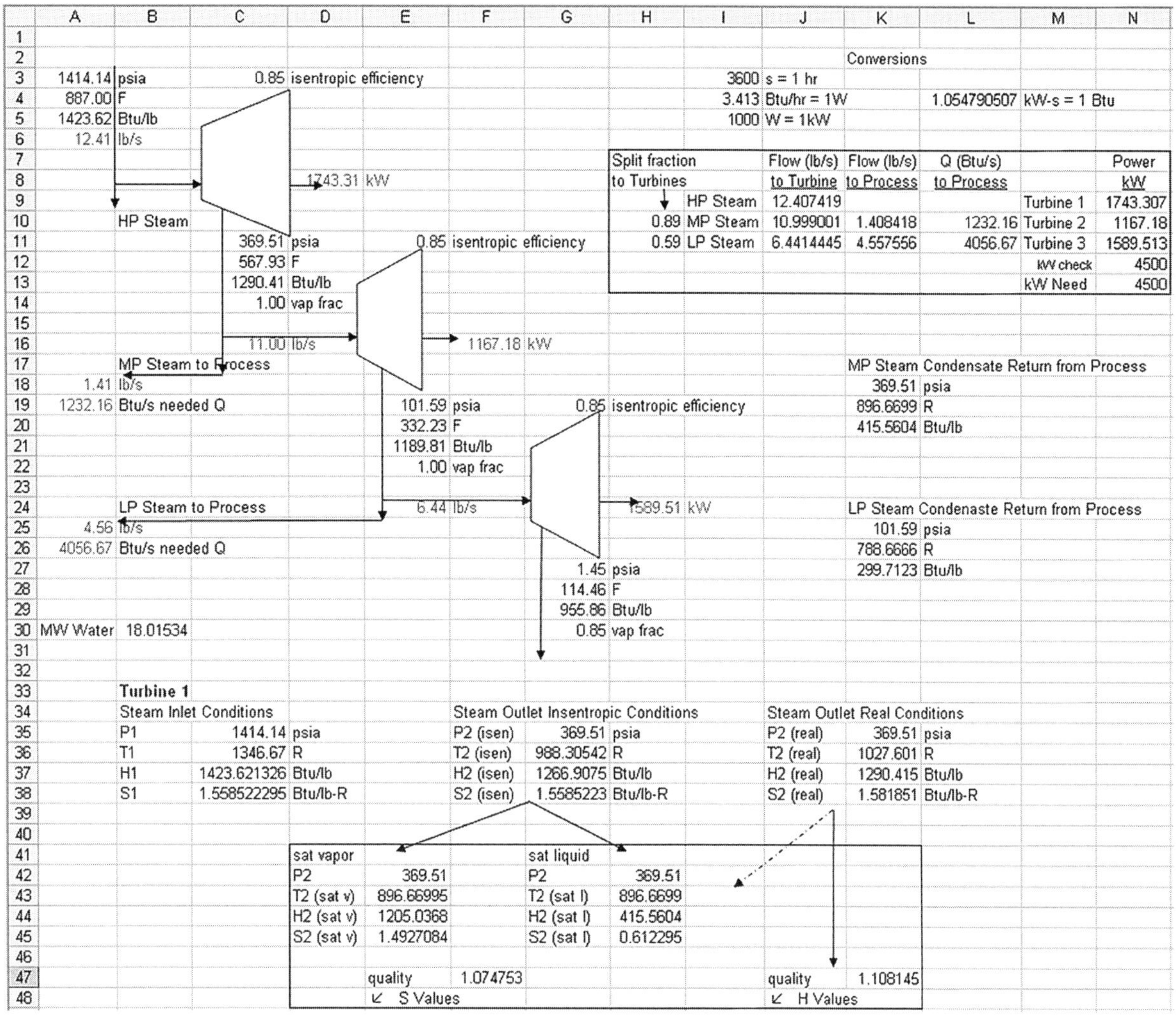

Figure 14.9 Steam power cycle with heat recovery Excel sheet solution using real steam properties.

as H_Steam(P_2, $T_2^{\text{isentropic}}$)/MW_H_2O). If $q_1^{\text{isentropic}}$ is <1, then the MP steam leaving the turbine is two phases. Here, $T_2^{\text{isentropic}} = T_2^{\text{Sat Vapor}}$ and the enthalpy of the two phase mixture is found as $\left(1-q_1^{\text{isentropic}}\right)\times \hat{h}_2^{\text{Sat Liquid}} + q_1^{\text{isentropic}} \times \hat{h}_2^{\text{Sat Vapor}}$.

Step 4: Determine HP turbine real outlet conditions using $\eta_{\text{HP Turbine}}^{\text{isentropic}} = 85\%$ isentropic efficiency

$P_2 = 369.51$ psia

$$\hat{h}_2^{\text{Real}} = \hat{h}_2 = \hat{h}_1 - \left(\hat{h}_1 - \hat{h}_2^{\text{isentropic}}\right)\times \eta_{\text{HP Turbine}}^{\text{Isentropic}} = 1290.41\ \text{Btu/lb}$$

$$q_1^{\text{Real}} = \frac{\hat{h}_2^{\text{Real}} - \hat{h}_2^{\text{Sat Liquid}}}{\hat{h}_2^{\text{Sat Vapor}} - \hat{h}_2^{\text{Sat Liquid}}} = 1.11 \rightarrow \text{steam leaving the HP}$$

turbine is all vapor as $q_1^{\text{Real}} \geq 1$

$T_2^{\text{Real}} = T_2 =$ If ($q_1^{\text{Real}} < 1.0$, $T_2^{\text{Sat Liquid}}$, T_HP_Steam($\hat{h}_2^{\text{Real}} \times$ MW_H_2O, P_2)) = 1027.60 R

$\hat{s}_2^{\text{Real}} = \hat{s}_2 =$ If ($q_1^{\text{Real}} < 1.0$, $\left(1-q_1^{\text{Real}}\right)\times \hat{s}_2^{\text{Sat Liquid}} + q_1^{\text{Real}} \times \hat{s}_2^{\text{Sat Vapor}}$, S_Steam($P_2$, T_2^{Real})/MW_H_2O) = 1.5819 Btu/lb-R

In step 4, we first determine the real enthalpy of the steam leaving the HP turbine $\hat{h}_2^{\text{Real}}$ using the isentropic efficiency:

$$\eta_{\text{HP Turbine}}^{\text{isentropic}} = \frac{\Delta \hat{h}^{\text{Real}}}{\Delta \hat{h}^{\text{isentropic}}} = \frac{\hat{h}_1 - \hat{h}_2^{\text{Real}}}{\hat{h}_1 - \hat{h}_2^{\text{isentropic}}}.$$

The steam quality leaving the HP turbine is determined as

$$q_1^{\text{Real}} = \frac{\hat{h}_2^{\text{Real}} - \hat{h}_2^{\text{Sat Liquid}}}{\hat{h}_2^{\text{Sat Vapor}} - \hat{h}_2^{\text{Sat Liquid}}}.$$

If q_1^{Real} is ≥1, the real condition of the MP steam leaving the HP turbine is all vapor. Here, T_2^{Real} can be found from the real steam enthalpy and pressure as T_HP_Steam($\hat{h}_2^{Real} \times$ MW_H_2O, P_2). The steam entropy is found as S_Steam(P_2, T_2^{Real})/MW_H_2O. If q_1^{Real} is <1, the MP steam leaving the turbine is actually two phases. Here, $T_2^{\text{Real}} = T_2^{\text{Sat Liquid}} = T_2^{\text{Sat Vapor}}$ and the real entropy of the two phase mixture is found as $\left(1-q_1^{\text{Real}}\right)\times \hat{s}_2^{\text{Sat Liquid}} + q_1^{\text{Real}} \times \hat{s}_2^{\text{Sat Vapor}}$.

These same procedures will be used for the MP and LP turbines. Part of the exhaust from the HP and MP turbines will be sent to the process for process heating. In some cases, HP steam may also be needed by the process. When steam is matched to the process, it will generally be condensed at the entering pressure. For the MP steam in this example, $\hat{h}_{MP}^{Sat\ Liquid}$ = H_Water(P_2, Tsat_Water (P_2))/MW_H_2O = 415.56 Btu/lb and $\Delta\hat{h}_{MP\ Steam}$ = (1290.41 − 415.56) = 874.85 Btu/lb. Similarly for the LP steam condensate, $\hat{h}_{LP}^{Sat\ Liquid}$ = 299.71 Btu/lb and $\Delta\hat{h}_{LP\ Steam}$ = (1189.81 − 299.71) = 890.1 Btu/lb.

We can next address the problem of determining the steam flows necessary to satisfy the heating and electric power requirements stated in the problem. Assembling the needed material and energy balances where

$F_{HP\_Turbine}$ = HP steam flow to turbine 1 (pound per second);
$F_{MP\_Turbine}$ = MP steam flow to turbine 2 (pound per second);
$F_{MP\_Process}$ = MP steam flow to the process (pound per second);
$F_{LP\_Turbine}$ = LP steam flow to turbine 3 (pound per second);
$F_{LP\_Process}$ = LP steam flow to process (pound per second);
$\dot{Q}_{MP}^{Process}$ = process heat load satisfied by MP steam (British thermal unit per second); and
$\dot{Q}_{LP}^{Process}$ = process heat load satisfied by LP steam (British thermal unit per second).

Turbine 1:

$$F_{HP\_Turbine} = F_{MP\_Turbine} + F_{MP\_Process} \tag{14.8}$$

$$\text{Power}_1 = F_{HP\_Turbine}\left(1423.62 - 1290.41\frac{\text{Btu}}{\text{s}}\right)\left(\frac{3600\ \text{s}}{\text{h}}\right)\left(\frac{1\ \text{W}}{3.413\ \text{Btu/h}}\right)\left(\frac{1\ \text{kW}}{1000\ \text{W}}\right) \tag{14.9}$$

Turbine 2:

$$F_{MP\_Turbine} = F_{LP\_Turbine} + F_{LP\_Process} \tag{14.10}$$

$$\text{Power}_2 = F_{MP\_Turbine}\left(1290.41 - 1189.81\frac{\text{Btu}}{\text{s}}\right)\left(\frac{3600\ \text{s}}{\text{h}}\right)\left(\frac{1\ \text{W}}{3.413\ \text{Btu/h}}\right)\left(\frac{1\ \text{kW}}{1000\ \text{W}}\right) \tag{14.11}$$

$$\dot{Q}_{MP}^{Process} = 1232.16\ \frac{\text{Btu}}{\text{s}} = (F_{MP\_Process})(\Delta\hat{h}_{MP\ Steam}) = F_{MP\_Process}\left(874.85\frac{\text{Btu}}{\text{lb}}\right) \tag{14.12}$$

Turbine 3:

$$F_{LP\_Turbine} = F_{LP\_Turbine} \tag{14.13}$$

$$\text{Power}_3 = F_{LP\_Turbine}\left(1189.81 - 955.86\frac{\text{Btu}}{\text{s}}\right)\left(\frac{3600\ \text{s}}{\text{h}}\right)\left(\frac{1\ \text{W}}{3.413\ \text{Btu/h}}\right)\left(\frac{1\ \text{kW}}{1000\ \text{W}}\right) \tag{14.14}$$

$$\dot{Q}_{LP}^{Process} = 4056.67\frac{\text{Btu}}{\text{s}} = (F_{LP\_Process})(\Delta\hat{h}_{LP\ Steam}) = F_{LP\_Process}\left(890.1\frac{\text{Btu}}{\text{lb}}\right) \tag{14.15}$$

Overall:

$$\text{Power}_1 + \text{Power}_2 + \text{Power}_3 = 4500\ \text{kW} \tag{14.16}$$

$$F_{HP\_Turbine} = F_{MP\_Process} + F_{LP\_Process} + F_{LP\_Turbine} \tag{14.17}$$

We have eight unknowns and eight independent equations. Solving,

From Equation (14.12), $F_{MP\_Process}$ = 1.408 lb/s;
From Equation (14.15), $F_{LP\_Process}$ = 4.558 lb/s; and
From Equation (14.16) using Equations (14.9), (14.11), and (14.14),

$$F_{HP\_Turbine}(140.5089) + F_{MP\_Turbine}(106.1119) + F_{LP\_Turbine}(246.7682) = 4500\ \text{kW}.$$

Simplifying using Equations (14.8) and (14.17) and solving,

$$F_{HP\_Turbine}(140.5089) + (F_{HP\_Turbine} - 1.408)(106.1119) + (F_{HP\_Turbine} - 5.966)(246.7682) = 4500\ \text{kW}$$
$$F_{HP\_Turbine} = 12.407\ \text{lb/s},$$

which allows determination of $F_{LP\_Turbine}$ = 6.441 lb/s and $F_{MP\_Turbine}$ = 10.999 lb/s.

These same calculations are performed in the Excel sheet. The final results in Figure 14.9 show that an HP steam flow of 12.41 lb/s will produce 1743.31 kW from turbine 1. The 12.41 lb/s of exhaust MP steam from turbine 1 will be distributed so that 11.00 lb/s go to turbine 2 and 1.41 lb/s go to the process. The MP steam to the process will supply the needed 1232.16 Btu/s of heating. The MP steam to turbine 2 will generate 1167.18 kW. The LP steam exhaust from turbine 2 will be distributed so that 6.44 lb/s goes to turbine 3 and 4.56 lb/s goes to the process. The LP steam to the process will supply 4056.67 Btu/s of heating and the LP steam to turbine 3 will generate 1589.51 kW. The total power produced from the three turbines meets the required 4500 kW needed by the process. The steam quality from turbines 1 and 2 is >1, which means the steam entering both turbine 2 and turbine 3 is superheated. The steam leaving turbine 3 is two phase with 85% quality.

We can easily explore the effect of changing the required power or process duty or turbine pressures—simply change any of the values in blue in the provided Excel file **Example 14.8.xls**, and the new system will be calculated. Of course, care must be taken that temperature of the steam being used by the process for heating meets ΔT_{min} requirements. ■

The intent of the examples explored in this chapter is to show how any cogeneration system, using real physical properties, can be constructed and evaluated to meet targeted process requirements. By developing general Excel-based solutions, we can easily explore the effect of changing processing conditions or costs.

14.3 SITE-WIDE UTILITY SYSTEM CONSIDERATIONS

In Chapters 13 and 14, we have been examining energy integration and utility system design for a distinct processing plant or area with clearly defined hot and cold streams and electric power requirements. This processing plant may produce a single product, or slate of similar products, through the same sequence of unit operations.

We now want to address the situation where our processing plant is one of several separate processing plants, or areas of processing integrity, all of which are part of a larger site (generally a corporate site). For example, a site may have both a vinyl chloride plant and a polyethylene plant with both plants sharing the intermediate, ethylene. There is no requirement that site processing plants share feeds or intermediates. However, a justification often made for multiple plants at a single site is that they can share the same central utility system. Our energy analysis can be extended from a single processing plant to a site-wide analysis—here, the possibility may exist for separate plants to exchange energy from existing hot and cold streams thereby reducing the overall utility load on the central utility system. In this section, we examine energy exchange between areas of integrity; here, modifications to the GCC for each processing plant can make site-wide energy integration evaluation straightforward.

EXAMPLE 14.9 *Direct Site Energy Integration*

Examine possible direct energy exchange between two processing plants, plant A and plant B, both located at the same site. These processing plants are currently sharing the same utility plant. Stream data for the two plants are supplied in Table 14.6. Here we use $\Delta T_{min} = 20°C$ for all matches (adapted from Shenoy, 1995).

SOLUTION It is instructive to first determine the maximum energy recovery (MER) network for each plant. This is the best starting point to then discuss possible energy exchange between plants. From our previous discussion (Chapter 13), we can solve the temperature interval analysis and the MER heat exchanger network design for each plant.

Stand-Alone Plant A: No Site-Wide Energy Integration

The temperature interval analysis (Figure 14.10a) shows plant A with a pinch temperature of 95°C, which is 105°C for hot streams and 85°C for cold streams. The heat exchanger network (Figure 14.10b) shows $N - 1$ heat exchangers above the pinch and $N - 1$ heat exchangers below the pinch. Target steam and cooling loads are met at 700-kW heating and 900-kW cooling.

Table 14.6 Stream Data for Plants A and B

Streams	*T* Supply (°C)	*T* Target (°C)	$F\hat{C}_p$ (kW/°C)
Plant A			
H1	175	45	10
H2	105	65	40
C3	20	155	20
Plant B			
H4	210	50	5
C5	140	170	10

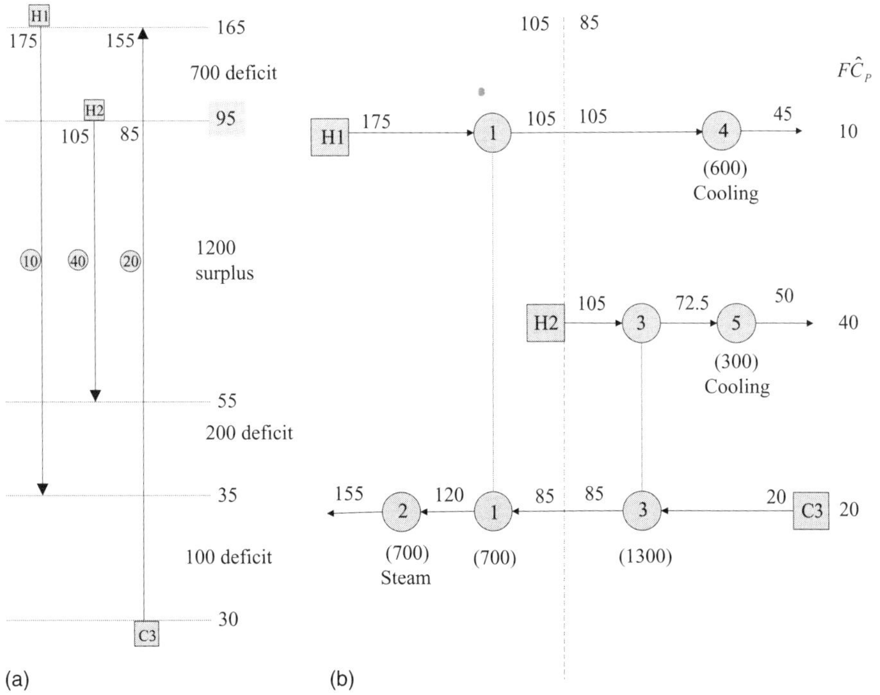

Figure 14.10 Plant A temperature interval analysis and MER heat exchanger network.

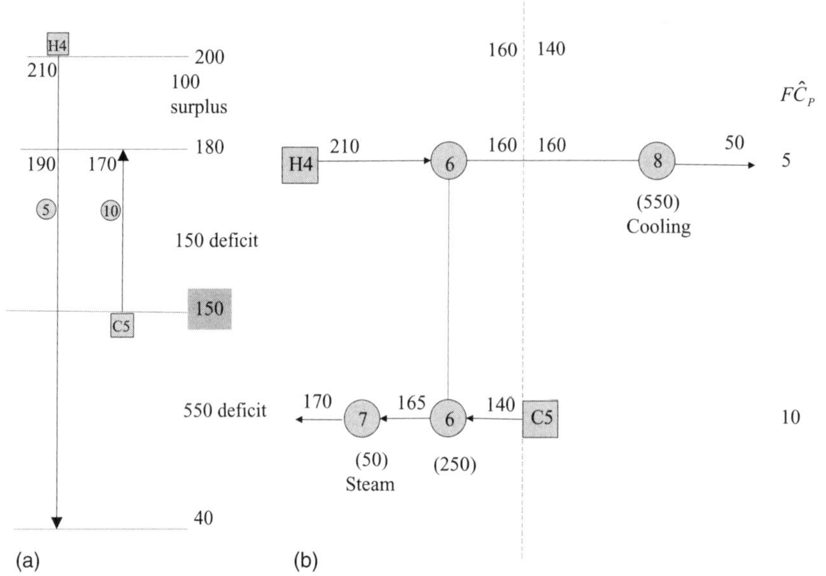

Figure 14.11 Plant B temperature interval analysis and MER heat exchanger network.

Stand-Alone Plant B: No Site-Wide Integration

The temperature interval analysis (Figure 14.11a) shows plant B with a pinch temperature of 150°C. The heat exchanger network (Figure 14.11b) shows target steam and cooling loads at 50-kW heating and 550-kW cooling, respectively. The heat exchanger network shows $N - 1$ heat exchangers above the pinch and $N - 1$ heat exchangers below the pinch; heat exchanger numbering is sequential from plant A to facilitate discussion of energy integration between the plants.

If plants A and B are treated as stand-alone areas, with no site-wide energy exchange, the utility plant will need to supply 700 + 50 kW = 750-kW heating and 900 + 550 kW = 1450-kW cooling. There may be many reasons to keep plant areas separate, including distance between plants making energy integration cost prohibitive, safety concerns, start-up, and control issues. If site-wide energy integration is to be considered, we can determine an upper bound on possible energy exchange by simply treating our two plants as one, in other words, we can allow direct integration of all the existing streams in the plants.

Direct Site Integration of Plants A and B: Upper Bound on Site-Wide Energy Integration

We can allow all streams in plants A and B to be considered for energy exchange. This would provide a maximum or upper limit for possible energy exchange. The temperature interval table in Figure 14.12a shows an overall pinch temperature of 95°C with 475 kW of heating and 1175 kW of cooling. Comparing this result to the results for stand-alone plant A (Figure 14.10) and stand-alone plant B (Figure 14.11), we see that plant A is controlling the location of the pinch. The needed utilities fall below that of a simple addition of the requirements of plants A and B.

In a retrofit application, we would want to preserve as much of the existing networks (Figures 14.10b and 14.11b) as possible. Figure 14.12b provides a possible network design meeting the target utilities in Figure 14.12a and where much of the structure of existing networks (Figures 14.10b and 14.11b) is preserved. Here, stream C3 in plant A must be split to preserve the match in heat exchanger 1 and to allow the addition of a new heat exchanger, heat exchanger 9, which brings H4 in plant B from 160°C to the hot stream pinch temperature of 105°C. Heat exchanger 9 allows the transfer of 275 kW from plant B hot stream H4 to plant A. This transfer lowers the cooling load in plant B from 550 kW (Figure 14.11b) to 275 kW and the transfer lowers the heating load of plant A from 700 kW (Figure 14.10b) to 425 kW. ■

As discussed, it may be impractical to allow direct energy integration between processing areas because of the distances involved, safety concerns, or start-up and control issues. Some of these concerns can be overcome by using indirect energy transfer via the steam system or a hot oil transfer loop.

EXAMPLE 14.10 *Indirect Site Energy Integration (Steam Loop)*

Examine using a steam loop for indirect energy exchange between the two processing plants of Example 14.9. Here again use $\Delta T_{\min} = 20$°C for all matches.

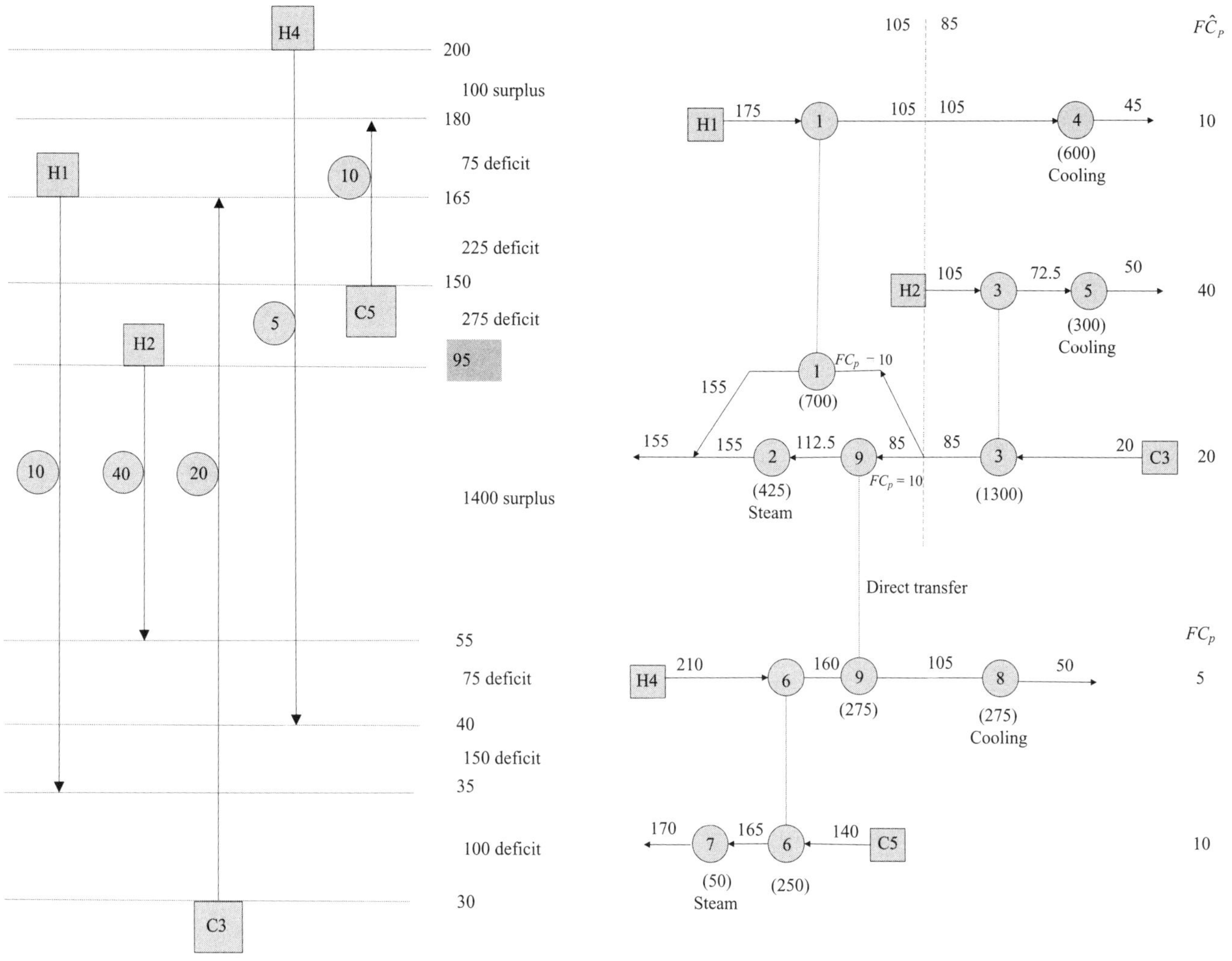

(a) (b)

Figure 14.12 (a) Plant A and plant B direct integration temperature interval analysis. (b) Plant A and plant B direct integration MER heat exchanger network.

SOLUTION Dhole and Linnhoff (1993) and Shenoy (1995) provide a straightforward procedure to determine the potential for indirect energy exchange between processing areas using steam.

Step 1: Remove pockets in the GCCs as indicated in Figure 14.13. Here, the motivation is that energy exchange in existing pockets (hot process and cold process stream energy exchange) is most economical in the local processing area.

GCC Plant A			GCC A Modified with Pocket Removal	
Temperature (°C)	Enthalpy (kW)	→	Temperature (°C)	Enthalpy (kW)
165	700		165	700
95	0		95	0
55	1200		65	900
35	1000			
30	900			

GCC Plant B			GCC B Modified with Pocket Removal	
Temperature (°C)	Enthalpy (kW)	→	Temperature (°C)	Enthalpy (kW)
200	500		160	50
180	0		150	0
150	1800		40	550
40	1600			

Step 2: Form a modified GCC with temperatures shifted $\Delta T_{min}/2$.

For step 2, points above the pinch (sink elements) are shifted upwards an additional $\Delta T_{min}/2$, and points below the pinch (source elements) are shifted downward an additional $\Delta T_{min}/2$.

Note that the pinch temperature in each plant is repeated—once for the start of the sink elements and once for

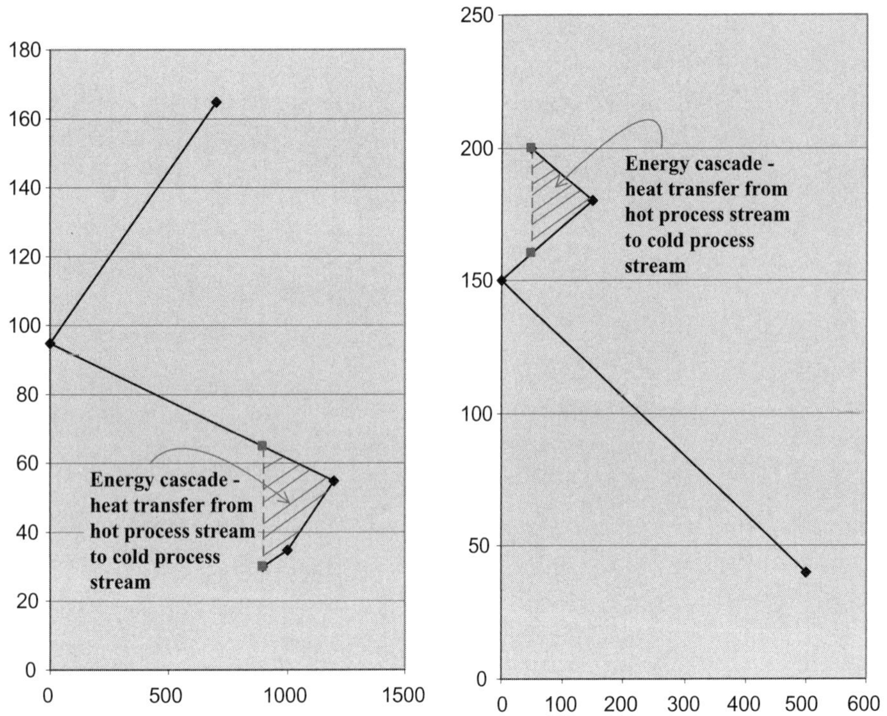

Figure 14.13 Plant A and plant B GCCs indicating pocket removal (hot process-to-cold process stream energy exchange).

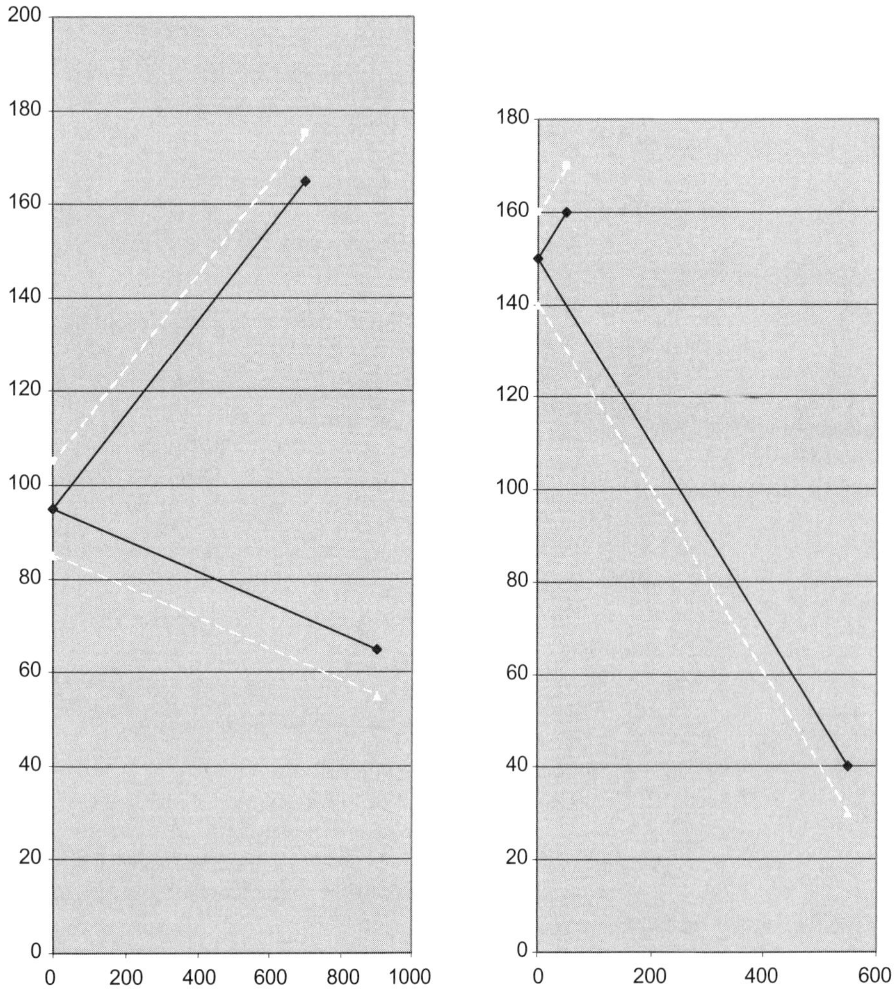

Figure 14.14 Plant A and plant B GCCs after pocket removal (in solid line) and modified GCC after shifting $\Delta T_{\min}/2$ both above and below the pinch (in dotted line).

the start of the source elements. The modified GCCs for plants A and B are shown in Figure 14.14.

Temperature (°C)	Enthalpy (kW)
Shifted GCC plant A	
175	700
105	0
85	0
55	900
Shifted GCC plant B	
170	50
160	0
140	0
30	550

Step 3: The shifted sink and source elements from each plant are combined to form the total site profiles (TSPs).

Temperature (C)	Enthalpy (kW)
Site sink profile	
105	0
160	550
170	700
175	750
Site source profile	
140	0
85	275
55	1325
30	1450

The site sink and source profiles are plotted on a T versus H diagram in Figure 14.15, with the site source profile plotted on a negative enthalpy axis.

The strength of this TSP construction is that energy transfer opportunities, from the source to the sink, can be directly read from the T versus H plot. This is a consequence of the "additional" $\Delta T_{\min}/2$ shifting in step 2. To see how Figure 14.15 can be used, allow that the site steam system has three pressure levels: an HP steam level, which may be used for some process heating as well as power generation; an intermediate pressure (IP) level, also used for power generation and process heating; and an LP level. From Figure 14.15, the IP steam level should be set at 175°C or higher. If the LP steam level is set at 125°C, there exists the possibility to transfer some energy from the source to the sink (the white line in Figure 14.15), thereby lowering the overall energy requirements in the site utility system. Steam that is generated by using waste heat from the process will not need to be generated by using fuel in the utility system. From Figure 14.15 the amount of steam at 125°C that can be raised from the source is $\dot{Q} = F\hat{C}_p(\Delta T) = 5(140 - 125) = 75$ kW. The total amount of LP steam needed by the sink is 10(125–105) = 200 kW. With energy exchange, this 200 kW is reduced to 125 kW. Figure 14.15 shows the steam loop network; here, for clarity, the steam profiles show only changes in latent heat. ■

EXAMPLE 14.11 *Indirect Site Energy Integration (Hot Oil Loop)*

An alternative to the energy transfer loop using steam is an energy transfer loop using oil. The use of an oil loop may increase energy exchange above that possible with a steam loop. There can be closer matching between $\Delta T_{\min}$ and the hot and cold side of the oil loop. Examine using a hot oil loop for indirect energy exchange between the two processing plants of Example 14.9. Here again use $\Delta T_{\min}$ = 20°C for all matches.

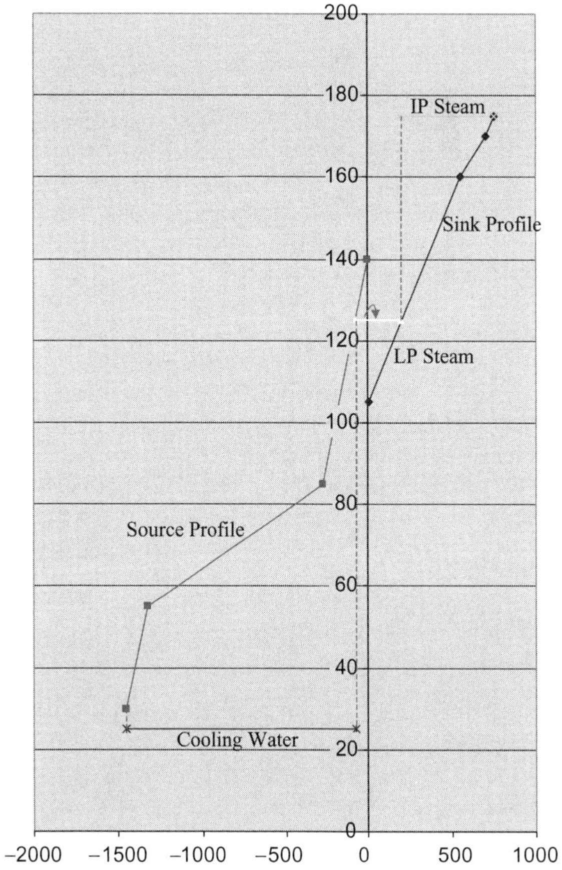

Figure 14.15 Plant A and plant B total site profiles (TSPs).

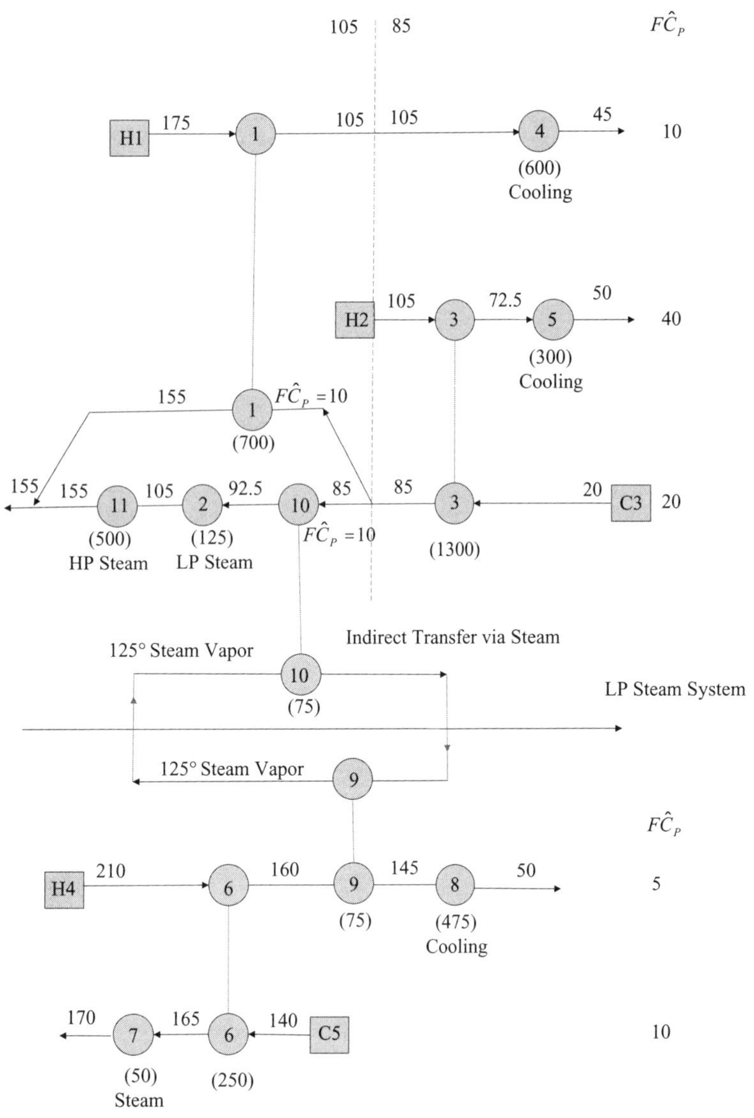

Figure 14.16 Plant A and plant B indirect steam loop based on TSP analysis.

SOLUTION If we examine Figure 14.16, we can visually think of the oil loop as a countercurrent heat exchanger located between the source and sink profiles. The oil will have a maximum temperature of 140°C as determined by the source and a minimum temperature of 105°C as determined by the sink. As $F\hat{C}_{P\text{source}} < F\hat{C}_{P\text{sink}}$ for the site, the source side controls the allowable energy exchange, $\dot{Q} = FC_{P\text{source}}\Delta T = 5(140 - 105) = 175$ kW. Recall the actual stream temperatures in the site source will be 160°C (for 140°C + $\Delta T_{\min}$) and 125 (for 105°C + $\Delta T_{\min}$).

In Figure 14.17, we show the oil loop to exchange energy between our two plants. Notice that the amount of energy that can be transferred has increased from 75 kW for the steam loop to 175 kW for the oil loop. More energy transfer is possible with the oil loop as an approach temperature of $\Delta T_{\min}$ is maintained on three sides of heat exchangers 9 and 10. The amount of energy transfer from either the steam or the oil loop falls below the maximum direct transfer of 275 kW. ■

14.4 CLOSING REMARKS

One of the expected outcomes from energy integration is a reduction in the required utilities and, consequently, a reduction in operating costs and emissions from utility system combustion processes—here also see Smith et al. (1990), Linnhoff and Dhole (1993), Klemes et al. (1997), Axelsson et al. (1999), and Varbanov et al. (2004a,b). In this chapter, we explored matching process energy needs, as determined by an energy integration analysis, with the design of the utility system. Here, process needs were matched to two fixed design cogeneration systems—a steam turbine system and gas turbine-based system. The use of real physical properties allowed determination of optimal sizing and operating conditions for the gas turbine

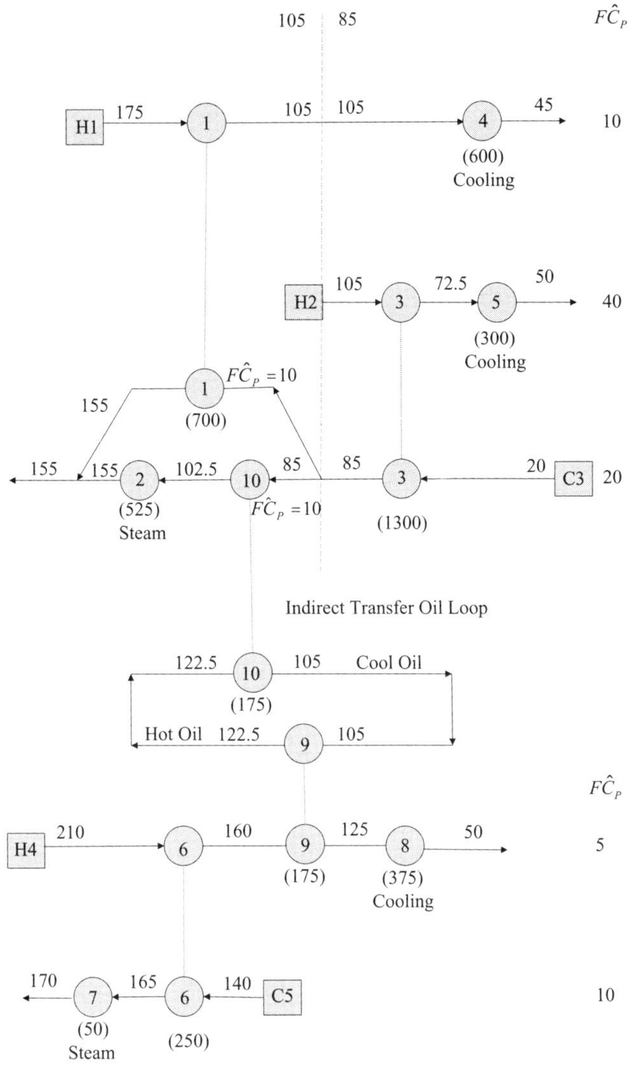

Figure 14.17 Plant A and plant B indirect hot oil loop based on TSP analysis.

system and also guided the design of the steam turbine system. We explored the possibility of heat integration between processing areas to reduce overall site utility requirements—here also see Ahmad and Hui (1991), Hui and Ahmad (1994a,b), Rodera and Bagajewicz (1999), and Kralj et al. (2002).

Optimization-based solution approaches, as discussed in the books by Floudas (1995) and Biegler et al. (1997), can be used to match process energy needs and utility system design. For example, Biegler et al. (1997) and Papoulias and Grossmann (1983) utilize a utility system superstructure to determine the optimal configuration for the utility system with known process demands—here see Problem 14.5.

REFERENCES

Ahmad, S. and D.C.W. Hui. 1991. Heat recovery between areas of integrity. *Comput. Chem. Eng.* 15(12): 809–832.

Axelsson, H., A. Asblad, and T. Berntsson. 1999. A new methodology for greenhouse gas reduction in industry through improved heat exchange and/or integration of combined heat and power. *Appl. Therm. Eng.* 19: 707–731.

Biegler, L.T., I.E. Grossmann, and A.W. Westerberg. 1997. *Systematic Methods of Chemical Process Design*. Prentice Hall, Upper Saddle River, NJ.

Dhole, V.R. and B. Linnhoff. 1993. Total site targets for fuel, cogeneration, emissions and cooling. *Comput. Chem. Eng.* 17(Suppl.): S101–S109.

Floudas, C.A. 1995. *Nonlinear and Mixed-Integer Optimization: Fundamentals and Application*. Oxford University Press, Oxford.

Hui, C.W. and S. Ahmad 1994a. Minimum cost heat recovery between separate plant regions. *Comput. Chem. Eng.* 18(8): 711–728.

Hui, C.W. and S. Ahmad. 1994b. Total site heat integration using the utility system. *Comput. Chem. Eng.* 18(8): 729–742.

Klemes, J., V.R. Dhole, K. Raissi, S.J. Perry, and L. Puigjaner. 1997. Targeting and design methodology for reduction of fuel, power and CO_2 on total sites. *Appl. Therm. Eng.* 17(8–10): 993–1003.

Kralj, A.K., P. Glavic, and M. Krajnc. 2002. Waste heat integration between processes. *Appl. Therm. Eng.* 22: 1259–1269.

Linnhoff, B. and V.R. Dhole. 1993. Targeting CO_2 emissions for total sites. *Chem. Eng. Technol.* 16: 252–259.

Papoulias, S.A. and I.E. Grossmann. 1983. A structural optimization approach in process synthesis—I. Utility systems. *Comput. Chem. Eng.* 7(6): 695–706.

Rodera, H. and M.J. Bagajewicz. 1999. Targeting procedures for energy savings by heat integration across plants. *AIChE J.* 45(8): 1721–1742.

Shenoy, U.V. 1995. *Heat Exchanger Network Synthesis: Process Optimization by Energy and Resource Analysis*. Gulf Publishing, Houston, TX.

Smith, R., E.A. Petela, and H.D. Spriggs. 1990. Minimization of environmental emissions through improved process integration. *Heat Recovery Syst. CHP* 10(4): 329–339.

Tensa Services and F.C. Knopf. 1990. *An Introduction to Pinch Technology and Industrial Energy Optimization*. Workbook for industrial short course. Tensa Services, Houston, TX.

Townsend, D.W. and B. Linnhoff. 1983a. Heat and power networks in process design. I. Criteria for placement of heat engines and heat pumps in process networks. *AIChE J.* 29: 742–748.

Townsend, D.W. and B. Linnhoff. 1983b. Heat and power networks in process design. II. Design procedure for equipment selection and process matching. *AIChE J.* 29: 748–771.

Varbanov, P., S. Doyle, and R. Smith. 2004a. Modeling and optimization of utility systems. *Trans. IChemE. Chem. Eng. Res. Des.* 82(A5): 561–578.

Varbanov, P., S. Perry, Y. Makwana, and R. Smith. 2004b. Top-level analysis of utility systems. *Trans. IChemE. Chem. Eng. Res. Des.* 82(A6): 4–795.

PROBLEMS

14.1 ***Modified GCC with Actual Temperatures*** Create the modified GCC with actual process stream temperatures above the pinch for Example 13.4 (Figure 13.10b) by

Step 1: Removing the pocket and

Step 2: Subtracting $\frac{1}{2}\Delta T_{\min}$ from the GCC temperature intervals; here, $-\frac{1}{2}\Delta T_{\min} = -5°C$.

Assume LP steam is available at 150°C. Indicate on this modified GCC the maximum amount of LP steam that can be utilized; again use a 10° approach. For comparison, also plot this LP steam on Figure 13.10.

14.2 ***Diesel Engine-Based Cogeneration System (A. Martin, pers. comm.)*** The process in Figure P14.2 will be modified by replacing the existing furnace with the exhaust gas stream from a diesel engine. The diesel engine will also supply electric power to the plant.

The reboiler (C1) uses 16 kW of sensible heat (150° → 190°C) and 104 kW of latent heat (190° → 190°C). The intercooler (R2) removes latent heat (95° → 95°C) from a column pump-around stream. The primary condenser (R1) will not be integrated in this analysis.

The exhaust gas from the diesel engine is available at 310°C and has a heat capacity of $\hat{C}_p = 1.102\ \mathrm{kJ/kg\text{-}°C}$.

This exhaust gas cannot be cooled below 128°C in order to prevent acid condensation in the heat recovery units. The net electric power generation is

$$\text{Power} = \left(0.72\left(\frac{\text{kW electicity}}{\text{kW}}\right)\times \text{hot utility load to the process (kW)}\right).$$

Cooling water is available at 20°C with a return temperature of 60°C. Cooling water has a heat capacity of $\hat{C}_p = 4.184\ \mathrm{kJ/kg\text{-}°C}$.

Determine the minimum hot and cold utility loads, the flow rate of the diesel exhaust gas, the electric power that can be generated, the energy loss in the diesel exhaust gas based on an ambient temperature of 25°C, and the water flow rate (based on 20°C water with a return temperature 60°C). Develop a heat exchanger network to meet the target utility requirements (an MER design). Where possible, try to preserve the existing network. Solve this problem maintaining a 20°C approach temperature ($\Delta T_{\min} = 20°$). Also, do remember to use $F\hat{C}_p$ (not just F or $\hat{C}_p$) for the water and exhaust gas streams in the heat exchanger design. In addition to the heat exchanger matches for the hot utility exhaust gas, show the heat exchanger matches for the cooling water (with 20°C water in and a return of 60°C).

14.3 ***Heat Recovery Exhaust Temperature*** Solve Example 14.6 with the requirement that $T_7 \geq 250°\mathrm{F}$. Compare results to Example 14.6.

14.4 ***Utility System Superstructure (Solution Provided)*** In a utility system superstructure (or any superstructure), we want to assemble all equipment that could realistically be

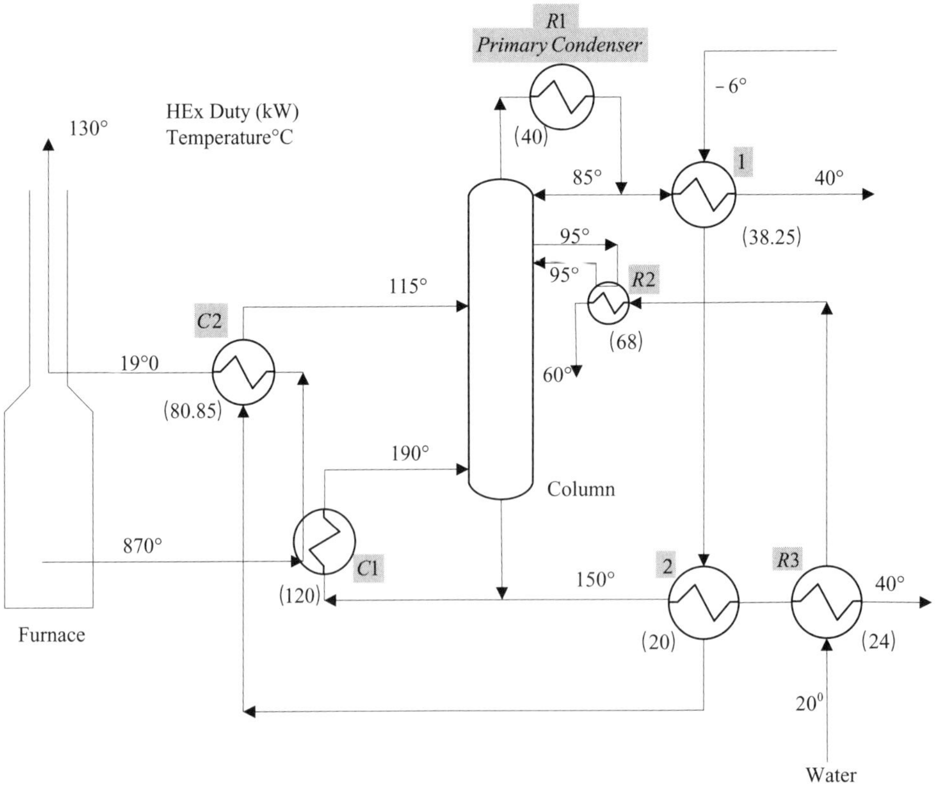

Figure P14.2 Column with furnace heater.

used to satisfy known process utility demands in a form allowing computer selection (mixed-integer program) of both the best configuration and operating conditions to minimize costs.

Biegler et al. (1997) and Papoulias and Grossmann (1983) provide a possible superstructure (Figure P14.4a) for a utility plant with the following process requirements: power A ≥ 7500 kW, a separate power B ≥ 4500 kW, MP steam ≥ 25 ton/h, and LP steam ≥ 85 ton/h. No HP steam is required. The superstructure allows for the possibility that steam can be generated from an HP or MP boiler and letdown valves can be used. Turbine enthalpy efficiency $\eta_{Turbine} = 65\%$. HP turbines can be expanded to MP or LP with the possibility of steam extraction at MP. The operation of the HP turbine is shown in more detail in Figure P14.4b. A final requirement is that each independent power demand (power A and power B) must be matched to a single unique turbine.

The following costing and physical property data are provided:

Cost Data

	Fixed Costs	Variable Costs
HP boiler	\$90,000 per year	\$9,600 per h/year-ton steam
MP boiler	\$40,000 per year	\$8,500 per h/year-ton steam
HP turbine	\$45,000 per year	\$25 per year-kW
MP turbine	\$25,000 per year	\$14.5 per year-kW
HP turbine steam extraction	\$20,000 per year	

If steam extraction from an HP turbine is used, for example, for HP turbine 1 if both $F_5 > 0$ and $F_6 > 0$, then an additional cost of \$20,000 per year is incurred.

Thermodynamic Data

$\Delta h_{(\text{HP to MP})} = 71$ kW-h/ton steam

$\Delta h_{(\text{MP to LP})} = 112$ kW-h/ton steam

Using the provided superstructure, we are asked to formulate and solve the mixed-integer linear programming (MILP) problem, which minimizes the utility system costs while meeting the required power and steam demands.

Formulation

We have been introduced to MILP problems in Chapter 12 where we developed the energy dispatch model for the cogeneration system. It is instructive to again show how these MILP problems can be formulated and then to discuss how solutions can be obtained. We begin by assigning an integer variable, δ_i (on/off), to each flow rate in Figure P14.4a. For example, we will use $\delta_5 F_5$ for the flow rate of stream 5; this allows F_5 to take on any value but, $\delta_5 = 1$ if flow in stream 5 is actually occurring and $\delta_5 = 0$ if flow is not occurring.

We can begin with energy balances to calculate the power from each turbine:

$$\text{Power}_1 = \left(F_3\ \eta_{\text{Turbine}}\Delta h_{(\text{HP to MP})}\right)\delta_3 + \left(F_6\ \eta_{\text{Turbine}}\Delta h_{(\text{MP to LP})}\right)\delta_6$$

$$\text{Power}_2 = \left(F_4\ \eta_{\text{Turbine}}\Delta h_{(\text{HP to MP})}\right)\delta_4 + \left(F_8\ \eta_{\text{Turbine}}\Delta h_{(\text{MP to LP})}\right)\delta_8$$

$$\text{Power}_3 = \left(F_9\ \eta_{\text{Turbine}}\Delta h_{(\text{MP to LP})}\right)\delta_9$$

$$\text{Power}_4 = \left(F_{10}\ \eta_{\text{Turbine}}\Delta h_{(\text{MP to LP})}\right)\delta_{10}$$

and the HP turbine mass balances (in–out):

HP turbine 1: $F_3\delta_3 - (F_5\delta_5 + F_6\delta_6)$
HP turbine 2: $F_4\delta_4 - (F_7\delta_7 + F_8\delta_8)$

The steam available to the process = steam produced – steam used:

$$\text{HP}_{\text{Process}} = F_1\delta_1 - \left(F_{11}\delta_{11} + F_5\delta_5 + F_6\delta_6 + F_7\delta_7 + F_8\delta_8\right)$$

$$\text{MP}_{\text{Process}} = \left(F_2\delta_2 + F_{11}\delta_{11} + F_5\delta_5 + F_7\delta_7\right) - \left(F_{12}\delta_{12} + F_9\delta_9 + F_{10}\delta_{10}\right)$$

$$\text{LP}_{\text{Process}} = \left(F_{12}\delta_{12} + F_6\delta_6 + F_8\delta_8 + F_9\delta_9 + F_{10}\delta_{10}\right)$$

We next add constraints to the formulation to match the process requirements. These constraints include

The two HP turbine mass balances: $F_3\delta_3 - (F_5\delta_5 + F_6\delta_6) = 0$ and $F_4\delta_4 - (F_7\delta_7 + F_8\delta_8) = 0$

The three process steam requirements: $HP_{\text{Process}} = 0$, $MP_{\text{Process}} \geq 25$ ton/h, $LP_{\text{Process}} \geq 85$ ton/h

The process power requirements:

Power A: $\text{Power}_1 + \text{Power}_3 \geq 7500$ kW
$\delta_1 + \delta_9 = 1$

Power B: $\text{Power}_2 + \text{Power}_4 \geq 4500$ kW
$\delta_4 + \delta_{10} = 1$

These power constraints ensure that power is generated using either an MP or HP turbine and that each power demand is met by a single turbine.

In order to check for extraction from the HP turbines, two additional integer variables are defined: $\delta_{T1\_Extract}$ and $\delta_{T2\_Extract}$, which represent yes/no for extraction from turbine 1 and turbine 2. Four additional constraints are used to check for extraction:

HP turbine 1: $\delta_5 + \delta_6 - \delta_{T1\_Extract} \leq 1$
$\delta_3 - \delta_{T1\_Extract} \geq 0$

HP turbine 2: $\delta_7 + \delta_8 - \delta_{T2\_Extract} \leq 1$
$\delta_4 - \delta_{T2\_Extract} \geq 0$

Solution Strategy (Using What's Best) and Solution Discussion

Recall that in What's Best you will need to highlight the integer variables; select WB! → Integer . . . → Integer Type • Binary-WBBIN. You can supply a name for the selected binary variables in the workbook. For example, supply the name Binary; What's Best will actually name the selected binary variables WBBINBinary.

Also, when using What's Best for all the MILP and MINLP problems in this text, you will need to select WB! → Options → General → Linearization → Degree

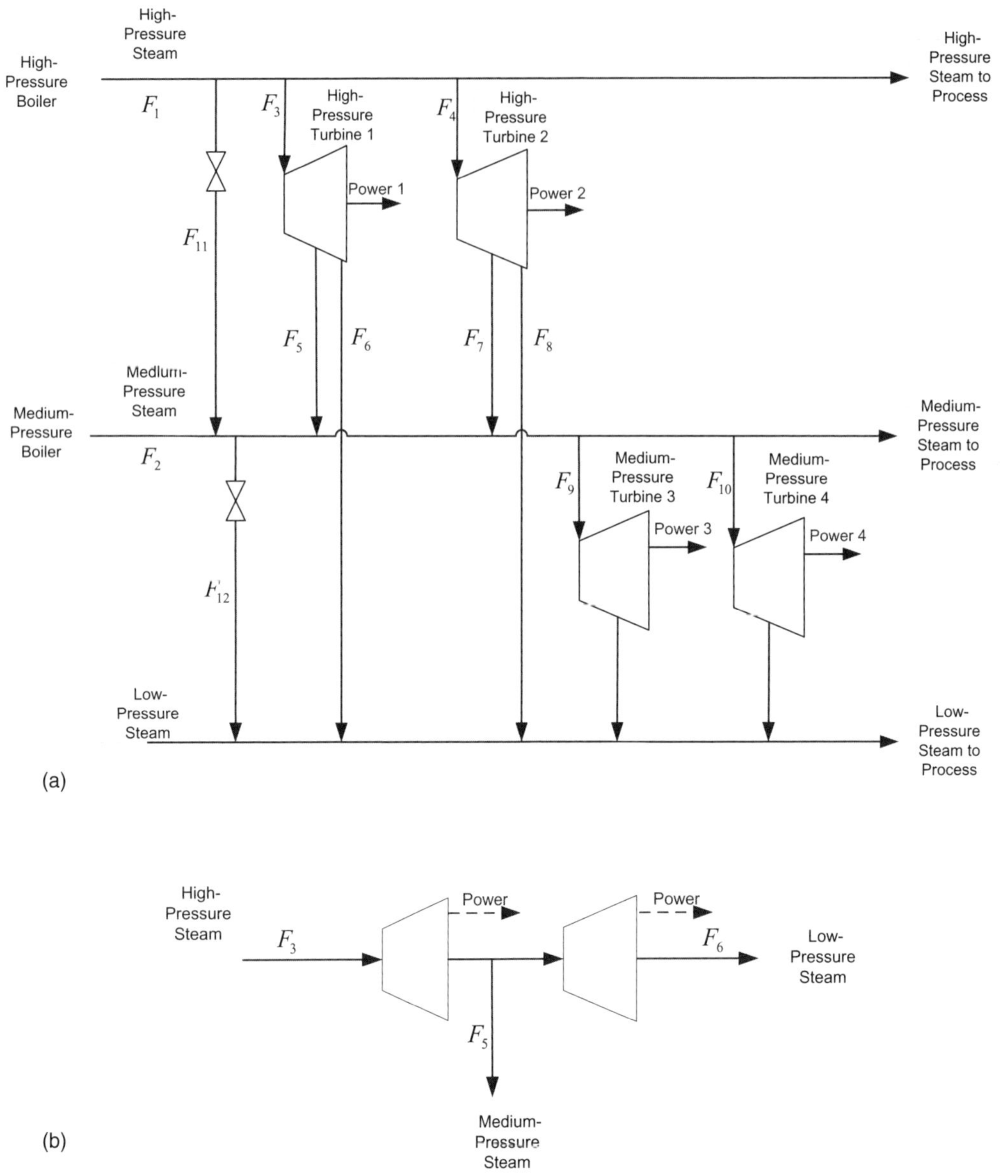

Figure P14.4 (a) Utility system superstructure (Biegler et al.,1997). (b) Expanded view of extraction turbine, here HP turbine 1 with total power output power 1.

. . . Math → OK. See the What's Best Manual for other options.

The final formulation consists of minimizing the objective function (left for the reader), subject to the process material and energy balances and 13 constraints. This formulation is correct, but the optimal solution may not be obtained within What's Best. As discussed in Biegler et al. (1997), there can be difficulties converging superstructure MILPs to the optimal solution. We can obtain an optimal solution to this problem by tightening the search region with the use of additional constraints. Here provide an upper bound to each flow rate; these 12 constraints can be $F_i \leq 130$ ton/h $i = 1, \ldots 12$. Also tighten the search region by converting the power requirements from inequality to equality constraints, $\text{Power}_1 + \text{Power}_3 = 7500$ kW and $\text{Power}_2 + \text{Power}_4 = 4500$ kW. The final solution is provided in **Problem 14.4.xls**. This optimal solution shows a system cost of $1,548,109 per year with use of an HP boiler, an HP turbine to meet power A requirements, and an MP turbine to meet power B requirements. Steam extraction is used with the HP turbine; MP steam requirements are met at 25 ton/h; and LP requirements are exceeded with 91.2 ton/h being delivered to the process.

14.5 ***Cogeneration Evaluation for the Hydrodealkylation of Toluene (HDA) Process*** In Problem 13.7, we solved the energy integration problem for the HDA process (Tensa Services and Knopf, 1990). In Problem 13.7, step 8, we determined needed utilities after energy integration (but without pressure shifting of the columns). You should have found the total hot utility need of the HDA process is 11.0 MMBtu / h.

Assume the HDA plant has an available 600-psia boiler capable of producing superheated steam at 680°F. Design a cogeneration system utilizing a steam turbine and where all the exhaust from the turbine will be used in the energy integrated HDA process as determined in step 8. The isentropic efficiency of the steam turbine is 0.75.

Hint: From the shape of the GCC (Problem 13.7, step 8), the exhaust steam from the turbine may be used to satisfy the benzene reboiler duty of 4.57 MMBtu / h.

Chapter 15

Site Utility Emissions

Emissions are broadly classified as either primary, including carbon dioxide, carbon monoxide, sulfur dioxide, particulates, and unburned hydrocarbons, or secondary, including nitrogen dioxide and sulfur trioxide. There are environmental concerns associated with each compound: CO_2 emissions with global warming, SO_x with acid rain, unburned hydrocarbons with smog, and particulate emissions with health effects. NO_x (representing both NO_2 and NO) is a precursor to ground-level smog.

There are several approaches to understanding emissions from combustion processes ranging from simple stoichiometric calculations to detailed simulations. Continuous emissions monitoring equipment can be used to quantify emissions. Emission predictions can involve varying levels of sophistication. A list of prediction techniques (with increasing sophistication) includes calculations based on

1. Stoichiometry alone;
2. Emission predictions based on manufacturer performance curves or manufacturer-provided calculation templates;
3. Thermodynamic equilibria;
4. Mass and energy balances using global (empirical) reaction kinetics;
5. The same, but with thermodynamically consistent elementary kinetics rate expressions;
6. The same, but solved on the microscale using computational fluid dynamics (CFD) combined with reduced set kinetics rate expressions.

In this chapter, we will examine the use of stoichiometric, thermodynamic equilibria and elementary kinetics rate expression calculations to predict emissions from utility systems.

We start our discussion by using stoichiometric calculations to predict emissions from a coal-fired power plant. Understanding coal-fired utility emissions is especially important as coal-fired utilities generate ~50% of the electricity in the United States. Carbon dioxide sequestration is often targeted at coal-fired utilities, and in Chapter 16 we predict how carbon dioxide sequestration would impact overall coal power plant efficiency and utility pricing.

Thermodynamic equilibrium calculations will be explored to estimate emissions from methane combustion in our cogeneration system of Example 14.4. Equilibrium calculations will require use of numerical methods we have developed in earlier chapters—particularly the use of the Newton–Raphson (NR) method we developed in Chapter 4.

A major emphasis in this chapter will be on understanding and using thermodynamically consistent elementary kinetics rate expressions particularly for modeling emissions from gas turbine systems. This will require numerical solution of rate expressions, which are ordinary differential equations (ODEs). We will find that these rate expressions (ODEs) cannot be solved by a simple Euler's method or Runge–Kutta method (fourth-order Runge–Kutta [RK4]) as developed in Chapter 5. Here we provide, as an Excel callable routine, CVODE from Lawrence Livermore National Laboratory (LLNL) (Collier et al., 2008) to allow solution of these stiff ODEs.

Modeling, Analysis and Optimization of Process and Energy Systems, First Edition. F. Carl Knopf.

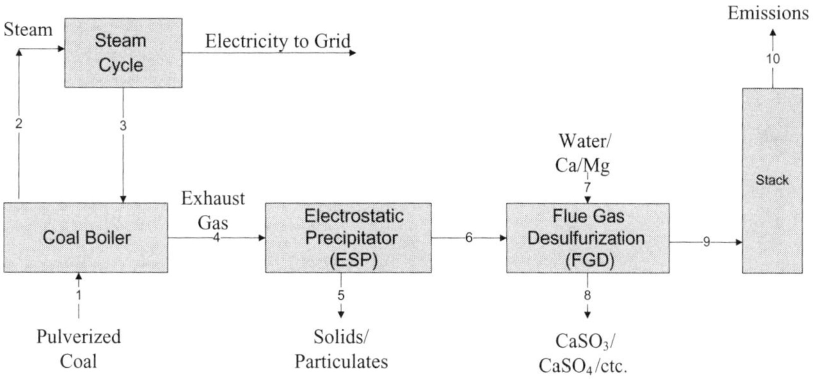

Figure 15.1 Simplified coal power plant process flow diagram (Conesville, Ohio, Unit #5).

15.1 EMISSIONS FROM STOICHIOMETRIC CONSIDERATIONS

Figure 15.1 shows a simplified process flow diagram based on the American Electric Power Conesville, Ohio, Unit #5 power plant (DOE/NETL-401/110907, US. DOE, 2007). This is a nominal 434-MW plant fired with bituminous coal. In the Conesville plant, an electrostatic participator (EP) removes particulates and a flue gas desulfurization (FGD) unit removes SO_2 in the exhaust gas prior to the stack.

EXAMPLE 15.1 *Coal Power Plant Emissions from Stoichiometric Considerations*

The bituminous coal used in the Conesville power plant has the following elemental analysis (in weight percent): C = 63.75, H = 4.50, O = 6.88, N = 1.25, S = 2.51, ash = 9.99, and water = 11.12. Develop the general equation to calculate the minimum or theoretical air required for complete combustion of this or any similar fuel when given the weight percent analysis. Apply this equation to the coal analysis to determine the minimum air flow needed for complete combustion. Assume air is 79% N_2 and 21% O_2 on a molar basis with an $MW = 28.84$. Also assume that the ash in the coal is inert and that the nitrogen in coal simply exits as N_2. In the following sections, we will discuss nitrogen conversion to NO_x, a process that can be both mass transfer and kinetically controlled in real combustors.

SOLUTION We want to develop a general equation for the theoretical air needed for combustion using coal as the representative fuel. The solution is also provided in the Excel file **Example 15.1.xls**. The amount of air needed for complete combustion of each element is determined as the following.

For $C + O_2 \rightarrow CO_2$, for each gram of carbon in the coal (here use 1 g), we would need

$$\left(\frac{1\text{ g C}}{12\text{ g C/mol}}\right)\left(\frac{1\text{ mol O}_2}{\text{mol C}}\right)\left(\frac{1\text{ mol air}}{0.21\text{ mol O}_2\text{ in air}}\right)\left(\frac{28.84\text{ g air}}{\text{mol air}}\right)$$
$$= 11.444\frac{\text{g air}}{\text{g C}}.$$

For $2H + \frac{1}{2}O_2 \rightarrow H_2O$, for each gram of H,

$$\left(\frac{1\text{ g H}}{1\text{ g H/mol}}\right)\left(\frac{0.5\text{ mol O}_2}{2\text{ mol H}}\right)\left(\frac{1\text{ mol air}}{0.21\text{ mol O}_2\text{ in air}}\right)\left(\frac{28.84\text{ g air}}{\text{mol air}}\right)$$
$$= 34.333\frac{\text{g air}}{\text{g H}}.$$

For O (this will reduce the amount of air needed) as

$$-\left(\frac{1\text{ g O}}{16\text{ g O/mol}}\right)\left(\frac{0.5\text{ mol O}_2}{\text{mol O}}\right)\left(\frac{1\text{ mol air}}{0.21\text{ mol O}_2\text{ in air}}\right)\left(\frac{28.84\text{ g air}}{\text{mol air}}\right)$$
$$= -4.292\frac{\text{g air}}{\text{g O}}.$$

For $S + O_2 \rightarrow SO_2$, for each gram of S,

$$\left(\frac{1\text{ g S}}{32\text{ g S/mol}}\right)\left(\frac{1\text{ mol O}_2}{\text{mol S}}\right)\left(\frac{1\text{ mol air}}{0.21\text{ mol O}_2\text{ in air}}\right)\left(\frac{28.84\text{ g air}}{\text{mol air}}\right)$$
$$= 4.292\frac{\text{g air}}{\text{g S}}.$$

We can write the general equation for the theoretical air needed for complete combustion (TACC) as

$$\text{TACC}_{\text{Air Min}} = 11.444\,C + 34.333\,\text{H} - 4.292\,\text{O} + 4.292\,\text{S},\ \frac{\text{g air}}{\text{g fuel}}. \tag{15.1}$$

For the coal analysis supplied here we would theoretically require

$$\left(11.444\frac{\text{g air}}{\text{g C}}\right)\left(\frac{0.6375\text{ g C}}{\text{g coal}}\right) + 34.333\left(\frac{0.045\text{ g H}}{\text{g coal}}\right)$$
$$-\ 4.292\left(\frac{0.0688\text{ g O}}{\text{g coal}}\right) + 4.292\left(\frac{0.0251\text{ g S}}{\text{g coal}}\right)$$
$$= 8.653\frac{\text{g air}}{\text{g of coal}}\text{ or }8.653\frac{\text{lb air}}{\text{lb of coal}}.$$

Coal-fired utilities are often cited as a major contributor to emission. It is important to appreciate the magnitude of coal-fired power plant emissions, which is explored in the next example. ■

EXAMPLE 15.2 ***Coal Power Plant Emissions: Concentration versus Flow Rate***

Using the results from Example 15.1, report CO_2, H_2O, O_2, N_2, SO_2, and particulate emissions for the theoretical air–coal combustion and coal combustion with 20% excess air. Assume the ash in the coal forms particulates and the electrostatic precipitator prior to the stack removes 100% of the particulates. Flue gas desulfurization removes 95% of the exhaust gas SO_2 prior to the stack. Based on plant data, for each mole of SO_2 removed by the flue gas desulfurization unit, 38 mol of water vapor are introduced into the stack exhaust stream (stream 9 in Figure 15.1). In order to deliver 434 MW to the transmission grid, the plant uses 374,455-lb raw coal/h.

Determine the emission concentrations in parts per million and flow rate (pound per hour), both directly after the boiler (stream 4) and at the stack (stream 9). In air emissions, parts per million gas phase is based on volume ratio or mole ratio. These (by volume or by mole) ratios are identical for an ideal gas and are generally taken as identical for emissions at atmospheric pressure. Here then, $y_i \times 106$ = ppm.

SOLUTION The solution is provided in the Excel file **Example 15.1.xls** for the exhaust stream after the boiler (stream 4 in Figure 15.1) and at the stack (stream 9). Both the theoretical air and 20% excess air cases are provided.

Assuming theoretical air and a basis of 1000 g coal/s after the boiler, we have

For the reaction $C + O_2 \rightarrow CO_2$,

$$\left(\frac{637.5 \text{ g carbon}}{12 \text{ g carbon/mol}}\right)\left(\frac{1 \text{ mol } CO_2}{\text{mol C}}\right) = 53.125 \text{ mol } CO_2.$$

For the moisture in the fuel and the hydrogen fuel reaction $2H + ½O_2 \rightarrow H_2O$,

$$\left(\frac{111.2 \text{ g } H_2O}{18 \text{ g } H_2O\text{/mol}}\right) + \left(\frac{45 \text{ g H}}{1 \text{ g H/mol}}\right)\left(\frac{1 \text{ mol } H_2O}{2 \text{ mol H}}\right) = 28.678 \text{ mol } H_2O.$$

For the reaction $S + O_2 \rightarrow SO_2$,

$$\left(\frac{25.1 \text{ g S}}{32 \text{ g S/mol}}\right)\left(\frac{1 \text{ mol } SO_2}{\text{mol S}}\right) = 0.784 \text{ mol } SO_2.$$

For nitrogen in the fuel and air,

$$\left(\frac{12.5 \text{ g N}}{14 \text{ g N/mol}}\right)\left(\frac{1 \text{ mol } N_2}{2 \text{ mol N}}\right) + \left(\frac{8653.2875 \text{ g air}}{28.84 \text{ g/mol air}}\right)\left(\frac{0.79 \text{ mol } N_2}{\text{mol air}}\right)$$
$$= 237.4817 \text{ mol } N_2.$$

Total gas-phase moles after combustion leaving the boiler

$$= 53.125 + 28.678 + 0.784 + 237.482 = 320.069 \text{ mol}$$

Gas-phase concentrations after combustion and leaving the boiler include CO_2 = 53.125/320.069 = 0.166 or 165,979.9 ppm and SO_2 = 0.784/320.069 = 2.450×10^{-3} or 2450 ppm.

The remaining calculations are developed in the Excel file **Example 15.1.xls** and are summarized in Table 15.1. In the Excel file, results are provided in both SI and field units.

Note in Table 15.1 that concentrations (as parts per million) change with the amount of excess air used. In contrast, emission flow rates (e.g., CO_2 and SO_2) remain constant even at different excess air levels. Emission flow rates often provide a better indication of both local and global environmental impact than concentration measurements. In a coal-fired power plant of this size (nominal 434 MW), actual particulate emissions may approach 100 lb/h (with total ESP collection efficiency, 99.3–99.8%) and NO_x emissions would be on the order of 2300 lb/h. ■

Near-term regulations may require CO_2 capture from coal-fired power plants. Cost analysis is often done assuming 90% CO_2 capture. With 90% CO_2 capture for this power plant (Table 15.1), 780,000 lb/h of pipeline quality CO_2 would be produced. We will examine costing of CO_2 capture as the case study of Chapter 16.

15.2 EMISSIONS FROM COMBUSTION EQUILIBRIUM CALCULATIONS

If the combustion process reaches equilibrium, we can calculate the equilibrium composition of the products. Here we

Table 15.1 Coal Power Plant Emissions—Gas-Phase Concentrations and Flow Rates

Species	Stream 4 with Theoretical Air (ppm)	Stream 4 with 20% Excess Air (ppm)	Stream 4 with Theoretical Air (lb/h)	Stream 4 with 20% Excess Air (lb/h)	Stream 9 with 20% Excess Air (lb/h)
CO_2	165,980	139,774	875,289	875,289	875,289
H_2O (total)	89,599	75,452	193,294	193,294	384,148
O_2 (air + fuel)	None	33,156	None	151,003	151,003
N_2 (air + fuel)	741,971	749,554	2,489,934	2,986,984	2,986,984
SO_2	2,451	2,064	18,798	18,798	940
Ash particulates	—	—	37,408	37,408	—
Total gas			3,577,314	4,225,367	4,398,364

will need physical properties of the reactants, products, and key radical species. Physical properties for species of interest in combustion processes are often obtained from the JANAF tables (thermochemical tables database of the National Institute of Standards Technology, see Chase et al., 1985). A convenient approach for computer implementation is provided by Prothero (1969)—here a polynomial expression for a species molar heat capacity, C_P, is fit to data from the JANAF tables. This heat capacity, when combined with the species standard heat of formation from its elements, $\Delta h^0_{f,298.15}$, and standard entropy of formation from its elements, $\Delta s^0_{f,298.15}$, allow determination of the necessary thermodynamic properties.

15.2.1 Equilibrium Reactions

When we combust any hydrocarbon or hydrocarbon mixture, the major reaction will be hydrocarbon + oxygen → carbon dioxide + water. As the combustion temperature increases above 1000°C, dissociation reactions of the gases become important. For methane combustion at high temperature, a simplified reaction set in Table 15.2 has been proposed (adapted from Rodriguez-Toral, 1999 and Gaydon and Wolfhard, 1979) and ξ variables are extents of reaction (mole reacted per second).

The methane combustion reaction 1 (R1) is considered irreversible. The dissociation reactions are reversible and the final product distribution will depend on the combustion temperature, combustion pressure, and feed composition to the combustor. It is possible to confirm that eight reactions are needed. Recall in Chapter 3, Problem 3.22 we established that the number of independent reactions is found as the number of species minus the rank of the atomic matrix (which is generally the number of elements). Here there are 13 species (CH_4, O_2, CO_2, H_2O, CO, N_2, H_2, OH, H, O, NO, Ar[in air], N_2O) and 5 elements (C, O, H, N, Ar), giving 13 − 5 = 8 reactions. As will be discussed in Section 15.3, often several hundred dissociation reactions are used to model combustion reactions; hundreds of reactions are possible as these kinetic equations are not be restricted to equilibrium compositions.

15.2.2 Combustion Chamber Material Balances

For the combustion chamber shown in Figure 15.2, the feed rate and composition (methane + air) are known.

We can assemble material balances (out − in = 0) for each species using the reactions in Table 15.2 as

$$CH_4\!: 0-(N_1y_{CH4,1}-\xi_1)=0 \tag{15.2a}$$

$$O_2\!: N_2y_{O2,2}-(N_1y_{O2,1}-2\xi_1+0.5\xi_2-\xi_3 + 0.5\xi_4-\xi_7-0.5\xi_8)=0 \tag{15.2b}$$

$$CO_2\!: N_2y_{CO2,2}-(\xi_1-\xi_2)=0 \tag{15.2c}$$

$$H_2O\!: N_2y_{H2O,2}-(2\xi_1-\xi_4-\xi_5)=0 \tag{15.2d}$$

$$CO\!: N_2y_{CO,2}-(\xi_2)=0 \tag{15.2e}$$

$$N_2\!: N_2y_{N2,2}-(N_1y_{N2,1}-\xi_3-\xi_8)=0 \tag{15.2f}$$

$$H_2\!: N_2y_{H2,2}-(\xi_4+0.5\xi_5-\xi_6)=0 \tag{15.2g}$$

$$OH\!: N_2y_{OH,2}-(\xi_5)=0 \tag{15.2h}$$

Table 15.2 Simplified Methane Combustion Reaction Set (Adapted from Gaydon and Wolfhard, 1979; Rodriguez-Toral, 1999; Rodriguez-Toral et al., 2000)

Reaction Number	Reaction	Reaction Extent	Reaction Extent Units
R1	$CH_4 + 2O_2 \rightarrow CO_2 + 2H_2O$	ξ_1	mol CH_4/s
R2	$CO_2 \leftrightarrow CO + \frac{1}{2}O_2$	ξ_2	mol CO_2/s
R3	$N_2 + O_2 \leftrightarrow 2NO$	ξ_3	mol N_2/s
R4	$H_2O \leftrightarrow H_2 + \frac{1}{2}O_2$	ξ_4	mol H_2O/s
R5	$H_2O \leftrightarrow \frac{1}{2}H_2 + OH$	ξ_5	mol H_2O/s
R6	$H_2 \leftrightarrow 2H$	ξ_6	mol H_2/s
R7	$O_2 \leftrightarrow 2O$	ξ_7	mol O_2/s
R8	$N_2 + \frac{1}{2}O_2 \leftrightarrow N_2O$	ξ_8	mol N_2/s

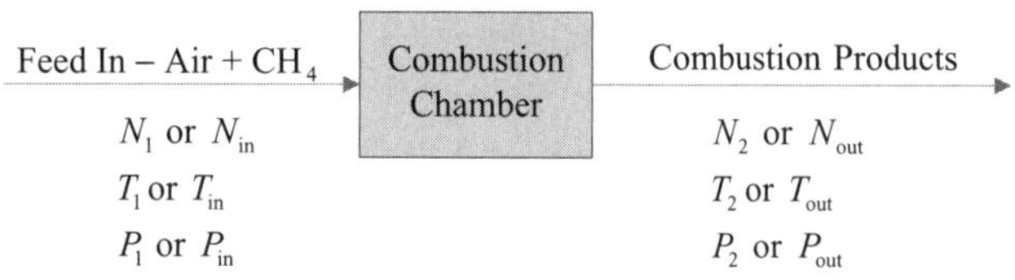

Figure 15.2 Combustion chamber with system nomenclature.

$$\text{H: } N_2 y_{\text{H},2} - (2\xi_6) = 0 \quad (15.2\text{i})$$

$$\text{O: } N_2 y_{\text{O},2} - (2\xi_7) = 0 \quad (15.2\text{j})$$

$$\text{NO: } N_2 y_{\text{NO},2} - (2\xi_3) = 0 \quad (15.2\text{k})$$

$$\text{Ar: } N_2 y_{\text{Ar},2} - (N_1 y_{\text{Ar},1}) = 0 \quad (15.2\text{l})$$

$$\text{N}_2\text{O: } N_2 y_{\text{N2O},2} - (\xi_8) = 0. \quad (15.2\text{m})$$

Here $y_{i,j}$ is the mole fraction of species i in stream j and N_j is the molar flow rate of stream j. We know that $\Sigma y_i = 1.0$, which for the combustion products provides

$$1 - (y_{O2} + y_{\text{CO2}} + y_{\text{H2O}} + y_{\text{CO}} + y_{\text{N2}} + y_{\text{H2}} + y_{\text{OH}} + y_{\text{H}} + y_{\text{O}} + y_{\text{NO}} + y_{\text{Ar}} + y_{\text{N2O}}) = 0. \quad (15.3)$$

Here we have dropped the subscripts on the outlet stream—recall we assume the feed composition and molar flow rate are known.

We can count the number of unknowns as the 13 possible product mole fractions, the exiting total molar flow rate, and the 8 extents of reaction—for a total of 22 unknowns. Above are assembled 14 equations; these equations are the 13 material balances and the $\Sigma y_i = 1.0$. We can quickly reduce the problem from 22 unknowns to 20 unknowns by noting that in Equation (15.2a) both N_1 and $y_{\text{CH4},1}$ are known, giving $\xi_1 = N_1 y_{\text{CH4},1}$, and with complete combustion, $y_{\text{CH4},2} = 0$. The problem then becomes 20 unknowns with 13 equations. The unknowns and variables ultimately used in the Visual Basic for Applications (VBA)-based solution approach (the VBA code is discussed in Example 15.5) are provided in Table 15.3.

Table 15.3 Combustion Calculation Unknowns and VBA Variable Assignments

Unknowns	Variable in VBA Code
N_2	$x(0)$
y_{O2}	$x(1)$
y_{CO2}	$x(2)$
y_{H2O}	$x(3)$
y_{CO}	$x(4)$
y_{N2}	$x(5)$
y_{H2}	$x(6)$
y_{OH}	$x(7)$
y_{H}	$x(8)$
y_{O}	$x(9)$
y_{NO}	$x(10)$
y_{Ar}	$x(11)$
y_{N2O}	$x(12)$
ξ_2	$x(13)$
ξ_3	$x(14)$
ξ_4	$x(15)$
ξ_5	$x(16)$
ξ_6	$x(17)$
ξ_7	$x(18)$
ξ_8	$x(19)$

15.2.3 Equilibrium Relations for Gas-Phase Reactions/Gas-Phase Combustors

The remaining seven equations needed to solve for the molar flow rates, species compositions, and extents of reaction come from equilibrium relationships—here it is assumed the seven dissociation reaction (R2–R8) are at equilibrium. In developing the equilibrium relations, we will try and be complete, but along the way, some of the equations and terms may be new to some readers (for additional background, see de Nevers, 2002).

For a reacting system, the Gibbs fundamental form for the molar free energy, G, can be expressed as

$$dG = -s dT + \upsilon dP + \sum \mu_i dn_i, \quad (15.4)$$

where s is the molar entropy, υ is the molar volume, μ_i is the chemical potential of species i (μ_i is defined next), and dn_i is the change in the moles of species i due to reaction. At equilibrium, with constant temperature and pressure, one can write

$$dG = \sum \mu_i dn_i. \quad (15.5)$$

Ultimately, we will develop equilibrium relations for the seven dissociation reactions (R2)–(R8).

For now, let us *consider just a single reaction*, say, reaction 2 (R2) in Table 15.2:

$$\text{CO}_2 \leftrightarrow \text{CO} + \frac{1}{2}\text{O}_2.$$

We can write (R2) as (products – reactants = 0):

$$\mathrm{CO}+\frac{1}{2}\mathrm{O}_2-\mathrm{CO}_2=0, \tag{15.6}$$

and if we number each species in (R2), $s_1 = CO_2$, $s_2 = CO$, $s_3 = O_2$, we can write Equation (15.6) as

$$\nu_1 s_1+\nu_2 s_2+\nu_3 s_3=0 \tag{15.7}$$

with $\nu_1 = -1$, $\nu_2 = 1$, and $\nu_3 = 1/2$.

The accepted convention in Equation (15.7) (with the right-hand side [RHS] = 0) is that the stoichiometric coefficients ν_i are positive for products and negative for reactants. Using our definition for the extent of reaction ξ, we can write

$$\frac{dn_1}{\nu_1}=\frac{dn_2}{\nu_2}=\frac{dn_3}{\nu_3}=d\xi, \tag{15.8}$$

where dn_1 is the change in the moles of s_1. For any species i, we can write

$$dn_i=\nu_i d\xi. \tag{15.9}$$

Using Equation (15.9) in Equation (15.5), we obtain

$$dG=\sum \mu_i dn_i=\sum \mu_i \nu_i d\xi \tag{15.10}$$

and

$$\frac{dG}{d\xi}=\sum \mu_i \nu_i. \tag{15.11}$$

For equilibrium, the extent of reaction (or simply the reaction) proceeds until the Gibbs free energy is a minimum; mathematically, $dG/d\xi = 0$, giving

$$\frac{dG}{d\xi}=0=\sum \mu_i \nu_i. \tag{15.12}$$

Equation (15.12) simply states that at equilibrium, the sum of the chemical potentials of the products must equal the sum of the chemical potential of the reactants. In Equation (15.12), we are considering a single reaction; there will be a unique extent of reaction associated with each reaction. The chemical potential for reactions is analogous to a voltage potential or voltage difference in electricity—a voltage potential is needed in order for electrons to flow.

Direct calculation of numerical values for the chemical potential is impossible. But the chemical potential μ_i can be written in terms of fugacity (Lewis) as

$$\mu_i=G_i^0+RT\ln\left(\frac{f_i}{f_i^0}\right), \tag{15.13}$$

where in Equation (15.13), as well as in subsequent equations in this section, the superscript ^{0} indicates evaluation at the standard state reference pressure, which here is 1 standard atmosphere; f_i is the fugacity of species i (f_i is defined next); f_i^0 is the standard state fugacity; and G_i^0 is the standard molar free energy of pure species i. We will see G_i^0 is a function only of temperature and the standard state fugacity $f_i^0 = 1$ atm. There can be significant differences in equilibrium calculations using 1 atm or 1 bar as the reference pressure. We do need to use the same units for both f_i and f_i^0 are used to ensure the ln ( ) is dimensionless. Using Equation (15.13) in Equation (15.12),

$$\sum(\nu_i)\left(G_i^0+RT\ln\left(\frac{f_i}{f_i^0}\right)\right)=0 \tag{15.14}$$

and

$$-RT\ln\prod_i\left(\frac{f_i}{f_i^0}\right)^{\nu_i}=\sum \nu_i G_i^0. \tag{15.15}$$

For gas-phase equilibrium calculations, the fugacity f_i is related to the partial pressure p_i ($p_i = y_iP$) by

$$f_i=\phi_i y_i P, \tag{15.16}$$

where ϕ_i is the fugacity coefficient for species i. For ideal gas mixtures, $\phi_i = 1$, and for nonideal systems, ϕ_i can be calculated from an equation of state for the mixture as we did in Chapter 8 (see Problems 8.5 and 8.6). For combustion processes of industrial interest (gas turbines, furnaces, etc.), the pressure is usually low enough to allow $\phi_i \cong 1$. For high-pressure systems, for example, our ammonia process of Chapter 5, $\phi_i \neq 1$, and an equation of state for the gas mixture must be used to determine ϕ_i.

At this point, it is convenient to define the Π term (the product term) in Equation (15.15) as a chemical equilibrium constant, K. To simplify the appearance of K, f_i^0 is omitted from the denominator ($f_i^0 = 1$ atm [or 1 bar]) giving

$$K=\prod_i(f_i[\text{atm}])^{\nu_i}=\prod_i(\phi_i y_i P[\text{atm}])^{\nu_i}=\prod_i(\phi_i y_i)^{\nu_i}P^{\Delta n}, \tag{15.17}$$

where here the change in moles $\Delta n = \nu_1 + \nu_2 + \nu_3$. For low-pressure systems of interest in industrial combustion processes, $\phi_i = 1$ (ideal gas) and the equilibrium constant can be written as

$$K = \prod_i (y_i)^{\nu_i} P^{\Delta n}. \quad (15.18)$$

The K for reaction 2, using Equation (15.18), can be written as

$$K_{R2} = \left(\frac{(y_{CO})^1 (y_{O_2})^{\frac{1}{2}}}{(y_{CO_2})^1} \right) P^{\frac{1}{2}}, \quad (15.19)$$

where P is in atmosphere.

Combining our definition of K and Equation (15.15), we obtain

$$-RT \ln K = \sum \nu_i G_i^0. \quad (15.20)$$

Our last remaining problem is determining G_i^0. By definition,

$$G_i^0 \equiv h_i^0 - Ts_i^0, \quad (15.21)$$

where G_i^0 is the molar free energy of species i at temperature T and at the standard state reference pressure of 1 atm (or 1 bar); h_i^0 is the molar enthalpy of species i; and s_i^0 is the molar entropy of species i, both at T and 1 atm. It is often useful to keep track of temperature, which, when needed, can be added as a subscript:

$$G_{i,T}^0 \equiv h_{i,T}^0 - Ts_{i,T}^0. \quad (15.22)$$

We can then write

$$h_i^0 = h_{i,T}^0 = \Delta h_{fi,T_0}^0 + \int_{T_0}^{T} C_{P,i} dT \quad (15.23)$$

and

$$s_i^0 = s_{i,T}^0 = \Delta s_{fi,T_0}^0 + \int_{T_0}^{T} \frac{C_{P,i}}{T} dT - R \ln\left(\frac{P}{P_0}\right).$$

Here with the requirement $P = P_0 = 1$ atm,

$$s_i^0 = s_{i,T}^0 = \Delta s_{fi,T_0}^0 + \int_{T_0}^{T} \frac{C_{P,i}}{T} dT, \quad (15.24)$$

where $\Delta h_{fi,T_0}^0$ and $\Delta s_{fi,T_0}^0$, respectively, are the species standard molar enthalpy of formation from its elements and the standard molar entropy of formation from its elements at the reference temperature T_0 ($T_0 = 298.18$ K); $C_{p,i}$ is the molar heat capacity of each species i, and the integration limits are T_0 to T. We can use Equations (15.23) and (15.24) to calculate G_i^0, giving us K from Equation (15.20). Equations (15.23) and (15.24) show that G_i^0 is only a function of temperature at the standard state reference pressure.

Before we continue, let us recall the statement made in the first paragraph of this section: "This heat capacity, when combined with the species standard heat of formation from its elements, $\Delta h_{f,298.15}^0$, and the standard entropy of formation from its elements, $\Delta s_{f,298.15}^0$, allow determination of the necessary thermodynamic properties." We can see now that the needed thermodynamic properties are h_i^0 and s_i^0.

The JANAF table fit of the molar heat capacity, at constant pressure, for a species i is provided by Prothero (1969)

$$C_{P,i} = a_1 + a_2 x + a_3 x^2 + a_4 x^3 + \ldots + a_7 x^6 \quad (0.3 \le x \le 2). \quad (15.25)$$

In Equation (15.25), $x = 10^{-3}\,T$, and for the coefficients, the temperature range 300–2000 K is appropriate for gas turbine and other industrial combustion applications. The coefficients for selected species of importance in combustion processes and their standard enthalpy of formation and entropy of formation are provided in Table 15.4.

EXAMPLE 15.3 *Integrate the Constant-Pressure Molar Heat Capacity Term $C_{p,i}$*

Show the integration for the heat capacity term (Eq. (15.25)) in the molar enthalpy (Eq. (15.23)) and entropy (Eq. (15.24)) terms.

SOLUTION

$$h_i^0 = h_{i,T}^0 = \Delta h_{f,298.15}^0 + \int_{T_o}^{T} C_{P,i}(T)\, dT$$

As $x = 10^{-3}\,T$, we can write $dx = 10^{-3}\,dT$ and

$$\int_{T_o}^{T} C_{P,i}(T)\, dT = \int_{x_o}^{x} \frac{C_{P,i}(x)}{0.001} dx$$

$$\int_{x_o}^{x} \frac{C_{P,i}(x)}{0.001} dx = \frac{a_1}{0.001}(x - x_0) + \frac{a_2}{(2)(0.001)}(x^2 - x_0^2) + \frac{a_3}{(3)(0.001)}(x^3 - x_0^3) + \ldots + \frac{a_7}{(7)(0.001)}(x^7 - x_0^7) \quad (15.26)$$

and

$$s_i^0 = s_{i,T}^0 = \Delta s_{f,298.15}^0 + \int_{T_o}^{T} \frac{C_{P,i}(T)}{T} dT \qquad \int_{T_o}^{T} \frac{C_{P,i}(T)}{T} dT = \int_{x_o}^{x} \frac{C_{P,i}(x)}{x} dx$$

$$\int_{x_o}^{x} \frac{C_{P,i}(x)}{x} dx = a_1 \ln\left(\frac{x}{x_0}\right) + \frac{a_2}{1}(x - x_0) + \frac{a_3}{2}(x^2 - x_0^2) + \cdots + \frac{a_7}{6}(x^6 - x_0^6) \quad (15.27)$$

■

Table 15.4 Species C_P Coefficients (for Eq. (15.25)), $\Delta h^0_{f,298.15}$ and $\Delta s^0_{f,298.15}$ (Prothero, 1969)

	a_1	a_2	a_3	a_4	a_5	a_6	a_7	$\Delta h^0_{f,298.15}$ (cal/mol)	$\Delta s^0_{f,298.15}$ (cal/mol-K)
CH_4	7.918404	−11.41722	63.73457	−75.25691	43.29269	−12.56732	1.469695	−17,895	44.49
C_2H_6	2.387968408	34.57751611	2.497260449	−18.36373457	10.38081242	−2.320482071	0.171401034	−20,033.4608	54.9713
C_3H_8	6.8008	28.71	62.349	−107.19	69.802	−21.101	2.4476	−24,891	64.355
n-C_4H_{10}	−39.772	412.9	−991.74	1325	−926.75	317.38	−42.031	−30,183	74.045
n-C_5H_{12}	−2.857971878	128.8347906	−90.84176048	48.23747357	−21.97105629	6.792072625	−0.922120358	−35,076.48184	83.1739
N_2	7.709928	−5.503897	13.12136	−11.67955	5.233997	−1.173185	0.103883	0	45.77
O_2	7.361141	−5.369589	20.54179	−25.86526	15.94566	−4.85889	0.5861501	0	49.004
CO_2	4.324933	20.80895	−22.9459	16.84483	−7.935665	2.121672	−0.2408713	−94,054	51.072
H_2O	7.98886	−1.506271	6.661376	−4.65597	1.696464	−0.3706212	0.03992444	−57,798	45.106
CO	7.812249	−6.668293	17.28296	−17.28709	8.860125	−2.314819	0.2447785	−26,417	47.214
NO	8.462334	−10.40669	27.54876	−30.28119	17.18511	−4.95726	0.5755281	21,580	50.347
H_2	6.183042	4.710657	−10.92135	12.54086	−7.016263	1.923395	−0.2084091	0	31.208
OH	7.6151	−1.936	0.877	2.6153	−2.6909	0.97789	−0.12695	9,432	43.88
H	4.968	0	0	0	0	0	0	52,100	27.392
O	5.974134	−4.241883	7.931254	−7.94423	4.403357	−1.271341	0.1491408	59,559	38.468
Ar	4.9681	0	0	0	0	0	0	0	36.983
N_2O	4.826714	20.13927	−22.13612	15.85518	−7.265313	1.897833	−0.2117446	19,610	52.546

EXAMPLE 15.4 ***Calculate the Equilibrium Constant for R2 Table 15.2, K_{R2}, at 1500 K***

Here we use the form for reaction 2 as provided in Equation (15.6):

$$CO + \frac{1}{2}O_2 - CO_2 = 0.$$

The data in Table 15.4 has been assembled in the Excel sheet in **Example 15.4a.xls**; you can rename this file and use it as the starting point to solve this example.

SOLUTION The solution is provided in **Example 15.4b.xls**. Here, with $T = 1500$ K, h_i^0 is determined from Equation (15.23), s_i^0 from Equation (15.24), and G_i^0 from Equation (15.22). The results are the following:

Species, i	h_i^0 (cal/mol)	s_i^0 (cal/mol-K)	G_i^0 (cal/mol)
CO_2	−79,305.22218	69.81219035	−184,023.5077
CO	−17,132.14205	59.34749256	−106,153.3809
O_2	9,705.187754	61.65525777	−82,777.69891

Next, we can solve for $\sum \nu_i G_i^0$:

$$\sum \nu_i G_i^0 = (-1)(G_{CO2}^0) + (1)(G_{CO}^0) + \left(\frac{1}{2}\right)(G_{O2}^0)$$
$$= 36{,}481.27737 \text{ cal/mol}$$

Finally, using Equation (15.20), $-RT\ln K = \sum \nu_i G_i^0$,

$$\ln K_{R2} = -\frac{\sum \nu_i G_i^0}{RT} = -\frac{\sum \nu_i G_i^0}{\left(1.987\dfrac{\text{cal}}{\text{mol-K}}\right)(1500\text{ K})} = -12.2399857$$

or $K_{R2} = 4.83328\text{E-}06$.

There are several important points to emphasize before we leave Example 15.4. Note that the system pressure used to calculate the equilibrium constant, K_{R2}, is $P = 1$ atm. Also, we did not need to know the feed composition or the equilibrium composition to calculate the equilibrium constant. To calculate any K_R, we only need the equilibrium reaction temperature, the reaction stoichiometry, and the thermodynamic properties as provided by Table 15.4. We also again want to comment on significant figures as we did in Chapter 1—we are reporting extra significant figures to allow the reader to quickly match the computer-generated solutions in the provided Excel files.

The careful reader may be concerned with our entropy term—recall in Chapter 8 we discussed entropy of mixing; there is no entropy of mixing used here (in Eq. (15.21) or (15.24)). In the calculations of Example 15.4, the chemical species are in stoichiometric proportion and each species is at the standard state reference pressure of 1 atm. The species are not at their partial pressure in a mixture; each species is separate and each is at the reference pressure.

Finally, the order of the reaction equation will influence the value of K_R, but ultimately, this will not affect the calculated equilibrium composition. For example, if R2 had been written in reverse,

$$CO + \frac{1}{2}O_2 \leftrightarrow CO_2,$$

then we would find for 1500 K

$$K = \frac{1}{4.83328\text{E-}06} = 206{,}898.83.$$

Also, if we had written R2 as $2CO_2 \leftrightarrow 2CO + O_2$, a different value for K_{R2} would be obtained ($K_{R2} = 2.33606\text{E-}11$). ■

15.2.4 Equilibrium Compositions from Equilibrium Constants

We can use equilibrium constants to determine the system composition at equilibrium. Here we use Equation (15.20) with the definition of K given by Equation (15.18). Using these equations, we can write the seven equilibrium relationships for R2–R8 in Table 15.2 as

$$\ln[K_{R2}] - \ln(y_{CO,2}) - \frac{1}{2}\ln(y_{O2,2}) + \ln(y_{CO2,2}) - \frac{1}{2}\ln(P_2) = 0, \quad (15.28)$$

$$\ln[K_{R3}] - 2\ln(y_{NO,2}) + \ln(y_{N2,2}) + \ln(y_{O2,2}) = 0, \quad (15.29)$$

$$\ln[K_{R4}] - \ln(y_{H2,2}) - \frac{1}{2}\ln(y_{O2,2}) + \ln(y_{H2O,2}) - \frac{1}{2}\ln(P_2) = 0, \quad (15.30)$$

$$\ln[K_{R5}] - \frac{1}{2}\ln(y_{H2,2}) - \ln(y_{OH,2}) + \ln(y_{H2O,2}) - \frac{1}{2}\ln(P_2) = 0, \quad (15.31)$$

$$\ln[K_{R6}] - 2\ln(y_{H,2}) + \ln(y_{H2,2}) - \ln(P_2) = 0, \quad (15.32)$$

$$\ln[K_{R7}] - 2\ln(y_{O,2}) + \ln(y_{O2,2}) - \ln(P_2) = 0, \quad (15.33)$$

and

$$\ln[K_{R8}] - \ln(y_{N2O,2}) + \ln(y_{N2,2}) + \frac{1}{2}\ln(y_{O2,2}) + \frac{1}{2}\ln(P_2) = 0. \quad (15.34)$$

Each K_R is determined at T_2, and we have indicated (P_2) and ($y_{i,2}$) to emphasize that we will calculate the equilibrium composition at condition 2, the combustion temperature and pressure. We can calculate each of the K_R values following the same process we used in Example 15.4. In fact, as part of the Excel sheet for **Example 15.4b.xls**, we calculated all seven K_R values.

With each K_R value determined, we now have 20 unknowns and 20 equations—the 12 material balances, the $\Sigma y_{i,2} = 1.0$, and the seven equilibrium relationships Equations (15.28)–(15.34).

EXAMPLE 15.5 *Equilibrium Composition: Simplified Methane Combustion Reaction Set*

In Example 14.4 (Chapter 14), we solved the energy balance assuming complete methane combustion with no dissociation reactions. The combustion chamber energy balance utilized the fuel lower heating value and a 2% combustion chamber energy loss was assumed. With a 5% pressure drop, results showed $T_{out} = 2011°F$ and $P_{out} = 4.197$ bar as indicated in Figure 15.3.

For Example 15.5, let us determine the equilibrium composition for a gas turbine combustion chamber with the same inlet fuel and air flow rates as in Example 14.4, but now with the methane reaction kinetics provided in Table 15.2. As indicated in Figure 15.4, we will simplify the problem slightly by neglecting the 2% combustion energy losses assumed in Example 14.4.

You will immediately notice in Figure 15.4 that the outlet temperature (T_{out} in our equilibrium relations) is not known—we must account for the dissociation reactions in order to determine T_{out}.

SOLUTION If you look ahead to Table 15.7, you will see that we can obtain results that are virtually identical to those found using the commercial code Chemkin-Pro. This problem is solved in three Excel files: (1) **Example 15.5a.xls**, (2) **Example 15.5b.xls**, and (3) **Example 15.5c.xls**. The use of three separate Excel files allows us to separate different calculation procedures but makes the calculations more time-consuming as information has to be moved between the files. As a quick overview, the *first* Excel file determines the outlet composition and temperature for complete combustion based solely on reaction R1 with this result serving as an initial guess to the solution; the *second* Excel file calculates needed equilibrium properties at any given temperature; finally, the *third* Excel file solves the material balance problem using the Newton–Raphson technique we developed in Chapter 4. The energy balance is solved after the material balance. As the temperature is varied to close the energy balance, the second Excel file must be used with each new temperature to determine equilibrium properties, and then these values must be copied to the third file to solve the nonlinear material balance problem.

Step 1: In the first Excel file, **Example 15.5a.xls**, we convert the mass flows of fuel and air to the combustor to molar flows and mole fractions for each species. For these calculations, the stream entering the combustor is noted as stream 1 (or in) and the stream leaving the combustor is stream 2 (or out). Based on the methane combustion reaction (R1), we calculated molar flows and species mole fractions for the combustion products shown in Figure 15.5.

An overview of the calculations preformed in **Example 15.5a.xls** is provided here:

Overview of the Excel File: **Example 15.5a.xls**
Convert the feed to the combustor in SI units and calculate $y_{i,1}$ values.
Determine products from the combustor, N_2 and $y_{i,2}$, assuming only reaction R1.

Step 2: In the second Excel file, **Example 15.5b.xls**, we first solve the energy balance around the combustor. As T_{in} is known, the enthalpy per mole of each species i can be determined using Equation (15.23). The species molar flow rates

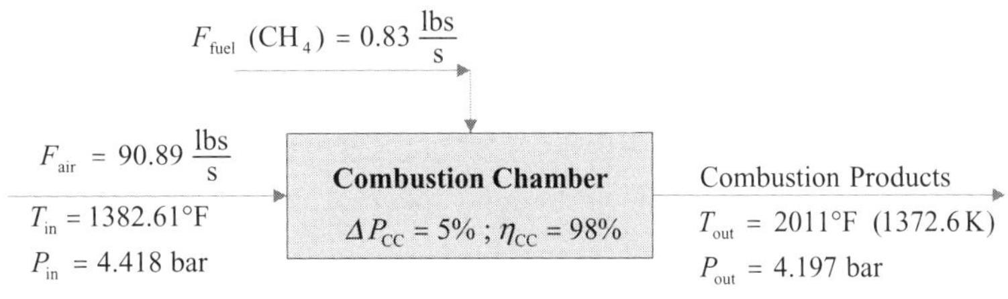

Figure 15.3 Combustion chamber results from Example 14.4.

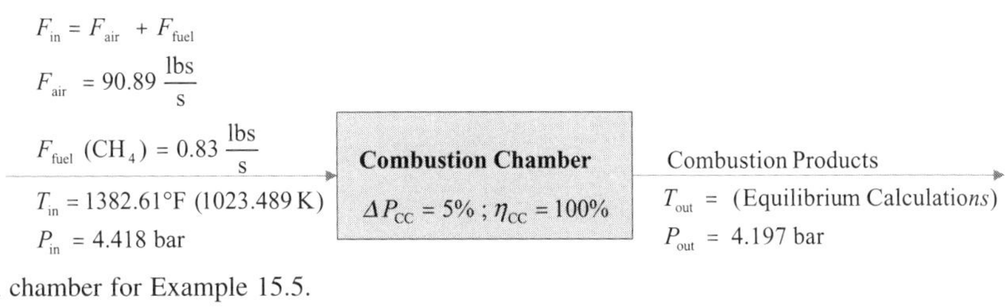

Figure 15.4 Combustion chamber for Example 15.5.

	A	B	C	D	E	F	G
25		In				Out Estimate	
26		in (g-mol/s)	yi		CH4+2O2=CO2+2H2O	out (g-mol/s) Estimate	yi Estimate
27	N Total In	1446.910344			ξ1 (g-mol/s)	23.46695211	
28	CH4 In	23.46695211	0.016218664		CH4 Out	0	0
29	N2 In	1111.566945	0.768234846		N2 Out	1111.566945	0.768234846
30	O2 In	298.2113906	0.20610219		O2 Out	251.2774864	0.173664863
31	Ar In	13.66505656	0.009444301		Ar Out	13.66505656	0.009444301
32	Sum Check		1		CO2 Out	23.46695211	0.016218664
33					H2O	46.93390421	0.032437327
34					N Total Out (Estimate)	1446.910344	

Figure 15.5 Composition leaving combustor based only on reaction R1.

Table 15.5 T_{out} Based on Reaction R1 and Species T_{out} Enthalpy Properties

	T_{in} Combustor	T_{out} Combustor
T initial (K)	298.15	298.15
T final (K)	1023.489	1398.227624
	h_i^0 (cal/mol)	h_i^0 (cal/mol)
CH_4	−8361.441035	−1291.518648
C_2H_6	−3947.425979	8056.010217
C3H8	−1641.972637	15,378.68268
n-C_4H_{10}	58.30174412	22,275.9065
n-C_5H_{12}	2299.425646	29,625.76685
N2	5,312.61292	8334.939924
O_2	5623.16849	8819.598739
CO_2	−85,765.7591	−80,718.49382
H_2O	−51,355.63511	−47,433.38136
CO	−21,047.13847	−17,985.15816
NO	27,084.273	30,208.79475
H_2	5112.248866	7888.859356
OH	14,608.26757	17,441.3838
H	55,703.48415	57,565.18564
O	63,228.45629	65,098.38943
Ar	3603.556686	5465.295644
N_2O	28,066.2165	33,140.57866

are known, allowing the energy rate in to be determined. From step 1 and based on R1, we know the outlet molar flow rates. With an assumed value for T_{out}, we can use Equation (15.23) to solve for the enthalpy of the combustion products. Next, a check if the energy rate in = the energy rate out (based on R1) can be made: $\Delta H_{R1} = \left(\sum y_{i,2} N_2 h^0_{i,T_{out}} - \sum y_{i,1} N_1 h^0_{i,T_{in}}\right) = 0$. T_{out} can be varied using Goal Seek until the energy balance closes. Table 15.5 provides the $h^0_{i,T_{in}}$ and $h^0_{i,T_{out}}$ values, which close the energy balance; $T_{out} = 1398.2$ K is the combustor chamber temperature considering methane combustion (R1) only.

This T_{out} represents a good initial temperature estimate for use with the dissociation reactions R2–R8. Using this T_{out}, we can calculate $h^0_{i,T_{out}}$, $s^0_{i,T_{out}}$, and $G^0_{i,T_{out}}$ for each species and the R2–R8 equilibrium constants, as we did in Example 15.4 and in the Excel file **Example 15.4b.xls**. These results are provided in the Excel file **Example 15.5b.xls** and shown in Table 15.6.

Table 15.6 Equilibrium Constants Based on T_{out} Value Determined from Reaction R1

	$\ln K_R = -\dfrac{\sum \nu_i G_i^0}{RT}$	K_R	Reaction
$CO_2 = CO + .5O_2$	−13.87829265	9.39148E-07	R2
$N_2 + O_2 = 2NO$	−12.51010418	3.68919E-06	R3
$H_2O = H_2 + .5O_2$	−14.64628467	4.35712E-07	R4
$H_2O = 0.5H_2 + OH$	−16.20224791	9.19291E-08	R5
$H_2 = 2H$	−24.52715461	2.22839E-11	R6
$O_2 = 2O$	−27.81465973	8.32237E-13	R7
$N_2 + 0.5O_2 = N_2O$	−15.93433341	1.20173E-07	R8

You will notice enthalpy values for some species (e.g., C_2H_6 and C_3H_8) that are not present in the problem are also calculated. It is easy to make a mistake copying the a_i values from Table 15.4 to the Excel sheet, so we have provided all these values on the sheet.

An overview of the calculations preformed in **Example 15.5b.xls** is provided here:

Overview of the Excel File: Example 15.5b.xls

Bring: N_1, $y_{i,1}$, N_2, $y_{i,2}$ values from **Example 15.5a.xls**. Temperature into combustor, T_{in}, is known.

Using data from Table 15.4 and Equation (15.23), calculate $h^0_{i,T_{in}}$ for each species i.

↓

Guess T_{out} from the combustor.

Using data from Table 15.4,calculate $h^0_{i,T_{out}}$ for each species i.

Iterations

Here, using only R1, check if $\Delta H_{R1} = \left(\sum y_{i,2} N_2 h^0_{i,T_{out}} - \sum y_{i,1} N_1 h^0_{i,T_{in}}\right) = 0$.

Using Goal Seek, adjust T_{out} until $\Delta H_{R1} = 0$.

↓

With converged T_{out}

At T_{out}, calculate all properties, $h^0_{i,T_{out}}$, $s^0_{i,T_{out}}$, $G^0_{i,T_{out}}$, for each species i and calculate ln (K) and K at T_{out} for reactions R2–R8.

At this point, we determined an estimate of the outlet composition (step 1) as well as an estimate of the outlet combustor temperature (step 2). These estimates are based solely on reaction (R1). But we can use the estimated outlet temperature to determine equilibrium constants for all the remaining dissociation reactions (step 2).

Before we move to the next Excel sheet, it is important to appreciate that in **Example 15.5b.xls**, we can supply *any* temperature (in cell B23) and find the resulting equilibrium constants for R2–R8 as well as the enthalpy of each reacting species. These data (the equilibrium constants and enthalpies at the selected T_{out}) will need to be supplied (via copy and paste) to the third Excel file.

Step 3: In the third Excel file, **Example 15.5c.xls**, we must first supply equilibrium constants for R2–R8 and species enthalpies at the selected T_{out}. This is done by using copy and paste from **Example 15.5b.xls** (cells B27–B52) to **Example 15.5c.xls** (cells G10–G35). In **Example 15.5c.xls**, we assemble the 20 unknowns (Table 15.3) and 20 equations—the 12 material balances, the $\Sigma y_{i,2} = 1.0$, and the seven equilibrium relationships (Eqs. (15.28)–(15.34)). In Equations (15.28–15.34), we account for the pressure P_{out}. We treat these as 20 nonlinear equations ($\Sigma y_{i,2} = 1.0$ is actually linear) and use the solution procedure we developed in Chapter 4—nonlinear Newton–Raphson. Recall the Newton–Raphson method we developed utilizes three matrices: The first matrix holds the starting guess, the second matrix holds the solution at each iteration and the third matrix contains the partial derivative terms evaluated at the current solution.

In general, for any chosen T_{out}, it takes 5–15 Newton–Raphson iterations to converge the nonlinear equation set. The energy balance is also solved as part of the VBA iteration code and reported at each step on the Excel sheet as $H_{out} - H_{in}$ (cell S85). Here we do want to be clear the energy balance can be solved for any converged species molar flows, but there will only be one unique T_{out} value such that the $H_{out} - H_{in} = 0$ (an adiabatic process). Actually, when using the converged species molar flow rates, we are looking for the T_{out} such that

$$\left|\frac{(H_{out} - H_{in})}{\text{mol fuel in}}\right| < 5\frac{\text{cal}}{\text{mol fuel}}.$$

If

$$\left|\frac{(H_{out} - H_{in})}{\text{mol fuel in}}\right| > 5\frac{\text{cal}}{\text{mol fuel}},$$

a new value for T_{out} is selected by trial and error. The energy balance will converge at $T_{out} = 1396.4$ K, with

$$\frac{(H_{out} - H_{in})}{\text{mol fuel in}} = -1.233\frac{\text{cal}}{\text{mol fuel}}.$$

The exiting combustion composition and extents of reaction are given in Table 15.7. This converged value for T_{out} is only 1.83 K from the initial guess, which was based solely on using R1 (step 2).

An overview of the calculations preformed in **Example 15.5c.xls** is provided here.

Overview of the Excel File: Example 15.5c.xls

Bring N_1, $y_{i,1}$, $h^0_{i,T_{in}}$, $h^0_{i,T_{out}}$, $s^0_{i,T_{out}}$, $G^0_{i,T_{out}}$; ln(K) and K at T_{out} from **Example 15.5b.xls**.

Assemble all material balances and all equilibrium equations (equations = unknowns).

Supply initial guess for all variables → all variables in Table 15.3.

Solve the equation set using Newton–Raphson elimination (Chapter 5) → obtain new values for variables in Table 15.3.

Check if

$$\Delta H_{\text{Combustion}} = \frac{\left(\Sigma y_{i,2} N_2 h^0_{i,T_{out}} - \Sigma y_{i,1} N_1 h^0_{i,T_{in}}\right)}{N_{\text{fuel\_in}}} < 5\frac{\text{cal}}{\text{mol fuel}}.$$

If $\Delta H_{\text{Combustion}} > 5\frac{\text{cal}}{\text{mol fuel}}$, then iteration

Iteration:

Guess new T_{out} and use this T_{out} in **Example 15.5b.xls** to get $h^0_{i,T_{out}}$, $s^0_{i,T_{out}}$, and $G^0_{i,T_{out}}$; ln (K) and K at new T_{out} for R2–R8.

Use Paste Special—Values to copy values to **Example 15.5c.xls** from **Example 15.5b.xls**.

Solve nonlinear equation set (Newton–Raphson method) in **Example 15.5c.xls**.

Check if $\Delta H_{\text{Combustion}} < 5\frac{\text{cal}}{\text{mol fuel}}$ then done, otherwise

At this point, you may be curious as to why we did not also make T_{out} a variable and then add the overall energy balance equation to the equation set (giving 21 unknowns and 21 equations). This addition would avoid the trial-and-error process to find T_{out}. However, the overall energy balance is highly nonlinear, making the trial-and-error process to find T_{out} more reliable. We may need a more sophisticated Newton–Raphson method to solve the combined material and energy balance problem, as will be discussed in the next section.

Table 15.7 provides the equilibrium composition leaving the gas turbine, when accounting for reactions R1–R8. From Table 15.7, 693.59 ppm NO and 0.018 ppm CO (for gases, $y_i \times 10^6$ = ppm) are expected to leave the gas turbine. However, NO_x concentrations from a methane-fired turbine typically fall in the range 60–100 ppm. Generally, equilibrium calculations overestimate NO_x concentration and underestimate CO concentration.

Also in Table 15.7, we show equilibrium calculation results using the commercial program Chemkin-Pro. Chemkin-Pro can also be used to predicted combustion emissions as will be detailed in the next section. The Chemkin-Pro equilibrium results are virtually identical with our equilibrium calculations. Notice that Chemkin-Pro does not require a feed rate—any basis can be used to solve the energy balance. ■

Table 15.7 Species Composition Leaving Combustor Using R1–R8 (Table 15.2) and Chemkin-Pro

Variables/Parameters	Final Values	Results from Chemkin
N_2 (g-mol/s)	1446.914568	–
T_{in} (K)	1023.489	1023.489
T_{out} (K)	1396.4	1396.4
P_{out} (bar)	4.197	4.197
y_{CH4}	0.0	3.68940E-39
y_{O2}	0.173314601	0.1733100
y_{CO2}	0.016218598	0.01621900
y_{H2O}	0.032431531	0.0324320
y_{CO}	1.81583E-08	1.77210E-08
y_{N2}	0.767885732	0.7678800
y_{H2}	1.69049E-08	1.64050E-08
y_{OH}	1.13715E-05	1.16820E-05
y_{H}	3.0656E-10	2.92980E-10
y_{O}	1.89064E-07	1.83910E-07
y_{NO}	0.000693595	0.00069530
y_{Ar}	0.009444273	0.0094440
y_{N2O}	7.42789E-08	7.76540E-08
ξ_2 (g-mol/s)	2.62736E-05	–
ξ_3 (g-mol/s)	0.501786025	–
ξ_4 (g-mol/s)	–0.008202121	–
ξ_5 (g-mol/s)	0.016453605	–
ξ_6 (g-mol/s)	2.21783E-07	–
ξ_7 (g-mol/s)	0.000136779	–
ξ_8 (g-mol/s)	0.000107475	–

Ultimately, the concentrations of NO_x, CO, and other pollutants depend on complex fluid flow, mass transfer, and kinetics effects. For example, actual residence times in different sections of the combustion zone may be uncertain due to back-mixing and other fluid flow processes. There are ongoing improvements in the physical models and detailed kinetic equations used to predict emissions, and we explore the use of these to predict emissions from gas turbines in the next section.

15.3 EMISSION PREDICTION USING ELEMENTARY KINETICS RATE EXPRESSIONS

15.3.1 Combustion Chemical Kinetics

There is a large ongoing research effort in assembling elementary kinetics rate expressions to predict the products of combustion. One example is the GRI mechanism (currently GRI-Mech 3.0) for methane and natural gas combustion (Smith et al.), which includes some 53 species and 325 elementary reactions. Elementary rate expressions are ones in which the kinetics are consistent with the stoichiometry. Consider the general bimolecular reaction

$$A + B \underset{k_b}{\overset{k_f}{\rightleftarrows}} C + D, \tag{15.35}$$

where A, B, C, and D are reacting species. In a bimolecular reaction, it is generally assumed that two molecules collide and react (with some probability) to form two different molecules. Here we are considering a single reaction, but we will develop our formulation using terms that will lend themselves to multiple reactions. We will also use and explain some of the nomenclature from the combustion literature, which is somewhat different from the chemical engineering reaction engineering literature.

We can write the rate expression r_1 for the reaction (reaction 1) in Equation (15.35) as the difference between the forward rate and backward (reverse) rate:

$$r_1 = r_{f1} - r_{b1} = k_{f1}(C_A)(C_B) - k_{b1}(C_C)(C_D), \quad \frac{\text{g-mol}}{\text{cm}^3\text{-s}}. \tag{15.36}$$

This rate expression depends on both the forward (k_f) and backward (reverse) (k_b) rate constants. We will use C_i to indicate species concentrations as g-mol/cm^3, but note that in many texts brackets, [ ] are used to indicate species concentration. A bimolecular rate constant has units cm^3/g-mol-s. Also note that the power on each concentration (here = 1) is the same as the stoichiometric coefficient in the reaction. The rate constant can be expressed as an Arrhenius expression:

$$k = Ae^{-E_a/RT}, \tag{15.37}$$

where A is termed the pre-exponential factor and E_a is the activation energy. To increase the usable temperature range, a three-parameter (A, β, E) rate constant can be used:

$$k = AT^{\beta}e^{-E_a/RT}. \tag{15.38}$$

At equilibrium, the forward and backward rates are equal, and we can write

$$k_{f1}(C_A)_{\text{eq}}(C_B)_{\text{eq}} - k_{b1}(C_C)_{\text{eq}}(C_D)_{\text{eq}} = 0, \tag{15.39}$$

where the subscript "eq" indicates equilibrium conditions. Rearranging,

$$\frac{k_{f1}}{k_{b1}} = \frac{(C_C)_{\text{eq}}(C_D)_{\text{eq}}}{(C_A)_{\text{eq}}(C_B)_{\text{eq}}} = K_c, \tag{15.40}$$

which shows that the ratio of the rate constants equals the equilibrium constant K_c based on concentration. The equilibrium constant can be related to the thermodynamic K used in the previous section and is fairly easy to calculate for ideal gases (Fogler, 2006). Also, as the equilibrium constant is considered more accurate than the rate constants, one generally finds in the literature k_f for each elementary reaction with k_b determined as $k_b = k_f/K_c$.

The forward and reverse rate constants for elementary reactions are said to be "thermodynamically consistent."

15.3.2 Compact Matrix Notation for the Species Net Generation (or Production) Rate

In combustion kinetics, an overall reaction mechanism is often based on a set of elementary reactions. For example, the Zeldovich mechanism (Zeldovich, 1946) explains the formation of nitric oxide from air at high temperature as a set of elementary reactions including

$$N + NO \rightleftarrows N_2 + O, \tag{15.41a}$$

$$N + O_2 \rightleftarrows NO + O, \tag{15.41b}$$

and

$$2O + M \rightleftarrows O_2 + M. \tag{15.41c}$$

The overall reaction mechanism is $N_2 + O_2 \rightleftarrows 2\,NO$. In the third elementary reaction, which accounts for oxygen dissociation, M represents all possible collision partners, which, under the ideal gas assumption, would be $C_M = P/RT$.

A general matrix form for a set of reactions can be written as

$$\nu' S \rightleftarrows \nu'' S, \tag{15.42}$$

where ν' is the reactant stoichiometric coefficient matrix, ν'' is a product stoichiometric coefficient matrix, and S is the species column vector. The elements of ν' are $\nu'_{m,i}$, the reactant stoichiometric coefficients, where we are using m for reactions as the row index and i for species as the column index. In ν'', the elements $\nu''_{m,i}$ are the stoichiometric coefficients for the products, all positive numbers.

EXAMPLE 15.6 ***Matrix Form for the Zeldovich Mechanism***

Develop the general matrix form (Eq. (15.42)) for the Zeldovich mechanism given by Equation (15.41).

SOLUTION

$$\begin{pmatrix} 0 & 0 & 1 & 0 & 1 & 0 \\ 1 & 0 & 0 & 0 & 1 & 0 \\ 0 & 0 & 0 & 2 & 0 & 1 \end{pmatrix} \begin{pmatrix} O_2 \\ N_2 \\ NO \\ O \\ N \\ M \end{pmatrix} \rightleftarrows \begin{pmatrix} 0 & 1 & 0 & 1 & 0 & 0 \\ 0 & 0 & 1 & 1 & 0 & 0 \\ 1 & 0 & 0 & 0 & 0 & 1 \end{pmatrix} \begin{pmatrix} O_2 \\ N_2 \\ NO \\ O \\ N \\ M \end{pmatrix}.$$

The rate expression r_m for each reaction can be written as

$$r_m = \left(k_{fm} \prod_{i=1}^{n_{\text{species}}} (C_i)^{\nu'_{m,i}} - k_{bm} \prod_{i=1}^{n_{\text{species}}} (C_i)^{\nu''_{m,i}} \right) \quad m = 1, 2, \ldots, n_{\text{reactions}}, \tag{15.43}$$

and the species net generation rate R_i can then be written as

$$R_i = \sum_{m=1}^{n_{\text{reactions}}} (\nu''_{m,i} - \nu'_{m,i})(r_m). \tag{15.44}$$

■

EXAMPLE 15.7 ***Species Generation Term for the Zeldovich Mechanism***

Determine the net generation rate R_i for each species in the Zeldovich mechanism using Equations (15.43) and (15.44).

SOLUTION Here, using our result from Example 15.6, we calculate $(\nu'' - \nu')$ as

$$\begin{pmatrix} 0 & 1 & 0 & 1 & 0 & 0 \\ 0 & 0 & 1 & 1 & 0 & 0 \\ 1 & 0 & 0 & 0 & 0 & 1 \end{pmatrix} - \begin{pmatrix} 0 & 0 & 1 & 0 & 1 & 0 \\ 1 & 0 & 0 & 0 & 1 & 0 \\ 0 & 0 & 0 & 2 & 0 & 1 \end{pmatrix} = \begin{pmatrix} 0 & 1 & -1 & 1 & -1 & 0 \\ -1 & 0 & 1 & 1 & -1 & 0 \\ 1 & 0 & 0 & -2 & 0 & 0 \end{pmatrix}.$$

The rate expressions from Equation (15.43) are

$$r_1 = \left(k_{f1} \prod_{i=1}^{n_{\text{species}}} (C_i)^{\nu'_{1,i}} - k_{b1} \prod_{i=1}^{n_{\text{species}}} (C_i)^{\nu''_{1,i}} \right) = k_{f1}(C_N)(C_{NO}) - k_{b1}(C_{N_2})(C_O)$$

$$r_2 = k_{f2}(C_N)(C_{O_2}) - k_{b2}(C_{NO})(C_O)$$

$$r_3 = k_{f3}(C_O)^2(C_M) - k_{b3}(C_{O_2})(C_M).$$

And with $R_i = \sum_{m=1}^{n_{\text{reactions}}} (\nu''_{m,i} - \nu'_{m,i})(r_m)$ from Equation (15.44), we find the species net generation or production rate as

$$R_{O_2} = (\nu''_{1,O_2} - \nu'_{1,O_2})(r_1) + (\nu''_{2,O_2} - \nu'_{2,O_2})(r_2) + (\nu''_{3,O_2} - \nu'_{3,O_2})(r_3)$$
$$= (0)(r_1) + (-1)(r_2) + (1)(r_3) = (-k_{f2}(C_N)(C_{O_2})$$
$$+ k_{b2}(C_{NO})(C_O)) + \left(+k_{f3}(C_O)^2(C_M) - k_{b3}(C_{O_2})(C_M)\right)$$

$$R_{N_2} = (k_{f1}(C_N)(C_{NO}) - k_{b1}(C_{N_2})(C_O))$$

$$R_{NO} = (-k_{f1}(C_N)(C_{NO}) + k_{b1}(C_{N_2})(C_O))$$
$$+ (k_{f2}(C_N)(C_{O_2}) - k_{b2}(C_{NO})(C_O))$$

$$R_O = (k_{f1}(C_N)(C_{NO}) - k_{b1}(C_{N_2})(C_O)) + (k_{f2}(C_N)(C_{O_2})$$
$$-k_{b2}(C_{NO})(C_O)) + \left(-2k_{f3}(C_O)^2(C_M) + 2k_{b3}(C_{O_2})(C_M)\right)$$

$$R_N = (-k_{f1}(C_N)(C_{NO}) + k_{b1}(C_{N_2})(C_O))$$
$$+ (-k_{f2}(C_N)(C_{O_2}) + k_{b2}(C_{NO})(C_O))$$

$$R_M = 0.$$

It is very easy to make mistakes when generating the rate expressions—note the coefficients and exponents associated with r_3 in the generation term for the oxygen radical R_O. The use of Equations (15.43) and (15.44) should help eliminate common mistakes. ■

In this section, we have used the Zeldovich or thermal NO formation mechanism for nitric oxide formation from air in combustion processes. This mechanism is important when air is at high temperature for extended times (see Example 15.8). Also important for NO formation from air are the prompt mechanism and the N_2O intermediate mechanism. When fuel contains nitrogen, the fuel-bound NO mechanism must also be included in the development of any comprehensive reaction mechanism. For a detailed discussion of NO mechanisms, see Miller and Bowman (1989) and Hill and Smoot (2000).

15.4 MODELS FOR PREDICTING EMISSIONS FROM GAS TURBINE COMBUSTORS

A gas turbine may be viewed as the combination of an air compressor, an annular combustor (Figure 15.6), and a turbine for power generation. Natural gas and compressed air are burned in the combustor, which delivers hot exhaust gases to the turbine. Roughly 50% of the total volume of air entering the diffuser mixes with fuel and burns in the primary zone. The other ~50% bypasses the primary zone and is used to help cool the combustor surfaces and turbine blades. The combustion products are expanded through the turbine, which supplies power for the compressor and produces electricity if the turbine shaft is coupled to a generator. In a cogeneration system, the exhaust gas from the power turbine is used to generate steam in a heat recovery steam generator (HRSG).

Detailed combustion kinetics mechanisms have been developed based on ideal experimental systems (e.g., premixed laminar flames). It is possible to utilize these mechanisms in real systems where they can serve as predictive tools for emissions and system performance. For gas turbines, this process begins by modeling the combustor as a series and/or parallel combinations of perfectly stirred reactors (PSRs) and plug flow reactors (PFR) (Andreini and Facchini, 2004). Figure 15.7 shows a possible reactor model of the gas turbine combustor in Figure 15.6, here using three PSRs and one PFR to represent the different zones.

We introduced material and energy balance equations for a PFR in Chapter 5 and in the following section, we will develop the PFR equations specific for combustion processes. Next, we develop the material and energy balance equations for a PSR.

15.4.1 Perfectly Stirred Reactor for Combustion Processes: The Material Balance Problem

A PSR or a continuously stirred reactor (CSTR) is depicted in Figure 15.8. In a PSR, the mixing of reactants and products is assumed to be instantaneous, and there is no variation of composition or temperature in the reactor. A PSR can be difficult to understand because there are at least three unusual concepts:

1. There is no variation in composition or temperature in the reactor
2. There are three timescales—a residence time associated with the flow in and out of the reactor, a "kinetics" time associated with the reaction events, and an elapsed time from an arbitrary initial condition.
3. Numerical integration of the unsteady-state PSR material balances over the elapsed time will lead to the material balance solution for both unsteady and steady-state PSR problems.

Addressing concept 1, one way to think of this is to say that the mixing time in the reactor is instantaneous, but there is still an average residence time for all molecules entering and exiting the reactor. A PSR is employed in combustion modeling to allow reactants and products to "instantaneously combust." For example, in Figure 15.7, PSR-2, which represents the primary zone, allows us to bring fuel, air, and products almost instantaneously to $T > 2000$ K. For a given residence time and feed composition, there will be generally one unique outlet composition allowing solution to the PSR material balance problem; however, in some cases, multiple steady-state solutions do exist.

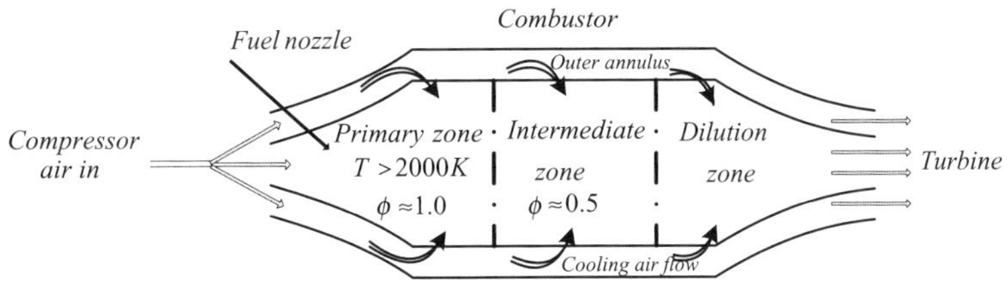

Figure 15.6 Schematic of a gas turbine annular combustor. ϕ = theoretical air for combustion/actual air used (adapted from Swithenbank et al., 1972; Yamamoto et al., 2002).

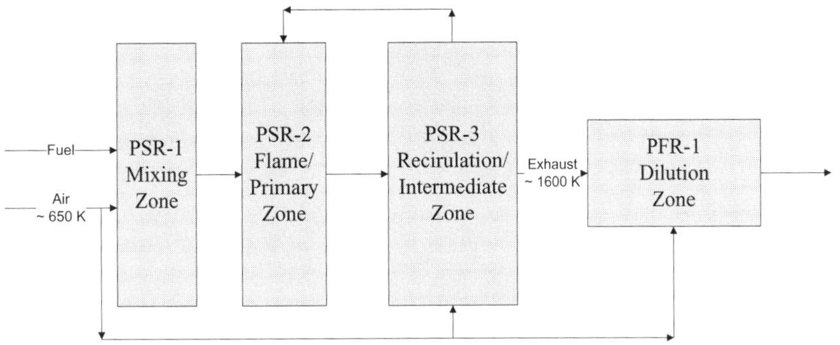

Figure 15.7 Conceptual model of gas turbine combustor using PSRs and a PFR.

The unsteady-state material balance for each species in the PSR (accumulation = flow in − flow out + generation by reaction) can be written as

$$\frac{d(C_iV_r)}{dt} = (Q_{\text{in}}C_{i,\text{in}}) - (Q_{\text{out}}C_i) + R_iV_r, \quad \frac{\text{g-mol}}{\text{s}}. \qquad (15.45)$$

In Equation (15.45), V_r is the known and fixed reactor volume (cm^3). Note that the outlet concentration C_i is the same as the species concentration in the reactor. Also, the volumetric flow rate into the reactor $Q_{\text{in}}(\text{cm}^3/\text{s})$ does not equal the volumetric flow rate from the reactor Q_{out}. For liquid phase processes, it is often a reasonable approximation to set $Q_{\text{in}} = Q_{\text{out}}$ (constant density approximation), but for vapor phase combustion processes, we must recognize that there will always be a change in the molar flow rates in and out and a large temperature change; consequently, there will be a change in Q_{out} from Q_{in}. Finally, in Equation (15.45), we have again used R_i for the generation or net production rate of species *s*. You will recall we defined the net generation of species *s* in our previous section (Section 15.3.2, "Compact Matrix Notation for the Species Net Generation [or Production] Rate") as $R_i = \sum_{m=1}^{n_{\text{reactions}}}(\nu''_{m,i} - \nu'_{m,i})(r_m)$. We want to make the distinction that time on the left-hand side (LHS) of Equation (15.45) is macroscopic or elapsed time and that time associated with a reaction event are not the same. For example, we can set the LHS of Equation (15.45) to be zero at steady state (the concentration leaving the PSR will not change with macroscopic time). But we cannot set $R_i = 0$ in our species net generation rate or we will simply obtain the blowout condition that (Q_{in} $C_{i,\text{in}}$) = (Q_{out} C_i)—this is a demonstration of concept 2. Molecules react because they remain in the PSR for a finite residence time.

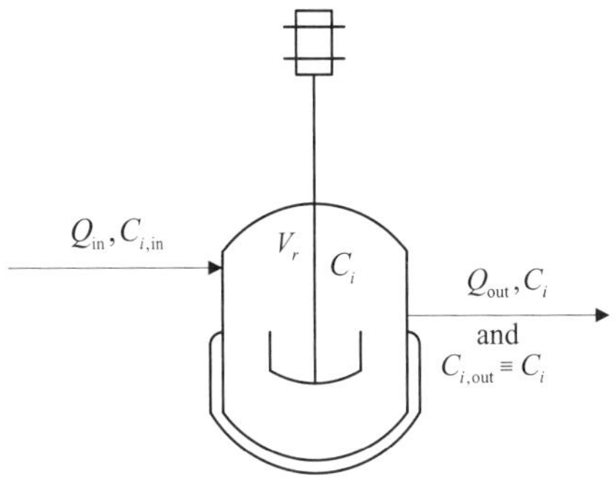

Figure 15.8 Perfectly stirred reactor (PSR) (for a PSR, $C_{i,\text{out}} \equiv C_i$).

In Equation (15.45), we can formulate the mean residence time τ of all molecules in the reactor as

$$\tau = \frac{V_r}{Q}[=]\frac{\text{cm}^3}{\text{cm}^3/\text{s}}[=]\text{s}. \qquad (15.46)$$

The residence time is the time required for one volume of product, V_r (measured at outlet conditions) to flow through the reactor. We do need to address the selection of Q in the residence time equation. In the chemical engineering literature, the residence time is generally based on Q_{in} or Q_{Ref} (volumetric flow rate at known reference conditions). In the combustion literature, the residence time is generally based on Q at outlet conditions; we use the combustion literature convention here.

In Equation (15.45), the ideal gas law can be used to express Q_{in} in terms of T_{in}, P_{in}, $\Sigma N_{i,in}$ and outlet Q, T, P, $\Sigma N_{i,out}$ as

$$Q_{in} = Q_{out}\left(\frac{T_{in}}{T}\right)\left(\frac{P}{P_{in}}\right)\left(\frac{\sum N_{i,in}}{\sum N_{i,out}}\right) = \left(\frac{V_r}{\tau}\right)\left(\frac{T_{in}}{T}\right)\left(\frac{P}{P_{in}}\right)\left(\frac{\sum N_{i,in}}{\sum N_{i,out}}\right). \quad (15.47)$$

With the reactor volume constant, our working material balance equation for the PSR can now be written as

$$\frac{dC_i}{dt} = \frac{1}{\tau}\left[\left(\frac{T_{in}}{T}\right)\left(\frac{P}{P_{in}}\right)\left(\frac{\sum N_{i,in}}{\sum N_{i,out}}\right)C_{i,in} - C_i\right] + R_i, \quad \frac{\text{g-mol}}{\text{s-cm}^3}. \quad (15.48)$$

Here, $\Sigma N_{i,in}$ and $\Sigma N_{i,out}$ are the sums of the molar species flow rates in and out of the PSR.

It is also common to write Equation (15.48) in terms of the mass flow rate of each species leaving the reactor. We again take advantage of the ideal gas law to write

$$\left(\frac{PV_r(MW_{mix})}{RT\tau}\right)[=]\frac{\text{g total}}{\text{s}},$$

where (MW_{mix}) is the molecular weight of the mixture leaving the PSR. Here then,

$$\frac{(MW_i)(V_r)dC_i}{dt} = \left(\frac{PV_r(MW_{mix})}{RT\tau}\right)[w_{i,in} - w_i] + R_i(MW_i)(V_r), \quad \frac{\text{g}}{\text{s}}, \quad (15.49)$$

where w_i is the weight fraction of species i leaving the PSR.

Solution Techniques for the Perfectly Stirred Reactor Material Balance

The solution of the PSR material balance problem is given by the simultaneous solution of $n_{species}$ (total number of species) equations defined in Equation (15.48). In these equations, the species net generation rate terms (the R_i terms in Example 15.7) are "stiff." A stiff equation implies that one or more terms (a single variable or grouping of variables) in the equation are changing very rapidly, while other terms are changing slowly. For example, reactions involving radical–radical recombinations (k[radical][radical]) are very slow because it is the product of two small concentrations.

Some PSR material balance problems, typically those with a small number of species, can be solved using nonlinear (NL)-optimization techniques—see, for example, Problem 15.2. The optimization approach often leads to multiple optimal solutions, so it cannot be considered a general solution approach.

As a first general solution approach, we can take advantage of the fact that the concentrations leaving the PSR will soon after start-up become constant with time, so we can set the LHS of Equation (15.48) = 0. This generates a set of $n_{species}$ nonlinear algebraic equations with $n_{species}$ unknown species. The Newton–Raphson method we developed in Chapter 5 can serve as a basis to solve this set of equations; however, experience has shown that the Newton–Raphson method must be suitably modified to solve these kinetic equations. Modifications include a line search (see Chapter 3) between the current accepted point $\left(C_i^{[k]}\right)$ and the Newton–Raphson solution $\left(C_i^{[k+1]}\right)$ to ensure that no mass fraction becomes less than zero or the sum of mass fractions becomes greater than 1. Additional modifications are often needed (Kelley, 2003).

A second general approach is to use numerical integration to solve the set of ODEs given by Equation (15.48)—this is concept 3. We can numerically integrate these unsteady-state equations until the concentrations leaving the PSR no longer change in value. Because the equations are stiff, the numerical ODE solution methods of Chapter 5, Euler's method and the RK4 method, will not work.

Lawrence Livermore National Laboratory has addressed the solution of these type problems (including nonlinear algebraic equation sets, ODEs, and ODEs/nonlinear algebraic equations) for several decades. They currently provide a state-of-the-art open source software suite, Suite of Nonlinear and Differential/Algebraic Equation Solvers (SUNDIALS) at https://computation.llnl.gov/casc/sundials/ for such problems. This suite includes the program CVODE for the solution of initial value ODE systems.

We have taken the code CVODE and created dynamic link libraries (DLLs) to allow CVODE to be called from Excel (Appendix A; J. Punuru, pers. comm.). Using these DLLs allows the solution to both PSR and PFR reactor problems utilizing elementary rate kinetics, as we will show in the next sections. There is also a modified Newton–Raphson code (KINSOL) available from LLNL that should allow solution of the nonlinear PSR material balance problem. However, we have chosen to use integration of the unsteady-state ODE equations for both the PSR and PFR because it provides a single solution method and it may also

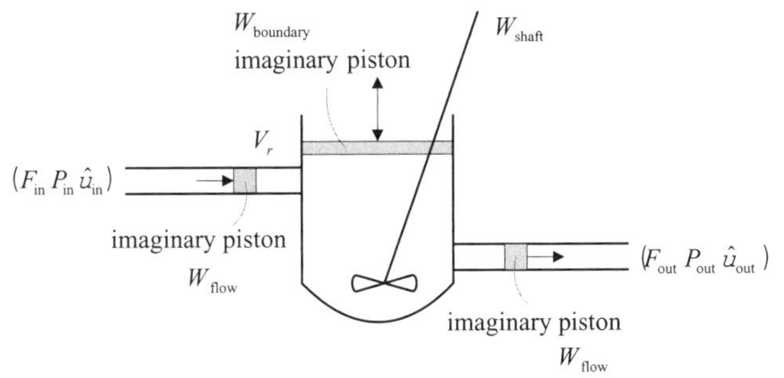

Figure 15.9 Open system for energy balance development.

offer the possibility that an oscillatory solution of the elementary rate equations may be detected.

15.4.2 The Energy Balance for an Open System with Reaction (Combustion)

An energy balance for the PSR (or the PFR) is needed to determine the energy input required to maintain a fixed outlet temperature (isothermal process) or to determine the outlet temperature if the combustion process is adiabatic. The energy balance (or first law of thermodynamics) can be written for a "black box" open system, as depicted in Figure 15.9, as

$$\begin{pmatrix}\text{Rate of energy}\\ \text{accumulation}\\ \text{in the system}\end{pmatrix} = \begin{pmatrix}\text{Rate of energy entering}\\ \text{by the inlet stream}\end{pmatrix} - \begin{pmatrix}\text{Rate of energy leaving}\\ \text{by the exit stream}\end{pmatrix} + \begin{pmatrix}\text{Rate of heat}\\ \text{added to the system}\end{pmatrix} + \begin{pmatrix}\text{Rate of work}\\ \text{done on the system}\end{pmatrix} \quad (15.50)$$

or

$$\frac{dU}{dt} = F_{\text{in}}(\hat{u}_{\text{in}}) - F_{\text{out}}(\hat{u}_{\text{out}}) + \dot{Q} + \dot{W}_T, \quad (15.51)$$

where U is the total internal energy (joule) of the system (the vessel contents), F_{out} is the flow rate out (kilogram per second), $\hat{u}_{\text{out}}$ is the specific internal energy at the exit (joule per kilogram), and $\dot{Q}$ and $\dot{W}_T$ are the heat and work rate terms. In writing Equation (15.51), we have assumed the system consists of a single phase, which for gas turbine combustion chambers will generally be vapor. Kinetic and potential energy terms, which are generally small in gas systems, have been neglected.

The total work rate can be written as

$$\dot{W}_T = \dot{W}_{\text{flow}} + \dot{W}_{\text{shaft}} + \dot{W}_{\text{boundary}}, \quad \frac{\text{J}}{\text{s}}, \quad (15.52)$$

where $\dot{W}_{\text{flow}}$ is flow work to move the fluid in and out of the reactor as $(F_{\text{in}}P_{\text{in}}\hat{\upsilon}_{\text{in}} - F_{\text{out}}P_{\text{out}}\hat{\upsilon}_{\text{out}})$ and where $\hat{\upsilon}_{\text{out}}$ is the exiting fluid specific volume; $\dot{W}_{\text{shaft}}$ is shaft work; and $\dot{W}_{\text{boundary}} = -PdV_r/dt$ is the rate of work due to a change in volume or expansion of the system.

Recalling that $\hat{u} = \hat{h} - P\hat{\upsilon}$, we can write Equation (15.51) as

$$\frac{dU}{dt} = F_{\text{in}}\left(\hat{h}_{\text{in}} - P_{\text{in}}\hat{\upsilon}_{\text{in}}\right) - F_{\text{out}}\left(\hat{h}_{\text{out}} - P_{\text{out}}\hat{\upsilon}_{\text{out}}\right) + \dot{Q} + \dot{W}_T, \quad (15.53)$$

where $\hat{h}_{\text{out}}$ is the specific enthalpy in the exiting stream. Incorporating the definition for flow work in Equation (15.53) and defining $\dot{W} = \dot{W}_{\text{shaft}} - PdV_r/dt$ as the combination of shaft and boundary work, we can write

$$\frac{dU}{dt} = F_{\text{in}}\left(\hat{h}_{\text{in}}\right) - F_{\text{out}}\left(\hat{h}_{\text{out}}\right) + \dot{Q} + \dot{W}. \quad (15.54)$$

For combustion systems, we simplify Equation (15.54) by neglecting the shaft work and PdV_r/dt terms as these are small compared to the flow-enthalpy terms. So far, our discussion has been completely general for any open system with reaction. We next want to particularize our discussion to PSR, and in a later section, we will return to Equation (15.50) and address energy balances for a PFR.

15.4.3 Perfectly Stirred Reactor Energy Balance

The transient energy balance can be added to the unsteady-state material balances we developed (Eq. (15.48)) and the equation set solved using CVODE. An equally valid alternative for the PSR is to solve the energy balance after the material balance problem is complete—we will use this latter approach.

Solution of the PSR material balances independent of the energy balance requires that the temperature in the reactor be known. As discussed next, iteration on the reactor temperature may be required, but the temperature in the reactor must be fixed before solution of the PSR material balance can begin.

At steady state, the composition and temperature in the PSR will not change with time and we can set the LHS of Equation (15.54) to zero.

Then, neglecting $\dot{W}$, the PSR energy balance for an *adiabatic system* $\left(\dot{Q}=0\right)$ is

$$F_{\text{out}}\left(\hat{h}_{\text{out}}\right)-F_{\text{in}}\left(\hat{h}_{\text{in}}\right)=0. \tag{15.55}$$

Equation (15.55) requires explanation as you may be wondering about the heat of reaction term. The heat of reaction is actually implicit in Equation (15.55). Recall in Equation (15.19) we wrote the energy balance for an adiabatic reactor as $N_{\text{out}}h_{\text{out}}+\sum_{m=1}^{n_{\text{reactions}}}\xi_m\Delta H_{r,m}-N_{\text{in}}h_{\text{in}}=0$. Here the enthalpy out is clearly a function of the enthalpy in and ξ_m $\Delta H_{r,m}$, where ξ_m were the extents of reaction for the $n_{\text{reactions}}$ reactions and $\Delta H_{r,m}$ are the heats of reaction at systems T and P.

What is important is that in combustion processes, accounting for the heats of reaction for each elementary reaction quickly becomes cumbersome. So, rather than trying to account for the many heats of reaction, it is more convenient to write our enthalpy balance terms using the standard enthalpy of formation for each species from it elements. Using the species enthalpy of formation, and all species with the same common reference state, eliminates the need for any heat of reaction information. The needed data are provided in the JANAF tables and in all similar thermochemical databases. Our enthalpy terms are

$$\hat{h}_i(T)=\hat{h}_{f,i}^0+\int_{T_{\text{ref}}}^{T}\hat{c}_{p,i}dT, \quad \frac{\text{J}}{\text{kg}}. \tag{15.56}$$

As we will see in Example 15.11, an adiabatic system will require an iterative solution to energy balance. We can guess an outlet temperature, solve the material balance problem, and check if Equation (15.55) equals 0. If Equation (15.55) does not equal 0, a new outlet temperature can be assumed, but here realize the material balance problem must be resolved as a new outlet temperature will change the rate constants in the material balances.

Neglecting $\dot{W}$, the PSR energy balance for an *isothermal system* can be written as

$$\dot{Q}=F_{\text{out}}\left(\hat{h}_{\text{out}}\right)-F_{\text{in}}\left(\hat{h}_{\text{in}}\right). \tag{15.57}$$

15.4.4 Solution of the Perfectly Stirred Reactor Material and Energy Balance Problem Using the Provided CVODE Code

Lawrence Livermore National Laboratory currently provides CVODE as an open source code for the solution of stiff ODEs. CVODE can be traced to the stiff ODE solver LSODE (ca. 1980) and GEAR (ca. 1970). Creating the DLLs to allow Excel to serve as a pre- and postprocessor for CVODE has been introduced in Appendix A. To use the Excel-based CVODE, we need to supply the unsteady-state material balance equations (Eq. (15.48)), including R_i for each species. The species net generation rate R_i includes both forward and reverse reactions, and here we compute k_b from the equilibrium constant K_c for the given PSR outlet temperature. The approach we have taken is to calculate K_c and k_b on the Excel sheet and to provide the material balance equations in a VBA program, which calls the DLLs for CVODE. The PSR energy balance is solved on the Excel Sheet after the material balance problem is completed.

CVODE is used to solve a PSR reactor problem in Example 15.8. Some readers may want first to try CVODE on simple problems. In Section 15.7, we provide such an example with two coupled ODEs.

EXAMPLE 15.8 ***Solve the Zeldovich Mechanism in a PSR Using CVODE Called from Excel***

Air at room temperature 298.15 K is instantaneously heated in a PSR of volume 100 cm^3 to a temperature of 2300 K as shown in Figure 15.10. The air is 21% oxygen and 79% nitrogen on a molar basis and the residence time in the reactor is 0.1 second. Determine the exiting mole fractions of all species if the Zeldovich mechanism (Eq. (15.41)) is applicable. Also, determine the mass flow rate for the given residence time and the energy required to heat the air to 2300 K.

The forward reaction rate constants $k_f=AT^{\beta}e^{-E_a/RT}$ are given in GRI-Mech 3.0 and are reported in Table 15.8.

The solution to this kinetics problem using CVODE is provided in the Excel file **Example 15.8.xls** with the macro named CVodeDenseCustom_Macro. We briefly detail the key VBA code (in `Sub fun`), which includes reading forward and backward (or reverse) rate constants from the Excel sheet:

```
k_1F = Sheet1.Cells(23, 8)

k_1Rev = Sheet1.Cells(23, 10)

k_2F = Sheet1.Cells(24, 8)

k_2Rev = Sheet1.Cells(24, 10)

k_3F = Sheet1.Cells(25, 8)

k_3Rev = Sheet1.Cells(25, 10)
```

and providing the Zeldovich mechanism species net production rate as determined in Example 15.7. Our species mass balance, Equation (15.48), here shown for O_2, gives

$$\frac{dC_{O_2}}{dt}=\frac{1}{\tau}\left[\left(\frac{T_{\text{in}}}{T}\right)\left(\frac{P}{P_{\text{in}}}\right)\left(\frac{\sum N_{i,\text{in}}}{\sum N_{i,\text{out}}}\right)C_{O_2,\text{in}}-C_{O_2}\right]+R_{O_2},$$

and with $R_{O_2}=\left(-k_{f2}(C_N)(C_{O_2})+k_{b2}(C_{NO})(C_O)\right)$
$+\left(k_{f3}(C_O)^2(C_M)-k_{b3}(C_{O_2})(C_M)\right)$,

we write the VBA code as

```
'balance for O2

' Also recall to continue a VBA line use a
space followed by an underscore

yd1 = (1 / Res_time) * ((T_in / T_out) *
(P_out / P_in) * (N_in / N_out)* O2in - y1)
_
 + (-k_2F * y5 * y1 + k_2Rev * y3 * y4 +
k_3F * y4 ^ 2 * M - k_3Rev * y1 * M)

Call set_Ith(ydot, 1, yd1)
```

where

$\frac{dC_{O_2}}{dt}$ = `yd1`; `y1` = C_{O_2}; `y2` = C_{N_2}; `y3` = C_{NO}; `y4` = C_O; `y5` = C_N; `k_2F` is the forward rate constant for reaction 2; `k_2Rev` is the backward (or reverse rate constant) for reaction 2; `O2in` is the oxygen concentration into the PSR as gram-mole per cubic centimeter. The statement `Call set_Ith(ydot, 1, yd1)` is used to keep track of the estimate for the exiting oxygen concentration.

The forward and backward rate constants are calculated on the Excel sheet. In Section 15.2, "Emissions from Combustion Equilibrium Calculations," we showed how data from the JANAF table could be used to calculate the molar enthalpy and molar entropy for provided species. These enthapies and entropies were calculated using species formation data and the integral of $C_p\,dT$. It was then possible to calculate the Gibbs free energy G as $G = H - TS$. On the provided Excel sheet, these calculations are in rows 3–19. For each of the three reactions (m = R1, R2 or R3), we can calculate

$$\ln(K_m)=-\frac{\sum_{i=1}^{n_{\text{species}}}(\nu''_{m,i}-\nu'_{m,i})G_i^0}{RT},$$

$$K_c=K_m\left(\frac{P^0}{RT}\right)^{\sum_{i=1}^{n_{\text{species}}}(\nu''_{m,i}-\nu'_{m,i})},$$

and k_b or $k_{\text{rev}}=k_f/K_c$; see rows 22–25. The outlet temperature is set on the Excel sheet in cell B11. For texts where the general reaction r is written as

$$aA+bB\underset{k_b}{\overset{k_f}{\rightleftarrows}}cC+dD,$$

you will find

$$K_c=K_r\left(\frac{P^0}{RT}\right)^{c+d-a-b}.$$

The results at 2300 K and with τ = 0.1 second are

	O_2	N_2	NO	O	N
gmol/cm$_3$	1.09403E-06	4.1684E-06	3.02177E-08	5.87269E-09	5.9069E -14
mol fractions	0.206477914	0.786710662	0.005703047	0.001108366	1.11482E -08

The mass flow rate is 0.1528 g/s and the energy balance (Excel sheet rows 56–60) shows $F_{\text{in}}(\hat{h}_{\text{in}})(T = 298.15\text{ K}) = 0$ cal/s; $F_{\text{out}}(\hat{h}_{\text{out}})(T = 2300\text{ K}) = 86.78$ cal/s, giving $\dot{Q}_{\text{added}} = 86.78$ cal/s.

Notes on using the macro CVodeDenseCustom_Macro: You can make changes in `Sub fun()` for variables and parameters. In `Sub CVodeDenseCustom_Macro()`, you must identify the row on the Excel sheet for `NEQ; NOUT; RTOL ; T0 ; T1 ; TMULT` or `set the values` in the subroutine (see rows 101–106). You must also set the initial conditions as shown here starting in row 109. The reader should download the available manuals for CVODE from LLNL (Collier et al., 2008; Hindmarsh and Serban, 2009). The parameters used in this example will solve most kinetics combustion problems.

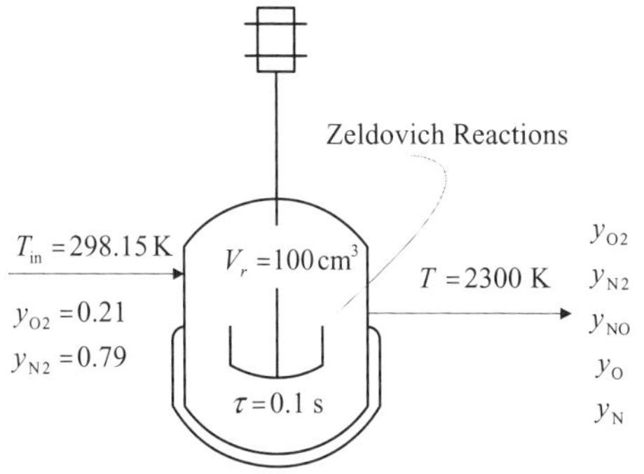

Figure 15.10 Perfectly stirred reactor for the Zeldovich Reactions.

Table 15.8 Zeldovich Mechanism Rate Constants

Reaction	A (g-mol, cm^3, s)	β (T in K)	E_a (cal/g-mol)
$N + NO \Leftrightarrow N_2 + O$	2.70 E + 13	0	355
$N + O_2 \Leftrightarrow NO + O$	9.00 E + 09	1.0	6500
$2O + M \Leftrightarrow O_2 + M$	1.20 E + 17	−1.0	0

EXAMPLE 15.9 ***Explore the Effect of Residence Time and Temperature on the Zeldovich Mechanism in a PSR Using CVODE Called from Excel***

Explore the effect of residence time and outlet temperature on the PSR of Example 15.8. Use outlet temperatures of 2300, 2100, and 1700 K with residence times of 0.01, 0.1, 1.0, and 10.0 seconds. Also, explore the effect of raising the inlet air temperature from 298.15 to 650 K (cell B69) and of changing the PSR volume from 100 to 200,000 cm^3 (cell B51).

The solution to this kinetics problem using CVODE is provided in the Excel file **Example 15.9.xls**. You will need to change the required parameters on the Excel sheet and run the macro CVodeDenseCustom_Macro. The outlet temperature is set in cell B11, the residence time in cell B30, the volume in cell B51, and the inlet feed temperature in cell B69.

Notice that changing either the volume or the inlet air temperature has no effect on the concentrations leaving the PSR. Changing either of these will, however, affect the energy balance. As shown in the provided solution in **Example 15.9.xls**, keeping the volume = 100 cm^3 and changing the inlet air temperature to 650 K, we find $F_{\text{in}}\left(\hat{h}_{\text{in}}\right)(T = 650\text{ K}) = 13.29\text{ cal/s}$; $F_{\text{out}}\left(\hat{h}_{\text{out}}\right)(T = 2300\text{ K}) = 86.78\text{ cal/s}$, giving $\dot{Q}_{\text{added}} = 73.49\text{ cal/s}$. To find $\dot{Q}_{\text{in}}$, set the PSR temperature to T_{in} (in cell B11) and determine $\left(\hat{h}_{\text{in}}\right)$ for the feed—this value is saved in cell H55.

In this example, we have allowed changes in the outlet temperature, volume, and residence time—with the mass flow then calculated. For gas turbine emission calculations, specifications will generally include the mass flow rate and the residence time—with the volume then calculated.

15.4.5 Plug Flow Reactor for Combustion Processes: The Material Balance Problem

If we look back at Figure 15.7, we see that a gas turbine combustor can be modeled as a series of PSRs and PFRs with recycle, in order to predict emissions. We have examined the solution to the PFR problem in Chapter 5 where a single global reaction rate was used to represent system kinetics. Here we want to develop the material balance equation for the PFR when elementary combustion reactions are available; we will also develop the PFR energy balance in a later section.

A PFR is depicted in Figure 15.11.

The unsteady-state material balance for each species in the PFR within the volume element (Figure 15.11) can be written (molar accumulation = flow in − flow out + generation by reaction) as

$$\frac{\partial (C_i \Delta V)}{\partial t} = QC_i\big|_z - QC_i\big|_{z+\Delta z} + R_i \Delta V. \qquad (15.58)$$

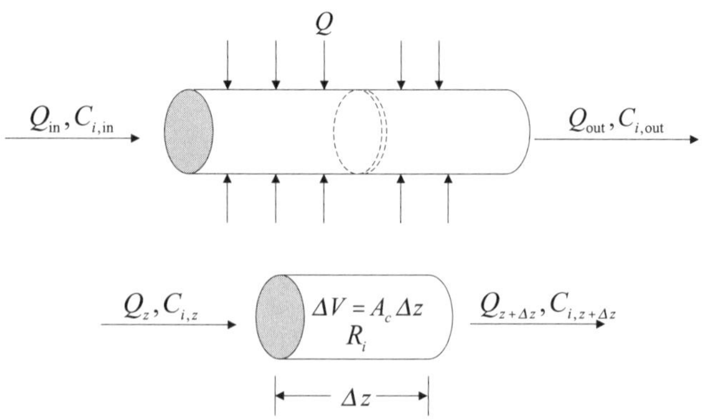

Figure 15.11 Plug flow reactor with volume element.

Dividing by ΔV (ΔV is not a function of time) and taking the limit as $\Delta V \to 0$ gives

$$\frac{\partial (C_i)}{\partial t} = -\frac{\partial (QC_i)}{\partial V} + R_i. \qquad (15.59)$$

For combustion within the gas turbine, the PFR is empty (no catalyst), and if the reactor has a constant cross-sectional area, A_c, the reactor length z can be expressed in terms of the reactor volume as $z = V/A_c$, giving $A_c dz = dV$:

$$\frac{\partial (C_i)}{\partial t} = -\frac{\partial (QC_i)}{A_c \partial z} + R_i. \qquad (15.60)$$

Recognizing that the fluid volumetric flow rate is related to the fluid velocity as $\nu = Q/A_c$, we can also write Equation (15.60) in a commonly found form:

$$\frac{\partial (C_i)}{\partial t} = -\frac{\partial (\nu C_i)}{\partial z} + R_i. \qquad (15.61)$$

We can set the time derivative = 0 in Equation (15.60), giving us a set of ODEs for the concentration of each species as

$$\frac{\partial (QC_i)}{A_c \partial z} = R_i. \qquad (15.62)$$

In order to use Equation (15.62), we can write the dependent variables, Q and C_i, as single variable N_i, as $N_i = QC_i$, giving our working material balance equations for the PFR as

$$\frac{\partial (N_i)}{\partial z} = (A_c) R_i. \qquad (15.63)$$

When using Equation (15.63), we will modify our concentration-based generation equations, R_i, by replacing C_i with N_i/Q.

Solving the PFR material balance is conceptually straightforward—we know the composition, T and P of the feed stream, and we can use Equation (15.63) to integrate in the z-direction until the end of the reactor or some specified exit condition is reached. We can use the ideal gas law to account for the change of Q at any z in terms of the known feed conditions as

$$Q_z = Q_{\text{in}} \left(\frac{T_z}{T_{\text{in}}}\right)\left(\frac{P_{\text{in}}}{P_z}\right)\left(\frac{\sum N_{i,z}}{\sum N_{i,\text{in}}}\right), \tag{15.64}$$

where the subscript "in" indicates feed conditions, which are known and constant.

15.4.6 Plug Flow Reactor for Combustion Processes: The Energy Balance Problem

We have discussed the energy balance for an open flowing system in Section 15.4.2. The energy balance for the PFR can begin with Equation (15.50), which can be written for the incremental volume element of the constant cross section in Figure 15.11 as

$$\begin{pmatrix}\text{Rate of energy}\\ \text{accumulation in}\\ \text{the volume element}\end{pmatrix} = \begin{pmatrix}\text{Rate of energy entering}\\ \text{the volume element by flow}\end{pmatrix} - \begin{pmatrix}\text{Rate of energy leaving}\\ \text{the volume element by flow}\end{pmatrix} + \begin{pmatrix}\text{Rate of heat added to}\\ \text{the volume element}\end{pmatrix} + \begin{pmatrix}\text{Rate of work added to}\\ \text{the volume element}\end{pmatrix} \tag{15.65}$$

or

$$A_c \Delta z \frac{d\hat{\rho}\hat{u}}{dt} = F_z(\hat{u}_z) - F_{z+\Delta z}(\hat{u}_{z+\Delta z}) + \dot{Q}\Delta z + \dot{W}_T \Delta z, \tag{15.66}$$

where $\hat{\rho}$ is the fluid specific density in the volume element, F is the flow rate (kilogram per second), $\hat{u}$ is the specific internal energy (joule per kilogram), and $\dot{Q}$ and $\dot{W}_T$ are the heat and work rate terms now per unit length of reactor. Following our same discussion of individual terms in Equation (15.66) as for the PSR energy balance, with $\hat{u} = \hat{h} - P\hat{\upsilon}$ and incorporating the definition for flow work and neglecting the combination of shaft and boundary work, we can write

$$A_c \Delta z \frac{d\hat{\rho}\hat{u}}{dt} = F_z\left(\hat{h}_z\right) - F_{z+\Delta z}\left(\hat{h}_{z+\Delta z}\right) + \dot{Q}\Delta z. \tag{15.67}$$

Dividing by A_c Δz and taking the limit as $\Delta z \to 0$ gives

$$\frac{d\hat{\rho}\hat{u}}{dt} = -\frac{1}{A_c}\frac{\partial}{\partial z}\left(F\hat{h}\right) + \frac{1}{A_c}\dot{Q}. \tag{15.68}$$

At steady state, the composition, temperature, and pressure at any point in the reactor do not change with time, so of accumulation of energy in the reactor (the LHS of Eq. (15.68)) can be set to zero and the energy balance for the PFR becomes

$$\dot{Q} = \frac{Fd\hat{h}}{dz}. \tag{15.69}$$

Here again, as discussed in the development of the PSR energy balance equation, the heats of reaction are implicit in the definition of the enthalpy. In the next two examples, we will assume the temperature profile (as a function of z) in the PFR is known so an overall energy balance can be performed after the PFR material balance problem is solved. This energy balance is simply $\dot{Q} = F\left(\hat{h}_{\text{out}} - \hat{h}_{\text{in}}\right)/\Delta z$.

For an adiabatic PFR ($\dot{Q} = 0$), the temperature profile in the PFR will not be known, and the outlet temperature for each dz step is generally determined using an iterative process. We show this iterative process for the PSR in Example 15.11 and in Problem 15.6, we develop the iterative process for the PFR using our bounding and interval refinement methods as developed in Chapter 3.

EXAMPLE 15.10 ***Methane Combustion and the Zeldovich Reactions in a PFR***

Methane and air at a flow rate of 28,385 g/s are added to a PFR. The PFR is isothermal and is at a temperature of 1600 K. Assume pressure drop is negligible. The reactor dimensions and feed composition are provided in Figure 15.12. Determine the outlet composition of the reacting species.

For the reacting species, assume that in addition to the Zeldovich reactions, four elementary methane combustion reactions (Table 15.9) occur (adapted from Tsuji et al., 2003).

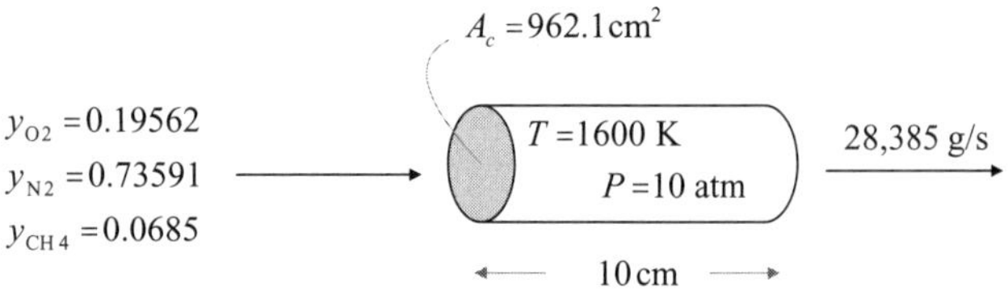

Figure 15.12 PFR for methane and air combustion.

Table 15.9 Combustion Reactions and Rate Constants for Use in Example 15.10

	Reactions	A (g-mol, cm^3, s)	β (T in K)	E_a (cal/g-mol)
r_1	$N + NO \Leftrightarrow N_2 + C$	2.70 E + 13	0	355
r_2	$N + O_2 \Leftrightarrow NO + C$	9.00 E + 09	1.0	6500
r_3	$2O + M \Leftrightarrow O_2 + M$	1.20 E + 17	−1.0	0
r_4	$CH_4 + H_2O \Leftrightarrow CO + 3H_2$	0.30 E + 9	2.0	30,000
r_5	$2CH_4 + O_2 \Leftrightarrow 2CO + 4H_2$	0.44 E + 12	3.0	30,000
r_6	$2H_2 + O_2 \Leftrightarrow 2H_2C$	0.68 E + 16	1.0	20,000
r_7	$CO + H_2O \Leftrightarrow CO_2 + H_2$	0.275 E + 10	0.5	20,000

SOLUTION We calculate $(\nu'' - \nu')$ (see Equation (15.42)) for the seven reactions as

	O_2	N_2	NO	O	N	M	CH_4	H_2O	CO	H_2	CO_2
r_1	0	1	−1	1	−1	0	0	0	0	0	0
r_2	−1	0	1	1	−1	0	0	0	0	0	0
r_3	1	0	0	−2	0	0	0	0	0	0	0
r_4	0	0	0	0	0	0	−1	−1	1	3	0
r_5	−1	0	0	0	0	0	−2	0	2	4	0
r_6	−1	0	0	0	0	0	0	2	0	−2	0
r_7	0	0	0	0	0	0	0	−1	−1	1	1

And with $R_i = \sum_{m=1}^{n_{\text{reactions}}} (\nu''_{m,i} - \nu'_{m,i})(r_m)$, we find the species net generation (or production) as

$$\begin{aligned} R_{O_2} &= (-1)(r_2) + (1)(r_3) + (-1)(r_5) + (-1)(r_6) \\ &= \left(-k_{f2}(C_N)(C_{O_2}) + k_{b2}(C_{NO})(C_O)\right) + \left(+k_{f3}(C_O)^2(C_M)\right. \\ &\quad \left.-\, k_{b3}(C_{O_2})(C_M)\right) + \left(-k_{f5}(C_{CH_4})^2(C_{O_2}) + k_{b5}(C_{CO})^2(C_{H_2})^4\right) \\ &\quad + \left(-k_{f6}(C_{H_2})^2(C_{O_2}) + k_{b6}(C_{H_2O})^2\right) \end{aligned}$$

$$R_{N_2} = \left(k_{f1}(C_N)(C_{NO}) - k_{b1}(C_{N_2})(C_O)\right)$$

$$\begin{aligned} R_{NO} &= \left(-k_{f1}(C_N)(C_{NO}) + k_{b1}(C_{N_2})(C_O)\right) \\ &\quad + \left(k_{f2}(C_N)(C_{O_2}) - k_{b2}(C_{NO})(C_O)\right) \end{aligned}$$

$$\begin{aligned} R_O &= \left(k_{f1}(C_N)(C_{NO}) - k_{b1}(C_{N_2})(C_O)\right) + \left(k_{f2}(C_N)(C_{O_2})\right. \\ &\quad \left.-k_{b2}(C_{NO})(C_O)\right) + \left(-2k_{f3}(C_O)^2(C_M) + 2k_{b3}(C_{O_2})(C_M)\right) \end{aligned}$$

$$\begin{aligned} R_N &= \left(-k_{f1}(C_N)(C_{NO}) + k_{b1}(C_{N_2})(C_O)\right) \\ &\quad + \left(-k_{f2}(C_N)(C_{O_2}) + k_{b2}(C_{NO})(C_O)\right) \end{aligned}$$

$$R_M = 0$$

$$\begin{aligned} R_{CH_4} &= \left(-k_{f4}(C_{CH_4})(C_{H_2O}) + k_{b4}(C_{CO})(C_{H_2})^3\right) \\ &\quad + \left(-2k_{f5}(C_{CH_4})^2(C_{O_2}) + 2k_{b5}(C_{CO})^2(C_{H_2})^4\right) \end{aligned}$$

$$\begin{aligned} R_{H_2O} &= \left(-k_{f4}(C_{CH_4})(C_{H_2O}) + k_{b4}(C_{CO})(C_{H_2})^3\right) \\ &\quad + \left(2k_{f6}(C_{H_2})^2(C_{O_2}) - 2k_{b6}(C_{H_2O})^2\right) \\ &\quad + \left(-k_{f7}(C_{CO})(C_{H_2O}) + k_{b7}(C_{CO_2})(C_{H_2})\right) \end{aligned}$$

$$\begin{aligned} R_{CO} &= \left(k_{f4}(C_{CH_4})(C_{H_2O}) - k_{b4}(C_{CO})(C_{H_2})^3\right) \\ &\quad + \left(2k_{f5}(C_{CH_4})^2(C_{O_2}) - 2k_{b5}(C_{CO})^2(C_{H_2})^4\right) \\ &\quad + \left(-k_{f7}(C_{CO})(C_{H_2O}) + k_{b7}(C_{CO_2})(C_{H_2})\right) \end{aligned}$$

$$\begin{aligned} R_{H_2} &= \left(3k_{f4}(C_{CH_4})(C_{H_2O}) - 3k_{b4}(C_{CO})(C_{H_2})^3\right) \\ &\quad + \left(4k_{f5}(C_{CH_4})^2(C_{O_2}) - 4k_{b5}(C_{CO})^2(C_{H_2})^4\right) \\ &\quad + \left(-2k_{f6}(C_{H_2})^2(C_{O_2}) + 2k_{b6}(C_{H_2O})^2\right) \\ &\quad + \left(k_{f7}(C_{CO})(C_{H_2O}) - k_{b7}(C_{CO_2})(C_{H_2})\right) \end{aligned}$$

$$R_{CO_2} = \left(k_{f7}(C_{CO})(C_{H_2O}) - k_{b7}(C_{CO_2})(C_{H_2})\right).$$

Table 15.10 Comparison of Simple PFR Solution with GRI-Mech 3.0 Complete Mechanism

Conditions/ Species (mol Fraction)	Solution from the Seven-Reaction Mechanism (Table 15.9)	Chemkin Solution Using Complete GRI-Mech 3.0
Temperature (K)	1600	1600
Pressure (atm)	10	10
F (g/s)	28,385	28,385
MW	27.7158	27.970
Q (cal/cm-s)	−234,384	
y_{O2}	0.067355	0.058716
y_{N2}	0.729133	0.73583
y_{NO}	2.3331E-13	9.50718E-07
y_O	9.8787E-10	3.33132E-06
y_N	3.2804E-17	1.23894E-13
y_{CH4}	5.4423E-14	−1.2593E-09
y_{H2O}	0.13422	0.13684
y_{CO}	0.01697	0.000144
y_{H2}	0.00145	6.8082E-06
y_{CO2}	0.05087	0.06831
Σy_i	1.0	0.99986

Our material balance equations for the PFR are the set of ODEs given by $d(N_i)/dz = (A_c)R_i$, where in R_i, C_i is replaced by N_i/Q and Q is determined from Equation (15.64). The solution is found in the Excel file **Example 15.10.xls** and in Table 15.10. The total enthalpy of the feed stream is 15,939.8 cal/s; the total enthalpy of the product stream leaving the PFR is −2,327,903 cal/s.

Table 15.10 shows that the results from the simple 10-species, 7-reaction model are similar to the detailed mechanism from Chemikin (GRI 3.0 with 53 species and 325 reactions). We emphasize that in order to accurately determine emissions from a combustion process, the full GRI mechanism should be used; this can be added to the Excel file we provide. ■

EXAMPLE 15.11 *Emissions from a Simplified Gas Turbine System*

As a final example, let us combine our PSR and PFR developments to predict emissions from the simplified gas turbine system shown in Figure 15.13. In Figure 15.13, the flows, temperatures, PSR retention times, and PFR cross section and length are based on a GE LM2500 aeroderivative gas turbine combustor. Here, combustion product gas recirculation has been eliminated.

Air and fuel at 10 atm and 650 K with mole fraction composition, $y_{O2} = 0.19562$, $y_{N2} = 0.73591$, and $y_{CH4} = 0.0685$, undergo combustion in PSR-1 with an estimated outlet temperature of ~2050 K and a residence time of 0.002 second. PSR-1 is assumed adiabatic, and we will need to determine the outlet temperature.

Dilution air with mole fraction composition, $y_{O2} = 0.21$ and $y_{N2} = 0.79$, is added after the PSR-1 flame zone. This single air addition to PSR-2 (via mixing point A) represents the air injection and subsequent cooling found in both the intermediate and dilution zones of the gas turbine. PSR-2 has a residence time of 0.001 second, and its outlet is fixed at 1613 K; here an overall energy balance will be required to determine the energy added. The temperature drop in PFR-1 (the dilution zone) is linear from 1613 to 1416 K over its known length of 35 cm and cross-sectional area of 962.1 cm2. Assume the methane combustion reactions are those given in Table 15.9.

SOLUTION The solution can be found in three Excel files representing PSR-1, **Example 15.11a.xls**; PSR-2, **Example 15.11b.xls**; and PFR-1, **Example 15.11c.xls**; the results from these Excel files are summarized in Table 15.11. Comparable results from Chemkin-Pro are provided in Table 15.12 (and are discussed next).

In the PSRs, we have now allowed the mass flow rate to be specified with the reactor volume calculated.

Table 15.11 Solution to the Gas Turbine Combustor System Using Table 15.9 Reaction Set

Conditions/ mol Fraction	PSR-1	Mixing Point A	PSR-2	PFR-1
T (K)	2055.77	1295.6556	1613	1416
P (atm)	10	10	10	10
F (g/s)	28,385	69,002.76	69,002.76	69,002.76
MW	27.8427	28.4271	28.4537	28.4667
$\dot{Q}$ (cal/s) or (cal/cm-s)	0	0	6,370,049.5	−123,160
y_{O2}	0.063091	0.148299	0.147497	0.147110
y_{N2}	0.732469	0.765837	0.766554	0.766902
y_{NO}	1.01315E-06	4.25516E-07	4.27683E-07	4.2849E-07
y_{O}	5.22163E-06	2.19306E-06	2.19455E-06	2.1908E-06
y_{N}	1.79757E-11	7.54973E-12	4.17680E-14	2.3671E-15
y_{CH4}	5.20293E-06	2.18521E-06	7.51754E-09	7.2395E-17
y_{H2O}	0.134599	0.056531	0.057069	0.05717
y_{CO}	0.007659	0.003217	0.001828	0.00100
y_{H2}	0.001686	0.000708	0.000228	0.00015
y_{CO2}	0.060484	0.025403	0.026821	0.02766
Σy_i	1.0	1.0	1.0	1.0

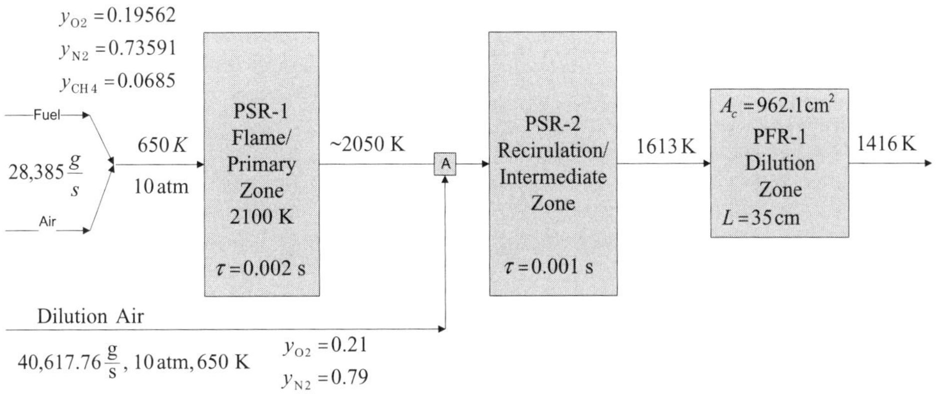

Figure 15.13 Simplified model of a gas turbine combustor.

Table 15.12 Chemkin-Pro Solution Using Complete GRI-Mech 3.0

Conditions/ mol Fraction	PSR-1	Mixing Point A	PSR-2	PFR-1
T (K)	2099.41	1316.8	1613	1416
P (atm)	10	10	10	10
F (g/s)	28,385	69,002.76	69,002.76	69,002.76
MW	27.937	28.468	28.481	28.483
$\dot{Q}$ (cal/s) or (cal/cm-s)	0	0	6,062,100	−126,990
y_{O2}	0.058693	0.146580	0.146590	0.146600
y_{N2}	0.73490	0.76691	0.76726	0.76732
y_{NO}	7.99580E-05	3.35162E-05	3.32022E-05	3.31080E-05
y_{O}	1.97410E-04	8.27462E-05	8.90378E-06	5.31015E-07
y_{N}	1.10290E-09	4.62320E-10	3.29684E-13	3.37889E-15
y_{CH4}	2.93240E-05	1.22917E-05	2.17471E-07	5.54138E-12
y_{H2O}	0.13540	0.05674	0.05727	0.05734
y_{CO}	0.00106	4.4339E-04	8.1704E-05	9.9624E-06
y_{H2}	0.00025104	0.000105229	3.4472E-06	1.1224E-07
y_{CO2}	0.067276	0.028200	0.028593	0.028668
Σy_i	0.997885	0.999110	0.999839	0.999973

PSR-1

From the Excel file **Example 15.11a.xls**, the air and fuel feed stream at $T = 650$ K and $P = 10$ atm has a total enthalpy of 1,391,820 cal/s. The temperature leaving PSR-1 must be varied (in cell B16) until the energy rate in the exiting stream is 1,391,820 cal/s; $\dot{Q}_{in} = \dot{Q}_{out}$ for an adiabatic process. As the temperature in cell B16 is varied, a new solution to the PSR problem must be generated. The temperature in PSR-1 (Table 15.11) is found to be 2055.77 K. The comparable result from Chemkin-Pro in Table 15.12 is 2099.41 K.

Mixing Point A

Adiabatic mixing is also assumed at mixing point A. The incoming dilution air at $T = 650$ K, $P = 10$ atm, and $F = 40{,}617.76$ g/s, contains 3,533,533.6 cal/s. This stream is adiabatically mixed with outlet from PSR-1. The molecular weight of the stream leaving A is $MW = 28.427$ with $T = 1295.655$ K; here see **Example 15.11b.xls**.

PSR-2

From the Excel file **Example 15.11b.xls**, the feed stream to PSR-2 shows $T = 1295.655$ K, $P = 10$ atm, and energy rate 4,925,354 cal/s. The temperature leaving PSR-2 is fixed at 1613 K with a retention time of 0.001 second. In order to maintain this outlet temperature, heat at a rate of 6,370,049.5 cal/s must be added to PSR-2.

PFR-1

From the Excel file **Example 15.11c.xls**, the feed stream to PFR-1 shows $T = 1613$ K, and $P = 10$ atm, and an energy rate of 11,295403.5 cal/s. The temperature leaving PFR-1 is fixed at 1413 K. The additional equation,

$$\frac{dT}{dz} = \frac{(1416 - 1613)}{35}$$

(with initial condition $T(z = 0) = 1613$ K), is added to the ODE reaction set to account for the linear temperature change along the PFR. The exiting stream has a total enthalpy of 6,984,810.5 cal/s; therefore, −4,310,593 cal/s or −123,160 cal/cm-s have been added. The negative quantity means heat has been removed from the PFR to the surroundings. It can often be easier to think of this negative quantity in terms of heat lost, for example, the heat lost in the PFR is 123,160 cal/cm-s.

Energy Balances for PSR-2 and PFR-1

The careful reader will have noticed that PSR-2 is endothermic and PFR-1 is exothermic; this is somewhat confusing for a combustor. Figure 15.13 (as well as Figure 15.7) is a simplification of the annular combustor. In actuality, some dilution air will be sent to PSR-2, some to PFR-1, and some used to directly cool the turbine blades. For simplicity, here we simply added all dilution air to PSR-2. The 6,370,050 cal/s added to PSR-2 to maintain an outlet temperature of 1613 K is roughly accounted for by the 4,310,593 cal/s removed from PFR-1 to obtain an exit temperature of 1415 K—so, overall the combustor is basically an adiabatic process. You can lower the temperature leaving PSR-2 (this temperature is actually an assumed value) to make the overall combustion process adiabatic.

Emission Calculations

We do want to comment on the emission calculations. The species mole fractions from the PSR-1 to PSR-2 to the PFR-1 show the same trend when using either the simplified kinetics (Table 15.9) or the full Chemkin-Pro results. However the values for the species mole fractions are very different. For example, a key emission species is NO_x. Results from the simplified kinetics for NO differ some one or two orders of magnitude from the full Chemkin-Pro results (which will also account for other NO_x species). This occurs, in part, because in the simplified kinetics, we have not accounted for all the NO formation mechanisms (see Hill and Smoot, 2000; Miller and Bowman, 1989). Some species mole fractions are close, for example, O_2 is within 10%, and N_2 and H_2O are within 1%. The important point here is the entire GRI-Mech 3.0 mechanism should be used if you are trying to quantify or predict emissions from natural gas-fired turbines. But the kinetics developed in this chapter provide a good start to appreciating the modeling of gas turbine combustors and emissions calculations. ■

15.5 CLOSING REMARKS

In this chapter, we explored equilibrium reaction calculations and particularized our discussion to equilibrium combustion calculations. The important feature of equilibrium calculations is that the species leaving any reaction process can be determined (the equilibrium composition) without knowledge of the actual reaction chemistry. The reader interested in additional details of equilibrium calculations is referred to the book by de Nevers (2002). For large equilibrium problems, you should consider downloading KINSOL from Lawrence Livermore National Laboratory.

We also introduced elementary reaction kinetics and saw how kinetic sets could be used in combination with PSRs and PFRs to model gas turbine combustor emissions. The kinetic sets for combustion processes were stiff ODEs and here we provided the CVODE from LLNL for problem solution.

The GRI mechanism (Smith et al./Web site) is available for methane and natural gas combustion. Detailed mechanisms for other species are available in the literature, and here the reader is referred to El-Mahallay and Habik (2002), Glassman and Yetter (2008), and Miller et al. (1990). An excellent overview of combustion kinetics and combustion modeling is also provided in Turns (2000).

REFERENCES

Andreini, A. and B. Facchini 2004. Gas turbines design and off-design performance analysis with emissions evaluation. *J. Eng. Gas Turbines Power* 126(1): 83–91.

Chapra, S.C. and R.P. Canale 2010. *Numerical Methods for Engineers* (6th edition). McGraw Hill, New York.

Chase, M.W., C.A. Davies, J.R. Davies, D.J. Fulrip, R.A. McDonald, and A.N. Syverud 1985. JANAF thermochemical tables (3rd edition). *J. Phys. Chem. Ref. Data* 14(Suppl. 1).

Collier, A.M., A.C. Hindmarsh, R. Serban, and C.S. Woodward 2008. User documentation for KINSOL v2.6.0. Technical Report UCRL-SM-208116, LLNL (Lawrence Livermore National Laboratory).

de Nevers, N. 2002. *Physical and Chemical Equilibrium for Chemical Engineers*. John Wiley and Sons, New York.

El-Mahallay, F. and S.E. Habik 2002. *Fundamentals and Technology of Combustion*. Elsevier, Oxford, UK.

Fogler, H.S. 2006. *Elements of Chemical Reaction Engineering* (4th edition). Prentice Hall, Englewood Cliffs, NJ.

Gaydon, A.G. and H.G. Wolfhard 1979. *Flames*. Chapman and Hall, London.

Glassman, I. and R.A. Yetter 2008. *Combustion* (4th edition). Academic Press/Elsevier, Burlington, MA.

Hill, S.C. and L.D. Smoot 2000. Modeling of nitrogen oxides formation and destruction in combustion systems. *Prog. Energy Combust. Sci.* 26: 417–458.

Hindmarsh, A.C. and R. Serban 2009. Example programs for CVODE v2.6.0. Technical Report UCRL-SM-208110, LLNL (Lawrence Livermore National Laboratory).

Kelley, C.T. 2003. *Solving Nonlinear Equations with Newton's Method*. SIAM, Philadelphia, PA.

Miller, J.A. and C.T. Bowman 1989. Mechanism and modeling of nitrogen chemistry in combustion. *Prog. Energy Combust. Sci.* 15: 287–338.

Miller, J.A., R.J. Kee, and C.K. Westbrook 1990. Chemical kinetics and combustion modeling. *Annu. Rev. Phys. Chem.* 41: 345–387.

Prothero, A. 1969. Computing with thermochemical data. *Combustion Flame*. 13: 399–408.

Rodriguez-Toral, M.A. 1999. Synthesis and Optimization of Large-Scale Utility Systems. Ph.D. dissertation, The University of Edinburgh.

Rodriguez-Toral, M.A., W. Morton, and D.R. Mitchell 2000. Using new packages for modeling, equation oriented simulation and optimization of a cogeneration plant. *Comput. Chem. Eng.* 24: 2667–2685.

Smith, G.P., D.M. Golden, M. Frenklach, N.W. Moriarty, B. Eiteneer, M. Goldenberg, C.T. Bowman, R.K. Hanson, S. Song, W.C. Gardiner, V.V. Lissianski, and Z. Qin. http://www.me.berkeley.edu/gri_mech/ (accessed November 2010).

Swithenbank, J., I. Poll, M.W. Vincent, and D.D. Wright 1972. Combustion Design Fundamentals. Fourteenth International Symposium on Combustion, The Combustion Institute.

Tsuji, H., A.K. Gupta, T. Hasegawa, M. Katsuki, K. Kishimoto, and M. Morita 2003. *High Temperature Air Combustion—From Energy Conservation to Pollution Reduction*. CRC Press, Boca Raton, FL.

Turns, S.R. 2000. *An Introduction to Combustion—Concepts and Applications* (2nd edition). McGraw Hill, Boston, MA.

U.S. Department of Energy 2007. Carbon dioxide capture from existing coal-fired power plants. Publication No. DOE/NETL-401/110907, DOE Office of Fossil Energy's National Energy Technology Laboratory, Pittsburgh, PA (revision date November 2007).

Yamamoto, T., T. Furuhata, N. Arai, and A.K. Gupta 2002. Prediction of NO_x emissions from high-temperature gas turbines: Numerical simulation for low-NO_x combustion. *JSME Int. J. B.* 45(2): 221–230.

Zeldovich, Y.B. 1946. The oxidation of nitrogen in combustion explosion. *Acta Physicochimica USSR* 21: 577–628.

CVODE TUTORIAL

Use the provided CVODE code to solve the following set of ODEs from Chapra and Canale (2010).

$$\frac{dy_1}{dx} = -0.5y_1$$

$$\frac{dy_2}{dx} = 4 - 0.3y_2 - 0.1y_1.$$

Integrate from $x = 0$ and at $x = 0$, $y_1 = 4$, and $y_2 = 6$. The solution can be found in the Excel file **Example 15.12xls**.

In `Sub fun`, you will need to supply the two ODEs as

```
y1 = get_Ith(y, 1)

y2 = get_Ith(y, 2)
```

```
'

 yd1 = -0.5 * y1

 Call set_Ith(ydot, 1, yd1)

'

 yd2 = 4 - 0.3 * y2 - 0.1 * y1

 Call set_Ith(ydot, 2, yd2)
```

and In Sub CVodeDenseCustom_Macro(), you will need to set for CVODE

```
' NEQ 2 /* number of equations */

' NOUT /* number of output times */

' RTOL 1.00E-06 /* scalar relative
tolerance */

' T0 /* initial time */

' T1 /* first output time */

' TMULT /* output time factor */

 NEQ = Sheet1.Cells(3, 2) 'Cell B2

 NOUT = Sheet1.Cells(4, 2) 'Cell B3

 RTOL = Sheet1.Cells(5, 2)

 T0 = Sheet1.Cells(6, 2)

 T1 = Sheet1.Cells(7, 2)

 TMULT = Sheet1.Cells(8, 2) 'Cell B8



'Starting Row on the Excel Sheet for
the Initial Guesses

 Start_Row_IG = 11
```

All these needed parameters and initial guesses are set on the Excel sheet. The CVODE code is run from the toolbar: Tools → Macro → Macros → CVodeDenseCustom_Macro.

The CVODE solution from the Excel sheet can be compared to Euler's method and the RK4 method (provided in Chapra and Canale, 2010 with a step size of 0.5):

x	y_1 Euler's	y_2 Euler's	y_1 RK4	y_2 RK4	y_1 CVODE	y_2 CVODE
0	4	6	4	6	4	6
0.5	3	6.9	3.115234	6.857670	3.1152	6.8577

PROBLEMS

15.1 ***Reactor Type from the Rate Expression*** Consider the general bimolecular reaction

$$A + B \underset{k_b}{\overset{k_f}{\rightleftarrows}} C + D,$$

where A, B, C, D are the reacting species. In some texts, you will see the elementary rate expression for the formation D as

$$\frac{d(C_D)}{dt} = k_f(C_A)(C_B) - k_b(C_C)(C_D), \quad \frac{\text{g-mol}}{\text{cm}^3\text{-s}}$$

or

$$\frac{d[D]}{dt} = k_f[A][B] - k_b[C][D], \quad \frac{\text{g-mol}}{\text{cm}^3\text{-s}}.$$

What type of reactor must be used for this rate expression to be true?

15.2 ***Optimization Solution for the PSR Zeldovich Reactions Material Balance Problem*** Solve Example 15.8 as an optimization problem. Air at room temperature 298.15 K is instantaneously heated in a PSR of volume 100 cm^3 to a temperature of 2300 K. The air is 21% oxygen and 79% nitrogen on a molar basis, and the residence time in the reactor is 0.1 second. The key reactions from Table 15.8 are

Reaction	A (g-mol,cm^3,s)	β (T in K)	E_a (cal/g-mol)
$N + NO \Leftrightarrow N_2 + O$	2.70 E + 13	0	355
$N + O_2 \Leftrightarrow NO + O$	9.00 E + 09	1.0	6500
$2O + M \Leftrightarrow O_2 + M$	1.20 E + 17	−1.0	0

Solution Hints

I actually tried many optimization formulations for this problem with the typical result being that multiple optimums were found—in other words, different starting points gave different solutions.

When using Solver, one formulation that gave the correct result from all starting points was to use Equation (15.48) at steady state:

$$\frac{1}{\tau}\left[\left(\frac{T_{\text{in}}}{T}\right)\left(\frac{P}{P_{\text{in}}}\right)\left(\frac{\sum N_{i,\text{in}}}{\sum N_{i,\text{out}}}\right)C_{i,in} - C_i\right] + R_i = 0. \qquad \text{(P15.2a)}$$

The objective function was formulated as

$$\min \sum_{i=1}^{n_{\text{species}}} \left(\frac{1}{\tau} \left[\left(\frac{T_{\text{in}}}{T} \right) \left(\frac{P}{P_{\text{in}}} \right) \left(\frac{\sum N_{i,\text{in}}}{\sum N_{i,\text{out}}} \right) C_{i,\text{in}} - C_i \right] + R_i \right)^2 . \tag{P15.2b}$$

In addition, as we actually want each term in the objective function to be zero, we add n_{species} species constraints to the problem as

$$\frac{1}{\tau} \left[\left(\frac{T_{\text{in}}}{T} \right) \left(\frac{P}{P_{\text{in}}} \right) \left(\frac{\sum N_{i,\text{in}}}{\sum N_{i,\text{out}}} \right) C_{i,\text{in}} - C_i \right] + R_i = 0 \quad i = 1, 2, \ldots, n_{\text{species}}. \tag{P15.2c}$$

To help further constrain the search region, atomic balances for both nitrogen and oxygen were added as well as the requirement that the sum of mole fractions out = 1:

$$\text{Atoms } N_{\text{in}} = \text{Atoms } N_{\text{out}};\ \text{Atoms } O_{\text{in}} = \text{Atoms } O_{\text{out}};\ \sum y_i = 1.0. \tag{P15.2d}$$

The problem variables were the species mole fractions out. When using Solver, you must also select within Solver → Options: Automatic Scaling; assume Nonnegative; and central derivatives.

Even when using this formulation (Eq. (P15.2b–d)), Solver may stop and indicate an optimal solution was not found. Here accept this nonoptimal solution and continue running Solver—the optimal solution will eventually be found. All these difficulties with the optimization formulation—even for this small kinetics problem—show why the ODE solution approach as developed in the chapter is a better general approach.

15.3 ***Temperature and Residence Time Dependence for the PSR Zeldovich Reactions*** Starting with the solution from Example 15.9 (air only), explore the effect of temperature and PSR residence on NO concentration. Make a log–log plot of NO concentration (in parts per million) versus residence time (from 0.001 to 10 seconds) at temperatures of 1700, 2100, and 2300 K. It is generally assumed that the Zeldovich reaction is not important at temperatures at or below 1700 K. Is that confirmed by your plots?

15.4 ***Stoichiometric Coefficient Matrices*** The following problem is from Turns (2000). Determine the coefficient matrices for the following reactions for the production of ozone from the heating of oxygen:

$$O_3 \underset{k_{1b}}{\overset{k_{1f}}{\rightleftarrows}} O_2 + O$$

$$O + O_3 \underset{k_{2b}}{\overset{k_{2f}}{\rightleftarrows}} 2O_2.$$

15.5 ***Work Energy Figure*** Establish the units for work in Figure 15.9 including $(FP\hat{\upsilon})$, where $\hat{\upsilon}$ is the exiting fluid specific volume; $\dot{W}_{\text{shaft}}$ is the shaft work; and $\dot{W}_{\text{boundary}} = -PdV_r/dt$ is the rate of work due to a change in volume of the system.

15.6 ***Adiabatic PFR (Solution Provided)*** In the PFR-1 of Example 15.11, the temperature profile in the reactor was assumed known. The profile was taken as linear with 1613 K at the reactor inlet and 1416 K at the reactor outlet. This allowed us to add an additional ODE, $dT/dz = (1416 - 1613)/35$, to the material balance set to account for temperature along the length of the 35-cm reactor.

For Problem 15.6, determine the outlet composition and outlet temperature for PFR-1 using the same inlet conditions as Example 15.11 but now assuming PFR-1 operates under adiabatic conditions. In developing the code for this problem, use the simplification that the temperature entering each differential volume element can be taken constant over the differential volume element. We will need to be sure that the temperature entering each element is the adiabatic temperature based on the composition leaving the previous volume element. This simplification allows solution of the material balance problem for each volume element independent of the energy balance. A more accurate solution can be obtained by adding a nonlinear algebraic energy balance equation, $T = T(C_i(z))$, for adiabatic reactor operation, to the ODE material balances.

Solution Hint: Even when using the simplification of constant temperature over each differential volume element, we still have to determine the adiabatic inlet temperature to that element. This inlet temperature, T_z, comes from the energy balance as indicated in Figure P15.6. An iterative process is required to find T_z ($h_{i,z} = f(T_z)$), which normally requires that T_z be bounded and then the interval refined as we did in Chapter 3 for single-variable problems.

For PFR-1, virtually all the methane has combusted and temperature changes in PFR-1 under adiabatic operation are expected to be small. Here then, we can skip the bounding step and simply set the minimum temperature into any volume element = 1550 K and the maximum temperature = 1650 K—this temperature range will cover all possible endothermic and exothermic kinetics. Using this range, we only need to provide an interval refinement strategy, and here we used interval halving.

The final solution can be found in the Excel file **Problem 15.6.xls**. For the objective function in the interval halving strategy, we are finding the minimum of $f(T_z) = ((N_{i,\text{in}})(h_{i,\text{in}}) - (N_{i,z})(h_{i,z}))^2$. Do note that the solution from CVODE for the reactor is slightly longer than 35 cm (here 38.765 cm) in order to maintain convergence properties. The adiabatic outlet temperature from the reactor is found to be ~1621.78 K. The solution for the adiabatic reactor outlet temperature from Chemkin-Pro, when using these same reaction kinetics, is 1621.24 K at 38.765 cm. This slight difference in temperature may be explained by adiabatic temperature convergence of 0.5 K for each differential element as used in the interval halving routine provided here. Do note that as temperature accuracy on the adiabatic energy balance over each differential element is increased, the solution times will

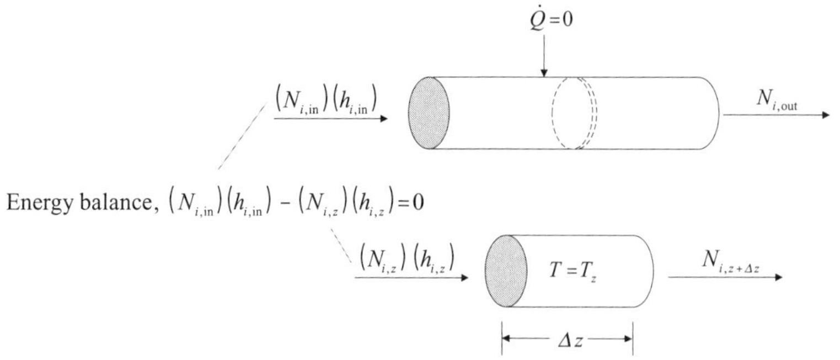

Figure P15.6 Adiabatic energy balance over differential element.

Table P15.7 Comparison of Gas Turbine Combustor Emissions with and without Water Injection; Chemkin-Pro Solution Using Complete GRI-Mech 3.0

Conditions/mol Flow	PSR-1	PSR-1	PSR-1(Chemkin)	PSR-1 (Chemkin)
	No Water Injection	With Water Injection	No Water Injection	With Water Injection
T (K)	2055.77		2099.41	2030.9234
P (atm)	10	10	10	10
F (g/s)	28,385	29,203.47	28,385	29,203.47
MW	27.8427		27.937	27.514
$\dot{Q}$ (cal/s)	0	0	0	0
N_{O2} (mol/s)	64.3201		59.634000	59.712000
N_{N2} (mol/s)	746.7368		746.69	746.7
N_{NO} (mol/s)	0.0010329		0.081240	0.037141
N_{O} (mol/s)	0.0053233		0.200570	0.141180
N_{N} (mol/s)	1.833E-08		1.12062E-06	5.31715E-07
N_{CH4} (mol/s)	0.0053043		2.97939E-02	3.88231E-02
N_{H2O} (mol/s)	137.2208		137.54000	183.15000
N_{CO} (mol/s)	7.80789		1.074700	0.996760
N_{H2} (mol/s)	1.7192		0.25507	0.26815
N_{CO2} (mol/s)	61.6621		68.354000	68.400000

increase from several minutes (with 0.5 K) to much longer run times.

15.7 ***Water Injection for Emissions Control*** Water injection directly into the primary zone of the gas turbine combustor can be used to reduce emissions. In the PSR-1 of Example 15.11, we introduced 28,385 g/s of an air–fuel mixture with composition $y_{O2} = 0.19562$, $y_{N2} = 0.73591$, and $y_{CH4} = 0.0685$ at 650 K and 10 atm. The outlet composition and temperature were determined assuming adiabatic operation and $\tau = 0.002$ second.

For this problem, see the effect of adding 818.47 g/s of water vapor at 650 K and 10 atm to the existing air–fuel mixture entering PSR-1 of Example 15.11. This water addition will give a new inlet flow rate of 29,203.47 g/s and inlet mole fractions of $y_{O2} = 0.1872$, $y_{N2} = 0.7043$, $y_{CH4} = 0.0655$, and $y_{H2O} = 0.0430$. When comparing results before and after water injection, it is instructive to use species molar flow rates (as opposed to parts per million, which can be difficult to interpret when the inlet flow rate changes). Complete Table P15.7 for the effect of water injection. For comparison, Table P15.7 provides combustor emissions with and without water injection from Chemkin-Pro using the complete GRI-Mech 3.0 and our results (without water injection) from Example 15.11.

Chapter 16

Coal-Fired Conventional Utility Plants with CO_2 Capture (Design and Off-Design Steam Turbine Performance)

This book provides a unified framework for coupling processing plants (or energy users) and cogeneration systems in order to improve energy efficiency and energy usage. In Chapter 1, we cautioned that even in the most straightforward case, where process electrical and steam demands are known and relatively constant, fluctuation in cogeneration fuel costs (e.g., the cost of natural gas) can quickly move a cogeneration project from being economically viable to nonviable. For example, many industrial natural gas fuel-based cogeneration projects are economically viable with natural gas prices in the \$4–\$6 per MMBtu range. However, as natural gas prices rise above \$6, the fuel mix for utility companies, which generally includes coal, natural gas, hydroelectric and perhaps nuclear energy, allows the sale of electricity at a price that eliminates many of the advantages for cogeneration. A key point noted in Chapter 1 was that this \$6 benchmark will certainly move upward if CO_2 capture and sequestration from conventional coal-fired power plants becomes a reality. Understanding the impact of CO_2 capture on utility pricing is important for (1) any industrial in discussion with utility companies on rates, (2) any industrial considering cogeneration installation, or (3) any industrial considering cogeneration expansion.

The goal of this chapter is to bring together many of the factors that will determine utility pricing under CO_2 capture. First, we will extend our modeling of steam turbines by utilizing the provided Excel-based thermodynamic functions (*TPSI+*) to predict the performance of a conventional coal-fired utility plant operating in both the design and off-design modes. To help understand off-design modeling, we will examine the case when steam throttling control is used to reduce utility plant power output. We will then predict plant off-design performance if a CO_2 capture technology is incorporated.

Postcombustion carbon dioxide capture is analogous to the incorporation of a chemical plant within the commercial utility's fence line; here the goal of the chemical plant will be to produce pipeline quality CO_2. Most technologies under study will require steam extraction from the power plant to help drive CO_2 separation processes (Ciferno et al., 2009). CO_2 capture technologies involving solvent or sorbents will require steam for regeneration. These additional steam extractions result in off-design operation of the power plant. In order to understand the impact of CO_2 capture and sequestration on utility pricing, we will utilize a levelized energy cost as dollar per kilowatt-hour. Consider a 600-MW power plant that after 90% CO_2 capture only produces 500 MW. The levelized energy cost will have three main components: First will be the cost to produce the 500 MW of delivered electricity; second will be recovery of capital and operating costs for the CO_2 capture and sequestration process; and, finally, makeup electricity at 100 MW will be needed to be purchased to keep available power from the plant constant.

Modeling, Analysis and Optimization of Process and Energy Systems, First Edition. F. Carl Knopf.

16.1 POWER PLANT DESIGN PERFORMANCE (USING OPERATIONAL DATA FOR FULL-LOAD OPERATION)

Utility company power plants are designed and optimized to produce electricity. Performance or efficiency is often reported as the turbine net heat rate or plant gross heat rate where the heat rate in British thermal unit per kilowatt-hour is the British thermal unit from the fuel required to produce a kilowatt-hour of electricity. When utility plant performance is based on normal full-load operating conditions, it is termed the "design case." It is also important to predict the performance of the power system in off-design or part load operation.

Steam turbine manufacturers can supply a "guaranteed" or expected performance for the turbine system operating under full-load conditions. They can also provide performance for the turbine under various operating conditions. This information comes from actual turbine tests and turbine tests in conjunction with detailed calculations often involving proprietary blade design data. General Electric (GE) has published a report (Spencer et al., 1963, 1974; also see Spencer note in references) that allows *calculation* of design and off-design performance for both reheat and non-reheat steam turbines.

We can also determine power plant design performance using *operational* plant data. Even with operational data, we will generally still want to use turbine exhaust end loss curves for part of the performance determination. Figure 16.1 shows a typical fossil fuel-fired utility plant configuration. Let us assume that the data (Li and Priddy, 1985) shown in Figure 16.1 represent normal full-load operating conditions. Actually, only the data highlighted in blue in the provided Excel file, **Example 16.1.xls** (and which are also given in Table 16.1a) are needed for the design case performance calculations.

The steam conditions from the boiler, $P = 3515$ psia and $T = 1000°F$, are supercritical (for water, $P_{Critical} \cong 3200.1$ psia and $T_{Critical} \cong 705.1°F$). The plant has steam reheat to 1000°F following the high-pressure turbine (HPT) section, and there are seven stages of regenerative feedwater heating. The feedwater heaters (FWHs) are numbered sequentially starting from the condenser. The condenser operates at 1.23 psia (2.5″ Hg). Table 16.1a,b summarizes the data available for design case calculations.

In Table 16.1a, steam pressures and temperatures at the turbine section inlets and exhaust are provided as well as the steam pressures and temperatures at each extraction point. When the exhaust from the turbine is single phase, its thermodynamic properties are completely specified by the pressure and temperature. When the exhaust is in two phases, we must specify the pressure or temperature and an independent quantity such as steam quality or steam enthalpy. In Table 16.1a, the steam from the low-pressure turbine (LPT) at extraction (1) and the outlet (expansion line end point [ELEP]) are two-phase flows; hence, in Table 16.1a, the enthalpies are provided.

Steam extraction is used to heat the condensed water returning to the boiler. The data in Table 16.1b provides approach temperature information for the FWHs. The terminal temperature difference (TD) is the TD between the condensate return stream leaving an FWH and the entering steam saturation temperature. The entering steam saturation temperature is determined at its extraction pressure:

$$T_{Out}^{Condensate} = T_{Saturation}^{Steam} - TD. \tag{16.1}$$

For example, for feedwater heater 7 (FWH-7) in Figure 16.1, the HP turbine extraction (7) steam is $P = 744$ psia and $T = 598.15°F$. At $P = 744$ psia, the steam saturation temperature using our Excel function calls = Tsat_Water (744) = 969.7 R = 510.1°F. Using Equation (16.1) with TD = 5°F, the temperature of the condensate return leaving FWH-7 and to the boiler would be 505.1°F.

In Table 16.1b, the drain cooler (DC) approach provides the temperature of the drain leaving the FWH as

$$T_{Out}^{Drain} = T_{In}^{Condensate} + DC. \tag{16.2}$$

Again, examining FWH-7 in Figure 16.1, $T_{In}^{Condensate} = 414.55°F$, giving $T_{Out}^{Drain} = 414.55 + 15 = 429.55°F$; $T_{In}^{Condensate}$ can be determined using Equation (16.1) for feedwater heater 6 (FWH-6).

We can now assemble the turbine and feedwater heater material and energy balances to determine the steam system design performance. These calculations are provided in the Excel sheet in **Example 16.1.xls**. In the Excel sheet, the design case calculations are on the left-hand side (LHS) of the sheet (columns A–M) and the off-design calculations (discussed next) are on the right-hand side (RHS) of the sheet (columns O–Z).

We have previously performed many of the needed steam turbine calculations—see Example 14.8. For the design case, there are really two separate problems being addressed on the Excel sheet. The first is determining the enthalpy, entropy, and steam quality in and out of each turbine section, as well as turbine section efficiency. This information allows the second problem of determining the steam extraction flows to meet the condensate return heating requirements to be solved. We can outline the calculation procedure as follows:

1. Use the known turbine inlet conditions and exhaust conditions at each extraction point to determine enthalpy, entropy, and steam quality in and out of each turbine section, as well as turbine section efficiency.
2. Use the known pressure drops in the reheater and boiler, terminal TD, and the DC approach to determine steam properties at appropriate locations.

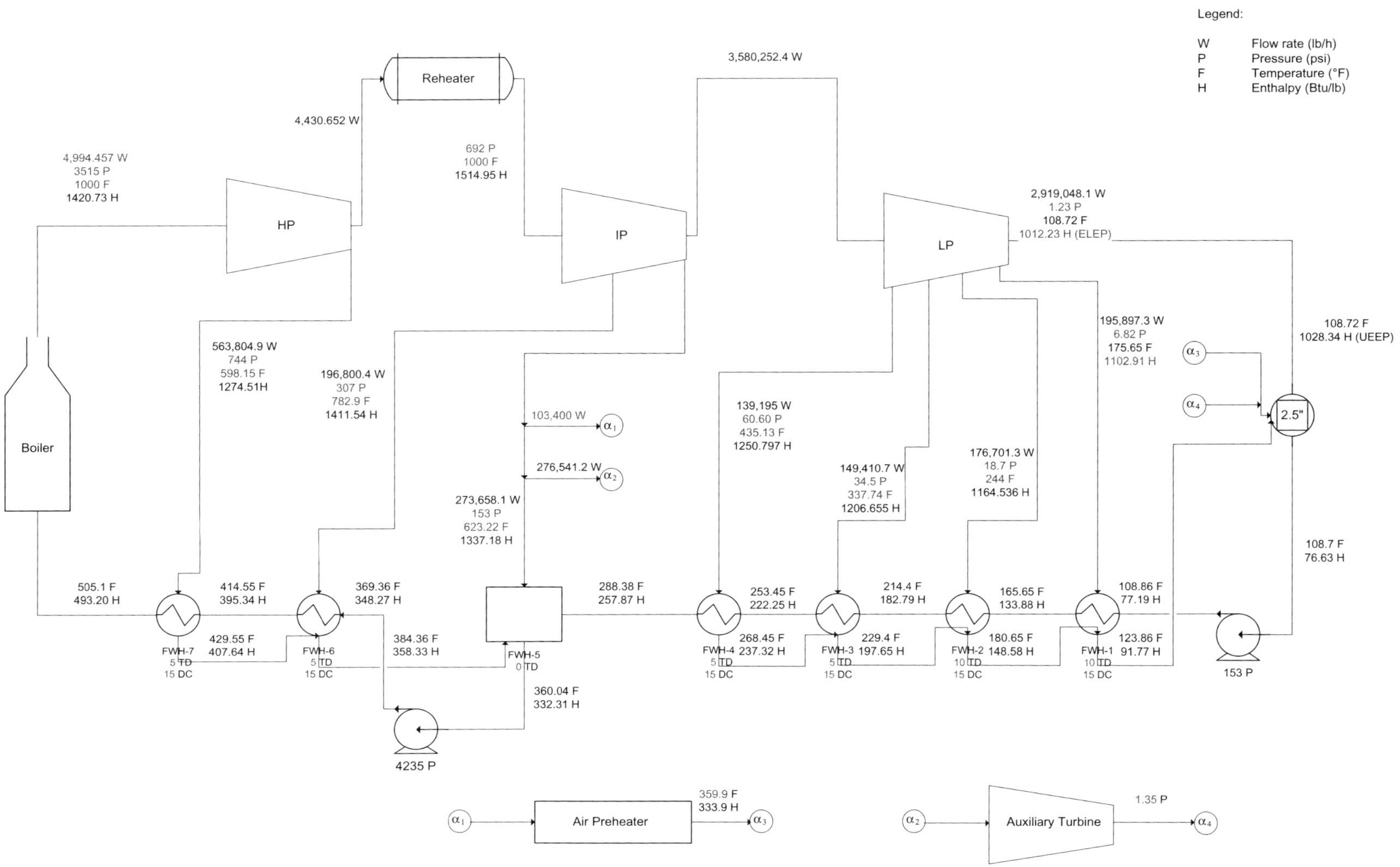

Figure 16.1 Heat balance diagram for the given design conditions (throttle steam flow ratio 1.0).

Table 16.1a Design Operating Data for Power Plant Heat Balance (Li and Priddy, 1985)

	Pressure (psia)	Temperature (F)	Enthalpy (Btu/lb)	Flow Rate (lb/h)	Pressure Drop (Dimensionless)	Efficiency/Size
1. Boiler					0.17	0.90
Outlet	3515.15	1000.00		4,994,457.0		
2. High-pressure turbine (HPT)						
Inlet HPT	3515.15	1000.00				
Extraction (7)	744.00	598.15				
Outlet P (psia)	744.00	598.15				
3. Reheater					0.07	
Outlet	691.92	1000.00				
4. Intermediate pressure turbine (IPT)						
Inlet (IPT)	691.92	1000.00				
Extraction (6)	307.00	782.91				
Extraction (5)	153.00	623.22				
Outlet (IPT)	153.00	623.22				
5. Low-pressure turbine (LPT)						
Inlet (LPT)	153.00	623.22	1102.91			
Extraction (4)	60.60	435.13	1012.23			
Extraction (3)	34.50	337.74				
Extraction (2)	18.70	244.00				
Extraction (1)	6.82	175.65				
Outlet (ELEP)	1.23	108.72				
6. Condenser in (UEEP)	1.23	108.72				
7. Condensate pump (motor driven)						0.82
8. Feedwater pump (auxiliary turbine driven)						0.85
9. Auxiliary turbine						0.80
Inlet from extraction 5	153.00	623.22				
Exhaust	1.35					
10. Air preheater						
Inlet from extraction 5	153.00	623.22		103,400.0		
Exhaust	(Sat. liquid)	359.9				
11. Generator						0.985
12. Low-pressure exhaust blading						30-in. with four flows

Table 16.1b Design Data for the Feedwater Heaters (Li and Priddy, 1985)

Feedwater Heater Number	Type of Heater	Drain Disposal	TD Terminal Temperature Difference (°F)	DC Drain Cooler Approach (°F)	Pressure Drop (Dimensionless)
1.	Surface heater with drain cooler	Charge to condenser	10.0	15.0	0.0
2.	Surface heater with drain cooler	Charge to next heater	10.0	15.0	0.0
3.	Surface heater with drain cooler	Charge to next heater	5.0	15.0	0.0
4.	Surface heater with drain cooler	Charge to next heater	5.0	15.0	0.0
5.	Contact heater	Contact heater	0.0		0.0
6.	Surface heater with drain cooler	Charge to next heater	5.0	15.0	0.0
7.	Surface heater with drain cooler	Charge to next heater	5.0	15.0	0.0

3. Determine the steam extraction flow rates starting with FWH-7 and continuing, FWH-6, feedwater heater 5 (FWH-5), . . . , until feedwater heater 1 (FWH-1) is reached.
4. Calculate the power used by the auxiliary turbine and low-pressure (LP) feedwater pump.
5. Calculate the turbine exhaust end loss using Spencer et al. (1963).
6. Sum the output from all the turbines; account for all losses and determine the turbine heat rate.

16.1.1 Turbine System: Design Case (See Example 16.1.xls)

Our thermodynamics database (*TPSI+*) provides the needed functions to determine steam properties and we are familiar (Chapters 7–9) with the use of Excel sheet functions to call these C functions. Here we detail the process for the HP turbine; the calculation procedures for intermediate pressure turbine (IPT) and LP turbine sections are identical. Values from the calculations and our thermodynamic functions are shown for the HP turbine section (recall all our developed thermodynamic functions contain an underscore; see Tables 8.1 and 8.2):

Step 1: Determine the HP turbine inlet enthalpy and entropy conditions.

$$P_{\text{HPT In}} = 3515.00 \text{ psia}$$

$$T_{\text{HPT In}} = 1000°\text{F} = 1459.67 \text{ R}$$

$$\begin{aligned}\hat{h}_{\text{HPT In}} &= \text{H\_Steam}(P_{\text{HPT In}}, T_{\text{HPT In}})/\text{MW\_H}_2\text{O}\\ &= \text{H\_Steam}(3515.00, 1459.67)/18.01534\\ &= 1420.73 \text{ Btu/lb}\end{aligned}$$

$$\begin{aligned}\hat{s}_{\text{HPT In}} &= \text{S\_Steam}(P_{\text{HPT In}}, T_{\text{HPT In}})/\text{MW\_H}_2\text{O}\\ &= \text{S\_Steam}(3515.00, 1459.67)/18.01534\\ &= 1.4690 \text{ Btu/lb-R}\end{aligned}$$

Step 2: Determine the HP turbine outlet saturation conditions (vapor and liquid states). These values will be needed to determine the steam quality.

Saturated Vapor

$$P_{\text{HPT Out}} = 744.00 \text{ psia}$$

$$\begin{aligned}T^{\text{Sat Vapor}}_{\text{HPT Out}} &= T_2^{\text{Sat Liquid}} = \text{Tsat\_Water}(P_{\text{HPT Out}})\\ &= 969.74 \text{ R}\end{aligned}$$

$$\begin{aligned}\hat{h}^{\text{Sat Vapor}}_{\text{HPT Out}} &= \text{H\_Steam}(P_2, T^{\text{Sat Vapor}}_{\text{HPT Out}})/\text{MW\_H}_2\text{O}\\ &= 1200.70 \text{ Btu/lb}\end{aligned}$$

$$\begin{aligned}\hat{s}^{\text{Sat Vapor}}_{\text{HPT Out}} &= \text{S\_Steam}(P_2, T^{\text{Sat Vapor}}_{\text{HPT Out}})/\text{MW\_H}_2\text{O}\\ &= 1.4238 \text{ Btu/lb-R}\end{aligned}$$

Saturated Liquid

$$P_{\text{HPT Out}} = 744.00 \text{ psia}$$

$$T^{\text{Sat Liquid}}_{\text{HPT Out}} = \text{TSat\_Water}(P_{\text{HPT Out}}) = 969.74 \text{ R}$$

$$\begin{aligned}\hat{h}^{\text{Sat Liquid}}_{\text{HPT Out}} &= \text{H\_Water}(P_{\text{HPT Out}}, T^{\text{Sat Liquid}}_{\text{HPT Out}})/\text{MW\_H}_2\text{O}\\ &= 499.61 \text{ Btu/lb}\end{aligned}$$

$$\begin{aligned}\hat{s}^{\text{Sat Liquid}}_{\text{HPT Out}} &= \text{S\_Water}(P_{\text{HPT Out}}, T^{\text{Sat Liquid}}_{\text{HPT Out}})/\text{MW\_H}_2\text{O}\\ &= 0.7009 \text{ Btu/lb-R}\end{aligned}$$

Step 3: Determine the turbine outlet conditions (entropy, $\hat{s}^{\text{Isentropic}}_{\text{HPT Out}}$; quality, $q^{\text{Isentropic}}_{\text{HPT Out}}$; temperature, $T^{\text{Isentropic}}_{\text{HPT Out}}$; and enthalpy, $\hat{h}^{\text{Isentropic}}_{\text{HPT Out}}$) assuming a 100% efficient isentropic expansion.

$$P_{\text{HPT Out}} = 744.00 \text{ psia}$$

$$\hat{s}^{\text{Isentropic}}_{\text{HPT Out}} = \hat{s}_1 = 1.4690 \text{ Btu/lb-R}$$

In step 3, we determine the steam quality leaving the HP turbine as

$$q^{\text{Isentropic}}_{\text{HPT Out}} = \frac{\hat{s}^{\text{Isentropic}}_{\text{HPT Out}} - \hat{s}^{\text{Sat Liquid}}_{\text{HPT Out}}}{\hat{s}^{\text{Sat Vapor}}_{\text{HPT Out}} - \hat{s}^{\text{Sat Liquid}}_{\text{HPT Out}}}.$$

Excel If statements (If (calculation, calculation true, calculation false)) are used to determine temperature and enthalpy values. If $q^{\text{Isentropic}}_{\text{HPT Out}}$ is ≥1, then the steam leaving the HP turbine is all vapor. Here, $T^{\text{Isentropic}}_{\text{HPT Out}}$ can be found from the known steam entropy and pressure as T_SP_Steam ($\hat{s}^{\text{Isentropic}}_{\text{HPT Out}} \times \text{MW\_H}_2\text{O}$, $P_{\text{HPT Out}}$). The steam enthalpy can be found as H_Steam ($P_{\text{HPT Out}}$, $T^{\text{Isentropic}}_{\text{HPT Out}}$)/MW_H$_2$O). If $q^{\text{Isentropic}}_{\text{HPT Out}}$ is <1, then the steam leaving the turbine is two phases. Here, $T^{\text{Isentropic}}_{\text{HPT Out}} = T^{\text{Sat Vapor}}_{\text{HPT Out}}$ and the enthalpy of the two-phase mixture is found as $(1 - q^{\text{Isentropic}}_{\text{HPT Out}}) \times \hat{h}^{\text{Sat Liquid}}_{\text{HPT Out}} + q^{\text{Isentropic}}_{\text{HPT Out}} \times \hat{h}^{\text{Sat Vapor}}_{\text{HPT Out}}$:

$$q^{\text{Isentropic}}_{\text{HPT Out}} = \frac{\hat{s}^{\text{Isentropic}}_{\text{HPT Out}} - \hat{s}^{\text{Sat Liquid}}_{\text{HPT Out}}}{\hat{s}^{\text{Sat Vapor}}_{\text{HPT Out}} - \hat{s}^{\text{Sat Liquid}}_{\text{HPT Out}}} = 1.062$$

$$\rightarrow \text{steam leaving HP turbine is all vapor as } q^{\text{Isentropic}}_{\text{HPT Out}} \geq 1$$

$$T^{\text{Isentropic}}_{\text{HPT Out}} = \text{If}\,(q^{\text{Isentropic}}_{\text{HPT Out}} < 1.0, T^{\text{Sat Vapor}}_{\text{HPT Out}}, \text{T\_SP\_Steam}\,(\hat{s}^{\text{Isentropic}}_{\text{HPT Out}} \times \text{MW\_H2O}, P_{\text{HPT Out}})) = 1019.66 \text{ R}$$

$$\hat{h}^{\text{Isentropic}}_{\text{HPT Out}} = \text{If}\,(q^{\text{Isentropic}}_{\text{HPT Out}} < 1.0, (1 - q^{\text{Isentropic}}_{\text{HPT Out}}) \times \hat{h}^{\text{Sat Liquid}}_{\text{HPT Out}} + q^{\text{Isentropic}}_{\text{HPT Out}} \times \hat{h}^{\text{Sat Vapor}}_{\text{HPT Out}}, \text{H\_Steam}\,(P_{\text{HPT Out}}, T^{\text{Isentropic}}_{\text{HPT Out}})/\text{MW\_H2O}) = 1245.56 \text{ Btu/lb.}$$

Step 4: Determine the HP turbine real outlet conditions and turbine section efficiency.

$$P_{\text{HPT Out}} = 744.00 \text{ psia}$$

$$T_{\text{HPT Out}} = 598.15°\text{F} = 1057.82 \text{ R}$$

In step 4, we determine the real enthalpy of the steam leaving the HP turbine $\hat{h}^{\text{Real}}_{\text{HPT Out}}$ using known $P_{\text{HPT Out}}$ and $T_{\text{HPT Out}}$ values. The steam quality leaving the HP turbine is determined as

$$q^{\text{Real}}_{\text{HPT Out}} = \frac{\hat{h}^{\text{Real}}_{\text{HPT Out}} - \hat{h}^{\text{Sat Liquid}}_{\text{HPT Out}}}{\hat{h}^{\text{Sat Vapor}}_{\text{HPT Out}} - \hat{h}^{\text{Sat Liquid}}_{\text{HPT Out}}}.$$

If $q^{\text{Real}}_{\text{HPT Out}}$ is ≥1, the real condition of the steam leaving the HP turbine is all vapor. If $q^{\text{Real}}_{\text{HPT Out}}$ is <1, the steam leaving the turbine is actually two phases. Here, $T_2^{\text{Real}} = T_{\text{HPT Out}} = T_{\text{HPT Out}}^{\text{Sat Liquid}} = T_{\text{HPT Out}}^{\text{Sat Vapor}}$, and in this formulation, the value $\hat{h}_{\text{HPT Out}}^{\text{Real}}$ must be supplied. The real entropy of the two-phase mixture is then found as $(1-q_1^{\text{Real}}) \times \hat{s}_2^{\text{Sat Liquid}} + q_1^{\text{Real}} \times \hat{s}_2^{\text{Sat Vapor}}$:

$$\hat{h}_{\text{HPT Out}}^{\text{Real}} = \text{H\_Steam}\,(P_{\text{HPT Out}}, T_{\text{HPT Out}})/\text{MW\_H2O}$$
$$= 1274.51 \text{ Btu/lb}$$

$$q_{\text{HPT Out}}^{\text{Real}} = \frac{\hat{h}_{\text{HPT Out}}^{\text{Real}} - \hat{h}_{\text{HPT Out}}^{\text{Sat Liquid}}}{\hat{h}_{\text{HPT Out}}^{\text{Sat Vapor}} - \hat{h}_{\text{HPT Out}}^{\text{Sat Liquid}}} = 1.105$$

$\rightarrow$ steam leaving HP turbine is all vapor as $q_{\text{HPT Out}}^{\text{Real}} \geq 1$.

For this problem, if $q_{\text{HPT Out}}^{\text{Real}} < 1$, $\hat{h}_{\text{HPT Out}}^{\text{Real}}$ must be provided (e.g., see Table 16.1a—LPT extraction (1) and outlet). Here then, $\hat{h}_{\text{HPT Out}}^{\text{Real}}$ could not be determined using H_Steam ($P_{\text{HPT Out}}$, $T_{\text{HPT Out}}$)/MW_H_2O as we would be in the two-phase region at the turbine outlet:

$$T_2^{\text{Real}} = T_{\text{HPT Out}}$$

$$\hat{s}_{\text{HPT Out}}^{\text{Real}} = \hat{s}_{\text{HPT Out}}$$
$$= \text{If }(q_{\text{HPT Out}}^{\text{Real}} < 1.0, (1-q_{\text{HPT Out}}^{\text{Real}}) \times \hat{s}_{\text{HPT Out}}^{\text{Sat Liquid}} + q_{\text{HPT Out}}^{\text{Real}}$$
$$\times \hat{s}_{\text{HPT Out}}^{\text{Sat Vapor}}, \text{S\_Steam}\,(\text{P}_{\text{HPT Out}}, T_{\text{HPT Out}}^{\text{Real}})/\text{MW\_H2O})$$
$$= 1.4969 \text{ Btu/lb-R.}$$

The real turbine efficiency is $\eta_{\text{HPT}}^{\text{Isentropic}} = \dfrac{\Delta\hat{h}^{\text{Real}}}{\hat{h}^{\text{Isentropic}}} = \dfrac{\hat{h}_{\text{HPT In}} - \hat{h}_{\text{HPT Out}}^{\text{Real}}}{\hat{h}_{\text{HPT In}} - \hat{h}_{\text{HPT Out}}^{\text{Isentropic}}}$.

These same procedures will be used for each IP turbine and LP turbine section. The turbine sections have steam extraction used for boiler feedwater heating. Also, part of the extracted steam from an IP turbine section is used to preheat air and to drive an auxiliary turbine used to compress the boiler feedwater.

16.1.2 Extraction Flow Rates and Feedwater Heaters

We can next address the problem of determining the extraction steam flows necessary to satisfy the feedwater heating requirements. These calculations must begin with FWH-7 (see Figure 16.2) and cascade until feedwater (1) is reached. We can assemble the needed material and energy balances for FWH-7; the other FWH will follow the same calculation procedure.

Feedwater Heater-7

Letting $F_{\text{FWH 7}}^{\text{Extraction 7}}$ = Extraction steam flow to FWH-7 (pound per hour);

$F_{\text{FWH 7}}^{\text{Drain 7}}$ = Drain flow (condensed steam) from FWH-7 (pound per hour);

$F_{\text{FWH 7}}^{\text{Condensate 7}}$ = Condensate flow from FWH-7 (to boiler) (pound per hour);

$F_{\text{FWH 6}}^{\text{Condenstae 6}}$ = Condensate flow from FWH-6 (pound per hour).

We can write the energy balance for FWH-7 as

$$\left(F_{\text{FWH 7}}^{\text{Extraction 7}}\right)\left(\hat{h}_{\text{HPT Out}}^{\text{Real}}\right) + \left(F_{\text{FWH 6}}^{\text{Condenstae 6}}\right)\left(\hat{h}_{\text{FWH 6}}^{\text{Condenstae 6}}\right)$$
$$= \left(F_{\text{FWH 7}}^{\text{Drain 7}}\right)\left(\hat{h}_{\text{FWH 7}}^{\text{Drain 7}}\right) + \left(F_{\text{FWH 7}}^{\text{Condensate 7}}\right)\left(\hat{h}_{\text{FWH 7}}^{\text{Condensate 7}}\right). \quad (16.3)$$

And using the material balance,

$$\left(F_{\text{FWH 7}}^{\text{Extraction 7}}\right) = \left(F_{\text{FWH 7}}^{\text{Drain 7}}\right) \quad (16.4)$$

gives

$$\left(F_{\text{FWH 7}}^{\text{Extraction 7}}\right)(1274.51) + (4{,}994{,}457)(395.34)$$
$$= \left(F_{\text{FWH 7}}^{\text{Extraction 7}}\right)(407.64) + (4{,}994{,}457)(493.20)$$

and

$$F_{\text{FWH 7}}^{\text{Extraction 7}} = 563804.9 \text{ lb/h.}$$

Results for each FWH are shown on Figure 16.1.

Feedwater Heater-5

The extraction at (5) also includes steam for air preheating, which is fixed at 103,400 lb/h and steam for an auxiliary turbine used to drive the high-pressure (HP) feedwater pump.

16.1.3 Auxiliary Turbine/ High-Pressure Feedwater Pump

For the auxiliary turbine, its efficiency ($\eta_{\text{AuxT}} = 0.80$), inlet steam conditions, and the turbine exhaust pressure are all known. An isentropic expansion calculation results in a two-phase exhaust with $\hat{h}_{\text{AuxT Out}}^{\text{Isentropic}} = 976.80$ Btu/lb. The actual enthalpy leaving is then found as

$$\hat{h}_{\text{AuxT Out}}^{\text{Real}} = \hat{h}_{\text{Extraction (5)}} - \eta_{\text{AuxT}}\left(\hat{h}_{\text{Extraction (5)}} - \hat{h}_{\text{AuxT Out}}^{\text{Isentropic}}\right)$$
$$= 1048.875 \text{ Btu/lb} \quad (16.5)$$

The power output of the auxiliary turbine is used by the HP feedwater pump to compress the boiler feedwater leaving FWH-5 from 153 psia to the boiler feed pressure of 4235 psia. The boiler feedwater leaves FWH-5 at 332.31 Btu/lb (153 psia, 360.04°F), and with a HP feedwater pump efficiency, $\eta_{\text{HP-Pump}} = 0.85$, the enthalpy out of the pump is

$$\hat{h}_{\text{HP-Pump Out}}^{\text{Real}} = \hat{h}_{\text{HP-Pump In}} + \frac{\left(\hat{h}_{\text{HP-Pump Out}}^{\text{Isentropic}} - \hat{h}_{\text{HP-Pump In}}\right)}{\eta_{\text{HP-Pump}}}$$
$$= 332.31 + \frac{(345.88 - 332.31)}{0.85} = 348.27 \frac{\text{Btu}}{\text{lb}}. \quad (16.6)$$

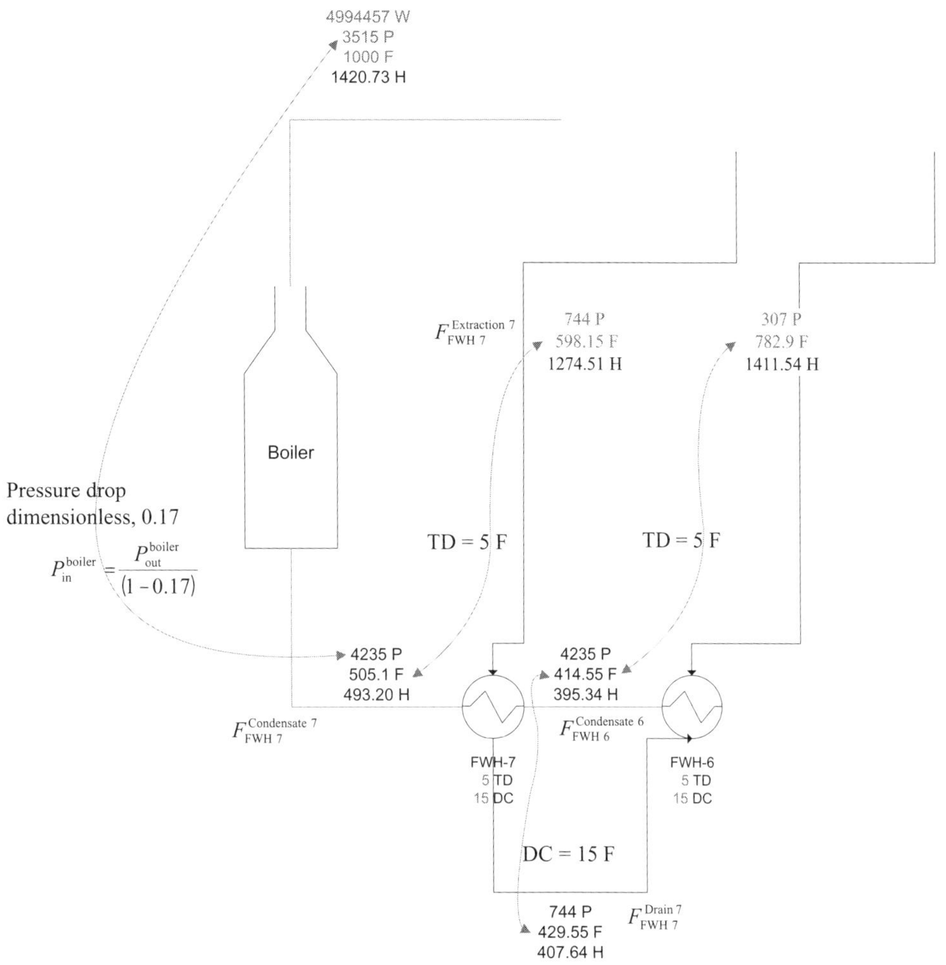

Figure 16.2 Figure showing nomenclature for feedwater heater 7 calculations.

This allows the steam demand of the auxiliary turbine to be found as

$$F_{AuxT}^{Steam} = \frac{F_{Feedwater}\left(\hat{h}_{HP\text{-}Pump\ Out}^{Real} - \hat{h}_{HP\text{-}Pump\ In}\right)}{\hat{h}_{Extraction\ (5)} - \hat{h}_{AuxT\ Out}^{Real}} = 276{,}541.16\frac{\text{lb}}{\text{h}}. \quad (16.7)$$

The power output of the auxiliary turbine would be

$$F_{AuxT}^{Steam}\left(\hat{h}_{Extraction\ (5)} - \hat{h}_{AuxT\ Out}^{Real}\right)\left(\frac{\text{Btu}}{\text{h}}\ \text{conversion to kW}\right)$$
$$= 23{,}360.29\ \text{kW}.$$

16.1.4 Low-Pressure Feedwater Pump

An electric drive motor is used to compress the condensate from the condenser from 1.23 psia to the pressure of FWH-5 at 153 psia. The LP pump efficiency $\eta_{HP\text{-}Pump} = 0.82$. The condensate leaves the condenser at 76.63 Btu/lb (1.23 psia, 108.7°F) and the enthalpy out of the LP pump is

$$\hat{h}_{LP\text{-}Pump\ Out}^{Real} = \hat{h}_{LP\text{-}Pump\ In} + \frac{\left(\hat{h}_{LP\text{-}Pump\ Out}^{Isentropic} - \hat{h}_{LP\text{-}Pump\ In}\right)}{\eta_{LP\text{-}Pump}}$$
$$= 76.63 + \frac{(77.09 - 76.63)}{0.82} = 77.19\frac{\text{Btu}}{\text{lb}}. \quad (16.8)$$

The power required by the LP pump is 642.13 kW.

16.1.5 Turbine Exhaust End Loss

The turbine exhaust end loss is the energy (the enthalpy) lost between the last stage of the LP turbine and the condenser inlet. The energy at the exit of LP turbine is the expansion line end point (ELEP) and the energy at the condenser inlet is the used energy end point (UEEP). These values, ELEP and UEEP, are not the same. Spencer et al. (1963) studied available GE turbines and provided curves for the turbine exhaust end loss as shown in Figure 16.3.

From Table 16.1a, the size of the LP turbine section exhaust blading is provided as 30 in. and there are four exhaust flows to the condenser. Figure 16.3 is for a turbine with a last-stage blade length of 30 in., a pitch diameter of 85 in., an annulus area of 55.6 ft^2, and an operating at 3600 rpm. Curves are available for different blade lengths and turbine speeds (see Spencer et al., 1963). In order to use Figure 16.3, we determine the annulus velocity as

$$V_{\text{an}} = \frac{Q_a \hat{\upsilon}(1-0.01Y)}{3600\ A_{\text{an}}}, \tag{16.9}$$

where

V_{an} = annulus velocity in foot per second;
Q_a = (total condenser flow/number of condenser flows) in pound per hour
= (2,919,048.06/with four condenser flows) = 729,762 lb/h;
$\hat{\upsilon}$ = saturated vapor specific volume at the end point in cubic foot per pound
= V_Steam (1.23 psia, Tsat)/MW_H_2O = 274.49 ft^3/lb;
Y = percent moisture of the ELEP (Y = 1 – quality), here Y = 9.34%; and
A_{an} = annulus area in ft^2 = 55.6 ft^2,

giving for this design case

$$V_{\text{an}} = \frac{\left(\dfrac{2919048.06}{4}\right)(274.49)(1-0.01(9.34))}{3600(55.6)} = 907.32\ \text{ft/s}.$$

Then from Figure 16.3, we find the exhaust loss is ≈21.74 Btu/lb of dry flow. In order to convert the exhaust loss in British thermal unit per pound of dry flow to British thermal unit per pound of steam flow, Spencer et al. (1963) provide

$$\begin{aligned}\text{Exhaust end loss per lb of steam flow} \\ &= (\text{Exhaust loss})(0.87)(1-0.01Y)(1-0.0065Y) \\ &= (21.74\ \text{Btu/lb of dry flow})(0.87)(1-0.01(9.34)) \\ &\quad (1-0.0065(9.34) \\ &= 16.108\ \text{Btu/lb of steam flow.}\end{aligned} \tag{16.10}$$

Then,

$$\begin{aligned}\text{UEEP} &= \text{ELEP} + (\text{Exhaust end loss/lb of steam flow}) \\ &= 1012.23 + 16.108 = 1028.338\ \text{Btu/lb.}\end{aligned} \tag{16.11}$$

In the Excel file **Example 16.1.xls**, we supply Figure 16.3 as a function, Function Exhaust_Loss (V_{an}), which contains three curves as

V_{an}, Annulus Velocity (ft/s)	Exhaust Loss, Btu/lb of Dry Flow
$V_{\text{an}} \le 400$	$= (196.71)(2.7182818)^{(-0.0054)(V_{\text{an}})}$
$400 \le V_{\text{an}} \le 850$	$= (-3.72463 \times 10^{-7})(V_{\text{an}})^3 + (0.0009103)(V_{\text{an}})^2 - (0.694268)(V_{\text{an}}) + 179.68773$
$V_{\text{an}} \ge 850$	$= (0.0527)(V_{\text{an}}) - 26.074$

For V_{an} values below 100 ft/s or above 1400 ft/s, see Spencer et al. (1963). There can be confusion over the ELEP and UEEP; notice that the enthalpy at the UEEP is larger in value compared to the enthalpy at the ELEP. The ELEP represents the turbine exhaust conditions if there are no losses at the LP turbine exit—this would be the maximum power the LP section could raise. Losses associated with the last stage (see Raj, 2008) decrease the amount of power that can be raised in the LP turbine section. We will calculate the maximum LP turbine work based on the ELEP and then account for work lost from the ELEP to the actual LP turbine exit conditions at the UEEP.

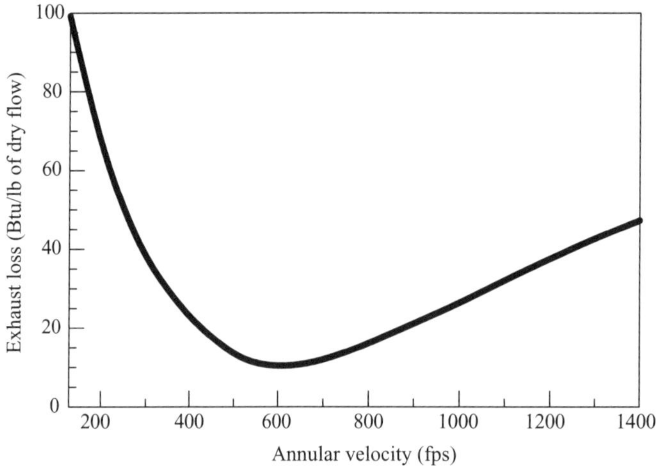

Figure 16.3 Turbine end exhaust loss versus annulus velocity—3600 rpm.

16.1.6 Steam Turbine System Heat Rate and Performance Parameters

We can now use the results from our material and energy balance calculations as assembled in the Excel file **Example 16.1.xls** to determine the steam plant heat rate. The steam turbine gross work output is the sum of work for each turbine section as

$$\dot{W} = \sum_{j'=\text{In}} F_{j'}\hat{h}_{j'} - \sum_{j=\text{Out}} F_j\hat{h}_j. \tag{16.12}$$

High-Pressure Turbine Section

$$\dot{W}_{\text{HPT}} = \sum_{j'=\text{HPT In}} F_{j'}\hat{h}_{j'} - \sum_{j=\text{HPT Out}} F_j\hat{h}_j$$

$$= \frac{(4{,}994{,}457(1420.73) - 563{,}804.9(1274.51) - 4{,}430{,}652.1(1274.51))}{3600\,\text{s/h}}$$

$$= 202{,}859.5\ \text{Btu/s} = 213{,}974.3\ \text{kW}$$

Intermediate-Pressure Turbine Section

$$\dot{W}_{\text{IPT}} = \frac{(4{,}430{,}652(1514.95) - 196{,}800.4(1411.54) - (103{,}400 + 276{,}541.2 + 273{,}658.1 + 3{,}580{,}252.4)(1337.18))}{3600\ \text{s/h}}$$

$$= 214719.28\ \text{Btu/sec} = 226483.85\ \text{kW}$$

Low-Pressure Turbine Section

$$\dot{W}_{\text{LPT}} = \frac{3{,}580{,}252.4(1337.18) - 139{,}196(1250.8) - 149{,}410.7(1206.66) - 176{,}701.3(1164.5) - 195{,}897.3(1102.9) - 2{,}919{,}048(1012.2)}{3600\ \text{s/h}}$$

$$= 293{,}465.5\ \text{Btu/s} = 309{,}544.6\ \text{kW}$$

The total (gross) turbine output is $\dot{W}_{\text{HPT}} + \dot{W}_{\text{IPT}} + \dot{W}_{\text{LPT}} = 750{,}002.79\ \text{kW}$

The *turbine net output* is found by subtracting the exhaust end losses from the turbine gross output. The exhaust end losses are

$$\text{Exhaust end losses} = \left(2{,}919{,}048.1\frac{\text{lb}}{\text{h}}\right)\left(1012.23 - 1028.338\frac{\text{Btu}}{\text{lb}}\right)\left(\frac{\text{h}}{3600\ \text{s}}\right)\left(\frac{1.05479\ \text{kW-s}}{\text{Btu}}\right) = -13{,}776.79\ \text{kW}$$

$$\text{Turbine net output} = (750{,}002.79 - 13{,}776.79\ \text{kW}) = 736{,}226.0\ \text{kW}.$$

The *generator net output* is found from the turbine net output and generator efficiency as

$$\text{Generator net output} = (736{,}226.0\ \text{kW})(0.985\ \text{generator efficiency}) = 725{,}182.6\ \text{kW}.$$

The *turbine net heat rate* is found as the amount of heat supplied to the turbine system (from the boiler including reheat) divided by the generator net output:

$$\text{Turbine net heat rate} = \frac{\text{Heat input to boiler including reheat (Btu/h)}}{\text{Generator net output (kW)}}$$

$$= \frac{4{,}994{,}457(1420.727 - 493.20) + 4{,}430{,}652.1(1514.948 - 1274.506)}{725{,}182.61}$$

$$= 7857\frac{\text{Btu}}{\text{kW-h}}. \tag{16.13}$$

The *turbine gross heat rate* is found as the amount of heat supplied to the turbine system divided by the generator net output + auxiliary turbine output:

$$\text{Turbine gross heat rate} = \frac{\text{Heat input to boiler including reheat (Btu/h)}}{\text{Generator net output} + \text{Auxiliary turbine output (kW)}}$$

$$= \frac{4{,}994{,}457(1420.727 - 493.20) + 4{,}430{,}652.1(1514.948 - 1274.506)}{725{,}182.61 + 23{,}360.29} = 7611.87\frac{\text{Btu}}{\text{kW-h}}. \tag{16.14}$$

The plant or system net heat rate is found as

$$\text{Plant net heat rate} = \frac{\text{Turbine net heat rate (Btu/kW-h)}}{\eta_{\text{Boiler}}\left(1 - \dfrac{\text{Auxiliary power (\%)}}{100}\right)}. \tag{16.15}$$

Here, Li and Priddy (1985) provide the plant auxiliary power consumption as 7.0% (including, e.g., coal handling), and with $\eta_{\text{Boiler}} = 0.90$, the plant net heat rate is found as 9387.2 Btu/kW-h. The net heat rate for a natural gas-fired power plant is typically 6800–7000 Btu/kW-h and a coal-fired power plant will be nominally 10,000 Btu/kW-h. Finally, the plant power output is found as

$$\text{Plant power output} = \text{Generator net output}\,(1 - \text{Auxiliary power consumption (\%)}) \tag{16.16}$$

Here the plant power output is 674,420 kW; this 674,420 kW would be the power delivered to the transmission line.

16.2 POWER PLANT OFF-DESIGN PERFORMANCE (PART LOAD WITH THROTTLING CONTROL OPERATION)

Typically, in response to reduced demand for power, steam power plants in the United States reduce steam flow in the boiler thereby producing less electricity. Reducing steam flow to the HP turbine section while maintaining boiler design outlet P and T is termed steam throttling or throttling control operation. Here we develop equations to predict turbine system performance under throttling control.

Once we have completed Section 16.2, we will find we only need to specify a change in the steam flow from the boiler to allow complete prediction of off-design performance. The complete off-design calculations are already provided in the Excel file **Example 16.1.xls**, columns O–Z.

The solution to the off-design problems involves

1. Supplying initial estimates for all pressures and efficiencies in the turbine system
2. Modifying pressures in the turbine system based on the off-design turbine inlet conditions
3. Modifying the efficiencies using the correlations in Spencer et al. (1963) or based on an approximate method utilizing the off-design velocity into each turbine section

16.2.1 Initial Estimates for All Pressures and Efficiencies: Sub Off_Design_Initial_Estimates ( )

In Section 16.1, we solved the design case material and energy balances for the turbine system using our thermodynamic functions and known $P$and T or P and $\hat{h}$ (for a two-phase steam) at each turbine section inlet, outlet, and extraction point. We begin the off-design case by "resolving" the design case using known P and calculated η for each turbine section. The efficiency for each turbine section is

$$\eta^{\text{Isentropic}}_{\text{Turbine Section}} = \frac{\Delta\hat{h}^{\text{Real}}_{\text{Turbine Section}}}{\hat{h}^{\text{Isentropic}}_{\text{Turbine Section}}} = \frac{\hat{h}_{\text{Turbine Section In}} - \hat{h}^{\text{Real}}_{\text{Turbine Section Out}}}{\hat{h}_{\text{Turbine Section In}} - \hat{h}^{\text{Isentropic}}_{\text{Turbine Section Out}}}. \tag{16.17}$$

In the Excel file **Example 16.1.xls**, these efficiency calculations are done as part of the design case calculations (column E). These $\eta^{\text{Isentropic}}_{\text{Turbine Section}}$ values are copied to the off-design case in the macro subroutine Off_Design_Initial_Estimates ( ).

Also in this subroutine, the pressure at each stage i is modified from the design pressure by multiplication with the flow ratio:

$$P_{i,\text{Off-Design}} = P_{i,\text{Design}} \times \text{Flow\_ratio}, \tag{16.18}$$

where the Flow_Ratiio = Off-design flow rate/Design flow rate.

For the demonstration calculations in Section 16.2, we will use a flow ratio = 0.75. The boiler pressure drop is also multiplied by the flow ratio. For the off-design case, the pressure and temperature leaving the boiler (3515 psia, 1000°F) and the pressure at the condenser (1.23 psia) remain at design conditions.

The off-design system performance can now be solved for the given off-design steam flow rate from the boiler using the design case efficiencies, estimated pressures, and assuming that all other conditions, including pump efficiencies and feedwater heater approach temperatures, remain at design conditions. The complete off-design calculations are provided in the Excel file **Example 16.1.xls**, columns O–Z.

16.2.2 Modify Pressures: Sub Pressure_Iteration ( )

We next enhance our off-design performance calculations by improving both our pressure and efficiency estimates. For off-design turbine performance, when controlled by steam throttling, we can begin with Stodola's ellipse law (Stodola and Lowenstein, 1927), which provides a relation between steam flow and pressure drop in a turbine section as

$$F_{\text{Turbine Section}} = K\sqrt{1-\frac{P_{\text{Out}}}{P_{\text{In}}}}, \tag{16.19}$$

where $F_{\text{Turbine Section}}$ is the steam flow in the turbine section (pound per hour); K is a proportionality constant; and P_{Out} = pressure out of the section and P_{In} = pressure in. But Equation (16.19) cannot be used for off-design calculations since it does not take into account the effect of varying inlet temperature or the effect of inlet pressure if the pressure ratio is constant. Erbes and Eustis (1986) utilize a modification of the ellipse law suggested by Sylvestri, which accounts for varying inlet conditions as

$$F_{\text{Turbine Section}} = K\sqrt{\frac{(P_{\text{In}})^2-(P_{\text{Out}})^2}{\hat{\upsilon}_{\text{In}}P_{\text{In}}}}, \tag{16.20}$$

where $\hat{\upsilon}_{\text{In}}$ = the stage inlet specific volume (cubic foot per pound) and the remaining terms are the same as Equation (16.19).

In order to use Equation (16.20), the constant K is first determined for each turbine section using design conditions. In the Excel file **Example 16.1.xls**, $K = K^{\text{Turbine Section}}_{\text{Design}}$ calculations are done as part of the design case calculations (column L); $\hat{\upsilon}_{\text{In}}$ values (column K) are found using our function = V_Steam (P, T)/MW_H_2O.

EXAMPLE 16.1 *Calculate K*

Calculate K ($K_{\text{Design}}^{\text{Turbine Section}}$) for the first stage of the IP turbine and the last stage of the LP turbine.

SOLUTION The first stage of the IP turbine is IPT In to Extraction (6).

From the Excel sheet, the design case conditions for the first stage of the IP turbine show $F_{\text{Turbine Section}} = 4{,}430{,}652.11$ lb/h = 1230.74 lb/s.

$P_{\text{In}} = 691.92$ psia
$P_{\text{Out}} = 307$ psia
$\hat{v}_{\text{In}}$ = V_Steam (691.92 psia, 1000 + 459.67 R)/MW_H_2O = 1.2182 ft^3/lb

$$K = K_{\text{Design}}^{\text{First Stage IPT}} = \frac{F_{\text{Turbine Section}}}{\sqrt{\dfrac{(P_{\text{In}})^2 - (P_{\text{Out}})^2}{\hat{v}_{\text{In}} P_{\text{In}}}}} = \frac{1230.74}{\sqrt{\dfrac{(691.92)^2 - (307)^2}{(1.2182)(691.92)}}}$$

$$= 57.62$$

The last stage of the LP turbine is Extraction (1) to LPT Out:

From the Excel sheet, the design case conditions for the LP turbine last stage show

$F_{\text{Turbine Section}} = 2{,}919{,}048.06$ lb/h = 810.85 lb/s
$P_{\text{In}} = 6.82$ psia
$P_{\text{Out}} = 1.23$ psia
q_{In} = steam quality in = 0.96639
V_Steam (6.82 psia, Tsat = 175.65 + 459.67 R)/MW_H_2O = 54.97797 ft^3/lb
V_Water (6.82 psia, Tsat = 175.65 + 459.67 R)/MW_H_2O = 0.016484 ft^3/lb

$\hat{v}_{\text{In}}$ = 0.96639 (54.97797) + (1 − 0.96639) (0.016484) = 53.1307 ft^3/lb

$$K = K_{\text{Design}}^{\text{Last Stage LPT}} = \frac{F_{\text{Turbine Section}}}{\sqrt{\dfrac{(P_{\text{In}})^2 - (P_{\text{Out}})^2}{\hat{v}_{\text{In}} P_{\text{In}}}}} = \frac{810.85}{\sqrt{\dfrac{(6.82)^2 - (1.23)^2}{(53.1307)(6.82)}}} = 2300.9$$

The results for all $K_{\text{Design}}^{\text{Turbine Section}}$ are given in Table 16.2.

With each $K_{\text{Design}}^{\text{Turbine Section}}$ determined, we can use Equation (16.20) in a reverse-order iterative calculation, starting at the LP turbine outlet, to update the off-design pressure distribution in the turbine system. Equation (16.20) can be rearranged to solve for an updated value for P_{In} as

$$P_{\text{In}}^{\text{Stage } i} = \frac{\left[\hat{v}_{\text{In}}^{\text{Stage } i}\left(\dfrac{F_{\text{Turbine Section}}^{\text{Stage } i}}{K_{\text{Design}}^{\text{Stage } i}}\right)^2 + \sqrt{\left(\hat{v}_{\text{In}}^{\text{Stage } i}\left(\dfrac{F_{\text{Turbine Section}}^{\text{Stage } i}}{K_{\text{Design}}^{\text{Stage } i}}\right)^2\right)^2 + 4\left(P_{\text{Out}}^{\text{Stage } i}\right)^2}\,\right]}{2}. \quad (16.21)$$

All the terms on the RHS of Equation (16.21) are known from the initial estimate of the off-design conditions. Do note that this modified form of Stodola's ellipse is suitable for predicting pressure in multiple-stage turbines; it is not intended for use with a single turbine stage and it does not apply when the inlet is throttled. The inlet pressure to the HP turbine is controlled using throttling to keep the boiler outlet pressure constant.

We have summarized the off-design initial estimates from Sub Off_Design_Initial_Estimates ( ) in Table 16.2. Starting at the exit of the LP turbine, where the pressure is fixed at $P_{\text{Out}} = 1.23$ psia, we can solve for P_{In} for each turbine section using Equation (16.12) (see Example 16.2); these results are also shown in the last column of Table 16.2. ■

Table 16.2 Initial Estimates for Off-Design Conditions (from Sub Off_Design_Initial_Estimates ( ))

	$K_{\text{Design}}^{\text{Turbine Section}}$ Fixed from Design Case	$F_{\text{Turbine Section}}^{\text{Stage i}}$ Off-Design Estimate (lb/s)	$P_{\text{In}}^{\text{Stage } i}$ Off-Design Estimate (psia)	$P_{\text{Out}}^{\text{Stage } i}$ Off-Design Estimate (psia)	$\hat{v}_{\text{In}}^{\text{Stage } i}$ (ft^3/lb) Based on Off-Design $P_{\text{In}}^{\text{Stage } i}$, $T_{\text{In}}^{\text{Stage } i}$	$P_{\text{In}}^{\text{Stage } i}$ Update by Equation (16.21) (psia)
HPT						
In to out	10.8553	1040.51	3515.00	558.00	0.2056	
Reheater			558.00	518.94		
IPT						
In to Ex(6)	57.6228	930.11	518.94	230.25	1.6373	529.44
Ex(6) to out	118.6722	892.07	230.25	114.75	3.1576	235.63
LPT						
In to Ex(4)	177.8415	754.20	114.75	45.45	5.5428	117.62
Ex(4) to Ex(3)	438.9098	726.99	45.45	25.88	11.6006	46.37
Ex(3) to Ex(2)	681.5781	697.67	25.88	14.03	18.1809	26.51
Ex(2) to Ex(1)	1009.3094	663.39	14.03	5.12	29.6189	14.57
Ex(1) to out	2300.9113	628.40	5.12	1.23	70.1381	5.54

EXAMPLE 16.2 ***Calculate Updated Values for P_{In}***

Calculate P_{In} for the last two stages of the LP turbine using Equation (16.21).

SOLUTION For Extraction (1) to LPT Out, using the data of Table 16.2,

$$P_{\text{In}}^{\text{Extraction (1)}} = \frac{\left[70.1381\left(\frac{628.40}{2300.9113}\right)^2 + \sqrt{\left(70.1381\left(\frac{628.40}{2300.9113}\right)^2\right)^2 + 4(1.23)^2}\right]}{2} = 5.54.$$

For Extraction (2) to Extraction (1), using the data of Table 16.2,

$$P_{\text{In}}^{\text{Extraction (2)}} = \frac{\left[29.6189\left(\frac{663.39}{1009.3094}\right)^2 + \sqrt{\left(29.6189\left(\frac{663.39}{1009.3094}\right)^2\right)^2 + 4(5.12)^2}\right]}{2} = 14.57.$$

Do note that the updated P values are not used in Equation (16.21); here, for example, $P_{\text{In}}^{\text{Extraction (1)}}$ is *not* used in the calculation of $P_{\text{In}}^{\text{Extraction (2)}}$.

When updated values for all the P_{In}'s have been determined, these values can replace the initial estimated $P_{\text{In}}^{\text{Stage } i}$ values found in Table 16.2. These new P values, when placed on the Excel sheet (off-design case), will generate new flow rates, temperatures, and steam specific volumes into each turbine section. This replacement (iteration process) continues until the pressure values remain unchanged. In Sub Pressure_Iteration ( ), 10 iterations are used, but it would be straightforward to add a convergence criteria. Also on the Excel sheet, we account for the possibility that two phases may be present in the last four stages of the LP turbine section.

With converged pressures for the IP turbine and LP turbine sections, we next update the HP turbine stage. The HP section is not included in the iteration process as the pressure drop in the reheater, $\Delta P_{\text{Reheater}}^{\text{Off-Design}}$, fixes the outlet pressure of the HP turbine as

$$P_{\text{Out}}^{\text{HPT}} = P_{\text{In}}^{\text{IPT}} + \Delta P_{\text{Reheater}}^{\text{Off-Design}}. \quad (16.22)$$

The pressure drop in the reheater can be determined by assuming a homogeneous flow model. This flow model provides the pressure in a pipe section as

$$\Delta P = \frac{4fLV_{\text{Avg}}^2\rho}{2d},$$

where f is the friction factor, L is the pipe length, V_{Avg} is the average fluid velocity, d is the pipe diameter, and ρ is the fluid density. Using the homogeneous flow model, we can determine the off-design pressure drop in the reheater in terms of the average fluid velocity for the design and off-design cases and the known design pressure drop as

$$\Delta P_{\text{Reheater}}^{\text{Off-Design}} = \frac{\left(\frac{V_{\text{In Reheater}}^{\text{Off-Design}} + V_{\text{In Reheater}}^{\text{Off-Design}}}{2}\right)^2}{\left(\frac{V_{\text{In Reheater}}^{\text{Design}} + V_{\text{In Reheater}}^{\text{Design}}}{2}\right)^2} \Delta P_{\text{Reheater}}^{\text{Design}}. \quad (16.23)$$

We also use an equation analogous to Equation (16.23) to find the pressure drop in the boiler. These calculations are assembled in the macro subroutine Sub Pressure_Iteration ( ). ■

16.2.3 Modify Efficiencies: Sub Update Efficiencies ( )

In the off-design calculations, we have utilized the $\eta_{\text{isentropic}}$ values found in the design case. We next want to update these efficiency values. Here a commonly used approach is to modify efficiencies using the correlations presented in Spencer et al. (1963). An alternative *approximate* approach, which we have found easy to implement, is based on turbine sections in actual operation showing both impulse and reactive characteristics. For our purposes, the actual details of steam expansion in the turbine are not needed, but discussion of impulse and reactive turbines can be found in Salisbury (1950).

It is reasonable to assume 50% reaction blading for each turbine section and from Salisbury (1950), stage efficiency can be found as

$$\eta_{\text{Design}}^{\text{Isentropic}} = 2y\left[(a-y)+\sqrt{(a-y)^2+1-a^2}\right], \quad (16.24)$$

where $a = \sqrt{1-x}$, x = fraction of the stage energy released in the bucket system, $y = W_{\text{Design}}/V_{\text{Design}}$, W_{Design} = turbine rotational speed, and V_{Design} = inlet steam velocity to the turbine section. For a 50% reaction stage, $x = 0.5$ and $a = 0.7071$. Furthermore, for a 50% reaction stage, the maximum efficiency

$$y_{\text{Optimal}} = \left(\frac{W_{\text{Design}}}{V_{\text{Design}}}\right)_{\text{Optimal}} = 0.7071,$$

giving $\eta_{\text{Design}}^{\text{Isentropic}} = 100\%$.

For power generation turbines, rotational speed will remain constant (3600 rpm) in both the design and off-design cases, allowing us to write

$$\frac{\left(\frac{W_{\text{Off-Desin}}}{V_{\text{Off-Design}}}\right)}{\left(\frac{W_{\text{Design}}}{V_{\text{Design}}}\right)_{\text{Optimal}}} = \frac{(V_{\text{Design}})_{\text{Optimal}}}{V_{\text{Off-Design}}}. \quad (16.25)$$

It is then possible to use Equation (16.24) to ratio the off-design and design efficiencies as

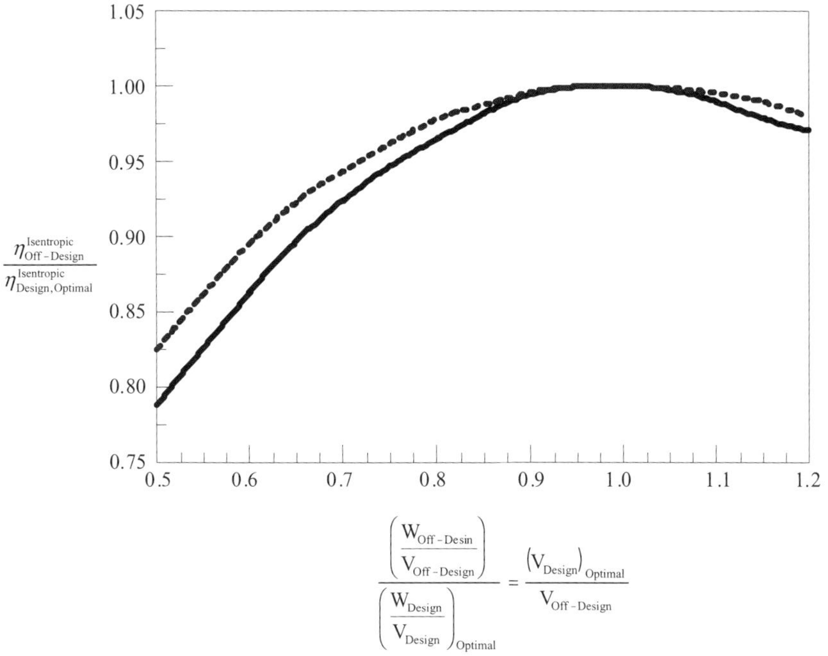

Figure 16.4 Variation of isentropic efficiency with velocity ratio for experimental data and ideal stage calculations (Eq. (16.26)) from Erbes and Eustis (1986).

$$\frac{\eta^{\text{Isentropic}}_{\text{Off-Design}}}{\eta^{\text{Isentropic}}_{\text{Design, Optimal}}} \cong 2\frac{a}{\left(\dfrac{(V_{\text{Design}})_{\text{Optimal}}}{V_{\text{Off-Design}}}\right)}\left[\left(a-\frac{a}{\left(\dfrac{(V_{\text{Design}})_{\text{Optimal}}}{V_{\text{Off-Design}}}\right)}\right)+\sqrt{\left(a-\frac{a}{\left(\dfrac{(V_{\text{Design}})_{\text{Optimal}}}{V_{\text{Off-Design}}}\right)}\right)^2+1-a^2}\,\right]. \tag{16.26}$$

In Equation (16.26), $a = 0.7071$. Equation (16.26) is plotted in Figure 16.4 and shows good agreement with experimental stage data over the expected

$$\frac{(V_{\text{Design}})_{\text{Optimal}}}{V_{\text{Off-Design}}}$$

range of 0.5–1.2.

Equation (16.26) allows us to modify the design efficiency to account for changes in steam velocity between the design and off-design calculations. Here, the assumption is made that $(V_{\text{Design}})_{\text{Optimal}} = (V_{\text{Design}})$ for each stage; in other words, it is assumed that for the design case, each stage of the turbine was optimally designed. The calculation using Equation (16.26) are found in macro subroutine Sub Update Efficiencies ( ).

The complete off-design calculation procedure can be implemented by calling the macro subroutine Sub Complete_Off_Design_Calculations ( ), which in turn calls Off_Design_Initial_Estimates, Pressure_Iteration, and Update_Efficiencies. Details for the exhaust end loss can be found in the Excel file **Example 16.1.xls**; basically, the part load expansion line is taken parallel to the design expansion line. The final results for the flow ratio = 0.75 are provided in Figure 16.5 and are summarized in Table 16.3.

The stage efficiencies using the approximate method (Sub Update Efficiencies ( )) show a ~4% reduction in HP turbine efficiency and virtually no efficiency change in the other stages, when comparing the design case to a 0.75 steam throttle. This result is in good agreement with calculations done following the method outlined in Spencer et al. (1963).

The off-design solution method developed here is robust. As noted by Erbes and Eustis (1986), "Since the pressure calculations (Sub Pressure_Iteration) are made in reverse order through the turbine, the effects of changes in condenser pressure, variations in extraction flow rates, the removal of feedwater heaters from service, and changes in feedwater-pump turbine extraction flow rate on the upstream pressure distribution are automatically taken care of."

16.3 LEVELIZED ECONOMICS FOR UTILITY PRICING

Concerns for global warming are increasing the probability that CO_2 emissions will be regulated. A most likely target will be coal-fired power plants, which generate about one-half of the 600+ GW of electricity used in the United States today. Continued reliance and expansion of coal-fired plants for electricity are anticipated for the next 20+ years (Energy Information Administration [EIA]). Therefore, any true

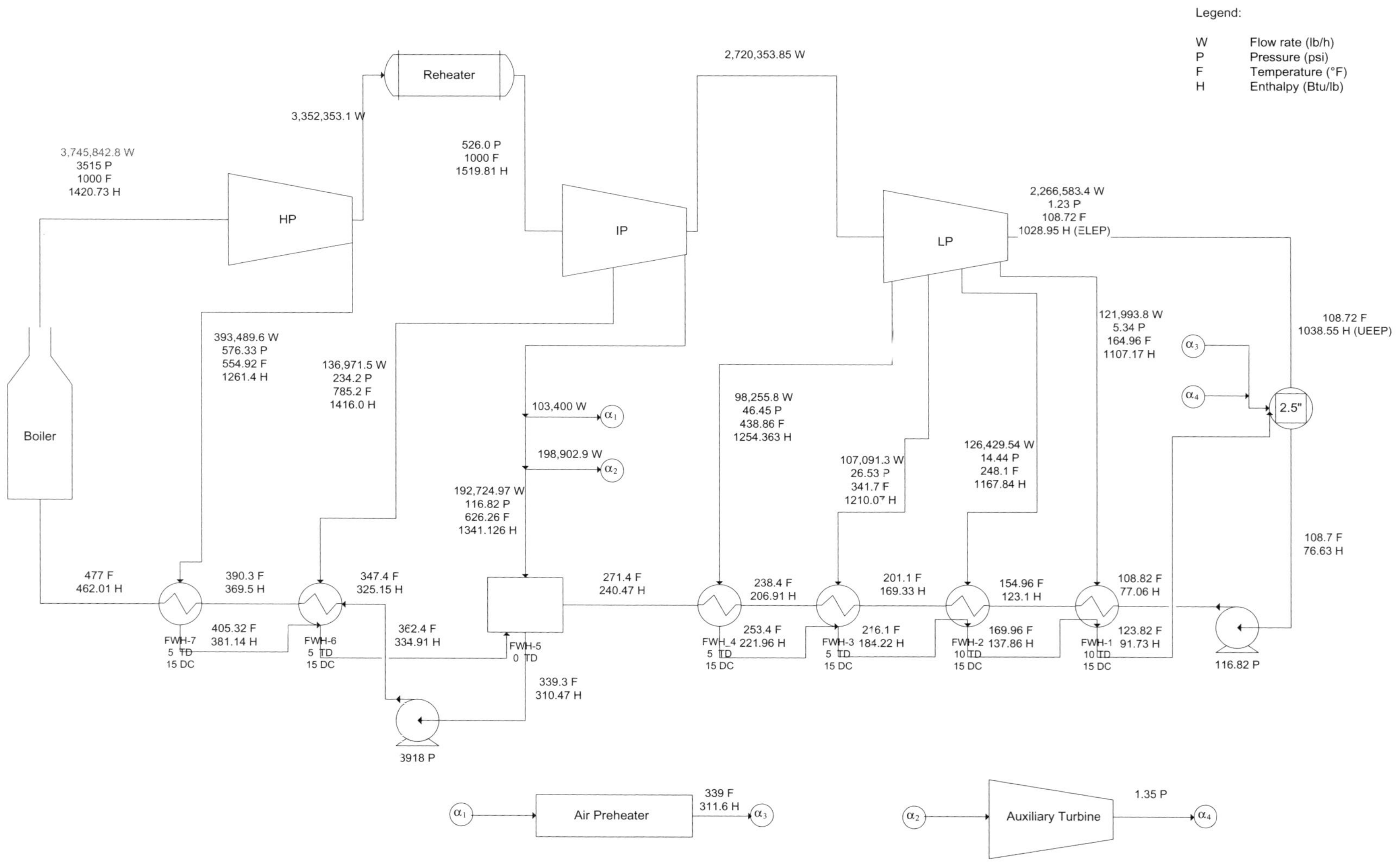

Figure 16.5 Heat balance diagram for off-design conditions with throttling steam flow ratio: 0.75.

Table 16.3 Comparison of Base Case and 0.75 Throttle Ratio on Power Plant Performance

	Base Case/ Design Case	Case 1 0.75 Flow Ratio
Throttle steam flow ratio	1.0	0.75
Steam turbine system ($F, P, T, \hat{h}$)		
Steam flow from boiler (lb/h)	4,994,457	3,745,843
Steam pressure from boiler (psia)	3515	3515
Steam temperature from boiler (°F)	1000	1000
Steam enthalpy from boiler (Btu/lb)	1,420.727	1,420.727
Water enthalpy into boiler (Btu/lb)	493.196	462.005
Steam pressure to IPT section (psia)	691.92	526.00
Steam pressure to LPT section (psia)	153.00	116.82
Stage efficiencies		
HPT	0.8348	0.8018
IPT to extraction (6)	0.8900	0.8900
IPT to extraction (5)	0.8640	0.8640
LPT to extraction (4)	0.8868	0.8868
LPT to extraction (3)	0.8640	0.8640
LPT to extraction (2)	0.8553	0.8553
LPT to extraction (1)	0.8657	0.8657
LPT to outlet ELEP	0.8659	0.8658
Steam turbine system heat rate		
Turbine net output (kW)	736,226	569,702
Generator efficiency (%)	98.5	98.5
Generator net output (kW)	725,183	561,156
Turbine net heat rate (Btu/kW-h)	7857	7943
Boiler efficiency (%)	90.0	90
Steam plant auxiliary power consumption (%)	7.0	7.0
Plant output (kW)	674,420	521,875
Plant net heat rate (Btu/kW-h)	9387	9490

emission reduction strategy must address CO_2 capture from both existing and future coal-fired plants.

Proper cost accounting is required for comparison of energy systems including fossil fuel-fired power plants, fossil fuel-fired power plants with capture and sequestration technologies, and all renewable energy systems. We have introduced the needed economic techniques in Chapter 2 and here we bring these techniques together to provide a levelized energy cost as dollar per kilowatt-hour. We can summarize the process as the following: (1) Bring all costs to a present value (PV) basis; (2) determine an equivalent or uniform annual cost (dollar per year); and (3) determine a levelized energy cost as the ratio of the equivalent annual cost (dollar per year) to the electricity generated (kilowatt-hour per year). Here an important consideration is accounting for escalating fuel costs. The steps needed to determine the levelized energy cost are shown in the next three examples and then are applied to a utility power plant incorporating CO_2 capture.

EXAMPLE 16.3 *Present Value of Pump Alternatives Both with and without Fuel Escalation*

Pump A has a total installed cost (TIC) of $1000 and, using current electricity prices (time = 0), has a "fuel" cost of $100 per year. A more energy efficient pump B is available, which has a $1200 installed cost with $50 per year electricity cost at current electricity prices. The pumps both have a 5-year life expectancy and electricity costs are taken as end-of-the-year costs. The discount rate $i = 10\%$. Determine the present value of each pump, first without electricity escalation and then if electricity costs are expected to escalate at $e_{\text{fuel}} = 6\%$ per year.

SOLUTION See the Excel file **Example 16.3.xls**. Using the present value equation developed in Chapter 2,

Without Electricity (Fuel) Escalation

$$PV_{\text{Pump A}} = 1000 + \frac{100}{(1.1)} + \frac{100}{(1.1)^2} + \frac{100}{(1.1)^3} + \frac{100}{(1.1)^4} + \frac{100}{(1.1)^5} = 1379.08 \quad (16.27)$$

$$PV_{\text{Pump B}} = 1200 + \frac{50}{(1.1)} + \frac{50}{(1.1)^2} + \frac{50}{(1.1)^3} + \frac{50}{(1.1)^4} + \frac{50}{(1.1)^5} = 1389.54 \quad (16.28)$$

With Electricity (Fuel) Escalation

$$PV_{\text{Pump A}} = 1000 + \frac{100(1.06)}{(1.1)} + \frac{100(1.06)^2}{(1.1)^2} + \frac{100(1.06)^3}{(1.1)^3} + \frac{100(1.06)^4}{(1.1)^4} + \frac{100(1.06)^5}{(1.1)^5} = 1448.03 \quad (16.29)$$

$$PV_{\text{Pump B}} = 1200 + \frac{50(1.06)}{(1.1)} + \frac{50(1.06)^2}{(1.1)^2} + \frac{50(1.06)^3}{(1.1)^3} + \frac{50(1.06)^4}{(1.1)^4} + \frac{50(1.06)^5}{(1.1)^5} = 1424.01 \quad (16.30)$$

Without fuel escalation, pump A would be a "better" (lower present value) purchase. With fuel escalation at 6% per year, pump B would be a better purchase. ■

When all costs have been brought to a present value basis, we can convert this present value to a uniform annual cost (an annuity), A, using Equation (2.7) from Chapter 2 as

$$A = \text{PV}\left(\frac{i(1+i)^n}{(1+i)^n - 1}\right). \quad (16.31)$$

Recall in Equation (16.31)

$$\left(\frac{i(1+i)^n}{(1+i)^n-1}\right)$$

is termed the capital recovery factor (CRF); Equation (16.31) can be written as $A = PV \times \text{CRF}(i, n)$. Equation (16.31) will immediately allow conversion of the total installed cost to an equivalent annual installed cost for the given discount rate i as $A_{\text{Installed Cost}} = \text{PV}_{\text{Installed Cost}} \times \text{CRF}(i, n)$. For example, for pump A, the $1000 total installed cost has an annualized installed cost of

$$A = \text{PV}\left(\frac{i(1+i)^n}{(1+i)^n-1}\right) = 1000\left(\frac{0.1(1.1)^5}{(1.1)^5-1}\right) = 1000(0.2638)$$
$$= 263.80\frac{\$}{\text{year}}.$$

The situation is a little more complex when converting an escalating cost, such as the fuel cost in Example 16.1, to an equivalent annual cost. For any discount rate, i, and escalation rate, e (fuel escalation, operation and maintenance [O&M] escalation, etc.), we can write

$$\frac{1+e}{1+i} = \frac{1}{1+i'}, \tag{16.32}$$

where the equivalent discount rate with escalation i' can be found as

$$i' = \frac{i-e}{1+e}. \tag{16.33}$$

For example, the equivalent discount rate with escalation i' in Example 16.1 would be

$$i' = \frac{i-e}{1+e} = \frac{0.1-0.06}{1+0.06} = 0.0377.$$

We can use the equivalent discount with escalation in place of i in the annuity present value calculation, which allows us to write

$$\text{PV} = A\left(\frac{(1+i')^n-1}{i'(1+i')^n}\right), \tag{16.34}$$

where in Equation (16.34)

$$\left(\frac{(1+i')^n-1}{i'(1+i')^n}\right)$$

can be considered a series present value factor (SPVF); Equation (16.34) can be written as $\text{PV} = A \times \text{SPVF}(i, e, n)$. Equation (16.34) allows conversion of an escalating cost (fuel cost, etc.) to the equivalent annual cost or levelized annual cost (LA) for the given discount rate i and escalation rate e as

$$\text{LA} = A\left(\frac{(1+i')^n-1}{i'(1+i')^n}\right)\left(\frac{i(1+i)^n}{(1+i)^n-1}\right), \tag{16.35}$$

where Equation (16.35) can be written as $\text{LA} = A \times \text{SPVF}(i, e, n) \times \text{CRF}(i, n)$. Here the product $\text{SPVF}(i, e, n) \times \text{CRF}(i, n)$ is termed the levelization factor (LF):

$$\text{LF}(i, e, n) = \left(\frac{(1+i')^n-1}{i'(1+i')^n}\right)\left(\frac{i(1+i)^n}{(1+i)^n-1}\right). \tag{16.36}$$

For example, for pump A with fuel escalation in Example 16.3, the $100 per year fuel costs would have a levelized annual fuel cost of

$$\text{LA} = 100\left(\frac{(1+i')^n-1}{i'(1+i')^n}\right)\left(\frac{i(1+i)^n}{(1+i)^n-1}\right)$$
$$= 100\left(\frac{(1+0.0377)^5-1}{0.0377(1.0377)^5}\right)\left(\frac{0.1(1.1)^5}{(1.1)^5-1}\right) = 100(1.1819)$$
$$= 118.19\frac{\$}{\text{year}}.$$

The equivalent annual installed cost and the levelized annual fuel cost can be added to determine the annual levelized cost as dollar per year.

EXAMPLE 16.4 *Levelized Annual Cost of Pump Alternatives (Dollar per Year)*

Determine the annual levelized cost as (dollar per year) for the pumps of Example 16.3.

SOLUTION See the Excel file **Example 16.4.xls**. We can determine the annual levelized cost by bringing all costs back to their present value and by using Equation (16.31). Alternatively, we can use Equation (16.31) with the total installed costs and Equation (16.35) with the fuel costs to determine the annual levelized cost. Note when there is no fuel escalation, $e = 0$ and $i' = i$, giving $\text{LF}(i, e, n) = 1$. ■

EXAMPLE 16.5 *Levelized Energy Cost of a Steam Turbine (Dollar per Kilowatt-Hour)*

A steam-driven turbine is used to directly drive the compressor of a refrigeration system. The compressor requires 300 kW (~400 hp). The installed cost of the turbine is $250 K and it has a 20-year life expectancy. The heat rate of the turbine is 12,000 Btu/kW-h and at the site, LP steam is currently valued at $4 per 1000 lb ($4/10^6 Btu). The discount rate $i = 10\%$ and steam (fuel) costs are expected to escalate at $e = 6\%$ per year. Determine the levelized energy cost as dollar per kilowatt-hour (adapted from Masters, 2005).

SOLUTION The solution is provided in the Excel file **Example 16.5.xls** and is shown in Figure 16.6.

Here we determine the equivalent annual installed cost (Eq. (16.31)) as

$$A = \text{PV}\left(\frac{i(1+i)^n}{(1+i)^n-1}\right) = 250{,}000\left(\frac{0.1(1.1)^{20}}{(1.1)^{20}-1}\right) = 250{,}000(0.1175)$$
$$= 29364.91\frac{\$}{\text{year}}.$$

	A	B	C	D	E	F	G
1							
2		Steam Turbine					
3							
4	Total Installed Cost	250000					
5	life (years)	20					
6	Compressor Power (kW)	300					
7	Turbine Heat Rate (Btu/kW-hr)	12000			Steam Turbine		Steam Turbine
8	Fuel Cost ($/10^{6}Btu)	4			Annual Costs		Levelized Costs
9	Operating hours/year	8000			($/yr)		($/kw-hr)
10	Yearly Fuel Cost	115200		Installed Cost ($/yr)	29364.91	Installed Cost ($/kW-hr)	0.0122
11	Discount Rate	0.1		Fuel ($/yr)	187638.04	Fuel ($/kW-hr)	0.0782
12	Escalation Rate	0.06		Annual Cost ($/yr)	217002.95	Energy Cost ($/kW-hr)	0.0904
13	Discount with Escalation	0.0377					
14	Capital Recovery Factor	0.1175					
15	Levelization Factor	1.6288					

Figure 16.6 Excel solution for the turbine levelized energy cost ($/kW-h).

The yearly fuel cost

$$= \text{Compressor power (kW)} \times \text{operating hours}\left(\frac{\text{h}}{\text{year}}\right) \times \text{Turbine heat rate}\left(\frac{\text{Btu}}{\text{kW-h}}\right) \times \text{Fuel cost}\left(\frac{\$}{\text{Btu}}\right) = 115{,}200\frac{\$}{\text{year}}$$

and the annual levelized fuel cost (Eq. (16.35)) is

$$\text{LA} = 115{,}200\left(\frac{(1+i')^n - 1}{i'(1+i')^n}\right)\left(\frac{i(1+i)^n}{(1+i)^n - 1}\right)$$

$$= 115{,}200\left(\frac{(1+0.0377)^{20} - 1}{0.0377(1.0377)^{20}}\right)\left(\frac{0.1(1.1)^{20}}{(1.1)^{20} - 1}\right)$$

$$= 115{,}200(1.6288) = 187{,}638.04\frac{\$}{\text{year}}.$$

We then determine the annual levelized cost as 29364.91 + 187638.04 = 217002.95 $/year.

The total power generated by the turbine is

$$300\text{ kW} \times 8000\frac{\text{h}}{\text{year}} = 2.4 \times 10^6\frac{\text{kW-h}}{\text{year}}$$

and

$$\text{the levelized energy cost} = \frac{217002.95\frac{\$}{\text{yr}}}{2.4 \times 10^6\frac{\text{kW-h}}{\text{year}}} = 0.0904\frac{\$}{\text{kW-h}}. \quad \blacksquare$$

16.4 CO_2 CAPTURE AND ITS IMPACT ON A CONVENTIONAL UTILITY POWER PLANT

For our economic analysis, we can take a very simplified view of the actual power plant and a postcombustion sequestration process as shown in Figure 16.7. There are several CO_2 separation processes in development (see Ciferno et al., 2009), including the use of amine solutions and solid sorbents. However, for existing fossil fuel-fired power plants, the most likely near-term technology for carbon dioxide capture is postcombustion flue gas scrubbing with monoethanolamine (MEA); this capture process is expensive (see Yang and Hoffman, 2009). After the CO_2 is removed from the boiler flue gas, it will be compressed for transport and possibly used in enhanced oil recovery or enhanced gas recovery operations or simply stored in underground facilities. The current target is 90% removal of generated CO_2 from a coal-fired power plant.

We want to evaluate the impact of adding an amine-based CO_2 capture system to the base case power plant we developed in Section 16.1. We will assume that the installed costs and utility needs of the CO_2 capture and sequestration process will be provided. The utility needs of this capture process will include steam, which will be taken from the power plant as well as electricity and possibly natural gas. We will need to determine the impact of steam removal from the power plant; generally, steam will be removed at or before the LP turbine section. This steam removal will lower the amount of generated electricity and also increase the fuel (coal) needed by the power plant, as less steam will be available for feedwater heating. We will first determine the impact of steam removal on power plant performance. We will then determine the levelized energy cost (dollar per kilowatt-hour) for CO_2 capture—here we must account not only for the capture process total installed casts but also the lower heat rate realized in the power plant as well as the costs for additional purchased electricity and natural gas.

EXAMPLE 16.6 ***Impact of CO_2 Capture on Power Plant Performance (Amine-Based Scrubbing)***

We want to evaluate the impact of adding CO_2 capture to the base case power plant we developed in Section 16.1. Here we utilize an amine-based scrubber with solvent regeneration as reported by the U.S. Department of Energy (DoE, 2007). The regeneration of the capture system when capturing 90% of the power plant base case-generated CO_2 will require 1,700,000 lb/h of LP steam. This steam flow rate is ~1/3 (34%) of the total HP steam generated in the power plant boiler and ~50% of the steam currently being fed to the LP

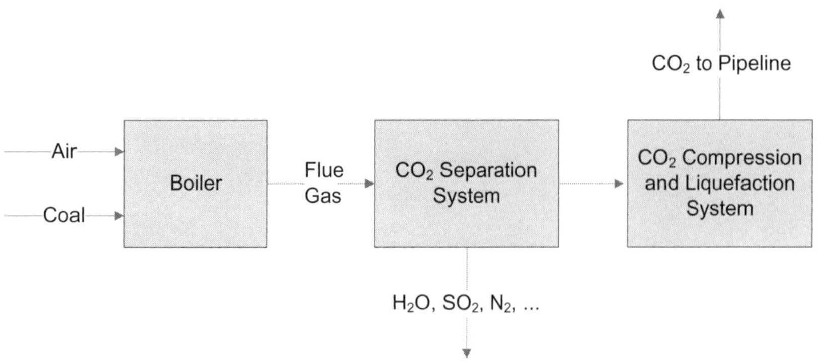

Figure 16.7 Post-Combustion CO_2 Capture and Sequestration Process.

turbine system. Determine the impact this parasitic steam load will have on power plant performance. Also, calculate the expected CO_2 emissions. For the power plant, again assume a generator efficiency of 98.5%, a boiler efficiency of 90%, and an auxiliary power consumption of 7%. The coal to the boiler has a carbon content of 71.72 wt %, with a lower heating value of 12,712 Btu/lb.

SOLUTION We need to extract 1,700,000 lb/h of LP steam from the turbine system for regeneration purposes in the capture process. In our base case power plant design, we extracted α_l = 103,400 lb/h of LP steam for air preheating. This extraction was just prior to the LP turbine section. We can modify the off-design performance model developed in Section 16.2 to account for increased steam extraction. Here we set the throttle steam flow ratio = 1.0 and we set α_l = 103,400 + 1,700,000 = 1,803,400 lb/h to account for both the air preheat and regeneration. Results from the Excel file **Example 16.6.xls** are shown in Figure 16.8 and are summarized in Table 16.4. Calculations for the steam turbine system heat rate and CO_2 emissions can be found in rows 349–365 of the Excel sheet.

The results from Table 16.4 show that the power plant will produce 106,326 kW less electricity due to steam extraction for the amine-based regeneration process. In order to properly determine the economics of the capture process, we will need to purchase 106,326 kW of makeup electricity. In addition, the regeneration process itself will have associated capital costs, O&M costs, and utility costs. All these costs will need to be brought to a levelized energy cost basis to determine the impact of carbon dioxide capture on utility pricing. ■

EXAMPLE 16.7 ***Impact of CO_2 Capture on Power Plant Economics***

An amine-base process DOE (2007) to capture 90% of the CO_2 produced in the base case power plant developed in Section 16.1 will have a total installed cost of $700,000,000. Annual O&M costs will be 5.33% of the total installed cost. The capture process itself will require 15,000 SCF/h of natural gas and 1000 kW. Determine the levelized energy cost for the process. Be sure to include the needed makeup electricity of 106,326 kW. Allow that currently, the utility or large industrial will pay 6.4¢ per kilowatt-hour for any purchased electricity and $6.35 per mSCF ($6.35 per 1000 SCF) for natural gas. For the capture process economics, use a discount rate, i = 10%, with fuel cost escalation (electricity and natural gas) at e_{fuel} = 1.98% per year and an O&M escalation at $e_{\text{O\&M}}$ = 1.89%. Assume a 20-year project life with 8000-h/year operation, and to account for taxes and debt (discussed next), use a capital recovery factor = 0.175 with the total installed cost.

SOLUTION Results are provided in the Excel file **Example 16.6.xls** and are summarized in Table 16.5. Economic calculations can be found in rows 369–402 of the Excel sheet.

The levelized energy cost for the capture process is 5.07¢ per kilowatt-hour. Using a levelized electricity cost of 7.46¢ per kilowatt-hour (7.46¢ per kilowatt-hour = 6.4¢ per kilowatt-hour × 1.1650 (fuel LF)), the amine-based capture process would increase utility costs by 68%. The current target for sequestration technology is a maximum of 35% increase in electricity cost (see Problem 16.2). Finally, in Example 16.5, we used a capital recovery factor of 0.175, which is larger than the captial recovery factor of 0.1175 determined in Example 16.3. In Example 16.5, the effect of income taxes, depreciation, debt, and stock are included within the capital recovery factor (here see Fleischer, 1994). ■

16.5 CLOSING COMMENTS

The power plant model developed here is general and will allow determination of system performance for design conditions and off-design operating conditions when using throttling control. There is also the possibility of sliding pressure control in power plants, which is not addressed here. The power plant model can serve as a basis for any steam power plant using any fuel (including nuclear). In the design case, we solved the full-load operation material and energy balances for the turbine system using our Excel-based thermodynamic functions and known P and T or P and $\hat{h}$ (for two-phase steam) at each turbine section inlet, outlet, and extraction point. This data can come from manufacturer data (see Spencer et al., 1963) or from the plant data acquisition system when the plant is operating under normal full-load conditions.

Given full-load data, standard commercial simulators including Aspen PLUS by Aspen Technology, HYSYS (as Aspen HYSYS by Aspen Technology or UniSim Design by Honeywell), and PRO/II by Simulations Sciences Inc. can

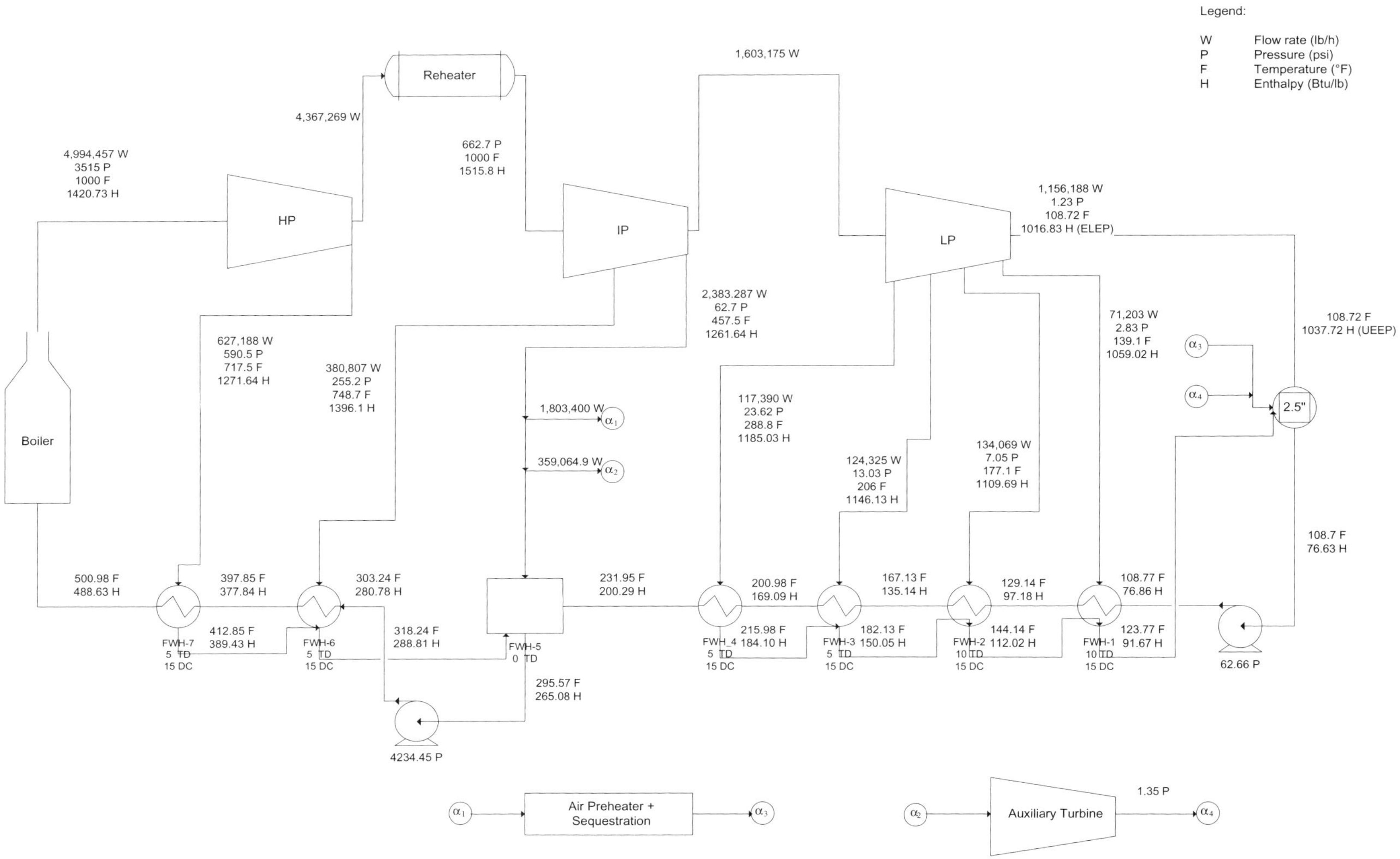

Figure 16.8 Heat balance with steam extraction for solvent regeneration (throttle steam flow ratio: 1.0).

Table 16.4 Impact of CO_2 Capture on Power Plant Performance

	Base Case	Case 1—90% Capture
Percent CO_2 capture	0%	90%
Steam turbine system (F, P, T, $\hat{h}$)		
Steam flow from boiler (lb/h)	4,994,457	4,994,457
Steam pressure from boiler (psia)	3515	3515
Steam temperature from boiler (°F)	1000	1000
Steam enthalpy from boiler (Btu/lb)	1420.727	1420.727
Water enthalpy into boiler (Btu/lb)	493.196	488.627
Steam pressure to IPT section (psia)	691.92	662.73
Steam pressure to LPT section (psia)	153.00	62.66
Steam extraction rate for regeneration (lb/h)	NA	1,700,000
Steam turbine system heat rate		
Turbine net output (kW)	736,226	620,156
Generator efficiency (%)	98.5	98.5
Generator net output (kW)	725,183	610,854
Turbine net heat rate (Btu/kW-h)	7857	9367
Boiler efficiency (%)	90.0	90.0
Steam plant auxiliary power consumption (%)	7.0	7.0
Plant output (kW)	674,420	568,094
Plant net heat rate (Btu/kW-h)	9,387	11,191
Steam turbine system CO_2 emissions		
Coal carbon wt %	71.72	71.72
Coal *LHV* (Btu/lb)	12,712	12,712
Coal required, *LHV* (lb/h)	498,027	500,111
CO_2 produced (lb CO_2/lb coal)	2.6281	2.6281
Total CO_2 produced (lb/h)	1,308,885	1,314,363

be used to calculate power plant performance. However, these packages do not include methodologies used for predicting off-design power plant equipment performance. User-written subroutines, as we developed in Chapter 16, must be added to account for off-design pressure and efficiency changes. There are more specialized commercial packages available for power plant analysis—GateCycle™ from GE or IPSEpro from SimTech Simulation

Table 16.5 Economic Impact of CO_2 Capture

	Base Case	Case 1—90% Capture
Percent CO_2 capture	0%	90%
Power generation		
Steam flow from boiler (lb/h)	4,994,457	4,994,457
Steam extraction rate for regeneration (lb/h)	NA	1,700,000
Plant output (kW)	674,420	568,094
Plant net heat rate (Btu/kW-h)	9,387	11,191
Total CO_2 produced (lb/h)	1,308,885	1,314,363
Economics		
Capital		
Total capital investment ($)		700,000,000
Capital recovery factor (year^{-1})		0.175
Annualized capital cost ($/year)		122,500,000
Annualized capital cost (¢/kW-h)		2.70
Operation and maintenance		
Annual O&M cost (5.33% TCI)		37,310,000
O&M levelization factor		1.1567
Levelized O&M cost ($/year)		43,157,013
Levelized O&M cost (¢/kW-h)		0.95
Natural gas + electricity		
Natural gas cost ($/year)		762,000
Capture process electricity (kW)		1,000
Makeup electricity (kW)		106,326
Electricity cost to industry ($/kW-h)		0.064
Electricity cost ($/year)		54,950,885
Fuel levelization factor		1.1650
Levelized NG + electricity cost ($/year)		64,904,199
Levelized NG + electricity cost (¢/kW-h)		1.43
Levelized capture cost ($/year)		230,561,212
Levelized capture cost (¢/kW-h)		5.07

Technology—that are designed to perform both design and off-design calculations for power systems.

Steam turbine technology is a mature topic and the reader interested in additional details should see the books by Salisbury (1950) and Cotton (1998). Additional details of turbine off-design modeling can be found in Spencer et al. (1963), Phillips (1986), Erbes (1987), and Eustis et al. (1987). For additional reading on levelized economics, see Masters (2005). Power plant design and operation is discussed in Li and Priddy (1985). A good overview book for combined cycle power plants is by Kehlhofer et al. (2009).

REFERENCES

CIFERNO, J.P., T.E. FOUT, A.P. JONES, and J.T. MURPHY. 2009. Capturing carbon from existing coal-fired power plants. *CEP* April: 33–41.

COTTON, K.C. 1998. *Evaluating and Improving Steam Turbine Performance* (2nd edition). Cotton Fact Inc, Rexford, NY.

ERBES, M.R. 1987. Phased construction of integrated coal gasification combined-cycle power plants. PhD Thesis, Stanford University.

ERBES, M.R. and R.H. EUSTIS. 1986. A computer methodology for predicting the design and off-design performance of utility steam turbine-generators. *Proc. Am. Power Conf.* 48: 318–324. Chicago, IL.

EUSTIS, R.H., M.R. ERBES, and J.N. PHILLIPS. 1987. Analysis of the Off-Design Performance and Phased Construction of Integrated-Gasification-Combined-Cycle Power Plants Volume 2: Models and Procedures. Electric Power Research Institute Final Report, Research Project 2029-12, February 1987.

FLEISCHER, G.A. 1994. *Introduction to Engineering Economy*. PWS Publishing Company, Boston.

KEHLHOFER, R., F. HANNEMANN, F. STIRNIMANN, and B. RUKES 2009. *Combined-Cycle Gas & Steam Power Plants* (3rd edition). PennWell Corporation, Tulsa, OK.

LI, K.W. and A.P. PRIDDY 1985. *Power Plant System Design*. John Wiley and Sons, New York.

MASTERS, G.M. 2005. *Renewable and Efficient Electric Power Systems*. John Wiley. Published on line January 28.

PHILLIPS, J.N. 1986. A study of the off-design performance of integrated coal gasification combined-cycle power plants. PhD Thesis, Stanford University.

RAJ, K.S.S. 2008. Last stage performance considerations in low-pressure turbines of power plants: A case study. *J. Eng. Gas Turbines Power* 130 March: 023004-1–023004-7.

SALISBURY, J.K. 1950. *Steam Turbines and Their Cycles*. Robert E. Krieger Publishing, Huntington, NY.

SPENCER, R.C., K.C. COTTON, and C.N. CANNON. 1974. A Method for Predicting the Performance of Steam Turbine-Generators. GER-2007C, General Electric Co. *Note:* The 1974 GE report by Spencer et al. is an updated version of the 1963 paper. The 1974 GE report can be difficult to obtain so we have chosen to reference the original 1963 paper in the text.

SPENCER, R.C., K.C. COTTON, and C.N. CANNON. 1963. A method for predicting the performance of steam turbine-generators—16,500 kW and larger. *J. Eng. Power* October: 249–301. Also see note after the next Spencer et al. reference.

STODOLA, A. and L.C. LOWENSTEIN (translation) 1927. *Steam and Gas Turbines*. Vol. 1. McGraw-Hill, New York.

U.S. DEPARTMENT OF ENERGY (DOE). 2007. Carbon Dioxide Capture from Existing Coal-Fired Power Plants, Publication No. DOE/NETL-401/110907, DOE Office of Fossil Energy's National Energy Technology Laboratory, Pittsburgh, PA (revision date November 2007).

YANG, W.-C. and J. HOFFMAN 2009. Exploratory design study on reactor configuration for carbon dioxide capture from conventional power plants employing regenerable solid sorbents. *Ind. Eng. Chem. Res.* 48: 341–351.

PROBLEMS

16.1 ***Solution Provided*** An important area for improvement in CO_2 capture technology is reducing the regeneration system steam requirements. This steam energy is often reported as British thermal unit (steam) for regeneration per pound-CO_2 captured. In Example 6.6, 1,700,000 lb/h of LP steam was required for regeneration of the amine-based solvent. Determine the British thermal unit (of steam) per pound-CO_2 captured.

Next, assume that the needed steam pressure for regeneration is lowered. Instead of steam being taken from extraction point (5) in Figure 16.8, assume the required 1,700,000 lb/h of steam for regeneration will be removed from extraction point (4). Determine the British thermal unit (of steam) per pound-CO_2 captured. For steam removal at extraction point (4), it will be useful to tune the model convergence by changing the number of pressure iterations. In addition, a macro subroutine that calls only Pressure_Iteration and Update_Efficiencies (without Off_Design_Initial_Estimates) should be included in the code to help achieve convergence.

SOLUTION The solution for extraction point (5) is found in **Example 16.6.xls**.

The solution for extraction point (4) is provided in the Excel file **Problem 16.1.xls**. Here we start with the code from Example 16.6/16.7 in **Example 16.6.xls**. The first change needed is to account for the steam extraction at extraction point (4). This is accomplished in the Excel sheet in the cell (row 51, column T). In the problem solution, a new macro subroutine Sub A_Slow_Convergence ( ) that calls only Pressure_Iteration and Update_Efficiencies has been added. This macro subroutine will need to be called several times to achieve convergence.

16.2 The nonpartisan Congressional Budget Office in 2006 estimated that if sequestration technology were fully implemented, the average household utility bill will increase $175 per year. Can you determine how the Budget Office arrived at this figure? The U.S. EIA supplies yearly electric sales, revenue, and average price data at the Web site http://www.eia.doe.gov/cneaf/electricity/esr/esr_sum.html

16.3 Sequestration economics can be presented in a number of ways, but the work we have accomplished in Examples 16.6 and 16.7 is a necessary first step. Use results from Examples 16.6 and 16.7 in the following problems:

(a) The cost estimates for sequestration in the literature range from $15 to $75 per metric ton (of CO_2 sequestered or CO_2 avoided). A metric ton = 2204.6 lb. What cost did we determine in Example 16.7?

(b) The cost of carbon (CO_2) credits is currently suggested at $20 per metric ton. This would be the cost you would have to pay if you chose not to implement any sequestration strategy and instead purchased carbon credits. How does this compare with our projected sequestration costs in part a?

(c) Fuel switching is often cited as a means of reducing greenhouse gases with natural gas as the preferred fuel choice. How would fuel switching impact the CO_2 emissions from our base case coal plant? Assume a methane lower heating value of 21,500 Btu/lb and that the plant net heat rate remains unchanged.

(d) In Examples 16.6 and 16.7, costs reflected producing liquid CO_2 at the power plant "fence line." This CO_2 must find a home. What impact would the following costs (*Source:* Intergovernmental Panel on Climate Change) have on the sequestration process we explored?

Process	Cost Range ($/ metric ton CO_2 Captured)	Comments
Transportation	$1.00–$8.00	Per 155 miles via pipeline
Geological storage	$0.50–$8.00	Not including any enhanced oil or natural gas recovery revenue
Monitoring of storage	$0.10–$0.30	Depending on regulations

16.4 In Example 16.6, we calculated the coal requirement (pound per hour) for the design case as

$$\frac{(\text{Plant output, kW})\left(\text{Plant net heat rate, } \frac{\text{Btu}}{\text{kW-h}}\right)}{\left(\text{Coal } LHV, \frac{\text{Btu}}{\text{lb}}\right)}$$

$$= \frac{(674{,}420 \text{ kW})\left(9387 \frac{\text{Btu}}{\text{kW-h}}\right)}{\left(12{,}712 \frac{\text{Btu}}{\text{lb}}\right)} = 498{,}027 \frac{\text{lb}}{\text{h}}.$$

Show the same result is obtained if heat input to the boiler is used to determine the coal requirement. Be sure to include the reheater heat input.

16.5 ***Power Plant Efficiency*** In Chapter 1, Equation (1.1), we defined a thermal or first law efficiency as

$$\frac{\text{Usable energy output from the system}}{\text{Energy supplied to the system}},$$

and in Chapter 1 the power plant (Figure 1.5) showed an efficiency of 38.08%. Determine the design efficiency of the power plant detailed in Chapter 16, Section 16.1; be sure to include steam reheat and boiler efficiency.

Chapter 17

Alternative Energy Systems

Chapters 1–16 provide a unified treatment of energy flows in a processing plant and the most efficient/economical means of supplying process energy needs. A focus in these chapters has been energy efficiency in the process and developing cogeneration-based systems for supplying needed energy; the best currently available large-scale conservation technology is cogeneration. We also introduced levelized economics and utility pricing as the cost of purchased electricity is a key factor in determining cogeneration economic viability. Understanding utility pricing when using coal as the fuel source is especially important as ~50% of the electricity generation in the United States is from coal. Our discussion of cogeneration and energy efficiency is important, but these are only one aspect of a national energy portfolio. This book can also be used as a basis to examine alternative energy systems. This is especially true for alternative systems where the energy inputs to these processes (chemical, from fuel, or radiation from the sun) can be used to produce steam (or to vaporize an organic) and can be used in a Rankine power cycle, or the chemical energy can be converted into a synthesis gas and used in gas turbines, or some combination of these. In this final chapter, we discuss levelized costing of alternative energy systems. We show the connection between levelized costs and the price we must pay for utilities if the alterative technology is implemented. In a chapter example, we detail the levelized costing process for an organic Rankine cycle (ORC). In the final example, we introduce nuclear power cycles. These examples help bring together material from many of the previous chapters.

17.1 LEVELIZED COSTS FOR ALTERNATIVE ENERGY SYSTEMS

Any discussion of alternatives energy systems should begin with an understanding of the actual cost of the technology, independent of incentives such as tax credits. Table 17.1 (D.E. Dismukes, pers. comm.) provides the break-even costs for various energy systems; this is the levelized price each technology must sell electricity in order to break even.

Table 17.1 provides a cross-section of the available alternative energy technologies. There will always be uncertainty in levelized costs as reported in Table 17.1 as calculations must be based on a number of economic assumptions. In Table 17.1 hydrokinetic is the use of the energy in a naturally flowing body of water to spin a generator—these are turbines and other devices that are in rivers or streams (or even tidal) that turn with the river's flow. In Problem 17.1 we outline some of the factors and considerations which must be included in any detailed costing analysis. The break-even costs in Table 17.1 do represent a starting point to begin a discussion of energy alternatives.

In Chapter 16 (Example 16.7) we determined a levelized selling price of electricity from a utility to a large industrial at 7.46 ¢/kW-h. The non-biomass alternatives in Table 17.1 all show a levelized cost above 7.46 ¢/kW-h. In order for these technologies to break even, they must sell generated electricity at a price above the current levelized market value for electricity. There must either be an accepted premium price paid for these technologies or some form of supplement (tax credits or environmental credit) must be made available to help offset losses. It is important to keep these (and other) technologies in our national energy portfolio as technological breakthroughs with each generation of development will help to lower the break-even cost.

Table 17.1 also indicates that biomass/renewable materials, when cofired with coal in existing coal utility plants, should actually lower electricity costs. Here, there are limits on the percentage of biomass materials that can economically be combined with the coal. There are also limits on how far biomass materials can be transported and remain cost-effective. Construction of new facilities to cofire

Modeling, Analysis and Optimization of Process and Energy Systems, First Edition. F. Carl Knopf.

Table 17.1 Break-Even Costs for Alternative Energy Technologies (D.E. Dismukes, Pers. Com.)

Technology	Levelized Cost (¢/kW-h)
Alternatives nonbiomass	
Geothermal	8.6
Solar photovoltaic	36.3
Low-temperature Rankine cycle	9.0
Wind	10.3
Fuel cells	17.0
Traditional hydroelectric from dams	5.3
Hydrokinetic	26.8
Solar thermal	20.4
Biomass cofire with existing coal generation	
Switchgrass	5.0
Wood	5.0
Corn stover	5.9
Biomass cofire with new coal generation	
Wood	12.7
Rice hulls	10.4
Next generation	
Biomass gasification	11.1
Integrated coal gasification combined cycle (IGCC)	9.2
Nuclear	10.9

biomass and coal will necessitate an increase in the cost of electricity. The next generation of technologies including gasification and clean coal processes would also require increased electricity costs if implemented.

Table 17.1 does not address specifically cogeneration systems, which has been an emphasis in the previous chapters. Decisions about cogeneration systems within processing plants often utilize a different cost analysis from that in Table 17.1. Generally, the electricity and steam generated by the cogeneration system is used within the process. There is opportunity to sell electricity at a qualifying cogeneration facility, but this electricity is generally purchased by the local utility at the local utility's avoided cost. The local utility's avoided cost may be defined as the lowest price electricity available in the purchase network of the utility. In many cases, the sale of electricity is not a key component in the analysis of the cogeneration system profitability. Process cogeneration systems are generally economically viable when natural gas prices are at or below the $4–6 per MMBtu range. As we discussed in Chapter 1, utility companies will produce and often purchase electricity from a number of sources using fuels including coal, natural gas, hydro and nuclear energy. As natural gas prices increase above the $4–6 range, the energy mix available to the utility company eliminates dollar savings possible by cogeneration.

We can use materials developed in this book to determine the efficiency and levelized cost of many alternative energy systems—especially those systems that involve Rankine power cycles, gas turbines, or some combination. We next examine organic Rankine cycles, which can be used to generate electricity from low-level or waste heat sources. Applications include geothermal (Hettiarachchi et al., 2007) and solar thermal systems (Schuster et al., 2007), diesel engines and gas turbines (Bohl, 2009; Leslie et al., 2009), and furnaces and ovens such as those found in cement plants (Lukawski, 2009; Obernberger et al., 2002). Organic Rankine cycle is even being studied for possible use within conventional power plants (Mohamed et al., 2009).

17.2 ORGANIC RANKINE CYCLE (ORC): DETERMINATION OF LEVELIZED COST

An ORC cycle involves the same steps as a steam Rankine cycle (Chapter 16) except an organic is used as the prime driver. Following Figure 17.1, (1) high-pressure liquid organic is vaporized by exchanging energy with a hot oil stream; (2) the organic is expanded through a turbine generating work; (3) the low-pressure vapor is then condensed; and (4) liquid organic is then compressed. An organic (here n-pentane) can be chosen, which has a lower freezing temperature when compared to water/steam. This opens the possibility for ORC use in remote locations, especially where water is generally not available or where cold weather can make water management difficult.

EXAMPLE 17.1 *Levelized Costing for an ORC System: Energy Recovery at a Natural Gas Compression Station*

There are over 1200 natural gas compression stations within the U.S. interstate natural gas pipeline system (Leslie et al., 2009). Most of the over 300 larger stations within the system use a gas turbine to directly drive a compressor. The exhaust gas from these gas turbines is often just released to the atmosphere. In these simple cycle gas turbine configurations, there is no heat recovery steam generator (HRSG) to help improve efficiency, as generally there is no need for steam at these locations. Generated steam could be used with a steam turbine to produce electricity for sale, but water/steam management in these often remote locations would be difficult. An area of active development is the use of an ORC to recover energy and to generate electricity from large compression stations.

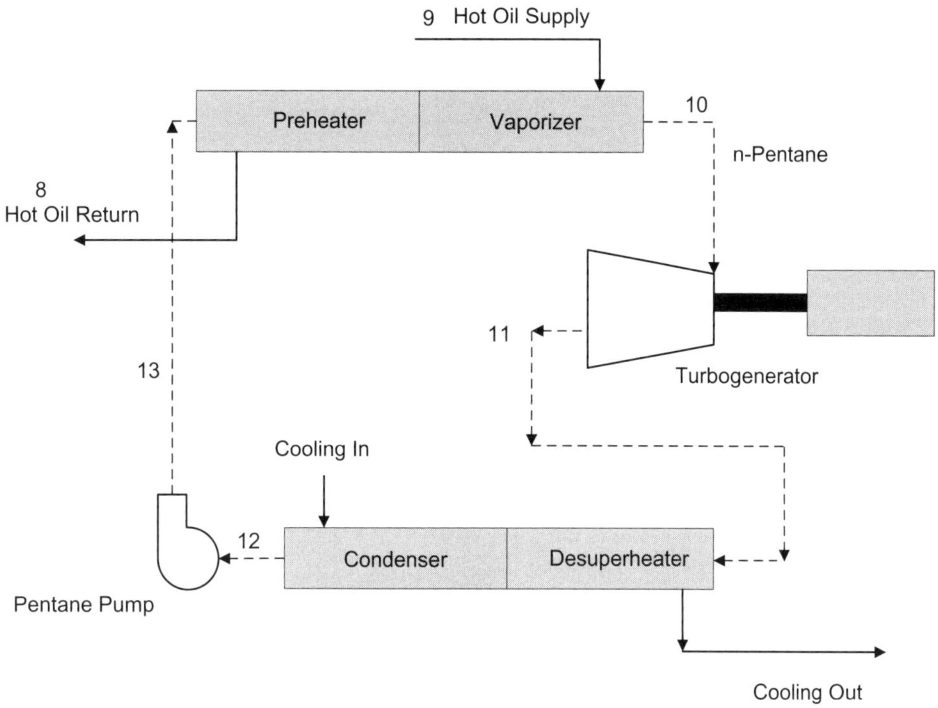

Figure 17.1 A simple organic Rankine (bottoming) cycle system.

A large gas compression station requires 27.9 MW to drive the compressor. Determine the levelized cost for electricity production if an n-pentane-based ORC is used to generate electricity from the gas turbine exhaust.

SOLUTION We can solve Example 17.1 by first utilizing our solution to Example 9.4 (gas turbine with regeneration). In place of the HRSG, energy will be recovered from the gas turbine exhaust gas with an ORC. The solution to Example 17.1 is provided in **Example 17.1a.xls** and key results are shown in Figure 17.2. For the provided solution, we have neglected pressure drop in the heat exchangers and assumed all units are adiabatic in the ORC system.

The solution begins by solving the material and energy balances for the ORC process. This will allow sizing and costing of the ORC system and then levelized economics can be used to determine the break-even cost of electricity.

Gas Turbine System

Solver is used to minimize the fuel flow rate with the requirements that the net power generated by the gas turbine is 27.9 MW, the temperature from the combustion chamber is 2500 R, and approach temperature in the air preheater is 20 R. The design variables used in Solver are the air flow rate, the fuel flow rate, and T_3, the air temperature leaving the air preheater.

Waste Heat Oil Heater (WHOH)

Hot oil is used to recover energy from the gas turbine exhaust gas. The temperature of the hot oil leaving the WHOH is 968.67 R and the hot oil return is 675.27 R; the oil has a maximum allowable working temperature of 650°F. The stack gas exhaust temperature is T_7 = 716.67 R. The heat capacity of the oil (British thermal unit per pound-Fahrenheit) is $\hat{C}_p = 0.3706 + (0.000397)T$.

The oil flow rate can be determined as

$$F_{\text{hot oil}} = \frac{\dot{Q}_{\text{turbine exhaust}}}{\left(\hat{h}_9 - \hat{h}_8\right)} = \frac{F_{\text{exhaust}}\left(\hat{h}_6 - \hat{h}_7\right)}{\left(\hat{h}_9 - \hat{h}_8\right)} \qquad (17.1)$$

$$= \frac{234.505(94.854)}{150.93} = 147.37\frac{\text{lb}}{\text{s}}.$$

Vaporizer

The vaporizer consists of two sections: a preheat section and the vaporizer section. In the preheat section, compressed liquid n-pentane from the pump is heated to near-boiling conditions with an approach temperature of 5 R. At 22.8 bar, the boiling temperature for n-pentane is 799.27 R. In the vaporizer section, liquid n-pentane is converted to superheated n-pentane at T_{10} = 858.87 R and P_{10} = 22.8 bar.

Until the n-pentane cycle is completed, we do not know the exact conditions into the preheater. Once the n-pentane conditions into the preheater are known, we can use an overall energy balance around both the preheater and vaporizer to determine the n-pentane

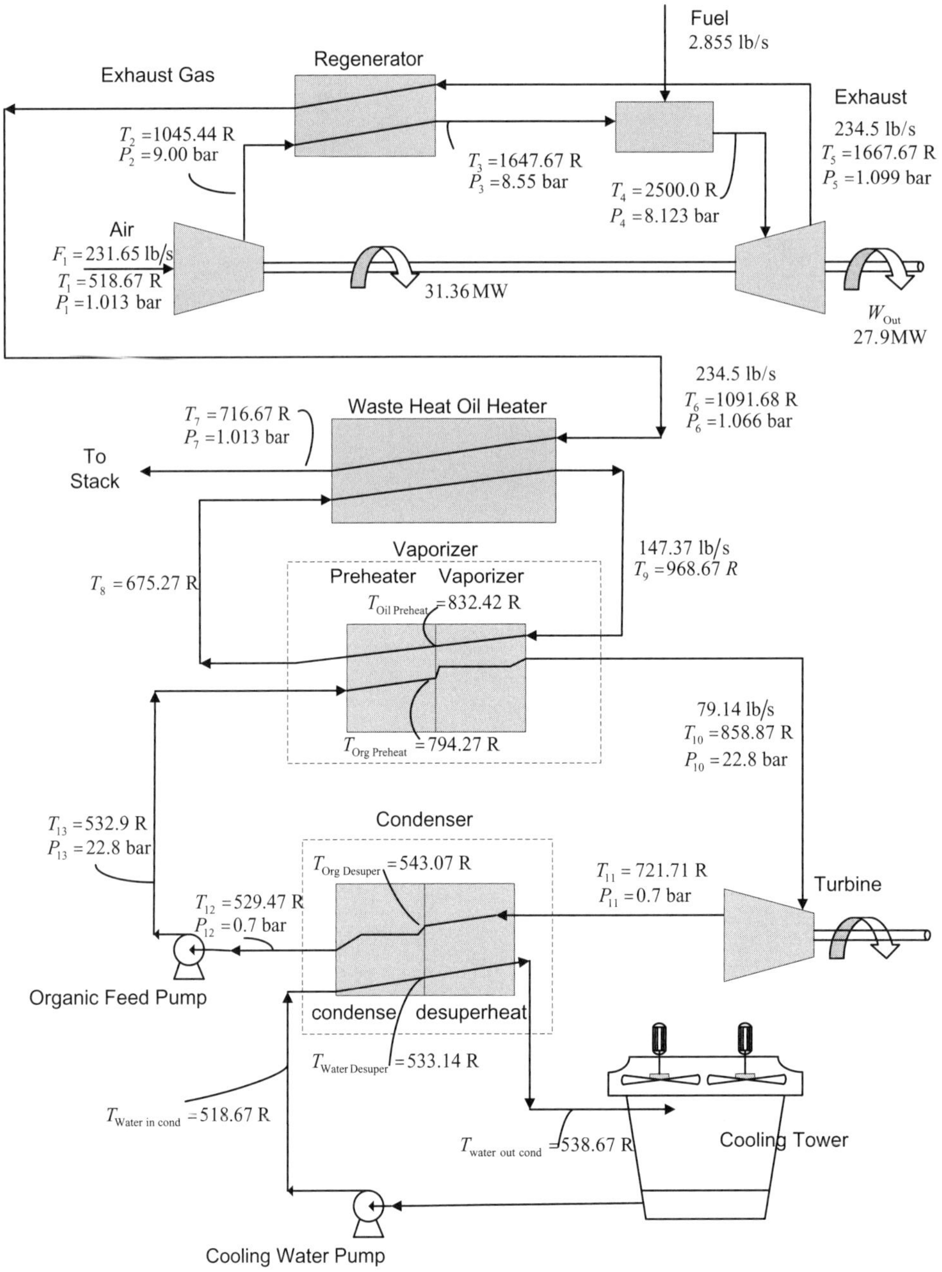

Figure 17.2 ORC for energy recovery from a natural gas compression station.

flow rate. For now, the calculations can proceed by assuming a flow rate, say, 1 lb/s of n-pentane from the vaporizer.

Organic Turbine

The organic driven turbine is assumed to have an isentropic efficiency of 0.85. Following Example 8.3 (Chapter 8), we can use the Redlich–Kwong equation of state (EOS) to estimate needed enthalpy and entropy values for n-pentane. The Excel file **Example 17.1b (vapor).xls** uses Equations (8.11) and (8.12) to provide vapor-phase properties. Here $h_0 = 6188.42$ cal/gmol and $s_0 = 3.9986$ cal/gmol-K with $T_0 = 250$ K to allow comparison with pure component n-pentane data provided by Reynolds (1979). For n-pentane C_υ^o,

$$C_\upsilon^o = -0.3692 + 0.1085T - \left(5.365\times10^{-5}\right)T^2 + \left(1.01\times10^{-8}\right)T^3, \quad \left(\frac{\text{cal}}{\text{gmol-K}}\right) \tag{17.2}$$

and $T_{critical} = 469.8$ K and $P_{critical} = 33.3$ atm (see Table 8.9). For n-pentane leaving the vaporizer, $T_{10} = 858.87$ R and $P_{10} = 22.8$ bar, and we solve for the molar volume, υ, using Equation (8.10). In the Excel file, Equation (8.10) is provided in cell 6J and Goal Seek is used to vary υ in cell 3L; here we find $\upsilon = 1277.43$ cm^3/g-mol. Then using Equations (8.11) and (8.12), $\hat{h}_{10} = 341.79$ Btu/lb and $\hat{s}_{10} = 0.52636$ Btu/lb-R.

Following an isentropic expansion to 0.7 bar, we find using the Excel file that $T_{11,\text{isen}} = 700.63$ R, and $\hat{h}_{11,\text{isen}} = 274.076$ Btu/lb. To use the Excel file, you can first solve for T, which gives $\hat{s} = 0.52636$ Btu/lb-R at 0.7 bar. Solving for υ will give a new T and this process continues (about three or four iterations) until υ and T remain unchanged. Using the isentropic efficiency, $\eta_{\text{organic turbine}} = 0.85$, we find

$$\hat{h}_{11,\text{actual}} = \hat{h}_{10} - \left(\hat{h}_{10} - \hat{h}_{11,\text{isen}}\right)\left(\eta_{\text{organic turbine}}\right)$$
$$= 341.79 - (341.79 - 274.076)(0.85) = 284.23\frac{\text{Btu}}{\text{lb}}. \tag{17.3}$$

With the actual enthalpy $\hat{h}_{11}$ (or $\hat{h}_{11,\text{actual}}$) and pressure P_{11} leaving the turbine, we find $T_{11} = 721.7$ R using the provided Excel file.

Condenser

The condenser consists of two sections, a desuperheat section and a condenser section. At 0.7 bar, the condensing temperature for n-pentane is 538.07 R. In the desuperheat section, superheated n-pentane from the turbine discharge (stream 11) is cooled to 543.07 R; this is an approach temperature of 5 R. In the condenser section, n-pentane is cooled to $T_{12} = 529.47$ R with $P_{12} = 0.7$ bar. Here we can use the provided Excel file Excel file **Example 17.1c (liquid).xls** to solve for υ_{12} and determine $\hat{h}_{12} = 58.087$ Btu/lb and $\hat{s}_{12} = 0.12802$ Btu/lb-R. When solving for υ (here you should find $\upsilon_{12} = 130.479$ cm^3/g-mol), use Excel Solver with υ bounded, as $105 \leq \upsilon \leq 150$ cm^3/g-mol, to find the liquid molar volume root.

Pump

Following an isentropic compression to $P_{13} = 22.8$ bar, we find, using the Excel file for liquid n-pentane, that $T_{13,\text{isen}} = 530.99$ R and $\hat{h}_{13,\text{isen}} = 59.7975$ Btu/lb. Again, to use the Excel file for liquid pentane, you can first solve for T, which gives $\hat{s} = 0.12802$ Btu/lb-R at 22.8 bar. Solving for υ will give a new T and this process continues (about three or four iterations) until υ and T remain unchanged. Then, using the pump isentropic efficiency, $\eta_{\text{organic pump}} = 0.65$, we find

$$\hat{h}_{13,\text{actual}} = \hat{h}_{12} + \left(\frac{\left(\hat{h}_{13,\text{isen}} - \hat{h}_{12}\right)}{\eta_{\text{organic turbine}}}\right)$$
$$= 58.087 + \left(\frac{(59.7975 - 58.087)}{(0.65)}\right) = 60.719\frac{\text{Btu}}{\text{lb}}. \tag{17.4}$$

With the actual enthalpy $\hat{h}_{13}$ (or $\hat{h}_{13,\text{actual}}$) and known pressure P_{13} leaving the pump, we can find $T_{13} = 532.85$ R.

Finally, using the enthalpy of n-pentane entering and leaving the vaporizer ($\hat{h}_{13}$ and $\hat{h}_{10}$), we can determine $F_{\text{n-pentane}}$:

$$F_{\text{n-pentane}} = \frac{F_{\text{hot oil}}\left(\hat{h}_9 - \hat{h}_8\right)}{\left(\hat{h}_{10} - \hat{h}_{13}\right)} = \frac{147.37(150.93)}{(281.071)} = 79.14\frac{\text{lb}}{\text{s}}. \tag{17.5}$$

The net power generated is

$$\text{Net power} = \text{Organic turbine}(kW) - \text{Pump(kW)}$$
$$= 4804.70 - 219.71 = 4584.99 \text{ kW}. \tag{17.6}$$

We next address the ORC economics, which begin with purchased equipment costs (PECs). The factors that should be included in determining the total capital investment (TCI) of a system have been discussed in Chapter 2, Section 2.5—"Plant Design Economics."

Equipment Capital Costs (Purchased Equipment Costs)

Equations to estimate equipment capital costs for the ORC (in 2010 dollar) are provided in Table 17.2. The capital cost equation for organic turbines is valid with power output 250 < kW < 5000. Heat exchangers costs assume carbon steel.

For preliminary cost estimates, heat exchangers area, can be determined using

$$\dot{Q} = UA\Delta T_{lm}, \tag{17.7}$$

where $\dot{Q}$ is the heat transferred (British thermal unit per second), U is the overall heat transfer coefficient (British thermal unit per second-square meter-Rankine), A is the area (square meter), and ΔT_{lm} is the log mean temperature difference (Rankine); the log mean temperature difference is detailed in Chapter 7 (Section 7.8). U values for the ORC system have been estimated (Lukawski, 2009) and are provided in Table 17.3.

U values can also be determined using the inside (h_i) and outside (h_o) tube heat transfer coefficient, where

$$\frac{1}{U} \approx \frac{1}{h_i} + \frac{1}{h_o}$$

when neglecting wall resistances and fouling. Values for the individual heat transfer coefficients can be determined for ORC system heat exchangers (example calculations can be found in Mohamed et al., 2010 and in Hettiarachchi et al., 2007).

Waste Heat Oil Heater: Capital Cost

For the WHOH,

$$\Delta T_{\text{lm}} = \frac{(T_6 - T_9) - (T_7 - T_8)}{\ln\left(\frac{(T_6 - T_9)}{(T_7 - T_8)}\right)}$$
$$= \frac{(1091.678 - 968.67) - (716.67 - 675.27)}{\ln\left(\frac{(1091.678 - 968.67)}{(716.67 - 675.27)}\right)} = 74.94 \text{ R}, \tag{17.8}$$

and using the provided overall heat transfer coefficient from Table 17.3,

Table 17.2 Equations to Estimate Purchased Equipment Costs (PECs) (Adapted from Lukawski, 2009)

Unit Operation	Characteristic in Cost Equation	Equation for Capital Cost ($)
Heat exchangers: waste heat oil heater, preheater, vaporizer, desuperheater, condenser	Area, A (m^2)	$1667(A)^{0.68}$
Organic turbine	Work, $\dot{W}$ (kW)	$5.691E06 - 2.06E06\ln(\dot{W}) + 0.2185E06(\ln(\dot{W}))^2$
Generator	Work, $\dot{W}$ (kW)	0.20 (cost of the turbine)
Pump	Work, $\dot{W}$ (kW)	$5160(\dot{W})^{0.48}$
Cooling tower	Heat transfer, $\dot{Q}$ (kW)	$915(\dot{Q})^{0.6}$

Table 17.3 Overall Heat Transfer Coefficient Values for the ORC System (Adapted from Lukawski, 2009)

Heat Exchanger	$U\left(\frac{\text{kW}}{\text{s-m}^2\text{-K}}\right)$	$U\left(\frac{\text{Btu}}{\text{s-m}^2\text{-R}}\right)$
Waste heat oil heater	0.06	0.03159
Preheater	1.1	0.57915
Vaporizer	2	1.053
Desuperheater	0.7	0.36855
Condenser	1.6	0.8424

$$A = \frac{\dot{Q}}{U\Delta T_{\text{lm}}} = \frac{F_{\text{exhaust}}\left(\hat{h}_6 - \hat{h}_7\right)}{(0.03159)(74.94)} = 9396\ \text{m}^2. \tag{17.9}$$

The capital cost of the WHOH is given in Table 17.4.

Table 17.4 Costing the ORC System

Unit Operation	Characteristic in Cost Equation	Cost ($)
Waste heat oil heater	9395.9 m^2	838,560
Preheater	188.5 m^2	58,777
Vaporizer	190.6 m^2	59,205
Organic turbine	4804.7 kW	3,930,262
Generator	4804.7 kW	786,052
Desuperheater	296.1 m^2	79,889
Condenser	1767.5 m^2	269,237
Pump	219.7 kW	68,665
Cooling tower	18,882 kW	336,549
Purchased equipment cost (PEC)		9,312,294
Fixed capital investment (FCI)		
Total capital investment (TCI)		13,968,441
Yearly operating and maintenance		200,000

Vaporizer and Condenser: Capital Costs

The vaporizer may be sized as three heat exchangers, one that heats liquid n-pentane to near-boiling conditions (the preheater), one in which the n-pentane is vaporized, and a final heat exchanger in which the n-pentane is slightly superheated. Here we treat the vaporizer as two heat exchangers. In the preheater section, liquid n-pentane is heated from inlet conditions $P_{13} = 22.8$ bar and $T_{13} = 532.85$ R to 794.27 R, which is 5 R below the normal boiling temperature of n-pentane at 22.8 bar (this is an approach of 5 R). In the vaporizer section, the n-pentane is heated to vapor conditions with $P_{10} = 22.8$ bar and $T_{10} = 858.87$ R. The areas and costs for the preheater and vaporizer are given in Table 17.4.

The condenser is also taken as two heat exchangers (rather than three heat exchangers); in the desuperheater, vapor-phase n-pentane at $P_{11} = 0.7$ bar and $T_{11} = 721.71$ R is cooled to 543.07 R, which is 5 R above the normal boiling temperature of n-pentane at 0.7 bar (this is an approach of 5 R). In the condenser section, the n-pentane is cooled to liquid conditions with $P_{12} = 0.7$ bar and $T_{12} = 529.47$ R. Costing for the condenser system is given in Table 17.4. In this example, the cold side of the condenser uses cooling water. In remote locations, air cooling would generally be used.

The total purchased equipment cost (PEC) is \$9.31 + MM. As discussed in Chapter 2, there are additional anticipated costs (see Douglas, 1988) for equipment installation, piping, the process control system, and various engineering costs. These costs, when added to the PEC, form the fixed capital investment (FCI). The total capital investment (TCI) then includes the FCI plus start-up costs and working capital. Here we simply estimate the TCI as TCI = 1.5(PEC). Finally, yearly operation and maintenance (O&Ms) costs for a nominal 5-MW n-pentane-based ORC are estimated a \$200,000 (Leslie et al., 2009).

Levelized Cost

Levelized costs were developed in Chapter 16 (Section 16.3) in order to account for a discount rate, i, and an escalation rate, e (e.g., a fuel escalation rate). For the total capital investment associated with this ORC system, there are no escalating costs to consider—we are simply recovering waste energy from the gas turbine exhaust and generating power. The annual levelized TCI can be determined using Equation (16.31):

$$\text{Levelized TCI} = \text{PV}\left(\frac{i(1+i)^n}{(1+i)^n - 1}\right) = 13{,}968{,}441(0.1524)$$
$$= \$2{,}128{,}898/\text{year}, \tag{17.10}$$

where a discount rate for money, $i = 0.085$, and project life, $n = 10$ years, have been used (Lukawski, 2009).

For the yearly O&Ms, the levelized cost assuming an O&M escalation rate, $e = 0.04$, is given by Equation (16.35):

$$\text{Levelized O\&M} = A\left(\frac{(1+i')^n - 1}{i'(1+i')^n}\right)\left(\frac{i(1+i)^n}{(1+i)^n - 1}\right)$$
$$= 200{,}000(1.2163) = \$243{,}258/\text{year}, \tag{17.11}$$

where from Equation (16.32)

$$i' = \frac{i-e}{1+e} = 0.04327.$$

The sum of the levelized TCI and O&M costs is $2,372,156 per year. If we assume 8000 h/year operation and a net organic turbine output of 4584.99 kW (from Eq. (17.6)), the levelized break-even cost for electricity production from the ORC system is 6.47¢ per kilowatt-hour. ■

In Table 17.1, the break-even cost for the ORC is estimated at 9¢ per kilowatt-hour as opposed to the 6.47¢ per kilowatt-hour determined in Example 17.1. There are clearly many factors that will impact a reported break-even cost. In Example 17.1, we assumed 8000 hours of operations per year, which is basically a 100% capacity/availability factor. A lower availability factor would increase the break-even cost. A more accurate determination of the total capital investment as well as the selected discount and escalation rates and years of operation will impact levelized costs. Ambient air conditions impact gas turbine performance and, consequently, the energy available in the gas turbine exhaust gas (gas turbine performance in off-design conditions has been detailed in Chapters 7 and 9). The final selected design will impact costs—for example, a recuperative heat exchanger can be used to exchange energy between stream 11 and stream 13 (in Figure 17.2), which may lower the break-even cost. The Monte Carlo simulation techniques developed in Chapter 2 (Section 2.7) can be useful in exploring uncertainty in economic parameters.

17.3 NUCLEAR POWER CYCLE

We can utilize the ODE numerical methods developed in Chapter 5 to introduce performance aspects of the next possible generation of nuclear reactors. We will determine the temperature profile in an individual reactor pebble (or pellet). As a term project (see Problem 17.3), provided reactor temperature and flow rate information can be combined with heat recovery (Chapters 7, 9, and 10) and the steam turbine developments (Chapter 16) to determine the performance of a nuclear reactor-based steam-driven power plant.

17.3.1 A High-Temperature Gas-Cooled Nuclear Reactor (HTGR)

A nuclear reactor is essentially a device to convert nuclear fission to heat, primarily by the conversion of fast neutrons to thermal neutrons, which are then absorbed by the surrounding media to generate heat. One type of nuclear fuel heavily studied in recent years is "TRISO" (tri-isotopic structural layers). A TRISO fuel consists of small spherical radioactive particles in a graphite matrix or shell (see Figure 17.3). The heat generated by the particles can then be transferred to either water, supercritical water, or a gas. One type of cylindrical TRISO-containing reactor uses spherical graphite "pebbles" as the graphite matrix, with heat transfer from the pebbles to flowing He. It is important to understand the temperature profiles in such a pebble to ensure that no part of it exceeds temperature limits at which stress cracking might occur. Such cracking could result in the release of heavy isotopes of Cs, Sr, and Ag, or radioactive gases, into the He, which if leaked from the reactor core would constitute a safety hazard.

A key metric for nuclear reactor beds is power density, which is defined as the generated power/volume of radioactive material (megawatt per cubic meter). Power density is an experimentally measured quantity. Here, for example, helium with a known flow rate can be passed through a test bed containing a known volume of radioactive material (encased within the graphite shells). At steady state, the MW generated would be $\dot{Q}_{\text{test\_bed}} = F_{\text{He}}\left(\hat{h}_{\text{He}}^{\text{out}} - \hat{h}_{\text{He}}^{\text{in}}\right)$ and the power density can be determined. The nuclear reactor examples and discussion in this section are from K.M. Dooley, personal communication, used with permission.

EXAMPLE 17.2 ***Numerical Solution for TRISO Pebble Center Temperature***

This example is adapted from Wang (2004; web.mit.edu/pebble-bed). A new TRISO fuel has an experimentally determined power density of 3.65 MW/m^3. The spherical graphite particles have a

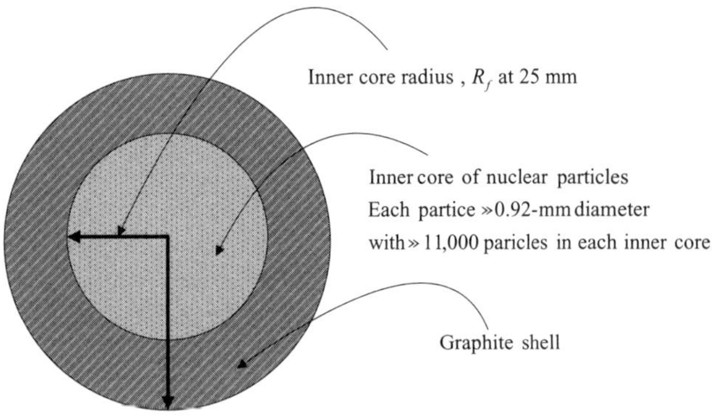

Figure 17.3 Nuclear reactor pebble—radioactive particles in graphite matrix (shell).

radius, R_p. The inner core (radius = R_f) of this graphite shell contains the small nuclear fuel particles, which themselves are only ~0.92 mm in diameter. The number of fuel particles in the inner core is ~11,000, so we can assume the heat is generated uniformly. The fuel particles are composites of several materials, but here assume the thermal conductivities of the outer and inner cores are that of pure graphite. Determine the maximum expected temperature at a pebble core if the temperature leaving the nuclear reactor is 850°C. For *safety purposes*, in these type calculations, the measured power density is multiplied by *five*.

SOLUTION Equations for heat conduction in the HRSG were developed in Chapter 10. For one-dimensional steady-state heat conduction in spherical coordinates (here you may need to refer to a standard transport text to see the derivation of these equations), temperature profiles can be determined from the following:

With No Heat Generation

$$\frac{1}{r^2}\frac{\partial}{\partial r}\left(r^2 k\frac{\partial T}{\partial r}\right)=0 \tag{17.12}$$

With Heat Generation and Where $\dot{e}_{generation}$ Is the Heat or Power Generated (MW/m^3)

$$\frac{1}{r^2}\frac{\partial}{\partial r}\left(r^2 k\frac{\partial T}{\partial r}\right)+\dot{e}_{\text{generation}}=0, \tag{17.13}$$

where r is the radius and k is the thermal conductivity (watt per meter-Celsius) and $\dot{e}_{\text{generation}}$ is the generation/power density term (watt per cubic meter). At steady state, the rate of heat generated in the inner core must equal the rate of heat conducted through the outer graphite shell. The equation without heat generation will be used to account for the temperature change in the graphite shell, and the equation with power generation will be used to account for the temperature change in the inner core.

Integrating each equation once (where C's are the constants of integration),

$$\left(r^2 k\frac{\partial T}{\partial r}\right)=C_1 \tag{17.14}$$

and

$$\left(r^2 k\frac{\partial T}{\partial r}\right)=\left(-\dot{e}_{\text{generation}}\right)\left(\frac{r^3}{3}\right)+C_2. \tag{17.15}$$

We can solve for C_2; at $r = 0$, $\dot{e}_{\text{generation}} = 0$ (no particles in zero volume) and symmetry requires $\partial T/\partial r = 0$, giving $C_2 = 0$. We can write Equation (17.15) as

$$\left(k\frac{\partial T}{\partial r}\right)=\left(-\dot{e}_{\text{generation}}\right)\left(\frac{r}{3}\right). \tag{17.16}$$

This allows us to determine a numerical solution for the *temperature in the inner core fuel zone* using Euler's method as

$$\left(\frac{T_r-T_{r-1}}{\Delta r}\right)=\left(\frac{\left(-\dot{e}_{\text{generation}}\right)(r)}{3k_{\text{fuel zone}}}\right), \tag{17.17}$$

where here we are numerically integrating from the outside in (from $r = R_f$ to $r = 0$) as

$$T_{r-1}=T_r+(\Delta r)\left(\frac{\left(\dot{e}_{\text{generation}}\right)(r)}{3k_{\text{fuel zone}}}\right). \tag{17.18}$$

To account for the graphite shell, we will need to determine C_1 in Equation (17.14). Here we can utilize the fact that the heat of transfer rate by conduction through the graphite shell must equal the rate of heat convection from the surface (at $r = R_p$):

$$\left(-k\frac{\partial T}{\partial r}\right)=h\left(T_{R_p}-T_\infty\right),\quad \text{at } r=R_p, \tag{17.19}$$

where h is the heat transfer coefficient (watt per square meter-Celsius) and T_∞ is the bulk fluid temperature around the pebble. We can solve for C_1 as

$$\left((R_p)^2 k\frac{\partial T}{\partial r}\right)=C_1=(R_p)^2 h\left(T_\infty-T_{R_p}\right), \tag{17.20}$$

allowing us to determine a numerical solution for the *temperature in the graphite shell* using Euler's method as

$$\left(\frac{T_r-T_{r-1}}{\Delta r}\right)=\left(\frac{(R_p)^2 h\left(T_\infty-T_{R_p}\right)}{k_{\text{graphite shell}}(r)^2}\right), \tag{17.21}$$

and we will numerically integrate from the outside in (from $r = R_p$ to $r = R_f$) as

$$T_{r-1}=T_r+(\Delta r)\left(\frac{(R_p)^2 h\left(T_{R_p}-T_\infty\right)}{k_{\text{graphite shell}}(r)^2}\right). \tag{17.22}$$

The solution to the numerical temperature determination is provided in the Excel file **Example 17.2.xls**. All of the needed physical properties and parameters for a possible nuclear are given in the spreadsheet. With a correlation for h and physical properties for the pebble, we can numerically integrate the two ordinary differential equation (ODEs) (Eqs. (17.22) and (17.18)) from the pebble surface to the center of the pebble. For now, assume h = 4296.00448 W/m^2-K and $\dot{e}_{\text{generation}} = (3.65\text{ MW/m}^3) =$ (3.65 MW/m^3), but recall for safety we must increase this value by a factor of 5, so we will set $\dot{e}_{\text{generation}} = (5)(3.65\text{ MW/m}^3) =$ (5)(3.65 MW/m^3). The correlation used to determine h can be found in Wang (2004).

We begin with the numerical integration for temperature distribution in the graphite shell. Here we use T_∞ = 850°C, which is the maximum observed bed temperature; this will give the highest pebble core temperature. It will be useful to also relate the rate of heat generation in the inner core with the rate of heat transfer from the pebble surface, $\dot{q}_p$ (watt per square meter); these two quantities must be equal:

$$\left(\frac{4}{3}\pi\left(R_f^3\right)\left(\dot{e}_{\text{generation}}\right)\right) = \left(4\pi\left(R_p^2\right)\dot{q}_p\right) \tag{17.23}$$

or

$$\dot{q}_p = \left(\frac{1}{3}\frac{R_f^3}{R_p^2}\left(\dot{e}_{\text{generation}}\right)\right). \tag{17.24}$$

We can determine the pebble surface temperature as

$$T_{r=R_p} = \frac{\dot{q}_p}{h} + 850,$$

where again for *safety* we must actually use $(5)q_p$. It is also convenient to use the expression for q_p in Equation (17.22) as

$$T_{r-1} = T_r + (\Delta r)\left(\frac{(R_p)^2(5)\dot{q}_p}{k_{\text{graphite shell}}(r)^2}\right). \tag{17.25}$$

The solution provided in the Excel file **Example 17.2.xls** shows that the pebble core temperature may reach 975.59°C. The long-term stability properties of graphite (e.g., resistance to cracking) at this temperature, and including the possibility of temperature cycling, would need to be evaluated. ■

REFERENCES

BOHL, R. 2009. Waste Heat Recovery from Existing Simple Cycle Gas Turbine Plants—A Case Study. IAGT paper 09-IAGT-205, October 2009.

DORF, R.C. 1978. *Energy, Resources, & Policy*. Addison-Wesley Publishing Company, Reading, MA.

DOUGLAS, J.M. 1988. *Conceptual Design of Chemical Processes*. McGraw Hill, Boston.

HETTIARACHCHI, H.D.M., M. GOLUBOVIC, W.M. WOREK, and Y. IKEGAMI 2007. Optimum design criteria for an organic Rankine cycle using low-temperature geothermal heat sources. *Energy* 32: 1698–1706.

LESLIE, N.P., R.S. SWEETSER, O. ZIMRON, and T.K. STOVALL 2009. Recovered energy generation using an organic Rankine cycle system. ASHRAE paper Ch-09-024.

LUKAWSKI, M. 2009. Design and Optimization of Standardized Organic Rankine Cycle Power Plant for European Conditions. MS Thesis, University of Iceland & the University of Akureyri, Akureyi, February 2010.

MOHAMED, K.M., M.C. BETTLE, A.G. GERBER, and J.W. HALL 2010. Optimization study of large-scale low-grade energy recovery from conventional Rankine cycle power plants. *Int. J. Energy Res.* 34: 1071–1087.

OBERNBERGER, I., P. THONHOFER, and E. REISENHOFER 2002. Description and evaluation of the new 1000 kW organic Rankine cycle process integrated in the biomass CHP plant in Lienz, Austria. *Euroheat Power* 10: 1–17.

REYNOLDS, W.C. 1979. *Thermodynamic Properties in SI: Graphs, Tables and Computational Equations for Forty Substances*. Department of Mechanical Engineering, Stanford University, Stanford, CA.

SCHUSTER, A., J. KARL, and S. KARELLAS 2007. Simulation of an innovative stand-alone desalination system using an organic Rankine cycle. *Int. J. Thermodyn.* 10(4): 155–163.

WANG, J. 2004. An integrated performance model for high temperature gas cooled reactor coated particle fuel. PhD Thesis, MIT.

PROBLEMS

17.1 ***(Term Project) Economic Evaluation: Conventional versus Alternative Energy Systems*** In this problem, which is best suited as a group term project, we want to compare common electricity generation methods with green and alternative energy technologies. A partial list of technologies that can be evaluated includes utility-based coal; utility-based nuclear energy; utility-based natural gas; hydroelectric dams; wind energy; cogeneration systems; biomass to electricity and biomass to liquid fuels; solar photovoltaic systems; and solar thermal, geothermal, and ocean energies. Comparison should include generation costs as, for example, dollar per kilowatt-year or cent per kilowatt-hour. Performance calculations for each technology should also be developed. The levelized break-even costs provided in Table 17.1 (cent per kilowatt-hour) provide reasonable values, but as shown in Example 17.1, you should develop your own analysis and levelized costs.

Some problem considerations are outlined here. A significant difficulty with this type comparison is the system size. There is an economy of scale that will affect all costs, especially capital costs. In order to allow solution of Problem 17.1, we can pick a "typical" size for each technology. For example, for wind turbines, it is convenient to examine a single turbine. Another consideration is the capacity factor, which provides the percent of time that a given utility plant with operate. For example, a typical utility company assumption for wind turbines is that only 10–15% of the rated capacity will be available. For hydroelectric dam power, 90% of the rated capacity is generally assumed available. Be sure to consider pollution abatement costs as well as fuel delivery and transmission costs. For example, the economics for biomass plants can become unfavorable when biomass feedstock is more than 50 miles (radius) from the plant.

17.2 ***(Term Project) Energy Future*** For those especially interested in alternative energy technologies/policies, and as a term project for an individual (rather than a group project), I do encourage comparing the energy concerns and planned energy directions of the late 1970s with those of today. A good place to start for the late 1970s is *Energy, Resources, & Policy* by Dorf (1978).

17.3 ***(Term Project) HTGR Nuclear Reactor-Based Steam Power Plant Performance*** The design of a nuclear reactor for use in power generation is extremely complex. Wang (2004) and web.mit.edu/pebble-bed showed that a cylindrical reactor bed with length ~10 m, radius ~1.48 m, void fraction ~0.4, and containing some 360,000 pebbles as those detailed in Example 17.2 can be used to heat helium flowing at 121 kg/s from 450°C entering the reactor bed to 850°C at the exit. The reactor can be assumed to operate at a nominal 1 atm. These data are also supplied in the Excel file **Example 17.2.xls**.

Determine the performance of a steam power plant using this reactor and helium as the heat transfer fluid. You will need to account for the HRSG system, which will use the energy available in the helium to produce steam (see Chapters 7, 9, and 10). The start of the steam turbine system to generate electricity is supplied in Chapter 16. Assume that parallel reactors may be used to increase the MW of electricity produced.

Appendix

Bridging Excel and C Codes

Microsoft Excel provides a flexible pre- and postprocessor for low-level languages such as C/C++. In this appendix, the details for passing single variables, vectors, and matrices from Excel ↔ Visual Basic for Applications (VBA) ↔ C programs are provided. The C programs will be connected to VBA as dynamic link libraries (DLLs, .dlls). The conduit for passing information between Excel and C programs is VBA. Passing single variables is straightforward. Passing vectors requires knowledge of pointers. Passing matrices requires an understanding of pointers and row-major and column-major storage strategies.

We assume the reader has some coding experience, but we also provide background for those not familiar with C. At the end of Appendix A we provide a tutorial on C programming, "Tutorial Microsoft C++ 2008 Express: Creating C Programs and DLLs," in which we utilize the discussion and examples from Appendix A to create C-based DLLs for examples from Chapter 2.

All the examples in Appendix A use Microsoft C++ 2008 Express as the development environment. The user must have Microsoft C++ 2008 Express installed to duplicate the examples here. Microsoft C++ 2008 Express is available as a free download from the Microsoft site (go to the Microsoft site → Downloads). You should download this file and allow the default installation.

Our intent in Appendix A is to use simple examples to show how the bridge between Excel and C is constructed. Example A.1 shows how simple math operations performed in C can be accessed from an Excel sheet. Example A.2 shows how vectors from an Excel sheet can be passed to a C code where basic operations are performed and the results are returned to the sheet.

We also introduce the construction of a basic C thermodynamics library. The library developed in Example A.3 shares many common coding developments with the rigorous thermodynamics code for cogeneration calculation, which is detailed in Chapter 8.

Finally, matrix transfer between Excel and C is demonstrated using the example of the solution of n linear equations with n unknowns. The solution for the n unknowns is obtained using the Gauss–Jordan matrix elimination method (written as a C program). This foundation allows exploration of the simultaneous or equation-based solution approach for both linear and nonlinear material balance problems in Chapter 4.

A.1 INTRODUCTION

Microsoft Excel is a widely used software package for engineering calculations (Coronell, 2005; de Moura, 2005; Ferreira et al., 2004). It provides an excellent interface for data input and storage and for the presentation of results. Many calculations can be performed directly on the spreadsheet. Macros may also be created, which simplify repetitive calculations and some complex calculations. Macros may be created by recording during Excel sheet operations or by coding directly in VBA as we did in Chapters 2 and 3; both features are provided within Excel (Walkenbach, 2001). As the level of complexity of the engineering problem increases, low-level languages such as Fortran or C/C++ are often required. Low-level languages provide more complete mathematical functions and a more structured coding format when compared to VBA. To summarize, simple engineering calculations can be performed on spreadsheets; more difficult calculations are performed using VBA; and when problems reach a certain complexity level, low-level languages are utilized.

Given the flexibility of Excel for data manipulation, data storage, and result presentation, it should be considered

Modeling, Analysis and Optimization of Process and Energy Systems, First Edition. F. Carl Knopf.

as the pre- and postprocessor for low-level languages (Rosen and Partin, 2000). Here we focus on how bridging between Excel and C/C++ can be accomplished. This bridging has been previously shown via the use of the __stdcall calling convention in C (Rosen, 2001a,b). Here an equivalent method is presented, which is easier to implement. Step-by-step procedures are shown to allow the transfer of variables, vectors, and matrices between Excel and C programs.[1]

The bridge between C and Excel is established by creating a DLL of the functions in C. These functions are loaded to the memory when the Excel application is invoked, and they remain in memory until the corresponding Excel program closes. The user can access these C-defined functions from the Excel sheet using Excel function calls or Excel macros. The overall problem then is to pass variables, constants, and data between Excel and a C program in a seamless fashion.

In this appendix, file location is somewhat complicated; this is in part due to locations determined by the C complier and in part because we do not want to move files out of the folder C:\POEA\Bridging Excel and C Codes. Figure A.1 shows file locations that will be used in Example A.1. For example, to access the DLL Simple_C_Math_dll.dll, we will use the complete path to this file, which is

C:\POEA\Bridging Excel and C Codes\Examples\Simple_C_Math_dll\Debug\Simple_C_Math_dll.dll.

Also, as shown in Figure A.1, the reader is provided with an empty folder, \New User to duplicate Appendix examples.

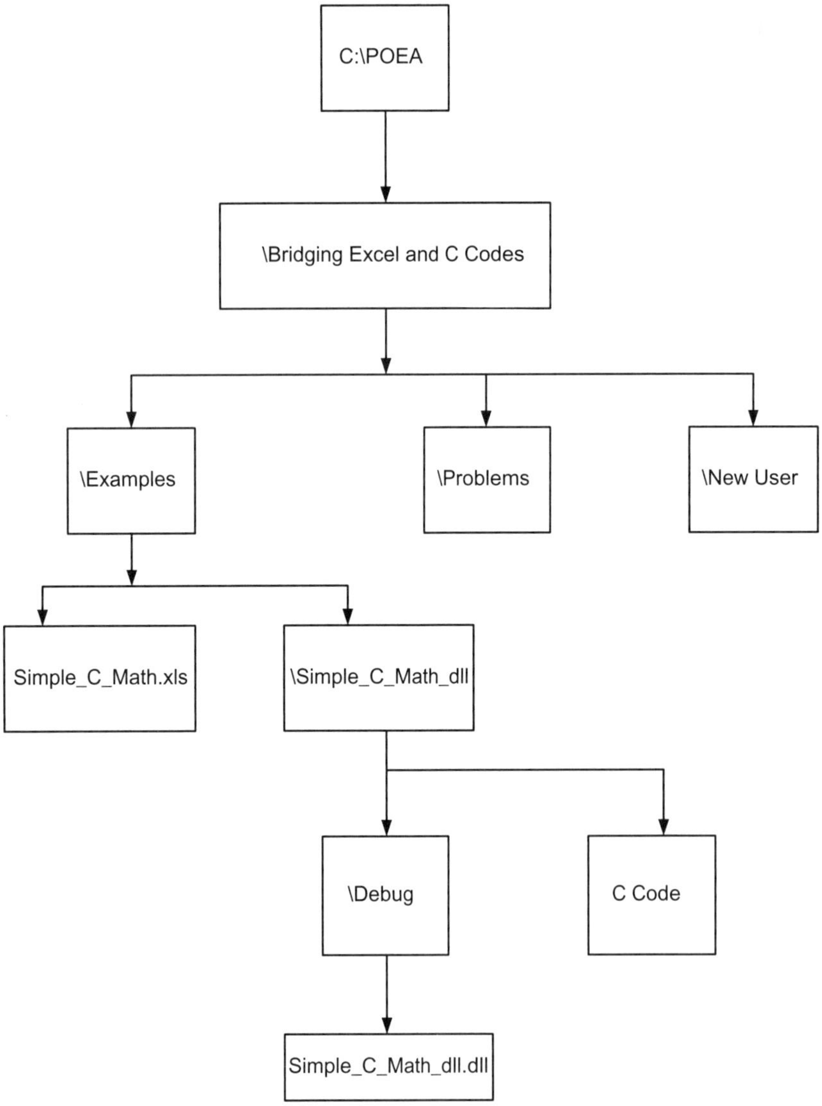

Figure A.1 Location of C files for use with Example A.1.

[1] This same procedure will work with C or C++ codes; the examples here use C code only.

A.2 WORKING WITH FUNCTIONS

EXAMPLE A.1 *Working with Simple Functions*

In this example, we address how simple C functions, such as exponential, difference, and sum, can be accessed from Excel. A stand-alone C code is available to perform these operations, and the code appears as shown in Figure A.2. This code is available in

C:\POEA\Bridging Excel and C Codes\Examples\Simple_C_Math_Stand_Alone\Simple_C_Math.dsw.

If you are not familiar with creating a stand-alone C program, see the Appendix Tutorials AT.1 and AT.2 *after* reading this example. Appendix Tutorials AT.1 and AT.2 provide a tutorial on creating a stand-alone C program.

If you double click on Simple_C_Math.dsw, the file shown in Figure A.2 will open (provided you have installed Microsoft C++

```
#include <stdio.h>
#include <math.h>

double calculate_sum (double z1, double z2);
double calculate_difference (double z1, double z2);
double calculate_exponential (double z1, double z2);

int main()
{
      double x1, x2, f_sum, f_diff, f_exp;

            x1 = 5.0;
            x2 = 3.0;

            f_sum =  calculate_sum  (x1, x2);
            f_diff = calculate_difference (x1, x2);
            f_exp =  calculate_exponential (x1, x2);

// Send Answers to Screen
            printf("sum = %lf\n", f_sum);
            printf("diff =  %lf\n", f_diff);
            printf("expon = %lf\n", f_exp);

            return 0;
}

double  calculate_sum (double z1, double z2)
{
      double sum;
      sum = z1 + z2;
      return sum;
}
double calculate_difference (double z1, double z2)
{
      double difference;
      difference = z1 - z2;
      return difference;
}


double calculate_exponential (double z1, double z2)
{
      double exponential;
      exponential = pow(z1,z2);
      return exponential;
}
```

Figure A.2 Stand-alone C functions for simple math calculations.

2008 Express). You can run this file from the toolbar buttons, Compile (or Ctrl + F7) → Build (or F7) → Execute (or Ctrl + F5), and the results for sum, difference, and exponential for x1 = 5 and x2 = 3 should appear on the screen.

The logic used in Figure A.2 should be clear after some comments. The include statements will load existing C libraries for math and input/output operations. The next three statements before main are termed prototypes. Placing these prototypes here (one for each function) allows compilation of the program without concern for the order in which the three functions are called. The main function assigns two variables, `x1 = 5.0` and `x2 = 3.0`, then the three C functions (`calculate_sum`, `calculate_difference`, `calculate_exponential`) are called and the results are printed on the screen.

In order to access these three C functions from an Excel spreadsheet, the following steps are needed:

Step 1: Open Microsoft Visual C++ 2008 Express Edition. You can close the start page window which will appear. To create a new project, from the toolbar, File → New → Project. From Project Types, Select: General; and from Templates, Select: Empty Project. For the location, I used `C:\POEA\Bridging Excel and C Codes\Examples` as the location and `Simple_C_Math_dll` as the project name.

If the reader wants to duplicate any example in this appendix, do not use the `\Examples` folder as this folder already holds working results. Instead use an empty folder, say, \New User. The path would then be `C:\POEA\Bridging Excel and C Codes\New User`; only this change in the folder location to `New User` is needed.

Step 2: To add the C code to the project, from the toolbar, Project → Add New Item. From Categories, Select: Code, and from Templates, Select: C++ File (.cpp). Use `C_Math.cpp` as the file name → Add, see Figure A.3. Copy the source code from Figure A.4 to the `C_Math.cpp` file. Here `"C_Math.h"` is a local header file, which will be created in step 4.

Step 3: To add the definition file to the project, from the toolbar, Project → Add New Item. From Categories, Select: Code, and from Templates, Select C++ File (.cpp). For the name, use `Simple_C_Math_dll.def` (this name must be the same as the project name with the file extension `.def`) → Add. Even though we started step 3 by adding a cpp file, the extension .def overrides the .cpp extension and creates the definition file in the project. The `Simple_C_Math_dll.def` file must be typed as

```
LIBRARY Simple_C_Math_dll
EXPORTS
 calculate_sum
 calculate_difference
 calculate_exponential
```

Do also recall the name following the key word `LIBRARY` must be the project name. This will be the name of the created DLL. All the functions callable from Excel must be listed following the key word `EXPORTS`.

Step 4: To add the header file to the project, from the toolbar, Project → Add New Item. From Categories, Select: Code, and from Templates, Select: Header File (.h). For the

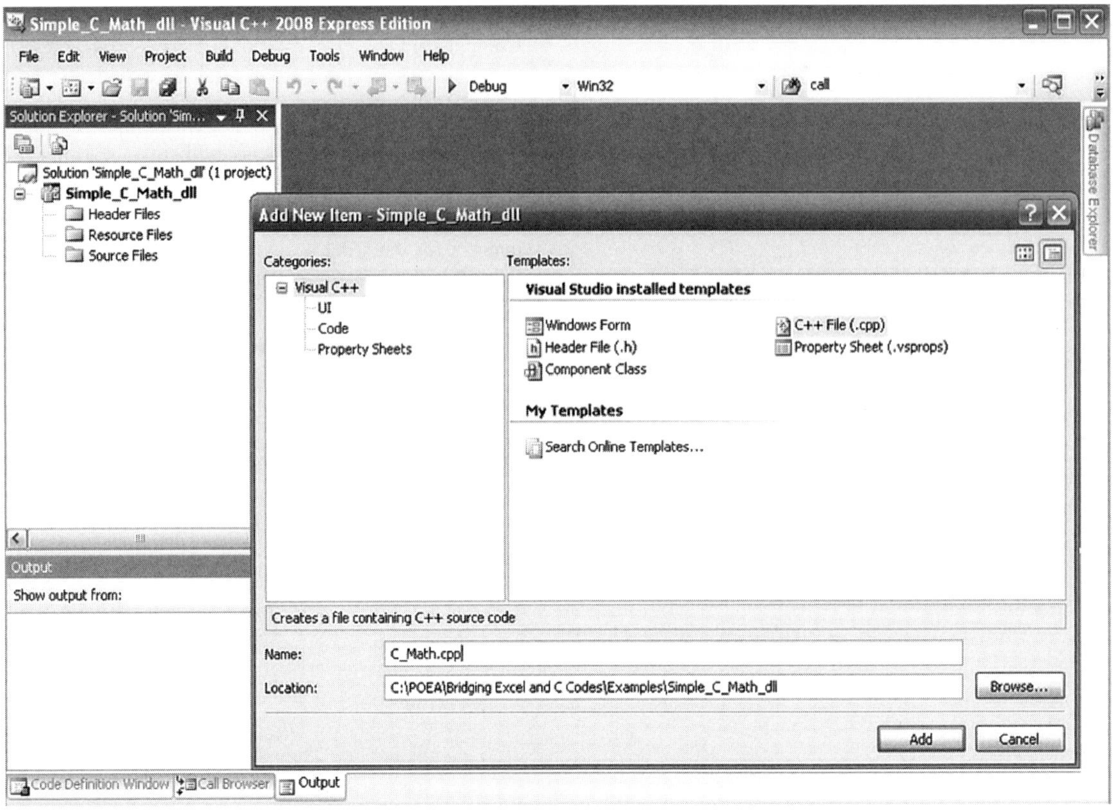

Figure A.3 Defining the C++ code name and location.

```
#include <math.h>
#include "C_Math.h"

double  calculate_sum (double z1, double z2)
{
      double sum;
      sum = z1 + z2;
      return sum;
}

double calculate_difference (double z1, double z2)
{
      double difference;
      difference = z1 - z2;
      return difference;
}

double calculate_exponential (double z1, double z2)
{
      double exponential;
      exponential = pow(z1,z2);
      return exponential;
}
```

Figure A.4 C functions to allow DLL math calculations.

name, use `C_Math.h` (the `.h` extension will be automatically generated if not supplied). The header file contains the prototypes of the functions and these are provided in the word document → Add. The `C_Math.h` file must be typed as

```
double calculate_sum (double z1, double z2);
double calculate_difference (double z1,
double z2);
double calculate_exponential (double z1,
double z2);
```

Step 5a: From the toolbar, select Project → Simple_C_Math_dll Properties (Alt + F7) → Configuration Properties. Expand the Configuration Properties window by highlighting the + sign next to Configuration Properties → General → Project Defaults → Configuration Type. The configuration type will currently be Application (.exe); use the drop-down window to select Dynamic Library (.dll) → Apply. Close the Configuration Manager window, which is shown in Figure A.5.

Step 5b: From the toolbar, select Project → Simple_C_Math_dll Properties (Alt + F7) → Configuration Properties → C/C++. Expand the C/C++ window by highlighting the + sign next to C/C++ → Advanced → Calling Convention. The current calling convention will be _cdecl (/Gd); use the drop-down window to select _stdcall(/Gz) → Apply. Close the Configuration Manager window. The use of /Gz is an alternative to using _stdcall prior to each of the C functions. The /Gz enables all the functions defined in the header files (step 4) to be accessed following the _stdcall calling convention → OK. The /Gz option will actually not remain as the last entry, but it will (and must) be in the option list.

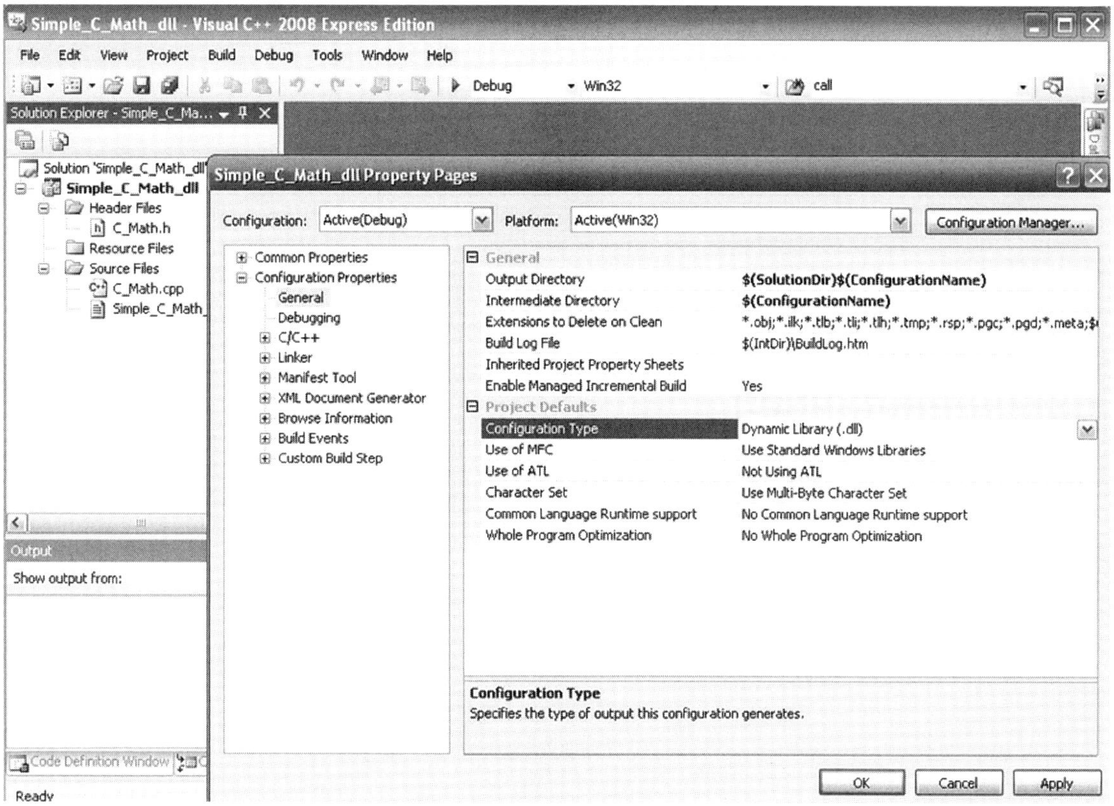

Figure A.5 Configuration manager—setting the project type as .dll.

Step 5c: From the toolbar, select Project → Simple_C_Math_dll Properties (Alt + F7) → Configuration Properties → Linker. Expand the Linker window by highlighting the + sign next to Linker → Input. The module definition file will be blank; add Simple_C_Math_dll.def → Apply. This name must be the same as the project name with the file extension `.def`, as shown in Figure A.6. Close the Configuration Manager window.

The changes and additions required in Microsoft C++ 2008 Express to create the DLL are completed. Now the successful build (F7) generates the `Simple_C_Math_dll.dll` in the debug folder of the project directory. You can close out of all C windows, and if you go to the \Examples or `\New Users` folder → Simple_C_Math_dll folder → Debug folder, you will find the DLL `Simple_C_Math_dll.dll`.

To call these three functions from the Excel worksheet, the following additions need to be made in the Excel file.

Step 6: Open the Excel file and go to Macros → Visual Basic Editor from the menu.

Create a new module and add the following lines:

```
Public Declare Function calculate_sum Lib
"C:\POEA\Bridging Excel and C Codes\ Exam-
ples \Simple_C_Math_dll\Debug\Simple_C_Math_
dll.dll" (ByVal z1 As Double, ByVal z2 As
Double) As Double
Public Declare Function calculate_difference
Lib"C:\POEA\Bridging Excel and C Codes\
Examples \Simple_C_Math_dll\Debug\Simple_C_
Math_dll.dll" (ByVal z1 As Double, ByVal z2
As Double) As Double
Public Declare Function calculate_exponen-
tial Lib "C:\POEA\Bridging Excel and Codes\
Examples \Simple_C_Math_dll\Debug\Simple_C_
Math_dll.dll" (ByVal z1 As Double, ByVal z2
As Double) As Double
```

If you have used the New User folder option (creating your own DLL), then in the declare statements, change `\Examples\` to `\New User\`.

The typical syntax for the external DLL function declaration is `Public Declare Function` *`function name`* `Lib` *`dll name with path`* `([Parameters]) [as return type]`.

Step 7: Test the C functions. The three C functions should appear in the User Defined function list in the current Excel worksheet. These functions will work the same as any Excel function.

The Excel sheet to test these functions is provided in **Example A.1.xls** The declare statements using the \Examples folder are provided in this file.

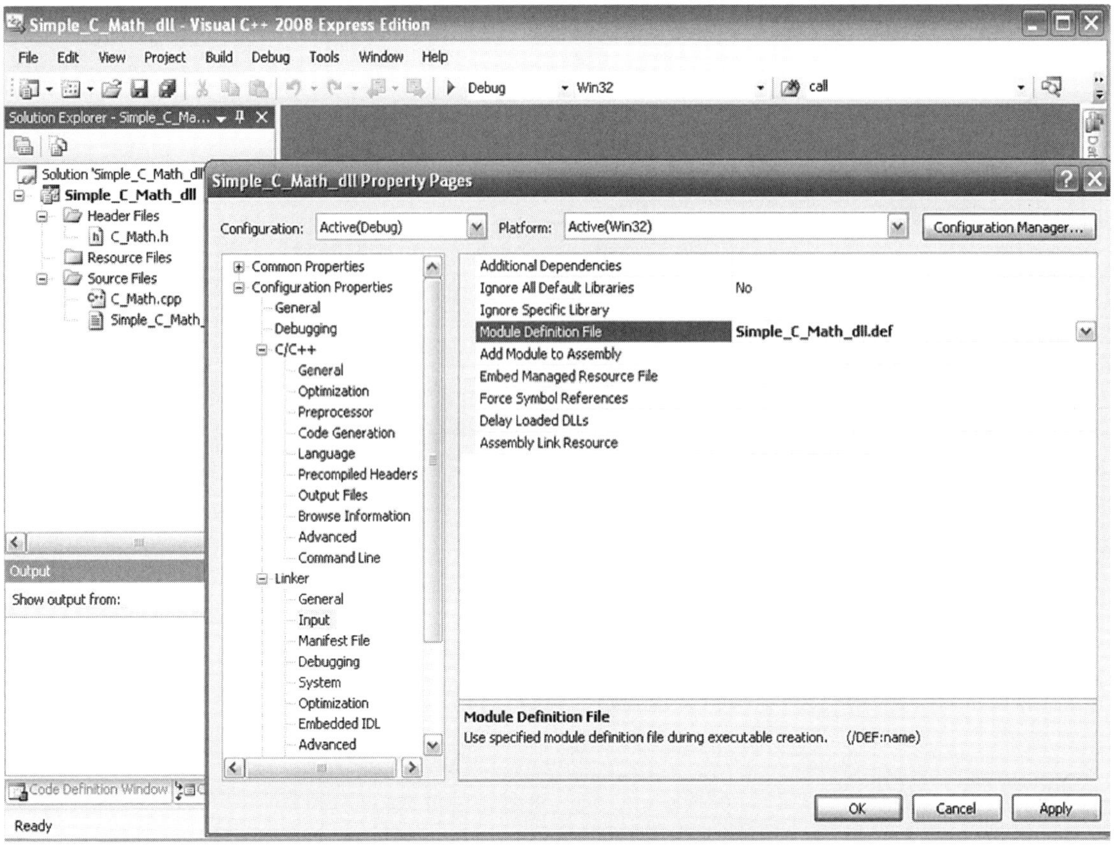

Figure A.6 Configuration manager—setting the project definition file.

A.3 WORKING WITH VECTORS

EXAMPLE A.2 *Working with Vectors*

Here, two vectors on an Excel sheet are summed in a C function and the resulting vector and its average are displayed on the Excel sheet. This is accomplished by VBA macros utilizing C functions.

In general, data in vectors (or matrices) are passed by reference, with the memory address to the first element in the vector passed in the parameter list. This process is transparent to the user if a stand-alone VBA or stand-alone C code is being used. However, to bridge Excel and C codes, we must understand this data transfer and specifically the use of pointers.

First, let us develop the stand-alone C code to sum two vectors, determine the average of the resulting vector, and display the output to the screen. This code is given in Figure A.7 and is available in

C:\POEA\Bridging Excel and C Codes\Examples\Simple_C_Vector_Stand_Alone_No_Pts\ Simple_C_Vector_Stand_Alone_No_Pts.dsw.

If you double click on Simple_C_Vector_Stand_Alone_No_Pts.dsw, the file in Figure A.7 will open. You can run this file from the toolbar buttons, Compile (or Ctrl + F7) → Build (or F7) → Execute (or Ctrl + F5), and the results for vector addition and averaging with `Col_A[4] = {1, 2, 3, 4}` and `Col_B[4] = {5, 6, 7, 8}` should appear on the screen.

The logic used in Figure A.7 should be clear after some comments. The stand-alone code begins by including the math and input/output libraries. Prototypes are then defined for `Calculate_Vector_Sum` and `Calculate_Vector_Average`. Prototypes allow compilation regardless of the calling order of the functions.

In function main, vectors `Col_A[4]` and `Col_B[4]` are defined, each containing four values. The vector `Col_C[4]` will also contain four values. It is important to recall that associated with any vector, for example, `Col_A[4]`, there will be numeric values, `{1, 2, 3, 4}`, and a memory address for each numeric value. The key to moving vectors (or matrices) between Excel and C is understanding the connection between the numeric values in elements and memory addresses.

It is instructive to next examine the parameters in the function call. In the function call `Calculate_Vector_Sum`, parameters `Col_A`, `Col_B`, and `Col_C` actually contain the memory addresses of the first element of the corresponding vectors. `Size` is a variable containing value = 4.

We next examine the function definition `void Calculate_Vector_Sum`. The function `Calculate_Vector_Sum` does not return a specific value as indicated by `void` before the function

```
#include <stdio.h>
#include <math.h>

void  Calculate_Vector_Sum (double Col_1[],  double Col_2[],
 double  Col_C1_C2[], long Nsize);
double  Calculate_Vector_Average (double Column[], long Nsize);


main()
{
      double Col_A[4] = {1, 2, 3, 4};
      double Col_B[4] = {5, 6, 7, 8};
      double Col_C[4];

      long Size = 4;

      double Vector_Average;

      Calculate_Vector_Sum ( Col_A,  Col_B, Col_C, Size);

      Vector_Average = Calculate_Vector_Average (Col_C, Size);

      // Send Answers to Screen also Note Index 0 to 3
            printf("Col_C[0] = %lf\n", Col_C[0]);
            printf("Col_C[1] = %lf\n", Col_C[1]);
            printf("Col_C[2] = %lf\n", Col_C[2]);
            printf("Col_C[3] = %lf\n", Col_C[3]);

            printf("Vector Average = %lf\n", Vector_Average);


      return 0;
}

void  Calculate_Vector_Sum (double Col_1[],  double Col_2[],
 double  Col_C1_C2[], long Nsize)
{
      int i;
      for(i =0; i<Nsize; ++i)
      {
            Col_C1_C2[i] = Col_1[i] + Col_2[i];
      }
      return;
}

double  Calculate_Vector_Average (double Column[], long Nsize)
{
      int i;
      double Sum =0;
      for(i =0; i<Nsize; ++i)
      {
            Sum = Sum +Column[i];
      }

      return Sum/Nsize;
}
```

Figure A.7 Stand-alone C functions for vector calculations.

name. Inside the `for` loop, *i*++ indicates that the variable *i* is to be incremented by 1 after each pass through the loop, starting with $i = 0$. In both C and VBA, vector storage begins in the zeroth element.

In `void Calculate_Vector_Sum`, the variable `Col_1[i]` is actually `*(Col_1 + i)`, where `(Col_1 + i)` is a memory address and `*(Col_1 + i)` is the numeric value in that memory address. Here `Col_1` will contain the same memory address value as `Col_A` and `*` acts as a pointer operator that converts the memory address to the numeric value in that address.

The C function `double Calculate_Vector_Average` determines the average value in a vector. In main, the parameter list in `Calculate_Vector_Average (Col_C, Size)` contains the starting address of the vector `Col_C[4]`, and `Size` again contains value 4. The `double` before the function declaration `Calculate_Vector_Average` is the data type for the return value for the function `Sum/Nsize`; `Sum/Nsize` returns the average of the vector to main as `Vector_Average`.

In Figure A.7, we are implicitly using pointers to solve the problem. The use of pointers or memory addresses in Figure A.7 is transparent to the user. However, to create a C-dll that can solve this example and that can be accessed in Excel, we must explicitly use pointers. We begin this process by explicitly using pointers in the stand-alone code. This is shown in Figure A.8; note the changes

```
#include <stdio.h>
#include <math.h>

void  Calculate_Vector_Sum (double* Col_1,  double* Col_2, double*  Col_C1_C2
, long Nsize);
double  Calculate_Vector_Average (double* Column, long Nsize);


main()
{
      double Col_A[4] = {1, 2, 3, 4};
      double Col_B[4] = {5, 6, 7, 8};
      double Col_C[4];

      long Size = 4;

      double Vector_Average;

      Calculate_Vector_Sum ( Col_A,  Col_B, Col_C, Size);

      Vector_Average = Calculate_Vector_Average (Col_C, Size);

      // Send Answers to Screen also Note Index 0 to 3
            printf("Col_C[0] = %lf\n", Col_C[0]);
            printf("Col_C[1] = %lf\n", Col_C[1]);
            printf("Col_C[2] = %lf\n", Col_C[2]);
            printf("Col_C[3] = %lf\n", Col_C[3]);

            printf("Vector Average = %lf\n", Vector_Average);


      return 0;
}


void  Calculate_Vector_Sum (double* Col_1,  double* Col_2, double*  Col_C1_C2
,long Nsize)
{
      int i;
      for(i =0; i<Nsize; ++i)
      {
            Col_C1_C2[i] = Col_1[i] + Col_2[i];
      }
      return;
}

double  Calculate_Vector_Average (double* Column, long Nsize)
{
      int i;
      double Sum =0;
      for(i =0; i<Nsize; ++i)
      {
            Sum = Sum +Column[i];
      }

      return Sum/Nsize;
}
```

Figure A.8 Stand-alone C functions for vector calculations using pointers.

in the prototypes and the C-function parameter lists. The code in Figure A.8 is available in

C:\POEA\Bridging Excel and C Codes\Examples\Simple_C_Vector_Stand_Alone\ Simple_C_Vector_Stand_Alone.dsw.

If you double click on Simple_C_Vector_Stand_Alone.dsw, the file shown in Figure A.8 will open. You can run this file from the toolbar buttons, Compile (or Ctrl + F7) → Build (or F7) → Execute (or Ctrl + F5), and the results for vector addition and averaging with `Col_A[4] = {1, 2, 3, 4}` and `Col_B[4] = {5, 6, 7, 8}` should appear on the screen.

The formulation in Figure A.8 emphasizes that memory addresses (e.g., `double* Col_1`) are being transferred from main to the C functions. The `double` indicates double-precision data and the * indiates that `Col_1` is a pointer variable. Again, vector elements are not passed directly, but rather a pointer variable is passed and the pointer variable stores the memory address of the first element in the vector. What is important about this change (compare Figure A.7 to Figure A.8) is that the C functions given

```
#include "C_Vector.h"

void  Calculate_Vector_Sum (double* Col_1,  double* Col_2, double*  Col_C1_C2
, long Nsize)
{
      int i;
      for(i =0; i<Nsize; ++i)
      {
            Col_C1_C2[i] = Col_1[i] + Col_2[i];
      }
      return;
}

double  Calculate_Vector_Average (double* Column, long Nsize)
{
      int i;
      double Sum =0;
      for(i =0; i<Nsize; ++i)
      {
            Sum = Sum +Column[i];
      }

      return Sum/Nsize;
}
```

Figure A.9 C Functions to allow DLL vector calculations.

in Figure A.8 allow direct conversion to a DLL, as shown in Figure A.9.

The procedure for creating the DLL project using Microsoft C++ 2008 Express is same as in Example A.1. The process is shown again here:

Step 1: Create a new project from the toolbar, File → New → Project. From Project Types, Select: General, and from Templates, Select: Empty Project. For the location, I used * `C:\POEA\Bridging Excel and C Codes\Examples` as the location and `Simple_C_Vector_dll` as the project name.

Step 2: To add the C code to the project, from the toolbar, Project → Add New Item. From Categories, Select: Code and from Templates, Select: C++ File (.cpp). Use `C_Vector.cpp` as the file name → Add. Copy and paste the source code from Figure A.9 to the `C_Vector.cpp` file. If the user is duplicating Example A.2, the location should be `\New User`.

Steps 3 and 4: Step 3 is to add the definition file (text file) `Simple_C_Vector_dll.def` and type the following lines:

```
LIBRARY Simple_C_Vector_dll
EXPORTS
 Calculate_Vector_Sum
 Calculate_Vector_Average
```

and step 4 is to add the header file `C_Vector.h` and type the following lines:

```
void   Calculate_Vector_Sum   (double*   Col_1,
double* Col_2, double* Col_C1_C2, long Nsize);
double    Calculate_Vector_Average    (double*
Column, long Nsize);
```

Step 5a: From the toolbar, select Project → Simple_C_Vector_dll Properties (Alt + F7) → Configuration Properties. Expand the Configuration Properties window by highlighting the + sign next to Configuration Properties → General → Project Defaults → Configuration Type. The configuration type will currently be Application (.exe); use the drop-down window to select Dynamic Library (.dll) → Apply.

Step 5b: Now expand the C/C++ window by highlighting the + sign next to C/C++ → Advanced → Calling Convention. The current calling convention will be _cdecl (/Gd); use the drop-down window to select _stdcall(/Gz) → Apply.

Step 5c: Now expand the Linker window by highlighting the + sign next to Linker → Input. The module definition file will be blank; add Simple_C_Vector_dll.def → Apply. This name must be the same as the project name with the file extension `.def`. Close the Configuration Manager window by selecting OK.

The changes and additions required in Microsoft C++ 2008 Express to create the DLL are completed. Now the successful build (F7) generates the `Simple_C_Vector_dll.dll` in the debug folder of the project directory. You can close out of all C work, and if you go to the `\Examples` or `\New Users` folder → Simple_C_Vector_dll folder → Debug folder, you will find the DLL `Simple_C_Vector_dll.dll`.

To call these functions from the Excel worksheet the following additions need to be made in Excel file.

Step 6: Open the Excel file and go to Macros → Visual Basic Editor from menu.
Create a new module and add the following lines.

```
Public Declare Sub Calculate_Vector_Sum Lib
" C:\POEA\Bridging Excel and C Codes\Examples
\Simple_C_Vector_dll\Debug\Simple_C_Vector_
dll.dll" (ByRef C1 As Double, ByRef C2 As
Double, ByRef C3 As Double, ByVal NSize As
Long)
Public Declare Function Calculate_Vector_
Average Lib " C:\POEA\Bridging Excel and C
Codes\ Examples \Simple_C_Vector_dll\Debug\
Simple_C_Vector_dll.dll" (ByRef C As Double,
ByVal NSize As Long) As Double
```

Note the use of key words `Sub` and `Function` in the VBA function declaration statements. The key word `Sub` (short for `Subroutine`) satisfies the VBA interpreter that the C function `Calculate_Vector_Sum` will return no value (recall our use of void in this C function). The key word `Function` requires that the C function `Calculate_Vector_Average` return a value, which will be `Sum/Nsize` (see the C function `Calculate_Vector_Average`); this value is returned as `As Double` at the end of the function declaration. Here C1, C2, and C3 are declared as `ByRef` to indicate that these variables are sent by location (or memory address) of the values rather than by sending the actual numerical values.

Unlike the C functions developed in Example A.1, the C functions developed in Example A.2 do not take individual variable values and return a single calculated value. In fact, the C function Calculate_Vector_Sum was declared a Sub to allow VBA interpretation; it will not appear as an available user-defined function in the Excel sheet. The function Calculate_Vector_Average does generate a single value and it does appear as a user-defined function in the Excel sheet; however, the needed vector input ByRef C As double cannot be accomplished directly because we do not have access to the starting address of the vector from the Excel sheet.

VBA macros must be used to access the two C functions, Calculate_Vector_Sum and Calculate_Vector_Average, from the Excel sheet. Figure A.10 provides a macro, Sub ADD_Vector_Macro(), for summing vectors on the Excel sheet and a macro, Sub AVG_Vector_Macro(), for determining the average of a vector on the Excel sheet; both these macros call C functions utilizing the VBA declarations developed in step 6. The macros in Figure A.10 are provided in the Excel file **Example A.2.xls**.

In Sub ADD_Vector_Macro(), three vectors, C1, C2, and C3, are defined and sized to the number of elements. Existing data from columns 1 and 2 on the Excel sheet are read into C1 and C2 and the C function Calculate_Vector_Sum is called. In the argument list, the vectors are passed as C1(0),

```
Public Sub ADD_Vector_Macro()
    Dim C1() As Double
    Dim C2() As Double
    Dim C3() As Double
    Dim NSize As Long

  'Determine the number of elements in each vector - here using second column
    NSize = Application.WorksheetFunction.CountA(Range("B:B"))

    ReDim C1(NSize)
    ReDim C2(NSize)
    ReDim C3(NSize)
    Dim i As Integer
    For i = 0 To NSize - 1
     ' VBA and C vector index from zero.  Excel columns index from one.
        C1(i) = Sheet1.Cells(i + 1, 1)
        C2(i) = Sheet1.Cells(i + 1, 2)
    Next i
    Calculate_Vector_Sum(C1(0), C2(0), C3(0), NSize)

    For i = 0 To NSize - 1
        Sheet1.Cells(i + 1, 3) = C3(i)
    Next i

End Sub

Public Sub AVG_Vector_Macro()
    Dim C3() As Double
    Dim AvgC3 As Double
    Dim NSize As Long

    NSize = Application.WorksheetFunction.CountA(Range("C:C"))

    ReDim C3(NSize)

    Dim i As Integer
    For i = 0 To NSize - 1
        C3(i) = Sheet1.Cells(i + 1, 3)
    Next i

    AvgC3 = Calculate_Vector_Average(C3(0), NSize)
    Sheet1.Cells(NSize + 1, 3) = AvgC3

End Sub
```

Figure A.10 VBA macros allowing vector calculations using C-dlls.

C2(0), C3(0). Recall that these parameters are defined by ByRef in the VBA declaration statement, so that the corresponding function call sends the address of the first (zero) element. The DLL function accesses the numerical values in the individual elements of the vectors from the corresponding memory locations. Once the execution of the DLL function is completed, it returns control to the VBA macro. The array C3 contains the computed values. The vector sum in C3 is printed to column 3 of the Excel sheet.

In Sub AVG_Vector_Macro(), the third column on the Excel sheet is read into C3 and this vector is used in the argument list of the call to the C function Calculate_Vector_Average. The C3(0) passes the memory address of the first element in the C3 vector. The column average is returned as AvgC3 and is printed on the Excel sheet.

Step 7: Test the C functions. The two C functions should be accessible from the Excel sheet using the macros created earlier. An Excel sheet to test these macros is provided in **Example A.2.xls**. Data should be provided in the first two columns of the sheet and then the VBA macro `ADD_Vector_Macro` run. This will add each element of the first and second columns and write the result to the third column. There is no need to specify the length of the vectors in columns 1 and 2. The length is determined in the VBA macro by the statement

```
NSize = Application.WorksheetFunction.
CountA(Range("B:B")),
```

which determines the length of the second column. Next, run the macro `AVG_Vector_Macro` and the average of the elements in the third column will be determined.

EXAMPLE A.3 *Creating a Simple Thermodynamics Library*

Our next example illustrates the construction of a thermodynamics library. Although this "library" will consist of only one simple function for the calculation of species enthalpy using heat capacity at constant pressure, the general construction of the library mirrors the construction of the rigorous library we provide in Chapter 8. Upon completion of this example, the user will have many of the tools necessary to modify or expand the existing rigorous combustion library.

It is not possible to report an absolute value for the enthalpy of a species—enthalpy must be reported relative to a reference state. This is discussed in more detail in Chapter 8. We can calculate a change in enthalpy for a species between two states. If the two states are at the same pressure and phase, we can use the species heat capacity to estimate the enthalpy change between two temperatures. The heat capacity of species i at constant pressure, $C_{P,i}$, of a vapor or liquid phase can be represented by

$$C_{P,i} = a_i + b_i T + c_i T^2, \tag{A.1}$$

where $C_{P,i}$ is in kilojoule per mole-kelvin or kilojoule per mole-Celsius; T is the temperature in kelvin or Celsius; and a_i, b_i, and c_i are species, phase, and temperature-dependent constants, respectively. We can estimate the change in enthalpy of a species, Δh_i, between (T_0, P and T_1, P) as

$$\Delta h_i = \int_{T_0}^{T_1} C_{P,i} dT = a_i(T_1 - T_0) + \frac{b_i}{2}(T_1^2 - T_0^2) + \frac{c_i}{3}(T_1^3 - T_0^3). \tag{A.2}$$

In the simple C library, we will supply, based on Equation (A.2), the reference temperature, $T_0 = T_{\text{ref}} = 298.15\,\text{K}$ or 25°C, and the constants (a, b, and c) for each species and phase.

For liquid- and vapor-phase water and for liquid- and vapor-phase ethanol, C_p constants provided in Felder and Rousseau (2005) are given in Table A.1.

We want this library to be called from the Excel sheet. From the sheet, we will specify the substance, its state, and the system temperature T_1 in Celsius.

Step 1: Create a new project from the toolbar, File → New → Project. From Project Types, Select: General, and from Templates, Select: Empty Project. For the location, I used `C:\POEA\Bridging Excel and C Codes\Examples` as the location, and as the project name, `CP_Lib`. `CP_Lib` is the folder (project) that will hold our library. Again, recall if the user is duplicating Example A.2; the location should be `\New User`.

Step 2: Figure A.11 shows the C code `CP_Code.cpp` for Equation (A.2). Here we include a header file and we define a function, `PureEnthalpy`, which accepts `T1`, `TRef`, and the C_p

Table A.1 Species C_p Constants a, b, and c, with T (°C)

	a	b	c
Water (liquid)	75.4E-03	0	0
Water (vapor)	33.46E-03	0.688E-05	0.7604E-08
Ethanol (liquid)	103.1E-03	0	0
Ethanol (vapor)	61.34E-03	15.72E-05	−8.749E-08

```
#include "CP_headers.h"

double PureEnthalpy(double T1, const double TRef, const double A[])
{

    return  A[0]*(T1 - TRef) + A[1]/2*(pow(T1, 2) -  pow(TRef, 2)) +
                               A[2]/3*(pow(T1, 3) -  pow(TRef, 3));
}
```

Figure A.11 C code for Equation (A.2)—line numbers added for discussion.

constants as `A[]`. The value for Δh is returned as a double-precision number from the function. Both `TRef` and `A[]` are declared as constants, which prevents their values from being changed by the code. We only need one equation, with the appropriate constants, to determine the change in enthalpy for any substance.

Do note in Figure A.11 and in some subsequent figures that we have added line numbers for discussion purposes. These line numbers should *not* appear in the actual working code.

The project window of Figure A.12 (source files and header files) indicates several files in this library that we have not yet added—this will be done in steps 3 and 4. We will compartmentalize different aspects of the project, which can make coding easier on larger projects, such as the rigorous combustion thermodynamics package.

Step 3 and Step 4: To add the C code `CP_Calls.cpp` to the DLL project, go to Project → Add New Item. From Categories, Select: Code, and from Templates, Select: C++ File (.cpp). and use `CP_Calls.cpp` as the file name (→OK). Add the file `CP_Calls.cpp` as shown in Figure A.13.

We can view `CP_Code.cpp` as the working equations and `CP_Calls.cpp` as the calling functions (each in a separate location). Notice the same construction in each calling function. For example, `lines 2-5` show a call to function `double H_L_Water(double T1)`, which in turn calls the function (working equation) `PureEnthalpy(T1, T0, WATER_L)`. `PureEnthalpy` is supplied with the system temperature T_1, the reference temperature T_0, and the C_p constants for liquid-phase water, `WATER_L`.

To add the header file `CP_Constants.h` to the DLL project from the toolbar, Project → Add New Item. From Categories, Select: Code, and from Templates, Select: Header File (.h). and use `CP_Constants.h` as the file name (→OK). Add the file `CP_Constants.h` as shown in Figure A.14.

The header file `CP_Constants.h` is a little unusual; here we have placed the constants (a, b, and c) for each substance. Note here the `//a`, `//b`, and so on, are equivalent to comments and have no impact on the one-dimensional array values. By placing all the constants in one header file, changes or additions to the constants can be quickly made.

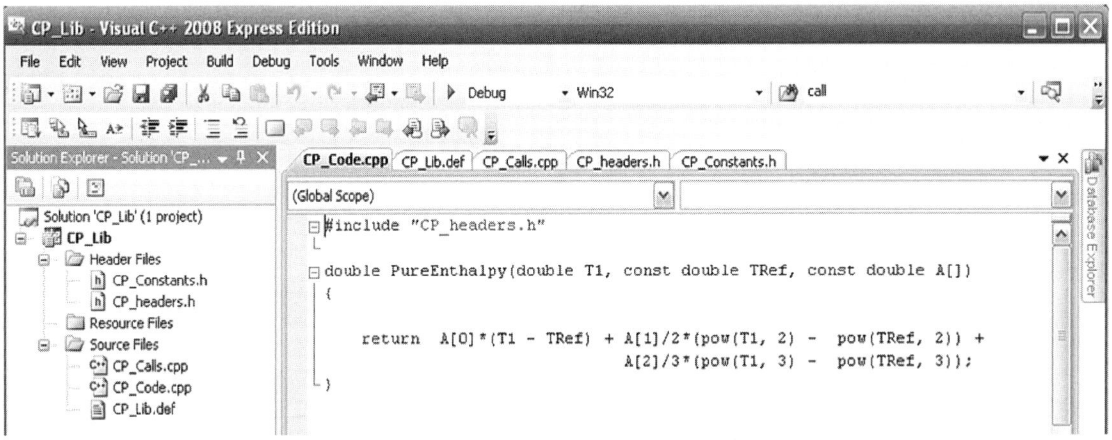

Figure A.12 Project window for enthalpy library.

```
line 1  #include "CP_headers.h"

line 2  double H_L_Water(double T1)
line 3  {
line 4        return PureEnthalpy(T1, T0, WATER_L);
line 5  }

line 6  double H_V_Water(double T1)
line 7  {
line 8        return PureEnthalpy(T1, T0, WATER_V);
line 9  }

line10  double H_L_Ethanol(double T1)
line11  {
line12        return PureEnthalpy(T1, T0, ETHANOL_L);
line13  }

line14  double H_V_Ethanol(double T1)
line15  {
line16        return PureEnthalpy(T1, T0, ETHANOL_V);
line17  }
```

Figure A.13 C code for `CP_Calls.cpp`—line numbers added for discussion.

```
const double WATER_L[] = {75.4E-03,  //a
                          0.0000, //b
                          0.0000  //c
                         };                                line 1

const double ETHANOL_L[] = {103.1E-03,  //a
                            0.00, //b
                            0.000 //c
                              };                           line 2

const double WATER_V[] = {33.46E-03,  //a
                          0.6880E-05, //b
                          0.7604E-08 //c
                         };                                line 3

const double ETHANOL_V[] = {61.34E-03,  //a
                            15.72E-05, //b
                            -8.749E-08 //c
                              };                           line 4

const double T0 = 25;                                      line 5
```

Figure A.14 Code for `CP_Constants.h`.

```
#include<stdio.h>                                          line 1
#include<math.h>                                           line 2
#include "CP_Constants.h"                                  line 3

double H_L_Water(double T1);                               line 4
double H_V_Water(double T1);                               line 5
double H_L_Ethanol(double T1);                             line 6
double H_V_Ethanol(double T1);                             line 7

double PureEnthalpy(double T1, const double T0, const double Generic_A[]);
                                                           line 8
```

Figure A.15 Code for `CP_Headers.h`.

```
LIBRARY CP_Lib                                             line 1
EXPORTS                                                    line 2
      H_L_Water                                            line 3
      H_V_Water                                            line 4
      H_L_Ethanol                                          line 5
      H_V_Ethanol                                          line 6
```

Figure A.16 Code for `CP_Lib.def`.

We are familiar with the construction of the header file `CP_Headers.h` shown in Figure A.15. In the header file, we must account for all functions in the C++ code. It is often convenient to group other header files in one central location, as done in `lines 1-3`. This simplifies the addition of header files to just the inclusion of just CP_Headers.h in other parts of the code. In previous header files, we have not included variable names in the argument list as done in `lines 4-8`. Inclusion of variable names is not required; for example, `line 4` could appear as `double H_L_Water(double)`.

The final file is the definition file `CP_Lib.def` shown in Figure A.16, which simply lists the files we want to export (allow to be called by programs)—in this case called by either VBA or the Excel sheet.

Step 5a: From the toolbar, select Project → CP_Lib Properties (Alt + F7) → Configuration Properties. Expand the Configuration Properties window by highlighting the + sign next to Configuration Properties → General → Project Defaults → Configuration Type. The configuration type will currently be Application (.exe); use the drop-down window to select Dynamic Library (.dll) → Apply.

Step 5b: Now expand the C/C++ window by highlighting the + sign next to C/C++ → Advanced → Calling Convention. The current calling convention will be _cdecl (/Gd); use the drop-down window to select _stdcall(/Gz) → Apply.

Step 5c: Now expand the Linker window by highlighting the + sign next to Linker → Input. The module definition file will be blank; add CP_Lib.def → Apply. This name must be the same as the project name with the file extension `.def`. Close the Configuration Manager window by selecting OK.

The changes and additions required in Microsoft C++ 2008 Express to create the DLL are completed. Now the successful build (F7) generates the `CP_Lib.dll` in the debug

```
Option Explicit                                                         line 1

Public Declare Function H_L_Water Lib " C:\POEA\Bridging Excel and C Codes\
Examples \CP_Lib\Debug\CP_Lib.dll" (ByVal T As Double) As Double        line 2

Public Declare Function H_V_Water Lib " C:\POEA\Bridging Excel and C Codes\
Examples \CP_Lib\Debug\CP_Lib.dll" (ByVal T As Double) As Double        line 3

Public Declare Function H_L_Ethanol Lib " C:\POEA\Bridging Excel and C Codes\
Examples \CP_Lib\Debug\CP_Lib.dll" (ByVal T As Double) As Double        line 4

Public Declare Function H_V_Ethanol Lib " C:\POEA\Bridging Excel and C Codes\
Examples \CP_Lib\Debug\CP_Lib.dll" (ByVal T As Double) As Double        line 5
```

Figure A.17 Excel VBA calls.

folder of the project directory. You can close out of all C work and if you go to the `\Examples` or `\New Users` folder → Simple_C_Vector_dll folder → Debug folder, you will find the DLL `CP_Lib.dll`.

Step 6: To access these functions (either from the Excel sheet or from VBA), the additions shown in Figure A.17 need to be made in the Excel file.

Here lines 2–5 all show the same construction, with one statement for each function we want to connect with in the C library. The form of the connect are the key words `Public Declare Function` (this establishes the function as a global function) followed by the function name, then the key word `Lib` and then the root directory location of the DLL. Next, we identify the names of the variables and the variable type in the function argument list; these variables will be connected to the function argument list in the C++ library. Here, the system temperature T will be passed to the C++ library `ByVal` as double type. The value (the change in enthalpy) returned from each function is specified `As Double` (note `As Double` at the very end of the call).

These four functions and declarations are available in the Excel sheet in **Example A.3.xls**. From the Excel sheet, click on Insert → Function → under "Or select a category:" → User Defined. You should find the following values for Δh (kilojoule per mole) if $T_1 = 46°C$ and $T_0 = T_{ref} = 25°C$.

Δh_L_Water ($T_1 = 46°C$)	1.5834	kJ/mol
Δh_V_Water ($T_1 = 46°C$)	0.707996	kJ/mol
Δh_L_Ethanol ($T_1 = 46°C$)	2.1651	kJ/mol
Δh_V_Ethanol ($T_1 = 46°C$)	1.40295	kJ/mol

The construction of this simple thermodynamics library shares many features with the library developed for combustion calculations. In Example A.3, calling C functions from Excel or VBA is straightforward, as only single variables are sent to the C program and a single result is returned. One-dimensional arrays (see Figure A.14) are used to store species-dependent constants, but this is all accomplished within the C library—here there is no direct use of pointers.

A.4 WORKING WITH MATRICES

Our final handshaking topic is moving matrices between Excel ↔ VBA ↔ C programs. This is an advanced coding topic.

Perhaps the most common use of matrices in engineering occurs in the solution of linear equation sets: systems of n linear equations and n unknowns. The general form of this problem is $Ax = b$ where A is the matrix of coefficients, x is the vector of unknowns, and b is the vector of right-hand side (RHS) (RHS values). The general solution for these problems is $x = A^{-1}b$. One solution approach to the system of n linear equations would be to invert A. A less time-consuming approach (in total mathematical operations) is direct algebraic substitution using Gauss–Jordan elimination.

A.4.1 Gauss–Jordan Matrix Elimination Method

We will use the Gauss–Jordan matrix elimination method as developed in Chapter 4 as our example of moving matrices between Excel ↔ VBA ↔ C programs. You will recall in Chapter 4 that we summarized the method as the following.

Outline of the Gauss–Jordan Method

$k = 0$ to number of rows

Step 1: $k = k + 1$. Select a_{kk} as the pivot element. Divide all elements of row k by a_{kk}. Here,

$$a_{kj} = \frac{a_{kj}}{a_{kk}}, \quad j = 1, 2, \ldots J. \tag{A.3}$$

Step 2: For each row i, where $i \neq k$, calculate new elements as

$$a_{ij} = a_{ij} - a_{ik}a_{kj}, \quad i = 1, 2, \ldots I \quad \text{for} \quad j = 1, 2, \ldots J. \tag{A.4}$$

Next k

EXAMPLE A.4 *The Gauss–Jordan Matrix Elimination Method*

Recall that in Chapter 4, Example 4.1 we solved the following equation set using the Gauss–Jordan method:

$$\begin{aligned} 2x_1 \quad -2x_2 \quad +5x_3 \quad &= 13 \\ 2x_1 \quad +3x_2 \quad +4x_3 \quad &= 20 \\ 3x_1 \quad -x_2 \quad +3x_3 \quad &= 10. \end{aligned} \tag{A.5}$$

With the final result, $x_1 = 1$, $x_2 = 2$, and $x_3 = 3$. For Example A.4, please review Example 4.1 and also review the general VBA code we developed for the Gauss–Jordan method.

In the next example, we develop a general Gauss–Jordan C program. We must ensure that the coefficient being normalized $a_{kk} \neq 0$; if $a_{kk} = 0$, we need to exchange rows with one of the following rows (one not yet normalized). We also will need to make adjustments with our indexing—in the C program, we will index starting from 0 for both rows and columns.

A.4.2 Coding the Gauss–Jordan Matrix Elimination Method

EXAMPLE A.5 *Working with Matrices (Multidimensional Arrays)*

We want to solve Example A.4 by writing a C function utilizing the Gauss–Jordan matrix elimination method. This method must be general to allow solution of any set of n linear independent equations with n unknowns. We also want to use Excel as the pre- and postprocessor. The equation set will be input to the Excel sheet. Coefficients from the equation set will be read by a VBA macro, which will then call the Gauss–Jordan algorithm a C function. Results from the function will be returned to the VBA macro and then displayed on the Excel sheet.

A very significant problem here is that VBA and C store multidimensional data differently. VBA stores multidimensional data in column-major order and C stores multidimensional data in row-major order. Figure A.18 illustrates differences between VBA column-major order and row-major order storages. Note, that VBA data storage is actually more complicated than simple column-major as VBA adds a zero element at the end of each column; we note these additional elements in bold. In Figure A.18, we also indicate a possible memory address for the first element in each matrix storage type. The VBA memory address will be used in the next discussions.

If we are coding entirely in VBA, or entirely in C, we do not need to be concerned with the internal details of multidimensional array storage—it is consistent within the programming language and transparent to the user. However, to move multidimensional arrays between VBA and C, we must convert between column-major and row-major storage strategies.

For Example A.5, we will use a VBA macro to read data from the Excel sheet—data will be stored by VBA in column-major order. This data must be converted to row-major order before the C function performing the Gauss–Jordan elimination can be called. Once the C function is completed, data must be converted to column-major order before returning to the VBA macro for results display on the sheet. Example A.5 involves three tasks:

1. Conversion of a column-major multidimensional array to a row-major
2. Implementation of Gauss–Jordan algorithm as a C function
3. Conversion of a row-major multidimensional array to a column-major

Figure A.19 addresses task 1 and provides a general C function to convert a VBA column-major multidimensional array to a row-major matrix as required by C.

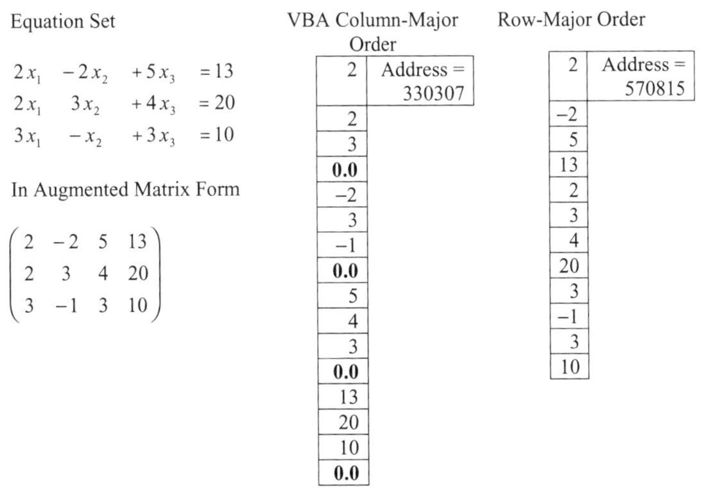

Figure A.18 Column-major and row-major storage comparison.

```
void ConvertVBA_Matrix_to_C_Matrix(double* Matrix, long Nrows, long Ncolumns
, double Matrix_Row_Major[100][100])
{
 //double Matrix_Row_Major[100][100];  it may be necessary to change the size

      int i;
      int j;

      for(i =0; i<Nrows; ++i)
      {
            int columnPos = 0;
            for(j =0; j<Ncolumns; ++j)
            {
                  Matrix_Row_Major[i][j] = *(Matrix+i+columnPos+j);
                  columnPos = columnPos + Nrows;
            }
      }

      return;
}
```

Figure A.19 Conversion from column-major to row-major storage.

```
void ConvertC_Matrix_to_VBA_Matrix(double Matrix_Row_Major[100][100],
 long Nrows,
long Ncolumns, double* Matrix_Col_Major)
{
      int i;
      int j;

      for(i =0; i<Nrows; ++i)
      {
            int columnPos = 0;
            for(j =0; j<Ncolumns; ++j)
            {
                  *(Matrix_Col_Major+i+columnPos+j) = Matrix_Row_Major[i][j];
                  columnPos = columnPos + Nrows;
            }
      }
      return;
}
```

Figure A.20 Conversion from row-major to column-major storage.

Examining the first line of the function in Figure A.19,

```
void ConvertVBA_Matrix_to_C_Matrix(double* Matrix, long Nrows, long Ncolumns, double Matrix_Row_Major[100][100]).
```

The memory address of the first element in the VBA-generated column-major matrix is stored in the variable `Matrix`; here, the `double*` indicates that `Matrix` is a pointer variable of type `double`. Regardless of the matrix storage order (row-major or column-major), all elements are stored in consecutive memory locations. A working matrix `double Matrix_Row_Major[100][100]` is also defined. Here, the size `[100][100]` has been arbitrarily selected to hold the converted matrix. Since the size of the input matrix is not known before the creation of the DLL, we assumed a maximum size of 100 × 100. This size declaration can be avoided by allocating memory dynamically using the malloc function in C (see Pururu and Knopf, 2008).

Applying two lines of the code to the example in Figure A.18,

```
Matrix_Row_Major[i][j] = *(Matrix+i+columnPos+j);

columnPos = columnPos + Nrows;
```

initially, `i =0`, `columnPos = 0`, `j =0` and `Matrix` is the address of the first element of the VBA matrix (`Matrix = 330307` from Figure A.18). `*(Matrix+i+columnPos+j)` gives the numeric value for that address, which is then assigned to `Matrix_Row_Major[0][0]` and = 2. The variable `columnPos` is then set = 3 and the counter on `j` is incremented to `j = 1`. `Matrix_Row_Major[0][1]` = `*(Matrix` + 0 + 3 + 1), which will be the fourth element in the VBA matrix, which here = −2; remember counting from the first element begins with zero. Matrix calculations in the C code will use `Matrix_Row_Major`.

Task 3 requires that matrix results from the C function be converted to column-major order before the matrix can be transferred to the Excel sheet. This procedure is shown in general form in Figure A.20.

Examining the first line of the function in Figure A.20,

```
void ConvertC_Matrix_to_VBA_Matrix(double Matrix_Row_Major[100][100], long Nrows, long Ncolumns, double* Matrix_Col_Major).
```

Here, `double Matrix_Row_Major[100][100]` is the C row-major matrix we want to convert to column-major. We will send the memory address of the first element in the column-major matrix using the pointer variable `double* Matrix_Col_Major`.

The matrix operation in Figure A.20 requires some explanation. Consider a variable, x, and a pointer variable, p, to the memory address of x. The variable x can be incremented by 1 using $x = x + 1$; x++; or $*p = *p + 1$;. Recall that $*p$ provides the value in memory address p.

Applying two lines of the code to the example in Figure A.18,

```
*(Matrix_Col_Major+i+columnPos+j) = Matrix_
Row_Major[i][j];

columnPos = columnPos + Nrows;
```

initially, `i` =0, `columnPos = 0`, `j` =0 and `Matrix_Col_Major` is the address of the first element of the column-major matrix (`Matrix_Col_Major` = 330307, see Figure A.18). This line can be interpreted as the numeric value in the memory address (330307+0+0+0) = `Matrix_Row_Major[0][0]`

and here = 2. The variable, `columnPos` is then set = 3 and the counter on `j` is incremented to `j` =1, giving `*(` 330307+0+3+1) = `Matrix_Row_Major[0][1]`, which sets the fourth memory address in the column-major matrix = `Matrix_Row_Major[0][1]` = −2.

Note that we do not create a new matrix to hold numeric values as we convert from row-major to column-major storage. Instead, every numerical value from the row-major matrix is assigned to a memory location. The memory locations are selected to give a column-major order if they are accessed sequentially. Therefore, all that is required in the VBA code is to send the memory address of the first element (`double* Matrix_Col_Major`) in the sequential column-major order list.

To avoid making changes in existing C codes, `ConvertVBA_Matrix_to_C_Matrix` and `ConvertC_Matrix_to_VBA_Matrix` may be used as "wrappers" on top of the existing codes. The calls to these wrapper functions and the C function for the Gauss–Jordan matrix elimination method are shown in Figure A.21.

```
void GJ_Elimination_Main(double* Matrix, long Nrows, long Ncols)      line 1
{                                                                      line 2
    double A[100][100];                                               line 3
    ConvertVBA_Matrix_to_C_Matrix(Matrix, Nrows, Ncols, A);          line 4
    GJ_Elimination(A, Nrows);                                         line 5
    ConvertC_Matrix_to_VBA_Matrix(A, Nrows, Ncols, Matrix);          line 6
}                                                                      line 7

void GJ_Elimination(double A[100][100], long row)                     line 8
{                                                                      line 9
    long col = row+1;                                                 line 10
    int nonZerIdx = 0;                                                line 11

    for(int i =0; i<row; ++i)                                         line 12
    {                                                                 line 13
        if(A[i][i] == 0.0)                                            line 14
        {                                                             line 15
            for (int i2 = i+1; i2<row; ++i2)                          line 16
            if(A[i2][i] != 0)                                         line 17
            {                                                         line 18
                nonZerIdx = i2;                                       line 19
                break;                                                line 20
            }                                                         line 21
            for(int j1 =0; j1<col; ++j1)                              line 22
            {                                                         line 23
                double tmp =  A[i][j1];                               line 24
                A[i][j1] = A[nonZerIdx][j1];                          line 25
                A[nonZerIdx][j1] =tmp;                                line 26
            }                                                         line 27
        }                                                             line 28

        double tmpAii = A[i][i];                                      line 29
        for(int j =0; j<col; ++j)                                     line 30
            A[i][j] = A[i][j]/tmpAii;                                 line 31

        for(int i1 =0; i1<row; ++i1)                                  line 32
        {                                                             line 33
            if(i1 != i)                                               line 34
            {                                                         line 35
                double tmpAi1 = A[i1][i];                             line 36
                for(int j =0; j<col; ++j)                             line 37
                    A[i1][j] = A[i1][j] - tmpAi1*A[i][j];             line 38
            }                                                         line 39
        }                                                             line 40
    }                                                                 line 41
}                                                                      line 42
```

Figure A.21 VBA matrix → C, Gauss–Jordan reduction, C matrix → VBA.

Our intent here is to convert a multidimensional array from VBA storage → C storage (`line 4`), then to execute a C function (`line 5`) using the properly modified array, and finally, to convert the resulting multidimensional array from C storage → VBA storage (`line 5`). Task 2, the C function performing the Gauss–Jordan algorithm, is provided in Figure A.21; it will also be discussed in Chapter 4. In the Gauss–Jordan subroutine procedure (`line 8`), the augmented matrix is named `A` and the number of rows is named `row`. In `lines 12-14`, the current diagonal element is examined; if zero, this must be corrected as a zero diagonal element cannot serve as a pivot element. Subsequent rows (`line 16`) are examined for the first nonzero element (`line 17`) in the pivot column. This row is marked as `nonZerIdx` in `line 19`. In `lines 22-27`, elements are exchanged, one at a time, between the current row (`i`) and the row with the nonzero pivot element (`nonZerIdx`). The process of checking the diagonal element and exchanging rows, if necessary, ends with `line 28`. `Lines 29-31` divide each element of the pivot row by the pivot element as shown in Equation (A.3). The pivot column in all rows except the pivot row (`line 34`) is made zero by lines `lines 32-40`. Equation (A.4) is implemented in `line 38`.

The procedure for creating the Empty Project using Microsoft Visual C++ 2008 Express Edition is the same as before. We are starting with **Step 2**. Combine the codes in Figures A.19–A.21 and add the include statement `#include "C_Matrix.h"`. Use `Simple_C_Matrix_dll` as the project file and the file holding the C code is `C_Matrix.cpp`. We next need **Steps 3 and 4**; step 3 is to add the definition file `Simple_C_Matrix_dll.def`,

```
LIBRARY Simple_C_Matrix_dll

EXPORTS

 GJ_Elimination_Main
```

and step 4 is to add the header file `C_Matrix.h`,

```
#include <stdio.h>

#include <math.h>

void ConvertVBA_Matrix_to_C_Matrix(double*
Matrix, long Nrows, long Ncolumns, double
Matrix_Row_Major[100][100]);

void ConvertC_Matrix_to_VBA_Matrix(double
Matrix_Row_Major[100][100], long Nrows,
long Ncolumns, double* Matrix_Col_Major);

void GJ_Elimination_Main(double* Matrix,
long Nrows, long Ncols);

void GJ_Elimination(double A[100][100],
long row);
```

Step 5a: From the toolbar, select Project → Simple_C_Matrix_dll Properties (Alt + F7) → Configuration Properties. Expand the Configuration Properties window by highlighting the + sign next to Configuration Properties → General → Project Defaults → Configuration Type. The configuration type will currently be Application (.exe); use the drop-down window to select Dynamic Library (.dll) → Apply.

Step 5b: Now expand the C/C++ window by highlighting the + sign next to C/C++ → Advanced → Calling Convention. The current calling convention will be _cdecl (/Gd); use the drop-down window to select _stdcall(/Gz) → Apply.

Step 5c: Now expand the Linker window by highlighting the + sign next to Linker → Input. The module definition file will be blank; add Simple_C_Matrix_dll.def → Apply. This name must be the same as the project name with the file extension `.def`. Close the Configuration Manager window by selecting OK.

The changes and additions required in Microsoft C++ 2008 Express to create the DLL are completed. Now the successful build (F7) generates the `Simple_C_Matrix_dll.dll` in the debug folder of the project directory. You can close out of all C work and if you go to the `\Tutorial C++ 2008 Express` or `\New Users` folder → Simple_C_Matrix_dll folder → Debug folder, you will find the DLL `Simple_C_Matrix_dll.dll`.

To access the Gauss–Jordan matrix elimination method, the following additions need to be made in Excel file.

Step 6: Open the Excel file and go to Macros → Visual Basic Editor from the menu.

Create a new module and add the VBA code in Figure A.22. The VBA procedure in Figure A.22 is provided in **Example A.5.xls**. It solves Example A.5.

Code Modifications Required for Each New Problem

The VBA macro calling the Gauss–Jordan method will need slight modification for each new problem. We must specify the number of rows and columns in the augmented matrix (`lines 6-7`), the first cell of the input matrix on the Excel sheet (`lines 13 and 16, here row = 8, column = 2`), and the starting cell on the Excel sheet for the results (`line 23, here row = 12, column = 2`). The Gauss–Jordan method is called in `line 20`, where the location of the first element in the matrix, `C(0, 0)`, is passed. As shown in Figure A.23, the results from the Gauss–Jordan elimination are $x_1 = 1$, $x_2 = 2$, and $x_3 = 3$.

A.5 CLOSING COMMENTS

A systematic approach has been presented to facilitate the transfer of data as single variables, vectors, and matrices between Excel and C programs. The transfer of matrices was complicated by the difference in matrix storage of each language; VBA uses column-major order and C uses row-major order. Two wrapper programs were developed, which allow multidimensional arrays to be transferred from VBA → C programs and from C → VBA programs without modification to existing VBA or C programs.

```
Public Declare Sub GJ_Elimination_Main Lib "C:\POEA\Bridging Excel and
C Codes\Examples \Simple_C_Matrix_dll\Debug\Simple_C_Matrix_dll.dll"
(ByRef Matrix As Double, ByVal Nrows As Long, ByVal Ncolumns As Long)

'comment

Public Sub Gauss_Jordan_Macro()
    Dim C() As Double
    Dim Nrows, Ncolumns As Long

    Nrows = 3
    Ncolumns = 4

    ReDim C(Nrows, Ncolumns)
    Dim i As Integer
    Dim j As Integer
    For i = 0 To Nrows - 1
        For j = 0 To Ncolumns - 1
            If Sheet1.Cells(i + 8, j + 2) = " " Then
                C(i, j) = 0
            Else
                C(i, j) = Sheet1.Cells(i + 8, j + 2)
            End If
        Next j
    Next i

    GJ_Elimination_Main C(0, 0), Nrows, Ncolumns


    For i = 0 To Nrows - 1
        For j = 0 To Ncolumns - 1
            Sheet1.Cells(i + 12, j + 2) = C(i, j)
        Next j
    Next i
End Sub
```

Figure A.22 VBA macro to call Gauss–Jordan matrix elimination method.

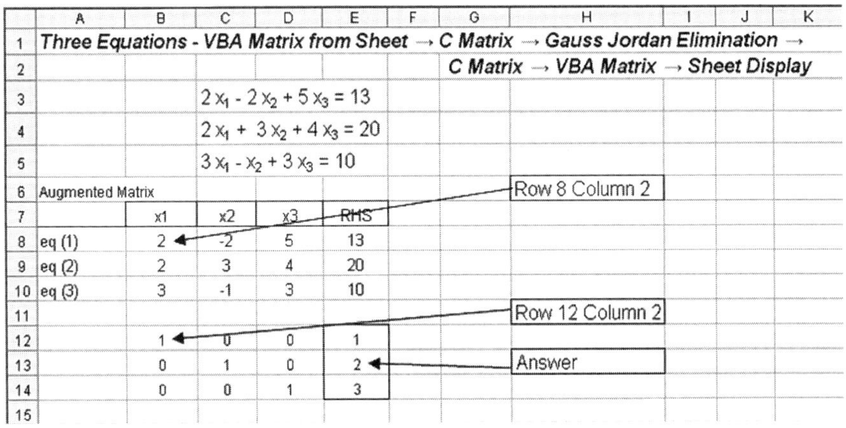

	A	B	C	D	E	F	G	H	I	J	K
1	Three Equations - VBA Matrix from Sheet → C Matrix → Gauss Jordan Elimination →										
2							C Matrix → VBA Matrix → Sheet Display				
3			$2x_1 - 2x_2 + 5x_3 = 13$								
4			$2x_1 + 3x_2 + 4x_3 = 20$								
5			$3x_1 - x_2 + 3x_3 = 10$								
6	Augmented Matrix							Row 8 Column 2			
7		x1	x2	x3	RHS						
8	eq (1)	2	-2	5	13						
9	eq (2)	2	3	4	20						
10	eq (3)	3	-1	3	10						
11								Row 12 Column 2			
12		1	0	0	1						
13		0	1	0	2			Answer			
14		0	0	1	3						
15											

Figure A.23 Results from the Gauss–Jordan elimination.

There are a few remaining issues that deserve special comment. First, integer declarations should be avoided in the VBA code; if integer data are being transferred to a C program. Integers in VBA are not equivalent in memory size to integers in C. Use long type in VBA to match with integers in C. There is no problem sending double (VBA) to double (C).

Before beginning the process of creating C functions and C-dlls, it is important to review allowable operations in VBA and Excel when using macros, functions, and subroutines as summarized in Table A.2.

Table A.2 Functionality of VBA Macros, VBA Subroutines, and VBA Functions

	Callable from Worksheet	Accept Arguments	Update Worksheet
VBA macro	Yes	No	Yes
VBA subroutine	No	Yes	Yes
VBA function	Yes	Yes	Yes/No*

*A function can only update the cell in which the function is called; this cell is updated with the return value of the function. Function parameter values (sent by ByRef) cannot be updated in the worksheet directly.

A *VBA macro* can be called from an Excel sheet, and it can change values on the sheet and all sheet values will be updated. A macro does not allow for an argument list. A *VBA subroutine procedure* allows for an argument list, but it cannot be called directly from an Excel sheet; it can be called from a macro. A VBA procedure can change values on the Excel sheet and the sheet values will be updated. A *VBA function* can be called directly from the Excel Sheet or from either a macro or a procedure. If a VBA function is called from the Excel sheet, it cannot change or write any value on the sheet. It should return a calculated value (from the function) to the sheet calling cell. If a VBA function is called from a macro or a procedure, it returns a calculated value and it can change or write values on the Excel sheet.

A VBA macro, a VBA subroutine procedure, or a VBA function can call a C/C++ function or a C/C++ subroutine procedure as a DLL. A C function allows for an argument list and it can also be called directly from an Excel sheet. A C function cannot change values on the Excel sheet. VBA expects a single value to be returned if a C function is called. A C subroutine procedure allows for an argument list, but it cannot be called directly from an Excel sheet. VBA does not expect a value to be returned from a C subroutine; however, calculations can be performed in the subroutine, for example, changing array values, which can then be accessed by VBA.

REFERENCES

CORONELL, D.G. 2005. Computer science or spreadsheet engineering. *Chem. Eng. Ed.* 39: 142–145.

DE MOURA, L.F. 2005. Microsoft Excel: A simple way of teaching complex tasks. *CACHE Winter 2005 Newsletter*, http://www.che.utexas.edu/cache/newsletters/fall2005_cover.html

FELDER, R.M. and R.W. ROUSSEAU. 2005. *Elementary Principles of Chemical Processes* (3rd edition). John Wiley & Sons, Hoboken, NJ.

FERREIRA, E.C., R. LIMA, and R. SALCEDO. 2004. Spreadsheets in chemical engineering education–A tool in process design and process integration. *Int. J. Eng. Ed.* 20(6): 928–938.

PUNURU, J.R. and F.C. KNOPF. 2008. Bridging Excel and C/C++ codes. *Comput. Appl. Eng. Educ.* 16(4): 289–304.

ROSEN, E.M. 2001a. Use of C++ DLLs in Visual Basic for Applications with Excel 2000. *CACHE Spring 2001 Newsletter*, http://www.che.utexas.edu/cache/newsletters/spring2001_cover.html

ROSEN, E.M. 2001b. On testing C++ DLLs, *CACHE Fall 2001 Newsletter*, http://www.che.utexas.edu/cache/newsletters/fall2001_cover.html

ROSEN, E.M. and L.R. PARTIN 2000. A perspective: The use of the spreadsheet for chemical engineering computations. *Ind. Eng. Chem. Res.* 39: 1612–1613.

WALKENBACH, J. 2001. *Excel 2002 Power Programming with VBA*. M&T Books, New York.

TUTORIAL

Microsoft C++ 2008 Express: Creating C Programs and DLLs

Our intent here is to provide a series of example problems, as tutorials, to reinforce the material discussed previously in Appendix A. In Problems A-T.1 and A-T.2, we use C to solve examples from Chapter 2. Here, the solution is output to the screen. In Problem A-T.3, we will combine these two C programming examples into one C program, and in Problems A-T.4 and A-T.5, we will connect this program to VBA and Excel as a DLL.

The solution to each example problem is provided in the \Problems folder (see Figure A-T.1). The user is encouraged to duplicate the tutorials by creating programs in the \New User folder.

Problem (Tutorial) A-T.1

Solve Example 2.6 using C programming. If you want to have $500,000 in 20 years, how much money must be set aside each month? Here the nominal interest rate is 10%.

Answer from Excel

$$A = FV\left[\frac{i}{(1+i)^n - 1}\right] = 500{,}000\left[\frac{\left(\frac{0.10}{12}\right)}{\left(1+\frac{0.10}{12}\right)^{12\times 20} - 1}\right] = 658.44. \tag{TA.1}$$

Solution Using C (Solution to Console–Computer Screen):

Set up the Project Folder

Step 1: Begin by checking that the folder New User exists in C:\POEA\Bridging Excel and C codes.

Step 2: Open Microsoft Visual C++ 2008 Express Edition, Start → Programs → Microsoft Visual C++ 2008 Express Edition → Microsoft Visual C++ 2008 Express Edition.

You can close the start page window which will appear.

Click on File → New → Project.

You should see a new window with various project types.

From Project Types, Select: General, and from Templates, Select: Empty Project.

Location → C:\POEA\Bridging Excel and C Codes\New User.

Project Name → Time_Value_Money_Annuity (this step will create a new folder named Time_Value_Money_Annuity in the folder New User).

Click on OK.

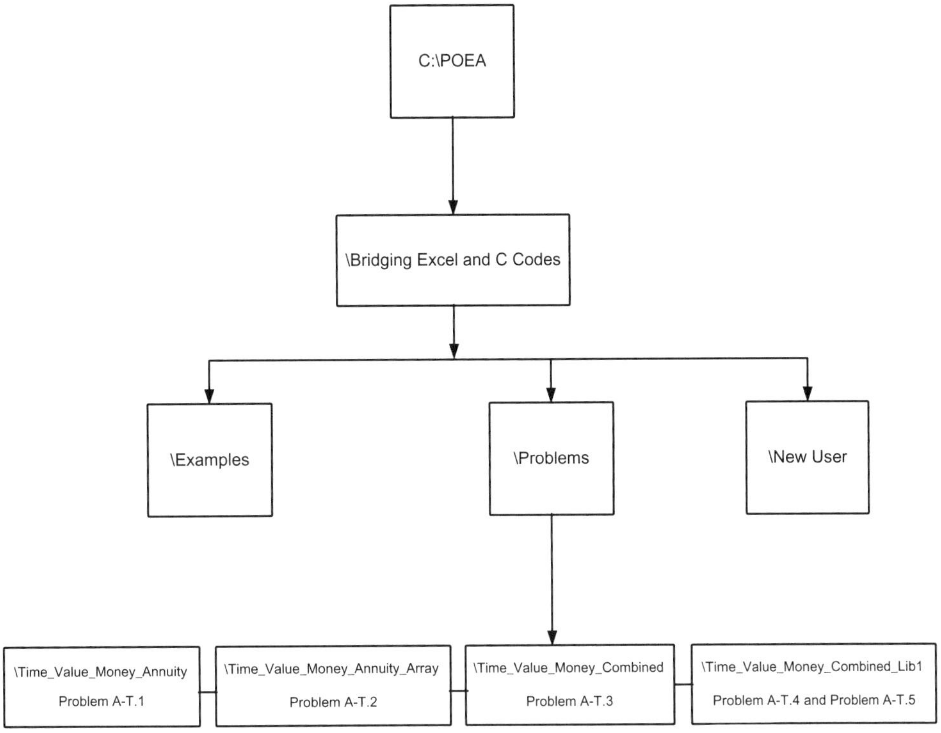

Figure A-T.1 Solution locations for Problems A-T.1–A-T.5.

Now the environment for developing the Time_Value_Money_Annuity project has been created. To add the actual C source code, follow the steps listed below:

To add the C code to the project, from the toolbar, Project → Add New Item.

You should see a new window with several types of files this project can take.

From Categories, Select: Code, and from Templates, Select: C++ File (.cpp).

Then give a file name: Annuity_Coding_Ex. The extension .cpp will be automatically added.

Click on OK.

You are ready to code!

This is a good time to save your work and to exit out so you can review what you have accomplished:

So far, we have in C:\POEA\Bridging Excel and C Codes\New User a folder (*the project name*),Time Vaue_Money_Annuity. Inside the Time_Value_Money_Annuity folder, there is an empty file, Annuity_Coding_Ex.cpp, which will eventually contain our C++ code.

Programming Problem A-T.1

Let us set up and solve this simple annuity problem in C. Return to the Time_Value_Money_Annuity folder and click on Time_Value_Money_Annuity.dsw to open the project workspace. Add the C++ code shown in Figure A-T.2, which we will discuss further.

Examining the code, we begin by including two header files (`line 1` and `line 2`)—`<stdio.h>` for standard input/output operations (e.g., printing) and the second, `<math.h>` for standard math operations (e.g., pow, a^b); these header files can be considered libraries of needed operations.

The next statement, `double Annuity_Amount (double, double, int);`, is termed a *prototype*. This prototype may seem a little unusual; however, if you look a little ahead in the code, you will see that the main calls the function `Annuity_Amount` in `line 10`. Without the prototype declaration, during the compilation of main, an error would result as the compilation will not recognize the function in `line 10`. A simple way to view the prototype is that it is telling the compiler to "trust me"; there is a function Annuity_Amount available; it returns a double value; and it will take three arguments (a double-precision variable, a double-precision variable, and an integer variable).

The next somewhat unusual convention is the use of a main program as `int main ()` in `line 5`, and `return 0;` in `line 12`; here simply use these statements as written—return 0 is a way of telling the operating system that the program executed successfully.

```
line 1   /* Annuity Coding Example */

line 2   # include <stdio.h>
line 3   # include <math.h>

line 4   double Annuity_Amount (double, double,  int);

line 5   int main ()
line 6   {
line 7         double Future_Value = 500000.;
line 8         double Interest_Rate = 0.10/12;
line 9         int Period = 20*12;

line10         double A_A = Annuity_Amount (Future_Value, Interest_Rate, Period);

line11         printf ("Monthly Payment or Annuity Amount = %7.3lf\n", A_A);

line12         return 0;
line13   }
line14   double Annuity_Amount (double F_V, double I_R, int Per)
line15   {
line16         return F_V*I_R/(pow((1+I_R),Per)-1);

line17   }
```

Figure A-T.2 Code for annuity example.

Following this we have a { indicator for the start of our code followed by standard variable declarations and call of the function Annuity_Amount as `double A_A = Annuity_Amount (Future_Value, Interest_Rate, Period);` from main. Do notice that after the statement main ( ), no semicolon is used to indicate the end of the line of code—here the brackets { code } provide the code associated with main ( ).

When the program is executed, the function call in line 10 moves control to line 14 as the function double Annuity_Amount (double F_V, double I_R, int Per), where the variables Future_Value, Interest_Rate, Period become double F_V, double I_R, and int Per. Again the { indicator is used for the start of the code in the function and one simple calculation is performed:

$$\frac{F\_V*I\_R}{(1+I\_R)^{Per}-1}, \quad \text{which is } FV\left[\frac{i}{(1+i)^n-1}\right]$$

(see Equation (TA.1)).

In C, the power calculation a^b is performed as pow (a,b). The result of this calculation is returned to main and assigned to `double A_A`. When using pow (a,b) in C++ 2008 Express, the parameters a and b cannot both be integers; in this problem, a is a double and b is an integer.

This value for the annuity is output (to the monitor) by `line 11`, with A_A occupying `7.3lf` (which is seven total number of spaces, with three after the decimal point—lf—indicates a floating point double-precision variable. The \n is used to return the cursor to a new line.

> In order to execute the code—after it is typed into Annuity_Coding_Ex.cpp, you must use the icon buttons → compile (or Ctrl + F7), then build (or F7) and then execute the program (Ctrl + F5). If needed, after the compile operation, the F4 key can be used to highlight errors.

Before we leave this example problem, it is instructive to examine the use of semicolons (;) indicating the end of each line. Do note that semicolons are not used after main or the function Annuity_Amount; the start and end of these functions are controlled by the {, and }.

Problem (Tutorial) A-T.2

Let us produce a table that shows how the account in Example 2.6 will grow each month. Recall that with this annuity formula, payment is made at the end of the period (monthly).

Month	End-of-Month Payment	Interest Earned for the Month	Total Funds Available (End of Month)
1	658.44	0	658.44
2	658.44	658.44 (.10/12) = 5.49	1322.37
3	658.44	11.02	1991.83
. . .			
239	658.44	4087.23	495213.60
240	658.44	1978.17	499998.82

There is some round error from the desired $500,000, as $658.44+ would need to be invested at the end of each month to produce the exact amount.

Again, Set Up the C++ Code to Run from the Console (Computer Screen)

Open Microsoft Visual C++ 2008 Express Edition, Start → Programs → Microsoft Visual C++ 2008 Express Edition → Microsoft Visual C++ 2008 Express Edition.

You can close the start page window which will appear.

Click on File → New → Project.

You should see a new window with various project types.

From Project Types, Select: General, and from Templates, Select: Empty Project.

Location → C:\POEA\Bridging Excel and C Codes\ New User.

Project Name → Time_Value_Money_Annuity_Array (this step will create a new folder named Time_Value_ Money_Annuity_Array in the folder New User).

Click on OK.

You should be at the main window.

On the toolbar → Project → Add New Item.

You should see a new window.

From Categories, Select: Code, and from Templates, Select: C++ File (.cpp).

Then give a file name: Annuity_Coding_Array_Ex. The extension .cpp will be automatically added.

Click on OK.

You are ready to code!

The C++ code to solve this example problem is given in Figure A-T.3.

```
/* Future Value of Monthly Annuity Payments*/          line 1

# include <stdio.h>                                     line 2
# include <math.h>                                      line 3

void Annuity_Payout(double, int, double, double[], double[]);   line 4

int main ()                                             line 5
{                                                       line 6
     int   i_counter;                                   line 7
     double Payment = 658.44;                           line 8
     double Interest_Rate = 0.10/12;                    line 9
     int Period = 20*12;                                line10

     double Future_Value_Interest[1000];                line11
     double Future_Value_Annuity_Payout[1000];          line12

     Annuity_Payout(Interest_Rate, Period, Payment, Future_Value_Interest,
                              Future_Value_Annuity_Payout);  line13

printf("Month      End of Month        Interest Earned      Total \n");  line14

printf("           Payment              for the month    Payment \n");   line15

     for (i_counter = 1; i_counter <=Period; i_counter++)    line16
     {                                                  line17
         printf (" %3d       %6.2f       %6.2f           %8.2f\n",
         i_counter, Payment, Future_Value_Interest[i_counter] ,
         Future_Value_Annuity_Payout[i_counter]);       line18
     }                                                  line19

     return 0;                                          line20
}                                                       line21


void Annuity_Payout(double I_R, int Per, double Pay, double FVI[1000]
, double  FVAP[1000])                                   line22
{                                                       line23
   int i;                                               line24

   FVAP[0] = 0.;                                        line25
   for (i = 1; i <=Per; i++)                            line26
   {                                                    line27
        FVI[i]  = FVAP[i-1]*I_R;                         line28
        FVAP[i] = FVAP[i-1] + Pay + FVI[i];             line29
   }                                                    line30

}                                                       line31
```

Figure A-T.3 Code for annuity example with arrays.

Examining the code, we again include the two standard libraries (`lines 2` and `3`). The next statement, `void Annuity_Payout(double, int, double, double[], double[]);`, is again a prototype. If you look a little ahead in the code, you will see that main accesses the function `Annuity_Payout` in `line 13`. Without this prototype, an error would result during the compilation of main, as `Annuity_Payout` would not yet be available (complied).

We can see in `line 4` that the function `Annuity_Payout` returns no value to the main program (`void`). The argument list indicates that five variables will be passed; we are familiar with `double` and `int` specifications; the specifications of `double[], double[]` indicate that the fourth and fifth variables will actually be arrays (both single dimension); notice we do not need to define the length of the arrays. The key word `void` requires additional explanation; it indicates that no value will be returned from `Annuity_Payout`. Recall that generally, we would expect to see something similar to `A_P = Annuity_Payout (.....)` and `A_P` would be a value calculated in the function `Annuity_Payout`. Even though no value is returned from function `Annuity_Payout`, calculations are performed in the function `Annuity_Payout` and these calculated values are placed in arrays—these values (arrays) are ultimately passed to and printed in main.

In the variable definitions, the arrays `double Future_Value_Interest[1000]` and `double Future_Value_Annuity_Payout[1000]` (`lines 11` and `12`) are both dimensioned to a length of 1000. It is not good coding practice to simply dimension arrays larger than needed, but for now, we will accept this and move on with coding efforts. For now, the `[1000]` sets up a single-dimension array (vector) with indices from 0 to 999; this can hold a maximum of 1000 real numbers.

We will often encounter the construction `for (i_counter = 1; i_counter <=Period; i_counter++)`. This statement sets `i_counter = 1` and checks if `i_counter` is less than or equal the variable `Period` (Period = 120). If true, the calculations enclosed in the trailing brackets { } are performed; here, this is simply a print statement. Then `i_counter` is incremented; here, `i_counter++` is equivalent to setting `i_counter = i_counter + 1`. The trailing operation (print statement here) is performed until the value of the variable Period is reached. Do note that a semicolon is not used at the end of the `for` statement. The `for` statement is applied to all operations in the immediately trailing {code}.

Next, we turn our attention to line 22, the function Annuity_Payout, which is coded as void Annuity_Payout(double I_R, int Per, double Pay, double FVI[1000], double FVAP[1000]). When this function is accessed, the variables Interest_Rate, Period, Payment, Future_Value_Interest, and Future_Value_Annuity_Payout become I_R, Per, Pay, FVI[1000], FVAP[1000]; here, FVI (Future Value of the Interest) and FVAP (Future Value of the Annuity Payout) are both set as double-precision single-dimension arrays of length 1000.

Initially in `line 25 FVAP [0]`, the zeroth element in the array `FVAP` is set = 0. The `for` statement starts the counter `i` at 1 and this is checked against `Per` (here `Per` = 120). In the loop

```
FVI[i] = FVAP[i-1]*I_R;
FVAP[i] = FVAP[i-1] + Pay + FVI[i];
```

which gives for `I` = 1,

```
FVI[1] = FVAP[0]*I_R = 0 *
0.10/12 = 0
FVAP[1] = FVAP[0] + Pay + FVI[1] = 0
+ 658.44 + 0 = 658.44
```

and for `i` = 2,

```
FVI[2] = FVAP[1]*I_R = 658.44 *
0.10/12 = 5.487
FVAP[2] = FVAP[1] + Pay +
FVI[2] = 658.44 + 658.44 +
5.487 = 1322.367.
```

The code is executed until `i` reaches 241; here, the loop is exited. Control is returned to main, where the values for each month (`i_counter`), payment, future interest value, and future total annuity value are printed to the screen `lines 14-19`.

Problem (Tutorial) A-T.3

Let us combine the two C++ codes developed in Problems A-T.1 and A-T.2 into one file. This is the start of the process to create a very simple C++ library.

Hopefully, these steps now seem familiar and little explanation is needed. Create a new folder (project) in \New User named Time_Value_Money_Combined andcreate a new C++ file Annuity_Calcs.cpp for the combined programs. Open Annuity_Calc.cpp and copy both programs into the file. This can be done by opening Annuity_Coding_Ex.cpp and selecting all the code (Ctrl + a), then paste. The same process can be followed with Annuity_Coding_Ex.cpp. Here then, both codes will be in the same file Annuity_Calcs.cpp. The only major modification required is that only one main program can be present. Try to rearrange the code (here it is easy to use Ctrl + c to copy followed by Ctrl + v to paste). The code should appear as shown in Figure A-T.4.

The only significant change in the code is that the variable `Payment` is no longer fixed at 658.44, but rather it is first set to the value `double A_A` calculated in function `Annuity_Amount` by `line 14` and then set to `double Payment = A_A;` (`line 16`).

At this point, it is important to run the code and to make sure all errors have been corrected. To improve the

```
/* Annuity Coding Examples */                                        line 1

# include <stdio.h>                                                  line 2
#include <math.h>                                                    line 3

double Annuity_Amount (double, double,  int);                        line 4
void Annuity_Payout(double, int, double, double[], double[]);        line 5

int main ()                                                          line 6
{                                                                    line 7
int   i_counter;                                                     line 8
double Future_Value = 500000.;                                       line 9
double Interest_Rate = 0.10/12;                                      line10
int Period = 20*12;                                                  line11
double Future_Value_Interest[1000];                                  line12
double Future_Value_Annuity_Payout[1000];                            line13

double A_A = Annuity_Amount (Future_Value, Interest_Rate, Period);   line14

printf ("Monthly Payment or Annuity Amount = %7.3lf\n\n", A_A);      line15

double Payment = A_A;                                                line16

Annuity_Payout(Interest_Rate, Period, Payment, Future_Value_Interest,
                              Future_Value_Annuity_Payout); line17

printf("Month      End of Month  Interest Earned    Total \n");      line18
printf("Payment     for the month          Payment \n");             line19

    for (i_counter = 1; i_counter <=Period; i_counter++)             line20
    {                                                                line21
    printf ("  %3d        %6.2f       %6.2f            %8.2f\n",
     i_counter, Payment, Future_Value_Interest[i_counter] ,
    Future_Value_Annuity_Payout[i_counter]);                         line22
    }                                                                line23


return 0;                                                            line24
}                                                                    line25



double Annuity_Amount (double F_V, double I_R, int Per)              line26
{                                                                    line27
   return F_V*I_R/(pow((1+I_R),Per)-1);                              line28
}                                                                    line29


void Annuity_Payout(double I_R, int Per, double Pay, double FVI[1000], double
FVAP[1000])                                                          line30
{                                                                    line31
   int i;                                                            line32

   FVAP[0] = 0.;                                                     line33
   for (i = 1; i <=Per; i++)                                         line34
   {                                                                 line35

        FVI[i]  = FVAP[i-1]*I_R;                                     line36
        FVAP[i] = FVAP[i-1] + Pay + FVI[i];                          line37
   }                                                                 line38

}                                                                    line39
```

Figure A-T.4 Code for annuity examples.

appearance of the output, an additional blank line was added after the printing of the annuity amount by adding a second "\n."

Problem (Tutorial) A-T.4

We next want to take our combined C++ code developed in Problem A-T.3 and convert it to a library, which can be accessed by VBA code, an Excel spreadsheet, or both.

The process of converting our combined code to a C++ library is conceptually straightforward. We will need to eliminate the main function, and we will directly call the working C functions either from Excel or VBA. We do emphasize that the procedure of developing a working code (Problem A-T.3), which is then converted to a C++ library (Problem A-T.4), is a procedure that should be followed.

In Problem A-T.4, we focus on calling the C++ library from VBA code, and in Problem A-T.5, we examine calling the library from the Excel spreadsheet.

Step 1: We begin by creating a new folder (project) in \New User named.

Click on File → New → Project.

You should see a new window with various project types.

From Project Types, Select: General, and from Templates, Select: Empty Project.

Location → C:\POEA\Bridging Excel and C Codes\New User.

Project Name → Time_Value_Money_Combined_Lib1 (this step will create a new folder named Time_Value_Money_Combined_Lib1 in the folder New User).

Click on OK.

You should be at the main window.

On the toolbar → Project → Add New Item.

You should see a new window.

From Categories, Select: Code, and from Templates, Select: C++ File (.cpp).

Create a C++ file named Annuity_Calcs_Lib1.cpp.

Step 2: Copy the source code Annuity_Calcs.cpp from Problem A-T.3 into Annuity_Calcs_Lib1.cpp.

We next need to comment out or remove (as done next) all the header files, prototypes, and the main function. The code, which is shown in Figure A-T.5, is basically reduced to the existing functions, with one new #include statement, which we will discuss further.

This process is fairly simple, but there are several more steps, and as noted earlier, we must make some changes to this code when it is being used in a DLL.

Step 3: A definition file is added to the project file: A definition file is used during compilation to identify functions that will be exported or made available to other programs. The .def file is created within the workspace (Time_Value_Money_Combined_Lib1.dsw).

To add the definition file to the project, from the toolbar, Project → Add New Item. From Categories, Select: Code, and from Templates, Select C++ File (.cpp). For the name, use Annuity_Calcs_Lib1.def (this name must be the same as the project name with the file extension `.def`) → Add. Even though we started step 3 by adding a cpp file, the extension .def overrides the .cpp extension and creates the definition file in the project.

The file (Annuity_Calcs_Lib1.def) should appear as

```
; if comments needed
LIBRARY Time_Value_Money_Combined_Lib1
EXPORTS
 Annuity_Amount
 Annuity_Payout
```

The semicolon allows comment statements. The key word `LIBRARY` allows identification of the project name, `Time_Value_Money_Combined_Lib1`. The key word `EXPORTS` is followed by the names of the

```
#include "Annuity_Calcs_Lib1.h"                                  line 1


/*System header files and function prototypes are moved to Annuity_Calcs_Lib.h
 which will serve as the library header file                     line 2

double Annuity_Amount (double F_V, double I_R, int Per)          line 3
{                                                                line 4
  return F_V*I_R/(pow((1+I_R),Per)-1);                           line 5
}                                                                line 6


void Annuity_Payout(double I_R, int Per, double Pay, double FVI[1000], double
FVAP[1000])                                                      line 7
{                                                                line 8
   int i;                                                        line 9

   FVAP[0] = 0.;                                                 line10
   for (i = 1; i <=Per; i++)                                     line11
   {                                                             line12
            FVI[i]  = FVAP[i-1]*I_R;                             line13
            FVAP[i] = FVAP[i-1] + Pay + FVI[i];                  line14
   }                                                             line15

}                                                                line16
```

Figure A-T.5 Code for annuity example as a library.

functions, which will be externally available. These functions (`Annuity_Amount`, `Annuity_Payout`) will be available to other codes.

Finally, we must make one change: In the C++ code Annuity_Calcs_Lib1.cpp `line 7`, change

```
void  Annuity_Payout(double  I_R,  int
Per, double Pay, double FVI[1000],
 double FVAP[1000])              line 7
```

to

```
void  Annuity_Payout(double  I_R,  int
Per, double Pay, double* FVI,
 double* FVAP)                   line 7
```

where the size of the arrays, here [1000], has been replaced with pointers.

Step 4: We next need to add a header file, which will take the place of the header files and prototypes we removed within the workspace (Time_Value_Money_Combined_Lib1.dsw),

To add the header file to the project, from the toolbar, Project → Add New Item. From Categories, Select: Code, and from Templates, Select: Header File (.h). For the name, use Annuity_Calcs_Lib1 (the `.h` extension will be automatically generated if not supplied).

In this header file (Annuity_Calcs_Lib1.h), add the code provided in Figure A-T.6.

We have encountered the first three prototype statements in the header file; the first two are libraries and the third is the function `Annuity_Payout` with arguments (double, double, integer). The fourth statement announcing the function `Annuity_Payout` with arguments `(double I_R, int Per, double Pay, double* FVI, double* FVAP)` looks familiar; however, here we may have expected the last two arguments associated with the arrays to appear as double[], double[]. The asterisks are actually pointers and they denote the location of the first element in each array. We also could have written the header for `Annuity_Payout` with arguments `(double, int, double, double*, double*)`; there is actually no need to name the variables.

Step 5a: From the toolbar, select Project → Time_Value_Money_Combined_Lib1 Properties (Alt + F7) → Configuration Properties. Expand the Configuration Properties window by highlighting the + sign next to Configuration Properties → General → Project Defaults → Configuration Type. The configuration type will currently be Application (.exe); use the drop-down window to select Dynamic Library (.dll) → Apply.

Step 5b: Now expand the C/C++ window by highlighting the + sign next to C/C++ → Advanced → Calling Convention. The current calling convention will be _cdecl (/Gd); use the drop-down window to select _stdcall(/Gz) → Apply.

Step 5c: Now expand the Linker window by highlighting the + sign next to Linker → Input. The module definition file will be blank; add Annuity_Calcs_Lib1.def → Apply. This name must be the same as the project name with the file extension `.def`. Close the Configuration Manager window by selecting OK.

Step 6: The next step is to set up the Excel worksheet, which will allow the C++ library to be accessed either directly from the worksheet or from the VBA code. In either case, the process begins by creating a module for the VBA code within the Excel worksheet. In the VBA module, the connection to the C++ library is made with the three lines of VBA code shown in Figure A-T.7. The code lines are wrapping below in the word processing package; again this is just three lines of code. Also recall, if necessary, in VBA a line continuation is a blank space followed by underscore.

```
# include <stdio.h>                                                     line 1
#include <math.h>                                                      line 2

double Annuity_Amount (double , double , int );                       line 3
void Annuity_Payout(double I_R,int Per, double Pay,double* FVI, double *FVAP);
                                                                       line 4
```

Figure A-T.6 Header file for annuity example.

```
Option Explicit                                                        line1

Public Declare Function Annuity_Amount Lib " C:\POEA\Bridging Excel and C
Codes\Problems\Time_Value_Money_Combined_Lib1\Debug\Time_Value_Money_Combined
_Lib1.dll" (ByVal Future_Value As Double, ByVal Interest_Rate As Double,
ByVal Period As Long) As Double                                        line2

Public Declare Sub Annuity_Payout Lib " C:\POEA\Bridging Excel and C
Codes\Problems\Time_Value_Money_Combined_Lib1\Debug\Time_Value_Money_Combined
_Lib1.dll" (ByVal Interest_Rate As Double, ByVal Period As Long, ByVal
Payment As Double, ByRef Future_Value_Interest As Double, ByRef
Future_Value_Annuity_Payout As Double)                                 line3
```

Figure A-T.7 Declare statements for the excel file.

Option Explicit—this requires all variables to be specified in type or an error message will result; this is good coding practice. There is then one statement for each and every function we want to connect with in the C++ library. The form of the connect are the key words `Public Declare Function` (this establishes the function as a global function) followed by the function name, then the key word `Lib`, and then the root directory location of the DLL. The DLL is initially created in the debug folder within the project space; often, experienced coders will move the working DLL out of the debug folder. Next, we identify the names of the variables and variable type in the function argument list; these variables will be connected to the function argument list in the C++ library. These exact names will, of course, be used in the VBA code. You will note that variables that are being passed to the C++ library (from the VBA code) are sent `ByVal` (by value) and variables returned from the C++ code for use in the VBA code are passed `ByRef` (by reference). Arrays are discussed next. Finally, the type of value returned by the function is specified `As Double` (note `As Double` at the very end of the call to `Annuity_Amount`).

It is important to examine the call to the function `Annuity_Amount` in a little more detail. After the function `Annuity_Amount`, we specify the library location and we identify as the calling parameters (argument list) `Future_Value, Interest_Rate, Period`. Because values for all these variables are being sent to the C++ function, each is specified as `ByVal`. The value for the annuity amount is returned `As Double`; the `As Double` at the end of the statement. We next must address another idiosyncrasy of the DLL library. The careful reader will note that the variable `Period`, which is an integer (both in the C++ code we complied as a DLL and the VBA code we will write shortly), is specified as a long in the argument list, `Period As Long`. This is simply ugly, but unfortunately, it is necessary. When the DLL is created in C++, integer variables are set to occupy full integer length. The integer length in VBA does not match the integer length in the C++ DLL and this creates major problems. The easy solution is to specify the integer variable Period as a long in the argument list, and the handshaking between the argument lists (VBA ↔ C++ DLL) is fine.

> *Note:* An integer variable in VBA is defined as an integer between −32,768 and 32,767, and a long variable is defined between −2,147,483,648 and 2,147,483,647. In C/C++, for integers or long integers, the range is −2,147,483,648 to 2,147,483,648. The range −32,768 and 32,767 is a short integer in C/C++.

We can next examine the construction of the second of our "functions," `Annuity_Payout`. We have discussed (preceding paragraph) the variable `Period` must be specified as a long. Our two arrays, `Future_Value_Interest` and `Future_Value_Annuity_Payout`, are specified `As Double` (with no length specified or brackets to indicate these are arrays). Arrays are always specified as `ByRef` regardless if we are passing values from the VBA code to the C++ code or vice versa. Here the `ByRef` key word allows the location of the first element in the array to be specified in the VBA code—this is shown next in the VBA code for Problem A-T.4.

Another idiosyncrasy caused by the creation of the DLL is the inclusion of the `As type` at the end of the function declare statement. The careful reader will recall that the function `Annuity_Payout` was complied as `void Annuity_Payout`, so no value for the function `Annuity_Payout` is expected. To avoid this problem, we declare `Annuity_Payout` to be a subroutine (`Sub`); a subroutine does not require a value to be returned.

We can now write the VBA code, which will access these two functions in the C++ library. The code, which is given in **Problem TA.4.xls**, would appear as shown in Figure A-T.8,

The VBA code begins with the `Option Explicit` and `Public Declare Function` statements we have discussed in the preceding paragraph. We next define a VBA subroutine `Public Sub Get_Annuity()`, which will call our two C++ functions. The structure of this subroutine `Public Sub Get_Annuity()` is identical to an Excel Spreadsheet macro; therefore, a macro, `Get_Annuity`, will also be available from the Excel spreadsheet.

Our variables are defined as expected. It is interesting to note that the variable `Period` could be defined as an integer in the VBA code; however, since this variable is being passed to the C-dll, it is declared a long in VBA to match with integer type in C. The function `Annuity_Amount` is called and the calculated annuity is stored in the variable `A_A`, which is then written on the spreadsheet in cell (1,1) in `line14`; here, the construct is Cell (row, column).

The variables to call `Annuity_Payout` are then defined. Do note that the two arrays are set (0–999) in parenthesis for VBA. The value for the payout is simply the `A_A` value found from the function call to `Annuity_Amount`.

For our two arrays, we must specify the starting location of the first element, `Future_Value_Interest(0)`, `Future_Value_Annuity_Payout(0)`, which in both cases is the zero element. The careful reader will note that these arrays are both identified as `ByRef` in the `Declare Function` statement, which implies values will

```
Option Explicit                                                     line 1

Public Declare Function Annuity_Amount Lib " C:\POEA\Bridging Excel and C
Codes\Problems
\Time_Value_Money_Combined_Lib1\Debug\Time_Value_Money_Combined_Lib1.dll"
(ByVal Future_Value As Double, ByVal Interest_Rate As Double, ByVal Period As
Long) As Double                                                     line 2

Public Declare Sub Annuity_Payout Lib " C:\POEA\Bridging Excel and C
Codes\Problems
\Time_Value_Money_Combined_Lib1\Debug\Time_Value_Money_Combined_Lib1.dll"
(ByVal Interest_Rate As Double, ByVal Period As Long, ByVal Payment As Double,
ByRef Future_Value_Interest As Double, ByRef Future_Value_Annuity_Payout As
Double)                                                             line 3

Public Sub Get_Annuity()                                            line 4
    Dim Future_Value As Double                                      line 5
    Dim Interest_Rate As Double                                     line 6
'In the Declare Function (above) Dim Period As Integer will not work bcz
allocation of size compatibility problems between C++ DLLs and VBA 7
    Dim Period As Long                                              line 8
    Dim A_A As Double                                               line 9

    Future_Value = 500000                                           line10
    Interest_Rate = 0.10/12                                         line11
    Period = 20 * 12                                                line12

    A_A = Annuity_Amount(Future_Value, Interest_Rate, Period)       line13
    Sheet1.Cells(1, 1) = A_A                                        line14

    Dim Payment As Double                                           line15
    Dim Future_Value_Interest(0 To 999) As Double                   line16
    Dim Future_Value_Annuity_Payout(0 To 999) As Double             line17
                                                                    line18
    Payment = A_A                                                   line19
                                                                    line20
    Annuity_Payout Interest_Rate, Period, Payment, Future_Value_Interest(0),
Future_Value_Annuity_Payout(0)                                      line21
                                                                    line22
    Dim i As Integer                                                line23
    For i = 1 To 240                                                line24
        Sheet1.Cells(i, 3) = Future_Value_Interest(i)               line25
        Sheet1.Cells(i, 5) = Future_Value_Annuity_Payout(i)         line26
    Next i                                                          line27

End Sub                                                             line28
```

Figure A-T.8 Excel module with VBA code for annuity example.

be returned from the C++ library. This is true, but we still must pass the starting element location (0) to the C++ library. In `line21`, the call to the C procedure `Annuity_Payout` can be made as `Call Annuity_Payout (parameter list)` or `Annuity_Payout parameter list`, a space without parentheses; the latter approach is used in `line21`. Array values are returned from the C++ library and we print these values on the Excel Spreadsheet in `lines 23-27`.

Problem (Tutorial) A-T.5

We want to call the C functions available in our library, `\Time_Value_Money_Combined_Lib1.dll`, directly from the Excel sheet.

The macro created in **Problem TA.4 .xls** provides Excel sheet connection to the library of our two functions, `Annuity_Amount` and `Annuity_Payout`. From the Excel sheet, we can use $f_x \rightarrow$ category: User Defined to look for available functions. Here we will find only `Annuity_Amount` is available. If you provided values (or cells) with Future_Value, Interest_Rate, and Period (500,000; = 0.10/12; = 20*12), the annuity amount will be calculated by the C function and the result displayed in the calling cell on the Excel sheet. Do notice that with any change in Future_Value, Interest_Rate, or Period, the annuity amount will be updated. This is important and will be needed for optimization strategies involving our combustion library.

The second function, `Annuity_Payout`, is not available from the Excel sheet. The Excel sheet ↔ VBA connection indicates `Sub Annuity_Payout` is a subroutine procedure. Subroutines cannot be called from the Excel sheet; see Table A.1 for allowable calls from the sheet.

Index

Modeling, Analysis and Optimization of Process and Energy Systems, First Edition. F. Carl Knopf.